A Handbook of SILICATE ROCK ANALYSIS

To Barbara, Edmund, Esther, Samuel and Tessa

A Handbook of SILICATE ROCK ANALYSIS

P. J. POTTS
Research Fellow in Earth Sciences
The Open University
Milton Keynes, UK

Blackie

Glasgow and London
Published in the USA by
Chapman and Hall
New York

Blackie & Son Limited
Bishopbriggs, Glasgow G64 2NZ
7 Leicester Place, London WC2H 7BP

Published in the USA by
Chapman and Hall
in association with Methuen, Inc.
29 West 35th Street, New York, NY 10001

First published 1987

British Library Cataloguing in Publication Data

Potts, P.J.
A handbook of silicate rock analysis.
1. Silicates—Analysis 2. Chemistry, Analytic
I. Title
549′.6 QD181.S6

ISBN 0-216-91794-8

Library of Congress Cataloging in Publication Data

Potts, P.J.
A handbook of silicate rock analysis.

Bibliography: p.
Includes index.
1. Rocks, Silicate—Analysis. I. Title.
QE38.P68 1987 549′.6 86-18868
ISBN 0-412-00881-5 (U.S.)

Printed in Great Britain by Bell & Bain (Glasgow) Ltd.

Contents

Preface

The techniques available for the chemical analysis of silicate rocks have undergone a revolution over the last 30 years. No longer is the analytical balance the only instrument used for quantitative measurement, as it was in the days of classical gravimetric procedures. A wide variety of instrumental techniques is now commonly used for silicate rock analysis, including some that incorporate excitation sources and detection systems that have been developed only in the last few years. These instrumental developments now permit a wide range of trace elements to be determined on a routine basis.

In parallel with these exciting advances, users have tended to become more remote from the data production process. This is, in part, an inevitable result of the widespread introduction of microcomputers for instrument control, and in part a consequence of the logistics of organizing a modern laboratory for rapid and efficient routine analysis. The resultant lack of interaction between user and machine leads to the danger of a 'black-box' attitude towards analytical chemistry—samples in at one end, result out at the other—without an appreciation of what happens in between.

However, to use an analytical technique most effectively, it is essential to understand its analytical characteristics, in particular the excitation mechanism and the response of the signal detection system. In this book, these characteristics have been described within a framework of practical analytical aplications, especially for the routine multi-element analysis of silicate rocks. All analytical techniques available for routine silicate rock analysis are discussed, including some more specialized procedures. Sufficient detail is included to provide practitioners of geochemistry with a firm base from which to assess current performance, and in some cases, future developments.

This manuscript could not have been completed without the constant help of my wife Barbara, the assistance of many friends and colleagues at the Open University and the patient encouragement of the publishers.

P.J.P.

SIEMENS

Advances in X-ray fluorescence analysis:

User-friendly operation with the SRS 300 wavelength dispersive X-ray spectrometer

With the SRS 300 spectrometer, Siemens has developed a system that answers the needs of today's analytical chemist. Its success in the user's laboratory has proven that this is a spectrometer system that you can believe in for reliability, user convenience, and analytical performance for not only today but the future.

The unique automation system of the SRS 300 puts you, the operator, in complete control – allowing you the advantage of end window X-ray optics for better sensitivity and low Z analysis, sample handling capabilities, trouble-free control of crystals, filters, collimator masks, and sample changers among numerous other high-performance features of the SRS 300.

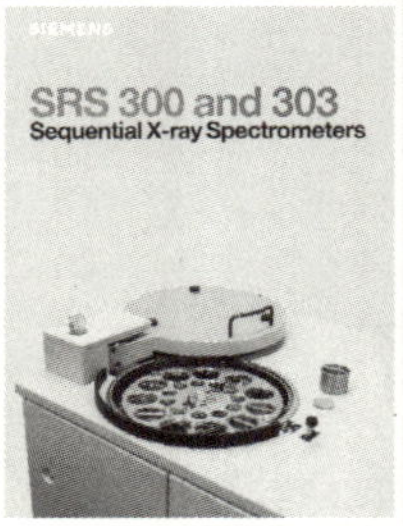

For further information, a quotation, consultation or a demonstration, please contact:

Siemens AG, Measuring and Process Control Systems Division, E 689
Postfach 21 12 62, D-7500 Karlsruhe 21,
Telephone (07 21) 5 95-47 30, Telex 78 255-69,
or a Siemens branch in your area.

The new SRS 300 X-ray spectrometer from Siemens

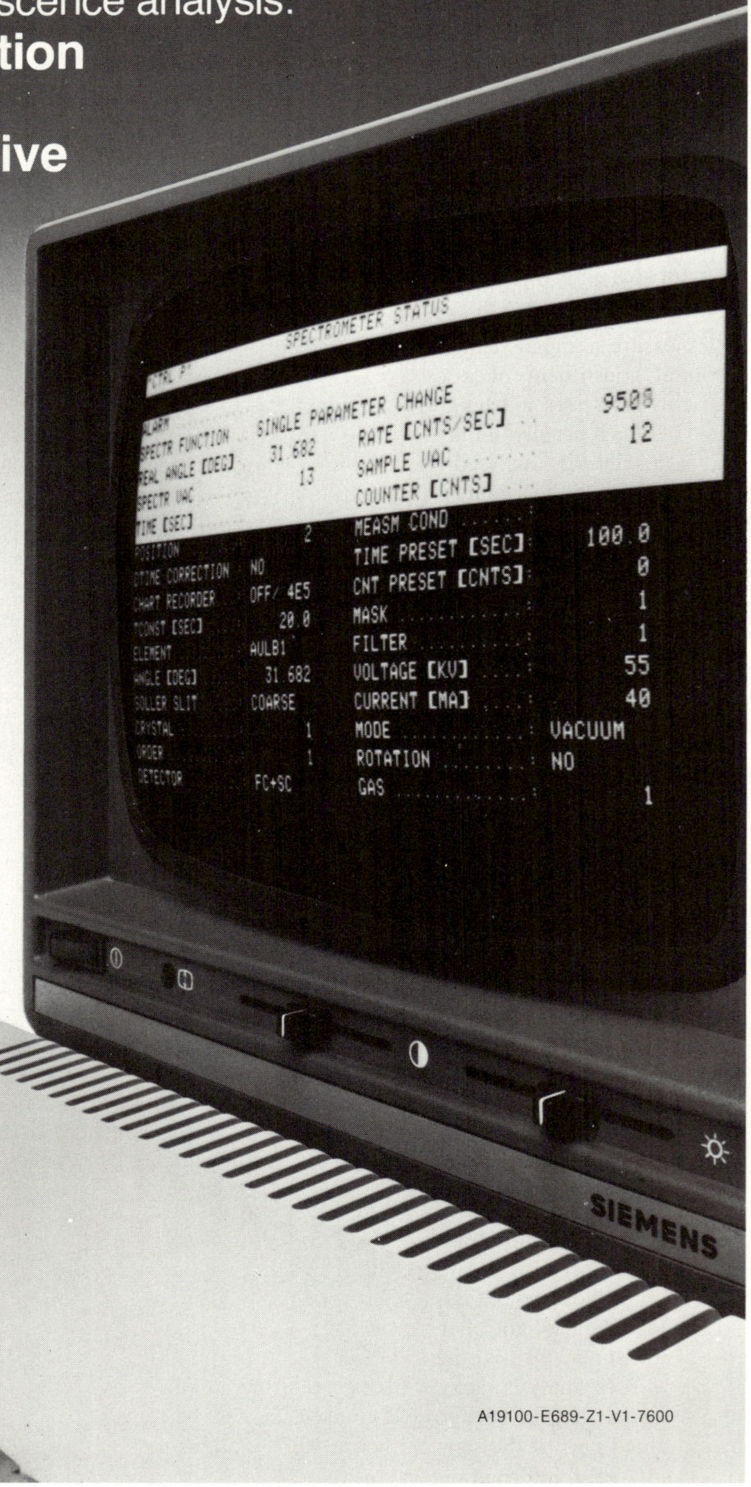

A19100-E689-Z1-V1-7600

1 Concepts in analytical chemistry

1.1 Introduction

The analytical chemistry of silicate rocks and minerals takes in a diverse range of techniques employing a wide variety of physical phenomena. Although the concepts on which many of these techniques are based appear to have little in common, a degree of uniformity does embrace each in terms of the underlying analytical principles. Thus the desired goal is to measure a signal related to the concentration of an elemental constituent of a sample. This analyte signal invariably possesses a background component, the magnitude of which usually limits the lowest concentration that can be satisfactorily analysed (the detection limit). The precision with which the signal can be measured depends on a variety of factors, both fundamental and instrumental in origin. These could include counting statistical errors, signal to background ratios, instrumental noise, drift and sensitivity. All techniques suffer interference effects and, in some, account must be taken of blank and contamination levels. It is generally necessary to prove the accuracy of measurements by the comparative analysis of international rock reference materials.

The purpose of this introductory chapter is to discuss some of these basic concepts so that it will then be possible to examine in more detail the characteristics of individual techniques in subsequent sections.

1.2 Terms and definitions in analytical chemistry

Several terms are in widespread use to describe phenomena affecting all analytical techniques. Some have specialized usage in analytical chemistry. The more important of these terms are as follows.

signal, background: The signal is the physical response containing analytical information related to the concentration of an elemental component of a sample. It is therefore the object of measurement. The magnitude of the signal is not, however, always linearly related to concentration (in some schemes of x-ray fluorescence analysis, for example, it is necessary to apply a non-linear iterative correction to raw count data to account for the influence of matrix composition on fluorescence x-ray emission). A variety of signals, including weight changes, voltage differences and special measurements, is used in analytical techniques applied to rock analysis: some of these are listed in Table 1.1. All signals have a background component, this being the unwanted part of the signal which usually contains no analytically useful information. In general, the background arises from two sources. The first is a natural background component (such as the x-ray continuum on which a characteristic x-ray line sits). The second is an instrumental background component characteristic of the measuring device. In the example outlined above, this could arise from fluorescence of the diffracting crystal or Compton scatter with an Si(Li) detector used for x-ray spectrum measurements.

detection limit: This refers to the limit down to which an analytical signal can be measured and distinguished at a specified confidence level from the background signal. This concept is considered in more detail in section 1.5. Unfortunately, the term 'detection limit' is one of the most abused in analytical chemistry, since without precise specification this term can be applied to a variety of signal levels.

sensitivity: This term is often qualitatively used to refer to the detection limit of a technique, although strictly this connection is secondary. The sensitivity of a technique refers to the gradient of a signal *v.* concentration calibration line and is measured as the signal strength per unit concentration. As a means of referring to the detection limit capability, account must be taken of the background intensity. However, for constant background, an element that shows high sensitivity must also possess a lower detection limit and vice versa (see Figure 1.1).

bias: In an analytical context, bias is the phenomenon whereby the average tendency of analysed results is either high or low compared with the expected composition. In electronic terminology, bias refers to an electrical potential applied to an electrode (for example, a bias of 1 kV).

drift: Drift arises when an instrument response changes with time, and is often associated with the effect of changes in ambient temperature on the stability of electronic circuits.

gain: Gain is the amplification factor applied to an electronic circuit.

noise: Short-term instability, which is often associated with

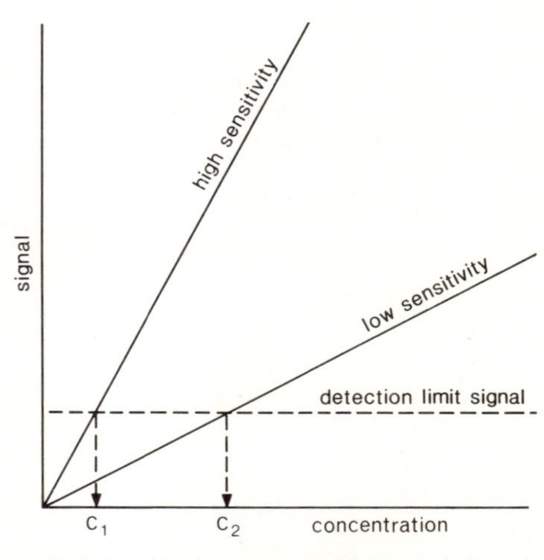

Figure 1.1 Relationship between sensitivity and detection limits. The diagram shows the signal *v.* concentration calibration plot for two elements, one having high sensitivity and the other having low sensitivity. Providing the magnitude of the detection signal limit for both is the same, the detection limit, converted to concentration, of the high-sensitivity element will be the lower.

Table 1.1 Analytical signals and instrumentation used in geochemical analysis.

Technique	Acronym	Analytical signal	Method of excitation	Instrumentation for detecting signal
Gravimetry		Weight loss or gain	Precipitation of element as insoluble compound or ignition of sample to release volatile phase (e.g. H_2O or CO_2)	Analytical balance
Titrimetry		Volume of titrant	Stoichiometric reaction of elemental species with specific reagent in solution	Burette
Colorimetry		Colour intensity as measured by optical absorption	Reaction of elemental species in solution with a chromophore (a reagent that forms a coloured complex with the element in question)	Optical monochromator or optical filters plus photomultiplier tube
X-ray fluorescence analysis	WD–XRF or ED–XRF	Count rate of secondary fluorescence x-ray line	Excitation of sample with x-ray spectrum from an x-ray tube	Bragg diffraction from a crystal spectrometer fitted with gas proportional or scintillation counter detectors or solid state Si (Li) detector coupled to a multichannel analyser
Electron probe micro-analysis	EPMA	As above	Bombardment of sample with a focused beam of electrons accelerated to 15 to 30kV	As above
Flame photometry		Intensity of emission of characteristic optical radiation from an atomized sample	Atomization and excitation of a sample solution in a flame (e.g. air/propane)	Optical monochromator or simple optical filters plus photomultiplier tube
Spark/arc discharge optical emission spectrometry		Intensity of emission of characteristic optical radiation from atoms and ions	Atomization (ionization), and excitation of a sample in an electrical discharge	Optical monochromator often with a large number (20 to 40) of fixed exit slits each with its own photomultiplier detector. Older instruments used a photographic plate for detection
Inductively coupled plasma optical emission spectrometry	ICP–OES or ICP–AES	As above	Atomization (ionization), and excitation of a sample solution in an argon plasma	Optical monochromator (scanning or fixed channel) plus photomultiplier tube
Atomic absorption spectrometry	AAS	Degree of absorption of characteristic optical radiation by atoms	Atomization of sample solution in a flame (e.g. air/acetylene) Absorption of monochromatic radiation from a hollow cathode lamp by atomic species in the flame	Scanning optical monochromator plus photomultiplier tube
Neutron activation analysis	INAA or RNAA	Intensity of gamma-ray emission from radioactive isotopes	Neutron capture by element nuclei to form radioactive daughter nuclei during irradiation by neutrons in a nuclear reactor	Solid state germanium gamma-ray detector coupled to a multichannel analyser
Delayed neutron fission		Intensity of neutron emission following fission of ^{235}U	Neutron-induced fission of ^{235}U during irradiation by neutrons in a nuclear reactor	Boron trifluoride neutron counter
Ion-selective electrodes	ISE	Voltage change across an ion-specific electrode	Diffusion of ions through an ion-selective membrane causing a change in the potential generated in an electro-chemical cell	Voltmeter
Spark source mass spectrometry	SSMS	Intensity of mass/charge signal	Atomization and ionization of sample in a radiofrequency electrical discharge	Ions dispersed using both electrostatic and electromagnetic sectors. Photographic plate detection
Thermal ionization mass spectrometry	TIMS	As above	Atomization and ionization of a chemically purified sample on an incandescent metal filament	Ions dispersed by a single magnetic sector and detected using Faraday cup, electron multiplier or Daly detectors

Table 1.1 (*Continued*)

Technique	Acronym	Analytical signal	Method of excitation	Instrumentation for detecting signal
Gas source mass spectrometry	GSMS	As above	Stepped combustion of sample to release gas (e.g. H_2, CO_2, SO_2 or noble gases) that are chemically purified and ionized by electron bombardment in the ion gun of mass spectrometer	Ions dispersed by a single electromagnetic sector and detected using Faraday cup or electron multiplier. Detectors often designed to operate in tandem or triplicate to permit simultaneous detection of the ion beams of interest
Inductively coupled plasma mass spectrometry	ICP-MS	As above	Atomization and ionization of a sample solution by aspirating into an inductively coupled argon plasma	Plasma gases sampled into a quadrupole mass analyser and detected using an electron channel multiplier coupled to a multichannel analyser

WD–XRF = wavelength-dispersive XRF.
ED–XRF = energy-dispersive XRF.
INAA = instrumental neutron activation analysis.
RNAA = radiochemical neutron activation analysis.
ICP–AES = ICP atomic emission analysis.
Acronyms generally follow those given by Delaey and Arkens (1981).

the random movement of electrons in individual electronic components, causes a random variation in the analytical signal called noise.

figure of merit: In the optimization of an analytical technique, measurements are normally made at a variety of instrumental settings. It is then desirable to plot some parameter as a function of those instrument settings in order to identify optimum operating conditions. The parameter selected is the figure of merit. The simplest form is the signal to background ratio, although this is not universally applicable.

precision and accuracy: Precision is a measure of the analytical reproducibility of a measurement. Accuracy is a measure of the closeness of this measurement to the 'true' value. In geochemical analysis precision is easy to measure: accuracy is occasionally impossible. In most cases values quoted for the precision of a technique are a grossly optimistic measure of accuracy. These concepts are discussed further in section 1.4.2.

reference material: Reference materials are the means by which the accuracy of analytical results can be assessed. They comprise silicate rock powders of proven homogeneity that have been distributed for analysis by a variety of techniques to as many laboratories as possible worldwide, usually on a voluntary co-operative basis. The assessment of resultant analytical data to estimate 'true' compositions is not trivial since data are usually affected by discrepancies arising from interlaboratory bias, particularly in trace elements analysis. For this reason rock reference materials never aspire to the name 'standards'. The reliability and status of rock reference materials are considered further in section 1.10.

standards: Standards are the samples selected by the analyst to calibrate an instrument. They may consist of rock reference materials or samples prepared from high-purity reagents. The use of this term when applied to reference materials in this context is not meant to imply that these samples can be used as an international basis for 'standard' measurements.

standard additions: A method of calibration in which different aliquots of a sample are 'spiked' with increasing amounts of an element. The graph of analysed signal *v.* concentration of element added, gives a calibration line, the intercept of which with the concentration axis gives the amount of that element present in the unspiked sample. This technique is sometimes used when analyses are subject to interferences that cannot be compensated easily in any other way. More details are given in section 1.9.

aliquot: this refers to that fraction of the sample taken for analysis ('5g sample aliquots').

analyse, determine: Samples are 'analysed' for elements but elements are 'determined' in a sample.

analyte: The species for which a determination is sought.

Interference: A non-linearity introduced into the analytical signal, the magnitude of which depends on the concentration of some other components in the sample.

matrix effect: An interference caused by the composition (or mineralogical structure) of the matrix in which the element to be analysed resides.

stoichiometry: A mineral or a chemical compound is said to be stoichiometric if the elemental constituents exist in the exact atomic proportions predicted by the chemical formula (for example, Eu_2O_3 is stoichiometric, $Eu_{2-x}O_3$ is not).

blank: The analytical blank is the concentration of an element in a sample that has been chemically prepared and analysed without the addition of a silicate rock sample. Blank concentrations of an element arise from impurities in chemical reagents used for preparative chemistry and from contamination from the laboratory environment. Measures to overcome these contamination effects are considered in section 1.7.

homogeneity: Homogeneity relates to the property that different aliquots of the same sample are identical in chemical composition. Since almost all silicate rocks are composed of discrete minerals, and therefore have bulk compositions that are intrinsically inhomogeneous at the mineral level, it is necessary to specify the mass at which the rock sample is expected to be homogeneous ('homogeneous at the 1g sampling level'). Sample homogeneity is discussed further in section 1.6.

Table 1.2 SI base units

Physical quantity	Name of SI unit	Symbol
length	metre	m
mass	kilogram	kg
time	second	s
electric current	ampere	A
thermodynamic temperature	kelvin	K
amount of substance	mole	mol
luminous intensity	candela	cd
Supplementary units:		
plane angle	radian	rad
solid angle	steradian	sr

Table 1.4 SI defined prefixes for multiples and fractions of 10

Fraction or multiple	SI prefix	Symbol
10^{-18}	atto	a
10^{-15}	femto	f
10^{-12}	pico	p
10^{-9}	nano	n
10^{-6}	micro	u
10^{-3}	milli	m
10^{-2}	centi	c
10^{-1}	deci	d
10	deca	da
10^{2}	hecto	h
10^{3}	kilo	k
10^{6}	mega	M
10^{9}	giga	G
10^{12}	tera	T

1.3 Units of measurement: the international system (SI) of units

1.3.1 *The SI system*

The present generation of scientists has seen some major changes in the units internationally adopted for scientific measurements. In the tussle between the proponents of the cgs system (based on the centimetre, gram and second) and the MKS system (based on the metre, kilogram and second), the latter prevailed. The ruling body on these matters is the General Conference on Weights and Measures (CIPM). In general, as far as analytical chemistry is concerned, these decisions are enacted through the influential International Union of Pure and Applied Chemistry (IUPAC).

The convention internationally adopted for the reporting of all scientific measurements is the SI system (the International System of units). This convention defines not only base units but also a range of compatible derived units together with rules for abbreviations and terminology. The system recognises seven base units and two additional supplementary units. These units form a unique and coherent system.

1.3.2 *SI base units*

The base units used in the SI system are all dimensionally independent. They are listed in Table 1.2, together with the two supplementary units.

Each of these units is precisely defined as shown in Table 1.3. Particular emphasis must be placed on distinguishing between the meaning of *mass* and *mole*. The mass of a substance is related to the amount of that substance present as normally measured in the laboratory by its weight. The number of moles present is related to the quantity of discrete entities present. In 0.012 kg (i.e. 12 g) of carbon-12, the number of carbon atoms present is given by the Avogadro constant, that is 6.022×10^{23}. This constitutes one mole of carbon-12 atoms. Similarly one mole of the reagent ferric chloride ($FeCl_3$) is equivalent to 162.206 g of reagent ($Fe = 55.847 + 3Cl = 3 \times 35.453$), one mole of Fe has a mass of 55.847 g and one mole of electrons e^- has a mass of 5.4859×10^{-10} g. Each of these masses contains the same number of specified entities (6.022×10^{23}).

Table 1.3 SI base units

Metre: the length equal to 1 650 763.73 wavelengths in vacuum of the radiation corresponding to the transition between the levels $2p_{10}$ and $5d_5$ of the krypton-86 atom.

Kilogram: the mass equal to the mass of the international prototype of the kilogram; that is a certain piece of platinum–iridium kept at the International Bureau of Weights and Measures, Sèvres, France.

Second: the duration of 9 192 631 770 periods of the radiation corresponding to the transition between two hyperfine levels of the ground state of the caesium-133 atom.

Ampere: that constant current which if maintained in two straight parallel conductors of infinite length of negligible circular cross-section, and placed one metre apart in vacuum, would produce between these conductors a force equal to 2×10^{-7} newton per metre of length.

Kelvin: unit of thermodynamic temperature being the fraction 1/273.16 of the thermodynamic temperature of the triple point of water. The triple point of water is the temperature at which ice, water and water vapour can coexist in equilibrium.

Mole: the amount of a substance of a system which contains as many elementary entities as there are atoms in 0.012 kilograms of carbon-12. When the mole is used, the elementary entities must be specified (e.g. atoms, molecules, ions, electrons etc).

Candela: the luminous intensity, in the perpendicular direction, of a surface of 1/600 000 square metre of a black body at the temperature of freezing platinum under a pressure of 101 325 newtons per square metre.

Supplementary SI units

Radian: the plane angle between two radii of a circle which cuts off on the circumference an arc of length equal to the radius.

Steradian: solid angle which, having its vertex at the centre of a sphere, cuts off an area of the surface of the sphere equal to that of a square with sides of length equal to the radius of the sphere.

Table 1.5 SI derived units

Physical quantity	Name of SI unit	Symbol	Definition in SI units
area	square metre		m^2
concentration	mole per cubic metre		$mol\,m^{-3}$
density	kilogram per cubic metre		$kg\,m^{-3}$
velocity	metre per second		$m\,s^{-1}$
volume	cubic metre		m^3
wavenumber	1 per metre		m^{-1}
frequency	hertz	Hz	s^{-1}
energy	joule	J	$kg\,m^2\,s^{-2}$
force	newton	N	$kg\,m\,s^{-2} = J\,m^{-1}$
power	watt	W	$kg\,m^2\,s^{-3} = J\,s^{-1}$
pressure	pascal	Pa	$kg\,m^{-1}\,s^{-2} = N\,m^{-2} = J\,m^{-3}$
electric charge	coulomb	C	A s
electrical potential difference	volt	V	$kg\,m^2\,s^{-3}\,A^{-1} = J\,A^{-1}\,s^{-1}$
electrical resistance	ohm	Ω	$kg\,m^2\,s^{-3}\,A^{-2} = VA^{-1} = S^{-1}$
electrical conductance	siemens	S	$kg^{-1}\,m^{-2}\,s^3\,A^2 = \Omega^{-1}$
electrical capacitance	farad	F	$A^2\,s^4\,kg^{-1}\,m^{-2} = A\,s\,V^{-1}$
magnetic flux	weber	Wb	$kg\,m^2\,s^{-2}\,A^{-1} = Vs$
inductance	henry	H	$kg\,m^2\,s^{-2}\,A^{-2} = V\,A^{-1}\,s$
magnetic flux density (magnetic induction)	tesla	T	$kg\,s^{-2}\,A^{-1} = V\,s\,m^{-2}$
radioactivity	becquerel	Bq	s^{-1}
radiation absorbed dose	gray	Gy	$J\,kg^{-1}$
radiation absorbed dose equivalent	sievert	Sv	$J\,kg^{-1}$ in tissue

1.3.3 *Fractions and multiples of ten*

The SI convention defines prefixes for specifying fractions and multiples of ten of the base units. These prefixes and symbols are listed in Table 1.4. When applied to the kilogram, these prefixes are used as if the base unit were the gram (that is 10^{-6} kg = milligram, not microkilogram).

1.3.4 *Derived units*

Having established a set of base units, the SI system then defines a set of dimensionally compatible derived units. These units are listed in Table 1.5. Thus the SI unit of pressure is the pascal. Note that the unit itself is written with a lowercase p (even though it might otherwise be assumed to represent a proper name) whereas the symbol is written in uppercase: Pa.

1.3.5 *Other SI units*

Due to popular usage or the lack of a suitable SI alternative a few 'rogue' units are tolerated under the SI system. These are listed in Tables 1.6 and 1.7. Table 1.6 lists two units that effectively represent 'nicknames' of legitimate SI units, the litre (instead of the cubic decimetre) and the tonne (instead of the megagram). Their use has been continued under the SI system due to their widespread adoption in science and commerce. In Table 1.7, some non-SI units, which nevertheless continue to be used under the SI system, are listed. These units include the hour and day and geometric degree, minute and second. It is interesting to note from the point of view of measuring the half-life of long-lived radioactive elements or the age of rocks, neither the month nor the year appear to be exactly defined under the SI system. Both the atomic mass unit and the electron volt continue to be used in scientific circles due to the lack of a suitable alternative. Neither is exactly defined under the SI system, since both are physical quantities, the measurement of which involves finite error.

All the units defined above are suitable for modern scientific usage and, with the exceptions mentioned, are fully coherent. The SI symbol is internationally accepted for all

Table 1.6 Common names for SI units accepted for use under the SI system

Physical quantity	Common name of unit	Symbol	SI name of unit	SI symbol
volume	litre	L	cubic decimetre	dm^3
mass	tonne	t	megagram	Mg

Table 1.7 Non-SI units accepted for use

Physical quantity	Name of unit	Symbol	SI definition of unit
time	minute	min	1 min = 60 s
	hour	h	1 h = 3600 s
	day	d	1 d = 86 400 s
geometric angle	degree	°	$1° = (\pi/180)$ rad
	minute	′	$1' = (\pi/10\,800)$ rad
	second	″	$1'' = (\pi/648\,000)$ rad
mass	atomic mass unit	amu	1/12 mass of a ^{12}C atom 1 amu ≈ 1.6605×10^{-27} kg
energy	electron volt	eV	1 eV ≈ 1.6022×10^{-19} J

scientific publications although some language variations exist in the spelling of the corresponding SI name. SI units may be used flexibly and usage is not strictly limited to dimensions specified in terms of base units. Thus density measurements quoted in traditional units of g cm^{-3} are acceptable under the SI system as an alternative to the more formal kg m^{-3}.

1.3.6 *Units not compatible with the SI system*

Several traditional units are still in widespread scientific use. All are unnecessary, since perfectly acceptable SI alternatives exist: the continued use of these units is therefore discouraged and is likely to be eventually outlawed in scientific publications. Units in this category are listed in Table 1.8. They can be divided into two categories: units which are exact multiples of 10 of equivalent units under the SI system, and secondly, units where conversion involves multiplication by some other factor. For the more important units in the latter category, the exact conversion factor has been agreed under the SI system and is listed in Table 1.8. It can be seen that one or two of these non-compatible units have very persistent scientific traditions, for example, the use of the barn to quote nuclear reaction cross-sections, the torr to measure vacuum pressure and the degree Celsius (or even worse, the degree Centigrade) to measure temperature. All are, however, doomed: use of the femtometre^2, pascal and kelvin will eventually prevail! Although discouraged, their use is perpetuated in this text where it reflects current scientific practice.

1.3.7 *Physical constants*

Many physical phenomena are used to make analytical measurements. In the mathematical treatment of these phenomena, various physical constants are used. The more important of these, expressed in SI units, are listed in Table 1.9.

Table 1.8 Non-SI units, use of which is discouraged

	Decimal fractions of SI units			Non-SI units exactly defined		
Physical quantity	Name of unit	Symbol	Definition	Name of unit	Symbol	Definition
length	angstrom	Å	10^{-10}m	inch	in	2.54×10^{-2}m
	micron	μ	10^{-6}m = 1 μm			
mass				pound	lb	0.453 592 37 kg
area	barn	b	10^{-28} m^2 = 100 fm^2			
force	dyne	dyn	10^{-5}N			
pressure	bar	bar	10^5Pa	atmosphere	atm	101 325Pa
				millimetre of mercury	mm Hg	13.5951 × 9.806 65Pa
				torr	Torr	(101 325/760)Pa
energy	erg	erg	10^{-7}J	kilowatt hour	kWh	3.6×10^6J
				thermochemical calorie	cal_{th}	4.184J
kinematic viscosity	stokes	St	10^{-4}m^{2}s^{-1}			
dynamic viscosity	poise	P	10^{-1}Pa s			
magnetic flux	maxwell	Mx	10^{-8}Wb			
magnetic flux density (magnetic induction)	gauss	G, Gs	10^{-4}T			
activity				curie	Ci	3.7×10^{10}s^{-1}
radioactive exposure				rontgen	R	2.58×10^{-4}Ckg^{-1} (in air)
radiation absorbed dose	rad	rad	10^{-2}Jkg^{-1} = 10^{-2}Gy			
radiation absorbed dose equivalent	rem	rem	10^{-2}J kg^{-1} = 10^{-2}Sv			
thermodynamic temperature interval	Celsius	°C	$T - 273.15$K			

By international agreement, conversion factors given in this table are exact. Data from McGlashan (1971).

Table 1.9 Recommended values of selected physical constants.

Name of physical constant	Symbol	Value
speed of light in vacuum	c	$2.997\,925\,0 \times 10^{8}\,\mathrm{m\,s^{-1}}$
unified atomic mass constant	$m_u = m_a(^{12}C)/12$	$1.660\,531 \times 10^{-27}\,\mathrm{kg}$
mass of a proton	m_p	$1.672\,614 \times 10^{-27}\,\mathrm{kg}$
mass of a neutron	m_n	$1.674\,920 \times 10^{-27}\,\mathrm{kg}$
mass of an electron	m_e	$9.109\,558 \times 10^{-31}\,\mathrm{kg}$
charge on an electron	e	$1.602\,191\,7 \times 10^{-19}\,\mathrm{C}$
electron radius	$r_e = e^2/4\pi e_o m_e c^2$	$2.817\,939 \times 10^{-15}\,\mathrm{m}$
Boltzmann constant	k	$1.380\,622 \times 10^{-23}\,\mathrm{J\,K^{-1}}$
Planck constant	h	$6.626\,196 \times 10^{-34}\,\mathrm{J\,s}$
Bohr magneton	$m_B = eh/4\pi m_e$	$9.274\,096 \times 10^{-24}\,\mathrm{A\,m^2}$
Avogadro constant	L or N_A	$6.022 \times 10^{23}\,\mathrm{mol^{-1}}$
gas constant	$R = Lk$	$8.314\,34\,\mathrm{J\,K^{-1}mol^{-1}}$
Faraday constant	$F = Le$	$9.648\,670 \times 10^{4}\,\mathrm{C\,mol^{-1}}$

Data from McGlashan (1971).

1.3.8 *Molarity, normality and concentrations*

When specifying the concentration of reagent in solution for use in a chemical reaction, strict SI conventions are not followed, and we enter an area where the tradition of the chemical literature is not supported by SI ratification. The formal unit of concentration under the SI system is the mole per cubic metre (or the mole per kg for solid solutions). However, this convention is not followed in the chemical literature. Concentration of reagents is normally expressed as molarity or molar concentration. This is the moles of reagent dissolved in one litre of solution (concentrations in grams per litre are also sometimes used).

Thus a one molar (1M) solution of HCl contains a concentration equivalent to 36.461g of anhydrous HCl (i.e. one mole) dissolved in one litre of solution. By the same reasoning, a 0.5M solution of H_2SO_4 contains the equivalent of $0.5 \times 98.074 = 49.04$g of anhydrous H_2SO_4 in one litre of solution.

In some circumstances it is desirable to specify the concentration of a reagent that is equivalent to that of another reagent participating in a specified chemical reaction. The unit selected to describe this phenomenon is the *normality*. This unit has meaning only if the participating reaction is specified. Such reactions could include acid–base neutralization, oxidation–reduction, complex formation or precipitation reactions.

In a simple example of an acid-base neutralization,

$$HCl + NaOH \rightarrow H_2O + NaCl \qquad (1.1)$$

for this reaction to go to completion one equivalent of hydrochloric acid solution must be mixed with one equivalent of sodium hydroxide solution. The equivalent weight in the context of this acid–base reaction is the weight of reagent (in grams) that reacts with, or contributes one mole of hydrogen ions to, the reaction. A one normal solution of the reagent contains one equivalent weight of the reagent dissolved in one litre of solution.

Thus, in the above example, a one normal (1N) solution of HCl represents a concentration equal to one equivalent weight of HCl (i.e. 36.461g) in one litre of solution. For both these reagents (HCl and NaOH) the concentration of a 1M solution is exactly equivalent to that of a 1N solution. However this is not the case for the acid–base reaction

$$H_2SO_4 + 2NaOH \rightarrow Na_2SO_4 + 2H_2O \qquad (1.2)$$

Sulphuric acid, although not fully ionized in solution, can nevertheless contribute two hydrogen ions to this reaction. A 1N solution is therefore now identical to a 0.5 M solution of H_2SO_4 and contains $0.5 \times 98.074 = 49.04$g of anhydrous sulphuric acid per litre of solution.

Similar definitions of equivalent weight and normality exist for solutions used in oxidation–reduction and complex-formation reactions. In each case, the molarity of a reagent refers to its concentration expressed as the proportion of formula weight (in grams) per litre of solution and the normality refers to the equivalent weight per litre identified in terms of a specific reaction.

In trace element solution chemistry it is frequently necessary to prepare or specify solutions at low concentrations where mg, μm or ng of element per mL of solution are the appropriate units. In this context the unit 'part per million' (ppm) is both unnecessary and ambiguous as an alternative to the μg mL^{-1}, since it is often misinterpreted as representing one part by weight of analyte per 10^6 parts by volume (not weight) of solution. This would only be the case on the unlikely asumption that the solution density was exactly 1 kg m^{-3}. Use of the ppm is equally unnecessary, though still widely used, in reporting trace element concentrations as a fraction of sample weight. The SI equivalent μg/g is more correct. Major elements are normally reported as % by weight calculated as the oxide.

1.4 Statistics

Two views are taken of statistics. One is a political condemnation as lies, damned lies and statistics, and the other is a more positive definition in terms of the art of determining the probable from the possible. Whatever the point of view, there is no doubt that statistics have an essential, though sometimes abused, role to play in analytical chemistry.

1.4.1 *Errors in analytical data*

All analytical measurements are subject to error and these can be classified into three groups.

(*a*) *Indeterminate or random errors*. Such errors are an inherent part of the operation of all analytical instrumentation and cannot be eliminated, though it is sometimes possible to reduce their magnitude. Indeterminate errors arise from electronic noise, the random emission of x-ray or gamma-ray photons, flame flicker effects, etc. The magnitude of these errors can be determined by repetitive measurement on the same sample. The principal use of statistical procedures is to predict the 'true' value from a group of individual measurements each containing random errors.

(*b*) *Determinate errors or systematic bias*. One of the more pragmatic laws of analytical chemistry is that all analytical techniques are subject to determinate error and that all analysts are subject to systematic bias. Often such bias remains unsuspected and undetected and is the major cause of interlaboratory discrepancies. Systematic bias can arise from instrumental drift effects, uncertainties in matrix correction procedures, or instrument misalignment. Questionable personal judgments, such as an inappropriate choice of reference materials to set up a calibration, or doubtful choice of operating conditions, also result in systematic bias. Personal bias also has the effect of influencing results so that they come closer to a preconceived notion of the true value for a measurement. The only guard against bias is a careful and systematic study of the performance of a technique in the analysis of as many reference materials as are available. Given the uncertainties in 'recommended' compositions of such materials (see section 1.10.7), this remains the ultimate challenge facing the analyst. Systematic bias does have one prominent feature—it normally affects results in one direction so that all data are either too high or too low.

(*c*) *Gross errors.* A third class of error affects analytical results. These are much less insidious than systematic bias, are usually readily detected, and lead to the abandonment of the analysis. Such errors arise from serious malfunction of equipment, mislabelling of samples or failure of the operator to press the correct buttons in the right order (i.e. finger trouble). Usually such errors are temporary and have a readily identifiable source. Errors in this category can be avoided by self-discipline, a systematic scheme of work with cross-checking, and sometimes by automation so that repetitive operations are carried out under computer control.

In these categories, statistics can only be used to predict the true value from sets of data subject to random error. Statistics can be used to test for the presence of, but not to correct for, the effects of systematic bias.

1.4.2 *Accuracy and precision*

The distinction between accuracy and precision often causes some confusion to new users of analytical techniques. The difference can be seen from data in Figure 1.2.

Precision is a measure of the analytical repeatability. A precise analysis is one where a set of replicate analyses forms a tight cluster about the average. The degree of precision is normally measured by the standard deviation of the analyses (see below), a measure of width of the data distribution peak.

Accuracy is a measure of how close the analysed data lie to the 'true' composition of the sample. One of the difficulties in silicate rock analysis is that the true composition even in reference material is sometimes poorly known. To charaterize the difference between accuracy and precision, data in Figure 1.2 show the hypothetical determination of cobalt by two techniques. The neutron activation data in Figure 1.2*a* are of high precision, cobalt being efficiently excited by INAA. Conversely, the x-ray fluorescence data are of rather poor precision (this particular element is usually inefficiently excited by XRF and suffers some spectrum interferences). However, neither set of data in Figure 1.2*a* can be considered to be accurate since each set clusters about an average which does not centre on the 'true' value.

Data obtained by INAA in Figure 1.2*b* can be considered to be both precise and accurate. The equivalent XRF data could not be considered to be accurate since the precision is

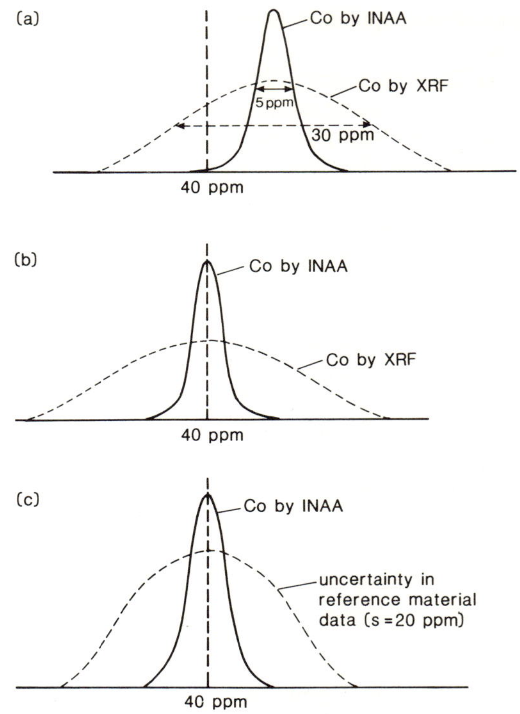

Figure 1.2 Accuracy and precision in the analysis of reference materials. Schematic data are shown for the repetetive analysis of cobalt by neutron activation analysis (one-sigma $s = 5$ ppm) and x-ray fluorescence (one-sigma $s = 30$ ppm) and the relationship of these results to the 'true' value 40 ppm (see text for details).

so low. However, analysts should be aware that the 'true' value is often a consensus estimate from quite disparate data. The uncertainty in consensus averages obtained from such data often significantly exceeds the precision attainable in modern instrumental techniques, particularly in the analysis of trace elements. In the light of this (Figure 1.2*c*) it is unwise to claim accuracy from the analysis of individual reference materials alone, an aspect considered further in section 1.10.

1.4.3 *Definitions in statistics*

Statistics are used to model the behaviour of sets of data. There are three distribution models encountered in analytical chemistry: the binomial distribution, the normal or gaussian distribution and the poisson distribution. Before considering these in turn it is necessary to define the statistical terms used in their evaluation.

A *frequency distribution* is effectively a histogram plotting the value of a datum (analysis) against its frequency of occurrence. Individual data points contributing to a distribution are usually denoted by the symbol x_1. Two cases are normally differentiated: an ideal distribution composed of an infinite number of measurements, called the *population distribution*, and that derived from a finite number of measurements, called the *sample distribution*.

The *population mean* (μ) is the arithmetic mean of the population distribution (infinite number of measurements).

The *sample mean* ($\bar{x}$) is the arithmetic mean of a finite number of measurements (n). The sample mean is thus an estimate of the population mean which is itself usually an unknown.

The *median* is the result which is midway in the frequency distribution and thus splits the distribution into two equal sets of data, one half being numerically smaller, the other half being numerically larger than the median.

The *mode* is an estimate of the most frequent result, there being several models by which modes are calculated.

For symmetrical distributions, mean, median and mode are indistinguishable. However, for skewed distributions, each of these values becomes distinct (Figure 1.3). The analysis of skewed distributions arising from interlaboratory cooperative analyses is discussed in section 1.10.5.

The *standard deviation* is an estimate of the width of the distribution and is calculated as the square root of the mean standard deviation. The standard deviation for a population distribution (σ) (that is for a large or infinite number of measurements) is given by

$$\sigma = \sqrt{\frac{\sum_{i=1}^{n} (\bar{x}_i - \mu)^2}{n}} \qquad (1.3)$$

where x_i is the ith data point, μ the population mean and n the number of data points. For a small number of points ($n < 20$) the sample standard deviation (s) is an estimate of the population standard deviation and is given by

$$s = \sqrt{\frac{\sum_{i=1}^{n} (\bar{x}_i - \bar{x})^2}{n-1}} = \sqrt{\frac{\left(\sum_{i=1}^{n} \bar{x}_i^2\right) - n\bar{x}^2}{n-1}} \qquad (1.4)$$

Substitution of $(n-1)$ (the *number of degrees of freedom*) for n gives a better estimate of σ for small data sets ($n < 20$).

The *coefficient of variation* (CV) is the relative standard deviation expressed as a percent of the mean and is calculated from

$$\text{CV} = 100 \,.\, \frac{s}{\bar{x}} (\%). \qquad (1.5)$$

The *variance* is also a measure of the width of a distribution and is equal to s^2. The variance is not often quoted as a measure of the precision of analytical results since the alternative (s) is in the same units as the mean ($\bar{x}$). However, there is one advantage from the use of variances. That is, if the error in an analysis originates from three independent (and random) sources, for example, the combined error is given by the sum of the variances.

$$s^2 = s_A^2 + s_B^2 + s_C^2. \qquad (1.6)$$

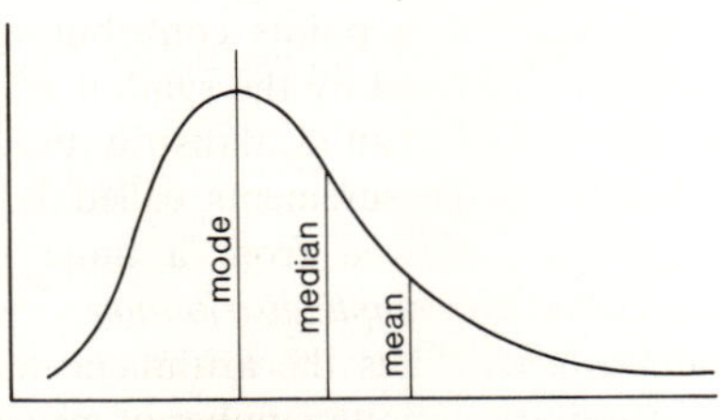

Figure 1.3 Relationship between mode, median and mean for a skewed data distribution set. (Lister, 1982.)

1.4.4 *Propagation of errors—absolute errors and standard deviations*

In mathematical calculations involving data, each of which has a known error, the combined error in the result of the calculation can be found as follows.

(*a*) *Operations involving addition and subtraction.* The error in the sum or difference is calculated from the *absolute errors* in the data to be added or subtracted. For example if Δa and Δb are the absolute errors in data values a and b, the absolute error Δc in the sum or difference is given as follows:

sum: $$c = a + b \qquad (1.7)$$

absolute error in sum: $\Delta c = \Delta a + \Delta b$ (note the + sign) (1.8)

difference: $$c = a - b \qquad (1.9)$$

absolute error in difference:

$$\Delta c = \Delta a - \Delta b \quad \text{(note the } - \text{ sign)} \qquad (1.10)$$

where the absolute errors Δa or Δb can take positive or negative values as appropriate.

(*b*) *Operations involving multiplication or division.* The error in the result of a multiplication or division is calculated from the *relative errors* of data used in the calculation. Thus:

multiplication: $$c = a \times b \qquad (1.11)$$

relative error in product:

$$\frac{\Delta c}{c} = \frac{\Delta a}{a} + \frac{\Delta b}{b} \quad \text{(note the + sign)} \qquad (1.12)$$

division: $$c = a/b \qquad (1.13)$$

relative error in quotient

$$\frac{\Delta c}{c} = \frac{\Delta a}{a} - \frac{\Delta b}{b} \quad \text{(note the } - \text{ sign)} \qquad (1.14)$$

where again the values of the absolute errors can take positive or negative values as appropriate.

For a fuller treatment of error propagation see, for example, Skoog and West (1982) pp.73–80.

Standard deviations propagate through mathematical calculations in a slightly different way from absolute errors.

(*c*) *Addition and subtraction.* If the standard deviation in a is s_a and in b is s_b, then the error in the sum or difference (s_c) can be found by summing variances, thus:

addition: $$c = a(\pm s_a) + b(\pm s_b) \qquad (1.15)$$

variance: $$s_c^2 = s_a^2 + s_b^2 \qquad (1.16)$$

subtraction: $$c = a(\pm s_a) - b(\pm s_b) \qquad (1.17)$$

variance: $$s_c^2 = s_a^2 + s_b^2. \qquad (1.18)$$

(*d*) *Multiplication or division.* The error in the product or quotient is calculated by summing the relative variances as follows:

multiplication: $$c = a \times b \qquad (1.19)$$

relative variance:

$$\left(\frac{s_c}{c}\right)^2 = \left(\frac{s_a}{a}\right)^2 + \left(\frac{s_b}{b}\right)^2 \qquad (1.20)$$

division: $c = a/b$ (1.21)

relative variance:

$$\left(\frac{s_c}{c}\right)^2 = \left(\frac{s_a}{a}\right)^2 + \left(\frac{s_b}{b}\right)^2. \quad (1.22)$$

1.4.5 *Frequency distributions*

When a large number of measurements is made of some physical phenomenon, individual data will be found to be distributed about a mean value in a characteristic way. A histogram of the value of the measurement versus the frequency of occurrence constitutes a frequency distribution, and the mathematical equation that describes the phenomenon is the distribution function. The three frequency distributions most commonly encountered in analytical chemistry are as follows.

(*a*) *Binomial distribution.* The binomial distribution is usually exemplified by flipping a coin. The result can have only two outcomes, heads or tails, each of which has equal probability. The probability of either outcome remains the same throughout the experiment. The binomial distribution therefore describes the probability that in a finite number of trials (coin flips), there will be a specified number of heads (successes) or a specified number of tails (failures).

The general form of the binomial distribution is

$$P(r) = \frac{n!}{r!(n-r)!} \cdot p^r.(1-p)^{n-r} \quad (1.23)$$

where n is the number of trials, r is the number of successes, p is the probability of success in any one trial, $n!$ is 'factorial' n $(= n.n-1.n-2\ldots1)$ and the entire function $P(r)$ measures the probability of getting exactly r successes in n trials.

A binomial distribution is plotted in Figure 1.4 for $n = 10$ trials and p (the probability of success) = 0.2. As expected, this distribution shows a maximum probability at two successes in ten and that, for the set parameters, the probability of 7, 8, 9 or 10 successes out of ten trials becomes vanishingly small.

(*b*) *Poisson distribution.* Like the binomial distribution, the poisson distribution is also a discrete frequency distribution (i.e., it is applicable to a finite number of measurements). The poisson distribution is used to describe the number of events in a given time interval providing those events occur randomly with time. The main criteria are that there should be a large number of occasions during which an event could occur and that on any individual occasion the probability that the event will occur is small but constant. For constant observation periods, the average rate of occurrence is thus the same, and the poisson distribution then gives the probability that during a specified period of time the event will occur a given number of times.

The form of the poisson distribution is as follows:

$$P(x) = \frac{\bar{x}^x.\mathrm{e}^{-\bar{x}}}{x!} \quad (1.24)$$

where x is the mean value of that variable, e is the base of natural logs = 2.7183 and $P(x)$ is the probability that a variable will take the value x.

One of the more colourful illustrations of poisson distributions is in predicting the probability of a cavalryman being killed by a horse kick during the Crimean War. The number of opportunities when this *could* take place is large (there being many cavalrymen and horses involved in the Crimean War). However, the probability of occurrence is very small. In analytical applications, poisson distributions are used to describe the rates of x-ray emission in XRF or microprobe analysis or of gamma-ray emission in neutron activation or gamma spectrometry techniques. In these applications, the poisson distribution (like the gaussian distribution to be described next) has one important property. The standard deviation of the distribution is closely estimated by the square root of the mean value of the variable, provided its value is sufficiently large. Thus if a total of x counts is accumulated in a given x-ray measurement, the standard deviation predicted by poisson counting statistics is $\sqrt{x}$ and the relative standard deviation (coefficient of variation) is given by

$$\mathrm{CV} = \frac{100\sqrt{x}}{x} = \frac{100}{\sqrt{x}} \quad (1.25)$$

The shape of the poisson distribution changes with the rate of occurrence. For small mean values of the variable ($x < 1$) the distribution is strongly asymmetric. For larger values ($x > 10$), the distribution becomes almost symmetric and approximates to gaussian in shape (Figure 1.5).

(*c*) *Gaussian or normal distribution.* Unlike the two distributions described above, the gaussian or normal distribution is a continuous (not discrete) distribution function. Many measurements of natural phenomena, when repeated a large number of times, are found to be distributed in a characteristic way. Values cluster around a central point and the frequency of measurements tails off exponentially on either side (Figure 1.6).

The equation of the gaussian distribution function is

$$f(x) = \frac{1}{\sigma\sqrt{2\pi}} \cdot \exp\left(\frac{-(\bar{x}-\mu)^2}{2\sigma^2}\right) \quad (1.26)$$

where μ is the population mean (i.e. the mean value derived from an infinite number of measurements, σ is the population standard deviation, x is the value of an individual measurement and $f(x)$ represents the frequency of occurrence of each value of $(x-\mu)$.

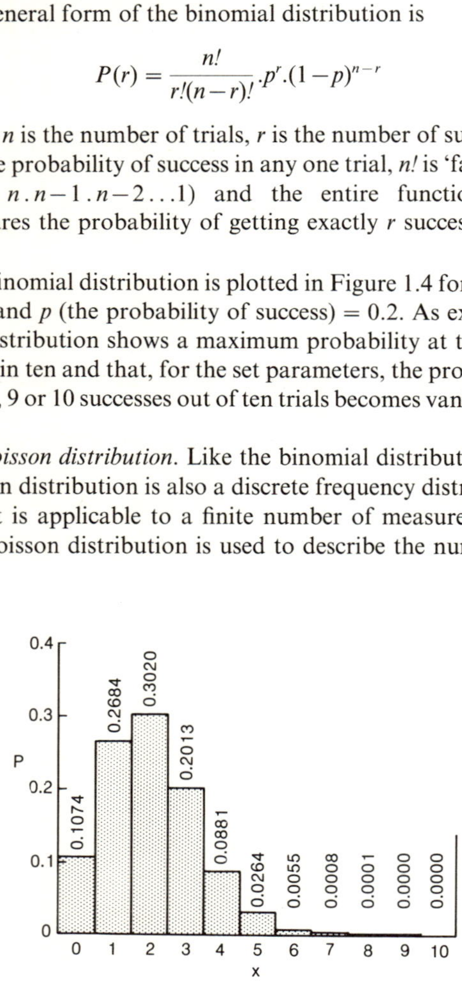

Figure 1.4 Binomial distribution for $n = 10$ trials where the probability (p) of success in any one trial is 0.2. (Zemany, 1978.)

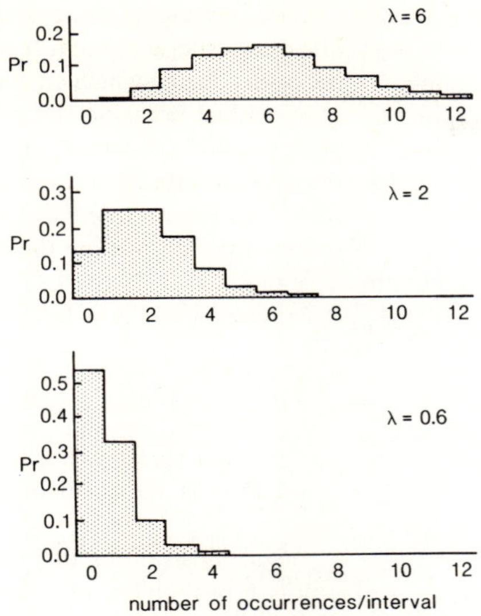

Figure 1.5 Poisson distribution: the shape of the distribution changes with the rate of occurrence, being highly unsymmetrical at a rate of occurrence $\lambda = 0.6$ and almost gaussian in shape at a rate of occurrence $\lambda = 6$, for example. (Davis, 1973).

This function extends from $-\infty$ to $+\infty$ and is symmetrical about the value $x = \mu$. Usually the values μ and σ are unknown. Estimates of these parameters from finite data sets are denoted by $\bar{x}$ and s, respectively.

From a practical viewpoint, one of the useful properties of the gaussian distribution is that it is possible to calculate the probability of an individual measurement lying within a specified range about the mean, as follows:

68.27% of measurements lie within the range $\mu \pm 1\sigma$

95.44% of measurements lie within the range $\mu \pm 2\sigma$

99.74% of measurements lie within the range $\mu \pm 3\sigma$

or, conversely:

50% of measurements lie within the range $\mu \pm 0.676\sigma$

Table 1.10 Correction factors that must be applied to standard deviation values calculated according to Equation 1.4 for small data sets ($n = 2, 3, 4$)

Number of measurements	Correction factor
2	1.253
3	1.128
4	1.085

90% of measurements lie within the range $\mu \pm 1.645\sigma$

95% of measurements lie within the range $\mu \pm 1.960\sigma$

99% of measurements lie within the range $\mu \pm 2.576\sigma$

99.9% of measurements lie within the range $\mu \pm 3.290\sigma$

1.4.6 *Standard deviations of small sets of data*

For small data sets (that is, less than about 20 values) standard deviations are normally calculated with the number of degrees of freedom $(n-1)$ substituted for the number of measurements (n) to give a better estimate of the population standard deviation (σ)—1 see equations (1.3) and (1.4). However, for very small data sets (less than 5) even this approximation gives a biased estimate of σ, as shown in Table 1.10.

An alternative approach to estimate the standard deviation from small data sets is to estimate s from the spread of results w (that is, the difference between the highest and lowest datum). The standard deviation is then estimated from

$$s = \frac{w}{k} \tag{1.27}$$

where k is a factor dependent on the number of observations, as shown in Table 1.11.

It must be emphasized that these conventions can only be

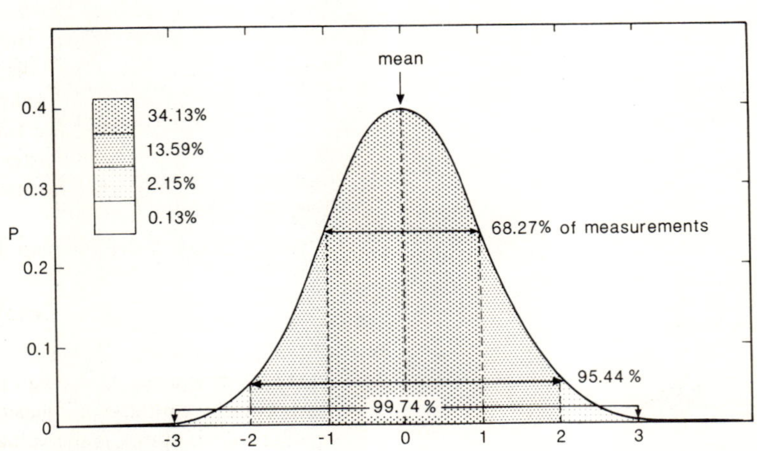

Figure 1.6 The gaussian distribution function. The proportion of data lying within the range ± one sigma, ± 2 sigma and ± 3 sigma of the mean is indicated. (Zemany, 1978.)

Table 1.11. Values of the factor k used to estimate standard deviations from the difference between maximum and minimum values in a set according to equation (1.27)

n	2	3	4	5	6	7	8
k	1.128	1.693	2.059	2.326	2.534	2.704	2.847
n	9	10	11	13	14	15	16
k	2.970	3.078	3.173	3.258	3.336	3.407	3.472

used to derive an approximate estimate of standard deviations from very small data sets.

1.4.7 *Tests on the reliability of statistical data*

A number of statistical tests can be carried out on sets of data to check their reliability. The most commonly used are Student's t test, the F-test and the chi-squared test, the most important features of which will now be discussed briefly.

(*a*) *Student's t test.* This test was devised by W. S. Gosset who published under the pen-name 'Student'. When small data sets are used to calculate a mean $\bar{x}$, there is uncertainty as to how good an estimate the value of $\bar{x}$ is to the population mean μ. The Student's t test can be used in two ways. The first is to test whether one set of data (of mean $\bar{x}$) belongs to another population (of mean μ) to within a specified confidence limit. The value of Student's t is calculated from

$$\pm t = (\bar{x}-\mu)\,.\frac{\sqrt{n}}{s} \tag{1.28}$$

For a population whose mean has a value of $\bar{x}$, to have a high probability of belonging to the population whose mean value is μ, the value of t calculated according to equation (1.28) must be smaller than the appropriate value listed in Table 1.12. This value of t must be selected for the appropriate number of degrees of freedom ($n-1$), and the desired confidence level. Thus, for example, if the value of t calculated from the mean ($\bar{x}$) and standard deviation (s) of 10 atomic absorption determinations of an element as compared with the expected element composition (μ) of a reference material according to equation (1.28) is less than 2.262 (Table 1.12), then at a confidence level of 95% there is no statistical difference between the AA results ($\bar{x}$) and the reference material data mean (μ).

Table 1.12 Student's t-distribution

Degrees of freedom	1	2	3	4	5	6	7	8	9
95% confidence	12.706	4.303	3.182	2.776	2.571	2.447	2.365	2.306	2.262
99% confidence	63.657	9.925	5.841	4.604	4.032	3.707	3.499	3.355	3.250

Degrees of freedom	10	11	12	13	14	15	20	30	∞
95% confidence	2.228	2.201	2.179	2.160	2.145	2.131	2.086	2.042	1.960
99% confidence	3.169	3.106	3.055	3.012	2.977	2.947	2.845	2.750	2.576

Data from Hayslett (1981).

The Student's t test can be used in a second manner: to calculate the error limits about a data-set mean ($\bar{x}$) within which the population mean (μ) must lie for a given confidence level.

$$\mu = \bar{x} \pm \frac{ts}{\sqrt{n}} \tag{1.29}$$

where s is the standard deviation of the data set, n is the number of measurements, and t is now the appropriate value read from Table 1.12.

Values of Student's t are listed in Table 1.12 for 95% and 90% confidence limits as a function of the number of degrees of freedom ($n-1$, where n is the number of observations (measurements)). The Student's t test is not restricted to 90% or 95% confidence limits, although these are the most commonly used. Tables are available with Student's t limits calculated for alternative confidence levels (see for example, Lister, 1982).

(*b*) *The F-test.* The F-test is a procedure whereby two independent sets of sample variances can be tested to see if they belong to the same population at a specified confidence level. One use of this test is in the analysis of reference samples. Analysts are supplied with two or more randomly selected bottles of the same reference material and asked to perform replicate analyses on each bottle. The F-test is used to predict, within specified confidence levels, if the variance of analyses from the same bottle is equivalent to the variance of analyses between bottles. If it is not and the between-bottle variance is greater than a specified limit, then inhomogeneity between different bottles of nominally identical samples is to be suspected.

In the example outlined above, the F ratio is computed from the measured variances s_1 (e.g. within-bottle variance) and s_2 (e.g. between-bottle variance) as follows:

$$F = \frac{s_1^2}{s_2^2} \tag{1.30}$$

where s_1 is the larger of the two values (i.e. $F > 1$).

The experimental F ratio is then compared with a calculated F ratio for a specified confidence level taken from tables. Some values of the F ratio calculated for a significance level of 5% are listed in Table 1.13 as a function of the number of degrees of freedom of both sets of measurements.

An example of the application of the F-test in checking the homogeneity of reference materials is given by Lister (1982) for the analysis of cobalt in an ore reference sample IGS-22. Samples from 10 bottles were analysed in duplicate. The variance (that is, s_2^2) between samples (obtained by comparing analyses of samples taken from one bottle with those from another) was 436.87 with 9 degrees of freedom. The variance within samples (calculated from duplicate analyses of samples taken from the same bottle) was 153.9 with 10 degrees of freedom. The experimental F ratio is therefore $436.87/153.9 = 2.84$. The critical value of F from Table 1.13, relevant for 9 and 10 degrees of freedom respectively, is 3.02. So the difference in variance within samples and between bottles could therefore be considered to be not significant and the sample to be chemically homogeneous in cobalt at the 95% confidence level. Details of the steps involved in the F-test calculation are given by Lister (1982).

Table 1.13 *F*-ratio for a significance level of 5%

n_1 \ n_2	1	2	3	4	5	6	7	8	9	10	20	30	∞
1	161.45	199.50	215.71	224.58	230.16	233.99	236.77	238.88	240.54	241.88	248.01	250.09	254.32
2	18.513	19.000	19.164	19.247	19.296	19.330	19.353	19.371	19.385	19.396	19.446	19.462	19.496
3	10.128	9.5521	9.2766	9.1172	9.0135	8.9406	8.8868	8.8452	8.8123	8.7855	8.6602	8.6166	8.5265
4	7.7086	6.9443	6.5914	6.3883	6.2560	6.1631	6.0942	6.0410	5.9988	5.9644	5.8025	5.7459	5.6281
5	6.6079	5.7861	5.4095	5.1922	5.0503	4.9503	4.8759	4.8183	4.7725	4.7351	4.5581	4.4957	4.3650
6	5.9874	5.1433	4.7571	4.5337	4.3874	4.2839	4.2066	4.1468	4.0990	4.0600	3.8742	3.8082	3.6688
7	5.5914	4.7374	4.3468	4.1203	3.9715	3.8660	3.7870	3.7257	3.6767	3.6365	3.4445	3.3758	3.2298
8	5.3177	4.4590	4.0662	3.8378	3.6875	3.5806	3.5005	3.4381	3.3881	3.3472	3.1503	3.0794	2.9276
9	5.1174	4.2565	3.8626	3.6331	3.4817	3.3738	3.2927	3.2296	3.1789	3.1373	2.9365	2.8637	2.7067
10	4.9646	4.1028	3.7083	3.4780	3.3258	3.2172	3.1355	3.0717	3.0204	2.9782	2.7740	2.6996	2.5379
20	4.3513	3.4928	3.0984	2.8661	2.7109	2.5990	2.5140	2.4471	2.3928	2.3479	2.1242	2.0391	1.8432
30	4.1709	3.3158	2.9223	2.6896	2.5336	2.4205	2.3343	2.2662	2.2107	2.1646	1.9317	1.8409	1.6223
∞	3.8415	2.9957	2.6049	2.3719	2.2141	2.0986	2.0096	1.9384	1.8799	1.8307	1.5705	1.4591	1.0000

n_1, n_2 = number of degrees of freedom.
Data from Hayslett (1981).

(*c*) *Chi-squared test.* The chi-squared test is a procedure that can be used to see if the degree of statistical fluctuation within a data set is compatible with statistical predictions. The test can be applied to any statistical model. The chi-squared test therefore answers questions such as: to what level of confidence does the variation in a particular data set follow that predicted by a gaussian distribution?

The chi-squared function is calculated from:

$$\chi^2 = \sum_{i=1}^{n} \frac{(O_i - E_i)^2}{E_i} \tag{1.31}$$

where O_i is the *i*th observed frequency (experimental datum) and where E_i is the expected frequency calculated from a specified model (for example a gaussian or poisson distribution).

The resultant value of χ^2 is compared with values such as those given in Table 1.14, relevant to the appropriate number of degrees of freedom, in order to evaluate the significance of the correlation. For example, if the chi-squared value calculated from a set of results having 4 degrees of freedom were less than 9.488 (Table 1.14), the discrepancy of the measured data set from the specified model distribution would not be significant at the 5% level (that is, there is a 95% confidence that the data set does follow the specified theoretical distribution).

Table 1.14 χ^2 distribution

	Significance level							
n	0.995	0.99	0.975	0.95	0.05	0.025	0.01	0.005
1	0.000039	0.00016	0.00098	0.0039	3.841	5.024	6.635	7.879
2	0.100	0.0201	0.0506	0.103	5.991	7.378	9.210	10.579
3	0.0072	0.115	0.216	0.352	7.815	9.348	11.345	12.838
4	0.207	0.297	0.484	0.711	9.488	11.143	13.277	14.860
5	0.412	0.554	0.831	1.145	11.070	12.832	15.086	16.750
6	0.676	0.872	1.237	1.635	12.592	14.449	16.812	18.548
7	0.989	1.239	1.690	2.167	14.067	16.013	18.475	20.278
8	1.344	1.646	2.180	2.733	15.507	17.535	20.090	21.955
9	1.735	2.088	2.700	3.325	16.919	19.023	21.666	23.589
10	2.156	2.558	3.247	3.940	18.307	20.483	23.209	25.188
15	4.601	5.229	6.262	7.261	24.996	27.488	30.578	32.801
20	7.434	8.260	9.591	10.851	31.410	34.170	37.566	39.997
30	13.787	14.953	16.791	18.493	43.773	46.979	50.892	53.672

n = number of degrees of freedom.
Data from Hayslett (1981).

1.4.8 *Calibration lines: least squares regression analysis*

Almost all analytical techniques are calibrated by measuring the signal obtained from a set of calibration standards. The assumption is normally then made that the measured signal is a linear function of the concentration of the appropriate element in the 'standard'. If calibrations are to be set up over extended concentration ranges, a quadratic function is sometimes assumed.

Under ideal conditions, measured data would form an exact linear function of concentration. However, errors are always superimposed on real data and a statistical procedure called regression analysis based on least squares minimization is normally used to calculate the best fit calibration line. Several assumptions are implicit in this approach, namely:

(i) An exact linear (or if selected, quadratic) relationship exists between measured signal and concentration
(ii) The concentrations of the standards are known exactly
(iii) The measured signal only is subject to error.

The calibration is therefore a plot of an independent variable (x), the concentration, and a dependent variable (y), the analytical signal. The linear least squares minimization technique is a method for calculating the best fit to data, for example, of the function

$$y = a + bx \tag{1.32}$$

based on minimizing the sum of the squares of the differences

$$\sum_{i=1}^{n} (\hat{y}_i - y_i)^2 \tag{1.33}$$

where $\hat{y}_i$ is the estimate of the analytical signal y_i derived for a specified concentration x_i according to equation (1.32) in the *i*th calibration standard.

A fuller statistical treatment of the least squares minimization procedure is given for example by Davis (1973), pp.192–221.

Various indicators of the goodness of fit are available to assess calibration data. One is to calculate the percentage difference between individual measurements and the values predicted by the calibration line. Another is the *correlation coefficient* (R) calculated as follows.

Correlation coefficient,

$$R = \sqrt{\left[\frac{\text{total sum of squares}}{\text{sum of squares due to regression}}\right]} \qquad (1.34)$$

where total sum of squares $= \sum_{i=1}^{n} (y_i - \bar{y})^2$, sum of squares due to regression $= \sum_{i=1}^{n} (\hat{y}_i - \bar{y})^2$ where $\bar{y}$ is the mean value of y and the other symbols are as defined above. For perfect fit, $R =$ unity. For a good calibration data set $R > 0.98$.

In routine analytical calibrations, numeric assessment of 'goodness of fit' rarely gives the analyst an adequate indication of the quality of a calibration data set. Experience indicates that it is *always* necessary to plot the calibration data graphically to obtain a satisfactory visual assessment. In the analysis of silicate rocks, the situation is often encountered where a single datum point can show poor agreement with the rest of the data set. Inclusion of such data can seriously bias the resultant calibration line and yet not unduly affect the magnitude of numeric indicators of calibration linearity. Such deviations are usually immediately apparent from a visual examination of a graphical plot of calibration data.

The linear regression procedure outlined above is widely used in all forms of analytical calibration. However, it cannot be used uncritically. One reason is outlined in the previous paragraph. Another reason is the weighting applied to individual points in data sets and arises from the assumption that the concentration of standards is known exactly. In the determination of trace elements by some techniques (XRF for example), international rock reference materials are often used to set up calibrations. The situation then arises that one of the calibration standards may contain significantly higher concentrations of an element than all the other samples in the calibration set. An extreme example is found in the use as a calibration standard for zirconium of lujavrite NIM-L which contains 14900 ppm ZrO_2. Most other silicate reference materials contain less than 1000 ppm Zr. Due to weighting inherent in the conventional least squares minimization procedure, experience has shown that the resultant calibration line will pass almost exactly through NIM-L data, even though measurements on this reference material may contain uncertainties that are as large as in the remaining data. In some circumstances, this can cause serious bias in the calibration line at lower concentrations (Figure 1.7). A similar situation arises in the calibration of

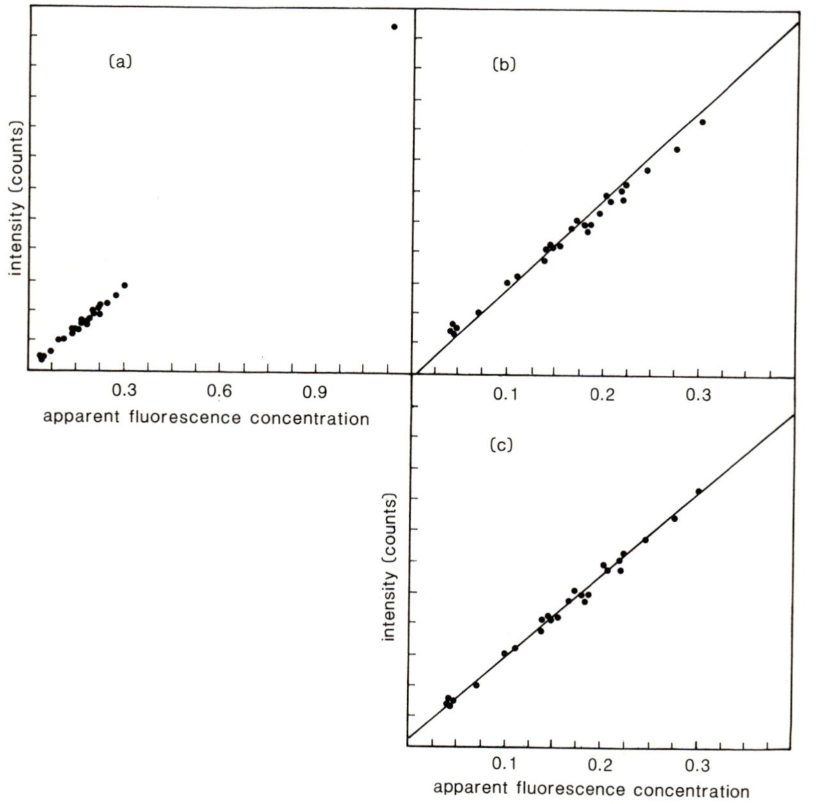

Figure 1.7 (*a*) Calibration data set for zirconium by energy dispersive XRF analysis. Data for 27 international rock reference samples are plotted as the intensity of the Zr x-ray emission signal *v.* the apparent fluorescence concentration (% oxide). (*b*) Linear regression of all data plotted in (*a*) showing lower section of calibration line. Due to the inherent weight placed on the highest datum (NIM-L = 1.49% ZrO_2) caused because it is present at over ten times the concentration of other values, some discrepancy is observed in fit to data at the lower end of the calibration. (*c*) Linear regression of the data set from which NIM-L has been removed. The balance of fit at the lower end of the calibration is significantly improved.

chromium. Inclusion of pyroxenite NIM-P (35100 ppm Cr_2O_3) in the data set will virtually guarantee that the resultant calibration will pass through this reference material data irrespective of the magnitude of relative errors. This is not to question the validity of the recommended analyses of these samples, rather to emphasize that the inclusion of calibration standards having more than 2–3 times the elemental abundance of the next highest sample in the calibration data set can, in some circumstances, cause an unexpected bias in calibration lines.

Thompson (1982) has reviewed some procedures to mitigate the influence of these effects. He compared the performance of three regression procedures: the simple linear regression, a weighted linear regression and a new method based on maximum likelihood estimation. In the weighted procedure, estimations of both x and y values were weighted by a factor inversely proportional to the variance of the y value. This produces a calibration line which passes closer to points with smaller variance. The maximum likelihood regression procedure takes account of variances independently in both x and y values of a calibration data set. Thompson concluded that the simple linear regression was satisfactory, providing:

(i) Ten or more sample are used to set up the calibration

(ii) These samples cover the concentration range from zero upwards *fairly uniformly*

(iii) The results from the method with the smaller variance are used as the independent variable (i.e. the x value).

If all these conditions could not be met, then the other regression procedures could show some advantage, though at the cost of requiring more data input.

Statistical procedures are particularly important in the assessment of isochron data for mass spectrometric determinations of Rb/Sr, Nd/Sm, U/Th/Pb and K/Ar isotope ratios. Such procedures are required so that an accurate assessment of errors can be made both in the slope of the isochron (related to the age of the sample suite) and in the intercept (related to the initial isotope ratio). Linear regression procedures, with particular relevance to mass spectrometric applications, have been discussed by York (1966, 1969) and Cumming *et al.* (1972).

1.4.9 *Quality assurance and secondary standards*

(*a*) *The analyst's viewpoint.* Several stages are involved in setting up and calibrating a new analytical technique. There is usually an intensive period of instrument optimization and 'debugging', including an investigation of interference effects. The procedures are then usually 'proved' by analysing rock reference materials to assess accuracy, precision and unsuspected interferences, etc. After this period of concentrated effort, the instrumentation is available for routine analysis of samples. The level of 'expert' supervision may then be relaxed considerably. It is essential, therefore, to have some means of monitoring continuously the quality of instrumental performance, both from batch to batch and from year to year. One of the most effective ways of doing this is to include one or more 'secondary standards' or 'quality assurance samples' within each analytical run. These materials are used to indicate whether the instrument is operating within a permitted error window. It is not essential that such sample should have well-characterized compositions before adoption on an instrument. It is important, however, that they are analysed as unknowns at the time the instrument calibration was investigated in detail. Their subsequent use is then to identify short-term and long-term calibration drift. Ideally, a range of secondary standards should be run to cover the concentration ranges of each element to be analysed. However, in practice, only one quality assurance sample is likely to be accommodated in a neutron activation analysis run, whereas several could be included in rapid techniques with low unit costs such as XRF and ICP–AES. The essential feature of the use of quality assurance samples is that the same sample is analysed as an unknown in every analytical run.

(*b*) *The user's viewpoint.* Circumstances often arise where a user visits several analytical installations or sub-contracts analyses to independent laboratories to obtain complete data for a project. In such situations, the user may find that he has no control over the accuracy or precision of the results obtained. When the same samples are analysed in more than one laboratory for the same range of elements, the results are likely to disagree. In collating such data sets, it is then desirable to be able to assess which is usable and which is suspect. The simple answer would be to include a rock reference material as an unknown in each set of samples submitted for analysis and to compare how individual laboratories perform. Bearing in mind the uncertainties in recommended compositions of rock reference materials, particularly in the case of the trace elements (see section 1.10.7), it is doubtful whether analysis of even a few reference materials is sufficient to assess the accuracy of a technique. It is suggested that a *minimum* of 10 'reliable' reference materials should be submitted for analysis as unknowns if a full assessment of accuracy is desired. If results are then plotted as a graph of expected composition *v.* analysed data, accuracy can readily be assessed. The most complete lists of silicate rock reference material data are those of Abbey (1983) and Govindaraju (1984). The results of such procedures can be ambiguous and should even be discouraged if an adequate number of the most reliable rock reference samples is not available.

For many users, such an elaborate assessment of accuracy may be impractical because of the expense or the time involved. If this is the case, it is still possible for the user to select one or two of his own samples, representative of the whole suite, and submit these as unknowns in each batch put through an instrument. This will allow the user to identify calibration drift from batch to batch over the entire time-span of the project and also to calculate a realistic analytical precision. Although there is a cost attached to such a procedure in terms of replicate analyses that must be undertaken, it is suggested that the value of the remaining analyses is enhanced due to the additional confidence that may be placed in these data.

1.5 Detection limits

The term 'detection limit' is perhaps one of the most widely abused in geochemical analysis. Not only is there a general lack of agreement in specifying the relative magnitude of detection-limit data, but no convention exists as to the name to be attached to such units.

Many users of analytical techniques regard the detection limit as a measure of the smallest concentration of an element that can be analysed. This is not always the case. Detection limits are often quoted in units representing the limit below which *qualitative* identification of analytical signal is no longer possible. This section attempts to resolve some of this confusion. The basic philosophy behind detection limit calculations is reviewed and a rationalized scheme of detection limit nomenclature is proposed. It must be emphasized that the terminology used here is presented to prevent ambiguity, but does not have international ratification.

Important reviews concerning detection limits include those of Long and Winefordner (1983), Bourmans (1978), Kaiser (1978), and Curry (1968). These show that detection limits are related to the confidence which may be placed on distinguishing a signal from the background measurement. If an infinite number of measurements of a background signal is made, then, providing errors are random, the set of measurements would obey a gaussian (normal) distribution characterized by a mean background signal μ_B and a standard deviation σ_B (Figure 1.8). Because real measurements are always limited to a finite number, these parameters are here denoted by $\bar{x}_B$ (an estimate of μ_B) and s_B (an estimate of σ_B).

One relevant property of a gaussian distribution is that the probability of any individual signal falling within the window $\bar{x}_B \pm 1s_B$ is 68.27%; $\bar{x}_B \pm 2s_B$ is 95.44% and $\bar{x}_B + s_B$ is 99.74%. Or, more appropriately for detection-limit calulations, the probability that a background signal will not *exceed* $\bar{x}_B \pm 1s_B$ is 84.14% (68.3 + .(100 − 68.3)/2); equivalent values for $\bar{x}_B + 2s_B$ and $\bar{x}_B + 3s_B$ are 97.72% and 99.87% respectively.

In practical analytical situations the standard deviation on the background is measured in various ways. In x-ray and gamma-ray methods it is usual to measure a background count rate or background area and then to estimate the standard deviation on the basis of ideal-counting statistical calculations. In optical emission methods a blank sample or solution is normally analysed ten times to estimate standard deviations. In practical situations it is unrealistic to assume that such measurements will follow an exact normal (gaussian) distribution. If this is the case, Long and Winefordner (1983) for example show that the confidence limit of $\bar{x}_B \pm 2s_B$ can fall from 95.44% (normal distribution) to 75% (non-normal distribution) and that of $\bar{x}_B \pm 3s_B$ from 99.74% (normal distribution) to 89% (non-normal distribution). On this basis, detection limits using $2s_B$ limits are not considered here to afford sufficient security of confidence for practical analytical measurements. The following detection-limit signals are therefore identified.

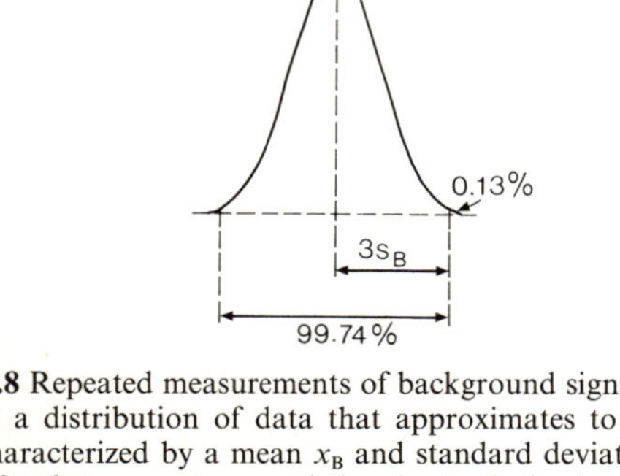

Figure 1.8 Repeated measurements of background signal (top) will result in a distribution of data that approximates to a gaussian shape characterized by a mean x_B and standard deviation s_B. The probability that a measurement belonging to the background population will exceed $x_B + 3s_B$ is 0.13% (gaussian statistics).

1.5.1 *Lower limit of detection* ($\bar{x}_{LLD}$)

Consider the signal lying at a level $3s_B$ above the mean background $\bar{x}$ (Figure 1.8). Based on normal distribution statistics, there is a 99.87% probability that this signal does *not* belong to the background distribution. This threshold is frequently taken as one measure of detection limit, normally called the *lower limit of detection*. It is important, however, to note that quantitative analysis is not possible at this signal level, and in fact this threshold represents the limit of qualitative analysis. This can be demonstrated as follows.

A signal ($\bar{x}_{LLD}$) having a mean value of $\bar{x}_B + 3s_B$ (Figure 1.9) will itself be characterized by a distribution. Assuming this distribution is normal, it will be characterized by a standard deviation s, which to all intents and purposes is identical to the standard deviation on the background distribution (s_B). From the form of such a distribution, 50% of all measurements of this signal will lie below the mean $\bar{x}_{LLD}$ and will be indistinguishable from the background population. The lower limit of detection therefore represents the threshold at which errors in a measured signal tend to infinity. It is important to emphasize that quantitative analysis is not possible at the concentration represented by the lower limit of detection.

1.5.2 *Limit of determination* ($\bar{x}_{LoD}$)

The question then arises: what is the smallest signal which can be quantitatively measured (as opposed to qualitatively

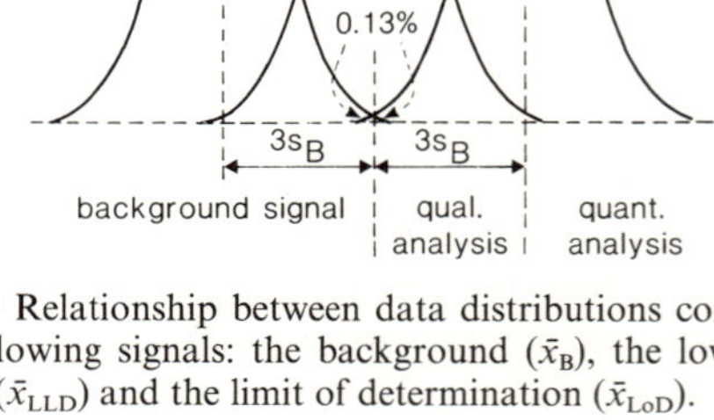

Figure 1.9 Relationship between data distributions corresponding to the following signals: the background ($\bar{x}_B$), the lower limit of detection ($\bar{x}_{LLD}$) and the limit of determination ($\bar{x}_{LoD}$).

recognized) above background. This signal, called here the 'limit of determination', is represented by a distribution having a mean lying $\bar{x}_B + 6s_B$ above the background. This distribution is selected because only the tail below $\bar{x}_{LoD} - 3s_B$ falls below the $\bar{x}_B + 3s_B$ background threshold (Figure 1.9). The limit of determination signal represents the smallest measure that can be confidently analysed (based on 3*s* error limits). The ideal error in measurements at this signal level is 16.7% relative (one sigma). The limit of determination represents the threshold below which measurements become increasingly qualitative. It is therefore suggested that this is the most appropriate unit to select for detection limit measurements since it is a direct estimate of the limit of quantitative analysis. Detection limit data in this work have accordingly been recalculated, where possible, to 'Limit of Determination' units (the six-sigma detection limit).

1.5.3 *Limit of quantitation* ($\bar{x}_{LoQ}$)

In many practical analytical situations, even the Limit of Determination can give a false measure of confidence as to the lowest signal that can be quantitatively measured. For this reason, one national body, the American Chemical Society Committee on Environmental Improvement (1980), has recommended a detection limit called the 'limit of quantitation' set at a level $10s_B$ above the mean background distribution. This unit was selected to ensure additional confidence in a unit designed to estimate the limit of quantitative analysis. The ideal error in measurements at this signal level is 10% relative (one sigma). It is suggested that the limit of quantitation is the most appropriate unit to use when there are legal, commercial or statutory implications placed on the interpretation of detection limit data.

1.5.4 *Summary of detection-limit signals*

The various threshold signals identified above are summarized as follows.

(i) $\bar{x}_B + 2s_B$; an over-optimistic estimate of the lower limit of detection favoured by some instrument manufacturers, use of which should be abandoned.

(ii) $\bar{x}_B + 3s_B$; the *lower limit of detection.* The limit to qualitative analysis; errors in determinations at this level are infinite.

(iii) $\bar{x}_B + 6s_B$; the *limit of determination.* The smallest signal that can be quantitatively measured. It is suggested that this unit is the most appropriate for adoption by geo-analysts to permit a realistic assessment of the lowest concentration levels that can be analysed.

(iv) $\bar{x}_B + 10s_B$; the *limit of quantitation.* This unit affords additional confidence as the smallest signal that can be quantitatively measured and is appropriate where there are legal, commercial or statutory implications placed on the interpretation of data.

(v) *Detection limit*; a term so abused that it has no specific meaning other than to refer in a general way to the performance of a technique towards measuring small signals (see also section 8.10).

1.5.5 *Interpretation of detection-limit data*

Apart from the variety of units that can be calculated for the estimation of detection limits, two further difficulties arise.

(*a*) *Matrix-dependence of detection limits.* The magnitude of the standard deviation on the background is, for most techniques, matrix-dependent. Detection limits for trace elements in distilled water by optical emission measurements will thus appear far more optimistic than data obtained from the same measurements on a rock solution containing high concentrations of calcium. Similarly, rare-earth detection limits obtained by instrumental neutron activation analysis would appear far worse if samples regularly had high cobalt contents. For these reasons it is important that detection-limit data should be accompanied by details of the matrix type and any relevant instrumental factors. It is at the discretion of the analyst to select a 'representative' sample for detection limit measurements. Given these variables, it is not sensible to place too much significance on differences in detection limits between two techniques unless the same units have been used for detection-limit calculation and these data differ by more than 100%.

(*b*) *Calibration-line effects.* A second factor influences whether ideal detection limits can be achieved in practice. This concerns the behaviour of the calibration line in the vicinity of the origin. It should be noted that the detection limit thresholds have been derived in terms of 'signals' (i.e. multiples of the standard deviation of a background measurement) rather than concentrations. Conversion of this detection limit signal to a concentration involves use of the analytical calibration function. Particular difficulties arise if the signal *v.* calibration line does not pass through the origin. Three situations can be envisaged.

(i) The calibration line passes through the origin (Figure 1.10*a*). The true detection limit is then indeed given by the concentration equivalent of the detection limit signal (e.g. $6s_B$) to which should be added the concentration uncertainty in the calibration function.

(ii) The calibration line gives a *negative signal intercept* (Figure 1.10*b*). Conversion of the detection limit signal will cause the equivalent concentration to be increased by an amount dependent on degree of offset from the origin of the calibration line.

(iii) The calibration line gives a positive signal intercept (Figure 1.10*c*). This situation could arise, for example, if measurements are affected by blank contamination levels. The measured background then comprises the true background signal plus a signal due to a contaminant. Standard-deviation measurements on this combined signal are likely to be significantly larger, leading to enhanced detection limits.

1.5.6 *Graphical representation of detection limit signals*

It is not often possible to view the actual signal from which detection limit data are calculated. To give some indication of the magnitude of the lower limit of detection and limit of determination signals these data have been artificially created in Figure 1.11. Spectra were synthesized on an ED-XRF system by recording a spectrum of Zr *Kα* for a short period (about 10s) and then adding tube continuum, scat-

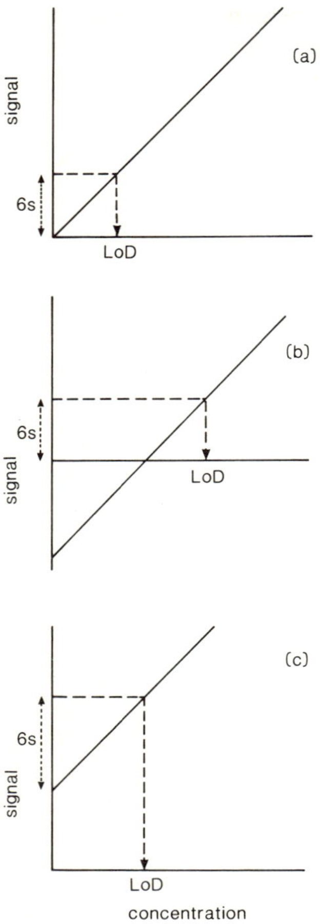

Figure 1.10 Conversion of limit of determination signal (six sigma) to concentration using calibration lines that (*a*) pass through the origin, (*b*) have a negative intercept with the signal axis, (*c*) have a positive intercept with the signal axis due, for example, to a significant blank component in the background signal.

tered off a graphite block. This caused the background spectrum under the Zr $K\alpha$ peak to increase progressively. Data acquisition was stopped when the magnitude of the background had reached the level equivalent to $3s_B$ and $6s_B$ signal levels respectively. Resultant spectra are shown in Figure 1.11 and give some indication of the relative magnitude of the signals from which detection limit data are calculated.

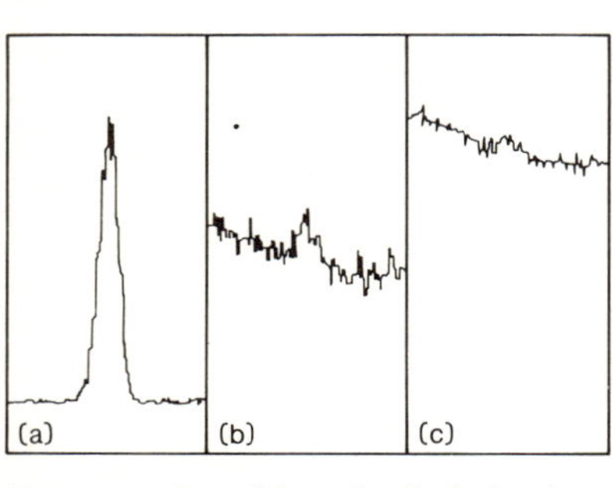

Figure 1.11 Representation of detection limit signals corresponding to (*b*) Limit of determination (six-sigma detection limit) and (*c*) Lower limit of detection (three-sigma detection limit) both constructed by adding the appropriate background component to a Zr $K\alpha$ signal excited by ED-XRF (*a*).

1.6 Sampling strategies: inhomogeneity effects

The collection and preparation of geological samples is an often overlooked aspect of silicate rock analysis. There are two aspects to the problem. The first is the strategy used to sample rocks in the field and to prepare a representative subsample suitable for chemical analysis. The second is, having submitted a ground sample for analysis, to identify the minimum mass on which physical measurements must be made to ensure that sample inhomogeneity effects are insignificant.

1.6.1 *Field sampling*

Field sampling is not an exact science. Although it is possible to devise sampling programs in the laboratory, modifications inevitably have to be made in their execution. Sampling programs may be dictated by the degree of exposure or logistical considerations such as accessibility of the field area and limitations on the weight of material that can be transported. However, unless some attempt is made to follow a systematic sampling program, an analysis on an individual hand specimen may be representative only of the material collected. Extrapolations as to the composition of the related outcrop, sampling locality, lava flow or mountain range can only be made with increasing uncertainty. There can be no universal scheme for the design of a sampling program. The number, distribution and weight of subsamples required will then depend on the size of the rock mass, whether it has a homogeneous matrix composition, and the mode of occurrence of the elements for which determinations are sought. Other factors would include the presence of any secondary effects, including weathering, chilled margins or metamorphism. Thus, criteria for a basaltic lava flow would be different from that for a differentiated magma sequence. Similarly, sampling strategy for major element analysis would be different from that required to obtain representative samples for a gold or platinum group element assay.

For conventional silicate rock analysis (major and trace elements), where there is no reason to suppose that there are gross inconsistencies in the composition of the material to be sampled, it is suggested that a target for the minimum weight of material collected for each sample should be 1kg. The grinding sequence to prepare a powdered sample for chemical analysis would then be as follows;

(i) The 1kg sample is split and reduced to grains 2mm across (8 to 10 mesh) by passing through a jaw crusher
(ii) This material is then passed through a disc grinder to further reduce the particle size to about 0.6mm (25 to 30 mesh)
(iii) This 25 to 30 mesh material is homogenized and split by coning and quartering to separate a 200g subsample
(iv) This subsample is ground in a planetary ball or swing mill to a particle size of 75μm (less than 200 mesh)
(v) This ground material is presented for analysis.

Sampling strategies are much more highly developed in the mineral exploration industry where accurate sampling is an important aspect of the economic assessment an ore body. A theory for the sampling of broken ores has been developed by Gy (1974). Practical aspects of sampling have been considered by Griffiths (1974), Jones and Beaven (1974) and Nicholls (1974) (all contributors to the conference proceedings, edited by Jones (1974)), and by Murphy (1982), for example. In the assessment of ore bodies, it would not be unusual to collect up to 5 tonnes of material, from which a representative sample aliquot for analysis must be separated.

1.6.2 *A note on mesh size*

The mesh size to which a powder has been ground is often used as a measure of the particle size distribution of a sample. It is normal to identify the proportion of a ground sample that will pass through a sieve of a given mesh size. Sieves are supplied according to mesh number. This defines the constructional parameters of the sieve including the size of mesh apertures, wire diameter and manufacturing tolerances. There are two standard specifications that are widely used to describe grades of sieve. One is defined by the British Standards Institution and the other by the American Society for Testing and Materials. Each specification is slightly different and the nominal aperture opening defined under each is listed in Table 1.15.

1.6.3 *Laboratory sampling procedures*

The analyst may have no control over field sampling procedures, reporting analytical results on an 'as received' basis. However, no amount of expense or effort in obtaining data of the highest accuracy and precision can compensate for samples that are not representative of the source material.

Silicate rocks are inherently heterogeneous materials as far as their bulk chemical composition is concerned, since they are composed of discrete minerals of diverse chemical composition and physical character. The ultimate limitation on analytical precision is therefore the homogeneity of the sample aliquot taken for analysis, that is, how representative it is of the bulk sample for which chemical analysis is required. Some of the uncertainties over sample homogeneity effects are overcome by modern standards of rock crushing which normally specify that the rock powders should be milled to pass a 200-mesh sieve. However, difficulties can arise in two areas. The first is in the analysis for major elements in coarse-grained materials, such as some granites, where there may be some difficulty in attaining complete homogeneity in the 0.1 to 0.5g samples often taken for analysis using modern instrumental techniques. The second area is in the analysis of certain elements at trace abundances, especially those elements that are not extensively partitioned into silicate mineral phases. These elements occur frequently as small minor-phase inclusions, and might include sulphur, phosphorus, titanium (rutile), chromium (chromite), Zr and Hf (zircon), Y, rare-earth elements, U and Th (all of which are often found as discrete phosphate, sulphide and oxide phases, for example, in some granite rocks), Au, Ag and platinum group elements (that occur almost exclusively as discrete alloy, sulphide, arsenide or antimony phases).

(*a*) *Studies of major-element homogeneity effects.* Several investigations of these problems have been carried out. Wilson (1964) presented a mathematical derivation of sampling precision, considering the variance of element distributions in terms of particle size, and concluded that 1g of rock sample crushed to a mesh size of 72 was adequate for the accurate determination of major elements in most rock types, but that smaller sampling aliquots were not recommended. Ingamells *et al.* (1972) considered that in some instances sampling error could be the controlling factor dictating analytical precision. Edelman (1962) highlighted several aspects of this problem relating to the major element analysis of coarse-grain materials. Kleeman (1967) considered that rock samples proposed as reference materials should be crushed to pass a 230-mesh sieve and pointed out difficulties in the coarse grain size of USGS: W-1 and G-1 as originally distributed.

Ridley *et al.* (1976) carried out a detailed investigation of sampling errors in the major element analysis of a medium- to coarse-grained biotite granodiorite, especially collected for the purpose. A comparison was carried out to relate the precision of analysis of six major elements (K_2O, Na_2O, CaO, Fe_2O_3(total), MgO and MnO) to three different sample preparation prcedures. One batch of sample was ground to pass a 100-mesh sieve, a second batch of powder was fused in a muffle furnace at 1450°C for 2h before crushing, and a third batch was fused with a flux (a lithium metaborate/tetraborate mixture) at 1200°C for 1h to ensure complete homogeneity prior to grinding. The grain size of the individual minerals in the original granodiorite was 1 to 7mm. Ridley *et al.* (1976) compared analytical precision of the six major elements listed above in 0.5g aliquots of the various samples. These data indicated that errors attributable to sample inhomogeneity could be distinguished in data from the unfused rock. Their work also indicated that optimum precision was obtained by grinding the rock to a mesh size of −140, +200, although the precise range is likely to be different for different rock samples.

Ellis *et al.* (1983) critically reviewed statistical tests that

Table 1.15 Nominal aperture sizes of standard sieves (in mm)

Mesh number	BS 410	ASTM E-11-61	Mesh number	BS 410	ASTM E-11-61
3	5.60	—	45	—	0.354
3½	4.75	5.66	50	—	0.297
4	4.00	4.76	52	0.300	—
5	3.35	4.00	60	0.250	0.250
6	2.80	3.36	70	—	0.210
7	2.36	2.83	72	0.212	—
8	2.00	2.38	80	—	0.177
10	1.70	2.00	85	0.180	—
12	1.40	1.68	100	0.150	0.149
14	1.18	1.41	120	0.125	0.125
16	1.00	1.19	140	—	0.105
18	0.850	1.00	150	0.106	—
20	—	0.841	170	0.090	0.088
22	0.710	—	200	0.075	0.074
25	0.600	0.707	230	—	0.063
30	0.500	0.595	240	0.063	—
35	—	0.500	270	—	0.053
36	0.425	—	300	0.053	—
40	—	0.420	325	—	0.044
44	0.355	—	400	—	0.037

BS = British Standards Institution.
ASTM = American Society for Testing and Materials.

can be carried out to measure variations in the state of homogeneity of powedered reference materials. Murphy (1982) reviewed the importance of representative sampling in mining and mineral exploration geology. Filby *et al.* (1985) investigated the problem of sample homogeneity at its extreme by investigating the consistency of replicate analyses of milligram aliquots of a coal fly ash reference material (SRM 1633a), a basalt (USGS: BCR-1) and a standard pottery sample. BCR-1 was found to be homogeneous at the 7mg level for the elements Na, K, Fe, Al, Ti, Mg, Ca, V, Mn, Co, Cs, Ba, La, Ce, Nd, Sm, Eu, Tb, Tm, Yb, Lu, Hf, Zr, Ta, and Th. Only Ni, Cr, Dy, and Ho gave poorer precision than expected.

It is apparent from this that sample homogeneity is unlikely to be a problem in the analysis of the major elements in fine-grain rocks such as basalts provided standard sample preparation techniques are followed. For more coarse-grained samples such as granites or granodiorites, the effect of sample inhomogeneity on the precision of major element determinations is more ambiguous; there is some evidence that inhomogeneity effects can be observed in the analysis of 0.5g aliquots of samples consisting of 2 to 7mm mineral grains that have been ground to 140 mesh (Ridley *et al.*, 1976).

(*b*) *Trace element homogeneity effects.* Sample inhomogeneity has always been important in the analysis of gold and platinum group elements, and hence fire assay procedures that are capable of fusing 50g sample aliquots (Chapter 15) are essential. However, it is suggested here that inhomogeneity effects may be a more widely overlooked problem in the analysis of certain other key trace elements including zirconium, chromium, rare-earth elements, uranium and thorium in some samples. The problem is related to the 'nugget effect', well known in the analysis of gold. If the mineral of interest is predominantly present in discrete minor-phase grains, inhomogeneity effects will be exaggerated. The larger the size of these grains and the lower the bulk concentration of the element of interest, the more serious will be the effects of inhomogeneity.

It is possible to carry out some crude calculations by assuming that an element is entirely present as a discrete mineral phase and that all the grains are spherical in shape and of uniform size. Providing these particles are randomly distributed in the sample, then their distribution will approximate to poisson statistics and the standard deviation of the number of particles (n) in any given sample aliquot will be $\sqrt{n}$. If these assumptions are accepted, it is possible to estimate the weight of sample required to achieve a specified sampling precision. This weight will vary as a function of the assumed diameter of the particle spheres (the larger the diameter, the smaller the number of discrete particles in a given sample aliquot) and also as a function of the bulk concentration of the element (the larger the concentration, the greater the number of discrete particles present in a given sample aliquot).

Such calculations have been carried out for gold grains of 1, 5 and 10μm diameter and for bulk concentrations of between 1ppb and 10ppm (Table 1.16). For each grain size, the weight of sample that must be taken for analysis to achieve a specified sampling precision of between 0.1 to 20% relative (one sigma) are listed. The dividing line indicates the limit corresponding to a 50g fire assay charge. Only sample compositions below the line can meet the specified precision requirements for a 50g aliquot. In some cases, of course, a far smaller sample weight is adequate. Thus if a 1% sampling precision is required the bulk concentration of gold is 100ppb, then 0.8g of sample must be analysed, assuming all the gold were present as 2μm diameter particles. If 10μm diameter gold grains are assumed, the corresponding sample aliquot is 760g. Clearly such data can only give a rather crude indication of sample inhomogeneity effects, since the statistical model does not take into account the spectrum of grain size present in real samples.

In Table 1.17, equivalent data has been calculated for the analysis of zirconium, this time for grain sizes of 5, 10 and 30μm and covering the bulk concentration range from 10ppb to 5000ppm. The dividing line this time has been drawn on the assumption that zirconium is determined on a 7g powder pellet by XRF. Let us take as an example a sample containing zirconium at the 100ppm level for which a 0.5% sampling error is considered acceptable. If the zirconium grain size were 5μm, 0.1g of sample is adequate for analysis. The corresponding figures for grain sizes of 10μm and 30μm are 0.7g and 39g sample aliquots respectively. These data are subject to the same limitations as those described for gold above. However, the implication of these calculations is that there may be some samples containing relatively coarse zircon or gold grains for which sampling errors may be the dominant contribution to the overall analytical precision. This conclusion must be judged in relation to the distribution of particle sizes present in a sample that has been ground to 200 mesh, for example (Figure 8.24). A prior knowledge of the grain size distribution is obviously invaluable in assessing the magnitude of this effect. In most routine analytical applications the additional time and cost in obtaining such information is rarely justified. It is then necessary either to rely on the specialized knowledge of the user, or to carry out analyses in duplicate on separate aliquots to check for errors in excess of those expected from instrumental factors.

1.7 Contamination effects

Rock samples become contaminated from three principal sources:

(i) Equipment used to split and crush materials down to a powder suitable for analysis

(ii) The laboratory environment in which chemical manipulations are carried out

(iii) Analytical reagents used in chemical preparation procedures in the course of an analysis.

All analytical techniques are therefore affected by contamination. These effects can normally be ignored in the bulk analysis of major and trace elements, but there are some areas where contamination effects become significant or even critical to successful analysis. One such area is in the analysis of a trace elements in samples that have been ground using equipment in which that element is a constituent of the grinding container. Another area is in the analysis of any trace element at concentrations significantly below the 1ppm level. A third area is in mass spectrometric determinations of isotope ratios where contamination blank

Table 1.16 Weight of sample (g) required for analysis to meet specified sampling precision. (1) Gold.

Diameter of grains = 1 μm

Sampling precision (% one sigma)	Bulk concentration of gold 1ppb	5	10	20	50	100	200	500	1ppm	2	5	10
	Weight of sample required for analysis											
0.1	>	>	760	380	150	75	40	15	8	4	1.5	0.8
0.2	>	380	190	95	40	20	10	4	2	1	0.4	0.2
0.5	300	60	30	15	6	3	1.5	0.6	0.3	0.15	<	<
1.0	80	15	8	4	2	0.8	0.4	0.15	<	<	<	<
5.0	3	1.6	0.3	0.15	<	<	<	<	<	<	<	<
10.0	0.8	0.15	0.1	<	<	<	<	<	<	<	<	<
20.0	0.2	<	<	<	<	<	<	<	<	<	<	<

Diameter of grains = 5 μm

Sampling precision (% one sigma)	Bulk concentration of gold 1ppb	5	10	20	50	100	200	500	1ppm	2	5	10
	Weight of sample required for analysis											
0.1	>	>	>	>	>	>	>	>	950	470	190	95
0.2	>	>	>	>	>	>	>	470	240	120	50	25
0.5	>	>	>	>	760	380	190	80	40	20	8	4
1.0	>	>	950	474	190	95	50	20	10	5	2	1
5.0	380	80	40	20	8	4	2	0.8	0.4	0.2	0.1	<
10.0	95	20	10	5	2	1	0.5	0.2	0.1	<	<	<
20.0	25	5	2.4	1.2	0.5	0.3	0.1	<	<	<	<	<

Diameter of grains = 10 μm

Sampling precision (% one sigma)	Bulk concentration of gold 1ppb	5	10	20	50	100	200	500	1ppm	2	5	10
	Weight of sample required for analysis											
0.1	>	>	>	>	>	>	>	>	>	>	>	760
0.2	>	>	>	>	>	>	>	>	>	950	380	190
0.5	>	>	>	>	>	>	>	600	300	150	60	30
1.0	>	>	>	>	>	760	380	150	80	40	15	8
5.0	>	600	300	150	60	30	15	6	3	1.5	0.6	0.3
10.0	760	150	80	40	15	8	4	1.5	0.8	0.4	0.15	<
20.0	190	40	20	10	4	2	1	0.4	0.2	0.1	<	<

Model: It is assumed that the sample is crushed so that all the gold is present as native gold (100% Au) in spheres of specified diameter (1, 5 or 10 μm) of density 19.3 kg/m^{-3}. The table lists the weight of sample that must be taken for analysis to achieve a sampling precision of 0.1—20% relative one sigma at a specified bulk concentration of gold. The line divides the table into samples which meet this specification if a 50 g fire assay charge is taken for analysis (data below the line) and those which do not (data above the line).

> means more than 1 kg would be required for analysis.

< means less than 0.1 kg is required for analysis.

levels become the limiting factor to successful analysis. It is only comparatively recently that problems and errors due to contamination have been examined in detail. Early analytical techniques had sensitivities only adequate for the analysis of the major elements and some minor elements and were relatively untroubled by contamination effects. However, with the development of more sensitive techniques for trace and ultra-trace element analysis (for example, neutron activation analysis, graphite furnace AA and thermal ionization and spark source mass spectrometry), contamination effects began to be investigated in more detail. Indeed, the sensitivity of some techniques such as thermal ionization mass spectrometry is so high that only laboratories having the highest standards of cleanliness are satisfactory for sample preparative chemistry. The sources of contamination listed above will now be examined in more detail.

1.7.1 *Sample grinding*

Two sources of contamination arise during sample crushing. The first is the abrasive action of the sample on grinding

Table 1.17 Weight of sample (g) required for analysis to meet a specified sampling precision. (2) Zirconium. Diameter of grains = 5 μm

Sampling precision (% one sigma)	Bulk concentration of zirconium 10ppb	50	100	500	1ppm	5	10	50	100	500	1000	5000
	Weight of sample required for analysis											
0.1	>	>	>	440	220	44	22	5	3	0.4	0.3	<
0.2	>	>	550	110	55	11	6	1	0.6	0.1	<	<
0.5	880	180	90	20	9	2	0.9	0.2	0.1	<	<	<
1.0	220	44	22	5	3	0.5	0.3	<	<	<	<	<
5.0	9	2	0.9	0.2	<	<	<	<	<	<	<	<
10.0	3	0.5	0.3	<	<	<	<	<	<	<	<	<
20.0	0.6	0.1	<	<	<	<	<	<	<	<	<	<

Diameter of grains = 10 μm

Sampling precision (% one sigma)	Bulk concentration of zirconium 10ppb	50	100	500	1ppm	5	10	50	100	500	1000	5000
	Weight of sample required for analysis											
0.1	>	>	>	>	>	360	180	35	18	4	2	0.4
0.2	>	>	>	890	450	90	45	9	5	0.9	0.5	0.1
0.5	>	>	710	140	70	14	7	1.4	0.7	0.2	<	<
1.0	>	360	180	36	18	4	2	0.4	0.2	<	<	<
5.0	70	14	7	1.4	0.7	0.2	0.1	<	<	<	<	<
10.0	20	4	2	0.4	0.2	<	<	<	<	<	<	<
20.0	5	1	0.5	0.1	<	<	<	<	<	<	<	<

Diameter of grains = 30 μm

Sampling precision (% one sigma)	Bulk concentration of zirconium 10ppb	50	100	500	1ppm	5	10	50	100	500	1000	5000
	Weight of sample required for analysis											
0.1	>	>	>	>	>	>	>	>	>	950	500	100
0.2	>	>	>	>	>	>	>	>	600	120	60	12
0.5	>	>	>	>	>	770	390	77	39	8	4	0.8
1.0	>	>	>	960	480	100	48	10	5	1	0.5	0.1
5.0	380	80	38	8	4	0.8	0.4	0.1	<	<	<	<
10.0	50	10	5	1	0.5	0.1	<	<	<	<	<	<
20.0	6	1.2	0.6	0.1	<	<	<	<	<	<	<	<

Model: It is assumed that the sample is crushed so that zirconium is entirely present as spherical grains of the specified diameter (5, 10 or 30μm) of density 4.5 kg/m^{-3}. The table lists the weight of sample that must be selected for analysis to meet the specified sampling precision (0.1 to 20% relative, one sigma) at the indicated bulk zirconium concentration. The line divides the table into those samples for which this specification is met for a 7 g powder pellet taken for XRF analysis (data below the line), and those that do not (data above the line).
> means more than 1 kg of sample would be required for analysis.
< means that less than 0.1 g of sample is required for analysis.

and splitting equipment. This can cause significant amounts of this material to be incorporated into the sample. Anyone who has attempted to grind pure white quartz in a tungsten carbide swing mill cannot fail to notice how the powder becomes tinged with grey. The second is the possibility of cross-contamination between samples, whereby small fractions of the previous sample embedded in the grinding equipment are incorporated into the current sample.

The effect of metallic iron, introduced into samples as splinters from steel jaw splitters, on the determination of Fe(II) will be considered in section 2.9. Thompson and Bankston (1970) undertook a detailed investigation of the effect of a variety of grinding and sieving materials on the contamination levels of trace elements determined by spark source–optical emission spectrometry. Aliquots (0.5g) of high purity SiO_2 were analysed after grinding in proprietary equipment made of tungsten carbide, artificial sapphire, boron carbide, aluminium ceramic, agate and lucite. Trace-element contaminants found after grinding twice in each of these devices are listed in Table 1.18. Significant levels of Co, Ti (and W and Ta) are expected after grinding in a tungsten carbide vial. The levels of contaminants in agate or lucite grinding mills are low. A boron carbide grinding vial imparted significant boron (as expected), and smaller amounts of Cu and Zn. The alumina and aluminium ceramic grinding apparatus would generally not be acceptable for

Table 1.18 Contamination during grinding and sieving

Grinding equipment	Contaminating elements	Concentrations in SiO_2 test samples Batch 1	Batch 2
Tungsten carbide	Co	32ppm	8.6ppm
	Ti	124	102
	Ta*	5	5
	W	not measured	
Al_2O_3 (diamonite) (artificial sapphire)	Al	>2000	>2000
	Cr	225	228
	Fe	8.7	26
	B	2.2	1.7
	Ga	2.5	3.3
	Mn	1.1	2.3
	Zr	?	?
Boron carbide	B	23	30
	Cu	2.7	2.1
	Ni	2.4	2.2
Aluminium ceramic	Al	>2000	>2000
	Cu	2.6	2.0
	Fe	34	55
	Ga	21	54
	Li	1	1
	Ti	11	31
	Zn	2.9	<2
Agate	B	1.8	2.3
	Cu	1.1	3.9
	Si	not measured	
Lucite	Mn	3.7	2.6
	Y	1.5	1.1

Sieve type	Contaminating elements							
Nylon sieve 200 mesh	none							
Brass sieve	B	Co	Cu	Fe	Mn	Ni	Sn	Zn
100 mesh	1.4	7.2	5.1	20	2.0	1.2	3.2	1.7
Stainless steel sieve	B	Co	Fe	M	Ni	Zn		
100 mesh	1.9	9.1	17	3.2	3.8	1.5		

Data from Thompson and Bankston (1970).
*See section 9.1.

routine trace analysis due to the range and level of contamination found (Table 1.18).

Thompson and Bankston found that sieving through nylon (200 mesh) caused no significant contamination, but that contamination from several elements was detected after using brass and stainless steel sieves (Table 1.18). They also detected significant levels of cross-contamination, even though equipment was cleaned using scouring powder, rinsed with hot tap water and soaked in concentrated nitric acid (where appropriate), distilled and then deionized water before use. The further use of proprietary scouring powder was discouraged since this material can itself contain zirconium and other alumino-silicate elements that could cause additional contamination. In order to reduce cross-contamination levels it was recommended that grinding equipment should be pre-cleaned by grinding and then discarding a batch of the current sample.

The precise data reported in Table 1.18 is obviously only directly applicable to the specific equipment on which tests were carried out. It is expected that the degree of contamination will also be affected significantly by the hardness of the sample. However, these data indicate in a general way the contaminants to be expected with various crushing equipment and confirm the reason for selecting agate in the most demanding applications.

1.7.2 *The laboratory environment*

The air in a 'typical' laboratory may contain up to 30 000 000 individual particles of between 0.5 and 100μm in size in each cubic metre of air (Moody, 1982*a*). In suburban areas, a significant proportion of these particles is likely to have originated from automobile exhausts and consist of lead material. In rural areas the laboratory air is also likely to contain siliceous materials from farming operations. Particulate matter from both sources will show diurnal and seasonal variations and will be significantly affected by weather conditions. In the analysis of silicate rocks for trace elements where the mass of element separated for analysis is less than 10 to 100μg, particulate fallout from the laboratory atmosphere is a serious source of contamination.

A further source of contamination within a chemical laboratory arises from the corrosion of metallic fixtures by acid fumes likely to be found in such an environment. Since the ultimate detection limit that can be attained by techniques such as mass spectrometry is limited by the procedural blank, great care must be taken over contamination levels in laboratories where preparative chemistry is carried out.

The specification normally used to set the maximum level of airborne contamination in a clean laboratory is the so-called 'class 100' standard of United States Federal Standard 209b (1973). Laboratories meeting the class 100 specification must have an airborne particle count of less than 100 particles of size 0.5μm in diameter per cubic foot of air (or pro-rata). The particles that are normally considered to contribute most to the analytical blank are those in the range 0.1 to 10μm in diameter. Smaller-sized particles constitute a rather small mass and do not therefore increase the blank appreciably, larger particles are relatively easy to filter from air supplies and do not remain in suspension in air as readily.

The principal means of achieving class 100 standards is by the use of high-efficiency particulate (HEPA) filters in the forced-air supply to the laboratory. Such filters have a minimum particulate efficiency of, typically, 99.99% as regards 0.3μm particles. However, of equal importance is the design of the laboratory itself, criteria for which have been considered in some detail by Moody (1982*b*). Such design considerations might include;

(i) Careful positioning of working benches in relation to filtered supply air to eliminate the possibility of cross-contamination between samples and to minimize contamination arising from the person of the analyst

(ii) Elimination of dust traps by for example, using continuous sheet floor coverings coved to the walls

(iii) Elimination, wherever possible, of metal fittings in the laboratory (e.g. door handles, water taps, steel trunking) and substituting plastic equivalents. If metal fittings must be used, then aluminium is preferred, provided it is sealed with polyurethane paint (Moody, 1982*b*)

(iv) Sealing of all porous surfaces with gloss finish paint,

chlorinated rubber paint being preferred in applications where lead is to be determined because of its very low inorganic lead content
(v) Provision of overshoes, tacky mat and protective clothing at the entrances to the laboratory to reduce ingress of contamination with the analyst
(vi) Design of the air supply system so that the clean laboratory is maintained at a small positive pressure compared with the surroundings—dust cannot then be sucked in.

For operations such as acid digestions where it is necessary to work in a fume hood, the standard design of turbulent flow fume hood (Figure 1.12*a*,*b*) is unlikely to be satisfactory. The reason is that in this design, air is sucked through the sash from the general laboratory environment, normally first flowing past the operator. This design does not therefore afford the cleanest working area. There are two means of overcoming these deficiencies. The first is to place small Teflon enclosures over each sample solution. Each enclosure is fed with a filtered air supply so that there is a locally clean environment for evaporation from individual solutions, for example (see section 17.1). The second option is to substitute *laminar flow cabinets* for normal fume hoods. The laminar flow cabinet consists of an enclosure in which 90% of the exhausted air is specially ducted to the cabinet, and only the remaining 10% is sucked in from the laboratory itself. This small proportion passing through the sash opening ensures that toxic fumes cannot diffuse back into the laboratory atmosphere. Laminar flow is normally achieved by delivering the supply air to a plenum above the working area. Air is then forced through a HEPA filter in the roof of the cabinet and a laminar curtain of air passses down through the working area and is exhausted through perforations in the floor of the cabinet (Figure 1.12*c*). Very low particulate counts can be achieved within the laminar flow cabinet. If lead is to be determined, it is usually desirable to pass the air supply through an activated-carbon filter bed before ducting to the cabinet to ensure complete removal of volatile organic lead compounds.

1.7.3 *Apparatus and reagents*

It goes without saying that providing a clean laboratory environment is a waste of resources unless preparatory chemistry is carried out in clean apparatus using highly purified reagents.

(*a*) *Preferred materials*. For ultra-trace element analysis conventional borosilicate glass apparatus is considered far from satisfactory (Murphy, 1976). The most widely used material for both apparatus and solution storage is polytetrafluoroethylene, marketed under the name 'Teflon' (a registered trade mark of DuPont de Nemours and Co. Inc.). Teflon is a fluorocarbon and is available in two forms, TFE (tetrafluoroethylene) and FEP (fluorinated ethylpropylene). Not only does Teflon have outstanding chemical resistance to acids, alkalis and organic solvents, but it can also be heated to temperatures of about 260°C without risk of thermal degradation. Moody (1982*a*) considers Teflon to be the cleanest of materials for ultra-trace analyses. It is, however, interesting to note that Dabeka *et al.* (1976) found that Teflon-FEP bottles constituted a persistent source of iron, nickel, manganese and chromium, presumably from contamination introduced into the polymer during manufacture using stainless-steel moulds. Not all batches of Teflon-FEP bottles are likely to be affected in this manner. Teflon is expensive, and other plastic materials have been used, particularly for solution storage during ultra-trace element analysis, including low-density polyethylene and polycarbonate.

Plastic substitutes for Teflon have lower maximum temperature tolerances and lower overall chemical resistance, as can be seen from data in Table 1.19. Moody (1982*a*) recommends low-density polyethylene as the most suitable alternative to Teflon for one-time storage of sample solutions, but re-use is not possible because of the low resistance of polyethylene to a vigorous cleaning regime.

Quartz glass is the only material to rival Teflon. Contamination levels from quartz containers are expected to be

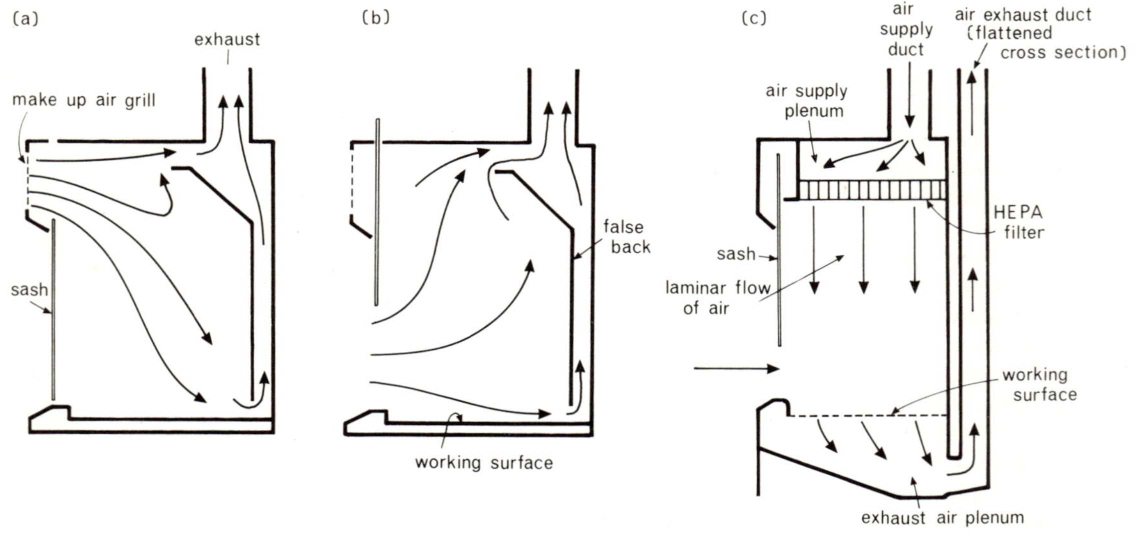

Figure 1.12 Cross-sectional design and air-flows in (*a*) fume cabinet with sash closed; (*b*) fume cabinet with sash open; (*c*) laminar flow fume cabinet in which 10% only of make-up air is sucked in through the sash opening.

Table 1.19 Physical properties of plastic and glass materials

	Polytetrafluoroethylene thermoplastic	Polyethylene low-density thermoplastic	Polypropylene thermoplastic	Polycarbonate thermoplastic	Quartz glass	Borosilicate glass
Specific gravity	2.1–2.3	0.91–0.925	0.90	1.2	2.21	2.32
Specific heat ($Jkg^{-1}K^{-1}$)	1046	2300	2090	1255	753	84
Coefficient of linear thermal expansion ($\times 10^5$)	10	10–20	8–10	6.6	0.055	0–33
Maximum continuous service temperature (°C)	260	60–77	120	110	900–1100	230–430
Chemical resistance to:						
strong mineral acids	excellent	good	excellent	fair	excellent (except HF)	excellent (except HF)
concentrated oxidizing acids	excellent	good to poor (attacked by oxidizing acids)	attacked slowly by oxidizing acids		excellent	excellent
strong alkalis	excellent	good	good	poor	poor	poor
alcohols	excellent	good below 60°C	good below 80°C	poor	excellent	excellent
aliphatic hydrocarbons	excellent	fair	good below 80°C	poor	excellent	excellent
bromatic hydrocarbons	excellent	fair	good below 80°C	poor	excellent	excellent
oils	excellent	good	good below 80°C	poor	excellent	excellent
Clarity	translucent	translucent	translucent	clear	clear	clear

Data from Weast (1973).

very low. Furthermore, this material can be safely heated to temperatures above 1000°C. However, quartz is a brittle material, and is not stable if it comes into contact with either hydrofluoric acid or concentrated solutions of alkalis.

(*b*) *Cleaning plastic containers*. The cleaning procedure proposed by Moody and Lindstrom (1977) for plastic containers is as follows;

(i) Fill with 1+1 HCl (analytical grade) and allow to stand for one week at room temperature (80°C for Teflon)

(ii) Empty and rinse with distilled water

(iii) Fill with 1+1 HNO_3 (analytical grade) and allow to stand for one week at room temperature (80°C for Teflon)

(iv) Empty and rinse with distilled water

(v) Fill with the purest available distilled water and allow to stand for several weeks or until needed, changing the water periodically

(vi) Rinse with purest water and allow to dry in a particle-free fume-free environment.

(*c*) *Losses of water and mercury*. If solutions are stored in a sealed plastic container, some loss of water by permeation through the walls of the container is to be expected, and this can approach 3% per year (Moody and Lindstrom, 1977). Equivalent losses from quartz containers should be negligible (less than 0.01% per year).

One cause for concern when storing solutions in any container is the long-term stability. Ions from solution can readily become adsorbed on to the walls of the container. This is normally eliminated by stabilizing solutions in dilute acid solution, 0.5M HNO_3 being adequate in most instances. Heydorn and Damsgaard (1982) have considered gains and losses of ultra-trace elements to the walls of polyethylene containers in detail. These workers have shown that, exceptionally, polyethylene will absorb mercury if exposed to the vapour in the laboratory environment. Furthermore, large losses of mercury occur when solutions of this element are stored in 0.1M HNO_3 in sealed polyethylene containers (Figure 1.13). The authors propose that this loss is caused by the reduction of mercury (II) ions in solution to mercury (0) by reaction with the polyethylene walls, mercury vapour then diffusing into the laboratory atmosphere.

Dabeka *et al.* (1976) investigated the effect on contamination of apparatus made of various plastic materials. They evaporated 250 mL high-purity hydrochloric acid from beakers made of Vycor, polypropylene, linear polyethylene, Teflon-FEP and poly (4-methyl 1-pentene) (TPX). All had been pre-cleaned in hot concentrated hydrochloric acid for several days. The elements titanium, aluminium and zinc were detected at low concentration in the solution taken from the linear polyethylene beaker. Silicon and zirconium were detected from the Vycor beaker. Polypropylene and TPX were relatively free of contaminants, and iron, nickel,

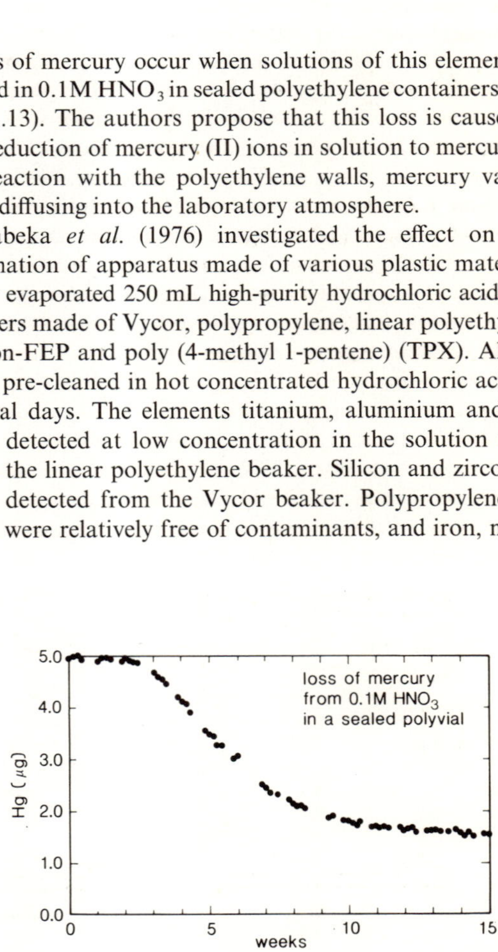

Figure 1.13 Loss of mercury as a function of time from a polyethylene container initially filled with 5 µg Hg as mercury (II) nitrate in 0.1 M nitric acid. (Heydorn and Damsgaard, 1982.)

manganese and chromium were found in the solution from the Teflon beaker as mentioned above.

(*d*) *Purification of water*. Conventions for purifying water vary from laboratory to laboratory. Many use water that has been distilled twice in a still made of quartz glass ('doubly distilled water'). Dabeka *et al.* (1976) compared trace-element contaminants in water that had been distilled from a conventional 'tin' still, with water that had been further purified by sub-boiling distillation in (i) a polypropylene still, (ii) a quartz still or (iii) by ion exchange. Data for eight elements found in these waters are listed in Table 1.20. Water produced by sub-boiling distillation in either still was superior to the original singly distilled water but the ion exchange water had even lower trace-element contamination levels. One further advantage of the ion exchange procedure is the rapid rate of purification (300mL min^{-1}) as compared with the sub-boiling method (about 2 to 5mL min^{-1}).

(*e*) *Sub-boiling distillation of reagents*. Sub-boiling distillation is a procedure widely adopted, especially in mass spectrometry sample preparation laboratories, to produce water and concentrated acid reagents (in particular, hydrofluoric, hydrochloric, and nitric acids) of the highest purity.

A very simple sub-boiling apparatus for purifying hydrofluoric acid was developed by Mattinson (1972), based on principles established by Coppola and Hughes (1952). The apparatus (Figure 1.14) consists of two FEP-Teflon bottles that screw into a TFE-Teflon connecting block. These bottles are held at right angles to one another. The reagent to be purified (e.g. concentrated hydrofluoric acid) is held in one bottle and heated to a temperture well below its boiling point by an infra-red lamp. With the collecting bottle held in position *A* (Figure 1.14), hydrofluoric acid evaporates out of the feed bottle and condenses on the inner surfaces of the connector block and collector bottle. These surfaces are cleaned, and condensed reagent falls back into the feed bottle. The collector bottle is then moved to position *B* and cooled in a basin of water. Purified acid is then collected at the rate of about 200mL of HF (of HCl) or 100–120mL of HNO_3 per day. Contamination can be reduced by ensuring that bubbles caused by excessive heating do not form in the feed bottle solution. When these rise to the surface and burst, tiny droplets of reagent are formed and can be carried over to the collecting bottle without participating in the distillation process.

Table 1.20 Trace-element contamination in distilled and deionized water.

Element	'Tin' still	Polypropylene still	Quartz still	Deionized water
Zn	33	0.1	0.4	0.02
Al	0.9	0.3	3	0.05
Mg	0.8	0.1	0.06	0.01
Cd	0.02	0.002	0.002	0.0002
Pb	2.5	0.02	0.003	0.01
Cu	12	0.2	0.007	0.004
Fe	0.3	0.3	0.3	0.03
Ni	0.07	0.04	0.002	0.001

Concentrations in ng mL^{-1}.
Data from Dabeka *et al.* (1976).

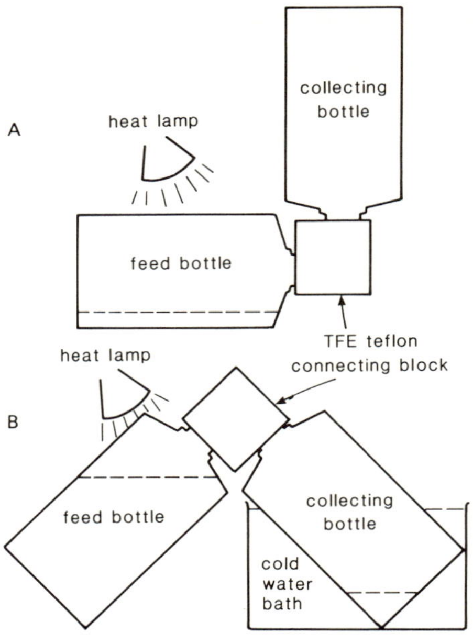

Figure 1.14 Simple sub-boiling Teflon still comprising two FEP Teflon bottles (one feed, the other collecting) joined by a TFE Teflon connecting block, as designed by Mattinson (1972).

An alternative apparatus has been described by Kuehner (1972) using sub-boiling distillation from a quartz still into a Teflon collecting bottle (Figure 1.15). A commercially available version of this still is claimed to give distillation rates of 5.8g min^{-1} (water), 7.5g min^{-1} (HNO_3) 7.1g min^{-1} (HCl) and 0.8g min^{-1} (H_2SO_4) using an 800W infra-red lamp. Maas and Dressing (1983) have described a simple full-boiling still constructed of pyrex glass suitable for producing up to 800ml day^{-1} of high-purity nitric acid. Dabeka *et al.* (1976) constructed a sub-boiling still from polypropylene.

Performance of sub-boiling distillation in reducing trace-element contamination is demonstrated by data for nitric acid in Table 1.21 and for hydrofluoric and hydrochloric acids in Table 1.22. These results indicate that this dis-

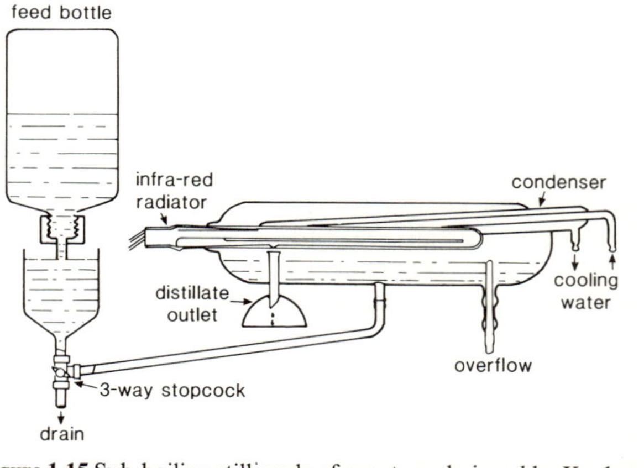

Figure 1.15 Sub-boiling still made of quartz as designed by Kuehner *et al.* (1972).

Table 1.21 Concentration of impurities in nitric acid (in ng mL^{-1})

Element	Kuehner *et al.* (1972)		Dabeka *et al.* (1976)	
	Reagent grade	Sub-boiling quartz still	Reagent grade	Sub-boiling quartz still
Ag	0.03	0.1		
Al			7	0.9
B				0.2
Ba	8	0.01		0.02
C	30	0.2		
Ca			0.2	0.4
Cd	0.1	0.01		<0.03
Co				<0.01
Cr	6	0.05		0.1
Cu	20	0.04		0.2
Fe	24	0.3	1300	0.8
In		0.01		
K	10	0.2		9
Mg	13	0.1	3	0.4
Mn				0.007
Na	80	1		0.5
Ni	20	0.05	0.7	0.03
P			0.8	0.5
Pb	0.2	0.02	0.05	0.04
S			0.6	15
Se	0.2	0.09		
Si				8
Sn	0.1	0.01		
Sr	2	0.01		
Te	0.1	0.01		
Ti				0.8
Tl	0.2			
V				<0.003
Zn	4	0.04		0.08
total	220	2.3	1312.4	37.0

tillation process reduces contamination levels by factors of, typically, between 10 and 100.

1.8 Reporting analytical data

A variety of conventions is used for reporting analysed results. The one adopted here follows broadly the protocol adopted by Abbey (e.g. 1983) and others. The major elements are listed as per cent oxide and reported in an order mainly controlled by decreasing cation valency; SiO_2, TiO_2, Al_2O_3, Fe_2O_3, FeO (or Fe_2O_3 total), MnO, MgO, CaO, Na_2O, K_2O, P_2O_5, H_2O+, CO_2, S, Cl, TOTAL. Results in this work have, in general, been quoted on a dry basis, that is after elimination of water evolved by heating the sample to 105°C for several hours (H_2O-). The H_2O+ listed above represents 'structural' water.

All other elements are reported as parts per million element, and one convention that has generally been used is to list these elements in alphabetic order.

Conversion factors used for converting element concentration to oxide concentration are listed in Table 1.23.

Data are usually reported to one more figure than can be quoted with confidence (significant figure convention). Less significant figures are therefore rounded up or down as appropriate. Problems can arise in the rounding up or down

Table 1.22 Concentration of impurities in HF and HCl acids (in ng mL^{-1})

Element	Hydrofluoric acid		Hydrochloric acid	
	Commercial high purity	Sub-boiling polypropylene still	Commercial high purity	Sub-boiling quartz still
Al	<30	0.5	100	0.8
B	1	2	0.9	0.06
Ba	<1	0.008	6	0.01
Ca	11	0.4	40	0.3
Cd	<1	<0.005	0.3	0.001
Ce			0.3	0.001
Co			0.3	0.001
Cr	4	0.06	2	0.008
Cu	<3	0.3	3	0.03
Fe	2	0.6	35	0.5
K	11	0.4	40	0.1
La			0.07	0.001
Mg	10	0.1	17	0.3
Mn	<1	0.03	3	0.001
Na	<20	0.6	30	0.2
Ni	<1	0.05	3	0.005
P	3	7	4	0.2
Pb	<2	0.03	10	0.006
Rb			0.2	0.0002
S	70	3	230	3
Si	34	4	90	1
Sn			0.3	0.02
Ti	60	0.5	2	0.06
V			0.1	0.08
Zn	3	0.1	7	0.03
Zr	<1	0.01	0.3	0.001

Data from Dabeka *et al.* (1976).

Table 1.23 Oxide ratios

AN	Oxide	Ratio	AN	Oxide	Ratio	AN	Oxide	Ratio
4	BeO	2.7758	34	SeO_3	1.6079	63	Eu_2O_3	1.1579
5	B_2O_3	3.2202	37	Rb_2O	1.0936	64	Gd_2O_3	1.1526
6	CO_2	3.6644	38	SrO	1.1826	65	Tb_2O_3	1.1510
7	N_2O_5	3.8551	39	Y_2O_3	1.2699		Tb_4O_7	1.1762
11	Na_2O	1.3480	40	ZrO_2	1.3508	66	Dy_2O_3	1.1477
12	MgO	1.6582	41	Nb_2O_5	1.4305	67	Ho_2O_3	1.1455
13	Al_2O_3	1.8895	42	MoO_3	1.5003	68	Er_2O_3	1.1435
14	SiO_2	2.1392	43	TcO	1.1649	69	Tm_2O_3	1.1421
15	P_2O_5	2.2916	44	RuO_2	1.3166	70	Yb_2O_3	1.1387
16	SO_3	2.4972	45	Rh_2O_3	1.1555	71	Lu_2O_3	1.1371
19	K_2O	1.2046	46	PdO	1.1504	72	HfO_2	1.1793
20	CaO	1.3992	47	Ag_2O	1.0741	73	Ta_2O_5	1.2211
21	Sc_2O_3	1.5338	48	CdO	1.1423	74	WO_3	1.2610
22	TiO_2	1.6681	49	In_2O_3	1.2091	75	ReO_2	1.1719
23	V_2O_5	1.7852	50	SnO_2	1.2696	76	OsO_4	1.3365
24	Cr_2O_3	1.4615	51	Sb_2O_5	1.3284	77	IrO_2	1.1665
25	MnO	1.2912	52	TeO_3	1.3762	78	PtO	1.0820
26	FeO	1.2865	55	Cs_2O	1.0602	79	Au_2O_3	1.1218
	Fe_2O_3	1.4297	56	BaO	1.1165	80	HgO	1.0798
27	CoO	1.2715	57	La_2O_3	1.1728	81	Tl_2O_3	1.1174
28	NiO	1.2725	58	Ce_2O_3	1.1713	82	PbO	1.0772
29	CuO	1.2518		CeO_2	1.2284	83	Bi_2O_5	1.1914
30	ZnO	1.2448	59	Pr_2O_3	1.1703	90	ThO_2	1.1379
31	Ga_2O_3	1.3442		Pr_6O_{11}	1.2082	92	UO_3	1.2015
32	GeO_2	1.4408	60	Nd_2O_3	1.1664			
33	As_2O_5	1.5339	62	Sm_2O_3	1.1596			

AN = atomic number of element.
The ratio is the number by which an elemental concentration must be multiplied to convert it into an element oxide concentration.

of 5. A convention that will avoid systematic errors in one direction is to round 5 to the nearest even number. Thus 15.45 rounds down to 15.4 whereas 15.55 rounds up to 15.6, where the figure after the decimal point is uncertain according to the analytical precision appropriate for this determination.

In assessing the reliability of analytical data for the major elements, summation to 100% is often examined. This is unsatisfactory from several points of view that arise in part from the fact that modern 'rapid' instrumental procedures (e.g. XRF, ICP–AES, AAS) are incapable of analysing all the major elements (in particular iron in its appropriate oxidation state), H_2O+ and CO_2. This leads to the following uncertainties in summation.

(i) Iron is often determined as total iron (regardless of oxidation state). Unless a separate (wet chemical) determination for FeO is carried out and Fe_2O_3 calculated by difference, ambiguity is introduced into the 100% summation. For example, some basalts can contain in excess of 10% FeO. If such analyses are reported as Fe_2O_3(total), the sum of major elements could legitimately add up to more than 101%.

(ii) Since modern instrumental techniques are incapable of determining H_2O+ and CO_2 individually, an estimate of their composition is often made by determining a combined 'loss on ignition'. The weight loss after fusing at 1000°C is taken as an estimate of H_2O+ plus CO_2. The shortcomings of this procedure are well known. A few minerals are resistant to releasing structural water (section 2.10). There is no control over the proportion of FeO in the sample that will simultaneously oxidize to Fe_2O_3 on ignition. In the case of the basalt mentioned above, a negative loss on ignition could be obtained if the weight increase due to FeO oxidation exceeds the weight loss due to loss of volatiles.

(iii) There is no reason why errors in individual determinations should exactly cancel to give a 100% total. Although it is very salutary when analyses do add up to 100%, this is an unsafe criterion for the routine assessment of analytical data.

1.9 Standard additions calibrations

Three methods are widely used to calibrate analytical techniques:
(i) From measurements on international rock reference materials (section 1.10)
(ii) By preparing calibration standards from known weights of analytically pure reagents
(iii) By standard additions.
The standard additions technique involves taking a natural silicate sample, dividing it into separate aliquots, and adding to each increasing quantities of a reagent containing the element under analysis. These increments are usually made equal and a minimum of five standard addition mixtures is desirable to set up a calibration set. The calibration set then consists of one unspiked sample and a set of spiked materials, all of which possess approximately identical matrices. These samples are analysed and a graph constructed of the concentration of spike added versus the analytical signal. This graph is extrapolated back to the x axis to determine the (negative) concentration of the element present in the unspiked sample. An example of a standard additions calibration for europium by neutron activation analysis is shown in Figure 1.16.

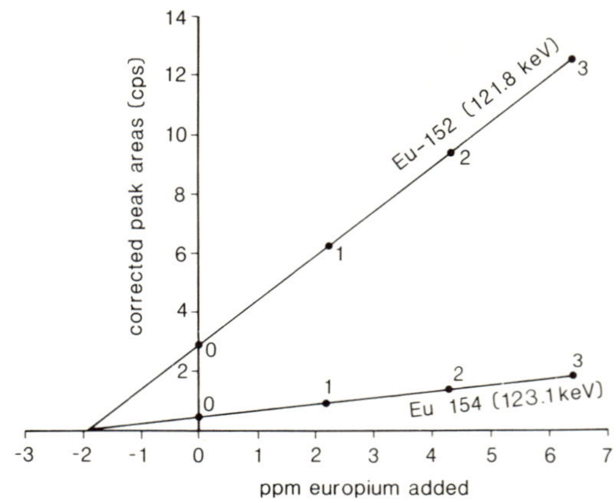

Figure 1.16 Standard additions calibration plot for determination of europium by neutron activation analysis. The gamma-ray signal from two europium isotopes (^{152}Eu and ^{154}Eu) activated in the samples by neutron irradiation is plotted as a function of the concentration of europium 'spike'. Extrapolation of data to concentration axis indicates that the unspiked sample contains 1.95 ppm Eu. (Potts *et al.*, 1981.)

Theoretical aspects of this procedure have been considered by Ratzlaff (1979), who has shown that to obtain optimum precision the standard addition increment should be at least equivalent to the expected concentration of the element in the unspiked sample so that excessive errors in extrapolation from analysed data are avoided.

Standard additions techniques are extensively used where it is important that calibration standards possess matrices identical to the samples to be analysed. Thus in graphite furnace AAS, the size of the analyte signal can be significantly affected by matrix differences between samples. Standard additions calibration offers a rapid and convenient way of compensating for these effects.

1.10 Rock reference materials

1.10.1 *G-1 and W-1: history*

It was not until the beginning of the 1950s that the first two rock reference materials were distributed for international co-operative analysis. H.W.Fairburn of the Massachusetts Institute of Technology and a group of colleagues at the Carnegie Institution and the US Geological Survey wished to develop dc arc-optical emission spectography as a rapid technique for determining as many elements as possible in silicate rocks. This technique showed clear advantages over traditional 'classical' methods then prevalent in terms of both speed and sensitivity of analysis. However, in order to evaluate this technique, some well-analysed reference materials were required for use in instrument calibration. Two rocks were selected and crushed in large quantities. These were a granite designated G-1 and a diabase designated W-1. These two samples were distributed to 34 laboratories of repute. As Ahrens (1977) emphasized, the intention of this exercise was simply to obtain reliable analyses of two natural rock samples and not to test the accuracy and precision of wet chemical methods of analysis. However, it is to the credit of Fairburn and his colleagues that this was indeed the outcome. The results of this interlaboratory co-operative investigation were published as a USGS Bulletin (Fairburn and others, 1951). The results reported by 34 laboratories in 10 countries were mainly major element analysis by classical wet chemical procedures, but a small number of trace-element results were also listed. These data are summarized in Table 1.24 for the sample W-1.

The impact of this report should not be underestimated—it has been described as a 'bombshell' (Abbey, 1983, p.2). Participating analysts from different laboratories, each expert in their field and using well-tried methodologies, reported data that showed large interlaboratory bias. Chayes (1951, p.67) commented that the normative variation was so extreme that in the absence of direct evidence to the contrary, there would be little reason to suppose that all the analyses submitted were of the same two rocks.

This first report was followed nine years later by a second report (Stevens and others, 1960) containing new analytical data for trace elements, as well as the majors. At that time, G-1 and W-1 were the only widely analysed reference materials available and there was considerable demand from geochemical laboratories for samples. This soon led to a rapid depletion of stocks, so that G-1 was completely used up by about 1965 and W-1 by 1972 (Flanagan, 1976). At the same time, there was considerable interest in assessing published analyses so that values for general use could be recommended. Notable among these assessments were the reports of Fleischer and Stevens (1962), Ingamells and Suhr

Table 1.24 Compilations of data for diabase USGS: W-1 over the period 1951–1983.

	Schlecht and Stevens (1951) x	(CV %)	Ahrens (1951)	Stevens and Niles (1960)	Fleischer and Stevens (1962)	Ingamells and Suhr (1963)	Fleischer (1965)	Flanagan (1973)	Abbey (1983)
SiO_2	52.41	(0.7)		52.54	52.72	52.62		52.72	52.72
TiO_2	1.04	(23)		1.07	1.07	1.08		1.07	1.07
Al_2O_3	15.19	(5.6)		15.05	14.87	14.95		15.02	15.02
Fe_2O_3	1.75	(55)		1.41	1.41	1.38		1.40	1.40
FeO	8.56	(8.1)		8.75	8.75	8.72		8.73	8.73
MnO	0.18	(41)		0.16	0.16	0.17		0.17	0.17
MgO	6.54	(7.2)		6.64	6.63	6.53		6.63	6.63
CaO	10.97	(1.6)		10.98	10.98	10.93		10.98	10.98
Na_2O	2.07	(10.2)		2.07	2.07	2.15		2.15	2.15
K_2O	0.70	(25)		0.64	0.64	0.63		0.64	0.64
H_2O	0.57	(41)		0.53	0.53	0.45		0.53	0.53
CO_2	0			0.05	0.05	0.07		0.06	0.06
P_2O_5	0.13	(36)		0.14	0.14	0.14		0.14	0.14
F					0.020?		0.025?	0.025	0.025
S	0.018	(118)			0.014		0.013	(0.012)	(0.012)
total	100.13			100.03	100.05	99.82		100.28	100.28
Ba (ppm)	340	(46)	300		225	130	180	160	160
Be			0		0.8	1	0.8	(0.8)	(0.8)
Ce							25?	(23)	(23)
Co			29		52?	44	50?	47?	47?
Cr	140	(50)	120		120?	110	120?	114?	115?
Cu			93		110?	110	110?	110	110
La			(11)		30		20?	9.8?	9.8?
Nb					10		10	9.5	9.5
Nd							17	15	15
Ni			90		80?	76	78	76?	76?
Pb			4		8?	6?	8?	7.8	7.8
Rb			39		22?	20	22	21	21
Sc			33		33?	34	34	35.1	35?
Sm					5.1		5	3.6?	3.6?
Sr	310	(127)	270		220? } 175? }	180	180?	190	190
Ta					0.7		0.7?	0.50	0.50
Th					2.4?		2.4?	2.42	2.4
U					0.52?		0.52?	0.58?	0.58?
V	400		210		240?	250	240?	264?	260?
Y					35?	19	25?	25	25
Yb					3	3.0	3	2.1	2.1
Zn			88		82?	110	82?	86	86
Zr			81		100?	160	100?	105	105

All data recalculated on a dry basis.

(1963), Fleischer (1965) and the widely-used report of Flanagan (1973). As G-1 and W-1 were the first rock reference materials, it is relevant to examine what changes have occurred in average values reported in the literature at various stages since distribution. Data for the major elements and selected trace elements are listed in Table 1.24. It must be emphasized that these results are averaged values assessed from data available up to the time of publication (with the exception of the first trace-element results of Ahrens, 1951).

For the first ten years or so after distribution, usable data were available for only the major elements. This is not surprising when it is remembered that many of the modern analytical techniques for trace determinations were either in their infancy, e.g. optical emission spectography, or largely unavailable for routine analysis (e.g. x-ray fluorescence, neutron activation, atomic absorption, ICP–optical emission). Of the major element results listed in Fairburn and others (1951) the spread of data reported for TiO_2, Fe_2O_3, MnO, K_2O, H_2O, P_2O_5 and S was particularly wide. As new analyses became available, changes in the 'proposed' values for Fe_2O_3, FeO, MgO, Na_2O and K_2O were particularly significant (Table 1.24). Sufficient data to recommend values for trace elements did not become available until the early 1960s (that is, at the time supplies of G-1 were almost exhausted). Recommended values for rare-earth elements were reported by Fleischer (1965), and a relatively full compilation of trace-element data was given by Flanagan (1973). Subsequently, virtually all of Flanagan's values have become 'fossilized', being re-adapted in Abbey's 1983 compilation almost without change.

Several salutary lessons were learnt from these studies on G-1 and W-1. Not only was attention focused on the problems of interlaboratory bias but also on the care that must be expended on the preparation of materials worthy of such detailed investigation. Flanagan (1976) reported that the samples prepared for distribution were originally crushed far too coarsely (G-1 is described as passing an 80-mesh screen, W-1 a 100-mesh screen). After their initial distribution, these samples were re-ground to less than about 300 mesh. Flanagan (1976) also reports that the method of homogenizing samples and distributing sample bottles in a random manner was not unambiguously documented. This makes it difficult to be dogmatic in judging the vigorous debate that occurred in the literature between Chayes (1969, 1970) and Vistelius (1970, 1971) concerning the importance of inter-analyst differences or sample homogeneity in explaining the systematic bias in contributed data for G-1 and W-1.

1.10.2 *Other reference materials*

The experience with G-1 and W-1 clearly established the need for natural silicate rock reference materials. From the early 1960s onwards, there occurred an explosion of reference materials distributed by various agencies and analysed to varying degrees of uncertainty.

Notable amongst these was a group of six samples often collectively called USGS II. These constituted an andesite, AGV-1; a basalt, BCR-1; a dunite, DTS-1; a granite, G-2; a granodiorite, GSP-1; and a peridotite, PCC-1. The distribution and compilation of data for these materials is closely associated with the work of Flanagan (1967, 1969, 1973, 1976). The USGS II reference materials represent some of the most reliably analysed samples to date and are the ones most commonly quoted when it is desired to establish the accuracy of a technique. None of these samples is now available for distribution. Further series of reference materials were distributed by the US Geological Survey. Other agencies also involved in this field of activity included two French research institutes, the Centre de Recherches Pétrographiques et Géochemiques (CRPG), and later the Association Nationale de la Recherche Technique (ANRT) (reference material distribution coordinated by H. de la Roche and later K.Govindaraju); the Geological Survey of Japan (A.Ando); the National Institute for Metallurgy, South Africa (T.W.Steele) and others.

Perhaps one of the difficulties of these early reference material programmes was that there was no easy means of disseminating analysed data to contributing laboratories, which by and large received only summary data. An important step in overcoming these difficulties occurred in 1977 with the publication of a new international journal, *Geostandards Newsletter*, under the editorship of K.Govindaraju. Not only did this journal serve as a forum for the discussion of all aspects of geochemical reference materials, but it also undertook publication of full lists of 'raw' data submitted as part of co-operative investigations of reference materials. For the first time this raw data became widely available and this naturally stimulated interest in the development of statistical (and subjective) techniques for evaluating 'consensus' values.

A further development associated with the publication of *Geostandards Newsletter* was the organization by Govindaraju of the Groupe Internationale de Travail–International Working Group (GIT–IWG) comprising geochemical laboratories worldwide willing to contribute data to new cooperative programmes of rock reference material characterization. The first of these programmes involved the cooperative analysis of three samples, an anorthosite AN-G, a basalt BE-N and a granite MA-N. A statistical assessment of data on these samples was published by Govindaraju in April 1980. The programme of analysis had been accomplished in nine months during which time 122 laboratories had contributed over 6000 individual elemental analyses, a commendable achievement.

During this period of development, several authors have undertaken the assessment of all available geochemical reference materials, summarizing analyses in a form which can be readily used by individual analysts in the evaluation of calibrations, etc. Notable among these compilations of data are the following references:

(i) *F.J.Flanagan* (1973): 1972 values for international geochemical reference samples. *Geochim. Cosmochim. Acta* **37**, 1189–1200.

(ii) *S.Abbey* (1983): Studies in 'standard samples' of silicate rocks and minerals. Geol. Survey Canada Paper 83-15, 114p.

(iii) *K.Govindaraju* (1984): 1984 compilation of working values for 170 international reference samples of mainly silicate rocks and minerals. *Geostand. Newslett. Spec. Issue* 8, July 1984.

A list of the more widely used silicate and associated reference materials is given in Table 1.25. Samples are listed

Table 1.25 Silicate and related reference materials

United States Geological Survey (*USGS*)			
USGS I:			
	GLA83a	G-1	granite (Rhode Island)
		W-1	diabase (Centreville, Virginia)
USGS II:			
		AGV-1	andesite (Guano Valley, Oregon)
	FLA76	BCR-1	basalt (Columbia River, Washington)
	GLA83a	DTS-1	dunite (Hamilton, Washington)
	ABB78a	G-2	granite (Rhode Island)
		GSP-1	granodiorite (Silver Plume, Colorado)
		PCC-1	peridotite (Sonoma County, California)
USGS III:			
	GLA81	BHVO-1	basalt (Kilauea caldera, Hawaii)
		MAG-1	marine mud (Gulf of Maine)
		QLO-1	quartz latite (Lake County, Oregon)
		RGM-1	rhyolite (Glass Mountain, California)
		SCo-1	shale (Cody shale, Wyoming)
		SDC-1	mica schist (Rock Creek Park, Washington DC)
		SGR-1	shale (Green River oil shale)
		STM-1	syenite (Table Mountain, Oregon)
USGS IV:			
	GLA83a	BIR-1	basalt (S.W. Iceland)
		DNC-1	diabase (Durham, North Carolina)
		W-2	diabase (Centreville, Virginia)
Others:			
		NOD-A-1	manganese nodule (Atlantic)
		NOD-P-1	manganese nodule (Pacific)
USGS with the Association of Exploration Geologists (*AEG*)			
		GXR-1	jasperoid soil (Drum Mountains, Utah)
		GXR-2	soil (0–18 cm depth) (Park City, Utah)
	GLA84	GXR-3	Fe–Mn–W rich hot spring deposit (Humboldt County, Nevada)
	GLA79	GXR-4	porphyry copper mill heads (Utah)
		GXR-5	soil (B horizon, 7.5–25 cm depth) (Somerset County, Maine)
		GXR-6	soil (B horizon, 15–45 cm depth) (Davidson County, North Carolina)
National Bureau of Standards (*NBS*)			
	URI81a	NBS-278	basalt (Clear Lake, Newberry Crater, Oregon)
	URI81b	NBS-688	obsidian (Jackpot, Nevada)
Centre de Recherches Pétrographiques et Géochimiques, France (*CRPG*)			
	GOV77	GA	granite (Andlau, Vosges, France)
		GH	granite (Hoggar, France)
		BR	basalt (Essey la Côte, Lorraine, France)
	GOV79	Mica-Fe	biotite (Massif Central, France)
		Mica-Mg	phlogopite (Bekily, Madagascar)
Association Nationale de la Recherche Technique, France (*ANRT*)			
ANRT I:			
	GOV82	DR-N	diorite (Rocher de Neuntelstein, Vosges, France)
		UB-N	serpentine (Bagnenelles, France)
		BX-N	bauxite (Brignoles, Vosges, France)
		DT-N	kyanite (Nanga-Eboko, Cameroun)
ANRT II:			
	GOV84a	GS-N	granite (Senones-Launois area, France)
		FK-N	potash feldspar (Madras, India)
	ROC76	GL-O	glauconite (Cauville, Basse-Normandie, France)
	ROC73	VS-N	synthetic glass
Groupe Internationale de Travail/International Working Group (*GIT/IWG*)			
GIT/IWG I:			
	GOV80	AN-G	anorthosite (Fiskenaesset, Greenland)
		BE-N	basalt (Essey-la-Côte, Meurthe-et-Moselle, France)
		MA-N	granite (Clermont-Ferrand, France)
GIT-IWG II:			
GOV84b	AL-I	albite	(Trente, Italy)
		IF-G	iron formation sample (West Greenland)

Table 1.25 (*Continued*)

Canadian Certified Reference Material Project (CCRMP)			
	ABB81	SY-2	syenite (Bancroft, Eastern Ontario, Canada)
		SY-3	syenite blende (as above, blended with radioactive concentrate)
		MRG-1	gabbro (Mount Royal, Montreal, Canada)
	ABB83 ABB85	FER-1	iron formation sample (Bathurst, New Brunswick, Canada)
		FER-2	iron formation sample (Bruce Lake, Ontario, Canada)
		FER-3	iron formation sample (Temegami, Ontario, Canada)
		FER-4	iron formation sample (Temegami, Ontario, Canada)
	GLA85	SO-1	regosolic clay soil (Hull, Quebec, Canada)
	BOW79	SO-2	podzolic B horizon soil (Montmorency Forest, Quebec, Canada)
		SO-3	calcareous C horizon soil (Guelph, Ontario, Canada)
		SO-4	chernozemic A horizon soil (Saskatoon, Saskatchewan, Canada)
Geological Survey of Japan			
GSJ I:			
	AND71	JB-1	basalt (Myokanji-toge, Japan)
	AND74, 75	JG-1	granite (Sori, Gunma-ken, Japan)
GSJ II:			
		JA-1	andesite (Hakone Volcano, Kanagawa, Japan)
		JB-2	basalt (Oshima Volcano, Japan)
		JB-3	basalt (Fuji Volcano, Yamanashi, Japan)
		JGb-1	gabbro (Utsukushigatake, Abukuma, Fukushima, Japan)
		JR-1	rhyolite (Wada-Toge, Nagano, Japan)
		JR-2	rhyolite (Wada-Toge, Nagano, Japan)
		JP-1	peridotite (Horoman, Hokkaido, Japan)
		JG-1a	granodiorite (Sori, Gunma, Japan)
		JB-1a	basalt (Sasebo, Nagasaki, Japan)
British Chemical Standards (UK)			
		BCS-375	soda feldspar (S.E. Sudan)
		BCS-376	potash feldspar (Glamsland, S. Norway)
National Institute for Metallurgy, South Africa (NIM)			
	STE79	NIM-D	dunite (Bushveld complex, Transvaal, South Africa)
	STE78	NIM-G	granite (Bushveld complex)
		NIM-L	lujavrite (Pilansberg complex, Transvaal)
		NIM-N	norite (Bushveld complex)
		NIM-P	pyroxenite (Bushveld complex)
		NIM-S	syenite (Phalaborwa complex, Transvaal)
Institute of Geochemistry, Irkutsk, USSR (IGI)			
	ABB78b	St-1A	trap (also known as 2001)
		SGD-1A	gabbro (also known as 2003)
		SG-1A	granite (also known as 2005)
Institute of Geophysical and Geochemical Exploration, China (IGGE)			
	XIE85	GSD-1	stream sediment (Shanxi Province, China)
		GSD-2	stream sediment (Jiangxi Province, China)
		GSD-3	stream sediment (Jiangxi Province, China)
		GSD-4	pond sediment (Anhui Province, China)
		GSD-5	pond sediment (Anhui Province, China)
		GSD-6	stream sediment (Qinghai Province, China)
		GSD-7	stream sediment (Liaoning Province, China)
		GSD-8	stream sediment (Quangxi, China)
International Atomic Energy Agency, Vienna (IAEA)			
	DYB79	SOIL-5	soil (Lima, Peru)
Geological and Mining Service, Suriname			
	BUR85	K-3	tropical soil composite (Western Suriname)
		CS-3	tropical soil composite (Central Suriname)
		SUR-1	gold-bearing soil (Central Suriname)
		BAR-1	lateritic bauxite (Western Suriname)

References:
ABB78a S. Abbey (1978) *Geostand. Newslett.* **2**, 141–146.
ABB78b S. Abbey and K. Govindaraju (1978) *Geostand. Newslett.* **2**, 15–22.
ABB81 S. Abbey (1981) *Geostand. Newslett.* **5**,13–26.
ABB83 S. Abbey, C. R. McLeod and Wan Liang-guo (1983) Geol. Surv. Canada, Paper 83–19.
ABB85 S. Abbey and R. M. Rousseau (1985) *Geostand. Newslett.* **9**, 1–16.
AND71 A. Ando, H. Kurasawa, T. Ohmori and E. Takeda (1971) *Geochem. J.* **5**, 151–164.
AND74 A. Ando, H. Kurawasa, T. Ohmori and E. Takeda (1974) *Geochem. J.* **8**, 175–192.
AND75 A. Ando, H. Kurawasa, T. Ohmori and S. Uchiumi (1975) *Geol. Surv. Japan Bull.* **26**, 17–(335) to 30–(348).
BOW79 W. S. Bowman, G. H. Faye, R. Sutarno, J. A. McKeague and H. Kodama (1979) *Geostand. Newslett.* **3**, 109–113.
BUR85 K. E. Burke (1985) *Geostand. Newslett.* **9**, 69–78.
DYB79 R. Dybczynski, A. Tugsavul and O. Suschny (1979) *Geostand. Newslett.* **3**, 61–87.
FLA76 F. J. Flanagan (1976) US Geol. Surv. Prof. Paper 840, 131–183.
GLA79 E. S. Gladney, D. R. Perrin, J. W. Owen and D. Knab (1979) *Anal. Chem.* **51**, 1557–1569.
GLA81 E. S. Gladney and W. E. Goode (1981) *Geostand. Newslett.* **5**, 31–64.
GLA83 E. S. Gladney, C. E. Burns and I. Roelandts (1983) *Geostand. Newslett.* **7**, 3–226.
GLA84 E. S. Gladney, C. E. Burns and I. Roelandts (1984) *Geostand. Newslett.* **8**, 119–154.
GLA85 E. S. Gladney, C. E. Burns and I. Roelandts (1985) *Geostand. Newslett.* **9**, 35–68.
GOV77 K. Govindaraju (1977) *Geostand. Newslett.* **1**, 67–98.
GOV79 K. Govindaraju (1979) *Geostand. Newslett.* **3**, 3–24.
GOV80 K. Govindaraju (1980) *Geostand. Newslett.* **4**, 49–138.
GOV82 K. Govindaraju (1982) *Geostand. Newslett.* **6**, 91–159.
GOV84a K. Govindaraju (1984) *Geostand. Newslett.* **8**, 173–206.
GOV84b K. Govindaraju (1985) *Geostand. Newslett.* **8**, 63–113.
ROC73 H. de la Roche and K. Govindaraju (1973) *Analusis (Paris)* **2**, 59–70.
ROC76 H. de la Roche, K. Govindaraju and G. S. Odin (1976) *Analusis (Paris)* **4**, 385–397.
STE78 T. W. Steele, A. Wilson, R. Goudvis, P. J. Ellis and A. J. Radford (1978) *Geostand. Newslett.* **2**, 71–106.
STE79 T. W. Steele and R. G. Hansen (1979) *Geostand. Newslett.* **3**, 135–172.
URI81a G. A. Uriano (1981) National Bureau of Standards Certificate: Standard Ref. Material 688, basalt rock.
URI81b G. A. Uriano (1981) National Bureau of Standards Certificate: Standard Ref. Material 278, obsidian rock.
XIE85 X. Xie, M. Yan, L. Li and H. Shen (1985) *Geostand. Newslett.* **9**, 83–159.

under the relevant distributing agency and details of rock type and origin are given. The type references listed in Table 1.25 were chosen to give details of contributed raw data and its statistical assessment in evaluating consensus values. These references are important since the accuracy of all bulk geochemical data rests ultimately on the confidence that may be placed in consensus values derived in these works. This point is elaborated further below.

1.10.3 *The role of silicate reference materials*

The importance of silicate rock reference materials in identifying inter-analyst bias has already been considered. In view of the very poor reproducibility of calcium and magnesium, for example, at low concentrations, highlighted in the first co-operative investigation of G-1 and W-1, Ahrens (1977) considered that the likelihood is high that several petrological mistakes may have arisen in earlier data due to analytical error.

The development of modern 'instrumental' analytical techniques has placed even more importance on the availability of reliable reference materials. It is generally considered that several modern techniques can only be adequately calibrated using reference materials (e.g. XRF trace analysis of powder pellets) and that others (e.g. ICP–optical emission) can only be checked satisfactorily for analytical bias by analysis of such materials. This is in contrast to early wet chemical procedures in which all calibrations were carried out against standard elemental solutions; here the arbiter of accuracy was generally considered to be the chemical purity of the standard reagent and the accuracy of the analytical balance. The analysis of rock reference materials is the only accepted way of establishing the accuracy of a modern analytical technique. For these reasons considerable effort has been expended in developing statistical (and subjective!) procedures for estimating 'true' compositions from reference material analyses. These will now be considered in more detail.

1.10.4 *Statistical analysis of reference material data*

In any co-operative programme for the analysis of reference materials, considerable effort is made by the distributors to ensure both that the sample selected for crushing is homogeneous and that systematic bias is avoided in all crushing, milling and bottling procedures. Some details of these procedures, including tests for homogeneity before distribution, are given by Govindaraju (1980). If there is reason to believe that the starting material is not homogeneous, then additional precautions must be taken. Steele *et al.* (1975), for example, describe the preparation of a precious metal ore reference material. It was necessary to collect 7500kg of rock so that sufficient material would be available for fire assay measurements using 100–200g sample aliquots. Furthermore, the starting material was collected from five separate localities. It was therefore necessary to devise an elaborate scheme of mixing and grinding, summarized in Figure 1.17. This scheme involved sixteen individual operations to homogenize the initial sample to the 240 3kg portions available for distribution.

This rather elaborate preparation procedure re-emphasizes the fact that any reference material distributed for co-operative analysis must have proven homogeneity. In most programmes, sample bottles are numbered and distributed in random sequence. As a further check on homogeneity, participants in some schemes of analysis are requested to analyse a minimum of two individual sample aliquots from each of two separate bottles so that checks can be made on the variance within and between bottles (section 1.4).

Participants are normally requested to analyse reference materials by well-tried methods appropriate to the sample type. In the absence of bias, all data submitted would be expected to be randomly distributed about a mean value. The data would in fact follow a normal distribution function and the assessment of an average value would be a trivial statistical operation. That is not generally so in practice, as is illustrated by data in Figure 1.18. This shows a numeric histogram of data for the elements Rb, Zr,Y in a granite reference material GIT-IWG: MA-N, submitted as part of Govindaraju's 1980 co-operative analysis programme. These results are by no means untypical. A large number of analyses (63) was submitted for rubidium, with the results distributed approximately symmetrically about the mean. The arithmetic mean of these data was 3582 ppm with a standard deviation of no less than 402 ppm. The coefficient of variation over all submitted results is thus 11.2%. A smaller but still statistically significant number of zirconium analyses was submitted. The 28 results varied from less than 4 to 150 ppm. The arithmetic mean was 33.6 ppm with a standard deviation of 28 ppm (coefficient of variation = 83%!). It is not obvious that this distribution is normal. The 12 analyses submitted for yttrium varied from 0.38 to 35 ppm and were distributed in such a way as to make the calculation of statistical parameters meaningless. The proposed values assigned to these elements by Govindaraju (1980) were Rb = 3600 ppm, Zr = 27 ppm and Y = 1 ppm (see also section 10.6.6). It must be emphasized that each of these elements is a common trace element routinely analysed in geochemical laboratories, and that all the data were obtained between June 1979 and March 1980. Although there are some unusual features in the sample itself, it is hard to avoid the conclusion that the uncertainties that surrounded the analysis of major elements in the first co-operative programme in 1951 are equally applicable to trace-element determinations in the early 1980s. It is now proposed to consider statistical procedures that have been developed in an attempt to overcome interlaboratory bias.

1.10.5 *Statistical assessment of reference material data*

Data in Figure 1.18 indicate that the most important feature of any statistical procedure designed to estimate the true composition of a trace-element is a method of recognizing

1. Samples taken from 5 deposits to give a total mass of 7500 kg.
2. Transferred to J.C.I. Mineral Research Laboratory where it was made into 8 batches.
3. Crushed in a jaw-crusher and then in a roller-crusher to about 0.5 cm in size.
4. Mixed 4 times by coning and flattening of the cone.
5. Crushed in a roller–crusher to 0.25 cm in size.
6. Ground in a ball mill (311 kg steel balls of 7 cm diameter) to 50 per cent larger than 200 mesh (Tyler).
7. Transferred to 34 numbered drums, and drums transported to NIM.
8. Drums divided into 4 distributed batches.

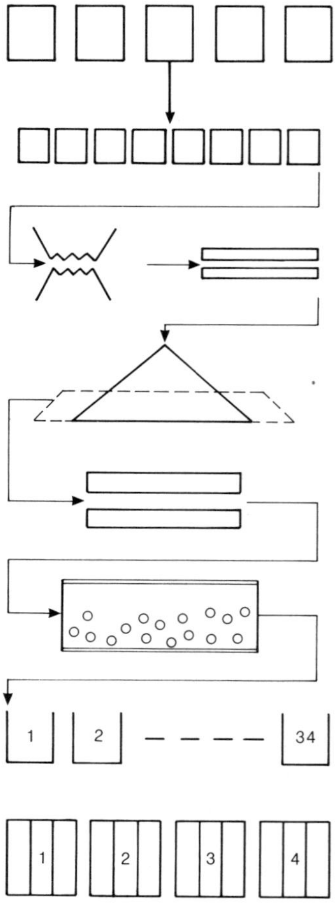

Figure 1.17

9. A flat heap of each batch formed.

10. Batch 1 divided into three portions by sequential shovels from the periphery (1 shovelful = 5 kg). Batches 2 to 4 each in turn treated similarly and ore superimposed on the previous batch (a mixing and dividing operation).

11. Steps 9 and 10 repeated on portion C.

12. Steps 9 and 10 repeated on portion F with a scoop holding 2 kg.

13. Batch 1 divided into 5 portions by feeding of the sample on to a rotating table (1st stage), and then division of the sample into five equal portions with an angular divider (a mixing and dividing operation).

14. Each portion in turn ground to material 95 per cent smaller than 200 mesh in a rubber-lined pebble-mill (a mixing operation).

15. Random (3 kg) samples taken from I, and placed in plastic pails.

16. Steps 13 to 16 repeated for batches G and H, to give total of 240 portions each weighing 3 kg.

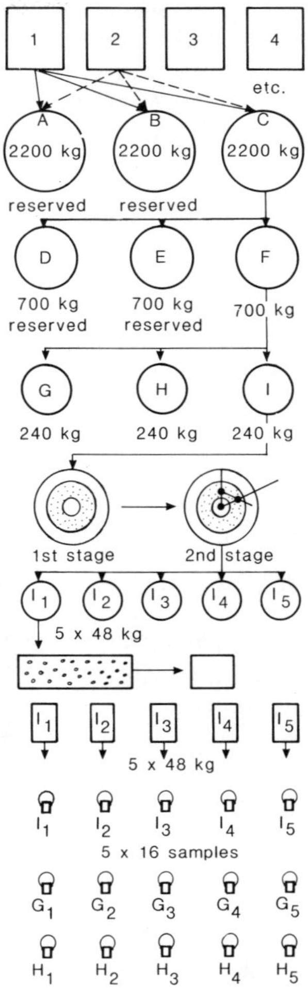

Figure 1.17 Steps in the procedure of Steele *et al.* (1975) for the crushing, grinding, homogenization and bottling of 7500 kg of precious metal ore collected from 5 separate deposits.

systematic bias, that is non-random error, in contributed data. Unfortunately, there are no statistically rigorous treatments that can fully compensate for non-random bias, although some procedures do permit the effective identification and subsequent rejection of outliers (data subject to gross systematic bias). One of the most successful data assessment procedures has included a degree of subjectivity, using the expertise of the evaluator in interpreting statistical and analytical reliability of data. Thus, Govindaraju (1980) used statistical indicators to evaluate data, but the final value was assigned on the basis of the perceived reliability of techniques used for making individual determinations. Conversely, Abbey (1983) used a 'selective laboratories approach' to assess data. This procedure took advantage of the fact that some laboratories could be shown to produce consistently more reliable data than others. These procedures are discussed in more detail in section 1.10.6.

Statistical techniques for the evaluation of reference material data have been reviewed by Lister (1982) and Abbey (1983). The most widely used statistical indicators used in such evaluations are as follows.

(i) *Arithmetic mean* $\bar{x}_a$: This is the simple arithmetic average of a data set. If individual results are denoted x_1, x_2, x_3, ... then the arithmetic mean is given by:

$$\bar{x}_a = \frac{\sum_{i=1}^{n} x_i}{n} \tag{1.35}$$

where n is the total number of results.
The arithmetic mean is an estimate of the 'true' value only when data follow a normal (random) distribution.

(ii) *Median M*: The data set is arranged in ascending order of magnitude. The median is the middle value of the distribution (odd number of results) or the average of the two middle values (even number of results). If the data set is skewed, the median is usually considered to be a more representative estimate of the 'true' value than the arithmetic mean and is much less sensitive to the presence of outliers.

(iii) *Geometric mean* $\bar{x}_{geo}$: An estimator for skewed data sets. The geometric mean is calculated as the square root of the sum of the squares of individual values in the data set. Sankar Das (1979) proposed that the geometric mean compared favourably with other estimators for highly skewed data sets.

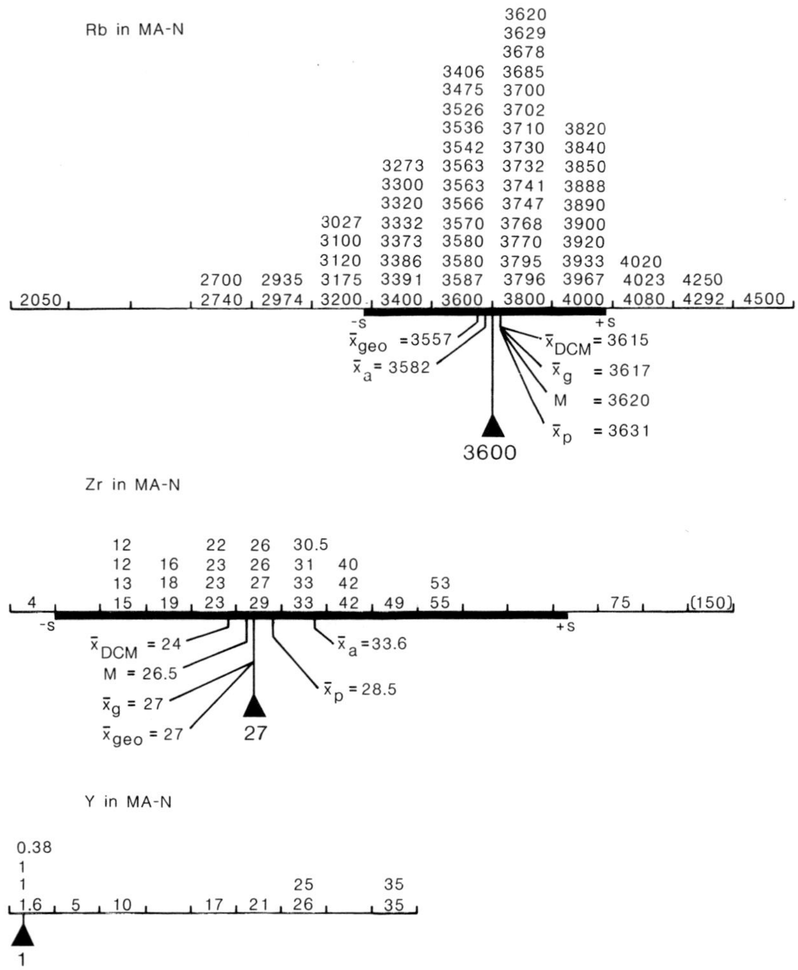

Figure 1.18 Numeric histograms of data for rubidium, zirconium and yttrium submitted during the 1980 cooperative analysis of granite GIT-IWG: MA-N. The proposed value (arrow) and 'indicators of central value' (see text) derived by Govindaraju (1980) are marked.

(iv) *Preferred mean* $\bar{x}_p$: The arithmetic mean and standard deviation are calculated. Outliers are then rejected if they fall outside the limits $\bar{x}_a \pm s$. The preferred mean is the arithmetic average of the residual data set. This is an example of trimming the data set. Several other conventions have been used by other workers, as summarized by Abbey (1983), pp.5–6.

(v) *Gamma central value* $\bar{x}_g$: An estimator for skewed distributions proposed by Christie and Alfsen (1977). A mathematical procedure in which data are transformed into a symmetrical distribution. Mean and standard deviations are then calculated and then back-transformed into the original data set. The technique has the advantage of permitting confidence limits to be calculated for a skewed distribution equivalent to those of a normal distribution. Colombo (1980) suggested that this procedure determined an estimate equivalent to the median.

(vi) *Dominant cluster mode* $\bar{x}_{DCM}$: Ellis *et al.* (1977) suggested a method of calculating the modal value that shows most agreement between analysts. The mean of a data set is calculated and values more than k standard deviations from the mean are discarded. The mean and standard deviation of residual data are calculated and the cycle repeated using a smaller value of k until only the dominant central cluster of data remains. The average of these data is calculated. Ellis *et al.* propose various conventions for performing the iterative calculation.

(vii) *Gastwirth median.* Gastwirth (1966) proposed an estimator calculated by adding (0.4 times the true median) to (0.3 times the sum of the two results), one lying one-third in from the lowest and the other one-third in from the highest datum.

(viii) *Skewness*: The coefficient of skewness is a measure of by how much the symmetry of a set of data deviates from a normal distribution and can be calculated from:

coefficient of skewness

$$= \sum_{i=1}^{n} (x_i - \bar{x})^3 \sqrt{\frac{n}{\left[\sum_{i=1}^{n} (x_i - \bar{x})^2\right]^3}}. \quad (1.36)$$

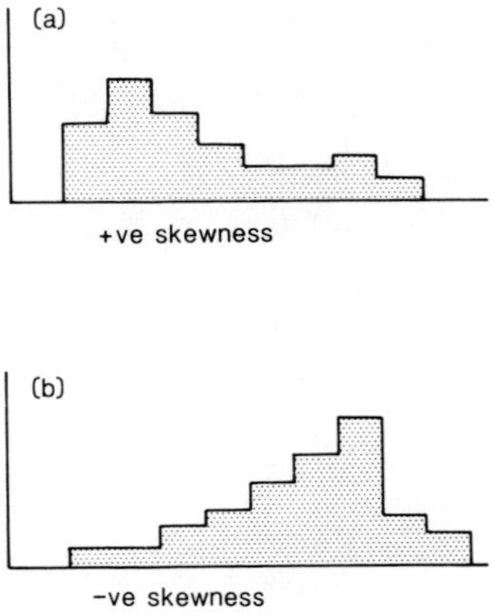

Figure 1.19 Skewed data distribution sets: (*a*) shows positive skew and (*b*) shows negative skew. (Lister, 1982.)

For a normal distribution this coefficient is zero. Distributions can show positive or negative skew (Figure 1.19).

(ix) *Kurtosis*: The coefficient of kurtosis is a measure of how well a set of data corresponds to the shape of a normal distribution and can be calculated from:

$$\text{coefficient of kurtosis} = x_i \frac{n \,.\, \sum_{i=1}^{n} (x_i - \bar{x})^4}{\left[\sum_{i=1}^{n} (x_i - \bar{x})^2\right]^2}. \qquad (1.37)$$

For a normal distribution this parameter has the value 3. If less than 3, the distribution has a flatter top than a normal distribution; greater than 3, the distribution is more pointed. The applicability of these tests for skewness and kurtosis is discussed by Lister (1982).

These parameters identify the more frequently used parameters calculated to evaluate data on rock reference materials.

1.10.6 *Practical schemes of reference material data assessment*

(*a*) *Selected laboratories method.* The selected laboratories method was developed by Abbey over a series of papers as a method of estimating 'usable' values from data of variable reliability (Abbey, 1970, 1972, 1975, 1977, 1982). The basis of this procedure is the observation that in any co-operative investigation, the reliability of data from some laboratories is considerably greater than that from others. Abbey's procedure is therefore designed to identify these 'select' laboratories so that their data alone can be used to calculate 'usable' values.

The procedure is applicable to constituents for which 10 or more results are available and the steps in the calculation are as follows.

(i) Each data point is classified as poor, fair or good according to its position in an ordinal distribution. Poor results are those with values $x_i > (\bar{x}+s)$ or $< (\bar{x}-s)$ where $\bar{x}$ and s are the arithmetic mean and standard deviation of the entire data set.

(ii) After eliminating poor results from the distribution, the mean and standard deviation of the residual data set $\bar{x}_1$, and s_1 are calculated.
Fair results are classified as those with values $x_i > (\bar{x}_1+s_1)$ or $< (\bar{x}_1 - s_1)$
Good results are those in the range $(\bar{x}_1 - s_1) < x_i < (\bar{x}_1+s_1)$.

(iii) Each contributing laboratory is then given a score rating (R) based on the numbers of good (N_G), fair (N_F) or poor (N_P) analyses as follows:

$$R = \frac{(N_G - N_P)}{(N_G + N_F + N_P)} \,.\, 100 \qquad (1.38)$$

(iv) Only results reported by laboratories having a rating that exceeds a specified level are used to calculate the usable value for each element. The specified level depends on several criteria, but in general is set to include those laboratories with the highest rating that contribute between one-third and one-half of the results identified as being 'good'. Full details of this methodology are listed by Abbey (1983), together with a history of the evolution of this procedure. The performance of Abbey's selective laboratories approach has been compared with the statistically rigorous procedures of Dybczynski (1980) by Abbey and Rousseau (1985).

(*b*) *Procedure of Dybczynski.* Dybczynski (1980) described a statistically rigorous procedure for assessing interlaboratory data that does not depend on subjective assumptions. This procedure assumes that analytical results are derived from the same normally distributed population characterized by a population average (μ) and a population standard deviation (σ) (μ and σ are unknowns, for which estimates, x and s, are sought). It is assumed that contributed data also contain a certain number of results (outliers) which come from some other population characterized by a shift in μ and σ. Dybczynski therefore selected five statistical tests designed to identify outliers. Data were treated as if only one outlier were present at any one time. If one of these tests identified the highest or lowest value in the distribution as an outlier, that value was eliminated and the test reiterated until no further outliers could be identified. Average value, standard deviation and variance were then calculated for the remaining population. The statistical tests applied to data were Dixon's test (Dixon, 1953; Grubb, 1969); Grubb's test (Grubb, 1950); coefficient of skewness (Grubb, 1969; Ferguson, 1961); and coefficient of kurtosis (1.10.5(ix)).

Where appropriate these tests were used to reject outliers at the 5% confidence level. Results indicated that concurrent use of all four criteria gave better results than could be expected from the use of any single test. Good selectivity was obtained in the rejection of outliers while avoiding excessive rejections (Dybczynski *et al.*, 1979).

In a paper addressing the approaches of pragmatism versus statistical rigour in assessing interlaboratory data, Abbey and Rousseau (1985) debated the advantages of the selective laboratories method and Dybczynski's method when applied to silicate rock data. Abbey proposed that the selective laboratories method emphasized the importance of understanding the shortcomings of analytical procedures rather than relying blindly on statistical parameters that cannot always be satisfactorily applied to disparate data submitted for co-operative analysis. Rousseau proposed that the selective laboratories procedure involved a series of empirical and subjective considerations and that only the originator is likely to use it correctly. Objective and rigorous statistical methods are to be preferred since they removed human bias from an assessment of data.

(*c*) *Govindaraju's procedure*. The procedures evolved by Govindaraju are neither statistically rigorous nor empirically subjective. They do, however, strike a happy medium between these extremes and have been successfully tested in a number of interlaboratory co-operative analyses. The first step in Govindaraju's procedure is to calculate some simple statistical parameters, namely:

$\bar{x}_a$ arithmetic mean
s standard deviation
s_k coefficient of skewness
M median
$\bar{x}_p$ preferred (trimmed) mean
$\bar{x}_g$ gamma central value
$\bar{x}_{DCM}$ dominant cluster mode
$\bar{x}_{geo}$ geometric mean

Govindaraju termed the last five parameters 'evaluators of the central value' of which the median and preferred mean are considered to be the most reliable. Govindaraju calculated these parameters for all submitted data but also for subsets comprising data analysed by individual techniques. Central values were then evaluated in relation to their internal consistency, and also taking into account any analytical difficulty expected in a particular technique.

Statistical data corresponding to the distributions for Rb and Zr plotted in Figure 1.18 are shown in Table 1.26. Govindaraju's (1980) interpretation of these results for granite GIT-IWG MA-N was as follows.

Rubidium: The mean of the five central values by atomic absorption, XRF, neutron activation and flame photometry was 3607 ppm and that for the two flame methods above (AAS and flame photometry) was 3602 ppm. Accordingly 3600 ppm was proposed as the consensus value for rubidium.

Zirconium: There was clearly some difficulty in evaluating these data since the means of the central values by emission spectrometry are systematically higher than those by XRF. However, the average of the central values by both techniques (27 ppm) was close to that of all the data taken together and this was accordingly adopted as the proposed value.

Yttrium: Govindaraju arbitrarily proposed a value of 1 ppm from these data. The distribution (Figure 1.18) is so scattered that meaningful statistical parameters cannot be calculated. However, it is relevant to note that one analysis by mass spectrometry and two by direct reading emission spectrometry cluster between 0.38 and 1ppm, whereas all the remaining data were obtained by XRF analysis and spread between 1.6 and 35 ppm. It is expected that a serious overlap interference of Rb $K\beta$ on Y $K\alpha$ will significantly interfere with XRF determinations in view of the very high rubidium content of this sample (3600 ppm).

(*d*) *Other schemes*. A detailed account of statistical assessment procedures as applied to ore reference material analysis has been given by Lister (1980). In evaluating the reliability of various procedures when applied to geochemical data, Lister (1984) suggests that the best estimator depends on the number of data, but that the most reliable in approximate

Table 1.26 International cooperative analysis of granite GIT-IWG: MA-N. Statistical analysis of submitted data.
Rubidium: proposed value = 3600ppm

Technique	n	$\bar{x}_a$	s	s_k	M	$\bar{x}_p$	$\bar{x}_g$	$\bar{x}_{DCM}$	$\bar{x}_{geo}$
All submitted data	63	3582	402	−1	3620	3631	3615	3715	3557
Atomic absorption	14	3524	427	0	3519	3478	3524	3522	3500
X-ray fluorescence	33	3609	397	−2	3685	3659	3669	3706	3583
Neutron activation	5	3209	358	0.2	3175	3192	3191	3209	3193
Emission spectrography	2	3120 3820							
Flame photometry	7	3705	119	0.1	3770	3700	3700	3649	3703
Photographic Spectrography	2	3820 4500							

Zirconium: proposed value = 27ppm

Technique	n	$\bar{x}_a$	s	s_k	M	$\bar{x}_p$	$\bar{x}_g$	$\bar{x}_{DCM}$	$\bar{x}_{geo}$
All submitted data	28	33.6	28	2.9	26.5	28.5	27	24	27
Atomic absorption	1	(75)							
X-ray fluorescence	20	25.9	11	0.2	24.5	24.3	25.4	23.8	22.8
Neutron activation	1	(150)							
Emission spectrography	5	34	19	0	33	35	34.2	34.4	29.4
Mass spectrometry	1	(27)							

Data from Govindajaru (1980).
n = number of submitted analyses; $\bar{x}_a$ = arithmetic mean; s = standard deviation; s_k = coefficient of skewness; M = median; $\bar{x}_p$ = 'preferred' mean; $\bar{x}_g$ = gamma central value; $\bar{x}_{DCM}$ = dominant cluster mode; $\bar{x}_{geo}$ = geometric mean.

order are Hampel's *M*-estimator, the median, Gastwirth's mean and the 25% trimmed mean. These are derived in more detail in the proceedings of a year-long symposium on robust estimators of location (Andrews *et al.*, 1972). As far as is known, procedures such as Hampel's *M*-estimator and the Gastwirth median have not yet been applied systematically to silicate reference material data.

1.10.7 *Reliability of rock reference material data*

The importance of 'reliable' analyses of rock reference materials in the calibration of modern instrumental techniques has already been stressed. In this respect, compilations of data for all available silicate samples such as those of Flanagan (1972), Abbey (1983) and Govindaraju (1984) are very valuable. However, these lists of data do have one drawback: they give little indication of the error limits in quoted values apart from a crude classification into 'usable', 'proposed' or 'recommended' values as opposed to 'for information' or 'order of magnitude values' depending on the favoured terminology of the compiler. The previous section has considered in some detail the problems facing compilers of reference material data in attempting to estimate central values from disparate data which may include gross outliers. The calculation of statistically meaningful uncertainty limits from such data is not simple since interlaboratory bias cannot readily be quantified on a statistical basis. And yet without some feel for the reliability of ref-

Table 1.27 Reliability of major element compositions (% oxide) of selected rock reference materials

	Basalt USGS: BCR-1			Granodiorite USGS: GSP-1			Basalt GIT-IWG: BE-N		Albite GIT-IWG: AL-I	
	Abbey (1983)	Gladney *et al.* (1983)		Abbey (1983)	Gladney *et al.* (1983)		Govindajaru (1980)		Govindajaru (1984)	
	$\bar{x}$	$\bar{x}$	(CV%)	$\bar{x}$	$\bar{x}$	(CV%)	$\bar{x}$	(CV%)	$\bar{x}$	(CV%)
SiO_2	54.53	54.78	(0.47)	67.32	67.73	(0.36)	38.39	(1.6)	69.38	(1.0)
TiO_2	2.26	2.24	(2.3)	0.66	0.659	(3.0)	2.62	(5.3)	0.12	(82)
Al_2O_3	13.72	13.74	(0.92)	15.28	15.24	(0.92)	10.12	(4.0)	18.60	(2.6)
Fe_2O_3	3.48	3.62	(3.9)	1.70	1.76	(5.1)	5.37	(8.6)	0.03	(75)
FeO	8.96	8.95	(0.90)	2.32	2.32	(2.4)	6.77	(4.5)	0.04	(40)
MnO	0.18	0.183	(3.3)	0.04	0.040	(6.3)	0.20	(10.0)	0.004	(100)
MgO	3.48	3.48	(2.5)	0.97	0.993	(3.9)	13.22	(3.3)	0.035	(56)
CaO	6.97	7.00	(1.1)	2.03	2.05	(2.5)	13.94	(2.7)	0.384	(20.7)
Na_2O	3.30	3.30	(1.7)	2.81	2.81	(1.6)	3.20	(7.2)	10.60	(4.3)
K_2O	1.70	1.70	(2.4)	5.51	5.54	(1.3)	1.40	(7.9)	0.14	(25)
H_2O+	0.67	0.79	(12.8)	0.58	0.53	(10.4)	2.25	(17.3)	0.29?	(39)
CO_2	0.02			0.12	0.11	(13.6)	0.74	(32)	0.20?	(114)
P_2O_5	0.36	0.37	(2.7)	0.28	0.28	(3.6)	1.06	(12.5)	0.038	(49)
F	0.05	0.048	(4.2)	0.37	0.36	(4.2)	0.10	(17.5)	0.005	(25.7)
S	0.04?	0.042	(1.6)	0.03?	0.035	(17.1)	0.03		0.009	(52)
Total	99.72	100.243		100.02	100.457		99.41		99.875	

Abbey (1983): 'usable' values.
Gladney *et al.* (1983): 'consensus' values. Coefficient of variation calculated for data accepted after statistical rejection of outliers.
Govindaraju (1980, 1984): Recommended and proposed values obtained after evaluation of statistical indicators and assessment of the reliability of the originating technique. Coefficient of variation is calculated from all submitted data without rejection of outliers.
All data recalculated, where necessary, on a dry basis.

erence material data, the cycle of interlaboratory bias is likely to be perpetuated due to the uncritical reliance on reference material calibration data sets.

In order to give some indication of the confidence limits, reference material data are listed in Table 1.27 (major elements) and Table 1.28 (trace elements) incorporating an indication of standard deviation uncertainties.

(*a*) *Major elements.* In Table 1.27 (major elements) data are listed for two USGS reference materials: basalt BCR-1 and granodiorite GSP-1. These results are taken from the

Table 1.28 Reliability of trace element compositions (ppm element) of selected rock reference materials

	Basalt USGS: BCR-1					Granodiorite USGS: GSP-1			Basalt CRPG: BR		Basalt GIT-IWG: BE-N	
	Abbey (1983)	Gladney *et al.* (1983)		IDMS		Abbey (1983)	Gladney *et al.* (1983)		Govindaraju and de la Roche (1977)		Govindaraju (1980)	
	$\bar{x}$	$\bar{x}$	(CV%)	$\bar{x}$	(CV%)	$\bar{x}$	$\bar{x}$	(CV%)	$\bar{x}$	(CV%)	$\bar{x}$	(CV%)
La	27	25.0	(0.16)	24.7	(3.6)	195	183	(3.6)	80	(40)	82	(15.7)
Ce	53	53.7	(0.74)	53.5	(2.0)	360	406	(2.5)	140	(12.9)	152	(21.6)
Pr	7?	6.9	(4.3)			(50)	51	(7.8)				
Nd	26?	28.7	(1.0)	28.9	(1.8)	190	190	(4.5)	60?		70	(21.3)
Sm	6.5	6.58	(1.3)	6.65	(2.4)	25?	26.8	(4.7)	12?		12	(16.0)
Eu	2.0	1.96	(1.3)	1.96	(2.2)	2.4?	2.36	(4.7)	3.7?		3.6	(24.1)
Gd	6.6	6.68	(1.0)	6.72	(2.3)	(15)	13	(7.7)			9	
Tb	1.0	1.05	(4.3)			1.4?	1.36	(5.1)			1.3	
Dy	7?	6.35	(0.94)	6.33	(1.9)	5.7?	5.4	(3.7)				
Ho	1.2?	1.25	(5.6)				1.2	(21)				
Er	3.5?	3.61	(1.2)	3.63	(2.5)	(3)	2.5	(8.0)				
Tm	0.6?	0.590	(3.0)				0.450	(13.3)				
Yb	3.4	3.39	(1.2)	3.41	(2.4)	1.9	1.7	(11.8)	2?		1.8	(36.8)
Lu	0.5?	0.512	(2.4)	0.528	(6.5)	0.2?	0.220	(11.4)			0.24	(52)
As	0.8?	0.64	(10.9)			(0.09)						
Ba	680	678	(1.2)			1300	1310	(0.38)	1050	(19.3)	1025	(14.1)
Co	36	36.3	(2.2)			7.8	6.5	(6.2)	50	(24.2)	61	(12.3)
Cr	15	16	(12.5)			12	13.0	(10.0)	380	(21.1)	360	(15.0)
Cs	0.96	0.97	(6.7)			1	0.95	(8.4)	1.3?		0.8	
Cu	16	19	(10.5)			33	34	(7.4)	72	(34.3)	72	(15.5)
F	500	480	(4.2)			3700	3600	(4.2)	1000		1000	(17.5)

(*Continued*)

Table 1.28 (*Continued*)

	Basalt USGS: BCR-1					Granodiorite USGS: GSP-1			Basalt CRPG: BR		Basalt GIT-IWG: BE-N	
	Abbey (1983)	Gladney *et al.* (1983)		IDMS		Abbey (1983)	Gladney *et al.* (1983)		Govindaraju and de la Roche (1977)		Govindaraju (1980)	
	$\bar{x}$	$\bar{x}$	(CV%)	$\bar{x}$	(CV%)	$\bar{x}$	$\bar{x}$	(CV%)	$\bar{x}$	(CV%)	$\bar{x}$	(CV%)
Ga	22	22	(4.5)			23	22	(6.8)	20	(76)	17	(35.5)
Hf	5	4.9	(3.1)			14?	15.0	(4.3)			5.4	(7.3)
Li	14	12.9	(1.6)			30	31	(6.5)	13	(38.5)	12	(23.1)
Mo	1.5?	1.2	(8.3)			1.5?			3?			
Nb	19?	14	(10.7)			23?	26	(7.7)	100	(8.1)	100	(24.5)
Ni	10	13	(15.4)			9	9.8	(16.3)	260	(28.4)	267	(10.9)
P	1570	1600	(2.8)			1220	1225	(3.3)				
Pb	14	13.56	(0.11)			54	54	(6.5)	8	(61.3)	4	(56)
Rb	47	47.1	(0.64)			250	254	(0.39)	47	(34.2)	47	(18.3)
S	400?	412	(1.6)			300	350	(17.1)	390?		300	
Sb	0.6	0.62	(8.1)			3.1?	3.2	(6.3)				
Sc	33	32.8	(2.6)			6.6	6.1	(4.1)	26?		22	(23.7)
Sn	2.5	2.1	(14.3)			5?	6.6	(10.6)	8?			
Sr	330	330	(0.76)			240	234	(0.64)	1320	(21.9)	1370	(7.9)
Ta	0.8?	0.79	(5.7)			1?	0.91	(7.7)			5.5	
Th	6.1	6.04	(5.0)			105	105	(2.4)	12?		11	(21.5)
U	1.7	1.71	(4.7)			2.1	2.2	(6.8)	3?		2.4	(16.0)
V	420	404	(4.9)			54	53	(6.6)	235	(38.0)	235	(16.1)
Y	40	39	(9.0)			29	29	(10.3)	30	(40.0)	30	(16.7)
Zn	125	129	(0.39)			105	103	(4.4)	150	(23.8)	120	(18.2)
Zr	185	191	(1.3)			500	530	(6.6)	250	(26.5)	265	(15.5)

Data of Abbey (1983) are 'usable' values: ? indicates less reliable data: ( ) indicates 'for information only'.
Data of Gladney *et al.* (1983) are 'consensus' values and the coefficient of variation was calculated for data accepted after statistical rejection of outliers.
IDMS (isotope dilution mass spectrometry): data represents arithmetic mean and coefficient of variation of 23 IDMS analyses listed in Table 17.17.
Data of Govindaraju and de la Roche (1977) and Govindaraju (1980) are mean values derived by evaluation of statistical indicators and a subjective assessment of techniques: coefficients of variation were calculated from *all* data submitted without rejection of outliers. Where analyses are listed without a corresponding CV, too few results were available to permit meaningful statistical analysis.

Table 1.29 Techniques used for major element analysis

	Basalt GIT-IWG: BE-N							Albite GIT-IWG: AL-I							
	% oxide	XRF	CH	AES	AA	NAA	FP	% oxide	XRF	CH	AES	AA	NAA	FP	IR
SiO_2	38.20	37	41	9	13			69.34	43	33	10	15			
TiO_2	2.61	37	37	12	13	1	1	0.12	52	16	14	16			
Al_2O_3	10.07	37	22	11	27	2	1	18.59	42	14	11	31		1	
Fe_2O_3	12.84	32	30	9	25	9		0.075	33	22	11	28	5		
MnO	0.20	36	17	12	34	1		0.004	26	5	16	49		2	
MgO	13.15	34	17	11	38			0.035	25	3	15	55			
CaO	13.87	36	17	11	34	1	2	0.384	36	5	12	42		1	
Na_2O	3.18	25	4	9	31	3	28	10.59	32		6	38	6	19	
K_2O	1.39	32	5	7	30	1	25	0.14	41		5	35		20	
P_2O_5	1.05	38	55	5			1	0.038	51	40	9				
H_2O+	2.24		97					0.29		91					9

BE-N: Data from Govindaraju (1980).
AL-I: Data from Govindaraju (1984).
Data presented as the number of analyses by specified techniques per 100 submitted.
Techniques: XRF, x-ray fluorescence analysis.
CH, chemical procedures (gravimetric, colorimetric, volumetric).
AES, direct reading emission spectrometry.
AA, atomic absorption spectrometry.
NAA, neutron activation analysis.
FP, flame photometry.
IR, infra-red spectrometry.

statistical analysis of Gladney *et al.* (1983) which was based on all the then available published data on these samples. Between 100 and 150 individual data points were available for most elements. Coefficients of variation listed with these analyses were calculated from residual data after elemination of outliers and therefore represent the standard deviation of data that were considered good enough to be used in the calculation of the 'consensus' values. These values are compared in Table 1.27 with Abbey's 'usable' values derived by applying the selected laboratory's procedure to data sets.

Results for these two USGS samples are contrasted with more recent data obtained by Govindaraju (1980, 1984) under the auspices of the International Working Group. Coefficients of variation for these samples were calculated from *all* submitted data without rejection of outliers and are therefore not strictly comparable with the more selective values of Gladney *et al.* These results indicate that there are few major elements whose values are known with a confidence better than 1% (one sigma). Furthermore, for several elements, coefficients of variation exceed 5%, sometimes substantially so, even though the concentration of the element is significantly above the expected detection limit of modern analytical techniques. And so we have the contradiction that many modern instrumental methods are capable of achieving instrument precisions often exceeding 0.1% relative. Uncertainties in analyses of individual reference materials used for calibrating instruments can to

Table 1.30 Techniques for the determination of trace elements: 1980 data for BE-N and AN-G

	Basalt GIT-IWG: BE-N								Anorthosite GIT-IWG: AN-G								Reference
	ppm	XRF	NAA	AES	AAS	MS	ISE	FP	ppm	XRF	NAA	AES	AAS	MS	ISE	FP	technique
Ba	1025	*	○	*	*				34	*		*	○				IDMS
Ce	10	*	*						4.7		*						NAA, IDMS
Cl	140																NAA
Co	61	*	*	*	*				25	*	*	*	○				NAA
Cr	360	*	○	*	*				50	*	*	*	*				
Cs	0.8		*						?								NAA
Cu	72	*		*	*				19	*		*	*				
Eu	3.6		*						0.37		*						NAA, IDMS
F	1000						*		120						*		ISE, colour
Ga	17	*		*					18	*		*					NAA
Gd	9		*						?								NAA, IDMS
Hf	5.4		*						0.38		*						NAA
La	82	*	*	○					2		*						NAA, IDMS
Li	12				*			○	13				*			○	IDMS
Lu	0.24		*						0.12		*						NAA, IDMS
Nb	100	*							2	*							colour
Nd	70	○	*						2		*						NAA, IDMS
Ni	267	*	○	*	*				35	*		*	○				
Pb	4	*			○	○			2					○			IDMS
Rb	47	*	○		*	○		○	1	○			○			○	IDMS
S	300	○							140	○							
Sc	22	○	*	*					10		*	*					NAA
Sm	12		*						0.7		*						NAA
Sr	1370	*		*	*				76	*		○	○				IDMS
Ta	5.5		*						0.2		*						NAA
Tb	1.3		*						0.2		*						NAA
Th	11	*	*	○					?								
U	2.4	○	*						?								
V	235	*		*	○				70	*		*					
W	29		*						90		*						NAA
Y	30	*		*					8	*		*					
Yb	1.8		*						0.85		*						NAA, IDMS
Zn	120	*		*	*				20	*		○	*				
Zr	265	*		*					15	*		*					

BE-N and AN-G: Data from Govindaraju (1980).
Techniques: XRF, x-ray fluorescence analysis.
NAA, neutron activation analysis.
AES, atomic emission spectrometry (direct reading AES and photographic spectrography).
AAS, atomic absorption spectrometry.
MS, mass spectrometry.
IDMS, isotope dilution mass spectrometry.
ISE, ion-selective electrodes.
FP, flame photometry.
Reference technique is the preferred analytical technique for determining reference values taken from Gladney *et al.* (1983) and following the guidelines of the National Bureau of Standards (Uriano and Gravatt, 1977).
* means a significant proportion of reliable data was contributed by the specified technique.
○ signifies a less widely used or less reliable technique.

Table 1.31 Techniques for determining trace elements: 1984 data for albite AL-I

Albite GIT-IWG: AL-I							
	ppm	XRF	NAA	ICP AES	AAS	FP	ISE
As	0.8				○		
Ba	85	○		*	*		
Be	2.7			*	*		
Ce	21	*	*	*			
Cl	110						
Co	1		*				
Cr	4		○	○			
Cs	0.34		*			*	
Cu	3	*		○	*		
Dy	1.4		*	*			
Er	0.59			*			
Eu	0.19		*	*			
F	45						*
Ga	20	*					
Gd	1.9		○	*			
Hf	2.6		*				
Ho	0.2		○	*			
La	9.9	*	*	*			
Li	1				*	○	
Lu	0.11		*	*			
Mo							
Nb	4	*					
Nd	10	*	*	*			
Ni	3	*			*		
Pb	4.5	*			*		
Pr				○			
Rb	5.8	*	○		○	○	
S	85						
Sc	1.74		*	○			
Sm	2.74		*	*			
Sr	80	*		*	○		
Ta	1.85		*				
Tb	0.3		*				
Th	9.5	*	*	○			
Tm			○				
U	5.8	*	*				
V	5.5	○		○	○		
Y	7	*		*			
Yb	0.78		*	*			
Zn	8	*		○	*		
Zr	43	*		○			

Data from Govindaraju (1984).
Techniques: XRF, x-ray fluorescence analysis.
NAA, neutron activation analysis.
ICP–AES, inductively coupled plasma–atomic emission spectrometry.
AAS, atomic absorption spectrometry.
FP, flame photometry.
ISE, ion-selective electrodes.
* means a significant proportion of reliable data was contributed by the specified technique.
○ signifies a less widely used or less reliable method.

some degree be overcome by incorporating a large number of such samples (often over 20) in the calibration data set so that discrepancies will cancel out. However, the only way in which the accuracy of a calibration can be satisfactorily tested is by the analysis of individual reference materials (preferably ones which have not been included in the calibration data set) and comparing analysed results with data such as that in Table 1.27. It is important to emphasize that the success of such a comparison will to some degree be limited by the confidence that can be placed in 'consensus' values of rock reference material data.

(*b*) *Trace elements*. The situation with regard to the confidence which may be placed in trace-element analysis of reference materials is shown in Table 1.28. The USGS basalt BCR-1 must represent one of the more extensively (and reliably) analysed materials available. The 'usable' values of Abbey (1983) calculated by the selected laboratories approach are compared with 'consensus' values calculated by Gladney *et al.* (1984) using a screening technique in which outliers are rejected on statistical criteria. The coefficients of variation from Gladney *et al.* (1984) were calculated from the residual (accepted) data set and these data for the rare-earth elements are compared with selected results by isotope dilution mass spectrometry taken from Table 17.17. Trace-element data for granodiorite USGS:GSP-1 are also listed in Table 1.28 where data of Abbey (1983) are compared with those of Gladney *et al.* (1983). Surprisingly large differences are apparent between some of these data (Ce for example). Results for basalt CRPG: BR in Table 1.28 were mainly obtained from earlier work as compiled by Govindaraju and de la Roche (1977). The reference material GIT-IWG: BE:N was collected from the same source as BR and more recent data from Govindaraju (1980) are also listed in Table 1.28. Coefficients of variation for BR and BE-N were calculated from *all* submitted results rather than the more selective approach used by Gladney *et al.* (1983).

A survey of these data indicates that, with a few notable exceptions, errors in the trace-element analyses of reference materials usually exceed 5% relative (one sigma) and that for some elements, confidence limits are an order of magnitude greater. The exceptions mainly involve certain elements in BCR-1 and GSP-1 for which there are a significant number of high-precision isotope dilution mass spectrometry results (including selected REE, Pb, Rb, and Sr). It is interesting to note that the errors in selected published analyses by isotope dilution mass spectrometry taken from Table 17.17 all exceed the coefficients of variation calculated by Gladney *et al.* (1983).

The problems mentioned above for setting up and assessing the accuracy of major element calibrations are even more serious for trace element data. An associated difficulty is that it is often necessary to determine these elements down to detection limit levels. Such data cannot be achieved unless the calibration line passes through the origin. In instruments that are calibrated directly from reference materials this is not always easy to achieve without a highly critical evaluation of the reliability of individual datum points.

1.11 Which technique for which element?

The technique selected to analyse a sample for a particular element depends upon a number of factors, not all related to analytical sensitivity. Such factors might include the availability of specialized facilities (such as a nuclear reactor for neutron activation analysis), capital cost of instrumentaion, and speed and convenience to user of undertaking the analysis. It is in fact desirable that a range of instrumental techniques is in widespread use since no one technique is ever likely to be completely free of bias. There are therefore two

Table 1.32 Comparative detection limits (in ppm) of selected analytical techniques

	chem	flame phot	flame AAS	hydride gen. AAS	ICP AES	ICP AES REE	ICP MS		arc OES	SSMS	XRF bead	XRF pellet	INAA	gamma	alpha		DN	ISE	TIMS	gas MS	fire assay	prompt gamma
Ac								Ac						*		Ac						
Al	*		12		9		3	Al	*		320					Al						
Ag								Ag	0.2							Ag			*		0.005	
Ar								Ar								Ar				*		
As				0.2	15			As				10	*			As						
Au								Au					0.05			Au					0.001	
B					2		4	B	20							B						0.1
Ba			10		1		2	Ba	2	0.01		24	60			Ba			*			
Be			0.6		0.1			Be	2							Be						
Bi								Bi								Bi						
Br								Br								Br						
C	*							C								C				*		
Ca	*		0.6		4			Ca	*		100	9				Ca						
Cd								Cd								Cd			*			
Ce					15	0.5	3	Ce	40	0.03		30	1.5			Ce			*			
Cl								Cl								Cl		*				
Co			2		2		2	Co	2			6	0.14			Co						
Cr			3		3		2	Cr	2			10	1.5			Cr						
Cs		*					2	Cs		0.005			0.17			Cs			*			
Cu			1		5		2	Cu	2			6				Cu						
Dy						0.3	2	Dy	6	0.02						Dy			*			
Er						0.12		Er		0.02						Er			*			
Eu					2	0.02		Eu	2	0.02			0.05			Eu			*			
F								F								F		50				
Fe	*		3		2.5		4	Fe	*		180	10	100			Fe						
Ga					16			Ga	4			6				Ga						
Gd					5	0.3	4	Gd	6	0.02			3.9			Gd			*			0.1
Ge								Ge								Ge						
H	*							H								H				*		
He								He								He				*		
Hf								Hf		0.02			0.17			Hf						
Hg				*				Hg								Hg						
Ho						0.06	1	Ho		0.02			(1.7)			Ho						
I								I								I						
Ir								Ir		0.05						Ir						0.0001
K		*	0.6		2400			K	7000		120	7		500		K			*			
Kr								Kr								Kr				*		
La					7	0.1	4	La	30	0.03		15	0.5			La			*			
Li		*	1		3		8	Li	0.8							Li						
Lu						0.02		Lu		*			0.10			Lu			*			

Table 1.32 (*Continued*)

	chem	flame phot	flame AAS	hydride gen. AAS	ICP AES	ICP AES REE	ICP MS		arc OES	SSMS	XRF bead	XRF pellet	INAA	gamma	alpha		DN	ISE	TIMS	gas MS	fire assay	prompt gamma
Mg	*		0.12		12		2	Mg	*		725					Mg						
Mn	*		1.2		15		4	Mn	*		180	13				Mn						
Mo							4	Mo	4			6				Mo						
N								N								N				*		
Na		*	0.2		12		4	Na	500		1780		1000			Na						
Nb					10		2	Nb	4	0.02		6				Nb						
Nd					30	0.3	4	Nd	30	0.03		36	4.6			Nd			*			
Ne								Ne								Ne				*		
Ni			3		6		4	Ni	14			6				Ni						
O								O								O				*		
Os								Os		0.05						Os					0.006	
P	*				30		600	P	2000		60	120				P						
Pa								Pa							*	Pa						
Pb			1		20		2	Pb	40	0.015		6		*		Pb						
Pd								Pd								Pd			*		0.0002	
Po								Po							*	Po						
Pr						1	2	Pr	4	0.03		30				Pr						
Pt								Pt		0.05						Pt					0.0016	
Ra								Ra						*	*	Ra						
Rb		*	1				2	Rb	6	0.02		3	10			Rb			*			
Re								Re		0.05						Re					*	
Rh								Rh								Rh					0.0001	
Rn								Rn							*	Rn						
Ru								Ru		0.025						Ru						
S								S			160	20				S				*	0.003	
Sb								Sb					4			Sb						
Sc					2		6	Sc	5			8	0.03			Sc						
Se				0.03				Se								Se						
Si	*		90		5			Si	*		470					Si						
Sm						0.5	4	Sm	10	0.02			0.1			Sm			*			0.1
Sn				*	6		4	Sn	6							Sn						
Sr			1		2		2	Sr	1	0.02		3				Sr			*			
Ta								Ta					0.07			Ta						
Tb						0.9		Tb		0.02			0.09			Tb						
Te				*				Te								Te			*			

Th			30		4	Th	45	0.025		9	0.22	0.5	*	Th	2		
Ti	*	42	1.5		4	Ti	*		140	4				Ti			
Tl						Tl		0.01						Tl		*	
Tm				0.08	1	Tm		0.02			0.34			Tm			
W					4	W					30			W			
U					2	U		0.02		6	0.5	0.2	*	U	0.2		
V		1	4		4	V	3			10				V			
Xe						Xe								Xe			
Y			2	0.03	2	Y	6	0.03		3				Y			*
Yb			1	0.04	2	Yb	5	0.02			0.14			Yb		*	
Zn		0.8	5		6	Zn	24			9	20			Zn			
Zr	*		6		2	Zr	6	0.02		6	100			Zr			

Techniques:

chem	Wet chemistry, colorimetric and gravimetric procedures
flame phot	Flame photometry
flame AAS	Flame atomic absorption spectrometry
hydride gen. AAS	Atomic absorption spectrometry using a hydride generator or cold vapour attachment (Hg)
ICP–AES	Inductively coupled argon plasma–atomic emission spectrometry
REE–ICP–AES	ICP–AES in which sample preparation scheme includes an ion exchange separation for rare earth analysis
ICP–MS	Inductively coupled plasma–mass spectrometry
arc–OES	Arc or spark discharge optical emission spectrometry
SSMS	Spark source mass spectrometry
XRF bead	x-ray fluorescence analysis in which samples are prepared for major element analysis as a lithium borate glass disc
XRF pellet	XRF analysis in which samples are prepared as powder pellets for trace element analysis
INAA	Instrumental neutron activation analysis
gamma	Gamma-ray spectrometry
alpha	Alpha spectrometry
DN	Delayed neutron counting
ISE	Ion-selective electrodes
TIMS	Thermal ionization mass spectrometry
gas MS	Gas source mass spectrometry
fire assay	Fire assay analysis using a lead or nickel sulphide fusion and grapite furnace AAS or neutron activation finish as appropriate
prompt gamma	Prompt gamma-ray neutron activation analysis

All detection limits calculated as six sigma limits of determination

*: detection limit data not available; technique may normally be used to determine indicated element in 'typical' silicate samples.

Flame AAS, ICP-AES, ICP-MS: sample assumed to be dissolved at a concentration of 1 g rock to 100 mL solution, with no chemical separation.

Gamma and alpha spectrometry: determinations refer to isotopes formed by the radioactive decay of ^{235}U, ^{238}U and ^{232}Th.

ICP-MS: sample assumed to be dissolved at a concentration of 0.2 g rock to 100 mL solution, with no chemical separation.

SSMS data are 'practical detection limits' by courtesy of S. R. Taylor (1985, pers. comm.).

ways of answering the question of which technique can be used for which elements. The first is to identify the techniques that are most widely used to determine individual elements. The second is to examine the analytical potential of individual techniques.

In order to identify the frequency of use of analytical techniques for silicate rock analysis, data have been abstracted from reports of Govindaraju (1980) and (1984) on the interlaboratory analysis of reference materials basalt GIT-IWG: BE-N, anorthosite GIT-IWG: AN-G and albite GIT-IWG: AL-I. These data are summarized in Tables 1.29, 1.30 and 1.31.

In Table 1.29, the frequency of use of selected analytical techniques has been listed for the determination of major elements in basalt BE-N and albite AL-I. These data indicate that x-ray fluorescence analysis and atomic absorption spectrometry were the most widely used techniques for the determination of the major elements as a group (excepting P_2O_5, which cannot be determined by AAS, and H_2O+, which cannot be determined by either AAS or XRF). Chemical procedures (including gravimetric, colorimetric and titrimetric) were also extensively used in the determination of selected major elements, including SiO_2, TiO_2, Al_2O_3, Fe_2O_3 and P_2O_5, and represent the only widely used procedures for the determination of Fe(II)O and H_2O+. At the time these analyses were carried out, only about 10% of the data was contributed from optical emission techniques including ICP–AES, although it is expected that this proportion will increase as this technique becomes better established. Flame photometry was used quite extensively for the determination of Na_2O and K_2O, presumably to complement wet chemical procedures used in the determination of the other major elements.

Data for trace-element determination are given in Tables 1.30 and 1.31. In Table 1.30, data from the 1980 report by Govindaraju are listed for the samples basalt BE-N and anorthosite AN-G. This is complemented by data in Table 1.31 from the 1984 report of Govindaraju on albite AL-1. In each case, a technique that contributed a significant proportion of consistently good data to the report is denoted by a solid circle and one that has contributed only a small proportion of the data, or for which results appeared somewhat variable, is denoted by an open circle. Data for the three rock types are listed because close examination indicates that not all techniques are capable of producing acceptable data down to the low concentrations of individual elements present in individual samples. Data for albite AL-1 are included because it is apparent that by 1984, ICP–AES had become a much better established technique for the analysis of trace elements, including the rare-earth elements.

Examination of data in Tables 1.30 and 1.31 demonstrates that four techniques are extensively used for routine multi-element trace analysis, namely, x-ray fluorescence, neutron activation analysis, atomic emission spectrometry (especially inductively coupled plasma spectrometry) and atomic absorption spectrometry. As might be expected, there is considerable overlap in the element coverage of ICP–AES and AAS, the former being more versatile. Data in these tables show that elements can be divided into the following categories.

(i) A core of trace elements that are routinely determined by several of these four techniques, including Ba, Co, Cr, Cu, Ni, Rb, Sr, Th, V and Zn.

(ii) The rare-earth elements that are normally determined by NAA or ICP–AES with XRF being used less frequently to analyse the light REEs (especially La, Ce, and Nd) at higher concentration levels.

(iii) A range of elements that are predominantly (or in some cases exclusively) routinely determined by one technique only. These are as follows:

XRF: Zr, Nb, Ga, Pb, U, Y
NAA: Ta, Hf, Cs, Sc, U
ICP–AES: Be, Y, Zr, Ga, Sc
ISE: F

(iv) A larger group of trace elements which, due either to very low natural abundances or to special analytical difficulties, are not routinely determined at all. These elements include Ge, Se, Br, Mo, platinum group elements, Ag, Cd, In, Sn, Sb, Te, I, W, Re, Au, Hg, Tl and Bi.

The capabilities of individual techniques used in geochemical analysis are indicated by data presented in Table 1.32. These data indicate 'typical' routine detection limits calculated as the six-sigma 'limit of determination'. Where precise data are not available or inappropriate (as in the case of some major elements), the applicability of a technique is indicated by a black spot. These results confirm the capability of XRF, INAA, AAS and ICP–AES as multi-element techniques but also demonstrate the value of delayed neutron fission and ion-selective electrodes, for example, for the determination of specific elements, U and F respectively, in this case. The principles and procedures whereby this performance can be achieved are discussed in the following chapters.

2 Classical and rapid methods of analysis

Chapters 2–7 are mainly concerned with the analysis of silicate rocks by schemes in which the sample is first taken up into solution. Although at first sight this appears to be a rather arbitrary classification, in many laboratories a progressive evolution has occurred in techniques based on rock solution analysis. The earliest schemes of analysis, now known as 'classical' techniques, were based principally on the use of the analytical balance in determining individual components. Such procedures, therefore, involved extensive chemical manipulations so that elements could be precipitated and assayed individually from starting samples. During the 1950s the 'classical' techniques evolved into 'rapid' schemes of analysis. These still relied on detailed solution chemistry but made use of flame photometry, spectrophotometry and complexometric titrations to eliminate some of the more time-consuming chemical manipulations. Rapid techniques have themselves now largely been superseded by more modern instrumental procedures: first atomic absorption spectrometry, followed by inductively coupled plasma–atomic emission spectrometry and now, in a few selected laboratories, inductively coupled plasma–mass spectrometry. All have one common requirement, that of taking the sample up into solution before analysis. In this chapter the techniques for dissolving rocks will first be considered and then the evolution of classical and rapid schemes of analysis.

2.1. Rock dissolution techniques: acid attack

2.1.1 *Properties of mineral acids*

Acid attack is the most widely used method of dissolving silicate rocks, and has evolved into two forms. In one form, the entire sample is taken up into solution so that the major elements and in particular silicon (and aluminium) may be determined. Solutions are often made up in this way to a concentration of 0.1–0.2g rock to 100ml solution. In the other, if silicon is not to be determined, this element is almost invariably removed as volatile silicon tetrafluoride (SiF_4) during sample attack by fuming in the presence of hydrofluoric acid. The resultant solution is more stable since fluorosilicic acid tends to hydrolyse and precipitate on standing or dilution. Such solutions may therefore be confidently prepared at higher rock concentration (up to 1g rock per 100mL solution) and are particularly valuable where the minor and trace elements are to be determined.

The only acid that will readily dissolve silicate material is hydrofluoric acid. This is in part due to the solubility of the hexafluorosilicate ion (SF_6^{2-}) in acidic aqueous solution. However, hydrofluoric acid is not normally considered suitable as a reagent alone, because some salts (those of potassium and calcium, for example), do not have high solubilities in this acid, and because the resultant solution is likely to contain some elements (iron, for example) in uncertain (mixed) oxidation states. Hydrofluoric acid is therefore normally mixed with one (or two) accompanying acids, most commonly perchloric and nitric acids. There is some mystery in the literature about the optimum combination of acid mixtures and indeed about the role of the second acid. Langmyhr and Sveen (1965), for example, questioned the role of the second mineral acid and suggested that, in some circumstances, HF alone was a more effective decomposition agent.

Some physical properties of mineral acids are listed in Table 2.1. The principal chemical properties are as follows:

(*a*)*Hydrofluoric acid.* The principal constituent of any acid decomposition procedure is hydrofluoric acid. This acid readily dissolves silica to form SiF_6^{2-} ions in acid solution. In some circumstances, this reaction can be vigorous, and it is preferable to wet a silicate rock powder with some other reagent before adding HF. HF forms a constant boiling point mixture (an azeotrope) with water, of composition 38.3% HF with a boiling point of 112°C.

Safety note: Hydrofluoric acid (48% HF), the reagent normally used for rock digestions, is one of the most hazardous mineral acids used in the laboratory. Any spills on the skin should be washed with copious cold water and the affected

Table 2.1 Physical properties of mineral acids

Acid	Chemical formula	Conc. reagent % acid	M	sg	b. pt.	
Hydrofluoric acid	HF	48%	29	1.15	112	(38.3% HF, constant boiling mixture)
Perchloric acid	$HClO_4$	70%	12	1.67	203	(72.4% $HClO_4$)
Nitric acid	HNO_3	70%	16	1.42	83 120	(HNO_3) (68% HNO_3, constant boiling mixture)
Hydrochloric acid	HCl	36%	6.8	1.18	110	(20.24% HCl, constant boiling mixture)
Sulphuric acid	H_2SO_4	98%	18	1.84	338	(98.3% H_2SO_4)
Orthophosphoric acid	H_3PO_4	85%	15	1.7	213	(decomposes to HPO_3)
Acetic acid	$CH_3.COOH$	99.5%	17.5	1.05	118	

M = molarity, sg = specific gravity, b. pt. = boiling point (°C).

Table 2.2 Volatile fluorides and chlorides

	b. pt.		b. pt.
SiF_4	−86	BCl_3	12.5
BF_3	−99.9	$AsCl_3$	130.2
AsF_3	−63	$GeCl_4$	84
GeF_4	sublimes	$SbCl_3$	283
SbF_3	319 (sublimes)	$HgCl_2$	302

b. pt. = boiling point (°C)

area treated with a gel containing monosodium glutamate. Failure to do so can cause agonizing pains an hour or more after the accident. HF appears to attack calcium in the body, and to form a complex with bone. One of the more vulnerable areas is under the fingernails. Full protective clothing is therefore essential for handling HF (as for any other concentrated mineral acid). All manipulations must be carried out in a fume cupboard. The material must never be dispensed by mouth pipette. A face visor, gloves and full laboratory coat should be considered as the *minimum* protective clothing.

Further difficulties arise because this acid can dissolve silica from laboratory glassware, even at low concentrations. Plastic laboratory apparatus manufactured from linear polypropylene (softening point 120°C) or PTFE (polytetrafluoroethylene) must be used instead. The latter can be heated as high as 280°C. However, above this temperature the polymer begins to break down, emitting highly toxic vapours. One of the advantages of hydrofluoric acid is the possibility of removing silicon as silicon tetrafluoride. However, in such procedures, losses of other volatile fluorides, including those of B, As, Ge, and Sb must be expected (Table 2.2).

As has been mentioned above, HF is rarely used by itself to dissolve silicate rocks, in part because the fluorides of the major elements do not have uniformly high solubilities, especially calcium fluoride (Table 2.3). The most common second acids include perchloric and nitric acids.

(*b*) *Perchloric acid.* Perchloric acid is one of the strongest mineral acids known. Furthermore, it is a powerful oxidizing agent, although this property is only observed when the acid is both concentrated and hot; the oxidizing properties are lost when the acid is cooled and diluted with water. The acid forms an azeotropic mixture with water containing 72.5% $HClO_4$ (i.e. $HClO_4.2H_2O$), which boils at 203°C (Table 2.1). This is the maximum concentration used as a laboratory reagent. Hot concentrated perchloric acid will react explosively with organic matter, especially alcohols and glycerol. If there is any possibility that organic matter may be present in a rock sample, it is preferable to add a small quantity of nitric acid as well as perchloric acid. Nitric acid reduces the potential for explosive oxidation by oxidizing the more reactive components of the sample at lower temperatures than does perchloric acid.

Salts of perchloric acid are usually extremely soluble, except for those of potassium (and Rb and Cs) (Table 2.3). Inorganic perchlorates are normally stable while moist or in aqueous solution. However, some of these salts become spontaneously flammable in the anhydrous form. For this reason, evaporations involving perchloric acid solutions must not be carried out in wooden fume cupboards. High density polypropylene is the preferred material for all fume cupboards and fume extraction pipework. Furthermore, such equipment must be provided with a washdown facility so that all inaccessible surfaces including the false back of the fume cupboard (if fitted) and the entire fume exhaust ductwork can be sprayed with water after use to prevent the build-up of any anhydrous perchloride deposits.

(*c*) *Nitric acid.* Concentrated nitric acid is a strong oxidizing agent. Its salts are generally very soluble in water (Table 2.3). The acid forms a constant boiling point mixture with water containing 68% HNO_3 with a boiling point of 120.5°C and this is the reagent strength normally used in the lab-

Table 2.3 Solubilities of common salts in aqueous solution

	Fluorides	Solubility in aqueous solution		*Chlorides*	Solubility in aqueous solution		*Perchlorates*	Solubility in aqueous solution	
		Cold	Hot		Cold	Hot		Cold	Hot
Na	NaF	$4.22^{(18)}$		NaCl	$35.7^{(0)}$	$39.12^{(100)}$	$NaClO_4$	s	vs
	Na_2SiF_6	$0.652^{(17)}$	$2.46^{(100)}$						
Mg	MgF_2	$0.0076^{(18)}$	i	$MgOl_2$	$54.25^{(20)}$	$72.7^{(100)}$	$Mg(ClO_4)_2$	$99.3^{(25)}$	vs
	$MgSiF_6$	65							
Al	AlF_3	$0.559^{(25)}$	s	$AlCl_3.6H_2O$	s	vs (HCl evolved)	$Al(ClO_4)_3.6H_2O$	s	s
Si	Na_2SiF_6	$0.652^{(17)}$	$2.46^{(100)}$						
K	KF	$92.3^{(18)}$	vs	KCl	$34.7^{(20)}$	$56.7^{(100)}$	$KClO_4$	$0.75^{(0)}$	$21.8^{(100)}$
	K_2SiF_6	$0.12^{(17.5)}$	$0.954^{(100)}$						
Ca	CaF_2	$0.0016^{(18)}$	$0.0017^{(26)}$	$CaCl_2$	$74.5^{(20)}$	$159^{(100)}$	$(CaClO_4)_2$	$188.6^{(25)}$	vs
	$CaSiF_6$	sls							
Ti	TiF_4	sd		$TiCl_4$	s	d			
Mn	MnF_2	$0.66^{(40)}$	$0.48^{(100)}$	$MnCl_2$	$72.3^{(25)}$	$123.8^{(100)}$			
	$MnSiF_6.6H_2O$	$140^{(18)}$	vs						
Fe (II)	FeF_2	sls		$FeCl_2$	$64.4^{(10)}$	$105.7^{(100)}$	$Fe(ClO_4)_2.6H_2O$	$97.8^{(0)}$	$116.1^{(60)}$
	$FeSiF_6.6H_2O$	128.2							
Fe(III)	FeF_3	sls	s	$FeCl_3$	$74.4^{(0)}$	$535.7^{(100)}$			

Table 2.3 (*Continued*)

	Nitrates	Solubility in aqueous solution		*Sulphates*	Solubility in aqueous solution		*Hydroxides*	Solubility in aqueous solution	
		Cold	Hot		Cold	Hot		Cold	Hot
Na	$NaNO_3$	$92.1^{(25)}$	$180^{(100)}$	Na_2SO_4	$4.76^{(0)}$	$42.7^{(100)}$	NaOH	$42^{(0)}$	$347^{(100)}$
Mg	$Mg(NO_3)_2.6H_2O$	125	vs	$MgSO_4$	$26^{(0)}$	$73.8^{(100)}$	$Mg(OH)_2$	$0.0009^{(18)}$	$0.004^{(106)}$
Al	$Al(NO_3)_3.9H_2O$	$63.7^{(25)}$	vs	$Al_2(SO_4)_3$	$31.3^{(0)}$	$98.1^{(100)}$	$Al(OH)_3$	i	i
							$NaAlO_2$	s	vs
Si							Na_4SiO_4	s	s
K	KNO_3	$13.3^{(0)}$	$247^{(100)}$	K_2SO_4	$12^{(25)}$	$24.1^{(100)}$	KOH	$107^{(15)}$	$178^{(100)}$
Ca	$Ca(NO_3)_2$	$121.2^{(18)}$	$376^{(100)}$	$CaSO_4.2H_2O$	0.241	$0.222^{(100)}$	$Ca(OH)_2$	$0.185^{(0)}$	$0.077^{(100)}$
Ti				$TiO(SO_4)$	d				
Mn	$Mn(NO_3)_2.4H_2O$	$426.4^{(0)}$	∞	$MnSO_4$	$52^{(5)}$	$70^{(70)}$	$Mn(OH)_2$	$0.0002^{(18)}$	
Fe (II)	$Fe(NO_3)_2.6H_2O$	$83.5^{(20)}$	$166.7^{(61)}$	$FeSO_4.7H_2O$	15.65	$48.6^{(50)}$	$Fe(OH)_2$	$0.00015^{(18)}$	
Fe (III)	$Fe(NO_3)_3.6H_2O$	$150^{(0)}$	∞	$Fe_2(SO_4)_3.9H_2O$	440	d			

Data from Weast (1972)
i = insoluble
sls = slightly soluble
s = soluble
vs = very soluble
d = decomposes

oratory. Nitric acid is a more moderate oxidizing agent than is perchloric acid, and is therefore, the preferred acid for attacking samples containing significant concentrations of organic matter. As with perchloric acid, the oxidizing properties of nitric acid are retained only in the concentrated form; when the acid is diluted below about 2M, these oxidizing properties are lost. Aqua regia is formed by mixing nitric with hydrochloric acid (1 : 3). It has specific applications in dissolving gold and some of the platinum group elements, and these are ascribed to the presence of both free chlorine and (ClNO) species in this mixture. A mixture of HF and aqua regia is one of the preferred acid mixtures for dissolving silicate rocks.

(*d*) *Hydrochloric acid.* Hydrochloric acid is normally available as a concentrated solution containing 36% HCl ($\sim$6.8M). Hydrochloric acid is not used alone to attack silicate rocks but is the preferred acid medium for subsequently taking up into solution samples for analysis by flame techniques, especially atomic absorption spectrometry. In chloride solutions interference effects are smaller than with sulphuric acid or nitric acid solutions. It should be noted that the chlorides of Ge, Hg, Sb, As and Sn are volatile (Table 2.2).

(e) *Sulphuric acid.* Sulphuric acid is not much used in the digestion of rocks. This is due to the low solubility of some inorganic sulphates (Table 2.3) but also to the effect it has in depressing the signal of samples to be analysed by atomic absorption spectrometry. Sulphuric acid is normally available as a near-anhydrous reagent comprising about 98% H_2SO_4. The anhydrous acid has an extremely high boiling point (338°C) and when hot and concentrated has oxidizing properties. Cold concentrated sulphuric acid has an extremely high affinity for water, much heat being evolved. When diluting, the concentrated acid must *always* be added slowly to excess water and not vice versa. Use can be made of this affinity when mixed with perchloric acid. The oxidizing properties of perchloric acid/sulphuric acid mixtures are almost equivalent to those of anhydrous perchloric acid.

2.1.2 *Procedures for acid dissolution: silicon to be retained in solution*

This sample preparation technique is normally used when some or all of the major elements (including silicon) are to be determined and has the advantage over fusion techniques (section 2.2) in that the salt content of the final solution is not artificially increased.

(*a*) *Typical procedure.* The method depends on the resistance of the sample to acid attack. Some silicates will dissolve by heating in hydrofluoric acid at temperatures below the azeotrope boiling point of HF–H_2O mixtures (112°C). Samples may then be contained in plastic bottles with screw-top lids. Many plastics deform at temperatures below 112°C (e.g. conventional polyethylene, polystyrene, polyvinyl chloride). Bottles made of linear polyethylene, polypropylene, polycarbonate or polytetrafluoroethylene (PTFE) are, however, suitable for such reactions (Langmyhr and Paus, 1968). A screw-top enclosure is necessary to prevent loss of silica. A typical method is as follows: a finely powdered rock sample (0.200g) is weighed into a suitable plastic bottle (100mL capacity). Nitric acid or aqua regia (0.5 to 1 mL) are added as wetting agents, then 5mL HF (40%). The mixture is heated on a water bath at 100°–110°C for 30 minutes. If a clear solution is obtained, saturated boric acid (50mL) is added and the solution diluted to 100mL, preferably in the reaction vessel.

(*b*) *Effect of boric acid.* The boric acid complexes with any free HF by a two-stage exothermic reaction:

$$H_3BO_3 + 3HF \rightleftharpoons HBF_3OH + 2H_2O \qquad (2.1)$$

$$HBF_3OH + HF \rightleftharpoons HBF_4 + H_2O. \qquad (2.2)$$

The resultant solution is far more stable and less hazardous to handle and may indeed be diluted to volume in glass volumetric flasks. Under certain conditions, however, fluoroboric acid hydrolyses to release HF. This reaction is promoted by increasing the hydrogen ion concentration. Calcium ions and probably other divalent ions have the same effect. The stability of $HF–H_3BO_3–H_2O$ mixtures towards laboratory glassware has been investigated by Bernas (1968). His data (Figure 2.1) show the concentration of silica found in 50mL solutions containing 1.5mL 48% HF and various concentrations of boric acid after two hours in contact with a glass container at ambient temperatures. These data indicate that solutions become contaminated by not more than 0.1mg silicon, providing 2.8g boric acid per 100 mL of solution are present. The stability imparted by boric acid to hydrofluoric acid solutions for periods of up to one hour has been confirmed by Buckley and Cranston (1971).

If a precipitate is formed after the acid attack stage, boric acid may be added to the reacting mixture, which is then reheated in the closed reaction vessel on a water bath for 15 to 30 minutes, allowed to cool, then diluted to 100mL with water.

(*c*) *Mineral phases resistant to acid attack.* Unfortunately, the scheme of analysis above cannot be adopted uncritically for routine general silicate analysis, since many samples contain mineral phases which are particularly resistant to acid attack. These include, for example: zircon ($ZrSiO_4$), rutile (TiO_2), corundum (Al_2O_3), spinel ($MgAl_2O_4$), tourmaline ($(Na(Mg, Fe, Mn, Li, Al)_3Al_6Si_6O_{18}(BO_3)_3(OH, F)_4)$), beryl ($Be_3Al_2Si_6O_{18}$), chromite ($FeCr_2O_4$) and cassiterite SnO_2. French and Adams (1973) investigated the problem of dissolving selected resistant minerals. Samples (0.1g) were attacked by 0.2mL aqua regia and 5mL 40% HF in 60mL screw-topped polypropylene bottles. These were heated on a steam bath for at least one hour. Many minerals dissolved readily under these conditions including alkali-feldspars, plagioclases, feldspathoids, quartz, biotite, muscovite, phlogopite, chlorite, hornblende, actinolite, epidote minerals, diopside, augite, olivine, calcite, dolomite, illite, kaolinite, montmorillonite, magnetite and ilmenite. However, some difficulty was experienced in achieving a satisfactory rate of dissolution for minerals such as pyrite, garnets, andalusite, sillimanite, kyanite, staurolite, tourmaline and hypersthene. Accordingly, 0.025g portions of these minerals were attacked by various HF–mineral acid combinations for periods of up to 10 hours. Their results are summarized in Figure 2.2. Pyrite was found to dissolve readily if the aqua regia content of HF solutions was increased to 1mL. In all cases, the rate of dissolution of minerals increased with finer grinding. The proportion of undissolved mineral was found to be reduced as a logarithmic function of time. However, the rate of attack on kyanite and staurolite was so low (Figure 2.2) that, for practical purposes, these minerals should be considered insoluble by this procedure. French and Adams (1973) reported that the rates of mineral decomposition increase by a factor of three if polycarbonate bottles are heated under pressure in domestic pressure cookers (see also French and Adams, 1973), though such procedures are slower and less convenient for routine batch analysis.

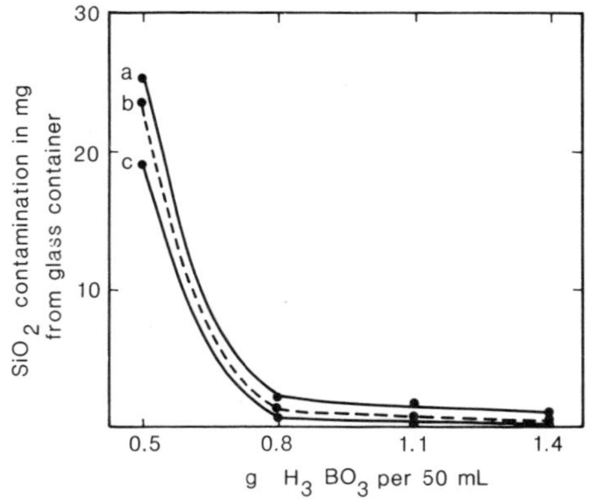

Figure 2.1 Effect of boric acid in suppressing the attack on glass containers by hydrofluoric acid solutions. Data were obtained by analysing the silica content of solutions containing 1.5 mL hydrofluoric acid (48%) diluted to 50 mL that had been in contact with borosilicate glass apparatus for 2 hours. Curve (*a*) represent the results from solutions containing HF and boric acid only. Curve (*b*): solutions as contained 0.25 mL aqua regia. Curve (*c*): solutions contained 0.25 mL aqua regia plus the ionic species obtained from 50 mg of a decomposed silicate-borosilicate glass sample. (Bernas, 1968.)

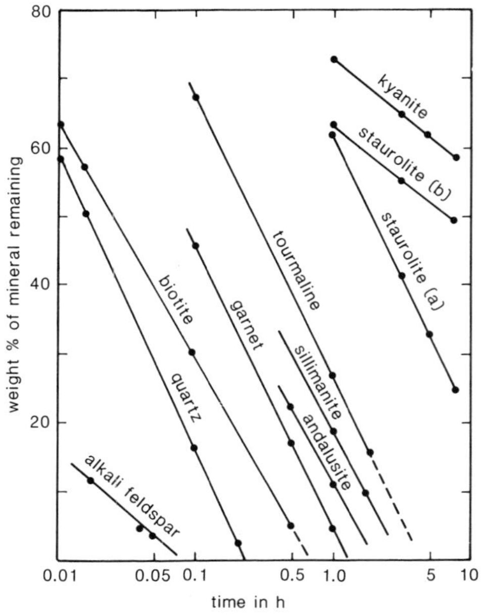

Figure 2.2 Approximate rates of dissolution of selected silicate minerals attacked by hydrofluoric acid (5 mL) and aqua regia (0.2 mL) in 60 mL polypropylene bottles. Sulphuric acid (5 mL) was added to the reagent used to dissolve staurolite (B). (French and Adams, 1973.)

(*d*) *Bomb techniques.* An important advance in acid dissolution procedures was made with the introduction of the acid digestion bomb popularized by Bernas (1968) and Langmyhr and Paus (1968). This consists of a PTFE beaker with lid, which fits tightly into an outer stainless steel (or monel) pressure jacket (Figure 2.3). This outer jacket has a screw-top lid (or, less satisfactorily, screw studs) which, when closed up, forms a gas-tight seal between PTFE beaker and lid. The charge, normally consisting of 200mg of powdered sample, is weighed into the inner Teflon beaker. A small volume of aqua regia (0.2mL) is then added. This acts in part as a wetting reagent, preventing excessive reaction in some samples when HF is added. Following the aqua regia, 5.0mL HF are added. The bomb is then sealed and heated in an oven to temperatures in the range 110°–150°C. Bombs

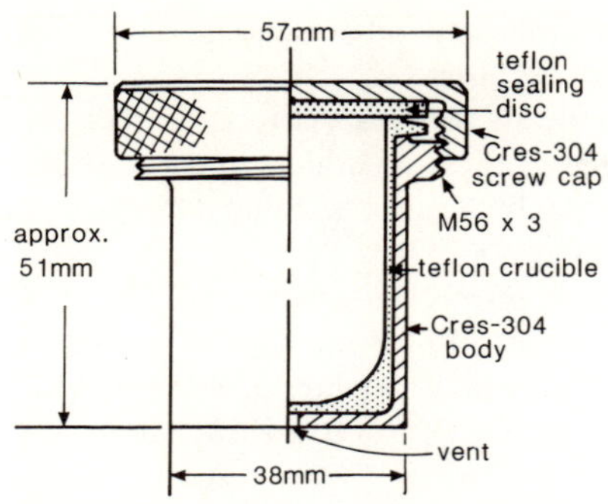

Figure 2.3 Teflon decomposition bomb used by Bernas, 1968.

must never be filled with liquid to more than 10% to 20% of their total free volume, nor should organic reagents and oxidizing agents be heated in a sealed bomb. In both cases the reaction may become explosive.

Calculations suggest that if all the reaction mixture vaporizes on heating, the pressure inside the bomb will rise to 70–100 atmospheres. Under these hostile conditions, many minerals, resistant to attack at normal pressures, will decompose rapidly. Much uncertainty as to whether a batch of samples will completely dissolve is thus eliminated in bomb techniques. After heating for up to one hour, the bomb is allowed to cool. It is essential that the bomb is not opened up until it has completely equilibrated to room temperature. The reaction mixture must then be examined to see if the sample has completely dissolved. Any precipitate formed on cooling may be redissolved by adding saturated boric acid and warming, as above, before finally adding water to make up a final volume of 100mL. The resultant solution contains rock at a total dissolved solids content of 0.2% (not including HF and H_3BO_3) and all the major elements have been retained in solution.

Because resistant minerals are successfully attacked, acid digestion bombs were soon adopted for dissolving zircons in the determination of lead isotope ratios by mass spectrometric techniques (see section 17.4).

Van Eenbergen and Bruninx (1978) investigated the potential loss of elements during attack in an acid digestion bomb. They concluded that virtually no losses occurred during decomposition, except for some elements (e.g. germanium) that might precipitate from solution in the bomb under unfavourable circumstances.

2.1.3 *Procedures for acid dissolution*: *silicon to be lost by volatilization*

If silicon is not to be determined in the resultant solution there are considerable advantages if silicon (and fluorine) are removed by evaporation during the acid attack procedure. These advantages are:

(i) Resultant solutions are far more stable (silica-bearing solutions tend to hydrolyse and precipitate on standing for long periods)

(ii) Silicon and fluorine cause interferences with some photometric determinations

(iii) The corrosive action of HF is removed, permitting manipulations with normal laboratory glassware

(iv) Such procedures reduce the salt content of resultant solutions; this is advantageous if instrumental procedures such as AAS or ICP-AES are to be used for determinations.

(*a*) *Typical procedure*. A typical procedure, developed from the work of Riley (1958*a*) is as follows. Rock powder (0.5g) is weighed into an open dish (normally made of platinum). To this powder is added 5 mL perchloric acid (60% vol/vol) and 15mL of 40% hydrofluoric acid. The mixture is heated on a sand-bath overnight at a temperature of 250–280°C to incipient dryness. During this evaporation, silicon and hydrogen fluoride are expelled. In some schemes of analysis, more perchloric acid is added to the residue and the evaporation repeated to ensure complete removal of volatile phases. The residue is taken up into solution in acid media and diluted to a final volume of 50mL. The choice of acid depends on the technique to be used for analysis. Walsh (1980) used 4mL perchloric acid, together with 15mL of water. The mixture was warmed gently to redissolve the residue. and diluted to 50mL after cooling. Samples were then analysed for selected major and trace elements by ICP–AES. However, perchloric acid 'poisons' some forms of cation exchange resins and must be removed if samples are to be subjected to such a chemical separation. Furthermore, perchloric acid causes some interference effects in atomic absorption spectrometry, and if this technique is to be used for determinations, samples after evaporation are normally taken up in hydrochloric acid (see e.g. Langmyhr and Paus, 1968).

(*b*) *The role of perchloric acid*. The main role of perchloric acid in the initial acid attack is to substantially increase the boiling temperature of the reaction mixture. (The boiling point of 72% perchloric acid is 203°C.) This has two results. First, fluorosilicic acid formed in solution by the attack of hydrofluoric acid on silicate materials, is efficiently decomposed at higher temperatures, leading to the volatilization of silica tetrafluoride:

$$H_2SiF_6 \rightarrow SiF_4\uparrow + 2HF. \qquad (2.3)$$

Secondly, hydrofluoric acid itself is progressively evaporated from the reaction mixture, since the HF–water azeotrope boils off at 112°C. During this evaporation, it is expected that other elements that form volatile fluorides will also be lost (Table 2.2).

Langmyhr and Graff (1965) commented that sulphuric acid with its higher boiling point (338°C) is more efficient than perchloric acid in this role. However, this acid is normally avoided in modern flame techniques because of the serious interferences that can then occur. Cohen *et al.* (1975) advocated the use of *m*-benzenedisulphonic acid, since this material has a much higher boiling point than perchloric acid and does not cause precipitation of salts such as potassium perchlorate, which have a low solubility. However, this acid has not found widespread use.

(*c*) *Resistant minerals*. As with closed vessel attack of samples with hydrofluoric acid (section 2.1.2), the HF/perchloric acid evaporation scheme fails to dissolve certain resistant minerals. This is of some concern since these solutions are routinely used to determine trace levels (by AAS and ICP–AES) of some elements that are extensively partitioned into resistant minor phases such as chromite, zircon and various rare earth element/U/Th minerals.

To overcome these difficulties, Walsh *et al.* (1981) recommended filtering the solution obtained from the HF/perchloric acid attack. Any insoluble residue was fused with NaOH or KHF_2. The resultant melt was dissolved in dilute hydrochloric acid and combined with the filtrate resulting from the initial acid attack. This procedure ensured that resistant rare earth element-bearing phases, including zircon, were brought completely into solution so that quantitative rare-earth element determinations could be carried out by ICP–AES.

Croudace (1980) reported difficulties in dissolving certain rock types (in particular those of tonalitic to alkali granite composition) using an HF/perchloric acid evaporation. An insoluble residue was occasionally formed subsequent to fuming off SiF_4 at 170°C. Attempts to dissolve this residue in 1M HCl failed. This residue was tentatively identified as having an x-ray diffraction pattern similar to that of the mineral ralstonite with a compositon $Na_{0.88}Mg_{0.88}Al_{1-12}(F,OH)_6.H_2O$. Croudace suggested that the most effective way of avoiding such insoluble residues was to prevent the reaction mixture from drying out completely. The first HF–perchloric acid evaporation should be carried out at 100°–110°C to a point where the sample is still damp. A few drops of perchloric acid are then added and the sample reheated to 160–180°C, but not to dryness. This is repeated up to four times to ensure that all traces of fluorine are eliminated. The residue can then be allowed to evaporate to dryness without risk of forming the insoluble fluoride salt.

It is apparent that hydrofluoric acid attack is a versatile method of dissolving silicate rocks, but that caution must be expressed in its application to some rock types that contain certain resistant minerals. In such circumstances, alternative procedures have been developed based on the fusion of the sample with a low melting point alkali salt.

2.2 Rock dissolution procedures: fusion with alkali salts

2.2.1 *Lithium metaborate fusions*

In rapid methods of analysis, it was necessary to prepare two sample solutions for the determination of all major elements, since silica and fluorine had to be removed from one solution (solution B) to prevent interferences affecting the determination of other constituents. However, after the introduction of atomic absorption spectrometry this limitation was no longer relevant and considerable interest arose in developing simple procedures for the determination of all major elements from one solution prepared by fusing samples with lithium metaborate ($LiBO_2$). Such procedures were first proposed by Ingamells (1964) and were rapidly adopted for schemes of analysis based on atomic absorption spectrometry (e.g. Suhr and Ingamells, 1966; Shapiro, 1967; van Loon and Parissis, 1969; Omang, 1969; Medlin et al., 1969). Sample decomposition procedures have been reviewed by Ingamells (1970).

The lithium metaborate fusion technique offers several advantages over acid attack procedures:

(i) Hazardous reagents such as HF and $HClO_4$ are not required

(ii) Specialized apparatus such as Teflon-lined digestion bombs is not required

(iii) Specifically with respect to atomic absorption analysis, the presence of excess lithium and borate ions in sample solutions has a buffering effect that suppresses some matrix interferences.

(*a*) *Fusion and dissolution procedures.* The rock powder is mixed with excess lithium metaborate powder in a crucible and fused for 15 to 30 minutes at 900°–1050°C in a muffle furnace or alternatively over a Meker burner (Figure 2.4). Van Loon and Parissis (1969) restricted temperatures to 900°C to avoid the risk of a slight loss of alkalis by volatilization. Non-wetting platinum crucibles are preferred for fusions. As an alternative, slightly less durable graphite crucibles may be used. Van Loon and Parissis (1969) recommended pre-igniting empty graphite crucibles for 30 minutes at 950°C before use. In early procedures, excess lithium metaborate was generally added to the sample in the proportion 5 to 7 parts of lithium metaborate to 1 part of rock powder. More recent work suggests that a flux-to-sample ratio of 3 to 1 is optimum (Crock *et al.*, 1984). After ignition, the melt is normally poured directly into a beaker containing dilute nitric acid (1M). In some earlier procedures, hydrochloric acid was used to dissolve lithium metaborate beads, although subsequent work (e.g. Crock *et al.*, 1984) suggests that boric acid will precipitate from solution under these conditions. The effect of pouring the molten bead directly into nitric acid solution is to shatter the sample into small fragments that will dissolve rapidly. Govindaraju *et al.* (1971) and Govindaraju and L'homel (1972) proposed the use of steel rollers to laminate the lithium metaborate melt. The molten sample was poured through the rollers and emerged as a thin (80μm thick) foil of solidified glass. This sheet was rapidly crushed in an agate mortar and, due to the large surface area then formed, the glass dissolved much more rapidly.

(*b*) *Interferences and resistant minerals.* Some difficulties have been encountered with the formation of insoluble polysilicic acid gels when dissolving the lithium metaborate melt in acid solution. To overcome this problem, Walsh and Howie (1980) allowed the crucible and melt to cool before immersing the crucible completely in about 200 mL of water containing 10 mL of concentrated nitric acid. A magnetic stirring

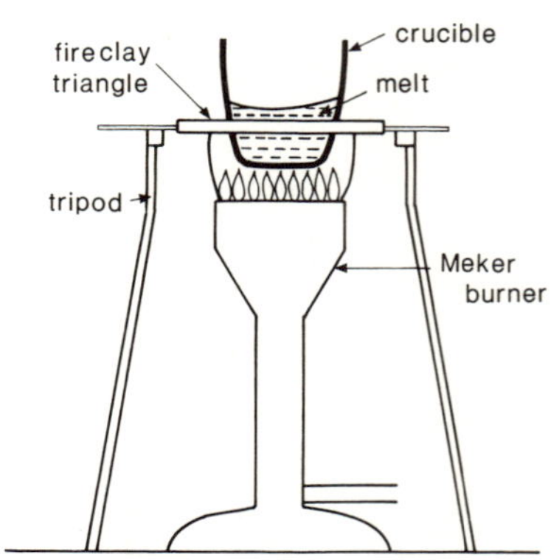

Figure 2.4 Apparatus for fusing a rock powder and alkali salt flux in a crucible over a Meker burner.

bar was then placed inside the crucible and stirring commenced immediately. The bead dissolved in 1–2 hours and the solution was made up to 250ml (from 0.5g of rock). Such solutions were found to be completely stable for several months.

Studies indicate that many common silicate minerals dissolve readily by lithium metaborate fusion; these include quartz, feldspars, amphiboles, pyroxenes and micas. Cremer and Schlocker (1976) carried out a detailed study of the success of this fusion method in dissolving a wide range of minerals that are sometimes present as accessory phases in silicate rocks. Samples of individual minerals, crushed to minus 100 mesh, were mixed with lithium metaborate in the proportions 0.1000g of mineral to 0.7000g of flux. The mixture was ignited in a graphite crucible in a muffle furnace at 950°C for 15 minutes. Molten beads were poured into 4% (by volume) nitric acid (100mL). Any insoluble residue was filtered, ignited for one hour at 600°C in a crucible, weighed and examined by x-ray diffraction and polarizing microscope. Experiments were repeated using a 2:1 mixture of lithium tetraborate/metaborate as flux. The results of this investigation are summarized in Table 2.4. This shows that a range of oxide and sulphide accessory phases including rare-earth-bearing minerals are not dissolved satisfactorily by the lithium metaborate fusion procedure.

Table 2.4 Minerals that leave insoluble residues on fusion with lithium metaborate or 2:1 lithium tetraborate/metaborate fluxes

Mineral or ore	Chemical formula	% residue after fusion in: lithium metaborate	2:1 lithium tetra/metaborate
Minerals			
Bastnaesite	(Ce, La)(CO_3)F	20	20
Chalcocite	Cu_2S	55–92	41–82
Chalcopyrite	$CuFeS_2$	39	44
Chromite	$FeCr_2O_4$	46	47
Corundum	Al_2O_3	<0.5	0.8
Fergusonite	Y(Nb, Ta)O_4	0.8	0.6
Fluocerite	(Ce, La, Y)F_3	5	6
Galena	PbS	35	13
Ilmenite	$FeTiO_3$	16	0.7
Monazite	(Ce, La, Th)PO_4	12	27
Pyrite	FeS_2	40	26
Wolframite	(Fe, Mn)WO_4	1.4	3
Xenotime	YPO_4	4	4
Zircon	$ZrSiO_4$	21	20
Ores			
Chrome	chromite	33	35
Refractory	(Fe, Mg) (Cr, Al)$_2O_4$		
Zinc ore	zincite ZnO sphalerite ZnS	34	79
Tin ore	cassiterite SnO_2	29–37	70–20
Titanium ore	rutile TiO_2	<0.5	29–32
Nickel oxide	bunsenite NiO	46	—

Minerals that dissolved or left a residue of less than 0.5–1%: actinolite, albite, allanite, anhydrite, ankerite, apatite, beryl, columbite, eudialyte, euxenite, ferberite, franklinite, gadolinite, garnet, goethite, graphite, magnetite, olivine, pyrolusite, quartz, scheelite, sphene, spinel, staurolite, tourmaline, turquoise, vesuvianite, vivianite, vonsenite, manganese ore, iron ore, phosphate rock, bauxite, calcium molybdate, burned refractory, fluorspar, lead–barium glass, clay, silica brick, burned magnesite, lithium ore, phosphate rock.
Fusion conditions: sample-to-flux ratio = 1:7; samples fused in graphite crucibles in a muffle furnace at 950°C for 15 minutes. Molten beads poured into 100 mL of 4% nitric acid.
All data from Cremer and Schlocker (1976).

One may conclude that a lithium metaborate fusion is a very effective technique for the major silicate-forming elements (with some reservation about some titanium and iron phases—Table 2.4) but that there must be some doubt about its effectiveness in systematically dissolving all minor and trace element phases.

2.2.2 *Alkali carbonate, oxide and hydroxide fusions*

Several alkali salts have been proposed for fusion of rock samples, including sodium carbonate (Na_2CO_3), sodium hydroxide (NaOH), sodium peroxide (Na_2O_2), the equivalent potassium salts, potassium pyrosulphate (a mixture of $K_2S_2O_7$ and $KHSO_4$), alkali fluorides (including KHF_2), and so on. Distinction must be drawn between *sintering* and *fusion* methods of sample decomposition. During the fusion process the flux melts and completely dissolves the powdered sample. In the sintering procedure, insufficient flux is added to dissolve the sample completely. Instead, on ignition, the flux breaks down the sample matrix, rendering it soluble when the mixture is cooled and dissolved in dilute acid solution. Many of the fluxes listed above have specific applications. However, in general the effectiveness of a flux in attacking silicate rocks increases in the order Na_2CO_3 < NaOH < Na_2O_2. The physical characteristics of individual fluxes are listed in Table 2.5; applications of individual mixtures are as follows.

(*a*) *Sodium carbonate* (Na_2CO_3). Sodium carbonate fusions are most widely used in the determination of fluorine by ion-selective electrodes. Indeed, for this application, Hopkins (1977) reports that analytical results are more consistent

Table 2.5 Salts used as fluxes to decompose silicate rocks

Salt		Melting point °C	Crucible for fusion
Lithium metaborate	$LiBO_3$	845	Platinum—
Lithium tetraborate	$Li_2B_4O_7$	930	5% gold or graphite
Sodium carbonate	Na_2CO_3	851	Platinum
Potassium carbonate	K_2CO_3	891	or nickel
Sodium hydroxide	NaOH	318	Zirconium
Potassium carbonate	KOH	360	(or gold, nickel or silver)
Sodium tetraborate (borax)	$Na_2B_2O_7$	741	
Sodium peroxide	Na_2O_2	480 (decomposes)	Zirconium
Potassium superoxide	KO_2	380	
Potassium fluoride	KF	846	
Potassium hydrogen difluoride	KHF_2	225 (decomposes)	
Potassium pyrosulphate	$K_2S_2O_7$	300	

Data from Weast (1972).

than those obtained by a sodium hydroxide fusion, for example. Anhydrous sodium carbonate is normally used as a flux by itself. However, other fusion mixtures have been reported for fluorine determination, including Na_2CO_3–K_2CO_3 (e.g. Troll et al., 1977) (this has a lower melting point than Na_2CO_3 alone); Na_2CO_3–ZnO (e.g. Ingram, 1970); and Na_2CO_3–KNO_3 (e.g. Haynes and Clarke, 1972), which is preferred for samples containing significant levels of sulphide. The properties of this melt cause the oxidation of sulphur to sulphate, so reducing interference caused by the sulphide ion.

Fusion mixtures are normally prepared with a minimum ratio of 4 parts of flux to 1 part of sample, and ignited at between 900 and 1000°C (depending on flux composition) for 20 minutes to 1 hour. Fusions may be carried out in platinum or nickel crucibles.

Jeffrey (1975, p.26) reports that 5 parts of flux to one part of rock will completely dissolve basic rocks, whereas some acidic rocks may require only 3 parts of flux to 1 of rock. If it is necessary to ensure complete digestion of some accessory phases such as zircon, rutile or chromite, fusion at 1000°C for 1 hour must be followed by an additional period at 1200°C for 10 minutes.

After cooling, the fusion mixture is normally dissolved in water (with heating on a steam bath if necessary). Hydrochloric acid (1M) is added to this solution, until slightly acidic, to release carbon dioxide.

Potassium carbonate is more hygroscopic than sodium carbonate and it may be desirable to preheat this flux to expel absorbed water before fusion with the sample.

(*b*) *Alkali hydroxides and peroxide fusions*. Alkali hydroxides and peroxides are extremely efficient at dissolving silicate rocks, and melt at much lower temperatures than do sodium carbonate fluxes. One important application of alkali hydroxide/peroxide fusions is in the determination of trace elements where it is essential that accessory phases such as zircon or rare-earth element minerals are quantitatively brought into solution. Sometimes sodium hydroxide is used alone for fusion, but more commonly NaOH–Na_2O_2 mixtures are used, especially if the rare-earth elements are to be determined. Both fluxes attack platinum crucibles, so gold, silver or nickel crucibles have been used, since these materials are more slowly attacked, and in more recent applications, zirconium crucibles have been preferred. Nevertheless, it is expected that a small fraction of the crucible material will always be dissolved in these melts. The resultant solutions cannot, therefore, be analysed for the crucible element (nor for the platinum group elements if nickel crucibles are used).

Sodium hydroxide has a long-standing history as a flux agent and was used, for example, in the 'rapid' schemes of analysis of Shapiro and Brannock (1956, 1962). Their solution A, used to determine Si and Al, was prepared as follows. Sodium hydroxide solution (5ml of 30% NaOH w/v) was added to a nickel crucible. The use of NaOH in solution form eliminated the problems caused by the very hygroscopic nature of NaOH pellets that makes reproducible weighing of NaOH in solid form difficult. However, concentrated NaOH solutions will readily attack glass, and the reagent must be stored in and dispensed from plastic apparatus. After evaporation of the NaOH solution to dryness, 50mg of silicate rock was added to the nickel crucible, which was then heated to dull red heat (about 800°C) for 5 minutes. About 50mL of water was then added to each crucible; these were left overnight to permit the melt to disintegrate. The crucible contents were then washed into a volumetric flask (1 litre) dissolved in dilute HCl and made up to the mark. Riley (1958*a*) reported more rapid fusions and consistently clear solutions if silver rather than nickel was used as the crucible material. Walsh *et al.* (1981) used a NaOH fusion (NaOH being added to the crucible in pellet form) in a silver crucible to dissolve the accessory mineral residue which remained after an HF–$HClO_4$ attack to ensure quantitative dissolution of resistant mineral phases, see 2.1.3(*c*) above.

Sodium peroxide is not normally used as a flux by itself, but is mixed with sodium hydroxide. Indeed Maxwell (1968, p.98), discussing the work of Theobald (1942), reports that the Na_2O_2–NaOH mixture is more effective than Na_2O_2 either alone or mixed with Na_2CO_3 in decomposing samples containing pyrite and chromite. Na_2O–NaOH fusions are extensively used in radiochemical neutron activation analysis group separation procedures where it is essential to obtain quantitative dissolution of all rare-earth element accessory phases. Normally samples are mixed in the ratio 4 parts of flux to 1 part of rock and fused for 15–30 minutes at about 800°C. After fusion, the melt is cooled and dissolved in water. Fe and Ti and some other elements will precipitate from this alkali solution as hydroxides and must be redissolved by acidifying the solution, usually with hydrochloric acid solution.

2.2.3 *Salt content of alkali fusion solutions*

Alkali salt fusion techniques have one principal disadvantage when used in conjunction with modern instrumental methods of analysis; namely the high salt content of the resultant solution. This is a disadvantage with AAS and ICP–AES techniques, for example, since solutions with high salt contents can cause:

(i) Erratic performance of nebulizers, including memory effects

(ii) Clogging of atomic absorption burner apertures

(iii) High background signals and scattered light effects in optical emission measurements.

These phenomena are further discussed in Chapters 4 and 5. Three procedures can reduce these effects.

(a) *Sample digestion by sintering*. Here the aim is to render the sample soluble in aqueous solution at minimum flux salt content. The sample is mixed with flux at a low ratio. Sodium peroxide is particularly suitable for this approach and sample-to-flux ratios can be as low as 1 : 1. The mixture is heated in a crucible to a temperature that is often not greatly above the melting point of the flux. The aim is to increase the activity of the flux to a point where it will attack the rock matrix, subsequently rendering it soluble in acidified solution.

Thus Jeffrey (1975, p.30) recommends fusing 0.5g of rock powder with 1g of Na_2O_2 at 500°C for 10 minutes (the melting point of Na_2O_2 is 480°C). Van Loon (1980, p.94) describes a similar procedure and reports that many minerals will decompose at 430°C but that 480°C for 2 hours is required to decompose the most resistant minerals. The cooled melt is normally extracted into cold water and any insoluble hydroxides dissolved by adding 5M HCl, excess HCl being added to a final concentration of 0.3 to 0.4M HCl.

(*b*) *Absorption of sodium by hydrated antimony* (*V*) *oxide.* In radiochemical neutron activation analysis procedures, high activity due to the isotope ^{24}Na is often found during the first two weeks after irradiation. This limits the sensitivity of data determined from spectra counted within this period. To overcome this problem, Girardi and Sabbioni (1968) showed that sodium could be removed by passing acidified sample solutions through exchange columns packed with hydrated antimony (V) oxide. It is possible therefore that the same technique could be adapted for reducing the salt content of samples fused with sodium-based fluxes.

(*c*) *Use of potassium superoxide.* Recently, Westland and Kantipuly (1983) have proposed the use of potassium superoxide (KO_2) as a fluxing agent. This material has the advantage that the salt content of resultant solutions can be reduced substantially by acidifying with perchloric acid. Potassium is precipitated and filtered off as the almost insoluble perchlorate. Rubidium and caesium perchlorates are also likely to be precipitated. Westland and Kantipuly used a flux containing 1 part of KO_2 to 3 parts KOH (the KOH was added to reduce the cost of individual fusions). Sample and flux were mixed in the ratio 1 : 3 and fused in a zirconium crucible to redness (nickel crucibles were substantially attacked by this reaction mixture). The melt was cooled and dissolved in dilute HCl. The effectiveness of this digestion procedure was reported to be equivalent to that of fusions using Na_2O_2. Of the four ions tested, quantitative recoveries of Eu and Th were obtained in the filtrate after one precipitation of potassium with perchloric acid. However, quantitative recoveries of calcium were obtained only after potassium perchlorate had been redissolved and reprecipitated. Silver could not be quantitatively recovered, and significantly contaminated the $KClO_4$ precipitate. The authors propose that this fusion procedure is particularly applicable to neutron activation procedures.

2.2.4 *Loss of silica during alkali fusion*

If silicon is to be determined following an alkali fusion, there is a possibility that a small proportion of silica can be lost during the fusion process in those samples with a significant content of fluorine if silicon and fluorine from the sample react to form volatile SiF_4. Harmer (1973) carried out some tests using Jeffrey and Wilson's (1960) method combining classical gravimetric and instrumental photometric procedures to determine silicon, and found that SiO_2 determinations in quartzose rocks could be low by as much as 0.5% for samples containing 1% to 3% F.

2.3 Classical methods of rock analysis

2.3.1 *Evolution of classical procedures*

The principles of classical analysis were laid down during the first decades of this century in two classical treatises by Washington (1918, 1932) and Hillebrand (1919). These schemes of analysis were mainly gravimetric, that is, they relied on the analytical balance for all measurements. Determinations could therefore only be carried out after extensive chemical manipulations designed to isolate quantitatively individual compounds. The simplicity of the instrumentation available for measurements was compensated for by the skill and experience of the analyst. Such was the specialized training of this discipline that when more modern instrumental procedures were introduced, their accuracy and reliability were mistrusted by some classical analysts. Few will regret the passing of this era, but many will admire the skill and ingenuity of those who sought to determine the major elements in silicate rocks using nothing more than an analytical balance. The classical analysis era lasted until the early 1950s. Stimulated by the need for more geochemical data, 'classical' procedures developed into 'rapid' methods of analysis that made use of the flame photometer, absorption measurements using simple optical spectrophotometers and element-specific complexometric titrations to avoid individual chemical separations. By the 1960s, these procedures had themselves been largely overtaken by atomic absorption spectrometry. This instrumental technique eliminated the need for almost all chemical separations. Atomic absorption analysis has in its turn been complemented by, and in some respects superseded by, atomic emission spectrometry using an inductively coupled argon plasma source. The introduction of this instrumentation has elevated wet chemical procedures to the status of a simultaneous multi-element technique, capable of determining many major and trace elements of geological interest.

2.3.2 *Classical schemes of analysis*

There is no one definitive classical scheme of analysis. Procedures evolved as new developments were made, and individual analysts retained personal preferences. However, the overall principle of analysis remained the same: that of chemically separating a rock sample dissolved in aqueous solution into individual elemental components that could then be determined gravimetrically. The components sought were SiO_2, Al_2O_3, Fe_2O_3, FeO, MgO, CaO, MnO, TiO_2, P_2O_5, Na_2O, K_2O, H_2O^+, H_2O^- and CO_2. H_2O^- refers to moisture absorbed from the atmosphere into the rock powder, which may be expelled by heating to 105°–110°C. H_2O^+ refers to structural water, that is water chemically incorporated into the minerals of a sample. In almost all schemes of analysis (including modern ones) Fe(II), H_2O and CO_2 are determined independently of the other major elements.

It is not proposed here to discuss in detail the practical procedures for classical analysis. It is, however, important to summarize the principles involved and the philosophy behind classical separation procedures, since these were the basis of all original geochemical studies. Detailed practical methodologies are discussed by Maxwell (1968) and Jeffery (1975, especially pp.34–47). Data that follow are based on details given by Jeffery, and describe a classical gravimetric scheme that incorporates some more modern photometric measurements designed to overcome some of the shortcomings of the strict gravimetric approach.

A summary of this scheme is shown in Figure 2.5. It is based on separating the constituents of a sample into four

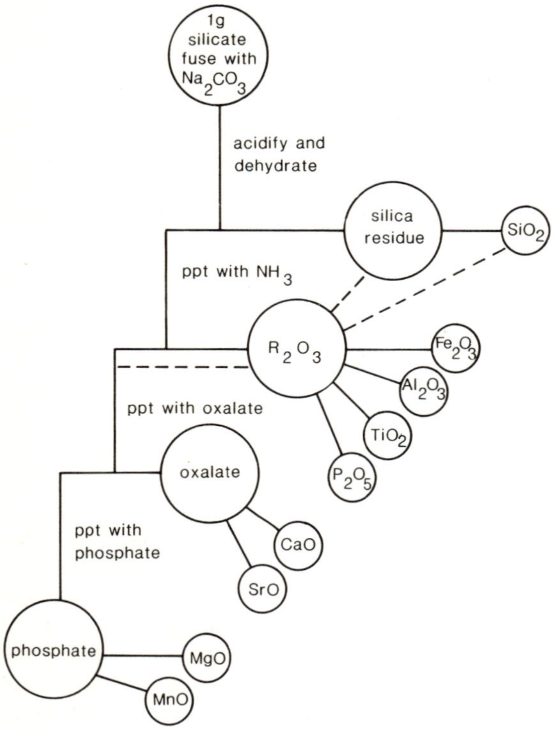

Figure 2.5 Schematic flow diagram for the chemical separation procedures involved in a classical scheme of analysis. (Jeffrey, 1975.)

groups of elements, each group being precipitated from the rock solution successively as follows:

(i) Silica is separated by dehydrating silicic acid H_2SiO_3 to insoluble SiO_2, accompanying elements are dissolved off in dilute acid solution

(ii) The so-called R_2O_3 group of elements, comprising Fe_2O_3, Al_2O_3, TiO_2 and P_2O_5, is precipitated as hydroxides when an acidic solution from which silica has been removed is neutralized with ammonia

(iii) Calcium and strontium are precipitated as insoluble oxalates

(iv) Magnesium and manganese are precipitated as insoluble phosphates.

2.3.3 *A typical procedure for the classical scheme of analysis*

(*a*) *Dissolution of rock.* The finely divided rock powder (1g) is fused with anhydrous sodium carbonate (3–5g) in a platinum crucible at temperatures increasing from 700° to 1200°C. After cooling, the melt is disintegrated with water (to which a few drops of ethanol must be added to reduce manganese, present as alkali manganate, to manganous(II) salts). Jeffrey (1975) suggests that this proportion of flux would now be considered excessive, a sinter of equal weight of sample and sodium carbonate being preferred. A small proportion of potassium nitrate can be added to oxidize samples that contain a large proportion of ferrous iron.

(*b*)*Determination of silica* (SiO_2). The aqueous mixture from (*a*) above is dissolved in hydrochloric acid and then evaporated to dryness until all hydrochloric acid fumes have been expelled. During this process silicic acid in solution is dehydrated to silica:

$$H_2SiO_3 \rightarrow H_2O + SiO_2 \qquad (2.4)$$

The residue after evaporation is attacked with hydrochloric acid solution. Soluble chloride salts are taken up into solution, leaving a gelatinous precipitate of SiO_2. This is filtered off and washed. However, the filtrate and washings still contain some dissolved silica. Part of this may be recovered by acidifying the filtrate with hydrochloric acid and evaporating to dryness a second time. The soluble chlorides are redissolved and the small silica residue filtered off and combined with the original residue. The soluble chloride filtrate is reserved for the next group separation.

The silica residue cannot be weighed directly since it contains small amounts of iron and aluminium contamination. Accordingly, this silica residue is moistened with concentrated sulphuric acid and ignited until the crucible and contents reach a constant weight. Silica is then fumed off by heating with hydrofluoric acid in the presence of a few drops of concentrated sulphuric acid.

The small residue in the crucible is ignited to constant weight. The weight loss of the crucible is the uncorrected silica content. Corrections must be applied for the silica content of filter papers ashed during the ignitions and the silica reagent blank (especially from hydrofluoric acid). Notwithstanding the two dehydrations for silica, a small fraction of this element is still carried over in the filtrate and this must be recovered in the next group separation for a 'traces silica' correction. The crucible and residue after silica volatilization (containing traces of iron and aluminium, in particular) are reserved for ignition of the ammonia precipitates.

(*c*) *Separation of mixed oxides* (*the* R_2O_3 *group*). The mixed oxides are recovered from the combined filtrates and washings from (*b*) above by adding ammonia to a hot solution acidified with hydrochloric acid until it becomes just alkaline ($pH = 7$). The precipitate contains iron, aluminium, titanium, phosphorus, zirconium, vanadium, chromium and some trace elements that form insoluble hydroxides. Traces of silica carried over from (*b*) above are also present. The hydroxide precipitate is filtered from the slightly alkaline solution. The filtrate is reserved for the next separation. A small proportion of calcium and magnesium will also be coprecipitated and these elements are removed by redissolving the precipitate after filtration and re-precipitating insoluble hydroxides with ammonia a second time. The filtrate is again combined with the first filtrate for the next separation. A small proportion of manganese is precipitated with the mixed oxides and the amount must be determined so that aluminium can be accurately determined gravimetrically by difference (see below). The main fraction of manganese is determined after precipitation with the magnesium (see (*e*) below) and total manganese determined independently on a separate sample aliquot. The difference represents manganese co-precipitated with the mixed oxides.

A small fraction of aluminium remains in the filtrate after the precipitation of mixed oxides. This can be removed by evaporating this filtrate to a small volume and re-precipitating with ammonia. The precipitate recovered in this way is added to the other mixed oxide residues.

These combined ammonia precipitates are then ignited at

temperatures up to 1200°C in the same crucible that contains the iron and aluminium residue after silica volatilization. The highly refractory residue remaining after ignition is brought into solution after fusion with potassium pyrosulphate. Any trace silica carried over from (*b*) above cannot be dissolved when the pyrosulphate melt is taken up in sulphuric acid solution. It is filtered off and silica determined by volatilization with HF and sulphuric acid as in (*b*) above. The pyrosulphate solution is made up to volume. Although individual gravimetric techniques are available, the elements iron, titanium, vanadium, phosphorus and zirconium would now normally be determined photometrically.

Aluminium is determined by difference. That is, the total weight of the mixed oxide residue remaining after ignition is calculated as a percentage of the sample taken for analysis. From this is subtracted the mixed oxide elements that are determined individually, calculated as the oxides Fe_2O_3, TiO_2, P_2O_3, V_2O_3, Cr_2O_3 and ZrO_2. Allowance must also be made for the proportion of co-precipitated manganese and trace silica. The difference is aluminium as Al_2O_3. The accuracy of aluminium determinations therefore depends on errors in the measurement of individual mixed oxide elements. For this reason, in any modern adaptation of classical procedures, these elements would normally be determined on a separate sample aliquot rather than on the appropriate group separation residue.

(*d*) *Oxalate precipitation of calcium and strontium.* The combined filtrates from (*c*) above are made just alkaline and heated to boiling. Hot ammonium oxalate solution is added. Calcium and strontium are precipitated as insoluble oxalates. Small amounts of magnesium are co-precipitated with calcium. This contamination can be removed by redissolving the calcium oxalate precipitate in hydrochloric acid and then re-precipitating it with ammonium oxalate under controlled conditions. The filtrates after both precipitations are combined and reserved for magnesium determination. The final calcium oxalate precipitate is ignited in a muffle furnace at about 1000°C and weighed as CaO or alternatively decomposed at 500°C in a muffle furnace and weighed as calcium carbonate. The latter route is preferred since calcium oxide is very hygroscopic and difficult to weigh in the laboratory atmosphere. The presence of strontium is normally ignored in the calcium residue as this element is unlikely to be at a sufficiently high relative concentration to cause an interference on calcium.

(*e*) *Insoluble phosphates–magnesium and manganese.* The combined filtrates from the oxalate separation are evaporated on a steam bath to a manageable volume and acidified with nitric acid. This solution is evaporated to dryness to remove ammonia salts from earlier separations that would interfere with the precipitation of magnesium. The residue is dissolved in dilute nitric acid, neutralized and, after digesting on a steam bath, filtered. Small residues, mainly of aluminium, may be recovered and determined by ignition for correction of aluminium determined by difference in (*c*) above. Magnesium is precipitated from the filtrate as magnesium ammonium phosphate hexahydrate by adding $(NH_4)_2HPO_4$ (diammonium hydrogen phosphate) and concentrated ammonia solution. After filtering and washing, this precipitate is dissolved in hydrochloric acid and then re-precipitated as above to remove any contamination from barium and traces of calcium and strontium not precipitated as oxalates. The combined filtrates can be used to determine nickel.

The magnesium ammonium phosphate hexahydrate precipitate and filter paper are heated on a bunsen burner at moderate heat to char the paper. The temperature is increased and the final ignition carried out in a muffle furnace at 1050°C. The residue is weighed as magnesium pyrophosphate ($Mg_2P_2O_7$):

$$2Mg(NH_4)PO_4.6H_2O \rightarrow 13H_2O + Mg_2P_2O_7 + 2NH_3 \quad (2.5)$$

This ignition is subject to some interferences since the magnesium ammonium phosphate hexahydrate precipitate can also contain some $Mg(NH_4)_4(PO_4)_2$, which decomposes to $Mg(PO_4)_2$ on ignition. This latter can be converted to the pyrophosphate only by heating to 1150°–1200°C. However, at these temperatures the pyrophosphate itself slowly loses P_2O_5.

In the determination of magnesium, manganese is precipitated as the ammonium phosphate. This forms manganese pyrophosphate ($Mn_2P_2O_7$) on ignition. The manganese content of the ignited residue is required both as a correction to the magnesium determination and also for the calculation of aluminium by difference. The ignited residue is dissolved in concentrated sulphuric acid solution to which is added nitric acid. Manganese can then be determined photometrically by oxidation to the permanganate using potassium periodate (section 2.4.5*a*).

(*f*) *Determination of sodium and potassium.* The alkali metals were determined in classical procedures by the so-called J. Lawrence Smith extraction. This involved igniting the sample with a mixture of ammonium chloride and calcium carbonate. The alkali metals were leached with water, sulphate salts converted to chlorides, and ammonium flux residues removed by ignition. Individual alkali metals were then determined gravimetrically after precipitation as insoluble salts: potassium as the chloroplatinate, perchlorate or cobaltinitrite complex and sodium by difference. Corrections were necessary to account for the solubility of the alkali salts. This procedure is essentially of historical interest and rapidly became obsolete when simple flame photometry instrumentation became widely available.

(*g*) *Other elements.* Classical gravimetric procedures were, in general, not sufficiently sensitive to allow determination of trace elements. However, procedures evolved for the determination of ferrous iron, water and carbon dioxide are still in widespread use in modern laboratories and are discussed in more detail in sections 2.9 and 2.10.

(*h*) *Limitations of classical gravimetric analysis.* Present-day analysts must be appalled by the prospect of the detailed chemical manipulations involved in classical chemical separation procedures. The time and professional skill required by such work clearly represents a serious limitation in relation to modern analytical practice and workloads. Perhaps a more fundamental limitation of classical procedures is that the analytical signal (a weight difference) is only characteristic of an element insofar as quantitative and uncontaminated recovery of that element has been achieved. The opportunities for small unsuspected discrepancies to arise in the hands of less experienced operators are clear. Further difficulties arise in the determination of elements such as aluminium and sodium. These were normally calculated by difference and were therefore subject to errors magnified by any discrepancies in the determination of other elements.

2.4 Evolution of rapid methods of analysis

Classical methods of analysis were revolutionized in the 1950s by three developments, two instrumental (photometric analysis, using uv/visible spectrophotometers, and flame photometry) and one chemical (the use of a unique complexing reagent, ethylenediaminetetra-acetic acid(EDTA) for element-specific titrations). The importance of these techniques was that schemes of major element analysis could be developed in which it was no longer necessary to separate each component chemically as was the case with gravimetric determinations. This led to the formulation of a range of new schemes of analysis, the best-known of which are those of Shapiro and Brannock (1956, 1962), Riley (1958*a*), Peck (1964) and Langmyhr and Graff (1965). These procedures are perpetuated today by the hybrid classical/rapid methods described by Kirschenbaum (1983). Although rapid schemes of analysis are still in use today, they have been superseded in many laboratories by instrumental procedures such as atomic absorption spectrometry, ICP–atomic emission spectrometry and x-ray fluorescence analysis. Photometric and gravimetric procedures are, however, still retained for some components, including Fe(II), CO_2 and H_2O, for which standard instrumental methods are not widely available.

Before discussing rapid schemes of analysis in detail, the three analytical developments will be considered.

2.5 Photometry

2.5.1 *Principles*

The solution to be analysed is mixed with a solution of a reagent called a chromophore. The chromophore is normally an organic 'ligand' that reacts specifically with the metal ions of the element to be determined. The complex formed by this reaction is coloured, showing strong absorption bands in the uv/visible region of the spectrum. The chromophore should itself be colourless, and should not react with any other elements in solution; if it does, the resultant absorption spectrum should not overlap that of the element complex to be determined. Determinations are then made by measuring the intensity of absorption of the element complex compared with that of a standard solution.

This technique is applicable to many of the major elements (see Table 2.6), especially to those in the first transition series, which often show a molar absorptivity greater than 10 000. This intense absorption is generally the result of charge transfer spectra.

2.5.2 *Charge transfer spectrum*

The reaction between metal ion and chromophore normally results in a complex in which the chromophore ligand is bonded, either in square planar or (more commonly) octahedral configuration (Figure 2.6). Bonding results in the formation of combined molecular orbitals between metal ion and ligand. This process is illustrated in a highly schematic manner in Figure 2.7. In this example, the energy levels of the 3*d*, 4*s* and 4*p* electron orbitals from the metal ion that participate in bonding are plotted. During bond formation these orbitals combine with the ligand (chromophore) orbitals to form a new set of molecular orbital energy levels. The molecular orbitals are classified as bonding, anti-bonding and non-bonding orbitals and their contribution to complex formation is as follows.

(i) Non-bonding orbitals are normally contributed by the metal ion and effectively take no part in molecular orbital formation, being unaffected in energy level.

(ii) Bonding orbitals are formed with lower energies than either of the original metal ion or ligand electron orbitals. These therefore contribute to the formation of stable metal ion–chromophore complexes.

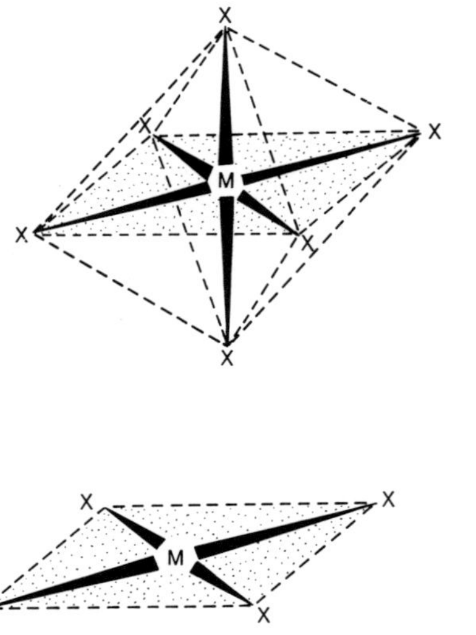

Figure 2.6 Complexation of metal ion (M) with ligand (X) in octahedral (MX_6) and square planar (MX_4) configuration.

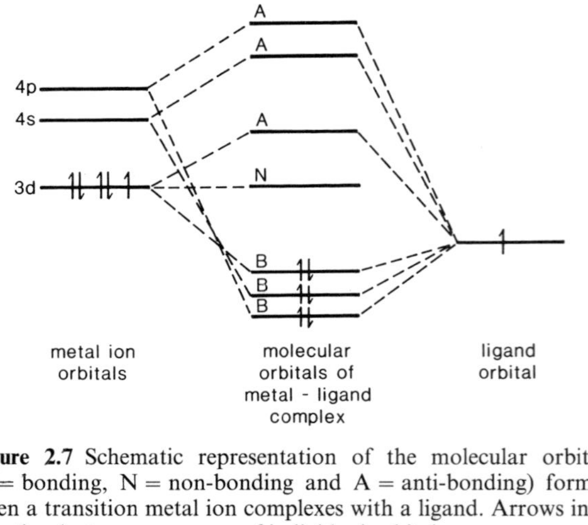

Figure 2.7 Schematic representation of the molecular orbitals (B = bonding, N = non-bonding and A = anti-bonding) formed when a transition metal ion complexes with a ligand. Arrows indicate the electron occupancy of individual orbitals.

Table 2.6 Selected reagents for the photometric determination of elements in rapid methods of analysis

Element	Photometric method	Chemical formula and conditions for complexation	Colour of complex	Absorption measurements in (nm)	Reference for details of application
Si	Molybdenum blue	$(NH_4)_6Mo_7O_{24}$ in a reducing solution.	Blue	812	Riley (1958)
Ti	Hydrogen peroxide	H_2O_2 in an acidic solution. Add phosphoric acid to suppress an interference from iron.	Yellow	400–410	Jeffrey (1975) Peck (1964)
Ti	Tiron	Disodium salt of 1, 2-hydroxy benzene-3, 5-disulphonic acid 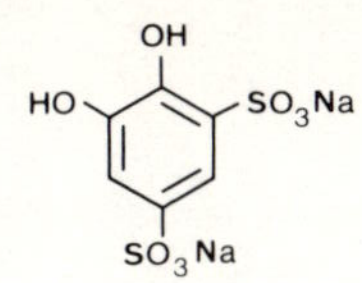	Yellow	380	Jeffrey (1975)
Al	Alizarin red S	$CO.C_6H_4.CO.C_6H(OH)_2.SO_3Na$	Yellow-orange	475	Shapiro and Brannock (1962)
Al	8-Hydroxyquinoline	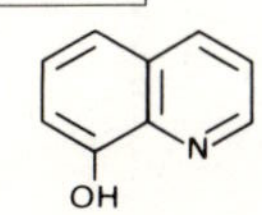	Yellow	410	Riley (1958)
Fe (total)	1, 10-Phenanthroline (orthophenanthroline)	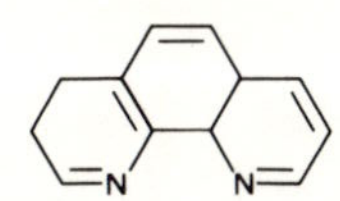 First reduce iron to Fe (II) with hydroxylammonium chloride	Red	508	Jeffrey (1975)
Fe (total)	Tiron	Disodium salt of 1, 2-hydroxy-benzene 3, 5-disulphonic acid. Iron first oxidized to Fe (III), measurements made in the presence of hydrogen peroxide.	Violet	560	Langmyhr and Graff (1965)
Fe	2,2′-dipyridyl	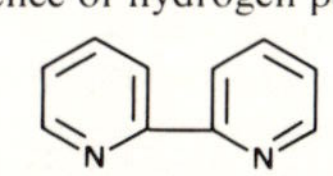 First reduce iron to Fe(II) with hydroxylammonium chloride	Red	522	Riley (1958)
Mn	Oxidize to permanganate	MnO_4^-	Red-purple	525	Shapiro and Brannock (1962)
Mg	Thiazole yellow		Yellow	545	Shapiro and Brannock (1962)
P	Convert to the molybdivanadophosphoric acid complex	$(NH_4)VO_3 + (NH_4)_6Mo_7O_{24} + KH_2PO_4$	Yellow	315 (maximum) 420–470 (to suppress an interference from iron)	Jeffrey (1975)
P	Heteropoly blue (or molybdenum blue)	$(NH_4)_6Mo_7O_{24}$ plus a reducing agent (e.g. ascorbic acid); remove silicon first.	Blue	830	Jeffrey (1975)

Table 2.6 (*Continued*)

Element	Photometric method	Chemical formula and conditions for complexation	Colour of complex	Absorption measurements in (nm)	Reference for details of application
Cr	*s*-Diphenyl carbazide		Reddish-violet	540	Maxwell (1968)
V	Phosphotungstate complex	Sodium tungstate (Na_2WO_4) plus phosphoric acid (H_3PO_4)	Yellow	absorption maximum in the uv; use 400 nm	Maxwell (1968)
Ni	Dimethylglyoxime	CH_3-C = NOH | CH_3-C = NOH	Deep red	530	Jeffrey (1975)

(iii) Anti-bonding orbitals are normally vacant and represent excited states to which molecular orbital electrons can be promoted on absorption of light energy. On the formation of the metal–ligand complex, electrons displaced from the original metal ion and ligand orbitals occupy the resultant molecular orbitals as shown in Figure 2.7, the lowest energy molecular orbitals being filled first. The metal–ligand complex is chemically stable if the sum of the energies of electrons occupying molecular orbitals is lower than that of metal ion and ligand separately.

If a molecular orbital has an energy closer in level to an original ligand electron orbital, for example, the character of that orbital will more closely resemble that of the original ligand and not those of the metal ion (and vice versa). Thus in Figure 2.7 the bonding orbitals can be considered to be more closely associated with those of the ligand whereas the anti-bonding orbitals can be considered to be associated with the metal ion. Thus if absorption of light causes excitation of a bonding electron to an anti-bonding molecular orbital, a notional transfer of electronic charge will have occurred from ligand to metal ion orbitals. Hence the term 'charge transfer spectra'.

For species dissolved in solution, collisions between metal–ligand complex molecules and solvent molecules will cause considerable random perturbation of the molecular orbital energy levels, disrupting the idealized discrete levels shown in Figure 2.7. Charge transfer spectra will therefore generally have rather broad, structureless shapes as shown in Figure 2.9 or 2.11, for example. The molecular orbital diagram in Figure 2.7 is appropriate for simple transition element ions in 6-fold (octahedral) co-ordination with simple (σ bonding) ligands. Some chromophores are based on more complex organic molecules containing 'conjugated' double bonds between alternate carbon atoms. These can contribute to another form of bonding with metal ions called π bonding. The molecular orbital diagram for such species is more complex, but the principles of charge transfer spectra remain the same.

The molecular orbital diagram in Figure 2.7 is typical of many transition-metal ion charge transfer complexes. There are, however, some exceptions, an important one being the complex between iron (in the ferrous state) and 1,10-phenanthroline. In this case, the energy levels of the metal ion orbitals that contribute to molecular orbital formation are *lower* than those of the 1,10-phenanthroline ligand (the opposite situation to that shown in Figure 2.7). On excitation, therefore, charge transfer notionally occurs in the opposite direction to that described above, from metal ion to ligand-associated orbitals. In the extreme this could be considered as an intra-molecular oxidation–reduction reaction stimulated by the absorption of light.

Many reagents have been proposed as chromophores for colorimetric analysis. Those most commonly encountered in rapid schemes of analysis are listed in Table 2.6 together with further analytical details.

2.5.3 *Instrumentation*

The design of a simple uv/visible absorption spectrometer as used in many early schemes of analysis is shown in Figure 2.8. The instrument consists of a light source (usually an

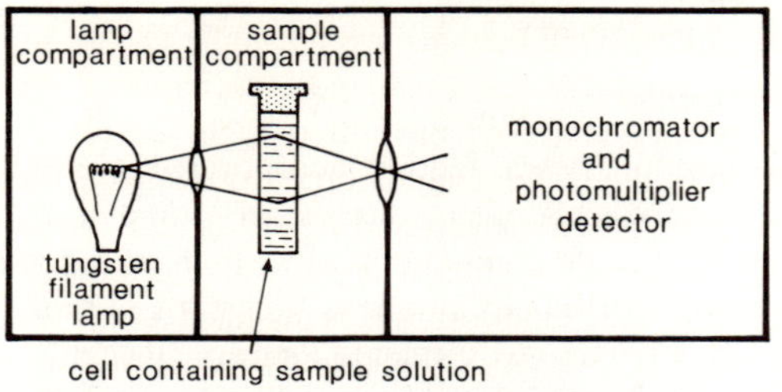

Figure 2.8 Schematic diagram of a simple single-beam spectrophotometer.

incandescent tungsten lamp), a sample cell compartment, an optical monochromator and a photomultiplier (or photocell) detector. The operation and characteristics of such devices are considered in more detail in Chapter 3. Early intruments were normally restricted to measurements in the 400 to 700 nm region, for several reasons. The tungsten lamp has a restricted output at lower wavelengths (for which a deuterium light source is required). Glass dispersion prisms used in early monochromators will not transmit radiation in the near ultraviolet spectrum. Early photomultiplier tubes or photocells had a limited response to radiation below 400 nm. These restrictions limited to some extent the chromophore complexes that could be satisfactorily measured. In early single-beam instruments, it was necessary to measure the absorption of a blank sample solution (comprising solvent and reagents and measuring cell) by substituting the blank sample in the light beam. On more modern double-beam instruments this comparison is carried out automatically, the light beam being split to pass through both sample and blank cells, the difference in absorption between the two light beams then being recorded directly. Quartz glass cells and optical components must be used if measurements are to be made below 400 to 300 nm. The internal path length of solutions within each sample cell is normally 10mm.

2.5.4 *Typical applications*

The role of photometry in rapid schemes of analysis is illustrated by two typical applications.

(*a*) *Determination of manganese as permanganate*. Manganese may be determined photometrically by taking a sample up into solution and oxidizing manganous(II) ions to permanganate. Permanganate ions have an intense charge transfer spectrum that shows an absorption maximum at about 525 nm (Figure 2.9). The degree of absorption of this purple-coloured complex is compared with that of a standard permanganate solution and the concentration of manganese in the unknown sample determined from the ratio of absorptions. Oxidation of Mn(II) ions is normally achieved by reaction either with a solution of ammonium persulphate, $(NH_4)_2S_2O_8$, in the presence of phosphoric and sulphuric acids containing a small amount of silver as catalyst (e.g. Riley, 1958*a*), or by boiling with a solution of potassium periodate (e.g. Shapiro and Brannock, 1956). The respective equations for these reactions are as follows:

$$2Mn^{2+} + 8H_2O + 5S_2O_8^{2-} \rightarrow 2MnO_4^- + 16H^+ + 10SO_4^{2-} \quad (2.6)$$

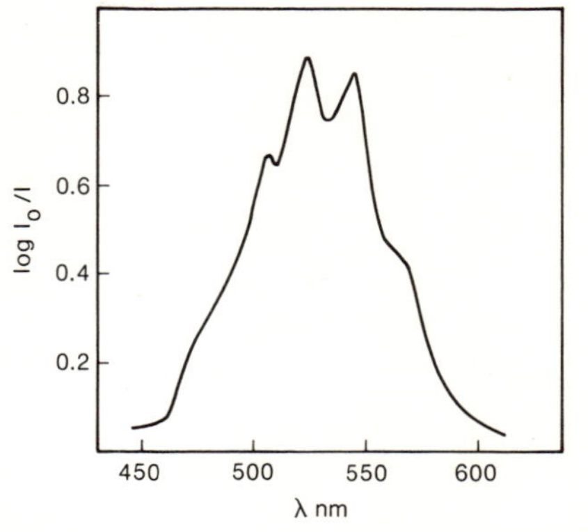

Figure 2.9 Absorption spectrum of the permanganate ion in solution. (Jeffrey, 1975.)

$$2Mn^{2+} + 5IO_4^- + 3H_2O \rightarrow 2MnO_4^- + 5IO_3^- + 6H^+ \quad (2.7)$$

In the reaction with ammonium persulphate, phosphoric acid must be added to eliminate a spectrometric interference at 525 nm that would otherwise occur due to uncomplexed ferric ions in solution. Unfortunately phosphoric acid can cause the precipitation of titanium phosphate. If the concentration of sulphuric acid is increased for such samples, this interference is eliminated (Jeffery, 1975, p.327). Hydrochloric acid must be avoided in the determination of permanganate, since the latter is sufficiently powerful to oxidize hydrochloric acid to chlorine. Similarly, preboiled water must be used to dilute all permanganate solutions, otherwise oxidizable contaminants may cause fading of the reagent colour.

The potassium periodate reaction appears to give similar results to persulphate. Absorption measurements of permanganate solutions suffer slight interference from the presence of chromium oxidized to chromate ions in solution. Jeffery (1975, p.327) recommends correcting this by measuring an absorption blank, not relative to a normal reagent blank, but relative to the permanganate solution to which sodium nitrite ($NaNO_2$) solution has been added after the initial absorption measurement. Sodium nitrite is a sufficiently strong reducing agent to reduce permanganate to manganous ions, but not to reduce chromate ions. Blank measurements will then automatically compensate for any chromate contribution to the permanganate absorption signal at 525 nm.

(*b*) *Photometric determination of total iron using* 1,10-*phenanthroline*. One widely used method of determining iron photometrically is by complexation as ferrous iron with the chromophore 1,10-phenanthroline. After dissolving the silicate sample and diluting it to volume, it is necessary to bring all the iron into the same oxidation state. Iron in the ferric state, Fe(III), can be reduced to Fe(II) using hydroxylammonium chloride solution ($HO.NH_3Cl$). The complexation reaction with 1,10-phenanthroline is buffered with sodium citrate or ammonium tartrate solution. The complex formed consists of an Fe(II) ion surrounded by three 1,10-phenanthroline ligands in an octahedral configuration (Figure 2.10). This complex exhibits an intense charge transfer spectrum that appears red in colour with an absorption maximum at 508 nm (Figure 2.11). It is important that a

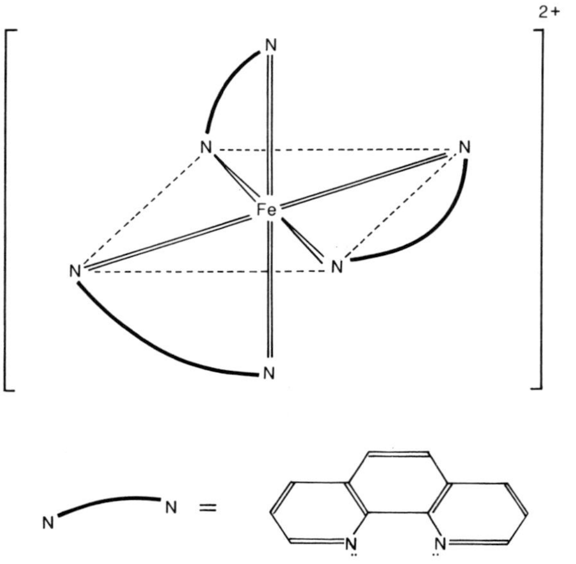

Figure 2.10 Schematic representation of the ferrous iron-(1,10-phenanthroline)$_3$ complex.

blank absorption measurement is subtracted from that of the sample solution. The blank sample is prepared in the same way but without addition of the sample. The resultant absorption measurements are compared with those from a standard solution prepared from pure iron filings or from ammonium ferrous sulphate $(NH_4)_2SO_4.FeSO_4.6H_2O$.

The success of this procedure rests on the complete nature of the reaction between Fe(II) and 1,10-phenanthroline, the intense colour of the complex formed and the absence of serious interferences. Skoog and West (1982) report that the stability constant for the reaction

$$Fe^{2+} + 3PhH^{+} \rightarrow Fe(Ph)_3^{2+} + 3H^{+} \qquad (2.8)$$

is 2.5×10^6 at 25°C. Quantitative formation of the ferrous complex is therefore obtained at solution pH values of between 2 and 9; pH 3.5 is normally maintained to prevent insoluble salts precipitating from solution.

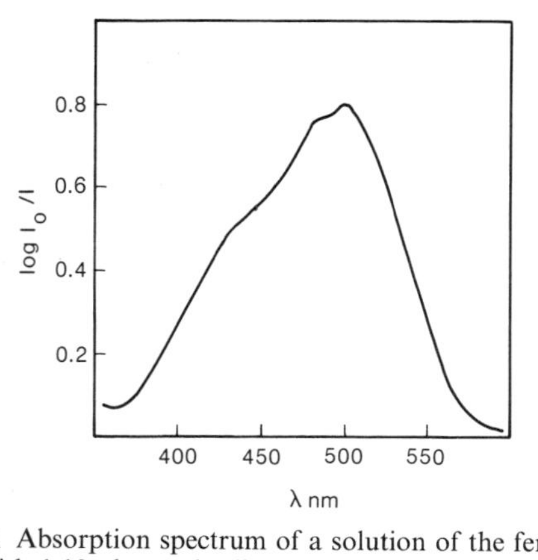

Figure 2.11 Absorption spectrum of a solution of the ferrous iron complex with 1,10-phenanthroline (1 cm cell, 0.6 μg Fe/100 mL). (Jeffrey, 1975.)

2.6 Flame photometry

2.6.1 *Principles*

Flame photometry was readily incorporated into rapid schemes of analysis to replace unsatisfactory gravimetric procedures for the determinations of sodium and potassium. The principle of this technique is that if a solution is aspirated into a flame, any elements that are efficiently excited by the thermal energy of the flame will, on de-excitation, emit light photons. These photons are detected at characteristic wavelengths and the intensity is proportional to the concentration of the element present in solution. Flame photometry instrumentation is thus a means of quantifying the qualitative observation that if sodium salts are heated in a simple gas flame, intense yellow-orange light is emitted. Potassium salts emit red-violet light, and so on. Flame photometry is an atomic emission spectrometric technique, the mechanisms of which are further discussed in Chapter 4.

The energy available in a natural gas or propane flame is quite low (the flame temperature is typically 1800–1900°C). A rather limited range of elements can therefore be excited effectively, notably the alkali metals.

2.6.2 *Instrumentation*

In many of the early rapid schemes of analysis, extremely simple instrumentation was used (Figure 2.12). This consisted of a burner assembly, into which the sample solution is aspirated, and a simple filter photometer. The latter isolates the wavelength range to be measured by means of selectable glass transmission filters. The intensity of light passing through the filter is measured using a photocell. Since the bandpass of the transmission filter is relatively wide compared with that of a prism monochromator, such intruments are suitable only for measuring well-separated lines such as the Na and K emissions (at 589 and 766 nm respectively) in a spectrum virtually free of any other significant emission.

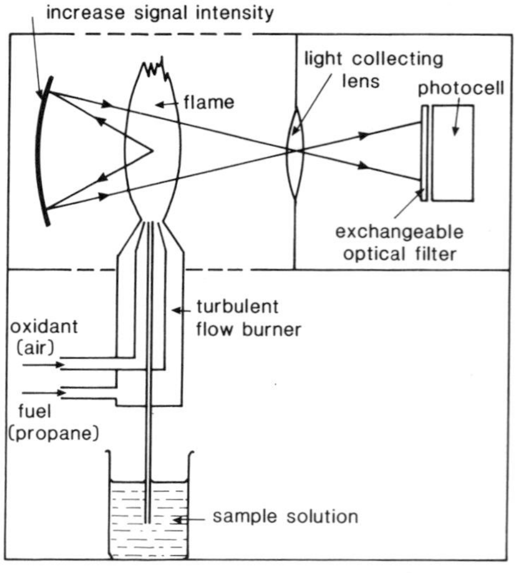

Figure 2.12 Schematic diagram of a simple flame photometer incorporating a total consumption burner and simple filter photometer and photocell for measurement of the flame emission signal.

The emission spectrum from a flame contains molecular emissions from the burning gas mixture that will to a small extent enhance the background under the sodium emission lines. The higher-wavelength potassium line is unlikely to be affected significantly. Due to the wide bandpass of filter photometers, a mutual interference is likely to be observed. A small proportion of the intense sodium emission, in particular, is transmitted by the potassium filter. It is necessary to apply a pre-calibrated correction to compensate for this effect by measuring the variation in the apparent concentration of potassium measured with the potassium filter on a set of solutions containing varying concentrations of sodium (Riley, 1958*a*).

(*a*) *Burners*. Many simple flame photometers employed a *total consumption burner* (Figure 2.13). This consists of an assembly in which the burner gases are mixed at the base of the flame. The sample solution is aspirated from a capillary directly into the flame by the venturi effect created by the flow of gases in the adjacent nozzle. This flame burns in a turbulent manner. The total consumption burner has the advantage that a large amount of sample solution reaches the flame (aspiration rates are usually 1–3 mL/min), but the considerable disadvantage that the emission signal observed from the flame is relatively noisy. This noise occurs because there is no control over the size of solution droplets entering the flame. This leads to varying rates of solvent evaporation and hence randomly varying atomic populations within the flame from which the atomic emission signal is observed.

(*b*) *Lundegardh burner*. These problems are significantly reduced in the Lundegardh burner (Figure 2.14). This design incorporates a premix chamber into which the sample solution is nebulized from a capillary by the venturi effect of the flow of oxidant gas through a nozzle. The solution droplets pass into the spray chamber where most of the large droplets condense on the chamber walls and drain to waste. Perhaps only 10% of the sample, comprising the smaller droplets, is swept into the flame. Desolvation and atomization of this aerosol occurs much more uniformly, resulting in a significant reduction of noise in the flame emission signal. Burner design and atomization processes are discussed in more detail in Chapter 4.

(*c*) *Monochromators*. A further refinement in more modern instruments is the substitution of grating (or prism) monochromators in place of the rather crude transmission filter. Such instruments have very much narrower bandpasses, permitting a further reduction in signal-to-noise ratio, since interferences due to molecular emissions that

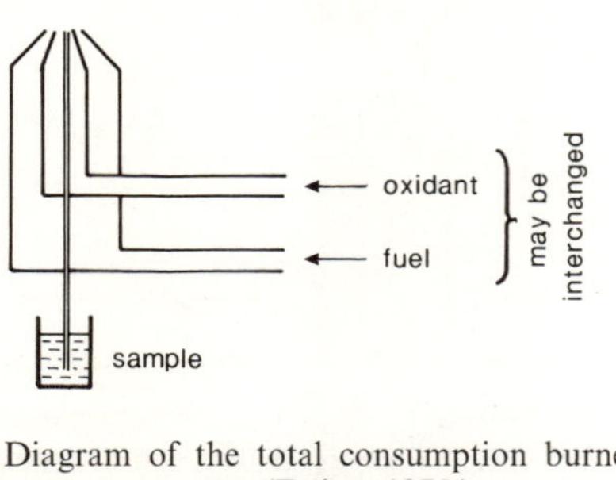

Figure 2.13 Diagram of the total consumption burner used for flame emission measurements. (Ewing, 1975.)

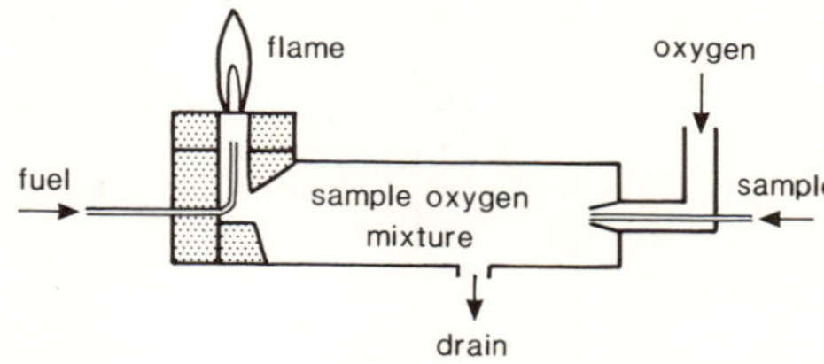

Figure 2.14 Diagram of the Lundegardh burner, incorporating a pre-mixing chamber that promotes quieter burning of the flame (cf. Figure 2.13). (Robinson, 1982.)

contribute to the analytical signal are reduced. It is then possible to extend applications to the determination of the other alkali metals such as lithium, caesium and rubidium at trace levels. The emission spectrum obtained from a flame photometer fitted with a scanning prism monochromator aspirated with a standard solution containing sodium, potassium, lithium and rubidium is shown in Figure 2.15.

(*d*) *Interferences*. Several effects influence the sodium and potassium signal measured on simple flame photometers. The most serious of these are changes in flame temperature and suppression of ionization. Firstly, the degree of excitation and therefore of emission of atoms is strongly influenced by flame temperature (see Chapter 4). Such changes can be caused, for example, by drift in gas-flow regulation or changes in the solution aspiration rate (due to differences in viscosity between standard and unknown solutions). Secondly, part of the atomic emission signal from the alkali metals is lost due to ionization of these elements in the flame; factors that change the degree of ionization will therefore affect the analytical signal. Thus the ionization of potassium is suppressed in solutions containing excess sodium, since sodium is more readily ionized than potassium. Varying concentrations of sodium will thus change the apparent potassium signal by varying amounts. Maxwell (1968, p.410) reports earlier work in which the presence of Mg^{2+} is shown to cause a depression in the potassium emission signal at low concentrations (0.005%–0.1%K). The magnitude of these effects is influenced by flame temperature; marked differences are expected between photometers using air–propane or air–natural gas (flame temperature 1800°–1900°) and those using oxygen–hydrogen mixtures (flame temperature 2800°). Several methods have been proposed to overcome these interference effects and to counter the influence of drift in response.

(*e*) *Internal standardization*. Instrumental drift may be effectively minimized by adding lithium solution at a constant concentration, typically 100 ppm Li, to all standard and unknown solutions analysed by flame photometry. Lithium atoms are also efficiently excited in the flame, emitting a strong emission line at 671nm. Measurements must therefore be made with a narrow bandpass filter monochronomator to avoid lithium interference on the adjacent sodium line at 589 nm. Determinations for sodium and potassium are then made by ratioing the respective signals against that of the lithium emission line. Providing the concentration of lithium is kept constant, any variations in instrumental conditions will simultaneously affect the intensity of analyte and inter-

C*

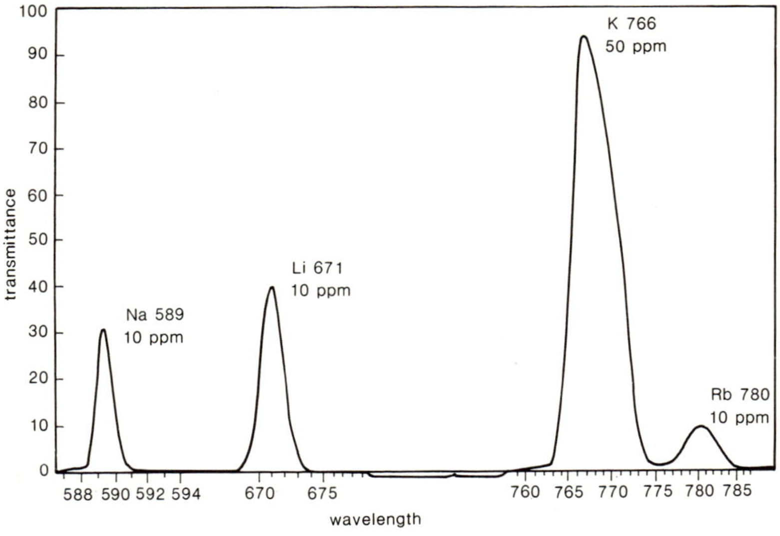

Figure 2.15 Example of a flame photometry emission spectrum obtained from a solution containing the indicated concentrations of sodium, potassium, lithium and rubidium. Wavelength in nm. (Maxwell, 1968.)

nal standard signals and will therefore be nullified in the ratio. The presence of excess lithium is also beneficial in that this element will effectively suppress the ionization of sodium and potassium in the flame, enhancing their emission signals and buffering differences in the salt content of sample solutions. Lithium internal standardization procedures were employed in schemes such as those of Shapiro and Brannock (1956) and Kirschenbaum (1983).

(*f*) *Spectroscopic buffer*. In the rapid method of Riley (1958*a*), a different approach was followed. From the solution to be analysed by flame photometry, Riley removed most of the major cations (Fe, Al and Ti) by anion exchange chromatography and then added ammonium sulphate solution as a spectroscopic buffer to give a final concentration of 0.4% thus overcoming interference effects on the sodium and potassium emission signals.

(*g*) *Matrix matching*. Langmyhr and Graff (1965) considered the effect of anions and cations on the determination of sodium and potassium by flame photometry; their conclusion, based in part on the observations of Dean (1960), was that the major interference arose from the presence of varying amounts of chloride and sulphate ions in solution and in particular from the effect these anions had on atomization mechanisms in the flame. Accordingly, the appropriate amounts of hydrochloric and sulphuric acids were added to all solutions analysed, including standard solutions, to compensate for these effects.

Maxwell (1968, p.403) reported the use of a 'radiation buffer'—a solution containing salts of Al, Fe, Mg and Ca. This radiation buffer solution was added both to rock solutions and to standard solutions. After an initial estimate of concentrations of Na, K (and Li) in the sample, standard solutions were made up so that the concentration of the alkali metals closely matched that estimated in the unknown solution. Samples were re-analysed against these matched standards, and inter-element effects were then minimized.

2.7 Titrations involving ethylenediaminetetra-acetic acid (EDTA)

Ethylenediaminetetra-acetic acid is a unique complexing agent. It is normally available as the disodium salt dihydrate and has the chemical structure shown in Figure 2.16. Its unusual properties stem from the fact that each EDTA molecule is capable of forming up to six bonds when it forms complexes with metal ions: one from each of the four carboxyl groups ($-COO^-$) and one from each of the two nitrogen atoms. Metal ions normally form complexes with EDTA in an octahedral configuration, the tentacles of the EDTA molecule wrapping themselves round the metal ion (Figure 2.17). The resultant complexes with a range of metal ions show high stability (Table 2.7). Apart from the electronic considerations, the effective stability depends on the ionic radius of the metal ion. Ions such as Mg that are too small to accomodate metal–EDTA bonds of optimum length form less stable EDTA complexes than calcium ions, for example, where the larger ionic size permits the formation of more stable bonds.

Stability constants of EDTA complexes (Table 2.7) indicate that a whole range of metal ions that may be present in rock solutions forms stable complexes. As well as Ca^{2+} and Mg^{2+}, other ions such as Sr^{2+}, Ba^{2+}, Mn^{2+}, Fe^{2+}, Fe^{3+}, Cu^{2+}, Zn^{2+}, Al^{3+}, for example, will all be effectively complexed with EDTA.

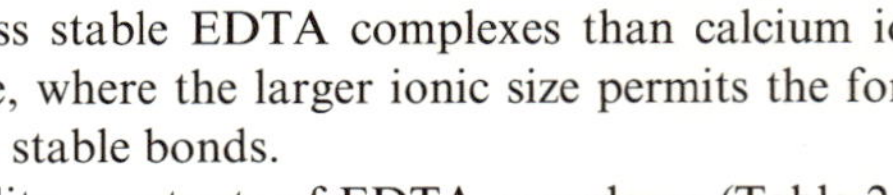

Figure 2.16 The chemical structure of ethylenediaminetetra-acetic acid (EDTA) (disodium salt).

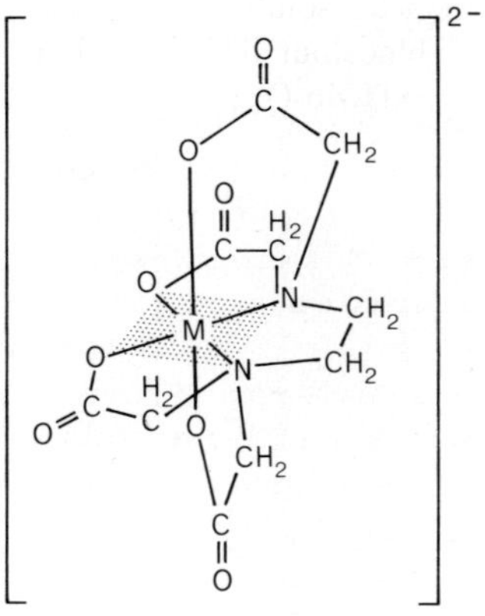

Figure 2.17 The structure of a metal ion–EDTA complex. (Skoog and West, 1982.)

Following the initial work of Schwarzenbach in 1945 on EDTA and related tertiary amines containing carboxylic acid groups, the analytical applications of these complexing reagents were widely investigated. In rapid schemes of analysis, EDTA was applied to the determination of Ca and Mg, elements that were not sufficiently sensitive to be analysed by flame photometry and for which suitable photometric analyses were difficult to perform reliably. The procedures first required the removal of interfering elements. Two titrations were then performed. The first, at pH greater than 12, was completely precipitated magnesium as the hydroxide. Results of this titration gave quantitative analyses of Ca. A second titration, at a lower pH, retained both calcium and magnesium in solution and results gave the total concentration of these elements. From these data, magnesium was determined by difference. Errors could, however, arise in such titrations, in particular in the determination of calcium in the presence of a large excess of magnesium, due to co-precipitation of some calcium hydroxide with $Mg(OH)_2$. Furthermore, if manganese was not effectively removed from starting solutions, a pre-calibrated correction was required to account for an associated interference effect. Strontium normally remained in solution and was determined simultaneously with calcium and magnesium. For rocks with ppm levels of strontium, this interference would normally be ignored.

A critical step in the determination of calcium and magnesium by EDTA titration was clearly the removal of interfering ions. Riley (1958*a*) used an extraction procedure in which interfering ions are reacted with 8-hydroxyquinoline (see Table 2.6 for structure) in aqueous solution at pH 5.0. The metal–chelate complexes are soluble in chloroform, and Riley (1959) developed a continuous extraction procedure to remove the interfering ions. Manganese remained in the aqueous phase after this separation and a correction has to be made for this on the Ca/Mg titration.

Shapiro and Brannock (1962) proposed the direct titration of calcium and magnesium in solutions to which had been added potassium cyanide and triethanolamine (2, 2′, 2″-hydroxytriethylamine). These reagents effectively react with iron, aluminium and other heavy metals as cyano- and triethanolamino- complexes that are not broken down during titration with EDTA. Calcium and magnesium may then be determined free of interference except from manganese, for which again a small correction must be made.

(*a*) *Calcium.* Having neutralized interfering ions, either by extraction as 8-hydroxyquinoline complexes into chloroform or by the addition of chelating reagents (potassium cyanide and triethanolamine), the solution is made alkaline ($pH > 12$) and a reducing agent such as ascorbic acid or hydroxyammonium chloride added. At high pH under these reducing conditions, magnesium and manganese are quantitatively precipitated as hydroxides (note the possibility of co-precipitation of $Ca(OH)_2$ if a large excess of Mg is present—see above). The resultant mixture is then titrated with EDTA solution using an indicator. The indicator changes colour at the point of equivalence when all calcium ions have just complexed with EDTA. Indicators used for this titration include murexide (Shapiro and Brannock, 1962), calcein (Riley, 1958) and solochrome dark blue (Langmyhr and Graff, 1965). Shapiro and Brannock used a filter photometer to detect the endpoint of murexide, which is otherwise difficult to detect visually. Maxwell (1968, p.157) overcame this difficulty by screening the murexide endpoint with naphthol green B; the combined indicators give a pinkish-brown to clear blue colour change. Calcein turns from light yellow to darkish brown under screened uv light. Solochrome dark blue turns from red to blue at the point of equivalence. The determination is quantified by repeating the titration using a standard solution of calcium.

(*b*) *Combined calcium and magnesium.* To an aliquot of the same solution in which the interfering ions have been removed or complexed, ammonium hydroxide buffer solution is added to increase the pH to about 10 (the magnesium–EDTA complex is unstable below $pH = 7.6$). This

Table 2.7 Formation constants for selected metal ion—EDTA complexes

Cation	K	log (K)	Cation	K	log (K)	Cation	K	log (K)
Ag^+	2.1×10^7	7.32	Fe^{2+}	2.1×10^{14}	14.33	Hg^{2+}	6.3×10^{21}	21.8
Mg^{2+}	4.9×10^8	8.69	Co^{2+}	2.0×10^{16}	16.31	Pb^{2+}	1.1×10^{18}	18.04
Ca^{2+}	5.0×10^{10}	10.70	Ni^{2+}	4.2×10^{18}	18.62	Al^{3+}	1.3×10^{16}	16.13
Sr^{2+}	4.3×10^8	8.63	Cu^{2+}	6.3×10^{18}	18.8	Fe^{3+}	1×10^{25}	25.1
Ba^{2+}	5.8×10^7	7.76	Zn^{2+}	3.2×10^{16}	16.5	V^{3+}	8×10^{25}	25.9
Mn^{2+}	6.2×10^{13}	13.79	Cd^{2+}	2.9×10^{16}	16.46	Th^{4+}	2×10^{23}	23.2

$K =$ (M.EDTA)/(M)(EDTA), where (M) is the activity of the metal ion, (EDTA) is the activity of uncomplexed ethylenediaminetetra-acetic acid and (M.EDTA) is the activity of the metal ion–EDTA complex.
Data from Skoog and West (1982), originally published in Schwartzenbach (1957).

solution is then titrated with EDTA solution. Most procedures use eriochrome black T as indicator; this gives a colour change from pink to clear blue-green. The interference from Mn must be accounted for by pre-calibrating the volume of EDTA solution required to react with a known amount of manganese. The concentration of manganese is determined independently and the calculated equivalent volume of EDTA solution subtracted from the sample titration value. Alternatively, in some schemes, manganese can be removed from the solution before titration by boiling with sodium chlorate solution and precipitating it as hydrated manganese dioxide (Riley, 1959).

2.8 A rapid scheme of analysis

There is no single rapid scheme of analysis—like classical techniques, rapid methods evolved through a continuous series of developments, starting with gravimetric procedures. During the 1950s and early 1960s it became fashionable for individual laboratories to publish their own schemes, each incorporating methodologies preferred at that institution. The most enduring of these include the schemes of Shapiro and Brannock (1956, 1962), Riley (1958), Langmyhr and Graff (1965), Peck (1964) and Kirschenbaum (1983).

There were many common threads running through these schemes. The most obvious was the near-absence of gravimetric techniques; instead, spectrometric, flame photometric and EDTA titration procedures, described in the sections above, were used. Furthermore, two solutions were commonly used for the determination of the major elements. Solution A was normally used to determine silica and alumina (silica only in the case of Riley's scheme). Solution B was subjected to a hydrofluoric/perchloric acid digestion in which silica was fumed off. Total iron, CaO, MgO, TiO_2, P_2O_5, MnO, Na_2O and K_2O were determined on this fraction. Determinations of Fe(II), S, CO_2 and H_2O were required to complete the analysis, and procedures for these will be discussed in later sections.

In view of its widespread influence on the development of rapid schemes of analysis, Shapiro and Brannock's 1962 procedure will be summarized here in outline.

(1) *Solution A* was prepared by taking 50mg of rock power and adding 5mL of 30% sodium hydroxide solution. The mixture was evaporated to dryness in a nickel crucible, then fused to dull red heat for about 5 minutes. The cooled melt was leached with water, acidified with HCl and diluted to 1 litre.

(2) *Solution B* was prepared by taking 0.5g of sample and digesting it with a mixture of HF (12mL), H_2SO_4 (2.5mL) and HNO_3 (0.6mL) on a steam bath overnight in a covered teflon beaker. The residue was transferred to a Vycor beaker using the minimum quantity of water, and reheated on a hot plate until SO_2 fumes were emitted. Four drops of $HClO_4$–HNO_3 mixture (1 : 1) were then added and the samples reheated until strong fumes were evolved. Any organic matter was destroyed at this stage. After cooling, water (225mL), concentrated HNO_3 (5mL) and hydrazine sulphate (1mL) were added and the solution boiled. The hydrazine sulphate reduces any insoluble manganese dioxide to soluble Mn(II) salts. After cooling the solution was made up to a final volume of 500mL.

(3) *Silica* was determined on solution A (8mL aliquot) photometrically by the blue colour developed by adding ammonium molybdate ($(NH_4)_6Mo_7O_{24}.2H_2O$) solution and a reducing agent (sodium sulphite/bisulphate solution). Maximum absorption is at 812 nm (Figure 2.18), although Shapiro and Brannock specified 640 nm since this wavelength is reported to give more reproducible determinations in the presence of iron (Jeffery and Wilson, 1960).

(4) *Alumina* was determined photometrically on a 5mL aliquot of solution A using the red colour developed with alizarin red-S, the absorption being measured at 475 nm. Unfortunately iron and titanium also form complexes with this reagent and cause interferences in absorption measurements. Accordingly, Shapiro and Brannock removed the iron interference by complexing with potassium ferricyanide and thioglycolic acid solution before the addition of alizarin red S. An empirical table was required to correct for the effect of the titanium interference. In view of these difficulties some workers preferred Riley's method in which aluminium was extracted from solution B. A complexing reagent, containing hydroxylammonium chloride (to reduce iron to Fe(II)), sodium citrate (a buffer) and beryllium sulphate (to complex residual traces of fluoride) was added to this solution. The iron interference was removed by complexing as the ferrous-dipyridyl complex. Aluminium was complexed with 8-hydroxyquinoline and extracted, free from iron, into chloroform. The absorption of the yellow aluminium–8-hydroxyquinoline complex was measured at 410 nm. A small correction for a titanium interference was still required. Care must be taken with this procedure as beryllium salts are highly toxic.

(5) *Total iron oxide* was determined photometrically on a 5mL aliquot of solution B. Iron was first reduced to Fe(II) using hydroxylammonium chloride solution buffered with sodium citrate, and the intensity of the orange colour developed with 1,10-phenanthroline measured at 555 nm. Jeffrey (1975, p.289) reports that the absorption maximum of this complex is at 508 nm. Details of this reaction are discussed further in section 2.5.4(*b*).

(6) *Titanium oxide* was determined photometrically on 5mL aliquots of solution B from the yellow colour developed

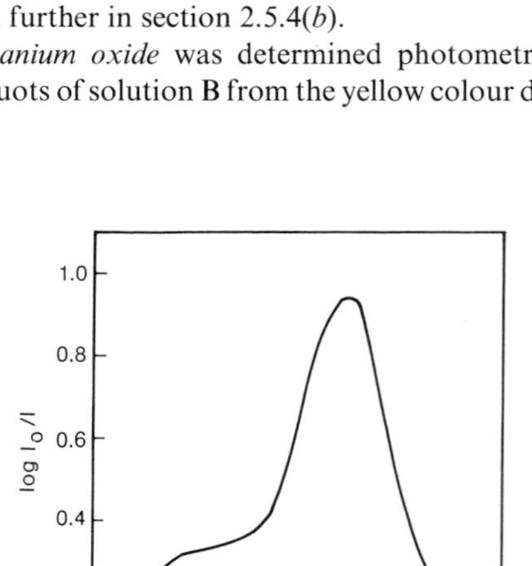

Figure 2.18 The absorption spectrum of a silicomolybdate solution after reduction to molybdenum blue (2 cm cells, 125 μg SiO_2/100 mL). (Jeffrey, 1975.)

with tiron (disodium 1,2-dihydroxybenzene-3,5-disulphonate). Iron interferes with the absorption measurements but can effectively be removed by reducing to Fe(II) with thioglycolic acid. In the procedure of Shapiro and Brannock, sodium dithionate ($Na_2S_2O_4$) was used as the reducing agent. However, Jeffrey (1975) reported that this reagent tends to deposit sulphur from solutions at low pH, the resultant turbidity interfering with absorption measurements. The absorption of the titanium–tiron complex was measured at 380 nm (Jeffrey, 1975, p.472). It was necessary for Shapiro and Brannock to make measurements at 430 nm since sodium dithionate absorbs strongly below 410 nm (Langmyhr and Graff, 1965, p.87).

(7) *Manganese oxide* was determined on 25mL aliquots of solution B by oxidizing Mn(II) ions in solution to permanganate using ammonium persulphite $(NH_4)_2S_2O_8$, catalysed by the presence of silver nitrate solution as described in section 2.5.4(*a*). The intense purple colour was measured at 525 nm. The correction of interferences from chromate ions has been discussed above.

(8) *Phosphorus pentoxide* was determined from 25mL aliquots of solution B, mixed with sodium metavanadate ($NaVO_3$) and sodium molybdate ($Na_2MoO_4.2H_2O$) solution. The absorption maximum of the molybdovanado–phosphoric acid complex is at 315 nm. However a wavelength of 420 nm was selected for measurement as a compromise between adequate sensitivity and minimizing and absorption interference from iron.

(9) *Sodium and potassium oxides* were determined by flame photometry using lithium as an internal standard (section 2.6).

(10) *Calcium oxide* was determined by EDTA titration in a strongly alkaline solution, using murexide as an indicator, of 25mL aliquots of solution B. Fe and Al and heavy metals were complexed with cyanide and triethanolamine before titration as described in section 2.7.

(11) *Magnesium oxide* was determined by difference between an EDTA titration for calcium plus magnesium and the calcium determination alone. A correction was required for the presence of manganese (section 2.7).

(12) *Water* was determined by igniting 1g of rock powder with fused sodium tungstate as flux in a pyrex test tube. The water released was absorbed on a pre-weighed filter paper held in the upper part of the test tube—see section 2.10 3(*a*).

(13) *Ferrous iron* was determined by decomposing 0.5g of rock powder with sulphuric acid (10mL of 1+3) and hydrofluoric acid (5mL, 48%). After boiling, the activity of the hydrofluoric acid was neutralized by adding boric acid crystals (10g). The liberated Fe(II) was titrated against potassium dichromate solution using diphenylamine sulphonate as indicator. A small (known) spike of ammonium ferrous sulphate was added to the rock solution to assure a definitive endpoint in those samples with very low Fe(II) contents. Iron(III) was then determined by difference against the total iron data. The determination of ferrous iron is described in section 2.9.

(14) *Carbon dioxide* was determined by a technique developed by Shapiro and Brannock in which CO_2 was liberated by the action of hydrochloric acid on 1g of rock powder using a specially designed tube and side-arm. The liberated gas was trapped in the side-arm and its volume estimated with a millimetre rule. The decomposition was carried out in the presence of mercuric chloride to prevent liberation of hydrogen from any 'tramp' iron that might have contaminated the sample during washing. Procedures for determining carbon dioxide are considered in more detail in section 2.10.

(15) Provision in the scheme was also made for the determination in phosphate rocks of both *total sulphur* (determined gravimetrically as $BaSO_4$) and *fluorine* (determined photometrically by the bleaching effect of the fluorine ion on the calcium–aluminium alizarin red S complex).

To give some indication of the accuracy and precision of analyses for the major elements by rapid schemes of analysis, results from three such procedures, those of Riley (1958), Langmyhr and Graff (1965) and Shapiro and Brannock (1956), are compared in Table 2.8 for granite USGS: G-1. These results are compared with average values compiled from the interlaboratory evaluation of G-1 and W-1 by Stevens and Niles (1960), derived mainly from 'rapid' techniques. Each set of data is compared with modern results from the work of Abbey (1983). These data indicate that analytical precisions of as low as 0.1 to 0.2% (relative, one sigma) can be obtained by rapid schemes under optimum conditions. However, analytical precisions are generally significantly worse than this and do not match instrumental precision expected of modern XRF or ICP–AES data.

Rapid methods of analysis survive today, not usually for the complete major element analysis of rock solutions, but for the determination of those constituents that are difficult or impossible to analyse by modern instrumental techniques such as XRF, AAS or ICP–AES. These components include Fe(II), H_2O, CO_2, Cl and F. The determinations of fluorine and chlorine are considered in Chapter 7 (ion-selective electrodes) and section 2.11 (auto-analyser techniques). Procedures for the other three constituents will now be considered in more detail.

2.9 Determination of ferrous iron

2.9.1 *Oxidation of rock powders during grinding*

The determination of ferrous iron causes some difficulty in modern schemes of analysis based on instrumentation (e.g. XRF, AAS, ICP–AES) that cannot distinguish the oxidation state of an analyte species. Such schemes must therefore be complemented by determinations for ferrous iron based on titrimetry. Conversely, there is no widely accepted method of determining ferric iron; this is normally calculated by difference from independent measurements of total iron and ferrous iron.

Any attempt to determine ferrous iron must take into account two potential interferences arising during sample preparation, which could invalidate any subsequent measurement. The first is the possibility of introducing splinters of iron ('tramp' iron) into the sample when splitting samples using steel crushing equipment. The second and more important possibility is that of atmospheric oxidation of ferrous iron during the final stages of grinding samples. Fitton and Gill (1970) investigated this problem in detail. Their results for the variation in the apparent FeO content

Table 2.8 Determination of the major elements in granite USGS: G-1 by selected rapid schemes of analysis

Scheme		SiO_2	TiO_2	Al_2O_3	Fe_2O_3	FeO	MnO
Riley (1958)	Expected	72.68	0.26	14.05	0.87	0.96	0.03
	Analysed	72.6	0.24	14.1	0.87	0.93	0.029
	CV ($n=5$)	0.12	6.1	0.9	9.0	1.9	26.4
	Method of determination	Photometric (molybdenum blue)	Photometric (H_2O_2)	Photometric (8-hydroxy-quinoline)	By difference	Titration (potassium dichromate)	Photometric (permanganate)
Langmyhr and Graff (1965)	Expected	72.68	0.26	14.05	0.87	0.96	0.03
	Analysed	72.73	0.237	13.81	0.66	0.96	0.034
	CV ($n=10$)	0.20	1.7	3.1	14.1	4.8	9.9
	Method of determination	Photometric (Molybdenum yellow)	Photometric (titron)	Titration (EDTA plus zinc)	By difference	Titration (potassium dichromate)	Photometric (permanganate)
Shapiro and Brannock (1956, 1962)	Expected	72.68	0.26	14.05	0.87	0.96	0.03
	Analysed ($n=2$)	72.4	0.24	14.5	0.86	0.99	0.02
	Method of determination	Photometric (molybdenum blue)	Photometric (tiron)	Photometric (calcium alizarin S)	By difference	Titration (potassium dichromate)	Photometric (permanganate)
Stevens and Miles (1960)	Expected	72.68	0.26	14.05	0.87	0.96	0.03
	Analysed	72.35	0.26	14.32	0.95	0.99	0.03
	CV ($n=48$ to 60)	0.66	16.6	2.6 (Interlaboratory compilation)	31.5	11.3	33.6

Expected compositions taken from Abbey (1983). Units: % oxide, CV: % relative (one sigma).
Data of Stevens and Niles is the arithmetic mean of contributed analyses; other statistical averages were calculated by these authors.

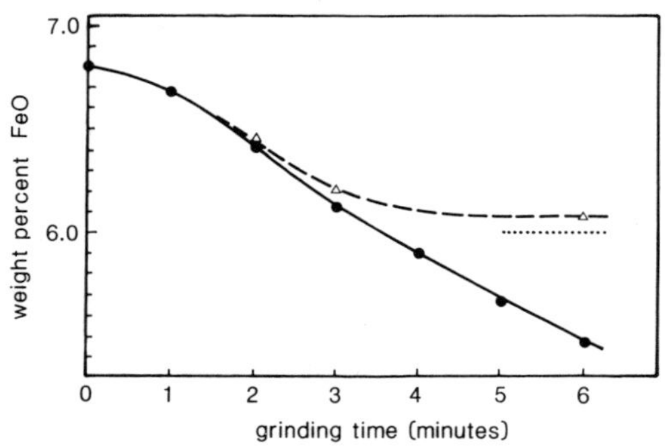

Figure 2.19 Oxidation of ferrous iron to ferric iron as a function of the grinding time of a basalt (Lake District) using a tungsten carbide 'Tema' swing mill. The solid line is the result of experiments in which sample grinding was interrupted to permit removal of some powder for analysis, simultaneously admitting fresh air. The broken line represents data from samples that were ground continuously for the time indicated in a sealed mill. The theoretical limit of oxidation in such circumstances is indicated by the dotted line. (Fitton and Gill, 1970.)

of a basalt ground for varying lengths of time in a tungsten carbide swing mill are shown in Figure 2.19, and indicate that considerable atmospheric oxidation can occur if samples are milled for longer than about one minute: in this example, the apparent FeO content of 6.8% falls to about 6.0% after 4 minutes grinding. This work showed that samples containing hydrated minerals such as biotite and chlorite were much more susceptible to oxidization than rocks in which the majority of the iron is held in fresh anhydrous minerals such as olivine or pyroxene, and Fitton and Gill therefore proposed that samples to be analysed for ferrous iron content should be mechanically ground for not longer than 30 seconds to reduce atmospheric oxidization, even though the particle size might then not be as fine as would normally be accepted for chemical analysis.

2.9.2 *Wilson's method*

Perhaps one of the most widely used methods of determining ferrous iron is that proposed by Wilson (1955). This procedure is based on the oxidation of ferrous to ferric iron using ammonium metavanadate (NH_4VO_3). In order to prevent atmospheric oxidation during chemical dissolution of the sample, ammonium metavanadate is added to the digestion mixture. As soon as ferrous iron is released from the rock matrix, it reacts as follows:

$$V^{V}O_2^+ + Fe^{2+} + 2H^+ \rightarrow V^{IV}O^{2+} + H_2O + Fe^{3+}. \quad (2.9)$$

The reduced form of vanadium ($V^{IV}O^{2+}$) is stable to atmospheric oxidation. The analysis procedure is as follows. Ammonium metavanadate crystals (0.1g) are accurately weighed into two polyethylene or Teflon beakers, one containing rock powder (0.5g) and the other empty (the blank). Hydrofluoric acid (8mL, 48%) is then added to each beaker to dissolve the rock. The beakers are sealed and left overnight or until the sample has dissolved. Saturated boric acid solution is then added to neutralize the hydrofluoric acid. The vanadium(IV) content of this solution is equivalent to the amount of ferrous iron liberated from the sample. This

CaO	MgO	Na_2O	K_2O	P_2O_5	H_2O+	CO_2	Fe_2O_3 (T)	Total
1.39	0.38	3.32	5.48	0.09	0.34	0.07	1.94	99.92
1.4	0.40	3.4	5.5	0.08	0.34	nd		99.89
4.2	2.5	1.4	2.1	5.7	6.3	—		
Titration (EDTA)	Titration (by difference)	Flame photometry	Flame photometry	Photometric (molybdo-vanado complex)	Gravimetric ignition	Gravimetric ignition	Photometric (dipyridyl)	
1.39	0.38	3.32	5.48	0.09	0.34	0.07	1.94	99.51
1.27	0.39	3.27	5.05	0.071			1.73	98.48
1.2	5.1	1.2	1.2	11.7			7.3	
Titration (EDTA)	Titration (by difference)	Flame photometry	Flame photometry	Photometric (molybdo-vanado complex)			Photometric (tiron)	
1.39	0.38	3.32	5.48	0.09	0.34	0.07	1.94	99.51
1.4	0.31	3.3	5.4	0.08	0.31	0.11	1.8	99.9
Titration (EDTA)	Titration (calcium precipitated as tungstate)	Flame photometry	Flame photometry	Photometric (molybdo-vanado complex)	Gravimetric (modified Penfield)	Volumetric ignition	Photometric (1,10-phenanthroline)	
1.39	0.38	3.32	5.48	0.09	0.34	0.07	1.94	99.51
1.40	0.40	3.31	5.42	0.10	0.36	0.08	2.04	101.57
9.0	32.2	7.1	7.3	55.1	48.7	13.0	13.9	
			(interlaboratory compilation)					

is determined by adding a known excess of Fe(II) solution (prepared from ammonium ferrous sulphate, $(NH_4)_2SO_4.FeSO_4.6H_2O$) to each beaker. All unreacted vanadium (V) ions are then reduced to vanadium (IV) as in equation (2.9). Each solution then contains excess ferrous ions, the concentration of which is determined by titration with a standard solution of potassium dichromate using sodium or barium diphenylamine *p*-sulphonic acid as indicator:

$$Cr_2O_7^{2-} + 6Fe^{2+} + 14H^+ \rightarrow 2Cr^{3+} + 6Fe^{3+} + 7H_2O. \quad (2.10)$$

The difference in titration volumes between sample and blank solutions is equivalent to the concentration of ferrous iron released from the rock sample.

Wilson's method is not free of interferences. Most importantly, it is unable to distinguish between the reducing powers of ferrous iron and of sulphide sulphur that may also be dissolved out of the sample. Systematic errors (which must be corrected for) will occur in samples containing acid-soluble sulphide, e.g. pyrrhotite (FeS). Similar problems are also likely to occur with samples containing organic matter (but not graphite, or manganese dioxide in some sedimentary rocks). These interferences are considered further in section 2.9.4. Whipple (1974) has undertaken a thorough chemical optimization of this procedure and has demonstrated several minor analytical interferences and ways of minimizing them.

2.9.3 *Pratt's method*

The other standard wet chemical procedure to determine ferrous ion is the Pratt method (1894). Conventionally, this involves dissolving a rock powder (0.5g) in sulphuric acid (10mL of 1+3) and hydrofluoric acid (5mL of 48%) in a large platinum crucible. The crucible is covered to exclude atmospheric oxygen and the mixture boiled (10 minutes). The hot crucible and contents are then submerged in cold boric acid solution. This solution is then titrated with potassium dichromate solution using barium diphenylamine sulphonate as indicator. In Shapiro and Brannock's adaptation of this technique (1962) a small known excess of ferrous ammonium sulphate solution was added before titration of the blanks to ensure that the colour change of the indicator could be observed more distinctly. To carry out this procedure successfully, air must be excluded during sample decomposition. Goldrich (1984) discussed a procedure in which decompositions are all carried out in a single silica flask using potassium permanganate for the $Fe^{II} \rightarrow Fe^{III}$ titration. The Pratt technique suffers from the same interferences as the Wilson method (section 2.9.2).

2.9.4 *Interference effects in Wilson's and Pratt's methods*

Langmyhr and Graff (1965) have summarized some analytical difficulties in the determination of ferrous iron related to the problem of achieving complete sample dissolution. Their work indicates that a fluoride precipitate is often formed after HF attack. This is usually assumed to consist of insoluble magnesium and calcium fluorides. However, Langmyhr and Kringstad (1966) found that fluoride precipitates present after HF attack on a diabase contained CaF_2, $MgAlF_5.xH_2O$, $NaAlF_4.xH_2O$ and Fe(II)Al-Fe(III)$F_5.xH_2O$. It is important, therefore, to bring into solution any precipitate observed after HF attack before commencing the Fe(II) titration. Furthermore, simple boiling of rock samples with hydrofluoric acid or HF–sulphuric acid mixtures will not dissolve the more resistant silicate minerals that may contain significant proportions of Fe(II)–see also section 2.1.2(*c*). A further difficulty arises because some iron sulphide minerals are likely to dissolve (e.g. chalcopyrite, $CuFeS_2$ and pyrrhotite, FeS) whereas pyrite (FeS_2) is likely to remain insoluble. These difficulties have not been completely resolved in wet chemical methods for the determination of ferrous iron.

2.9.5 *Other procedures*

Few instrumental procedures have been described for the determination of iron in a specified oxidation state. Some consideration has been given to x-ray analysis, since the energy of the iron *L*-lines is sensitive to oxidation state (section 10.9). These lines occur at very low x-ray energies (0.6–0.7keV) and although some results have been obtained for individual minerals by electron probe microanalysis, the technique has not been proven for bulk sample analysis. Kiss (1977) described a potentiometric method in which, following a Pratt-type sample decomposition, the reaction mixture is titrated against potassium dichromate solution using an automatic titrator. The signal from a platinum–20% rhodium bimetallic electrode is used to detect the endpoint. Precision of about 0.35% (relative standard deviation) was reported, and the technique could readily be modified to permit determination of total iron. The presence of sulphide minerals was claimed not to interfere with the determination.

2.10 The determination of water and carbon dioxide

The determination of structural water and carbon dioxide is an integral part of all major-element determinations of silicate rocks. However, this procedure has recently tended to be overlooked in some laboratories, because a separate sample analysis is necessary to complement modern instrumental schemes available for the other major elements.

In this section, the more common procedures for determining water and carbon dioxide are considered. Those who would attach detailed geochemical significance to the results of such determinations should first study Chapter 18. Gas source mass spectrometer measurements of samples such as ocean floor basalts by stepped combustion procedures suggest that only gases released above 600°C are unambiguously representative of the primary composition. Gases released at lower temperatures appear to be associated with various sources of contamination from both laboratory and geological environments.

2.10.1 *Loss on ignition*

In many x-ray fluorescence procedures for major elements, it has become conventional to estimate the proportion of water and carbon dioxide in a sample by a simple weight loss on ignition. Samples are ignited in an unglazed porcelain or vitreous silica crucible at 1000°–1050°C for about one hour.

Such determinations have little geochemical value for several reasons.

(i) The maximum temperature used in such procedures is not sufficiently high to decompose all hydrated minerals. In particular, temperatures above 1200°C are required to decompose minerals such as talc, topaz, staurolite, cordierite and epidote.

(ii) In some samples a gain in weight is observed after ignition (i.e. a negative loss on ignition) due to the oxidation of ferrous iron in the sample:

$$2Fe^{II}O + \tfrac{1}{2}O_2 \rightarrow Fe_2^{III}O_3 \qquad (2.11)$$

The gain in weight caused by this reaction, if it were to go to completion, is equivalent to a negative loss on ignition of 1.13% for each 10% FeO in the sample. Unfortunately, there is no reason to assume that complete oxidation occurs under the conditions routinely used for ignition (see Table 2.10).

(iii) Some loss by volatilization of alkali metals, fluorine and oxides of sulphur may also occur. Loss on ignition data can only therefore be used as a general indication of the volatile element content of a sample.

2.10.2 *Hydroscopic or absorbed water* (H_2O-)

Before a determination of structural water can take place, the amount of water absorbed on to the sample powder must be determined. By convention this is measured by heating the powdered sample at 105°–110°C in an oven for at least 2–3 hours (or preferably to constant weight). However, such determinations are of doubtful geochemical significance. Errors occur, as a result of the slow oxidation of Fe(II) to Fe(III) even in powdered samples heated to this low temperature, and the decomposition of some minerals (e.g. clay minerals), which evolve structural water at this low temperature. Furthermore, Jeffrey (1975, p.262) has shown that the apparent adsorbed water content of powdered samples can change significantly when exposed to varying relative-humidity conditions in the laboratory atmosphere. Samples at risk from this phenomenon include those containing hydrated phases such as chlorite (Figure 2.20). Absorbed water is often reported as part of any major

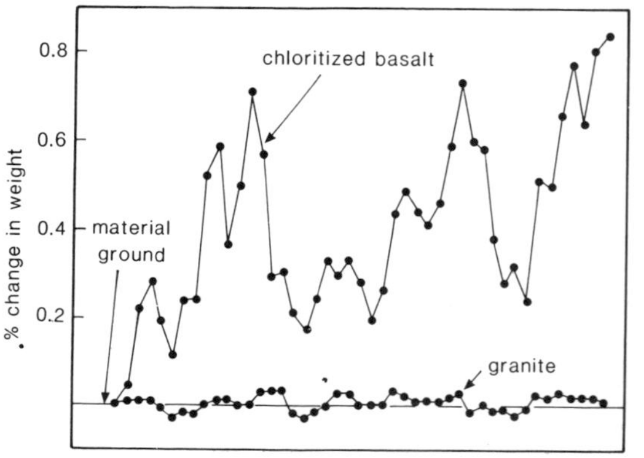

Figure 2.20 Day-to-day weight changes of two silicate rock powders on exposure to air. (Jeffrey, 1975.)

element analysis and is normally designated H_2O- to distinguish the determination from that of structural water (H_2O+).

Some workers, for example Abbey (1983), suggest that it is not meaningful to report absorbed water as part of a major-element analysis, preferring results to be reported on a dry basis. The uncertainties and potential variabilities outlined above provide considerable support for this point of view. The convention of reporting major-element data on a dry basis has been adopted in this work.

2.10.3 *Ignition methods for the determination of water and carbon dioxide*

The rapid methods incorporated into the scheme of Shapiro and Brannock (1962) together with methods based on the selective absorption of gaseous components, are considered here as representative. It must be noted that these techniques do not distinguish the origin of evolved gases (structural water, hydroxyl ions, organic constituents, carbonates). Although, by convention, measurements are reported as CO_2 and H_2O+, determinations are really representative total hydrogen and total carbon, irrespective of source.

(*a*) *The Penfield method for water.* Penfield (1894) originally proposed a rapid method for the determination of total water. As incorporated into Shapiro and Brannock's rapid scheme of analysis, the method is as follows. The rock powder (1g) is mixed with a flux, sodium tungstate (2g), which has been pre-fused to evolve water of crystallization from $Na_2WO_4.2H_2O$. The mixture is placed in a pyrex test tube. A 2×2 inch piece of filter paper is pre-weighed and then inserted into the top of the test tube, which is then stoppered (the stopper has a small hole in it). The test tube is then inserted into a cooling jacket (Figure 2.21). The sample end of the test tube is heated for 5 minutes using a Fisher burner. The flux melts and assists in breaking down the structure of thermally resistant minerals. Water evolved is condensed on the filter paper. After cooling for one minute, the filter paper is rotated within the test tube to mop up all the condensed water, and is weighed in a stoppered bottle. Delong and Lyman (1982) have described a modification to the design of cooling jacket to ensure uniform condensation of water vapour. This simple and rapid method suffers from the disadvantage that not all the evolved water can be collected on the filter paper, and so a small empirical correction must be applied. This is estimated by Jeffery (1975) to be equivalent to 10% of the recorded weight of water if less than 20mg of H_2O are collected, or 2mg more if the evolved water exceeds this figure.

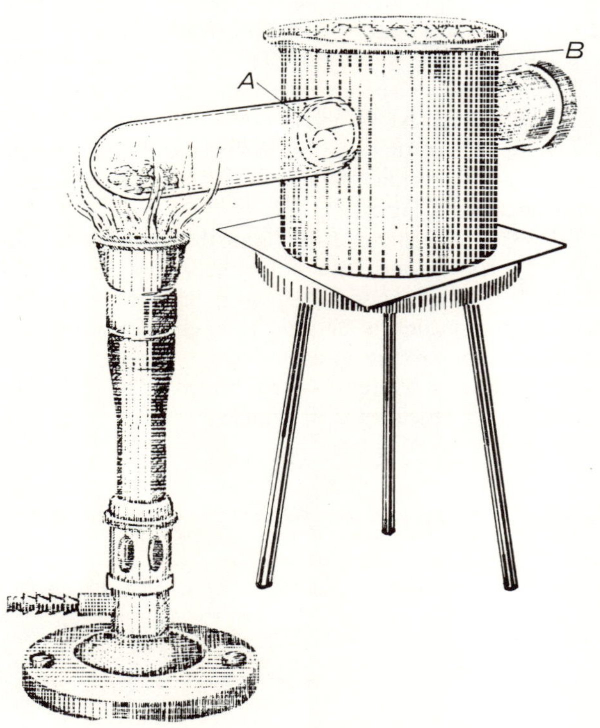

Figure 2.21 The determination of water using Shapiro and Brannock's apparatus. *A* is a coil of filter paper used to absorb water; *B* is a polyethylene container filled with crushed ice plus sodium chloride crystals. (Shapiro and Brannock, 1962; by permission of the copyright holder.)

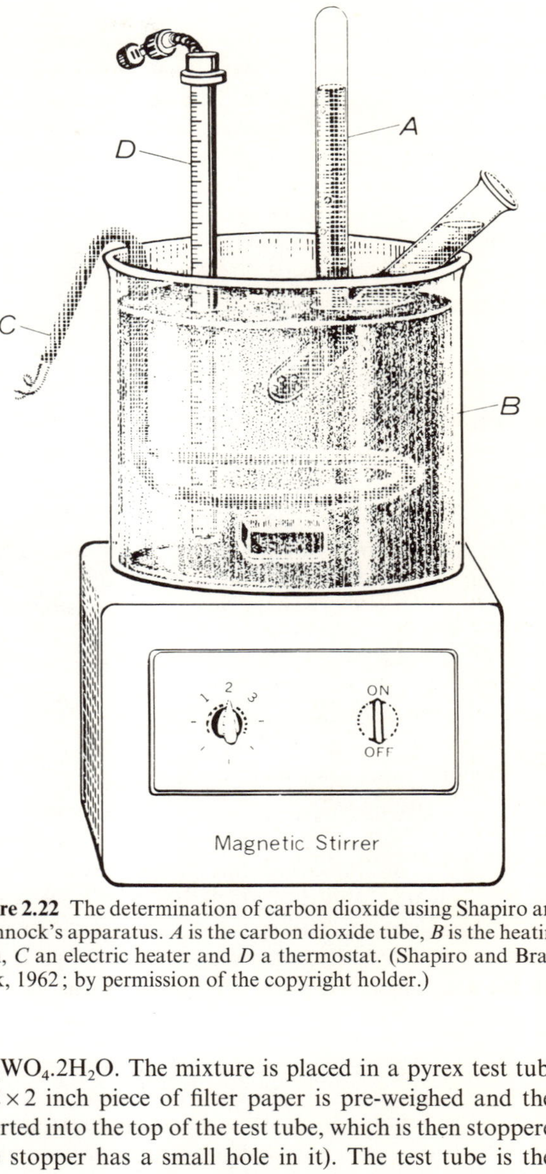

Figure 2.22 The determination of carbon dioxide using Shapiro and Brannock's apparatus. *A* is the carbon dioxide tube, *B* is the heating bath, *C* an electric heater and *D* a thermostat. (Shapiro and Brannock, 1962; by permission of the copyright holder.)

(*b*) *Rapid determination of carbon dioxide.* Shapiro and Brannock (1962) adapted a technique, originally proposed by Fahey (1946), to liberate CO_2, suitable for samples yielding less than about 6% of the gas. The apparatus (Figure 2.22) consists of a boiling tube on to which is fused a long,

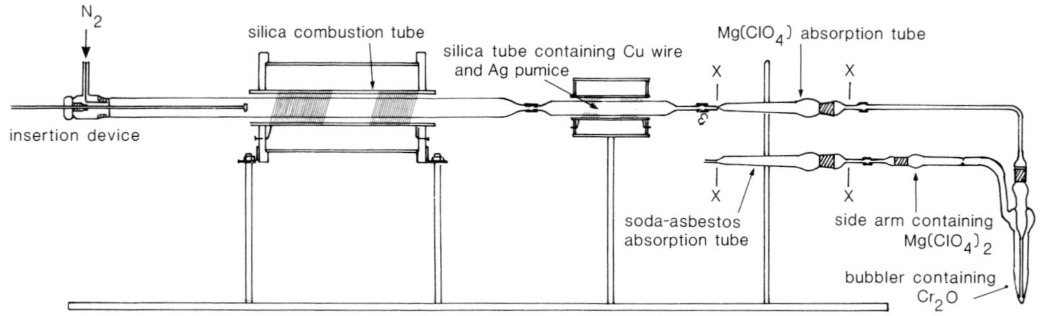

Figure 2.23 Riley's apparatus for the simultaneous determination of water and carbon dioxide by thermal decomposition of the sample in a silica tube furnace at 1000°C. (Riley, 1985*b*.)

relatively narrow side-arm, the end of which is sealed. The powdered sample (1g) is poured into the boiling tube together with solid mercuric chloride $HgCl_2$ (1g) and saturated mercuric chloride solution (2mL). The boiling tube is filled to the mark with paraffin oil and rotated to displace all air from the side-arm. The boiling tube is then positioned so that the side-arm is vertical. Hydrochloric acid (1+1, 2mL) is added to the sample and the mixture heated to 110°–120°C in an oil bath for at least 5 minutes. The liberated CO_2 gas displaces oil from the side-arm. When cooled to room temperature, the amount of oil displaced is proportional to the concentration of CO_2 in the original sample. The apparatus is calibrated using samples of known CO_2 content. The principal interference arises from the presence of any metallic-iron flakes introduced into the sample during splitting and grinding. When attacked by acid, these would release hydrogen gas. The presence of mercuric chloride overcomes this difficulty, since metallic iron is taken into solution without the evolution of hydrogen by reaction with this salt:

$$Fe^{O} + 2Hg^{2+} \rightarrow Fe^{2+} + 2Hg^{+} \qquad (2.12)$$

Jeffrey (1975, p.177) summarizes earlier work that indicates that certain minerals (e.g. scapolites) do not release all their carbon dioxide when attacked with hydrochloric acid alone and that a mixture of HCl and HF is then required to break down the silicate matrix. This modification does not appear to have been widely adopted.

(*c*) *Riley's method for the simultaneous determination of CO_2 and H_2O*. The apparatus described by Riley (1958*b*) was designed to permit the simultaneous gravimetric determination of water and carbon dioxide by thermal decomposition (Figure 2.23). The powdered sample (0.5–1.5g) is placed in a 2 mL aluminium boat lined with a piece of nickel foil. This boat is then ignited at 1100°C in a silica tube furnace for 30 to 40 minutes. During ignition a gentle flow of nitrogen gas (3 litres per hour) sweeps any evolved gases through two absorption tubes, one packed with magnesium perchlorate ($Mg(ClO_4)_2$), which quantatively absorbs water vapour, and one packed with soda-asbestos to absorb CO_2. If the sample contains significant concentrations of sulphur (<0.5%), a bubbler containing a saturated solution of chromium trioxide in syrupy phosphoric acid must be placed between the two absorption tubes to prevent the absorption of SO_2 by the soda-asbestos tube.

A systematic interference in the determination of carbon dioxide can occur due to the formation of oxides of nitrogen in the carrier gas during passage through the ignition furnace. These are decomposed by passing gases first through a tube furnace packed with copper turnings heated to 700°–750°C.

Blank levels are minimized by passing the nitrogen supply through absorption columns containing soda lime (to absorb CO_2), fused calcium chloride and magnesium perchlorate (to absorb H_2O), before the gas is delivered to the ignition furnace. After ignition, the increase in weight of the main absorption tubes is measured to determine the weight of CO_2 and H_2O evolved by the sample. With care, blank levels of 0.1mg per hour of water and 0.2mg per hour of CO_2 can be achieved.

Riley (1958*b*) reported that some minerals such as talc, topaz and epidote do not release all their water at temperatures below 1100°C; but found that the loss of water was complete after igniting at 1200°C for 30 minutes. This was confirmed by comparing decompositions of these minerals by the proposed method at 1200°C with measurements in which a sodium tungstate flux had been added to promote sample decomposition. Data by the two methods were in good agreement (Table 2.9).

Riley (1985*b*) compared results of the determination of water by his method with other recognized procedures (Table 2.10). He analysed a variety of samples containing varying proportions of ferrous iron and compared the apparent water contents measured by:

(i) Simple loss on ignition (sample heated to 1000°C in an open platinum crucible for 30 minutes)

Table 2.9 Determination of water content (%) of minerals resistant to thermal decomposition

Mineral	Epidote	Staurolite	Topaz	Talc	Phlogopite
Fusion with sodium tungstate flux (Groves, 1951, p.97)	1.96	1.83	1.69	4.82	2.72
Thermal decomposition at 1200°C (Riley, 1958*b*)	1.98	1.87	1.66	4.88	2.77

Data from Riley (1958*b*).

Table2.10 Comparison of methods for the determination of water in rocks

Rock type	% FeO	Residual % FeO after ignition	Loss on ignition: Uncorrected %	Loss on ignition: Corrected A %	Loss on ignition: Corrected B %	Penfield method %	Penfield method with PbO %	Shapiro & Brannock method %	Riley method %
Biotite–garnet schist	9.42	1.32	2.35	3.39	3.24	2.32	2.88	2.48	2.98
Garnet–sillimanite–sericite schist	10.29	1.93	4.48	5.62	5.41	5.19	5.54	4.91	5.47
Garnet–cordierite hornfels	11.91	3.01	2.70	4.02	3.69	2.87	2.83	2.94	3.27
Garnet–cordierite hornfels	9.58	2.22	4.17	5.84	5.59	4.38	4.37	4.47	4.67
Cordierite–garnet–biotite hornfels	8.92	2.29	3.80	4.79	4.50	3.92	3.94	3.86	4.10
Biotite-rich cordierite hornfels	11.13	1.50	3.08	4.33	4.16	3.25	3.65	3.41	3.70
Biotite–sericite schist	8.70	0.18	3.58	4.55	4.53	3.93	4.22	3.99	4.50
Hornblendite	7.15	0.16	1.19	1.98	1.96	1.84	2.42	1.44	2.57
Garnet–sillimanite–biotite schist	9.82	2.28	3.57	4.65	4.40	4.03	4.08	3.80	4.32

Data from Riley (1958*b*).
A: corrected on the assumption of total oxidation of ferrous iron.
B: corrected on the assumption of partial oxidation of ferrous iron.

(ii) The Penfield method (sample decomposed in a pyrex glass bulb for 20 minutes at about 800°C)
(iii) The Penfield method as above but with a lead oxide flux added
(iv) The Shapiro and Brannock (1955) method (sodium tungstate added as a flux)
(v) Riley's own method (sample decomposed at 1100°C for 30 to 40 minutes).

The results of these determinations (Table 2.10) show the proportion of ferrous iron not oxidized to ferric iron in samples after simple loss on ignition, and the effect of a correction for the associated gain in weight on the determination for water.

Riley suggested that these data indicate that the higher temperature specified in his procedure was more effective in releasing water than the other established procedures, which tended to give systematically low results. No correction was applied by Riley for the water vapour not absorbed on the filter paper in the standard Penfield procedure. The recommended empirical correction factor was, however, applied to results by Shapiro and Brannock's method; this factor was considered by Riley to be too low. The shortcomings of the loss on ignition analyses are apparent, since the proportion of ferrous iron remaining after ignition appears to be both significant and variable.

(d) The C–H–N elemental analyser. In the quantative characterization of organic compounds, routine determinations are required of bulk carbon, hydrogen and nitrogen. The C–H–N elemental analyser (Figure 2.24) was designed to fulfil this role and has been adapted by a small number of laboratories for the analysis of silicate rocks for water and carbon dioxide (e.g. Vin and Jones, 1978). The principles on which this instrumentation are based are in some respects similar to those used for Riley's apparatus. The sample is pre-dried at 105–110°C for 1 to 3 hours before analysis. Sample powder (25mg) and flux (100mg) are then weighed into a nickel boat. The flux both causes breakdown of the silicate matrix and acts as a source of oxygen, ensuring that

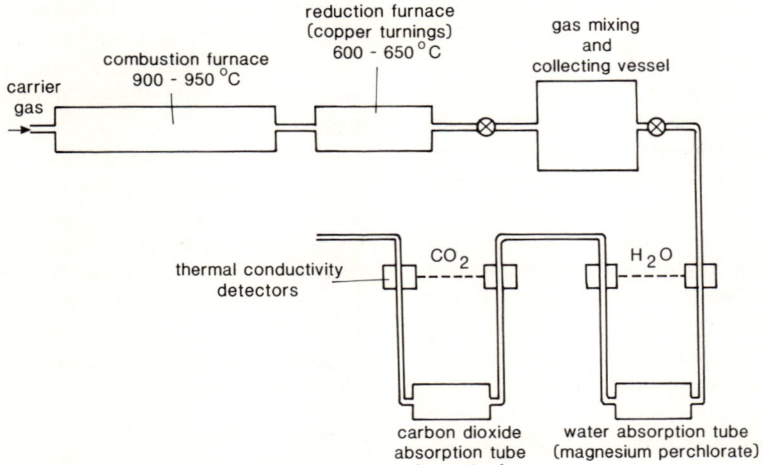

Figure 2.24 Highly schematic representation of a C–H–N elemental analyser for the instrumental determination of carbon as CO_2 and hydrogen as H_2O.

Table 2.11 Determination of H_2O+ and CO_2 in selected rock reference materials using an elemental analyser

Reference material	Expected % CO_2 (Abbey, 1983)	Analysed total C as % CO_2	Expected % H_2O (Abbey, 1983)	Analysed total H as % H_2O+
GSJ:				
JB-1	0.18?	0.13	1.01	0.92
JG-1	0.08?	0.08	0.54	0.50
USGS:				
AGV-1	0.02	0.17	0.78	0.87
BCR-1	0.02	0.20	0.67	0.93
DTS-1	0.07	0.15	0.42	0.42
PCC-1	0.18	0.12	4.70	4.76

Analysed data from Din and Jones (1978).

all carbon for example is oxidized to CO_2. Din and Jones investigated the efficiency of various fluxes (PbO, Na_2WO_4, $PbCrO_4$, CuO) and found that a mixture consisting of PbO, CuO and $PbCrO_4$ in the proportions 10 : 1 : 1 was the most effective. The sample is loaded into the ignition furnace and any atmospheric gases introduced at the same time are purged from the instrument with helium. Samples were then ignited at 900°–950°C, and evolved gases flushed through a reduction furnace containing copper turnings heated to 600°–650°C. Nitrogen oxides formed during ignition are reduced to nitrogen, and excess oxygen absorbed by passing combustion gases over copper turnings heated to 600°–650°C.

The gases are then passed through tubes containing (i) magnesium perchloride (to absorb water vapour), and (ii) colorcarb (to absorb CO_2). A thermal conductivity bridge senses the gas flow on the inlet side of each absorption tube in comparison with that on the outlet side. Absorption of gas within the tube causes the thermal conductivity meter to generate a signal, the magnitude of which depends on the difference in composition between inlet and outlet gases. This is itself dependent on the amount of gas (H_2O or CO_2) absorbed on the column. The imbalance of the thermal conductivity signal is therefore a measure of the total hydrogen or carbon contents of the sample, measured as H_2O or CO_2 respectively.

The instrument is calibrated by igniting a sample, preferably a rock reference material, of known composition. Data from Din and Jones (Table 2.11) demonstrates the reliability of this technique. Detection limits for a 25mg sample were about 0.02wt% CO_2 or H_2O. Particular care must be taken to ensure that the analysed aliquot (25mg) is homogeneous and representative of the bulk sample. The technique can be extended to permit the determination of sulphur evolved during combustion as SO_2. After removal of impurities from the ignited gas stream, sulphur dioxide is absorbed on a silver oxide column, the signal again being measured using a thermal conductivity bridge (e.g. Culmo, 1972).

(*e*) *Other techniques*. Other instrumental techniques have been developed for the determination of water and carbon dioxide in silicate rocks. *Gas chromatography* has been described by Jeffery and Kipping (1962) and Carpenter (1962) for the determination of carbon dioxide. In Jeffery and Kipping's procedure, the gas was liberated from powdered samples by attack with orthophosphoric acid. Hydrogen was used as the carrier gas. Carpenter claimed that gas chromatography was about 100 times more sensitive than gravimetric or volumetric procedures. Le Guern *et al.* (1982) has used a field-based chromatograph for the analysis of high-temperature volcanic gases. Barker and Chatten (1982) have discussed the application of gas chromatography to the determination of low concentrations of total carbonate in geological materials. *Infra-red spectrometry* in the determination of carbonate and non-carbonate carbon has been described by Terashima (1978, 1979). Carbonate carbon was released by heating the powdered sample to about 150°C and then adding hydrochloric acid (6M). The mixture was shaken and then allowed to stand until no more CO_2 was evolved. Nitrogen carrier gas was used to sweep carbon dioxide to the infra-red detector. Non-carbonate carbon was determined after first removing carbonate carbon by heating to 130°C with hydrochloric acid (11M). The mixture was evaporated to dryness to remove all traces of HCl. The residue was covered with iron powder, chips of tungsten were added and the mixture ignited at about 1800°C for 35 to 60 seconds to release non-carbonate carbon. Results for some rock reference materials are listed in Table 2.12. *Karl Fischer titrations* were used by Turek *et al.* (1976) to determine combined water in silicate rocks. Dried rock powder was mixed with about twice its weight of litharge as flux in a porcelain boat. The sample was ignited in a nichrome wire furnace at 850°C for 30 minutes. Evolved gases were swept with oxygen-free nitrogen carrier gas into a titration vessel containing methanol that had previously been brought to the end-point with Karl Fischer reagent. The volume of Karl Fischer reagent required to re-establish the end-point was proportional to the amount of water released from the sample. Karl Fischer reagent consists of a solution of iodine, sulphur dioxide and pyridine (C_5H_5N) in methanol. This reacts with water as follows:

$$C_5H_5N.I_2+C_5H_5N.SO_2+C_5H_5N+H_2O \rightarrow 2C_5H_5N.HI+C_5H_5N.SO_3 \quad (2.13)$$

The end-point can be detected visually by a change from

Table 2.12 Determination of carbonate carbon and non-carbonate carbon in selected reference materials by infra-red spectrometry

Reference material	Carbonate carbon (ppm)	Non-carbonate carbon (ppm)	Total carbon (ppm)	Expected total carbon (ppm)
GSJ:				
JB-1	353	114	467	(491?)
JG-1	172	43	215	(218?)
USGS:				
AGV-1	9	15	24	(55)
BCR-1	10	20	30	(55)
DTS-1	243	90	333	(191)
G-2	214	24	238	(218)
GSP-1	297	33	330	(327)
PCC-1	360	112	472	(491)

Analysed data from Terashima (1979).
Expected total carbon recalculated from Abbey (1983). Data are quoted in brackets to indicate uncertainties that may have arisen due to rounding erors in figures originally quoted as % CO_2.

yellow to brown caused by the appearance of the pyridine/iodine complex, signalling that all the water has reacted. Potentiometric methods of end-point detection are, however, preferred. Data from Turek *et al.* for structural water determined by Karl Fischer titration are compared with results by the Penfield method in Table 2.13. The authors claim that the former procedure is superior in terms of accuracy, precision and rapidity to the established Penfield method.

2.11 The auto-analyser

If rapid methods of analysis had remained within the mainstream of geochemical data-gathering, instrumentation such as auto-analysers would be far more widely used today. Although extensively used in clinical and some industrial laboratories, auto-analysers have found only restricted application in geochemical laboratories in the photometric determination of 'difficult' elements such as fluorine and chlorine (e.g. Fuge, 1976).

Table 2.13 Determination of structural water (H_2O+) in selected reference materials by the Penfield method and by Karl Fischer titration

Reference material	Technique	Analysed results mean (%)	s (%)	CV (%)	Expected composition (%)
USGS:					
BCR-1	Penfield	—	0.35	48.0	0.67
	Karl Fischer	1.00	0.09	9.0	
CCRMP:					
SY-2	Penfield	0.31	0.08	25.8	0.43
	Karl Fischer	0.59	0.14	23.7	
CCRMP:					
SY-3	Penfield	0.52	0.14	27.7	0.42
	Karl Fischer	0.70	0.07	10.0	

s = standard deviation (units: weight %).
CV = coefficient of variation = 100.s/mean (units: % relative).
Analysed data taken from Turek *et al.* (1976).
Expected composition taken from Abbey (1983).

The auto-analyser is an instrument capable of the automated analysis of samples taken up into solution by procedures based on spectrophotometry or flame photometry. The heart of the instrument is a multichannel peristaltic pump. This consists of a manifold across which are stretched a number of plastic capillary tubes (Figure 2.25). Solutions within each capillary are pumped by the action of a set of rotating rollers. These squash the capillary tube walls against a platen and squeeze the solution forward during their rotational movement. The rate at which solutions are pumped depends on the rate of rotation of the rollers and the internal diameter selected for a particular capillary line. The peristaltic pump therefore permits a continuous flow of sample and reagent solutions and avoids the need for pipetting to mix the appropriate proportion of reactants.

However, cross-contamination and the uncontrolled diffusion of reagents would be a serious problem in auto-analyser operation if it were not for the scavenging action of air bubbles introduced into each capillary line. In this way sample and reagent solutions are segmented into small packets, each about 10 mm long, that are permitted to mix and react in a controlled manner. Samples may then be introduced into the system at a rate of about 20 to 40 per hour, each separated by a water blank.

Fuge (1976) described apparatus for the photometric determination of fluoride by its fading action on the zirconium–xylenol orange complex and of chlorine by the ferric thiocyanate method. The experimental arrangement for this latter procedure is shown in Figure 2.26. The sample solution was prepared by a sodium carbonate fusion in which the rock powder (0.2 g) was mixed with Na_2CO_3 flux (1.2 g) and fused at 900°C for 30 minutes. After cooling, 2 to 3 drops of ethanol were added to reduce manganese, and the fused sample was slurried with water and leached with a solution to which ammonium carbonate had been added. This mixture was then filtered, acidified with nitric acid to expel CO_2, and made up to volume (50 mL).

The reagents used in the determination for chlorine were mercuric thiocyanate solution (dissolved in a 1 : 1 mixture of water and methanol) and ferric ammonium sulphate solution (prepared in a solution acidified with nitric acid). These reagents were pumped through manifolds 1 and 2 (Figure 2.26), mixed and segmented with air. This combined reagent

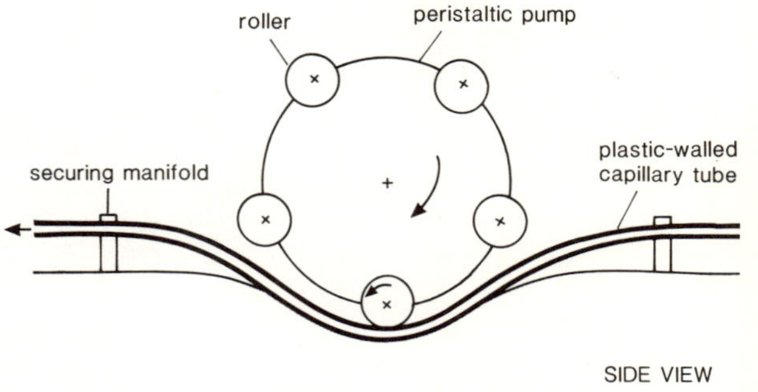

Figure 2.25 Schematic diagram of peristaltic pump.

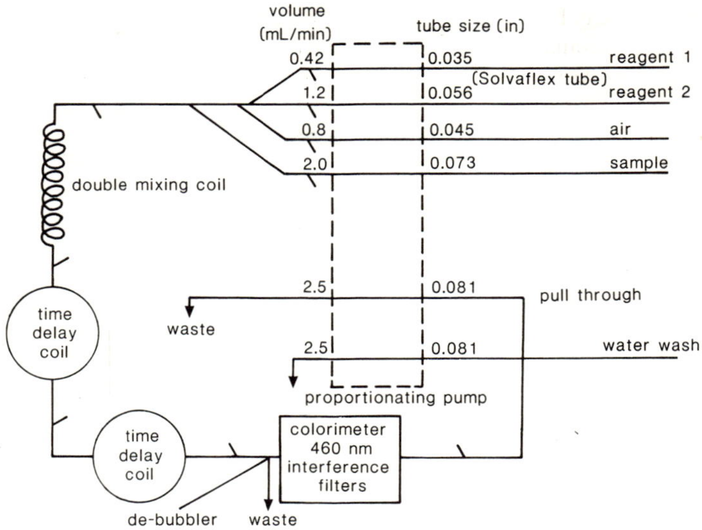

Figure 2.26 Manifold and 'circuit' diagram of capillary tubing used for the photometric determination of chlorine using an auto-analyser. (Fuge, 1976.)

was then mixed with the sample solution. In the absence of Cl^- ions, the mercuric ions complex with thiocyanate leaving free (hydrated) Fe(III) ions in solution. However, on mixing the sample solution, any Cl^- ions attack the mercuric thiocyanate complex to form stable mercuric chloride species in solution. The displaced thiocyanate ions are then free to form a complex with Fe^{3+}, the intense red-coloured ferric thiocyanate ion ($Fe(SCN)_6{}^{3-}$).

The remainder of the auto-analyser circuit consists of a mixing coil to ensure complete mixing of reagents, and time delay coils, which ensure both that the reaction goes to completion and that reactants equilibrate to laboratory temperature, if necessary. Solutions then pass through a device to remove air bubbles and then on into a flow cell situated in a filter photometer. This continually monitors the absorption of reagent solutions at 460 nm, the absorption maximum of the ferric thiocyanate complex.

Fuge (1976) undertook analyses in duplicate and the sampler was run a rate of 40 solutions per hour (for chlorine). Determinations down to 10 ppm Cl could be made by this technique, and repetitive analyses of a granodiorite sample gave a mean value of 220 ppm Cl with a coefficient of variation of 5.6%. Interferences could occur in samples containing more than 1000 ppm Cr due to the yellow colour of the $CrO_4{}^{2-}$ ion in the sample solution. This was removed by adding a further manifold through which hydrogen peroxide (3%) solution was pumped to reduce this species to Cr^{3+}. Some results for chlorine by the above method and for fluorine by the zirconium–xylenol orange method are listed in Table 2.14. Fuge has also adapted the auto-analyser technique for the photometric determination of other trace elements such as chromium and vanadium (Fuge, 1967, 1973).

Table 2.14 Determination of chlorine and fluorine in selected reference materials by photometry using an auto-analyser

	Chlorine		Fluorine	
Reference	Expected	Analysed	Expected	Analysed
USGS:				
AGV-1	185	110	(400)	420
BCR-1	58?	62	(500)	460
GSP-1	340	338	(3700)	3610

All data in ppm.
Analysed data taken from Fuge (1976).
Expected compositions taken from Abbey (1983).
Data in brackets originally reported as % fluorine.

3 Optical spectrometry: principles and instrumentation

3.1 Principles

Optical spectrometry is the technique of measuring the intensity of absorption or emission of radiation in the ultraviolet–visible region of the spectrum. In analytical applications, these measurements are made by exciting, in various ways, transitions of electrons between outer orbitals of atoms, ions or molecular species. In most applications, it is necessary to dissolve the sample before analysis.

The uv–visible region of the electromagnetic spectrum lies between the infra-red and x-ray regions (Figure 3.1). From an instrumental point of view, it is convenient to define the wavelengths bracketing this region by two phenomena that effectively restrict spectrometer performance. The upper wavelength limit is about 850 nm and represents the point beyond which conventional photomultiplier tubes cease to respond satisfactorily. Higher wavelengths constitute part of the near infra-red region. The lower limiting wavelength is about 180 nm and represents the point at which absorption of radiation by air becomes excessive. Wavelengths below this limit belong to the vacuum ultraviolet spectrum.

Optical absorption or emission measurements are useful in analytical applications only if some means is used to excite samples, and the resultant optical absorption or emission signal must be linearly related to the concentration of a specified analyte in the sample. Several different techniques have been devised; those most commonly used in geological applications (Figure 3.2) are as follows.

(*a*) *Uv–visible spectrophotometry*. This was one of the earliest optical techniques, adapted extensively for rapid silicate rock analysis, and made use of the fact that some element ions react in solution with certain ion-specific reagents to form intensely coloured complexes. Iron, for example, reacts in the ferrous state with 1,10-phenanthroline to form a red-coloured complex with a broad absorption spectrum having a maximum at 508 nm (section 2.5). It is thus possible to determine iron by measuring the absorption of this complex using a suitable uv–visible spectrophotometer (Figure 3.2*a*). The principles and applications of such determinations are considered in more detail in Chapter 2.

(*b*) *Flame photometry*. If a solution is aspirated into a suitable flame, optical emission lines are observed from certain elements (e.g. sodium and potassium). These are the elements in which electron transitions can readily be excited by the thermal energy of the flame. The emission lines are observed at characteristic wavelengths and the intensity may be measured using a simple flame photometer (Figure 3.2*b*). The design and operation of such instruments is considered in more detail in Chapter 2.

(*c*) *Atomic absorption analysis*. During the 1950s, a new and revolutionary technique was devised by Walsh and his colleagues in Australia (Walsh, 1955). This was based on the principle that strong absorption of optical radiation by atoms of an element could be induced if a sample were excited with the atomic radiation of that element. Such atomic line spectra could conveniently be excited in a hollow cathode lamp. The sample in solution was nebulized and the resultant aerosol mixed with gases and combusted in a flame. The flame, therefore, both atomized the sample and served as the sample cell. The unique aspect of this technique was that the atomic absorption mechanism was strongly element-specific, eliminating virtually all spectrum overlap interferences that afflicted other contemporary optical emission techniques such as arc/spark discharge–atomic emission spectrometry. Time-consuming chemical separations, an integral part of some previous schemes of analysis, were for the first time no longer necessary. Atomic absorption instrumentation, outlined in Figure 3.2*c*, is discussed in more detail in Chapter 4.

(*d*) *Inductively coupled plasma–atomic emission spectrometry*. Although element-specific, atomic absorption measurements were restricted by some matrix interference effects as well as by inadequate sensitivity towards some trace elements of geochemical importance. Many of these problems were overcome during the 1970s by the development of emission spectrometry, based on aspirating a sample into the fireball of an inductively coupled argon plasma (Figure 3.2*d*). At the very high temperatures to which sample aerosols are heated in the plasma (about 8000K maximum), atomization interferences are virtually eliminated. Strong optical emission signals can be observed from both atomic and ionic species. With the appropriate instrumentation (Chapter 5), the technique is capable of operating in a simultaneous multi-element data acquisition mode. Some success has also been achieved with the dc coupled argon plasma (Chapter 5), although this source has not proved as versatile as the inductively coupled plasma.

(*e*) *Arc/spark emission spectrometry*. Emission spectrometry was revolutionized by the inductively coupled plasma source. This overcame many of the difficulties encountered in earlier instrumentation, developed during the 1950s, based on electrical discharge excitation of solid samples (Chapter 6). Although arc or spark emission spectrometry is highly sensitive to a wide range of trace elements, the technique suffers from signal instability problems inherent in the excitation mechanism. The technique is not therefore thought to provide trace-element data of the highest precision, but has found valuable applications in the semi-quantitative analysis of samples for geochemical survey and exploration studies.

Although a range of excitation mechanisms is used in the techniques outlined above, they have in common the

Energy		Wavenumber σ	Wavelength λ	Frequency ν
kcal/mol	Electron volts, eV	cm^{-1}	cm	Hz
9.4×10^{7}	4.1×10^{6}	3.3×10^{10}	3×10^{-11}	10^{21}
9.4×10^{5}	4.1×10^{4}	3.3×10^{8}	3×10^{-9}	10^{19}
9.4×10^{3}	4.1×10^{2}	3.3×10^{6}	3×10^{-7}	10^{17}
9.4×10^{1}	4.1×10^{0}	3.3×10^{4}	3×10^{-5}	10^{15}
9.4×10^{-1}	4.1×10^{-2}	3.3×10^{2}	3×10^{-3}	10^{13}
9.4×10^{-3}	4.1×10^{-4}	3.3×10^{0}	3×10^{-1}	10^{11}
9.4×10^{-5}	4.1×10^{-6}	3.3×10^{-2}	3×10^{1}	10^{9}
9.4×10^{-7}	4.1×10^{-8}	3.3×10^{-4}	3×10^{3}	10^{7}

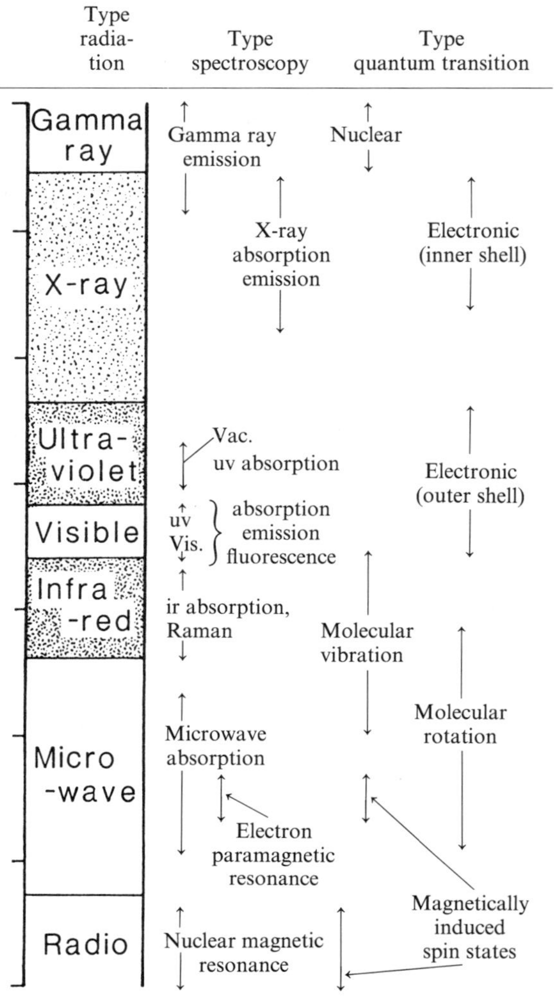

Figure 3.1 Subdivisions and excitation characteristics of radiation in the electromagnetic spectrum. (Skoog and West, 1980.)

necessity of making absorption or emission measurements of light in the uv–visible spectrum. The principles and instrumentation used for such measurements will now be considered.

3.2 The nature of light

(*a*) *Energy and wavelength.* Light is an electromagnetic radiation and may be considered to consist of two sinusoidal waveforms propagated along the same axis in planes perpendicular to one another. One waveform comprises an electric vector, the other a magnetic vector (Figure 3.3). Since the latter plays no part in conventional optical spectroscopy, it will not be considered further here. (Note, however, the splitting effect on light of magnetic fields, considered further in section 4.13). According to quantum theory, each light photon corresponds to a discrete quantum of energy (E) related to the wavelength of the electromagnetic wave (λ) by the equation:

$$E = \frac{hc}{\lambda} = h\nu = hc\bar{\nu} \qquad (3.1)$$

where h is the Planck constant (6.63×10^{-34} Js), c is the velocity of light in vacuo (2.9979×10^{8} ms^{-1}, ν is the frequency of radiation (hertz), and $\bar{\nu}$ is the reciprocal wavelength = wavenumber (cm^{-1}).

(*b*) *Units.* The fundamental property of light radiation is its discrete quantum energy. However, although the electron

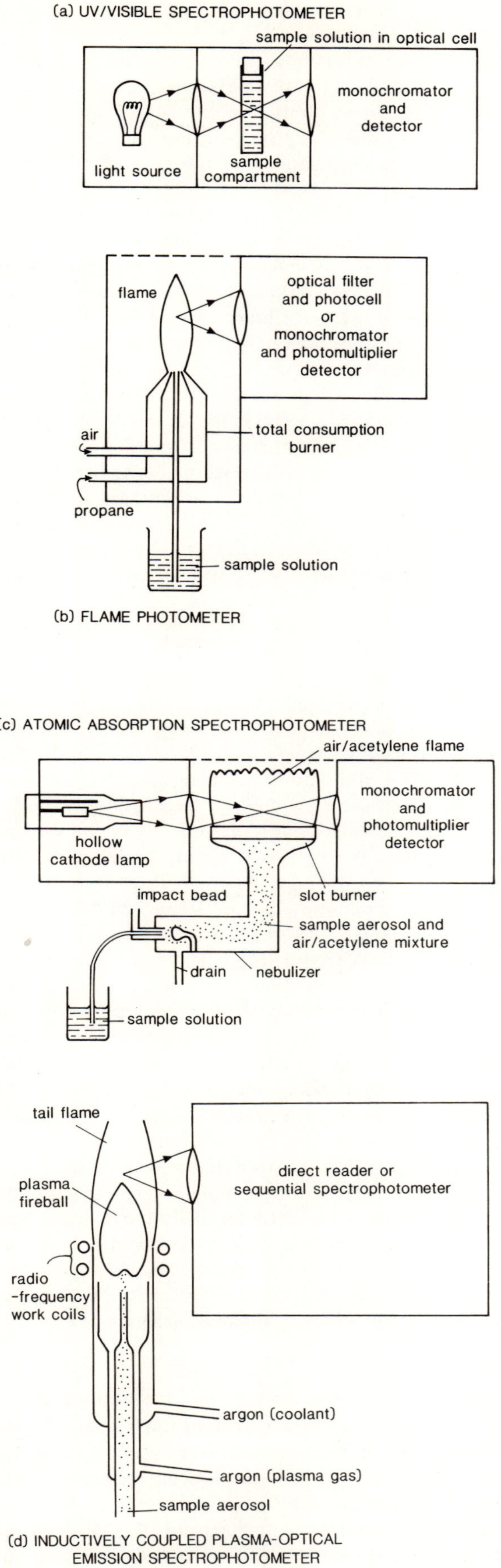

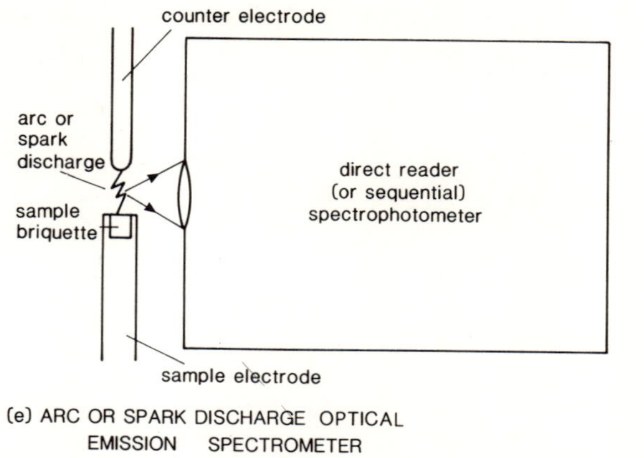

Figure 3.2 A comparison of spectrometer designs: (*a*) uv–visible absorption spectrometer; (*b*) flame photometer; (*c*) atomic absorption spectrometer; (*d*) inductively coupled plasma–atomic emission spectrometer; (*e*) arc or spark discharge optical emission spectrometer.

volt is used by convention as the unit of energy in the x-ray region of the electromagnetic spectrum, it is more common to refer to uv–visible light photons in terms of their wavelength (expressed in nm). The units of wavenumber (in cm^{-1}) are normally reserved for the lower-energy infra-red region of the spectrum. Due to the reciprocal relationship between energy and wavelength (equation 3.1), higher-energy light photons correspond to shorter wavelengths (see also Figure 3.3).

(*c*) *Colour*. Radiation in the visible region of the spectrum, unlike radiation of other wavelengths, can be sensitively detected by the human eye. Light of different wavelengths (energy) can be discriminated by the eye as different colours. These are listed in Table 3.1. It is important to distinguish between colours perceived by absorption and emission. If a sample *emits* yellow light in the range 580–595 nm (as in flame photometry for example), the flame appears yellow in colour. If, however, a solution containing a metal ion–chromophore complex *absorbs* yellow light in the same

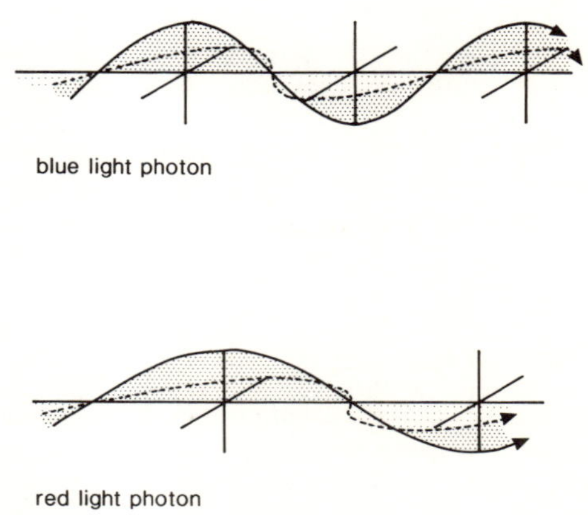

Figure 3.3 The propagation of light photons. The solid line delineates the electric vector, the dotted line the magnetic vector component. Blue light corresponds to a higher photon energy than red, and in consequence has a shorter wavelength.

Table 3.1 Colours of light in the visible spectrum

Approximate wavelength range (nm)	Colour	Complementary colour
400–435	Violet	Yellow-green
435–480	Blue	Yellow
480–490	Green-blue	Orange
490–500	Blue-green	Red
500–560	Green	Purple
560–580	Yellow-green	Violet
580–595	Yellow	Blue
595–650	Orange	Green-blue
650–750	Red	Blue-green

Data from Skoog and West (1982).

wavelength range, the solution will appear blue in colour. Blue is the *complementary* colour perceived by the eye for white light from which yellow has been absorbed. Complementary colours are also listed in Table 3.1.

3.3 Atomic spectroscopy

Atomic spectra are derived from transitions of electrons from one discrete electron orbital in an atom to another. Optical spectroscopy takes its name from the fact that the difference in energy between respective orbitals corresponds to the energy of electromagnetic radiation in the uv–visible region of the spectrum. As outlined above, analytical applications involve two forms of measurement, absorption and emission.

In *absorption measurements*, a sample is excited by electromagnetic radiation (light photons) causing the excitation of electrons from a lower orbital energy state to a higher one (Figure 3.4*a*). The intensity of absorbed light is then proportional to the concentration of an element species in the sample cell. Absorption measurements are used both in atomic absorption spectrometry, where the analyte species are atoms of an element formed in a flame, and uv–visible spectrometry, where the absorbing species is an element ion–chromophore complex formed in solution. In order to induce significant absorption, incident light radiation must have exactly the same energy as the difference in energy between the two electron orbital states involved in the transition. Hence the importance of the hollow cathode lamp in atomic absorption spectrometry, since this permits an

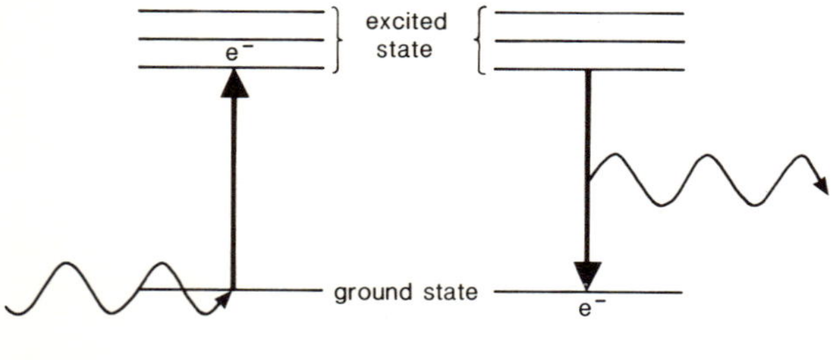

Figure 3.4 Electron orbital transitions associated with (*a*) optical absorption and (*b*) optical emission.

atomized sample to be excited with an atomic line spectrum of precisely the correct wavelength.

Optical *emission* is induced when some external source of energy (such as an argon plasma) is used to excite electron transitions within an atom. The intensity of light emitted when electrons relax back to their ground state is then a measure of the atomic population of the sample cell (Figure 3.4*b*).

Samples can be excited by several means, including the thermal energy of a flame (as in flame photometry), an argon plasma (as in ICP–AES), or an electrical discharge (arc/spark discharge–AES).

Optical fluorescence is a branch of atomic spectroscopy that has, to date, not been fully accepted for practical analytical applications. The sample solution is atomized and excited by an intense light source. Optical emission is then observed from atomic species in the sample. The technique offers the prospect of simultaneous multi-element determination. However, analytical performance has not yet been unambiguously shown to offer advantages over conventional absorption or emission procedures. Experimentally, excitation has been achieved using a laser or the intense continuum emission from an inductively coupled argon plasma.

The energy (or wavelength) of light involved in absorption or emission measurements is characteristic of an element since it is derived from the difference in energy between a unique pair of electron orbitals participating in the transition. An understanding of the origin of atomic spectra is, therefore, closely related to an understanding of the electronic structure of atoms. This is comparatively simple for only one atom, namely hydrogen. Since hydrogen consists of a nucleus surrounded by one orbital electron, the relevant wave equations can be solved exactly to define electron orbital structure. Solutions for multi-electron atoms can only be approximations. Indeed, atomic spectra of such elements can be complex. As is well known, the emission spectrum of iron as measured by ICP–atomic emission spectrometry contains hundreds of discrete lines, each caused by the transition of an electron between a unique pair of electron orbital energy levels. It is important for atomic spectroscopists to be able to identify unambiguously the electronic states corresponding to each of these transitions. In the following sections, the electronic structure of atoms in the periodic table is discussed briefly in terms of simple quantum theory. The way in which this model is used in the Russell–Saunders spin-orbital coupling scheme to derive the spectroscopic terms used to identify atomic states participating in electron transitions is then considered.

3.4 The electronic structure of atoms: quantum theory

3.4.1 *Wave mechanics and quantum numbers*

It is not proposed here to discuss the derivation of atomic orbital structure based on a rigorous application of wave mechanics. It is sufficient to describe an atom as consisting of a nucleus, which for a given isotope of an element contains a unique number of neutrons and protons. Each proton possesses a notional positive charge of one unit, and the total number of protons on the nucleus defines the atomic

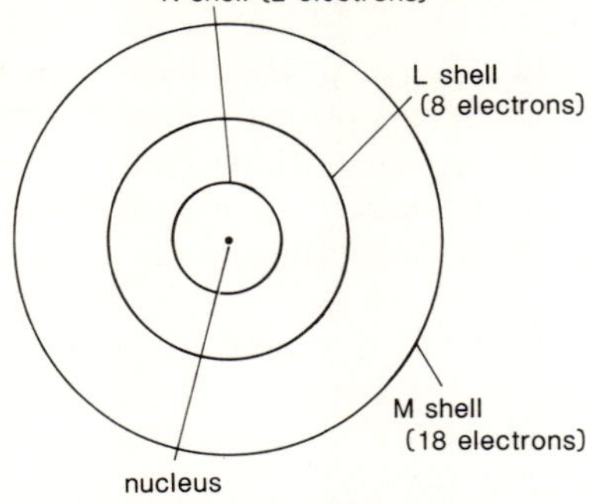

Figure 3.5 Highly simplified 'planetary' model of the atom showing nucleus and *K*, *L* and *M* electron orbital shells.

number of the element. Conversely, the total number of protons plus neutrons defines the atomic weight of the isotope. In a neutral atom, the nucleus is surrounded by negatively charged electrons, the total number of which exactly balances the number of positively charged protons on the nucleus. For some analytical techniques (x-ray fluorescence, for example), the simple 'planetary' model of the atom (Figure 3.5) is adequate to explain emission phenomena. However, for atomic spectroscopy, it is necessary to develop this model further. Wave mechanics is used to describe mathematically the electron orbitals available for occupancy around the atom. Electron orbitals characterize the space around the nucleus in which the probability of finding a specific electron is significant. Electrons are conceived, therefore, not as occupying simple planetary trajectories, but more diffuse spatial areas round the nucleus, called orbitals. The exact position of the electron in an orbital can be defined only in terms of statistical probabilities. The shape of such orbitals can usually be visualized by drawing in the space enclosing regions in which the integrated probability of finding the electron approaches unity (Figure 3.6). Characteristic shapes of orbitals are considered further below.

An integral part of the wave-mechanical approach is that each electron orbital can be characterized by four quantum numbers.

The principal quantum number (n) defines the overall energy level of an orbital electron shell and corresponds to the *K,L,M*,... electron orbital notation used to describe electron transitions giving rise to the emission of characteristic x-ray lines (Chapter 8). The correspondence between principal quantum number and orbital shell notation is as follows:

principal quantum number n : 1 2 3 4 5 ....

electron orbital shell: *K L M N O*....

Of these, *K*-shell electrons ($n = 1$) are most closely bound to the nucleus. Put very simply, the principal quantum number describes the overall size of an electron orbital, a *K*-shell orbital being much more compactly distributed about the nucleus than an equivalent *L*-shell orbital.

Each orbital shell does not represent a single energy level, but may be split into several subshells. These subshells are classified by the *azimuthal quantum number* (l). For each principal quantum number n, the azimuthal quantum number l may take integral values between 0 and $n-1$. Thus for an *M*-shell orbital ($n = 3$) three subshells may exist, having values $l = 0$, 1 and 2. For historical reasons associated with early measurements on the atomic spectrum of hydrogen, these subshells are normally identified by a letter of the alphabet, as shown in Table 3.2.

In physical terms, the azimuthal quantum number defines the angular momentum of the electron orbital. As a practical guide, this quantum number, together with the magnetic quantum number described next, defines the characteristic 'shape' of an orbital (see Figure 3.6).

Each sub-orbital shell is further characterized by a *magnetic quantum number* (m). The magnetic quantum number describes the behaviour of the electron sub-orbital under the influence of an externally applied magnetic field. In such circumstances a group of 'degenerate' sub-orbitals (that is, several individual orbitals all having identical energies) can be split into several different energy levels. For each value of the azimuthal quantum number l, the magnetic quantum number m may take the following values:

$l = 0 \quad m = 0$

$l = 1 \quad m = -1, 0, 1$

$l = 2 \quad m = -2, -1, 0, 1, 2$, etc.

As mentioned above, visually speaking, the azimuthal and magnetic quantum numbers can together be considered to define the 'shape' of an individual electron orbital. Examples of these shapes are shown schematically in Figure 3.6 for the following combination of quantum numbers.

1*s* ($n = 1$, $l = 0$, $m = 0$), for which one only orbital is defined;

2*p* comprising a subshell of three orbitals,

$2p_y$ ($n = 2, l = 1, m = -1$)

$2p_z$ ($n = 2, l = 1, m = 0$)

$2p_x$ ($n = 2, l = 1, m = +1$)

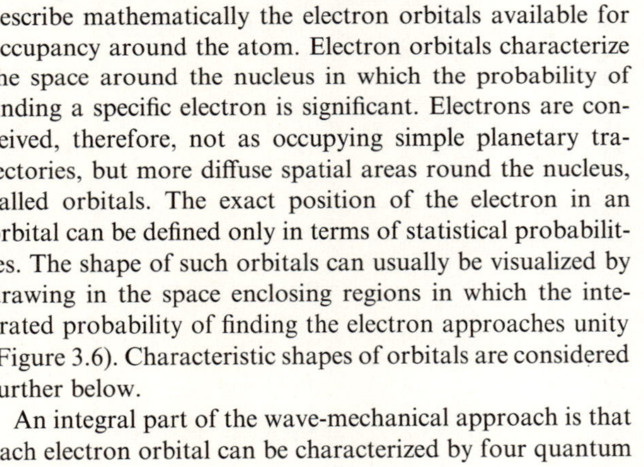

Figure 3.6 Balloon diagrams showing the shapes of 1*s*, 2*p* (p_x, p_y and p_z) and 3*d* (d_{z^2}, $d_{x^2-y^2}$, d_{xy}, d_{yz} and d_{zx}) electron orbitals. (Cotton and Wilkinson, 1966.)

Table 3.2 Notation for the azimuthal quantum number, l

Azimuthal quantum number (l)	Symbol	Historical derivation of symbol
0	*s*	sharp
1	*p*	principal
2	*d*	diffuse
3	*f*	fundamental
4	*g*	—
5	*h*	—

3*d* comprising a subshell of five orbitals:

$3d_{xy}$ ($n = 3, l = 2, m = -2$)
$3d_{yz}$ ($n = 3, l = 2, m = -1$)
$3d_{z^2}$ ($n = 3, l = 2, m = 0$)
$3d_{xz}$ ($n = 3, l = 2, m = +1$)
$3d_{x^2-y^2}$ ($n = 3, l = 2, m = +2$)

Finally, each individual electron possesses the property of '*spin*' characterized by the quantum number *s*. The spin quantum number may take the values $+\frac{1}{2}$ or $-\frac{1}{2}$ only.

3.4.2 *The Pauli exclusion principle*

The Pauli exclusion principle states that no two electrons occupying an orbital in an atom may possess identical values of all four quantum numbers *n*, *l*, *m* and *s*. By implication, therefore, electrons will generally occupy individual electron orbitals (identified by specific values of *n*, *l* and *m*) in pairs, each electron having opposite values of spin.

3.4.3 *The aufbau principle*

'Aufbau' (German: 'building up') refers to the principle that with a knowledge of the relative energy levels of electron orbitals it is possible to build up the electronic structure of each element in the Periodic Table. The principle embodies a few rules and is subject to a few exceptions. Each atom possesses a set of electron orbitals, not all of which are necessarily occupied. The electronic structure of the atom is formulated by filling these orbitals with the requisite number of electrons, the most stable orbitals being filled first. The order of orbital stability derived for the hydrogen atom is as follows, the most stable orbitals being listed first:

1*s*, 2*s*, 2*p*, 3*s*, 3*p*, 3*d*, 4*s*, 4*p*, 4*d*, 4*f*, 5*s*, 5*p*, 5*d*, 5*f*, 5*g*, ...

Taking into account both quantum theory and the Pauli exclusion principle, the maximum number of electrons required to fill each orbital type is as follows:

s orbitals: 2 electrons
p orbitals: 6 electrons
d orbitals: 10 electrons
f orbitals: 14 electrons
g orbitals: 18 electrons

In shorthand notation, orbital occupancy is defined by means of superscripts. Thus $3d^8$ signifies a set of 3*d* orbitals containing, in total, eight electrons.

The aufbau principle thus works as follows. The hydrogen atom, comprising a nucleus surrounded by one orbital electron, has an electronic structure of $1s^1$, the electron occupying the lowest-level orbital in the ground state atom. The helium atom has two orbital electrons that fill the 1*s* shell as $1s^2$. Lithium has the electronic structure $1s^2\ 2s^1$, and so on. The ground state electronic configurations of each atom in the periodic table are listed in Table 3.3. As can be seen, the simple building rules break down in respect of the outer electrons of heavier atoms. Iron, for example, with 26 orbital electrons, has the orbital structure $1s^2\ 2s^2\ 2p^6\ 3s^2\ 3p^6\ 4s^2$. In this instance the stability level of the 4*s* orbitals falls below that of some of the 3*d* levels because of complex interactions between electrons in occupied orbitals. The last two electrons therefore fill the 4*s* orbitals, in preference to occupying the 3*d* orbitals. These anomalies are in part influenced by the additional stability conferred on an atom if the outer orbitals are either exactly half filled, or fully filled, with electrons.

Table 3.3 Electronic configuration of atoms in the Periodic Table

Element		1*s*	2*s*	2*p*	3*s*	3*p*	3*d*	4*s*	4*p*	4*d*	4*f*	5*s*	5*p*	5*d*	5*f*	5*g*
1.	H	1														
2.	He	2														
3.	Li	2	1													
4.	Be	2	2													
5.	B	2	2	1												
6.	C	2	2	2												
7.	N	2	2	3												
8.	O	2	2	4												
9.	F	2	2	5												
10.	Ne	2	2	6												
11.	Na	2	2	6	1											
12.	Mg	2	2	6	2											
13.	Al	2	2	6	2	1										
14.	Si	2	2	6	2	2										
15.	P	2	2	6	2	3										
16.	S	2	2	6	2	4										
17.	Cl	2	2	6	2	5										
18.	Ar	2	2	6	2	6										
19.	K	2	2	6	2	6		1								
20.	Ca	2	2	6	2	6		2								
21.	Sc	2	2	6	2	6	1	2								
22.	Ti	2	2	6	2	6	2	2								
23.	V	2	2	6	2	6	3	2								
24.	Cr	2	2	6	2	6	5	1								
25.	Mn	2	2	6	2	6	5	2								
26.	Fe	2	2	6	2	6	6	2								
27.	Co	2	2	6	2	6	7	2								
28.	Ni	2	2	6	2	6	8	2								
29.	Cu	2	2	6	2	6	10	1								
30.	Zn	2	2	6	2	6	10	2								
31.	Ga	2	2	6	2	6	10	2	1							
32.	Ge	2	2	6	2	6	10	2	2							
33.	As	2	2	6	2	6	10	2	3							
34.	Se	2	2	6	2	6	10	2	4							
35.	Br	2	2	6	2	6	10	2	5							
36.	Kr	2	2	6	2	6	10	2	6							
37.	Rb	2	2	6	2	6	10	2	6			1				
38.	Sr	2	2	6	2	6	10	2	6			2				
39.	Y	2	2	6	2	6	10	2	6	1		2				
40.	Zr	2	2	6	2	6	10	2	6	2		2				
41.	Nb	2	2	6	2	6	10	2	6	4		1				
42.	Mo	2	2	6	2	6	10	2	6	5		1				
43.	Te	2	2	6	2	6	10	2	6	6		1				
44.	Ru	2	2	6	2	6	10	2	6	7		1				
45.	Rh	2	2	6	2	6	10	2	6	8		1				
46.	Pd	2	2	6	2	6	10	2	6	10						
47.	Ag	2	2	6	2	6	10	2	6	10		1				
48.	Cd	2	2	6	2	6	10	2	6	10		2				
49.	In	2	2	6	2	6	10	2	6	10		2	1			
50.	Sn	2	2	6	2	6	10	2	6	10		2	2			
51.	Sb	2	2	6	2	6	10	2	6	10		2	3			
52.	Te	2	2	6	2	6	10	2	6	10		2	4			
53.	I	2	2	6	2	6	10	2	6	10		2	5			
54.	Xe	2	2	6	2	6	10	2	6	10		2	6			

3.5 Spectroscopic notation for electron orbital configurations: the Russell–Saunders coupling scheme

3.5.1 *Spin-orbital coupling*

It has been assumed thus far that individual electrons in individual orbitals behave essentially independently of one

Table 3.3 (*Continued*)

Element		*K*	*L*	*M.*	4*s*	4*p*	4*d*	4*f*	5*s*	5*p*	5*d*	5*f*	5*g*	6*s*	6*p*	6*d*	6*f*	6*g*	6*h*	7
55.	Cs	2	8	18	2	6	10		2	6				1						
56.	Ba	2	8	18	2	6	10		2	6				2						
57.	La	2	8	18	2	6	10		2	6	1			2						
58.	Ce	2	8	18	2	6	10	2	2	6				2						
59.	Pr	2	8	18	2	6	10	3	2	6				2						
60.	Nd	2	8	18	2	6	10	4	2	6				2						
61.	Pm	2	8	18	2	6	10	5	2	6				2						
62.	Sm	2	8	18	2	6	10	6	2	6				2						
63.	Eu	2	8	18	2	6	10	7	2	6				2						
64.	Gd	2	8	18	2	6	10	7	2	6	1			2						
65.	Tb	2	8	18	2	6	10	9	2	6				2						
66.	Dy	2	8	18	2	6	10	10	2	6				2						
67.	Ho	2	8	18	2	6	10	11	2	6				2						
68.	Er	2	8	18	2	6	10	12	2	6				2						
69.	Tm	2	8	18	2	6	10	13	2	6				2						
70.	Yb	2	8	18	2	6	10	14	2	6				2						
71.	Lu	2	8	18	2	6	10	14	2	6	1			2						
72.	Hf	2	8	18	2	6	10	14	2	6	2			2						
73.	Ta	2	8	18	2	6	10	14	2	6	3			2						
74.	W	2	8	18	2	6	10	14	2	6	4			2						
75.	Re	2	8	18	2	6	10	14	2	6	5			2						
76.	Os	2	8	18	2	6	10	14	2	6	6			2						
77.	Ir	2	8	18	2	6	10	14	2	6	7			2						
78.	Pt	2	8	18	2	6	10	14	2	6	9			1						
79.	Au	2	8	18	2	6	10	14	2	6	10			1						
80.	Hg	2	8	18	2	6	10	14	2	6	10			2						
81.	Tl	2	8	18	2	6	10	14	2	6	10			2	1					
82.	Pb	2	8	18	2	6	10	14	2	6	10			2	2					
83.	Bi	2	8	18	2	6	10	14	2	6	10			2	3					
84.	Po	2	8	18	2	6	10	14	2	6	10			2	4					
85.	At	2	8	18	2	6	10	14	2	6	10			2	5					
86.	Rn	2	8	18	2	6	10	14	2	6	10			2	6					
87.	Fr	2	8	18	2	6	10	14	2	6	10			2	6				1	
88.	Ra	2	8	18	2	6	10	14	2	6	10			2	6				2	
89.	Ac	2	8	18	2	6	10	14	2	6	10			2	6	1			2	
90.	Th	2	8	18	2	6	10	14	2	6	10			2	6	2			2	
91.	Pa	2	8	18	2	6	10	14	2	6	10	2		2	6	1			2	
92.	U	2	8	18	2	6	10	14	2	6	10	3		2	6	1			2	
93.	Np	2	8	18	2	6	10	14	2	6	10	5		2	6				2	
94.	Pu	2	8	18	2	6	10	14	2	6	10	6		2	6				2	
95.	Am	2	8	18	2	6	10	14	2	6	10	7		2	6				2	
96.	Cm	2	8	18	2	6	10	14	2	6	10	7		2	6	1			2	
97.	Bk	2	8	18	2	6	10	14	2	6	10	8		2	6	1			2	
98.	Cf	2	8	18	2	6	10	14	2	6	10	10		2	6				2	
99.	Es	2	8	18	2	6	10	14	2	6	10	11		2	6				2	
100.	Fm	2	8	18	2	6	10	14	2	6	10	12		2	6				2	
101.	Md	2	8	18	2	6	10	14	2	6	10	13		2	6				2	
102.	—	2	8	18	2	6	10	14	2	6	10	14		2	6				2	
103.	Lw	2	8	18	2	6	10	14	2	6	10	14		2	6	1			2	

Data from Cotton and Wilkinson (1966).

another. This is not so. In interpreting atomic spectra arising from the transitions of electrons between various orbital quantum levels, it is not satisfactory simply to identify the two individual orbitals participating in the transition. Rather, it is necessary to identify the overall state of the atom corresponding to each of the participating electron orbital configurations. The state of the atom is characterized by the total orbital energy, total orbital angular momentum and total electron spin. These parameters describe the way in which the electric and magnetic fields of different occupied electron orbitals interact with one another to produce the final energy-level pattern.

The convention used to identify the overall electron orbital energy state is the *Russell–Saunders spin-orbital coupling scheme*. This scheme assumes that it is possible to determine the total angular momentum of an atom by vector addition of the angular momenta of individual orbital electrons. By this means, the way in which the entire atom interacts with an applied external field can be described. Angular momentum arises both from the intrinsic spin of individual electrons and from the motion of electrons around the nucleus (orbital angular momentum).

A simple visual description (Moore, 1966) of the way in which the angular momentum vectors of two electrons

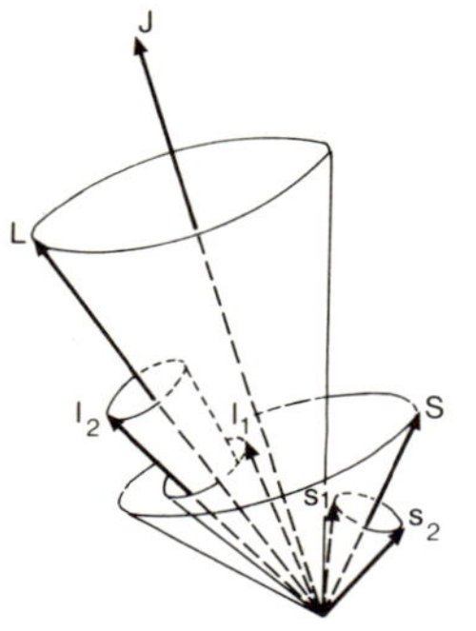

Figure 3.7 Schematic representation of Russell–Saunders spin-orbital coupling. The orbital angular momentum vectors l_1 and l_2 couple to form a resultant L. The spin angular momentum vectors s_1 and s_2 couple to form a resultant S. The L and S terms couple and precess about their resultant J. (Moore, 1966.)

combine and interact with an external field is shown in Figure 3.7. The angular momenta due to the spin of individual electrons (s_1 and s_2) are assumed to combine by vector addition to form a resultant vector S. Individual spin vectors s_1 and s_2 will precess (i.e. spin like a top) round the resultant S. The orbital angular momenta, l_1 and l_2 are assumed to combine by vector addition to form a resultant vector L about which the individual vectors l_1 and l_2 will again precess. The L and S vectors are then assumed to combine to form a total angular momentum vector J. In the absence of an external field, the total spin and orbital angular momentum vectors will precess round the resultant J.

The Russell–Saunders vector terms for multi-electron atoms are derived as follows.

The *total spin vector* S for an atom is determined from the sum of the spin vectors (i.e.quantum numbers) of individual electrons:

$$S = \sum_i s_i \tag{3.2}$$

The resultant S can have only integral or half-integral values, since s_i can only take the values $+\frac{1}{2}$ or $-\frac{1}{2}$. The value of S is effectively determined only from (outer) partially-filled orbitals, since summation over all electrons in a filled orbital must cancel to zero (Pauli exclusion principle).

The *total orbital angular momentum vector* (L) may be calculated from the orbital angular momentum quantum numbers of individual electrons (l_i) as follows:

$$L = \sum_i l_i \tag{3.3}$$

The value of L must take an integral value and once again, summation over filled orbitals will cancel.

The *total angular momentum vector* (J) is derived from the total spin (S) and total orbital (L) angular momentum quantum numbers:

$$J = L + S \tag{3.4}$$

In the Russell–Saunders notation, these total angular momentum quantum numbers are summarized in the *spectroscopic term symbol*.

$$^{2S+1}L_J$$

where the values of $L = 0, 1, 2, 3, \ldots$ are represented by the symbols $S, P, D, F, \ldots$.

The term $2S+1$ is referred to as the *multiplicity* of the spectral state. Singlet, doublet and triplet multiplicity states correspond to the $2S+1$ values of 1 (i.e. $S = 0$), 2 (i.e. $S = \frac{1}{2}$) and 3 (i.e. $S = 1$), respectively.

The Russell–Saunders coupling scheme can be applied to those atoms in which the interaction between electron spin and orbital vectors is stronger than that between individual electrons and the positively charged nucleus. This assumption breaks down when heavier atoms are considered.

3.5.2 *Allowed and forbidden transitions*

The Russell–Saunders scheme is extensively used by spectroscopists to identify emission lines from an atom by specifying the Russell–Saunders terms describing the excited and ground state electronic configurations involved in the respective transition. However, the rules of quantum theory do not permit indiscriminate transition between any electronic state. Some transitions are allowed, some are forbidden. Selection rules must be applied to identify if a transition is allowed. Translated into the Russell–Saunders convention, these selection rules state that for an allowed transition, there must be a change in the J and L values of the spectroscopic terms representing the end states of the transition as follows:

$$J = 0 \text{ or } \pm 1 \tag{3.5}$$

$$L = \pm 1. \tag{3.6}$$

3.5.3 *Spectroscopic terms applied to the sodium atom*

As an illustration of these concepts, the electron orbital energy level diagram for a sodium atom is shown in Figure 3.8. The ground-state electronic configuration of sodium is $1s^2\,2s^22p^63s^1$. The inner $1s$, $2s$ and $2p$ electron orbitals are thus filled, with one unpaired electron in the $3s$ orbital. The Russell–Saunders spectroscopic term for this ground-state is thus $^2S_{1/2}$ since $S = \frac{1}{2}$ and $L = 0$ (the value $S = -\frac{1}{2}$ is in this instance considered trivial and is ignored). Atomic spectroscopy is concerned principally with the excitation of the outer $3s$ electron to higher energy vacant orbitals. Those available for excitation of the $3s$ electron include the $3p$, $3d$, $4s$ and $4p$ orbitals. The Russell–Saunders terms for the electronic configurations representing the atom in which the $4s$ electron has been excited to each of these orbital levels are derived in Table 3.4. These data demonstrate that the $3s$ and $4s$ electron configurations have identical spectroscopic terms as do $3p$ and $4p$, and $3d$ and $4d$, etc. In order to distinguish which orbital is involved in a transition, it is conventional to add the corresponding principal quantum number to the term:

$3\,^2S_{1/2}$ refers to the $3s$ electron configuration
$4\,^2S_{1/2}$ refers to the $4s$ electron configuration.

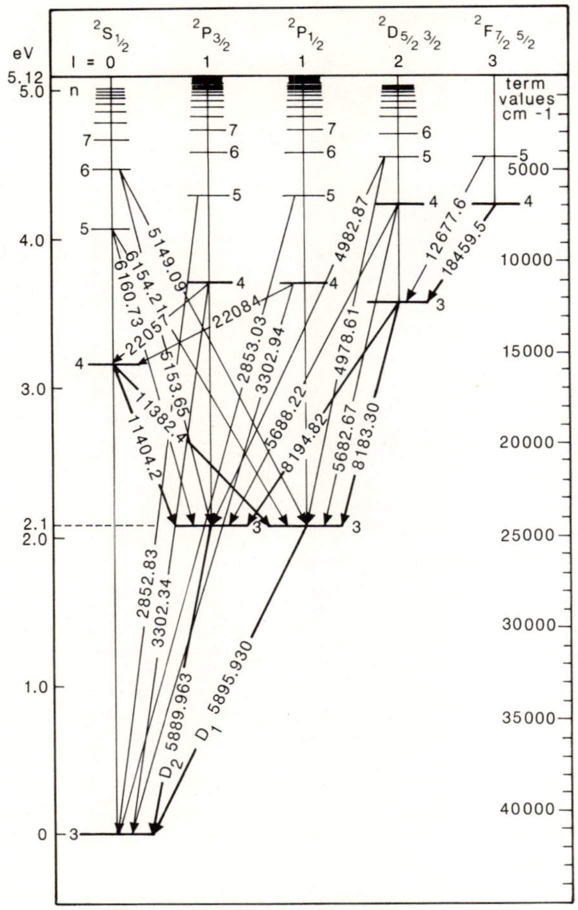

Figure 3.8 Grotian diagram of the sodium atom showing orbital energy levels and electron transitions observed in the atomic emission spectrum of sodium. Energy levels are defined relative to that of the 3*s* orbital. (Walker and Straw, 1966.)

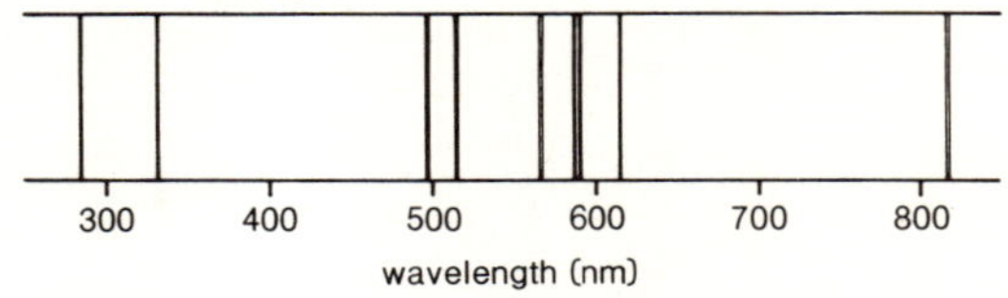

Figure 3.9 Atomic emission line spectrum of the sodium atom.

In all the arguments presented above, it is assumed that 1*s*, 2*s* and 2*p* orbitals remain fully occupied and are unaffected by any excitation process.

When selection rules are applied to the spectroscopic terms in Table 3.4, it is apparent that transitions are permitted between

3*s* level ($^2S_{1/2}$) and 3*p*, 4*p*, 5*p* etc. ($^2P_{1/2,3/2}$)

3*p* level ($^2P_{1/2,3/2}$) and (i) 4*s*, 5*s* etc. ($^2S_{1/2}$), and (ii) 3*d*, 4*d* etc. ($^2D_{3/2,5/2}$)

3*d* level ($^2D_{3/2,5/2}$) and 4*f*, 5*f* etc. ($^2F_{5/2,7/2}$).

For historical reasons related to discoveries on the simple spectrum of the hydrogen atom, these series of lines are named as follows:

- transitions involving the *s* to 3*p* levels: the sharp series
- transitions involving the *p* to 3*s* levels: the principal series
- transitions involving the *d* to 3*p* levels: the diffuse series
- transitions involving the *f* to 3*d* levels: the fundamental series.

Atomic lines involving the ground-state (in this case the 3*s* orbital) are referred to as *resonance lines*: these are normally the most prominent in atomic spectra.

A further point to note from Table 3.4 is that the spectroscopic term for the 3*p*, 3*d* and 4*f* subshells of the sodium atom has two components (e.g. $^2P_{1/2}$ and $^2P_{3/2}$ for the 3*p* level). This indicates that the 3*p* subshell comprise orbitals that possess two distinct angular momentum components. In an externally applied field, these components will split in energy to a small extent. This phenomenon is observed when the emission spectrum of atomic sodium (Figure 3.9) is related to transitions on the orbital energy level diagram (Figure 3.8). The prominent yellow emission lines in this spectrum may be correlated with transitions involving the 3*p* and 3*s* orbitals. Due to the influence of fields within the sodium atom, the 3*p* levels are split in energy slightly. The doublet at 589 nm corresponds to the following transitions:

558.9963 nm: $3^2P_{3/2} \rightarrow 3^2S_{1/2}$

589.5930 nm: $3^2P_{1/2} \rightarrow 3^2S_{1/2}$.

A second emission doublet is observed at 330.234 and 330.294 nm, represented by exactly the same spectroscopic terms but this time corresponding to transitions between 4*p* and 3*s* levels.

3.5.4 *Grotian diagrams and energy level conventions*

Orbital energy level diagrams such as those in Figure 3.8 are called Grotian diagrams after their originator. A unique orbital diagram may be drawn for each element. As the number of unpaired electrons in partially filled orbitals increases, the number of allowed transitions increases in complexity. The example discussed above (sodium) represents one of the simpler atomic spectra and Grotian diagrams, and was selected to model the principles of atomic spectroscopy and Russell–Saunders coupling. More spe-

Table 3.4 Spectroscopic terms for selected excited states of the sodium atom

Electronic configuration	L	S	J	Russell-Saunders spectroscopic term	
$1s^22s^22p^63s^1$	0	$+\frac{1}{2}, (-\frac{1}{2})$	$\frac{1}{2}$	$^2S_{1/2}$	Ground state
$1s^22s^22p^63p^1$	1	$+\frac{1}{2}, -\frac{1}{2}$	$\frac{1}{2}, \frac{3}{2}$	$^2P_{1/2}, {}^2P_{3/2}$	
$1s^22s^22p^63d^1$	2	$+\frac{1}{2}, -\frac{1}{2}$	$\frac{3}{2}, \frac{5}{2}$	$^2D_{3/2}, {}^2D_{5/2}$	Excited states
$1s^22s^22p^64f^1$	3	$+\frac{1}{2}, -\frac{1}{2}$	$\frac{5}{2}, \frac{7}{2}$	$^2F_{5/2}, {}^2F_{7/2}$	

For the ground state configuration ($3s^1$), the value $^2S_{-1/2}$ is considered trivial.

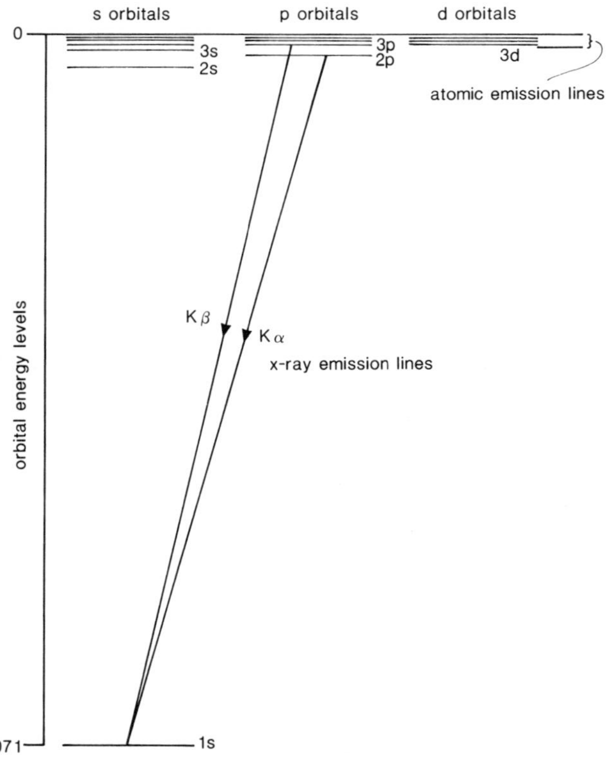

Figure 3.10 Electron orbital energy level diagram of the sodium atom showing the relationship between transitions that give rise to x-ray emission lines and atomic (optical) emission lines. The difference in energy (eV) between outer electron orbitals has been exaggerated.

cialized monographs discuss the rigorous interpretation of these principles (e.g. Walker and Straw, 1961).

The energy scale in Figure 3.8 is defined, for convenience, relative to the lowest energy orbital under consideration (the 3*s* orbital). However, the 'absolute' energy of an electron orbital is, by convention, defined as the ionization energy of an electron occupying that orbital, that is the energy required to remove the electron out of the orbital to infinite distance from the atom. Since work must be done to remove an electron in this manner, the potential energy of an orbital takes a negative value. This convention has been adopted in Figure 3.10 (see also next section). Thus we have the apparent paradox that the more tightly bound electron orbitals, in which an electron confers higher stability to the atom, possess higher (negative) potential energies. Excitation of such electrons by absorption of energy by the atom causes a transition to an orbital of lower (negative) potential energy. Using this convention, the potential energy of an ionized electron is zero.

3.5.5 *Comparison between optical and x-ray emission spectra*

X-ray fluorescence spectrometry also involves the measurement of radiation (this time x-ray radiation) caused by electron transitions within an atom. The relationship between this and atomic spectrometry is illustrated in Figure 3.10, again taking the sodium atom as an example. X-ray fluorescence first involves the ionization of an inner 'core' electron, the 1*s* electron in the case of the *K* series of x-ray lines. This highly excited state rapidly falls back to the ground state by rearrangement of orbital electrons involving $2p \rightarrow 1s$ orbitals (the $K\alpha$ lines) or $3p \rightarrow s$ orbitals (the $K\beta$ lines). The energy difference between these electron orbital states is balanced by the emission of an x-ray photon. Such transitions are normally described in terms of the Siegbahn notation, making use of the simple shell model of electronic orbitals (section 8.3). However, rigorous Russell–Saunders spectroscopic states can be used to identify these transitions (Table 3.5). By contrast, atomic spectroscopy involves very much smaller energy transitions between outer electron orbitals, such transitions being induced by the excitation of outer electrons to a higher energy state, without ionization.

3.6 The absorption of light

3.6.1 *Transmittance and absorbance*

When light is passed through a sample cell, a fraction of it is absorbed (Figure 3.11). The power (P) of the emerging beam can be measured in relation to the power (P_0) of the incident beam in two ways.

The *transmittance* (T) is the fraction of light emerging from the sample:

$$T = P/P_0 \tag{3.7}$$

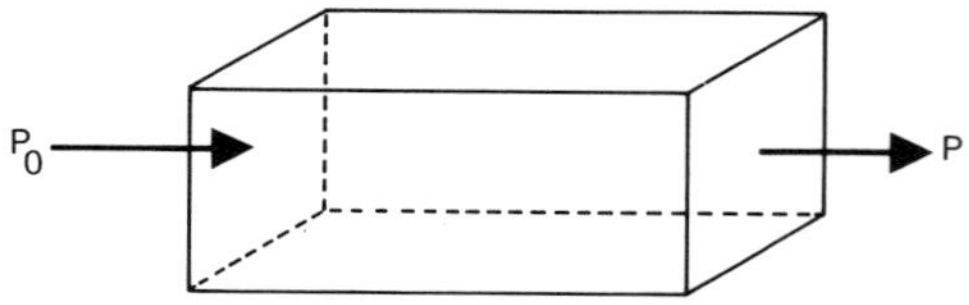

Figure 3.11 Absorption cell used to model Beer's Law. *P* represents the power of light transmitted through the cell.

Table 3.5 Comparison of Siegbahn and Russell–Saunders spectroscopic notation for x-ray emission from the sodium atom

x-ray line	keV	Excited-state electronic configuration	Ground-state electronic configuration	Transition notation: x-ray	Transition notation: spectroscopic
$K\alpha_1$	1041	$1s^12s^22p^63s^1$	$1s^22s^22p^53s^1$	$L_{III} \rightarrow K$	$^2P_{3/2} \rightarrow {}^1S_0$
$K\alpha_2$	1041	$1s^12s^22p^63s^1$	$1s^22s^22p^53s^1$	$L_{II} \rightarrow K$	$^2P_{1/2} \rightarrow {}^1S_0$

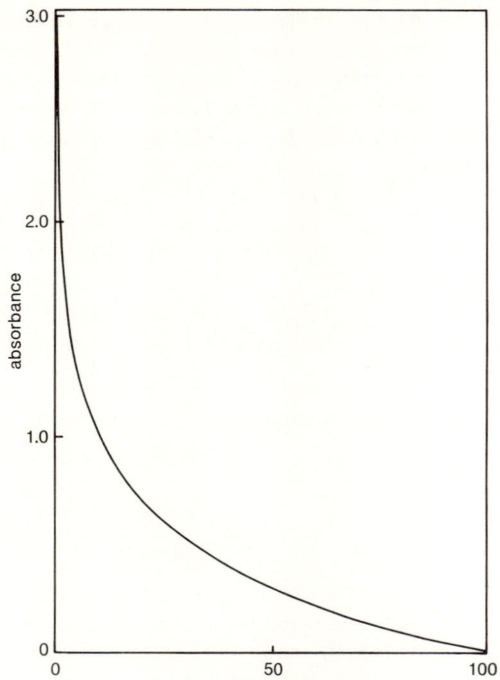

Figure 3.12 Graphical relationship between transmittance and absorbance.

and is normally expressed as a percentage. 100% transmittance means that no light is absorbed within the sample, 0% transmittance means that the sample is opaque to light.

Conversely, the *absorbance* (A) is a measure of the light power absorbed within the sample:

$$A = -\log_{10}(T) = \log_{10}(P_0/P). \qquad (3.8)$$

Absorbance and transmittance are expressed in terms of the power of light beams, rather than their intensity, since in practical use it is the energy per unit area reaching the detector that is measured.

The relationship between transmittance and absorbance is plotted in Figure 3.12. If optical measurements are to be made to the highest precision, the range is normally restricted to 15% to 65% transmittance (0.8 to 0.2 absorbance units, see section 3.15). Conversion factors between absorbance and transmittance are listed in Table 3.6. It is interesting to note that modern instruments, capable of making absorbance measurements down to 3.0 units, are detecting only 0.1% of the incident light.

The older term 'optical density' is no longer used scientifically; it is synonymous with, and has been superseded by, the term 'absorbance'.

3.6.2 *Beer's Law*

The laws governing the absorbance of light within a sample were developed by Beer, with contributions from Lambert and Bouguer, and are here referred to as Beer's Law. In its simplest form, this law states that successive increments in the number of identical absorbing molecules in the path of a beam of monochromatic radiation absorb equal fractions of the radiant energy traversing them (Ewing, 1975, p.36).

One form for the mathematical expression of this law is:

$$P = P_0.\exp(-k.b) \qquad (3.9)$$

where P_0 is the power of the incident radiation, P is the power of the transmitted radiation, b is the path length of radiation within the sample, and k is a constant *for a given sample* at the specific frequency of radiation.

Further development of this expression to describe samples containing different amounts of absorbing species shows that the constant k incorporates a term directly proportional to the concentration of the absorbing species. Setting absorbance $A = \log_{10}(P_0/P)$ and incorporating a term equal to $\log_{10}e$ into the constant, an analytically useful form of Beer's Law is obtained:

$$A = \log_{10}(P_0/P) = abc \qquad (3.10)$$

where a is the absorptivity (or absorption coefficient), b is the path length within the sample, and c is the concentration of the absorbing species.

If the path length b is expressed in cm and the concentration c is expressed as the number of moles of absorbing species (as appropriate for solution analysis) then the constant a is written as C, the *molar absorptivity*.

A more complete mathematical derivation of these expressions is given, for example, by Skoog and West (1982, pp.489–491) or Ewing (1975, pp.35–37).

3.6.3 *Limitations of Beer's Law*

Beer's Law is the basis of all photometric and atomic absorption determinations. However, in practical applications there are some significant limitations. The most important of these is that in relating absorbance to concentration, Beer's Law assumes that all absorbing species behave independently of one another so that each has an equal opportunity of interacting with transmitted radiation. In the analysis of solutions by spectrophotometry this is normally only true at lower concentrations (<0.01M). At higher concentrations, several types of interaction can occur, causing deviations from strict linearity. Those most commonly encountered include the following.

Table 3.6 Conversion of transmittance to absorbance

Transmittance (%)	Absorbance
0.1	3.000
0.2	2.699
0.5	2.301
1	2.0
2	1.699
3.16	1.5
5	1.301
10	1.0
20	0.699
31.62	0.5
50	0.301
80	0.097
90	0.046
100	0

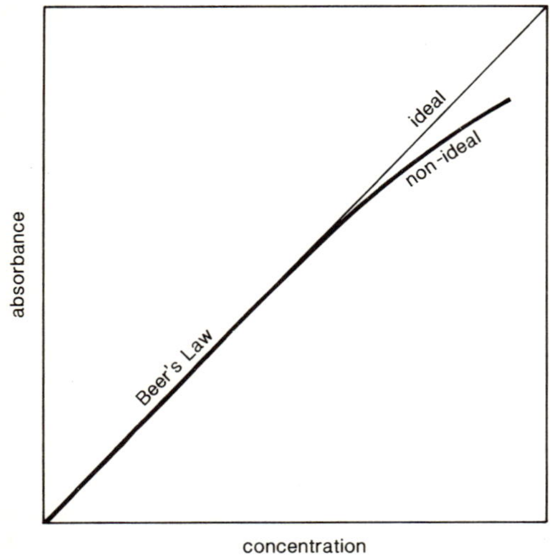

Figure 3.13 Plot of concentration *v.* absorbance showing ideal and non-ideal behaviour in relation to Beer's Law.

(i) At high concentrations, the mean distance between absorbing species and other species, such as electrolyte ions in solution, is reduced. If this distance diminishes to the point where the electrostatic fields due to ionic species begin to perturb the energy levels within the absorbing atom, the effective absorptivity constant will be changed and Beer's Law will no longer be obeyed.

(ii) In solution, the position of some chemical equilibria change as the concentration of participating species is altered. The example often quoted is that of a solution of potassium dichromate:

$$\begin{array}{cc} Cr_2O_7^{2+} + H_2O \rightleftharpoons & 2CrO_4^{2+} + 2H^+ \\ \text{orange} & \text{yellow} \\ E_{max} = 350 \text{ nm} & E_{max} = 375 \text{ nm} \end{array} \qquad (3.11)$$

This equilibrium shifts with concentration (the greater the dilution, the further the equilibrium shifts to the right). Serious deviations from Beer's Law are therefore expected with changes in the concentration of potassium dichromate solutions.

The net result of these effects is shown in Figure 3.13. At low concentrations, Beer's Law is expected to be followed exactly, giving a linear concentration *v.* absorbance calibration. However, at higher concentrations a point will be reached where non-linearity effects are observed resulting from the type of interferences described above. It is important, therefore, that instrument calibrations that rely on Beer's Law for quantification are set up with a range of standard samples that exceed the concentration range expected for unknown samples. In this way, non-linearities such as those shown in Figure 3.13 may be detected and if necessary accounted for.

3.7 The emission of light

3.7.1 *Emission mechanisms*

In emission spectroscopy, a sample is normally excited by the thermal energy of a flame (flame photometry), an argon plasma (ICP-AES) or an electrical discharge (arc/spark spectrometry). Atoms in the sample absorb thermal energy, causing outer orbital electrons to become excited from their ground state to a higher orbital energy level. After a very short lifetime (typically 10^{-6} to 10^{-9} s) the excited atoms decay back to the ground state. Simultaneously, electromagnetic radiation, normally in the form of light in the uv–visible region, is emitted. The intensity of this emitted light is governed by the fraction of atoms excited to the higher energy state. This will in turn depend both on the thermal energy available in the excitation source and the difference in energy between excited and ground-state levels.

3.7.2 *The Boltzmann equation*

The degree of excitation by a thermal energy source is described by the Boltzmann distribution equation. If N_1 is the number of atoms in a specified excited state and N_0 the number in the ground state, then the fraction excited is given by

$$\frac{N_1}{N_0} = \frac{g_1}{g_0}.\exp\frac{(-E)}{(kT)} \qquad (3.12)$$

where E is the energy difference between ground and excited states, T is the temperature (K), k is the Boltzmann constant, equal to 1.380622×10^{-23} J K^{-1}, and g_0 and g_1 are weighting factors so that allowance may be made if either state corresponds to more than one quantum level. $g = 2J + 1$, where J is the combined spin and orbital angular momentum vector of the ground and excited electron states as appropriate; see section 3.5 above.

The proportion of atoms in specified excited states has been calculated as a function of temperature by Walsh (1955). These data have been recalculated and extended up to temperatures of 10 000K (Table 3.7). It can be seen from these data that the fraction of excited atoms is very strongly dependent on the energy difference between ground and excited states. Thus at any given temperature, the number of excited caesium atoms (band gap 1.46 eV) is much higher than that of zinc (band gap 5.80 eV). At the flame temperatures likely to be reached in atomic absorption analysis (2000–3000K) the proportion of excited species in the flame is likely to be very low. In fact, Ebdon (1982, p.29) estimates that in an air-propane flame into which is aspirated 1mg l^{-1} of zinc solution, there will exist at any one time only one excited atom of zinc.

Conversely, when considering atomic emission techniques, data in Table 3.7 indicate that any phenomena causing a change in the excitation temperature will have a significant effect on the intensity of atomic emission lines. At the temperature of the fireball of an inductively coupled argon plasma (say 8000K), the Boltzmann equation predicts that about 24% of the caesium atoms will be in an atomic excited state, but only 0.07% of the zinc atoms.

3.7.3 *Self-absorption effects*

If atoms are efficiently excited in an emission source (e.g. an argon plasma), there is an increasing probability that a significant proportion of the emission lines originating from excited atoms in the hot core of the plasma will undergo a

Table 3.7 Proportion of atomic population residing in specified excited state at selected tempertures

	Atomic emission line nm	eV	Russell-Saunders terms for the transition	g_1/g_0	Ratio of excited state to ground state atoms at specified temperature 2000K	3000K	4000K	5000K	6000K	8000K	10 000K
Cs	852.1	1.455	$6^2P_{3/2} \rightarrow 6^2S_{1/2}$	2	4.31×10^{-4}	7.19×10^{-3}	2.94×10^{-2}	6.83×10^{-2}	1.20×10^{-1}	2.42×10^{-1}	3.70×10^{-1}
Na	589.1	2.104	$3^2P_{3/2} \rightarrow 3^2S_{1/2}$	2	9.97×10^{-6}	5.84×10^{-4}	4.47×10^{-3}	1.51×10^{-2}	3.42×10^{-2}	9.45×10^{-2}	1.74×10^{-1}
Ca	422.7	2.933	$4^1P_1 \rightarrow 4^1S_0$	3	1.22×10^{-7}	3.55×10^{-5}	6.05×10^{-4}	3.32×10^{-3}	1.03×10^{-2}	4.26×10^{-2}	9.98×10^{-2}
Zn	213.9	5.795	$4^1P_1 \rightarrow 4^1S_0$	3	7.46×10^{-15}	5.51×10^{-10}	1.50×10^{-7}	4.32×10^{-6}	4.07×10^{-5}	6.70×10^{-4}	3.60×10^{-3}

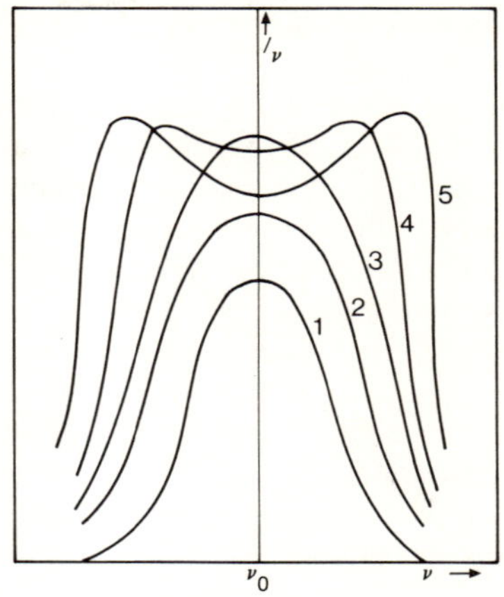

Figure 3.14 The effect of self-absorption on the shape of an atomic emission line. The notional atomic population increases in concentration from 1 (no self-absorption) to 5 (strong self-absorption). (Ebdon, 1982.)

resonance absorption by atoms of the same element in the cooler outer layers of the plasma. The intensity of the emission line reaching the spectrometer will then be lower than expected. This phenomenon is called self-absorption. The practical effect is that non-linearities may be observed in calibration lines at high element concentrations when self-absorption effects are likely to be most serious, in a manner similar to Beer's Law discrepancy effects shown in Figure 3.13.

Self-absorption also affects the shape of the emission lines. Atoms at the centre of a flame/plasma are relatively hot and therefore possess higher thermal energy. The width of their emission lines is therefore significantly broadened by the Doppler effect (see 3.7.4(i) below). Conversely, atoms on the periphery of the plasma are colder and possess a smaller degree of thermal energy. These atoms therefore absorb predominantly the central resonance frequencies of the emission profile. The resultant line profile changes as shown in Figure 3.14 as the degree of self-absorption increases. In analytical applications, self-absorption effects are avoided if at all possible. The one exception to this is a novel background correction technique applicable to atomic absorption analysis that makes use of self-broadening effects observed when the current fed to a hollow cathode lamp is pulsed (section 4.13).

3.7.4 *Line-broadening effects*

The natural width (measured as the full linewidth at half maximum) of an atomic absorption or emission line is very narrow—about 10^{-5}nm. However, in practical applications absorption and emission lines are broadened by several phenomena, including the following.

(i) *Doppler broadening* arises from the motion of the emitting atoms relative to the fixed measuring spectrometer. Atoms travelling towards the spectrometer entrance slits will apparently emit radiation at slightly higher frequencies and vice versa, due to the Doppler effect. As the velocity of atoms is controlled by their thermal energy, this broadening effect will increase with temperature.

(ii) *Pressure broadening* (also known as Lorentz or collisional broadening) occurs when the energy levels of atoms are perturbed to a small extent by collisions with other atoms. As a result of these random perturbations, the apparent width of an atomic spectral line is broadened. Since the number of collisions increases with pressure, line widths are expected to be broader for observations made at higher gas pressures.

(iii) *Zeeman broadening* is caused by the influence of magnetic fields (see section 4.8), and *self-absorption broadening* effects have been considered in the previous section.

To give some indication of the relative magnitude of these effects, Price (1979, p.15) estimates that the Doppler width of the sodium 589 nm spectral line is 0.00048 nm at 3000K, but that if this line is observed after atomization in a flame, the pressure effect will broaden it to about 0.008 nm. This is still a relatively narrow line width compared with the resolution of modern instrumentation, which is usually 0.01–0.2 nm.

3.8 Instrumentation for optical spectroscopy

3.8.1 *Nomenclature*

Specialized terms are used in optical spectrometry, including some that denote specific forms of instrumentation.

(i) A *spectrum* is the result obtained when light (electromagnetic radiation) is resolved into its constituent wavelengths, being effectively a plot of the power (intensity) of light absorption or emission *v.* wavelength (or energy).

(ii) A *monochromator* is an instrument designed to disperse radiation into its constituent frequencies. The source radiation passes through an entrance slit and is dispersed by a combination of optical components including glass prisms or diffraction gratings and mirrors. The dispersed radiation is focused on an exit slit so that only the desired wavelength is transmitted. Monochromator design is discussed in more detail in section 3.9 below.

(iii) A *colorimeter* is an instrument in which the degree of absorption of a sample solution is measured by comparison with the 'colour' or a set of standard solutions made up to bracket the absorption of the sample. Simple colorimeters were used in early rapid schemes of analysis (section 2.4) and used the human eye as detector.

(iv) A *photometer* is an instrument for measuring the intensity of photons emitted from a light source. Simple photometers consist of a light source, optical filter and photoelectric detector. Wavelength can only be changed by changing the filter (see section 2.5).

(v) A *spectroscope* is an instrument incorporating a monochromator, the dispersed spectrum from which is viewed with the human eye.

(vi) A *spectrograph* is an instrument incorporating a monochromator and a photographic plate for detection. The configuration was used for example on early spark/arc source optical emission instruments.

(vii) A *spectrometer* is an instrument that incorporates a monochromator to measure the energy distribution of radiation in a spectrum.

(viii) A *spectrophotometer* is a spectrometer in which radiation dispersed by the monochromator is detected using a photoelectric device such as a photomultiplier tube.

3.8.2 *Properties of materials used in the construction of optical instruments*

The design and construction of spectrometers has been strongly influenced by the optical transmission properties of materials available for the construction of optical components and by the absorption properties of air. These phenomena are compared in Figure 3.15.

Although *air* transmits light in the visible region of the spectrum, strong absorption occurs in the ultraviolet region below about 180 nm. Measurements below this limit must be performed under vacuum (hence the term vacuum uv) or with light paths purged with non-absorbing gases such as argon or nitrogen. Air also has strong absorption bands in the infra-red region of the spectrum. Most optical systems incorporate lenses and sometimes prisms made of glass. '*Normal*' *silicate glass* transmits radiation in the visible

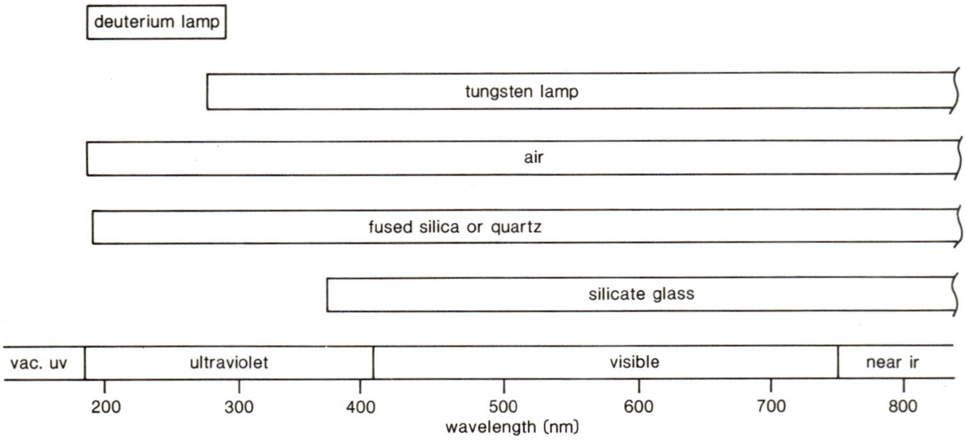

Figure 3.15 Optical transmission ranges of selected comonents used in the construction of an optical spectrometer. (Adapted from Skoog and West, 1982, p.503.)

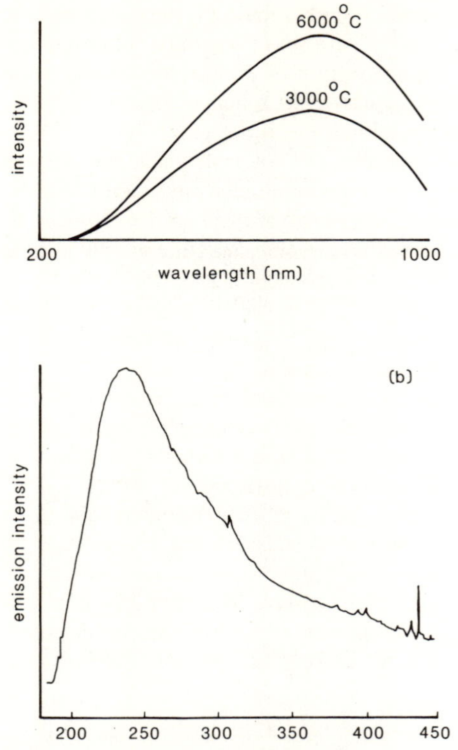

Figure 3.16 Lamp source outputs as a function of wavelength: (*a*) incandescent tungsten lamp at 3000° and 6000°C (Robinson, 1982); (*b*) deuterium arc lamp (Sotera and Kahn, 1982). The output of the latter becomes discontinuous at wavelengths above 350 nm.

region but starts to absorb strongly at the blue end at wavelengths below about 360 nm. For measurements below 360 nm, *quartz glass* may be substituted for optical components, as it transmits light down to about 180 nm. The incandescent *tungsten filament* is an efficient source of radiation in the visible and infra-red region of the spectrum. However, its intensity output diminishes significantly below about 300 nm and it is an ineffective source of radiation in the blue and uv regions of the spectrum (Figure 3.16). Conversely, the *deuterium discharge tube* efficiently emits radiation in the uv–visible region of the spectrum down to below 200 nm; however, its light output becomes discontinuous above 350 nm (Figure 3.16). The sensitivity of detection devices, including *photomultiplier tubes*, also varies with the wavelength of incident light. Normally the coating of the photocathode is optimized for detecting radiation in either the uv–blue region or the visible–red region of the spectrum.

These response characteristics must be taken into account in spectrometer design.

3.8.3 *Spectrometer design: light sources and sample configurations*

A spectrometer consists of a source of light, a sample, a wavelength selector (the monochromator), a photon detector and some form of read-out system. The configuration of source and sample depends on the excitation mechanism appropriate to a specific technique. Typical combinations include tungsten lamp–uv–visible absorption in a sample solution cell (Chapter 2), hollow cathode lamp–absorption by a sample atomized in a flame (atomic absorption spectrometry, Chapter 4), argon plasma–emission from an atomized sample (ICP–atomic emission spectrometry, Chapter 5) and electrical discharge excitation of a sample prepared as a solid electrode (arc/spark emission spectrometry, Chapter 6). The characteristics of these sources and excitation mechanisms have been discussed briefly above, but are considered in more detail in the chapters listed.

3.9 Monochromator

The function of the monochromator is to disperse incident polychromatic radiation into its constituent wavelengths, the intensity of which can then be measured. There are two basic designs. In the spectrograph, the entire spectrum is dispersed and brought into focus along a plane where its intensity may be detected by exposing a photographic plate. In conventional spectrophotometers, however, the dispersed beam is brought into focus on the exit slit of the monochromator. The dispersing element(s) may be rotated so that either the entire spectrum is scanned across the exit slit or, more usually, the monochromator adjusted to transmit the specified wavelength required for measurement. The intensity of the transmitted beam is normally detected using a photomultiplier tube.

3.9.1 *Prism monochromators*

(*a*) *The Bunsen (prism) monochromator*. The earliest monochromators were based on refraction of light in a glass prism (Figure 3.17). Triangular prisms with an apex angle of 60° were normally used. Dispersion of light occurs in a prism

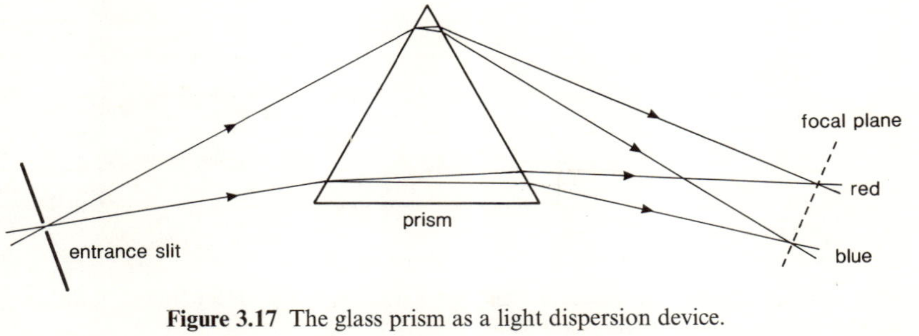

Figure 3.17 The glass prism as a light dispersion device.

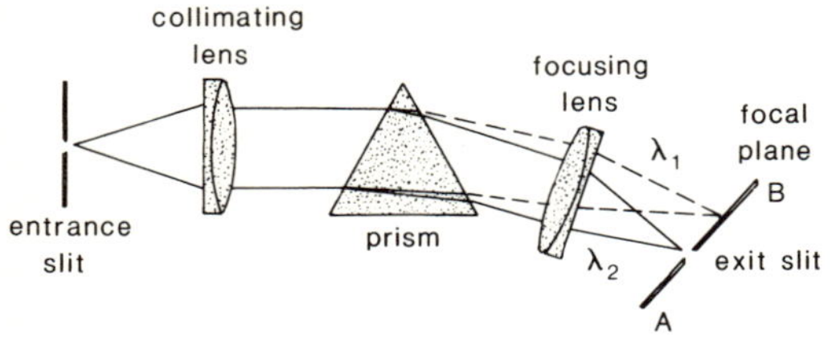

Figure 3.18 The simple Bunsen monochromator based on 60° glass prism. (Skoog and West, 1982.)

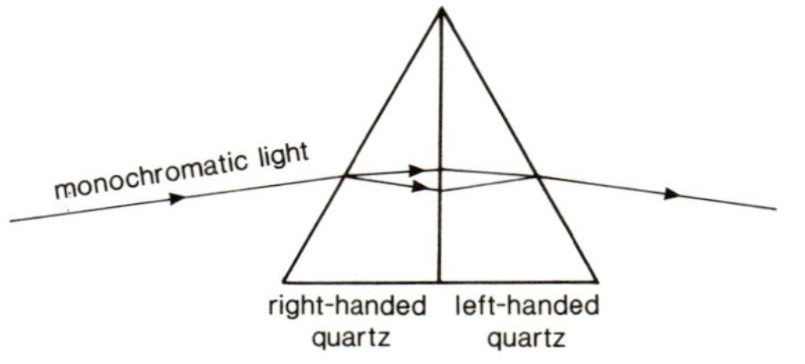

Figure 3.20 The Cornu prism made up of two 30° prism segments, one comprising left-handed rotating quartz, the other right-handed rotating quartz. This combination overcomes aberration effects due to optical activity.

because the refractive index of light varies slightly as a function of wavelength. At the air–glass boundary, red light is refracted least and blue light is refracted to the greatest extent. Incident white light is therefore dispersed into its constituent wavelengths. The simple prism monochromator designed by Bunsen is shown in Figure 3.18. The lenses are designed to gather source radiation and focus refracted light on the exit slit. They do not directly contribute to the dispersion of incident light.

(*b*) *Aberration effects.* Standard glass prism monochromators suffer from a serious aberration called birefringence. If a monochromatic beam of light is refracted through the prism, it cannot be brought into exact point focus at the exit slit. This is because unpolarized light consists of electric vectors that oscillate in all planes relative to the vertical and horizontal axes. The refractive index of vertically polarized light entering the prism is slightly different from that of horizontally polarized light. A monochromatic beam is therefore focused as a band, the extremes of which correspond to vertically and horizontally polarized components of the incident beam (Figure 3.19).

This aberration effect may be overcome in two ways. In the Cornu design (Figure 3.20), the 60° prism is fabricated from two 30° segments of quartz glass. One segment comprises right-handed quartz having positive optical activity, the other left-handed quartz. In this way the aberration effect is cancelled out. In the Littrow design (Figure 3.21) the prism comprises a single 30° segment, the rear surface of which is silvered. Incident light is therefore reflected back through the prism. Aberrations due to optical polarization are cancelled out. The Littrow prism can form a very economical monochromator design since the same optical components used for focusing the incident beam may also be used for the refracted beams.

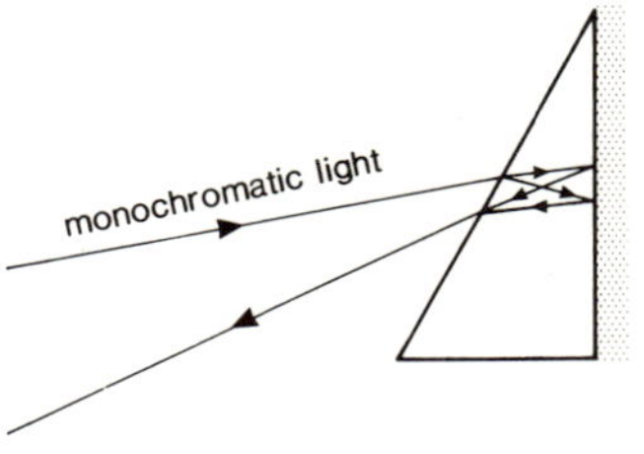

Figure 3.21 Littrow prism used to overcome optical activity aberration effects.

(*c*) *Advantages and disadvantages of prism monochromators.* The principal advantage of prism monochromators is the relative ease and therefore cheapness of manufacture. However, in modern spectrometer designs, glass prisms have some serious drawbacks. One is that light must be transmitted through the prism. Thus quartz must be used if measurements are to be made in the blue–uv region, although this material also suffers a cut-off at about 185 nm, since quartz strongly absorbs light of lower wavelengths. A further difficulty is that the light dispersion through a glass prism is not linear with wavelength—the resolution of lines at the blue end of the spectrum is significantly better than that for lines at the red end after refraction through a glass prism. Also, operating problems arise in recording spectral scans using prism monochromators. If the prism is rotated at a linear rate, the dispersed spectrum crossing the monochromator exit slits is detected on a reciprocal wavelength scale (see Figure 3.25). To compensate for this effect, an accurately machined snail-shaped cam can be used to rotate the prism so that wavelength scans may be recorded which are linear in wavelength. For these reasons, modern high-

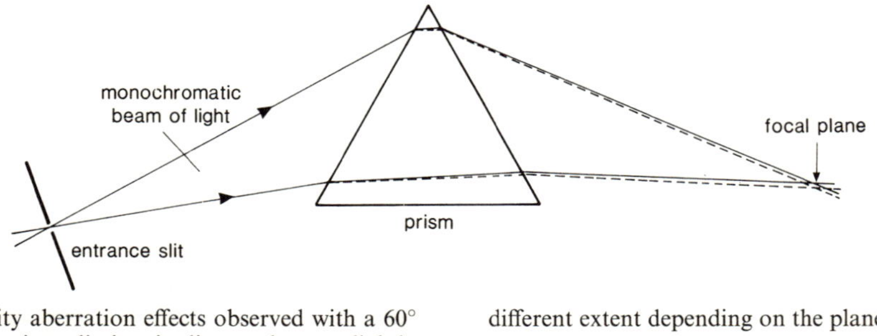

Figure 3.19 Optical activity aberration effects observed with a 60° glass prism. Monochromatic radiation is dispersed to a slightly different extent depending on the plane of polarization.

performance spectrometers depend on the *diffraction* of light by a *grating* rather than the *refraction* of light by a quartz *prism*.

3.9.2 *Diffraction gratings*

When light is incident on a substrate on which a large number of parallel lines has been ruled, interference effects are observed such that specific wavelengths are strongly diffracted only at specific angles with respect to the incident beam. Diffraction gratings exploit this phenomenon.

Two types of grating are available:

(i) *Transmission gratings*, the earliest type of grating, consisted of a quartz glass substrate on which a large number of lines was ruled. Diffraction occurs as the light beam is transmitted through the grating. Transmission gratings suffer a number of disadvantages, largely overcome with the developments of reflection gratings.

(ii) *Reflection gratings* are made by ruling a large number of parallel grooves on a highly reflective substrate. Diffraction occurs in the beam reflected from the grating.

In either type of grating, diffraction will be observed only if the frequency of line spacing is of the same order of magnitude as the frequency of incident radiation. Conventional gratings must, therefore, be ruled at a density of 1500–4000 lines per mm in order to diffract uv–visible light.

3.9.3 *Transmission grating*

The behaviour of light incident on a transmission grating is shown in Figure 3.22. Each of the lines ruled on the quartz substrate scatters incident radiation, effectively behaving as a blanking strip. Light is, however, transmitted through the non-ruled areas. The net effect is that the grating behaves as if it formed a large number of slits, each splitting the incident radiation into individual beams. Each beam is propagated, according to Huygen's principle, through the unruled portion of the slit as a hemispherical wavefront shown in semicircular cross-section in Figure 3.23. Interference will occur between wavefronts from adjacent slits. At most angles relative to the grating these interferences will be destructive, since adjacent wavefronts will be out of phase. However, at specific angles, constructive interference of specific frequencies will occur, the waveforms from each secondary source then being exactly in phase. The angle θ (with respect

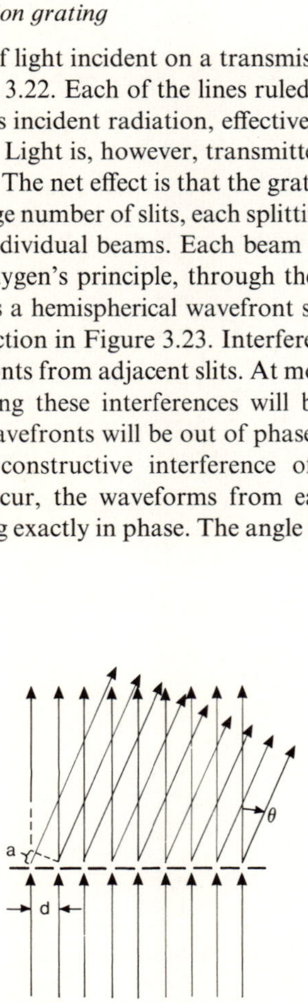

Figure 3.22 Diffraction at a transmission grating. (Ewing, 1975.)

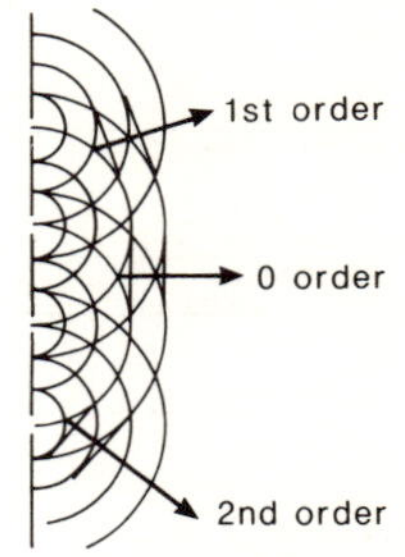

Figure 3.23 Huygens' principle. This diagram shows the propagation of monochromatic light through adjacent slits. Constructive interference will occur in the transmitted beam of light at certain angles of diffraction. (Moore, 1966.)

to the incident beam) at which constructive interference will be observed depends on both the wavelength of light λ and the spacing d between the lines ruled on the grating, and is given by

$$n\lambda = d \, . \sin\theta \tag{3.13}$$

where n is an integer $n = 0$ for zero-order diffraction (i.e. light passing straight through the grating, undiffracted, $n = 1$ for first-order diffraction, $n = 2$ for second-order diffraction, etc.

Thus a polychromatic beam of light incident on a transmission grating will be diffracted into its wavelength components. These will be observed as zero-, first-, second-order etc. beams that will be diffracted symmetrically on either side of the zero-order line (Figure 3.24).

At any particular angle of diffraction, not only will first-order lines appear ($n = 1$ in equation 3.13) but also multiple-order lines. Thus red light at 650 nm is diffracted at the same angle as uv light at 325 and 216.7 nm. If necessary, it is possible to remove such interferences by filtering the diffracted beam with an optical filter that permits transmission of radiation in the waveband of interest only. Alter-

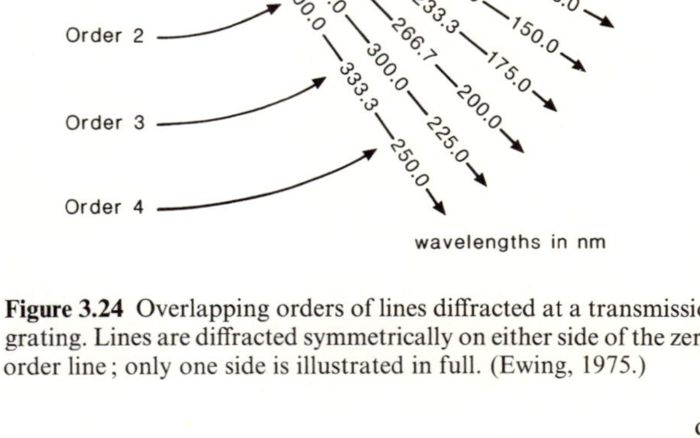

Figure 3.24 Overlapping orders of lines diffracted at a transmission grating. Lines are diffracted symmetrically on either side of the zero-order line; only one side is illustrated in full. (Ewing, 1975.)

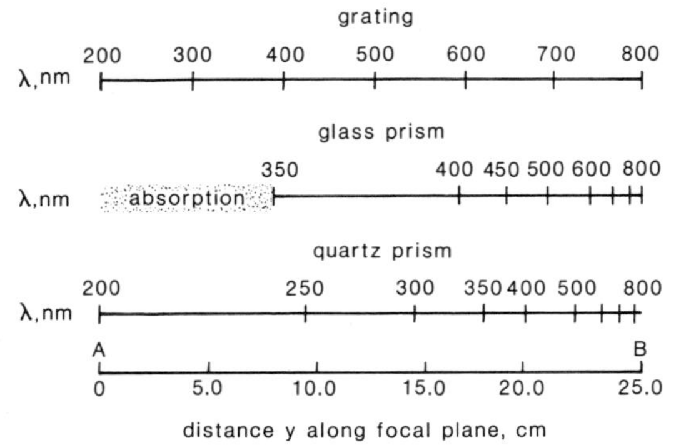

Figure 3.25 Dispersion of wavelengths along the focal plane of a grating monochromator and two prism monochromators of the Littrow design, one using a glass prism, the other using a quartz prism. (Skoog and West, 1982.)

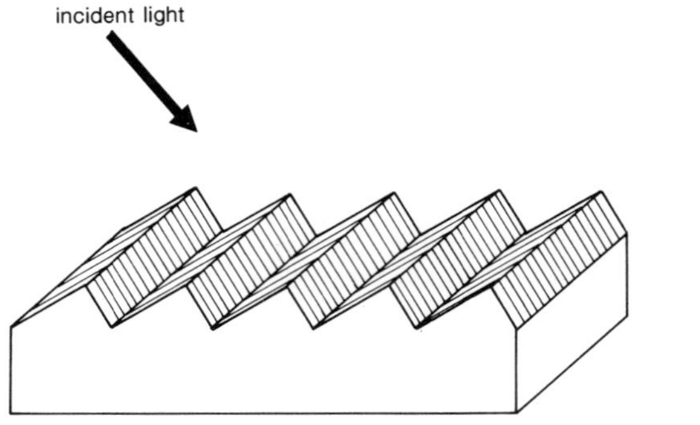

Figure 3.26 Pattern of grooves on a reflection grating. The arrow shows the direction of incident radiation,

natively, a secondary quartz prism can be placed in the diffracted beam. This prism acts as an 'order sorter'. Multiple-order lines are refracted to different degrees so that overlap interferençes can be distinguished. An example of such an application is given in section 3.9.7.

Transmission gratings (like reflection gratings) have an important advantage over prism monochromators; the angular dispersion of a diffraction grating is essentially linear over the entire uv–visible wavelength range (Figure 3.25). Diffraction gratings are, therefore, as effective at separating lines at the red end of the spectrum as at the blue. This is not the case with glass prisms.

The simple transmission grating has a number of disadvantages, however, that preclude its widespread use. The most important is that radiation must be transmitted through the grating medium with the attendant problems of unfavourable absorption characteristcs in the uv region, as with quartz prisms. Moreover transmission gratings are not very efficient, giving an unwanted intense zero-order line. Furthermore, 50% of the diffracted spectrum (on one side of the zero-order line) is normally unused. These problems were largely overcome with the development of reflection gratings.

3.9.4 *Reflection gratings*

Reflection gratings consist of a glass or metal substrate on which is ruled a large number of equally spaced parallel *grooves*. The face of each of these grooves is precisely aligned at a specified angle to the surface of the substrate (Figure 3.26). The angle that each face makes with respect to the substrate surface is called the *blaze angle*.

Diffraction occurs at a reflection grating in a manner similar to that described for a transmission grating above. Incident radiation is reflected from the face of each groove, each of which effectively behaves as a secondary point source of light (Figure 3.27). Destructive interference will occur at most angles in this multiply-reflected beam. Constructive interference will only occur at specific angles, providing the distance between beams reflected off successive groove faces equals an exact number of integral wavelengths. Mathematically, this condition is given by the equation

$$n\lambda = d(\sin i \pm \sin r) \qquad (3.14)$$

where d is the groove spacing, i is the angle of incident radiation, and r is the angle of diffracted radiation (Figure 3.27).

(*a*) *Blaze angle*. The blaze angle of a grating is not arbitrarily selected. It is desirable to concentrate as much of the diffracted beam energy as possible into first-order rather than multiple-order lines and to optimize the performance of the grating for a particular spectral range (see Figure 3.32). These objectives can be attained by careful selection of blaze angle. The monochromator specifications normally include the wavelength for which the grating is blazed. This is the wavelength, in a specified order (normally the first), for which the angle of reflection from the groove face equals the angle of diffraction from the grating. To cover the entire uv–visible range with optimum efficiency it is sometimes desirable to employ two gratings, one blazed at about 220–250 nm for the uv region from 180 up to 400 nm and a second blazed at 500–550 nm to cover the visible region from 400 up to 800 nm. The optical efficiencies of two grating monochromators, one blazed at 250 nm and the other blazed at 500 nm, are compared in Figure 3.28.

(*b*) *Replication*. The reflection grating described above is called an *echellette grating*. Its widespread adoption in

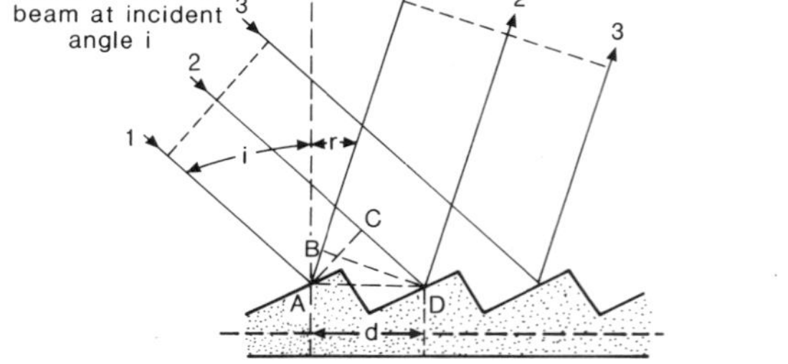

Figure 3.27 Mechanism of diffraction from an echellette type reflection grating; (*d*) is the distance between adjacent grooves (i.e. grating periodicity). (Bauman, 1962; reproduced by Skoog and West, 1980.)

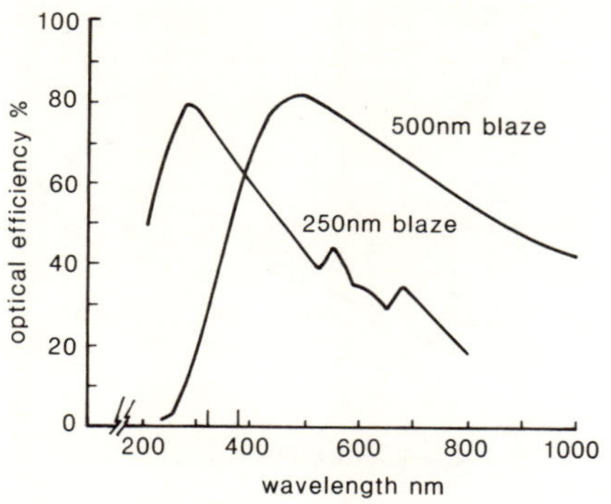

Figure 3.28 Optical efficiency of two grating monochromators, one with a grating blazed at 250 nm, the other at 500 nm. (Courtesy of Instrumentation Laboratory.)

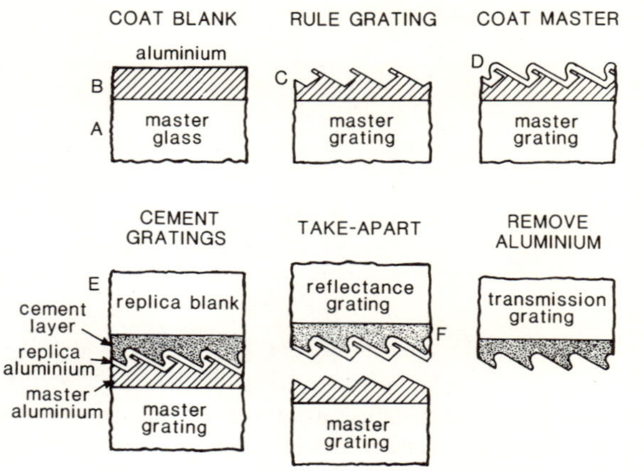

Figure 3.29 Manufacture of master gratings using the replication process. Replication removes the overhanging burr that significantly reduces diffraction efficiency since it intercepts part of the reflected beam. This process has now largely been superseded by the holgraphic process. (Richardson, 1969.)

optical instruments rested on overcoming several technical difficulties in achieving its extremely high accuracy in groove periodicity and linearity. Richardson (1969) has highlighted these technical difficulties, summarizing the achievements of various workers over the years in increasing the number of grooves that could be ruled on a grating (Table 3.8).

Originally each echellette grating was ruled individually, using an instrument known as a ruling engine. However, such mechanical instruments can blaze only a relatively small number of gratings per year. To overcome this problem a replication technique was developed. A master grating was ruled on an aluminum-coated glass substrate. A plastic material was then cast on to this master grating and, when set, stripped off and cemented on to a rigid substrate (Figure 3.29). The surface of this replica grating was then coated with aluminium to give a highly reflective surface. However, even these techniques have now been superseded by recent developments in holographically ruled gratings. The appropriate grating is created from the interference patterns generated from two synchronous lasers. This pattern is etched into a photosensitive resin. The grating pattern created in this way can be over an order of magnitude higher in accuracy and linearity than those ruled by mechanical means.

3.9.5 *Angular dispersion of prisms and gratings*

Angular dispersion is a measure of the ability of a prism or grating to separate polychromatic radiation into individual frequencies and is measured as either the rate of change of refraction angle (or diffraction as appropriate) with wavelength (the angular dispersion), or the number of wavelength units per unit distance measure along the focal plane of the monochromator (nm per mm)—the linear dispersion

(*a*) *Prisms.* For a prism the angular dispersion depends on: (i) the refraction index (η) of the prism material, which itself varies with wavelength of light (λ); (ii) the angle of incidence of light (θ); and (iii) the prism angle a. For a 60° glass prism, angular dispersion ($d\theta/d\eta$) is given by

$$d\theta/d\eta = (1-\eta^2/4)^{-1/2}. \quad (3.15)$$

The net effect is that the angular dispersion of glass prisms is approximately linear with wavenumber (i.e. reciprocal in wavelength). This means that such prisms give good dispersion in the blue–uv region, but lines at the red end of the spectrum are crowded together (see Figure 3.25).

(*b*) *Diffraction gratings.* The angular dispersion of a diffraction grating is given by

$$\frac{dr}{d\lambda} = \frac{n}{d\cos r} \quad (3.16)$$

where n is the order of diffracted radiation, d is the distance between successive grooves on the grating, and r is the angle of diffraction.

From this equation it is apparent that the angular dispersion of a grating increases as the number of grooves per nm increases. Furthermore, since the angle of diffraction r does not vary a great deal over narrow wavelength ranges, cos r is expected to be approximately constant and the dispersion of a grating will be almost linear with wavelength.

Table 3.8 Historical development of the diffraction grating

Year	Spectroscopist	Grooves per mm
1785	Rittenhouse	4.2
1823	Fraunhofer	315
1846	Norbert	236
1870	Rutherford	689
Present	Modern uv–visible spectrophotometers	1800 to 3600 (typical)

Data from Richardson (1969).

3.9.6 *Resolving power*

The resolving power of a prism or grating is a measure of its ability to separate two lines at adjacent wavelengths λ_1 and λ_2 as distinct images. Resolving power R is normally expressed by

$$R = \frac{\lambda}{(\lambda_1 - \lambda_2)} = \frac{\lambda}{\delta\lambda} \quad (3.17)$$

where λ is the average wavelength.

Thus the resolving power required to separate two lines at 400.0 and 400.2 nm is 400.1/(400.2 − 400.0) = 2000.

(*a*) *Prism*. For a prism the resolving power is given by

$$R = b(\mathrm{d}\eta/\mathrm{d}\lambda) \quad (3.18)$$

where b is the base length of the triangular prism and $\mathrm{d}\eta/\mathrm{d}\lambda$ is related to the angular dispersion, (as described above) being the rate of change of refraction index (η) with wavelength. Thus for a given angular dispersion, high resolving powers are observed from large prisms.

(*b*) *Diffraction grating*. The resolving power of a diffraction grating is given by

$$R = nN \quad (3.19)$$

where n is the order of the diffracted line, and N is the number of grooves illuminated by the incident beam.

Thus the larger the physical size (area) of the grating, the higher the resolving power.

Simple calculations show that the resolving power required to distinguish two lines at about 500 nm separated by small wavelength differences are as follows:

Separation between lines (nm):	10	5	1	0.5	0.1	0.05	0.01
Required resolving power:	50	100	500	1000	5000	10 000	50 000

3.9.7 *Echelle grating*

A normal echellette grating is ruled at several thousand grooves per mm and is designed to operate using first- (and second-) order diffracted lines. Multiple-order lines are normally unwanted and considered to be interferences. In contrast, an echelle grating is normally ruled at between 30 and 300 grooves per mm and designed specifically for measurements on multiple-order lines, often in the range $n = 30$ to $n = 120$. The grating itself consists of step-like grooves (Figure 3.30) of typical height 5.6 μm and width 13 μm (Richardson, 1969). These dimensions should be compared with those of conventional echellette gratings, which are normally in the tenths of μm range. The technical difficulties in manufacturing echelle gratings are severe because of the very high degree of accuracy required to maintain periodicity and linearity, without which ghost images and distortion seriously degrade the performance of the grating. Because the echelle grating is designed to operate over such a large range of orders, steps must be taken to separate one order of diffraction from another so that multiple-order line overlaps can be distinguished from one another. This is normally achieved by placing a 30° quartz prism in front of the diffraction grating. The function of the prism is to act as an '*order sorter*' (Figure 3.31). Successive orders of lines are diffracted to slightly different degrees so that the diffracted spectrum appears as a two-dimensional array of wavelength

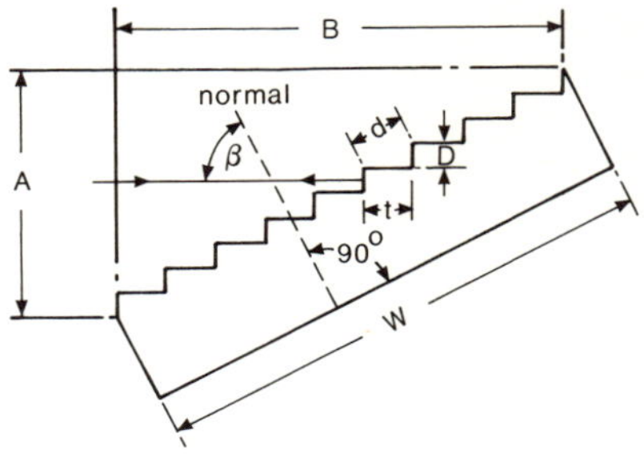

Figure 3.30 Cross-sectional profile of an echelle grating. (Richardson, 1969.)

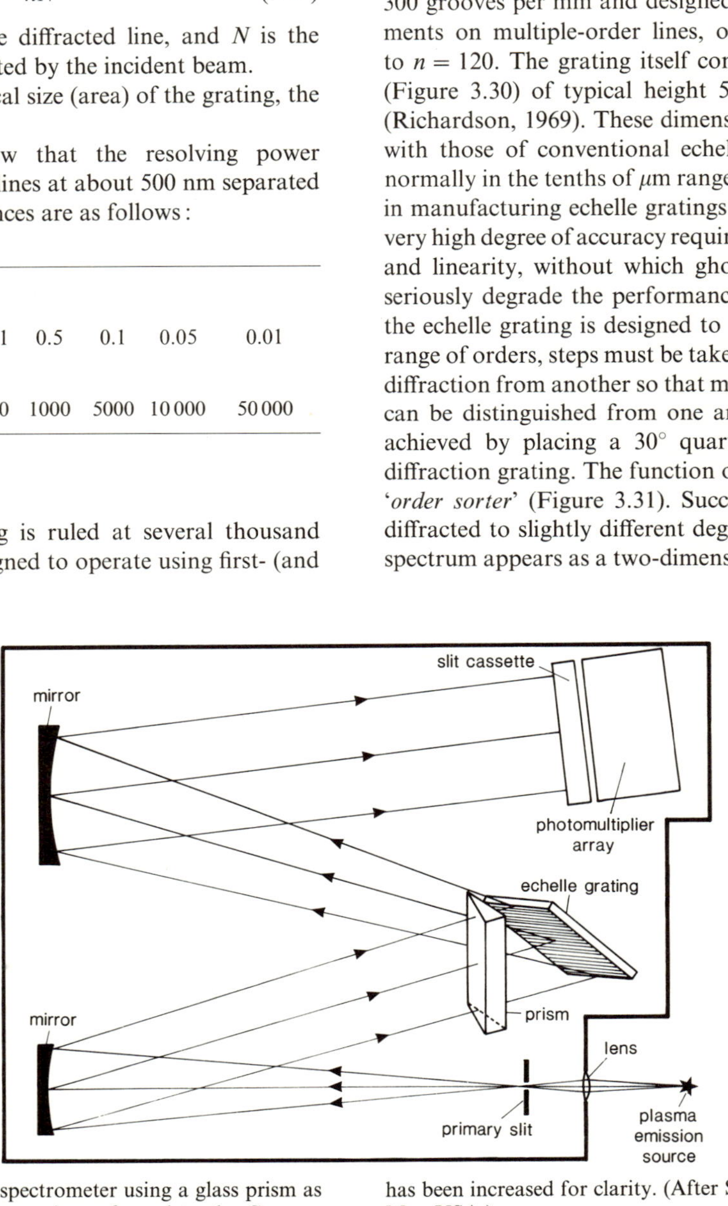

Figure 3.31 An echelle grating spectrometer using a glass prism as an 'order sorter'. The spectrometer is configured to the Czerny–Turner design. The angle of incident of light beam on the grating has been increased for clarity. (After Spectrometrics Inc, Andover, Ma., USA.)

v. order number. Clearly there is some duplication in the wavelength range covered by successive orders of line. This is called the *free spectral range*, (F_λ) defined as the range of wavelengths from λ to $(\lambda + \delta\lambda)$ for which the mth order of wavelength $(\lambda + \delta\lambda)$ coincides with the $(m+1)$th order of wavelength (λ) (Richardson, 1969, p.42). For a typical echelle configuration, F_λ may be calculated from:

$$F_\lambda = \delta\lambda = \lambda^2/2t \qquad (3.20)$$

where t is the width of the echelle groove (Figure 3.30).

Several advantages are claimed for a spectrometer based on the echelle grating.

(i) *Resolving power*. Because echelle monochromators are designed to operate at high orders of diffraction ($n = 30$ to 120 in equation 3.19) the resolution of these instruments is extremely high, often 5–10 times better than equivalent echellette monochromators. Thus commercial monochromators can readily achieve resolutions of the order of 0.003 nm, whereas 0.02 nm is a more typical figure for a conventional grating spectrometer. The resolving power of an echelle spectrometer can therefore be as high as two million.

(ii) *Angular dispersion*. Echelle spectrometers also show a very high angular dispersion: 0.12 nm/mm is claimed for one commercial design. Furthermore, because one order can be stacked on top of another, spectra can be recorded as spectrographs in a very compact manner. One commercial instrument records the entire 190–800 nm range on a 10 × 12.5 cm photograph at an angular dispersion that would otherwise require a linear distance of about 2 m using a conventional instrument.

(iii) By selecting the appropriate order of line, the grating is effectively *blazed for all wavelengths*, whereas the performance of a conventional echellette grating is optimized for a particular wavelength range (Figure 3.32).

The performance of an echelle spectrometer in an ICP–AES application is further discussed in section 5.13. As a result of commercial interests and patent protection, these devices have not been as widely used as might be expected from their technical performance.

3.10 Optical filters

Two forms of optical filter are available.

(i) *Absorption filters* are often used to eliminate multiple-order interferences in spectrometer design. In this respect, the 30° order sorter prism discussed above is a rather specialized design for overcoming the same effect. In flame techniques, such as atomic absorption spectrometry, the filter has a second function. The flame itself is a source of light that can degrade the atomic absorption signal through stray light effects. Depending on the wavelength of the atomic line, an optical filter can be used to remove this unwanted effect. Optical absorption filters are normally made of coloured glass that transmits a relatively wide band of wavelengths (Figure 3.33). These filters are adequate for removing multiple-order line interferences that are in any case widely separated in wavelength. The narrowest bandwidth available using an absorption filter is about 40 nm. Transmission filters may be used as wavelength selectors without the expense of a monochromator in simple flame photometry (section 2.6).

(ii) The *interference filter* works on a different principle and offers a much narrower bandpass, often in the range 1 to 10 nm. It consists of a substrate (often magnesium fluoride), coated with two semi-transparent silver films. Some of the incident radiation is transmitted straight through the filter. However, much is internally reflected several times within the substrate before onward transmission (Figure 3.34). The transmitted beam consists, therefore, of directly transmitted incident radiation together with radiation that has been internally reflected 2, 4, 6, ... times within the filter substrate. At most wavelengths, destructive interference will occur between the various components in this transmitted beam. Only at those wavelengths where the thickness of substrate satisfies equation (3.21) will constructive interference occur:

$$n\lambda = 2t \qquad (3.21)$$

where n is the order of radiation and t is the thickness of the

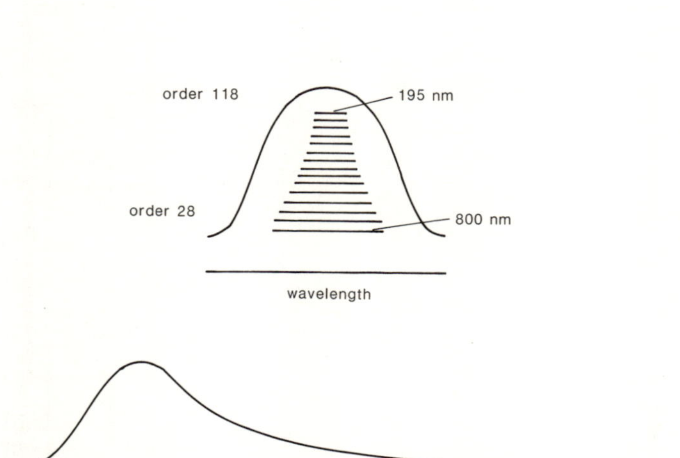

Figure 3.32 Comparison of diffraction efficiency between an echelle-type grating and a conventional echellette reflection grating blazed for optimum efficiency at 300 nm. Due to the characteristics of the echelle grating, an order of diffraction can always be selected at the optimum of the grating blaze function. (Spectrometrics Inc, Andover, Ma., USA.)

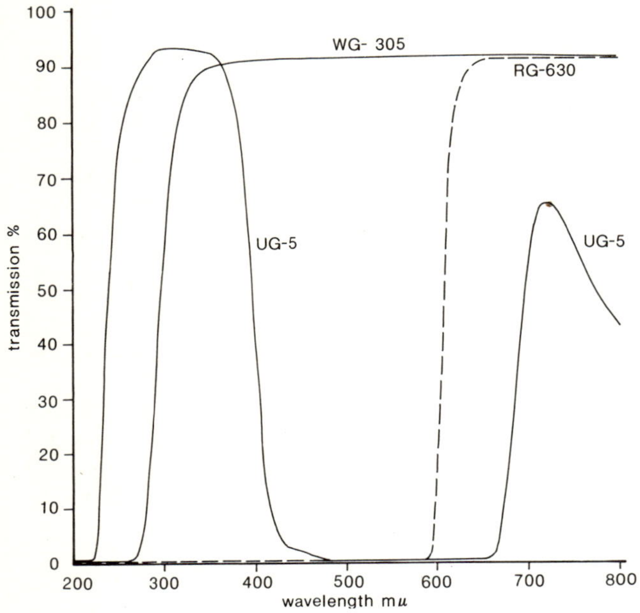

Figure 3.33 Transmission characteristics of absorption filters used in atomic absorption spectrometry. UG-5 is designed to improve signal to noise measurements of lines in the uv region. WG-305 and RG-630 are for eliminating second-order interferences on lines in the visible and red region of the spectrum respectively. (Varian Techtron, Victoria, Australia.)

filter and it is assumed that the light beam is normal to the filter.

In designing transmission filters, it is necessary to select a compromise between a narrow bandwidth and adequate transmission intensity. A complete range of interference filters can be manufactured covering the uv–visible region by adjusting the filter thickness to the appropriate value. Manufacturing tolerances on maintaining constant filter thickness are critical.

3.11 Slits

Slits are an essential, though sometimes overlooked, component of monochromator design. They normally consist of a pair of knife-edges designed so that the edges are parallel and in the same plane. The distance between the knife-edges is usually adjustable, with entrance and exit slits identical in width. The entrance slit effectively defines the radiation source. The optical path through the monochromator normally incorporates focusing lenses or concave mirrors (of specified focal length) so that an image of the entrance slit, after diffraction at a grating, is focused on the exit slit.

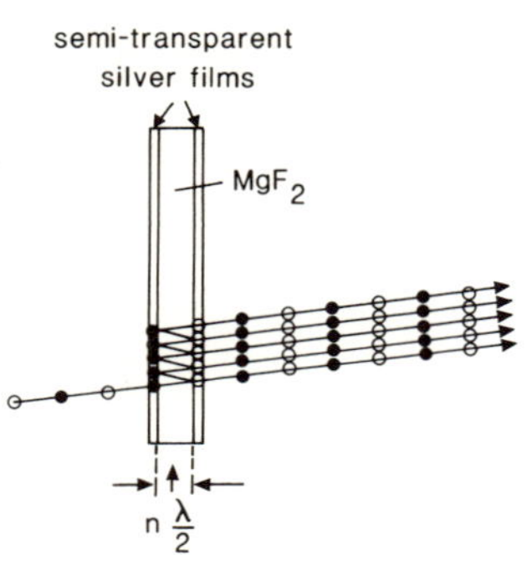

Figure 3.34 Transmission mechanism in an interference filter. Black circles represent troughs and open circles peaks of radiation that undergo constructive interference through this filter. (Ewing, 1975.)

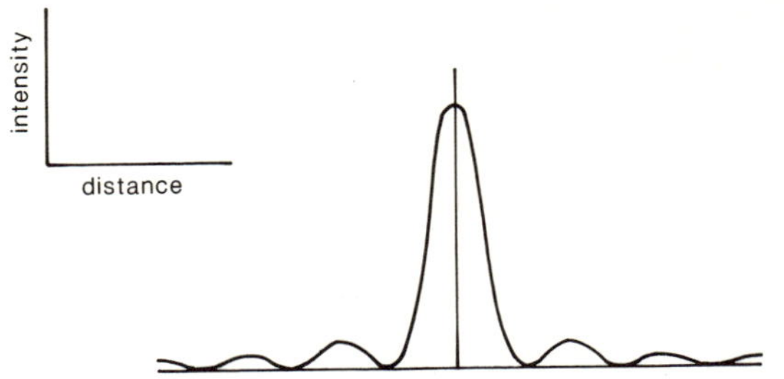

Figure 3.35 Diffraction pattern across a single slit plotted as a function of intensity *v.* distance across the diffracted beam parallel to the slit opening. (Ewing, 1975.)

The behaviour of a light beam passing through a slit depends on the slit's relative width. If the width of a slit approaches the wavelength of incident light, interference fringes are formed so that the central image of the slit is surrounded symmetrically by subsidiary interference bands (Figure 3.35). The width of the central band ($\delta\lambda$), measured between adjacent minima is given by

$$\delta\lambda = 2f\lambda/d \qquad (3.22)$$

where d is the slit width, λ is the wavelength, and f is the focal length of the focusing lens or mirror.

At very narrow slit widths, the intensity of the interference fringes is significant and the transmitted beam is of low intensity and diffuse. As the slits are gradually increased in width, the central interference maxima become sharper and more intense, a condition under which optimum resolution is observed. As the slits become even wider, the slit image broadens as more energy is transmitted and the interference

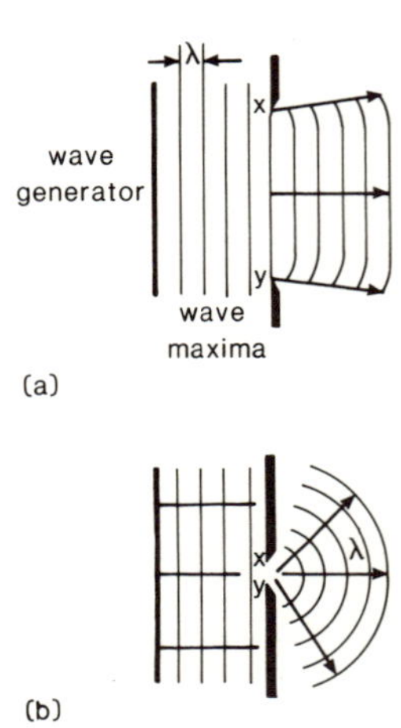

Figure 3.36 Propagation of monochromatic light waves through (*a*) wide ($xy \gg \lambda$) and (*b*) narrow ($xy \simeq \lambda$) slits. (Skoog and West, 1980.)

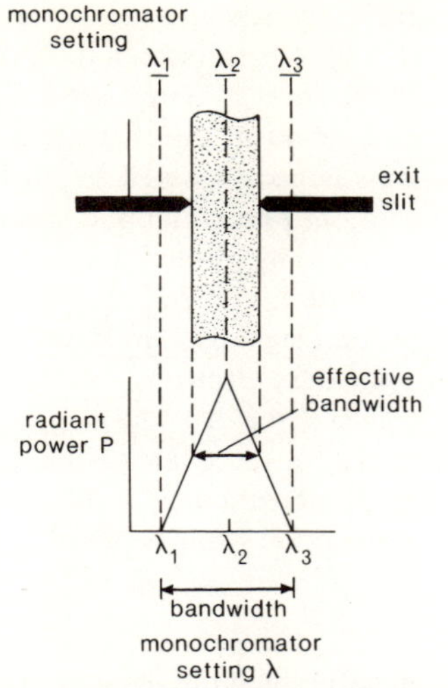

Figure 3.37 Illumination of an exit slit by monochromatic radiation (λ_2) at various monochromator settings, showing the relationship between monochromator bandwidth and effective bandwidth. Entrance and exit slits are assumed to be identical in width. (Skoog and West, 1980.)

fringes effectively disappear (Figure 3.36). The slit width (*d*) under which optimum resolution is observed is given by

$$d = 2f\lambda/w \tag{3.23}$$

where *w* is the diameter of the focusing lens or mirror, *f* is its focal length, and λ is the appropriate wavelength.

For practical measurements, it is usual to select slit widths wider than the optimum to increase the intensity of the transmitted beam.

Under these conditions, monochromatic radiation emerging from the exit slit can be represented approximately by a triangular function (Figure 3.37). The wavelength difference corresponding to the full-width half maximum of this function is known as the *effective bandwidth* of the monochromator, which should not be confused with the (total) *bandwidth* representing the width (in nm) of the base of the triangle.

Selection of the appropriate slit width is an important aspect of monochromator operation for the following reasons.

(i) If the monochromator uses a prism as the dispersion element, the angular dispersion varies significantly with wavelength (section 3.9.1(c)). A much narrower slit width is therefore required for measurements in the red region of the spectrum to achieve satisfactory resolution in comparison with performance in the uv (Figure 3.38). This problem does not affect grating monochromator operation where angular dispersion is almost constant.

(ii) Selection of slit width has an important effect on the severity of interferences between adjacent atomic emission lines in complex spectra observed from ICP sources, for example.

(iii) In many optical methods it is desirable to minimize slit widths to reduce the magnitude of the background signal reaching the detector and hence contribute to noise (shot noise, see section 3.15) in the absorption measurements.

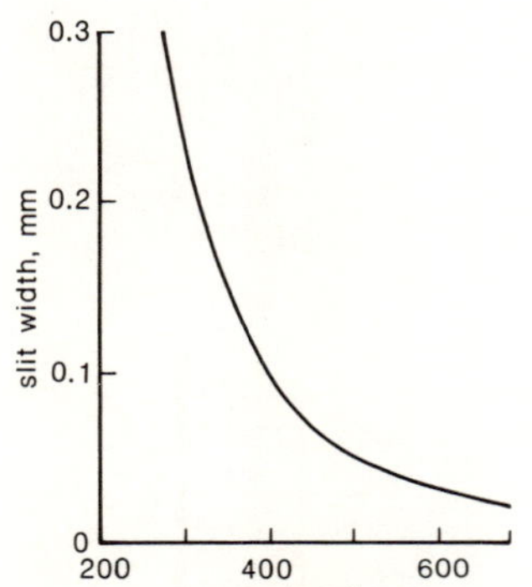

Figure 3.38 The slit width required to maintain an effective bandwidth of 1 nm using a Littrow quartz prism monochromator. (Skoog and West, 1980.)

3.12 Photon detectors

The final element of spectrometer instrumentation to be considered here is the detector used for measuring the intensity of radiation transmitted by the monochromator. Although photocells are used on some simple instruments, most modern spectrometers use the photomultiplier tube. This is similar in design to those used to measure scintillation radiation in sodium iodide gamma-ray spectrometers (section 13.6).

The photomultiplier tube consists of a photodynode and a series of dynode elements, all of which are enclosed under high vacuum in a glass envelope (Figure 3.39). Each dynode

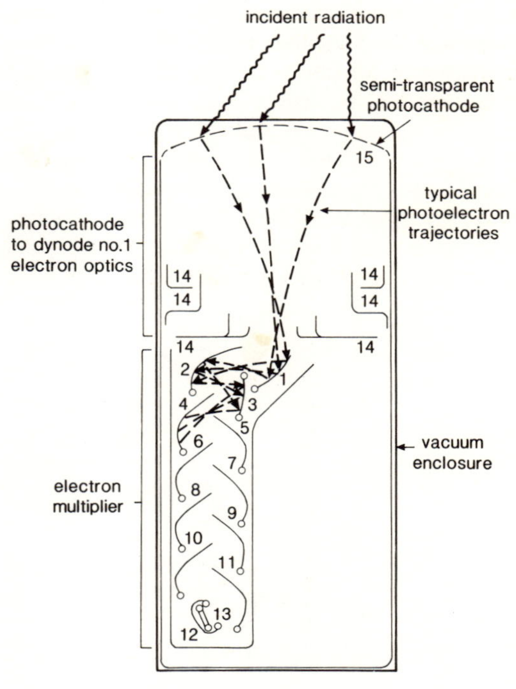

Figure 3.39 Schematic diagram of a photomultiplier showing photocathode (15), electron focusing electrodes (14), dynodes (1–11) and anode (13). (RCA Corporation.)

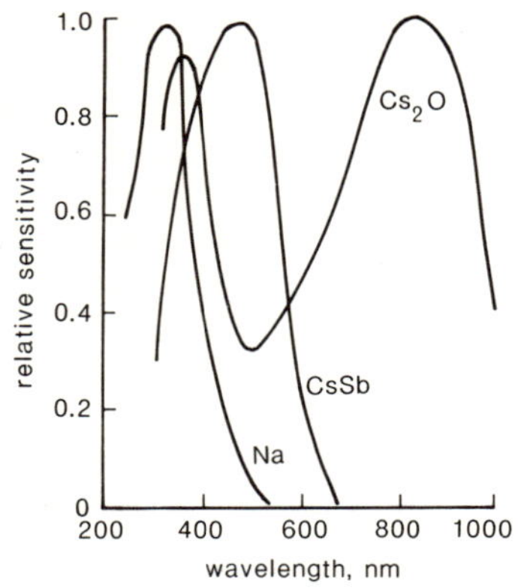

Figure 3.40 Spectral responses of three photoemissive surfaces used to coat photocathodes. (Skoog and West, 1980.)

is biased at a successively higher positive voltage relative to the photodynode. Light photons impinging on the photodynode cause electrons to be ejected by the photoelectric effect. The electrostatic fields between successive dynodes cause these electrons to be accelerated and focused on to the next element in the dynode chain. Between two and four secondary electrons are ejected at each electron–dynode impact. The electron multiplication factor across 12 dynodes may be 10^6 to 10^7. That is, for each photoelectron emitted at the photodynode, a cascade of over 10^6 secondary electrons is delivered to the anode. The precise value of the gain depends, in fact, on the number of dynode elements (n) in the photomultiplier and the voltage difference (V) maintained between each element such that

$$\text{photomultiplier gain} \propto V^{0.7n}. \tag{3.24}$$

The selection of phodynode material is critical to successful operation of the photomultiplier as there is no one coating that possesses a uniformly high photoelectric response throughout the entire spectral range from 180 to 850 nm. The materials available to coat the photodynode include Cs_3Sb, K_2CsSb, Na_2KSb (Cs doped), NaKCsSb, GaAs and Cs_2O. The spectral responses of three of these materials are plotted in Figure 3.40. As can be seen from equation (3.24), if the photomultiplier bias V is increased, a corresponding increase will be observed in the magnitude of the signal. Unfortunately, this is accompanied by an increase in the *dark current*: that is, the photomultiplier output signal produced in the absence of any photon input. An increase in V also causes an increase in the degree of *shot noise*. This is the variation in output signal due to statistical variations in the number of secondary electrons emitted by the dynode elements. For a given value of V the magnitude of the shot noise is proportional to the square root of the intensity of light falling on the photocathode. The influence of shot noise on errors in the absorption signal will be discussed in section 3.15.

3.13 Classical monochromator designs

A variety of arrangements of mirrors/lenses and prisms/gratings are used in modern spectrometer designs. Each has its own operating characteristics, the design arrangement normally being called after the originator. Two factors, in particular, have influenced the designs adopted for present-day spectrometers: glass and quartz lenses absorb radiation at the uv end of the spectrum and can cause aberration effects, and concave (or toroidal) mirrors are therefore often preferred as focusing elements to avoid these interference effects; secondly, glass prisms can suffer the same deficiency, as well as causing a non-linear dispersion response. Gratings are therefore usually the preferred dispersion element.

The more important designs incorporated in modern instruments are as follows.

(*a*) *Littrow monochronomator*. The Littrow design makes use of a 30° prism, silvered on one face to overcome chromatic aberration effects (as discussed in section 3.9.1(*b*). This basic geometry is often adopted in modern designs, with a grating replacing the glass prism, and a concave mirror rather than lenses to focus the incident and diffracted beams (Figure 3.41). The diffraction grating must be rotated so that the desired wavelength can be focused on the exit slit. This design has the advantage that it uses the minimum

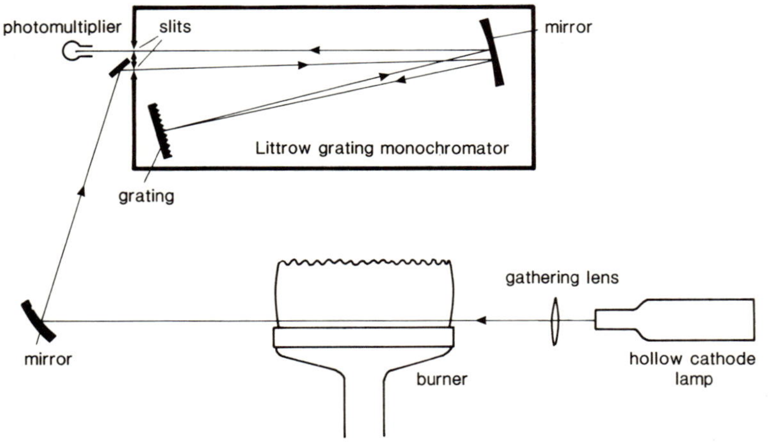

Figure 3.41 Littrow grating monochromator used for flame atomic absorption spectrometry. (After Perkin Elmer Corp. Norwalk, Connecticut, USA.)

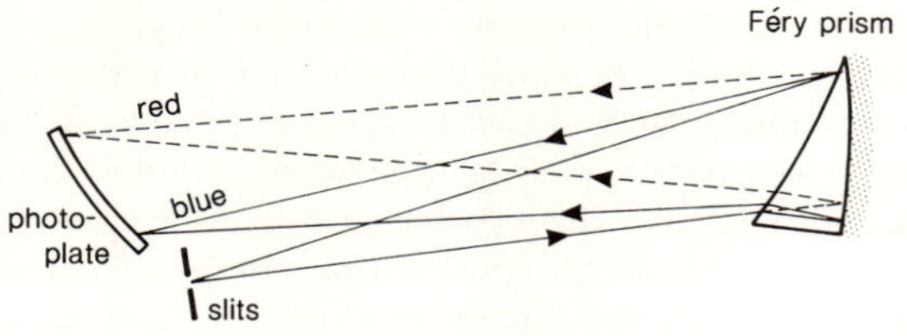

Figure 3.42 The Féry prism monochromator configured as a spectrograph.

of optical components and makes for a compact instrument. In Figure 3.41 a Littrow design monochronomator is shown coupled to a flame AA spectrometer.

(*b*) *The Féry spectrograph.* In early Littrow monochromator designs, using glass prism dispersion, performance in the uv was poor due to attenuation caused by both the prism and an additional glass focusing lens. The latter was dispensed with in the Féry design by both refracting and focusing the incident beam of light using the prism with a curved and silvered rear face (Figure 3.42). The spectrograph shown in this diagram suffered from severe astigmatism, although there was some improvement in uv measurements.

(*c*) *Ebert–Fastie monochromator.* The Ebert–Fastie monochromator uses a single spherical mirror to reflect and focus incident and diffracted beams (Figure 3.43). Wavelengths are again selected by rotating the diffraction grating. The original Ebert design used prism dispersion. However, this was reoptimized for grating diffraction by Fastie (1952) in the design used widely today. Fastie showed that parallel-jaw slits do not give optimum resolution with this monochromator design and that for optimum performance circular curving slits should be used (Meltzer, 1969). In Figure 3.43, an Ebert–Fastie monochromator is shown coupled to a uv–visible absorption spectrometer.

(*d*) *Czerny–Turner monochromator.* The Czerny–Turner design is similar to the Fastie arrangement. However, the single large collimating mirror is replaced by two smaller mirrors. This design saves the expense of manufacturing the large Ebert mirror. Wavelength selection is again achieved by rotating the grating. In Figure 3.44, a Czerny–Turner monochromator is coupled to a flame atomic absorption spectrometer.

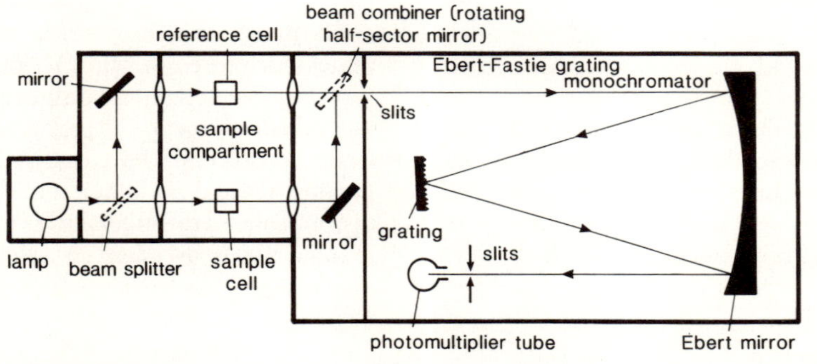

Figure 3.43 An Ebert–Fastie grating monochromator configured for uv–visible absorption spectrometry.

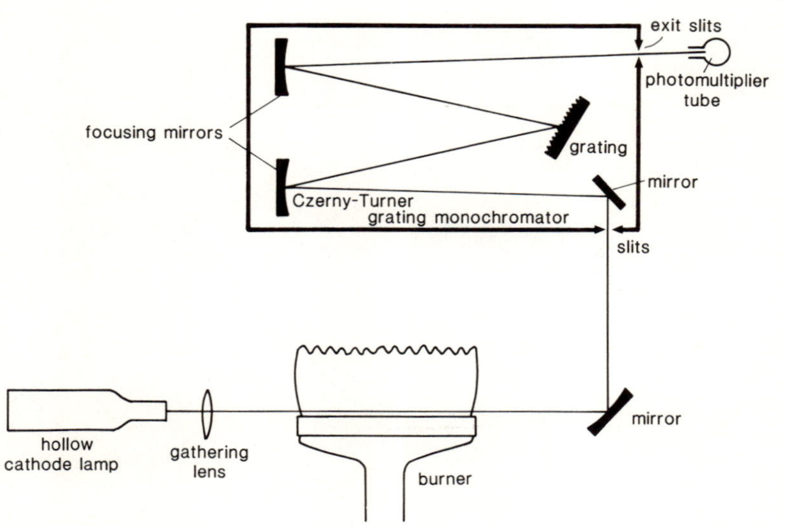

Figure 3.44 A Czerny–Turner grating monochromator configured for atomic absorption spectrometry. (After Varian Techtron, Victoria, Australia.)

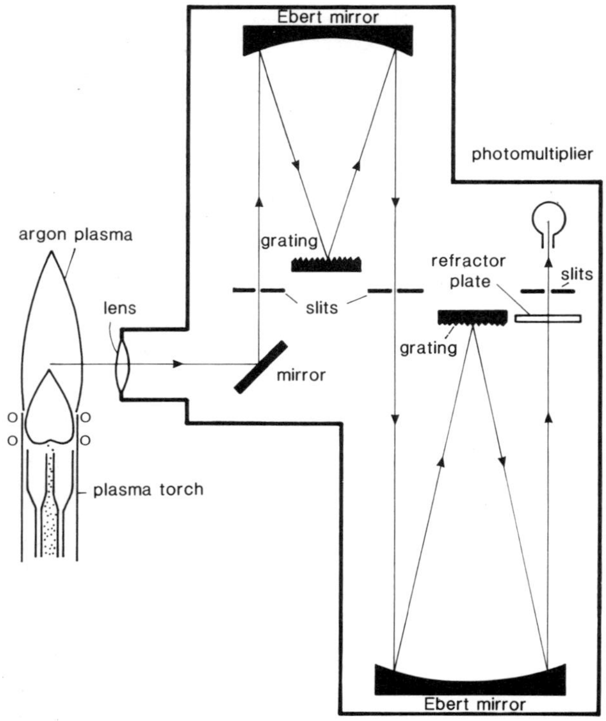

Figure 3.45 A double Ebert monochromator used for ICP–atomic emission measurements. (After Instrumentation Laboratory, Andover, Ma., USA.)

(*e*) *Double monochromators.* In more demanding applications, the performance of a single monochromator is not always adequate. A double Ebert–Fastie monochromator design is shown in Figure 3.45, used in conjunction with an ICP emission source. This arrangement is claimed to overcome two particular problems associated with ICP emission spectrometry. The first is that stray light effects from the intense emission lines of elements such as calcium are minimized; the second, that multiple-order interferences are effectively removed. This a particular advantage in analysing the complex spectra observed from emission sources. The first grating effectively behaves as a line order sorter and the second is designed to optimize resolution to about 0.02 nm.

(*f*) *Direct reading spectrometers.* In each of the monochromator designs discussed above, wavelength is selected by rotating the dispersing element (grating or prism). In routine operation, these instruments must be used in sequential mode, the monochromator being scanned from wavelength to wavelength, one after another, to complete a multi-element analysis. However, an alternative class of instrument exists, permitting simultaneous multi-element operation. This is based on principles set out by Rowland in 1882. Rowland disovered that if a concave grating is placed on the circumference of a circle (the *Rowland's circle*) on which are also placed the entrance slits of the monochromator, then all diffracted lines are also brought into focus on the Rowland's circle. The radius of curvature of the grating must be twice that of Rowland's circle. The concave grating thus disperses and focuses the diffracted beam and eliminates the need for collimating mirrors or lenses.

The most common arrangement of these principles used in optical spectroscopy is the *Paschen–Runge mounting* (Figure 3.46). In this design the grating and entrance slits are fixed and effectively define the circumference of the Rowland's circle. Individual wavelengths are then diffracted and focused at fixed points around the Rowland's circle. Such instruments were originally designed for use with photographic plates as detectors. However, in modern instru-

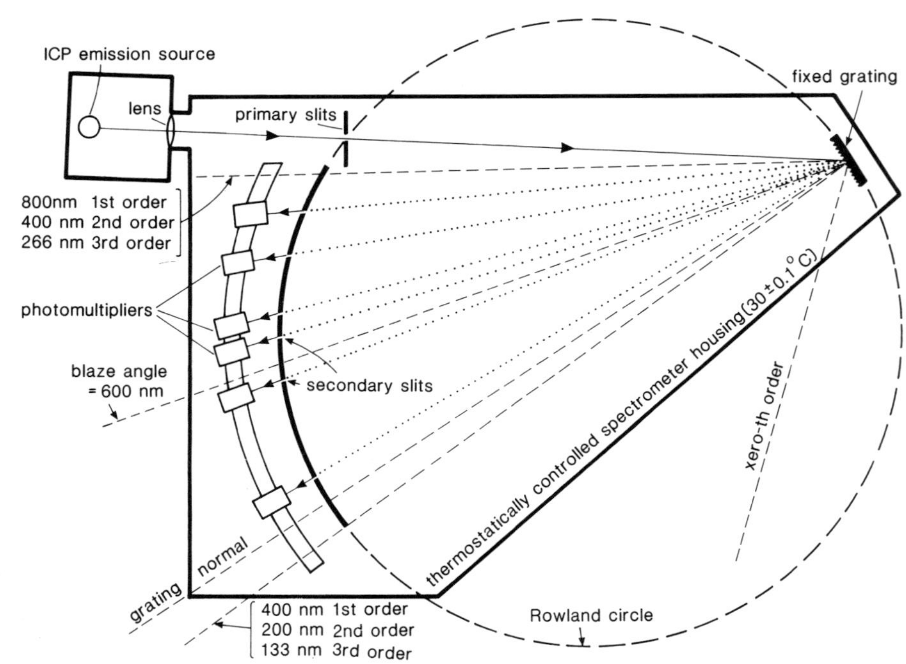

Figure 3.46 Paschen–Runge grating monochromator configured for ICP–atomic emission measurements. The spectrometer is designed as a direct-reader instrument and the diagram shows six photomultipliers adjacent to appropriate slits. (Based on Thompson and Walsh, 1983, p.40.)

ments, this design has been adapted for simultaneous multi-element analysis by replacing the photographic plate by a series of fixed exit slits, each placed on the Rowland's circle at the precise position of an atomic emission line of interest. Each exit slit has an associated photomultiplier tube. The number of analytical lines that may be measured simultaneously is then only limited by the number of fixed channels installed on the instrument. This arrangement is often referred to a 'direct' reader spectrometer and is widely used on ICP emission and arc/spark emission instruments. The fixed exit slits are sometimes complemented by a single movable slit and photomultiplier assembly to permit more flexible operation.

The diameter of the Rowland's circle used on most instruments is between 1 and 1.5m. Any distortion in the dimension of the monochromator, especially in one that incorporates multiple fixed exit slits, will cause a serious defocusing effect and loss in intensity. Such monochromators are, therefore, usually constructed on a rigid vibration-resistant optical bench and installed in a thermostatically controlled cabinet that permits control of temperature to a tolerance of no more than 30 ± 0.1°C.

The most serious technical deficiency of the concave grating monochromator is *astigmatism.* A point source at the entrance slit is diffracted and focused as a line on the Rowland's Circle. Diffracted beams are not, however, focused in the plane perpendicular to the optical axis. Inevitably, some loss in intensity occurs due to masking of the top and bottom of each diffracted line due to the finite height of the exit slits. This effect can to some extent be compensated by reflecting the source beam on to the grating using a concave mirror. This solution is a cheaper alternative to manufacturing a toroidal-shaped diffraction grating.

A further consequence of Rowland's-circle geometry is that if the central groove of the grating is cut at a blaze angle optimized for a particular wavelength, the edge grooves will be optimized for a slightly different angle, due to the divergent nature of the incident beam. This has the effect of spreading out the optimum wavelength response of the grating.

3.14 Stray light effects

Due to imperfections in the finish of optical components used in a monochromator, a small fraction of the source spectrum, irrespective of wavelength, is transmitted to the exit slits. This unwanted light therefore consists of wavelengths outside the spectral bandpass for which the monochromator grating has been set and can constitute a serious error in absorbance measurements. The difference between true and measured absorbance is plotted for a spectrometer having stray light contributions of 1%, 0.1% and 0.01% in Figure 3.47. As can be seen, the effect of 1% stray light causes increasingly large errors in absorbance measurements significantly above 1.0 units.

Stray light originates from imperfections in the grating, unwanted reflections from collimating mirrors and/or lenses and sometimes from light scattered off the walls of the monochromator housing. The amount of stray light can be reduced by inserting light baffles where appropriate inside

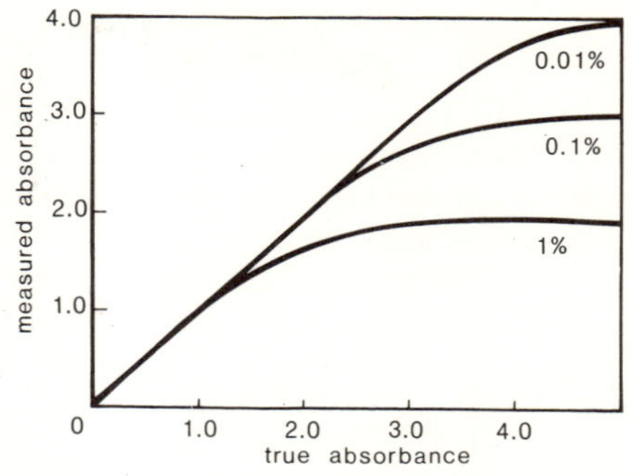

Figure 3.47 The effect of stray light on absorbance measurements assuming a contribution of 1%, 1.0% and 0.01% from stray light to the analyte signal. (Sharpe, 1984.)

the monochromator to intercept unwanted reflections, and by improvements in the manufacturing tolerances of optical components. Mirrors and gratings are generally coated with aluminium which, when new, has an extremely high reflectance. However, aluminium slowly oxidizes, increasing the potential for light scatter especially of radiation in the uv region. Sharpe (1984) has discussed this problem in some detail. Modern optical components are normally coated with magnesium fluoride (which is soft and not entirely chemically resistant) or with silica or fused quartz to overcome these oxidation effects. After this treatment, aluminium-coated surfaces show virtually no deterioration with age.

3.15 Errors in spectrometric measurements

In many analytical measurements (x-ray fluorescence spectrometry, for example), the higher the intensity of the signal, the smaller the relative statistical error in the recorded signal. This is, however, not the case in spectrophotometric absorbance measurements.

In order to understand this apparent contradiction, it is necessary to examine the sequence of operations that must be followed in setting up an instrument to make an absorbance measurement:

(i) Adjust the instrument to read 100% transmission
(ii) Adjust the instrument to read 0% transmission
(iii) Measure the transmission with the sample placed in the light path.

Instrumental errors are associated with each of these three operations.

3.15.1 *Sources of noise*

(*a*) *Shot noise.* At 100% transmission (absorbance = 0) the maximum intensity of radiation is transmitted by the monochromator and falls on the photomultiplier tube. Under these conditions, the precision of a measurement is limited by *shot noise* in the photomultiplier. As mentioned above, shot noise arises due to statistical variations in the number of electrons emitted by each dynode stage. In fact, these errors are dominated by fluctuations in emission from the photocathode and first dynode stage. Shot noise is a function of transmittance:

$$s = k_1\sqrt{(T^2+T)} \qquad (3.25)$$

where T is the % transmittance and k_1 a constant.
Errors in this reading will most seriously affect the measurement of low absorbance values.

(*b*) *Flicker effects*. A further source of noise that will also affect low-absorbance measurements is source flicker. Flicker noise arises from fluctuations in the intensity of the light source. This could be a tungsten lamp in a uv–visible spectrometer or a hollow cathode lamp in an atomic absorption spectrometer. Rothman *et. al.* (1975) estimate that flicker noise is proportional to the transmittance (T):

$$s = k_2T \qquad (3.26)$$

where k_2 is a constant.

The influence of flicker noise will be most serious in low-absorbance measurements; its influence is negligible for high-absorbance readings.

(*c*) *Other sources of noise*. Several other sources of noise can influence spectrophotometric measurements. Some of these have been discussed by Skoog and West (1980, p.159). The most serious of these are signal amplifier noise and limited readout resolution. Both can influence absorption measurements at low transmittance values. The effect of limited readout resolution is further considered below.

3.15.2 *Combined effect of noise*

Skoog and West (1980) have estimated the relative influence of these sources of noise (Table 3.9). This table itemizes the relative concentration uncertainty in absorbance or transmittance measurements subject to 0% T or Johnson noise, signal shot noise and flicker noise.

0% T or Johnson noise arises from uncertainties in making the 0% transmittance adjustment arising from drift in the dark current reading of the detector and the effect of temperature changes on detector and amplifier circuit response. Johnson noise arises from random fluctuations in the movement of electrons in electronic circuits.

These two effects are independent of transmittance reading, and a absolute uncertainty of $\pm 0.003\%$ transmittance has been assumed in calculating data in Table 3.9 (see Skoog and West, 1980, pp.158–159, for further details).

Signal shot noise arises from random fluctuations in signal gain in a photomultiplier detector, as discussed above. Data in Table 3.9 have been calculated using a value $k_1 = \pm 0.003$ in equation (3.25). Flicker noise arises from random fluctuations in the output of the light source, as discussed above. Data in Table 3.9 has been calculated assuming a value of $k_2 = \pm 0.003$ in equation (3.26).

The relative effects of these noise sources can be judged from data plotted in Figure 3.48. The combined effect of these sources of error is to give a U-shaped plot of relative error in measured signal *v*. absorbance reading, errors being at a minimum between absorbance readings in the range 0.2 to about 0.8.

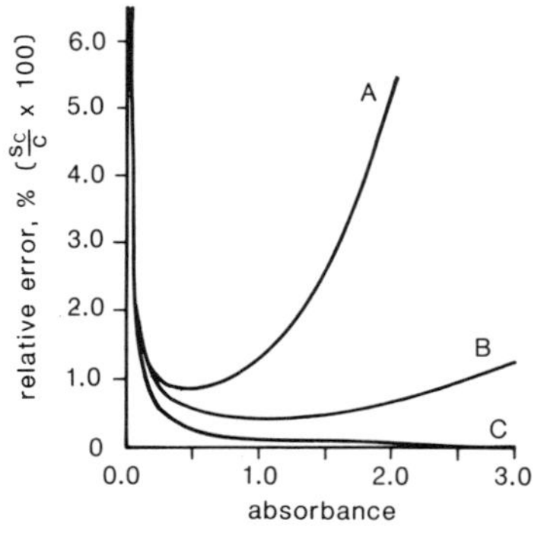

Figure 3.48 Error in concentration as a function of absorbance plotted from data listed in Table 3.9. Curve A: 0%T and Johnson noise; curve B: signal shot noise; curve C: flicker noise. (Skoog and West, 1980.)

Table 3.9 Error in concentration measurement caused by the effect of various sources of noise on transmittance and absorbance readings

Transmittance	Absorbance	Relative error in concentration (%) for signals limited by:		
		0% T or Johnson noise	Signal shot noise	Flicker noise
0.95	0.022	±6.2	±8.4	±5.8
0.90	0.046	±3.2	±4.1	±2.8
0.80	0.097	±1.7	±2.0	±1.3
0.60	0.222	±0.98	±0.96	±0.59
0.40	0.398	±0.82	±0.61	±0.33
0.20	0.699	±0.93	±0.46	±0.18
0.10	1.00	±1.3	±0.43	±0.13
0.032	1.50	±2.7	±0.50	±0.09
0.010	2.00	±6.5	±0.65	±0.06
0.0032	2.50	±16.3	±0.92	±0.05
0.0010	3.00	±43.4	±1.4	±0.04

Data from Skoog and West (1980).
See text for details of revised calculations.

3.15.3 *Limited readout capability*

The shape of this error curve (Figure 3.48) can be further compounded if the instrument readout system is not sufficiently sensitive. Rearranging Beer's Law (equation 3.10) in terms of concentration,

$$c = -\frac{1}{Cb}.\log(T) = -\frac{0.434}{Cb}.\ln(T) \qquad (3.27)$$

where C is the molar absorptivity, b is the path length of the sample cell, and T is the transmittance.

By differentiating this equation, it is possible to determine the rate of change in concentration with respect to changes in transmittance as follows:

$$\frac{1}{C}\cdot\frac{dc}{dT} = \frac{1}{T\ln(T)}. \qquad (3.28)$$

This function is plotted relative to concentration in Figure 3.49. At both low and high transmittance, a relatively large change in concentration is required to effect unit change in

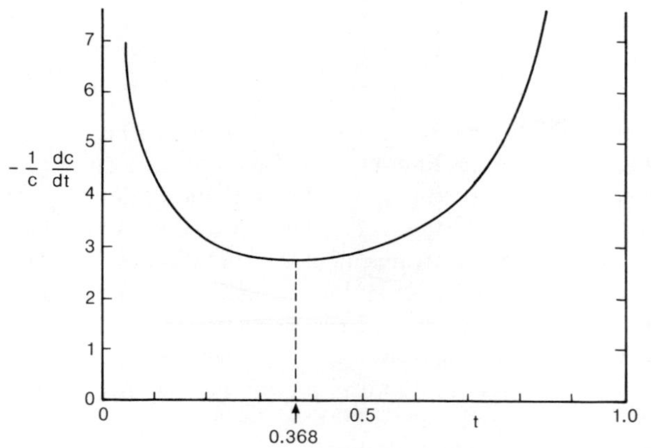

Figure 3.49 Relative error in concentration derived as a function of transmittance by differentiation of Beer's Law. (Walton and Reyes, 1973.)

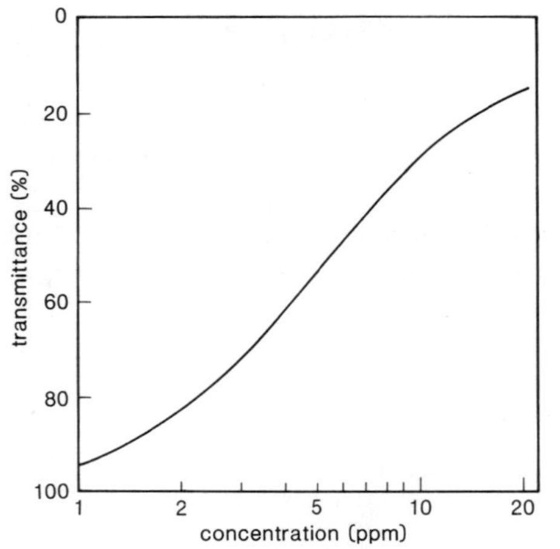

Figure 3.50 A Ringbom plot of concentration (log ppm) *v.* transmittance. The preferred analytical range is represented by the linear portion of the curve of maximum gradient.

transmittance. Put another way, the sensitivity of a spectrophotometric measurement (that is, the change in transmittance induced by unit change in concentration) is greatest at intermediate concentrations and achieves a maximum at 36.8% transmittance (0.434 absorbance units). In general it is therefore desirable to perform all absorption measurements in the relatively flat central region of the plot in Figure 3.49 between limiting values of about 15 and 65% transmittance (0.8 to 0.2 units of absorbance).

On some spectrophotometers, the uncertainty in measuring the readout signal is almost constant over the entire transmittance range and depends on the ability to discern small differences in the reading of a meter needle relative to a linear transmission scale. Under these circumstances the uncertainty in this reading will have a much larger effect on errors in absorbance and therefore concentration measurements at high and transmission values as indicated by the general trend of the error curve in Figure 3.49.

3.15.4 *Ringbom plot*

Analytical calibrations are normally plotted as dictated by Beer's Law, using a concentration *v.* absorbance scale (Figure 3.13). However, the variation in signal sensitivity indicated by equation (3.28) can be demonstrated if calibrations are plotted, as suggested by Ringbom (1939), using log (concentration) *v.* transmittance coordinates (Figure 3.50). This 'Ringbom' plot shows that optimum sensitivity occurs at the linear part of the calibration (steepest gradient). If Beer's Law is obeyed exactly, this linear calibration will correspond to the minimum error in absorbance measurement predicted by equation (3.28) to be between 15 and 65% transmittance (0.2 to 0.8 absorbance units).

In summary, therefore, noise in an absorbance signal arises from a combination of sources. First, the inherent instrument sensitivity is at a maximum over a range of absorbances from about 0.2 to 0.8 units (15–65% transmission). Secondly, unlike x-ray measurements, for example, the precision of high absorbance readings is significantly degraded by the effects of electronic noise in the amplifier and by the limited resolution of some readout systems. Thirdly, low absorbance readings are degraded by shot noise in the detector and flicker in the intensity output of the light source.

4 Atomic absorption spectrometry

4.1 Introduction

The phenomenon of atomic absorption has been known for a long time. As early as the first quarter of the nineteenth century, Wollaston and Fraunhofer observed lines in the spectrum of the Sun now known to be caused by the absorption of light by atomic vapour in the Sun's atmosphere.

However, there was considerable delay in the development of a practical instrument that could apply the phenomenon of atomic absorption to analyses in the laboratory.

Difficulties were caused by the narrowness of the line over which absorption measurements must be made. The peak width of an atomic absorption line is extremely small—approximately 0.002 to 0.005 nm. Absorption phenomena can be induced in atoms only by incident radiation of precisely the wavelength that falls within this very narrow envelope. Thus, when an atomic vapour is excited by a continuous light source, as in most other forms of uv–visible spectrometry, only a minute fraction of the source output is absorbed. Measurements would then have to be made using a highly tuned monochromator having a bandpass equivalent to the width of the atomic absorption line. Otherwise, the signal due to the small fraction of light absorbed would be swamped by the intense continum simultaneously transmitted by the monochromator (Figure 4.1). These problems defeated attempts to make practical atomic absorption measurements until the work of Walsh (1955) and, independently, of Alkemade and Milatz (1955).

Walsh's proposal was to replace a continuous light source with an atomic line source. Atomic spectra can be conveniently excited in a hollow cathode lamp, the cathode being coated with the element of interest. The sample was atomized from a prepared aqueous solution by nebulization in which the resultant sample aerosol was premixed with air and acetylene gases that were then burnt as a flame. Radiation from the hollow cathode lamp was passed through the flame and the degree of atomic absorption detected using a monochromator tuned to the wavelength of the atomic emission line (Figure 4.2). Using this arrangement, the monochromator need have only relatively low resolution. Indeed, the resolution must be adequate to separate adjacent lines in the atomic spectrum of the hollow cathode lamp, since, unlike a continuous radiation source, a large proportion of the exciting atomic line radiation falling within the bandpass of the monochromator now suffers absorption (Figure 4.3). This arrangement has been called the 'lock and key effect' since the hollow cathode lamp emits atomic emission radiation of a specific element which is only able to induce absorption by atoms of that element in the sample. The extremely high monochromator resolution otherwise required for atomic absorption measurements is effectively transferred to the light source.

Walsh's account, describing the early development of this form of instrumentation (Walsh, 1980), contains lessons for all who seek recognition for new analytical techniques. After a gestation period of several years, during which the practical applications of the idea were virtually ignored by the scientific community, its potential was recognized by a large instrument manufacturer. The technique then rapidly became established as perhaps one of the most widely applied methods of elemental analysis in solution. Atomic absorption spectrometry is indeed still one of the standard instrumental methods of analysing waters and process control samples. After commercial instrumentation became widely available in the early 1960s, atomic absorption was rapidly adapted for geological applications, virtually replacing classical and rapid schemes for the determination of the major and selected trace elements. However, in comparison with other modern instrumental methods, the technique suffers one important limitation: in general, only one element may be determined at once, since individual hollow cathode lamps are required for each element. This limitation has been overcome to some extent by the provision of multiple-lamp turrets, where several lamps can be powered simultaneously, permitting rapid exchange of one lamp for another in the analysis position. More sophisticated instruments have also been designed, incorporating two monochromators that permit simultaneous detection of absorption signals from two lamps. Despite these developments,

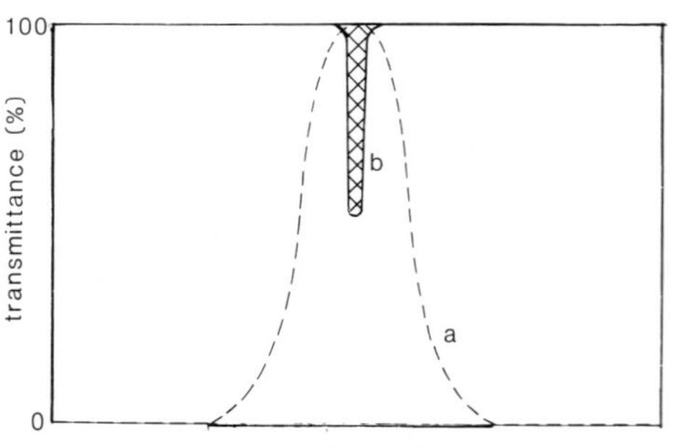

Figure 4.1 Atomic absorption using a continuum source. Curve *a* represents the range of wavelengths transmitted by a typical AA monochromator (FWHM = 0.2 to 1 nm). Curve *b* represents the small fraction of this wavelength range capable of inducing atomic absorption (FWHM typically 0.002 to 0.004 nm). The width of curve *b* has been exaggerated for clarity.

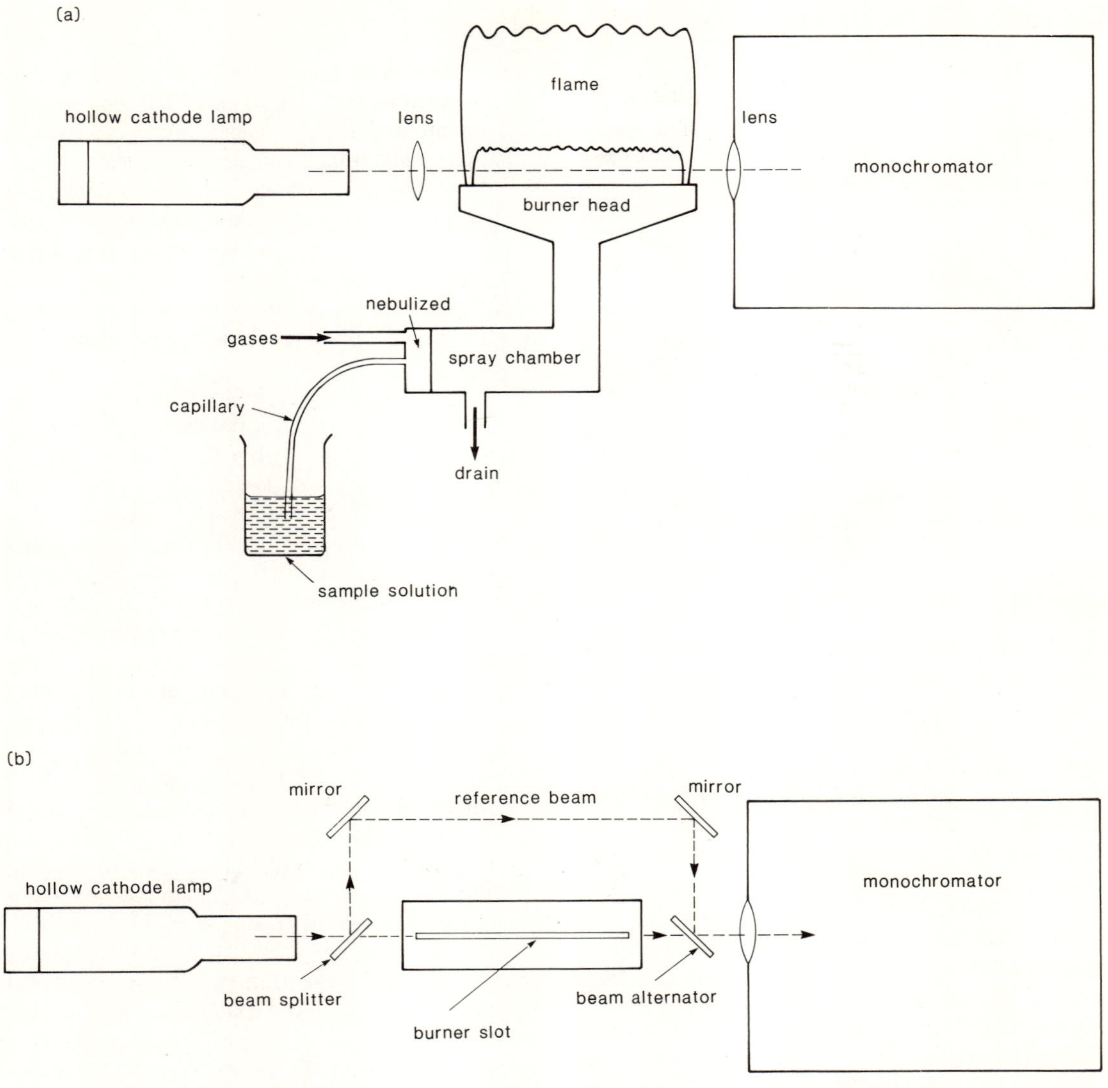

Figure 4.2 Typical instrumental configurations for atomic absorption spectrophotometry: (*a*) conventional flame AAS, (*b*) a double-beam flame AAS system viewed from above.

atomic absorption spectrometry cannot generally compete in speed with multi-element techniques such as ICP–AES or XRF analysis. The principal advantages of AAS remain the simplicity, and therefore low capital cost, of instrumentation and the relative freedom from spectral overlap interferences, a benefit resulting from the method of excitation using element-specific hollow cathode lamps. The most important geological application of AAS is probably not in routine major and selected trace-element analysis, but rather in more specialized applications in which flame atomization is replaced by graphite furnace or vapour generation attachments. In this way, exceptionally low detection limits of elements that are difficult to determine by other procedures may be achieved. These elements included arsenic, selenium and tellurium using a hydride generation device, mercury using a cold vapour generator and selected trace elements by graphite furnace atomization. These aspects of atomic absorption spectrometry are considered in more detail in sections 4.12–4.16.

4.2 Instrumentation

4.2.1 *General considerations*

An atomic absorption spectrometer consists of an atomic line source (normally a hollow cathode lamp but sometimes an electrodeless discharge lamp), a means of atomizing a sample and containing it in an atom cell (see below), a monochromator, a photomultiplier detector, and a readout system (Figure 4.2). The most widely used way of atomizing a sample solution and confining it as an atomic vapour for absorption measurements is by means of a flame. The flame is therefore used both to atomize the sample (using both the thermal and chemical energy of combusting gases) and to maintain a cell of atomic vapour within the light path of the instrument.

In electrothermal atomization using a hollow graphite furnace, samples are atomized by thermal energy caused by electrical resistance heating of a graphite tube. The resultant atomic vapour is held in the light path passing through the

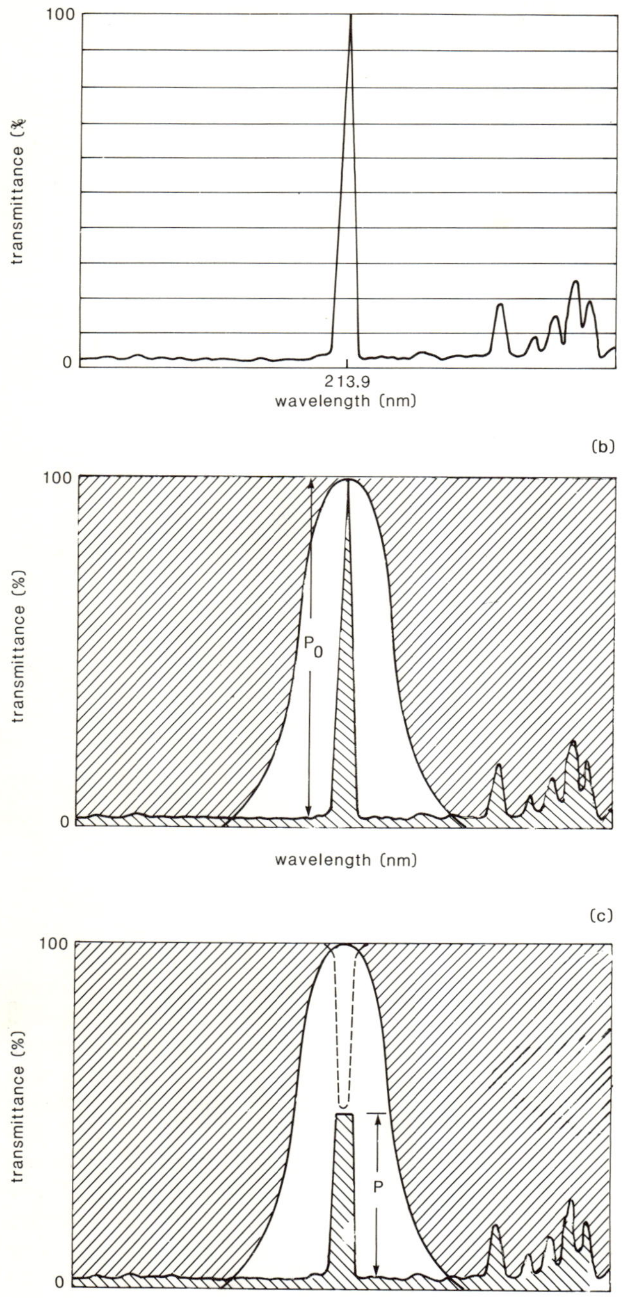

Figure 4.3 The lock and key effect in atomic absorption spectrophotometry. (*a*) The emission spectrum of a zinc hollow cathode lamp. (*b*) Wavelength bandpass of monochromator superimposed on zinc emission spectrum (from a hollow cathode lamp) showing how extraneous atomic emission lines are excluded from measurements. (*c*) Atomic absorption of the zinc emission line which causes the intensity of lamp radiation to be significantly reduced. (Ewing, 1975.)

centre of the hollow graphite tube for a short length of time and the transient absorbance signal measured before vapours are swept to waste. The components of an atomic absorption spectrometer required to make these measurements will now be discussed in more detail.

4.2.2 *Light sources*

(*a*) *The hollow cathode lamp.* The hollow cathode lamp (Figure 4.4) is designed to emit an intense atomic line spectrum of a specified element and must have operational characteristics that include a short warm-up time and stable light output over extended periods. Although not the only possiblility, the hollow cathode lamp is the most widely used light source for atomic absorption measurements, electrodeless discharge lamps having been developed for 'difficult' elements (see (*b*) below) where adequate light intensity cannot be obtained from an equivalent hollow cathode lamp.

The hollow cathode lamp consists of a glass envelope containing an inert gas at relatively low pressure. Two electrodes are arranged within the glass envelope; one is the anode, the other the cathode, which is machined as a hollow cylinder. The inside surfaces of the cathode are coated with a compound of the element from which the atomic emission spectrum is to be excited. The filling gas is normally neon or argon at a pressure of 5 to 10 Torr. When an electrical potential of several hundred volts is applied across the electrodes, ionization of the filling gas occurs. Cations of this gas migrate towards the cathode and, if the applied potential is sufficiently large, collide violently with the inner surface of the hollow cathode. This causes the element compound to be sputtered into the lamp atmosphere. In this way a cloud of atoms is created within the hollow cathode. Atomic emission of these atoms is excited by the energy of the electrical discharge. Since most of this discharge occurs within the hollow cathode, the lamp behaves as a compact and localized source of atomic emission radiation. Part of the atomic emission spectrum output from an iron hollow cathode lamp is shown in Figure 4.5. As optical emission depends, in part, on the sputtering action of the discharge, the hollow cathode coating will be progressively eroded and redeposited on cooler surfaces within the glass envelope. To prolong the effective life of these lamps, it is therefore desirable to limit the electrical discharge current to the lowest value that will give a stable light output of sufficient intensity. There is a further advantage in minimizing lamp current. At high discharge currents, the concentration of atomic vapour within the hollow cathode increases to such an extent that severe self-absorption can occur (section 3.7). That is, characteristic radiation emitted by excited atoms within the hollow cathode is re-absorbed by unexcited atoms in the same atomic vapour. The light output of the lamp can then be seriously diminished (Figure 3.14).

Hollow cathode lamps normally operate at a current of between 1 and 20 mA and an electrode potential of 100–300 V. The glass envelope is normally constructed of borosilicate glass but since this material does not effectively transmit radiation below 300 nm, a quartz glass end-window must be substituted for elements that emit atomic lines below this limit.

(*b*) *Electrodeless discharge lamp.* The hollow cathode lamp is a satisfactory light source for the majority of elements that may be determined by AAS. However, for a limited number of elements the intensity of atomic emission output can be significantly increased by using an electrodeless discharge lamp (Figure 4.6). The electrodeless discharge lamp

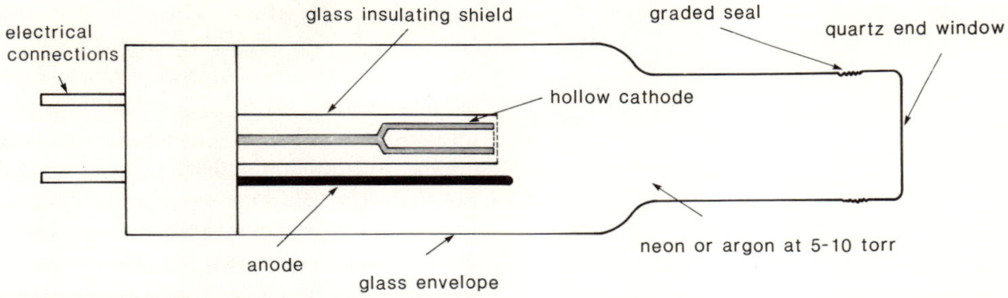

Figure 4.4 Schematic arrangement of a hollow cathode lamp. Only lamps emitting lines below 300 nm would be manufactured with a quartz glass end-window; borosilicate glass is satisfactory for higher wavelength lines.

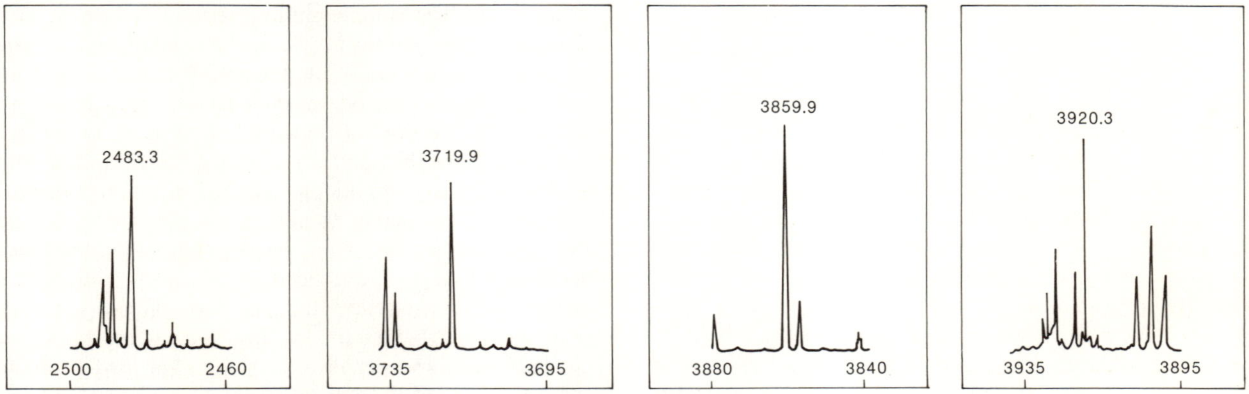

Figure 4.5 Some prominent lines in the iron emission spectrum from a hollow cathode lamp. (Varian Techtron Pty.)

consists of a sealed bulb containing a salt (often the iodide) of the element of interest in an atmosphere of argon at low pressure. This bulb is surrounded by a ceramic former around which is wound a copper coil. A radiofrequency signal (normally in the industrial frequency waveband at 27.12 MHz) is fed into this induction coil which is designed to deliver electrical power of up to 30 watts to the glass bulb. Under these conditions the salt in the bulb is atomized and excited and emits its characteristic atomic line spectrum. The radiofrequency coil must be enclosed in a metal envelope (with a quartz end-window) to minimize leakage of radiofrequency radiation.

The emission output from an electrodeless discharge lamp is in general more intense than that from an equivalent hollow cathode lamp, the former having a longer potential operating life. However, at the present stage of development electrodeless discharge lamps are notably less stable than hollow cathode lamps, requiring warm-up times of up to 30 minutes.

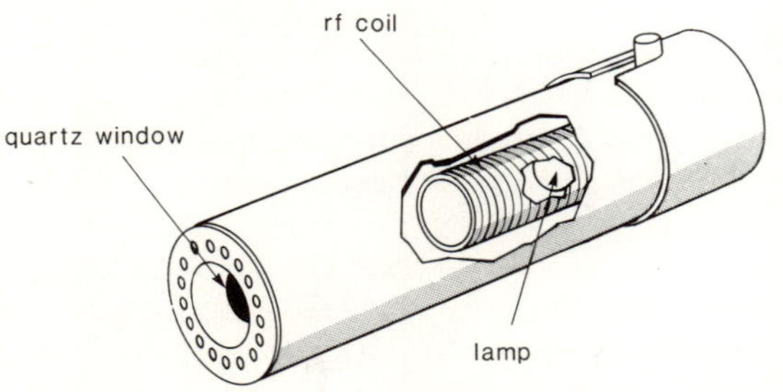

Figure 4.6 Schematic diagram of an electrodeless discharge lamp. (Barnett, 1973.)

Elements for which electrodeless discharge lamps are available include As, Bi, Cd, Hg, P, Pb, Sb, Se, Sn, Te, Tl and Zn. Estimates of the increased light output of these devices compared with hollow cathode lamps are shown in Table 4.1. In atomic absorption applications, a more intense light source will in general reduce baseline noise and therefore improve potential detection limits. However, notable improvements in sensitivity using electrodeless discharge lamps have been observed only in the atomic absorption determination of As, Hg and Se; increases in sensitivity over hollow cathode lamp excitation are reported by one manufacturer (Varian Techtron) to be about >100, 50 and 20% respectively. Electrodeless discharge lamps also have applications as light sources in atomic fluorescence analysis.

4.2.3 *Nebulizer design*

The function of the nebulizer is to convert a sample solution into an aerosol mist, and mix it with the oxidant and fuel gases suitable for burning as a flame. In other applications, this process might be called atomization. However, in atomic spectrometry, *nebulization* is the term used to describe the

Table 4.1 Relative intensities of electrodeless discharge lamps

Element	Power (W)	Approximate gain in intensity relative to a hollow cathode lamp
As	7	10
Bi	8	5
Cd	7	20
Hg	4	100
P	6	—
Pb	11	10
Sb	10	3
Se	5	10
Sn	8	10
Te	7	2
Tl	11	20
Zn	7	30

Data from Varian catalogue, 'Light sources for atomic absorption spectrometry', 1985.

physical process whereby a solution is broken up into aerosol droplets, whereas *atomization* is reserved for release of atoms resulting from the chemical dissociation of salt particles that remain after evaporation of solvent from these aerosol droplets in the flame. Nebulizers are not used exclusively for AA instrumentation, being also an integral part of some flame photometers and inductively coupled plasmas (used for both atomic emission and mass spectrometry measurements). Different designs are appropriate for different instruments and are discussed further under the appropriate section headings in Chapter 2 and 5.

(*a*) *Development of the AA nebulizer*. Early flame photometers used the *total consumption burner* (section 2.6). The tip of this burner was designed so that the venturi effect caused by the flow of fuel and oxidant gases through the adjacent orifice caused the sample solution to be sucked from a capillary tube directly into the flame. The atomic emission signal observed from this flame was very noisy, mainly because of the turbulent flow of the gas mixture and because there was little control over the size of droplets entering the flame. This led to erratic rates of desolvation and atomization. The *Lundegardh premix burner* (section 2.6) was an important advance in overcoming these difficulties. This design incorporated a spray chamber into which the sample solution was nebulized by the venturi effect of the oxidant gas. A large proportion of the droplets so formed (often 95%) coalesced, condensed on the walls of the spray chamber and drained away to waste. Those droplets that were left were small and of much more uniform size. They were swept into the flame and atomized at a more uniform rate. As a result, the signal from a premix burner was considerably more stable and free of noise, more than compensating for the lower nebulization efficiency compared with the total consumption burner.

(*b*) *Modern premix burners: the pneumatic nebulizer*. The principles of the Lundegardh burner are retained in the pneumatic nebulizer used on modern AA instruments (Figures 4.7, 4.8). The sample solution is nebulized into the spray chamber by the venturi action of the oxidant gas. In order to promote formation of an aerosol of uniform size, some nebulizer designs include a glass impact bead positioned in line with the flow of nebulized solution. This has two functions. First, the impact bead shatters some of the droplets into a fine aerosol (secondary fragmentation); secondly the impact bead serves to collect the larger droplets and efficiently remove them from the gas mixture. In other pneumatic nebulizer designs, the impact bead is omitted and baffles are inserted into the gas flow within the spray chamber. These baffles are designed to cause the gases flowing through the spray chamber to rotate. This both promotes efficient mixing of the aerosol and creates centrifugal forces that cause the larger aerosol droplets to be thrown against the walls of the chamber, so removing them from the gas mixture. In either design the efficiency of nebulization is only about 10% maximum, 90% of the aspirated solution draining to waste.

The process of nebulization is not well understood. The sample solution is shattered by the effect of a high-velocity jet of gas in an adjacent venturi (Figure 4.7). Browner and Boorn (1985) suggest that the mechanism may involve some form of surface stripping such that successive thin surface films of liquid sucked out of the capillary are removed by the gas flow and then spontaneously collapse into droplets under the influence of surface-tension forces. The resultant size distribution of aerosol droplets is shown in Figure 4.9.

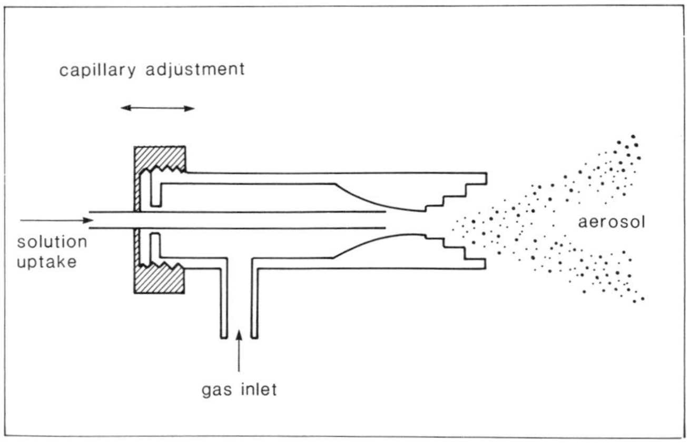

Figure 4.7 Aerosol production by a pneumatic nebulizer. (Browner and Boorn, 1984.)

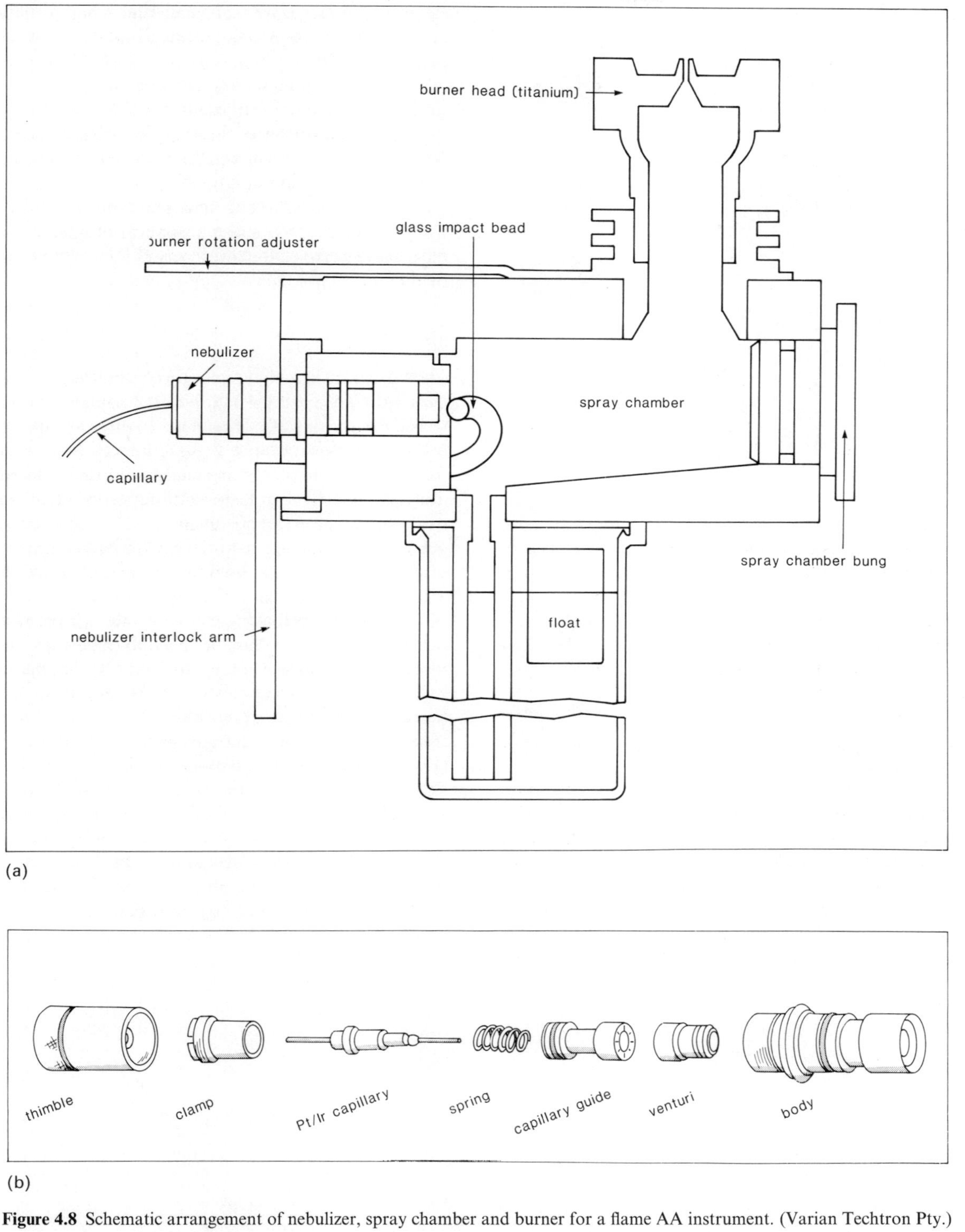

Figure 4.8 Schematic arrangement of nebulizer, spray chamber and burner for a flame AA instrument. (Varian Techtron Pty.)

In early AA instruments, a rather coarse range of particle size was generated by the nebulizer and it is now apparent that this was the cause of some of the matrix interference effects (e.g. silicon–aluminium) that affected earlier schemes of analysis. Such interferences can be virtually eliminated by appropriate modifications to nebulizer design, shifting the average aerosol particle size to smaller values (Browner and Boorn, 1984). Cresser and Browner (1980) described a method for measuring the size distribution of droplets in the range 0.5. to 10 μm produced by a pneumatic nebulizer, and quote earlier work by Willis (1967), which suggested that more than 90% of the droplets reaching burners have a size less than 5 μm in diameter.

One of the important analytical characteristics of a nebu-

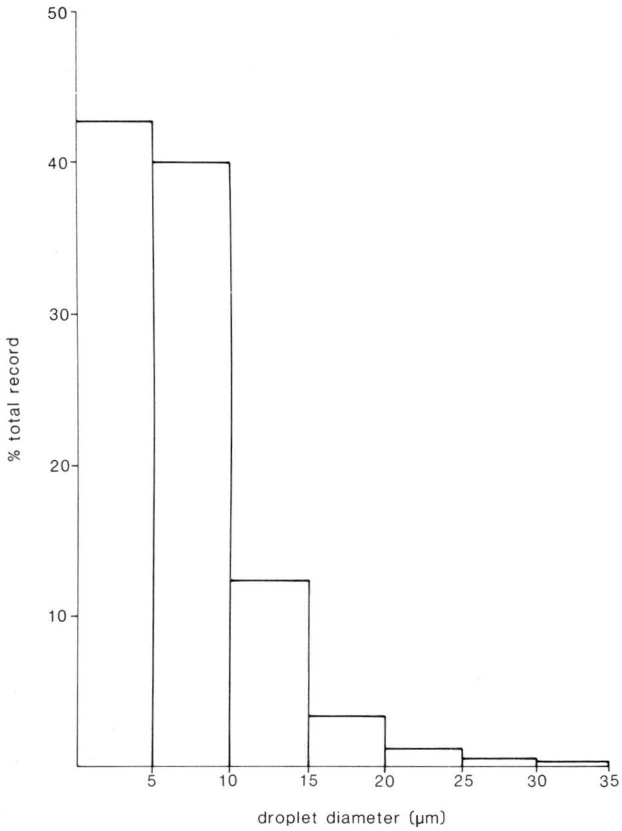

Figure 4.9 Distribution of aerosol droplets produced by a pneumatic nebulizer fitted with an impact bead but no spray chamber spoilers. (Morton, 1985.)

lizer is the effect of variations in sample uptake rate on the magnitude of the absorbance signal. The data in Figure 4.10 indicate that for a typical atomic absorption nebulizer there is an optimum sample uptake rate of about 4 mL per minute, but with a relatively flat response between 3 and 6 mL per minute. This suggests that nebulizer adjustment is not critical for optimum operation, and that small variations in sample uptake rate will not critically affect the accuracy of calibrations.

Samples for analysis are prepared in a variety of acid solutions, and so it is important that the nebulizer assembly should be resistant to corrosion. The nebulizer capillary is normally made of platinum–iridium alloy and the spray chamber from an inert plastic polymer.

Alternative nebulizer designs are appropriate for inductively coupled argon plasma applications because of the different gas flow characteristics, and these will be discussed further in Chapter 5.

4.2.4 *Burner design*

The burner most widely used in AA applications is the *slot burner* design (Figure 4.11). This consists of a narrow slot whose dimensions are determined by the gas mixture to be burnt. For slow-burning gases (air–acetylene or air–propane) a 100 mm slot is normally used (0.3 mm wide for acetylene flames). For faster-burning gases, such as nitrous oxide–acetylene, a slot 50 mm long by 0.5 mm wide provides greater protection against blowback. The burning velocities of the most commonly used flame gases are listed in Table 4.2. If the velocity at which gases are expelled from the burner slot falls below these critical values, a blowback will occur. Burner/spray chamber assemblies always, therefore incorporate blow-out plugs to prevent the build-up of explosive gas pressures under such circumstances. For similar reasons, nebulizer drains must be provided with a water trap to prevent the escape of flammable nebulizer gases (Figure 4.8). Particular attention is paid to by manufacturers to the design of the burner jaws. It is desirable to obtain a stable flame even when analysing solutions of high salt content (up to perhaps 10%). However, turbulent flow through the burner jaws, particularly at high gas flow rates, together with the effect of any manufacturing imperfections, tends to cause nucleation of salt particles that can then clog the

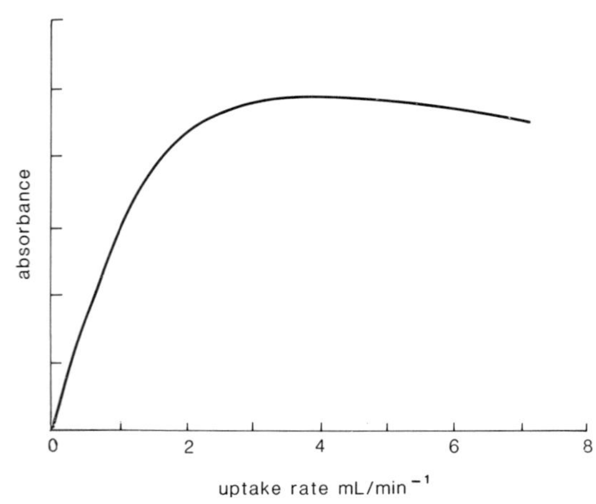

Figure 4.10 Absorbance signal from a pneumatic nebulizer as a function of solution uptake rate. (Price, 1979.)

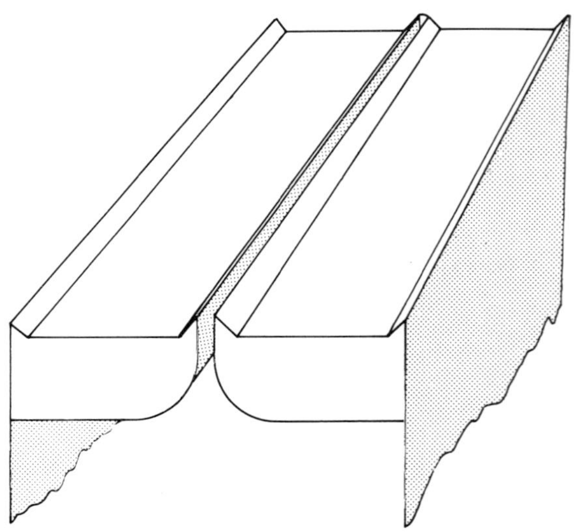

Figure 4.11 Cut-away diagram of an AA burner showing the profile of the slot and raised jaws. (Ebdon, 1982.)

Table 4.2 Details of flames used for atomic absorption analysis

Gas mixture	Flow rates (L/min) Fuel	Oxidant	Approximate temperature (K)	Flame speed (cm/s)
Air–propane				
fuel-lean	0.3	8		
stoichiometric	0.3–0.45	8	2200	45
fuel-rich	0.45	8		
Air–acetylene				
fuel-lean	1.2	8		
stoichiometric	1.2–1.5	8	2450	160
luminous	1.5–1.7	8		
fuel-rich	1.7–2.2	8	2300	
Nitrous oxide–acetylene				
fuel-lean	3.5	10		
stoichiometric	3.5–4.5	10	3200	285
fuel-rich	4.5	10		
Air–hydrogen				
stoichiometric	6	8	2300	320
Nitrous oxide–hydrogen				
stoichiometric	10	10	2900	380
Nitrous oxide–propane				
stoichiometric	4	10	2900	250

Data from Gaydon and Wolfard (1960); Aldous *et al.* (1972); Willis (1968); as reproduced by Price (1979, p. 29).

burner. Turbulent gas flow can be eliminated both by streamlining the base of the burner to promote laminar flow of gases through the slot and by manufacturing the jaw assembly to a polished finish. In order to prevent corrosion during use, the burner jaws are often manufactured from tantalum. One problem sometimes encountered when using a nitrous oxide–acetylene flame is the partial decomposition and oxidation of acetylene at the base of the flame, resulting in the clogging of the slot with carbon particles. In one design (Figure 4.12) these problems have been overcome by careful attention to burner design. The jaws of the slot are raised to promote circulation of air round the base of the flame so that any carbon deposits will be readily oxidized. The mass of metal within the jaws is made as small as possible, so that the high temperatures attained during operation will tend both to burn away incipient carbon deposits and to preheat the gas mixture before it enters the flame. The body of the burner is designed with cooling fins, to prevent overheating and promote stable operation.

Alternative burner designs have been developed for more specialized applications, the most important being the *Bowling burner*. This design incorporates three parallel slots and has a greatly reduced tendency to clog when solutions of high salt content are aspirated.

4.2.5 *Overall spectrometer design*

Several other features must be considered in the overall design of an atomic absorption spectrometer.

(*a*) *Modulation of lamp output.* In uv–visible instruments, sample solutions are normally measured in a light-tight sample compartment to prevent interference from stray light. This is clearly not practical in flame atomic absorption spectrometry. Not only ambient light, but also radiation emitted by the flame itself, enter the monochromator from the burner compartment. It is obviously necessary to distinguish between these unwanted sources of radiation and the atomic signal derived from absorption of the hollow cathode lamp emission. This is usually achieved be modulating the lamp output by an ac signal. This is easily obtained using the 50 or 60 Hz mains signal, although in some instruments 285 Hz is preferred since this is not a harmonic of the mains frequency and will not cause potential electronic interferences. The signal used to pulse the lamp output is also used to trigger the photomultiplier signal readout circuits. In this way it is possible to discriminate between the ac modulated photomultiplier signal (the desired atomic absorption signal) and a dc component (due to ambient light and flame emission). The direct influence of this dc component is thus eliminated. It is, however, desirable to minimize excess ambient illumination, since this increases

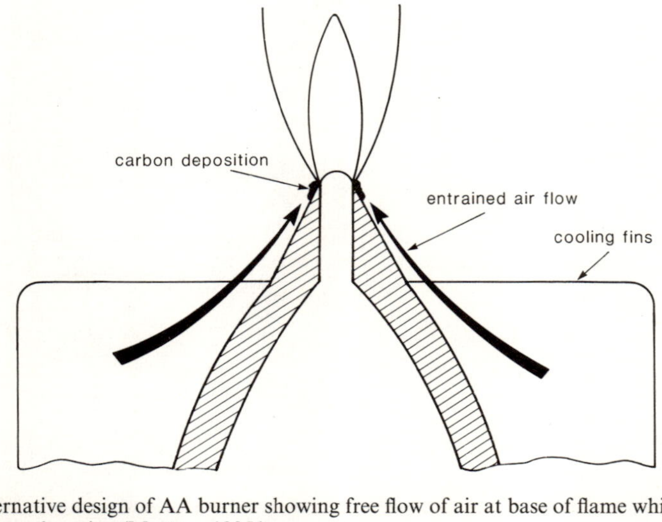

Figure 4.12 Alternative design of AA burner showing free flow of air at base of flame which minimizes the build-up of carbon deposits. (Morton, 1985.)

the shot noise uncertainty in the photomultiplier output (see section 3.15).

(*b*) *Double- or single-beam instruments*. Early AA instruments were designed with single optical beams so that the entire output from the lamp source passed through the flame and was measured by the monochromator readout system. However, subsequent improvements in the nebulizer/burner design, particularly those resulting from the introduction of support gas to stiffen the flame, meant that the most noisy component in the measurement chain became short-term drift in the hollow cathode lamp output. This could with advantage be reduced by a double-beam instrument design (Figure 4.2). In such instruments the lamp source beam is split, one beam passing through the flame as in conventional instruments, the other bypassing the flame before being recombined with the first for measurement by the monochromator/readout system. The atomic absorption signal is then measured relative to the reference beam. Since drift in the lamp output affects both beams this source of error is eliminated. Thus, Kahn (1968) showed that within a minute of applying power to a hollow cathode lamp, stable AA measurements could be made with a double-beam instrument even though the lamp output was not then fully stable.

However, double-beam instruments are not without disadvantages. In particular, (i) beam splitting causes a 50% reduction in the useful radiation passing through the flame, and (ii) noise in atomic absorption measurements due to flicker in the flame itself cannot be compensated for in double-beam instruments. Indeed, in modern instruments that incorporate lamp power supplies of high stability, flame flicker is probably the main source of signal noise. The superiority of double-beam instruments is not as clearcut. Although double-beam instruments will eliminate instrumental electronic drift and have the potential for yielding data of higher precision, single-beam instruments are optically more efficient, have the potential for yielding better detection limits, and are preferable for graphite furnace atomization where small transient signals must be measured.

4.3 Properties of flames

(*a*) *Gas mixtures*. The properties of gas mixtures in flame atomic absorption analysis are listed in Table 4.2. The most widely used gas mixtures are air–acetylene and nitrous oxide–acetylene. The air–propane flame burns at a lower temperature than the acetylene flames and has been proposed as being suitable for the determination of readily ionized alkali metals. However air–propane is rarely used in modern schemes of analysis, not the least because this flame is very prone to flicker in even slight draughts and is too cool to promote efficient atomization of the sample aerosol. The air–hydrogen flame finds a more specialized use in atomic fluorescence analysis.

(*b*) *The air–acetylene flame*. The air–acetylene flame was originally proposed for atomic absorption measurements by Walsh (1955) and is still one of the most widely used gas mixtures. The flame temperature is relatively high, so that the sample is efficiently atomized and burns in a stable blue flame. The gas mixture may be adjusted so that *stoichiometric* proportions of gases are burnt in the flame, exactly balancing the equation

$$2C_2H_2 + 5O_2 \rightarrow 4CO_2 + 2H_2O + \text{heat}. \qquad (4.1)$$

The stoichiometric flame burns as a stiff blue flame with a pale blue inner cone 2–4 mm high (Figure 4.13). This flame is widely used for the determination of many elements analysed by AAS. If the proportion of fuel to air is increased, the flame structure changes slightly. The inner cone increases in length and the flame becomes luminescent due to the presence of incandescent particles of carbon burning in the outer in the outer layers of the flame. These *fuel-rich* conditions are particularly favourable for elements that form stable oxides, such as Si, Cr, Mo, Al and Sn, although the presence of excess acetylene causes some reduction in flame temperature. If the proportion of acetylene in the flame is reduced, oxidizing conditions prevail. Under such *fuel-lean* conditions, elements that form refractory oxides do not

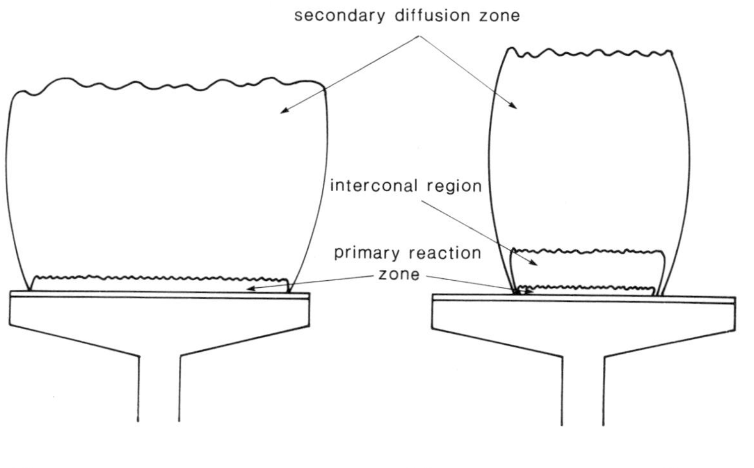

Figure 4.13 Regions in the flame produced by combustion of (*a*) air–acetylene (100 mm burner slot) and (*b*) nitrous oxide–acetylene (50 mm burner slot).

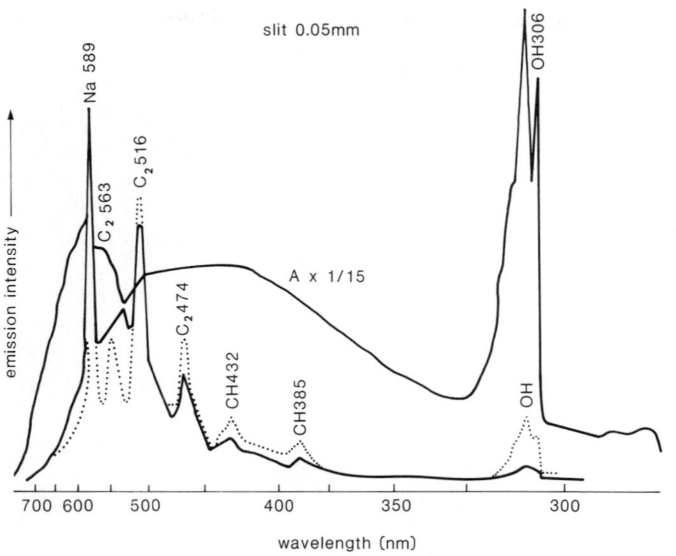

Figure 4.14 Background emission from an air–acetylene flame. Curve *A* represents a conventional flame, curve *B* represents a separated flame in which the emission from the secondary diffusion zone has been suppressed. The intensity of curve *A* has been reduced by a factor of 15 to bring it onto the same scale. The dotted line represents the effect of reducing the acetylene flow rate on the separated flame. (Price, 1979.)

break down into atomized particles efficiently. If the acetylene gas flow is reduced further, insufficient fuel is available to continue flame propagation and the flame lifts off from the burner slot and is extinguished.

(*c*) *Flame structure of an air–acetylene flame*. Two regions can be distinguished in an air–acetylene flame: an inner cone comprising the primary reaction zone and an outer cone comprising the secondary reaction or diffusion zone (Figure 4.13). In the *primary reaction zone*, acetylene and oxygen react by a series of chemical reactions that involve the breakdown of acetylene (C_2H_2) and oxygen (O_2) molecules into a series of *free radicals* including $H^{\bullet}$, $O^{\bullet}$ and $HO^{\bullet}$. A free radical is a highly reactive fragment of a molecule that has a transient lifetime in the course of a chemical reaction, but which is incapable of independent stable existence. The principal reaction products in this primary reaction zone are H_2, CO and H_2O.

In the *secondary reaction zone*, these reaction products combust further with oxygen that diffuses into the flame from the surrounding atmosphere to form the final reaction products H_2O and CO_2. These reactions cause a strong background emission signal, which is undesirable since it contributes significantly to noise in the atomic absorption signal (Figure 4.14). If the gas mixture is adjusted so that excess acetylene is present (a fuel-rich flame), $C_2^{\bullet}$ and $CH^{\bullet}$ radicals are formed within the primary reaction zone and may be detected by their respective emission bands. This can be seen in Figure 4.14, where the emission spectrum from a normal air–acetylene flame is compared with that from a 'sheathed' flame in which the secondary diffusion zone is suppressed. The contribution that this latter zone makes to overall flame emission can be seen to be significant.

The structure of a typical flame can be judged by analogy with the temperature contour profile of a natural gas–air flame (Figure 4.15). The relatively cool inner plume is surrounded by a much hotter secondary diffusion mantle. An understanding of the flame chemistry and temperatures is important in evaluating the sample atomization processes within the flame. Rann and Hambly (1965) have measured the distribution of selected atoms within both fuel-rich and fuel-lean air–acetylene flames. Their data for copper and calcium are shown in Figure 4.16. As can be seen from these data, gas composition as well as temperature profile within the flame can significantly influence atomic distributions, and therefore optimum analysis conditions, in the flame.

(*d*) *The nitrous oxide–acetylene flame*. The nitrous oxide–acetylene flame was developed by Willis (1965) to overcome some of the interference effects observed in an air–acetylene flame. This gas mixture burns at a much higher velocity and significantly higher temperature than the air–acetylene flame (Table 4.2). This causes some operating difficulties, since it is possible unintentionally to select gas flow conditions under which the speed of flame propagation exceeds the speed at which gases are expelled from the burner slot. The flame will then flash back into the nebulizer chamber. For this reason, nitrous oxide–acetylene flames are normally burnt using a burner with a 50 mm (not a 100 mm) slot. Furthermore, the ignition (and extinguishing) procedure must be carefully followed. A normal air–acetylene flame is first ignited at the burner. The acetylene flow is then increased until soot particles are observed in the flame. The oxidant gas supply is then switched over from compressed air to nitrous oxide. The flame is extinguished by following the reverse sequence.

Nitrous oxide and acetylene react in a stoichiometric flame as follows:

$$2C_2H_2 + 10N_2O \rightarrow 4CO_2 + 2H_2O + 5N_2. \quad (4.2)$$

Three regions can be distinguished in the flame (Figure 4.13). The *primary reaction zone* is pale blue in colour and 2–3 mm

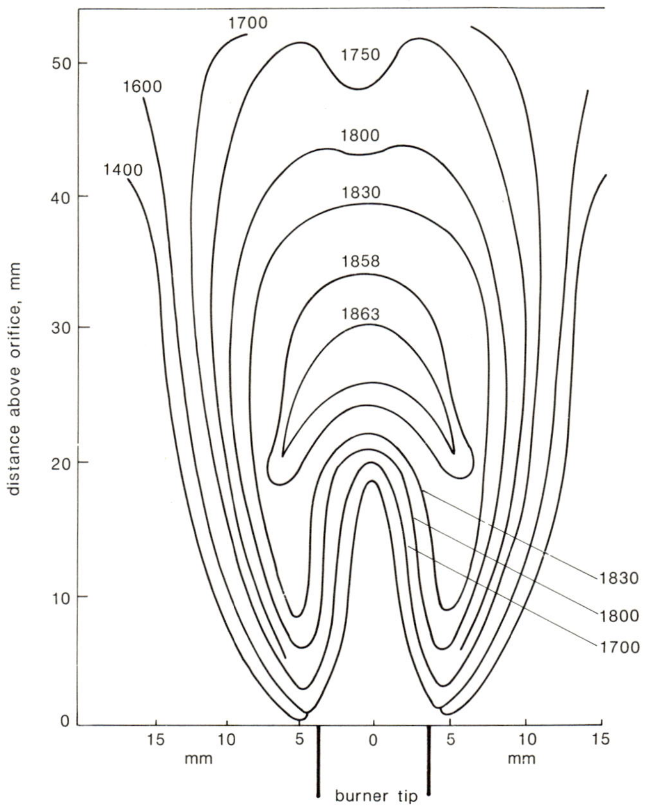

Figure 4.15 Distribution of temperatures in a cross-section through a natural gas–air flame (temperatures in °C). (From Lewis and van Elbe, 1943, as reproduced by Skoog and West, 1980.)

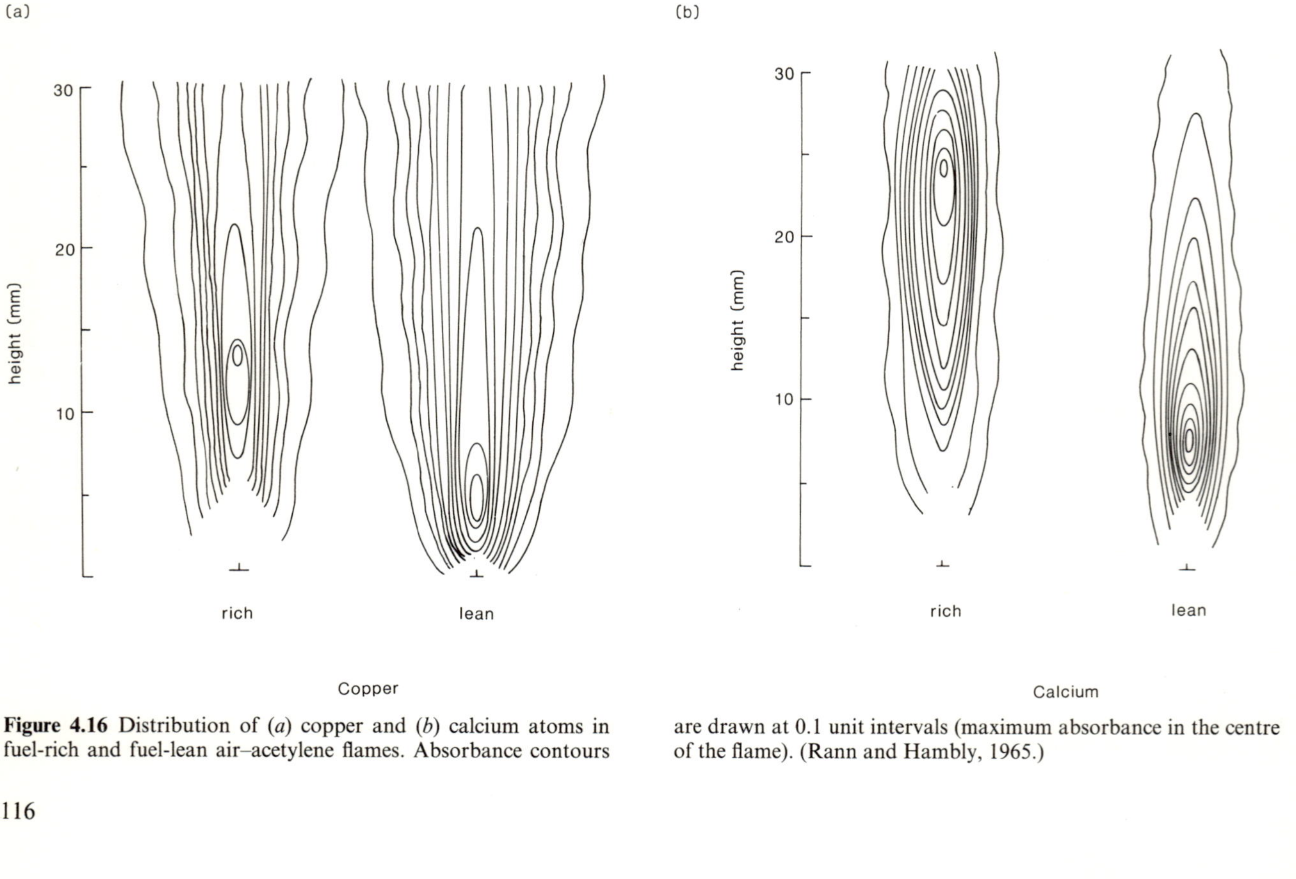

Figure 4.16 Distribution of (*a*) copper and (*b*) calcium atoms in fuel-rich and fuel-lean air–acetylene flames. Absorbance contours are drawn at 0.1 unit intervals (maximum absorbance in the centre of the flame). (Rann and Hambly, 1965.)

high. However, in contrast with the air–acetylene flame, this is surrounded by a red *interconal region* (the 'red feather') which increases in height as the fuel ratio is increased. This in its turn is surrounded by a blue *secondary diffusion zone*. By studying the emission bands found in the red interconal region, the presence of the $CN^{\cdot}$ and $HN^{\cdot}$ radicals has been established. It has been proposed that one of the reasons for the success of the nitrous oxide–acetylene in the determination of some elements is the highly reducing enviroment caused by the presence of these radicals. The elements that form refractory oxides, e.g. silicon, titanium, etc., are efficiently atomized in a nitrous oxide–acetylene flame due both to the high flame temperature and the scavenging action of these radicals towards oxygen. As can be seen from Figure 4.17, strong emission bands are observed from the nitrous oxide–acetylene flame.

4.4 Flame chemistry and atomization interferences in the flame

The key to successful atomic absorption is an understanding of the processes and reactions that occur when a nebulized solution is swept into the flame. Unfortunately, flame chemistry is rather complex, involving as it does free radical reactions of the flame gases. The flame is in fact a dynamic reaction cell, the residence time of individual atoms passing through the optical path within the flame being very short (about 10^{-4}s). The sequence of processes affecting a sample aerosol as it enters the flame may be visualized as shown in Figure 4.18. As the nebulized sample enters the flame, the rise in temperature first causes the solvent to evaporate ('desolvation') leaving solid microparticles of inorganic salt. The rapid rise in temperature then causes these salts either to sublime or to evaporate into molecular fragments that rapidly dissociate into atoms ('atomization').

The rate at which atomization processes occur is strongly dependent on the initial particle size distribution of aerosol droplets. Optimum atomic absorption signals are observed when these droplets are of uniform size and very small ($>5\mu$m).

The conventional theory of atomization involves the assumption of *dissociation equilibrium*. The generation of free atoms in the flame is controlled by a dissociation equilibrium involving oxide, hydroxide, hydride or halide:

$$MA \rightleftharpoons M^{\cdot} + A^{\cdot} \qquad (4.3)$$

where $M^{\cdot}$ is the metal atom and $A^{\cdot}$ is the oxide, hydroxide, etc., radical.

Halls (1977*a*) has critically examined this theory and shown that its conclusions are not always in agreement with actual observations made in an air–acetylene flame. Halls considered that reactions with flame constituents including CO, H_2 and the radicals $H^{\cdot}$, $C_2^{\cdot}$, $C^{\cdot}$ and $CH^{\cdot}$. Experimental evidence supported the view that $H^{\cdot}$ radicals were the species involved in reduction, and that atom production was controlled by fast binary reactions involving $H^{\cdot}$ radicals such as

$$MO + H^{\cdot} \rightleftharpoons M^{\cdot} + HO^{\cdot}. \qquad (4.4)$$

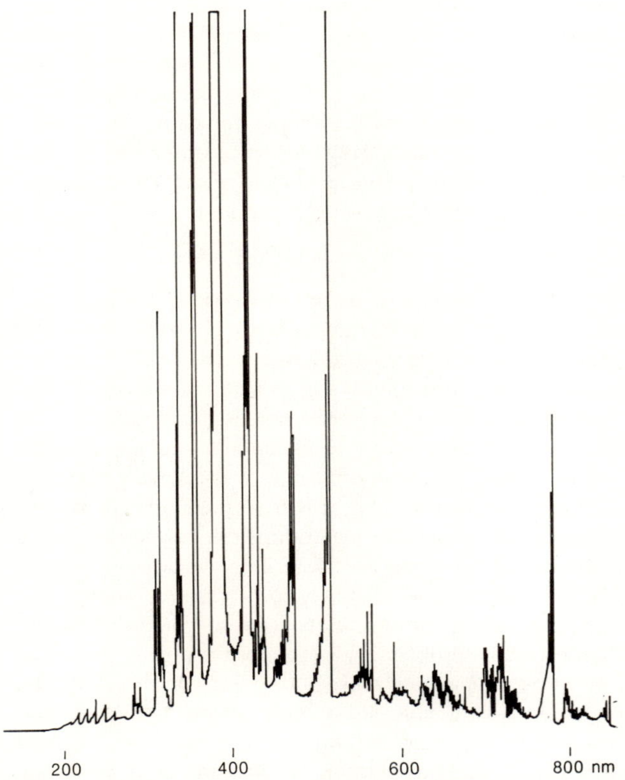

Figure 4.17 The emission spectrum from a nitrous oxide–acetylene flame (Pecsok *et al.*, 1976).

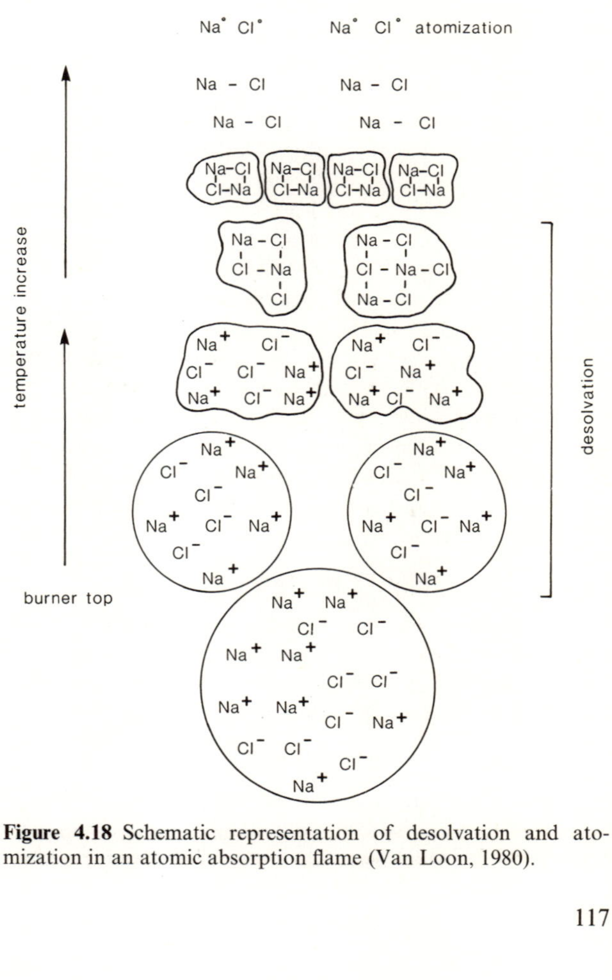

Figure 4.18 Schematic representation of desolvation and atomization in an atomic absorption flame (Van Loon, 1980).

The precise mechanism of atomization clearly depends on a number of factors, including the chemical reactivity of the element and any other accompanying components of the sample solution, flame temperature and reactivity and composition of flame gases. Thus some elements will form compounds that readily dissociate into atoms, some may first form more refractory salts (e.g. NaCl), or monoxides or hydroxides (e.g. CaO), before atomization. These reactions are complex and not fully characterized. However, many of the intermediate species formed in the flame will subsequently dissociate to atoms either by thermal effects or by reaction with free radicals. It is possible to generalize some typical behaviours.

(*a*) *Alkali metals*. The favoured medium for preparing solutions for atomic absorption analysis is dilute hydrochloric acid, since element chlorides are the most volatile of the common inorganic acid salts. It is likely that during atomization, alkali metal chloride solutions desolvate and then dissociate as discrete alkali chloride molecules. Subsequent reaction with free radicals in the flame will rapidly cause these molecules to break down, yielding atomic particles. Ebdon (1982, p.19) proposed that the following free radical reactions can occur in the simple hydrogen–oxygen flame into which a solution containing sodium and chloride ions is aspirated:

$$H^{\cdot} + NaCl \rightarrow Na^{\cdot} + HCl$$

$$H^{\cdot} + HO^{\cdot} + NaCl \rightarrow H_2O + Na^{\cdot} + Cl^{\cdot} + Q$$

$$H^{\cdot} + HO^{\cdot} + NaCl \rightarrow H_2O + Na^{\cdot *} + Cl^{\cdot} \qquad (4.5)$$

where Q represents the release of energy and $Na^{\cdot *}$ represents a sodium atom excited by chemiluminescence.

It is likely that these reactions are to some extent reversible, and that by the laws of mass balance, the degree to which atomic sodium is released is influenced by the chlorine concentration in the flame and therefore the acidity (HCl content) of the original solution. The atomic sodium signal will therefore be depressed by the presence of excess HCl, since this inhibits the liberation of free sodium in the flame.

The atomization efficiency of Na, K and Li in an air–acetylene flame has been considered in detail by Halls (1977*b*). Sample solutions containing the alkali metal in nitric acid solution in the presence of excess Cs or K as ionization suppressor were investigated. The atomization efficiencies of Na, K, and Li were found to be 98%, 85% and 26% respectively. The principal compound formed in the flame was assumed to be the hydroxide, no evidence being found for the presence of oxide or carbide species. The reaction that controlled the production of free alkali metal atoms was shown to be

$$MOH + H^{\cdot} \rightleftharpoons M^{\cdot} + H_2O. \qquad (4.6)$$

(*b*) *Ionization suppression*. One particular characteristic of the alkali metals is the low ionization potential. Indeed, the thermal energy of the flame is sufficient to cause a significant degree of ionization of elements such as sodium:

$$Na^{\cdot} + \text{energy} \rightarrow Na^{+} + e^{-}. \qquad (4.7)$$

In consequence, a reduction in the atomic absorption signal is observed, since sodium *ions* exhibit quite a different absorption spectrum from sodium *atoms*. One proposal for reducing this effect was to determine the alkali metals in a cooler air–propane flame to suppress thermal ionization effects. However, it is now more usual to use a conventional air–acetylene flame and to add a large excess of a salt that is more readily ionized than sodium. Thus if excess caesium is added to the sample solution, electrons are readily formed in the flame by ionization of caesium:

$$Cs^{\cdot} \rightleftharpoons Cs^{+} + e^{-}. \qquad (4.8)$$

The sodium ionization interference is suppressed:

$$Na^{+} + e^{-} \rightleftharpoons Na^{\cdot}. \qquad (4.9)$$

Although caesium is the favoured ionization suppressant for the determination of Na and K, care must be taken to ensure that it is free from sodium (or potassium) impurities that could significantly increase blank levels.

(*c*) *Calcium, magnesium and rare-earth elements*. These elements tend to form monoxides and oxy-salts in the flame. However, in this case, these intermediate complexes are broken down relatively easily in the high-temperature air–acetylene flame. The atomic absorption signal due to these elements can be depressed if solutions also contain significant concentrations of other elements such as Al, Be, P, Si, Ti, V and Zr. This depression is caused by removal of Ca, or Mg atoms in the flame due to the formation of compounds (mainly oxy-salts) of low volatility. Thus for example, if phosphate is present in solution, the resultant Ca-phosphate salt is rather more difficult to break down in an air–acetylene flame than, say, calcium chloride. Similarly, aluminium tends to form an aluminium–oxygen–magnesium compound in the flame, resulting in a depressed magnesium atomic absorption signal. The most widely used method of overcoming these interferences is to add a '*releasing agent*' to the solution to be analysed. Lanthanum (as 1% or 2% lanthanum chloride solution) is widely used in geological applications. Lanthanum reacts in a similar manner to calcium, magnesium, etc., with oxy-anions in the flame to form stable compounds. However, since lanthanum is added in a large excess, its presence tends to release the element of interest from stable compound formation:

$$Ca\text{–}O\text{–}X + La^{\cdot} \rightleftharpoons Ca^{\cdot} + La\text{–}O\text{–}X. \qquad (4.10)$$

The effect of lanthanum as releasing agent may be judged from Figure 4.19, taken from the work of Kahn (1966). This shows the effect of varying concentrations of aluminium and silicon on the absorbance signal from the alkaline earth element, strontium. This interference is eliminated at all but the highest concentrations of silicon and aluminium by the addition of a 1% lanthanum solution as releasing agent.

These interferences can be virtually eliminated without using a releasing agent if determinations are carried out in a hotter nitrous oxide–acetylene flame. However, calcium and magnesium have relatively low ionization energies, so that reductions in the atomic absorption signal by ionization then become a problem. It is therefore necessary to add an ionization suppressing reagent to overcome this difficulty, 0.1% potassium chloride solution being widely recommended. In some applications, uranium has been proposed as releasing agent in place of lanthanum due to its strong affinity for oxygen and its low atomic absorption sensitivity.

(*d*) *Titanium and aluminium*. These metals represent a group

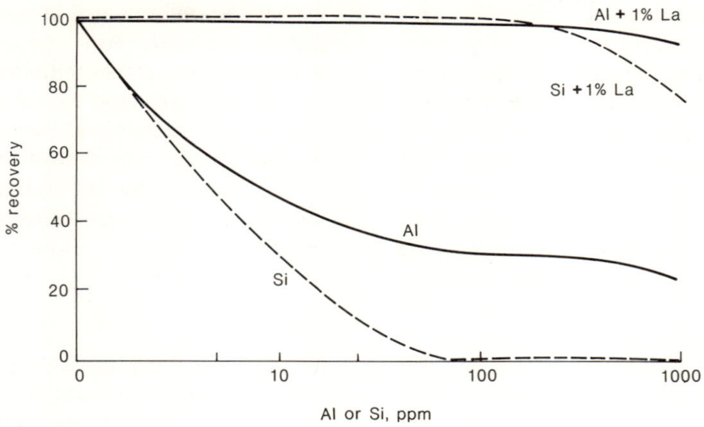

Figure 4.19 Example of the suppression of chemical interferences in an atomic absorption flame using lanthanum as releasing agent. The absorbance signal from strontium is plotted as a function of the concentration of aluminium and silica in solution. In the presence of 1% La solution, the interferences on the Sr signal are suppressed up to concentrations of several hundred ppm Si or Al. (Ewing, 1975.)

of elements that form very stable monoxides in the flame. As their dissociation energies are high (>5 eV) measurements can only be satisfactorily made in a hot nitrous oxide–acetylene flame. It is likely that the reducing properties of the CN˙ radical present in this flame greatly assist in the release of the atomic species (see 4.3(*d*) above).

Several ions are reported to interfere with the atomization of titanium, in particular aluminium, iron and fluoroboric acid. In many schemes of analysis, these interferences are overcome by matching the concentrations of interfering elements in standard and sample solutions. Even though the ionization potential of titanium and aluminium is much higher than that of calcium and magnesium, some ionization of the former elements occurs in the hot nitrous oxide–acetylene flame. Depending on the composition of the sample matrix, ionization may be effectively suppressed by adding a solution of lanthanum or potassium chloride.

(*e*) *Silicon.* Silicon is considered separately here because it poses some difficulty in accurate and stable determination. Indeed, in routine major element analysis by AAS, silica determinations are required to have a higher precision than any other element, since it accounts for over 40%–50% of the sample. There are, however, particular difficulties in achieving such standards. Silicon forms a highly refractory monoxide (SiO˙) with a high dissociation energy (8.3 eV). Silicon must, therefore be determined in a hot nitrous oxide–acetylene flame. However, the magnitude of the atomic absorption signal is highly sensitive to changes in the fuel and oxidant flow rate. In Table 4.3 the atomic absorption signal from silicon is listed as a function of nitrous oxide and acetylene gas flows. It can be seen that the optimum signal is obtained in a very rich flame. Under these conditions the flame has a tendency to luminesce, increasing noise in the atomic absorption measurement, and to clog the burner, possibly due to formation of silicon carbide in the burner jaws.

Gill and Kronberg (1975) investigated conditions under which precise determinations could be made of silicon in fluoroborate solutions of silicate rocks, using a nitrous oxide–acetylene flame. The most stable operation was found at slightly leaner flame mixtures than the fuel-rich conditions required to give the maximum absorbance signal. Sample solutions were adjusted to give silicon concentrations in the range 100–200 μg/mL, since a significant reduction in noise was observed compared with the 200–400 μg/mL solutions normally used for routine silicate analysis. The order in which standard and sample solutions were run was stringently controlled in order to negate the effect of drift in instrument response. With these precautions, precisions of 0.35 weight % (one sigma) could be achieved in the determination of silica in granite USGS:G-2 containing 69.2% SiO_2.

Barreda and Diez (1980) devised a method for the accurate determination of silica. Samples were decomposed by a $LiCO_3/H_3BO_3$ fusion, silica being stabilized in solution as the fluoride. Even so, some difficulties were encountered with the stability of sample solutions, deposits of lithium fluoride and potassium fluoroborate sometimes being precipitated in the AA sampling capillary. These difficulties were partially overcome by using sodium rather than potassium salts to buffer sample solutions. The reproducibility

Table 4.3 Change in absorbance signal from silicon as a function of flame stoichiometry

Flow rates (L/min) acetylene	nitrous oxide	Si absorbance
4.50	7.7	0.004
4.65	7.7	0.040
4.85	7.7	0.090
5.00	7.7	0.120
5.15	7.7	0.140
5.30	7.7	0.147
5.50	7.7	0.135

Data from Scobbie (1973).
Analyses performed on a silicon solution of concentration 50 μg/mL.

of measurements was reported to be ± 0.3 weight % SiO_2 in granite USGS: G-2 at the one sigma level.

Kiss (1982) reviewed in detail previous procedures for determining silica and proposed that considerable simplifications could be made by decomposing samples with HF under heat in a closed container. The presence of excess free HF, alone, resulted in excellent stability of sample solutions. Careful adjustment and periodic monitoring of flame stoichiometry was essential, but other interference effects were small. Reproducibility was typically in the range 0.1–0.2 weight % (one sigma) at the 50–60 weight % SiO_2 level.

(*f*) *Atomization interferences: general summary*. The examples above illustrate the principal interferences observed in flame AAS, namely those arising from ionization and stable compound formation effects. As has been discussed above, the principal methods of suppressing these are:
(i) By addition of an ionization suppressor or releasing agent, as appropriate
(ii) By matching matrix of standards to samples
(iii) By using a hotter nitrous oxide flame.
The efficiency with which atoms are formed in the flame clearly depends on a complex series of physical and chemical reactions, as well as being affected by the chemical properties of the analysed species itself. The practical implication of these effects is that the maximum atomic absorption signal will only be observed after optimization of several parameters, including instrument observation height in the flame as well as flame stoichiometry.

4.5 Instrumental and spectral interferences

Apart from influences on the atomization process, other interferences can affect the atomic absorption signal.

4.5.1 *Nebulizer flow rate*

For the most precise work, it is desirable to minimize the effect on atomic absorption signal of changes in the rate of nebulization caused by:
(i) Viscosity differences of standard and sample solutions due to differences in total salt content, pH etc.
(ii) The effect of changes in the depth to which the nebulizer capillary is immersed in the sample solution
(iii) The length of this capillary.
It has already been shown that, under optimum settings, modern pneumatic nebulizers are relatively insensitive to changes in solution uptake rate (Figure 4.10), providing the uptake rate is optimized to about 4 mL/min. This is in direct contrast with the performance of total consumption burners where small variations in depth of immersion of the sampling capillary can significantly influence the resultant emission signal.

4.5.2 *Flame temperature*

When a solution is aspirated into a flame, the temperature of the flame is lowered, by an amount that is influenced to a small degree by the composition of the sample solution. However, small changes in the flame temperature do not significantly change the ground-state atom population, since the proportion of excited atoms is always very small at AA flame temperatures (see Table 3.7). This is in contrast with ICP–AES where the magnitude of the atomic emission signal is much more sensitive to changes in plasma temperature. However, atomic absorption signals may be influenced to a small degree by temperature-related changes in atomization efficiencies that alter, for example, the rate at which refractory oxides are dissociated to atoms or the proportions of species that undergo ionization.

4.5.3 *Spectral interferences*

(*a*) *Direct line overlap*. Atomic absorption spectrometry is remarkably free of spectral overlap interferences. Due to the mechanism of atomic absorption (section 4.1), direct line overlap interferences can only occur if an atomic line due to a second element occurs at exactly the same wavelength as the resonance line of the element to be determined. Since the bandwidth of these lines is very narrow, this situation rarely occurs; the few interferences reported in the literature are listed in Table 4.4. If one of these interferences is likely to occur, it is usually possible to select an alternative lamp line, sometimes with a loss in sensitivity.

(*b*) *Non-specific absorbance*. Atomic absorption measurements can be affected either by absorption due to molecular species in the flame, or by scatter of the hollow cathode lamp radiation from particles in the flame. In both cases the atomic absorption signal is artificially enhanced and a correction for background absorbance must be made. This effect particularly influences measurements of lines at low wavelengths (especially below 300 nm).

One example of this effect is the interference on the Ba atomic absorption signal due to the absorption band spectra of CaOH as observed in an air–acetylene flame (Figure 4.20). This interference can be removed by using a hotter nitrous oxide–acetylene flame, since the higher temperature can break down the CaOH molecular fragment and hence remove the absorption band spectrum. However, other

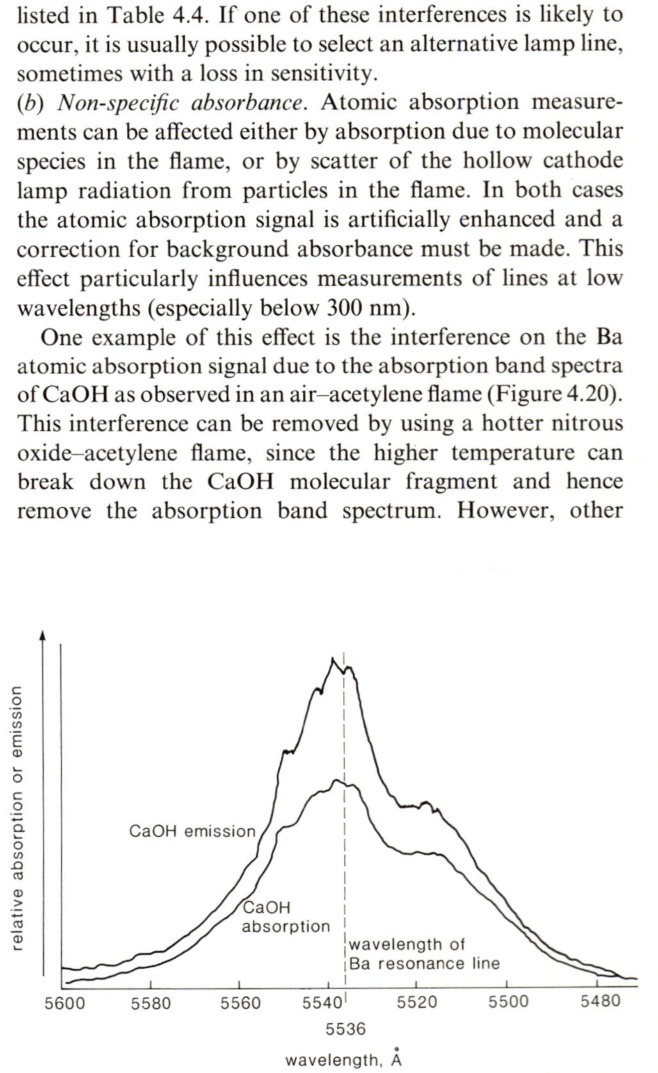

Figure 4.20 Example of a rare spectrum interference in AA spectroscopy between the molecular absorption bands of CaOH and the atomic absorption resonance signal of Ba at 553.6 nm. (Adapted from Capacho-Delgado and Sprague, 1965, by Skoog and West, 1980.)

Table 4.4 Spectral overlap interferences in atomic spectrometry

Lamp source line element	nm	Analyte	Absorption line (nm)	Separation (nm)	Sensitivity (ppm)
Aluminium	308.215	Vanadium	308.211	0.004	800
Antimony	217.023	Lead	216.999	0.024	250
Antimony	217.919	Copper	217.894	0.025	100
Antimony	*231.147*	Nickel	231.095	0.052	35
Antimony	323.252	Lithium	323.261	0.009	200
Arsenic	228.812	Cadmium	228.802	0.010	45
Copper	*324.754*	Europium	324.753	0.001	75
Gallium	*403.298*	Manganese	403.307	0.009	15
Germanium	422.657	Calcium	422.673	0.016	6
Iodine	206.163	Bismuth	206.170	0.007	10
Iron	*271.903*	Platinum	271.904	0.001	40
Iron	279.470	Manganese	279.482	0.012	0.04
Iron	285.213	Magnesium	285.213	<0.001	10
Iron	*287.417*	Gallium	287.424	0.007	250
Iron	324.728	Copper	324.754	0.026	0.8
Iron	327.445	Copper	327.396	0.049	1.1
Iron	338.241	Silver	338.289	0.048	150
Iron	352.424	Nickel	352.454	0.030	0.1
Iron	396.114	Aluminium	396.153	0.039	50
Iron	460.765	Strontium	460.733	0.032	20
Lead	241.173	Cobalt	241.162	0.011	15
Lead	247.638	Palladium	247.643	0.005	3.5
Manganese	*403.307*	Gallium	403.298	0.009	25
Mercury	*253.652*	Cobalt	253.649	0.003	100
Mercury	285.242	Magnesium	285.213	0.029	200
Mercury	359.348	Chromium	359.349	0.001	250
Neon	359.352	Chromium	359.349	0.003	0.1
Silicon	*250.690*	Vanadium	250.690	<0.001	65
Zinc	*213.856*	Iron	213.859	0.003	200

Italicized data are resonance lines.
The table lists those lamp source lines that suffer absorbance from an analyte line other than that of the element being sought. Thus the zinc 213.856 nm line from a hollow cathode lamp can be absorbed by iron in a sample (at 213.859 nm) as well as zinc. As an indication of the severity of this interference, the sensitivity of the analyte line (defined as the concentration in solution required to give an absorbance of 0.0044 units) is also listed.
Data from Price (1979, p.132) wherein the source references are listed.

background interferences are more intractable and require correction by direct measurement. The principal techniques available to carry out these corrections using flame atomization are the deuterium arc (continuum) correction procedure or the Smith–Hieftje hollow cathode lamp reversal method. These correction procedures are discussed in more detail in section 4.13.

4.6 Instrument optimization for routine analysis

4.6.1 *Instrument optimization*

A number of adjustments must be made to atomic absorption instrumentation before use. Assuming that a manual single-beam instrument is to be used and that the alignment of optical and burner assemblies is uncertain, the following series of adjustments would be required.

(i) Switch on the instrument to warm it up. Mount the appropriate hollow cathode lamp in the lamp turret and set the current to the value recommended by the manufacturer. Check that the nebulizer water trap is filled with water (Figure 4.8) and light the appropriate flame. A fume extraction hood is required above the burner to remove toxic fumes from the laboratory, emission from the nitrous oxide–acetylene flame being particularly corrosive.

(ii) Select the appropriate slit width for the monochromator (normally as recommended by the manufacturer) and optimize the wavelength setting of the monochromator to the chosen atomic emission line from the hollow cathode to obtain the maximum detected radiant power.

(iii) Adjust the mechanical alignment of the hollow cathode lamp to maximize the radiant power entering the monochromator.

(iv) Adjust photomultiplier gain so that the transmission readout is set to 100%.

(v) Switch readout to absorption and readjust the photomultiplier fine gain to 0.0 absorbance units.

(vi) Aspirate a solution containing the element of interest at about the concentration in the samples to be measured, and adjust the burner to give the maximum absorption signal as follows:

—horizontal position of burner to centralize on light beam
—vertical position of burner to obtain most sensitive viewing height
—stoichiometry of the flame to obtain maximum absorbance signal.

Since each of these adjustments is to some degree interrelated, it is necessary to reoptimize each a second time.

(vii) If desired, set scale expansion to the required value so that the measured signal is increased to distinguish small differences in absorbance between samples more readily. High scale expansion settings increase the inherent noise in the signal.

(viii) Select the appropriate value for the amplifier time constant. The time constant circuits are designed to reduce random fluctuations in the observed signal, at the cost of slowing down the response time of the meter readout. Whereas a small degree of damping is beneficial, it is essential to prevent over-damping of the output signal so that the readout is too sluggish to respond readily to changes in absorbance.

(ix) Aspirate distilled water and reset the absorbance signal to 0.00.

4.6.2 *Analysis sequence—manual instruments*

The instrument can now be used to make measurements of sample solutions. If data of the highest precision are required, a strategy for running sample and standard solutions is needed that will negate the effect of any drift in instrument response. Langmyhr and Paus (1968) proposed the following sequence of analysis.

(i) The sample solutions, and a set of standard solutions whose concentrations bracket the expected range of signal responses, are run rapidly one after another, each solution being interspersed with a distilled-water blank, to avoid memory effects and to allow the signal to reset to the baseline before the next sample is run.

(ii) The absorbance output is inspected and the two standard solutions whose absorbance signals lie on either side of each unknown solution are identified. Each sample solution is then re-run in the following order:

—distilled water to set zero
—low standard
—sample solution
—high standard
—distilled water to reset zero
—high standard
—sample solution
—low standard.

The complete sequence is then repeated.

If E_1, E_2 and E_x are the absorption measurements of low standard, high standard and sample solution, C is the concentration equivalent of the element in the low standard (% weight equivalent) and ΔC the difference in concentration between high and low standards, then the concentration of the element in the unknown solution is given by

$$\text{concentration in sample} = C+\Delta C.\left[\frac{E_x-E_1}{E_2-E_1}\right]. \quad (4.11)$$

The above series of measurements gives a total of four individual estimates of the concentration in the sample, which can be averaged to yield a value from which drift errors have effectively been eliminated. If the upper and lower standard solutions are selected to bracket the sample solution closely enough, any non-linearities in the calibration curve are also allowed for.

It is assumed that compensation has been made for any chemical interference differences between sample and standard solutions.

4.6.3 *Analysis sequence—automated instruments*

Although suitable for older manual instruments, the sequence of analysis above is not particularly efficient for more modern instruments in which short-term drift is expected to be small. Such instruments are often required to analyse large numbers of samples under microprocessor control in conjuction with an automatic sample changer. Repetitive measurement of standard solutions is both inconvenient and unnecessary. It is far more effective to precalibrate the instrument using a set of standard sloutions, and to compute and store the calibration equation for future use. The function can be linear or curved. At the beginning of an operating session, the absorbance of a blank (e.g. distilled water) and a high standard are measured. The stored calibration shape function is recalled and these data fitted to it to recalculate the gradient and intercept of the calibration function appropriate to the current operating session. In this way it is not necessary to re-run the entire standard solution set. The absorbances of sample solutions may then be measured one after the other, with only occasional checks on blank and high standard.

4.6.4 *Blank measurements*

The term 'blank' may be used in several contexts. The *instrument blank* normally refers to a sample of distilled water used to measure the zero absorbance signal. However, in any practical scheme of analysis, it is essential to run a *procedural blank*. This is a solution that has undergone the entire sample dissolution procedure (digestion, evaporation, dilution, addition of ionization suppressors or releasing agents, etc.) except that no rock powder was added to the reaction mixture. The procedural blank is then analysed as an unknown. Normally the procedural blank will also give a zero absorbance reading. If it does not, then the reagents, water or apparatus used in the solution preparation procedure are contaminated with the element of interest, or the results are affected by a non-specific (or molecular) absorbance interference. In the former case, careful examination of water and reagent contamination levels is required. In the latter case, a background absorbance correction may be necessary (section 4.13).

4.7 Schemes of analysis using flame atomic absorption

4.7.1 *Major elements*

The introduction of commercial atomic absorption instruments in the early 1960s had an important impact on the rapid schemes of analysis that were then common in many laboratories. Many of the alternative time-consuming chemical separation steps were made redundant. Magnesium could be determined by a technique (AAS) specific to that element, instead of by difference in an EDTA titration with calcium (section 2.7). Interferences that affected photometric determinations were by-passed due to the high element specificity of atomic absorption analysis. It is not surprising, therefore, that many laboratories abandoned their existing rapid procedures and adapted AA spectrometry for major-element determinations, other than for the elements P, S, Cl, F, and for H_2O/C_2O, for which gravimetric or photometric procedures still had to be retained. It was soon found, however, that interference effects could not be completely ignored and that careful matrix matching of sample and standard solutions was in some cases necessary.

Early applications of atomic absorption spectrometry to major-element analysis were described by Trent and Slavin (1964), Billings and Adams (1964), Ingamells (1966) and Suhr and Ingamells (1966). In the procedure of Shapiro (1967), the elements Ca, Mg, Na, K and Mn were determined by atomic absorption, the other major elements being determined by Shapiro and Brannock's traditional 'rapid' methods (Shapiro and Brannock, 1962). Shapiro found that this hybrid scheme halved the time required to undertake an analysis, all determinations being made from one sample solution based on a lithium metaborate fusion. Lanthanum was used as a releasing agent in the determination of calcium and magnesium.

However, this application was soon followed by more comprehensive schemes for the determination of all major elements (except P, H_2O, CO_2, F and Cl) by atomic absorption (Langmyhr and Paus, 1968; Van Loon and Parissis, 1969; Omang, 1969; Yule and Swanson, 1969; Medlin *et al.*, 1969; Boar and Ingram, 1970). Previous rapid schemes of analysis had involved the preparation of two separate sample solutions. The stimulus derived from the introduction of AAS to develop reliable single-solution dissolution procedures, based in particular on a fusion with lithium metaborate, has been discussed in Chapter 2.

Bernas (1968) advocated the use of an HF/aqua regia attack using a Teflon-lined digestion bomb. Boric acid was subsequently added to sample solution, both to neutralize the activity of HF towards glassware and to promote the dissolution of insoluble fluorides by the action of fluoroboric acid, H_3BF_4, then formed in sample solutions (see section 2.2). Bernas claimed that an important aspect of this dissolution procedure was that the presence of fluoroboric acid either significantly diminished or entirely eliminated chemical, ionization and matrix interference effects in atomic absorption measurements.

Langmyhr and Paus (1968) described a comprehensive method of major-element analysis based on the HF/boric acid system, and used lanthanum, potassium or caesium as ionization buffers or releasing agents to remove residual interferences. Up to 10 elements (Si, Ti, Al, Fe, Mn, Mg, Ca, Na, K, Cr) were determined from a single sample solution.

Van Loon and Parissis (1969) advocated the lithium metaborate fusion technique followed by nitric acid attack, sample solutions then being compatible with previously developed colorimetric and spectrographic procedures for the elements which could not be analysed by AA. These workers carried out a comprehensive study of potential interference effects. Omang (1969) and Boar and Ingram (1970) used a similar fusion technique, the former reporting further details of interference effects in the AA analysis of silicates, the latter applying the technique in particular to coal ash and mineral silicates.

4.7.2 *Trace elements*

The detection limit data in Table 4.9 indicate that flame AA spectrometry may also be used to determine a range of geologically important trace elements, including Ba, Be, Co, Cr, Cu, Li, Ni, Pb, Rb, Sr, V and Zn, following simple sample digestion.

Buckley and Cranston (1971) described a comprehensive scheme of determining 18 elements by AAS, using the HF/aqua regia attack followed by addition of boric acid. Interferences were minimized by matching the matrix of standard solutions with that of unknowns. In this way determinations could be made of the major elements plus Co, Cr, Cu, Li, Ni, Rb, Sr, V and Zn. Rantala and Loring (1975) extended this range to include Ba, Be and Pb as well. Samples were digested to give a final dilution of 0.1g to 100mL for major-element determination, or 1g to 100mL for trace elements.

Warren and Carter (1975) determined trace levels of Co, V, Cr, Ni, Co and Ba using a low-pressure HF attack in which 2g of sample was dissolved to give a final solution of 200 mL. Interferences were minimized by matching the matrices of sample and synthetic standard solutions. For this purpose, the concentrations of the major elements in these synthetic standards was set to be equivalent to 8% Fe_2O_3, 2% MgO, 4% Na_2O, 5% Al_2O_3, 5% CaO and 3% K_2O.

4.8 Interference suppression

Before examining details of how interference effects have been suppressed in practical schemes of analysis, the comments of Browner and Boorn (1984) must be taken into account. Quoting the work of Smith (1983), Browner and Boorn suggested that it is possible virtually to eliminate all matrix-induced vaporization interference by shifting the particle-size distribution produced in the nebulizer aerosol to smaller values. The implications of this are that the severity of some interferences are a function of the nebulizer design and efficiency, and that data reported in the literature for measurements made on earlier instrumentation may not be appropriate for modern spectrometers fitted with more efficient nebulizers. These factors must be taken into consideration when evaluating the following data.

4.8.1 *Major elements*

The precautions taken in various schemes of analysis to suppress flame AA interferences vary from full matrix matching to the addition of specific ionization suppression or releasing agents. There is general consensus that the presence of excess fluoro-boric acid or lithium metaborate in solution (depending on which solution preparation procedure has been used) is beneficial in providing a near-constant solution matrix and in suppressing some interferences. However, this still leaves some residual effects for which compensation must be made. Langmyhr and Paus (1968) carried out a detailed investigation of interferences in the AA determination of major elements using a hydrofluoric acid attack.

Table 4.5 Atomic absorption analysis of major elements, interference effects and their correction

Analysis conditions			Langmyhr and Paus (1968)		Van Loon and Parissis (1969)		Boar and Ingram (1970)	
Element	Line (nm)	Oxidizing gas	Interferences	Action taken	Interferences	Action taken	Interferences	Action taken
Silicon	251.6	N_2O	None		None		Al	Mask by adding tartaric acid (1%) to sample solution; add Al to standard solution
Titanium	364.3	N_2O	Al, $(H_3BO_3 + HF)$	Add Al, H_3BO_3 and HF to standard solutions	Complex	Adjust Al concentration in standard to solutions to between 700–1000 ppm and acidify to 4N HCl	Al, Si	Add excess Al to sample solution; add Si to standard solution
Aluminium	309.3	N_2O	$(H_3BO_3 + HF)$	H_3BO_3 and HF added to standard solutions	$LiBO_2$	Add Li to standard solutions	None	
Iron	248.3	Air	None		None		None	More reliable results obtained using the nitrous oxide flame
Manganese	279.5	Air	None		$LiBO_2$	Add Li to standard solutions	Not investigated	
Magnesium	285.2	Air	Si, Al, P	Add La to sample and standard solutions	Si, Al, $LiBO_2$	Add 1% La to sample and standard solutions; add Li to standard solutions	Al	Use the nitrous oxide flame
		N_2O	None		Not investigated		None	
Calcium	422.7	Air	Si, Al, P	Add La to sample and standard solutions	Si, Al, $LiBO_2$, Ti	Add 1% La to sample and standard solutions; add Li to standard solutions; keep Si concentration below 170 ppm	Al, SO_4^{2-}	Use the nitrous oxide flame
		N_2O	Ionization	Add K to sample and standard solutions	Not investigated		Ionization	Add excess K to sample and standard solutions
Sodium	589.0	Air	Alkalis	Add Cs to sample and standard solutions	$LiBO_2$	Add Li to standard solutions	Ionization	Add excess K to sample and standard solutions
Potassium	766.5	Air	Alkalis	Add Cs to sample and standard solutions	Na, $LiBO_2$	Add Li to standard solutions; add Na to standard solutions when required (i.e. when Na > 5 ppm)	None reported	Low and variable results obtained from an air–propane flame

Sample dissolution procedures
Langmyhr and Paus (1968):
0.2 g powdered sample + HF (5 mL); digest with heating; dissolve residue in 50 mL saturated boric acid and make up to 100 mL.
Van Loon and Parissis (1969):
Fuse 0.2 g powdered sample with 1.4 g lithium metaborate at 900°C; dissolve melt in 100.0 mL of dilute nitric acid.
Boar and Ingram (1970):
Fuse 0.5 g sample with 2 g lithium metaborate at 900°C; dissolve melt in dilute nitric acid; make up to 250 mL with 2.5 g tartaric acid (buffer). Matching concentrations of tartaric acid and lithium metaborate added to all standard solutions.

Boric acid was added to suppress the activity of hydrofluoric acid, and residual interferences were compensated for by adding either lanthanum as releasing agent or potassium/caesium as ionization suppressant. Their findings are summarized in Table 4.5. These data are compared with the results of equivalent studies by van Loon and Parissis (1969) and Boar and Ingram (1970). In the latter two schemes, samples were attacked by fusion with lithium metaborate. Boar and Ingram (1970) added lithium metaborate, nitric acid and tartaric acid to all standard solutions to match the matrix of unknowns. Potassium was added to sample and standard solutions as appropriate to suppress ionization and the hotter nitrous oxide flame was favoured to remove interferences in the determination of Ca and Mg. Van Loon and Parissis (1969) used lanthanum as ionization suppressor and added lithium to standard solutions to suppress ionization when appropriate. In some cases, there is agreement between these three schemes of analysis as to the precise source of an interference and the action to be taken. In other cases, opinions diverge. This is no doubt to some degree a reflection of (i) instrumentation used (particularly the design of the nebulizer, see above); (ii) the concentration of analysed solutions; and (iii) the precise height in the burner flame at which observations were made. Nevertheless it is clear from these data that titanium is a particularly difficult element to determine reliably by flame AA.

A detailed investigation of the effect of interferences arising from silicon and aluminium in the AA determination of other major elements has been carried out by Lee and Guven (1975). Sample solutions were again buffered by the fluoroboric acid system. Interference effects were sought on the elements Na, K, Ca, Mg, Fe and Li. Silicon, aluminium and calcium were measured using a nitrous oxide–acetylene flame, the other elements (including magnesium) using a conventional air–acetylene flame. Their results are summarized in Table 4.6. Serious interferences, chiefly caused by the simultaneous presence of both silicon and aluminium, were reported in determining the elements calcium, magnesium and iron. Lee and Gurven recommended matching the concentration of silicon and aluminium in all standard solutions to that of sample solutions in order to overcome this effect, and demonstrated that serious errors in quantitative analysis could arise if this precaution was not followed (Table 4.7).

Table 4.6 Interferences of silicon and aluminium on the determination of major elements by AAS

Element	Oxidizing gas	Line (nm)	Results of interference tests
Si	N_2O	251.6	Al, Na, K, Ca, Mg, Fe, Li do not cause interferences
Al	N_2O	309.3	As above
Na	air	589.0	Si and Al do not cause an interference
K	air	766.5	
Li	air	670.8	
Mg	air	285.2	Minor interference from Si, large interference from Al
Ca	N_2O	422.7	Si and Al cause a significant interference
Fe	air	248.3	Si and Al cause a significant interference when present together

Data from Lee and Guven (1975).
Dissolution procedure: Standard solutions prepared with 6 mL HF and 5.6 g H_3BO_3 in 500mL solution. Interference effects studied over a range of concentrations expected after dissolving 0.1 g silicate powder to 500mL solution. Releasing agents or interference suppressants were *not* added to any of these solutions.

Table 4.7 Chemical interferences in flame AAS: effect of matching Si and Al content of standard solutions on the analysis of granite, USGS: G-1

	SiO_2	Al_2O_3	Fe_2O_3	CaO	MgO	Na_2O	K_2O	total
Expected	72.68	14.05	1.94	1.39	0.38	3.32	5.48	99.24
Analysed (with matching)	72.63	14.04	1.83	1.35	0.37	3.52	5.59	99.33
Analysed (without matching)	72.63	14.04	1.58	1.16	0.20	3.52	5.59	98.72

Data from Lee and Guven (1975).
Expected compositions taken from Abbey (1983).
Sample preparation procedure: 0.1g rock powder dissolved in 6 mL HF by heating in a pressure vessel for 1hr at 110°C. After cooling, the sample was dissolved with 5.6 g boric acid to give a solution of volume 500mL.
Sample solutions were analysed against standard solutions (i) with silicon and aluminium added to Ca, Mg and Fe standard solutions to compensate for interference effects; and (ii) with no added Si or Al. Ionizing or releasing agents were *not* added to any solutions.

4.8.2 *Trace elements*

The interference effects expected in the determination of selected trace elements are listed in Table 4.8. This is taken from the work of Rantala and Loring (1975), and shows the action recommended to suppress interferences. In general, determinations of the elements Be and Cr (in a nitrous oxide flame) and Cu, Li and Zn (in an air–acetylene flame) are free from interferences. Barium and strontium suffer from ionization in the flame, and potassium was added in this scheme of analysis to overcome this effect. The use of lanthanum as releasing agent in the determination of strontium is illustrated in Figure 4.19.

Background correction was used by Rantala and Loring in the determination of Co (240.7 nm), Ni (232.0 nm) and Pb (217.0 nm). If chromium is determined in an air–acetylene flame, care must be taken in preparing standard solutions, since the atomic absorption response of chromium in the 6^+ oxidation state (as prepared from standard potassium dichromate reagent) is different from that in the 3^+ oxidation state (as is likely to be present in the sample solution). The oxidation state of a 6^+ standard chromium solution can be reduced to the 3^+ state by the addition of a mixture of perchloric acid and hydrogen peroxide (Rantala and Loring, 1975). Vanadium suffered enhancement from Al and K; these elements were added to standard solutions to match interferences. The response of rubidium depended on the potassium concentration. Below 600 ppm K, the Rb signal was enhanced but above 600 ppm K, the Rb signal was

Table 4.8 Trace element analysis by atomic absorption: interference effects and their correction

Element	Line (nm)	Oxidizing gas	Interference	Action taken
Ba	553.6	N_2O	Ionization	Add 1500 ppm K to sample and standard solutions
Be	234.9	N_2O	None	
Co	240.7	air	Molecular absorption	Background correction required
Cr	357.9	air	Cr^{6+}/Cr^{3+}, Fe	Standard solutions prepared from $K_2Cr_2O_7$ must be reduced to Cr^{3+} to avoid low absorbance; add Fe to standard solutions
		N_2O	None	Above interference not observed but Cr sensitivity is reduced
Cu	324.7	air	None	
Li	670.8	air	None	
Ni	232.0	air	Molecular absorption	Background correction required
Pb	217.0	air	Molecular absorption	Background correction required
Rb	780.0	air	K, variable	K $<$ 600 ppm: signal enhanced; K $>$ 600 ppm: signal depressed; match K content of standard with that of sample solution; effect of Na much smaller
Sr	460.7	N_2O	Ionization	Add 1500 ppm K to sample and standard solutions
V	318.4	N_2O	K, Al enhances	Add 100 ppm Al, 100 ppm K to standard solutions to swamp interferences
Zn	213.9	air	None	

Data from Rantala and Loring (1975).
Sample preparation: 1 g rock powder attacked by 1 mL aqua regia and 6 mL HF at 90–100°C in a covered Teflon vessel on a water bath. After cooling the contents are diluted to 100 mL together with 5.6 g boric acid.
Fe interference on Cr (air–acetylene) was reported by Langmyhr and Paus (1968).

Table 4.9 Detection limit data for flame and graphite furnace atomic absorption spectrometry

Element	Line (nm)	Relative sensitivity	Spectral bandwidth (nm)	Whole rock detection limits		Ideal solution detection limits	
				Walsh 1986 (ppm)	calculated (ppm)	Flame AAS (μg/mL)	Graphite furnace AAS (μg/mL)
Ag	328.07	1	0.5	1		0.006	0.000006
	338.28	2	0.5				
Al	309.27	1.0	0.5		**12**	0.06	0.00003
	396.15	1.5	0.5				
	237.34	5.1	0.5				
B	249.77	1.0	0.2	90		2.1	0.045
	208.95	2.3	0.2				
Ba	553.55	1.0	0.5	**10**		0.06	0.00012
	350.11	572	0.5				
Be	234.86		1.0		**0.6**	0.003	0.000006
Bi	223.06	1.0	0.2	5		0.12	0.00003
	222.83	3.6	0.2				
	306.77	4.3	0.5				
Ca	422.67	1	0.5		**0.6**	0.003	0.00003
	239.86	200	0.2				
Cd	228.80	1	0.5	0.8		0.003	0.0000006
	326.11	410	0.5				
Co	240.73	1.0	0.2	**2**		0.015	0.000024
	304.40	16	0.5				
Cr	357.87	1.0	0.2	**3**		0.009	0.000015
	428.97	7.5	0.5				
Cu	324.75	1	0.5	**1**		0.009	0.000021
	217.89	8.3	0.2				
Fe	248.33	1.0	0.2		**3**	0.015	0.00003
	371.99	9	0.2				
K	766.49	1.0	1.0		**0.6**	0.003	0.000009
	769.90	2.0	1.0				
Li	670.78	1.0	1.0	**1**		0.003	0.00003
	323.26	400	0.2				
Mg	285.21	1.0	0.5		**0.12**	0.0006	0.0000006
	202.50	30	1.0				
Mn	279.48	1.0	0.2		**1.2**	0.006	0.000002
	403.08	12.5	0.2				
Mo	313.26	1.0	0.5	6		0.06	0.00009

Table 4.9 (*Continued*)

Element	Line (nm)	Relative sensitivity	Spectral bandwidth (nm)	Whole rock detection limits		Ideal solution detection limits	
				Walsh 1986 (ppm)	calculated (ppm)	Flame AAS (μg/mL)	Graphite furnace AAS (μg/mL)
Na	589.00	1.0	0.5		**0.2**	0.001	0.000009
	589.59	2.0	1.0				
	330.23	550	0.5				
Nb	334.91	1.0	0.2	90		6.0	—
	358.03	1.1	0.5				
Ni	232.00	1.0	0.2	**3**		0.015	0.00015
	352.45	5.1	0.5				
Pb	217.00	1.0	1.0	**1**		0.06	0.000021
	283.30	2.0	0.5				
Rb	780.02	1.0	0.2	**1**		0.015	—
	794.76	2.8	0.2				
Sb	217.58	1.0	0.2	5		0.12	0.00024
	231.18	2.0	0.5				
Sc	391.18	1.0	0.2	10		0.15	—
	327.36	2.8	0.2				
Si	251.61	1.0	0.2		**90**	0.45	0.0006
	250.69	2.9	0.5				
Sn	235.48	1.0	0.5	20		0.3	0.00018
	286.33	1.3	0.5				
Sr	460.73		0.5	**1**		0.015	0.00003
Ti	364.27	1.0	0.5		**42**	0.21	0.0009
	365.35	1.1	0.2				
U	358.49	1.0	0.2	90		90	0.09
	356.66	1.5	0.2				
V	318.54	1.0	0.2	**1**		0.15	0.0003
	318.40	1.4	0.2				
W	255.14	1.0	0.2	90		3.6	—
	400.88	3.8	0.5				
Y	410.24	1.0	0.5	50		0.6	0.03
	414.29	1.5	0.5				
Zn	213.86	1.0	1.0	**0.8**		0.003	0.000006
Zr	360.12	1.0	0.2	90		4.5	—
	468.72	8.2	0.2				

Sensitivity: most sensitive line = 1.0; less sensitive lines are quoted in multiples relative to this value (e.g. '3' means 'three times less sensitive').
Spectral band width: equivalent to recommended monochromator slit width.
Absorption line, sensitivity and spectral band width data from Varian: 'Light sources for atomic absorption spectroscopy', 1985.
Whole rock detection limits data taken from Walsh (1986) ('working detection limits'); calculated whole rock and ideal solution detection limits taken from Table 5.8.
Flame AAS detection limits that fall within the range of concentrations found in 'typical' silicate rocks are in boldface.

depressed. It was essential, therefore, to pre-analyse samples for potassium and then to accurately match standard solution composition. Rubidium was probably the most difficult trace element to determine reliably in this scheme of analysis.

4.9 Detection limits

Detection limits of the elements that may be determined by flame AA in silicate rocks are listed in Table 4.9. Although these data are expected to be representative of multicomponent rock solutions, specific interference problems encountered in individual elements have not been allowed for.

These data show that flame AA has extremely low detection limits for some of the major elements (especially, for example, Na, K, Mg and Ca). Because of this high sensitivity, it is normally necessary to prepare solutions for major-element analysis to a dilution of 0.1g rock to 100mL solution (compared with 0.5 or 1g rock to 100mL for trace-element determination). Even at this lower rock concentration, the atomic absorption signal for several major elements often significantly exceeds the sensitive range for determination (>0.65 absorbance units, section 3.15). The absorbance signal can be reduced by one of the following methods.

(i) Dilute further separate aliquots of the sample solution to bring the absorbance reading back into the analytical range.

(ii) Select a less sensitive atomic absorption line from the hollow cathode lamp.

(iii) Rotate the burner head so that it is no longer in line with the optical axis of the instrument. This effectively shortens the path length of the atomic cell thus reducing atomic absorption signal (Figure 4.21). Kahn *et al.* (1981)

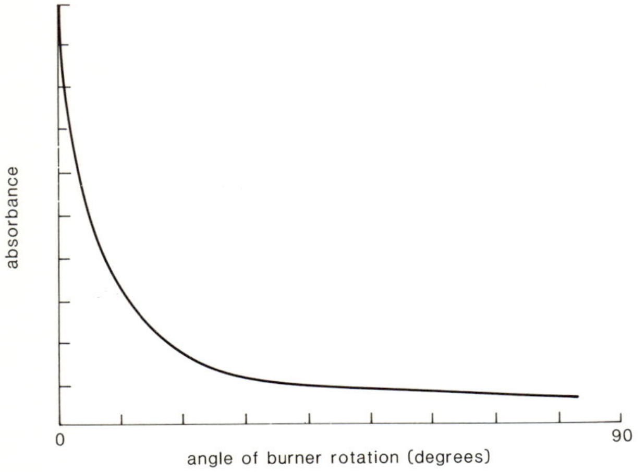

Figure 4.21 Absorbance signal as a function of burner head rotation (0° represents the burner axis parallel with the optical axis of instrument). (Price, 1979.)

used this device to analyse samples containing metals at high concentration. Their data showed that in the determination of copper and zinc, a 10° rotation of the burner head reduced the atomic absorption signal by a factor of 16, and a 90° rotation caused a reduction of about 75 times compared with normal operation.

Data in Table 4.9 show that the elements that can normally be determined by flame atomic absorption include the major elements Si, Ti, Al, Fe, Mn, Mg, Ca, Na, K and the trace elements Ba, Be, Co, Cr, Cu, Li, Ni, Pb, Rb, Sr, V and Zn. Other geochemically important trace elements cannot normally be determined without preconcentration. The detection limit performance of atomic absorption is compared with other spectroscopic techniques based on solution analysis (graphite furnace AAS, ICP–AES, ICP–MS) in Table 5.8.

4.10 Routine performance

The accuracy and precision of flame AA in major element analysis is demonstrated by data in Table 4.10. Data on USGS: W-1 (diabase) from Buckley and Cranston (1971) is compared with analyses taken from Langmyhr and Paus (1968) and Rantala and Loring (1975). Buckley and Cranston's data are particularly interesting, showing the precision that may be achieved in analyses obtained over a period of several months. Under controlled conditions, these results indicate that long-term precision of most major elements determined by flame AAS is likely to be within the limits 3 to 5% (coefficient of variation), short-term precision normally being within the range 1 to 3%.

Precision data for trace elements are given in Table 4.11. The results of Cranston and Buckley (1971) again represent the performance for 20 analyses carried out over several months. Other data represent short-term analytical precision calculated from smaller data sets. As might be expected, precision figures for trace elements are more variable, depending to a greater extent on instrument sensitivity and relative concentration. For many elements, errors of less than 5% relative (one sigma) were recorded, the longer-term precision data of Buckley and Cranston (1971) generally being in the range 10–15% (coefficient of variation).

4.11 Electrothermal atomization

4.11.1 *Principles and status*

Flame atomization, as demonstrated for atomic absorption measurements by Walsh (1955), is still extensively used with modern instruments. However, an important development was proposed by L'vov (1958, 1961) who suggested replacing the flame with a *graphite furnace* as an atom cell. This normally consists of a hollow graphite tube which can be heated by passing a low voltage dc current through the graphite (Figure 4.22). The sample is deposited as a solution of 1 to 100μL in volume on to the inner wall of the graphite tube. The furnace is then heated through a pre-programmed temperature cycle as follows.

(i) *Drying stage*: the furnace is heated to between 100° and 150°C for 45 to 60s to remove the solvent by evaporation.

(ii) *Ashing stage*: the residue is heated to intermediate temperatures (400°–1000°C for 30 to 45 s) to break down the salt matrix and drive off volatile components. Considerable molecular absorption may occur at this stage due to the emission of smoke.

(iii) *Atomization stage*: the residue after ashing is heated to a high temperature (900°–3000°C for 5 to 8 s) to atomize

Table 4.10 Major element analysis of diabase, USGS: W-1 by flame atomic absorption spectrophotometry

USGS: W-1		SiO_2	TiO_2	Al_2O_3	Fe_2O_3	MnO	MgO	CaO	Na_2O	K_2O	Total
	expected:	52.72	1.07	15.02	11.11	0.17	6.63	10.98	2.15	0.64	100.49
Buckley and	analysed:	52.62	1.05	15.44	11.04	0.169	6.96	10.72	2.10	0.650	100.75
Cranston (1971)	CV ($n = 5$):	4.8	4.8	3.5	3.5	3.8	6.9	10.0	7.3	3.7	
Langmyhr and	analysed:	52.65	1.08	15.27	11.20	0.167	6.59	10.83	2.11	0.598	100.50
Paus (1968)	CV ($n = 5$):	0.2	0.9	1.0	2.0	1.8	1.4	1.9	1.0	1.5	
Rantala and	analysed:	52.62	1.12	14.51	11.38	0.16	6.40	10.96	2.16	0.67	99.98
Loring (1975)	CV ($n = 6$):	2.6	1.5	1.5	0.4	1.1	0.1	0.7	1.2	4.7	

CV = coefficient of variation = 100 (standard deviation)/(mean analysis).
Concentrations in weight % oxide.
Buckley and Cranston (1971): precision represents reproducibility in 20 replicate analyses carried out over several months.
Langmyhr and Paus (1968): data derived from 5 analyses and presumed to represent the within-run precision.
Rantala and Loring (1975): data derived from 6 analyses and presumed to represent the within-run precision.
Expected composition from Abbey (1983).

Table 4.11 Trace-element analysis of selected reference materials by flame atomic absorption spectrophotometry

		Ba	Be	Co	Cr	Cu	Li	Ni	Pb	Rb	Sr	V	Zn
USGS: AGV-1													
	expected:	1200	2?	16	10	59	12	15	33	67	660	125	86
Rantala and	analysed:	1140	2.6	18	9	59	12	12	40	67	698	128	90
Loring (1975)	CV (n = 3)	2.0	2.8	14	5.6	4.2	2.6	—	4.3	2.3	2.2	1.2	1.3
Warren and	analysed:	1330		16	11	62		19				129	
Carter (1975)	CV (n < 6)	0.7		15	8	1.2		—				1.6	
USGS: W-1													
	expected:	160	(0.8)	47?	115?	110	14.5?	76?	7.8?	21	190	260?	86
Rantala and	analysed:	156	0.9	41	116	122	14	66	<12	21	194	285	86
Loring (1975)	CV (n = 3):	3.2	4.0	2.4	0.5	3.7	2.2	4.5	—	4.1	5.1	0.7	0.7
Buckley and	analysed:			68	128	105	14	81	<50		167	249	88
Cranston (1971)	CV (n = 20):			26	6	16	21	15	—		11	14	9

CV = coefficient of variation = 100 (standard deviation)/(mean analysis).
Concentrations in ppm.
Expected concentrations taken from Abbey (1975).
Dilution factors:
Rantala and Loring (1975): 1 g rock to 100 mL.
Buckley and Cranston (1971): 0.1 g rock to 100 mL.
Warren and Carter (1975): 2 g rock to 200 mL.

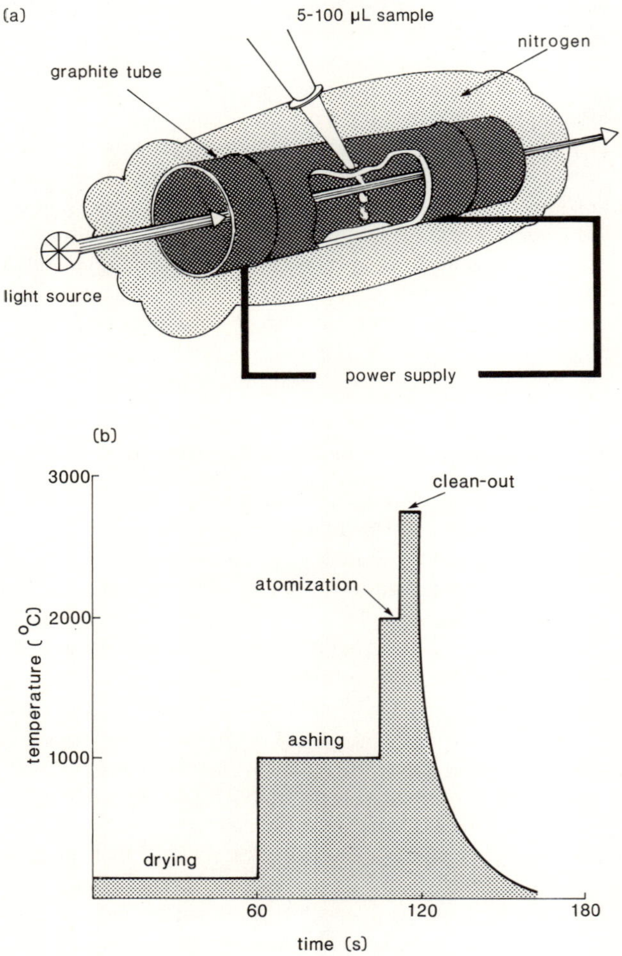

Figure 4.22 Schematic diagram of a hollow graphite furnace used for electrothermal atomization of samples. (*a*) A sample solution being pipetted into the furnace. In modern furnaces argon would normally be used as flushing gas. (*b*) A typical furnace heating cycle used to dry, ash and then atomize samples. ((*a*): Walsh, 1982.)

the sample. The transient atomic absorption signal detected during this stage is a measure of the analyte concentration.

Precise values of furnace temperature and heating times depend on the volatility of both the analyte and accompanying matrix.

The furnace itself is designed to replace the conventional burner head so that radiation from the lamp source passes through the hollow longitudinal axis of the graphite tube (Figure 4.35). During the heating cycle, the graphite tube must be flushed with argon (or nitrogen) to prevent rapid atmospheric oxidation. The current status of this technique has been comprehensively reviewed by Slavin (1982) and Slavin and Manning (1982).

4.11.2 *Analytical performance*

Interest in substituting the hollow graphite furnace for carrying out atomic absorption measurements centres on the extremely low solution detection limits that may be achieved by this technique (Table 4.9, 5.8). Indeed, excluding mass spectrometry techniques, graphite furnace AAS is the most sensitive technique for solution analysis. Thus, solution detection limits exceed those attained by flame AA analysis, often by more than two to three orders of magnitude. In comparison with inductively coupled plasma—AES, solution detection limits by graphite furnace AA are often lower by one to two orders of magnitude. Such low detection limits can be achieved because during atomization, the atomic species are concentrated for a period of up to 0.5 s within the optical path of the instrument confined inside the graphite tube. This long residence time leads to a very high specific absorbance signal.

4.11.3 *Interference effects*

This high-performance specification is accompanied by some serious analytical difficulties associated with the mechanism of atomization.

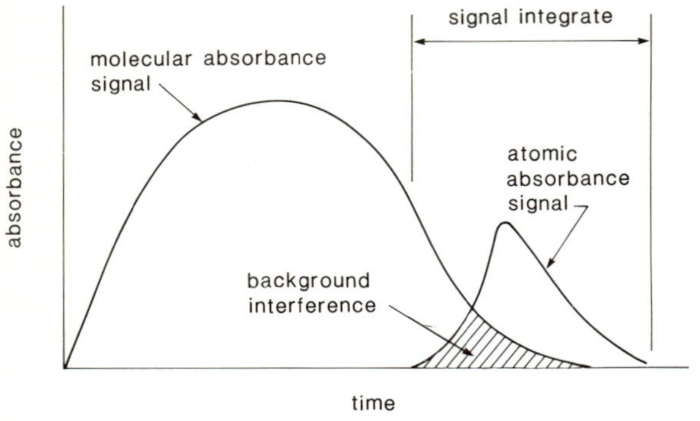

Figure 4.23 Atomic absorption signals observed during ashing and atomization cycles. The molecular absorption signal is produced from smoke and other molecular species emitted during ashing. The tail of this signal (shaded area) overlaps the desired atomic absorption signal observed during atomization and must be subtracted using a suitable background correction procedure.

(i) The ashing stage is usually accompanied by the emission of smoke, which causes a large non-specific molecular absorbance. This absorbance signal is often much larger than the desired atomic absorption signal. It is usually difficult to completely eliminate the molecular absorption signal during the critical atomization stage. The accuracy of atomic absorption measurement depends therefore on minimizing the molecular absorption by removing as much of the matrix as possible (but none of the analyte) from the sample solution before atomization (Figure 4.23). Any residual molecular interference must then be removed using a background correction technique (section 4.13).

(ii) As outlined above, unlike that of flame AA, the graphite furnace absorbance signal is transient. At the high temperatures reached during the atomization cycle, the element of interest is rapidly boiled off as an atomic vapour. The resultant absorption signal sharply increases in intensity, reaches a maximum and then begins to decline as the atomic vapour is swept by the purge gas out of the hollow graphite tube (Figure 4.23). The entire signal lasts for less than about 4 s. It is very important that measuring and readout instrumentation should be fast enough to follow these transient changes in absorbance.

(iii) Serious chemical affects can influence the atomization mechanism. These are discussed in more detail in section 4.12.

4.11.4 *Limitations of early instrumentation*

Following the development of commercial graphite furnace attachments in the early 1970s, this technique was rapidly taken up by many groups to solve analytical problems for which no other technique was readily available. This stimulated a flood of analytical method development research papers. However, with the benefit of hindsight, it is clear that some of the technical deficiencies of early instrumentation were not fully appreciated by either instrument manufacturers or users at the time of introduction. In particular:

(i) Many of the original AA instruments had measuring and readout systems that were designed for the measurement of steady-state absorbance signals originating from flame atomization. These instruments were too slow to follow the rather rapid changes in the transient graphite furnace signal. This made the recognition, interpretation and correction of interferences unreliable, as has been discussed, for example, by Lundberger and Frech (1981).

(ii) The earliest instruments employed a hollow graphite tube in which sample solutions were deposited directly on to the inner wall. Samples were then atomized into the atmosphere within the tube that was rapidly changing in temperature. This phenomenon can cause some serious chemical interferences during atomization, but these were largely eliminated with the introduction of the L'vov platform (see 4.12.3(*c*)).

(iii) Techniques for the accurate correction of molecular absorption were not fully developed until well after the introduction of the first graphite furnace atomic absorption instruments.

These instrumental deficiencies, together with an incomplete understanding of the atomization process, meant that early analytical results obtained by graphite furnace AA gained the reputation of poor precision and erratic interference effects. This reputation is no longer deserved, and many analysts would consider that the technique now has equivalent or even superior precision to flame AA, though retaining interference effects that require careful management if consistent results are to be achieved. In geological applications these interferences usually preclude the analysis of sample solutions that have not undergone some form of pre-concentration to remove matrix elements. For this reason the detection limits listed in Table 5.8 are for elements in solution only, not whole rock, since the latter depend on the sample dilution factor, which may vary depending on the pre-concentration procedure. Typical applications, therefore, are in the determination of trace elements that cannot readily be determined 'instrumentally' by other techniques such as flame AA, ICP–AES or XRF analysis. Examples of applications are discussed further in section 4.14.

4.11.5 *Characteristics of modern instrumentation*

In order to achieve consistent and reliable performance, the following must be satisfied where appropriate:

(i) Detection and readout instrumentation must operate at short time constants to follow transient absorbance signals accurately
(ii) Detection of absorbance must be based on the integrated signal area, not peak height
(iii) If appropriate, the sample solution must be atomized off a L'vov platform into an isothermal atmosphere
(iv) The sample must be treated with matrix modifiers to remove chemical interferences, especially premature vaporization during the ashing cycle
(v) The absorbance signal requires accurate background correction.

These aspects of electrothermal atomization will be considered further in the following sections.

4.12 Atomization in the hollow graphite furnace

The rather complex processes leading up to atomization in a hollow graphite furnace are the critical aspect of furnace

AA analysis. The processess that occur during the three heating cycles (drying, ashing and atomization) will now be considered. Details are included of interference effects and how they may be overcome.

4.12.1 *Drying cycle*

(*a*) *Procedure.* A small volume of the sample solution (normally 5–50 μL) is pipetted into the graphite furnace, through a small hole drilled in its wall (Figure 4.22). The furnace is then heated to above the boiling point temperature of the solvent, and the solution is evaporated to dryness, forming an anhydrous residue of inorganic salts. However, the drying temperature must not be too high, otherwise spitting will occur, causing non-reproducible atomization. In fact the drying cycle is critical in ensuring reproducible sample deposition within the graphite furnace.

Several factors influence this process.

(*b*) *Electrolytic and pyrolytic graphite.* Normal electrolytic graphite has a rather porous structure and behaves towards many samples rather as though it were a sponge. Solutions can therefore soak into the graphite wall. During atomization, the atoms are released at an erratic rate, causing poor reproducibility in the atomic absorption measurement. This problem can largely be overcome by substituting pyrolytically coated tube. These are formed from conventional graphite tubes that are heated to temperatures of around 2500°C in an atmosphere normally containing methane or other hydrocarbons. The methane is reduced to carbon at the surface of the tube, causing a dense imperious layer to be formed. This 'pyrolytic' graphite layer is normally 30 to 50 μm thick, and is much more impervious and unreactive towards the sample solution and subsequently-formed vapours than is standard electrolytic graphite. Recently, tubes made entirely of pyrolytic graphite have become available. These are technically superior to pyrolytically coated tubes in several respects. Pyrolytic graphite has a greater strength-to-weight ratio than coated tubes. This material can therefore be used to manufacture tubes with thinner walls than coated graphite (Dymott, 1985). Thinner walls permit higher rates of heating, an important aspect in minimizing vapour-phase interferences during the atomization stage.

(*c*) *Properties of graphite tubes–tube ageing effect.* It is convenient here to consider other characteristics of graphite tubes, particularly those related to the deterioration in performance observed with use. The effect of tube ageing on the absorbance signal shape as a function of the number firings for a 2 ng sample of aluminium injected on to a pyrolytically coated platform in a pyrolytically coated tube is shown in Figure 4.24. These data demonstrate the value of determining the transient absorbance signal by peak integration rather than peak height measurement. The shape of absorbance pulse, shown in Figure 4.24, changes with age due to small variations in the rate at which atoms are released from the walls of the tube. However, using a modern pyrolytically coated tube, the total number of atoms released, as represented by the integrated signal, does not change significantly. This conclusion is valid only if the readout and measurement electronics are sufficiently fast to measure this transient signal, as has been mentioned. Data in Figure 4.25 show the change in absorbance as a function of the number of tube firings for a pyrolytically coated tube and a totally pyrolytic graphite tube. The latter is clearly more durable.

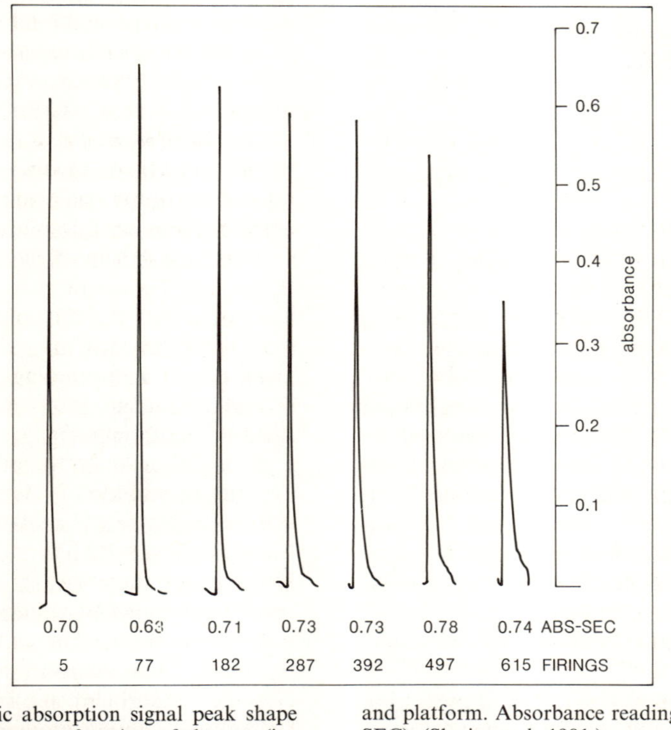

Figure 4.24 Change in the atomic absorption signal peak shape from a 2 ng sample of aluminium as a function of the age (i.e. number of firings) of a pyrolytically-coated graphite tube furnace and platform. Absorbance readings are in peak area mode (ABS–SEC). (Slavin *et al*., 1981.)

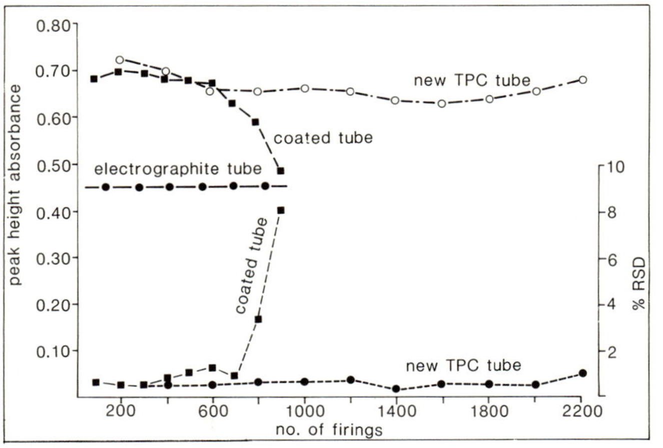

Figure 4.25 Comparison of the ageing effects between a pyrolytically-coated and a totally pyrolytic graphite furnace as a function of the number of firings for (i) 10 μL 10 ppb Mn solution (squares) in a coated tube, and (ii) 10 μL 8 ppb Mn solution (open circles) in a totally pyrolytic graphite tube. Lower curves show the respective relative standard deviations in the measurements. All data derived from peak height measurements. (Dymott, 1985.)

Data in Figure 4.26 from Slavin *et al.* (1981) demonstrate the effect of tube ageing on the magnitude of atomization interference effects. The change in absorbance of a 2 ng sample of aluminium in a solution that also contained increasing amounts of $CaCl_2$ (an interfering matrix) is shown as a function of the number of firings. The results from two types of tube are shown; one with a modern impervious pyrolytic coating and the other, of older manufacture, with an imperfect pyrolytic coating structure. Nor only is the modern pyrolytically coated tube considerably less susceptible to ageing effects, but the influence of the calcium chloride interference is also considerably suppressed. Slavin *et al.* (1981), attribute this to the fact that small amounts of calcium chloride become trapped in the porous structure of the older design of tube (a process called *intercalation*).

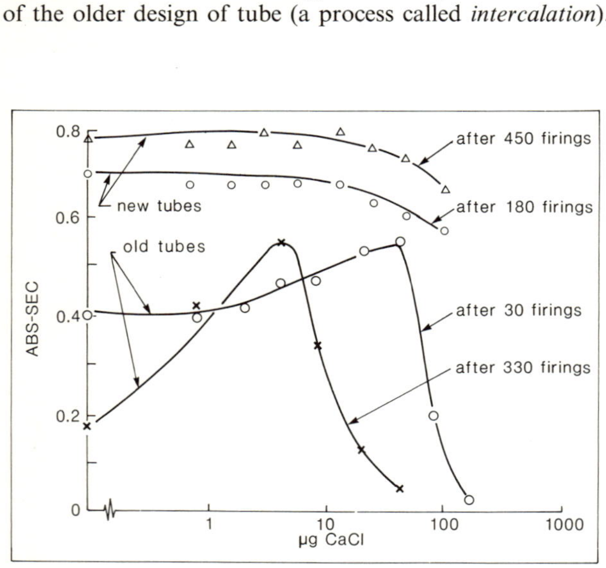

Figure 4.26 Ageing effect comparison between old pyrolytic graphite tubes (with ineffective coating) and newer pyrolytic graphite tubes. The absorbance signal from 2 ng Al is plotted as a function of $CaCl_2$ concentrations for tubes that have been used for the specified number of firings. Absorbance measurements in peak area mode (ABS–SEC). (Slavin *et al.*, 1981.)

Instead of being driven off during the ashing stage, this calcium chloride is retained in small amounts and emitted, together with aluminium, during subsequent atomization. Atomized aluminium is able to react in the vapour phase to form aluminium chloride. The corresponding aluminium atomic absorption signal is then depressed to an erratic degree.

(*d*) *Sample loading*. It is generally accepted that the way in which a sample solution is loaded into the graphite tube is an important source of errors in reproducibility. Manual pipetting is the least precise method, because of the small variations that are likely to occur in the exact position, size and shape of the solution drop deposited within the furnace. Automatic sample loading devices may reduce this problem. However, an alternative approach has been described by Matousek (1977) based on aerosol deposition in the furnace. Sotera *et al.* (1983) described a similar system, used in conjunction with an automatic sampler that deposited the sample as an aerosol in the furnace after the solution had been aspirated through a conventional pneumatic nebulizer (Figure 4.27). During this process the furnace is heated to a sufficiently high temperature to ensure flash vaporization of the solvent. The sample is therefore deposited in the furnace as an extremely fine mist of inorganic salt particles, rather than the haphazard range of crystal sizes expected after conventional sample drying. This system is claimed to offer several advantages. First, analytical precision is improved, since the uniformly small crystals will atomize in a more reproducible manner. Second, Sotera *et al.* (1983) showed that the magnitude of some molecular absorbance interferences could be considerably reduced by aerosol sample deposition. Their data in the determination of tellurium in nickel based alloys are shown in Figure 4.28, where the absorbance signal for tellurium is shown as a function of time in seconds since the start of atomization. The solid line represents tellurium absorbance, the dotted line tellurium plus the background absorbance. The magnitude of the background component is considerably reduced using aerosol sample deposition techniques.

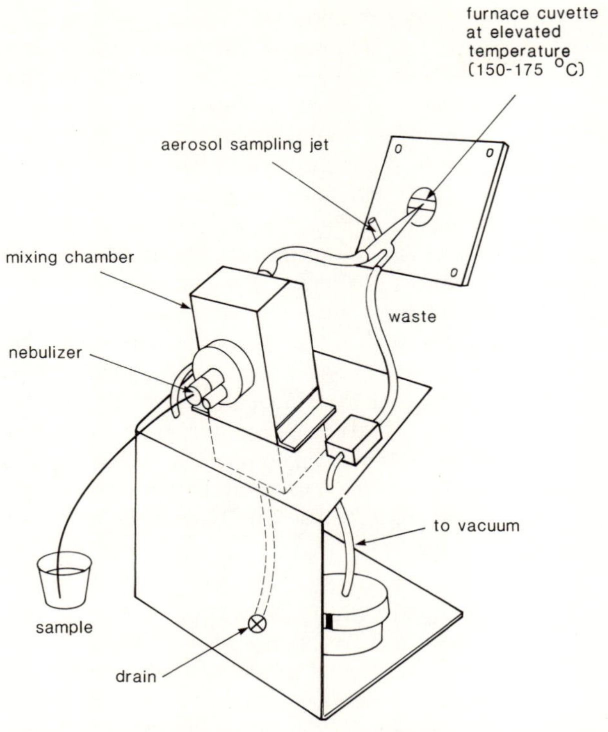

Figure 4.27 Schematic diagram of a device that deposits the sample solution as an aerosol into the graphite furnace after nebulization. (Sotera *et al.*, 1983.)

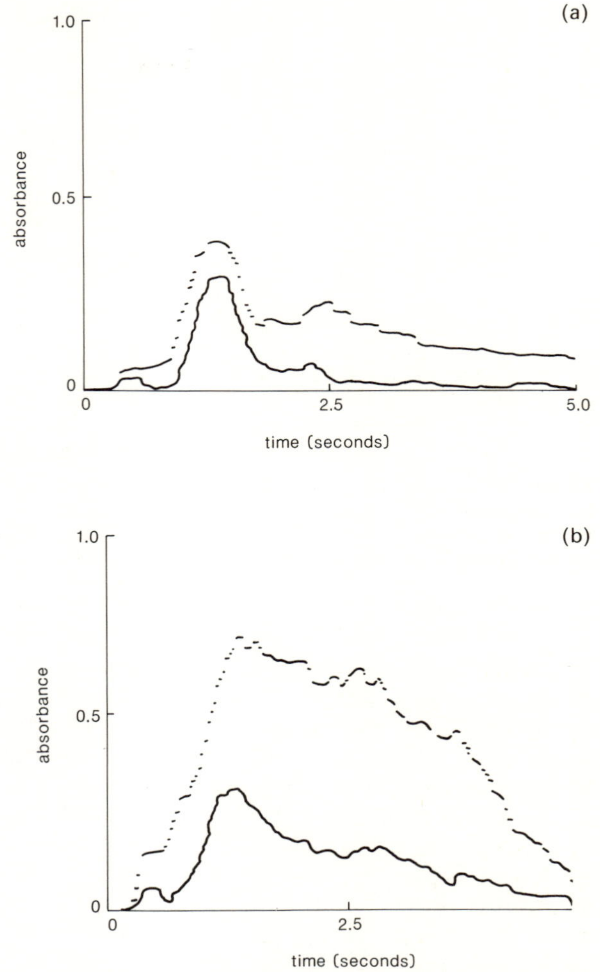

Figure 4.28 Reduction in background absorbance resulting in the determination of tellurium in nickel alloys using aerosol sample deposition. (*a*) The absorbance signals during atomization after manual pipetting of 25 μL sample solution. (*b*) The same data after aerosol deposition. The solid line plots the tellurium atomic absorption signal, the dotted line the molecular absorbance interference. (Sotera *et al.*, 1983.)

4.12.2 *Ashing/charring cycle*

(*a*) *Mechanisms.* In the analysis of aqueous solutions of single salts, the ashing stage is not particularly critical. However, if sample solutions contain an excess of matrix components, then ashing becomes critical in removing both chemical interferences that could depress the atomic absorption signal and molecular absorbance effects that could swamp the desired atomic absorption signal.

During ashing, the inorganic matrix of the sample is broken down, usually leaving the element of interest as a simple metallic salt (for example the oxide). At the same time, unwanted matrix components are driven off.

Ideally, ashing should be carried out at a temperature just below that at which the element of interest or its salts are volatilized. Premature vaporization during ashing is one of the causes of low or erratic absorbance signals. The presence of chloride ions in sample solution has been associated with this phenomenon.

(*b*) *Interferences from chloride ions.* Two problems have been identified. The first affects the ashing cycle. Many metal chlorides are volatile, so that there is a high probability that significant proportions of these elements will be prematurely lost during the ashing cycle. The magnitude of the residual atomic absorption signal detected during atomization is thus reduced. This is demonstrated by the increased loss of lead during low-temperature ashing in the presence of excess magnesium chloride, compared with solutions containing lead alone, as shown in Figure 4.29. The second affects the atomization stage. Any chloride residue remaining in the

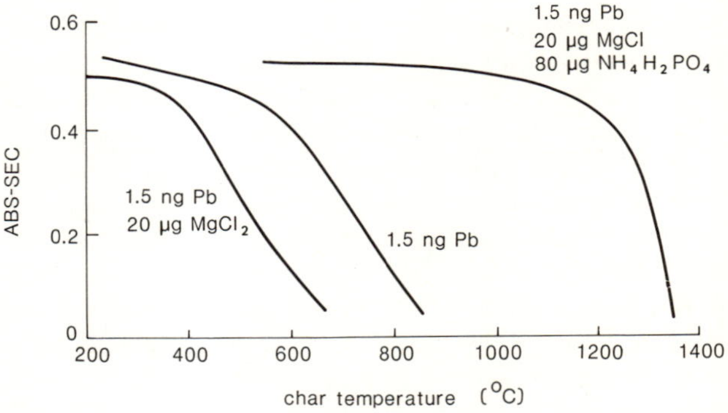

Figure 4.29 Optimizing ashing temperature in the hollow graphite furnace determination of lead. A 1.5 ng sample of lead can be ashed at temperatures up to 600°C without premature loss of lead. In the presence of 20 μg $MgCl_2$, this temperature is reduced to about 350°C due to the formation of volatile lead chlorides. If 80 μg $(NH_4)H_2PO_4$ is added as a matrix modifier, ashing temperatures can be safely increased to over 1000°C. (Slavin *et al.*, 1984.)

Table 4.12 Dissociation energies of selected metal monochlorides

Element	kJ/mol	Element	kJ/mol	Element	kJ/mol
Hg	97	Ni	367	In	427
Cd	205	Tl	368	Ba	444
Zn	288	Mg	377	Si	452
Pb	297	Be	385	V	472
Bi	302	Ca	393	Ga	473
Mn	309	Co	393	Li	473
Ag	310	Sr	402	Sc	490
Se	318	Ge	406	Ti	490
Fe	347	Na	410	Al	498
Cu	355	Sn	411	Y	524
Sb	356	K	423	B	545
Cr	363	H	427		

Data from Slavin and Manning (1980), quoting B. V. L'vov and L. A. Pelieva (1980), *Prog. Anal. Atomic Spectrosc.* **3,** 65, and B. V. L'vov (1978), *Spectrochim. Acta* **33B,**153–193.

furnace after ashing will subsequently become vaporized during atomization. Reactions can then occur in the vapour phase whereby molecular metal chloride species are formed. The resultant atomic absorption signal is depressed. One such interference has already been mentioned above: that on aluminium resulting from the presence of $CaCl_2$ that may become intercalated into the wall of the graphite tube. At the high furnace temperatures reached during atomization, chlorine atoms are released and react in the vapour phase to form aluminium chloride species, depressing the atomic aluminium signal. Elements most likely to be affected by the formation of stable molecular chlorides in the vapour phase can be judged from data in Table 4.12. This lists the dissociation energies of monochlorides: the higher the dissociation energy, the greater the stability of the corresponding metal–chloride complex. As can be seen, the elements most likely to suffer from interferences through this phenomenon include B, Y, Al, Sc, Li, Ti, Ga and V.

Unless other action is taken (see (*c*) below) it is desirable to avoid the use of hydrochloric acid in sample dissolution procedures for graphite furnace AA analysis (or other halide-based inorganic acids). Sulphuric, nitric or phosphoric acids are preferred, and in some cases contribute in a positive manner to the suppression of matrix interferences. This must be contrasted with flame atomic absorption procedures where the reverse is true, hydrochloric acid being the preferred mineral acid for sample dissolution. The problem of chloride vapour phase interferences has been investigated in detail by Slavin *et al.*(1984), who described a modified graphite tube ('curtain tube') to minimize this interference effect. Since chloride ion interferences can be associated with absorption of sample solutions in imperfect wall linings, it is likely that the magnitude of these phenomena will be reduced for measurements made using totally pyrolytic graphite tubes.

(*c*) *Matrix modifiers.* In addition to the problems of loss of volatile chlorides, some elements are intrinsically volatile, such as the elements and salts of Hg, Cd, Tl, Pb, etc. There is then a high probability that substantial proportions of these elements will be volatilized even during low-temperature ashing, and that it will then be difficult to separate the very large molecular absorbance signal from a (possibly small) atomic absorption signal. To overcome this problem, Ediger *et al.* (1974) and Ediger (1975) proposed the addition of a *matrix modifier* to sample solutions, a concept that is now accepted as an essential component of graphite furnace AA analysis for many elements. The matrix modifier is normally an inorganic salt added in excess to the sample solution. During the ashing stage this salt reacts with the inorganic residue, suppressing the premature loss of by volatilization of elements of interest. One often-quoted example which illustrates the suppression of loss of volatile chlorides comes from the work of Ediger *et al.* (1974) in the analysis of trace-elements in sea water. Sea water contains excess sodium chloride, so that there is a high probability of premature loss of volatile metal chlorides during ashing. However, by the addition of excess ammonium nitrate as a matrix modifier, the chloride can effectively be boiled off during ashing at temperatures below 500°C, removing this interference:

$NaCl$	+	NH_4NO_3	→	$NaNO_3$	+	$NH_4Cl\uparrow$	
boils at 1413°C		decomposes at 210°C		decomposes at 380°C		sublimes at 335°C	(4.12)

Further examples of the use of matrix modifiers include the addition of nickel to samples to be analysed for arsenic or selenium. During ashing, nickel arsenide or selenide salts are formed, and are not prematurely volatilized until the higher-temperature atomization stage.

Although a universal matrix modifier does not exist, magnesium nitrate or ammonium hydrogen phosphate have a wide range of applications. The action of ammonium dihydrogen phosphate on the atomization of lead is shown in Figure 4.29, taken from the work of Slavin *et al.* (1984). This shows the absorbance signal due to lead after charring at the specified temperatures. As can be seen, premature loss of lead occurs at charring temperatures above about 600°C. In the presence of excess magnesium chloride, lead losses become unacceptable at even lower temperatures (about 300°C), due to the formation of volatile lead chloride. However, if excess ammonium dihydrogen phosphate is added as matrix modifier, charring can be safely carried out at a much higher temperature, up to 1000°C, without significant loss of lead. Under these conditions, it is possible to remove the molecular absorbance interference before atomization far more successully. A list of appropriate matrix modifiers in the determination of selected elements using platform atomization (to be discussed below) is shown in Table 4.13.

Under ideal circumstances, all molecular species are completely volatilized and removed from the furnace during the ashing cycle. Inevitably, with real samples, this process is seldom complete (Figure 4.23). The tail of the molecular absorbance curve carries over to interfere with the atomic absorption signal during the atomization cycle. There are almost no circumstances where it is not essential to correct for this background signal and indeed the overall success of the graphite furnace AA often depends on the effectiveness of both the ashing stage and the subsequent background correction applicable during atomization. Background correction techniques are discussed further in section 4.13.

Table 4.13 Details of analysis by atomic absorption using hollow graphite furnace atomization

Element	Matrix modifier	Furnace temperature		Characteristic amount (pg)	Line (nm)	Zeeman sensitivity loss
		Ash °C	Atomize °C			
Ag	200 μg PO_4^{3-}	650	1600	1.4	328.1	6
Al	50 μg $Mg(NO_3)_2$	1700	2400	10	309.3	10
As	20 μg Ni^{2+}	1500	2500	17	193.7	11
Au	50 μg Ni^{2+}	1000	2200	13	242.8	16
B	5 μg Ca^{2+}	1000	2700	700	249.7	40
Ba		1200	2700	6.5	553.6	4
Be	50 μg $Mg(NO_3)_2$	1500	2500	1.2	234.9	50
Bi	20 μg Ni^{2+}	900	1900	23	223.1	35
Ca		1100	2600	0.8	422.7	5
Cd	200 μg PO_4^{3-} + 10 μg $Mg(NO_3)_2$	900	1600	0.35	228.8	2
Co	50 μg $Mg(NO_3)_2$	1400	2400	6	240.7	15
Cr	50 μg $Mg(NO_3)_2$	1650	2500	3.5	357.9	12
Cs		900	1900	5	852.1	—
Cu		1200	2300	8	324.7	45
Fe	50 μg $Mg(NO_3)_2$	1400	2400	2	248.3	8
Hg	$K_2Cr_2O_7$	250	1000	—	253.7	40
K					766.5	40
Li		900	2600	1.4	670.8	15
Mg		900	1700	0.35	285.2	9
Mn	50 μg $Mg(NO_3)_2$	1400	2200	2	279.5	9
Mo		1800	2700	9	313.3	5
Na					589.5	5
Ni	50 μg $Mg(NO_3)_2$	1400	2400	9	232.0	9
P	50 μg $La(NO_3)_3$	1400	2700	3000	213.6	30
Pb	200 μg PO_4^{3-} + 10 μg $Mg(NO_3)_2$	600	900	11	283.3	17
Pd		900	2300	24	247.6	9
Pt		1300	2500	17	265.9	20
Rb		800	1900	2.3	780.0	
Sb	20 μg Ni^{2+}	1100	2400	38	217.6	5
Se	20 μg Ni^{2+} + 25 μg $Mg(NO_3)_2$	900	2000	30	196.0	12
Si		1400	2700	40	251.6	2
Sn	200 μg PO_4^{3-} + 20 μg $Mg(NO_3)_2$	1000	2100	23	286.4	6
Sr		1300	2600	1.4	460.7	2
Te	20 μg Ni^{2+}	900	2000	15	214.3	7
Ti		1400	2700	43	365.3	4
Tl	1% H_2SO_4	600	1500	10	276.8	35
V		1500	2700	30	318.4	25
Zn	6 μg $Mg(NO_3)_2$	600	1800	0.04	213.9	12

Data from Slavin *et al.* (1984), wherein the source references giving details of matrix modifiers are listed.
Where PO_4^{3-} is specified, $(NH_4)_2HPO_4$ or $(NH_4)H_2PO_4$ can be used, whichever has the lower analyte contamination level.
Atomization details are as follows: sample size: 20 μL deposited on a L'vov type platform (except as indicated below) in a pyrolytically coated graphite furnace; drying cycle: 60 s at 160°C with a 1s ramp; ashing cycle: 45 s at indicated temperature with a 1 s ramp; atomization cycle: 4 to 6 s at the indicated temperature, integrated absorbance signal; clean-out cycle: 2600°C for 6 s; electrodeless discharge lamps are used where available; data is appropriate to Perkin–Elmer Zeeman corrected atomic absorption instruments. The refractory elements B, Ba, Ca, Mo, Pd, Pt, Sr, Ti, V must be atomized from the furnace wall, the others from a L'vov platform.
Characteristic amount is the mass in pg of analyte required to give an integrated signal of 0.0044 absorbance units (i.e. 99% transmission).
Zeeman sensitivity loss data lists the % reduction in sensitivity using the Zeeman background correction system compared with conventional deuterium arc background correction.
Sn would normally be determined from the 224.6 nm line using deuterium arc background correction.

4.12.3 *Atomization cycle*

(*a*) *Mechanism.* After ashing the sample, the temperature of the furnace is sharply increased, causing the residual sample to volatilize and dissociate into atoms. In some schemes of analysis, it is normal to switch off temporarily the argon purge supply (used to preserve the operating life of the graphite against atmospheric oxidation). The residence time of atomic species in the furnace is then prolonged, so maximizing the atomic absorption signal.

There are two main requirements to ensure optimum atomization. Firstly, the temperature of the furnace should be increased as rapidly as possible to its atomization temperature to ensure rapid and efficient generation of an atomic vapour. Secondly, this atomic vapour should be emitted into a furnace atmosphere that has reached thermal equilibrium.

Failure to achieve both these conditions can lead to unpredictable atomization interferences that were the source of many of the problems associated with early furnace designs.

(*b*) *The effect of furnace design on atomization interferences.* The original furnace design of L'vov (1961) (Figure 4.30*a*)

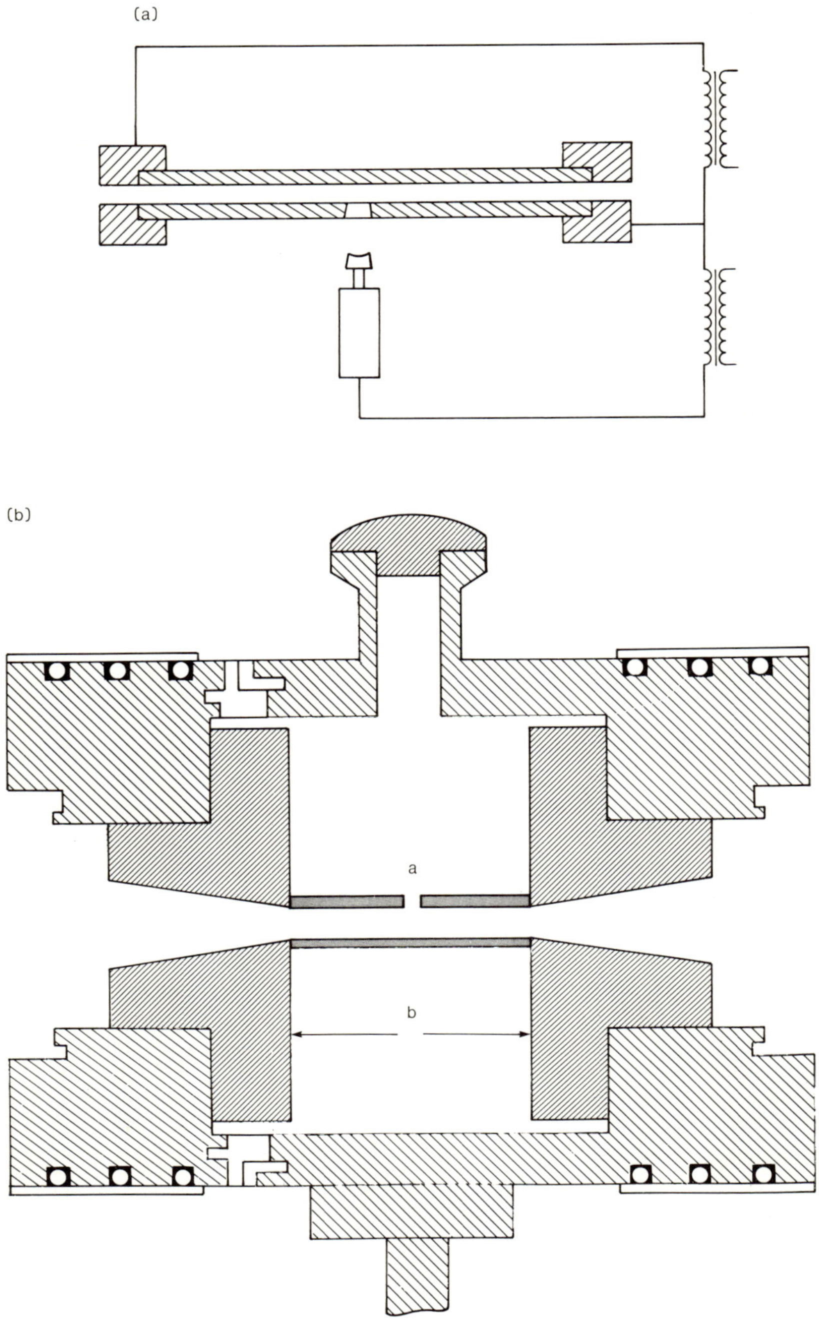

Figure 4.30 (*a*) The graphite furnace originally proposed by L'vov. (*b*) The graphite furnace developed by Massmann showing the graphite tube (*a*) and the end cones (*b*). (Price, 1979.)

was considered too complex for commercial exploitation. Instrument manufacturers therefore adopted a much simpler design due to Massmann (1968). This consists of a simple graphite tube clamped at either end to two steel contacts (Figure 4.30*b*). The furnace is heated by passing a dc current of up to 400 A at 10 V through the graphite. Samples are introduced through a small tube drilled at the centre of the tube. Commercial variations include designs in which the walls of the tube are profiled to ensure maximum and uniform heating at the centre where the sample is deposited. However, this design has an important analytical limitation. The walls of the furnace are heated due to their electrical resistance. The atmosphere inside the furnace, however, is heated principally by the radiation effect. Thus a sample deposited on the inner wall of the furnace is almost always atomized *before* the temperature of the furnace atmosphere has reached its equilibrium level (Figure 4.31). It is now recognized that this discrepancy in temperature is the source of a considerable number of interferences reported in the early literature, many of which were repeatable only with the temperature profiles obtained on individual furnace designs. In its original form, the simple graph-

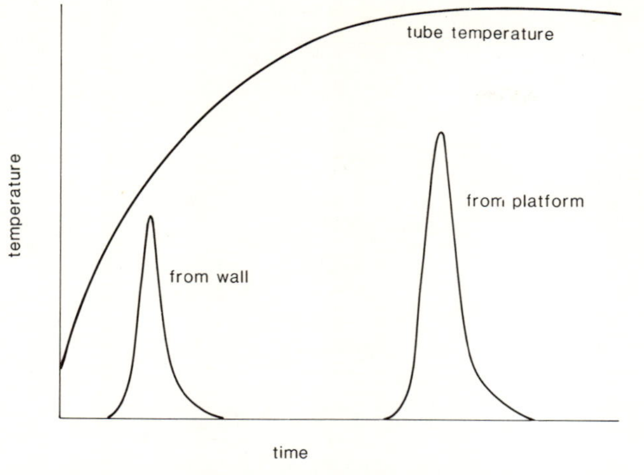

Figure 4.31 Schematic representation of the atomic absorption signal observed from wall atomization and platform atomization. The latter is delayed to a time when the temperature of the graphite furnace has almost reached equilibrium. (Slavin and Manning, 1980.)

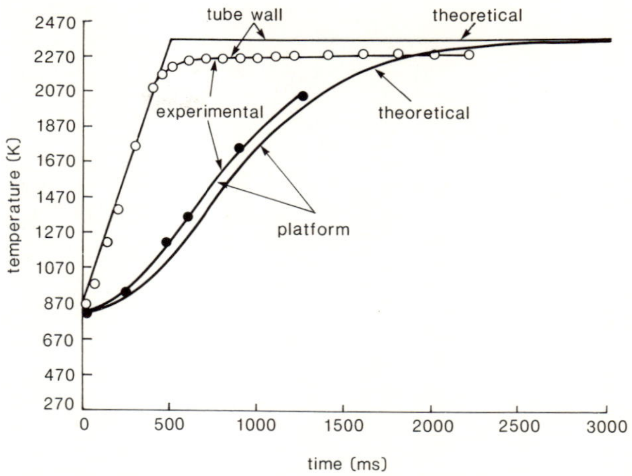

Figure 4.33 Theoretically and experimentally measured changes in temperature of tube wall and platform of a hollow graphite furnace heated at rates of 3 K ms^{-1} from an initial temperature of 870 K. (Chakrabarti *et al.*, 1983.)

ite tube furnace is a very difficult instrument to optimize and operate reliably, particularly if samples are prepared in a variety of matrices.

(*c*) *The L'vov platform.* In order to overcome these difficulties, L'vov *et al.* (1977) and L'vov (1978) proposed a remarkably simple modification now known as the L'vov platform (Figure 4.32). The L'vov platform consists of a small platform of pyrolytically coated graphite inserted into a Massmann-type tube furnace. The sample solution is deposited on to this platform. The platform makes rather poor thermal contact with the walls of the furnace. During atomization, the temperature of the platform and therefore that of the ashed sample increases more slowly than that of the furnace walls (Figure 4.33). This thermal lag is sufficient to ensure that atomization from the platform is delayed, so that atoms are released into an atmosphere that has essentially reached its equilibrium temperature (Figure 4.31). This so-called *isothermal* atomization has been investigated in detail by several groups of workers (for example Chakrabarti *et al.*, 1983; Kaiser *et al.*, 1981; Slavin and Manning, 1980; Slavin *et al.*, 1981). Experimental data indicate that many of the interferences encountered in using the conventional Massmann furnace either disappear

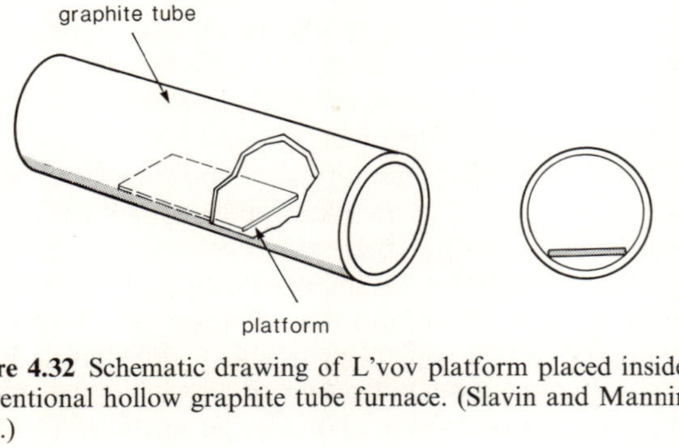

Figure 4.32 Schematic drawing of L'vov platform placed inside a conventional hollow graphite tube furnace. (Slavin and Manning, 1982.)

or are considerably reduced in the isothermal atomization procedure. Thus Kaiser *et al.* (1981) studied the atomization behaviour, from the L'vov platform and tube wall, of Pb and Cd (representative of volatile elements), Co and Cr (representing elements of intermediate volatility) and Sn and Al (representing elements that form stable oxides). In each case, a significant matrix effect was observed during atomization from the tube wall. Platform atomization, with and without matrix modifier, gave improved performance both in terms of a reduction in the interference effect and an improvement in the precision of absorbance measurement. Similar results were obtained by Chakrabarti *et al.* (1983), who observed greatly reduced interferences from chloride in the determination of Pb and Cd by platform atomization. This was attributed to the following benefits of isothermal atomization.

(i) Atomization occurs at a time when furnace tube wall and atmosphere have virtually reached thermal equilibrium. Turbulence caused by the rapid expansion of the purge gas will therefore have also subsided, permitting more precise measurement of the vapour-phase atomic absorption signal.

(ii) Atomization occurs into an atmosphere that has reached a higher (equilibrium) temperature. This higher temperature will promote dissociation of any molecular chloride species, giving a higher atomic absorbance signal. The proportion of any undissociated species is likely to remain constant, whilst absorption measurements are being made, reducing variations due to any residual matrix effects.

A further consideration in the atomization of samples from the L'vov platform is that it is often possible to operate at lower temperatures than would normally be required for wall atomization. The reason is seen from Figure 4.34, taken from Slavin *et al.* (1981). This shows the absorbance signal due to manganese in the presence of excess $Mg(NO_3)_2$ as matrix modifier. As the atomization temperature is reduced from 2300° to 1900°C, more of the atomization profile falls under that section of the temperature profile that indicates the furnace has reached thermal equilibrium. A temperature

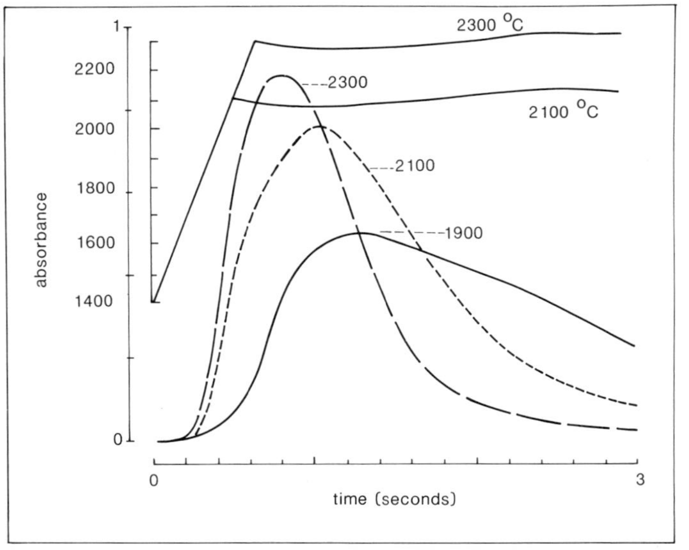

Figure 4.34 Delay in atomization of 0.4 ng Mn (+0.07% $Mg(NO_3)_2$ matrix modifier) as the temperature of atomization is reduced. Absorbance profile of Mn and temperature profile of furnace are plotted at selected temperatures. (Slavin *et al.*, 1981.)

of 2200°C would be selected as the best compromise for atomization in this example.

(*d*) *Stable compound formation.* One group of elements suffers interferences on atomization other than those already discussed above. These are the elements that tend to form refractory stable compounds that resist volatilization and dissociation, and include Ba, V, W, Mo, and Ta, which can react with the furnace lining to form stable carbides. This effect is minimized by modern pyrolytically coated graphite components that reduce close contact between sample and carbon. Slavin *et al.* (1981) reported that a significant improvement in the determination of Mo, Ti and V was not observed during atomization from a L'vov platform as opposed to the tube wall. It was suggested that the maximum temperature to which the furnace can be heated is rather too close to the atomization temperature of these refractory metals, and a platform cannot then be heated rapidly enough to obtain effective atomization.

4.12.4 *Cleaning cycle*

The atomization cycle is normally followed by a clean-out cycle in which the temperature of the furnace is increased to a few hundred degrees above atomization temperature to ensure that all the analyte is burnt off the furnace and to reduce memory effects. A delay of at least 20 s is necessary before injection of the next sample to permit the furnace to cool down.

Overall analysis time is therefore about 3 minutes per sample, and is broken down into its individual cycles as shown in Figure 4.22. This is longer than the analysis cycle for flame AA measurements, where samples would be aspirated for perhaps 15 s, followed by 15 s of blank solution. However, with the appropriate microprocessor control and an automatic sample-changing device, graphite furnace AA is suited for pre-programmed, unattended, overnight operation, a benefit that is normally not available for flame AAS on safety grounds.

4.13 Background correction

As has been mentioned in preceding sections, the ultimate success of furnace AA measurements in the analysis of complex matrices depends on the accuracy of the correction for non-specific molecular absorbance. This facility is also required in flame AA work, although for a rather limited number of elements—particularly those determined from lines close to the uv limit of the instrument.

Unlike flame applications, one difficulty that affects furnace corrections is that the proportion of background absorbance changes during the course of the atomic absorption measurement.

Three forms of background correction are available in atomic absorption spectroscopy: the traditional deuterium arc background correction; corrections based on the Zeeman splitting effect; and the Smith–Hieftje hollow cathode lamp reversal system. All three are capable of correcting background signals arising from absorption by molecular species or the scattering effect of particles (e.g. from smoke) in the light beam, providing the corresponding absorption band is relatively broad compared with the narrow atomic absorption line. These phenomena are always more serious for measurements in the uv region, and are often negligible for atomic absorption lines in the visible region above 430 nm. In some unusual furnace AA applications, it has become apparent that short-range variations in background absorption can be observed within the bandpass of the monochromator. Of the three techniques above, only the Zeeman and Smith–Hieftje correctors can cope with such non-linear background absorption effects. The characteristics of these correction methods will now be considered.

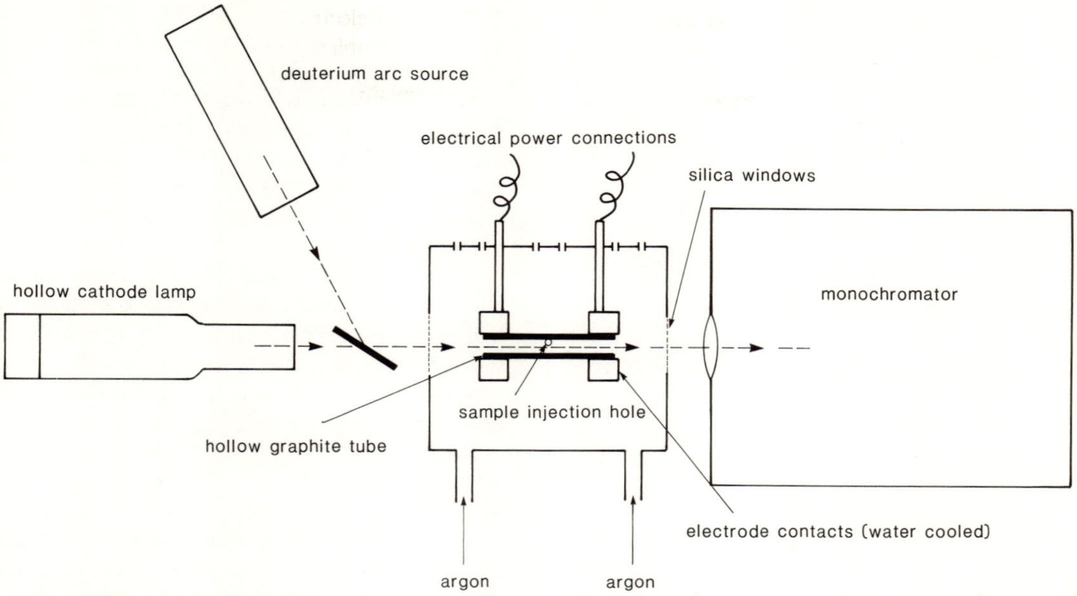

Figure 4.35 Instrumentation for electrothermal atomization using a hollow graphite furnace and deuterium arc background correction.

4.13.1 *Deuterium arc correction*

The deuterium arc lamp gives an intense continuous emission spectrum in the uv region (Figure 3.16), lamp emission becoming weak and discontinuous above 350 to 400 nm, setting an upper wavelength limit for the range of useful application of this source.

Deuterium arcs were proposed for background correction by Koirtyohann and Picket (1965) (using the hydrogen continuum). The principle of the technique is as follows.

The atomic emission line from a hollow cathode lamp alternates with the continuum output from a deuterium arc arranged to follow the same light path as in normal atomic absorption measurements. A rather wide wavelength range of continuum radiation (0.2 to 1nm, depending on monochromator slit selection) falls within the bandpass of the monochromator. Continuum radiation within this wavelength range will suffer both absorption by atomic species (as with a conventional hollow cathode lamp) and non-specific background absorption. However, the fraction of continuum wavelengths capable of inducing atomic absorption is very small compared with the total wavelength range transmitted by the monochromator that is susceptible to background absorption (Figure 4.1). Monochromator signal will, therefore, be a direct measure of the background absorbance averaged over the wavelength range that fills the bandpass of the monochromator.

In practice the instrumentation is arranged as shown in Figure 4.35. Radiation from the hollow cathode lamp is chopped with that from the deuterium arc, using a half-silvered mirror at a frequency of at least 10 times per second, so that each alternately passes through the AA furnace or flame. The photomultiplier output is synchronized with the chopper frequency using a suitable lock-on amplifier. If the absorbance readout is zeroed with respect to the deuterium arc signal (background absorption), the hollow cathode absorbance signal represents the background corrected atomic absorption signal.

The deuterium arc system is simple and effective, within the following limitations:

(i) Up to 50% of the useful signal is lost due to the necessity of chopping the hollow cathode lamp output.

(ii) The technique cannot adequately correct structured backgrounds, that is, background absorbances that change within the bandpass of the monochromator (see Figure 4.40), the correction signal being the average value for the wavelength range transmitted by the monochromator.

(iii) The technique is ineffective above about 350 to 400 nm, though corrections are not usually required in the visible region, especially in flame AA measurements; if necessary, a tungsten filament lamp could be substituted.

(iv) The magnitude of background absorbance that can be corrected depends on optical design of the monochromator, but is usually in the range 0.5 to 1.5 absorbance units.

Despite these limitations the deuterium arc is adequate for the correction of background interferences in almost all flame and many graphite furnace applications.

4.13.2 *The Zeeman effect*

Atomic absorption transitions normally arise from the excitation of electrons from a ground state orbital energy to an excited state (section 3.5). These 'resonance' absorbance lines are normally single lines of very narrow half width (<0.005 nm). However, if an atom is subjected to an intense magnetic field, the orbital energy levels become perturbed. In these circumstances, sub-orbital levels of nominally identical energy can become split into individual discrete energy levels (see section 3.5). In the simplest case, the excited orbital level splits into three discrete energy states labelled π, σ^+ and σ^-

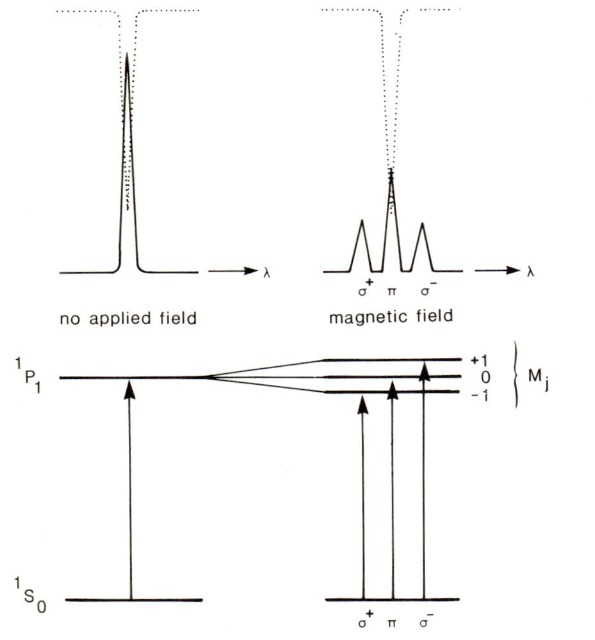

Figure 4.36 Electron transitions and atomic absorption spectra observed in the absence and presence of a magnetic field. The latter data shows the normal Zeeman effect. The dotted profile represents the atomic emission signal from a hollow cathode lamp.

(Figure 4.36). Transitions involving the central π band retain the original atomic absorption wavelength (though at one-half the original intensity). Transitions represented by the two side bands, σ^+ and σ^-, correspond to absorption lines that occur symmetrically on either side of the π band. The electron transitions that give rise to these bands are shown in Figure 4.36.

It is important to emphasize that this phenomenon occurs when *atoms* are subjected to the magnetic field, and is not simply due to the transmission of light through such a field. The magnitude of the wavelength difference between central and side bands depends on the intensity of the applied magnetic field.

The application of the Zeeman effect as a background corrector depends on one further phenomenon. Excitation or de-excitation of electrons to or from π as opposed to σ orbitals involves radiation polarized in different planes. Thus, radiation emitted in the central π component is polarized in a plane parallel to the magnetic field, that in the σ sidebands in a plane perpendicular to this. If an optical polarizer is placed in the light path of the instrument, these two components can thus be distinguished.

The Zeeman effect can be applied in one of two ways: either by applying a magnetic field to atoms in the hollow cathode lamp or alternatively to atomized samples in a hollow graphite furnace. It is not easy to adapt the technique to polarize atoms created in a conventional atomic absorption flame.

(*a*) *Normal Zeeman effect*. The line spectrum from a hollow cathode lamp is normally unpolarized and emitted at sharp atomic emission frequencies. If a strong magnetic field is applied to the lamp, atomic emission lines are split by the Zeeman effect due to the perturbation of electronic orbital energy levels described above. In the simplest case, each emission line is split into the three components, the π and σ lines, polarized at right angles to one another. Although background correctors can make use of this phenomenon (e.g. Koizumi and Yasuda, 1975, 1976*a*; Oishi *et al.*, 1980) such instruments have a reputation for instability and drift. Most practical instruments make use of the inverse Zeeman effect.

(*b*) *Inverse Zeeman effect*. Other considerations come into play if the magnetic field is applied to atoms generated in a hollow graphite furnace. Only atomic emission photons from the lamp that are orientated in a specific plane of polarization can induce atomic absorption phenomena. Atomic orbital energy levels are split by the magnetic field, as before. However, because of the way in which electron orbitals in atoms of the sample orientate themselves in the externally applied magnetic field, absorption can occur at three frequencies, corresponding to the wavelengths of the π, σ^+ and π^- bands (taking the simplest case). If light from a conventional (unpolarized) hollow cathode lamp is passed through the furnace, only atoms in the π orientation can be excited, and only that fraction of the lamp output that is polarized in a plane parallel to the magnetic field suffer absorption. Conversely, all planes of orientation of the lamp line can suffer non-specific (background) absorption.

Atoms with electron orbitals split into the σ sidebands cannot contribute to the atomic absorption signal because their absorption frequency has been shifted by the magnetic field outside the frequency range of the atomic emission line from the hollow cathode lamp. This phenomenon is known as the inverse Zeeman effect and is the basis of most practical background correction techniques applied to hollow graphite furnace AA.

(*c*) *Zeeman background correction instrumentation*. Two forms of instrumentation are available, using the *permanent magnet* and the *modulated magnet field*. The simplest form of instrumentation is shown in Figure 4.37. The hollow graphite furnace is surrounded by a permanent (dc) magnet capable of generating a large magnetic field (perhaps 10 kilogauss). A rotating polarizer is placed in the optical path between lamp source and furnace. The atomized sample in the furnace is thus excited by plane polarized radiation, the direction of which is controlled by the position of the polarizer. When this polarized light beam is parallel with the magnetic field, atomic absorption occurs, since exciting radiation is then in the correct plane to induce excitation of electrons to the π band. This absorption signal is superimposed on a background absorption. When the polarizer is rotated through 90°, atomic absorption cannot occur, since the lamp line no longer lies in the correct phase to induce atomic absorption, even though it possesses exactly the correct wavelength. However, in this orientation, non-specific background absorption will still occur. The difference between the absorption signal in these two limiting positions of the rotating polarizer is, therefore, the true background corrected atomic absorption signal. In practice, the provision of a rotating polarizer has not been entirely satisfactory. In some circumstances such a polarizer can cause unwanted reflections and represent an additonal source of noise. An alternative approach has been described,

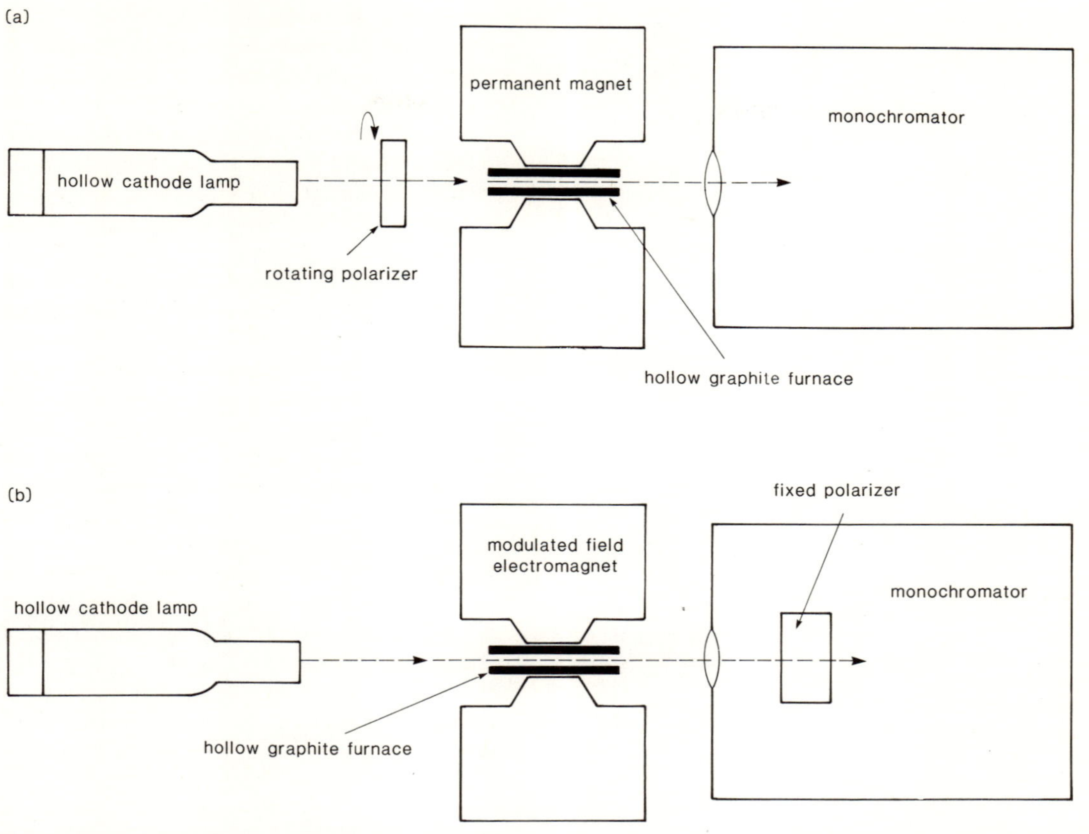

Figure 4.37 Instrumentation for background correction using the inverse Zeeman effect. (*a*) The graphite furnace is surrounded by a large dc magnet and a rotating optical polarizer is placed between hollow cathode lamp and furnace. (*b*) In an alternative design, the magnetic field is generated by an electromagnetic field excited by a modulated current and a fixed optical polarizer is now placed between furnace and monochromator.

for example by Liddell and Brodie (1980) in which the field created by the magnet is modulated.

In the modulated magnetic field, the rotating polarizer is replaced by a fixed polarizer now placed between furnace and monochromator. The polarizer itself is constructed from synthetic crystalline quartz. The permanent (dc) magnet is replaced by an electromagnet excited by current modulated at 50/60 Hz but still surrounding the furnace (Figure 4.37). When the electromagnet is switched off, unpolarized light from the hollow cathode lamp suffers both atomic and background absorption and the appropriate polarized phase can pass through the static polarizer and be detected. When the electromagnet is switched on, atomic absorption can no longer be detected, since the direction of the magnetic field is arranged such that atoms in the furnace will only absorb that component of the central π frequency perpendicular to the orientation of transmission of the polarizer. Absorption by atoms could be detected at the wing σ frequencies, since these have the appropriate orientation to be transmitted by the fixed polarizer. However, the wavelength required to induce absorption in these wing lines is again shifted out of the bandpass of the atomic emission line of the hollow cathode lamp. The only absorption signal detected, therefore, is that component of the hollow cathode lamp line, corresponding to the central π frequency, but perpendicular to the component that suffers atomic absorption, since this is the only radiation that can be transmitted by the fixed polarizer. This component will therefore suffer background absorption but not atomic absorption. The difference in absorption signals between magnetic field 'off' and 'on' is therefore a measure of the true background corrected absorption.

(*d*) *Normal and anomalous Zeeman splitting patterns.* The above is a rather simplified description of Zeeman background correction phenomena. The simple splitting pattern (Figure 4.36) is representative of only a fairly small number of atomic lines in which the transition of ground state to first excited state can be represented by the Russell–Saunders transition $^1S_0 \rightarrow {}^1P_1$ (section 3.5). Such atomic lines are said to display the *normal* Zeeman splitting effect. If other spectral terms are involved then the corresponding energy levels can be split by a magnetic field into a much more complex pattern in which a larger proportion of radiation is concentrated in side bands. Examples of such *anomalous* Zeeman splitting patterns are shown in Figure 4.38, from Koizumi and Yasuda (1976*b*). A further difficulty arises in that the magnetic field is not usually strong enough to shift the side bands completely out of the bandpass of the hollow cathode atomic emission line. A significant component due to atomic absorption will then contribute to the nominal background only absorption signal. This effect will reduce the sensitivity of the atomic absorption signal.

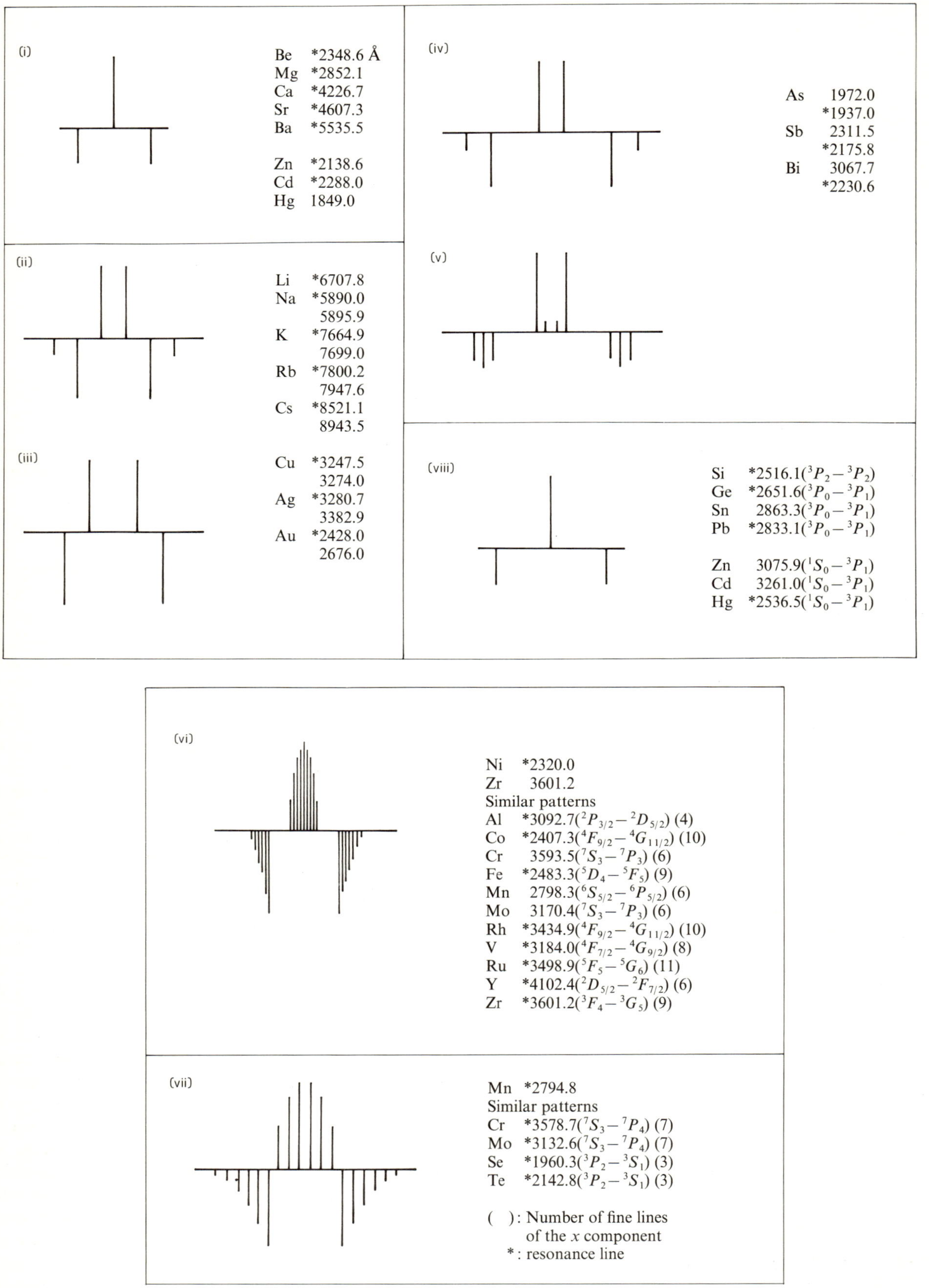

Figure 4.38 Splitting patterns produced by the normal (i) and anomalous (ii–vii) Zeeman effects. (i), $^1S_0-^1P_1$; (ii), $^2S_{1/2}-^2P_{3/2}$; (iii), $^2S_{1/2}-^2P_{1/2}$; (iv), $^4S_{3/2}-^4P_{1/2}$; (v), $^4S_{3/2}-^4P_{3/2}$; (vi), $^3F_4-^3G_5$; (vii), $^6S_{5/2}-^6P_{7/2}$; (viii), $^3P_0-^3P_1$, $^1S_0-^3P_1$, $^3P_2-^3P_2$. (Koizumi and Yasuda, 1976.)

It must further be emphasized that the type of Zeeman splitting depends on the characteristics of the two electron orbitals involved in the transition. Different lines in the atomic emission or absorption spectrum from the same element are expected to show different splitting patterns.

In summary, therefore, the effects of both complex band structure observed in the anomalous Zeeman splitting patterns and the incomplete separation of Zeeman side bands from the central band (as mentioned above) cause a loss in element sensitivity compared with deuterium arc background correction. The magnitude of this loss in sensitivity depends on the atomic absorption line selected for analysis and the corresponding Zeeman splitting characteristics of this line. This reduction in sensitivity is often less than 10–15% but can approach 40–50% for some elements (Table 4.13).

(*e*) *Calibration curve roll-over*. A serious problem can affect Zeeman corrected absorbance data. At high analyte concentrations the Zeeman side bands begin to broaden markedly. This causes an apparent increase in the background signal since an increasing proportion of atoms that can absorb radiation corresponding to the σ side band transitions, then become excited by the hollow cathode atomic emission line wavelength. The net background corrected atomic absorption signal is then progressively reduced in sensitivity as analyte concentration is increased. All calibration plots corrected by the inverse Zeeman effect, therefore, become curved at high absorbance values. Indeed at limiting high analyte concentrations, a roll-over effect is observed where the atomic absorption signal actually starts to decrease with increasing analyte concentration (Figure 4.39). Calibrations then become double-valued, since the same atomic absorption signal represents two concentrations. It is important to establish this roll-over point for individual elements and to restrict operation to lower concentrations.

(*f*) *Applications and advantages of Zeeman background correction*. For the reasons outlined above, the principal application of Zeeman background correction is in hollow graphite furnace measurements. The main advantage of Zeeman background correction over the deuterium arc method is that the background correction is performed at exactly the same wavelength as the absorption measurement. The accuracy of this procedure is not, therefore, affected by adjacent 'structured' background even if such fluctuations occur within the bandpass of the monochromator (Figure 4.40). Instrumentation for Zeeman background correction is, however, relatively bulky and expensive, although there are no lamp alignment problems. The main disadvantage of the technique is that correction is accompanied by some loss in sensitivity, and furthermore, calibration curves can become double-valued at high analyte concentrations. Nevertheless, Zeeman background correction is widely used to overcome intractable interference problems in some demanding furnace AA applications, and is capable of performing accurate background measurements up to at least two absorbance units.

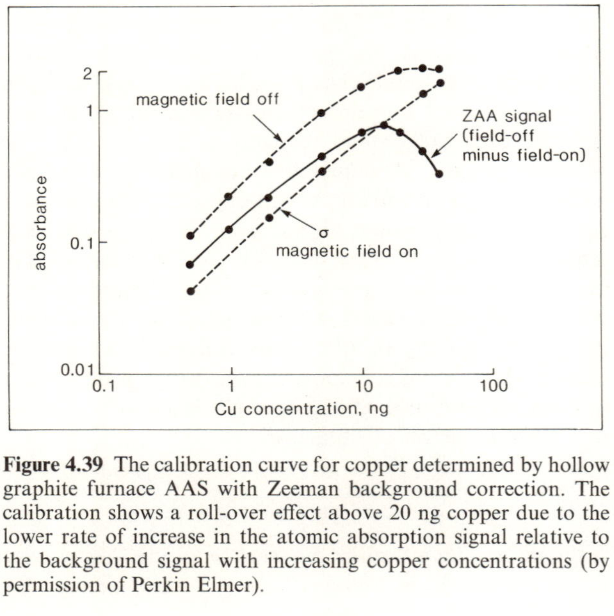

Figure 4.39 The calibration curve for copper determined by hollow graphite furnace AAS with Zeeman background correction. The calibration shows a roll-over effect above 20 ng copper due to the lower rate of increase in the atomic absorption signal relative to the background signal with increasing copper concentrations (by permission of Perkin Elmer).

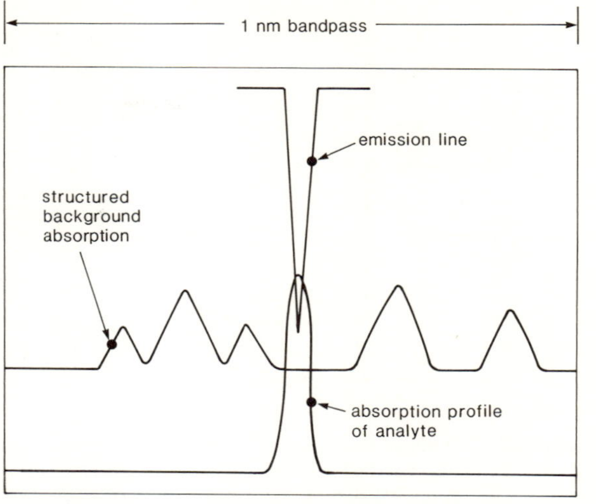

Figure 4.40 Structured atomic absorption backgrounds. This schematic diagram shows that the background absorption signal has peaks in it that are within the wavelength bandpass of the monochromator. Such background signals cannot be corrected adequately by the deuterium continuum method. (Sotera and Kahn, 1982.)

4.13.3 *Smith–Hieftje background correction*

The Smith–Hieftje background correction system operates on quite a different principle. From the earliest days of atomic absorption measurements, it was known that the operating current of hollow cathode lamps had to be carefully controlled. If too high a current was selected, the phenomenon of self-reversal was observed, in which the width of the atomic emission line broadened appreciably, but with a depression in the intensity at the peak centre (section 3.7). This degrades the sensitivity of atomic absorption measurements, and was avoided.

However, Smith, Schleicher and Hieftje (1982) developed a method of background correction based on this very phenomenon. Their system is based on conventional instrumentation using hollow cathode lamp excitation. The cur-

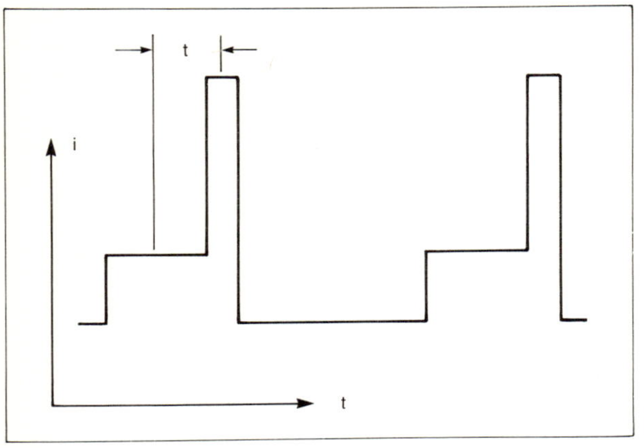

Figure 4.41 Modulated current used to excite the hollow graphite lamp for the Smith–Hieftje background correction system. The saw-tooth pattern represents the conventional lamp current. The short current pulses induce self-reversal in the hollow cathode lamp output for use in the background correction scheme; i = current, t = time. (By permission of Instrumentation Laboratory.)

rent to the hollow cathode lamp is modulated as in a normal instrument to distinguish the atomic absorption signal from ambient light and flame/furnace emission effects (section 4.2.5). However, during each modulation cycle, the current to the hollow cathode lamp is pulsed to induce self-reversal in the atomic emission line output (Figure 4.41). The atomic emission line profile broadens considerably with a corresponding reduction in intensity at the centre of the profile (Figure 4.42). Under conditions of self-reversal, the wings of the emission line fall outside the bandpass necessary to induce atomic absorption in the sample. These wings contribute only to the background absorption signal. At the same time, the relative proportion of energy at the centre of the atomic emission line profile is significantly reduced, causing a corresponding reduction in the relative magnitude of the true atomic absorption signal. The net effect is that during self-reversal the atomic absorption signal is much reduced but the background absorbs the same proportion of light as during normal hollow cathode lamp operation

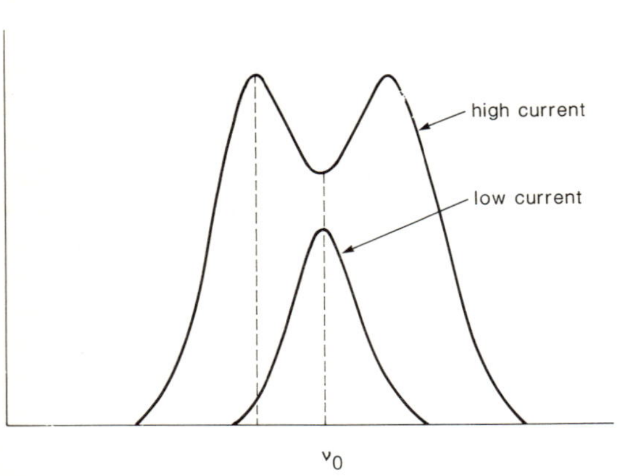

Figure 4.42 Wavelength emission profile as a function of operating current as used in the Smith–Hieftje background correction system. When pulsed with a high current, self-reversal is induced in the lamp output. (Sotera and Kahn, 1982.)

(Sotera and Kahn, 1982). The difference between signals obtained in the normal operating state and under conditions of self-reversal is therefore a measure of the true background corrected signal. However, since self-reversal is never complete and some atomic absorption always contributes to the background absorption signal, the overall sensitivity of the corrected signal is reduced. At the time of writing, no data are available to the author to compare the magnitude of this loss in sensitivity with that also experienced in the Zeeman background correction system. It seems likely that different elements will have sensitivities that are reduced by different amounts by each of the correction systems.

The Smith–Hieftje system does not suffer the same roll-over problem in calibration curves as affects Zeeman corrections. In consequence, it is claimed that this system can be applied over a wide range of concentrations and is capable of background correction for absorbances up to three units (0.1% transmittance). The Smith–Hieftje technique is applicable to flame as well as furnace atomization, can be used over the entire spectral range and requires no special instrumentation except a well-insulated hollow cathode lamp and associated power supplies. The technique is optically more efficient than either Zeeman or deuterium arc correction systems, although a reduction in sensitivity is expected compared with deuterium arc correction due to incomplete self-reversal in the hollow cathode lamp output. Arguably, the range of wavelengths over which background absorbance is sampled in the Smith–Hieftje system is slightly wider than that using the Zeeman effect.

4.13.4 *Choice of background correctors*

There is no universal choice of instrumentation for background correction, but it is possible to set out some general principles. For the majority of flame measurements, background correction is not required at all. The few cases where it is necessary generally correspond to measurements on lines in the uv below 250 nm. There is then little evidence that the traditional deuterium arc technique is not adequate, although, if competitively priced, the Smith–Hieftje corrector may be equally applicable for some elements.

For the majority of furnace meaurements, background correction is essential, the exception being for a few elements determined from atomic absorption lines in the visible region above say 450 nm. If optimum sensitivity is important, then the choice of background corrector is influenced by the specific AA line to be used for measurement. There are groups of elements for which each of the three background corrector systems would be preferred (including the deuterium arc system for elements such as Si, Hg, Al and Ti). However, if such measurements are affected by structured background (see Figure 4.40), the choice lies between the Zeeman and Smith–Hieftje systems. Selection is likely to be made on grounds of economy, element range and on other instrumental features of the spectrometers to which the respective correctors are attached.

4.14 Geological applications of furnace AAS

Although graphite furnace AA analysis is used extensively in environmental, hydrological and biological studies, this

Table 4.14 Trace element analysis by furnace atomic absorption spectrophotometry

Element	Line (nm)	Furnace temperatures Ash (°C)	Atomize (°C)	Clean-out (°C)	Blank level in flux (ppm)	Estimated detection limit (ppm)	NBS river sediment Certified value (ppm)	Analysed value (ppm)	CV (%)
As	193.7	1000	2700	2800	0.8–2.4	5	(66)	81	4.9
Cd	228.8	500	2000	2500	0.05–0.15	0.5	10.2	9.1	4.4
Cr	357.9	1200	2700	2800	0.3–1.0	1.4	2.96%	2.82%	2.5
Cu	324.7	1000	2500	2800	0.7–1.8	1.7	109	106	7.1
Co	240.7	1000	2700	2800	0.2–0.5	1.5	(8)	7.8	2.6
Mn	279.5	1100	2700	2800	0.2–1.5	1.8	785	791	0.8
Ni	232.0	1000	2700	2800	0.6–2.5	2.5	45.8	43.8	8
Pb	283.3	1000	2500	2800	0.8–2.5	3.3	714	699	3.4
V	318.4	1500	2700	2800	0.7–2.6	2.5	23.5	27.7	3.6
Zn	213.9	700	2300	2500	1.5–3.6	4.2	1720	1659	3.7

Data from Bettinelli (1983).
Reference material is NBS/SRM 1645.
Analysis conditions: drying cycle: 110°C for 45 s; ashing cycle: 60 s; atomization cycle: 5 s; clean-out cycle: 5 s; volume of sample solution injected: 20 μL; absorption measurements taken in a continuous argon flow mode.
The blank composition of flux represents the range of concentrations found in different batches of lithium tetraborate from four suppliers.
Detection limit is an estimate of the six sigma detection limit derived from Bettinelli's data. The original values have been multiplied by three (on the assumption that they were quoted as two-sigma limits) and recalculated as ppm rock assuming that 0.3 g sample is taken up into 250 mL solution.
Certified data for NBS/SRM-1645 are National Bureau of Standards certified values, data in brackets represent reported values only.
CV = coefficient of variation (% relative) derived from four analyses.
Chromium data was determined by flame AAS.

technique has found much more specialized application to geological samples, mainly because it is almost always necessary to separate the elements of interest from silicate matrix to avoid complex interference effects. Some details of atomization conditions and matrix modifiers used in the determination of selected trace-elements using L'vov platform are listed in Table 4.13.

However, in spite of these reservations, Bettinelli (1983) described a scheme of analysis in which eleven elements (As, Cd, Co, Cr, Cu, Fe, Mn, Ni, Pb, V and Zn) were determined in siliceous materials without chemical separation. Samples were fused in a platinum crucible with lithium tetraborate at 1000°C for 1 hour (0.3 g of sample to 1.5 g of flux). The cooled residue was then dissolved in hydrochloric acid and diluted to a final volume of 250 mL. In order to retain arsenic qualitatively, 78 mg of nickel nitrate was added to separate fusion mixtures from which this element was to be determined. Interferences that affect atomization were investigated. The presence of nickel stabilized the atomization of arsenic. The addition of phosphoric acid was found to reduce the interference of lithium tetraborate on lead. The presence of lithium tetraborate in all analysed solutions appeared to have buffered interference effects in other elements. The resultant analysis conditions are listed in Table 4.14, together with the troublesome blank levels found in the lithium tetraborate flux and estimates of detection limits. In the analysis of an NBS river sediment (SRM-1645) precision errors are generally within the range 1–5% (coefficient of variation) although at relatively high trace concentrations.

This work, unlike most published applications, indicates the performance of graphite furnace AA without a chemical separation for the analyte, but Sighinolfi (1973) determined thallium by acid digestion of the sample. The resultant residue was taken up in hydrobromic acid and thallium (as well as cadmium and possible small amounts of other components) extracted as bromides into di-isopropyl ether. Detection limits of the order of 50 ppb Tl using 0.2 g of sample material were claimed for this method. Elson *et al.* (1978) used a solvent extraction procedure followed by flameless AA to determine Ag, Cd, and Pb in rock reference materials. In this procedure the appropriate complex with dithiocarbamate was extracted into 4-methylpentan-2-one (methyl isobutyl ketone). This organic solution was injected directly into the furnace, calibration being achieved by the method of standard additions. Schnepfe (1979) determined germanium by solvent extraction (into toluene) and Barredo *et al.* (1978) determined strontium without a chemical separation step. Matrix interferences were serious, so that geochemical standards that matched the composition of the analyte were recommended. Detection limits of the order of 0.2 ppm Sr in rock were claimed.

Of particular interest is the determination of REEs by graphite furnace AA. Sen Gupta (1981) described results of such analyses. 1–3 g of sample were dissolved by acid attack (HF, HNO_3 and $HClO_4$). Any acid-resistant residue was filtered off and fused with potassium pyrosulphate. After dissolving the fusion products in dilute sulphuric acid, the resultant solution was combined with the main batch. The rare-earth elements were separated from matrix components by precipitation using methyl oxalate in the presence of calcium as carrier. After filtration and washing, the residue was redissolved in nitric acid, and calcium removed by precipitation as the hydroxide, using hydroxylamine hydrochloride and ammonia solution and in the presence of iron as carrier. The residue containing REEs, Sc, Y, Th and added Fe was taken up into solution and diluted into appropriate aliquots equivalent to 0.1 to 0.004 g of the original rock per mL (see Table 4.15). Furnace atomization conditions and an estimate of detection limits are also listed in Table 4.15, together with analytical data for two reference materials USGS:BIR-1 and W-2. Cerium and lanthanum were not determined in this scheme of analysis because furnace atomization of these elements gave strong negative absorbances. Subsequently, Sen Gupta revised these procedures, investigating the use of a pyrolytically coated graphite furnace lined with tantalum foil to enhance the sensitivity of cerium (Sen Gupta, 1984), and a combined ion-exchange separation plus tantalum-lined graphite furnace AA procedure for the determination of Sc, Y and the rare earths (Sen Gupta, 1985). In this latter work, the elements La,Ce, Pr, Gd, Tb and Lu were determined by atomization from a tantalum-lined furnace, the other elements being determined from a pyrolytically coated graphite furnace. Interferences were observed from minor concentrations of the elements Ca, Al, Mg, Ba, and Sr that were incompletely separated from the REE solution. These could be eliminated by heating the sample to 1800° to 2000°C before atomization at 2700°C (in a pyrolytically coated graphite furnace) or at 2600°C in a tantalum-lined furnace (tantalum foil degrades rapidly at 2700°C). Other rare-earth elements were found to cause a serious interference in the determination of cerium, and satisfactory measurements could be made only after chemical separation of cerium. This was achieved by co-precipitation of cerium from the ion exchange REE fraction, first with calcium oxalate and then with ferric oxide. Rare-earth element analyses by these procedures for granite NIM-G and basalt USGS: BHVO-1 are listed in Table 4.15. Horsky and Fletcher (1981) also combined an ion-exchange separation with graphite furnace AA in the determination of La, Sm, Eu, Yb and Lu in silicate rocks. Although satisfactory atomization of Sm, Eu and Yb could be achieved at 2800°C, the optimum temperature for La and Lu was above the 3000°C limit of the furnace; at this temperature, sensitivities were poor and the graphite furnace deteriorated rapidly.

4.15 Cold vapour and hydride generators

4.15.1 *Determination of mercury by cold vapour generation*

The atomic spectrum of mercury shows two prominent lines which are of use for analytical determinations. The principal atomic absorption line is at 184.9 nm, but at this wavelength both air and combustion products of flames absorb radiation so heavily that practical determinations cannot normally be made using flame atomization. Instead the 253.7 nm mercury line must be selected with an accompanying fall in sensitivity of between 20 and 40 times. This in fact means that mercury cannot normally be detected at the ultra-trace levels found in many silicate rocks. To overcome these difficulties, advantage has been taken of the unique property of metallic mercury, that of its high vapour pressure at room temperature. This is the basis of the cold vapour atomic absorption determination of mercury.

The principle of the technique is that the sample is taken into solution and, in a suitable reaction vessel, mercury ions in solution are reduced (normally with alkaline solution of sodium borohydride) to metallic mercury. The mercury vapour is bubbled out of the reaction solution by hydrogen that is also formed during the reduction reaction and passed to an absorption cell with quartz end-windows. The longitudinal axis of this cold vapour cell is arranged to coincide with the optical path of an AA spectrometer, replacing the conventional burner flame. Atomic lines from a mercury lamp suffer absorption when the mercury vapour is swept through the cell. The transient mercury atomic absorption signal is recorded. It is normally necessary to place a desiccant trap between reaction vessel and absorption cell to absorb water vapour and prevent condensation on the quartz end-windows. A high-intensity mercury electrodeless discharge lamp must normally be used to compensate for the absorption in air of the 253.7 nm mercury line. The general arrangement of reaction vessel and cold vapour cell is shown in Figure 4.43.

Thus, Sighinolfi *et al.* (1980) digested silicate samples (0.1–0.5 g) with a sulphuric acid–nitric acid mixture in a closed Teflon bomb at 180°C for 90 minutes. The closed reaction vessel was required to prevent premature losses of mercury vapour. The H_2SO_4–HNO_3 acid mixture was selected to ensure that all mercury would be leached out of sulphide minerals and other micro-inclusions even though the entire host silicate matrix was not taken into solution. After cooling, the reaction mixture and any unattacked residue were transferred to a polypropylene bottle and the volume adjusted to exactly 10 mL. This mixture was then transferred to the reaction vessel and mercuric ions reduced to the metal using 10 mL of sodium borohydride (3% $NaBH_4$) dissolved in 1% sodium hydroxide solution:

$$Hg^{2+} + 2BH_4^- \rightarrow Hg^0 + H_2\uparrow + B_2H_6 \qquad (4.13)$$

Hydrogen gas is liberated during this reaction and drives mercury vapour out of the reaction solution into the quartz absorption cell for measurement.

In the analysis of geological materials, vapour phase interferences in the absorption cell are probably negligible. Of more concern is the presence of gold, platinum group elements, selenium and tellurium in the sample solution. These elements can interfere with the release of mercury vapour by competing reactions involving the formation of amalgams, selenides, tellurides or other stable mercury compounds. Checks made by Sighinolfi *et al.* (1980) suggest that at the very low concentration at which these interfering elements normally occur in silicate rocks, these interferences can be ignored. In general the performance of the technique is limited not by analytical sensitivity but by the reagent blank. Even high-purity reagents can contain detectable amounts of mercury, not the least caused by absorption of vapour from some laboratory atmospheres. Sighinolfi *et al.* (1980) measured reagent blanks of 2.5 ng of mercury per 10 mL reagent solution. The derived detection limit (six sigma) for a 0.5 g sample aliquot is then 30 ppb Hg in whole rock. Analytical precision was found to be 2 to 3% (coefficient of variation) at the 100–200 ppb level. There is some concern over the stability of mercury sample and standard solutions. Mercury is very mobile and can escape by diffusion or absorption on to the walls of some kinds of plastic containers (section 1.7). Sighinolfi *et al.* (1980) daily prepared a range of working standard solutions from the 1000 μg/mL stock solution. The concentration of H_2SO_4 and HNO_3 was

Table 4.15 Analysis of rare-earth elements by graphite furnace AAS

Element	Line (nm)	Sen Gupta (1981)							Sen Gupta (1985)				
		Atomization temperature (°C)	Estimated detection (ppm rock)	Analysed concentration (g rock/mL)	USGS: BIR-1 Expected (ppm)	USGS: BIR-1 Analysed (ppm)	USGS: W-2 Expected (ppm)	USGS: W-2 Analysed (ppm)	Atomization temperature (°C)	NIM-G Expected (ppm)	NIM-G Analysed (ppm)	USGS: BHVO-1 Expected (ppm)	USGS: BHVO-1 Analysed (ppm)
La	550.1								2600 (Ta)	105?	120	17?	19
Ce	567.0								2600 (Ta)	200	200	39	49
Pr	495.1	2300	3	0.05–*0.2*	—	1	—	12	2600 (Ta)	—	18	—	7
Nd	492.5	2500	2	0.05–*0.1*	—	4.2	13.5?	24	2700 (PG)	68?	70	24	29
Sm	429.7	2700	0.4	0.05–*0.1*	1.0?	1	3.3?	3	2700 (PG)	16?	16	6.1	6
Eu	459.4	2500	0.2	*0.05*	0.55?	0.6	1.1	1.2	2700 (PG)	0.4?	0.5	2.0	2
Gd	368.4	2700	1	0.05–*0.1*	—	5	—	12	2600 (Ta)	—	12.4	6.0	6.6
Tb	432.7	2700	1.5	0.05–*0.1*	—	0.8	0.65?	1	2600 (Ta)	3?	3	1.0?	1
Dy	421.2	2500	0.05	*0.05*	3.7?	3.0	—	3.8	2700 (PG)	(15)	17	5?	5
Ho	410.4	2500	0.05	*0.05*	—	0.5	—	0.8	2700 (PG)	—	4.6	—	1
Er	400.8	2500	0.1	*0.05*	—	2.3	—	3.1	2700 (PG)	—	12	—	2.4
Tm	371.8	2500	0.01	*0.05*	—	0.4	0.4?	0.5	2700 (PG)	(2)	2	0.3?	0.3
Yb	398.8	2500	0.05	*0.004*	1.7?	1.9	2.1	2.2	2700 (PG)	14	15	1.9	2.4
Lu	336.0	2700	0.2	0.05–*0.2*	0.29?	0.7	0.33?	0.8	2600 (Ta)	2?	2	—	0.4
Y	410.2	2500	0.4	0.05–*0.1*	16?	17	23?	25	2700 (PG)	145	122	27?	27
Sc	391.2								2700 (PG)	1?	1	31	28

Data from Sen Gupta (1981, 1985).
The estimated detection limit is recalculated from the mass of element that gives an absorbance of 0.0044 units (Sen Gupta, 1981), assuming that 10 μL of solution is injected into the furnace at the rock dilution indicated in the table above (italicized). The resultant data have been doubled to estimate the six-sigma detection limit.
Expected compositions taken from Abbey (1983).
Ta = sample atomized from a tantalum foil-lined furnace.
PG = sample atomized from a pyrolytic graphite furnace.
Ce was determined after individual ion-exchange separation from other rare-earth elements.

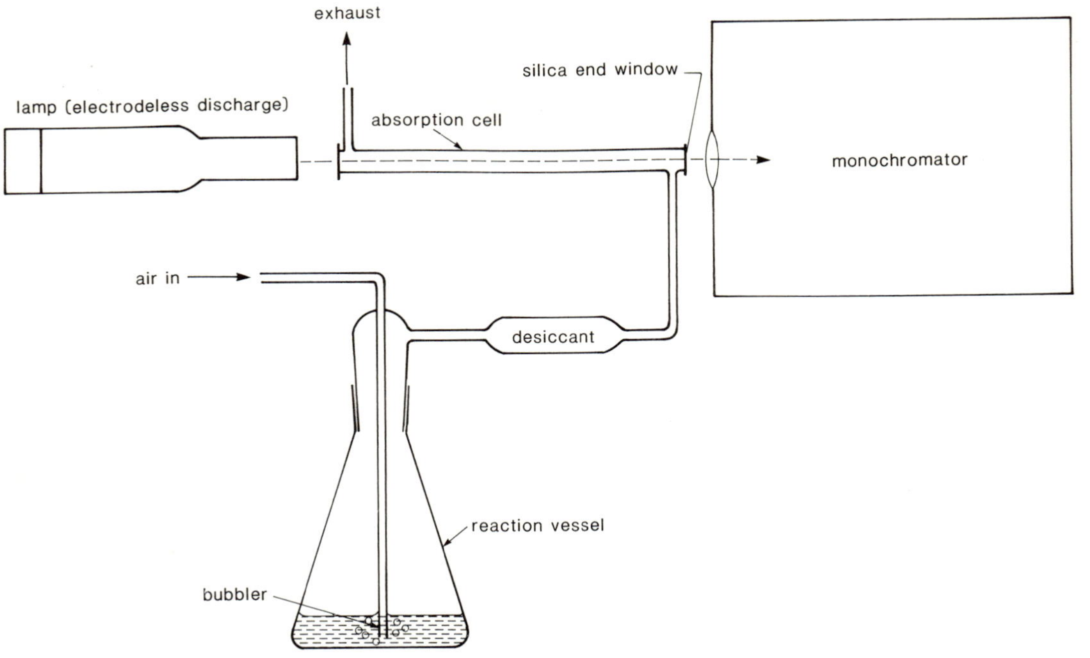

Figure 4.43 Apparatus for the cold vapour generation method of determining mercury. Mercury vapour, liberated in the reaction vessel, is swept into a quartz window absorption cell for measurement of the mercury atomic absorption signal.

adjusted in each of these to match the sample solutions. All standard solutions were stabilized by the addition of a few drops of 5% $KMnO_4$ or $K_2Cr_2O_7$ solution to prevent the premature formation of metallic vapour.

An alternative method of determining mercury has been described by Flanagan *et al.* (1982). Mercury was driven off from 500 mg samples by combustion at 1000°C and absorbed on gold chips by amalgam formation. These gold chips are subsequently ignited in an induction furnace and the evolved mercury vapour passed to a quartz absorption cell for measurement. These workers carried out a detailed study of the homogeneity of mercury in unopened bottles of six USGS rock reference materials, there being some doubt as to whether the wide scatter of data reported for this element was due to the inhomogeneous distribution between various splits.

Within uncertainties in the analytical procedure, mercury was found to be homogeneously distributed. However, it was reported that the high and variable mercury content of diabase USGS: W-1 probably resulted from the use of mercury fulminate detonators in the quarry before collection of this material. Such are the hazards of rock analysis!

4.15.2 *Hydride generators*

Arsenic and selenium are difficult elements to determine by flame AA because their absorption lines (at 193.7 nm and 196.0 nm respectively) are heavily absorbed by air and normal flame gases. Early workers in this field, including Holak (1969), Fernandez and Manning (1971) and Manning (1971) overcame these analytical difficulties by first converting the elements into volatile hydrides. In current practice, these hydrides are liberated by reduction with either stannous chloride or sodium borohydride solution, and may then be swept into an *argon–hydrogen flame* for conventional atomic absorption analysis. A special gas nebulizer is used to premix analyte gases with argon. As the sample is introduced into the nebulizer in gaseous form, the efficiency of nebulization is 100% compared with about 10% for the conventional pneumatic nebulizer. This leads to a useful improvement in sensitivity. The argon–hydrogen flame is much cooler than a normal air–acetylene flame and is therefore more appropriate for the determination of these volatile elements. The argon–hydrogen flame is also far more transparent to light at 190 nm than flames formed from more conventional gas mixtures.

Alternatively, the hydride gases are swept into a long narrow-bore *quartz tube* which is positioned in and is heated by an air–acetylene flame using conventional slot burner. As hydride gases pass through this heated tube, thermal decomposition occurs, releasing atoms of the element. The atomic absorption signal is measured by light passing down the longitudinal axis of this absorption cell. The tube must be of narrow bore to prevent dilution of analyte by expansion into a wide tube. Since the lamp output does not pass through the flame, but through the absorption cell itself, transmission is high. These instrumental arrangements are shown in Figure 4.44.

Hydride generation techniques may be applied to the elements As, Bi, Ge, Pb, Sb, Te and Sn. Sensitivities of all except Pb (and Ge) are improved using argon–hydrogen flame atomization compared with more conventional flame analysis.

As an example of hydride generator operation, Rubeska

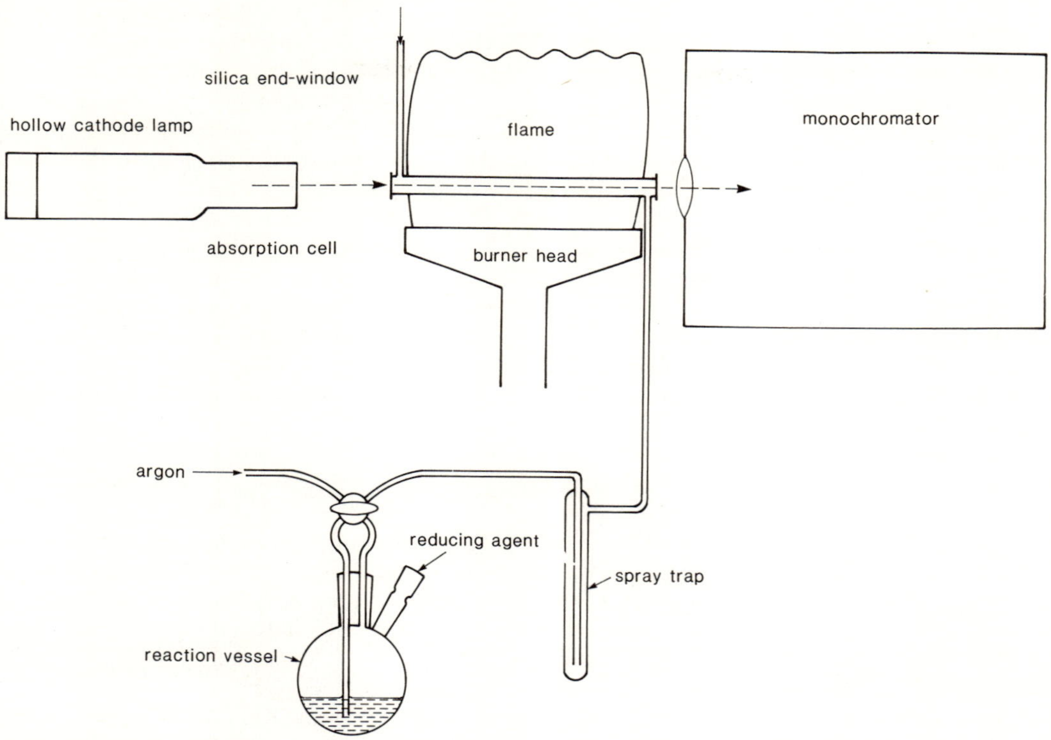

Figure 4.44 Apparatus for the determination of As, Se, Te, etc., by hydride generation. The volatile hydride formed in the reaction vessel is swept into the heated silicon tube where decomposition occurs, releasing atomic species for absorption measurement.

and Hlavinkova (1979) determined arsenic in rocks and soils and reviewed sample decomposition procedures and interference effects. Samples of 0.5 g were digested with aqua regia (10 mL), standing first at room temperature for one hour and then at 90°C for two hours in an open beaker. When almost dry the residue was taken up in hydrochloric acid (10 mL, 1 : 1) and diluted to 50 mL. In samples containing high concentrations of sulphide minerals, low arsenic results were obtained, probably due to the occlusion of arsenic in precipitated sulphur. To attack these samples, inverted aqua regia ($HCl:HNO_3 = 1:3$) was substituted to ensure complete oxidation of sulphur. Although this acid attack procedure does not dissolve silicate phases in the sample, investigations suggest that arsenic is quantitatively recovered.

After dissolution of the sample, an appropriate volume of this solution (containing 20 to 200 ng As) was added to the reduction vessel. Because the rate of hydride formation from As(III) is different from that of As(IV), it was necessary to pre-reduce the sample solution using a solution of potassium iodide (10%) in 3% HCl+ascorbic acid (1%) for stabilization. All the arsenic was then converted to the As^{3+} form, a reaction time of at least 5 minutes being required to ensure completion. Arsenic was then reduced to the hydride using a 10% solution of sodium borohydride ($NaBH_4$) in NaOH (2.5%). Evolved gases were swept into a heated quartz tube cell for atomic absorption measurements.

The principal interferences in this technique arise from elements in solution that interfere with the reduction of arsenic(III) to arsine, AsH_3. In geological samples these include Ni, Co, Cu and possibly Fe. The mechanism of this interference may involve the simultaneous reduction of ions of these elements to the free metal, which can then form arsenides or can possibly chemically decompose to arsine. Rubeska and Hlavinkova investigated the effect of these metals and found that interferences occurred only at concentrations exceeding 10 μg Ni, 100 μg Co and >1 mg Cu or Fe. As these levels are unlikely to be exceeded in normal silicate rock analysis, they were ignored. Care must also be taken in the storage of all reagent solutions, especially $NaBH_4$, since this alkaline solution can leach arsenic out of some types of glass bottle. Detection limits using this technique were significantly better than 0.2 ppm in rock.

A similar analytical procedure was used by Imai *et al.* (1984) to determine selenium in geological reference materials. The sample was digested by heating with a mixture of $HF/HNO_3/HClO_4$, which must not be allowed to evaporate to dryness, otherwise losses of selenium will occur. Interferences in the generation of selenium hydride were observed due to the presence of Cu, Ni, As and Sn. These interferences were suppressed by the addition of ferric chloride solution to reaction mixtures. If high concentrations of these interfering elements are present, a solvent extraction and graphite furnace AA determination was preferred to hydride generation. The silica absorption cell was heated electrically rather than in an air–acetylene flame. The detection limit (six sigma) was established to be 30 ppb, with a precision error in the order of 12% (one sigma) at the 100 ppb level.

4.16 Solid sampling and novel atomization devices

4.16.1 *Solid sampling*

One area of current active research in atomic absorption analysis is the atomization of samples that have been introduced into a graphite furnace in the solid form. The advantage of such procedures is, of course, that virtually no sample pretreatment is required (apart from grinding to a uniformly small sample size). The general application of this technique has not yet been proved, although some success has been obtained in the analysis of specific elements in simple sample matrices.

Several workers have reported analyses of silicate rocks by solid sampling techniques. For example, Langmyhr and Thomassen (1973) determined rubidium and caesium by solid sample introduction into a specially adapted graphite furnace. The sample mass presented for analysis was 1 to 10 mg, ground to pass a 325 mesh sieve in order to minimize sampling errors. The powder was deposited in a hollow graphite furnace, dried for 60 s at 180°C and then atomized at about 2000°C for 120 s. The appropriate atomic absorption signal was integrated over the atomization period. The authors were concerned that the analysed mass should be representative of the bulk sample. They considered that since Rb and Cs were expected to follow K in silicate rocks, it was reasonable to assume that these elements were homogeneously distributed between matrix phases. A 10 mg sample should, therefore, be adequate. Results indicated that detection limits down to the 0.5 ppm level could be obtained for Rb and Cs by atomizing, typically, 1 mg sample aliquots. Precisions were only slightly inferior to those expected by flame AA.

Govindaraju *et al.* (1973, 1977) determined caesium (1973) and silver (1977) by a flame atomic absorption technique using solid powder introduction. In this unusual technique the sample, milled to a particle size of 30 to 70 μm, was coated uniformly on to an iron screw rod before being atomized in the flame. The rod was first wetted with ethanol and then dipped into the sample powder. With the appropriate manipulations a uniform powder coating could be obtained. The sample rod was then burnt in an air–acetylene flame and the integrated atomic absorption signal recorded. Calibration was achieved by means of suitable solid samples. In the determination of silver, Govindaraju *et al.* (1977) successfully carried out a homogeneity check on a proposed reference material using only 50 mg sample aliquots, coefficients of variations of 10% being obtained at the 2 ppm level.

4.16.2 *Alternative devices*

The examples above illustrate two diverse procedures for solid sample introduction in AA analysis. Several research groups have proposed alternative methods of sample intoduction for both solids and solutions using both flame and electrothermal atomization. Examples of such proposals include the following.

(*a*) *The Kahn boat.* Kahn *et al.* (1968) described an atomization procedure in which the sample (solution or solid) was loaded into a tantalum boat (about 50 mm long by 1–2 mm wide). Solution samples could be dried by placing the boat close to a conventional air–acetylene flame before rapidly heating the boat in the flame itself. Volatile elements (including As, Cd, In, Pb, Hg, Se, Ag and Tl) are atomized

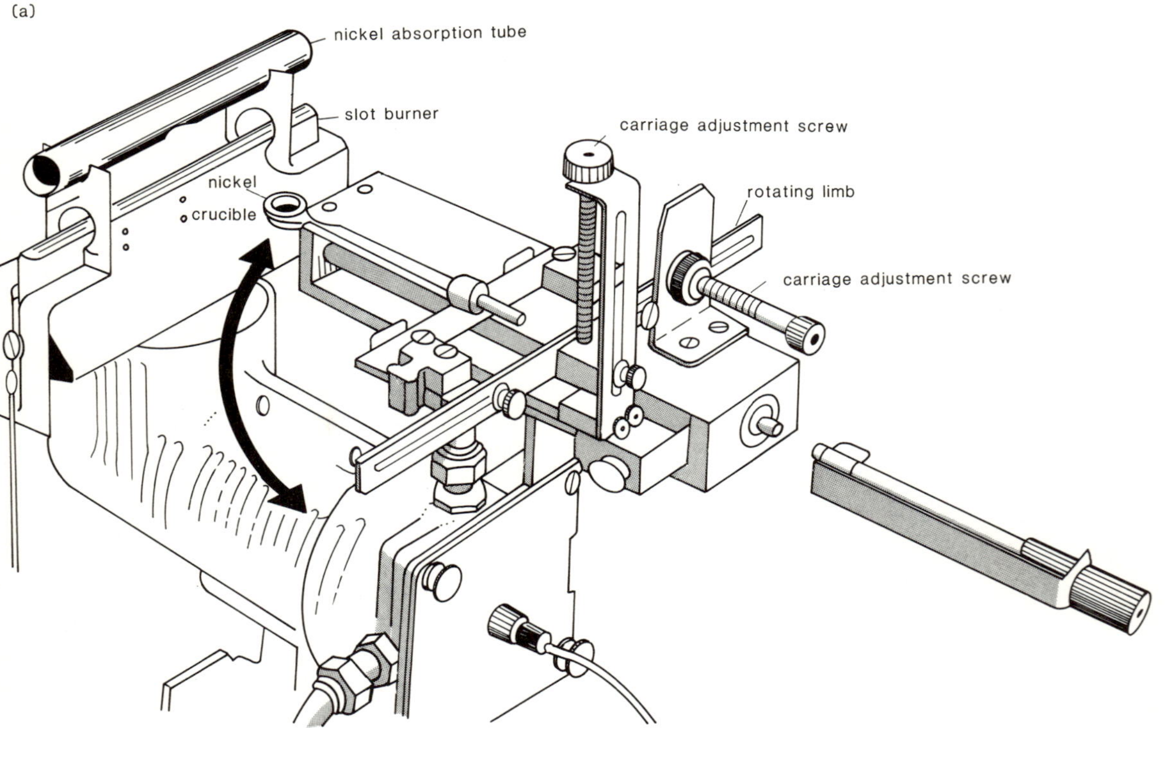

and the transient atomic absorption signal measured. Detection limits are significantly better than conventional flame atomization, although the technique has been mainly used to analyse biological and environmental samples.

(*b*) *The Delves cup*. This consists of a micro-crucible that may be heated in a flame in conjunction with a nickel or ceramic tube through which the hollow cathode lamp radiation passes, as shown in Figure 4.45 (Delves, 1970). The sample, loaded in the Delves cup, is first ashed close to the flame. It is then heated in the flame itself. The sample is partially atomized. The atomic vapour passes through a hole in the nickel or ceramic tube where it is retained in the optical path for a longer period than would be expected during normal flame atomization. High sensitivity can be achieved for volatile elements such as Pb, Cd and Tl, although reproducibility is often reported to be poor. This atomization technique was developed for the analysis of blood (and environmental) samples.

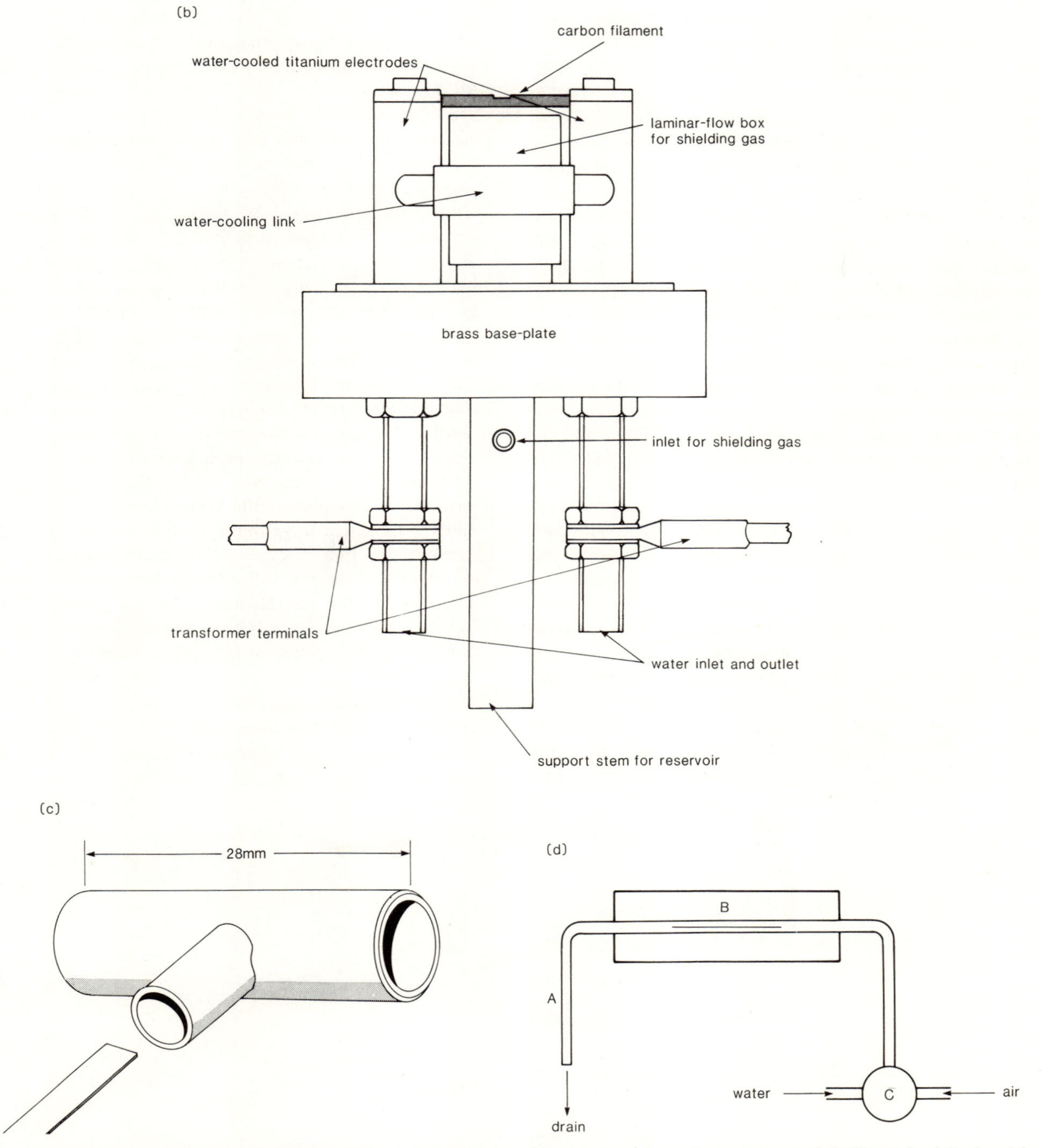

Figure 4.45 Novel atomization arrangements. (*a*) The Delves cup (Delves, 1970); (*b*) the West rod (taken from Ebdon, 1982); (*c*) the graphite probe and tee-tube furnace (Slavin and Manning, 1982); (*d*) atom trapping using a water-cooled silica tube (*A*) placed in flame over a conventional slot burner (*B*). (*C*) is a water/air valve. (Khalighie *et al.*, 1979.)

(*c*) *The West rod.* West and Williams (1969) reported an alternative method of electrothermal atomization in which the sample was mounted on a carbon rod 40 mm long and 1–2 mm in diameter (Figure 4.45). This filament was heated electrically in a manner similar to the hollow graphite furnace, using a stepped heating program. In the original design, the carbon rod was enclosed in a quartz tube flushed with argon to prevent oxidation. Later versions surrounded the rod with a sheath of argon. The transient atomic absorption signal is measured in the cloud of atoms formed immediately above the surface of the filament. Carbon rod atomizers have been reported to give good analytical sensitivity, although detection limits are not appreciably improved compared with conventional hollow carbon tube atomization.

Further developments of these principles include the substitution of a tantalum wire loop for the carbon rod (e.g. West *et al.*, 1979) and the development of a field, portable AA instrument based on atomization from tantalum wire and incorporating a Zeeman background correction system (Casteldine *et al.*, 1983).

(*d*) *The graphite probe constant temperature furnace.* Slavin and Manning (1982) sought to overcome some of the deficiencies of the conventional hollow graphite tube furnace caused when samples are atomized into an atmosphere that is still varying in temperature. They described a system whereby the sample was mounted on a strip of graphite sheet which could be introduced through a side tube into a conventional graphite furnace (Figure 4.45). The sample, mounted on the graphite probe, was dried and ashed (charred) outside the modified tube furnace. The latter was then pre-heated and when the atomization temperature had been reached, the sample probe was inserted through the side tube. Atomization thus occurred into an atmosphere that had already reached constant temperature. This new furnace showed significant improvement over wall atomization from a conventional tube furnace in reducing matrix interferences and the premature atomization of analyte. These advantages were not as clearcut when compared with conventional platform atomization.

(*e*) *Other atomization devices.* These include a slotted quartz tube in conjunction with an air–acetylene or air–hydrogen flame to determine elements such as As, Se, Sb, and Hg without the necessity of hydride or cold vapour generating apparatus (Watling, 1977). Khalighie *et al.* (1979, 1982) described a water-cooled silica tube situated above the primary reaction zone of an air–acetylene flame to condense out certain elements (e.g. Cu, Cd and Se) after conventional nebulization and flame atomization (Figure 4.45). After a suitable collection period, the cooling water is ejected with compressed air, the skin of the silica tube heats up, and releases the atomic species to the flame. The transient atomic absorption signal is measured immediately above the silica tube permitting detection limits to be achieved that are one to two orders of magnitude lower than by conventional flame atomization. Walsh (1982) discussed the merits of atomization by cathodic sputtering in a glow discharge.

(*f*) *Atomic fluorescence.* This technique has not yet proved its potential for routine analytical measurements. The technique involves exciting an atomic vapour, generated by one of the methods described above, with a polychromatic source of radiation. Atoms of individual elements fluoresce, emitting their characteristic line spectra. The technique, therefore, offers the potential of determining several elements simultaneously, provided the atoms of each element can be excited efficiently and the corresponding fluorescence line detected individually. Samples are usually atomized in either an argon-separated air–acetylene flame burnt from a burner designed to give a circular flame, or an inductively coupled argon plasma. Several sources of excitation have been used to induce a sufficiently intense atomic fluorescence signal, including the hollow cathode lamp (providing the output is pulsed), electrodeless discharge lamps, vapour discharge lamps, lasers (including the tunable dye laser), and even the emission from a second inductively coupled argon plasma. Such systems have not yet found practical analytical applications mainly because refractory compounds in the flame scatter into the detector light of the same wavelength as that of the atomic fluorescence radiation. The amount of scatter is variable, being matrix-dependent. It is difficult to devise an accurate method of correcting for this phenomenon. Atomic fluorescence techniques have been reviewed by Walsh (1982) and are discussed in more detail for example by Price (1979, pp.103–111).

The techniques described above are by no means an exhaustive review of developments in atomic spectroscopy but have been selected to give some indication of the direction developments in this field are taking.

5 Inductively coupled plasma–atomic emission spectrometry

5.1 Historic development and analytical capabilities

Optical emission spectrometry is a well proven analytical technique, in widespread use for the last 30 to 40 years. However, early devices available as optical emission sources (arc, spark or dc electrical discharges) had a number of disadvantages that prevented their widespread use for fully quantitative analysis of silicate samples. This situation changed dramatically during the mid-1960s with the development of inductively coupled argon plasma as an emission source for spectroscopic measurements.

Early work in establishing the formation and properties of an inductively coupled plasma that could be formed at atmospheric presure was carried out by Babat (1947). Further important advances in techniques to form and sustain inductively coupled argon plasma discharges were reported by Reed (1961, 1962, 1963). Based on this work, two research groups, one led by Greenfield (Greenfield *et al.*, 1964) and the other by Fassel (Wendt and Fassel, 1965), demonstrated the analytical potential of this plasma as an optical emission source. By viewing the appropriate region in the argon plasma tail flame, atomic and ionic emission lines could be measured against very low background emission intensities. Very high temperatures (10 000 K maximum) are attained in modern ICP plasmas. Samples must be prepared in solution form and are introduced into the plasma by nebulization. Aerosol droplets are heated to temperatures of between 6000 and 8000 K in the plasma, and under these conditions efficient atomization occurs. Indeed, in contrast to flame atomization used in atomic absorption spectrometry, chemical interference effects in ICP plasmas are normally very small. Furthermore, ICP–atomic emission sources benefit from a very wide linear response range, often quoted to be more than six orders of magnitude. These properties give the ICP–AES source some very powerful analytical characteristics.

However, the atomic and ionic line spectra emitted by samples excited in an ICP plasma are not simple; a very large number of individual lines are present in the spectra of 'typical' rock samples. One of the principal concerns in the routine application of this technique, especially for the determination of trace elements, is to account adequately for all existing and potential spectrum overlap interferences. ICP–atomic emission spectrometry does not benefit from the 'lock and key' effects that make spectrum overlap interferences rare in atomic absorption spectrometry (section 4.1).

However, perhaps the outstanding advantage of ICP–AES over atomic absorption methods is that instruments can be designed with direct reader spectrometers (Figure 3.46) that permit simultaneous multi-element determination. The number of elements that can be measured simultaneously is essentially limited by the number of fixed exit slits and photomultiplier assemblies fitted to the spectrometer, between 20 and 60 such channels being common. Some users prefer the flexible operation possible on instruments of alternative design incorporating sequential spectrometers, whose movement can be pre-programmed to measure selected peak and background wavelengths.

In terms of sensitivity, speed of analysis and capability for both trace and major element determinations, the inductively coupled plasma–atomic emission spectrometer has brought wet chemical schemes of analysis into the league of other established instrumental techniques such as x-ray fluorescence analysis. If there is a drawback to ICP–AES, it is that all samples must be reproducibly dissolved—compare the very simple powder pelleting procedures required for XRF trace element analysis. However, some would argue that such solution preparation procedures afford additional flexibility to ICP–AES, permitting, for example, the determination of the rare-earth elements by a procedure that incorporates an ion exchange separation.

The first commercial instrument became available in 1975, and since then ICP–AES has become the instrument of choice for routine major and trace-element determinations in many laboratories, especially those with a tradition of wet chemical analysis. Not the least attraction is the possibility of determining anything between 20 and 60 elements simultaneously in a cycle time of 2–3 minutes on a modern instrument fitted with a direct reader polychromator. The technique has become so well established that it is often referred to simply as 'ICP analysis'; this does *not* avoid confusion with recent analytical instrumentation in which the ICP source is used to generate ions for mass spectrometric determination. The ICP–AES technique is sometimes referred to as ICP–OES ('optical emission spectrometry'). The former terminology has been adopted here, following common usage, despite the fact that optical emission from ionic species is just as important as atomic emissions in routine analytical applications.

In parallel with the development of ICP sources, in recent years interest has increased in the direct current plasma as an optical emission source for solution analysis. Although not as versatile as the ICP, direct current plasmas have found some role in silicate rock analysis, and the analytical characteristics of this source are discussed in section 5.12.

5.2 The inductively coupled argon plasma

A *plasma* is any luminous volume of gas having a fraction of its atoms or molecules ionized. As indicated by Ebdon

(1982, p.36), this definition applies equally to a flame. However, in analytical spectroscopy, the term 'plasma' is normally reserved for an electrically excited discharge. A plasma source designed for emission spectrometry is shown schematically in Figure 5.1. It consists of a glass torch surrounded by a copper work coil. The functions of the components are as follows.

(*a*) *Torch assembly*. The torch is normally fabricated from glass in the form of three concentric glass tubes, the upper part of which is surrounded by the rf work coil. The torch assembly is designed to deliver gases so that a stable argon plasma is formed at the open end through which the sample aerosol can be injected. The two pioneering research groups developed separate torch designs: the Greenfield torch, which is now not widely used because of its need for a high gas flow, and the Fassel (or Scott) torch, which after progressive evolution, is now the industry standard.

The Greenfield torch (Figure 5.2*a*) was designed for use with conventional AA nebulizers that operate at a much higher gas flow (10L/min) than later ICP-type nebulizers (1L/min). The Greenfield torch consists of three concentric tubes, the outer one of which is 28 mm in internal diameter. Nitrogen at a flow rate of 20 to 70L/min is used as coolant gas, argon at a flow rate of 12 to 38L/min as injector and auxiliary gas (Thompson and Walsh, 1983). In order to sustain a stable plasma, relatively high power inputs are required (up to 8 kW).

The Fassel torch has evolved over a number of years to the design shown in Figure 5.2*b*. It is a more economical design, the outer tube having an internal diameter of 18 mm. The torch is designed to operate with nebulizers that require flow rates of only 1L/min of argon. Argon is also used as coolant gas and is supplied at a flow rate of 15 to 20L/min to the outer concentric tube. The function of the coolant gas is to supply argon to the plasma, to stabilize the plasma centrally in the torch and to cool the outer glass jacket, preventing it from fusing or distorting. Carrier gas and sample aerosol are injected into the plasma through the axial capillary at a flow rate of 0.5 to 1.5L/min. An auxiliary supply of argon can be delivered to the plasma through the intermediate tube. In some schemes of analysis, no auxiliary supply is required. In others, a small flow of argon is used (0–1.5L/min) to lift the plasma off the injector, preventing overheating of the glass capillary and sooting of the aperture, especially if organic solvents are to be nebulized. The gas flow to the intermediate and outer tubes is supplied tangentially, the vortex created by the spiralling motion of gases improving the stability of the plasma. These outer tubes are usually made of quartz glass because of its greater heat stability. Plasmas are sustained at operating powers of about 0.7 to 2 kW. The Fassel torch is, therefore, designed to form a stable toroidal plasma, through the centre of which the sample aerosol can be injected. Design features include a tulip-shaped intermediate tube and a capillary injector tube. Manufacturing tolerances are much tighter than for the Greenfield torch in respect of both concentricity and annular spacing of torch components.

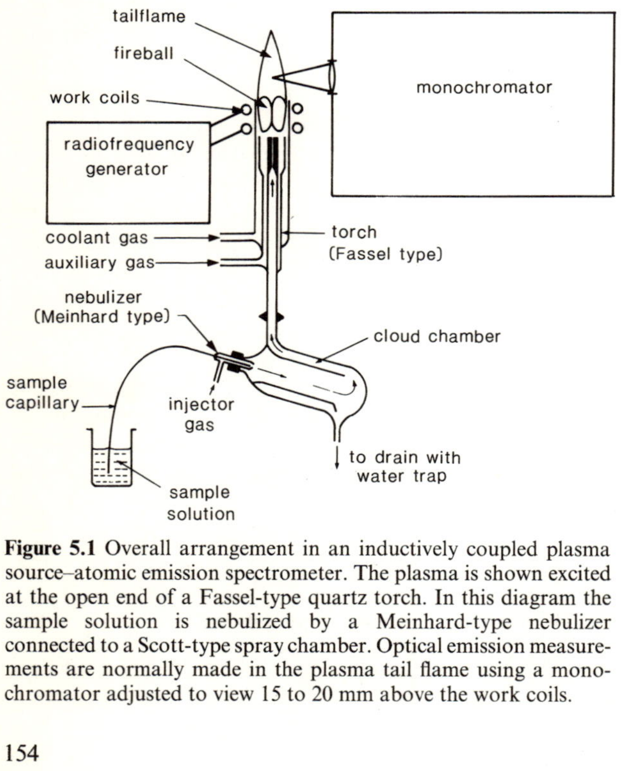

Figure 5.1 Overall arrangement in an inductively coupled plasma source–atomic emission spectrometer. The plasma is shown excited at the open end of a Fassel-type quartz torch. In this diagram the sample solution is nebulized by a Meinhard-type nebulizer connected to a Scott-type spray chamber. Optical emission measurements are normally made in the plasma tail flame using a monochromator adjusted to view 15 to 20 mm above the work coils.

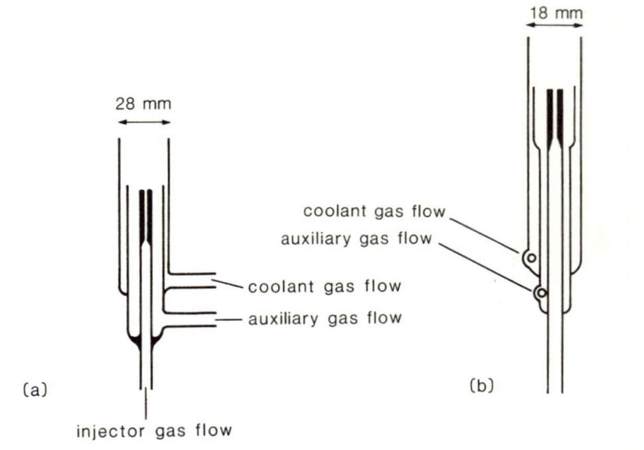

Figure 5.2 Torch designs for ICP atomic emission measurements. (*a*) The Greenfield torch having an overall internal diameter of 28 mm, an annular gap for coolant gases of 0.5 mm and an injector tip bore of 2 mm. (*b*) The Fassel torch as further developed by Scott *et al.* (1974) and others. The overall internal diameter is 18 mm, the annular gap for coolant gas 1.0 mm and the injector tip bore 1.5 mm. Current design includes a tulip-shaped intermediate tube and capillary injector tube. (Thompson and Walsh, 1983.)

(*b*) *Work coil: mechanism of plasma formation*. The work coil consists of two or three turns of copper tube down the centre of which passes cooling water. This copper coil is wrapped around the outer tube of the torch, and the spacing between coil and torch, the distance between individual copper coils and the concentricity of both coil and torch are important parameters in the formation of a stable plasma.

The copper coil is connected to a radiofrequency power generator. This creates oscillating rf magnetic fields within the argon gas flowing through the torch, as shown in Figure 5.3. However, power cannot be transferred to the torch until the argon gas has been '*seeded*' with a few electrons from a Tesla coil. These electrons are then subjected to intense

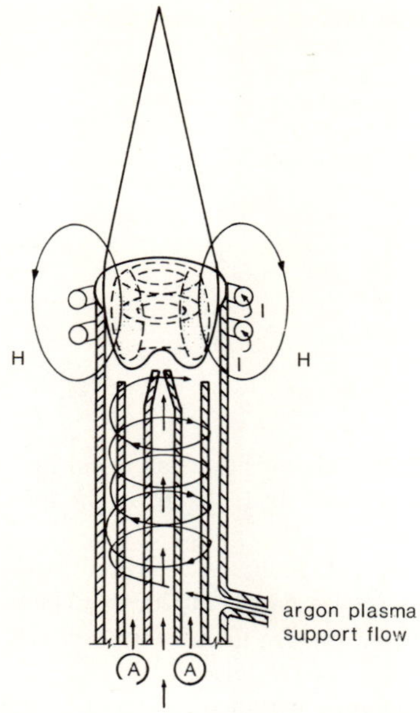

Figure 5.3 Structure of an inductively coupled plasma showing plasma fireball and tail flame. The magnetic fields (*H*) generated by the current (*I*) passing though the work coils are shown, together with the circular paths of electrons and argon ions induced in the plasma fireball. (Fassel, 1978.)

oscillations by the radiofrequency magnetic fields generated by the work coils. Collisions that occur between the free electrons and atoms of argon gas cause further ionization. Each charged particle formed in the argon gas is subjected to magnetic fields that cause it to travel in closed annular paths within the torch. However, other argon atoms impede the movement of this current of charge particles, causing ohmic heating of the gas. In consequence, its temperature rapidly rises to a maximum of around 10 000 K. An important parameter in sustaining stable plasma operation is that the density of current formed within the plasma should be sufficient to deliver enough energy to raise the temperature of the gas faster than the heat lost by conduction or radiation. The transfer of energy by this inductively coupling mechanism is unusual in that there are no electrical contacts within the plasma gas. The work coil can be considered to represent the primary winding of a transformer, the secondary winding of which is formed by the annular currents formed within the plasma. The efficiency with which power is delivered to the plasma depends on the electrical coupling between work coil and gas. In particular, the *impedance* between the two must be carefully matched to ensure that the electrical oscillator properties of plasma and work coil are tuned to ensure maximum power transfer. These coupling problems have been discussed by Greenfield *et al.* (1975), who estimated that the electrical properties of a plasma gas at 10 000 K are approximately 30 ohm/cm inductance and a few ohms resistance.

(*c*) *Radiofrequency power transfer.* The radiofrequency at which the torch operates is also an important parameter in sustaining plasma operation. National restrictions normally limit the choice of frequencies available for plasma torch operations; these prevent interferences being transmitted accidentally into telecommunications systems. The majority of crystal-controlled rf generators are designed to operate within the industrial frequency bands at 27.12 or 40.68 MHz, although plasma generators have been described operating at between 5 and 100 MHz. At the lower frequency range it is necessary to deliver up to 30 kW to sustain the plasma. At 27.12 MHz, generators are normally designed with maximum power ratings of between 2 and 5 kW, and are often used routinely to deliver about 1 kW to the plasma.

It is important to appreciate that at 27.12 MHz, power is delivered principally to the outer skin of the plasma fireball (Figure 5.3), the coupling decreasing exponentially towards the centre of the plasma. The thickness of this coupling layer is normally characterized by the *skin depth*, this being the thickness of plasma from its outer boundary by which the energy coupling has fallen to 0.37 of its maximum value. Skin depth for the ICP is normally estimated to be in the range 2 to 3 mm (see Hieftje, 1983). However, the skin depth increases significantly as the frequency of the radiofrequency field is reduced. Higher generator power is therefore required from a lower-frequency generator to compensate for the larger volume of plasma gas over which input energy is dissipated. In a conventional Greenfield or Fassel torch, very little power is therefore delivered to the centre of the plasma fireball, the region in which the plasma is injected.

(*d*) *Radiofrequency power generators.* An important consideration in the design of radiofrequency power supplies is that in order to achieve precise and repeatable data in analytical applications, the forward power delivered to the plasma must be regulated. Account must be taken of the fact that the impedance of the plasma will change slightly when samples of varying composition are injected, and that simultaneous changes must be effected in the rf power supply to ensure that constant power transfer occurs.

Many plasmas are excited by *free-running* generators designed to operate at the resonance frequency of the work coil and plasma. Small changes in impedance cause corresponding changes in the nominal rf frequency that automatically regulate the power delivered to the plasma. Greenfield *et al.* (1975) found that when concentrated mineral acids (HCl, HNO_3, $HClO_4$ or 50% H_3PO_4) were injected into a plasma operated at powers of 5.5 kW and 2 kW, the nominal rf value of 7 MHz changed by less than 1 and 1.5 kHz respectively. Since the work coil forms part of the oscillator circuit, the free-running generators must be mounted close to the torch to prevent power losses and electrical interference effects. *Crystal controlled* rf generators use the precise frequency obtained from a crystal oscillator to synchronize plasma power input. This type of generator may be mounted remote from the torch with an impedance matching circuit mounted with the work coil in the plasma source enclosure. Feedback is provided from plasma source to generator to achieve accurate power regulation. Using either generator, the actual power delivered to the plasma is typically 50% of the energy dissipated by the rf generator.

(*e*) *Low gas-flow torch designs and alternative plasma gases.* One of the disadvantages of conventional argon plasma torches is the high consumption of argon gas required for routine operation. Indeed, a conventional torch uses between 13 and 20 litres of argon per minute, mainly to ensure that the outer quartz glass tubing adjacent to the plasma is adequately cooled. Argon consumption, therefore, represents a significant cost in routine plasma operation and is often most economically satisfied by installing a liquid argon supply dewar rather than relying on conventional compressed gas cylinders. Hieftje (1983) estimated that each additional litre per minute increment in argon consumption increases the annual running cost of an ICP–AES in North America by $US1000 (1983 prices).

It is not, therefore, surprising that active development work is being undertaken to design plasma torches that will operate at much lower rates of argon consumption. For example, Montaser *et al.* (1984) report data on the operating characteristics of a conventional torch (internal diameter 18 mm) run at a gas flow of 5 L/min at a forward power of 720 W. Under optimized conditions, signal to background ratios of spectral lines for 20 elements were investigated. Some were equivalent to those observed in a conventional plasma, others were worse by a factor of 1.5 to 2.

Data for a novel torch designed to operate at 1L/min argon have been described by Ripson *et al.* (1982). This low flow rate was achieved by using compressed air to cool the outside of the torch. Specially designed work coils were used, in which cooling water was circulated to increase the rate of external cooling (Figure 5.4). Preliminary tests were carried out with conventional three-tube torches of internal diameter about 18 mm. However, a two-tube torch configuration (with the intermediate tube removed) appeared to give the most stable plasma operation. Results, in terms of detection limits and interference effects, were promising, being only slightly inferior to conventional torch performance. Hieftje (1983) has reviewed the development of mini-, micro- and high-efficiency torches for the ICP, and has concluded that miniaturized torches (for example of diameter reduced to 13 mm) offer the best immediate solution for operation at reduced rf power and coolant gas flow. In the longer term, dramatic reductions in power or argon gas flow may be achievable using external forced-air cooling in conjunction with outer torch tubes of materials of a higher thermal conductivity such as beryllium oxide or certain high-temperature ceramics.

(*f*) *Plasma gas.* Argon is selected as the plasma for conventional ICP sources for several reasons:

(i) Argon is an inert gas and is therefore likely to suppress chemical interferences when sample solutions are atomized in the plasma (compared with the rather more complex free radical gas phase reactions that can occur in AA flames)

(ii) Argon is optically transparent in the uv–visible region of the spectrum

(iii) Argon has a high first ionization energy (15.75 eV) which permits determination of almost all elements that can be excited to emit lines in the uv–visible region

(iv) Argon has a moderately low thermal conductivity, so that heat is retained within the plasma fireball, sustaining stable operation at moderate power inputs.

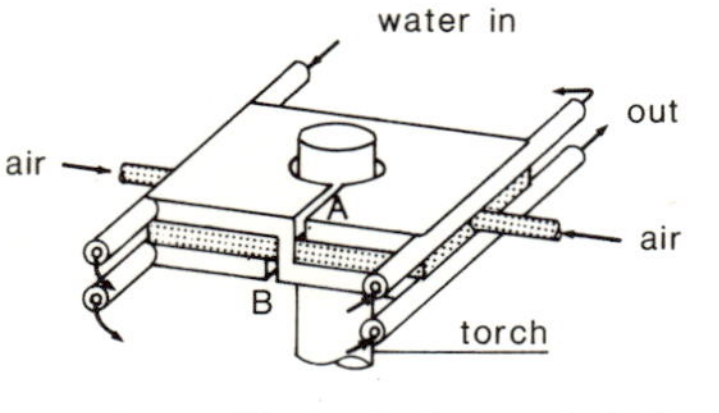

Figure 5.4 An experimental inductively coupled plasma torch configuration designed to operate at 1 L/min argon consumption. The work coil is water-cooled, the plasma torch is forced-air-cooled. (Ripson *et al.*, 1982.)

Table 5.1 Physical properties of plasma gases

Gas	Electrical resistivity at 25 °C $10^{-2}\Omega$cm	Thermal conductivity at 25 °C 10^{-6}cal/s.cm^2 °C	Specific heat at 25 °C cal/g
Helium	5	340	1.25
Argon	2	39	0.13
Nitrogen	10	57	0.25

Data from Robin (1982).

The practicalities of substituting alternative plasma gases for argon have been considered by Robin (1982). Only three gases have higher ionization energies than argon and would therefore be suitable for exciting difficult elements such as F, Cl, Br, or S. These are fluorine (17.4 eV), neon (21.6 eV) and helium (24.6 eV). In addition, nitrogen has been proposed as a cheap alternative plasma gas. The relevant physical properties of He, Ne and N_2 as summarized by Robin are listed in Table 5.1. Nitrogen has been used as plasma gas, either alone or mixed with argon. However, excited nitrogen emits a complex band spectrum that can cause many undesirable spectrum interferences and has not therefore been widely adopted. Some investigations have been carried out using helium plasmas. Due to the high thermal conductivity of this gas, very much higher power inputs are required. The resultant helium plasma can be propagated as a thin, bright needle-shaped discharge concentrated along the axis of the torch (Figure 5.5). At the current state of development, it has proved difficult to inject sample aerosols into this plasma.

Thus at the time of writing, the conventional argon plasma represents the yardstick (in terms of both detection limits and freedom from chemical interferences) against which new designs must be measured. None has yet been shown to be superior in routine analytical performance, though considerable active research and development work is currently being carried out in this area.

5.3 Nebulizers and spray chambers

As in atomic absorption spectrometry, samples are injected into the plasma as an aerosol. Sample solutions must, therefore, first be nebulized. However, there are important differences in the design criteria relevant to ICPs as opposed to AA spectrometers. In particular, the flow rate of oxidant gas available to operate the nebulizer in AA instruments is high—often 10 or more litres per minute. However, optimum ICP plasma conditions are usually observed when the argon injector flow rate, used to nebulize the sample

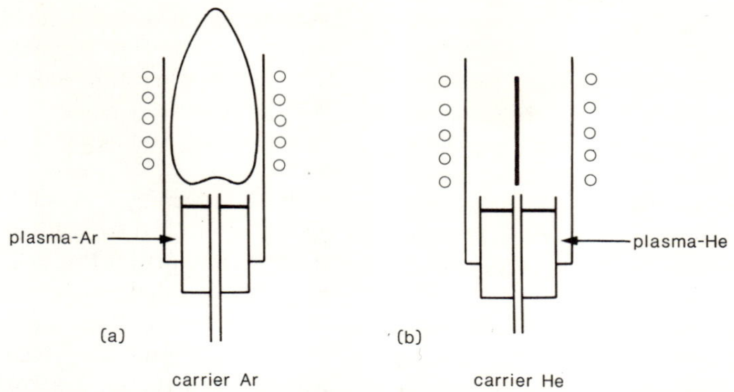

Figure 5.5 Comparison of the shape of inductively coupled plasmas excited using (*a*) argon and (*b*) helium. The latter discharge forms a bright needle-shaped plasma region. (Robin, 1982.)

solution, is adjusted to about one litre per minute (or even less). This reduction in gas flow rate causes two problems that significantly affect ICP nebulizer operation.

The first is that the efficiency of nebulization is much lower. That is, the fraction of sample solution aspirated into the nebulizer that is injected into the plasma is often as low as 2% (in comparison with 10% efficiency often quoted for pneumatic nebulizers used on AA instruments).

Secondly, the much lower gas flow passing through the nebulizer cloud chamber also reduces the efficiency with which the last sample can be washed out. Thus the ICP nebulizer can suffer from serious memory effects, a problem exacerbated by the characteristically higher sensitivities compared with AA instruments.

Whereas the concentric pneumatic nebulizer is almost universally adopted for AA spectrometers, several different designs have been tried for ICP instruments. However, the design criteria remain the same: that of efficient aerosol production of a uniformly small droplet size (and in any case less than 5 to 10 μm in diameter). In some cases a high tolerance against clogging and the ability to nebulize sample slurries is also important. As in AA instruments, the design of the cloud chamber and any associated baffles used to remove larger aerosol droplets is important in promoting reliable analytical operation.

Three types of nebulizer are employed on instruments used for practical analytical measurements: the concentric tube, cross-flow and Babington type nebulizers. The operational characteristics of these devices will now be considered.

5.3.1 *Concentric glass nebulizer*

The tolerances demanded in the construction of a pneumatic nebulizer suitable for ICP applications are much more stringent than for AA instruments, because of the lower gas flow rates involved. Whereas the gap between annular spaces in an AA nebulizer may be in the range 150 to 250 μm, successful aerosol generation in an equivalent ICP nebulizer requires gaps in the range 10 to 20 μm (Browner and Boorn, 1984). Earlier attempts to machine such devices using conventional workshop practices were not particularly successful. However, a breakthrough occurred with the development by Meinhard (1976) of a concentric-tube pneumatic nebulizer constructed of borosilicate glass (Figure 5.6). This all-glass device is constructed by standard glassblowing techniques and has been widely adopted on many ICP instruments. Argon gas at a flow rate of about 1L/min is passed through the outer glass tube, and the venturi effect created at the nozzle of this tube causes the sample solution to be sucked out of the axial capillary and simultaneously shattered into aerosol droplets as described in section 4.2. The efficiency and reproducibility of aerosol production is critically affected by the configuration and concentricity of the nozzle, especially in view of the extremely small annular spaces referred to above (Figure 5.7).

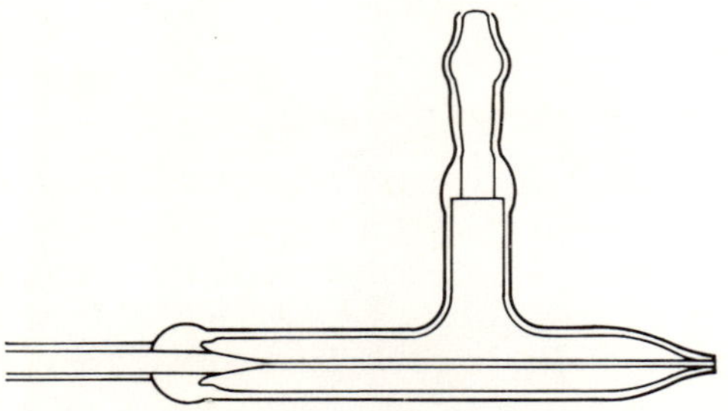

Figure 5.6 The Meinhard concentric glass nebulizer. Argon injector gas is introduced through the side arm, the sample solution through the axial capillary. (Garbarino and Taylor, 1980.)

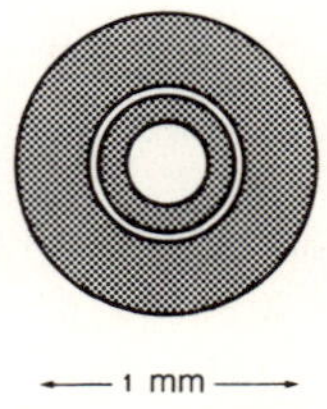

Figure 5.7 End view of the tip of a Meinhard nebulizer showing the central solution capillary and the very narrow annular gap (10–20 μm) through which the nebulizer gas flows. (Thompson and Walsh, 1983.)

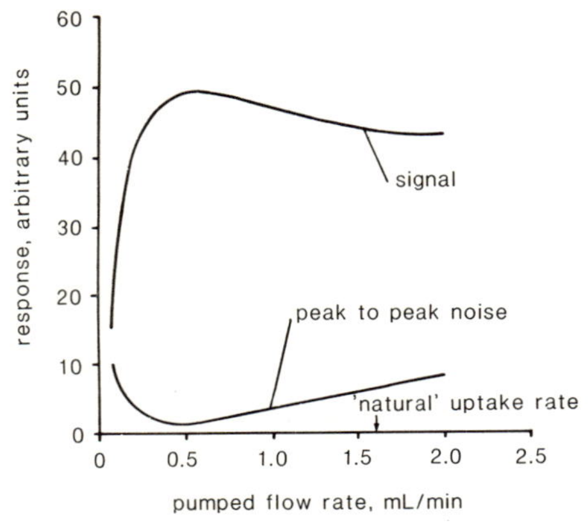

Figure 5.8 Variations in ICP signal intensity and signal to noise ratio as a function of solution uptake rate using a Meinhard nebulizer. Solution flow rate was controlled by a peristaltic pump. The signal response is relatively constant for uptake rates of between 0.3 and 2.0 L/min. The natural uptake rate is representative of a nebulizer with no sample capillary to throttle flow rates. (Thompson and Walsh, 1983.)

One of the advantages of the Meinhard nebulizer is that it is self-priming, the venturi effect being adequate to lift the sample solution up the input capillary. A peristaltic pump is not required for routine operation. The natural solution uptake rate in the free-running mode of operation (that is, not pumped) is typically 1.5m L/min with no flexible capillary tubing to throttle the solution supply rate. The effect of changing this uptake rate on signal response and signal to background ratio can be measured by controlling the solution feed rate with a peristaltic pump. Thompson and Walsh (1983) showed that the analytical response followed the curve shown in Figure 5.8. These data indicate that increasing solution input rate above the free-running rate detracts from analytical performance. However, significant improvements are observed if the uptake rate is reduced to about 0.5 mL/min, below which a serious deterioration in performance is observed. In order to avoid the unnecessary complexity of routine operation with a peristaltic pump, Thompson and Walsh (1983) suggest that a reasonable compromise is achieved by throttling input rate, using a 0.5 mm internal diameter flexible capillary 300 mm long which gives a flow rate of about 0.7 mL/min.

The Meinhard nebulizer offers a highly reproducible and reliable performance for a large number of routine applications involving the analysis of dilute aqueous solutions. However, it does have some limitations, in part associated with the susceptibility to attack of the glass construction material by some chemical reagents, and in part due to the tendency to clogging of the very narrow annular nozzle openings. These limitations may be summarized as follows.

(i) Glass nebulizers cannot be used to aspirate solutions made up with hydrofluoric acid, or even hydrofluoric acid neutralized with boric acid, nor concentrated solutions of alkalis (Thompson and Walsh, 1983). Such reagents cause rapid erosion of the nebulizer tip.

(ii) Because of the very narrow spacing in the tip of the Meinhard nebulizer, these devices are prone to clogging from particulate matter in solution or gas supplies. Slurried samples cannot be nebulized. However, other matter, including dust particles, lint, or cellulose fibres (from filter papers), for example, can cause blockages. It is therefore essential to filter the argon gas supply and to take precautions to avoid particulate contamination during solution preparation procedures.

(iii) The Meinhard nebulizer has a low tolerance to solutions with total dissolved salt contents at the 1%w/v level in the short term or 0.4% for continuous operation. If more concentrated solutions are nebulized, salts condense out on the tip of the nebulizer, causing partial blockage and progressively reducing analytical sensitivity. Thompson and Walsh (1983), quoting the work of Sharpe, suggest that the mechanism for this process is that sample aerosol droplets, recirculating within eddy currents, fall back on to the face of the nebulizer tip, where they evaporate to dryness, progressively building up a deposit that clogs the aperture.

Two procedures have been proposed to minimize this effect and permit reliable nebulization of solutions in the 0.4%w/v concentration range. The first is to 'humidify' the argon gas supply to the nebulizer to minimize the evaporation effect. The second is to clean the nebulizer after each sample has been run by injecting about 0.2 mL of water into the injector gas supply, so removing any incipient deposits from the tip.

5.3.2 *Spray chambers*

The aerosol cloud chamber plays an important role in limiting the size of aerosol droplets that can reach the plasma, as it does in flame atomic absorption instruments. Most ICP nebulizer cloud chambers incorporate baffles that induce sharp changes in direction of the gas flow. Only the smallest aerosol particles can follow such flow paths. The higher momentum of larger particles causes them to be thrown against the walls of the spray chamber, where they condense and drain to waste.

Meinhard nebulizers are often used in conjunction with

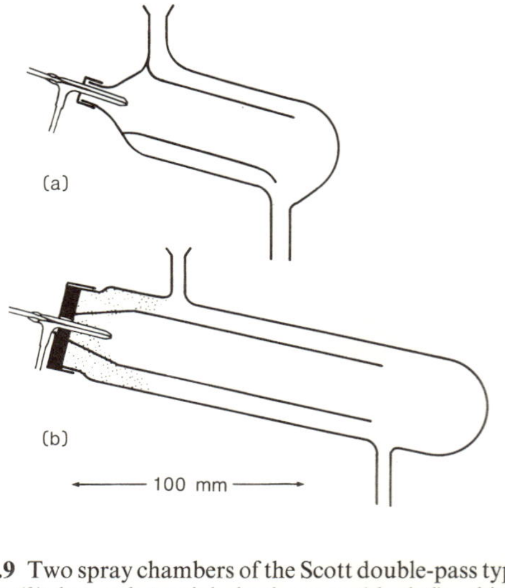

Figure 5.9 Two spray chambers of the Scott double-pass type. Spray chamber (*b*) has substantial dead space (shaded) which is not efficiently swept by the aerosol flow. (Ramsey *et al.*, 1983.)

the Scott double-pass type cloud chamber, a design originally described by Scott *et al.* (1974) (Figure 5.9). The aerosol gas flow is caused to double back on itself before entering the ICP plasma. The relatively long residence time of the aerosol in this chamber and the abrupt change of direction of aerosol gases cause efficient removal of larger droplets from the aerosol cloud.

The performance of this nebulizer/cloud chamber combination has been investigated in detail by Ramsey *et al.* (1983). These workers were particularly concerned with performance in routine analytical applications where it is important to maintain rapid rates of sample throughput. It is then important that the ICP–AES signal rapidly stabilizes when substituting sample for blank solution and then decays back to its background level after re-substituting the blank (1M HCl) solution. Ramsey *et al.* (1983) demonstrated that the time taken for a sample signal to stabilize is characterized by an onset time (the delay before the first traces of sample aerosol reach the plasma). Thereafter the signal exponentially increases to its equilibrium value (Figure 5.10). Furthermore, the total signal stabilization time was significantly affected by the existence of dead volume within the nebulizer. Four different capillary–nebulizer connections were investigated by Ramsey *et al.* (Figure 5.11). The connection with the largest solution dead volume gave a total stabilization time of 19 s (Type B); the optimized design in which dead volume was eliminated gave a signal stabilization time of 8.5 s (Type D).

Nebulizer design was also found to have a significant effect on the signal clean-out time, measured as the time delay, after substitution of a blank solution, before the sample signal has fallen to 1% of its steady-state background value. The influence of two spray chambers of the Scott design (shown in Figure 5.9) was investigated. The rate of fall of analyte signal using the larger spray chamber was found to be considerably lower than that for the smaller chamber (Figure 5.12), a fact attributed to the large proportion of dead space in the former design (shown shaded in Figure 5.9). When the smaller spray chamber was coupled to the various nebulizer-uptake capillary connection arrangements shown in Figure 5.11, clean-out time after substituting blank for sample solution varied from over 300 s in the worst case (Type B connection) to less than 20 s for the optimized connection (Type D). These data show that small differences in detail in nebulizer and cloud-chamber arrangements can have important affects on routine analytical performance.

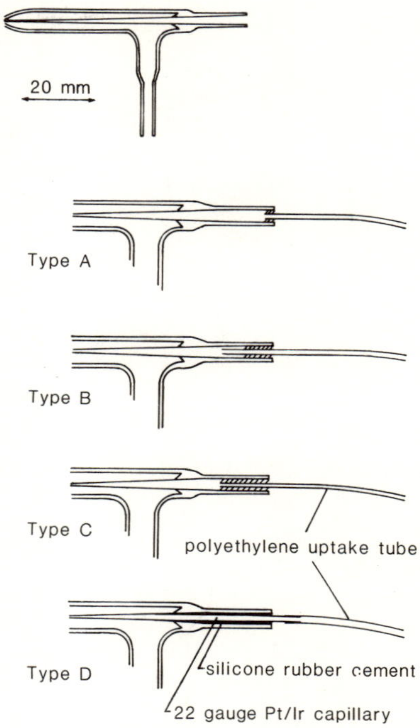

Figure 5.11 Four different ways of connecting the sample capillary to a Meinhard nebulizer. In types A, B and C, the capillary is connected with a short length of silicone rubber tubing (shaded); in type D, a platinum–iridium capillary connecting tube has been cemented to the nebulizer. (Ramsey *et al.*, 1983.)

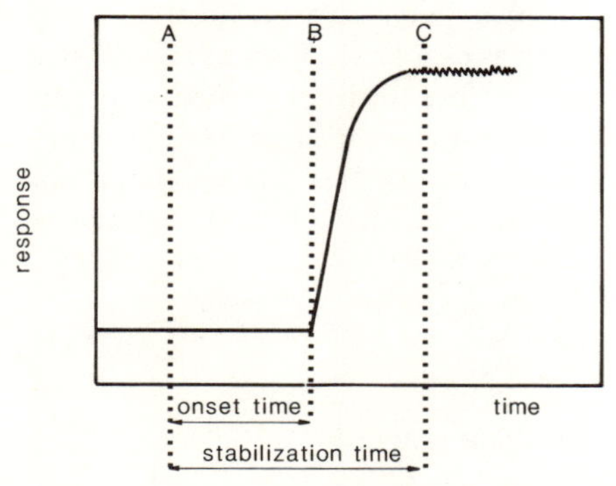

Figure 5.10 Typical delay time in ICP signal response when a sample solution is substituted for the blank. Change-over occurred at point *A*. *A–B* is the onset time for sample aerosol to reach the plasma; *A–C* is the total signal stabilization time. At point *C*, the signal has reached 99% of its equilibrium value. (Ramsey *et al.*, 1983.)

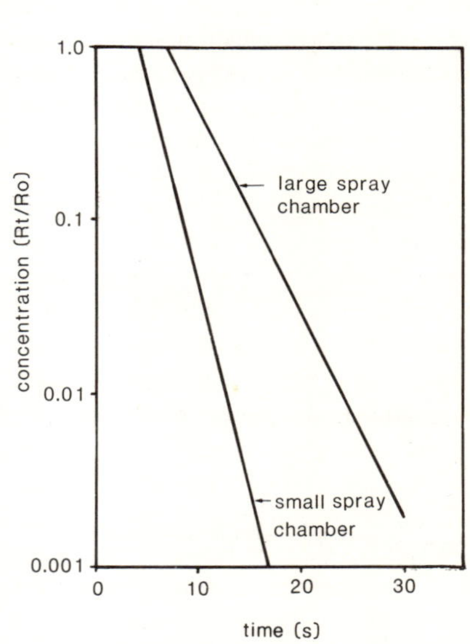

Figure 5.12 Effect of spray chamber design on clean-out characteristics using a Meinhard nebulizer. The larger spray chamber (*b* in Figure 5.9) has a larger clean-out time due to the mixing of new and old sample aerosols that can occur in the dead volume. (Ramsey *et al.*, 1983.)

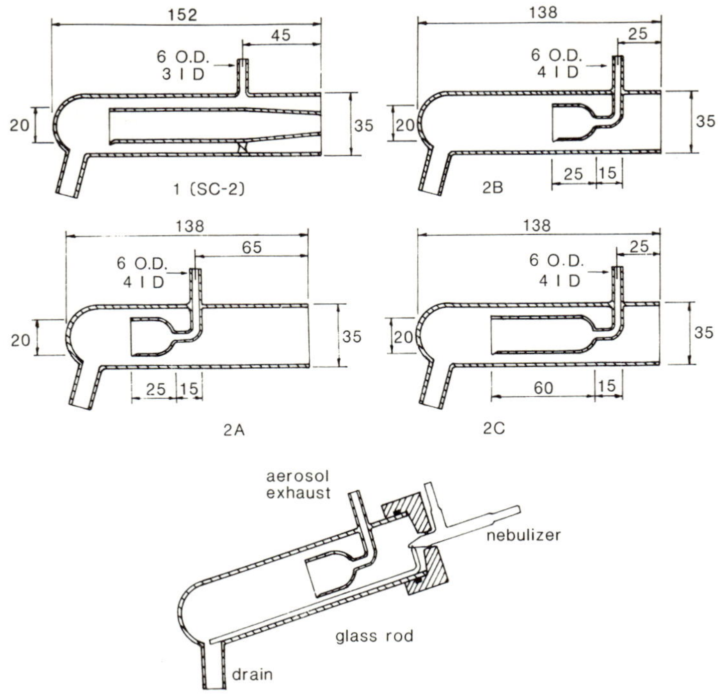

Figure 5.13 Experimental double-pass spray chambers tested by Schutyser and Janssens (1979). (Schutyser and Janssens, 1979.)

Using a combination of optimized nebulizer and cloud chamber, Ramsey *et al.* were able to analyse one sample every 45 seconds, the analysis interval composing 9 s pre-flush with sample solution, three 5 s signal integrations and a 21 s clean-out period nebulizing a 1M HCl blank solution before the next sample is run.

A study of the influence of spray-chamber design on the performance of the Meinhard concentric nebulizer was carried out by Schutyser and Janssens (1979). These workers investigated four cloud chamber designs, shown in Figure 5.13. The signal to background ratio in the determination of Li (670.78 nm) and Cd (226.50 nm) was used as the criterion for selecting optimum design. Their results indicated that it was desirable to position a glass rod with its tip adjacent to the nebulizer orifice, since otherwise discrete drops of solution condensed on the nebulizer nozzle, impeded the carrier gas flow and increased signal noise. An extension to this glass rod (Figure 5.15) ensured uniform collection of condensed sample solution inside the cloud chamber and removed spikes caused when drops of solution fell into the drain.

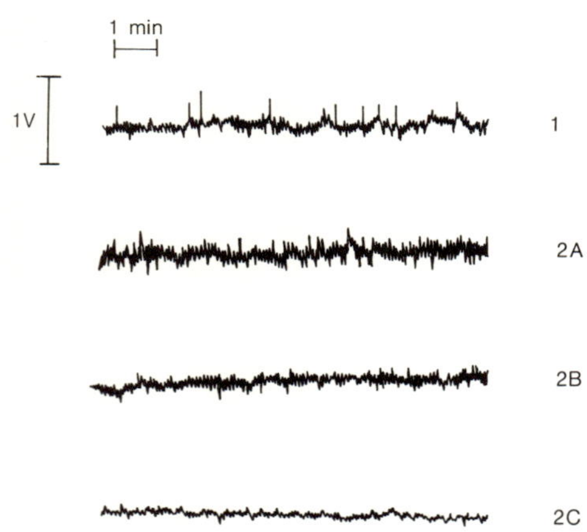

Figure 5.14 ICP emission signals for a 20 mg/L solution of Li (at 670.78 nm) using spray chambers of design 1, 2A, 2B and 2C respectively shown in Figure 5.13. (Schutyser and Janssens, 1979.)

The performance of each spray chamber in terms of noise in the emission reading from a 20 mg/L solution of Li is shown in Figure 5.14. Model 2B was favoured as the best compromise between optimum signal to background ratio and long-term signal stability. It was suggested that part of the success of this design stems from the fact that larger aerosol droplets impact on the back of the bell-shaped glass rod and were therefore efficiently removed from the aerosol stream. In view of the large apparent dead space in this spray chamber, it is not clear whether the clean-out characteristics of this design would be favourable, though this work contributes useful additional data on optimizing the performance of these devices.

5.3.3 *The cross-flow nebulizer*

Development of the cross-flow nebulizer by Valente and Schrenk (1970) was one attempt to overcome the limitations of the standard pneumatic nebulizer when analysing solutions of higher total salt contents. The principle of the cross-flow nebulizer is shown in Figure 5.15. A jet of argon

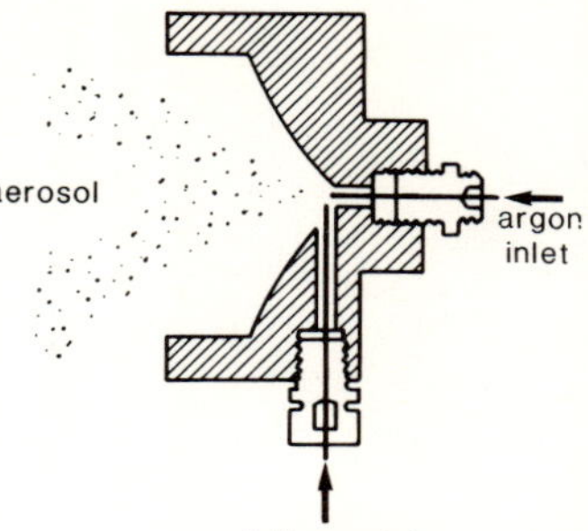

Figure 5.15 Schematic design of a cross-flow nebulizer showing aerosol production. (Browner and Boorn, 1984.)

gas is directed at a capillary held at right angles and containing the sample solution. The shearing action of this gas jet as it breaks over the sample capillary causes the liquid column to shatter into an aerosol. Not all cross-flow nebulizers are self-priming; some require the use of a peristaltic pump. Since the capillary apertures are significantly larger than those in a concentric glass nebulizer, the cross-flow nebulizer is less susceptible to blocking. However, the relative configuration of the two capillary tubes is critical for successful operation. Some manufacturers fabricate the gas and sample orifices from sapphire to reduce any tendency to clog and to provide a surface that is resistant to corrosive acids or other solvents. It is relatively easy to manufacture this nebulizer and associated cloud chamber from HF-resistant materials.

An early cross-flow design was devised by Kniseley *et al.* (1974) incorporating adjustable gas and solution capillaries. However, cross-flow nebulizers for routine analysis are normally of fixed construction, so that reproducible operation is guaranteed by a high degree of accuracy in alignment during the manufacturing process. Notable among these fixed nebulizers is the all-glass MAK cross-flow nebulizer (Figure 5.16), described by Anderson *et al.* (1981) and named after its originators, Meddings, Anderson and Kaiser. This nebulizer was designed to achieve the highest long-term precision in ICP measurements. Tests by Anderson *et al.* (1981) showed that their instrument was capable of restraining calibration drift to within 5% over five working days, and that within-run reproducibility for selected elements (Cu, Zr and Ti) was extremely low: within 0.06 to 0.09% relative standard deviation. These figures indicated that analytical precision was no longer controlled by noise during nebulization but rather was approaching the electronic stability limits of the instrument. In general use, analytical precisions, relative to an internal reference standard, of less than 0.5% relative standard deviation were claimed for this device.

5.3.4 *The Babington nebulizer*

Neither of the two nebulizer systems described above can be used to nebulize samples containing particulate material as is found in slurries, for example. However, several nebulizer designs now available, based on the concept described by Babington (1973) and developed originally for paint-spraying applications, can overcome these restrictions. This has been adapted for nebulization of viscous solutions and slurries in flame atomic absorption applications by Fry and Denton (1977, 1979) and for ICP applications by Garbarino and Taylor (1980).

The Babington nebulizer consists of a hollow glass or metal sphere into which a small orifice is drilled. The argon carrier gas supply is led into the sphere and arranged to escape as a jet of gas through the orifice. The sample solution is pumped so that if flows over the outside of the sphere. The jet of argon, escaping through the orifice, shatters the film of liquid pasing over it, forming an aerosol (Figure 5.17). Garbarino and Taylor (1980) machined the body of the nebulizer of PTFE and made the Babington sphere of 4

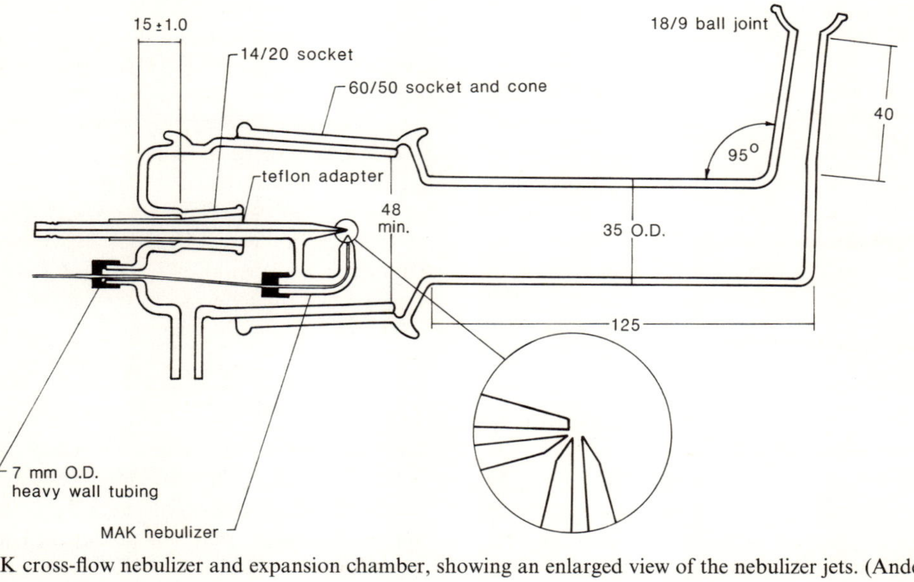

Figure 5.16 The MAK cross-flow nebulizer and expansion chamber, showing an enlarged view of the nebulizer jets. (Anderson *et al.*, 1981.)

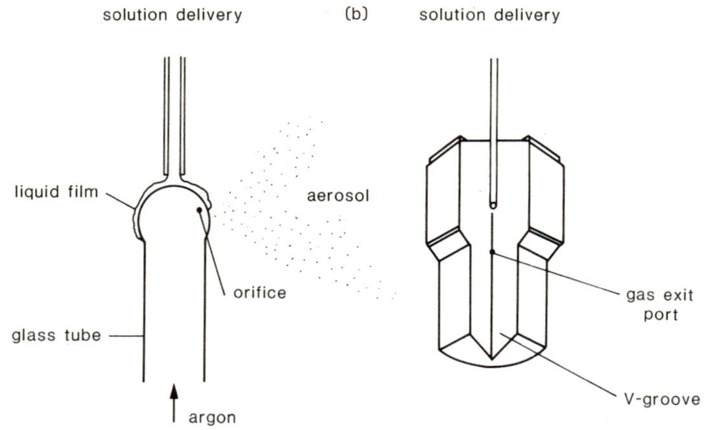

Figure 5.17 (*a*) The Babington nebulizer; (*b*), the V-groove nebulizer, both capable of nebulizing slurries and viscous solutions. (Browner and Boorn, 1984.)

mm outside diameter pyrex tubing with a 0.15 mm diameter orifice, required to deliver 1 litre of argon gas at 40 psi supply pressure. The nebulized aerosol was directed at a glass impactor bead positioned directly in front of the orifice to remove the larger aerosol droplets. Performance or the device was found to be equivalent to or superior in sensitivity and detection limits to a conventional cross-flow nebulizer, but without the troublesome maintenance and particulate clogging problems associated with that latter device.

An alternative form of nebulizer based on the Babington principle has been described by Suddendorf and Boyer (1978, 1981), designed specifically to overcome problems with conventional nebulizers that clog when aspirating solutions of high salt content. This nebulizer was fabricated from stainless steel that was subsequently gold-plated. A block of this material 50 mm long by 8 mm wide was machined with a 'V'-groove down one face (Figure 5.17). A hole (0.23 mm diameter) was drilled in the base of the 'V' about half-way down the block. A jet of argon gas (flowing at 1L/min) escaped through this orifice. The sample solution was pumped through a capillary and discharged into the top of the groove, running down over the hole. The argon jet shattered the liquid film and the resultant aerosol was directed at an impactor rod to remove the coarser droplets. Tests indicated that the 'V'-groove nebulizer was free from the drift and clogging associated with the conventional pneumatic nebulizer. Detection limits between the two devices were essentially equivalent, although the signal from the 'V' groove nebulizer was slightly more noisy and was affected by the ability of the peristaltic pump to deliver a constant and stable flow of solution.

A 'V'-groove nebulizer based on similar principles has been described by McKinnon *et al.* (1981). This is generally referred to as the GMK nebulizer, after its originators Giess, McKinnon and Knight. This nebulizer was developed to analyse solutions of high salt content, free of drift and slurried samples containing particulate matter up to 1 mm in diameter without clogging. The general arrangement is shown in Figure 5.18. The sample solution is pumped through an injector tube constructed from 1.5 mm glass tubing and shaped as shown in Figure 5.18. The minimum bore of the injector tube is constricted to 0.15 mm to give an argon flow rate of 1 L/min at a back pressure of 380 kPa. The aerosol was generated as the gas jet ruptured the liquid film running down the groove and was directed at a glass impactor bead placed about 1mm from the gas nozzle. In operation, the entire assembly was tilted at 20° so that the plasma gas and drain tubes were vertical. Various tests were carried out by McKinnon *et al.* (1981) to characterize precision, detection limits, memory effects, salt effects and slurry handling capacity. Results indicated that this nebulizer was not only equivalent to cross-flow and concentric glass nebulizers, but had additional advantages in accepting solutions with high salt contents and varying amounts of suspended matter.

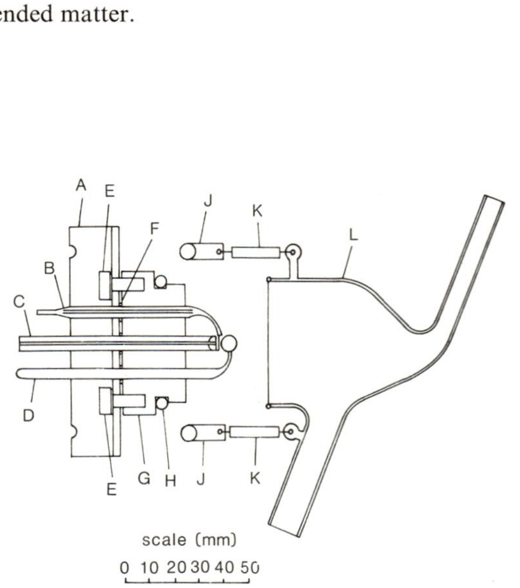

Figure 5.18 The GMK cross-flow nebulizer and expansion chamber, showing *A*, aluminium support bracket; *B*, glass solution injector capillary; *C*, gas injector tube; *D*, glass impactor bead; *E*, nylon screws; *F*, seal; *G*, Teflon support block; *H*, 'O' ring; *J*, *K*, retaining bars and springs; *L*, spray chamber. In operation the assembly is rotated through 20° so that the plasma gas and drain tubes are held vertically. (McKinnon *et al.*, 1981.)

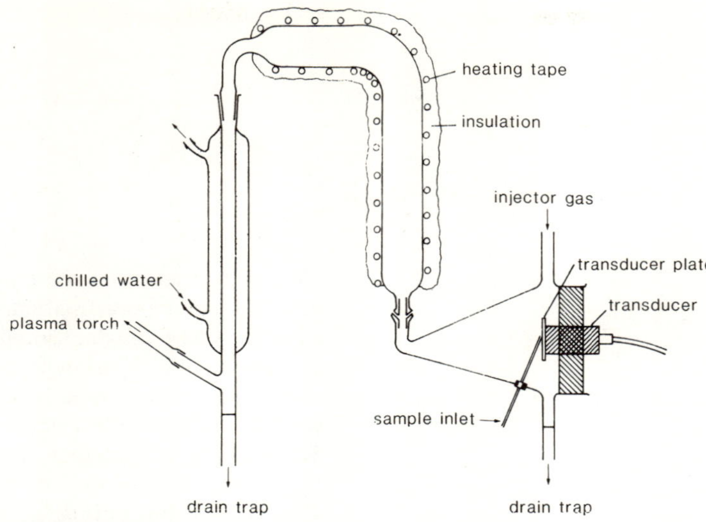

Figure 5.19 Schematic arrangement of an ultrasonic nebulizer and desolvation apparatus. (Thompson and Walsh, 1983.)

5.3.5 *Ultrasonic nebulization*

The main drawback of pneumatic nebulizers operated at the low argon flow rates required for ICP analysis is their low efficiency. Only 2–3% of the aspirated sample solution reaches the plasma; the rest is drained to waste from the cloud chamber. Ultrasonic nebulization has been proposed as a technique for overcoming some of these difficulties. Various designs have been proposed, and Robin (1982) has summarized recent practices.

The principles of an ultrasonic nebulizer are shown in Figure 5.19. The sample solution is loaded on to a thin membrane which is excited by ultrasonic waves (of frequency about 1–2 MHz) using a suitable piezoelectric transducer and waveguide. In order to promote efficient nebulization, the sample solution on the membrane should be maintained at a constant thickness. Furthermore, it is desirable that this film thickness is equivalent to the wavelength of the ultrasonic waves so that they are efficiently propagated through the sample solution. These aspects are critical for successful and reproducible operation by ultrasonic nebulization, since aerosols are formed by geyser formation at the liquid surface.

The ultrasonic nebulizer is extremely efficient in generating very fine aerosol particles, and for small volumes of solution, proportions of 40% or more may be converted into an aerosol. These droplets are transported to the plasma by the argon carrier gas, the flow rate of which is therefore independent of the rate of aerosol production. The plasma must normally be operated at higher power in order to prevent plasma extinction due to the much larger amounts of sample aerosol now injected. However, this is not completely effective, since the large quantities of water vapour that enter the plama cause an excessive cooling effect that significantly reduces sensitivities. The benefits of ultrasonic nebulization can only be obtained using a desolvation system to remove water vapour from the aerosol particles before they enter the plasma. Then detection limits can be improved by perhaps one order of magnitude in comparison with conventional pneumatic nebulization. Olson *et al.* (1977) suggested that this improved performance was almost entirely caused by a tenfold increase in the rate of sample delivery to the plasma torch. Ultrasonic nebulizers have not found widespread application in routine schemes of analysis, partly because of the additional complexity (and cost) of instrumentation, and partly because of the long clean-out times (2 to 5 minutes) required between samples.

5.4 Physical structure of the plasma

The shape and structure of the argon plasma are shown in Figure 5.3. The design of the torch, work coil and gas flows are such that the plasma region has a toroidal configuration centred at the open end of the quartz tube. The sample aerosol is injected through the centre of this plasma region. It has already been shown (section 5.2(*c*)) that most of the plasma power is induced in the surface skin of the plasma fireball, so that introduction of the carrier gas stream through the centre affects plasma power transfer to a remarkably small extent. The sample aerosol, therefore, passes through the centre of what is effectively a high-temperature tube furnace (that is the outer plasma fireball). There is little intermixing of sample aerosol and gases in the plasma generation region and this has an important effect on the analytical properties of this spectroscopic source, as will be discussed further below. Three regions within the plasma (plasma fireball, secondary region and tail flame) may be distinguished.

(i) *The plasma fireball* takes the form of a doughnut-shaped region in which energy is transferred into the plasma from the radiofrequency magnetic field created by the work coil by the mechanism discussed in section 5.2(*b*). It is generally accepted that temperatures in the region of 10 000 K are attained within this region. Under these conditions, argon atoms can be excited in several ways.

First, argon atoms are ionized by collision with rapidly

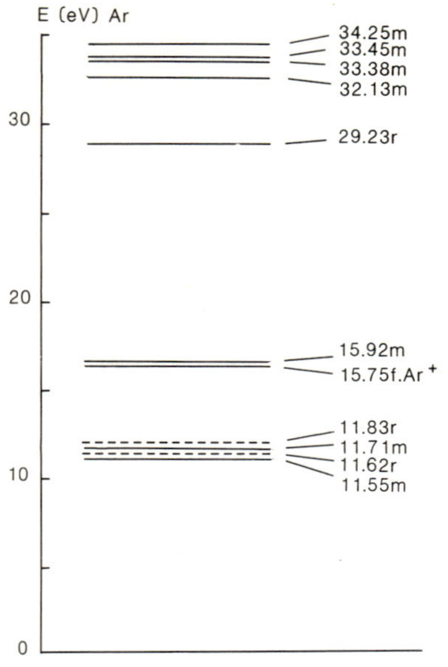

Figure 5.20 Simplified diagram of the electron orbital energy levels in argon: *r*, resonance level; *m*, metastable level; *f*, Ar^+ = ground state level of argon ions. (Robin, 1982.)

moving electrons:

$$Ar + e^- \rightarrow Ar^+ + 2e^- \quad (5.1)$$

This ionization process is essential for sustaining the plasma by the generation of Ar^+ ions.

However, the energy input to the plasma is sufficiently high that many highly excited argon species can be formed. The atomic energy level diagram for argon (Figure 5.20) shows some of the available excited energy levels in the argon atom. Robin (1982) proposed that the following excitation mechanisms can occur in the plasma fireball, all of which lead to the emission of light:

$$Ar^+ + e^- \rightarrow Ar^* + h\nu(\text{continuum}) \quad (5.2)$$

$$Ar^* \rightarrow Ar' + h\nu' \quad (5.3)$$

$$Ar'^* \rightarrow Ar + h\nu'' \quad (5.4)$$

$$Ar^* \rightarrow Ar^m + h\nu''' \quad (5.5)$$

$$Ar'^* \rightarrow Ar^m + h\nu'''' \quad (5.6)$$

where Ar^* and Ar'^* denote argon atoms in different excited states and Ar^m represents argon atoms in metastable excited states. As will be seen, the concept of metastable argon species has important consequences for sample excitation mechanisms. Metastable argon is considered to comprise argon atoms in a highly excited state (of energy 11.55 to 11.72 eV) that possess a finite lifetime of about 10^{-3}s (Mermet, 1975). This should be compared with 'normal' atomic excited states which have lifetimes of the order of 10^{-8}s.

The equations above indicate that the plasma fireball itself is an intense light source and is quite unsuitable for spectroscopic measurements. As well as the discrete emission lines from argon atoms described above, the fireball emits an intense continuum of radiation extending up to about 500 nm. This is caused by ion–electron recombination reactions (equation 5.2) and bremsstrahlung radiation generated within the plasma. The bremsstrahlung is a continuum of light radiation emitted when fast-moving electrons suffer a 'breaking' effect (that is, a stepwise loss of energy) by interaction with the loosely bound outer electrons of atoms or ions in the plasma.

(ii) The primary plasma fireball is surrounded by a *secondary region* which is also luminous but slightly transparent. The intensity of continuum emission from this secondary region is sharply reduced. This region then gives way to the tail flame.

(iii) The *tail flame* is almost invisible when deionized water is nebulized, though it takes on the familiar emission colours when inorganic ions such as sodium are injected into the plasma. The tail flame itself emits an astonishingly low emission spectrum, as can be seen from Figure 5.21. When

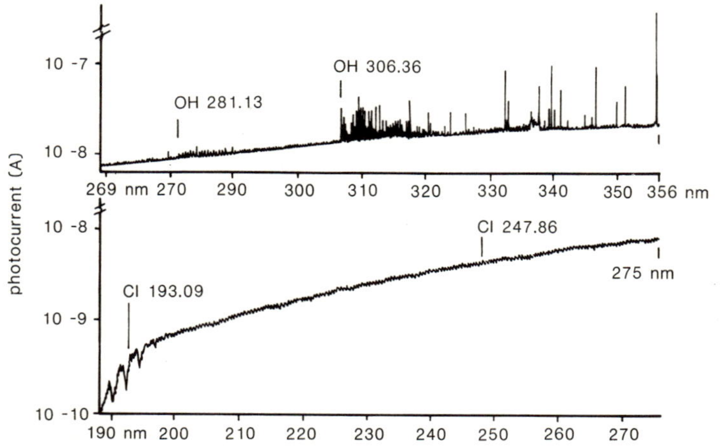

Figure 5.21 Background spectrum of an argon plasma in the range 188–356 nm. Emission lines due to C and Ar (above 320 nm) and emission bands due to OH are the only significant features in this spectrum. (Fassel, 1978.)

water is aspirated into the plasma, emission lines* due to the following species may be detected:

Ar I Forster *et al.* (1982) report the presence of 205 atomic argon lines in the region 317 to 600 nm. No argon emissions, were, however, detected between 200 and 300 nm and of those reported, the most intense were observed above 430 nm.

C I 193.1 and 247.9 nm (a contaminant from industrial-grade argon)

Si I A series of 15 lines between 212.4 and 288.2 nm may be observed due to attack of the silica torch by solutions containing HF, for example

OH Series of bands between 281.0–294.5 nm and 306.0–324.5 nm from the dissociation of water molecules

O_2, N_2, NH, N_2+ } Band emissions caused by the entrainment of atmospheric gases in the plasma tail flame

In addition the following lines may be observed in the fireball and secondary zones:

Ar II lines

H: the Balmer band series of lines at 388.9, 397.0, 410.2, 434.0 and 486.1 nm.

In relation to the intense emission lines observed when analyte species are injected into the plasma, many of these background lines are rather weak. All spectroscopic measurements are made in the tail flame, usually at a viewing height of between 15 and 20 mm above the work coil.

5.5 Temperature distribution in the plasma

In order to understand the analytical potential of the plasma, it is desirable to know the temperature of various regions within this source. This has been the subject of experimental work (for example by Kalnicky *et al.*, 1975) and of computer simulation modelling (for example Barnes and Schleicher, 1975; Boulos and Barnes, 1981).

The latter authors (Boulos and Barnes, 1981) have reviewed the development of computer simulation modelling of ICP plasmas, and some of their results are presented in Figure 5.22. These data were calculated for a Fassel-type spectrochemical torch having an internal diameter of 18 mm. The plasma was excited at 26.3 MHz and argon flow rates were: coolant 9 L/min; auxiliary: 1L/min; and injector: 1L/min. Calculated data are plotted as a function of distance from the axis of the torch (R) for the parameters (a) gas flow streamline function, (b) temperature, (c) swirl velocity of gases and (d) analyte concentration ratio. The main conclusions of these computer model results are as follows:

(i) *Gas flow rates*: the value $\psi = 0.045$ represents the streamline boundary of the central carrier gas. Mass transfer of material across this boundary can only occur by diffusion.

(ii) *Temperature*: injection of the carrier gas causes considerable local cooling of the plasma although the carrier gas temperature still reaches almost 8000 K. Despite injection of the carrier gas, strong *swirl velocities* are retained in the torch due to tangential introduction of coolant gas referred to above. The inlet swirl velocity in this model was assumed to be 13.3 m/s.

(iii) Calculated data for the *distribution of an analyte species* indicates that although it is predominantly concentrated in the plume of carrier gas along the central axis of the discharge, the analyte reaches most of the regions in the discharge.

Clearly the value of such data is limited by the reliability of the computer model selected by the authors and the assumptions made during its execution. Nevertheless, these diagrams are useful in indicating the general behaviour and are broadly confirmed by independent experimental measurements. Indeed, Kalnicky *et al.* (1975) measured the temperature distribution in the plasma from the relative intensity of selected lines in the atomic emission spectrum of iron. If the transition probabilities of these lines are accurately known, then by determining the relative populations of excited iron levels in various regions of the plasma, the corresponding temperature can be predicted from the Boltzmann distribution equation (see section 3.7). Based in part on this work, Fassel (1977) generated a slightly idealized model of the temperature distribution in a plasma shown in Figure 5.23. The plasma fireball is shown to achieve a temperature equivalent to 10 000 K. The core of the plasma jet is considerably cooler due to the injection of sample and carrier gas. However, by the effects of radiation, conduction and convection, the injected gas stream reaches a temperature of about 8000 K by the time it emerges above the top of the fireball. At the normal observation height for spectroscopic measurements (15–20 mm above the work coil) the sample gas stream is predicted to have cooled to between 6200 and 6500 K. This relatively small rate of cooling occurs because the central gas stream is surrounded by a hot jacket of gas emanating from the fireball itself. This insulates the sample stream from more rapid cooling effects.

5.6 Atomization and excitation processes

From the data above, it is apparent that the sample aerosol undergoes rapid heating up to a maximum temperature of 8000 K by conduction, convection and radiation effects as it passes through the core of the plasma fireball. Under these conditions, aerosol droplets rapidly suffer desolvation, volatilization and then atomization. The thermal energy available for atomization is at least twice that available in the hottest nitrous oxide–acetylene flame. Efficient atomization gives the ICP plasma one of its valuable analytical properties: a high degree of freedom from chemical interference effects.

Above the fireball, the carrier gas stream begins to fan out slightly (Figure 5.23) and mixing occurs between the atomized sample and argon atoms and ions from the fireball, some of which are still in a highly energetic state.

The precise mechanism of excitation is not yet fully understood and is, at the time of writing, an area of intense research interest. However, any model of excitation must account for several phenomena observed in the argon plasma that at first glance appear contradictory.

(i) When the intensity of many atomic lines is plotted

* The notation used to distinguish atomic and ionic emission lines is as follows:
Ar I signifies an emission line from atomic argon
Ar II signifies an emission line from the first ionized state of argon (Ar^+).

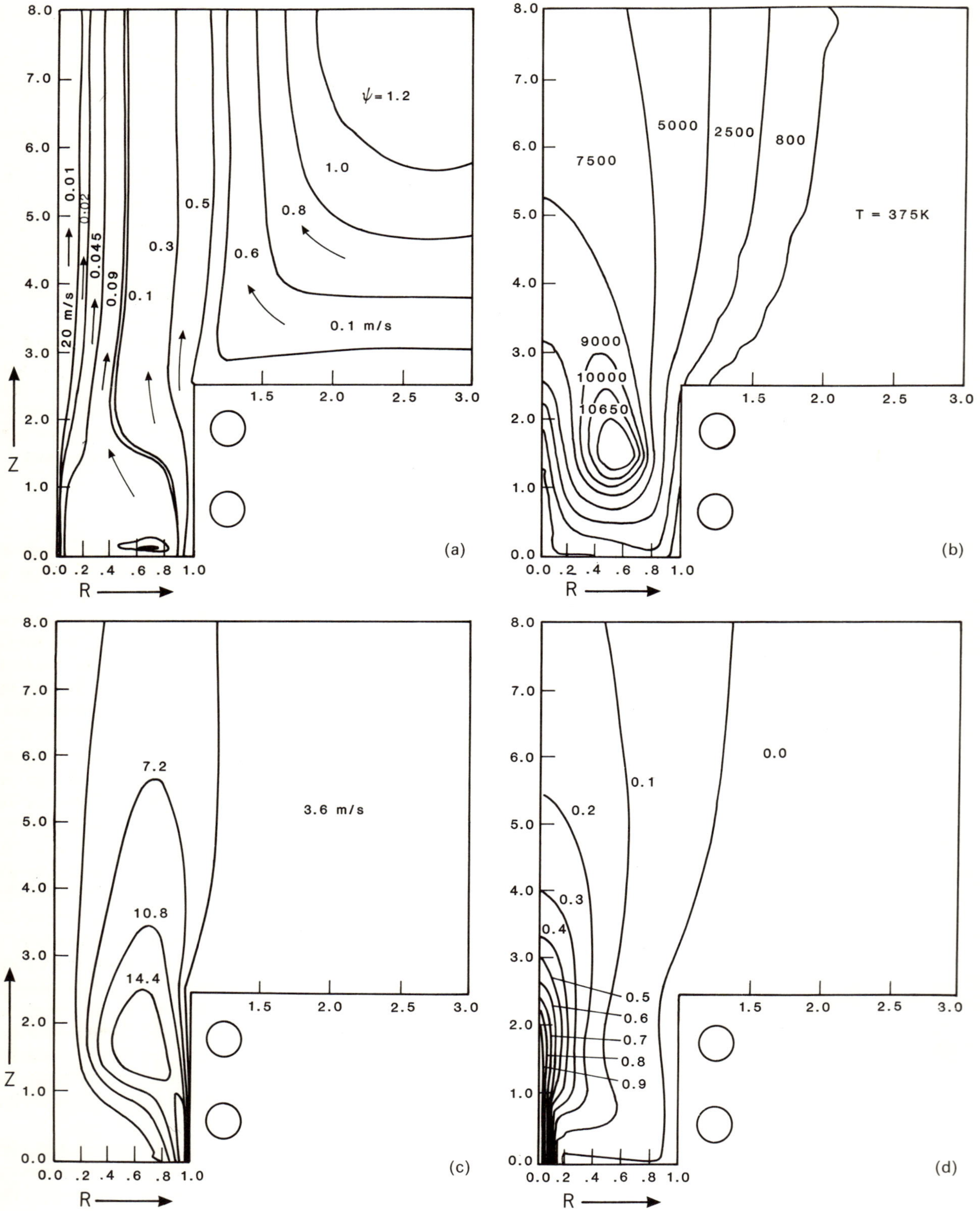

Figure 5.22 Computer-simulated structure of the argon plasma using a spectrochemical torch (18 mm internal diameter) operated at 26.3 MHz, forward power 1.0 kW, coolant flow 9 L/min, injector flow rate 1 L/min, auxiliary flow rate 1 L/min, initial coolant swirl velocity 13.3 m/s. The parameters plotted are: (*a*) gas mass flow rate (stream function ψ); (*b*) temperature T (K); (*c*) swirl velocity η_B (m/s); (*d*), analyte concentration ratio (X). (Boulos and Barnes, 1981.)

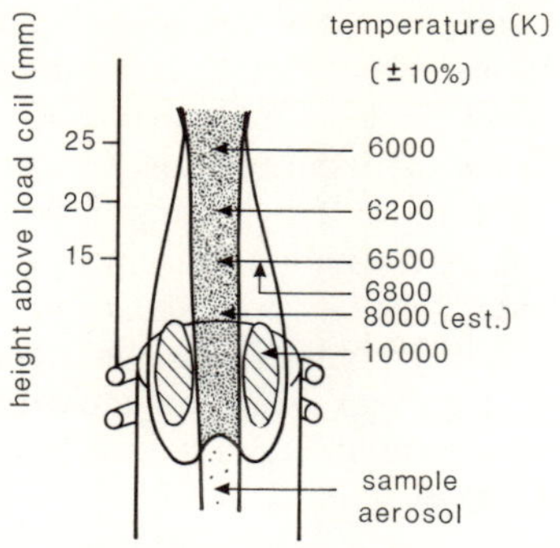

Figure 5.23 Temperatures in the argon plasma as measured by the spectroscopic method. (Fassel, 1977.)

as a function of observation height in the argon plasma, a maximum intensity is often observed in the region around the top of the fireball where the temperature of the injected sample stream reaches a maximum.

(ii) Optical emission is also observed from sample species that have undergone ionization. However, the intensity of many of these ion lines often reaches a maximum much higher up in the tail flame (usually between 15 and 20 mm above the work coil) in a region where it might be expected that the energy available for a thermal excitation mechanism is much reduced.

(iii) Recent observations (Blades and Horlock, 1981) indicate that the variation in intensity with observation height in the plasma of some species such as Cd I and Cd II emission lines is related, suggesting an interrelationship between the corresponding excitation mechanisms.

In general terms these phenomena can be explained as follows. The appearance of a maximum in the atomic line signal around the top of the fireball is explained on the basis of a model which assumes that the species in this region are in *local thermal equilibrium*. The thermal energy of the gases in this region is sufficient to excite atomic species according to the Boltzmann distribution equation (section 3.7). Lines excited in this region are sometimes termed *soft* lines and involve transition energies of about 3 to 10 eV.

However, in the tail flame 15 to 20 mm above the work coil, local thermal equilibrium conditions cannot exist, as transitions are observed in this region considerably above the nominal thermal energy available. The concept of gas 'temperature' then becomes meaningless. It is now thought that the highly energetic species formed in the plasma fireball are responsible for some excitation phenomena in this region of the tail flame. Following the suggestion of Mermet (1975), these species are considered to be metastable argon atoms. The number of excited argon atoms in this region of the plasma greatly exceeds that which could be supported by the notional temperature of plasma gases. The mechanism of excitation involves the turbulent mixing of 'energetically hot' plasma gases with 'cooler' sample gases. Energy transfer to sample atoms (X) can then occur according to Boumans (1982) via the following sequence of reaction mechanisms:

(i) *Penning ionization–excitation*:

$$Ar^* + X \rightarrow Ar + X^{+*} + e^- \qquad (5.7)$$

where Ar^* is a highly excited argon atom and X^{+*} is an excited *ion* of the sample species X.

(ii) *Spontaneous emission from the ionic level*:

$$X^{+*} \rightarrow X^+ + h\nu_{ion} \qquad (5.8)$$

where $h\nu_{ion}$ is an optical emission photon characteristic of the *ion* X^+.*

(iii) *Recombination reaction involving three-bodied collisions*:

$$X^+ + e^- + e^- \rightarrow X^* + e^- \qquad (5.9)$$

where X^* is an excited *atom* of X.

(iv) *Spontaneous emission from the atomic level*:

$$X^* \rightarrow X + h\nu_{atom} \qquad (5.10)$$

where $h\nu_{atom}$ is an optical emission photon characteristic of the *atom* X.

This hypothesis explains why ion lines are formed predominantly higher up in the tail flame (this being the region where highly excited argon species are able to mix with the atomized sample), and also why the emissions observed from different species can be related (since the formation of X^* is related to that of X^{+*} in the above reaction sequence).

5.7 Interferences in the argon plasma

(*a*) *Physical interferences*. Physical interferences are caused by variations in the rate of atomization of samples and are normally related to changes in the efficiency of nebulization caused by differences in the physical properties of solutions. Such effects may be caused by differences in viscosity or surface tension between sample solutions related, for example, to the total salt content or acid strength. Such factors affect the size distribution of droplets formed during nebulization. Many of these effects are likely to be small. However, for the most precise work it is necessary to match salt content and acidity of sample and standard solutions.

McQuaker *et al.* (1979) investigated the effect of acid strength on the uptake rate of various salts dissolved in perchloric acid. They found that changes in acid concentration caused differences in viscosity affecting the take-up rate of solutions by a coaxal pneumatic nebulizer. Their results are plotted in Figure 5.24. Sample uptake rate of a solution containing 10% perchloric acid was reduced to about 92% of its value when nebulizing acid-free solutions. The corresponding emission signal for a wide range of elements dropped by about 15% over the same concentration range. To obtain reproducible results in analytical applications, McQuaker *et al.* controlled the acid strength of all sample and standard solutions to $3.5 \pm 0.4\%$ $HClO_4$, and estimated that variations of $\pm 0.4\%$ in $HClO_4$ content could introduce analytical errors of up to 1.3% relative.

(*b*) *Chemical interferences*. Chemical interferences normally occur when the release of atoms during atomization is affected by the formation of some stable chemical species in the plasma. Many chemical interferences have been reported in flame AA analysis. The most serious are often caused by the formation of stable molecular compounds that have a low volatility. For example, a large depression occurs in the atomic absorption signal from calcium in the presence of phosphate due to the formation of calcium phosphate that resists atomization. Similarly, the absorption of magnesium

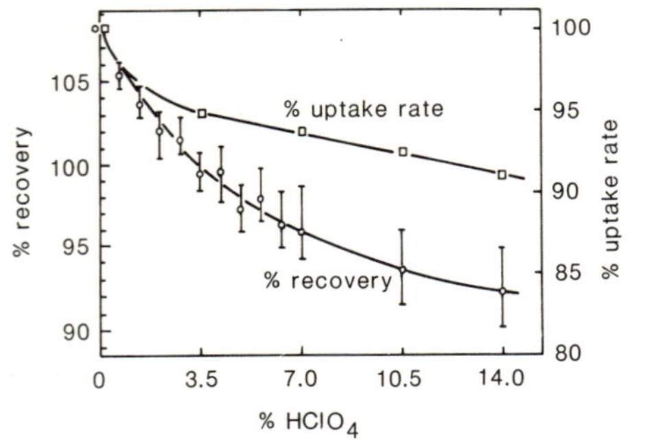

Figure 5.24 The effect of changes in the concentration of perchloric acid on the uptake rate and signal intensity (% recovery relative to 3.5% $HClO_4$) for an ICP system fitted with a coaxial pneumatic nebulizer. Signal recovery is the mean of measurements on selected lines of Cd, Cr, Cu, Fe, Mg, Mn, Mo, Pb and Zn, the bars indicating the spread of results. (McQuaker *et al.*, 1979.)

or calcium atoms is reduced in the presence of aluminium due to the formation of stable magnesium or calcium–aluminium–oxide complexes (see section 4.4 for more details of AA interferences). Due to the very much higher temperatures available in the argon plasma (up to 8000 K) these interferences are greatly reduced and under normal operating conditions are virtually negligible. As an indication of these effects, the calcium emission signal observed from solutions containing varying excess quantities of aluminium is plotted in Figure 5.25. Even when a hundredfold excess Al is present, the calcium atomic emission signal is depressed by no more than 1–2%.

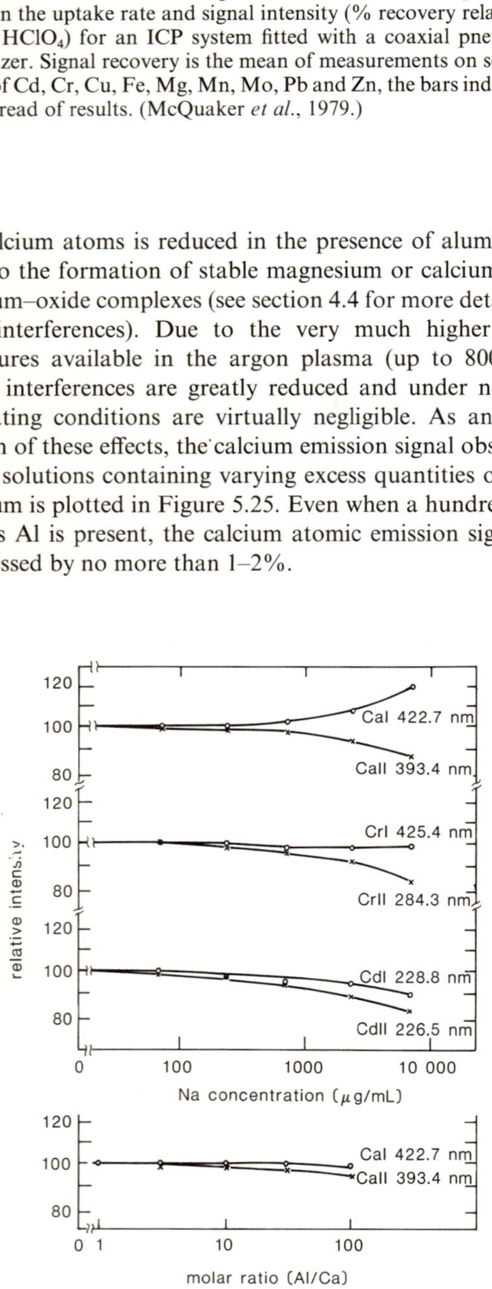

Figure 5.25 (*a*) Effect of changes in the concentration of sodium on the ICP emission signals of neutral atom and ion lines of Ca, Cr, Cd. (*b*) Effect of changes in the concentration of aluminium on the ICP emission intensities of calcium atom and ion lines. (Fassel, 1977.)

(*c*) *Ionization interference.* A significant depression in the atomic absorption signal of some elements (the alkali metals, for example) occurs in flame AA measurements due to ionization in the flame. Steps must usually be taken to suppress this effect, as far as possible, by adding an ionization buffer.

The situation in emission spectrometry from an ICP plasma is different. As has been shown in section 5.6(*a*) above, the observation of ion emission lines from the plasma tail flame is related to an ionization and excitation mechanism involving metastable argon species. Indeed, for many elements, optical emission of selected ion lines is more intense than that from corresponding atomic lines. The concern is not then to suppress ionization as such, but rather to ensure that the intensity of optical emission by either atomic or ionic species does not vary when samples that may contain varying concentrations of more readily ionizable elements are introduced into the plasma. It is perhaps a surprising observation that ionization interferences are generally low in the argon plasma, since it might be expected that the introduction of a more readily analysable element would increase the free electron population of the plasma and significantly influence ionization interferences:

$$M^0 \rightleftharpoons M^+ + e^- \quad (5.11)$$

That this is not so implies that the free electron density of the plasma is buffered by plasma excitation products.

Thus data presented by Fassel (1977) in Figure 5.25 show the change in the atomic and ionic emission signals from calcium, chromium and cadmium in solutions containing varying excesses of sodium. Only at very high sodium concentrations (often >1000 μg/mL) is a significant enhancement or depression in the analyte emission signal observed. It is certain that these results are not entirely caused by ionization interference effects; changes in the total salt content of analysed solutions also influence these data. The introduction of readily ionized species may therefore have a small effect on atomic and ionic emission signals, though not as large as might be expected from atomic absorption studies.

(*d*) *Plasma interferences—general.* Although the details of interferences above indicate that these effects are reassuringly small, such data must be treated with caution as they are only relevant to observations made at the 'normal' viewing height in the plasma. Several workers have reported that the magnitude of plasma interferences changes sig-

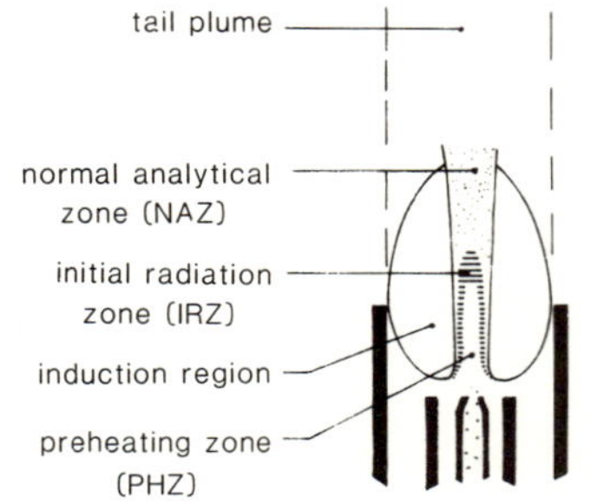

Figure 5.26 Spectroscopic radiation zones in the ICP argon plasma based on behaviour of sample aerosol. (Koirtyohann *et al.*, 1981.)

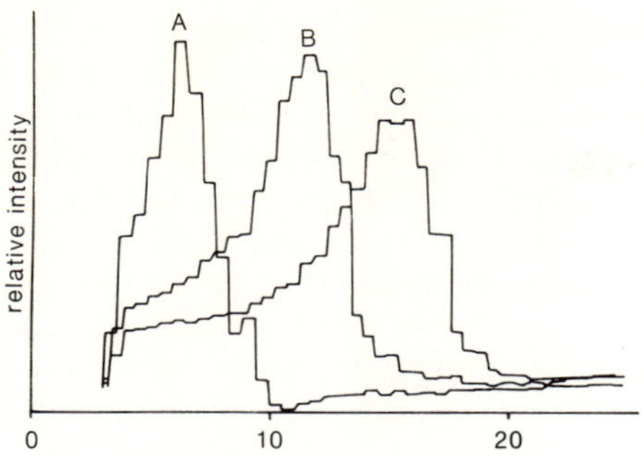

Fig. 5.27 Difference in emission intensity of a 20 μg/mL solution of calcium at 422.7 nm with and without the addition of 1000 μg/mL of lithium. Data are plotted for three injector gas flows: (*A*) 1.0 L/min, (*B*) 1.2 L/min and (*C*) 1.4 L/min. In each case the apex of the initial radiation zone was observed at 6, 11 and 16 mm respectively above the load coil, corresponding to the maximum enhancement in the interference effect plotted here. (Koirtyohann *et al.*, 1981.)

nificantly, depending on the zone from which measurements are being made. To this end Koirtyohann *et al.* (1981) proposed that interference effects should be measured relative to an internal reference point within the plasma rather than relative to the fixed work coils. These workers divided the plasma into regions based on the behaviour of the central sample aerosol channel which is of primary analytical interest (Figure 5.26). These zones are as follows.

The *preheat zone* (PHZ) is followed by the *initial radiation zone* (IRZ). This zone corresponds to the region where analyte atoms begin to emit radiation. The *normal analytical zone* corresponds to the region where spectral emission for many elements is from ions and represents the point at which most analytical meausurement are made. The position of the initial radiation zone can readily be viewed by nebulizing a solution of a few hundred μg/mL of sodium.

Koirtyohann *et al.* (1981) demonstrated the importance of these zones by measuring the interference effect of lithium on atomic and ionic emission lines from calcium as a function of height within the plasma. Some results from these studies are shown in Figure 5.27. This diagram shows the *difference* in emission intensity from a 20 μg/mL solution of Ca with and without 1000 μg/mL Li. Data are plotted for the plasma operated at three carrier gas flows: 1.0 L/min (*A*), 1.2 L/min (*B*) and 1.4 L/min (*C*). In each case, the apex of the initial radiation was observed at 6, 11 and 16 mm respectively above the load coil. These results indicate that the magnitude of the interference follows the position of the initial radiation zone. If analytical measurements were made at 15 mm above the work coils, what would be a negligible interference with a carrier gas flow of 1 L/min could represent a serious interference at 1.4 L/min. The mechanism for these interferences is related to plasma chemical reactions and is an area of active research interest.

5.8 Measurement and analysis of emission spectra

(*a*) *Spectral interferences.* By far the most serious potential interferences in ICP–atomic emission spectrometry arise from spectral overlaps between emission lines. The complexity of emission spectra from an ICP source can be illustrated by a small section of the spectrum of iron recorded over the range 287.10–288.85 nm by Michaud and Mermet (1982) (Figure 5.28). In this rather narrow 1.8 nm portion of the spectrum these authors detected 22 emission lines from both atomic and ionic iron. Indeed, in their study of the iron spectrum from 200 to 300 nm, over 1000 lines were observed, including 100 that had not previously been reported. At the time of writing, the emission lines and overlap interferences observed in an argon plasma have not yet been fully characterized. Several compilations of emission line data exist (see section (*e*) below) and some of these tables include data on potential interferences. However, much of this work depends on optical emissions observed from an electrical discharge emission source, and although wavelength data are accurate, the relative intensity data are not relevant for ICP measurements. Spectral overlaps therefore currently offer one of the more serious sources of potential analytical bias in ICP–AES.

(*b*) *Spectrum line widths.* The efficiency with which overlap interferences can be resolved depends in part on the natural line width of emission lines themselves and in part on the specification of the spectrometer available to make the measurement. It might be expected that the natural line width of emissions from an ICP source might be significantly enhanced by both Doppler and collisional broadening due to the high temperatures encountered inthe plasma (section 3.7). However, in a detailed study of this problem, Human and Scott (1976) observed that the collisional broadening component was considerably less than that observed in flames and that the predominant effect was Doppler broadening.

Doppler broadening is due to changes in the frequency of the emission line caused by the random movement of the emitting species with respect to the observer (i.e. the spectrometer); collisional broadening is caused by perturbations in the electron orbital energy levels of atoms or ions due to collisions with other species (see section 3.7).

Robin (1982) estimated that the FWHM of emission lines was increased from about 10^{-5}nm (the natural line width) to between 0.002 and 0.02 nm in an ICP source. Human and

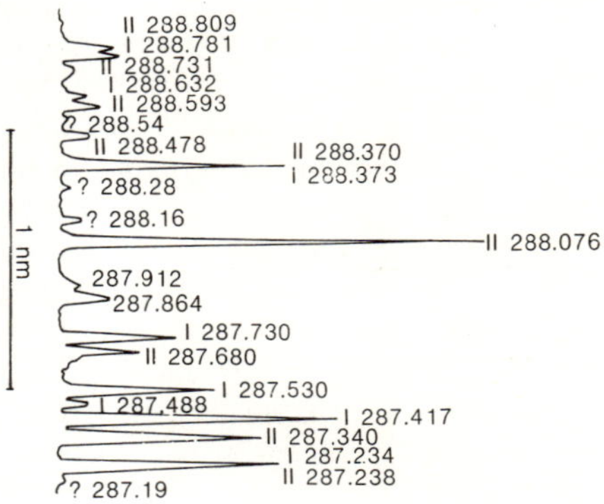

Figure 5.28 Section of the ICP emission spectrum of iron between 287.1 and 288.8 nm. (Michaud and Mermet, 1982.)

Scott (1976) estimated that plasma spectra line half-widths were of the order of 0.005 nm. This places constraints on spectrometer design. If the resolution specification of the spectrometer significantly exceeds these values, the degree of overlap interference will be controlled by intrumental factors. If spectrometer resolution were significantly better, then no benefit would result since the degree of overlap interference would controlled by the observed FWHM emission width caused by Doppler broadening. Therefore, there is considerable advantage in making emission measurements with a spectrometer with FWHM resolution towards the lower end of the range 0.02 to 0.005 nm so that interferences are not increased by instrumental effects. This is a very much higher specification than that required for routine atomic absorption measurements where wavelength selection is effectively provided by the line spectrum emitted by the hollow cathode lamp.

(*c*) *Background emissions.* A second spectrum interference problem is encountered in ICP–AES analysis—that of background emission. Background emission in an AA spectrometer can be readily distinguished from the atomic absorption signal by chopping or pulsing the output from the hollow cathode lamp source. This technique is obviously not possible with the continuous emission signal observed from an ICP. Nor is a background signal observed only from the plasma itself. An analytical signal can be enhanced by stray light effects and these are most serious where a line of low intensity is being measured in the presence of an intense emission due to another component. Particularly troublesome in silicate rock analysis are the intense emissions due to Ca, Fe and Al. The light entering the spectrometer is reflected off components into the detector. The magnitude of this stray light interference is affected by spectrometer design and is virtually eliminated using a double monochromator system (section 3.13, 3.14). Direct reader spectrometers are most seriously affected, and Walters *et al.* (1983) have shown that substitution of a holographic grating for a classically ruled grating can significantly reduce the magnitude of stray light effects.

(*d*) *Line selection and interference correction.* Line selection and the identification and correction of overlap interferences are critical aspects of successful ICP–AES measurements. Until comparatively recently, no definitive tables existed that listed relative line intensities appropriate to the inductively coupled argon plasma source. Older optical emission tables drawn up from arc and spark emission spectrography data have proved misleading, both in identifying the most sensitive ICP lines for analysis and in assessing the relative magnitude of any interference. The problem is compounded by the large range of potential interfering elements that can coexist in silicate rock solutions and the potential for multiple-order overlap interferences, caused by higher order diffractions from a grating monochromator. Faced with these problems, many analysts have resorted to a detailed investigation of individual overlap interferences by step scanning spectrometers over selected analytical lines to identify any potential interferences that could occur from the various combinations of elements that are present in samples. Examples of two such step scans are shown in Figure 5.29, taken from the work of McLaren *et al.* (1981) and relevant for the analysis of marine sediments. Figure

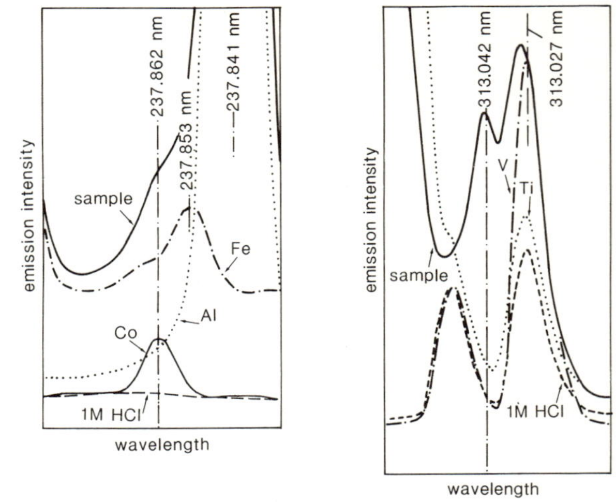

Figure 5.29 ICP spectrum wavelength scans recorded to identify spectral interferences on (*a*) the Co II 237.862 nm line and (*b*) the Be II 313.042 nm line. (McLaren *et al.*, 1981.)

5.29*a* shows the interferences observed on the Co(II) 237.862 nm line, indicating that significant spectral overlap occurs from the emissions of Al (237.841 nm) and Fe (237.853 nm). Similar problems beset the Be(II) 313.042 nm line as can be seen in Figure 5.29*b*. This suffers spectral overlap interferences from the HCl blank solution as well as from V and Ti. In such cases it is essential to make an accurate correction of background emission intensity and to remove any residual interference by calculating the intensity of the interfering element at the line of interest from an adjacent line of that interferent which does not itself suffer interference.

The above data indicate that the measurement and correction of background increases the complexity involved in any general scheme of analysis. If a sequential spectrometer is used for analysis, individual measurements of peak *and* background intensities are necessary. The number of measurements then required is illustrated by data in Table 5.2 which lists the peak and background wavelengths selected by McLaren *et al.* (1981) in the analysis of marine sediments. For some elements background correction was judged not to be necessary. For others it was necessary to record background measurements on either side of the analyte line to account for any slope in the background emission. Data in Table 5.2 also indicate the very accurate wavelength control required on an ICP spectrometer and the large number of spectroscopic measurements required to undertake a successful analysis. Criteria for deciding whether to select background measurements are similar to those outlined in XRF analysis in section 8.11. If the intensity of the peak measurement is expected to be sufficiently high relative to the background (p.279) the error in ignoring background measurements is negligible. In consequence, it is not always necessary to correct background emission in the determination of the major elements, but it is almost always necessary to make such corrections in the analysis of trace-element concentrations, especially those approaching detection limit levels.

Table 5.2 Line and background wavelengths used in the analysis of marine sediments

		Emission line (nm)	Background measurement positions (nm)	
B	I	249.773	—	—
Mo	II	202.030	202.003	202.057
As	I	193.696	193.678	193.714
Cd	I	228.802	228.778	228.826
Zn	I	213.856	—	—
P	I	214.914	214.895	214.931
Pb	II	220.353	220.329	220.376
Co	II	237.862	—	—
Ni	II	231.604	231.585	231.625
Mn	II	259.373	—	—
Al	I	237.335	—	—
Fe	II	259.940	—	—
V	II	292.402	—	—
Cr	II	267.716	267.688	267.745
Mg	II	279.079	—	—
Cu	I	324.724	324.719	324.789
Na	I	330.237	—	—
Ti	II	336.121	—	—
Be	II	313.042	312.988	312.988
Ca	II	317.933	—	—

Data from McLaren *et al.* (1981). A dash indicates that no background correction was performed.

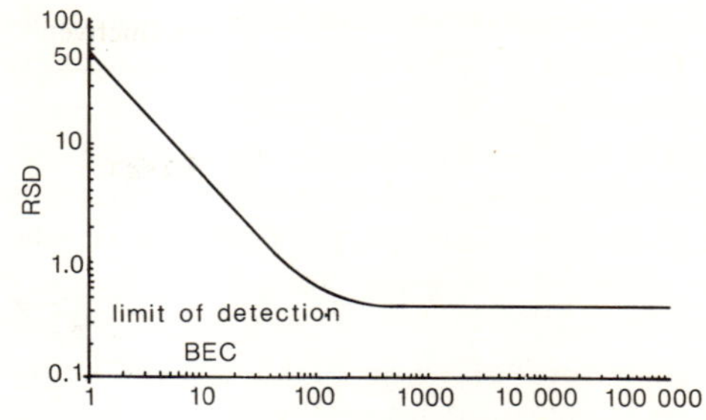

Figure 5.30 Diagram showing the theoretical analytical precision (relative standard deviation) plotted as a function of concentration expressed as multiples of the background equivalent concentration (BEC). The relative standard deviation flattens off at concentrations exceeding 100 BEC. (Church, 1981.)

The most intense lines are not always automatically selected for analysis. If both major and trace elements are to be determined from the same sample solution without intermediate dilution, some major element lines are found to be so intense that they saturate the photomultiplier and cause non-linear calibration curves. In such circumstances it is desirable to select a less intense line for analysis. Thus, Walsh (1980) selected the Ca(II) 315.887 nm and the Mg(II) 383.826 nm lines for analysis because the most intense lines at 399.3 and 279.553 nm respectively were considered to be far too intense for successful analysis.

In this respect it should be noted that the relative error in an optical emission measurement does not continue to fall with increasing emission intensity. The choice of a less sensitive line does not, therefore, necessarily reduce the precision of a measurement. Thus in Figure 5.30, the theoretical variation in the relative standard deviation of an emission intensity measurement is plotted as a function of concentration, calculated in multiples of the background equivalent concentration (BEC), as presented by Church (1981). The relative error in a measurement sharply falls as concentrations increase from detection limit level up to 100 times the background equivalent concentration. However, at higher concentrations, little improvement is observed in the relative standard deviation. All emission lines are expected to show similar curves; the precise value of the limiting error depends on a number of factors including plasma stability and nebulizer noise levels. A similar relative standard deviation relationship has been measured experimentally, for example in the determination of lead, by Watters (1983).

Taking these factors into account, the lines most commonly selected in ICP schemes of analysis biased towards the characterization of geological materials are listed in Table 5.3. It must be emphasized that the optimum choice of line depends on a number of factors, including the optical efficiency of the spectrometer used for the measurement (e.g. the blaze angle of the grating) and the magnitude of any interference effect. Data in Table 5.3 are therefore offered as a guide only, the lines selected depending on both application and instrumental factors.

Similarly, the listed interferences must be interpreted with caution. Some of these interferences are direct spectral overlaps. Others disappear if a line is measured in the first order instead of second order and vice versa. Still others cause background perturbations due to their presence as an intense emission adjacent to a weak trace-element line.

(*e*) *A note on line selection tables.* Line spectra for the optical emission of elements are well characterized due to the historic interest there has been in the study and analytical applications of electrical discharges as optical emission sources. This work has led to the compilation of two sets of emission tables normally referred to as the MIT tables (Harrison, 1969) and the NBS tables (Meggers *et al.*, 1975). These tables are specifically appropriate to electrical discharges, and although the tabulated wavelengths of emission lines can be regarded as accurate their relative intensities are not unambiguously appropriate to the inductively coupled argon plasma source. This makes it difficult to assess the most sensitive lines suitable for ICP analysis from these compilations alone. However these earlier data have been adapted for ICP–AES and have been used as a basis for predicting the presence of potential interferences in ICP spectra by Larson and Fassel (1970) (data based on the MIT tables); Boumans (1980, 1984) (data based principally on the NBS tables); and Parsons *et al.* (1980) (compiled from all available data, especially from arc/spark sources and in particular the MIT tables). For the reasons outlined above, these compilations were not entirely satisfactory. The deficiency in line emission data based on direct experimental measurements from the inductively coupled argon plasma has been progressively filled, in particular by Winge *et al.* (1979) who published a list of prominent lines observed in ICP–AES. This has been followed by further studies,

Table 5.3 Optical emission lines selected for the analysis of geological materials using an inductively coupled argon plasma

Element	Principal lines and interferences (nm) (I_n/I_b)		Other lines and interferences (nm) (I_n/I_b)	
Ag	I 328.068 (38.0)	Mg, Ca, Al, Fe, Ti, Zr, Mn		
Al	I 309.271 (13.0)	Mg, V, Be, OH band	I 237.312 (10.0)	Mg, Ca, Fe, Mn
	I 308.215 (6.6)	V	I 396.152 (10.5)	Mo, Ca, Zr
As	I 193.696 (56.0)	Mg, Fe, Al, V, Cr, Zr	I 189.042 (22.0)	Mg, Ca, Fe, Al, Ti, V, Cr, La
			I 197.197 (39.0)	
Au	I 242.795 (170.0)	Fe, Mn, V, Sr		
B	I 249.773 (63.0)	Fe		
Ba	II 455.403 (230.0)	V, Zr, Sr, Sc, Cr	II 493.409 (130.0)	
Be	II 313.042 (110.0)	V, OH band		
Bi	I 223.061 (87.0)	Mg, Ca, Fe, Al, Ti, V, Cr La, Ni	I 306.772 (40.0)	OH band
Ca	I 422.673 (1.5)	Sr	II 315.887	
	II 317.933 (1.5)		II 393.366 (89.0)	
			399.3	
Cd	II 226.502 (89.0)	Fe, Al, Ti, Ni, Zn	II 214.438 (120.0)	Fe
			I 228.802 (110.0)	
Co	II 228.616 (43.0)	Fe, Ti, V, W, Ni, Ba	II 237.862 (31.0)	Fe, Al
			II 238.892 (50.0)	Fe, Ti, V
			314.26	
			I 345.350	
Cr	II 283.536 (42.0)	Mg, Fe, Ti, V	I 425.435	Nb, Sr
	II 267.716 (42.0)	Mn, Mo, U, V		
	II 205.552 (49.0)	Al, Mn		
Cu	I 324.754 (56.0)	Fe, Ti, Be, Nb, Ar	II 224.700	Fe, Pb
		Mn, Mo, OH band		
Fe	II 259.940 (48.0)	Ta, Mo, Mn	II 263.105 (19.0)	
			302.06	
			II 238.204 (65.0)	
			271.4	
Ga	I 294.364 (64.0)		I 417.206 (45.0)	Fe, Ti
Ge	I 265.118 (62.0)		I 209.426 (75.0)	
Hf	241.5		II 277.336 (190.0)	
K	I 766.490		I 404.721 (0.7)	
Li	I 670.784	Ti, Sc	I 460.286 (3.5)	
Mg	II 279.079 (1.0)	Fe, Mn, Nb, Cu	II 279.553 (195.0)	Fe, Cu
	I 383.826 (0.9)		I 285.213	
Mn	II 257.610 (220.0)	W	II 259.373 (190.0)	Fe
Mo	II 202.030 (38.0)	Ca, Fe, Al, As	313.62	Fe, Cr, W, Zr
			II 281.615 (21.0)	Al, Mg, Mn
			I 386.411	
Na	I 588.995 (101.0)		I 330.237 (1.6)	
	I 589.592 (43.0)			
Nb	II 309.418 (83.0)	Al, Mg, Ti, V, Fe		
Ni	II 231.604 (19.0)	Fe, Zn, Co, Sr, Sc	I 341.476 (6.2)	Zr
	(221.647 (29.0) is most intense)			
P	I 213.618 (39.0)	Fe, Al, Cu, Zn, V, Mg	(178.29)	Al
			I 214.914 (39.0)	
Pb	II 220.353 (70.0)	Mg, Fe, Al, Ti, V, As, La, Zn, Nb	I 405.783 (11.0)	
Sb	I 206.833 (91.0)	Mg, Ca, Ti, Fe, Al, Cr, V, Zn		
	I 217.581 (68.0)			
Sc	II 361.384 (200.0)	Ti, Ce		
	II 424.683 (110.0)			
Se	I 196.026 (40.0)	Fe, Al, Mn, V		
Si	I 251.611 (250.0)		I 252.851 (95.0)	
	I 288.158 (110.0)			
Sn	II 189.989 (120.0)	Al, Mn, Ti, Zn, Mg, Ca, V, Zr, As	187.980	
			I 235.484 (31.0)	
Sr	II 407.771 (72.0)	Ba, Cu, Sn, Ta, Mo, Cr	II 421.552 (39.0)	
Ta	II 240.063 (105.0)		II 301.254	Be, Cr, Fe, Mn, Co
	II 226.230 (120.0)			
Te	I 238.578 (17.0)	Fe, Mn, Ta, W	I 214.281 (73.0)	
Th	II 353.959 (30.0)	Mg, Ca, Fe, Mn, V, Zr, Ce	II 283.730 (46.0)	
	II 401.913 (36.0)			
Ti	II 334.941 (79.0)	Cu, Co, Au, W	II 336.121 (57.0)	
	II 337.280 (45.0)		II 368.520 (26.0)	Fe, Mn
Tl	II 190.864 (74.0)		I 377.572 (13.0)	
	I 351.924 (15.0)			
U	II 409.014 (8.9)	Ca, Mn		
	II 385.958 (12.0)	Mg, Ca, Fe, Al, Ti, V, Ce, Cr, La		

Table 5.3 (*Continued*)

Element	Principal lines and interferences (nm) (I_n/I_b)		Other lines and interferences (nm) (I_n/I_b)	
V	II 311.071 (30.0)	Ti, Mn, Cr, Zr	II 292.402 (40.0)	Fe, Ti, Mn, La, Mg, Al, Cr
	II 268.796 (29.0)	Cr	II 310.230 (47.0)	OH band
	II 290.882 (34.0)	Fe, Nb	307.311	
	(309.311 (60.0) is most intense)			
W	II 239.709 (54.0)	Fe, Al	400.875	Al, Ca, Fe, Ti, Sr
	II 207.911 (100.0)			
Y	II 371.030 (86.0)		II 321.669 (38.0)	
Zn	II 202.548 (75.0)	Mg, Al, Cu, W, Fe, Nb, Zr, La, Cr		
	I 213.856 (170.0)	Mg, Fe, Al, Ni, Cu		
Zr	II 339.198 (39.0)		II 349.621 (30.0)	Y
	II 343.823 (42.0)	Ca		

(i) Selected lines and interferences compiled from data presented in Church (1981); Walsh and Howie (1980); Crock *et al.* (1983) (all silicate rock analysis); Walsh (1982) (geochemical exploration); Walters (1983) (geological samples); McLaren *et al.* (1981) (marine sediments); McQuaker *et al.* (1979) (environmental materials); Schramel (1983) (biomedical/environmental samples); Boumans and Bosveld (1979) (plasma interferences).
(ii) I indicates an atomic emission line; II signifies an ion emission line; I_n/I_b represents the net line intensity ratioed to the background signal intensity as measured by Winge *et al.* (1979) and is, therefore, a measure of the relative signal strength.
(iii) The elements listed as interferences should be taken as a guide, only, since their presence and relative severity depends on order of line measured, plasma operating conditions and design of spectrometer.
(iv) The distinction between 'principal line' and 'other line' is based on the most frequent use in the references indicated. However, choice of line is also affected by instrument/spectrometer design and will not always follow the ranking indicated above.

including, for example, Winge *et al.* (1982) and Anderson *et al.* (1982). However, the most complete work to date is that of Winge, Fassel, Peterson and Floyd (1984) who have listed relevant spectral data in *Inductively Coupled Plasma–Atomic Emission Spectroscopy: an Atlas of Spectral Information.*

5.9 Some instrument considerations–simultaneous *v.* sequential monochromators

The overall design of an ICP atomic emission instrument affects the operating strategy. Two distinct classes of equipment are currently in use (Figure 5.31).

(i) The *simultaneous spectrometer* is a direct reader design and is usually based on the Paschen–Runge mounting, discussed in detail in section 3.13. Light from the ICP source is focused on a fixed diffraction grating placed on the circumference of a Rowland's circle. Fixed exit slits are arranged on this Rowland's circle to coincide with the diffracted lines of a pre-selected range of elements. Light passing through these exit slits is focused on individual photomultiplier tubes. The number of elements that may be determined simultaneously is essentially limited by the physical constraints of siting photomultipliers around the Rowland's circle and the cost associated with the installation of additional channels. Polychromators fitted with 20–60 channels are available. The diameter of the Rowland's circle is often about 1m, and optical filters may be placed in front of individual photomultipliers to eliminate multiple order diffraction interferences. The inflexibility of this system (it is not generally a trivial matter to change element lines) is often ameliorated by providing a movable exit slit to permit the occasional determination of lines that have not been pre-programmed and to offer a spectral scanning facility. The strength of such instruments lies in the extremely rapid analysis times that may be achieved in the routine determination of fixed range of major and/or trace elements.

(ii) The *sequential spectrometer* (Figure 5.31*b*) is based on a conventional scanning monochromator which on a modern system can be pre-programmed to scan rapidly from line to line under computer control. The analyst, therefore, has the flexibility of being able to specify the sequence of lines to be measured and the versatility to amend this analysis programme according to changing requirements. The sequential nature of measurements means that analysis times are longer than with a simultaneous spectrometer. However, major instrumental alterations are not required to change the sequence of analysis as is often demanded in a research laboratory. The performance of a sequential spectrometer system has been compared with that of a direct reader polychromator instrument by Floyd *et al.* (1980). They concluded that in routine applications in which 30 elements are to be determined in a batch of 30 samples, total instrument analysis time was 2 hours on a simultaneous instrument and 6 hours on a sequential spectrometer system.

(iii) *Background measurements*: whichever spectrometer is used, corrections for background emission must normally be made. In the case of the simultaneous polychromator it is expensive and impracticable to provide individual photomultiplier channels to measure the background for each analytical line. Two instrumental modifications have been proposed to overcome these difficulties.

The first device involves the concept of moving the monochromator entrance slits. The arrangement is particularly suitable for direct reader polychromators with fixed exit slits and is shown in Figure 5.32. The entrance slit can be moved transversely by a stepper motor, normally under computer

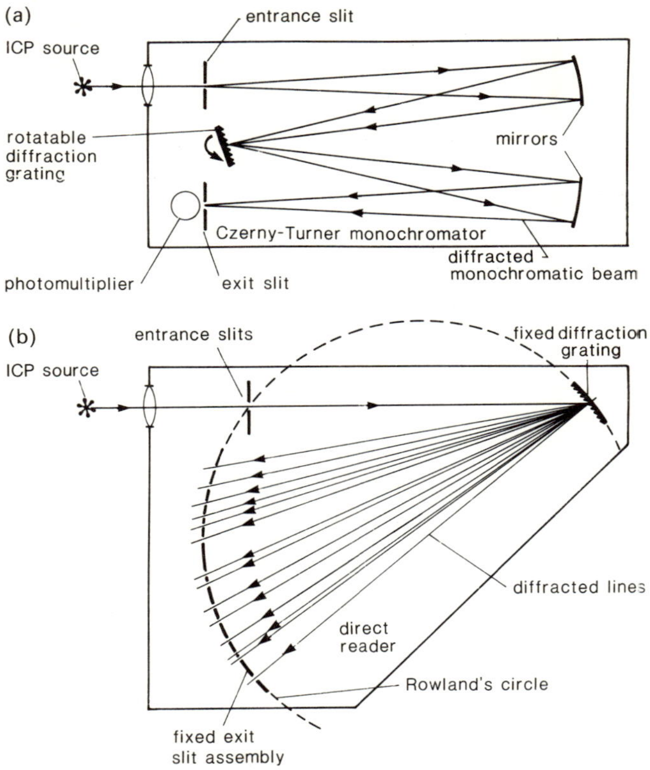

Figure 5.31 Schematic arrangement of (*a*) a sequential ICP spectrometer based on a single Czerny–Turner monochromator, and (*b*) a simultaneous ICP spectrometer based on a direct reader monochromator system. For clarity, individual photomultiplier assemblies behind the exit slits have been omitted.

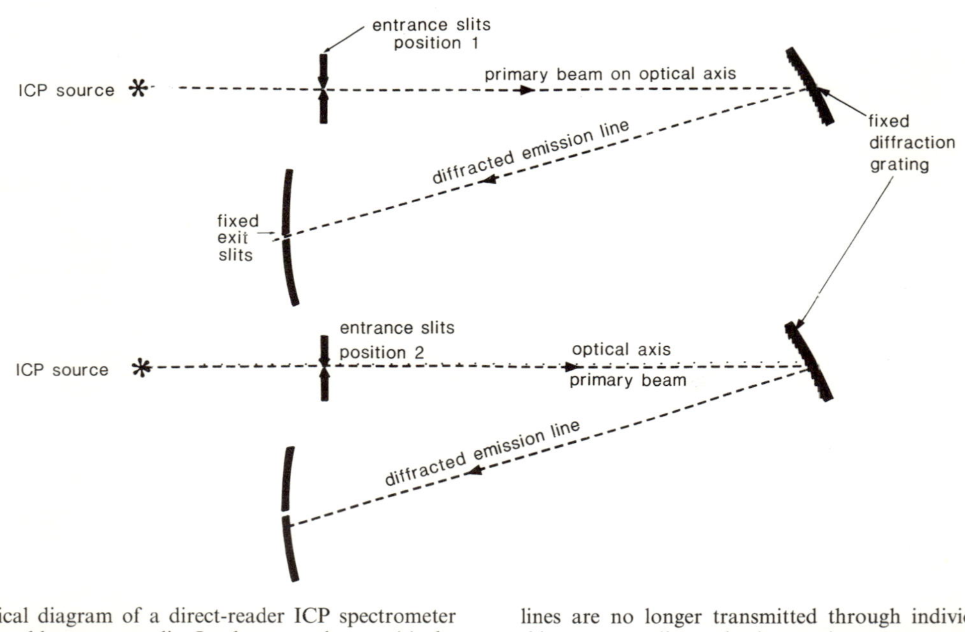

Figure 5.32 Optical diagram of a direct-reader ICP spectrometer fitted with a movable entrance slit. In the normal state (*a*) the monochromator is optimized so that diffracted emission lines are centred on fixed exit slits. With the entrance slit moved sideways, the diffracted spectrum is also shifted sideways so that the emission lines are no longer transmitted through individual exit slits. In this state, an adjacent background measurement can be made. The entrance slit would normally be operated under computer control in an automated spectrometer system.

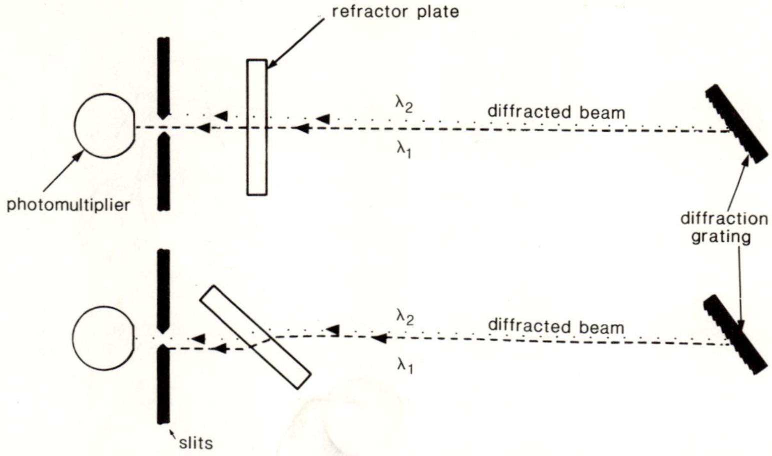

Figure 5.33 Use of a quartz refractor plate in optimizing wavelength setting on a sequential spectrometer. The grating is first rotated coarsely to find the approximate intensity maximum of a selected emission line (dashed line). The refractor plate is then rotated to fine-tune the monochromator setting to the precise peak maximum of that emission line (dotted line).

control. Background correction is obtained as follows. With the entrance slits in their 'normal' position, emission line intensities are measured as usual. The polychromator entrance slit is then moved through a predetermined distance in one direction. This has the effect of shifting the entire spectrum diffracted from the grating with respect to the Rowland's circle, so that each fixed exit slit now measures an off-peak (one hopes a background) wavelength. After a further integration the entrance slits can be moved in the opposite direction, if necessary, to permit background measurement on the other side of the emission line. These data permit accurate background correction of the emission signal as has been discussed, for example, in the analysis of geological samples by Walters *et al.* (1983). Wavelength profiles can also be measured using this system by progressively stepping the entrance slits so that the required range is scanned across the fixed exit slits. The total range that can be scanned in this way is usually about 1nm.

The second device uses a *quartz refractor plate* placed behind the primary entrance slits. On a sequential monochromator available from one instrument manufacturer, the refractor plate is positioned in front of the exit slits to improve wavelength optimization (Figure 5.33). The refractor consists of a plate of quartz glass. When positioned at right angles, a beam of light is transmitted without deviation. However, when the refractor plate is rotated, the incident beam is refracted, effectively causing a sideways shift in the transmitted spectrum. Fitted behind the entrance slits of a direct reader monochromator, the refractor plate if rotated has a similar effect to that of moving the primary slits as described above. Fitted in front of the exit slits of a sequential monochromator, the refractor plate can improve the speed and accuracy of monochromator movement. In resetting to a predetermined wavelength, the monochromator diffraction grating is slewed rapidly by a stepper motor designed to move in relatively coarse increments of 0.005 nm. Final adjustment to the precise wavelength can then be achieved by rotating the refractor plate, so achieving movements of as little as 0.001 nm. Programmed rotation of the refractor plate can also assist in searching for the intensity maximum of emission lines.

5.10 Optimizing operating parameters

In optimizing an inductively coupled argon plasma torch, the analyst has several parameters under his control. These include:

(i) Argon gas flows (coolant gas supply, typically between 12 and 20 L/min; auxiliary gas supply, typically 0 to 0.5 L/min; and injector gas supply, that is, the sample aerosol carrier gas, typically 0.8 to 1.6 L/min)
(ii) Forward power delivered to plasma
(iii) Observation height in plasma tail flame for spectroscopic measurement.

Of these parameters, the injector gas flow rate (and observation height) are particularly important.

Some examples of the influence that individual parameters have on the analytical signal will now be considered. Some of these data are characteristic of the individual designs of instrument and are not necessarily representative of the behaviour of all ICP spectrometers.

The influence of forward power delivered to the plasma on the signal to background ratio for selected element lines is shown in Figure 5.34. These data are taken from Boumans (1978) using a Philips designed torch and a cross-flow nebulizer. The highest signal to background ratios were observed at lower power inputs for measurements made on this system under constant gas flow conditions and constant observation height.

Figure 5.35 shows the effect of changes in the injector gas flow rate to the nebulizer on net peak intensity and signal to background ratio for various rare-earth element lines (Broekaert *et al.*, 1979*b*). The rare-earth element signal is strongly influenced by carrier gas flow rate. The authors considered that when the carrier gas flow rate is increased from 0.4 to 0.5 L/min, the efficiency of the Meinhard concentric glass nebulizer used for determinations increased. Solution uptake rate was simultaneously increased from

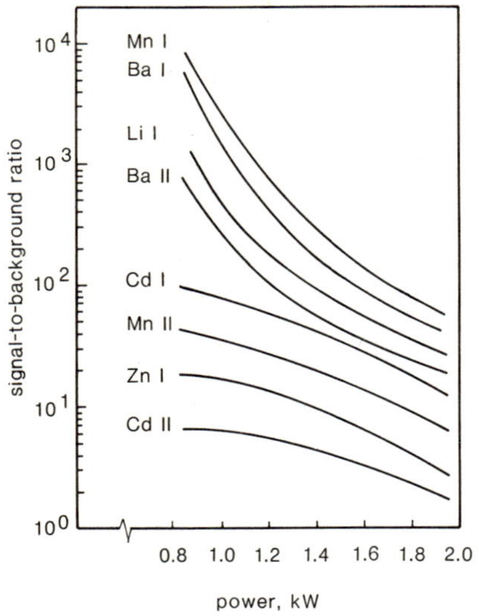

Figure 5.34 Effect of forward power delivered to plasma on the signal-to-background ratio of selected emission lines. Using the instrument on which these measurements were obtained, optimum signals are obtained at lower generator power settings. (Boumans, 1978.)

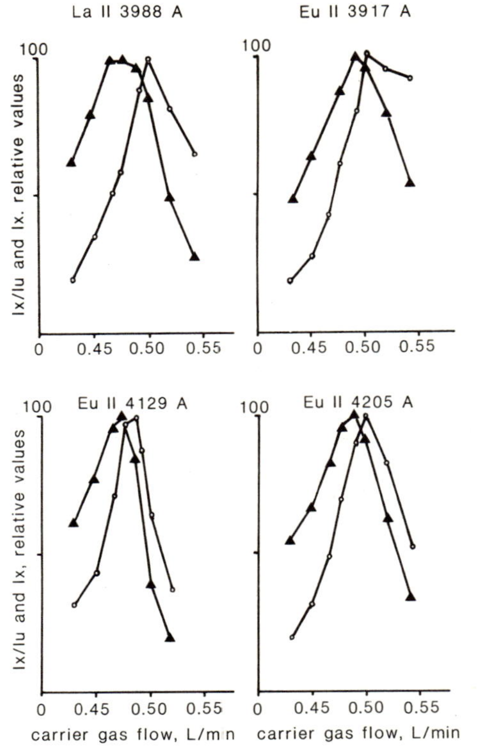

Figure 5.35 Effect of varying injector (carrier) gas flow on line to background intensity ratio (○) and net emission line intensity (▲) of 0.5 g/mL solutions of lanthanum and europium. (Broekaert *et al.*, 1979*b*).

0.08 to 0.15 mL/min and a smaller aerosol particle size was expected. The authors proposed that this led to more efficient desolvation and atomization, causing the sharp increase in sensitivities in Figure 5.35. The corresponding decrease in sensitivity at flow rates in excess of 0.5 L/min was ascribed to a reduction in the excitation temperature of the central part of the plasma. It seems possible that changes in plasma structure identified by the radiation zones of Koirtyohann *et al.* (1981) (section 5.7*d*) may in part account for these observations.

Brenner *et al.* (1981) studied the influence of carrier gas flow on the net signal intensity of selected REE lines as a function of observation height in the plasma. Samples were analysed on a medium power argon–nitrogen ICP using a cross-flow pneumatic nebulizer. Their data are shown in Figure 5.36. Using this instrumentation, optimum signal was obtained at a carrier gas flow rate of 2.0 to 2.4 L/min and observation heights of 12–14 mm.

Investigations by Nygaard *et al.* (undated) indicate that the emission signal of sodium varies significantly with observation height in the plasma. Although the emission signal was observed to reach a maximum 10 mm above the work coil (Figure 5.37) (the initial radiation zone), the precision of measurements was better at 20 mm above the work coil (the normal observation zone). This latter observation height was preferred for analysis in normal circumstances.

The results summarized above indicate the sort of studies carried out to optimize excitation conditions for individual elements. In optimizing conditions for simultaneous multi-element analysis, two additional factors must be taken into account. The first is that a unique set of operating conditions exists for optimum determination of each element. For practical schemes of analysis, therefore, compromise conditions must be selected. The second is that operating parameters tend to be interrelated, a change in one causing the optimum setting of others to alter. Two procedures have been used to overcome these difficulties.

(*a*) *Univariant optimization.* One parameter is varied while all others are kept constant at pre-determined 'reasonable' settings. An analytical signal is then measured as a function of that parameter. The most common 'figure of merit' used to identify optimum conditions is the signal to background ratio (Greenfield and Burns, 1980). When one parameter has been optimized, the same procedure is followed for the next. Under ideal circumstances, this procedure is carried out iteratively until no change results from one iteration to the next. In practice, assuming that reasonable values have been selected for the initial settings, one iteration is adequate. The sort of results from which optimum analytical conditions are evaluated are those plotted in Figures 5.34 to 5.37.

(*b*) *Simplex optimization procedure.* An alternative approach is the so-called simplex optimization procedure; its application to argon plasmas has been discussed, for example, by Ebdon *et al.* (1980*a*) (to compare the performance of low-power argon cooled plasma with high powered nitrogen cooled plasmas); Montaser *et al.* (1984) (to optimize operating conditions of a low gas flow argon plasma torch) and Terblanche *et al.* (1981) (to optimize six parameters on a high-power argon plasma torch). The simplex procedure as devised by Spendley *et al.* (1962) and modified by Nelder and Mead (1965) is the one most widely applied, and its application in general analytical chemistry

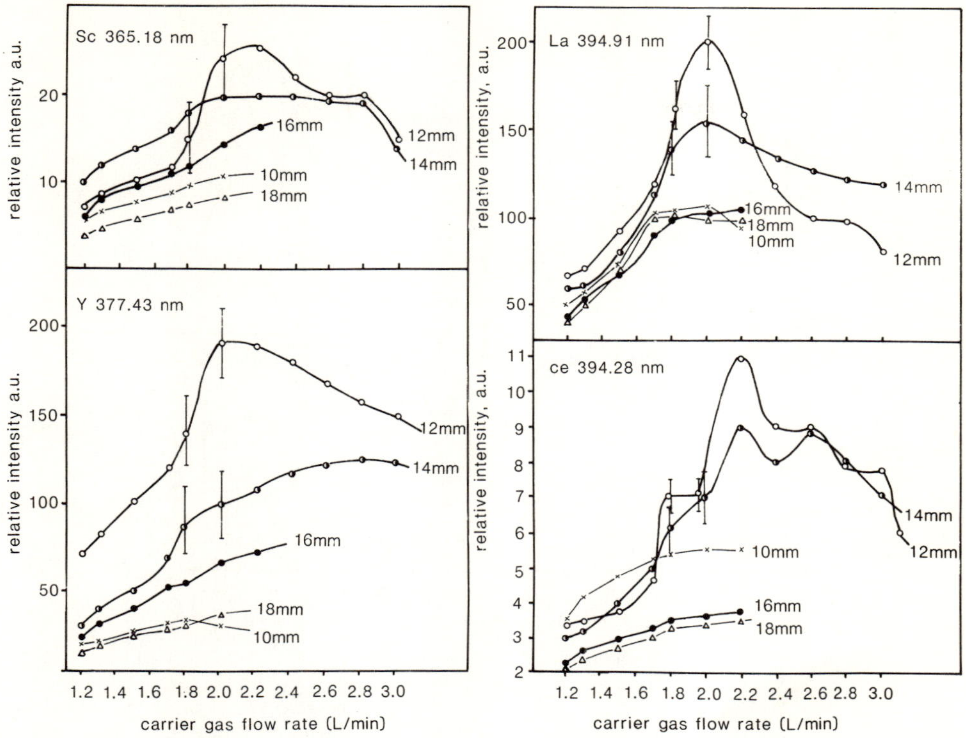

Figure 5.36 Effect of varying injector (carrier) gas flow rate on the net intensity of scandium and selected rare-earth elements as a function of observation height in the plasma tail flame. Measurements were made on a medium power argon–nitrogen ICP operated at 4.5 kW and coupled to cross-flow pneumatic nebulizer. (Brenner *et al.*, 1981.)

has been reviewed by Deming and Parker (1978) and Deming and Morgan (1983).

The procedure involves systematically varying all the parameters according to the rules of the Nelder and Mead algorithm. At each step, a 'figure of merit', normally again the signal to background ratio, is recorded. In order to ensure unambiguous optimization it is necessary to identify the maximum and minimum values of each parameter and to ensure that a large initial step is selected to cover this range (Yarbro and Deming, 1974). For example, the minimum gas flow is the smallest flow capable of sustaining a stable plasma; the maximum is the largest flow the equipment is capable of withstanding. In a simple manner, this optimization can be illustrated as in Figure 5.38. The contours represent equal values of signal to background ratio in factor space. The solid circles represent the different operating conditions obtained by changing all torch parameters simultaneously. The Nelder and Meads (1965) algorithm is designed to create experimental conditions that will identify the optimum value of each parameter that corresponds to maximum signal to noise ratios represented by the shaded

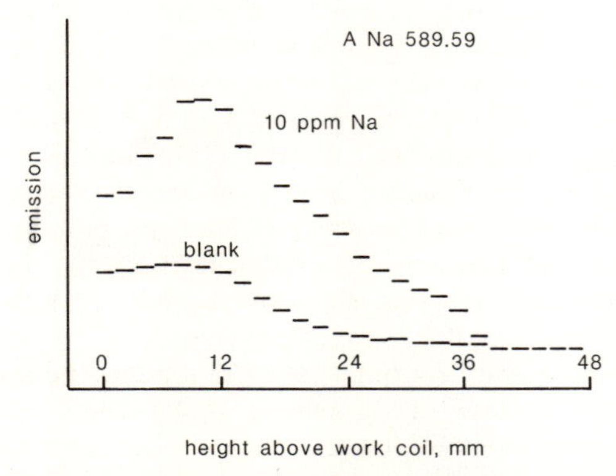

Figure 5.37 Variation in the sodium 589.59 nm emission as a function of observation height in argon ICP. (Nygaard *et al.*, by courtesy of Instrumentation Laboratory.)

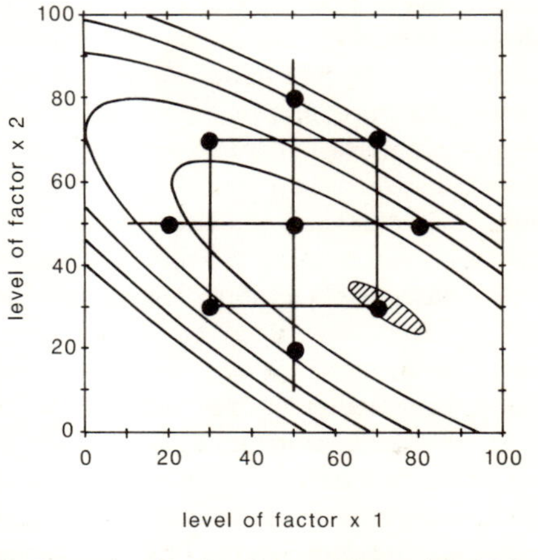

Figure 5.38 Example of a simplex experiment in which the experimental observations are plotted in factorial space. This is known as a two-factor central composite experimental design. (Deming and Morgan, 1983.)

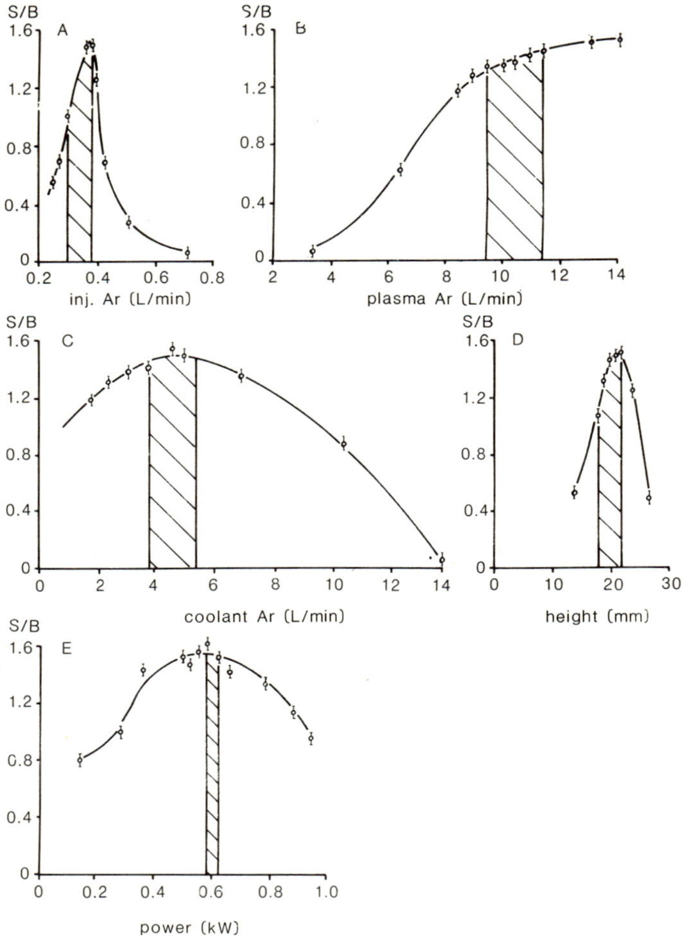

Figure 5.39 Confirmation of the simplex optimization of five variables: (*A*), injector gas flow rate; (*B*), auxiliary (plasma) gas flow rate; (*C*), coolant gas flow rate; (*D*), observation height in tail flame (mm above load coil); (*E*), power coupled into the plasma (kW). In each case, the signal to background ratio of the manganese 257.6 nm line is plotted as a function of the variable. The shaded areas indicate the regions identified as optimal by the simplex method. (Ebdon *et al.*, 1980.)

area in Figure 5.38. The data of Ebdon *et al.* (1980*a*) for the optimization of five operating variables are shown in Figure 5.39. The variables were injector, auxiliary and coolant argon flow rates, observation height and plasma power for the determination of the Mn(II) 257.6 nm ion line using a modified argon cooled torch described in Ebdon *et al.* (1980*b*). Each diagram in Figure 5.39 represents the results of a univariant optimization for individual operating variables. The shaded areas indicate the values identified as optimal by the simplex method. The sharp curves for both injector gas flow and observation height indicate the critical nature of these parameters in obtaining optimum analytical sensitivity.

(*c*) *Compromise operating conditions.* As has been mentioned above, a unique set of operating parameters can be derived by univariant or simplex optimization procedures for each individual element. However, in circumstances where ICP–AES instruments are used for simultaneous multi-element analysis, it is necessary to make a judgement on compromise conditions under which adequate excitation of a wide range of elements is achieved.

Boumans and de Boer (1975, 1976) have studied compromise conditions for the operation of a torch of similar design to that used by Fassel. The instrument was fitted with an ultrasonic nebulizer and desolvation apparatus and excited by a 50 MHz rf generator. Compromise torch operating conditions were:

Power input to plasma	0.7kW
Coolant gas flow	19L/min
Injector gas flow	1.3L/min
Observation height	15mm above work coil.

In a study of interference effects under compromise analysis conditions, Boumans and de Boer (1976) identified some interference effects caused in particular by the desolvation apparatus, and the complexity of this apparatus in conjunction with ultrasonic nebulization has discouraged manufacturers from adopting such equipment in routine applications.

Of particular concern in optimization procedures designed to identify compromise operating conditions is the selection of a suitable 'object function'. Thus, in a scheme of analysis designed to determine both major and trace elements, a greater weight should be placed on optimizing excitation conditions for those trace elements present at concentrations approaching the limits of detection of the instrument. Leary *et al.* (1982) considered several composite objective functions and used the sequential simplex technique to evaluate their success in simultaneous multi-element analysis by ICP–AES. The operating parameters identified by the final simplex in a test of this method did not correspond to the best conditions for any single element, but were judged to give a realistic compromise in the simultaneous determination of all elements.

5.11 Calibrations for ICP–AES

5.11.1 *Analytical advantages of ICP–AES*

The inductively coupled argon plasma has been widely adopted as an excitation source for emission spectrometry for several reasons.

(i) Low detection limits can be achieved largely because the background emission from the tail plume at the normal observation height is very low

(ii) The source can be used for simultaneous multi-element analysis

(iii) Chemical and ionization interference effects, which are of special concern in AA analysis, are small in the ICP plasma

(iv) Calibrations are linear over concentration ranges of between 3 and 5 orders of magnitude so that both major and trace elements can be determined from the same sample solution. This extended calibration linearity occurs because at the normal observation height the plasma behaves as an optically thin source. The sample emitting region is surrounded by a sheath of hot argon originating from the fireball. Self-absorption effects are, therefore, minimized in contrast to AA analysis, where this phenomenon is an important cause of non-linearities in calibrations at high analyte concentrations. Human and Scott (1976) reported that at observation heights of 10 mm above the work coil, self-reversal of spectral lines was not observed but that lines were considerably broadened. The authors considered that the core of the plasma appeared to behave as an absorbing region, whereas emission originated from an annulus around the core. However, at an observation height of 20 mm and above self-reversal effects were observed at the highest analyte concentrations. This appeared to be caused by a layer of cooler absorbing atoms surrounding the emission region. In spite of this problem, linear calibrations for Ca and Sr were observed over 3 orders of magnitude, self-reversal only being apparent at the highest concentrations studied.

The advantages outlined above can only be exploited if calibration procedures compensate for the sample interference effects in ICP–AES, particularly variations in nebulizer efficiency as a function of salt content and acidity of sample solutions.

5.11.2 *Preparation of standard solutions*

The introduction of ICP–AES stimulated the requirement, perhaps for the first time, for the preparation in solution form of a large number of elemental standards. Not only was it necessary to prepare standard solutions of major and minor elements, but also for a range of trace elements that had not previously been determined extensively by wet chemical methods.

The stability and compatibility of elemental standard solutions is perhaps a neglected area of analytical chemistry. The principle is simple enough: a known weight of the element or compound is dissoved to a preselected volume to prepare a stock solution (stock solutions of many elements are commercially available at concentrations of 1000 μg/mL). A range of working standard solutions is then prepared from these stock solutions. In this section it is proposed to examine some of the problems associated with the preparation of standard solutions.

The first concern is the selection of a suitable compound. Since this compound must normally be weighed (in air) on an analytical balance, several stringent requirements must be satisfied:

(i) The element or compound must be stoichiometric: thus ammonium ferrous sulphate, $(NH_4)_2SO_4.FeSO_4.6H_2O$, is suitable, whereas ferric oxide, $Fe_{(2+x)}O_{(3-y)}$, where x and y are likely to vary according to sample preparation and heating history, is not

(ii) The element or compound must be capable of being manufactured to a high purity

(iii) It must be inert and stable in air (thus NaCl is a suitable Na standard, NaOH is not, since it is very deliquescent, rapidly absorbing water vapour from air)

(iv) It must be readily soluble in water or dilute acids without requiring extensive pre-chemistry

(v) Aqueous solutions (in dilute acid if required) should be stable and resistant to precipitation/decomposition during storage prior to use.

Given these requirements, a rather limited number of elements or compounds is generally used to prepare standard solutions. Wherever possible the element itself is selected. If not, a stable stoichiometric compound must be substituted. A list of chemical materials widely used for the preparation of solution standards is listed in Table 5.4, together with a guide to dissolution procedures.

Some additional factors that are of concern if standards are to be mixed to prepare multi-element calibration solutions are now discussed.

5.11.3 *Stability of multi-element standard solutions*

Some combinations of ions, when mixed, will react to form insoluble compounds. This problem has been examined by McQuaker *et al.* (1979) in a scheme for the multi-element analysis of environmental samples. The principal difficulties encountered by these workers in the preparation of multi-element standard solutions were as follows.

(i) Chromium in chromate form was found to precipitate other cations as insoluble chromates. Chromium nitrate was therefore used to prepare the stock solution instead of the more common potassium dichromate.

(ii) For the same reason, sulphate salts were not used in solution preparation to avoid the precipitation of insoluble sulphates.

(iii) Some salts were found to polymerize on standing or in combination with other reagents and are therefore unsuitable for long-term storage. In particular, McQuaker

Table 5.4 Preparation of standard elemental stock solutions

	Compound	Drying conditions	*W* (g)	Suggested dissolution procedure
H	$CuSo_4.5H_2O$			Gravimetric methods
Li	$LiCO_3$		10.6454	50 mL 2M HCl (care: effervesces)
	$Li_2SO_4.H_2O$		18.4339	Water
Be	Elemental Be		1.0000	20 mL 1:1 sulphuric acid plus trace HCl !highly toxic!
	$BeSO_4.4H_2O$		19.6545	Dissolve in 2N sulphuric acid !highly toxic!
B	H_3BO_3		5.7197	Water
C	$CaCO_3$		(8.3339)	Gravimetric analysis
	$CH_3COO.Na$		(3.4147)	Gravimetric analysis
N				
O				
F	NaF	640/1–2hr	2.2102	Water
Na	NaCl	550/	2.5420	Water
	$Na_2CO_3.10H_2O$		6.2232	Dilute HCl; (care: effervesces)
Mg	Mg ribbon		1.0000	50 mL 1:1 HCl
	MgO	900/1hr	1.6585	Dilute HCl
Al	Al metal		1.0000	50 mL 1:1 HCl + few drops HNO_3 (mercuric salts catalyse dissolution)
	Al_2O_3	900/1hr	1.8894	50 mL 1:1 HCl + few drops HNO_3
Si	SiO_2	900/1hr	2.1395	50 mL HF then when dissolved, stabilize with saturated boric acid solution
	SiO_2	900/1hr	2.1395	Lithium metaborate fusion then dissolve in dilute nitric acid
P	KH_2PO_4	110/1hr	4.3937	Water
	$(NH_4)H_2PO_4$		3.7138	Water
S	$BaSO_4$		7.2801	For pyrometric determination of SO_2
	$CaSO_4$		4.2464	as above
	K_2SO_4		5.4351	Water (standard sulphate solution)
Cl	NaCl	550/	1.6485	Water
K	KCl	550/	1.9067	Water
Ca	$CaCO_3$	110/	2.4973	25 mL 1M HCl (care: effervesces)
Sc	Sc_2O_3	900/1hr	1.5338	10 mL 1:1 HCl
Ti	Ti metal		1.0000	50 mL water + 50 mL HCl; heat, dilute when dissolved with 1% HCl
	$(NH_4)_2TiO(C_2O_4)_2.H_2O$		6.1406	Water, salt may not be perfectly stoichiometric, may need independent standardization
V	NH_4VO_3		2.2964	Water (+ 100 volume H_2O_2)
	V metal	d	1.0000	40 mL HNO_3
Cr	Cr metal		1.0000	50 mL 1:1 HCl with gentle heat
	$K_2Cr_2O_7$		2.8289	Water
Mn	Mn flakes		1.0000	20 mL 1:1 HNO_3 or 50 mL conc. HCl
	$MnCl_4.4H_2O$		3.6022	50 mL conc. HCl
Fe	Fe metal	d	1.0000	20 mL 1:1 HCl
Fe(II):	$(NH_4)_2SO_4.FeSO_4.6H_2O$		7.0217	1M H_2SO_4
Co	Co metal	d	1.0000	50 mL 1:1 HNO_3
	$CoCl_2.6H_2O$		4.0373	Water + 2mL conc. HCl
Ni	Ni metal	d	1.0000	50 mL 1:1 HNO_3
	$NiCl_2.6H_2O$		4.0503	Water acidified with HCl
Cu	Cu metal	d	1.0000	50 mL HNO_3
	$CuSO_4.5H_2O$		3.9291	Water acidified with H_2SO_4
Zn	Zn metal		1.0000	40 mL 1:1 HCl or 40 mL 1:1 HNO_3
Ga	Elemental Ga		1.0000	50 mL 1:1 HCl
Ge	Elemental Ge		1.0000	30 mL aqua regia
As	Elemental powder		1.0000	50 mL HCl or 50 mL HNO_3
	As_2O_3		1.3203	50 mL 5M NaOH then just acidify with dilute HCl
Se	Elemental Se		1.0000	20 mL aqua regia
Br	KBr	110/1hr	1.4893	Water
Rb	RbCl	110/1hr	1.4148	Water
Sr	$SrCO_3$		1.6849	20 mL 1M HCl (or nitric acid)
Y	Y_2O_3	900/1hr	1.2699	20 mL HCl
Zr	Zr metal		1.0000	50 mL HF
	ZrO_2		1.3508	50 mL H_2SO_4
Nb	Nb metal		1.0000	10 mL HF + 5mL HNO_3 added dropwise
	Nb_2O_5		1.4305	10 mL HF + 5mL HNO_3 added dropwise
Mo	MoO_3		1.5003	10 mL conc HCl
	$(NH_4)_6Mo_7O_{24}.4H_2O$		1.8402	Water containing 1% ammonia
Tc				
Ru	$(NH_4)_2(RuCl_5.H_2O)$		3.2891	Water acidified with HCl

Table 5.4 (*Continued*)

	Compound	Drying conditions	*W* (g)	Suggested dissolution procedure
Rh	$(NH_4)_3(RhCl_6)$		3.5930	10% HCl
Pd	$(NH_4)_2PdCl_4$		2.6716	Water
	Pd metal		1.0000	20 mL aqua regia
Ag	$AgNO_3$		1.5748	Water; store in an amber glass bottle
	Ag metal		1.0000	20 mL 1:1 HNO_3
Cd	CdO		1.1422	20 mL 1:1 HNO_3
	$CdCl_2.2.5H_2O$		2.0313	Water
In	Elemental In		1.0000	20 mL 1:1 HNO_3
Sn	Metal granules		1.0000	100 mL HCl (warm); when dissolved, dilute to 1L with 10% HCl
	$(NH_4)_2SnCl_6$		3.0962	10% HCl
Sb	$KSbO.C_4H_4O_6$ (antimony potassium tartrate)		2.6687	Water acidified with H_2SO_4
	Elemental Sb		1.0000	10 mL conc HCl + few drops HNO_3 + 10 g tartaric acid
Te	Elemental Te		1.0000	20 mL aqua regia
	TeO_2		1.2509	20 mL aqua regia
I	KI		1.3081	Water
Cs	CsCl	110/1hr	1.2668	Water
Ba	$BaCO_3$		1.4370	20 mL 1M HCl (care: effervesces)
	$BaCl_2.2H_2O$		1.7788	Water
La	La_2O_3	900/1hr	1.1728	20 mL HCl
Ce	CeO_2	600/1hr	1.2284	20 mL 1:1 HCl
Pr	Pr_6O_{11}	900/1hr	1.2082	10 mL HCl
Nd	Nd_2O_3	900/1hr	1.1664	10 mL HCl
Sm	Sm_2O_3	900/1hr	1.1596	10 mL HCl
Eu	Eu_2O_3		1.1582	10 mL 1:1 HNO_3
Gd	Gd_2O_3	900/1hr	1.1526	20 mL 1:1 HCl
Tb	$Tb_2^{III}Tb_2^{IV}O_7$	900/1hr	1.1762	20 mL 1:1 HCl
Dy	Dy_2O_3	900/1hr	1.1477	20 mL 1:1 HCl
Ho	Ho_2O_3	900/1hr	1.1455	20 mL 1:1 HCl
Er	Er_2O_3	900/1hr	1.1435	20 mL 1:1 HCl
Tm	Tm_2O_3	900/1hr	1.1421	10 mL HCl
Yb	Yb_2O_3	900/1hr	1.1387	20 mL 1:1 HCl
Lu	Lu_2O_3	900/1hr	1.1372	20 mL 1:1 HCl
Hf	Hf metal		1.0000	5 mL HF + 5 mL water, then dropwise addition of 10 mL HNO_3 (care: vigorous reaction)
	HfO_2		1.1793	As above
Ta	Ta metal		1.0000	As above
	Ta_2O_5		1.2211	As above
W	WO_3		1.2611	10% NaOH solution
	W metal		1.0000	20 mL HNO_3 + 10 mL HF
Re	$(NH_4)_2ReO_4$		1.4406	Water
	$KReO_4$		1.5537	Water
Os	$(NH_4)_2(OsCl_6)$		2.3081	Water acidified with HCl (highly toxic, volatile if oxidized)
Ir	$(NH_4)_2IrCl_6$		2.2943	Water
Pt	$(NH_4)_2PtCl_6$		2.2753	Water
	Pt metal		1.0000	20 mL aqua regia
Au	Au metal		1.0000	20 mL aqua regia
Hg	Hg metal		1.0000	20 mL 1:1 HNO_3
	HgO		1.0798	20 mL 1:1 HNO_3
Tl	Tl_2SO_4		1.2349	Water
Pb	Pb metal		1.0000	50 mL 1:4 HNO_3
	$Pb(NO_3)_2$		1.5985	Water
Bi	Bi metal		1.0000	50 mL conc HNO_3
Th	ThO_2		1.1379	Fuse with KOH, leach with water, acidify with HCl
U	$UO_2(NO_3)_2.6H_2O$		2.1095	Water
	U_3O_8		1.1793	Evaporate to dryness with a few mL HNO_3 then with dilute HCl. Redissolve chloride residue in water.

Drying conditions list first the temperature (°C) and then the time (where known) for which reagents should be heated before weighing.
d: these metal powders oxidize rapidly on exposure to moist air. Reagent bottles should be opened in an inert dry atmosphere.
W is the weight of the listed reagent that must be weighed to prepare 1 litre of solution containing 1000 μg of the element.
Note that where the reagent is soluble in water alone, it would be normal to acidify such solutions to ensure stability during storage.
Mercury solutions should not be stored in plastic bottles.
Data have been compiled from Jeffery (1975); Van Loon (1980); Langmyhr and Paus (1968); Johnson and Matthey (1978); BDH (1983).
Acids are concentrated laboratory reagents unless otherwise stated; 1:1 means one part by volume of concentrated acid to one part of water.

et al. (1979) identified the trace elements Nb, Ta, Mo and W as falling into this category. Stock solutions of these elements were stable in basic or weakly acid media. However, when aliquots were added to more acidic multi-element calibration solutions, polymerization and co-precipitation could occur. These authors found that at all concentration levels, mixtures of Nb/Ta/Mo/W produced a precipitate. Accordingly, it was necessary to remove Ta and W from the analysis scheme altogether and to add Nb and Mo to multi-element standard solutions only at reduced concentrations.

(iv) A particular difficulty was encountered with the precipitation of thallium with a barium salt.

5.11.4 *Calibration using matrix-matched standard solutions*

The second concern in standard solution preparation is the desirability of eliminating analytical errors by matching the matrix composition of standard solutions with that of sample solutions. Church (1981) considered this problem in some detail and his scheme of multi-element standard solution preparation is listed in Table 5.5. The criteria used by Church were as follows:

(i) The *acid content* of all samples and standard solutions must remain constant to avoid analytical errors due to the pH dependence of solution uptake rate during nebulization (section 5.6*a*). This is not a simple matter when digesting a variety of samples during which the rate of heating and evaporation of acid reagents are expected to vary. Church overcame this problem as follows. Silicate samples (0.8–1.2 g) were digested in a Teflon beaker by attack with HF (10 mL), $HClO_4$ (3 mL) and HNO_3 (2 mL). After refluxing overnight, the HF (and silicon as SiF_4) were driven off. The residue was taken up with concentrated HCl (5 mL) and evaporated to dryness. 40 mL of 6.2M HCl was then added and the sample heated until it had completely dissolved. The importance of 6.2M HCl is that it forms a constant boiling point mixture (azeotrope) with water. Solutions of this composition boil to give off HCl and H_2O in constant proportions, the solution remaining at 6.2M in HCl. After heating the residue, Church readjusted the acid volume in

Table 5.5 Standard solutions used in a matrix-matching calibration scheme for the determination of major and trace elements by ICP-AES

	Acid Blank	std 2	std 3	std 4	std 5	std 6	std 7	std 8	std 9	std 10	std 11	std 12	std 13	std 14	
Na	0.05	—	500	1500	1000	250	50	100	10	—	—	—	—	—	Na
K	0.20	—	1500	1000	500	250	50	100	10	—	—	—	—	—	K
Mg	0.02	—	50	10	—	—	300	1000	—	1500	—	—	2500		Mg
Ca	0.01	—	50	10	500	1500	1000	250	100	—	2500	—	—	—	Ca
Fe	0.20	—	50	10	1000	500	—	250	1500	100	—	—	—	3600	Fe
Al	0.04	—	50	10	100	250	1500	—	1000	—	500	3000	—	—	Al
Ti	0	—	200	—	—	—	—	—	—	—	—	—	—		Ti
P	0	—	—	—	—	—	—	—	—	325	—	—	—	—	P
Ba	0	—	50	—	—	—	—	—	—	—	—	—	—	—	Ba
La	0	—	50	—	—	—	—	—	—	—	—	—	—	—	La
Ce	0	—	50	—	—	—	—	—	—	—	—	—	—	—	Ce
Zr	0	—	50	—	—	—	—	—	—	—	—	—	—	—	Zr
U	0	—	—	50	—	—	—	—	—	—	—	—	—	—	U
Th	0	—	—	50	—	—	—	—	—	—	—	—	—	—	Th
V	0	—	50	—	—	—	—	—	—	—	—	—	—	—	V
Cr	0	—	50	—	—	—	—	—	—	—	—	—	—	—	Cr
Mn	0	—	50	10	—	—	100	1000	250	500	—	—	—	—	Mn
Co	0	—	50	—	—	—	—	—	—	—	—	—	—	—	Co
Ni	0	—	50	—	—	—	—	—	—	—	—	—	—	—	Ni
Cu	0	—	50	—	—	—	—	—	—	—	—	—	—	—	Cu
Zn	0	—	50	—	—	—	—	—	—	—	—	—	—	—	Zn
Mo	0	—	—	—	—	—	—	—	—	50	—	—	—	—	Mo
Ag	0	10	—	—	—	—	—	—	—	—	—	—	—	—	Ag
Cd	0	—	—	10	—	—	—	—	—	—	—	—	—	—	Cd
Sn	0	—	—	50	—	—	—	—	—	—	—	—	—	—	Sn
Pb	0	—	—	50	—	—	—	—	—	—	—	—	—	—	Pb
Bi	0	10	—	—	—	—	—	—	—	—	—	—	—	—	Bi
As	0	—	—	50	—	—	—	—	—	—	—	—	—	—	As
Sb	0	—	—	50	—	—	—	—	—	—	—	—	—	—	Sb
Sum	—	20	2900	2850	3100	2750	3000	2700	2870	(2888)	3000	3000	2500	3600	
HCl		—	20	20	20	20	20	20	20	20	20	20	20	20	
$HClO_4$		—	5	5	5	5	5	5	5	5	5	5	5	5	
HNO_3		25	—	—	—	—	—	—	—	—	—	—	—		

Data from Church (1981).
Element concentrations in $\mu g/mL$; acid compositions in % (volume) of concentrated acid.
Sum is the total dissolved salts in each solution, std 10 includes potassium present in the phosphate salt used to make up this solution.
Data have been rounded to the nearest 10 $\mu g/mL$.

each sample solution to 40 mL using 6.2M HCl. The sample solutions were then made up to 100 mL, including 5 mL of $HClO_4$ solution to ensure oxidation state stability in the final solution. Church could therefore be confident that all sample solutions were made up to the same acid content, namely 20% HCl+5% $HClO_4$ and this, therefore, defined the acid composition of all standard solutions used in his scheme of analysis.

(ii) Variations in *nebulizer efficiency* have also been reported due to variations in the total salt content of sample solutions. After loss by evaporation of silica during sample digestion, silicate rock samples will leave different proportions of residual salts. To overcome this problem, Church (1981) took different weights of sample depending on rock type. Thus approximately 1.2 g of granite or about 0.9 g of basalt will both give a total salt content of 2500 ± 400 ppm when taken up in 100 mL solution using the sample preparation procedure outlined above. All multi-element standard solutions were also prepared to give a total salt content of 2500 to 3000 ppm. The 14 standard solutions covering 29 major and trace elements, designed so as to ensure solution matrix matching, are listed in Table 5.5. Working curves for the trace elements were found to be linear. However, second- and third-order polynomials were required to fit calibration curves for the major elements at the higher concentrations

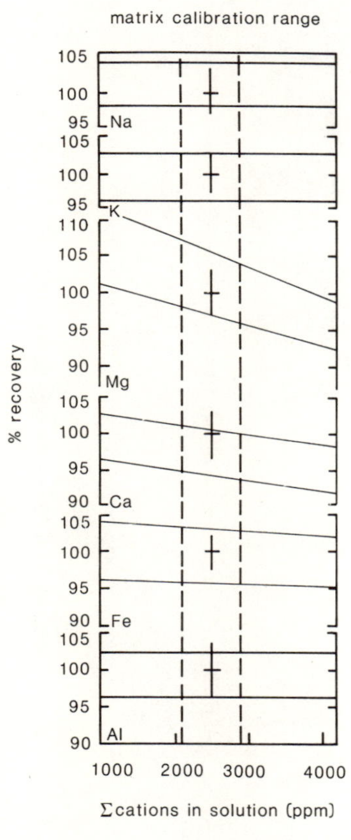

Figure 5.40 Effect of changes in the total salt content of analysed solutions (measured as the sum of cations present) on the emission signals of atomic lines of Na, K and Al and ion lines of Mg, Ca and Fe. Emission data are normalized to their values at 2500 ppm total dissolved cations. (Church, 1981.)

encountered in this work. The standard solutions for Ag and Bi were prepared in 25% nitric acid since precipitates would form in the corresponding HCl medium.

The importance of maintaining at a constant value the total salt content can be seen from Figure 5.40. The band enclosing the spread of data for the normalized intensity of selected major elements is plotted as a function of total cation content of analysed solutions. Data for the *atom* emission lines of Na, K and Al show negligible variation. Data for the *ion* emission lines of Mg, Ca and Fe show a significant variation.

A different approach was used by McQuaker *et al.* (1979) in a application concerning the analysis of environmental samples. Due to the nature of the sample, it was not necessary to employ matrix matching. However, it was necessary to calibrate the instrument from a set of multi-element standard solutions. It was not considered desirable to use multi-element solutions in which high concentrations of one element are mixed with low concentrations of another. Under these conditions, inter-element interference corrections (e.g. due to spectral line overlap) can be serious. In order to overcome these difficulties, McQuaker *et al.* grouped their calibration standards at equivalent concentration levels, as listed in Table 5.6.

5.11.5 *Calibration using rock reference materials*

An alternative approach in ICP–AES analysis is to calibrate instruments using solutions prepared directly from rock reference materials. The reference material is digested following the same procedures as for unknown samples permitting automatic matching of the matrix. This approach has been advocated by Walsh and Howie (1980) and Walsh (1982), who argued that, on balance, it was simpler to use standard reference materials than to follow more complex matrix matching procedures using synthetic solutions. Walsh and Howie (1980) prepared calibration solutions from USGS: W-1, BCR-1, AGV-1 and G-2 in the determination of the major and a range of trace elements. The principal disadvantage of this approach is the uncertain confidence that must be placed on 'recommended' values for rock reference materials particularly in relation to some trace elements.

5.12 Silicate rock analysis

5.12.1 *Sample preparation for simultaneous multi-element analysis*

Two approaches to sample dissolution have generally been followed depending on whether silica is to be determined or not. In many simultaneous schemes of analysis, silica is not determined and an acid attack using hydrofluoric–perchloric acid is used. After initial sample attack, the solution is evaporated to incipient dryness to permit all the silica to be lost by volatilization as silicon tetrafluoride. The residue is then taken up in dilute hydrochloric acid. Church (1981) added perchloric acid to this final solution to impart oxidation state stability. Walsh (1980) took this solution up in perchloric acid alone. Overall dilutions would normally be 1g of rock powder taken up to 100 mL final solution. This sample dissolution procedure will not dissolve some resistant

Table 5.6 Standard solutions used in the analysis of environmental materials by ICP–AES

	Spectrum	Wavelength (nm)	std 1	std 2	std 3	std 4	std 5	std 6	std 7	std 8	std 9	std 10	std 11	std 12	std 13	std 14	std 15	std 16	std 17	std 18	std 19	std 20	
Al	I	396.152	—	—	—	—	500	200	100	50	20	10	5	2	1	0.5	0.2	0.1	0.05	0.02	—	0	Al
As	I	193.77	—	—	—	—	—	—	—	—	20	10	5	2	1	0.5	0.2	0.1	0.05	—	—	0	As
B	I	249.773	—	—	—	—	—	—	—	—	20	10	5	2	1	0.5	0.2	0.1	0.05	0.02	0.01	0	B
Ba	II	455.403	—	—	—	—	—	—	—	—	20	10	5	2	1	0.5	0.2	0.1	0.05	0.02	0.01	0	Ba
Be	II	313.042	—	—	—	—	—	—	—	—	—	—	5	2	1	0.5	0.2	0.1	0.05	0.02	0.01	0	Be
Ca	I	422.673	—	—	—	—	500	200	100	50	20	10	5	2	1	0.5	0.2	0.1	0.05	0.02	—	0	Ca
Cd	II	226.502	—	—	—	—	—	—	—	—	20	10	5	2	1	0.5	0.2	0.1	0.05	0.02	0.01	0	Cd
Co	I	345.350	—	—	—	—	—	—	—	—	20	10	5	2	1	0.5	0.2	0.1	—	—	—	0	Co
Cr	II	283.563	—	—	—	—	—	—	—	—	20	10	5	2	1	0.5	0.2	0.1	0.05	0.02	0.01	0	Cr
Cu	I	324.754	—	—	—	—	—	—	—	—	20	10	5	2	1	0.5	0.2	0.1	0.05	0.02	0.01	0	Cu
Fe	II	259.940	—	—	—	—	500	200	100	50	20	10	5	2	1	0.5	0.2	0.1	0.05	0.02	0.01	0	Fe
Ga	I	417.206	—	—	—	—	—	—	—	—	20	10	5	2	1	0.5	0.2	0.1	—	—	—	0	Ga
In	I	451.131	—	—	—	—	—	—	—	—	20	10	5	2	1	0.5	0.2	—	—	—	—	0	In
Mg	II	279.079	—	—	—	—	—	200	100	50	20	10	5	2	1	0.5	0.2	0.1	—	—	—	0	Mg
Mn	II	257.610	—	—	—	—	—	—	—	50	20	10	5	2	1	0.5	0.2	0.1	0.05	0.02	0.01	0	Mn
Mo	II	281.615	—	—	—	—	—	—	—	—	—	—	5	2	1	0.5	0.2	0.1	0.05	0.02	0.01	0	Mo
Nb	II	309.418	—	—	—	—	—	—	—	—	—	—	5	2	1	0.5	0.2	0.1	0.05	0.02	0.01	0	Nb
Ni	II	231.604	—	—	—	—	—	—	—	—	20	10	5	2	1	0.5	0.2	0.1	0.05	—	—	0	Ni
P	I	213.618	200	100	—	—	—	—	—	50	20	10	5	2	1	0.5	—	—	—	—	—	0	P
Pb	II	220.353	—	—	—	—	—	—	—	—	20	10	5	2	1	0.5	0.2	0.1	—	—	—	0	Pb
Sb	I	206.833	—	—	—	—	—	—	—	—	20	10	5	2	1	0.5	0.2	0.1	—	—	—	0	Sb
Se	I	196.06	—	—	—	—	—	—	—	—	20	10	5	2	1	0.5	0.2	0.1	—	—	—	0	Se
Si	I	288.158	—	—	—	—	—	—	—	—	20	10	5	2	1	0.5	0.2	0.1	0.05	0.02	—	0	Si
Sn	I	189.99	—	—	—	—	—	—	—	—	—	—	5	2	1	0.5	0.2	0.1	0.05	—	—	0	Sn
Sr	II	407.771	—	—	—	—	—	—	—	—	20	10	5	2	1	0.5	0.2	0.1	0.05	0.02	0.01	0	Sr
Te	I	238.578	—	—	—	—	—	—	—	—	20	10	5	2	1	0.5	0.2	—	—	—	—	0	Te
Ti	II	337.280	—	—	50	10	—	—	—	—	—	—	5	2	1	0.5	0.2	0.1	0.05	0.02	0.01	0	Ti
Tl	I	377.572	—	—	—	—	—	—	—	—	—	—	5	2	1	0.5	0.2	—	—	—	—	0	Tl
V	II	292.402	—	—	—	—	—	—	—	—	20	10	5	2	1	0.5	0.2	0.1	0.05	0.02	0.01	0	V
Zn	I	213.856	—	—	—	—	—	—	—	50	20	10	5	2	1	0.5	0.2	0.1	0.05	0.02	0.01	0	Zn

Data from McQuaker *et al.* (1979).
All concentrations in μg/mL.
For data normalization, the low standard is std 20 and the high standards are std 1, 3, 6, 9 and 11 depending on group of elements to be determined.

mineral such as chromite, zircon, cassiterite, REE minerals etc. (see section 2.1). If the elements that concentrate in these minerals are to be determined, more elaborate precautions are required. Crock *et al.* (1983) carried out the initial acid attack in a thick-walled PTFE screw-capped bottle. On heating gently on a hotplate, pressure built up inside the bottle, permitting more effective attack on resistant minerals. Walsh *et al.* (1981) described a scheme specifically for the analysis of rare-earth elements. Sample powders were attacked with hydrofluoric–perchloric acid mixtures and heated to incipient dryness as usual. The residue was taken up in hydrochloric acid and if any insoluble resistant materials were observed, this solution was filtered and the residue fused with either NaOH (0.5 g) at 800°C in a silver crucible for 20 minutes, or alternatively with KHF_2 in a platinum crucible. The cooled residues were dissolved in hydrochloric acid and recombined with the main solution. This procedure was designed in particular to ensure complete dissolution of zircon. Crock and Lichte (1982) used a lithium metaborate fusion in place of acid attack if zircon was suspected in samples.

If silica is to be determined, an alternative procedure must be employed. For example, Walsh and Howie (1980) used a lithium metaborate fusion, the fused residue being taken up in nitric acid. As this solution was to be used for major element determination for which the ICP–AES technique is, if anything, too sensitive, a larger sample dilution was used, 0.5 g of sample being used with 1.5 g $LiBO_2$ and taken up to a final volume of 250 mL.

5.12.2 *Internal standardization*

To compensate for variations in instrument sensitivity, electronic drift, nebulizer efficiency, total salt content and other matrix effects, the technique of internal standardization is extensively used in ICP–AES. All solutions are spiked with identical concentrations of an internal standard element. This element must not be one for which determination is required nor present in samples at significant background concentrations. The ratio of all measurements to the intensity of the internal standard is then determined to compensate for any drift in the operating response of the plasma.

Thus, in the determination of rare-earth elements, Brenner *et al.* (1981) selected scandium as internal standard since this element was expected to respond in the same manner as the REE in the plasma. The authors considered that this internal standardization procedure adequately compensated for variations in excitation of the REE ion lines between different sample solutions and precluded the necessity of precise match of acid and total dissolved salts content. In more general analysis schemes, Walsh and Howie (1980) used La, Crock *et al.* (1983) used Lu as internal standards, and Crock and Lichte (1982) selected Cd in determination of REEs.

5.12.3 *Performance in routine analysis: accuracy, precision and detection limits*

Typical analytical results by ICP–AES are shown in Table 5.7. These data indicate that this technique is capable of determining 10 of the major elements (including phosphorus) and over 20 trace elements from samples prepared simply by dissolving 1g of rock to 100 mL of solution with no further preconcentration. From the same solution it is often possible to determine seven rare-earth elements (La, Ce, Nd, Sm, Eu, Gd, Yb) although detection limits are not adequate for routine application to all samples (see next section).

A comparison of detection limits (Table 5.8) indicates that ICP–AES is a much more powerful technique in the analysis of geological samples than flame AAS with limits of determination for many trace elements in the 1 to 10 ppm range. In comparison with ICP–mass spectrometry, detection limits for some elements, especially the lighter ones, are of the same order of magnitude. However, ICP–MS has generally far superior detection limits for the heavier elements. Neither technique can approach the ideal solution detection limits that can be achieved by graphite furnace atomic absorption spectrometry. In routine geochemical applications, the inability to determine rubidium by ICP–AES is a disadvantage.

The precision of major element data in Table 5.7 is often in the range 1 to 2% relative (one sigma). Church (1981) estimated that the corresponding accuracy was in the range ±3.6% or better in his scheme of analysis. As might be expected, the precision of trace element analysis is a little more variable, data in Table 5.7 often being in the range 1 to 5% relative (one sigma). Church (1981) estimated the corresponding accuracy to be in the range 5 to 10% relative.

5.12.4 *Rare-earth element analysis by ICP–AES*

Considerable interest has been shown in the development of schemes of analysis for the rare-earth elements by ICP–AES. This has been stimulated by the fact that alternative techniques such as INAA or isotope dilution mass spectrometry are time-consuming and require either expensive equipment or access to specialized reactor facilities.

The schemes of analysis developed for REE determination by ICP usually employ a simple ion-exchange separation. This has three advantages: the total dissolved salts content of the final solution is considerably reduced; many potential overlap interferences of matrix elements on REE lines are removed and the opportunity is created to reduce considerably the volume of the final analysed solution, thus minimizing the overall dilution factor.

There is some divergence of opinion over the most appropriate emission lines to use for REE analysis. Some of those proposed in the literature are listed in Table 5.9, together with reported interfering elements. It must be emphasized that some spectral interferences are dependent on the instrument used (for example order of spectral line measured, plasma torch operating conditions, etc.). Furthermore, many interferences listed in Table 5.9 disappear following an ion exchange separation.

As has been mentioned above, some REEs (La, Ce, Nd, Sm, Eu, Gd, and Yb) may be determined without ion exchange separation at higher concentration levels. However, for a general analysis programme, removal of matrix and sample preconcentration by ion exchange are essential, and the effect this has on analytical performance can be judged from the detection limit data presented in Table 5.10. Here the

Table 5.7 Simultaneous multi-element analysis of major and trace elements by ICP-AES

	Granodiorite USGS: GSP-1 Church (1981)			Basalt NBS: SRM-688 Crock *et al.* (1983)			Diorite ANRT: DR-N Borsier and Garcia (1983)		Norite NIM-N Walsh (1980)		
	Expected (% oxide)	Analysed (% oxide)	CV (%)	Expected (% oxide)	Analysed (% oxide)	CV (%) $n=9$	Expected (% oxide)	Analysed (% oxide)	Expected (% oxide)	Analysed (% oxide)	
SiO_2	67.32	—	—	48.4	—	—	52.98	54	52.64	53.6	SiO_2
TiO_2	0.66	0.657	1.2	1.17	1.23	1.2	1.09	1.16	0.20	0.19	TiO_2
Al_2O_3	15.28	15.05	1.8	17.36	17.08	0.6	17.56	18	16.50	16.37	Al_2O_3
Fe_2O_3	4.30	4.34	1.2	10.35	10.49	0.4	9.72	9.9	8.91	8.76	Fe_2O_3
MnO	0.04	0.0394	1.5	0.167	0.160	1.6	0.22	0.24	0.18	0.17	MnO
MgO	0.97	0.989	1.5	8.4?	8.42	0.4	4.41	4.4	7.50	7.37	MgO
CaO	2.03	2.01	1.5	12.17?	12.34	0.2	7.07	7.2	11.50	11.46	CaO
Na_2O	2.81	2.75	1.6	2.15	2.20	3.1	3.00	—	2.46	2.38	Na_2O
K_2O	5.51	5.58	1.3	0.187	0.20	1.9	1.70	2.3	0.25	0.47	K_2O
P_2O_5	0.28	0.277	1.3	0.134	0.128	3.6	0.25	0.176	0.03?	0.03	P_2O_5
TOTAL	(31.88)	(31.69)		(52.09)	(52.25)		(95.0)	(97.4)	100.17	100.80	TOTAL
	(ppm)	(ppm)	(%)	(ppm)	(ppm)	(%)	(ppm)	(ppm)			
B	—	—	—	—	—	—	—	<10			B
Ba	1300	1275	1.0	200?	178	1.1	390	401			Ba
Be	1?	1.32	1.1	—	0.20	25	1.8?	<2			Be
Cd	(0.06)	<2	—	—	—	—	0.09?	<2			Cd
Co	7.8	6.8	2.9	50?	47	2.1	35	35			Co
Cr	12	11.8	2.5	330	260	7.7	42	39			Cr
Cu	33	30.7	0.8	96?	90	1.1	50	51			Cu
Ga	23	—	—	—	17	41	22?	—			Ga
Li	30	28.4	2.5	—	7	14	42	10			Li
Mo	1.5?	<3	—	—	—	—	—	<2			Mo
Nb	23?	—	—	—	5	20	6?	<20			Nb
Ni	9	10.0	5.0	150?	143	1.4	16	24			Ni
Pb	54	52	9.6	3.3	<4	—	55	55			Pb
Sb	3.1?	<18	—	—	—	—	0.3?	<2			Sb
Sc	6.6	—	—	38?	43.3	1.2	28?	—			Sc
Sn	5?	6.8	15.4	—	—	—	—	<10			Sn
Sr	240	233	1.0	169	170	0.06	400	397			Sr
Th	105	80	5.0	0.3	<4	—	5	—			Th
U	2.1	<20	—	0.37	<80	—	1.5	—			U
V	54	50.1	1.5	250?	248	0.4	230	156			V
W	(0.1)	<7	—	—	—	—	—	101			W
Y	29	—	—	—	19.5	0.5	30?	23			Y
Zn	105	104	1.5	58?	79	1.3	145	120			Zn
Zr	500	—	—	—	63	6.3	125	116			Zr

Expected compositions from Abbey (1983).
CV = coefficient of variation = 100 × (standard deviation)/mean.

Table 5.8 Comparison of detection limit performance of atomic absorption, ICP–AES and ICP–mass spectrometry techniques.

Element	Flame atomic absorption				Graphite furnace atomic absorption	Inductively coupled plasma atomic emission spectrometry				Inductively coupled plasma mass spectrometry			
	Wavelength (nm)	Analysis conditions	Limit of determination Ideal solution (μg/mL)	Limit of determination Whole rock (ppm)	Limit of determination ideal solution (μg/mL)		ICP emission line (nm)	Limit of determination Ideal solution (μg/mL)	Limit of determination Whole rock (ppm)	Mass no. and charge	Limit of determination Ideal solution (μg/mL)	Limit of determination Whole rock (ppm)	
Al	309.3	no	0.06	**12**	0.00003	Al	309.271	0.046	**9***	27+	0.003	**(3)**	Al
							308.215	0.090					
Sb	217.6a	air	0.12	*5	0.00024	Sb	206.833	0.064	20			Sb	
	206.8	air					217.581	0.088					
As	193.7	no,bc,edl	0.3	60	0.0003	As	193.696	0.106	15	75+	0.060	(60)	As
							189.042v	0.272					
Ba	553.5b	no	0.06	*10	0.00012	Ba	455.403	0.0026	**1**	138+	0.002	**2**	Ba
	455.4	no					493.409	0.0046					Be
Be	234.9	no	0.003	**0.6**	0.000006	Be	313.042	0.00054	**0.1**				
Bi	223.1	air	0.12	*5	0.00003	Bi	223.061	0.068	20	209+	0.002	2	Bi
B	249.7	no	2.1	*90	0.045	B	249.773	0.0096	**2***	11+	0.004	**(4)**	B
Cd	228.8	air	0.003	*0.8	0.0000006	Cd	226.502	0.0068	5	114+	0.002	2	Cd
							214.438	0.0050					
Ca	422.7	(no)	0.003	**0.6**	0.00003	Ca	422.673	0.020	**(4*)**				Ca
							317.933	0.020					
C						C	247.856	0.35	70*		0.1		C
Ce	520.0					Ce	418.660	0.104	**15**	140+	0.003	**3**	Ce
							394.275	0.136					
Cs	852.1	air	0.06	12	0.00006	Cs	452.673	83.0	**16600***	133+	0.002	**2**	Cs
Cr	357.9	air	0.009	*3	0.000015	Cr	283.563	0.014	**3**	52+	0.002	**(2)**	Cr
							267.716	0.014					
Co	240.7	air,bc	0.015	*2	0.000024	Co	228.616	0.014	**2**	59+	0.002	**(2)**	Co
							238.892	0.012					
Cu	324.7	air	0.009	*1	0.000021	Cu	324.754	0.0108	**5**	63+	0.002	**(2)**	Cu
							234.75						
Dy	421.2	no	0.15	30		Dy	340.780	0.054	15		0.002	**2**	Dy
Er	400.8	no	0.12	21	0.0009	Er	369.265	0.036	15		0.004	4	Er
Eu	459.4	no	0.09	18	0.0003	Eu	381.967	0.0054	**2**		0.004	4	Eu
Gd	368.4	no	6.0	1200	0.024	Gd	376.839	0.050	**5**		0.004	**4**	Gd
	407.9	no					303.284	0.054					
Ga	287.4	air	0.21	42	0.00006	Ga	294.364	0.092	**16**				Ga
Ge	265.1	no	0.3	60	0.0003	Ge	265.118	0.096	40	72+	0.004	4	Ge
Au	242.8	air	0.03	6	0.00003	Au	242.795	0.034	10	197+	0.001	1	Au
	267.6	air											
Hf	307.3	no	6.0	1200		Hf	241.5	0.030	6*				Hf
							277.336						
Ho	410.4	no	0.18	36	0.0021	Ho	345.600	0.0114	15		0.001	**1**	Ho
In	303.9	air	0.12	24	0.00006	In	230.606	0.126	25*	115+	0.002		In
Ir	208.8	air	1.5	300	0.0015	Ir	224.368	0.054	11*		0.004	4	Ir
	263.9	air											
Fe	248.3	air	0 .015	**3**	0.00003	Fe	259.940	0.0124	**2.5***	56+	0.004	**(4)**	Fe
							263.105	0.030					
La	550.1	no	6.0	1200	0.0015	La	398.852	0.022	**7**	139++	0.004	**4**	La
							379.478	0.020					
Pb	217.0	bc,edl,air	0.06	*1	0.000021	Pb	220.353	0.084	**20**	208+	0.002	**2**	Pb
	283.3	edl,air											
Li	670.8	air	0.003	*1	0.00003	Li	670.78	0.045	**3**	7+	0.008	**(8)**	Li
Lu	336.0	no	0.9	180		Lu	261.542	0.0020	0.4*				Lu

Table 5.8 (*Continued*)

Element	Flame atomic absorption				Graphite furnace atomic absorption	Inductively coupled plasma atomic emission spectrometry				Inductively coupled plasma mass spectrometry			
	Wavelength (nm)	Analysis conditions	Limit of determination Ideal solution (μg/mL)	Limit of determination Whole rock (ppm)	Limit of determination ideal solution (μg/mL)		ICP emission line (nm)	Limit of determination Ideal solution (μg/mL)	Limit of determination Whole rock (ppm)	Mass No. and charge	Limit of determination Ideal solution (μg/mL)	Limit of determination Whole rock (ppm)	
Mg	285.2	(no)	0.0006	**0.12**	0.0000006	Mg	279.079	0.060	**12***	24+	0.002	**(2)**	Mg
							383.826	0.064					
Mn	279.5	air	0.006	**1.2**	0.000002	Mn	257.610	0.0028	**15**	55+	0.004	**(4)**	Mn
Hg	253.7	air	0.51	102	0.0006	Hg	194.227	0.050	10*	202+	0.004	4	Hg
	184.9	p,no					253.652	0.122					
Mo	313.3	no	0.06	*6	0.00009	Mo	313.62		10		0.004	**4**	Mo
							202.030	0.0158					
Nd	492.5c	no	3.0	600		Nd	417.732	0.272	**30**		0.004	**4**	Nd
	463.4	no					430.358	0.150					
Ni	232.0d	air,bc	0.015	*3	0.00015	Ni	231.604	0.030	**6**		0.004	**4**	Ni
	341.5	air											
Nb	334.9	no	6.0	*90		Nb	309.418	0.072	**10**		0.002	**(2)**	Nb
	334.4	no											
Os	290.9	no	0.3	60	0.006	Os	225.585	0.0007	0.14*		0.002	2	Os
Pd	247.6	air	0.045	9	0.0003	Pd	340.458	0.088	18*		0.001	1	Pd
P	213.6	no,edl	120.0	24000	0.06	P	213.618	0.152	**30***		0.6	**(600)**	P
							178.29v						
Pt	265.9	air	0.21	42	0.0006	Pt	203.646	0.11	22*		0.002	2	Pt
K	766.5	air	0.003	**0.6**	0.000009	K	766.49	12.0	**2400***				K
Pr	495.1	no	15.0	3000		Pr	422.293	0.094	40		0.002	**2**	Pr
Re	346.1	no	2.4	480	0.03	Re	221.426	0.012	2.4*		0.004	4	Re
Rh	343.5	(no)	0.015	3	0.0003	Rh	233.477	0.088	18*		0.002	2	Rh
Rb	780.0	air	0.015	***1**		Rb	420.185	75	15000*	85+	0.002	**(2)**	Rb
Ru	349.9	air	0.9	180		Ru	240.272	0.060	12*		0.008	8	Ru
Sm	429.7	no	3.0	600		Sm	443.432	0.166	15		0.004	**4**	Sm
							359.260	0.086					
Sc	391.2	no	0.15	*10		Sc	361.384	0.0030	**2**		0.006	**(6)**	Sc
							424.683	0.0054					
Se	196.0	no,bc,edl	0.6	120	0.0003	Se	196.026	0.15	50	78+	0.1	100	Se
Si	251.6	no	0.45	**90**	0.0006	Si	251.611	0.024	**5***				
							288.158	0.054					Si
Ag	328.1	air	0.006	*1	0.000006	Ag	328.068	0.014	5	107+	0.002	2	Ag
Na	589.0	air	0.001	**0.2**	0.000009	Na	588.995	0.058	**12***		0.004	**(4)**	Na
Sr	460.7	(no)	0.015	***1**	0.00003	Sr	407.771	0.00084	**2**		0.002	**(2)**	Sr
							421.552	0.00154					
Ta	271.5	no	4.5	900		Ta	240.063	0.056	10*				Ta
							226.230	0.050					
Te	214.3	air	0.15	30	0.00015	Te	238.578	0.352	16*	130+	0.002	2	Te
							214.281	0.082					
Tb	432.7	no	3.0	600		Tb	367.635	0.120	80		0.004	4	Tb
							350.917	0.056					
Tl	276.8	air	0.06	12	0.00003	Tl	190.864v	0.080	16*		0.002	2	Tl
							351.924	0.400					
Th	324.6	no				Th	353.959	0.200	30	232++	0.004	**4**	Th
							401.913	0.166					
Tm	371.8	no	0.3	60		Tm	313.126	0.0104	2*		0.001	**1**	Tm
							346.220	0.0162					
Sn	235.41	no,edl	0.3	*20	0.00018	Sn	189.99v	0.09	**6**		0.004	**4**	Sn
	286.3	no,edl					235.484	0.192					

Ti	364.3	no	0.21	**42**	0.0009	Ti	334.941	0.0076	**1.5***	48+	0.004	**(4)**	Ti
							337.280	0.0134					
W	255.1f	no	3.6	*90		W	239.709	0.110	15	184+	0.004	**4**	W
	400.9	no					207.911	0.060					
U	358.5g	no	90.0	*90	0.09	U	409.014	0.674	80	238+	0.002	**2**	U
	356.7	no					385.958	0.500					
V	318.4	no	0.15	***1**	0.0003	V	311.071	0.020	**4**	51+	0.004	**(4)**	V
							292.402	0.0150					
Yb	398.8	no	0.12	24	0.00003	Yb	328.937	0.0036	**1**		0.002	**2**	Yb
							369.419	0.0060					
Y	410.2	no	0.6	*50	0.03	Y	371.030	0.0070	**2**		0.002	**(2)**	Y
							321.669	0.0158					
Zn	213.9	air	0.003	**0.8**	0.000006	Zn	202.548	0.0080	**5**	66+	0.006	**(6)**	Zn
							213.856	0.0036					
Zr	360.1	no	4.5	*90		Zr	343.823	0.0142	**6**		0.002	**(2)**	Zr
							339.198	0.0154					

AA line overlaps

(a) Spectral line overlaps from copper and lead affect the antimony 217.6 nm line; 206.8 nm line is preferred for determinations in the presence of these elements, with some reduction in sensitivity.

(b) Molecular absorption by CaOH affects the barium 553.5 nm line; 455.4 nm *ion* line may be used instead providing ionization is not suppressed.

(c) Praesodymium may interfere with 492.5 nm atomic absorption neodymium line.

(d) Calibrations using the nickel 232.0 nm atomic absorption line may be non-linear due to an adjacent nickel 232.14 nm non-resonance line. At higher concentrations, Ni 341.5 nm is preferred with some loss in sensitivity.

(e) The tin 235.4 nm is preferred when using a nitrous oxide flame due to high sensitivity and good linearity of calibrations. However the tin 386.3 nm line is used when background absorbance at 235 nm becomes excessive.

(f) The tungsten 255.1 nm line is the most sensitive but W 400.9 nm is sometimes preferred for routine analysis due to better signal to noise ratios.

(g) The uranium 358.5 nm line is the most sensitive but is affected by noise from the nitrous oxide–acetylene flame cyanogen emission. The U 356.7 or 351.5 nm lines are sometimes preferred.

Analysis conditions (*atomic absorption*):

no: nitrous oxide–acetylene flame.

air: air–acetylene flame.

(no): nitrous oxide or air–acetylene flame have been used for analysis.

bc: background correction required.

edl: high intensity electrodeless discharge lamp required.

p: purge of optical path with nitrogen is required to prevent excessive absorption of the Hg 184.9 nm line in air.

Flame atomic absorption analysis:

(1) Atomic absorption lines and analysis conditions based on Price (1979) and selected instrument manufacturer's leaflets.

(2) Ideal solution detection limit is an estimate of the six-sigma detection limit of single element solutions and represents 'typical' values from instrument manufacturer's leaflets.

(3) Whole rock detection limit is an estimate of the six-sigma limit of determination calculated as the elemental concentration in whole rock using the ideal solution detection limits assuming 1g rock were taken up into 100 mL solution. These data have then been arbitarily increased by a factor of 2 to allow for non-ideal performance when analysing real sample solutions. Data marked by asterisk are taken from Walsh and Howie (1986) ('working detection limits'). Values in bold represent data within the range of concentrations encountered in 'normal' silicate rocks.

Graphite furnace AAS:

(4) Ideal solution detection limits are estimates of six-sigma detection limits for single element solutions compiled from selected instrument manufacturers' data on 100 μL sample injections.

Inductively coupled plasma–atomic emission spectrometry:

(5) The most sensitive emission lines are listed from the references listed in (7) below; v = vacuum monochromator essential to prevent unacceptable absorption of radiation below 190 nm in air.

(6) Ideal solution detection limits are the six-sigma detection limits taken from the work of Winge *et al.* (1979).

(7) Whole rock determination limit is an estimate of the six-sigma limit of determination calculated as the elemental concentration in the original rock sample assuming 1g of rock has been taken up into 100 mL solution, with no pre-concentration procedure. Where available, data has been compiled from the work of Church (1981); Walsh and Howie (1980, 1986); Crock *et al.* (1983); Bolton *et al.* (1980); and Floyd *et al.* (1980). Data marked by asterisk were calculated from ideal solution detection limits as described in (3) above. Values in bold again represent data that fall within the concentration range of 'normal' silicate rocks.

Inductively coupled plasma-mass spectrometry:

(8) Ideal solution detection limits are an estimate of six-sigma detection limits for single aqueous solutions compiled from Date and Gray (1981, 1982); Cantle and Goddard (1983); and Sciex Ltd.

(9) Whole rock data are an estimate of the six-sigma detection limits calculated as the elemental concentration in whole rock assuming that 0.2 g of sample has been taken up into 100 mL solution, with no sample preconcentration. Data are calculated from ideal solution detection limits, resultant values have been arbitrarily increased by a factor of 2 to allow for non-ideal behaviour when analysing real samples. Due to uncertainty in mass overlaps and matrix interferences at the light end of the spectrum, all data for elements lighter than molybdenum have been put in brackets. Additional solution detection limits for ICP–MS: S: 2000, F: 60000, Br: 200 and I: 4 μ/mL.

limits of determination obtained by Bolton *et al.* (1983) and Crock *et al.* (1982, 1983) are compared for solutions prepared directly from the sample and for those that have undergone an ion exchange separation. An improvement in detection limits by a factor of between 5 and 20 is normally observed following ion exchange separation.

Ion exchange performance is discussed in more detail in Chapter 14. Schemes applied to REE analysis by ICP–AES normally use a simple cation exchange separation. Thus Walsh *et al.* (1981) used a Dowex AG 50W-X8 cation exchange resin. The sample solution was loaded on to the ion exchange column and eluted with 400 mL 1.7N HCl. The REEs and some of the trace elements (all Ba, some Sr, Zr, Hf) were retained quantitatively on the column; most of the major elements were washed off and discarded. The REEs were then eluted from the column with 500 mL 4N HCl, evaporated to dryness and redissolved in 5 mL 10% HCl for analysis by ICP–AES. This analysed fraction also contained

Table 5.9 Rare-earth element emission lines and spectral interferences reported in ICP–AES

Element	Spectrum	Wavelength (nm)	I_n/I_b	Reported interferences	Literature reference
La	II	398.852	27.0	Ca,Zr	2,3,4,5,6
	II	379.478	30.0	Ca,Fe,Al,V,Ba Mn,Ti,Ce,Nd	1,4
	II	408.672	30.0		6
	II	394.910		Ti,Fe,Ca	5,7
Ce	II	418.660	5.7	Zr,Y,Fe,Dy,Sc	2,3,5
	II	413.765	6.2	Mg,Ca,Fe,Al,Ti,V, Cu,Cr,Ba,Mn,Sr	1,4,7
	II	394.275	4.4	Ca,Al,V,Eu,Sm,Gd Nd,Mn	4.5
	II	413.380	6.0	Dy	5
	II	401.239	4.0		6
Pr	II	422.293	6.3	Ce,Pr,Sr,Ca	2,3
	II	422.535	7.0	Sm	3,6
	II	410.075		Fe	5
	II	417.939	7.2	Zr	5
	II	390.844	8.1		7
Nd	II	406.109	31.0	Zr	3,5
	II	430.358	40.0	Ba,Zr,Fe	2,3,5,7
	II	417.732	22.0	Al,Ce,Eu,Y	4,6
		401.609		Ni,Ca,Fe,Ti,Sr,Sm	1
Sm	II	359.260	69.0	W,Ni,Mn,Fe,V,Ca Ti,Nd,Gd,Y	1,3,5,7
	II	442.434	55.0		2
	II	443.432	36.0	Ca,Ti,V,Ce,Nd	4
		373.92		Pr	3
	II	356.827	39.0	Sc	5
		392.8			6
Eu	II	381.967	110.0	Ca,Fe,Nd,Gd	1,2,3,4
	II	420.505	70.0	Gd	5,6,7
	II	412.970	70.0	Gd,Dy	5,6
	II	390.710	39.0	Sc,Fe	5
	II	393.048	53.0	Fe	5
	II	397.196	32.0	Zr	5
Gd	II	342.247	21.0	Ce,Ni,Fe,Dy,Cr,Sm	1,3,7
	II	335.047	14.0	Ti,Ca	3,5
	II	376.839	12.0	Ca,Fe,Ti,Mn,V,Nd, Ce,Th	3,4
	II	336.223	15.0	Sc,Y	3,5
	II	364.619	10.0		6
	II	303.284	11.0		2
Tb	II	350.917	130.0	Zr	5,6,7
	II	367.635	50.0		2
Dy	II	353.170		Mn,Ti,Sm,Nd	1,3,5,7
	II	400.035		Fe	5,6
	II	340.780			2
Ho	II	345.600	53.0	Ti,Zr	2,3,5,6
	II	389.102	18.0	Zr,Fe	7,5
Er	II	369.265	16.0	Y	2,3,5,6
	II	390.631	14.0		3
	II	337.271	29.0	Ti	5,7
Tm	II	376.133	26.0	Ti	5,6
	II	313.126	58.0		2
	II	346.220	37.0		7

Table 5.9 (*Continued*)

Element	Spectrum	Wavelength (nm)	I_n/I_b	Reported interferences	Literature reference
Yb	II	328.937	170.0	V,Fe	1,2,3,5,7
	II	369.419	100.0	Ti,Fe	4,5,6
Lu	II	261.542	150.0		2,3,7
	II	347.248	10.0		6
	II	339.707	14.0	Zr	5
Y	II	371.030	86.0	Ca,Ti	1,5,6,7
	II	321.669	38.0		2
	II	377.433	57.0		5
Sc	II	255.237	65.0	Mn,Fe	1
	II	361.384	200.0		6,7
	II	365.180	31.0	Fe	5
	II	431.409	38.0	Gd	5

Data from 1: Bolton *et al.* (1983); 2: Crock and Lichte (1982), Crock *et al.* (1983); 3: Walsh *et al.* (1981), Walsh and Howie (1980); 4: Church (1981); 5: Brenner *et al.* (1981); 6: Broekaert *et al.* (1979*a*); 7: Floyd *et al.* (1980).
I_n/I_b represents net analyte/background intensity and is a measure of relative sensitivity of emission lines, date being taken from Winge *et al.* (1979).
Some identified interferences may not be observed on all instruments. The absence of any listed interference is probably best interpreted as meaning 'not investigated'.
If a sample has undergone ion exchange separation, the only elements likely to be present in the REE fraction are Y, Sc, Ca, Sr and Ba. The other interferences listed above will be absent from analysed solutions.

Table 5.10 Limits of detection in the determination of rare-earth elements by ICP–AES

	Bolton *et al.* (1983)			Crock *et al.* (1982, 1983)		
Element	Wavelength (nm)	Limit of determination: No separation	Limit of determination: Ion exchange separation	Wavelength (nm)	Limit of determination: No separation	Limit of determination: Ion exchange separation
La	379.478	2.4	0.3	398.852	4	0.07
Ce	413.765	21	3.6	418.660	8	0.47
Pr				422.293	20	0.96
Nd	401.609	18	1.2	430.358	8	0.33
Sm	359.260	9	0.6	442.434	100	0.45
Eu	381.967	0.6	0.06	381.967	4	0.014
Gd	342.247	3	0.6	303.284	20	0.27
Tb				367.635	40	0.87
Dy	353.170	2.4	0.3	340.780	8	0.38
Ho				345.600	8	0.054
Er				369.265	8	0.12
Tm				313.126	—	0.078
Yb	328.937	0.6	0.06	328.937	2	0.033
Lu				261.542	—	0.015
Y	371.030	1.2	0.12	321.669	4	0.027
Sc	255.237	0.6	0.12	424.683	4	
Sample weight (g)		0.5	1		1	1
Sample preparation		Acid decomposition Si fumed off	Acid decomposition Ion exchange separation of REEs		Acid decomposition Si fumed off	Acid decomposition Double ion exchange for REEs
Final volume of analysed solution (mL)		50	15		50	5
Dilution factor		1:100	1:15		1:50	1:5

Limit of determination represents the concentration equivalent to six standard deviations on the background or blank signal, whichever is the greater.
Units of concentration = ppm

all the barium, yttrium, scandium and some Ca, Sr, Zr and Hf from the original sample solution.

Crock and Lichte (1982) used a similar procedure, but included a cation exchange in nitric acid solution followed by an ion exchange in hydrochloric acid solution to remove Fe. The final solution contained the REEs and Y plus minor amounts of other elements. Similar ion exchange techniques have been used for REE separations by Bolton *et al.* (1983), Brenner *et al.* (1981, 1984) and Broekaert *et al.* (1979). Detailed investigations of the separation of REEs by ion exchange appropriate for ICP–AES schemes of analysis have been reported by Broekaert and Hormann (1981) (Dowex 50W-X8 in HCl medium) and Crock *et al.* (1984) (Biorad AG50-X8 using both HCl and HNO_3 media), and are discussed in more detail in Chapter 14.

The accuracy and precision of REE analysis by ICP–AES may be judged by data in Table 5.11 (REE analysis on sample solution with no chemical separation) and Table 5.12 (REE analysis following ion exchange separation). The former level (no preconcentration) is satisfactory for an approximate survey analysis of some samples but is not sufficiently sensitive for routine research-grade determinations. Results of analysis following ion exchange separation in Table 5.12 demonstrate the excellent precision that may be achieved by this technique, data for most elements being in the range 1 to 4% relative.

Table 5.11 Rare-earth element determination by ICP–AES. Direct analysis of sample solution

	Andesite USGS: AGV-1			
	Abbey (1983) Expected	Church (1981) Analysed	CV (%)	Bolton *et al.* (1983) Analysed
La	36	38	1.6	51
Ce	71	60	1.7	63
Pr	(7)			
Nd	37?	34	10	43
Sm	5.9	6.1	15	<15
Eu	1.6?	1.8	2.8	1.4
Gd	(5.5)	5.2	4.8	<5
Tb	0.7?			
Dy	3.5?			4
Ho	(0.6)			
Er	(1.2)			
Tm	—			
Yb	1.9	1.7	1.5	3
Lu	0.3?			
Y	19			23
Sc	12.5			13

Units: ppm.
CV = coefficient of variation = 100 × standard deviation/mean analysis.
Sample preparation procedures do not include a chemical separation stage.
Overall dilutions: Church (1981): 1 g to 100 mL; Bolton *et al.* (1983): 0.5 g to 50 mL.

5.13 Direct current plasma–optical emission spectrometry

The direct current plasma is created by initiating an electrical discharge within a stream of argon gas between two electrodes. The electrodes are then drawn apart so that a continous plasma discharge is maintained within the argon gas stream. A sample that has previously been taken up into solution is nebulized and the sample aerosol directed at the plasma. In the vicinity of the plasma, the aerosol particles rapidly undergo desolvation, vaporization and atomization. The atomic (or ionic) optical emission signal is observed

Table 5.12 Rare-earth element analysis by ICP–AES following an ion exchange separation.

	Basalt USGS: BCR-1				Test sample	
	Abbey (1983) Expected	Crock and Lichte (1982) Analysed	CV $n = 3$	Walsh *et al.* (1981) Analysed	Walsh *et al.* (1981) Analysed	CV $n = 15$
La	27	26.6	0.4	26	5.83	6.5
Ce	53	53.8	0.2	53.2	12.3	3.5
Pr	7?	7.29	0.4	6.8	1.89	4.3
Nd	26?	29.7	0.5	33.5	9.64	3.5
Sm	6.5	6.7	2.2	7.1	2.48	2.2
Eu	2.0	1.98	2.0	2.09	0.90	1.1
Gd	6.6	6.9	1.4	7.0	2.59	2.7
Tb	1.0	1.0	5.0			
Dy	7?	6.72	0.6	6.4	2.92	0.9
Ho	1.2?	1.40	0.7	1.42		
Er	3.5?	3.80	0.4	3.87	1.75	2.3
Tm	0.6?	0.57	1.8			
Yb	3.4	3.70	0.4	3.50	1.79	1.1
Lu	0.5?	0.524	0.4	0.55	0.28	1.8
Y	40	35.3	0.4			

Units: ppm.
CV = coefficient of variation = 100 × standard deviation/mean analysis (%).
Overall dilution factors: Crock and Lichte, 1 g to 5–10 mL; Walsh *et al.* (1981), 0.5 g to 5 mL.

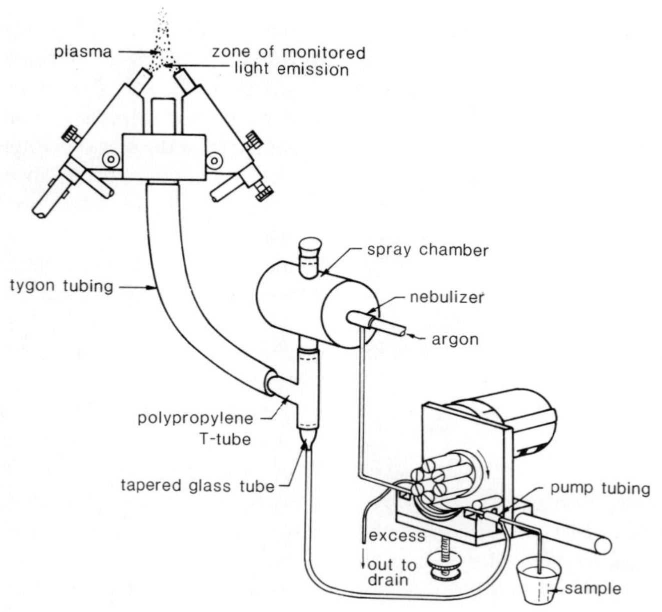

Figure 5.41 Schematic arrangement of a two-electrode direct current plasma–atomic emission spectrometer showing the configuration of electrodes, peristaltic pump, nebulizer and spray chamber. (Bankston and Fisher, 1977.)

using a suitable spectrometer. Dc plasma–atomic emission spectrometry is thus a technique for the analysis of samples prepared as solutions. The technique suffers from similar spectral overlap interferences to those described for inductively coupled plasma emission spectrometry. However, chemical interference effects are somewhat more serious in the dc plasma because the mechanism of atomization is not as efficient. The analytical characteristics of dc plasma jets have been reviewed by Greenfield *et al.* (1975). For some elements, dc plasma–AES can achieve detection limits approaching those attainable by ICP–AES. However, in general, performance falls some way below this goal. Instrumentation for dc plasma–AES benefits from a significant advantage both in lower initial capital cost and lower running costs compared with ICP–AES. However, perhaps because of the inability to match or exceed the ultimate performance of the latter technique, DCP–AES has not found as widespread an application in silicate rock analysis as might have been expected.

(*a*) *Instrumentation and plasma characteristics*. Earlier instrumentation consisted of two electrodes (Figure 5.41) which were set at an angle of about 75° to one another. The electrodes were made of tungsten and a stream of argon was directed through a sleeve surrounding each electrode. The argon served both to sustain the plasma and to cool the electrodes and to prevent their ceramic sleeves from melting. In service, this arrangement was found to be a relatively noisy optical emission source, mainly because of the poor positional stability of the plasma (see later). This problem was overcome by introducing a third electrode to stabilize the discharge (Figure 5.42). The two original electrodes both became anodes and a third common cathode was added above the plasma.

The plasma is struck by arcing a high-voltage spark from cathode to anode. The plasma is then sustained by passing a low voltage (40–50 V) current regulated to about 7 A through each of the anodes. The resultant plasma discharge forms an inverted 'Y' configuration. The argon gas flow rates required to maintain the plasma total 8 L/min and comprise 4 L/min for the nebulizer, 1 L/min for each anode and 2 L/min for the cathode (Decker, 1980). Welding-grade argon is reported to be adequate but in some analytical applications, special grade argon (99.9% pure) is preferred. During operation, small amounts of the electrodes are continuously ablated into the plasma, resulting in an intense background optical emission from this material. Conventional tungsten electrodes cause many potential emission

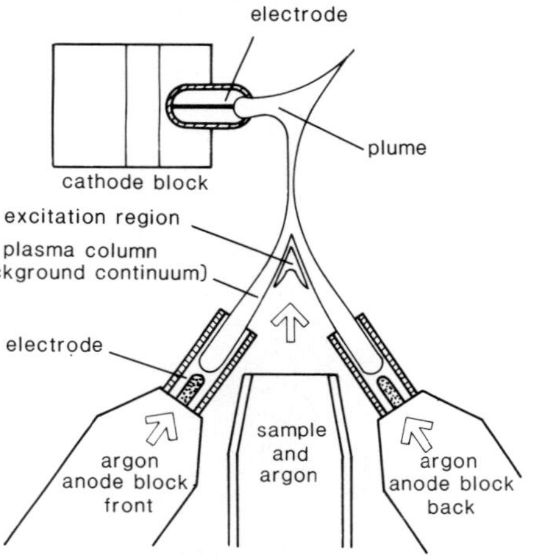

Figure 5.42 Schematic diagram of the direct current plasma source generated from three electrodes. (J. Reednick, by courtesy of Spectrometrics Inc.)

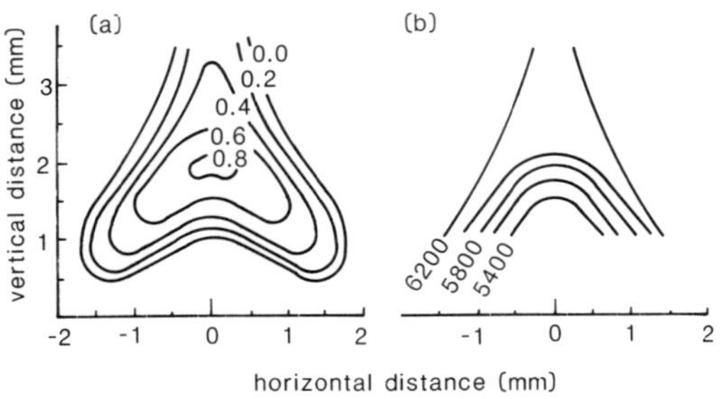

Figure 5.43 Contour map of (*a*) the logarithm of the line-to-background ratio of the zinc 307.6 nm line, and (*b*) temperature in the excitation region of a three-electrode direct current plasma atomic emission source. (Decker, 1980.)

interferences, and substitution of carbon anodes is preferred to overcome this problem. Decker (1980) reported that the solution uptake rate of the nebulizer, controlled by a peristaltic pump, was 1.6 mL/min, of which 0.26 mL formed an aerosol directed at the plasma. Sample nebulization procedures and limitations are as discussed under AAS and ICP analysis.

The sample aerosol and argon carrier gas flow are directed underneath the arms of the inverted 'Y' of the plasma. In this region, aerosol particles are dried and lose most of thier vertical component of velocity. These particles are then deflected laterally around the plasma core due to the thermal barrier in the excitation region of this source.

Decker (1980) has investigated the temperatures generated in this region of the plasma. Data for the signal to background ratio for the 307.6 nm line of zinc (Figure 5.43) show that in the region where signal to background ratio achieves a maximum, sample temperature reaches values of 5700–6000 K. In this region adjacent to the plasma the sample aerosol rapidly undergoes desolvation, atomization and excitation. The optimum emission signal is found just below the apex of the plasma but outside the plasma boundary beyond which an intense background continuum signal degrades spectroscopic observations. In comparison with ICP excitation, samples atomized in a dc plasma do not benefit from the much higher atomization temperatures (about 8000 K) experienced in the inductively coupled argon plasma. The thermal energy which is available for atomization of samples during dc plasma excitation is thus somewhat lower. Conversely, the temperatures to which samples are subjected in the dc plasma are considerably higher than the highest temperature encountered in a nitrous oxide–acetylene atomic absorption flame.

Data in Figure 5.43 also demonstrate the importance of positional stability in the dc plasma. The area from which maximum emission is observed is very small, perhaps 0.5 mm across. The emission source must, therefore, be maintained at the exact focal point of the associated monochromator if precise and reproducible measurements are to be made.

Data taken from Decker (1980) and plotted in Figure 5.44 show that some compromise is required in the optimization of observation height within the plasma for multi-element analysis. The maximum intensity of selected ion lines occurs at approximately the same observation height in the plasma. However, some variation is observed in the optimum height for measuring selected atomic emission line signals (Figure 5.44*b*). Decker correlated this variation with differences in the 'norm' temperature associated with individual lines.

(*b*) *Chemical interferences*. Several workers have noted that the emission signal of many elements from a dc plasma is affected by the presence of more readily ionized elements (e.g. Bankston *et al.*, 1979; Decker, 1980; Cantillo *et al.*, 1983). Bankston *et al.* reported that if sample solutions contained high concentrations of an easily ionized element, the analytical sensitivity of other elements excited by the dc plasma source was increased. Moreover, this enhancement effect was directly related to the concentration of the readily ionized element and the relative ease with which it could be ionized. In the determination of major, minor and trace elements in silicate rocks, Bankston *et al.* (1979) selected a lithium metaborate digestion (0.2 g sample + 1.4 g lithium metaborate) taken up into 100 mL of solution to swamp any ionization effects. Even at this relatively high concentration of lithium ions, variations in sample compositions were sufficient to cause non-linear calibration curves, those of the major elements Na and K being most severely affected. Bankston *et al.* concluded that the addition of excess Li from the flux was not in itself sufficient to suppress interference effects and that it was necessary to add caesium as

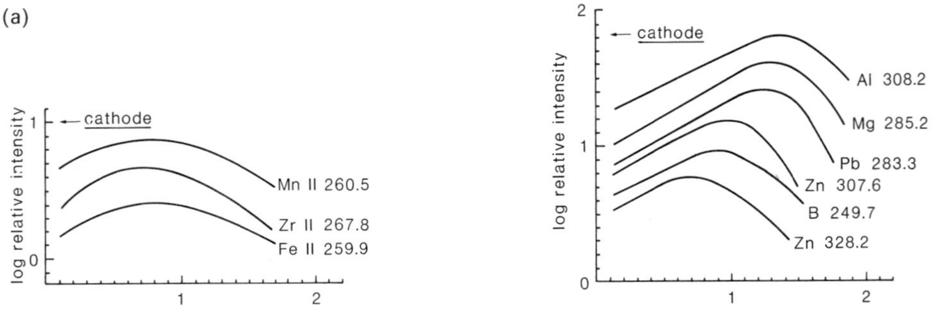

Figure 5.44 Distribution of the emission intensity of (*a*) the ion lines of Mn, Zr and Fe, and (*b*) the atom lines of Al, Mg, Pb, An and B as a function of vertical observation height in a three-electrode direct current plasma source. (Decker, 1979.)

Table 5.13 Apparent first ionization potentials of selected elements at 5000–6000 K

Cs	K	Ba	Sr	Na	Li	Ca	Ti	V	Al
4.4	4.74	5.03	5.47	5.49	5.75	5.82	6.55	6.79	6.8
Cr	**Mn**	**Mg**	**Fe**	**Ni**	**Cu**	**Si**	**Zn**	**P**	
6.95	7.33	7.33	7.67	8.11	8.13	8.41	9.06	10.17	

Units: eV.
Data from Bankston *et al.* (1979).

an auxiliary ionization buffer. Caesium concentrations of 2.2 mg/mL were sufficient to overcome the non-linear calibration of potassium but 3.16 mg/mL were required before a linear calibration for sodium was obtained. Caesium was selected not only because determinations were not required for this element, but also because its apparent first ionization potential at temperatures of 5000–6000 K is lower than that of every other element (Table 5.13). It may not be unrelated to the problems encountered above that the ionization potential of lithium (5.75 eV) is lower than all other elements analysed in this work apart from sodium and potassium.

Decker (1980) investigated ionization interference effects in the dc plasma and observed that solutions containing up to 0.5% potassium (as KNO_3) had a negligible effect on the emission intensity of a range of analytes present at the 100 μg/mL level. However, above this concentration level, marked enhancement effects were observed. It was suggested that high concentrations of potassium (5%) cause significant changes in the structure of the plasma. Specifically, the high concentration of an element that is easily analysed causes the following:

(i) The plasma expands its conducting column, weakening the plasma jets from the anodes which are reduced in intensity and length
(ii) Expansion of the plasma jet cores results in an overall expansion of the plasma—this expansion can result in melting of the ceramic sleeves adjacent to the cathodes
(iii) The small drop in potential between anode and cathode observed at the same time suggests that only the outer layers of the plasma are being affected, not the plasma core.

Table 5.14 A scheme of analysis for silicate rocks by direct current plasma–AES

Element	Spectrum series	Analyte line (nm)	Spectral interferences	Background line (nm)	Spectral interferences	Basalt USGS: BHVO-1			Detection limit
						Expected % oxide	Analysed % oxide	CV (%) $n=9$	
SiO_2	I	288.16	none	288.07	Fe II 288.08	49.90	48.36	4.8	
					W I 288.06				
TiO_2	II	334.90	none	334.80	Fe I 334.79	2.69	2.51	2.3	
Al_2O_3	I	394.40	none	394.28	Ca 394.28	13.85	14.11	3.1	
					Mn 394.29				
Fe_2O_3	II	259.94	none	259.86	W II 259.87	12.23	11.85	1.5	
MnO	II	257.61	W 257.62	257.53	Fe 257.53	0.17	0.172	2.2	
MgO	I	383.83	Ca 383.83	383.71	P 383.714	7.31	7.36	2.0	
			Mn 383.82		Fe I 383.714				
CaO	II	396.85	W I 396.86	396.73	none	11.33	11.40	0.99	
Na_2O	I	589.00	none	588.82	none	2.29	2.26	1.8	
K_2O	I	404.41	none	404.29	none	0.54	0.45	19	
P_2O_5	I	214.91	W 214.91	214.85	Fe II 214.85	0.28	0.28	21	
			FeI 214.917		Fe I 214.84				
						ppm	ppm		ppm
Ba	II	455.40	none	455.27	Ca 455.264	135	151	3.2	20
					Fe I 455.255				
					W I 455.253				
					Si III 455.251				
Cr	I	425.44	W I 425.43	425.31	none	300	285	3.3	20
Cu	I	327.40	Ca 327.396	327.30	Mn 327.30	140	111	1.9	4
					Ti 327.405				
Ni	I	305.08	none	304.99	W II 304.99	120	138	16	6
Sr	II	346.45	Fe II 346.45	346.34	Mn II 346.333	420	492	3.9	6
			W 346.44		Fe I 346.330				
V	I	437.92	none	437.79	none	320?	328	5.5	20
Zn	I	213.86	Fe I 213.86	213.79	none	105	106	19	30

Data from Bankston *et al.* (1979).
Expected data from Abbey (1983).
Analyses were undertaken on a two-electrode dc plasma optical emission system fitted with an echelle spectrometer. Samples were digested by lithium metaborate fusion and analysed solutions were buffered with 2 mg/mL Cs.
The authors' quoted detection limits have been increased by a factor of 2 and are representative of six-sigma limit of determination.
CV = coefficient of variation $\simeq 100 \times$ standard deviation/mean analysis (%).
Data represents mean of 9 determinations.

These physical changes in the plasma at high potassium concentrations also affect conditions in the sample excitation region:

(i) There is a marked increase in both intensity and signal to background ratios for both atom and ion lines—the enhancement of atom lines cannot be related to an ionization effect and Decker considers that high potassium contents may induce more efficient volatilization of the sample

(ii) The background is also increased in the presence of excess potassium (although the signal intensity increases at a faster rate)

(iii) There is a reduction of 200–300 K in the excitation temperature of the plasma near the core and a reduction in the temperature gradient at the edge of the plasma in the analytically useful region

(iv) The position of maximum intensity of lines with high 'norm' temperatures tends to move higher towards the plasma core, as might be expected from the observed fall in temperature gradients within the analytically useful region.

These results indicate that the mechanism of the ionization enhancement effect is not as straightforward as for flame AAS, and may be related to increases in the rate of volatilization of samples due to the presence of high concentrations of alkali metals.

(*c*) *Spectral interferences.* As in ICP spectroscopy, the emission spectrum observed from a dc argon plasma is complex. Excitation mechanisms in both these plasmas show some similarities (Decker, 1980) and the comments and precautions for investigating and correcting spectral line overlap interferences in ICP analysis are relevant here. Data in Table 5.14 indicate the lines selected for both peak and background measurement in a scheme of analysis devised by Bankston *et al.* (1979) for selected major and trace elements in silicate rocks. The interferences reported by the authors are listed. Tungsten line overlaps were of particular concern due to the use of tungsten electrodes to maintain the dc plasma.

One advantage some dc plasma instruments of one commercial design have over the ICP equivalents is that the dc plasma source is coupled to an echelle type monochromator (section 3.9). Such spectrometers both possess inherently high resolution (usually better than 0.005 nm) and disperse multiple-order diffractions. Both these characteristics help to minimize (or in some cases eliminate) spectral overlap interferences. A further advantage of these instruments is that of recording a spectrum on a photographic plate for rapid qualitative identification of the elements present. The echelle spectrometer can of course be used in conjunction

Table 5.15 Sediment analysis by direct current plasma–atomic emission spectrometry.

	Wavelength (nm)	Limit of detection (ppm)	Marine mud USGS MAG-1				
			Expected	Analysed same batch	CV (%) $n = 6$	Analysed 10 months	CV (%) $n = 15$
		(% oxide)	(% oxide)	(% oxide)		(% oxide)	
SiO_2	288.2	1.925	51.19	51.3	<2	49.2	5
TiO_2	334.9	0.020	0.75	0.80	4	0.82	6
Al_2O_3	396.2	1.700	16.46	14.4	<1	13.2	10
Fe_2O_3	373.7	0.00017	6.98	6.29	2	5.86	8
MnO	403.1	0.0039	0.10	0.089	3	0.087	3
		ppm	ppm	ppm		ppm	
Co	340.5	21	20	26	12	28	24
Cr	357.9	6	105	89	1	94	6
Cu	324.8	6	27	33	10	32	9
Ga	417.2	6	21?	28	3	27	11
Mo	379.8	—	1?				
Nb	408.0	—	9.6?				
Ni	341.5	6	54	53	4	53	6
Tl	351.9	—	—				
V	347.9	15	140	140	<4	140	4
W	400.9	—	—				
Y	371.0	6	27?	11	7	13	19
Zn	213.9	15	135	120	4	120	8
Zr	339.2	18	130	98	2	110	7
La	394.9	15	41?	41	2	42	17
Ce	418.7	30	86?	130	6	110	13
Dy	353.2	3	—	3.0	10	3.6	22
Yb	369.4	3	2.6	1.6	6	2.3	23

Data from Cantillo *et al.* (1983).
Expected compositions taken from Abbey (1983).
Same-batch data is average of 6 samples analysed one after another during the same operating session.
10-month data is average of 15 samples analysed periodically over a 10-month period.
CV = coefficient of variation = 100 × standard deviation/mean analysis (%).
0.2 g sample fused with 1 g $LiBO_2$ and finally dissolved in 100 mL solution.
Detection limits represent six-sigma limits of determination in sediments.

with conventional ICP sources, and Fernando (1982) has described figures of merit for the use of this instrument combination.

(*d*) *Applications to silicate rock analysis.* Few reports have appeared in the literature on the application of the dc plasma–AES in silicate rock analysis. However, Bankston *et al.* (1979) described a scheme for the determination of selected major and trace elements in silicate rocks, and Cantillo *et al.* (1983) have reported on the analysis of marine sediments by DCP–AES. Because of the interference effects associated with readily ionizable elements mentioned above, it is desirable to use a lithium metaborate fusion technique to bring the sample into solution. The resultant solution is then swamped with lithium ions. Thus Bankston *et al.* (1979) mixed 0.2 g of powdered sample with 1.4 g lithium metaborate, and fused the mixture in pre-ignited graphite crucibles at 950°C for 14 mins. After fusion, the melt was dissolved in nitric acid and made up to a final volume of 100 mL. To suppress additional chemical interferences, 0.2% (w/v) caesium was added to all solutions to give a final concentration of 0.2% (w/v) (0.32% w/v Cs in those aliquots to be analysed for sodium and potassium—see above).

Cantillo *et al.* (1983) used a similar preparation procedure in the analysis of marine sediments. Caesium was not added to sample solutions, but the lithium content was kept constant for both standard and sample solutions. Linear calibrations were obtained for all elements up to concentrations of 1000 μg/L, except for Mn, Al, Si, Ti and Fe, which were linear up to 200 μg/L, 200, 800, 10 and 200 mg/L respectively; Na and K were not determined in this scheme of analysis.

Results for the determination of basalt USGS:BHVO-1 by Bankston *et al.* (1979) are listed in Table 5.14, and for the analysis of marine mud USGS:MAG-1 by Cantillo *et al.* (1983) in Table 5.15. Coefficients of variation for the major elements in BHVO-1 vary from 1% (CaO) to 21% P_2O_5, and for the trace elements from 2% (Cu) to 19% (Zn). These results were obtained on a source fitted with only two electrodes and did not benefit from the additional precision attainable on a three-electrode stabilized plasma. Limits of determination (six sigma) reported by Bankston *et al.* for selected trace elements vary from 4 to 30 ppm in whole rock.

Cantillo *et al.* (1983) investigated both short-term and long-term precision of their method for the analysis of the marine sediment MAG-1. Short-term precision was demonstrated by the results of six samples analysed together as a batch (listed in Table 5.15). Long-term precision was assessed from 15 samples of MAG-1 analysed periodically over a 10-month period. As might be expected, most elements from this second batch show coefficients of variation that are significantly higher. Detection limits (six sigma) for most trace elements were in the range 3 to 20 ppm (whole rock).

(*e*) *Conclusion.* Direct current plasma–atomic emission spectrometry occupies a position intermediate between flame AA analysis and ICP–AES. It suffers from chemical interference effects which require careful attention as in flame AAS. Spectral overlap interferences are also of concern as in ICP–AES. The direct current plasma shares with ICP a powerful simultaneous multi-element analytical capability and versatile range of elements that may be determined directly in solution. The technique has not yet been widely adopted in silicate rock analysis and has untapped potential in this application.

6 Arc and spark source optical emission spectrometry

6.1 Historical perspective

Optical emission spectrometry using an electrical discharge generated across two electrodes was one of the first instrumental procedures to challenge classical wet chemical techniques in the spectrochemical analysis of silicate samples. In the late 1940s and early 1950s such instruments offered some exciting analytical possiblilities, including:

(i) The elimination of extensive chemical separation procedures

(ii) Extremely high sensitivity for a wide range of trace elements (down to ppm detection limits for many)

(iii) Simultaneous multi-element analysis capability

(iv) Extremely rapid analysis times (one sample every 3–4 minutes).

It was a result of the interest in exploiting this potential, and the consequent necessity to obtain reliable natural reference materials, that stimulated the first collaborative interlaboratory reference material programme involving two samples, granite G-1 and diabase W-1, in the early 1950s (Ahrens, 1977).

The passage of time has vindicated all the advantages of the techniques listed above. However, one problem that has never been solved satisfactorily is the inherently poor precision and repeatability of measurements observed using this excitation source. In consequence, arc or spark source optical emission spectrometry has not been widely accepted as a routine quantitative analytical procedure for silicate rock samples. This is in contrast to metallurgical applications, where emission spectrometry is often the technique of choice. Despite these reservations, arc/spark source emission spectrometry still has a role to play in geochemical analysis today. However, applications are in the area of general reconnaissance analysis or regional geochemical surveys, where the ability to estimate the major and trace element composition of samples to precisions that are rarely better than 10% relative is not a disadvantage. Indeed, for such studies, optical emission spectrometry is one of the few techniques capable of determining simultaneously all the metallic and semi-metallic elements in the Periodic Table, including Si, As, Sb, and Se.

The technique was subject to considerable research and development interest from the time when commercial instrumentation became widely available in the 1940s up to about 1975. However, since then further development work for applications involving geochemical analysis has been largely transferred to optical emission spectrometry using the inductively coupled argon plasma as excitation source. There is a broad degree of compatibility between the optical components of these two techniques—the first ICP emission spectrometers consisted of earlier instrumentation with the plasma source substituted for an arc/spark stand.

6.2 Instrumentation

6.2.1 *Overall arrangement*

The overall arrangement of instrumentation is shown in Figure 6.1 and consists of an arc/spark stand into which are mounted two electrodes. One of these electrodes contains the sample powder to be analysed. Optical emission from the electrical discharge is measured using a suitable mono-

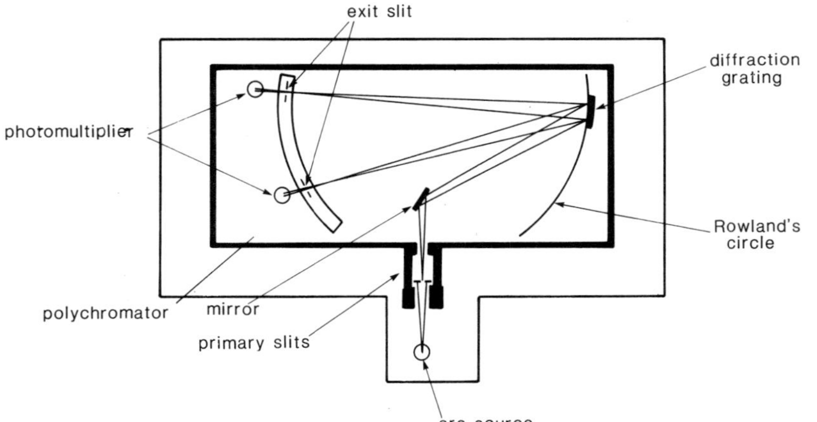

Figure 6.1 Schematic arrangement of an arc/spark source atomic emission spectrometer using a direct reading polychromator. For simplicity, only two detection channels are shown. (Based on a Philips PV 8250 design.)

chromator and read-out system. Almost all early instruments used the photographic plate for detection and were therefore known as emission spectro*graphs*. To exploit further the potential for simultaneous multi-element analysis, instrumentation was developed using a direct-reader polychromator system with electronic detection. Such systems can be fitted with a large number of exit slits and individual photomultipliers, as has been discussed in sections 5.9 and 3.13, making it possible to detect between 20 and 60 elements simultaneously (depending on instrument configuration).

Samples for analysis are prepared as pellets or lightly compressed powders loaded into one of a pair of electrodes across which either a dc current (an arc) or an ac current (a spark) is passed. In normal operation, this electrical discharge causes some of the sample to be vaporized. This atomic vapour is transported into the gap which is arranged at the point of focus of the monochromator. The intensity of selected emission lines from atomic and ionic species excited in the discharge is measured.

The poor precision referred to above arises largely from difficulty in controlling the electrical discharge on the sample electrode, since silicate rocks are composed of discrete heterogeneous particles that are non-conducting. These problems do not affect the analysis of metallurgical samples to the same degree.

6.2.2 *Arc and spark discharges*

Three methods are widely used to excite an electrical discharge: the dc arc, the ac arc and ac spark. Each has slightly different analytical characteristics.

(*a*) *The dc arc*. The dc arc is created by striking a dc current between two electrodes. One electrode comprises the sample itself. The other, the counter-electrode, is positioned between 2 and 10 mm away in normal operation. The discharge is initiated either by momentarily bringing the two electrodes into contact or by temporarily creating a high-voltage discharge across them. Electrical heating of the tips of the electrodes causes some sample material to be vaporized and ionized. As the electrodes are drawn apart, the number of these ions multiplies and they become capable of sustaining an electrical discharge across the gap. A current of between 1 and 30 A at 220 V (open circuit) is normally required (8–14 A in most analytical applications). Under these conditions, thermal energy created by the electrical resistance to movement of electrons and ions within the gap is sufficient to vaporize continuously material from the tip of the sample electrode. This material is broken down into atoms and ions within the electrical plasma discharge where species can become excited to higher electronic orbital states. It is the discrete line spectra emitted when atoms in these excited states fall back to their ground states that are detected by the optical spectrometer.

The dc arc is a very sensitive optical emission source but gives poor analytical precision in the excitation of silicate rock powders. The reason for this is that the dc discharge follows the line of least resistance, usually from some high point on the surface of the sample electrode. This high point (an individual mineral grain) heats up, becomes vaporized and is rapidly eroded away. The dc arc then wanders to an adjacent grain that now offers the line of least resistance. In practice the discharge wanders quite rapidly across the surface of the sample electrode, causing large fluctuations in the corresponding optical emission signal.

Temperatures in the dc arc are estimated to reach 6000–6500 K. The principal application of the technique is in the determination of trace elements where the highest sensitivity is required, although admittedly at inferior analytical precision.

(*b*) *The ac arc*. The ac arc is effectively a dc arc which is interrupted at a rate of 100 to 120 times per second. An electrical current of 1 to 5 A is required at an open circuit voltage of 2000 to 5000 V. The electrical plasma is continually formed and then interrupted and then reformed. In this way the formation of local hot spots is avoided and a more reproducible emission signal observed, though with some loss in sensitivity.

(*c*) *The ac spark*. The ac spark is generated by much higher voltages (15 000 to 40 000 V) but lower average currents than either of the two arc discharges. Spark formation is controlled by a circuit such as that shown in Figure 6.2. The primary coil of the step-up transformer is connected to the ac mains power supply. The secondary winding is connected to a circuit consisting of a capacitance, an inductor, the analytical spark gap and an interruptor spark gap. The interruptor is rotated by an electrical motor which revolves in synchronization with the mains line voltage. The circuit is designed to ensure that an electrical discharge occurs across the analytical gap only when the highest voltage is stored in the capacitor. In order to achieve this, the electrical discharge is not controlled by breakdown across the analytical gap, but rather by the position of the interruptor. In this way much higher precisions can be observed in optical emission measurements, since the intensity and duration of the discharge no longer depend on the properties of the sample or local variations in plasma discharge conditions. This higher precision is achieved at the expense of some loss in sensitivity compared with dc arc discharges.

The average current flowing during spark source excitation is low. However, the instantaneous current at the moment of discharge is high, perhaps up to 1000 A (Skoog and West, 1980, p.339). Under these conditions very high transient temperatures are reached in the associated plasma–perhaps equivalent to 40 000 K. The energy then available is much more favourable for the production of ions and excited ions. The ion emission spectrum observed from an ac spark is therefore much more prominent than the rather weak ion emissions observed from a dc arc.

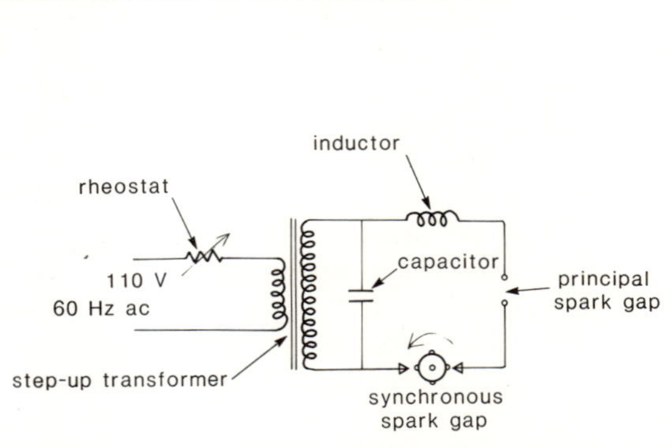

Figure 6.2 Electronic circuit (the Feussner circuit) used to generate a synchronous spark source. (Ewing, 1975).

6.2.3 *Electrodes*

(*a*) *Advantages of graphite.* The composition and surface texture of electrodes are critical aspects of arc/spark emission spectrometry. For the analysis of geological samples, it is necessary to mix a sample with the powder of a conducting element to promote electrical stability of the discharge (see section 6.3 below).

Electrodes are normally manufactured from carbon rod for several reasons:
(i) Carbon is an excellent electrical conductor and has mechanical properties suitable for the machining of electrodes
(ii) Carbon is an element for which determinations are not normally required in silicate rock samples
(iii) Inevitably, electrode material is ablated into the electrical discharge, but the emission bands of carbon do not cause serious interferences with sample spectra.

(*b*) *Electrode sheathing gas.* Carbon does have one drawback in emission spectrometry. At the high temperatures attained in the electrical discharge, carbon reacts with nitrogen, if the discharge is struck in air, to form cyanogen radicals ($CN^{\cdot}$). This species emits an intense spectrum in the region 360 to 420 nm which both enhances backgrounds and causes unwanted spectrum interferences. Molecular bands due to N_2 and NO from air entrained into the plasma may also be observed. These difficulties are normally overcome by sheathing the electrodes during discharge with a spectroscopically inert gas. This has two benefits. First, air is excluded from the plasma discharge region, suppressing cyanogen band emission. Second, the electrodes are efficiently cooled, preventing local overheating. This is of particular benefit in dc arc discharges and reduces the tendency of the electrical discharge to wander over the electrode surfaces. Stallwood (1954) designed a spark stand in which electrodes were cooled by a forced-air flow. This basic design has been widely adopted in later instruments. Various workers have described gas jet devices that both cool the electrodes and sheath the electrical discharge from air.

A mixture of argon with up to 30% oxygen is normally used for this purpose. Helz (1964) described a gas jet device that sheathed the electrodes in an argon–20% oxygen mixture, and did away with cumbersome enclosures round the arc. Cyanogen bands were absent from spectra recorded at sheathing gas flow rates of about 7 L/min. Boumans and Maessen (1969) described a double-flow device in which argon–oxygen mixtures were used as the electrode stabilizing gas, and pure argon as sheathing gas. These workers undertook a detailed investigation of changes in composition and flow rate of the argon–oxygen gas on the stability of a dc arc discharge. Criteria used were the degree of wander and excitation efficiency, evaluated in particular from emission intensities of lithium. Both composition and flow rate of sheathing gas were shown to be critical to efficient and stable operation of the dc discharge. The proportion of oxygen determined the rate of consumption of sample and electrode, whereas flow rate influenced the stability of the arc and mass transport of sample material into the excitation region. The field of stable operation of the dc discharge as a function of flow rate and oxygen gas composition is shown in Figure 6.3. Rather than the commercially available mixtures of argon–20% argon, Boumans and Maessen (1969) obtained

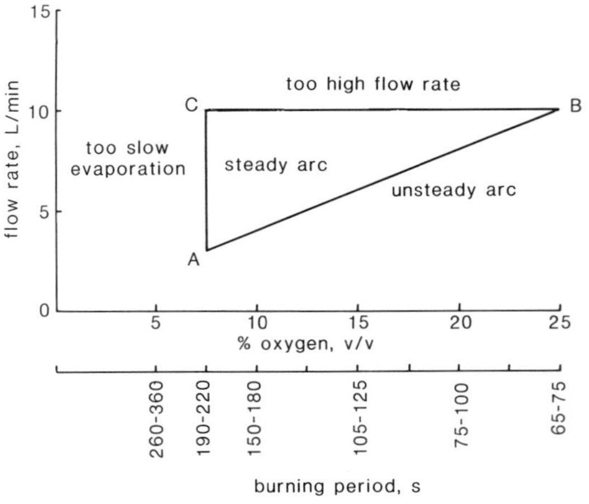

Figure 6.3 Plot of gas flow rate *v.* percent oxygen composition for argon–oxygen sheathing gas used to stabilize the dc arc discharge. The area of stable arc operation, as determined by Boumans and Maessen (1969) is delineated.

more reliable operation using 1 : 9 oxygen–argon mixtures. In the determination of trace elements in geological materials using the double-flow excitation stand, Maessen *et al.* (1976) used excitation conditions optimized as follows: stabilizing gas, 1 : 9 oxygen–argon at a flow rate of 10 L/min; sheathing gas, pure argon at a flow rate of 3.7 L/min; discharge current, 10 A dc.

(*c*) *Collimation of source radiation.* A further difficulty with carbon electrodes is that when electrical discharge is struck, the tips become incandescent, emitting a background continuum. Radiation entering the spectrometer must, therefore, be carefully collimated to the central region of the electrical discharge to prevent unwanted enhancement of background intensities.

6.3 Sample preparation

(*a*) *Basic procedures.* One of the advantages of arc/spark source emission spectroscopy is that samples may be analysed as powders with little pre-treatment. However, this convenience is also one of the limitations of the technique when applied to geological materials for the reasons outlined above, namely that silicate powders are non-conducting and are composed of discrete particles that are intrinsically inhomogeneous at the mineralogical level.

The first requirement, therefore, is that the sample must be ground to a fine particle size—200 mesh being the minimum acceptable. Samples must then be homogenized with powdered graphite to ensure good electrical conductivity and suppress some matrix effects. However, in some circumstances, electrodes prepared from such mixtures are still prone to both erratic emission intensities and mineralogical effects due to their particulate nature. These effects can be further suppressed to some extent by adding a 'spectroscopic buffer' to the sample mixture. The spectroscopic buffer is normally a low-melting-point alkali salt mixture that has two principal functions. The first is to promote local fusion of the sample during arc formation so that adjacent mineral grains are homogenized. The second is to increase the vola-

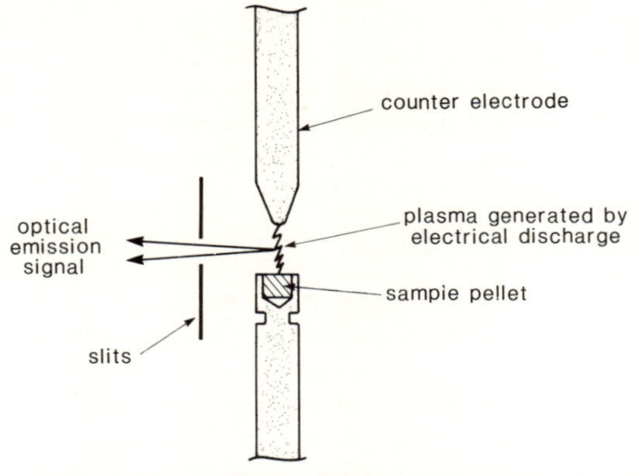

Figure 6.4 Schematic arrangement of electrode assembly and electrical discharge designed for the excitation of pelletted samples.

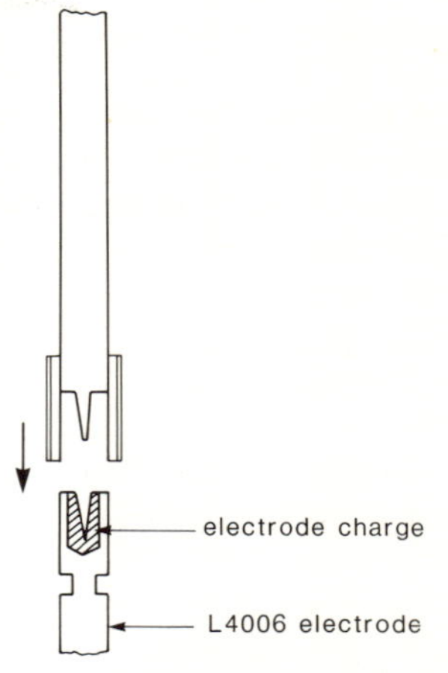

Figure 6.5 Tamping tool designed by Watson and Russell (1978) to pierce a gas vent in sample charges loaded into cupped graphite electrodes.

tility of the sample so that analyte species are more efficiently vaporized and released to the plasma discharge excitation source.

The choice of spectroscopic buffer will be considered further in the next section.

The most widely used form of electrode preparation is to compress the sample mixture in the form of a pellet or of a 'briquette'. Typically, 50 mg of rock powder is taken to form a pellet compressed in a die under a pressure of 2 to 4 tonnes to form a disc 4.5 mm diameter by about 4 mm thick. This pellet must be a press fit in a cup-shaped graphite electrode (Figure 6.4) to ensure both good electrical contact and thermal conductivity. To achieve these aims, Ordelman *et al.* (1966) reported that tolerance on die diameter used to make pellets was 4.5 ± 0.012 mm.

An alternative simplified sample preparation procedure was developed by Watson and Russell (1978) to avoid the contamination from Cu, Zn, Cr and Ni that can occur from briquetting equipment. Loose powders were loaded into cup-shaped electrodes and lightly compressed with a tamping tool. Initial experiments indicated that during excitation, a proportion of the sample was ejected from the electrode due to the explosive evolution of gases formed during heating. A special tamping tool was designed (Figure 6.5) that formed a gas vent in the electrode to eliminate this effect. Results by this invert packing procedure were not as reproducible as by the briquette method, though they were free of contamination effects and offered advantages in terms of speed of preparation and convenience.

The counter-electrode is normally made of carbon rod with a tapered end and rounded tip (Figure 6.4) to promote stable discharge formation. An improved burn is obtained from amorphous carbon cathodes (Watson and Russell, 1978).

(*b*) *Spectroscopic buffer*. The function of the spectroscopic buffer is (i) to enhance emission intensities and (ii) to suppress differences in excitation efficiencies between both silicate samples of varying matrix composition and calibration standards which have been made up of synthetic mixtures of oxides. The most effective salts to use as buffer materials are selected low-melting-point compounds of the alkali metals Na, K and Li, all elements of low ionization potential.

If accuracy and precision are of paramount importance in an analytical scheme, then by far the most satisfactory sample preparation procedure is to fuse samples and standards with an alkali metal flux such as lithium metaborate. After cooling, the melt is reground, blended with graphite and prepared as a briquette for arcing. Maessen *et al.* (1976) examined in detail the efficiency and reproducibility of excitation of selected elements in samples that were both fused and unfused. A substantial improvement in accuracy was observed for elements of intermediate volatility in pre-fused samples; the effect of elements of low or high volatility was not as marked.

However, in view of the other factors that affect precision, the inconvenience of fusing all samples prior to analysis would not always be considered acceptable. The alternative procedure is therefore to homogenize sample, flux and graphite and compress the dry mixture to form a briquette as described above. The flux and composition of such mixtures has varied. Maessen *et al.* (1976) prepared 35 mg briquettes containing 8 mg rock powder, 2 mg SiO_2 and internal standard elements, and 25 mg of flux containing 30% Li_2CO_3, 5% LiF and 65% graphite. Watson and Russell (1978) investigated samples prepared with fluxes of Li_2CO_3, $LiBO_2$, $Li_2B_4O_7$ and LiF. When various mixtures of these compounds and graphite were analysed, the optimum composition was found to be 25% silicate rock powder, 15% LiF and 60% graphite. This conclusion was generally in agreement with the work of Boumans and Maessen (1969) who reported that substitution of LiF for Li_2CO_3 had an appreciable effect on the excitation efficiency, particularly of refractory elements, presumably due to the greater volatility of fluoride salts. Avni *et al.* (1972) omitted the flux altogether and mixed silicate samples with graphite in the ratio 1 : 5. Ordelman *et al.* (1966) mixed 50 mg of sample with 450 mg of flux containing 12% LiF and 88% graphite.

6.4 Behaviour of elements in an arc discharge

6.4.1 *Chemical effects—fused and unfused samples*

The accuracy and precision of arc/spark emission spectrometry is closely related to mechanisms that occur in the electrical plasma discharge involving both volatilization and excitation of elements. This is a complex area that is not completely understood. However, Boumans and Maessen (1969) and Maessen *et al.* (1976) have systematically examined excitation effects. They considered that discrepancies in emission signals when samples of varying composition are excited can be caused by:

(i) Changes in the plasma discharge, in particular, temperature and electron pressure
(ii) Self-absorption effects, for example, that can reduce the intensity of emission lines generated within the plasma
(iii) Variations in the efficiency with which elements are volatilized from the sample electrode and transported into the plasma.

Under controlled analytical conditions, in which the sample is pelletted with a suitable buffer (e.g. LiF) and graphite, Maessen *et al.* (1976) considered that (iii) made the principal contribution to errors.

In the case of geological samples, the chemical environment surrounding a particular element oxide will vary slightly from sample to sample. The rate at which this element will then be released into the plasma discharge is, therefore, expected to vary as a function of:

(i) The mineralogical (and crystallographic) characteristics of the host matrix
(ii) The melting point and boiling point of the oxide
(iii) The boiling point of the element
(iv) The likely reduction of the oxide to the element by reaction with carbon from the graphite electrodes
(v) The propensity for the element to form stable (involatile) carbides
(vi) Differences in physical and chemical interactions due to the size of host particles
(vii) Differences in chemical interactions with the flux.

These factors are the cause of 'matrix effects', that is differences in the sensitivity of elements caused by variations in mineralogical composition between a set of samples.

Based on these factors, Maessen *et al.* (1976) were able to rationalize the benefits of pre-fusing samples with a lithium metaborate flux prior to briquetting, particularly for analytical procedures in which calibration is carried out relative to synthetic standards.

If, during arcing, the products formed by the chemical reaction of the analyte oxide with host matrix and flux differ appreciably in volatility from the original oxide, it is expected that evaporation of the element from a natural silicate sample will differ substantially from that occurring from a synthetic standard. If this occurs, significant calibration discrepancies will occur. Maessen *et al.* (1976) identified elements of moderate volatility such as Ga, Zn, Mn, Ba and V as falling into this category. In these circumstances pre-fusion of sample and synthetic standards causes a significant improvement in accuracy because the chemical environment of all samples presented for arcing is then equalized.

If reaction products formed during arcing have approximately the same volatility as the original oxides, little variation will be observed in the rates of evaporation of the analyte into the plasma. Maessen *et al.* (1976) identified elements of very high volatility (such as Pb) and low volatility (such as La, Ce, Zr, and Be) as falling into this category. In this case there is little advantage to be gained from pre-fusing samples. Although the efficiency with which elements enter the plasma is improved after fusion, there is little difference in the rates of evaporation between natural silicate or synthetic oxide samples, whether fused or not.

6.4.2 *Volatility of elements*

The work described above is important in characterizing the limitations of calibration procedures using synthetic oxide standards. However, it also highlights the problem that not all elements are expected to be evaporated into the plasma at the same rate during arcing. This has significant implications in devising a satisfactory programme of measurement.

The volatility of selected elements has been investigated in some detail by Boumans and Maessen (1969). They studied the variation in intensity of selected emission lines and the rate of evaporation of selected elements as a function of the burn time. Data were obtained for samples that had been pre-fused with lithium metaborate (1 part of sample to 1 part of lithium metaborate, fused at 1000°C for 5 minutes in a graphite crucible) and unfused samples. Each sample was made up of quartz and calcium carbonate (10% CaO w/w in mixture) to which was added selected element oxides. In each case the powdered sample was mixed with a 14:1 graphite–LiF mixture before arcing. The results of these experiments are plotted in Figures 6.6 (fused samples) and 6.7 (unfused samples). Samples were burnt using a 10A dc arc under conditions of slow, medium and fast evaporation (corresponding to 5, 10 and 15% oxygen content of the argon stabilizing gas). Results indicated a large range in both signal intensity and rate of evaporation between volatile elements such as Cd, Bi and Zn and refractory elements such as Y, REEs and W. Volatilization appeared to occur relatively smoothly from fused samples but more erratically from unfused samples, although the general trends remained the same. The intensity change of atom lines closely followed variations in the true rate of volatilization of elements, although this does not appear to be the case with ion lines (compare Mn (I) and Mn (II)).

6.4.3 *Studies of matrix effects*

A further insight into optimizing excitation conditions in the dc plasma arc is found in the work of Avni *et al.* (1972). These workers investigated variations in several analytical parameters as a function of axial distance within the plasma between anode and cathode. Samples were excited with a 14A dc arc using a 40 mg sample charge mounted in graphite electrodes placed 6.0 mm apart. The samples tested were pure SiO_2, $SiO_2+5\%$ each of Ca, Mg, Al and Fe, and $SiO_2+(Na+K)$, both with and without graphite. A spectroscopic buffer (flux) was not used in this work. Some of the results, plotted as a function of axial distance in the plasma from the anode, are shown in Figures 6.8 and 6.9.

In Figure 6.8 the variation in plasma temperature with axial distance from the anode is plotted for selected test samples. The addition of 5% each of Ca, Mg, Al and Fe did not markedly influence the temperature distribution along the plasma compared with SiO_2 alone, which averaged 6000

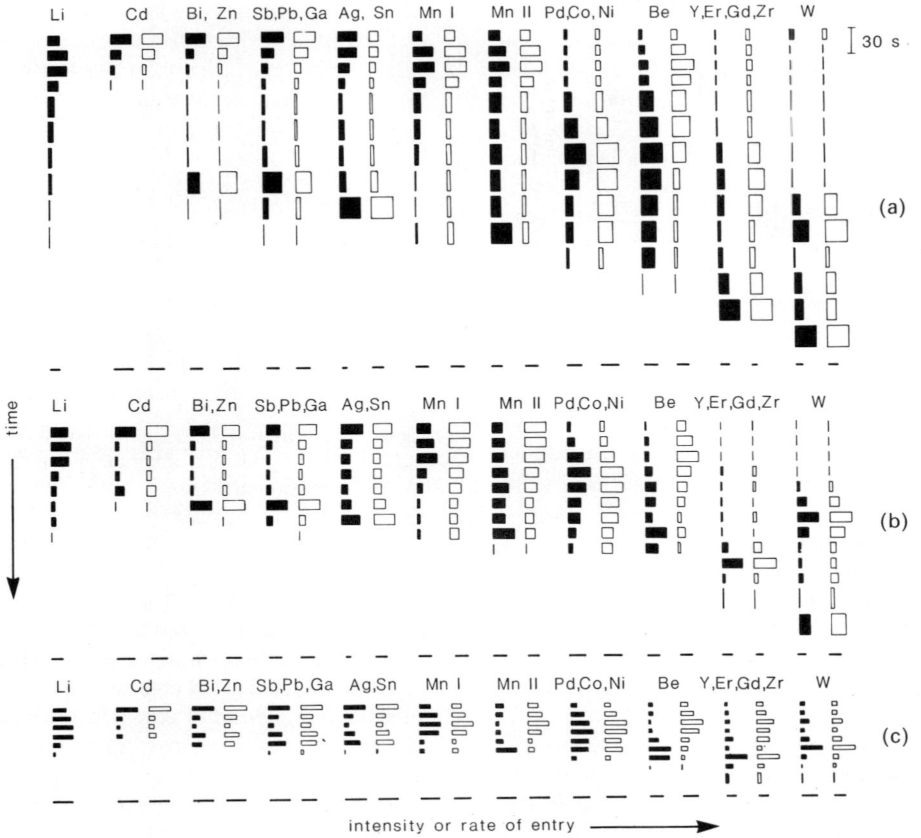

Figure 6.6 Evaporation patterns for selected elements measured as a function of arc burn time. Data are plotted for test samples that have been fused with lithium metaborate flux prior to analysis, and include (i) variation in intensity of emission line corrected for background (black rectangles) and (ii) variation in rate of evaporation of element (open rectangles). Results are plotted for (*a*) slow, (*b*) medium and (*c*) fast evaporation rates which corresponded to oxygen contents of the argon–oxygen arc stabilizing gas of 5, 10 and 25% respectively. The scale bar below each plot relates to the factor by which corresponding intensities must by multiplied to obtain comparable scales for the three evaporation modes. (Boumans and Maessen, 1969.)

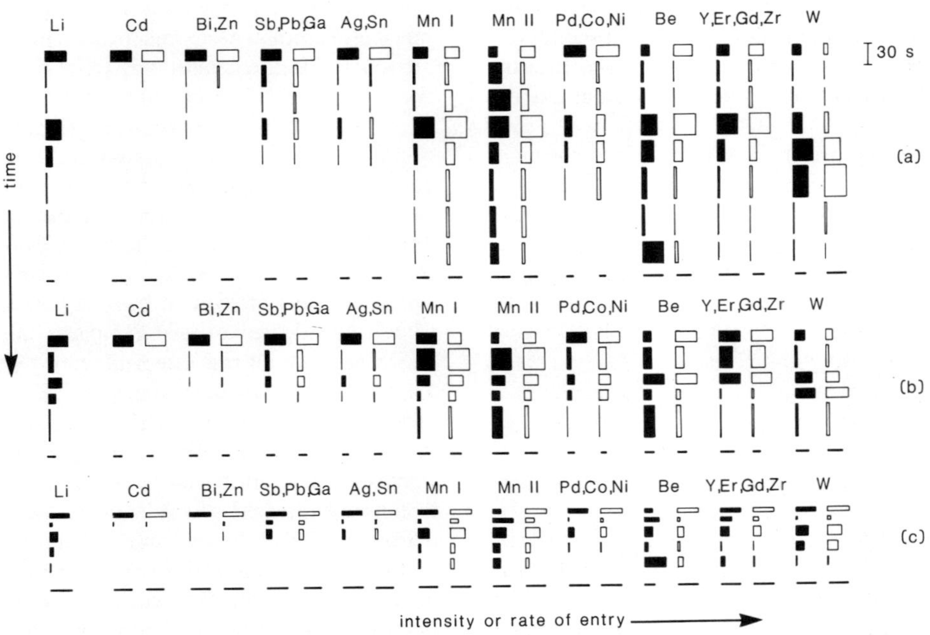

Figure 6.7 As for Figure 6.6, but data on the variation in emission intensity and evaporation rates is now plotted for test samples that were not fused with lithium metaborate flux prior to analysis. (Boumans and Maessen, 1969.)

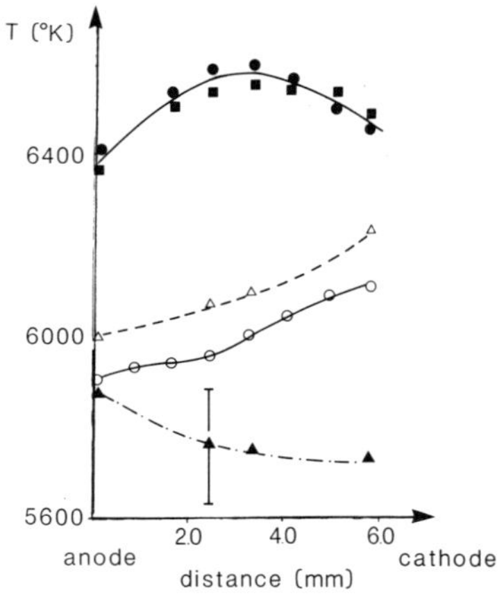

Figure 6.8 Axial variation of temperature in a dc arc plasma (14A, gap distance: 6mm) as a function of test sample composition as follows:
○ SiO_2:
△ SiO_2+5% each of Ca, Mg, Al, Fe
▲ SiO_2+5% (Na+K)
■ SiO_2+12% (Na+K)+500% graphite
● SiO_2+12% (Na+K)+300% graphite.
(Avni *et al.*, 1972.)

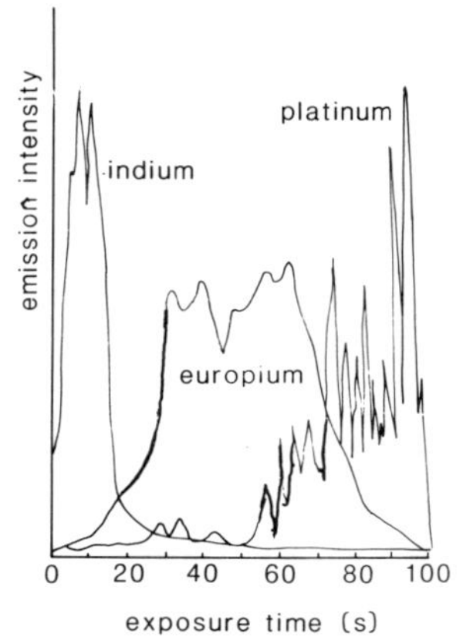

Figure 6.10 Variation in emission intensity of indium, europium and platinum as a function of burn time in a 12.5 A dc arc discharge. (British Geological Survey, Keyworth, Notts.)

K. However, addition of 2.5% each of Na and K to SiO_2 caused a significant depression in plasma temperature. Presumably this was caused by the low ionization potential of these elements which affected plasma characteristics. Samples to which graphite was added (SiO_2+12% (Na+K)+500% graphite) and (SiO_2+12%(Na+K) +300% graphite) show a marked increase in plasma temperature up to a maximum of about 6600 K in the central region.

Data on the axial variation in electron density within the plasma are shown in Figure 6.9. These results show that samples containing SiO_2+5% each of Ca, Mg, Al and Fe caused a marked reduction in electron density in the centre of the plasma although this is the area where the rate of change in electron density was lowest. The addition of

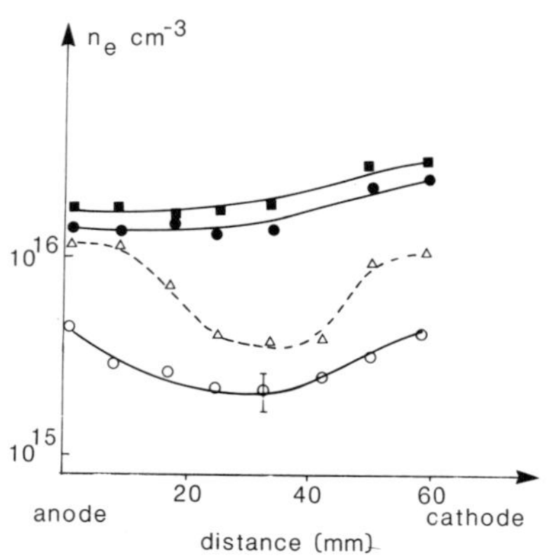

Figure 6.9 Axial variation of electron density (n_e) in a dc arc plasma (14 A; gap distance, 6 mm). Symbols correspond to identical test samples to those listed in Figure 6.8. (Avni *et al.*, 1972.)

graphite to samples caused a significant increase in electron density and a reduction in the spatial variation.

Avni *al.* (1972) also presented data to indicate that the intensity of emission of common trace elements was independent of sample matrix, providing measurements are restricted to the central region of the plasma discharge. They concluded that optimum results were obtained with samples diluted with graphite in the ratio 1 : 5 and that only slight improvements in analytical accuracy could be obtained by adding palladium to samples as an internal standard. This last conclusion is contrary to the general practice of other workers in arc source emission spectrometry, particularly in the determination of trace elements in silicate rocks.

6.4.4 *Rates of evaporation during arcing*

The data above indicate some of the factors that must be taken into account in setting up an analytical programme for arc source spectrochemical analysis. Of particular concern is the large variation in volatility between different elements. Data in Figure 6.10 demonstrate the variation in emission signals due to indium, europium and platinum during the course of a 100-second burn. Indium is completely burnt off within about 20 s. However, the europium signal becomes significant only after 20 s and has decayed to background by about 80 s. The emission intensity of platinum becomes significant only after 60 s. Date (1984) has suggested that these three elements are particularly suitable for internal standardization of trace elements of high, intermediate and low volatility. Measured intensities are normalizing to the emission signal of In, Eu or Pt, as appropriate.

A further consideration is the time over which emission signals are integrated. Data in Figure 6.11 show the change in emission intensity of the lead 405.78 nm line as the burn period proceeds. The lead signal effectively decays to background after 20 to 25 s. If the emission signal were integrated for any longer than 25 s, not only would the background intensity be needlessly increased, but also a potential interference may be encountered from manganese. Data in Figure 6.12 show that a similar problem can arise in the measure-

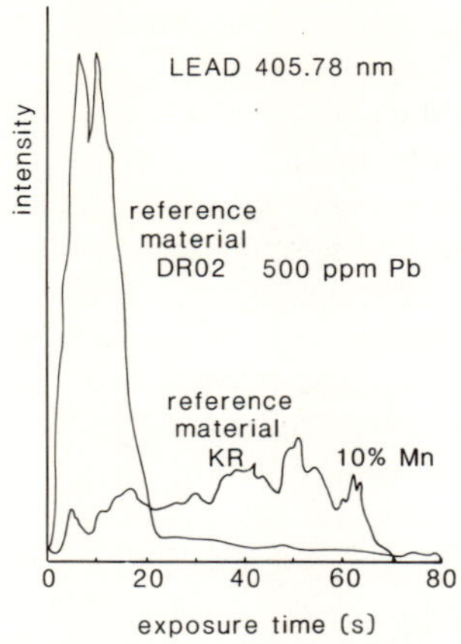

Figure 6.11 Variation in emission intensity of the lead 405.78 nm line as a function of dc arc burn time, demonstrating a potential interference from manganese if signal integration times are extended significantly beyond 20 s. (British Geological Survey, Keyworth, Notts.)

ment of the zinc line at 334.50 nm. This time, if signal integration is extended beyond about 20 s a significant interference is likely to be encountered from calcium.

Some other elements for which it is desirable to restrict integration times during arcing are listed in Table 6.1.

6.4.5 *Internal standardization*

In most routine schemes of analysis, especially using the dc arc source, it has proved necessary to normalize all intensity data against that of an internal standard so that the sometimes large fluctuations in emission can be corrected. The choice of three internal standards, In, Eu and Pt, as described above, is an unusual but clearly effective way of modelling variations in the emission intensity of individual elements according to their characteristic rates of evaporation. Most other workers have selected only one internal standard element. The most popular have included scandium (Bowditch and Watson, 1973) palladium (Galan, 1974) and germanium (Watson and Russell, 1978). Maessen *et al.* (1976) investigated the use of palladium and bismuth as reference elements for atom lines and erbium for ion lines. Their results were ambiguous and they reported that the use of reference elements gave some tendency to an improvement in precision when samples were arced in an unfused state. However, precision tended to deteriorate when samples had been prefused with lithium metaborate flux.

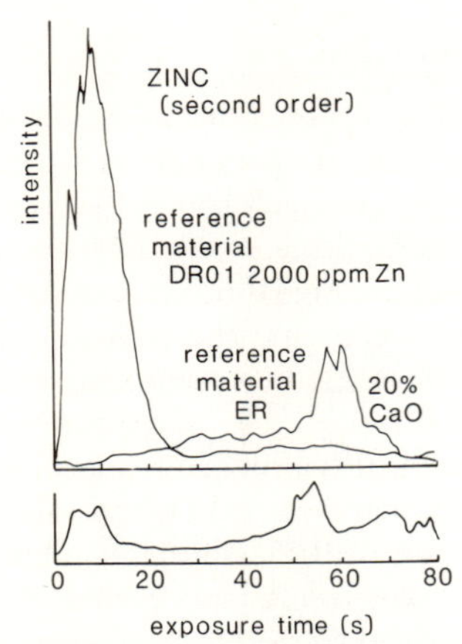

Figure 6.12 Variation in emission intensity of the zinc 334.50 nm line as a function of dc arc burn time, showing a potential interference from calcium if signal integration times are extended significantly beyond 20 s. (British Geological Survey, Keyworth, Notts.)

Table 6.1 Emission integration times for volatile elements

Element	Emission line (nm)	Maximum emission integration time after arcing initiated (s)
Ag	338.29	40
Be	234.86	60
Bi	306.77	25
Cd	228.80	10
Ga	294.36	35
P	255.33	30
Pb	283.31	25
	287.33	
Zn	213.86	20
	334.50	20

Data from Holmes (1982).
Analysis conditions: sample mixed in the ratio 1:1 with a spectroscopic buffer containing 50% graphite and 50% NaF. 30 mg pellet arced at 12.5 A dc for 100 s total burn time. Data normalized relative to Pd, In, Eu, Hf or Pt as internal standards.
Integration of emission intensities of the above elements was prematurely stopped after the indicated burn times to prevent enhanced background and inter-element interferences.

6.5 Simultaneous multi-element analysis

6.5.1 *Method development*

Due to the short burn time (usually less than 2 minutes per sample) and the large fluctuations that can occur in emission intensities, it is not appropriate to couple arc/spark sources to sequential spectrometers. Direct reader spectrometers are almost always used on modern instruments, fitted with a large number of fixed channels for simultaneous measurement of up to about 60 emission lines.

In setting up an analytical procedure for routine multi-element analysis, consideration must be given to the following points, derived from instrumental excitation factors described above:

(i) Optimization of sample/buffer composition to ensure efficient release of elements of interest to the plasma discharge (section 6.3 above)
(ii) Evaluation of matrix effects both between samples and relative to synthetic calibration standards (if used) to determine if pre-fusion of samples with lithium metaborate flux is desirable
(iii) Selection of most sensitive emission lines that avoid spectral interference effects
(iv) Evaluation of internal standardization procedures and, if appropriate, selection of elements for internal stan-

dardization that will adequately follow the behaviour of analytes having different rates of evaporation
(v) Optimization of emission integration interval to avoid the accumulation of unnecessary background counts or interferences
(vi) Optimization of instrumental conditions including dc arc *v.* ac spark excitation, electrode gap etc.
Emission lines and estimates of relative intensities from arc discharges are well documented in tables such as those of the NBS (Harrison, 1969) and MIT (Meggers *et al.*, 1975). However, since the degree of ionization in an arc spark is much greater than that in a dc arc, selection of the most sensitive ion or atom line will therefore depend on the type of discharge and energy input into the plasma.

A further consideration in line selection is that the arc/spark plasma cannot always be considered as an optically thin source. In the determination of some major elements, in particular, the atom/ion concentration in the plasma can be sufficiently high to cause self-absorption effects, particularly if resonance lines are used. Selection of less sensitive lines can usually avoid this difficulty. On older instruments that use spectrographic detection involving exposure of a photographic plate, a further difficulty can arise. If plasma conditions are optimized for the measurement of trace elements, major element lines will saturate the photographic emulsion and be indeterminable. In such circumstances it is often necessary to carry out multiple exposures of the photographic plate for varying integration times to compensate for the small dynamic range of this detection device.

6.5.2 *Major-element analysis*

If instrumentation is to be calibrated specifically for the determination of major and selected minor elements but not trace elements, ac spark conditions of excitation are most appropriate, since data of higher precision can be achieved. Thus, Bowditch and Watson (1973) described the commissioning of a direct-reading spectrometer calibrated for the determination of seven major elements (SiO_2, TiO_2, Al_2O_3, Fe_2O_3, MnO, MgO and CaO) and zirconium. Sample powders (0.25 g) were mixed with flux (2 g) (which contained scandium as internal standard) and fused in platinum–5% gold crucibles at 1000–1050°C for 15 to 20 minutes. After cooling, the melt was ground, homogenized with 3 g of graphite and pressed into a 1-inch diameter disc at 25 tonnes. The surface of the sample to be analysed was dressed with 600 grit paper.

Spark excitation was investigated using both a condensed arc (also called a low voltage spark) generated at 1 kV (nominal), 5 microfarad capacitance, 100 microhenry inductance and a more conventional ac spark generated at 20 kV (nominal), 0.0175 microfarad capacitance and residual inductance. Silica determinations under conditions of condensed arc excitation were not satisfactory due to interference from an oxide interference band. This problem could be eliminated by exciting samples in an argon atmosphere. However, spark excitation was proposed for all subsequent measurements. Optimum analytical conditions were selected as follows: pre-burn, 20 s (to allow excitation conditions to stabilize); integration period, 4×5 s (20 s total) relative to scandium internal standard; gap distance, 5.5 mm (checked after every four or five burnings).

The instrument was calibrated by analysing a range of 25 rock reference materials that covered the composition range required for the analysis of unknown samples. Sample pellets were then analysed by selecting suitable standards that gave emission intensities straddling the intensity of each element line in the unknown. Standards and samples were then sparked and the average intensity data calculated. If the mean data from the standard pellets (L_0 and H_0) were different from the values obtained during calibration, intensities were adjusted as follows:

$$\text{corrected counts} = C_u - L_0 \cdot \left(\frac{H_s - L_s}{H_0 - L_0}\right) + L_s + FE_1 \quad (6.1)$$

where C_u = counts of unknown sample, H_s = high standard counts obtained during calibration, L_s = low standard counts obtained during calibration, F = interference factor (equivalent to the number of counts given by 1% of the interfering element), and E_1 = concentration (weight %) of interfering element.

Some results from this calibration procedure are listed in Table 6.2. Accuracy of the method is indicated by the analyses of four reference materials listed in Table 6.2 taken from the calibration standard set.

In comparison with alternative procedures for silicate rock analysis, the lack of data for Na_2O, K_2O and P_2O_5 in the present scheme of analysis would be considered a disadvantage, as acknowledged by the original authors. The precision of data listed in Table 6.2, although adequate for general survey applications, is unfavourable compared with alternative modern instrumental techniques such as ICP–atomic emission or XRF analysis. Nevertheless the spark source–optical emission technique has in its favour the avoidance both of dissolution problems that might affect ICP analyses of samples of unusual composition (such as peridote PCC-1) and the uncertainties due to inter-element effects in samples which fall outside the range of normal silicate compositions that can reduce the accuracy of XRF data.

6.5.3 *Trace-element analysis*

In general, the role of spectrochemical analysis by arc/spark discharge is not in routine major element determination but rather in the analysis of geochemical survey samples for a wide range of trace elements. This can be achieved both because of favourable signal to noise ratios in the arc discharge emission source and the possibility of installing a large number of channels on polychromators of direct reader design.

Some indication of the range of elements and detection limits that can be achieved is shown in data listed in Table 6.3. These data include the wavelength and relative intensity of the two most sensitive lines as evaluated from the NBS copper arc data listed by Boumans and Bosveld (1979) and analytical detection limits evaluated from the data of Walthall (1976). These data should be considered as a guide only to analytical procedures. The choice of lines depends both on the optical characteristics of the spectrometer (many

Table 6.2 Analysis of major elements in reference materials by high-voltage spark discharge–optical emission spectrometry

Element oxide	Emission line (nm)	Calibration range (% oxide)	Coefficient of variation (% relative)	Basalt BCR-1		Peridotite PCC-1		Disthene DTS-1		Granite NIM-G	
				Expected	Analysed	Expected	Analysed	Expected	Analysed	Expected	Analysed
SiO_2	390.5	30–80	2.65	54.53	53.3	42.10	40.9	40.61	40.45	75.70	73.35
	288.1	30–80	4.0								
TiO_2	337.2	0.01–2.6	5.5	2.26	2.18	0.01	0.02	0.00?	0.02	0.09	0.09
Al_2O_3	308.2	0.4–20	2.05	13.72	13.7	0.73	0.86	0.25	0.42	12.08	12.10
	396.1	0.2–16	1.8								
Fe_2O_3	259.9	0.1–17	4.5	13.41	13.15	8.28	8.40	8.70	8.96	2.02	2.18
MnO	257.6	0.01–0.76	2.8	0.18	0.18	0.12	0.12	0.12	0.125	0.021	0.015
MgO	285.2	0.01–7	6.2	3.48	3.43	43.50	43.6	49.80	49.8	(0.06)	0.13
	518.3	0.5–50	2.2								
CaO	393.3	0.01–0.6	14.5	6.97	7.03	0.55	0.54	0.14	0.13	0.78	0.77
	317.9*	0.2–25	2.9								
ZrO_2	339.1*	0.02–1.6	9.0	0.02	0.03	0.00	<0.02	0.00	<0.02	0.04	0.03
Sc	402.0	internal std									

Data from Bowditch and Watson (1973).
Expected data from Abbey (1983).
Analyses represent best fit of emission intensity of four selected reference samples to practical calibration line set up from 25 reference materials, date being quoted in % oxide.
Samples (0.25 g) fused with lithium tetraborate (2.00 g), cooled, ground, homogenized with graphite (3.00 g) and compressed into a 1.25-inch disc for analysis.
Analysis conditions: High voltage spark discharge (20 kV), 5.5 mm electrode gap, 20 s pre-burn, 20 s emission intensity integration.
System calibrated against rock reference materials.
* Indicates lines were measured in the second order.

Table 6.3 Emission lines and detection limits for trace element analysis by arc source–optical emission spectrometry

Element	Spectrum	Wavelength (nm)	Sensitivity (NBS units)	Detection limit (ppm)
Ag	I	328.068	55000	0.2
	I	338.289	28000	
Al	I	396.152	9000	m
	I	309.271	6500	
As	I	197.197	65000	200
	I	200.334	44000	
	I	234.984	2600	
Au	I	201.200	11000	13
	I	267.595	3400	
B	I	208.959	6500	20
	I	249.773	4800	
Ba	II	455.403	65000	2*[1]
	II	233.527	2000	
Be	I	234.861	9500	2
	II	313.042	4800	
Bi	I	195.389	48000	2
	I	206.170	44000	
	I	306.772	36000	
Ca	II	393.366	42000	m
	II	396.847	22000	
	I	422.673	11000	
Cd	I	228.802	15000	30
	II	214.438	19000	
Ce	II	418.660	3500	40
	II	395.254	3100	
Co	I	345.350	21000	2
	I	340.598	11000	
Cr	I	425.435	20000	2
	I	357.869	19000	
Cs	I	852.124	15000	200*[2]
	I	455.531	400	
Cu	I	324.754	50000	2*[1]
	I	327.396	25000	
Dy	II	353.170	22000	6
	II	396.839	14000	
Er	II	390.631	11000	5
	II	369.265	7900	
Eu	II	420.505	60000	2
	II	381.967	39000	
Fe	I	373.487	7000	m
	II	259.940	2000	
Ga	I	417.206	20000	4
	I	294.364	9500	
Gd	II	376.839	8700	6
	II	342.247	6900	
Ge	I	199.824	42000	2
	I	265.118	12000	
Hf	II	202.818	8500	20
	II	339.980	2300	
Hg	I	253.652	15000	10000*[2]
	I	435.835	4000	
Ho	II	345.600	16000	2
	II	389.102	13000	
In	I	451.131	18000	10
	I	325.609	13000	
Ir	I	254.397	7900	14
	I	322.078	5100	
K	I	766.490	18000	7000*[2]
	I	404.414	320	
La	II	394.910	9000	30
	II	408.672	5500	
Li	I	670.784	36000	0.8*[1]
	I	323.263	170	
Lu	II	261.542	18000	6
	II	350.739	8300	
Mg	I	285.213	6000	m
	I	383.826	5000	

Table 6.3 (*Continued*)

Element	Spectrum	Wavelength (nm)	Sensitivity (NBS units)	Detection limit (ppm)
Mn	I	403.076	27000	m
	II	257.610	12000	
Mo	II	202.030	40000	4
	I	386.411	29000	
Na	I	588.995	20000	500*[2]
	I	330.237	300	
Nb	II	202.932	3300	4
	II	309.418	1800	
Nd	II	430.358	5400	30
	II	406.109	4700	
Ni	I	341.476	8200	14*[1]
	I	352.454	8200	
Os	I	201.814	29000	44
	I	290.906	9600	
P	I	213.618	3400	2000*[2]
	I	253.565	600	
Pb	I	405.783	34000	40*[1]
	I	368.348	14000	
Pd	I	340.458	24000	0.5
	I	360.955	20000	
Pr	II	417.938	5200	4
	II	422.298	3800	
Pt	I	204.937	5500	14
	I	306.471	3200	
Rb	I	780.023	30000	6*[1]
	I	420.185	320	
Re	I	346.046	55000	20
	II	227.525	5200	
Rh	I	369.236	9400	2
	I	352.802	8800	
Ru	I	372.803	11000	1
	I	349.894	8300	
Sb	I	206.833	42000	135
	I	259.805	6000	
Sc	II	361.384	28000	5*[2]
	II	363.075	20000	
Se	I	196.026	90000	
	I	203.985	40000	
Si	I	251.611	3600	m
	I	288.158	2600	
Sm	II	356.827	4200	10
	II	359.260	4200	
Sn	I	197.080	26000	6
	I	283.999	14000	
Sr	II	407.771	46000	1
	II	421.552	32000	
Ta	II	240.063	2400	640
	II	301.254	1800	
Tb	II	350.917	5700	95
	II	370.286	4700	
Te	I	200.202	26000	640
	I	238.578	1500	
Th	II	401.913	4200	45
	II	346.992	1300	
Ti	II	334.941	12000	m
	II	336.121	7200	
Tl	I	351.924	20000	6
	I	535.046	18000	
Tm	II	384.802	8900	6
	II	346.220	8500	
U	II	385.958	4900	430
	II	367.007	2800	
V	I	437.924	12000	3
	II	309.311	3800	
W	II	202.998	15000	20
	II	200.807	13000	
Y	II	371.030	13000	6
	II	437.494	12000	

Table 6.3 (*Continued*)

Element	Spectrum	Wavelength (nm)	Sensitivity (NBS units)	Detection limit (ppm)
Yb	II	369.419	32000	5*[2]
	II	398.799	32000	
Zn	I	213.856	10000	24*[1]
	I	334.502	1400	
Zr	II	339.198	5700	6
	II	257.139	2100	

Emission line and intensity data are taken from the NBS tables (Meggers *et al.*, 1975), as listed by Boumans and Bosveld (1979), relevant for the copper arc. In general the two most intense lines are listed, although when both these lines occur in the same spectral region, an alternative line has been listed.
I denotes atom emission lines; II denotes ion emission lines.
Detection limit data has been evaluated from the work of Walthall *et al.* (1976) and is estimated to represent the six-sigma limit of determination in whole rock for samples excited by arc source using photographic detection. Additional data (*) are taken from Holmes (1982)(*[1]) or Myers *et al.* (1975) (*[2]).
m = element normally present as a major element.

spectrometers being optimized for detection in the range 230 to 350 nm) and an adequate evaluation of spectral overlap effects. Detection limits are also affected by plasma conditions and the sensitivity of the detector available to measure the intensity of diffracted emission lines. Data in Table 6.3 were obtained by photo-plate detection.

In order to achieve the sensitivities indicated in Table 6.3, it is necessary to excite samples with a dc arc discharge. In view of the uncertain composition in rock reference materials of some trace elements for which determinations may be required, it is usually necessary to calibrate instruments against synthetic standards prepared from high-purity analytical reagents. Concern then arises over the influence and magnitude of matrix interference effects and the degree to which they can be overcome using a spectroscopic buffer, as discussed in section 6.4.

Watson and Russell (1978) investigated the effect of variation in major element composition on the excitation of trace elements by dc arc. The compositions investigated are listed in Table 6.4. The influence of changing matrix compositions on emission intensities of lines from Si, Al, Ca, Mg, Hg, Mn, Mo, Cu, Zn, Ni, Co, Pb, Cr, V, Sn and Zr (plus Sc and Ru internal standard elements) was investigated. Over the range of compositions examined, changes in major element concentrations affected all elements by less than 10% (except lead). However, large concentrations of iron, sodium and potassium caused some enhancement of all lines. As a result of these tests, it was considered satisfactory to prepare a matrix of intermediate major-element composition (Table 6.4) and to spike aliquots of this sample with high-purity oxides of the trace elements to be determined to produce a range of calibration standards containing 1000, 500, 300, 100, 30, 10 and 3 ppm of each of the elements.

The analysis procedure was then as follows. Samples were first ignited to remove moisture and CO_2 that might cause eruption of the sample during arcing. The highest temperature at which samples could reliably be ignited without risking significant losses of volatile elements such as zinc was found to be 750°C. Ignited samples were then blended with a spectroscopic buffer containing 80% graphite and 20% LiF in the ratio one part of sample to three parts of buffer. The buffer contained 0.03% GeO_2 as internal standard. Calibration procedures were investigated both by compressing samples as briquettes 4 mm in diameter and containing 50 mg of sample or by tamping loose powder into a cup-shaped electrode as described in section 6.3. Pellets gave the greater arc stability and improved analytical reproducibility, although sample preparation was more inconvenient and contamination from pelletting equipment was detected.

Table 6.4 Compositions investigated for major-element interference effects in arc source–optical emission spectrometry (units: wt %)

	Range of compositions investigated for matrix effects	Selected composition of an intermediate rock matrix
SiO_2	42 to 72	60
TiO_2	0.3 to 10.5	0.5
Al_2O_3	7 to 28	10
Fe_2O_3	7 to 28	7.5
MgO	4 to 34	6
$CaCO_3$	7 to 28	10
Na_2CO_3	2 to 13	3
K_2CO_3	2 to 13	3

Data from Watson and Russell (1978).

Samples were excited by a dc arc of 12 A using emission integration times of 90 s. Variations in these conditions of ± 1 A or ± 5 s did not affect results. Background emission intensities were corrected by measurements at the mercury channel, an element that was expected to be volatilized completely from samples during ignition. The results of some elements that had emission lines in the vicinity of mercury, but not others, were improved, and for general analysis, background correction in this manner was discontinued.

Analytical results for trace elements and selected major elements in rock reference materials determined by this technique are listed in Table 6.5. These data are compared with similar analyses taken from the work of Walthall *et al.* (1976), Brenner and Harel (1976) and Galan (1974). These data indicate that arc source spectrochemical determinations of trace elements often give acceptable, though sometimes variable, agreement with recommended values. Coefficients of variation are often in the range 8–20%. Some difficulty is encountered in determining elements such as Rb, Cs, Li, K and Na since the most sensitive lines of these elements lie above 550 nm, outside the sensitive range of detection of many emission source spectrometers. Similarly, difficulties can be encountered in the routine determination of As, Au, Hf, Os, P, Sb, Se, Sn, Te, W since the most sensitive lines of these elements lie below the range 210 to 200 nm, where absorption in air and quartz glass optical components (below 190 nm) becomes significant.

6.5.4 *Photoplate detection*

In modern instrumentation, detection of emission line signals is largely electronic, using photomultiplier tubes. How-

Table 6.5 Simultaneous multi-element analysis of selected reference materials by dc arc–excited optical emission spectroscopy

	Granodiorite USGS:GSP-1					Basalt USGS BCR-1					Shale USGS: SCo-1			
	Abbey (1983) Expected	Walthall *et al.* (1976) Analysed	CV $n = 6$	Galan (1974) Analysed	Watson and Russell (1978) Analysed	Abbey (1983) Expected	Walthall *et al.* (1976) Analysed	CV $n = 6$	Galan (1974) Analysed	Watson and Russell (1978) Analysed	Abbey (1983) Expected	Brenner and Harel (1976) Analysed	CV $n = 12$	
SiO_2	67.32			>40.36		54.53	54.3	7.8	>40.36					SiO_2
TiO_2	0.66	0.64	9.4	—		2.26	2.30	14			0.62	0.752	5.7	TiO_2
Al_2O_3	15.28	16.8	15	11.80		13.72	14.9	17	15.20					Al_2O_3
Fe_2O_3	4.30	5.98	15	3.45		13.41	15.3	3.4	16.45					Fe_2O_3
MnO	0.04	0.060	15	0.037	0.045	0.18	0.26	4.2	0.165		0.05	0.053	8.8	MnO
MgO	0.97	1.50	6.5	0.83		3.48	6.93	11	3.31					MgO
CaO	2.03	3.41	15	2.32		6.97	7.33	14	6.38					CaO
Na_2O	2.81	>0.43	—	2.74		3.30	>0.32		3.32					Na_2O
K_2O	5.51	7.07	2.5	4.35		1.70	>1.0		2.00					K_2O
P_2O_5	0.28	0.34	13	—		0.36	0.56	9.3						P_2O_5
Ag	0.083?	<0.1		0		0.034?	0.2	28	0					Ag
B	—	<10				4	<10				66?	64.4	11	B
Ba	1300	1165	6.5			680	571	7.1			590	744	8.3	Ba
Be	1?	1.3	8.5			1.6?	1.66	9.8						Be
Bi	0.037	<1.0				0.047?	<1.0							Bi
Cd				0					0					Cd
Co	7.8	6.61	7.3		<10	36	41.6	5.9		34	11	11.3	24	Co
Cr	12	11.5	18		<10	15	12.4	17		<10	71	65.2	13	Cr
Cu	33	<46		30	34	16	<46		20	16	28	30.6	12	Cu
Ga	23	25.4	7.3			22	22.9	10.7			14?	13.9	16	Ga
Li	30	—	—			14	—							Li
Mo	1.5?	1.43	26	0		1.5?	<2.2		0					Mo
Nb	23?	20.5	16			19?	12.2	39						Nb
Ni	9	7.5	8.5		5	10	11.7	10.8		12	30	29.8	9.3	Ni
Pb	54	68	4.7	45	64	14	17.7	14	24	40	28?	27.6	7.0	Pb
Rb	250	—	—			47	—							Rb
Sc	6.6	6.9	12			33	>68							Sc
Sn	5?	9.5	7.5	2		2.5	<3.2		0					Sn
Sr	240	318	10			330	425	6.0			170	223.6	8.9	Sr
Th	105	142	12			6.1	<22							Th
V	54	73	8.1		70	420	>215			400	135?	108.8	6.5	V
Y	29	31.4	13			40	39.8	9.1						Y
Zn	105	89	9.2	100	130	125	160.3	8.1	160	290				Zn
Zr	500	352	16		480	185	208.3	22		185	135	132.5	10.4	Zr
La	195	285	14			27	43.4	15						La
Ce	360	684	16			53	128	6.5						Ce
Pr	(50)	40	16			7?	<15							Pr
Nd	190?	195	22			26?	36.2	45						Nd
Sm	25?	27.9	16			6.5	<4.6							Sm
Eu	2.4?	2.38	14			2.0	2.16	7.6						Eu
Gd	(15)	16.4	11			6.6	9.70	20						Gd
Dy	5.7?	<3.2				7?	<3.2							Dy

Ho	—	—		1.2?	<1.0		Ho
Er	(3)	—		3.5?	<2.2		Er
Yb	1.9	2.26	9.3	3.4	5.70	15	Yb

Major-element oxide concentrations in weight % oxide.
Trace-element concentration in ppm element.
CV = coefficient of variation = 100 × standard deviation/mean analysis (%).
Analysis conditions
Walthall et al. (1976): Sample mixed with 2 parts of graphite and excited at 12 to 13 A dc (300 V) using a 5 to 6 mm gap in an argon–30% oxygen atmosphere. Spectra integrated for 120 s using photoplate detection. Semi-quantitative calibration against synthetic standards prepared from quartz to which selected major element oxides were added.
Galan (1974): Sample mixed with graphite and lithium borate in the proportions 20 : 30 : 50 and excited at 8 A dc. Spectra integrated for 40 s after a 5 s pre-burn and detected on a 28 channel direct reader polychromator. Calibration against synthetic standards with major-element matrix matching using palladium as reference element.
Watson and Russell (1978): Sample mixed in the ratio 1 : 3 with a spectroscopic buffer (80% graphite + 20% lithium fluoride) and excited at 12 A dc in an atmosphere of argon–10% oxygen with an outer argon sheath. Spectra were integrated for 90 s and detected on a 24 channel direct reader polychromator. Calibration was against synthetic standards prepared to match an intermediate rock composition using germanium as reference element.
Brenner and Harel (1976): Samples mixed in the ratio 1 : 5 with graphite and excited at 14 A dc using a 6 mm arc gap. Spectra were integrated for 50 s using photoplate detection. Calibration standards were prepared from silica using palladium as internal standard.

ever, on some present-day instruments, and almost all early ones, photographic plates were used for measuring the intensity of emission lines. It is then necessary to measure the optical absorbance of a large number of lines on a photographic emulsion to permit quantification of results. The general procedures for carrying out this process are discussed in the context of spark source mass spectrometry in section 19.5. Helz *et al.* (1969) described a procedure for the computer analysis of data obtained from a digital microphotometer used to measure transmission intensities of optical emission lines recorded on a photographic plate. Other workers, including van Hoyle *et al.* (1975), Vanderborght and van Grieken (1978) and Stüwer (1976) have considered various aspects of the automated evaluation of photoplate data for spark source mass spectrometry, aspects of which are also relevant to optical emission spectrometry.

6.5.5 *Semi-quantitative reconnaissance analysis*

One of the great advantages of photoplate detection is the possibility of rapid and systematic semi-quantitative analysis in which the presence of unsuspected elements can be identified. Myers *et al.* (1961) considered this aspect of dc arc optical emission spectrometry for application to rapid reconnaissance studies. Powdered samples (10 mg) were mixed with graphite (20 mg) and transferred to the cavity of a crater electrode. Samples were then excited using a dc arc of 12 to 13 A at 300 V generated across an analytical gap of 5 to 6 mm. Emission spectra (either 225–475 or 605–855 nm in the first order) were recorded on photoplates for arc times of 120 s (the time taken to burn most samples to completion). Processed plates were calibrated against pre-recorded standard plates using an optical comparator by visually identifying the two reference spectra that bracketed the emission intensity of the unknown sample. Semi-quantitative results were reported by the convention shown in Table 6.6, the quoted concentration indicating that the analysis lay within

Table 6.6 Concentration ranges for reporting semi-quantitative arc emission spectrographic data based on one-third order of magnitude

Limits of concentration defined by standards (%)	Reported concentration (%)
10.0–4.6	7
4.6–2.2	3
2.2–1.0	1.5
1.0–0.46	0.7
0.46–0.22	0.3
0.22–0.10	0.15
ppm	ppm
1000–460	700
460–220	300
220–100	150
100–46	70
46–22	30
22–10	15
10–4.6	7
4.6–2.2	3
2.2–1.0	1.5

Data from Myers *et al.* (1961).

the listed concentration range. Detection limits that could be achieved by this visual comparator technique are listed in Table 6.7.

Table 6.7 Approximate visual detection limit for semi-quantitative arc source–optical emission spectrography involving visual comparison of photoplates.

	ppm		ppm		ppm		ppm
Si	20	Cd	50	La	20	Sc	5
Al	10	Ce	200	Li	200	Sn	10
Fe	8	Co	5	Lu	100	Sr	2
Mg	5	Cr	1	Mo	5	Sm	100
Ca	50	Cs	20 000	Nb	10	Ta	200
Na	500	Cu	1	Nd	100	Tb	1000
K	7000	Dy	50	Ni	3	Te	1000
Ti	2	Er	50	Os	100	Th	200
P	2000	Eu	500	Pb	10	Tl	100
Mn	2	Ga	2	Pd	3	Tm	100
Ag	1	Gd	50	Pr	500	U	500
AS	1000	Ge	10	Pt	30	V	10
Au	20	Hf	100	Rb	100 000	W	100
B	20	Hg	10 000	Re	50	Y	10
Ba	2	Ho	100	Rh	50	Yb	5
Be	1	In	10	Ru	100	Zn	200
Bi	10	Ir	100	Sb	100	Zr	10

Data from Myers *et al* (1961).
Detection limits of the following elements can be reduced to the indicated values by a second exposure involving 20 mg of sample: Na: 5, K: 20, As: 100, Cs: 200, Hg: 20, Li: 0.6, Rb: 60 and Te: 100 ppm.

6.6 Conclusions

Arc/spark emission spectrometry as applied to geological samples is one of the first instrumental procedures that has run the whole gamut of analytical development. The technique has passed through three stages of use:
(i) A new experimental technique that was largely unproven in practical applications
(ii) The mainstay of some geochemical laboratories, being the only instrumental technique then available capable of rapid routine determinations of a range of trace elements
(iii) A technique that has largely been abandoned as far as the mainstream of instrumental analysis is concerned, superseded by the much more predictable excitation properties of the inductively coupled argon plasma source.

The legacy of spectrographic analysis using an arc/spark emission source can be summarized as follows:
(i) The realization of the enormous analytical potential of instrumental techniques capable of rapid simultaneous determinations of trace elements down to the ppm level
(ii) A systematic study of atomic and ionic emission lines subsequently of value in ICP–AES
(iii) Instrumental developments in terms of multi-channel direct reader spectrometers which have formed a firm basis for the first ICP–optical emission instruments.

Although no longer in the mainstream of trace element analysis, arc/spark emission spectrometry retains a direct application in geochemical survey and reconnaissance studies where trace element data of the highest precision is not a priority.

7 Ion-selective electrodes

7.1 Analytical perspective

The ion-selective electrode represents a technique in electrochemistry in which the simple pH meter has been adapted to element-specific analytical applications. In a conventional pH meter, an electrode measures changes in the hydrogen ion concentration of a solution to determine its acidity/alkalinity. The ion-selective electrode carries out a similar function, except that the electrode is designed to respond to changes in the activity of specific ions in solution. Such electrodes are now available for a wide range of ions, including Na^+, K^+, Ca^{2+}, NO_3^-, Cl^-, F^-, S^{2-}, etc. In geochemical analysis, the outstanding application is in the determination of fluoride ions. Indeed, due to the lack of any simple alternative, ion-selective electrodes now represent the standard instrumental technique for determining this element. They are, therefore, widely used, not least because of the importance of fluorine measurements in geochemical exploration and mineralization studies.

The only other element that has been extensively analysed by ion-selective electrodes in silicate samples is chlorine. However, the sensitivity and selectivity of specific electrodes towards chlorine is not as high as for fluorine, leading to some analytical difficulties in routine applications.

A typical analytical procedure involves a simple fusion of the sample, normally with a sodium carbonate flux. After cooling, the melt is taken into solution. Suitable buffer solutions are then added and the fluoride activity is measured from the electrochemical potential created by a fluoride-specific electrode relative to a standard electrochemical cell (the silver–silver chloride electrode). With corrections, the reading is directly related to the free fluoride content of the sample solution. The technique benefits from the very low capital cost of instrumentation, simple sample preparation and a relatively rapid rate of sample throughput.

7.2 Instrumentation

7.2.1 *Circuit description*

The instrumentation required for ion-selective electrode determination is shown in Figure 7.1. Two electrodes are immersed in the solution to be measured; one is the ion-selective electrode, the other a standard reference electrode. Each is connected to a pH (or rather a pIon) meter capable of measuring very small changes in electrical potential between the two. The circuit is completed by the electrolytic behaviour of the sample solution itself. Under equilibrium conditions, extremely small currents pass through this circuit. The direction in which this current passes depends of the choice of electrodes and their characteristic standard potential. In order for the technique to succeed, changes in potential must be caused only by the ion-selective electrode, and the response must relate directly to changes in the activity of the specific element to be measured.

7.2.2 *Activity*

The activity a of an ion is determined from

$$a = \gamma[\text{CONC}] \tag{7.1}$$

where CONC is the element concentration, and γ is the activity coefficient.

As the dilution of a solution increases, γ tends to unity and activity and concentration become identical. However, in more concentrated solutions, γ becomes increasingly smaller than unity. The concept of activity is therefore required to model the fact that in more concentrated solutions, the physical and chemical behaviour of ions is no longer linearly related to concentration; the proportionality factor for this non-ideal behaviour is the activity coefficient.

In the following sections, the components of the ion-selective electrode circuit will be considered individually before discussing analytical applications of this technique.

7.2.3 *The fluoride ion-selective electrode*

The fluoride ion-selective electrode (Figure 7.2) became commercially available in 1966 (Frant and Ross, 1966). It con-

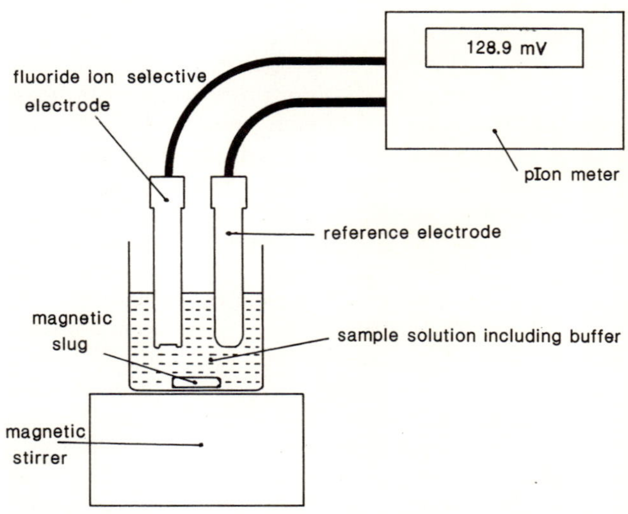

Figure 7.1 Schematic arrangement of instrumentation used for the determination of fluorine using a fluoride ion-selective electrode.

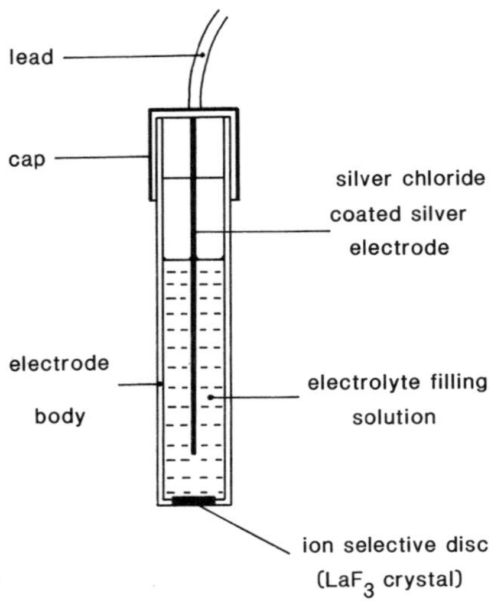

Figure 7.2 The fluoride ion-selective electrode showing the schematic arrangements of components.

sists of an outer tubular envelope containing an electrolyte solution in which is immersed a metal electrode contact. Plastics are normally used for the outer envelope since plastic material is chemically inert, resists the surface absorption of ions from sample solutions and is a good electrical insulator. The lower end of this tube is closed by an ion-selective membrane. This membrane represents the diffusion contact between electrolyte solution within the electrode and the sample solution. This membrane is selected so that it is impervious to all ions except the one of interest. In the case of the fluoride ion-selective electrode, the electrode is made of silver wire coated with silver chloride; the electrolyte is a chloride solution that also contains a low concentration of fluoride ions (at about 0.001M) and the membrane consists of a single crystal of lanthanum fluoride doped with europium to improve its electrical conductivity. The lanthanum fluoride matrix is essentially insoluble in aqueous solution. However, fluoride ions in this crystal are relatively mobile. When the electrode is immersed in a sample solution, fluoride ions can effectively diffuse through channels in the lanthanum fluoride crystal. These channels are too small to permit any other ion to pass through (except for the hydroxyl ion, OH^-; see section 7.2.4). Thus the electrode can respond only to changes in fluoride ion activity in sample solutions.

If the solution to be analysed contains a higher concentration of F^- ions than the electrode filling solution, F^- ions will rapidly migrate through the crystal membrane, leaving a surplus of positively charged counter-ions on the other side. Equilibrium will rapidly be established at the point where the rate of diffusion of F^- ions into the electrode filling solution is identical to the rate of diffusion in the reverse direction. The actual number of ions involved in this transfer is so small that the effect on analyte concentration in the sample solution is infinitesimal. At equilibrium a potential difference is created across the ion-selective membrane. This potential difference is proportional to the activity of fluoride in the sample solution and must be measured relative to a standard reference electrode.

7.2.4 *Interference effects: the influence of OH^-*

Hydroxyl ions (OH^-) are almost exactly the same size as F^- ions and can therefore also migrate through the ion-selective membrane. The response of the fluoride ion-selective electrode will therefore be affected by changes in the pH of sample solutions. In order to minimize this interference, the concentration of OH^- must be maintained at less than 10% of the F^- concentration. It is a simple matter to do this by making all sample solutions sufficiently acidic. Unfortunately, if solution pH is reduced below pH 5, the following reactions occur, reducing the free F^- ion concentration in solution:

$$H^+ + F^- \rightleftharpoons HF \tag{7.2}$$

$$HF + F^- \rightleftharpoons HF_2^- \tag{7.3}$$

Since neither HF nor HF_2^- can diffuse through the ion-selective membrane, measured activities of fluoride will be systematically low under these pH conditions. In reviewing the problem, Nicholson (1982) recommended that all measurements should be made under compromise conditions at pH 5.5.

One important advantage of the fluoride ion-selective electrode is that it has a selectivity over other common anions (except hydroxyl) of several orders of magnitude. Under ideal conditions, the sensitive operating range is from solutions saturated in fluoride down to 10^{-6}M F^-. The lower limit is set by the very low solubility of the lanthanum fluoride membrane in aqueous solutions.

7.2.5 *Standard reference electrodes*

A standard reference electrode is designed to produce a standard electrical potential, unaffected by changes in the composition of sample solutions, relative to which changes in the potential of the ion-selective electrode may be measured. There are two types of standard reference electrode in widespread use, the traditional calomel electrode and the silver–silver chloride electrode.

(*a*) *The calomel reference electrode*. The design of a calomel electrode is shown in Figure 7.3. It consists of an inner tube containing a paste of mercury metal, mercurous chloride (i.e. calomel) and usually saturated potassium chloride solution. Electrical contact to the pH meter is made by a metal electrode that dips into this paste. Electrical contact to the outer tube which contains saturated potassium chloride solution is made either via a small hole or a porous plug. To complete the circuit to a sample solution, the lower end of the outer tube normally has a porous disc or fibre plug through which ions can migrate. This plug offers a relatively large electrical resistance (2000–3000 Ω; Skoog and West, 1980, p.533) and limits the current-carrying capacity of the electrode. Only very small quantities of electrolyte solution leak through this plug and contaminate the sample.

In electrochemical terms, the electrode is liquid mercury. Mercurous chloride (calomel) has a very small solubility in water, about 0.00020 g per 100 mL at 25°C (Weast, 1973):

$$Hg_2Cl_{2(solid)} \rightleftharpoons Hg_2Cl_{2(solution)} \tag{7.4}$$

In solution, a small fraction of calomel dissociates as fol-

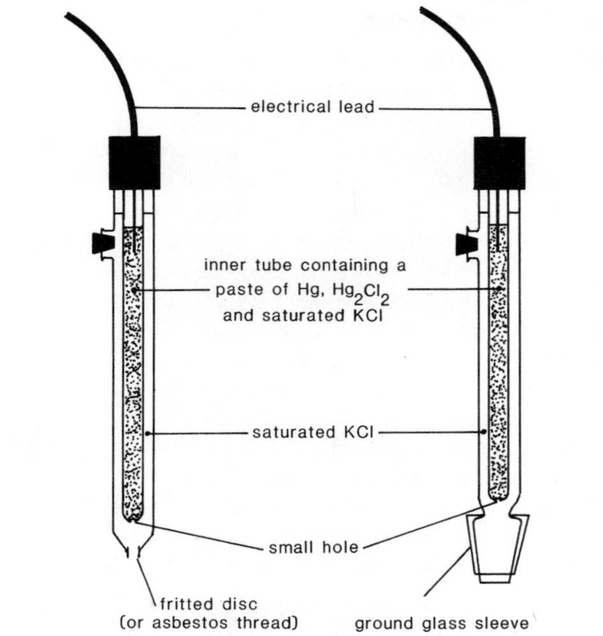

Figure 7.3 Typical designs of calomel reference electrodes in which electrolytic contact with the sample solution is made (*a*) by a porous fibre seal and (*b*) by the KCl solution film formed across a conical ground-glass joint. The latter has a much lower junction resistance, but tends to leak small quantities of KCl solution into the sample.

lows:

$$Hg_2Cl_{2(solution)} \rightleftharpoons Hg_2^{2+} + 2Cl^- \quad (7.5)$$

Mercury(I) ions exist in solution as dimers (Hg_2^{2+}), not as simple singly-charged species. The position of the equilibrium represented by equation (7.5) will therefore depend on the chloride ion concentration, since electrolyte solutions are saturated in Hg_2Cl_2. However, the chloride ion concentration is itself buffered by the presence of saturated KCl solution. The concentration of $Hg_2{}^{2-}$ ions in the calomel electrode is therefore maintained constant at a given temperature. These ions contribute to the electrochemical reaction:

$$Hg_2{}^{2+} + 2e^- \rightleftharpoons 2Hg_{(liquid)} \quad (7.6)$$

The overall electrochemical reaction can be considered to be:

$$Hg_2Cl_{2(solid)} + 2e^- \rightleftharpoons 2Hg_{(liquid)} + 2Cl^- \quad (7.7)$$

Using a saturated KCl buffer, this reaction creates an electrochemical potential of about 0.241V at 25°C independent of the composition of the solution into which the electrode tip is immersed. The precise potential depends on the potential drop (usually about 2 mV) across the porous plug connecting electrode filling solution and sample solution (the liquid junction potential). Calomel electrodes are available with different chloride filling solution concentrations, as listed in Table 7.1. This affects the electrode potential to a small degree. Data in Table 7.1 also indicate changes in potential expected due to changes in ambient temperature. The saturated calomel electrode is more sensitive to temperature than the silver–silver chloride electrode.

Table 7.1 Details of calomel reference eletrodes

Name	Concentration of Hg_2Cl_2	Concentration of KCl	Electrode potential (V) relative to the standard hydrogen electrode
Saturated	Saturated	Saturated	$+0.241-6.6\times10^{-4}\times(t-25)$
Normal	Saturated	1.0*F*	$+0.280-2.8\times10^{-4}\times(t-25)$
Decinormal	Saturated	0.1*F*	$+0.334-8.8\times10^{-5}\times(t-25)$

Data from Skoog and West (1980).
t = temperature in °C; F = molar activity.
Electrode reaction is Hg_2Cl_2 (s) $+ 2e^- \rightleftharpoons 2Hg(l) + 2Cl^-$.

Although widely used as a reference electrode in pH measurements, the calomel electrode has some disadvantages in ion-selective electrode applications. Calomel electrodes are relatively slow to re-equilibrate after changes in temperature and can suffer disequilibrium effects, since Cl^- ions are able to diffuse out of the porous junction more rapidly than K^+ ions. The precision of potential measurements using this device is generally considered to be ±1mV. For more precise work, the silver–silver chloride reference electrode is preferred.

(*b*) *Silver–silver chloride reference electrode*. The silver–silver chloride reference electrode consists of a silver wire, coated with silver chloride and immersed in an electrolyte solution saturated in silver chloride and containing excess chloride ions. As with mercury, silver chloride has a low solubility in aqueous solution, 0.000089 g per 100 mL at 10°C (Weast, 1973). The electrolyte solution is contained in a suitable outer tube connected to the sample solution via a porous plug. The schematic arrangement of a silver reference electrode, configured with a double liquid junction (see below), is shown in Figure 7.4.

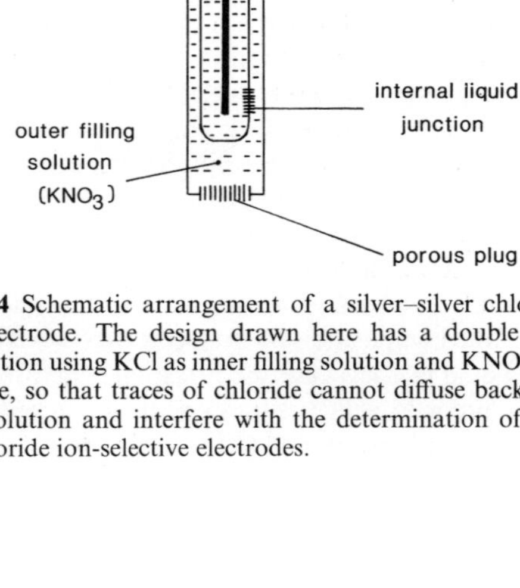

Figure 7.4 Schematic arrangement of a silver–silver chloride reference electrode. The design drawn here has a double junction configuration using KCl as inner filling solution and KNO_3 as outer electrolyte, so that traces of chloride cannot diffuse back into the sample solution and interfere with the determination of that ion using chloride ion-selective electrodes.

The mechanism of the electrode is similar to that described for the calomel electrode. A small concentration of silver ions is maintained in the electrolyte solution and is buffered at constant concentration by both the low solubility of AgCl and the presence of excess chloride ions. Constant electrochemical potential is therefore created by the following series of reactions:

$$AgCl_{(solid)} \rightleftharpoons AgCl_{(solution)} \quad (7.8)$$

$$AgCl_{(solution)} \rightleftharpoons Ag^+ + Cl^- \quad (7.9)$$

$$Ag^+ + e^- \rightleftharpoons Ag_{(solid\text{-}electrode)} \quad (7.10)$$

$$AgCl_{(solid)} + e^- \rightleftharpoons Ag_{(electrode)} + Cl^- \quad (7.11)$$

The electrochemical potential generated by this reaction again depends on the concentration of chloride used in the electrolyte filling solution. Using a saturated solution of potassium chloride, the potential is 0.197 V relative to the standard hydrogen electrode.

In many applications where the highest electrode stability is required, substitution of an *equitransferent* electrolyte filling solution is preferred. This solution contains KCl, NaCl and KNO_3 present at a total ionic strength about ten times higher than that of the solution to be analysed. The composition of the equitransferent solution is formulated such that the rate of diffusion of both positive and negative charged ions through the porous junction is equivalent. In this way disequilibrium effects that create unwanted liquid-junction effects are minimized.

The silver–silver chloride electrode is far less sensitive to variations in temperature than the calomel electrode and under optimum conditions can give stabilities of the order of ± 0.1mV.

Selective-ion electrodes are available for the determination of chloride in sample solution. Some difficulty then arises from the use of a silver–silver chloride reference electrode due to the small amount of potassium chloride electrolyte that can diffuse out into the sample solution, limiting the sensitivity of measurements. For such applications, the silver–silver chloride electrolyte solution is normally buffered with an additional electrolyte comprising 10% potassium nitrate (Figure 7.4). In this way electrical connection is maintained in this double junction reference electrode design without the possiblity of contaminating sample solutions with chloride ions.

7.2.6 *The pIon meter*

The final component in the ion-selective electrode circuit is the pIon meter, used to measure small changes in the electrical potential between reference and selective ion electrode. The pIon meter can have an analogue or digital readout which must be resolvable to at least 0.1mV. The input circuits must be designed to have a large impedance to prevent a significant current being drawn from the electrodes. If this did occur, the composition of the electrolyte would change and this would in itself alter the potential generated by the circuit. In order to achieve this performance, meters must have an internal resistance of 10^{12} Ω or more.

7.3 The Nernst equation

The electrical potential (E) created across a pair of electrodes is given by the Nernst equation:

$$E = E_0 + \frac{2.303\,RT}{nF}\log(a) \quad (7.12)$$

where E_0 is the standard electrode potential and is a constant at a particular temperature—the magnitude of E_0 depends on a choice of ion selective and reference electrode, R is the gas constant = 8.314 J K^{-1} mol^{-1}, T is the temperature (K), F is the Faraday constant = 9.648×10^4 C mol^{-1}, n is the charge of the ion, including its sign ($n = -1$ for F^-), and a is the activity of the ion that is causing the potential change.

This equation is in fact a simplification, in which it is assumed that the activity of all other species contributing to the electrochemical reaction is unity.

If all measurements are made at 25°C, then the term $2.303\,RT/F$ is a constant equivalent to 59.16 mV. The Nernst equation for the determination of fluoride then becomes

$$E = E_0 - 59.16 \log a_{F^-} \quad (7.13)$$

where E and E_0 are measured in millivolts.

A plot of the electrode potential difference versus log of the fluoride ion activity is, therefore, a straight line with an ideal gradient of -59.16 mV mol^{-1}. A typical calibration, taken from the work of Troll *et al.* (1977), is shown in Figure 7.5. This demonstrates the linear response obtained at higher activity levels. Nicholson (1983) estimated that activities of F^- can be measured from about 10^{-6} M (0.019 mg F^- L^{-1}) up to the point of saturation of a fluoride solution. However, particular difficulty occurs at low ion activities where the linear response breaks down. As fluoride ion activity is reduced further, the electrode potential measurements tend

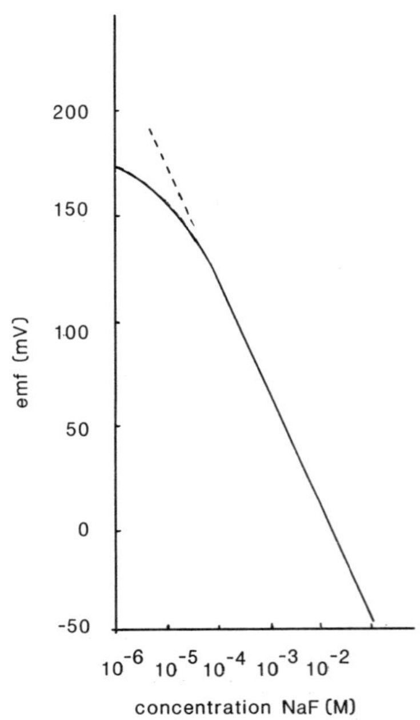

Figure 7.5 Calibration curve for NaF measured with a fluoride ion-selective electrode. (Troll *et al.*, 1977.)

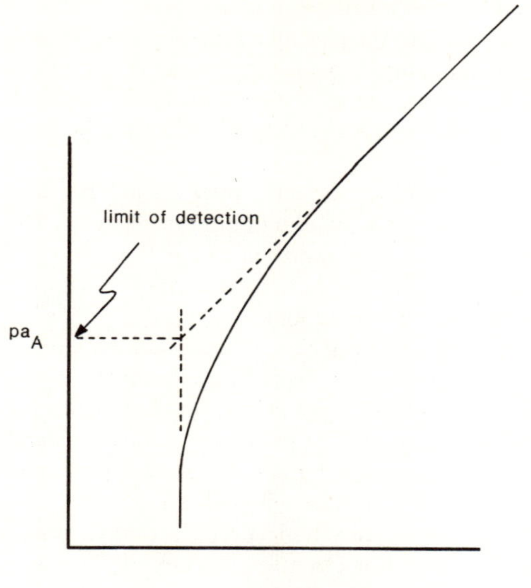

Figure 7.6 A definition of detection limit for use with ion-selective electrode calibrations as recommended by IUPAC (1976). (Nicholson, 1983.)

to approach asymptotically a constant value. Normal criteria cannot then be used to define the detection limit of measurements. To overcome this difficulty, the International Union of Pure and Applied Chemistry (IUPAC, 1976) recommended that a practical limit of detection should be taken as the activity (concentration) of analyte equivalent to the point at which the two linear segments of the calibration curve intersect (Figure 7.6).

7.4 Interference effects: non-ideal Nernst behaviour

Two types of interferences can influence the determination of fluoride using ion-selective electrodes: chemical and instrumental effects. In a review of the application of this technique to silicate rock analysis, Nicholson (1983) has evaluated these effects in detail.

7.4.1 *Chemical interferences*

Chemical interferences may be related to either ionic strength or to complex formation.

(*a*) *Ionic strength.* As outlined above, the Nernst equation predicts that changes in the potential difference between selective ion and reference electrode are linearly related to the log (activity) of the analyte (F^-). However, in relating measured activity to concentration (the desired measure) account must be taken of the effect of changes in ionic strength of sample solution on fluoride ion activity.

Ionic strength is calculated from

$$I = \frac{1}{2} \sum_{i=1}^{n} m_i z_i^2 \tag{7.14}$$

where m_i is the concentration of the ith ion, and z_i is its charge. The summation is carried out over all ions present.

Thus, errors in determinations will occur if the electrode response from a multi-element sample solution is compared with that from a standard fluoride solution unless steps are taken to match the respective ionic strengths. It is therefore important to buffer all samples with a standard amount of a solution of high ionic strength to swamp any differences. The formulation of the buffer solution depends on the second phenomenon.

(*b*) *Complex formation.* Certain elements in a sample solution may react with fluoride ions to form stable complexes that effectively remove fluoride from analyte solution. Unless this effect is suppressed, systematically low measurements would result. Of particular concern in geological applications is the presence of both iron and aluminium. If aluminium concentrations are sufficiently high, Nicholson and Duff (1981) suggested that it may be desirable to remove this element completely from sample solutions prior to ion-selective electrode measurements. In any case, it is necessary to add a complexing agent to sample solutions, which will selectively react with aluminium and iron, preventing the formation of stable fluoride species. There is some controversy in the literature concerning the most effective masking reagent (see for example Nicholson, 1982; Boyle, 1982). However, in silicate rock applications, Nicholson (1983) recommended a solution containing tri-ammonium citrate and CDTA (trans-1-,2-diaminocyclohexane-NNN′N′-tetra-acetic acid) (Figure 7.7). Taking into account the necessity to buffer the effect of differences in both pH and ionic strength as well as complex formation, Nicholson (1983) recommended the following composition for a *total ionic strength adjustment buffer* (TISAB):

NaCl:	58 g
glacial acetic acid:	57 mL
CDTA:	4 g
tri-ammonium citrate:	24 g

per litre of solution with a pH adjusted to 5.5 using 10M NaOH solution.

7.4.2 *Instrumental interferences*

Instrumental interference effects are mainly associated with disequilibrium phenomena that cause delay before a stable electrochemical potential measurement can be made. These interferences have been reviewed by Nicholson (1983). The most serious are caused by changes in temperature or the intensity of ambient illumination. Memory effects can also be observed in which the elapsed time before a stable

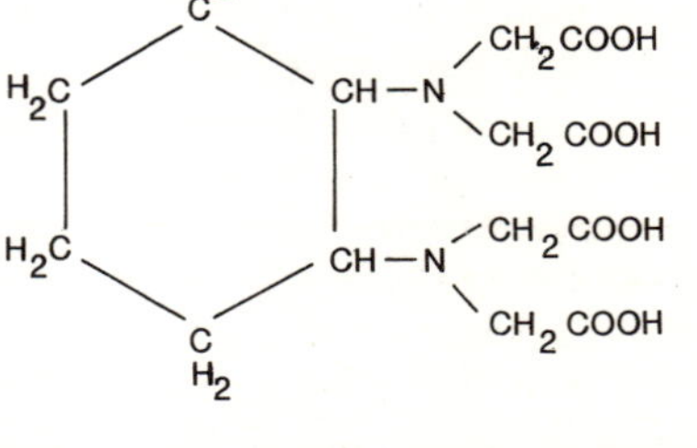

Figure 7.7 The structure of CDTA (i.e. 1,2-diaminocyclohexane-NNN′N′-tetra-acetic acid).

response is observed depends on the concentration of fluoride in the previous sample. This effect can be minimized if the last sample was of lower concentration than the present one. It is, therefore, desirable to analyse samples in order of increasing fluoride content. Response failure can occur due to loss of wet contact between the lanthanum fluoride sensing membrane and internal electrolyte solution due to spillage or evaporation of this solution. Bixler and Solomon (1984) have described a way restoring of electrode performance based on the creation of a solid-state internal $Ag/AgF/LaF_3$ interfacial structure which replaces the internal reference system.

7.5 Schemes for the analysis of geological samples for fluorine

Several schemes of analysis for fluorine by ion-selective electrodes have been described, for example,by Troll *et al.* (1977), Hopkins (1977), Olade (1976), Jagner and Pavlova (1972), Ingram (1970) and Bodkin (1977).

A typical scheme (after Troll *et al.*, 1977) is to fuse 0.05 to 1.0 g of rock powder (the precise weight depends on fluoride content) with 4 g of $Na_2CO_3+K_2CO_3$ flux in a platinum crucible. The crucible is placed in a furnace that has been pre-heated to 500°C. The temperature is then raised to 950°C and after one hour is further increased to 1000°C for half an hour. After cooling, the melt is dissolved in dilute hydrochloric acid and made up to a final volume of 250 mL. An aliquot of this solution (5 mL) is then mixed with 50 mL of buffer solution and left for one hour to ensure complete complexation of iron and aluminium. Calibration is achieved by standard additions; that is the change in potential resulting from the addition of a known volume of standard fluoride solution to the sample solution is used to determine the fluoride content of the sample (Figure 7.8).

Troll *et al.* (1977) investigated interference effects and optimized experimental parameters. They reported that careful attention to the following details was required if accuracies of better than 4% relative were required.

(i) Fusion: the sample should be carefully mixed with flux and loaded into the platinum crucible so that the minimum amount of sample touches the walls of the crucible. The sample mixture should be covered by an additional layer of carbonate and sintered by slow heating before the melt appears.

(ii) Dissolution: the cold melt should be dissolved at room temperature avoiding addition of excess HCl. After adding the conditioning (buffer) solution, sample aliquots for analysis must be left for one hour to complete complexation of interfering ions.

(iii) Analysis: electrolyte filling solutions should be similar in composition to that of measuring solutions. Sample solutions should be analysed in non-glass beakers with temperatures constant. The minimum volume of standard solution that causes a potential change of approximately −30 mV should be added for self-calibration.

Variations on this basic procedure have been used by other workers. For example: Bodkin (1977) fused samples (200 mg) with lithium metaborate (800 mg) in pre-ignited graphite crucibles at 1050°C for ten minutes. The melt was poured directly into dilute nitric acid and made up to 100 mL. Preliminary experiments in which the sample fusion was carried out using a gas burner gave low fluoride recoveries. It was suggested that this may have been caused by losses due to volatilization. This effect was not observed for fusions carried out in a muffle furnace using either graphite or platinum crucibles and temperatures of up to 1050°C for 25 minutes. Olade (1976) used a flux consisting of two parts of anhydrous Na_2CO_3 plus one part of KNO_3 and determined both chloride and fluoride by ion-selective electrodes. Hopkins (1977) compared a $(Na,K)_2CO_3$ fusion method with one using (i) NaOH alone and (ii) Na_2CO_3–KNO_3 flux mixtures. Results in the analysis of fluorspar and opal glass are listed in Table 7.2. These data show that only the $(Na,K)_2CO_3$ and NaOH fusions decomposed these high-fluoride samples completely. Comparison of analyses of geochemical exploration samples using carbonate and hydroxide fusions indicated that the $(Na,K)_2CO_3$ fusion procedure gave more precise results at lower abundance levels. Jagner and Pavlova (1977) fused samples with sodium carbonate and zinc oxide in the ratio 1 : 5 : 1 and investigated whether losses of fluorine

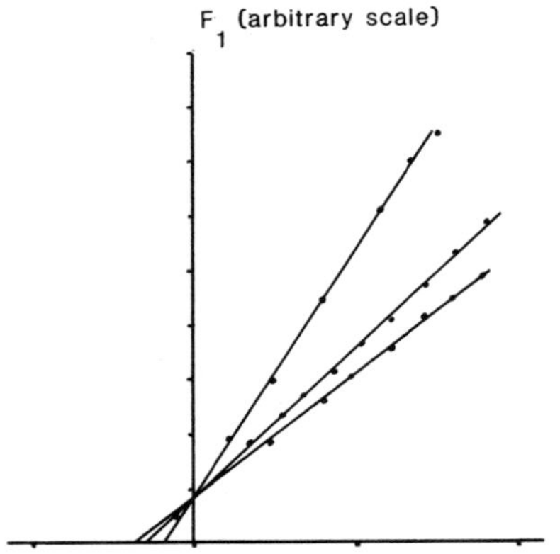

Figure 7.8 Standard additions determination of fluoride in 100, 150 and 200 mg aliquots of a granite reference sample by titration against v mL of standard 20μg/mL sodium fluoride solution. (Jagner and Pavlova, 1972.)

Table 7.2 Comparison of fluorine determinations using different fusion mixtures

Sample	Certificated composition	Na_2CO_3/ K_2CO_3 800°C/ 15 min	NaOH 620°C/ 15–20 min	Na_2CO_3/ KNO_3
	expected	analysed	analysed	analysed
NBS 91 opal glass	5.72	5.6	5.7	3.3
NBS 79 fluorspar	46.3*	45.0	48.0	23.0

Data from Hopkins (1977).
Concentrations in weight %; certified compositions are NBS values, except * which is calculated as F in CaF_2.
Temperature and time of fusion are indicated, where known.

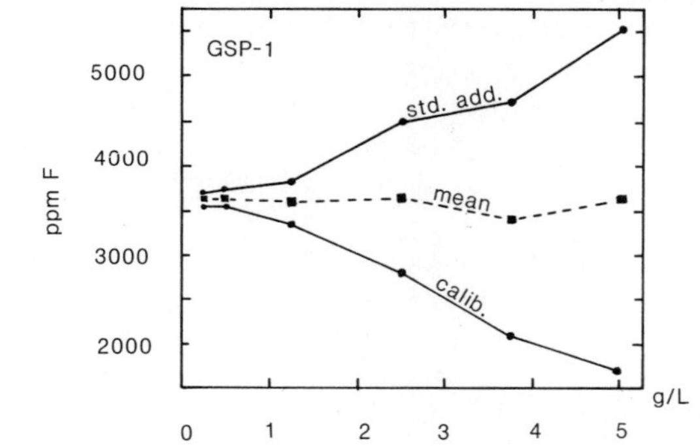

Figure 7.9 Fluorine analysis obtained from different ratios of sample weight to total dissolved volume by ion-selective electrodes using (*a*) standard additions and (*b*) comparison with standard fluoride solutions as calibration procedures. (Stecher, 1983.)

occurred whilst sintering samples at 900°C for 30 minutes. No losses could be detected.

Stecher (1983) determined fluorine using a conventional carbonate fusion following the general procedures of Troll *et al.* (1977) and compared the results by calibrating (i) by standard additions in which sufficient fluoride was added to sample solutions to induce a 30 mV difference in electrode potential and (ii) relative to independent 10 ppm fluoride calibration solutions. Results indicated that data by standard additions were systematically high, whereas data by standard solution calibration were systematically low by approximately the same proportion. This effect appeared to be a function of sample dilution factor, as shown in Figure 7.9. At sample concentrations of less than 0.5 g/L, these effects were small but still significant. However, large discrepancies were observed at high concentrations and were correlated by Stecher with matrix interference effects that had not been fully compensated by the total ionic strength buffer. The ionic strength buffer used in this work was NaCl (58 g/L), sodium citrate dihydrate ($Na_3C_6H_5O_7.2H_2O$, 294 g/L) and CDTA (100 mL of 0.5M CDTA solution per litre) made up to pH 6.0. As an expedient, Stecher averaged the results from both calibration procedures as individual discrepancies appeared to cancel out. Precisions of determinations were reported to decrease rapidly below 100 ppm F and the effective detection limit was about 30 ppm (six sigma). Data from Troll *et al.* (1977) indicated that whole rock detection limit by their procedure was about 50 ppm F.

Selected results for the determination of fluorine by ion-selective electrodes are listed in Table 7.3. These data indicate that analytical precisions are generally in the range 3 to 10% relative (one sigma).

7.6 Determination of chlorine by ion-selective electrodes

7.6.1 *The chloride ion-selective electrode*

The chloride ion-selective electrode incorporates a chloride-selective membrane and is similar in design to the fluoride ion-selective electrode (Figure 7.2), although the response reaction is different. The chloride ion-selective electrode consists of a silver-chloride-coated silver wire immersed in an aqueous solution saturated in silver chloride. The ion-specific membrane consists of either a pellet cast from silver chloride (sometimes mixed with silver sulphide) or alternatively a membrane in which finely divided particles of silver chloride (5–10 μm across) are suspended in a silicone rubber matrix. If the electrode is immersed in a solution containing a higher concentration of chloride ions than that in the electrolyte solution, exchange through the membrane causes chloride ions to diffuse into the electrode filling solution, thus changing the electrical potential across the membrane. An equilibrium is rapidly set up which prevents further net ion migration across the membrane. If the analyte solution contains a lower chloride concentration, the diffusion of ions takes place in the opposite direction. In either case, the change in chloride content of the electrolyte solution causes a shift in the equilibrium

$$Ag^+ + Cl^- \rightleftharpoons AgCl_{(solution)} \quad (7.15)$$

since the solubility product expressed in ion activities

$$\text{solubility product} = \{Ag^+\}.\{Cl^-\} \quad (7.16)$$

is effectively constant at 1.56×10^{-10} $(mol/L)^2$ at 25°C. Thus, if the chloride ion concentration of the electrolyte increases (sample solution is higher in Cl^-), the silver ion concentration of the electrolyte is reduced and vice versa. The electrode is therefore a silver–silver chloride electrode that responds to changes in the silver ion content of the electrolyte solution. These changes are directly proportional to the chloride concentration of the sample solution.

7.6.2 *Applications to silicate rock analysis*

The chloride selective-ion electrode is not as generally accepted as a method for the routine analysis of silicate rocks, because the chloride ion-specific electrode does not possess as high a degree of selectivity or sensitivity.

However, Olade (1976) has described the simultaneous determination of both fluorine and chlorine in geochemical exploration samples by ion-specific electrode. The powdered sample (0.25 g) was fused with 1 g of flux (consisting of 2 parts by weight of Na_2CO_3 and 1 part of KNO_3) at 900°C for 20 minutes in a muffle furnace. After cooling, the melt was digested in boiling deionized water, acidified with nitric acid and made up to 100 mL. Chloride was determined on a 10 mL aliquot using a double junction reference electrode (see section 7.2.5(*b*)) with calibration by standard additions. Fluoride was determined on a 10 mL aliquot of the same solution to which 10 mL of buffer solution was added (the buffer contained sodium citrate and potassium nitrate). Fluoride was calibrated relative to standard fluoride solutions, the matrix of which was matched with that of sample solutions in terms of both flux composition and acid strength. Fluorine data obtained by this procedure are listed in Table 7.3. Chlorine data are presented in Table 7.4. Precisions are in the range 1.2 to 9.3% (coefficient of variation). Some discrepancy was observed, particularly in the analysis of andesite USGS: AGV-1, although it must be emphasized that the technique was designed for rapid geochemical exploration applications rather than to obtain research-grade data of the highest accuracy and precision.

In a similar procedure, Haynes and Clark (1972) deter-

Table 7.3 Fluorine determinations in selected reference materials using ion-selective electrodes

		Abbey (1983) Expected	Stecher (1983) Expected	Hopkins (1977) Analysed	CV	Troll *et al* (1977) Analysed	CV	Bodkin (1977) Analysed	CV	Ingram (1970) Analysed	Stecher (1983) Analysed	Troll and Farzaneh (1978) Analysed	CV	Olade (1976) Analysed	CV
Diabase	W-1	250	235					248	7.3	240	206				
Andesite	AGV-1	400	435	546	3.8			430			422	442	3.1	613	5.9
Basalt	BCR-1	500	490	740	16.9	480	3.7	517	4.6		472	477	2.3		
Disthene	DTS-1	0?						20			<15	5	17		
Granite	G-2	1200	1295	1410	7.4			1272	4.2	1480	1363	1283	3.1	1235	2.3
Granodiorite	GSP-1	3700	3600			3800	2.9	3592	2.3	3863	3594	3659	1.1	3000	4.9
Peridotite	PCC-1	0?						20			<15	4	11		
Opal glass	NBS-91	5.72%		5.6	2.9	5.68	2.6			5.72		5.57	0.6		
n						15 to 48		10		1 to 2	1 to 4	5			
Flux				$Na_2CO_3+K_2CO_3$		$Na_2CO_3+K_2CO_3$		$LiBO_2$		Na_2CO_3+ ZnO	Na_2CO_3+ K_2CO_3	$(SiO_2)^*+Fe^*$		$2Na_2CO_3+KNO_3$	
Fusion				800°C for 15 min		950–1000°C for 1.5 hr (total)		1050°C for 10 min		900°C for 30 min	950–1000°C for 1.5 hr	1200°C for 15 min		900°C for 20 min	
Dilution				50 mg to 100 mL		50/1000 mg to 250 mL		200 mg to 100 mL		100 mg to 100 mL	500 mg to 100 mL	500–2000 mg to 250 mL		250 mg to 100 mL	
Buffer				NaCit + Cit		NaCl + NaCit + CDTA		NaCl + NaCit + DCTA		NaCit + KNO_3	NaCl + NaCit + CDTA	Cit		NaCit + KNO_3	

* pyrohydrolysis.
CV = coefficient of determination = 100.standard deviation/mean analysis (%).
Concentrations in ppm rock except NBS-91 (%F rock).
NaCit = sodium citrate; Cit = citric acid; DCTA = CDTA = trans-1-2-diaminocyclohexane-NNN′N′-tetra-acetic acid; *n* = number of analyses.

Table 7.4 Determination of chlorine in selected reference materials using ion-selective electrodes

	Abbey (1983) Expected	Aruscavage and Campbell (1983) Analysed	CV	Troll and Farzaneh (1980) Analysed	Haynes and Clark (1972) Analysed	CV	Olade (1976) Analysed	CV
Andesite AGV-1	185	119	4.2				266	1.2
Basalt BCR-1	58?	66	6.1					
Disthene DTS-1	11?	31	16					
Granite G-2	100	50	6.0		(a)84	13.2	60	6.0
					(b)75	22.4		
Granodiorite GSP-1	340	339	2.4		(a)347	7.1	350	9.3
					(b)351	15.7		
Peridotite PCC-1	80?	80	5.0					
Basalt BHVO-1	94?	102	4.9	119				
Quartz Latite QLO-1	220?	255	5.1	226				
Rhyolite RGM-1	540	530	3.2	492				
Shale SCo-1	51?	64	7.6	363(a)				
				295(b)				
Mica Schist SDC-1	35?	42	6.4	40				

Units of concentration: ppm rock.
CV = coefficient of variation = 100 × standard deviation/mean analysis (%).
Aruscavage and Campbell (1983): chloride oxidized to chlorine in a 'gas diffusion' cell before being reduced with an alkaline solution of sodium sulphite for determination by ion-selective electrode. Data represents mean of a total of 9 determinations on three sample bottles.
Troll and Farzaneh (1980): chlorine released by hydropyrolysis and determined (a) photometrically using ammonium ferric sulphate and mercuric (II) thiocyanate and (b) ion-selective electrodes.
Haynes and Clark (1972): samples (250 mg) fused with a flux (1 g) consisting of two parts of Na_2CO_3 plus one part of KNO_3 or ZnO over a Meker burner for 15 minutes. The cooled melt was dissolved in water, acidified with nitric acid and chloride determined by (a) spectrophotometry using the ferric thiocyanate method and (b) chloride ion-selective electrodes ($Cl^- > 5$ μg/mL) or silver nitrate titration using silver ion-selective electrodes to detect the end point ($Cl^- < 5$ μg/mL). Data represent mean of 4 determinations.
Olade (1976): samples (250 mg) fused in flux (1 g) consisting of a 2:1 sodium carbonate: potassium nitrate at 900°C for 20 min, disintegrated with boiling water, acidified with nitric acid. Chloride was then determined by ion-selective eletrodes. Data represent mean of 4 determinations.

mined chlorine by fusing samples (0.25 g) with 1g of either 2:1 Na_2CO_3/KNO_3 or 2:1 Na_2CO_3/ZnO at 900°C in glass tubes or platinum crucibles for 20 minutes. After cooling, melts were digested in boiling water acidified with nitric acid and made up to a final volume of 100 mL. Chloride ion activity was measured using an ion-selective electrode. If this activity was greater than 5 μg Cl^-/mL (≡2000 ppm whole rock), the technique was calibrated by standard additions by adding 1 mL of 1000 μg/mL standard chloride solution to sample solutions and remeasuring with a chloride ion-selective electrode. If the chloride activity fell below 5 μg Cl^-/mL, the electrode reading fell on the non-linear part of the Nernst response graph and analyses were carried out using a silver ion-selective electrode to detect the end-point of a titration with silver nitrate solution. Detection limit by the direct chloride ion-selective electrode procedure without silver nitrate titration was expected to be 1.8 μg Cl^-/mL (≡720 ppm Cl^- whole rock).

Haynes and Clark (1972) considered interferences from iodide, bromide and thiocyanate, all of which react by precipitation with silver during silver nitrate titration. These were not considered to be serious in rocks of 'normal' silicate composition. Of more concern was the possible presence of sulphide ions in sample solutions, since these can react with silver chloride in the sensing membrane of the chloride ion-selective electrode and form precipitates with silver nitrate. It was important therefore to oxidize sulphide-rich samples before analysis. This could be achieved by adding hydrogen peroxide to solutions during disaggregation of Na_2CO_3/ZnO melts. If a Na_2CO_3/KNO_3 flux was used, no further oxidation was required, although using this latter procedure, a significant drop in precision at low levels of chloride was observed. This occurred mainly because the pyrex glass culture tubes used for Na_2CO_3/KNO_3 fusions gave high chloride blank values of around 1.2 μg Cl^-/mL (≡480 ppm Cl^- whole rock in place of 0.02 μg Cl^-/mL) (≡8 ppm Cl^- whole rock) using a platinum crucible fusion. Data for reference materials are listed in Table 7.4.

Aruscavage and Campbell (1983) adopted a different approach to the determination of chlorine by ion-selective electrodes. Their procedure was based on the principle that in order to achieve uniform accuracy, chlorine must be quantitatively separated from the sample matrix. They designed an elegant solution to this problem—the gas diffusion cell. The powdered sample (200 mg) was weighed into the outer chamber of the diffusion cell (Figure 7.10). A reducing solution (2.5 mL) comprising 16% KOH and 0.8% anhydrous sodium sulphite was added to the central chamber. The sample was then attacked by 3.0 mL of an oxidizing solution containing 5% potassium permanganate in 15%w/v (sulphuric acid plus 4 parts of concentrated 48% HF). The gas diffusion cell was sealed with its lid and left on a gently oscillating table overnight. The HF attacked the rock matrix, releasing chloride ions into solution. These ions were then

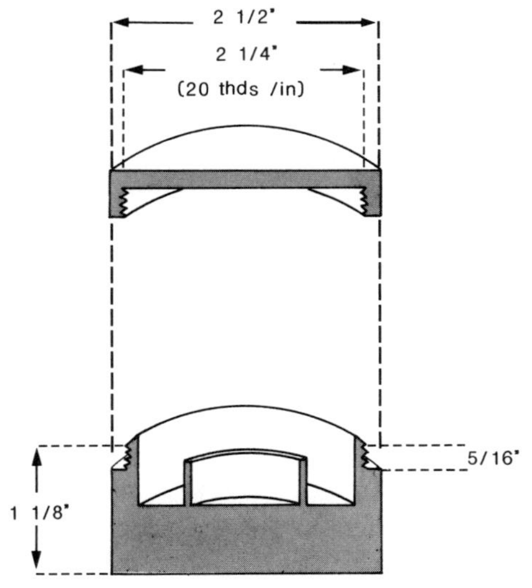

Figure 7.10 Cross-section through a gas diffusion cell used for the chemical separation of chlorine. The sample and an oxidizing solution are placed in the outer chamber, a reducing solution to collect evolved chlorine gas is placed in the inner chamber. (Aruscavage and Campbell, 1983.)

oxidized by permanganate to chlorine gas, which diffused into the central chamber where it reacted with the alkaline solution of sodium sulphite to form chloride again. Radiotracer studies by Aruscavage and Campbell indicated that about 90% of the chlorine was recovered in the reducing solution, 1% remaining in the oxidizing solution and about 10% being lost in the vapour phase. However, this low recovery did not significantly affect analytical results, since the proportion of chloride recovered from all samples remained constant to $\pm 2\%$. The chloride content of the reducing solution was determined directly using a chloride ion-selective electrode. The technique was calibrated by analysing standard solutions of sodium chloride with each batch of samples. Some interferences could occur from the simultaneous release of bromine and iodine although discrepancies were not expected to be important in samples of 'normal' silicate composition. The precision of the method was found to be better than 8%, and detection limits of the order of 10–20 ppm rock were expected in routine analysis. Selected results of Aruscavage and Campbell are listed in Table 7.4.

7.7 Other techniques for the determination of chlorine and fluorine

7.7.1 *Pyrohydrolysis*

An alternative procedure for releasing fluorine and chlorine for analysis from silicate rocks and minerals has been adapted by Farzaneh and Troll (1977, 1978) from earlier work (e.g. Shapiro, 1976; Bloxham, 1968). The technique, pyrohydrolysis, involves first drying the sample at 105–110°C for 2 hours. Depending on fluorine content, between 0.5 and 2 g of powder were mixed with silica and then iron metal in the ratio 1 : 2 : 20 in a quartz crucible. This mixture was pyrolysed by heating in an induction furnace in a stream of superheated steam. Fluorine was evolved as SiF_4 and HF which were absorbed in a 0.025 M solution of NaOH. This solution was subsequently mixed with a citrate buffer solution, diluted to 250 mL, and fluorine determined directly using fluoride ion-selective electrodes and a standard additions calibration technique. The precision of fluorine determinations was often less than 5% (one sigma relative) as can be judged from selected results taken from Troll and Farzaneh (1978) and listed in Table 7.3.

A similar procedure was used for the determination of chlorine. Samples (0.5 to 2.0 g) were mixed with equal weights of silica and pyrohydrolysed in an induction furnace for 10 minutes. Chloride was, however, determined photometrically from the reaction with ammonium ferric sulphate and mercury (II) thiocyanate solution using a filter photometer. Results for selected reference materials taken from Troll and Farzaneh (1980) are compared with conventional determinations by ion-selective electrodes in Table 7.4. The photometric procedure can suffer interferences in samples that contain high sulphide concentrations, for which chloride ion-selective electrode determination may be preferred.

7.7.2 *Ion chromatography*

(*a*) *General capabilities.* Ion chromatography is the term applied to the ion exchange separation of ionic species in solution in which detection of separated species is accomplished electronically using a conductivity detector. The technique is particularly associated with the separation of anions such as F^-, Cl^-, Br^-, PO_4^{3-}, NO_3^- and SO_4^{2-}, although other anionic species (such as the organic acids and cations) can also be separated and detected using suitable exchange columns.

In geological studies, ion chromatography has obvious applications in the detection and determination of anions in ground waters and brines. Interest in silicate rock analysis is in the determination of chlorine, fluorine and sulphur using instrumental separation and detection of respective anions.

(*b*) *Instrumental developments.* Highly ionic species such as F^- and Cl^- cannot be satisfactorily separated by conventional ion exchange procedures (Chapter 14) because they cause swelling of the resin and have very low mass transport rates. Nor can their presence normally be detected using a conductivity detector, because the signal is swamped by the high background conductivity of accompanying ions (including eluent) in sample solutions.

The pioneering work of Small *et al.* (1975) overcame these problems and led to the development of a new analytical technique, ion chromatography. Small *et al.* (1975) combined the three components shown schematically in Figure 7.11: an ion exchange column suitable for anions and other species, a separator column to remove unwanted ionic background and a conductivity detector. Their particular achievements were as follows:

(i) The development of special low-capacity ion exchange resins suitable for efficient separation of ions such as F^-, Cl^- and Br^-

(ii) The use of a second suppressor ion exchange column that reduced the high background conductivity

(iii) The application of a conductivity detector to measure separated ion peaks electronically.

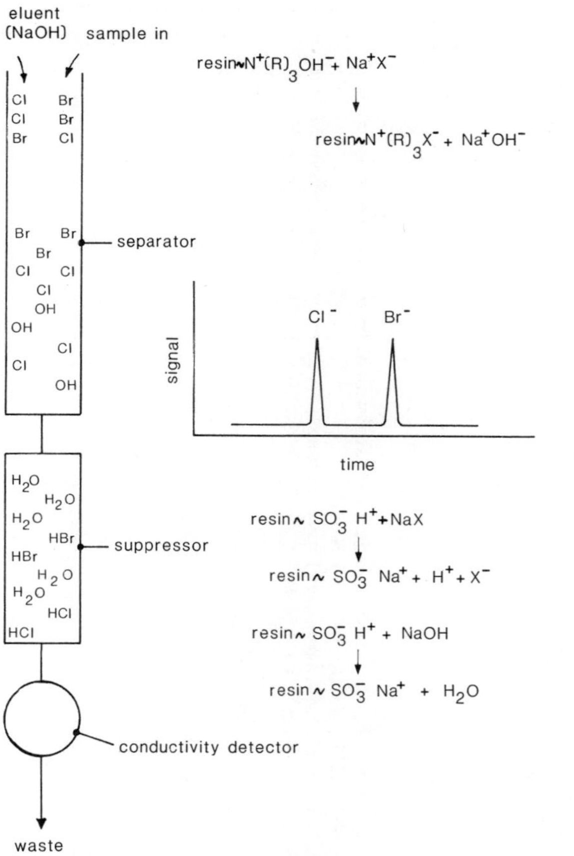

Figure 7.11 Schematic arrangement of ion chromatography instrumentation used here in the determination of Cl^- and Br^-. The arrangement of the separator and suppressor ion exchange columns and conductivity detector is shown. Dilute sodium hydroxide solution is used here as eluent. The resin exchange reactions are indicated. (Pohl and Johnson, 1980.)

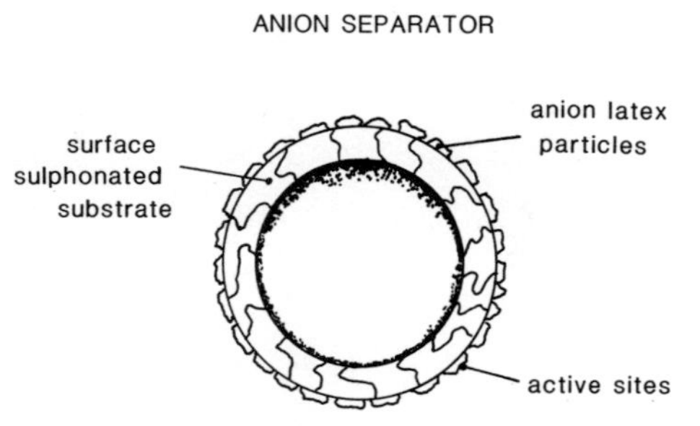

Figure 7.12 Diagram of an anion resin bead developed for ion chromatography showing the styrene–divinylbenzene core, the surface sulphonated outer layer and the anion exchange latex particles agglomerated on to the surface of the bead. (Pohl and Johnson, 1980.)

The early development of ion chromatography and subsequent exploitation has been reviewed in detail by Pohl and Johnson (1980), on which most of the following description is based.

Small *et al.* (1975) developed an *ion exchange resin* suitable for ion chromatography from conventional polystyrene-divinylbenzene substrates. The surface of beads of this substrate was sulphonated to provide cation exchange sites. The core of each polystyrene-divinylbenzene bead was, therefore, inert, providing a rigid structure impervious to highly ionic species. Ion exchange resins of this material were, therefore, both more resistant to swelling effects and permitted higher flow rates at moderate back pressures than were possible with conventional macroporous gel resins that were sulphonated throughout their structure. The same principle was used in the manufacture of anion exchange resins, based on surface-sulphonated styrene-divinyl benzene beads as described above. However, small anion exchange particles that incorporated the tetramethyl-ammonium functional group were then agglomerated on to the surface of individual beads.

In the original work of Small *et al.* (1975), the resultant ion exchange particles had a structure as shown in Figure 7.12. The anion functional groups are attached to a latex base that was then coated as a monolayer on to the surface of the sulphonated substrate. Anion exchange properties are, therefore, again restricted to the surface of these resin beads.

Anion exchange resins made in this way have very low anion capacities (in the order of 0.005 to 0.05 meq/g) and can be satisfactorily eluted using very low molar concentration solutions (e.g. 0.003M $NaHCO_3$/0.0024M Na_2CO_3).

The function of the *suppressor column* is to chemically modify the ionic composition of the eluent solution, removing ion background, so that selected species may be sen-

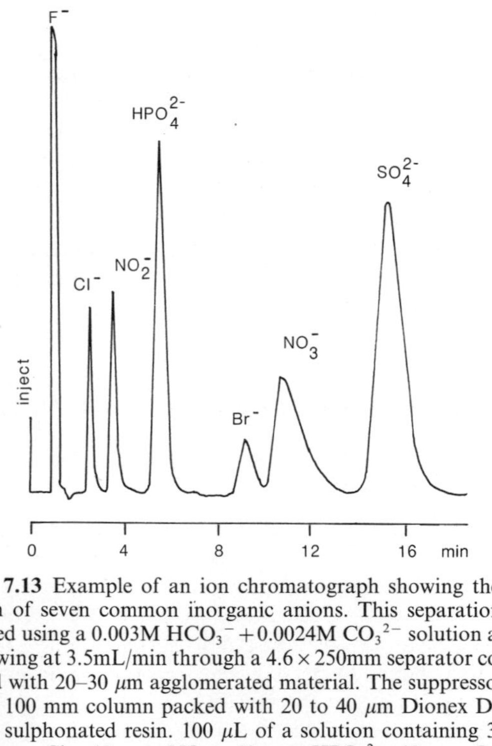

Figure 7.13 Example of an ion chromatograph showing the separation of seven common inorganic anions. This separation was achieved using a 0.003M HCO_3^- + 0.0024M CO_3^{2-} solution as eluent flowing at 3.5mL/min through a 4.6 × 250mm separator column packed with 20–30 μm agglomerated material. The suppressor was a 4.6 × 100 mm column packed with 20 to 40 μm Dionex DC–X8 totally sulphonated resin. 100 μL of a solution containing 3 ppm F^-, 4 ppm Cl^-, 10 ppm NO_2^-, 50 ppm HPO_4^{2-}, 10 ppm Br^-, 30 ppm NO_3^- and 50 ppm SO_4^{2-} was injected. (Pohl and Johnson, 1980.)

Table 7.5 Determination of chlorine, fluorine and sulphur in selected reference materials by ion chromatography

	CHLORINE				FLUORINE				SULPHUR			
	Abbey (1983)	Wilson and Gent (1983)	Evans *et al.* (1981)		Abbey (1983)	Stecher (1983)	Evans *et al.* (1981)		Abbey (1983)	Terashima (1979)	Evans *et al.* (1981)	
	Expected	Analysed	Analysed	CV	Expected	Expected	Analysed	CV	Expected	Analysed	Analysed	CV
Andesite AGV-1	185	170			400	435			100?			
Basalt BCR-1	58?	110			500	490			400?			
Granite G-2	100	81			1200	1295			100?			
Granodiorite GSP-1	340	330			3700	3600			300?			
Peridotite PCC-1	80?	120			0?	<15			100?			
Basalt BHVO-1	94?	98	77	3.9	380	390	393	1.2	110?	101	112	1.3
Marine mud MAG-1	3.09%	2.93%	2.97%		900?	800	882		4300?	3943	3700	
Quartz latite QLO-1	220?		193	3.2	280?	265	282	3.5	40?	13	40	6.5
Rhyolite RGM-1	540	550	479	2.8	340?	345	326	4.3	100?	10	32.5	11
Shale SCo-1	51?		49		770?	750	778		660?	608	616	
Mica schist SDC-1	35?		43.5		620?	600	630		670?	671	672	
Shale SGR-1	50?		92		1900?	1800	2285		15600?	16075	16400	
Nepheline syenite STM-1	450?	480	424		910?	900	955		?	22	62	

Units of concentration: ppm rock.

CV = coefficient of variation = 100 × standard deviation/mean analysis (%).

Wilson and Gent (1983): sample (100 mg) fused with sodium carbonate (200 mg) at red heat over a Meker burner for 5 min and disintegrated with water to make a total volume of 50 mL. Chlorine determined by ion chromatography using a Dionex S-2 column and elution with 0.003M Na_2CO_3+0.0015M NaOH.

Evans et al. (1981): Sample (200 mg) ignited at 1200°C for 20 min with 1 g accelerator (V_2O_3+Fe+Sn) in an atmosphere of oxygen saturated with water vapour (pyrohydrolytic release). Evolved gases were trapped in ice-cooled eluent and analysed by ion chromatography, eluting with 0.003M $NaHCO_3$+0.0024M Na_2CO_3. Sulphur was determined on a separate aliquot to which H_2O_2 was added to ensure complete oxidation to sulphate.

Terashima (1979): Sulphur determined by combustion and infra-red spectrometry and included for comparison with the ion chromatography data.

sitively measured using an ion conductivity detector. This column is designed to have minimal effect on the ionic separation achieved in the first column. In the case of anion analysis, a conventional strongly acidic cation exchange resin in the hydrogen form is used. This resin is fully sulphonated and it is designed to remove cations (e.g. Na^+) from eluted solutions as follows:

$$\text{Resin-}SO_3^-.H^+ + Na^+ + HCO_3^- \rightarrow \text{Resin-}SO_3^-.Na^+ + H_2CO_3 \quad (7.17)$$

Elution with $NaHCO_3/Na_2CO_3$ solutions is particularly favoured because sodium ions are efficiently absorbed by strongly acidic cation exchange resins and the product of the resin suppression reaction, H_2CO_3, (equation 7.17) is only weakly ionic and therefore offers a low conductivity background. Pohl and Johnson (1980) listed the optimum physical properties of the suppression resin:

(i) Low surface area (i.e. microporous resin)
(ii) Small bead diameter to minimize peak dispersion of ion peaks
(iii) Moderate cross-linking to minimize absorption and swelling effects
(iv) Minimal total volume to minimize the effects of absorption, dispersion and the retention of weak electrolytes.

The capacity for anion separation by ion chromatography is demonstrated by the spectrum in Figure 7.13, which shows the elution pattern for a sample solution containing several inorganic cations (F^-, Cl^-, Br^-, NO_2^-, NO_3^-, HPO_4^{2-} and SO_4^{2-}) in the concentration range 3 to 50 ppm.

(*c*) *Geological applications of ion chromatography*. The application of ion chromatography in the determination of fluorine, chlorine and sulphur in silicate samples has been described by Evans *et al.* (1981). Much of this work centred on the optimization of pyrohydrolysis conditions used to release these elements from rock powders. Samples (200 mg) were placed in a ceramic crucible and covered with V_2O_5 (0.2 g), low-sulphur iron powder (0.7 g) and tin accelerator (0.9 g). A porous quartz wafer was placed over crucibles to prevent spattering. Samples were then ignited at 1200°C for 20 minutes under an oxygen gas flow of 1.4L/min. Recoveries of fluorine were significantly increased if the oxygen was first saturated with water vapour by bubbling the oxygen carrier gas through distilled water at 60°C. Evolved gases were trapped in 50 mL of the standard eluent solution (0.0003M $NaHCO_3$, 0.0024M Na_2CO_3). Under conditions optimized for release of fluorine, recoveries of Cl and S were $100 \pm 5\%$ and $85 \pm 6\%$ respectively.

F and Cl were determined by ion chromatography. 0.2 mL of sample solution was injected and eluted with the standard eluent solution at a flow rate of 138 mL/min. Calibration was against a series of standard solutions prepared from NaF and NaCl. Sulphur was oxidized in a separate aliquot of sample solution to sulphate SO_4^{2-} using sufficient 3.3% H_2O_2 and run individually, since the presence of H_2O_2 caused interferences in ion chromatography that affected the determination of both fluorine and chlorine. Detection limits commonly averaged 1, 0.9 and 5 μg absolute for F, Cl and S (corresponding to concentrations of 5, 4.5 and 25 ppm whole rock assuming a 200 mg sample aliquot). Accuracy may be judged from data listed in Table 7.5 for determinations of F, Cl and S in selected reference materials.

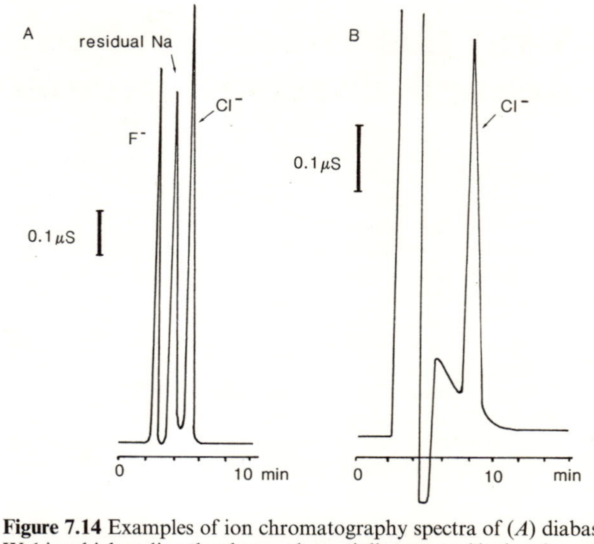

Figure 7.14 Examples of ion chromatography spectra of (*A*) diabase W-1 in which sodium has been substantially removed by batch mode ion exchange, and (*B*) granodiorite GSP-1 of a solution obtained directly by disaggregation with water of a sodium carbonate fusion of the sample. The response of the conductivity detector is plotted along the vertical axis in units of microsiemens (μS). (Wilson and Gent, 1983.)

Ion chromatography was also used by Wilson and Gent (1983) to determine chloride in geological samples. Samples (100 mg) were fused with sodium carbonate (200 mg) in a platinum crucible over a Meker burner at red heat for 5 minutes. After cooling, fusion mixtures were disaggregated with water and made up to 50 mL. Aliquots of 10 mL were filtered and anlysed by ion chromatography using two types of exchange column. In one type of column (Dionex S3) it was necessary to remove much of the sodium ion content of samples by shaking solutions with 3 mL AG-50W-X12 ion exchange resin for 4 hours. Samples were eluted with 0.003M $NaHCO_3$/0.0024M Na_2CO_3 solutions. In an alternative column (Dionex S2), chloride could be determined directly using 0.003M $NaHCO_3$/0.0015M NaOH for elution. A comparison of chromatograms by these two procedures is shown in Figure 7.14. In this latter procedure a potential interference from nitrate was not observed and instrument time from chromatography was 28 minutes per sample, permitting a total of 30 determinations to be made in a 24-hour period (including sample preparation). Calibration of the technique was relative to chloride standard solutions prepared in a matrix matched with sodium carbonate. Detection limits were reported to be 7 ppm chloride in whole rock at the 90% confidence level. Typical analyses of selected reference materials by this procedure are shown in Table 7.5.

Ion chromatography is clearly a powerful technique in the determination of anion concentrations in aqueous samples. In application to geological samples, the higher sensitivity, particularly for chlorine and the capability of multi-element determinations must be balanced against the high capital cost of instrumentation, especially in comparison with equipment for ion-selective electrode analysis.

8 X-ray fluorescence analysis: principles and practice of wavelength dispersive spectrometry

8.1 Analytical characteristics

X-ray fluorescence spectrometry is a technique for the analysis of bulk specimens. Samples are prepared as compressed powder pellets or fused glass discs and excited with x-ray radiation, normally generated by an x-ray tube operated at a potential of between 10 and 100 kV. Interaction of this primary radiation with atoms of the sample causes ionization of discrete orbital electrons. During the subsequent electronic rearrangement by which the atom then de-excites back to the ground state, fluorescence x-rays of energy characteristic of that element are emitted. The emission intensity of this characteristic radiation is measured with a suitable x-ray spectrometer and compared with that from a standard sample. The technique is one of the most widely used routine instrumental methods of analysing rock samples both for the major elements Na, Mg, Al, Si, P, K, Ca, Ti, Mn, Fe and selected trace elements, including Rb, Sr, Y, Nb, Zr, Cr, Ni, Cu, Zn, Ga, Ba, Pb, Th, and U. Detection limits for many of these trace elements lie in the range 1 to 10 ppm rock under routine operating conditions.

Trace element geochemistry came of age when XRF spectrometers became routinely available for silicate rock analysis. After the initial development of commercial instrumentation in the 1950s and 1960s, XRF was adopted as the standard instrumental technique for rock analysis by many laboratories following Norrish and Hutton (1969) and Leake *et al.* (1969) for example. This resulted, for the first time, in the widespread availability of routine major and trace-element analyses and stimulated the development of important geochemical modelling and discriminatory procedures, especially those based on the elements Sr, Rb, Y, Nb and Zr that are particularly well suited to analysis by this technique.

The successful adaptation of XRF as a routine analytical tool was not immediate and had to await two important developments. The first was the commercial availability of electronic instrumentation of exceptionally high stability, particularly with respect to the elimination of drift in the emission intensity of high-power x-ray tubes. The second was the development of suitable mathematical correction procedures to account for the discrepancy that observed fluorescent x-ray counts are not directly proportional to atomic concentration but must first be corrected for absorption-enhancement effects caused by the influence of all other elements in the sample.

These difficulties have now largely been overcome, and modern instrumentation is capable of achieving such a high degree of precision that it is generally accepted that analytical statistics in silicate rock analysis are principally limited by sample inhomogeneity effects. In fact the very high order of statistical precision that can be obtained from modern instrumentation has tended to mask the fact that analytical *accuracy* depends both on the reliability of the matrix correction program and the procedures used for calibration. This touches on one of the limitations of x-ray spectrometry: the fact that it is a comparative technique in which calibrations must generally be carried out relative to assumed compositions in silicate rock reference materials. Accuracy is therefore critically affected both by the selection of samples to be included in the calibration set and the confidence that may be placed in compiled compositions of these materials. There have been serious shortcomings in compilations of some elements (particularly trace elements) in some reference rocks, as is discussed in Chapter 1. If x-ray fluorescence spectrometry has one shortcoming, it is the confusion that can arise between accuracy and instrumental precision and the complacency that can then result when critically assessing analytical data.

Conventional instrumentation for x-ray fluorescence analysis is based on wavelength dispersive spectrometers. In consequence, the origin of x-ray spectra are normally considered in terms of the wavelength (in Ångstrom or nm) of the corresponding emission lines. However, the use of such a notation does not facilitate an understanding of mechanisms of x-ray emission, derived from transitions between electron orbitals of discrete *energy*. Nor is it compatible with x-ray instrumentation based on the energy dispersive spectrometers that are becoming more widely used. The following sections therefore consider the physics of x-ray production and interaction with matter on the basis of energy differences. Analytical applications in this chapter are restricted to conventional instrumentation using high power x-ray tubes (3–5 kW) and wavelength dispersive spectrometers. The principles and analytical characteristics of energy dispersive spectrometers will be discussed in Chapter 9, together with other recent developments in ED–XRF instrumentation and excitation geometries.

Much common ground exists, both in the physics of x-ray production and the instrumental characteristics of x-ray spectrometers, between the x-ray fluorescence and electron microprobe techiques, considered further in Chapter 10, for which the following sections constitute preliminary reading.

8.2 Energy and wavelength of x-rays

An x-ray is a high-frequency electromagnetic radiation of energy intermediate between the far ultraviolet and gamma-ray regions of the spectrum. In atomic spectroscopy, the

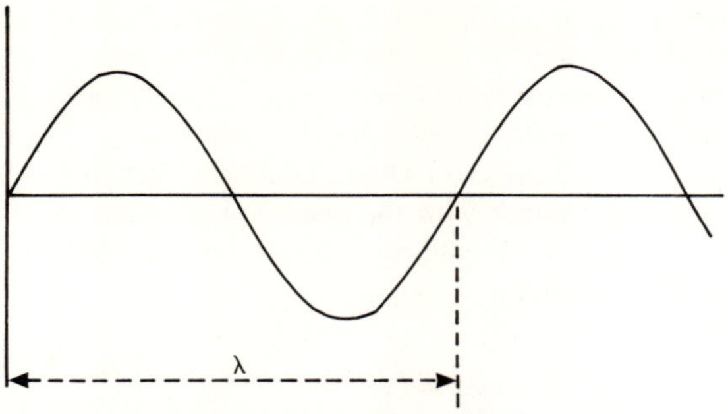

Figure 8.1 Representation of the sinusoidal waveform of an x-ray (λ equals one wavelength).

particular distinction of an x-ray is that it arises from electron transitions between discrete orbital shells of an atom (by convention, gamma-rays are emitted by the nucleus itself). The energy of an x-ray is conveniently measured in units of the electron volt; one electron volt being the increase in energy imparted to an electron after acceleration through a potential field of one volt. However, like all other electromagnetic radiations, x-rays possess a sinusoidal waveform (Figure 8.1) so that an x-ray may also be characterized by its wavelength; energy (E) and wavelength (λ) are related by the equation

$$E = h\nu = \frac{hc}{\lambda} \tag{8.1}$$

where h is the Planck constant $= 6.626 \times 10^{-34}$ joules s^{-1}, c is the velocity of light in vacuum $= 2.998 \times 10^8$ ms^{-1}, and ν is the frequency of the waveform (cycles per second).

Incorporating the conversion factor (1keV $= 1.602 \times 10^{-8}$ joule) then equation (8.1) may be more conveniently expressed as

$$E\,(\text{in keV}) = \frac{12.396}{\lambda} \tag{8.2}$$

where λ is the wavelength expressed in Ångstrom units (1 Å $= 10^{-10}$ m). Ångstrom units are traditionally used to express wavelengths in x-ray spectrometry, although their continued use is discouraged under modern SI conventions in favour of the nanometre. Unfortunately, the numeric conversion of fluorescence x-ray lines to nanometres generates numbers that are not as convenient for everyday use (the Rb $K\alpha_1$ line changes from 0.926 Å to 0.0926 nm).

Table 8.1 Conversion table, for keV *v*. Ångstrom units

Units A	0.2	0.3	0.4	0.5	0.6	0.7	0.8	0.9	1.0
Units B	62.0	41.3	31.0	24.8	20.7	17.7	15.5	13.8	12.4
Units A	1.2	1.5	2.0	2.5	3.0	4.0	6.0	8.0	10.0
Units B	10.3	8.26	6.20	4.96	4.13	3.10	2.07	1.55	1.24
Units A	12.0	15.0	20.0	30.0	40.0	50.0	60.0	80.0	100.0
Units B	1.03	0.83	0.62	0.41	0.31	0.25	0.21	0.15	0.12

If units A represent wavelength in Å, then units B represent energy in keV, and vice versa.

The x-ray region of the spectrum can be considered to extend from 0.1 to about 100 keV (120 – 0.12 Å). However, the region of interest in routine analytical x-ray spectrometry is much more restricted than this, usually being 1 to 20 keV (12–0.6 Å). Table 8.1 converts energy (in keV) to wavelength (in Å).

8.3 The origin of x-ray spectra

8.3.1 *Moseley's Law*

In comparison with other analytical techniques, x-ray fluorescence analysis is unusual in that there is a consistent and predictable variation in the energy of x-ray emissions from atom to atom. This unexpected observation was formulated by Moseley (1913, 1914) who found that the wavelength of an x-ray emission (λ) could be related to the atomic number of an element (Z) by the formula

$$\frac{1}{\lambda} = k(Z-\sigma)^2 \tag{8.3}$$

where k is a constant for a particular series of lines (K, L, M, etc), and σ is a shielding constant.

The remarkable correlation afforded by this equation is demonstrated in Figure 8.2 which shows the variation with atomic number of the wavelength of the $K\alpha$ x-ray emissions for the elements lithium to molybdenum in the Periodic Table.

The origin of Moseley's discovery, in terms of the electronic structure of atoms, was not fully understood at the time. However, it was subsequently used to confirm the atomic number of new elements in the Periodic Table (from

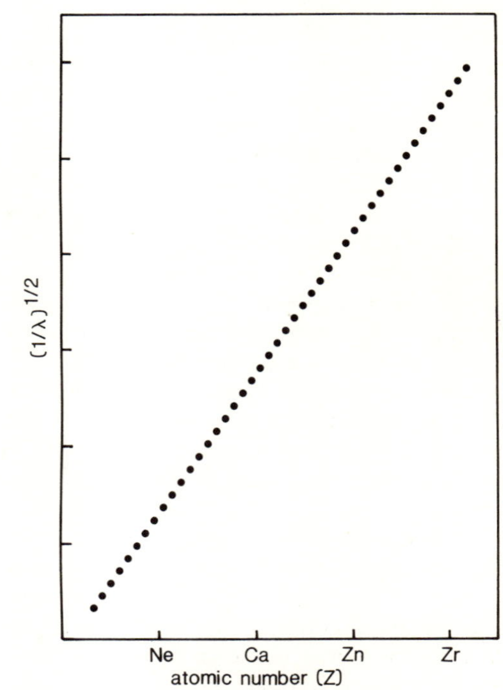

Figure 8.2 Moseley's Law, demonstrating the near-linear relationship between atomic number (Z) and $(1/\lambda)^{1/2}$ using modern values for the $K\alpha$ wavelength of the elements Li to Mo.

their appropriate x-ray signature) and can be considered as the origin of analytical x-ray spectrometry.

8.3.2 *Electronic structure of atoms*

The phenomenon observed by Moseley is in fact related to transitions of electrons between orbitals of discrete energy in an atom. To explain x-ray spectra, it is adequate to consider the structure of an atom as represented by the highly simplified model after Bohr (Figure 8.3). An atom is considered to consist of a nucleus composed of both protons and neutrons in which the number of protons (each of which has a notional single positive charge) defines the atomic number of the atom; the sum of protons and neutrons defines the atomic weight of the atom. The nucleus is surrounded by electrons occupying electron orbitals of discrete energy which for convenience may be grouped together as electron orbital *shells* designated by the letters *K*, *L*, *M*, *N*, *O*, etc. Following the rules of quantum theory (Chapter 3), each shell can be filled with electrons up to a maximum number as follows: *K*: 2, *L*: 8, *M*: 18, *N*: 32, etc. In a neutral atom, the positive charge on the nucleus (equal to the number of protons) is exactly balanced by an equivalent number of negatively charged orbital electrons. Since the *K* orbital shell has the lowest energy (see next paragraph), this shell is filled with electrons first, then the *L*-shell and so on, until charge balance with the nucleus is attained. A more detailed consideration of the electronic structure of the atom is given in Chapter 3.

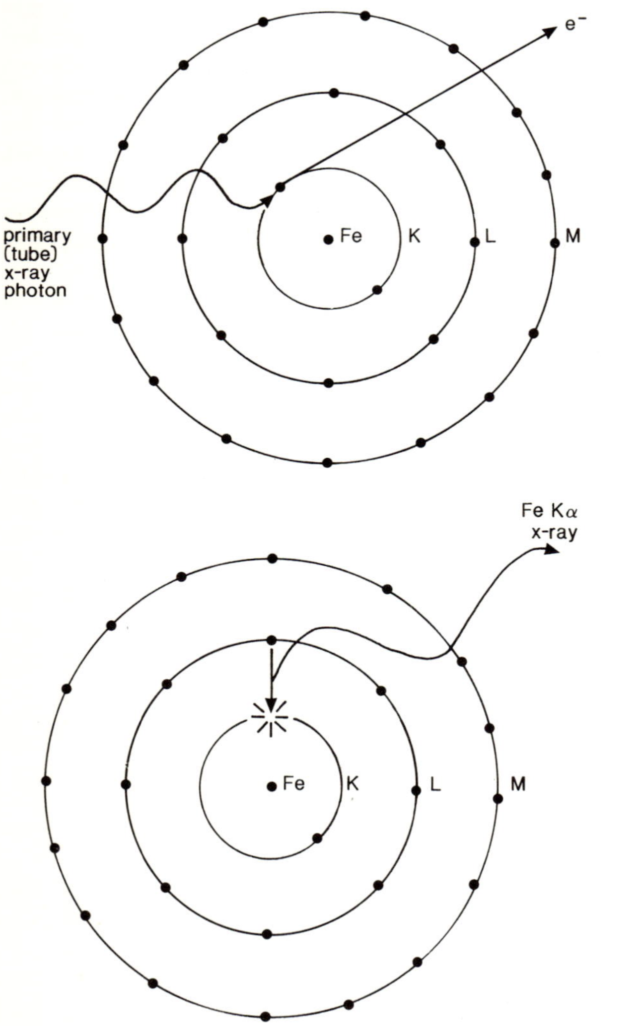

Figure 8.3 Schematic representation of the mechanism of x-ray fluorescence of an iron atom leading to the emission of an Fe *Kα* x-ray.

8.3.3 *Convention on orbital energy levels*

By convention, the energy of an electron orbital is defined as the energy required to remove that electron from the influence of the atom and so oppose the electrostatic attraction between negatively charged electron and positively charged nucleus. This energy is measured by the *ionization potential* of the orbital electron and thus electrons occupying orbitals of the highest ionization potential represent states that are energetically the most stable. Since work must be done to ionize an electron, the orbital energy is quoted with a notional negative sign. In the Bohr model, *K*-shell orbitals are the closest to the nucleus, are therefore subject to the largest electrostatic interactions between positively charged nucleus and negatively charged electron and will therefore possess the largest (negative) ionization potential. Electrons occupying outer *L*-, *M*- and *N*-shells are progressively more weakly bound to the nucleus.

As an illustration of these concepts, the simple electron orbital structure of an iron atom is drawn in Figure 8.4. The nucleus, comprising 30 neutrons and 26 protons (atomic weight = 56) is surrounded by 26 orbital electrons, 2 of which occupy the *K*-shell, 8 the *L*-shell and the remaining 16 the *M*-shell.

8.3.4 *X-ray emission*

Based on this simple electron orbital shell model of an atom, x-ray fluorescence may be simply represented as shown in Figure 8.3. A high-energy x-ray photon (from the x-ray tube) interacts with an orbital electron, causing it to be removed from the electron orbital by ionization. An electron can be removed from any of the occupied shells by this process. In the case of a *K*-shell electron, a vacancy is created in the *K*-shell (Figure 8.3) leaving the atom in a highly unstable state. In order to regain stability, an electron from the *L*- or *M*-shell falls down the potential energy gradient to refill this vacancy. The surplus energy which must be lost for this transition to take place is emitted as an x-ray having an energy characteristic of the difference between the two orbital levels.

The resultant x-ray spectra are complicated by a further phenomenon. In real atoms, orbital electrons interact with one another to split the energy levels of each shell into individual subshells. The *K*-shell is always mono-energetic. However, the *L*-shell can split into three subshells designated L_I, L_{II}, and L_{III}, the *M*-shell into five levels (M_I to M_V) and so on. Transitions from each subshell to a vacant orbital are governed by selection rules that are precisely defined by quantum theory (discussed in Chapter 3). Some orbital transitions are allowed, others are forbidden. As will be seen shortly, from a detailed knowledge of the electronic structure

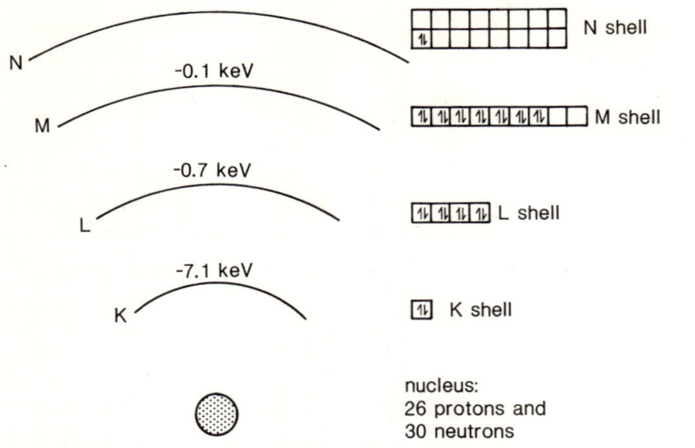

Figure 8.4 Diagram of the electron orbital structure of an iron atom. The number of electrons occupying each shell is indicated, together with the corresponding ionization energy (in keV). Interations between orbital electrons cause individual subshells to split in energy. In these circumstances, it is energetically more favourable for the last two electrons to occupy the *N*-shell rather than the *M*-shell, as might otherwise be expected.

of atoms, it is possible to correlate observed x-ray emissions in fluorescence spectra with electron transitions between individual orbital subshells.

Referring back to the phenomena embodied in Moseley's law, the overall factor in controlling the potential energy levels of orbitals from one atom to another is the magnitude of the positive charge on the nucleus. This sets the overall potential energy field with which individual orbital electrons interact. Hence Moseley's strong correlation between the energy of x-ray emission and atomic number. Furthermore, the effective nuclear charge available for an outer electron orbital will be diluted by the presence of intervening electrons—hence the shielding factor (σ) in equation (8.3).

A further consideration in the interpretation of x-ray spectra is that the magnitude of the split into sub-orbitals increases as the number of electrons on the atom increases. Thus low-atomic-number atoms will have relatively simple x-ray spectra, whereas a larger number of x-ray lines are likely to be distinguishable in the spectra of the heavier elements. The spectra of heavier elements are likely to be more complex due to the larger number of orbital electrons that are available to participate in transitions.

Some of these phenomena are illustrated in Figure 8.5, which shows schematically the more important x-ray lines observed in x-ray fluorescence spectra of silicate rocks in the 0–20 keV region of the spectrum. The $K\alpha$ transition is clearly resolved into two separate lines, $K\alpha_1$ and $K\alpha_2$ in the spectrum of rubidium, for example. However, these lines cannot be distinguished in the sodium spectrum. Comparing the lines visible in the *L*-spectrum of barium and lead, more transitions are clearly available to the latter element.

8.3.5 *Notation for x-ray emission*

The convention most commonly used to identify x-ray lines by analytical spectroscopists is the *Siegbahn notation.* In spite of the fact that this notation is neither systematic nor complete, it is almost universally accepted to designate emission lines in analytical applications. The Siegbahn notation was based on the relative intensity of lines observed in x-ray spectra and was formulated at a time when the corresponding electron orbital transitions were not fully understood.

Three series of analytically important lines are identified: the *K*-series, the *L*-series and the *M*-series. Each series corresponds to an x-ray emission derived from a transition required to fill a vacancy in the *K*-, *L*- or *M*-shells respectively. Within each series, x-ray lines are subclassified according to their observed relative intensity, $L\alpha$ lines being more intense than $L\beta$ lines which are more intense than $L\gamma$ lines. Splitting of these lines in turn is then denoted by a further subscript 1, 2, 3, etc. The full notation for an individual line is, for example: $K\alpha_1$, $L\beta_3$, etc.

Unfortunately, as will be seen below, when x-ray lines are correlated with their respective electron orbital transitions, this scheme appears to be rather arbitrary. Spectroscopists have their own transition notation based on quantum theory and the principle of spin-orbital coupling as discussed in outline in Chapter 3. However, although this rigorous approach defines precisely the electronic states involved in a transition, it has been neglected in the context of x-ray analysis.

8.3.6 *Typical x-ray spectra*

The electron orbital energy diagrams of two elements will now be examined and correlated with observed x-ray spectra.

(*a*) *Iron.* The electron orbital structure of an iron atom has already been considered schematically in Figure 8.4. However, as we have already seen, splitting of energy levels makes the orbital structure of the atom more complex. In Figure 8.6 the transitions that follow the ionization of a *K*- and an *L*-shell electron are marked on the orbital energy level diagram for an iron atom. The vertical scale is a potential energy scale, the zero level of which corresponds to an ionized electron at infinity (no longer bound to the atom). Individual transitions are annotated by the energy (keV),

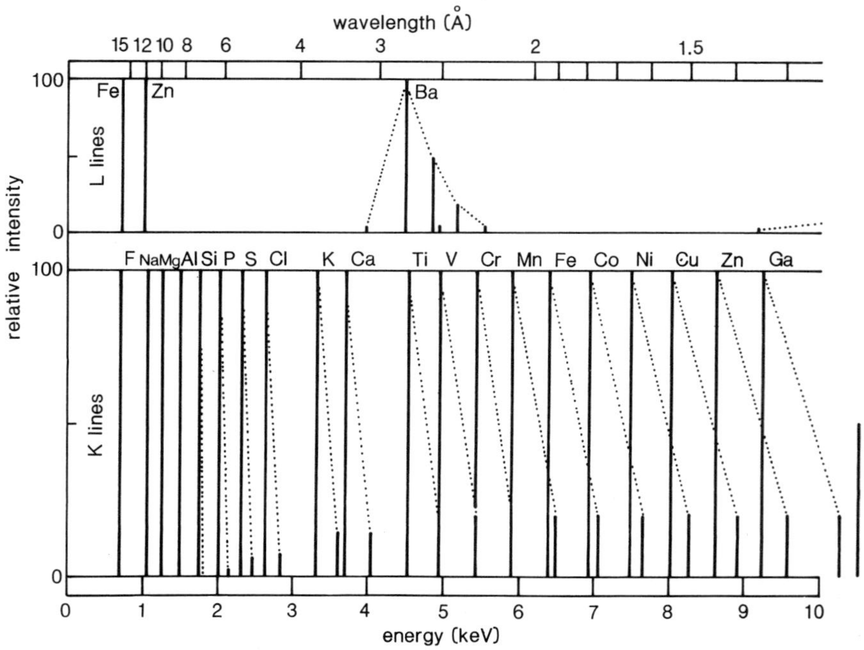

Figure 8.5 Diagram of the more important x-ray lines found in the spectrum of a rock sample in the 'analytical' region from 0 to 20

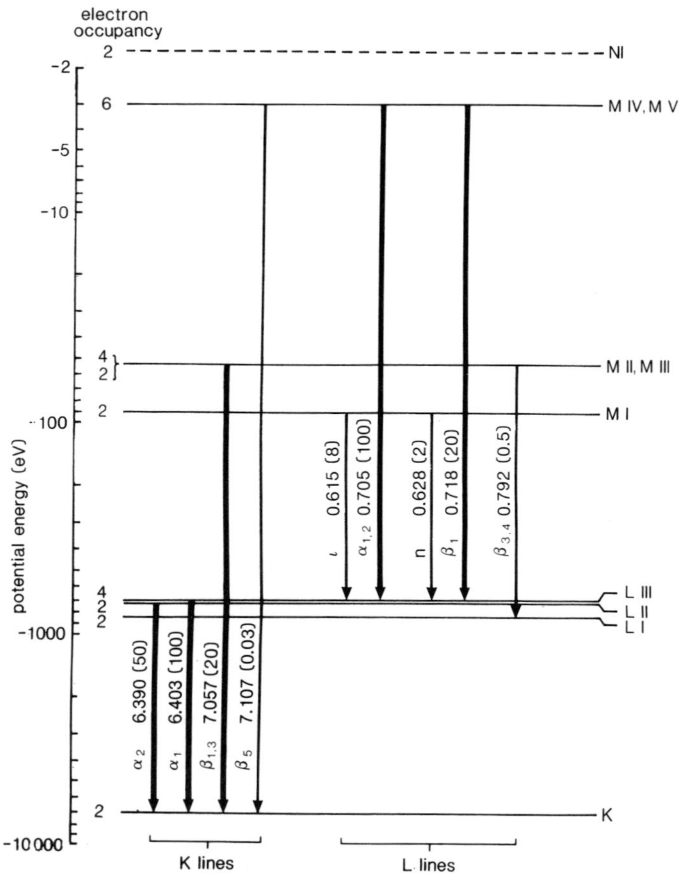

Figure 8.6 Orbital energy level diagram for an iron atom. Observed electron transitions are identified by the energy, relative intensity and Siegbahn notation for the corresponding x-ray fluorescence line.

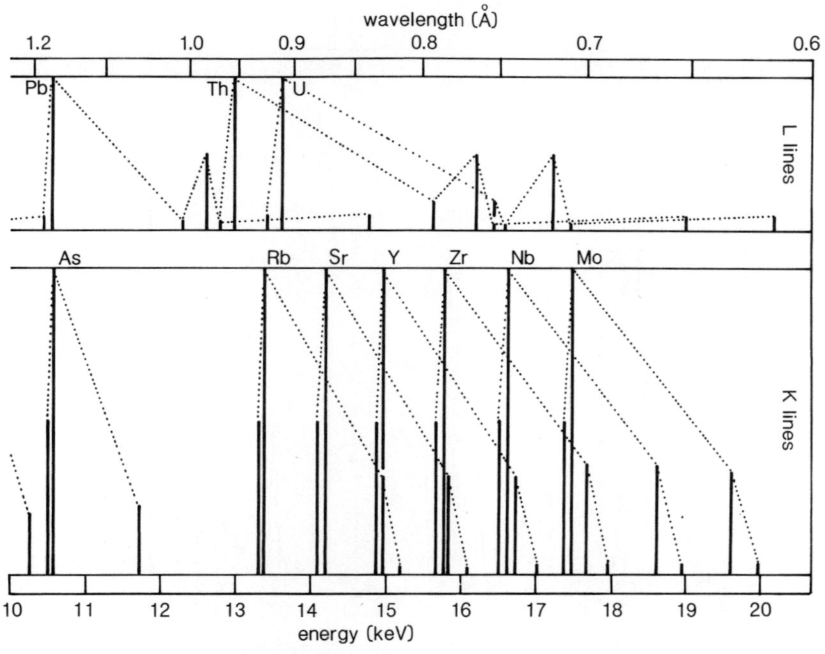

keV. Tie lines are used to identify the lines of an element within the particular *K*- or *L*-series.

relative intensity within a line series and Siegbahn notation for the accompanying x-ray photon. Ionization of a *K*-shell electron causes an electron transition that results in the emission of one of three x-ray photons: the $K\alpha_1$ and $K\alpha_2$ lines at 6.403 and 6.390 keV (which are not fully resolved by most x-ray spectrometers) and the $K\beta_{1,3}$ line at 7.057 keV. Transitions causing these x-ray emissions involve electrons falling from the L_{III}, L_{II} and $M_{II}M_{III}$ shell levels respectively. A fourth transition may be observed at 7.107 keV (the $K\beta_5$ line). This transition is forbidden according to strict quantum mechanical selection rules and its spectral intensity is 2 to 3 orders of magnitude less than the allowed lines, as may be seen in the measured x-ray spectrum derived from iron in Figure 8.7*a*.

Ionization of an electron from one of the three *L* subshell levels gives rise to the opportunity for a more complex series of transitions shown in Figure 8.6. Principal among these are the $L\alpha_{1,2}$ and $L\beta_1$ transitions. However, the corresponding x-ray emissions occur at such low energy (0.615 to 0.792 keV) that they cannot be detected efficiently on most x-ray spectrometers and are subject to serious attenuation effects within the sample. These lines are not used for quantitative analysis. The *L*-shell spectrum of iron is shown in Figure 8.7*b*. The overall intensity of individual *L*-lines is over an order of magnitude smaller than that of the *K*-lines. For these reasons the iron $K\alpha_{1,2}$ line would normally be selected for quantitative analysis.

(*b*) *Barium*. The orbital energy level diagram for barium is shown in Figure 8.8. The ionization potential of a barium *K*-shell electron is 37.45 keV. If the energy of incident (excitation) photons substantially exceeds this value, then the *K*-shell spectrum of barium may be excited as well as *L*- and *M*-shell emissions. The orbital transitions that would follow ionization of electrons from these shells are shown in Figure 8.8. As more orbital levels are occupied by electrons, a larger number of x-ray lines is observed in the barium spectrum (Figure 8.9). The $K\alpha/K\beta$ lines are emitted in the range 31–38 keV. These emissions lie at too high an energy for either efficient excitation or detection on most conventional XRF spectrometers. The *K*-lines of barium cannot normally be used for quantitative analysis.

The *L*-shell spectrum lies in the range 3.9–4.6 keV (Figure 8.9*b*), the most intense line of which (the $L\alpha_1$ transition) would normally be selected for analysis. However, in the case of silicate rocks, the Ba $L\alpha_1$ line suffers a severe overlap interference from Ti $K\alpha_{1,2}$, titanium being an element likely to be present at much larger abundances. In consequence, the Ba $L\beta_1$ line is normally selected for the analysis of this element using wavelength dispersive spectrometers even though sensitivities are lower.

The barium *M*-shell spectrum is observed in the range 0.6–1.0 keV (Figure 8.9*c*), below the level of efficient detection on many analytical instruments.

8.3.7 *Selection of lines for analysis*

These examples show that the selection of lines for quantitative analysis is much more straightforward than for optical emission spectroscopy, for example. The principal criteria are as follows.

(i) Selection of an emission line in the 1 to 20 keV (12 – 0.6 Å) region, since this is the range over which lines can be most efficiently excited and detected using conventional instrumentation.

(ii) Within a *K*-, *L*-, or *M*-line series, selection of the most intense emission line. Thus in XRF analysis, the $K\alpha_{1,2}$ would

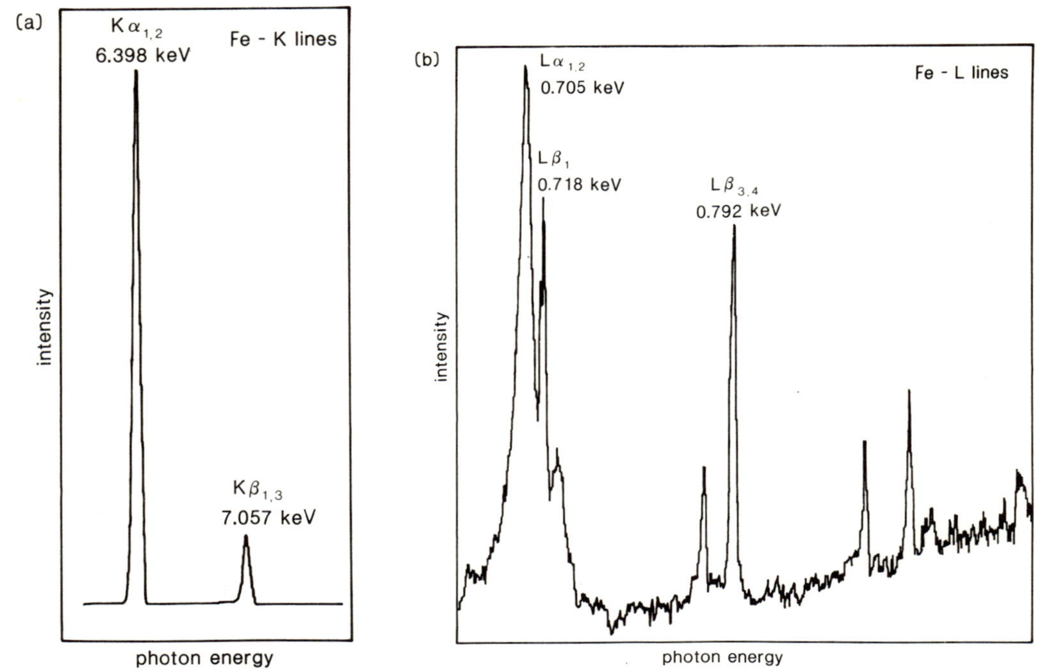

Figure 8.7 X-ray emission spectrum of iron showing (*a*) *K*-lines (6.39–7.06 keV), and (*b*) *L*-lines (0.705–0.792 keV). Principal lines attributable to iron are marked. Data were recorded from a wavelength-dispersive spectrometer scan of iron metal excited by electron probe.

normally be selected for the elements sodium to molybdenum and the $L\alpha_{1,2}$ line for elements such as barium, the rare-earth elements, lead, thorium and uranium.

(iii) Occasionally in silicate rock analysis, account must be taken of any serious spectral overlap interference (e.g. Ti on Ba, as described above) that makes it desirable to select a subsidiary line for analysis.

Line selection and spectrum interferences are considered further in section 8.10.2.

8.3.8 *Oxidation state and chemical bonding effects*

It is apparent from data plotted in Figures 8.6 and 8.8 that electron transitions corresponding to lines normally used for quantitative analysis involve inner orbital shells. These are least affected by chemical bonding or oxidation state effects which are expected to perturb only the outermost electron orbital levels—namely the N_I and $M_{IV}+M_V$ shells of iron (Figure 8.6) and the P_I and O levels of barium (Figure 8.8).

Thus, XRF analysis yields data that are normally independent of the chemical state of an element with respect either to oxidation or co-ordination number. Analyses are a measure of the atomic concentration of an element. As will be seen in section 10.10, this observation breaks down when considering, for example, electron probe microanalysis of individual minerals from the *K*-lines of light elements such as Na through to S. In such circumstances, wavelength shifts may be observed that result from the perturbation of *L*- and *M*-shell levels due to chemical bonding effects.

8.3.9 *Relative intensities*

The relative intensities of lines within an x-ray series (*K*, *L* or *M*) are not identical for all elements but depend on quantum mechanical selection rules, the magnitude of the difference in energy between the orbital levels involved and

Table 8.2 Relative intensity of *K*-, *L*- and *M*-lines of selected elements

	$K\alpha_1$	$K\alpha_2$	$K\beta_1$	$K\beta_2$	$L\alpha_1$	$L\alpha_2$	$L\beta_1$	$L\beta_2$	$L\beta_3$	$L\beta_4$	$L\gamma_1$	L_λ	L_η	$M\alpha_1$	$M\alpha_2$	$M\beta_1$	$M\gamma_1$
Si	100	50	2														
Ca	100	50	15		100		10					1	1				
Fe	100	50	20		100		20			0.5		8	2				
Rb	100	50	16	3	100	10	45		3	3		3	1				
Mo	100	50	17	4	100	10	45	1	3	3	1	3	1				
Ba	100	50	21	7	100	10	50	20	6	5	5	2	1				
W	100	50	26	10	100	10	50	20	6	4	10	3	1	100	100	45	1
Pb	100	50	88	11	100	10	50	20	6	4	10	3	1	100	100	60	5
U	100	50	31	11	100	10	50	20	6	4	10	3	1	100	100	60	5

Data from White and Johnson (1970).

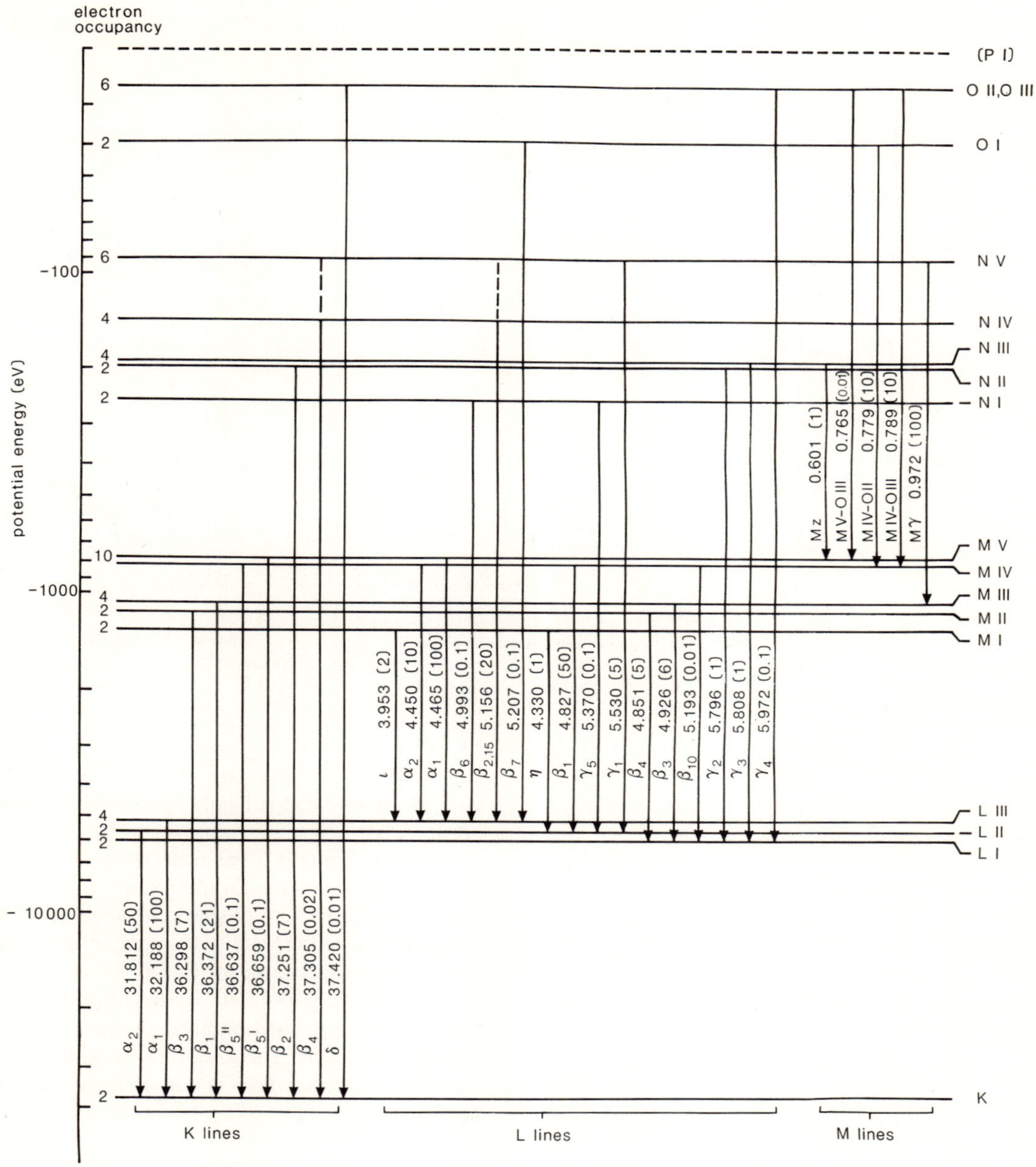

Figure 8.8 Orbital energy level diagram for a barium atom. Observed electron transitions are identified as in Figure 8.6.

orbital electron populations. Relative intensities are also influenced by the probability of internal conversion of the x-ray by the Auger effect, as will be discussed in the next section. Fundamental considerations of selection rules dictate that the $K\alpha_1 : K\alpha_2$ intensity ratio is always 100 : 50. However, the relative intensity of other lines varies, there being some correlation with atomic weight, as indicated by data for selected elements listed in Table 8.2.

8.4 Competing de-excitation routes

8.4.1 *The Auger effect*

The discussion so far has concerned the x-ray spectrum originating from the electron transitions which follow ionization of a *K*-, *L*-, or *M*-shell electron. However there is a second competing route by which an atom may de-excite. There is a certain probability that for example a $K\alpha$ x-ray, emitted when an *L*-shell electron fills a vacancy in the *K*-shell, can itself interact with an outer orbital electron before it has escaped from the influence of the atom. The energy of the x-ray is totally re-absorbed by the atom, resulting in the ionization of the second electron and causing a second orbital vacancy in the atom. This mechanism is shown schematically in Figure 8.10.

The process is similar to the internal conversion of gamma-rays described in section 12.3 and is known in x-ray spectrometry as the Auger effect. As the Auger electron has a discrete kinetic energy (being the difference between the energy of the original x-ray photon and the ionization poten-

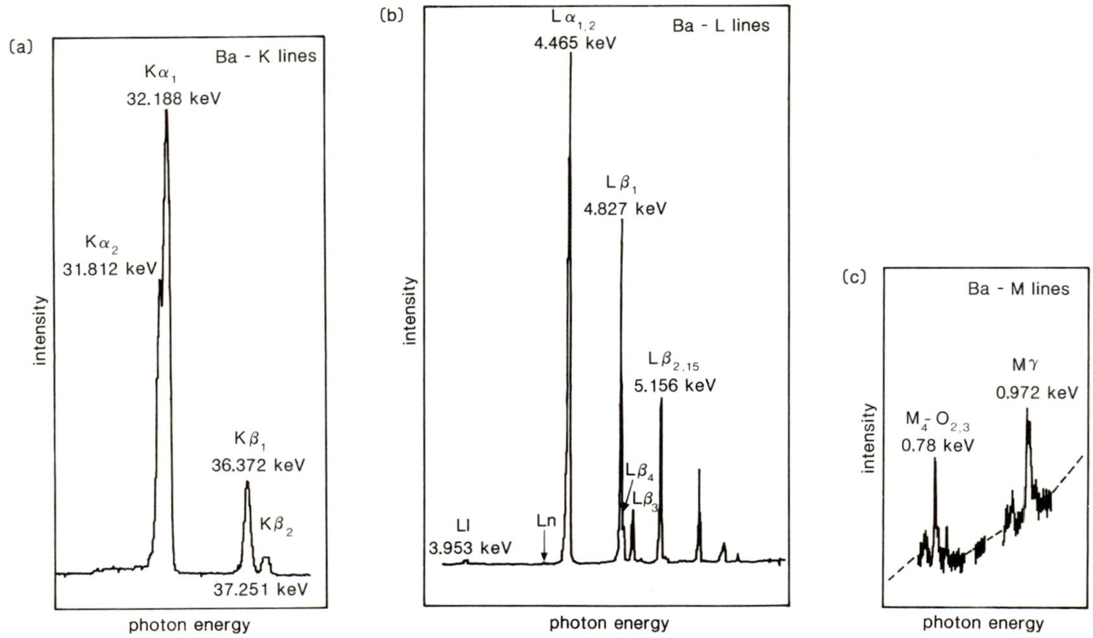

Figure 8.9 X-ray emission spectrum of barium showing (*a*) *K*-lines (31.8–37.2 keV); (*b*) *L*-lines (3.95–5.16 keV); and (*c*) *M*-lines (0.78–0.97 keV). *L*- and *M*-line spectra were recorded from a wavelength-dispersive scan of barite excited by electron probe. The *K*-lines were recorded from the radioactive decay of a ^{137}Cs source using a Si(Li) detector.

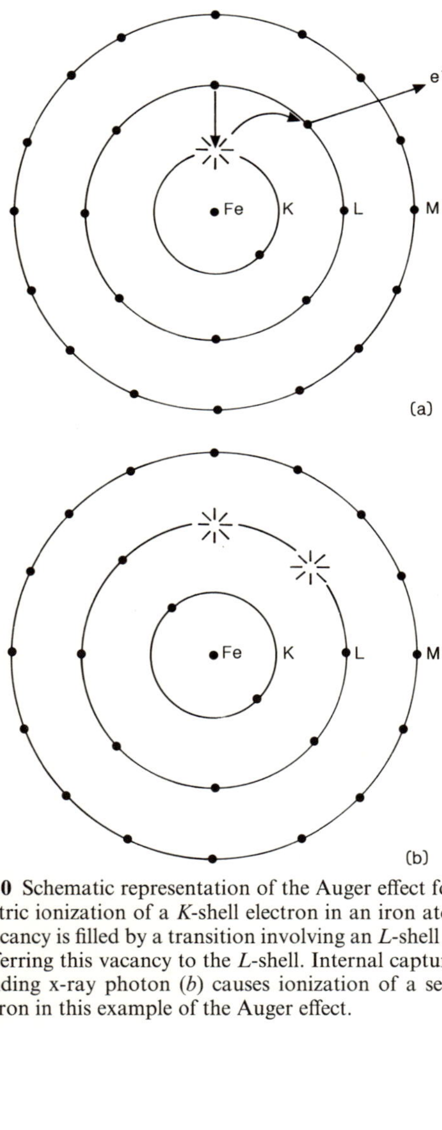

Figure 8.10 Schematic representation of the Auger effect following photoelectric ionization of a *K*-shell electron in an iron atom. The *K*-shell vacancy is filled by a transition involving an *L*-shell electron (*a*), transferring this vacancy to the *L*-shell. Internal capture of the corresponding x-ray photon (*b*) causes ionization of a second *L*-shell electron in this example of the Auger effect.

tial of the respective electron orbital), it is possible to use this phenomenon in a highly specialized form of electron spectroscopy used to measure surface effects in solid samples. Riviere (1982) has reviewed this and other related surface/thin film analytical techniques.

From an XRF point of view, the Auger effect is an undesirable phenomenon, as it reduces the intensity of observed x-ray emissions. The most serious losses occur when for example the $K\alpha$ photon energy lies close to the ionization potential of an *L*-shell electron. Auger energy transfer is then most efficient. This condition is most closely met for the light elements, where in consequence Auger losses are significant.

8.4.2 *Fluorescence yields*

The probability that an atom will de-excite by the emission of a characteristic x-ray is an experimentally measured parameter called the *fluorescence yield.* Thus, ω_K is the fluorescent yield of all *K*-shell x-rays, being the fraction of those atoms in which a *K*-shell vacancy has been produced that emits any *K*-line x-ray (i.e. $K\alpha + K\beta$). Conversely $(1-\omega_K)$ represents the fractional probability that de-excitation will involve internal conversion of these x-rays by the Auger effect. The data in Table 8.3 demonstrate the variation of fluorescent yield for selected elements with atomic number. The Auger effect is very high for the light elements but becomes a minor decay route for *K*-lines of the heaviest elements (e.g. U). When comparing the emission probabilities of *K*-, *L*- and *M*-shell x-rays for the same element, the fluorescent yield decreases markedly from *K*- to *M*-shells. For most sensitive determinations, it is therefore desirable to measure a *K*-line rather than an *L*- or *M*-line, *providing* excitation can be achieved efficiently.

Table 8.3 Fluorescent yields of selected elements for *K*-, *L*- and *M*-shell x-rays

	ω_K	ω_L	ω_M
Na	0.024		
Si	0.047		
Ca	0.163		
Fe	0.347		
Rb	0.669	0.01	
Mo	0.764	0.07	
Ba	0.901	0.09	
W	0.957	0.30	
Pb	0.968	0.39	0.029
U	0.976	0.53	0.06

Data from Bambynek *et al.* (1972).

Table 8.4 Relative intensity of selected satellite lines

	$K\alpha_2$	$K\alpha_2$	$SK\alpha_3$	$SK\alpha_4$	$K\beta_1$
Na	1.041 (100)		1.050 (10)	1.052 (10)	1.067 (0.5)
Si	1.739 (50)	1.740 (100)	1.752 (6)	1.754 (3)	1.829 (2)
Ca	3.687 (50)	3.691 (100)	3.711 (2)	3.715 (2)	4.012 (15)

Data from White and Johnson (1970).
$SK\alpha_3$ and $SK\alpha_4$ are satellite line equivalents to the $K\alpha_1$ and $K\alpha_2$ transitions.
Data listed are energy of line (in keV) and relative intensity (in brackets).

8.4.3 *Satellite lines*

The Auger effect has a further consequence concerning the appearance of satellite lines in the x-ray spectrum. This is shown schematically in Figure 8.11. The vacancy created in the *K*-shell by photo-electric ionization is filled by the transfer of an electron from say the *L*-shell. At this stage the atom has a single electron orbital vacancy in the *L*-shell. If the accompanying *Kα* x-ray is internally converted by another *L*-shell electron, the net effect is to leave the excited atom with two electron vacancies in its *L*-shell. As has already been noted, the overall control on the orbital energy of an electron is the magnitude of the attractive force between electron and nucleus. In a doubly-ionized atom, this potential energy will increase significantly, since the self-shielding effect on orbital electrons is reduced. The energy of x-rays emitted during electron transitions to de-excite this dual ionized species will, therefore, all be slightly increased compared with their singly-ionized counterparts resulting in the appearance of satellite lines in the x-ray spectrum.

Whereas this reasoning gives a satisfactory explanation for the presence of *L*, *M* satellite lines, it does not account for the prominent *K*-line satellite peaks in the spectra of many light elements. These must arise from the double ionization of *K*-shell electrons by primary x-ray photons from the x-ray tube. The relative importance of these *K* x-ray satellite lines may be judged from the data in Table 8.4, which shows that for elements lighter than about calcium,

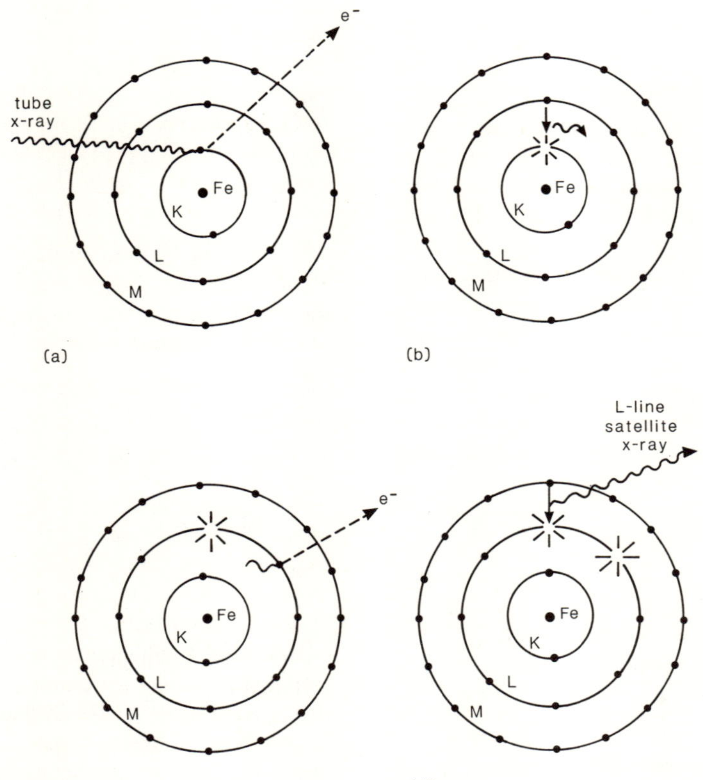

Figure 8.11 Mechanism for the generation of satellite x-ray lines in an iron atom, showing (*a*) photoelectric ionization of a *K*-shell electron; (*b*) $L \rightarrow K$ electron transition with generation of *K*-line x-ray; (*c*) internal capture (Auger effect) of *K*-line x-ray leading to ionization of an *L*-shell electron; (*d*) the atom is left with two electron vacancies in its *L*-shell which perturbs the orbital energy levels such that the resultant $M \rightarrow L$ electron transition generates an satellite *L*-line x-ray.

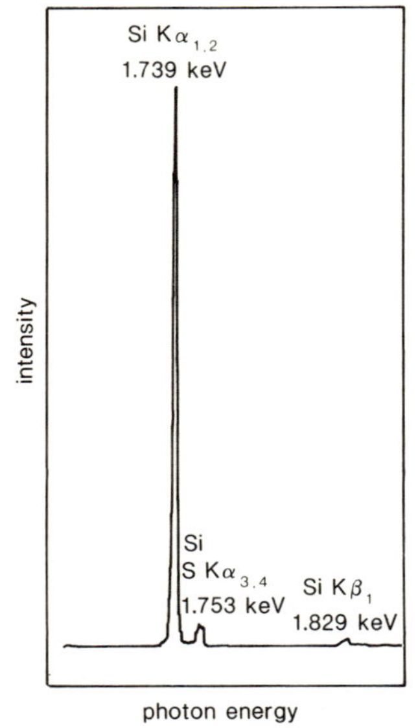

Figure 8.12 X-ray emission spectrum of silica showing the $K\alpha$, $K\beta$ and satellite $K\alpha_{3,4}$ lines.

the $SK\alpha_3$ and $SK\alpha_4$ satellite lines (the dual-ionized analogues to the $K\alpha_1$ and $K\alpha_2$ transitions) become increasingly prominent. To illustrate this phenomenon, the K-line spectrum of silicon is plotted in Figure 8.12. Satellite lines can be seen adjacent to their parent $K\alpha_{1,2}$ peak.

8.5 Excitation of x-ray spectra

8.5.1 *The x-ray tube—principles of operation*

In x-ray spectrometry, several sources have been used in routine analytical applications to excite samples, but by far the most widely used is by the x-ray tube. The traditional design is based on the Coolridge tube (Figure 8.13). This consists of an evacuated glass envelope containing a cathode which is held at a high negative potential and an anode which is earthed. The cathode comprises a cup-shaped focusing electrode and a tungsten filament through which sufficient current is passed to cause the wire to become incandescent. Under these conditions, thermionic emission of electrons occurs. Under the influence of the electrostatic fields created between the cathode and anode, electrons are accelerated

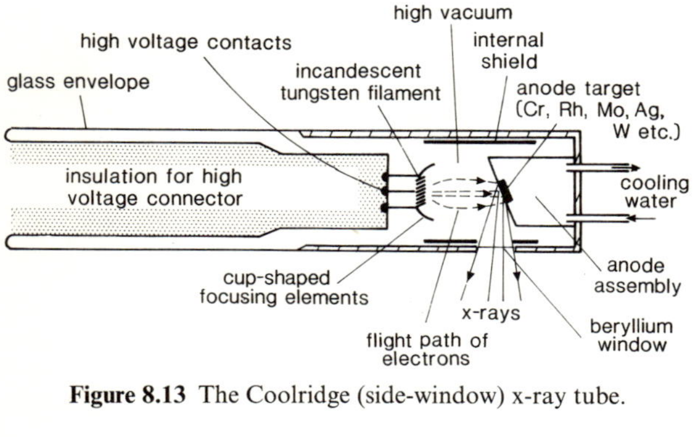

Figure 8.13 The Coolridge (side-window) x-ray tube.

through a potential gradient, normally selected by the analyst to be between 10 and 100 kV. Energetic electrons striking the anode cause the emission of x-rays that can escape from the tube through a window made of beryllium foil typically 200 to 500 μm thick. Beryllium is selected as window material since it is the lowest-atomic-number material that is available as a foil with the appropriate mechanical properties to form a vacuum-tight seal. The tube and power supplies are designed to give a long life and a highly stable beam of electrons impinging on the target.

There are three principal interactions between the electron beam and target:

(*a*) *Excitation of target atoms*. Energetic electrons cause ionization of target atoms, resulting in the emission of the characteristic K, L and M spectrum of the target material (see Figure 8.16). Since these characteristic photons play an important role in the efficient excitation of samples, the choice of target material (and therefore tube type) is an important consideration indetermining analytical strategy. Materials most commonly used for analytical work include chromium, molybdenum, rhodium, silver, gold, tungsten and scandium.

(*b*) *Tube continuum spectrum*. Characteristic line spectra sit on an x-ray continuum spectrum which is also excited within the target, having energies ranging from zero up to the operating potential of the x-ray tube (see Figure 8.16). This is called *bremsstrahlung* (or breaking) radiation and is generated when primary electrons are decelerated by interaction with the outer electrons of atoms of the target material. The shape of the continuum spectrum may be modelled mathematically from *Kramers' formula*, as a function of energy:

$$I = k[E_0 - E]/E \tag{8.4}$$

or of wavelength:

$$I = k'\left[\frac{1}{\lambda.\lambda_{\min}} - \frac{1}{\lambda^2}\right] \tag{8.5}$$

where I is the continuum intensity, E is the corresponding x-ray energy, λ is the corresponding x-ray wavelength, E_0 is the operating potential of the x-ray tube giving rise to a minimum continuum wavelength of $\lambda_{\min}$, and k and k' are constants.

It can be shown by differentiating Kramers' expressions that the continuum distribution has a maximum intensity at an x-ray energy/wavelength corresponding to

$$E = \tfrac{1}{2}V \tag{8.6}$$

or

$$\lambda = 2\lambda_{\min} \tag{8.7}$$

However, Jenkins (1976, p.16) suggests that various factors that attenuate the continuum spectrum generally shift this maximum to $E = 0.6$ V ($\lambda = 1.6\lambda_{\min}$). The importance of the tube continuum spectrum is that it considerably extends the useful range of excitation of an x-ray tube and is especially important in adequately exciting the light elements, though at the cost of some degradation in signal-to-background ratios.

(*c*) *Backscattered electrons*. The third important interaction

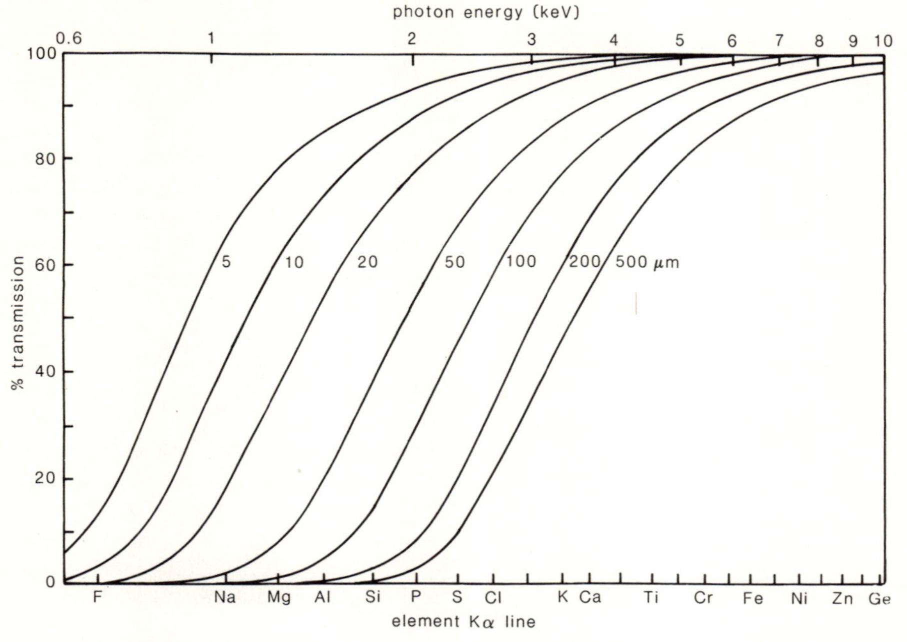

Figure 8.14 The attenuation properties of beryllium windows of thickness 5 to 500 μm as a function of x-ray energy.

at the anode is that a certain fraction of electrons is back-scattered out of the target. These energetic electrons interact with the walls and more importantly the adjacent beryllium window of the x-ray tube causing a heating effect. It is desirable that the target is positioned as close to the window as possible so that the maximum solid angle of x-rays is available to excite the sample. However, in order to withstand the mechanical strains created by this heating effect, windows must be made of relatively thick beryllium foil (often $>200\ \mu$m). This has the undesirable effect of seriously attenuating the low-energy continuum output from the tube, as can be seen from Figure 8.14. It was because of their thick (typically 1 mm) beryllium windows that older instruments were unable to excite adequately elements below magnesium in the Periodic Table.

(*d*) *Reverse bias tube*. One solution to this problem is to fit x-ray tubes with dished beryllium windows; that is windows that are concave in section, having a relatively thin centre to transmit low-energy continuum but retaining a thick outer circumference to maintain mechanical strength. However, by changes in the design of tube, this problem can be substantially overcome. In the reverse bias x-ray (Figure 8.15) the target anode is held at high positive potential; the rest of the tube including the walls and beryllium window are held at ground potential. Electrons are generated by thermionic emission from an incandescent filament, circular in shape. These electrons, under the influence of electrostatic fields created in the tube, impinge on the anode. However, since the cathode and the body of the tube including the beryllium window are at the same (ground) potential as the anode, electrons are deflected away from the window, reducing the heating effect in this material and permitting the use of thinner gauge beryllium foil.

(*e*) *Tube power*. The excitation of atoms using an x-ray tube source is an inefficient process, as is the detection of fluorescence x-rays using a wavelength dispersive spectrometer. In order to produce adequate count rates in analytical applications, it is necessary to excite samples with a high primary x-ray flux produced from a tube operated at a power rating of usually at least 3 kW on modern instru-

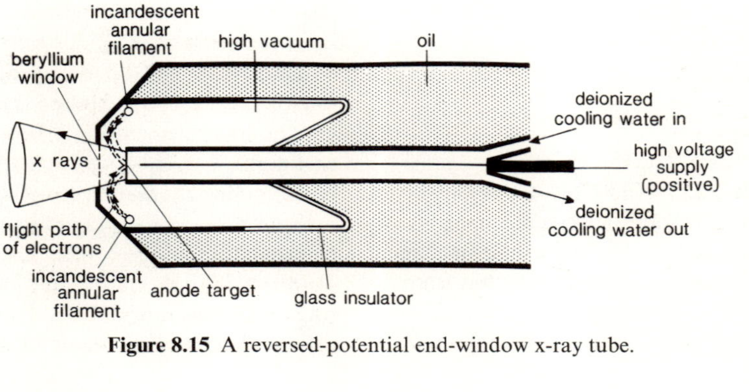

Figure 8.15 A reversed-potential end-window x-ray tube.

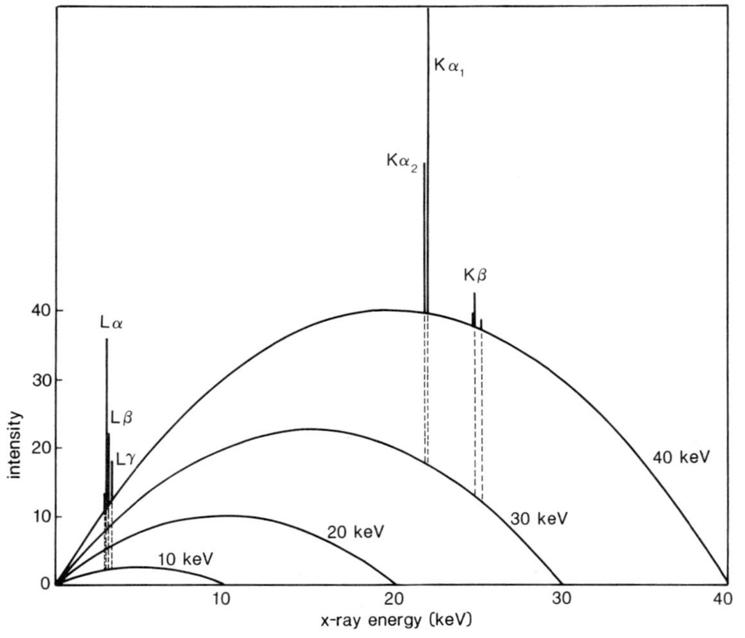

Figure 8.16 Schematic diagram of the spectrum from a silver-anode x-ray tube operated at 10, 20, 30 and 40 kV. The continuum has been calculated according to Kramers' formula and characteristic silver *K*- and *L*-lines are marked on the spectrum excited at 40 kV. The relative intensity of the latter have been considerably reduced in order to demonstrate their position with respect to the continuum. Dotted lines indicate other spectra in which characteristic tube lines appear. The FWHM of these tube lines is in the order of 1 eV.

ments. The tube power (in watts) can be calculated from the product of the operating potential (in kilovolts) and the current (in milliamps) of electrons emitted from the incandescent filament. To prevent self-destruction of the tube, it is essential that cooling water is circulated through the anode to dissipate the large amount of heat generated in these high-power tubes.

(*f*) *X-ray tube spectrum.* An x-ray spectrum from a silver anode tube is shown schematically in Figure 8.16 to illustrate the relationship between operating potential and emission characteristics.

If the tube is operated at between 10 and 20 kV, only the characteristic silver *L*-lines (at between 2.6 and 3.7 keV) are excited. Under these conditions, the sample is excited by these and a continuum component. At 30 kV, the silver *K*-lines at 22 to 25.5 keV are also excited but not very efficiently. At tube potentials above 40 kV, the silver tube *K*-lines are excited with increasing efficiency. Note also that there is a significant change in the distribution of energy in the tube continuum as a function of operating potential.

8.5.2 *Other excitation sources*

(*a*) *Radioisotope sources.* The x-ray and gamma-ray emissions from artificially-produced radioactive isotopes may be used as excitation sources for x-ray fluorescence analysis. However, the photon flux available from such sources is normally several orders of magnitude lower than that from a high-power x-ray tube. This shortcoming may only effectively be overcome by using such sources in conjunction with the more efficient detection characteristics of an energy dispersive spectrometer, and this mode of excitation is considered further in section 9.12.

(*b*) *Direct electron excitation.* In direct electron excitation, the sample is excited by a broad beam of electrons. In such instruments, the sample could be considered to behave as the anode of an x-ray tube. On a micrometre scale, this is of course the basis for electron microprobe analysis. However, in the analysis of bulk materials, direct electron excitation suffers the considerable disadvantage compared with x-ray tube excitation in that bremsstrahlung radiation is generated within the sample itself. This background radiation, therefore, competes for detection directly with the fluorescent lines of the elements of analytical interest. Thus, in general, peak-to-background ratios are considerably degraded, resulting in best detection limits in the 100 ppm range instead of 1–10 ppm. Moreover, rock samples must be both mixed with a suitable conducting binder to earth the electron charge and also analysed under high vacuum to minimize the absorption of exciting electrons.

Balanced against these operating difficulties is the fact that direct electron excitation is much more efficient in exciting the light elements (say boron to silicon) than photon excitation. This enhanced excitation efficiency outweighs the background disadvantages for these elements. This form of excitation has thus found some limited application in the determination of these light elements (as reviewed by Lucas-Tooth, 1978) but is not widely used in the analysis of silicate rocks. Robinson and Davis (1977) have described applications in the analysis of limestones.

8.6 Interaction of x-rays with matter

An understanding of the way in which the polychromatic beam of photons from an x-ray tube interacts with a sample to give the secondary x-ray spectrum is fundamental to x-ray fluorescence analysis.

8.6.1 *Beer's Equation*

When a beam of x-rays passes through a sample of finite thickness (x), the transmitted beam intensity (I) is related to the incident beam intensity (I_0) by the well-known Beer's Law (also known as the Beer–Lambert Law):

$$I = I_0 \exp(-\mu x) \tag{8.8}$$

where μ is the linear attenuation coefficient at the x-ray energy under consideration (normal units: cm^{-1}), and I represents the intensity of the beam transmitted at the same energy and direction as the incident beam I_o.

For the purposes of x-ray spectrometry, it is much more convenient to substitute for μ the *mass attenuation coefficient*, equal to μ/ρ (ρ is the sample density). The mass attenuation coefficient is thus the attenuation of unit area per unit mass of the sample and the attenuation equation becomes

$$I = I_0 \exp\left[-\left(\frac{\mu}{\rho}\right)\rho x\right]. \tag{8.9}$$

It should be noted that the term 'mass attenuation coefficient' is preferred to 'mass absorption coefficient' since the former implies that scatter of the incident beam as well as absorption within the sample are included.

The radiation of intensity ($I_o - I$), which is 'lost' within the sample, is absorbed or scattered by three principal phenomena.

(*a*) *Photoelectric absorption*. The photoelectric effect is the dominant form of interaction between x-ray beam and sample in the energy range under consideration, and is the process that gives rise to the fluorescent x-ray spectra already discussed in earlier sections. The incident x-ray causes the ionization of one of the inner orbital electrons with complete transfer of photon energy to electron. The subsequent electronic rearrangement of the excited atom results in the emission of characteristic fluorescent x-rays or, if internal capture occurs, Auger electrons. Tertian and Caisse (1982, p.17) estimated that photoelectric absorption usually accounts for more than 95% of all x-ray–sample interactions.

(*b*) *Rayleigh scatter*. Rayleigh scatter, also known as *elastic* or *coherent* scatter, results from the interaction of radiation with inner orbital electrons that are so strongly bound to the nucleus that electron excitation is not possible. Absorbed radiation is thus re-irradiated from the atom with its energy unmodified but with a random angular correlation with respect to the incident photon. The analytical significance of this effect is that a small fraction of the entire x-ray tube spectrum will be scattered off the sample into the detection system as well as any characteristic x-rays arising from photoelectric absorption within the sample itself. Thus the spectrum from samples fluoresced with a silver tube will contain both the scattered silver tube lines and continuum. High-atomic-number elements have a larger proportion of electrons tightly bound to a nucleus of high positive charge and will, therefore, be more efficient in promoting Rayleigh scatter than lighter elements. Similarly, low-energy tube radiation is more likely to fall below the threshold for exciting inner orbital electrons and will also therefore be more effectively scattered.

(*c*) *Compton scatter*. Compton scatter, also known as *inelastic* or *incoherent* scatter, results from the interaction of radiation with the more loosely bound outer electrons of an atom. Part of the energy of the incident photon is used in ionizing the electron, the remainder is re-irradiated as a photon of lower energy. The principles of conservation of energy and momentum apply to this interaction from which it can be shown that if E is the energy of the incident photon and θ the angle through which photon is scattered (see Figure 8.17), then the energy of the scattered photon E' is given by

$$E' = \frac{E}{1 + 0.001957(1 - \cos\theta)} \tag{8.10}$$

where E and E' are units of keV.

Conversely, in terms of wavelength shift ($\Delta\lambda$) (in Å),

$$\Delta\lambda = 0.0243\,(1 - \cos\theta). \tag{8.11}$$

In XRF instruments, the value of θ is fixed by the design of the spectrometer, being the angle subtended by lines drawn from the x-ray tube and spectrometer entrance slit to the surface of the analysed sample.

The intensity of Compton scatter photons varies as a function of scatter angle, as shown in Figure 8.18. Data in this diagram show that there is a broad minimum in intensity when scatter has occurred through an angle of 90–100° relative to the incident photon beam. This observation controls the design of all analytical x-ray spectrometers, since if the x-ray tube and spectrometer view the sample at about right angles, that component of the detected spectrum due to Compton scatter of the tube spectrum will be minimized. For $\theta = 90°$, $\cos\theta = 0$ and equations (8.9) and (8.10) reduce to

$$E' = \frac{E}{1 + 0.001957E}\,\text{keV} \tag{8.12}$$

and

$$\Delta\lambda = 0.0243\ \text{Å}. \tag{8.13}$$

Note that the wavelength shift in the scattered photon is independent of incident photon energy.

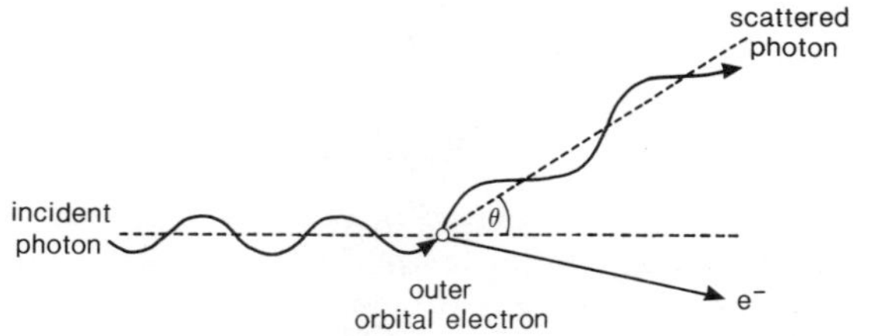

Figure 8.17 Mechanism of Compton scatter, showing the inelastic interaction of an incident photon with an outer orbital electron of an atom.

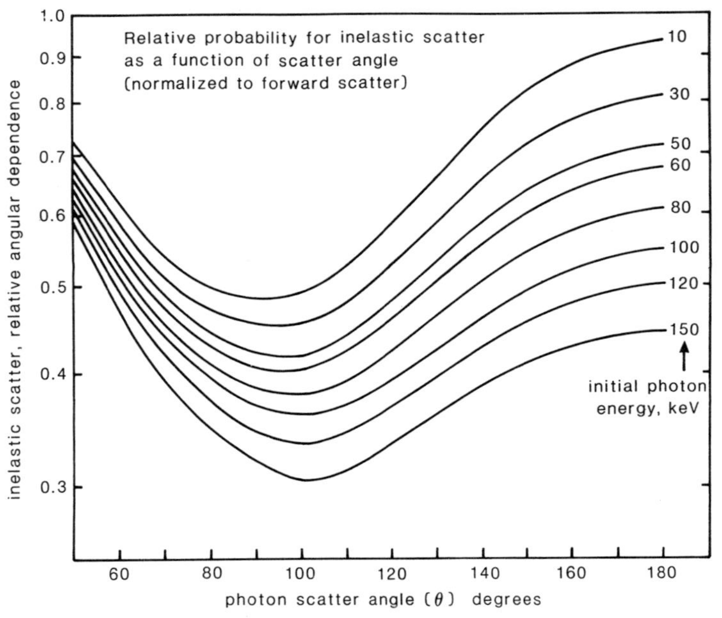

Figure 8.18 The relative probability of Compton (inelastic) scatter as a function of scatter angle. (Woldseth, 1973, p. 1.16.)

The analytical consequence of Compton scatter is that a modified peak of any characteristic or continuum line from the x-ray tube can be scattered from the sample into the detector, as shown in Figure 8.19. The energy of this scattered photon depends on spectrometer design (i.e. the value of θ) but will normally be modified, as given by equation

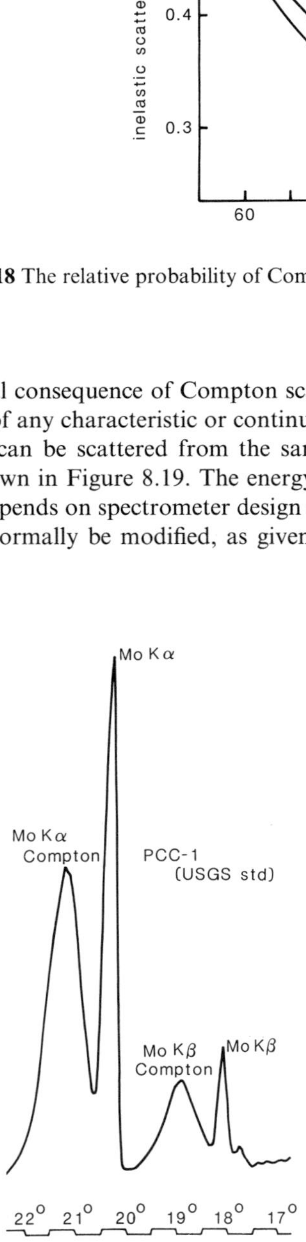

Figure 8.19 Wavelength scan of part of the spectrum scattered off a peridotite USGS: PCC-1, excited with a molybdenum-anode x-ray tube. The molybdenum *Kα* and *Kβ* lines are caused by Rayleigh scatter of the tube lines off the sample. The Compton peaks are caused by inelastic scatter of the tube lines off the sample. (Sumartojo and Paris, 1980.)

(8.12). The intensity of the Compton scattered spectrum bears a strong relationship to the mean atomic number of sample. Samples with light matrices (such as silicate rocks) are much more effective than heavier samples (such as metals and alloys) in scattering tube spectra. This greater effectiveness occurs because the binding energy between outer electrons and the nucleus is lower in low-atomic-number elements (due to the smaller positive charge on the nucleus) so enhancing the probability of Compton scatter. This phenomenon is the basis of a matrix correction procedure discussed further in section 8.8.4.

X-ray tube radiation scattered from the sample into the spectrometer by either Rayleigh or Compton processes is the main fundamental limitation to the detection limits that may be achieved by XRF analysis.

It must be emphasized that the Compton scatter peak shown in Figure 8.19 is not present in the spectrum emitted by the x-ray tube but is entirely a function of scatter in the sample. The way in which the relative proportions of Rayleigh and Compton scatter vary as a function of photon energy may be judged from the attenuation coefficient diagram for silicon (Figure 9.17).

8.6.2 *Absorption edges*

The variation of the mass attenuation coefficient (including contributions from scatter) for the elements Si, Ca and Fe with x-ray photon energy is shown in Figure 8.20 for the range 1 to 20 keV. There is an overall trend for the mass attenuation coefficient to decrease with increasing x-ray energy, as might be expected from the general increase in the penetrating power of higher energy x-rays. However, each plot shows a sharp discontinuity, known as an absorption edge. In these examples, the energy of the absorption edge corresponds to precisely the energy required to ionize a *K*-shell electron from the atom. This phenomenon has

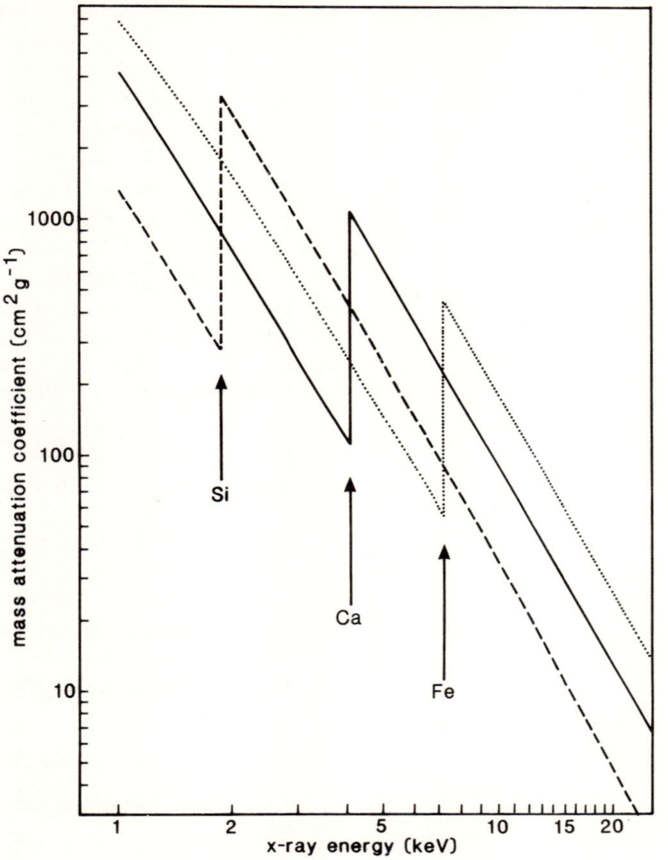

Figure 8.20 Mass attenuation coefficients of the elements silicon, calcium and iron plotted as a function of x-ray energy in the range 1–20 keV. The sharp discontinuities (absorption edges) correspond to the critical excitation energies of the respective *K*-shell orbital electrons.

important analytical consequences. An x-ray photon of energy just below the absorption edge is efficiently transmitted by the element; however one just above the absorption edge is strongly absorbed by the sample since it can then efficiently promote photo-electric ionization of a *K*-shell electron. As the energies of the characteristic $K\alpha$ and $K\beta$ photons always lie just below the *K*-line absorption edge, these lines are not efficiently absorbed by the element itself.

A further consequence relates to the analytical strategy in exciting a sample. Figure 8.20 shows that the most effective x-ray tube photons for exciting a sample are those with energies just above the appropriate absorption edge. Thus an x-ray tube with a chromium anode that emits characteristic lines at 5.41 keV (Cr $K\alpha_{1,2}$) and 5.95 keV (Cr $K\beta$) will effectively excite calcium (absorption edge 4.04 keV). However, silicon is likely to be excited predominantly by the low-energy continuum as its absorption edge (at 1.84 keV) is remote from the characteristic chromium tube lines. Conversely, the chromium tube lines cannot excite iron at all, since its absorption edge is at 7.11 keV. This element will only be excited, rather inefficiently, by tube continuum radiation above this threshold.

The magnitude of the step in attenuation at the absorption edge is an experimentally determined parameter called the *absorption jump ratio* (r) and is determined from the ratio of the mass attenuation coefficient at the top divided by that at the bottom of the absorption edge. The absorption jump ratio for a *K*-line edge (r_K) is a measure of the contribution that photoelectric absorption makes to the total attentuation at that energy.

Conversely, the *absorption jump factor* (J) is defined as that fraction of the total absorption which results in the ionization of *K*-shell electrons and is determined from

$$J = 1 - \frac{1}{r_K} = \frac{r_K - 1}{r_K}. \tag{8.14}$$

Absorption edges associated with *L*-shell electrons are more complex as these shells can split into three discrete energy levels (section 8.3.4) each of which therefore has an individual absorption edge associated with it. As an example, the variation in mass attenuation coefficient of barium as shown over the energy range 1–20 keV as plotted in Figure 8.21. The three distinct absorption edges at 5.99, 5.62 and 5.25 keV are associated with the $L_{\rm I}$, $L_{\rm II}$ and $L_{\rm III}$ shell ionization energies respectively. The *K*-shell absorption edge of barium lies at much higher energy (37.45 keV).

In the case of composite samples, the effective mass attenuation coefficient will be made up of contributions from each individual element in proportion to its concentration. The *bulk mass attenuation coefficient* ($\bar{\mu}$) can be calculated

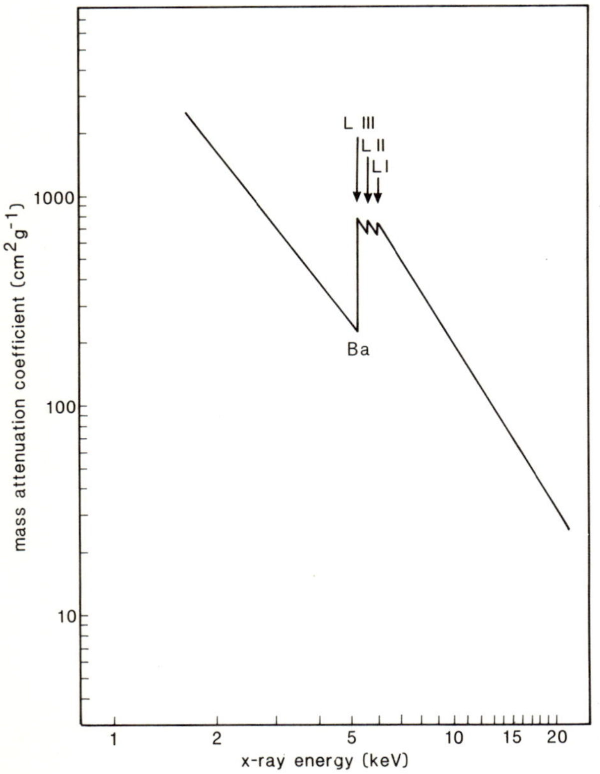

Figure 8.21 Mass attenuation coefficient of barium plotted as a function of x-ray energy. The sharp discontinuities correspond to the critical excitation energies of the L_I, L_{II} and L_{III} orbital subshells.

by summing each individual contribution as follows:

$$\bar{\mu} = \sum_{\substack{\text{all} \\ \text{elements}}} \left(\frac{\mu}{\rho}\right) . C \qquad (8.15)$$

where (μ/ρ) is the mass attenuation coefficient of an individual element at a given energy, and C is the weight fraction of that element in the sample such that:

$$\sum_{\substack{\text{all} \\ \text{elements}}} C = 1. \qquad (8.16)$$

As an example of the calculation of bulk mass attenuation coefficients, individual stages for a rock of assumed composition of 60% SiO_2, 20% CaO, 20% Fe_2O_3 and 2000 ppm BaO are shown in Table 8.5 for measurements corresponding to the Ca $K\alpha$ line at 3.690 keV. Note that the attenuation contribution of oxygen must be included in the calculation and that barium, as a trace element, makes a very small (almost insignificant) contribution to the overall summation.

In order to give some indication of the variation with energy of the bulk mass attenuation coefficient of a rock sample, this parameter has been plotted in Figure 8.22 for a sample having the composition of a basalt USGS: BCR-1. As in most silicate rocks, the most important contributions to the bulk mass attenuation coefficient are from the elements Fe, Ti, Ca, K, Si, Al, Mg, Na, and O, the absorption edges of which are clearly visible in Figure 8.22. However, if the sample is prepared for analysis as a fused glass disc, the attenuation contribution of the flux elements must also be taken into account. Tables of mass attenuation coefficients are used as the basis in some procedures for the correction of absorption interferences in quantitative analysis and published tabulations are available (e.g. Heinrich, 1966; Leroux and Thinh, 1977).

8.7 Matrix effects in geological samples

8.7.1 *Absorption-enhancement effects*

One of the most serious drawbacks to x-ray fluorescence analysis is that the intensity of an observed fluorescence line is not directly proportional to concentration but is affected by the concentrations of all other elements present in the sample. These interferences are called matrix, inter-element or more specifically absorption-enhancement effects and have stimulated a plethora of mathematical models designed to account for and correct these phenomena. In spite of the superficial algebraic complexity of mathematical correction procedures, the concepts on which most individual models are based are relatively straightforward. It should be noted that the colloquialism 'ZAF' should not strictly be applied to correction models for XRF analysis as it is in electron probe microanalysis, since XRF procedures are not as

Table 8.5 Example of calculation of bulk mass attenuation coefficient at the energy of the Ca $K\alpha$ line

	Si	Ca	Fe	Ba	O
oxide concentration	60%	20%	20%	2000ppm	
weight fraction element	0.280	0.143	0.140	0.0018	0.435
μ/ρ at 3.690keV (Ca $K\alpha_{1,2}$ line)	665	168	375	545	142
$\left(\frac{\mu}{\rho}\right) \cdot C$	186	24	53	1	62
$\bar{\mu} = \sum\left(\frac{\mu}{\rho}\right) \cdot C$	326 cm^2 g^{-1}				

Mass attenuation data (μ/ρ) taken from Jenkins and de Vries (1978).

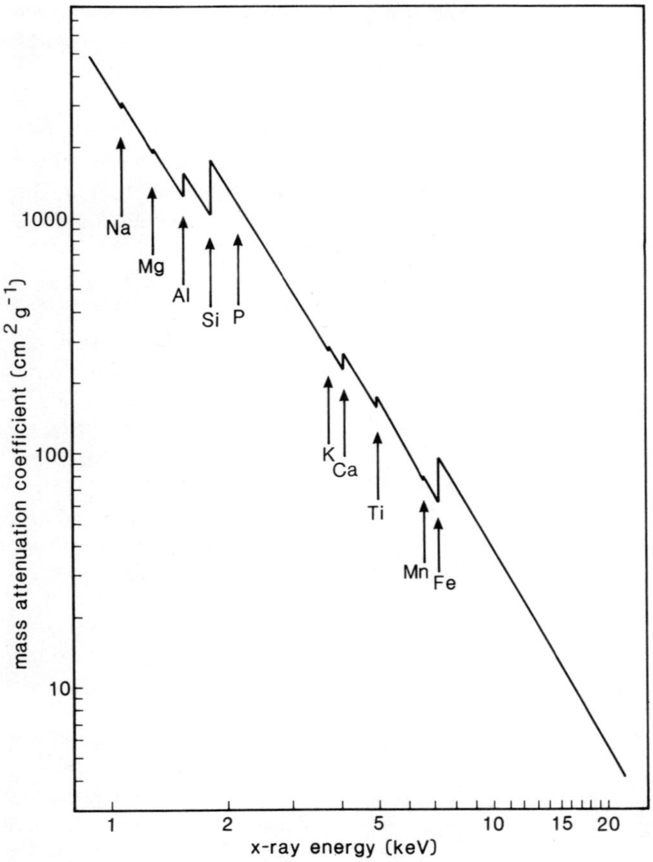

Figure 8.22 Bulk mass attenuation coefficient of an andesite rock (USGS: AGV-1) plotted as a function of x-ray energy, demonstrating the absorption edges caused by the major constituents of the sample.

directly related to atomic number, atomic weight and fluorescence effects.

Absorption-enhancement corrections must take into account the geometric configuration of sample excitation, as shown in Figure 8.23. This represents the excitation of a layer of the sample, of thickness dx, lying at a distance x beneath the surface. An element in this layer DE is excited by a cone of primary x-rays DEF from the x-ray tube. Fluorescence x-rays are emitted at all angles but only those within the cone BGH of solid angle Ω will enter the entrance slits (or collimator) of the spectrometer from segment B. The mean path length of primary radiation within the sample is represented by AB and depends on the mean take-off angle (ψ_1) of the x-ray tube with respect to the sample. Similarly

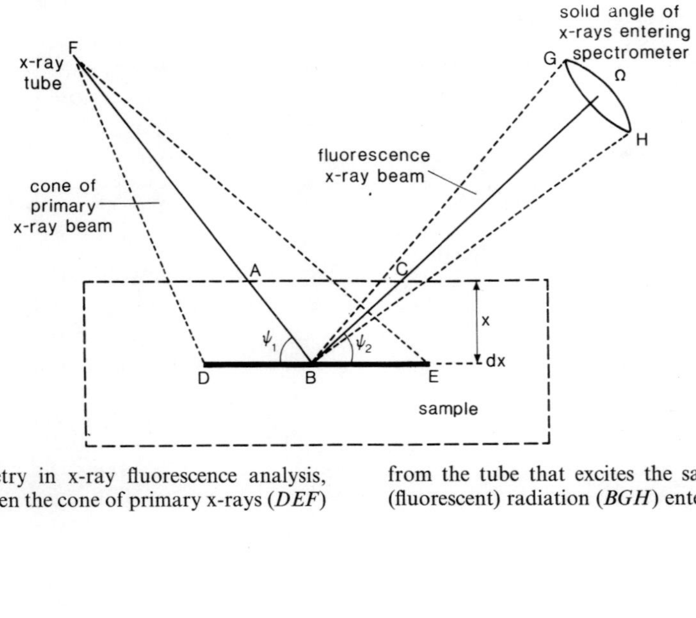

Figure 8.23 Excitation geometry in x-ray fluorescence analysis, showing the relationship between the cone of primary x-rays (DEF) from the tube that excites the sample, and a cone of secondary (fluorescent) radiation (BGH) entering the spectrometer.

the path length of fluorescence radiation within the sample (BC) depends on the take-off angle (ψ_2) of the x-ray spectrometer. The values of ψ_1 and ψ_2 are fixed by design so that ψ_1 is often greater than ψ_2.

The effects that must be included in an interference correction model include the following.

(i) *Attenuation of the primary x-ray beam* as it traverses path length AB to reach the layer DE in Figure 8.23. This effect is known as primary absorption.

(ii) *Attenuation of the fluorescence radiation* of each element as it traverses the sample along path length BC (secondary absorption).

(iii) *Secondary fluorescence* or second-element effects, where the observed fluorescence intensity depends on the presence of a second element. The secondary fluorescence effect occurs as follows, taking silicon as an example. The silicon $K\alpha$ line is likely to be intensely excited in rock samples. The energy of this line (1.739 keV) lies above the aluminium K-line absorption edge at 1.560 keV. There is therefore a possibility that a significant fraction of Si $K\alpha$ x-rays will suffer absorption along path length BC by aluminium atoms, causing them to fluoresce. The apparent count rate of Al $K\alpha$ will thus be enhanced by this process and that of Si $K\alpha$ will be diminished.

(iv) *Tertiary fluorescence* or third-element effects results form an extension of secondary fluorescence phenomena. In some rocks having high calcium concentrations, Ca $K\alpha$ radiation may be absorbed by silicon atoms within the sample, causing fluorescence of Si $K\alpha$ x-rays. These may themselves cause the secondary fluorescence of aluminium as described in (iii) above. The apparent count rate of aluminium $K\alpha$ is hence influenced by the concentration of calcium as well as silicon in this example of a third-element effect.

Much of the investigation of these effects has been carried out on the rather specialized Fe–Cr–Ni alloy system, the results of which do not necessarily assume the same importance when applied to silicate rock analysis.

Of the four effects described above, the secondary absorption effect (ii) is dominant; primary absorption (i) is somewhat less significant. Practical correction procedures applied to silicate rock analysis often ignore the contribution from second- and third-element effects ((iii) and (iv) above). Indeed, it has not been unambiguously demonstrated that these latter effects (especially (iv)) are large enough to warrant correction in silicate samples.

8.7.2 *Fundamental equation of excitation*

Ignoring interferences due to secondary and tertiary fluorescence, it is possible to set down the equation representing the apparent fluorescence intensity of the ith element (I_i) excited by polychromatic tube x-rays as follows.

If the sample is infinitely thick with respect to x-rays (see section 8.6.3) then

$$I_i = G.EX_i.C_i \int_{E_{\mathrm{ABS}_i}}^{E_{\max}} \frac{\left(\dfrac{\mu}{\rho}\right)_i \dfrac{I_{\mathrm{tube}}}{\sin\psi_1}}{\dfrac{\bar{\mu}_{\mathrm{tube}}}{\sin\psi_1}+\dfrac{\bar{\mu}_{K\alpha}}{\sin\psi_2}}\,\mathrm{d}E \qquad (8.17)$$

(i) G is a geometric factor to account for the solid angle of fluorescent x-rays entering the spectrometer:

$$G = \frac{\sin\psi_1}{\sin\psi_2}\cdot\frac{\Omega}{4\pi} \qquad (8.18)$$

(ii) Ex_i is an excitation factor for the element i defining the probability that an absorbed primary x-ray photon will result in the emission of a $K\alpha$ x-ray:

$$EX_i = \frac{r_K - 1}{r_K}\,.\,\omega_K \cdot g_{K\alpha} \qquad (8.19)$$

r_K = absorption jump ratio (section 8.6.2), ω_K = K-shell fluorescent yield (section 8.4.2), $g_{K\alpha}$ = proportion of K-line x-rays that appear as $K\alpha$ lines.

(iii) C_i is the true concentration of the element i.

(iv) The integration term is carried out over all energies from the absorption edge energy of the element i (E_{ABS_i}) to the operating potential of the x-ray tube ($E_{\max}$).

(v) The probability that tube radiation of intensity I_{tube} will excite the element i is given by the term

$$\left(\frac{\mu}{\rho}\right)_i \cdot \frac{I_{\mathrm{tube}}}{\sin\psi_1}$$

where $(\mu/\rho)_i$ is the mass attenuation coefficient of the element i. Both this term and I_{tube} are functions of energy and must be integrated.

(vi) It is also necessary to correct this expression for two absorption effects. The reduction in intensity of primary x-ray beam due to attenuation within the sample (cf. equation (8.15)) is given by

$$\bar{\mu}_{\mathrm{tube}} = \int_E \sum_{\substack{\text{all}\\ \text{elements}}} \left(\frac{\mu}{\rho}\right)_E . C_i. \qquad (8.20)$$

This term must also be integrated over the range of tube energies that excite the sample. If, however, an equivalent tube energy E_1 can be selected (perhaps the energy of a characteristic tube $K\alpha$ line), then this term effectively becomes a constant for a particular element and the expression reduces to a more simple summation:

$$\bar{\mu}_{\mathrm{tube}} = \sum_{\substack{\text{all}\\ \text{elements}}} \left[\left(\frac{\mu}{\rho}\right)_{E_1} . C_i\right] \qquad (8.21)$$

where mass attenuation coefficients $(\mu/\rho)E_1$ are selected at this equivalent energy (E_1).

(vii) Conversely, the attenuation of the fluorescence x-ray intensity is given by

$$\bar{\mu}_{K\alpha} = \sum_{\substack{\text{all}\\ \text{elements}}} \left(\frac{\mu}{\rho}\right)_{E_2} . C \qquad (8.22)$$

where mass attenuation coefficients are now those for energy E_2 corresponding to the $K\alpha$ fluorescent radiation.

(viii) The net effect of these two attenuation terms must take into account the difference in path length of primary and secondary radiation due to the take-off angles ψ_1, (x-ray tube) and ψ_2 (spectrometer):

$$\bar{\mu}_{\mathrm{net}} = \frac{\bar{\mu}_{\mathrm{tube}}}{\sin\psi_1} + \frac{\bar{\mu}_{K\alpha}}{\sin\psi_2} \qquad (8.23)$$

In the above derivation, it has been assumed for convenience and clarity that measurements are being made for the $K\alpha$ line of element *i*. However, the above expressions are equally valid on substituting data for any other x-ray line.

An important simplification can be made to the fundamental intensity equation (8.17). Assuming that it is possible to substitute an equivalent tube energy for excitation (see (vi) above), then for a given x-ray tube on a given instrument

$$C_i \propto I_i \left[\frac{\bar{\mu}_{\text{tube}}}{\sin \psi_1} + \frac{\bar{\mu}_{K\alpha}}{\sin \psi_2} \right] \tag{8.24}$$

which may also be written as:

$$C_i \propto I_i \left[\bar{\mu}_{\text{tube}} + \bar{\mu}_{K\alpha} \frac{\sin \psi_1}{\sin \psi_2} \right] \tag{8.25}$$

This form of the equation is the starting point for several practical correction procedures, as will be discussed further in later sections.

8.7.3 *Mineralogical effects*

The choice of correction procedures for absorption-enhancement effects in rock samples is influenced by mineralogical effects and sample preparation procedures designed to reduce the magnitude of these phenomena. It is appropriate, therefore, to examine sample preparation techniques before discussing mathematical correction models further.

As with almost all other techniques for bulk analysis, samples must first be ground to a flour-like consistency to ensure that the sample presented for analysis is homogeneous and representative of the original specimen. The normal specification for analytical work is that this powder must pass a 200-mesh sieve having a nominal sieve opening size of 74 μm. It is in fact impractical to crush powders routinely to a much finer grain size, because of the reduction in efficiency of the milling equipment and the increased degree of contamination introduced into the sample.

Early analytical procedures proposed the analysis of samples for all elements on simple powder pellets prepared by mixing rock powder with a suitable binding agent and compressing into a circular briquette. Whereas this is still the preferred technique for the heavy elements (especially those above iron in the Periodic Table), analytical discrepancies were apparent in the determination of the light elements, especially Na, Mg, Al and Si. Reliable calibrations could be carried out only over very narrow composition ranges. In order to understand the source of these errors, it is necessary to examine the critical depth to which a sample is excited for individual elements.

(*a*) *Critical depth of penetration—powder pellets.* As has been discussed above, fluorescence x-rays are significantly attenuated within a sample. There must, therefore, be some critical depth below the surface of a sample beyond which a fluorescent x-ray can no longer penetrate for detection by the spectrometer. This critical depth will vary from element to element, being dependent on both the penetrating power (i.e. energy of the corresponding x-ray photon) and sample composition.

It can be shown (see, for example, Bertin 1975, p.624) that if I_∞ is the intensity of an x-ray emission from a sample that is infinitely thick and I_x is the corresponding intensity from a sample of thickness x, then

$$I_x = I_\infty [1 - \exp\{-\bar{\mu}_{\text{net}} \rho x\}] \tag{8.26}$$

where ρ is the sample density and $\bar{\mu}_{\text{net}}$ is the net bulk mass attenuation coefficient as defined in equation (8.23).

If the ratio I_x/I_∞ is set at 0.99 (i.e. the sample is sufficiently thick to emit 99% of the radiation of an infinitely thick sample) then, solving equation (8.26) for x:

$$x = \frac{4.605}{\rho \cdot \bar{\mu}_{\text{net}}} \tag{8.27}$$

If $I_x/I_\infty = 0.999$ is chosen as the limit, then the term 4.605 in equation (8.27) must be replaced by 6.908.

This expression has been solved for various fluorescence x-ray lines for a pressed powder pellet of assumed effective density 2.1 kg m^{-3} and taking $\psi_1 = \psi_2 = 45°$ (Table 8.6). A further simplification has been made in assuming that that sample is excited exclusively by the $K\alpha$ line of the indicated x-ray tube (not necessarily valid for Na, Si, etc.). Calculations are based on the composition of andesite USGS: AGV-1.

Data in Table 8.6 show that 99% of all sodium $K\alpha$ radiation is detected from within 5 μm of the surface of the sample. The corresponding figures for Si and Ca are 13 μm and 36 μm respectively. In fact, if the pellet had an overall thickness of about 4 mm, as would be the case for a 6 g sample compressed to a 31.5 mm disc, then only 0.12% of the sample mass would be analysed for sodium, 0.3% for Si and 0.9% for Ca. For heavier elements, critical penetration depths increase from 0.18 mm for Fe, to 1.4 mm for Nb to over 10 mm for the REE *K*-lines.

(*b*) *Particle size distribution in powder pellets.* These data must be interpreted in relation to the range of particle sizes obtained on grinding a rock. Data from Leake *et al.* (1969) are reproduced in Figure 8.24 and show the particle size histogram for various rock types ground and sieved through a BS 250-mesh sieve (of aperture 68 μm). Particle sizes range from 10 to 55 μm with a maximum frequency at about 35 μm.

Taken together, these results indicate that the thickness of a compressed powder pellet for which analyses for light elements such as Na, Mg, Al and Si are representative corresponds to less than one monolayer of rock particles. Only in unusual circumstances is this layer likely to be completely representative of the bulk sample. In routine applications it is not always acceptable to sieve all samples, because of the additional time that must then be committed to sample preparation and also the uncertainties as to whether sieving will cause mineral fractionation effects. Without sieving, it is likely that the particle size distribution will be extended to an even wider range than that shown in Figure 8.24. This sample preparation problem is likely to be particularly serious in samples that contain flaky minerals such as biotite, which tend to align preferentially parallel with the pellet surface during compression.

Not only is the volume of sample analysed for the light elements Na to Si mineralogically unrepresentative of the

Table 8.6 Critical penetration depth of characteristic $K\alpha$ x-rays in powder pellets and glass fusion discs

				Critical penetration depth				
						Glass discs		
Element	Line	keV	X-ray tube	Powder pellet μm	Powder pellet %*1	1:6 Lithium borate (μm)	1:6 Heavy absorber (μm)	1:2 Low dilution (μm)
Na	$K\alpha$	1.04	Cr	4.8	0.12	4.8	4.2	4.8
Si	$K\alpha$	1.74	Cr	13	0.33	18	12	16
Ca	$K\alpha$	3.69	Cr	36	0.9	94	64	66
Fe	$K\alpha$	6.40	Rh	180	4.5	550	146	370
Rb	$K\alpha_1$	13.39	Rh	900	22.5	3000	749	1890
Nb	$K\alpha_1$	16.61	Rh	1400	35.0	4500	1090	2870
La	$K\alpha_1$	33.44	W	10600	—	23100	600	17700
Eu	$K\alpha_1$	41.53	W	15400	—	28800	35100	23400
Rh	$K\alpha_1$	20.21	tube	3930	—	12400	3080	8020

Calculations are based on the equations presented in the text using the composition of andesite USGS: AGV-1.
To simplify calculation of the primary beam absorption effect, it has been assumed that each element is exclusively excited by the $K\alpha$ line of the x-ray tube indicated above.
The effective density of a powder pellet has been taken as 2.1 kg m^{-3} and of glass discs, 2.3 kg m^{-3} (from laboratory measurements).
*1: Data in this column represent the proportion of powder pellet from which fluorescent x-rays can escape assuming that its total thickness is 4 mm.
Glass disc calculations are based on rock to flux dilution ratios of 1:6 or 1:2 as indicated, using the following flux compositions:
lithium borate: 80% lithium metaborate–20% dilithium tetraborate eutectic mixture.
heavy absorber: 85% dilithium tetraborate–15% lanthanum oxide.
low dilution: 100% dilithium tetraborate.

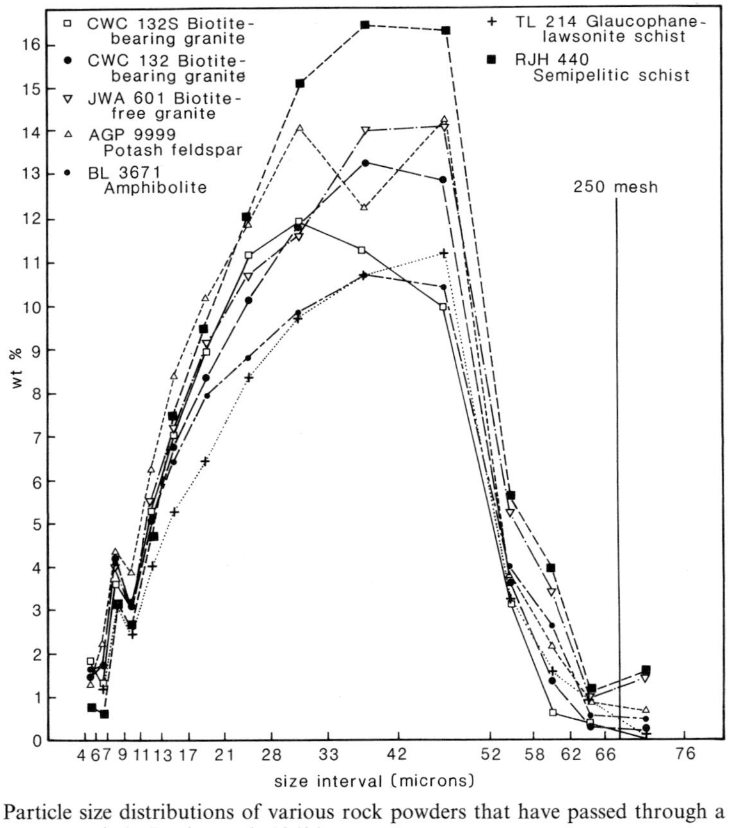

Figure 8.24 Particle size distributions of various rock powders that have passed through a BS-250 mesh sieve (68 μm aperture size). (Leake *et al.*, 1969.)

bulk, but also attenuation corrections can no longer be estimated with any degree of certainty. In these circumstances, attenuation occurs within discrete mineral phases, but the attenuation correction is calculated from the bulk composition of the sample. Errors will arise, since corrections based on the bulk sample cannot accurately model attenuation in individual phases that may have quite diverse chemical compositions. These arguments do not apply with the same force to elements significantly heavier than calcium, because fluorescent x-rays can escape from a more representative volume of sample. Although powder pellets are the standard method of determining the heavier major and trace elements, reliable data cannot normally be obtained for Si, Al, Mg, and Na unless meticulous precautions are taken over powder preparation and calibration.

(*c*) *Elimination of matrix effects by lithium borate fusions.* The principal method of overcoming mineralogical effects to permit reliable determination of light elements is to break down the rock matrix by fusion with a suitable flux prior to analysis. The sample and flux are mixed and heated in a crucible. The sample is dissolved in the molten mixure which is then cast as an amorphous glass disc suitable for analysis without further treatment. The use of such techniques in XRF analysis was first proposed by Claisse (1956, 1957) who fused samples with a flux based on sodium borate (borax). In modern techniques in which sodium is one of the elements sought, samples are usually fused with fluxes consisting of dilithium tetraborate ($Li_2B_4O_7$) and/or lithium metaborate ($LiBO_2$), following developments by Andermann and Allen (1961). Mixtures of sample and flux must be heated to temperatures of between 950° and 1200° to ensure complete dissolution of the silicate sample. When quenched and annealed as a disc, lithium borate fluxes form clear glasses that are durable and resist the absorption of water, unlike their sodium counterparts. As rock powders are diluted in constant ratio with flux, samples analysed as glass discs have a further advantage, namely mass attenuation *differences* between samples of similar composition are reduced. In this respect, Claisse (1956) found that matrix effects were reduced considerably by the addition of BaO_2, $BaSO_4$ or $K_2S_2O_7$ to his fusion mixtures. These compounds were primarily used to promote oxidation of sulphides in ore samples for which the analytical program was being developed. However, Rose *et al.* (1962) made use of this observation by developing a heavy absorber technique in which lanthanum oxide (La_2O_3) was added to dilithium tetraborate flux to suppress matrix effects even further. Lanthanum oxide was selected because it is not an element that is normally analysed with the major elements, causes the least spectrum line overlaps and promotes glass formation in cast discs. Heavy absorber fluxes incorporating lanthanum oxide are now the standard sample preparation technique for major element analysis by wavelength dispersive XRF.

The properties and applications of various lithium borate flux mixtures have been reviewed by Bennett and Oliver (1976). The principal properties of individual materials are as follows. *Dilithium tetraborate* (melting point 920°C) is a basic flux which readily dissolves alumina but which attacks silica more slowly. As it melts to form a relatively viscous liquid, fusion mixtures must be heated to 1200°C to ensure homogeneity of the melt. In contrast, *lithium metaborate* has a relatively low melting point (845°C), is more mobile when fused and will efficiently dissolve acidic oxides.

As a 'universal' flux, Bennett and Oliver (1976) proposed a *eutectic mixture* of 20% dilithium tetraborate to 80% lithium metaborate (melting point 840°C) to combine the advantages of both components. This flux is capable of dissolving a wide range of alumino-silicates varying from 100% SiO_2 to 100% Al_2O_3, providing it is mixed with the sample in the proportion 1 part of sample to 5 of flux.

When lanthanum oxide is added to fluxes, it increases the basicity of the melt. In the formulation of *heavy absorber fluxes*, this material is normally mixed with dilithium tetraborate in the proportion 15–20% La_2O_3 to 85–80% $Li_2B_4O_7$. As mentioned above, lanthanum oxide fluxes have the further advantage that like silicon and phosphorus, lanthanum is a network-forming element and so promotes glass formation. Heavy absorber fluxes may thus be used with advantage to form stable glass fusion discs of samples containing unusually high Mg or Ca concentrations—samples which tend to devitrify if prepared with some other fluxes.

(*d*) *Optimizing sample-to-flux ratios.* In choosing the appropriate flux-to-sample ratio, a balance must be reached between (i) minimizing (or even eliminating) matrix attenuation effects at high dilution ratios on the one hand, and (ii) ameliorating the loss in analytical performance due both to the reduced sensitivity (analytes being diluted by flux in the sample analysed) and higher background counts (due to the larger proportion of low atomic number matrix in the sample) that will then occur.

It is of interest to estimate the degree of dilution required to eliminate effectively mass attenuation diffences between samples. Very approximately, the limit appears to be one part of rock to 50 parts of a lithium borate flux, or one part of rock to between 10 and 15 parts of a lanthanum heavy absorber flux. However, in most schemes of analysis, the compromise posed in the first paragraph of this section is met at sample to flux ratios of one to five or one to six. Under these conditions, the mineralogical effect is eliminated, adequate sensitivity is retained, but some form of absorption-enhancement correction must still be applied to measured intensities. Although lanthanum-based heavy absorber fluxes are preferred for wavelength dispersive instruments (Norrish and Hutton, 1969), they are not appropriate for energy dispersive XRF techniques (see section 9.7). Furthermore, in wavelength dispersive spectra the characteristic lines of lanthanum do cause some minor overlap interferences as is discussed further in section 8.10.2.

Recently, 'low dilution' fusion mixtures with sample-to-flux ratios of 1 to 2 have been used (for example, Thomas and Haukka, 1978; Hutton and Elliott, 1980; Lee and McConchie, 1982). These mixtures both eliminate the mineralogical effect, allowing determination of the light major elements, but also retain adequate sensitivity to facilitate the determination of some trace elements such as the Rb–Sr–Y–Zr–Nb group of elements. Such procedures also obviate the necessity of preparing two samples for analysis (powder pellet for trace elements and glass disc for major elements). It has not been unambiguously established that all accessory minerals (e.g. zircon) dissolve satisfactorily in such flux mixtures (see, for example, Cremer and Schlocker, 1976).

The practical problems of preparing glass discs and powder pellets will be discussed further in section 8.10. However, it is of particular relevance here to consider the x-ray absorption properties of these materials in comparison with powder pellets.

(*e*) *Critical penetration depths in glass discs.* The critical penetration depths of x-rays from selected elements for three glass disc mixtures are listed in Table 8.6. Calculations have been carried out for a 1 : 6 dilution with a lithium tetraborate/metaborate eutectic flux, an equivalent dilution with a lanthanum heavy absorber flux and a 1 :2 low dilution mixture, using the compositions listed in Table 8.8.

As might be expected, the critical penetration depth of x-rays from the heavier elements in the 1 : 6 and 1 : 2 glass discs is significantly greater than that for a powder pellet. Thus, data for iron vary from 180 μm (powder pellet) to 550 μm (1 : 6 glass disc) to 370 μm (1 : 2 glass disc). However, perhaps surprisingly, the penetration depth of sodium $K\alpha$ is virtually identical, irrespective of sample preparation procedure and that for Si $K\alpha$ not very different. The reason for this is that these light element x-rays are heavily attenuated by oxygen either from the sample itself (powder pellet) or contributed by the flux.

The behaviour of elements in the lanthanum heavy absorber flux is interesting. Below the lanthanum *L*-shell absorption edge (5.5 keV), the presence of lanthanum slightly increases the attenuation of characteristic x-rays, though not markedly so, critical penetration depths being of the same order as for the 1 : 2 lithium borate glass discs. However, for characteristic x-rays of energies higher than that of the lanthanum absorption edge, much stronger attenuation occurs. Thus the critical penetration depth of iron is reduced to 146 μm, a value lower than that for a compressed powder pellet (180 μm).

Unless particularly thick samples are prepared, neither of the samples prepared as glass discs at dilutions of 1 : 6 or 1 : 2 with lithium borate-based fluxes can be considered infinitely thick with respect to Rh $K\alpha$ radiation. This precludes the use of scattered tube radiation for estimating mass attenuation corrections (see section 8.8.4). Furthermore, data in Table 8.6 would suggest that 1 : 2 'low dilution' glass discs cannot be considered infinitely thick to niobium $K\alpha$ radiation unless they are thicker than about 3 mm.

8.7.4 *The magnitude of inter-element effects in silicate rock samples*

(*a*) *Powder pellets.* In order to gain some insight into the relative magnitude of the x-ray absorption of one element on another, the mass attenuation contributions from individual elements on selected fluorescent x-ray lines are listed in Table 8.7. These calculations have been carried out for a powder pellet of a rock having the composition of an andesite USGS : AGV-1. Each entry represents the value of (μ/ρ). C, where C represents the weight fraction of attenuating element and (μ/ρ) the mass attenuation coefficient at the energy of the indicated $K\alpha$ line. The sum term at the foot of each column therefore represents $\bar{\mu}_{K\alpha}$ (see equation (8.22)). These data show that the major absorber of Na $K\alpha$ radiation is oxygen, that oxygen and aluminium are major absorbers of Si $K\alpha$, and that silicon is the dominant absorber of both Ca $K\alpha$ and Fe $K\alpha$ in a rock of this composition. The absorption contribution of the trace elements on these major element lines is small and can be ignored. Conversely, the absorption effect of the major elements on the Rb $K\alpha$ and Nb $K\alpha$ trace element lines is small but significant. Although data in this table are representative only of the identified rock, the trends

Table 8.7 Relative contributions of individual elements to the mass attenuation coefficient for the $K\alpha$ lines of selected elements in andesite USGS : AGV-1

Absorbing element			X-ray line suffering attenuation					
			Na $K\alpha$	Si $K\alpha$	Ca $K\alpha$	Fe $K\alpha$	Rb $K\alpha$	Nb $K\alpha$
Element	Weight %	keV:	1.041	1.739	3.690	6.398	13.373	16.581
		Å:	11.91	7.126	3.359	1.938	0.927	0.748
O	47.23		605	146	17	3.2	0.4	0.2
Na	3.15		82	21	1.8	0.6	0.07	0.04
Mg	0.92		2	8	1.1	0.2	0.03	0.01
Al	9.07		25	92	13.1	2.7	0.3	0.2
Si	27.67		110	28	49.2	10.3	1.6	0.7
P	0.21		1	0.3	0.4	0.1	0.01	0.006
K	2.41		26	6	8.8	2.4	0.3	0.1
Ca	3.53		43	11	1.6	3.8	0.5	0.2
Ti	0.63		9	3	0.4	0.9	0.1	0.05
Mn	0.07		1	0.5	0.1	0.02	0.02	0.09
Fe	4.73		95	31	4.8	1.1	1.2	0.7
traces	0.25		1	0.4	2.1	0.33	0.06	0.06
sum			1000	347	100	25.7	4.6	2.4

Tabulated data represent the product of the weight fraction of element times the mass attenuation coefficient at the energy of indicated $K\alpha$ lines.
In order to facilitate comparison between elements, all data has been arbitrarily normalized such that sum for sodium $K\alpha$ equals 1000. Original coefficients may be recalculated by multiplying all data by 3.125.
Mass attenuation coefficients taken from Jenkins and de Vries (1978).

Table 8.8 Composition and x-ray attenuation characteristics of glass discs

Rock : flux ratio Flux type		1:6 Meta/tetra-borate eutectic	1:6 Heavy absorber	1:2 Tetra-borate
Composition of glass disc				
% rock		14.3	14.3	33.3
% lithium metaborate		68.6	0	66.7
% dilithium tetraborate		17.1	72.9	0
% lanthanum oxide		0	12.8	0
Elemental contribution of flux to composition of glass disc				
% O		55.4	50.1	42.9
% Li		11.0	6.0	9.3
% B		19.3	18.6	14.5
% La		0	11.0	0
Percent contribution of flux to bulk mass attenuation coefficient for selected element Kα lines				
Na	1.041keV	84.4	86.7	64.3
Si	1.739keV	79.0	86.5	55.4
Ca	3.359keV	60.3	72.2	33.5
Fe	6.398keV	53.1	87.6	27.2
Rb	13.373keV	46.7	86.7	22.5
Nb	16.581keV	47.4	87.5	23.0

illustrated here are expected to be followed by other rocks of 'typical' silicate composition.

(*b*) *Glass discs.* An estimate of the influence of flux on the attenuation of selected x-ray lines is shown in Table 8.8. Here, the compositions of the the three flux types selected for consideration in Table 8.6 are listed together with the percentage contribution that each flux makes to the overall mass attenuation coefficient of selected $K\alpha$ lines. In the case of glass discs made up from the two lithium borate fluxes, the contribution of flux to the overall x-ray attenuation coefficient varies from 84% (for Na $K\alpha$) in the 1:6 eutectic down to 23% (for Nb $K\alpha$) in the 1:2 'low dilution' mixture. As might be expected, the flux contribution for the 1:6 heavy absorber mixture is higher, at about 87% for all x-ray lines except for Ca $K\alpha$ and other lines that fall just below the La L absorption edge. In this case the rock contributes only about 13% to the total sample attenuation. *Differences* in rock composition will not greatly influence the overall absorption of x-rays in 1:6 heavy absorber samples, so minimizing the magnitude (and uncertainty) in interference corrections.

8.8 Mathematical procedures for the correction of absorption-enhancement effects

8.8.1 *Fundamental parameter procedures*

Fundamental parameter models use the basic equation of excitation (equation (8.17)) and compatible equations for secondary fluorescence (see Tertian and Claisse, 1982, p.60) to calculate the magnitude of absorption-enhancement effects. These procedures require an accurate knowledge of various experimentally determined parameters including the fluorescent yield (ω_K), the absorption jump ratio (r_K), element mass attenuation coefficients and x-ray tube intensity, both the latter being functions of x-ray energy.

The best known algorithm is that of Criss and Birks (1968), which uses pure elements as standards. In this form fundamental parameter calculations may be applied to any sample matrix. These procedures have not, however, found widespread use in quantitative geological applications due to errors resulting from uncertainties in the fundamental parameters themselves.

8.8.2 *Iterative procedures based on the calculation of matrix attenuation*

(*a*) *Background.* Procedures used in geological applications include those that seek to correct inter-element attenuation effects according to equation (8.25) which relates the concentration of element i (C_i) to its x-ray intensity I_i by the relationship

$$C_i \propto I_i\left[\bar{\mu}_{\text{tube}}+\bar{\mu}_{K\alpha}\left(\frac{\sin\psi_1}{\sin\psi_2}\right)\right]$$

In applying this model to silicate rocks, the following assumptions are normally made:

(i) Secondary (and tertiary) fluorescence effects may be ignored.

(ii) Samples will be calibrated against standards of similar composition and calibrations are expected to be valid over restricted composition ranges (a silicate calibration would not be valid for the analysis of carbonate samples, for example).

(iii) The attenuation of tube radiation ($\bar{\mu}_{\text{tube}}$ in the equation above) is often assumed to be negligible in comparison with $\bar{\mu}_{K\alpha}$ (or alternatively, experimentally determined mass attenuation parameters are assumed to compensate for the small effect of this tube term (see below)).

(iv) Intensity data is normalized against a monitor of similar composition to unknown samples so that drift in the spectral distribution of the x-ray tube can be accounted for.

Procedures based on these principles have been described by Norrish and Hutton (1969) and further developed by Harvey *et al.* (1973) for glass discs, and Holland and Brindle (1966) and Brown *et al.* (1973) for powder pellets.

In simple terms, the algorithms described by these workers seek first to compare a measured intensity with the intensity *v.* concentration calibration line to determine an approximate (uncorrected) concentration. When all major constituents have been analysed in this manner, the set of apparent concentration data is used to calculate the inter-element mass attenuation data for each x-ray line measured using either published coefficients (e.g. Heinrich, 1966; Leroux and Thinh, 1977) or experimentally determined data (Norrish and Hutton, 1969). The mass attenuation terms so derived are used to revise the uncorrected concentrations. The procedure is then repeated, but using the revised concentrations as initial estimates until differences between values obtained from successive iterations are insignificant. Usually less than five interations are necessary. If more than ten are required, the correction is probably not converging and errors in data input should be suspected. The method is therefore an 'iterative self-consistent mass attenuation correction procedure'. An example (Table 8.9) of the output from one such algorithm taken from Holland and Brindle

Table 8.9 Example of an iterative self-consistent mass attenuation correction in the XRF analysis of granite USGS: G-1 s

Iteration number	1	2	3	4	5	6	7
SiO_2	74.227	73.124	72.958	73.020	73.011	73.012	73.012
Al_2O_3	15.099	14.316	14.480	14.457	14.460	14.460	14.460
Fe_2O_3	1.799	1.867	1.933	1.927	1.927	1.927	1.927
MgO	0.378	0.406	0.410	0.409	0.409	0.409	0.409
CaO	1.008	1.406	1.436	1.430	1.430	1.430	1.430
Na_2O	3.281	3.450	3.476	3.468	3.469	3.469	3.469
K_2O	3.971	5.087	5.044	5.028	5.031	5.031	5.031
TiO_2	0.214	0.231	0.238	0.237	0.237	0.237	0.237
MnO	0.0229	0.0235	0.0243	0.0242	0.0242	0.0242	0.0242

Data from Holland and Brindle (1966).

(1965) shows the convergent nature for iterative corrections in the analysis of a granite, USGS: G-1.

(*b*) *Glass discs—the method of Norrish and Hutton.* Special considerations are required in the analysis of glass discs to take into account the mass attenuation properties of the flux. The approach developed by Norrish and Hutton (1969) but incorporating some modifications by Harvey *et al.* (1973) will now be described, since it was developed specifically for the analysis of silicate rocks.

Ignoring primary absorption and secondary fluorescence, the concentration of an element (C) is given by:

$$C = kI\bar{\mu} \tag{8.28}$$

where k is a calibration constant,
I is the net intensity (peak minus background) of the measured line of the element, and μ is the bulk mass attenuation coefficient of the sample prepared as a glass disc at the energy of the appropriate x-ray line.

In this application, it is convenient to calculate $\bar{\mu}$ in terms of $\bar{\mu}_{\text{flux}}$, the bulk mass attenuation coefficient of a glass disc as if it were composed entirely of flux, to which must be added the increase in attenuation resulting from the substitution of rock for flux as given by

$$\cdot \sum_{\substack{\text{all}\\ \text{elements}}} (\mu_i - \bar{\mu}_{\text{flux}}) \,.\, C_i$$

where μ_i and C_i are the mass attenuation coefficient and concentration fraction of the ith element. This additional attenuation must be added in proportion to the ratio of rock to flux defined by the *dilution factor f* where

$$f = \frac{\text{weight of rock}}{\text{weight of rock} + \text{flux}}. \tag{8.29}$$

The bulk mass attenuation coefficient of the glass disc is then given by

$$\bar{\mu} = \bar{\mu}_{\text{flux}} + f. \sum_{\substack{\text{all}\\ \text{elements}}} (\mu_i - \bar{\mu}_{\text{flux}}) \,.\, C_i \tag{8.30}$$

If (i) the intensity–concentration calibration is linear and passes through the origin, and (ii) all intensity data is ratioed against the intensity of that element in a monitor standard (I_{mon}) at element concentration C_{mon}, then from equation (8.28) the concentration of the element in question can be written as

$$C = C_{\text{mon}} \cdot \frac{I}{I_{\text{mon}}} \cdot \frac{\bar{\mu}}{\bar{\mu}_{\text{mon}}} \tag{8.31}$$

Furthermore, the term $C_{\text{mon}} . I/I_{\text{mon}}$ is equivalent to the apparent concentration of the specified element (C'), uncorrected for mass attenuation effects. Combining equation (8.31) with (8.30), the full expression for C becomes

$$C = C'\left[\frac{\bar{\mu}_{\text{flux}}}{\bar{\mu}_{\text{mon}}} + f. \sum_{\substack{\text{all}\\ \text{elements}}} \left(\frac{\mu_i - \bar{\mu}_{\text{flux}}}{\bar{\mu}_{\text{mon}}}\right) C_i\right] \tag{8.32}$$

$$= C'(A + f. \sum_{\substack{\text{all}\\ \text{elements}}} B_i \,.\, C_i) \tag{8.33}$$

where A and B_i are constants for a given fusion mixture which may be determined either from published tables of mass attenuation data (Heinrich, 1966; Leroux and Thinh, 1977) or measured experimentally (Norrish and Hutton, 1969).

The equation was developed in this form so that correction coefficients would approximate to unity and could further be readily transferred from instrument to instrument.

Several further points must be considered in the successful application of this correction procedure.

(i) As C_i in equations (8.32) and (8.33) is not accurately known before applying this correction (since only apparent concentrations may be determined initially) it is necessary to apply this correction iteratively, as described earlier in this section.

(ii) Calibrations are normally set up by analysing a range of silicate reference materials to cover the compositions to be analysed and then applying a least squares regression to the resultant concentration–intensity calibration data set. However, since the intensities of reference materials are themselves affected by mass attenuation effects, it is necessary to apply a reverse mass attenuation correction using the known compositions of these 'standards' before an adequate calibration plot may be made. The most convenient way of doing this is to convert the known element concentrations into apparent fluorescent values (Harvey *et al.*, 1973) from

$$C'_{\text{RM}} = \frac{C_{\text{RM}}}{A + f \sum_i B_i C_{i(\text{RM})}} \tag{8.34}$$

where RM now refers to the known composition of the reference material. The calibration line is then constructed as a plot of intensity *v.* apparent fluorescent concentration.

(iii) It is essential that an accurate dilution factor (f) is used in this correction procedure. The ratio of flux to rock powder will change during fusion of the sample and flux due to the loss of absorbed water from the flux and the loss of both this and other volatile structural components (water, carbon dioxide etc.) from the rock powder. There are several ways of compensating for these effects. The flux can be stored at 400–600°C prior to sample preparation to prevent absorption of water vapour (or CO_2 in the case of La_2O_3-bearing fluxes). Loss on ignition of rock powder must be accounted for. Loss on ignition data for both flux and sample powder may be used to adjust the weight of flux to ensure a constant dilution factor in the glass disc after fusion.

Uncertainties arise in loss on ignition measurements due to weight *increases* that can occur during the ignition of samples at 1000°C due to the atmospheric oxidation of iron (II) (FeO) to iron (III) (Fe_2O_3). For each 1% FeO, an increase of 0.111% relative in weight will occur, assuming complete oxidation. This discrepancy is often ignored in routine schemes of analysis due to the relatively crude nature of the ignition loss measurement, uncertainty as to whether complete oxidation of Fe(II) occurs during ignition (see section 2.10) and the inconvenience of carrying out an independent non-instrumental determination of Fe(II). In order to minimize errors that can arise from these effects, some analysts prefer to prepare glass discs from samples that have been ignited rather than just dried at 105°C. Summation to 100% should then be improved although the reliability of such XRF analyses on an 'as received' rather than 'ignited' basis still depends on the accuracy of the loss on ignition measurement.

The procedures described above have been considered in some detail as they were the first to be widely adopted for the accurate correction of glass disc data. The inability of these procedures to account for secondary fluorescence and primary absorption has not caused apparent error, partially due to the dilution effect on the sample in the glass disc. Furthermore, if corrections make use of experimentally determined mass attenuation coefficients, as recommended by Norrish and Hutton (1969), rather than data from published tables, any residual errors are likely to be compensated.

One drawback of correction procedures based on self-consistent mass attenuation calculations is in the analysis of the heavier trace elements (e.g. Rb–Nb). It is sometimes inconvenient to pre-determine the major element abundances required for attenuation calculations since these would normally be determined on a different sample type (glass disc in place of a powder pellet) and under different instrumental conditions.

8.8.3 *Influence coefficient procedures*

(*a*) *Lachance–Traill alpha coefficients.* Influence coefficient procedures are empirical methods of correcting absorption-enhancement effects based on the principle that the attenuation of the x-ray line due to an element is influenced by the concentration of all the other elements present in the sample. Lachance and Traill (1966) (and Traill and Lachance, 1966) introduced the concept of the alpha coefficient based on a development of earlier work (see for example Tertian and Claisse, 1982, Chapter 10). If a_{ij} is the coefficient that quantifies the effect of element j on the element i under consideration, then the true concentration of element i (C_i) can be determined from the apparent concentration C_i' by the equation

$$C_i = C'_i\left(1+\sum_{\substack{\text{all}\\ \text{elements but } j\neq i}} a_{ij}C'_j\right) \tag{8.35}$$

which expands to

$$C_i = C'_i(1+a_{ij}C'_j+a_{ik}C'_k+\ldots) \tag{8.36}$$

In all these expressions, a prime represents an apparent elemental concentration.

In many practical applications, an additional coefficient describing the influence of the element i on itself is added to this algorithm (the modified Lachance–Traill equation):

$$C_i = C'_i(1+a_{ii}C'_i+a_{ij}C'_j+a_{ik}C'_k+\ldots) \tag{8.37}$$

The alpha coefficients may be positive or negative in value. A positive coefficient indicates that the overall effect of the element j on element i is absorption. Conversely, a negative coefficient suggests that enhancement dominates. An implicit assumption is that for a given system, the alpha coefficient is a constant and does not vary with sample composition.

(*b*) *Rasberry–Heinrich algorithm.* In a close examination of the Fe–Cr–Ni alloy system, third-element effects are observed. That is, the Lachance–Traill alpha coefficients are not completely independent of sample composition for such samples. To account for this deficiency in the basic Lachance–Traill algorithm, Rasberry and Heinrich (1974) proposed the addition of a further term to the Lachance–Traill expression. The Rasberry–Heinrich algorithm is thus:

$$C_i = C'_i\left[1+\sum_{\substack{j\neq i\\ j\neq k}} A_{ij}C'_j+\sum_{\substack{k\neq i\\ k\neq j}}\frac{B_{ik}C'_k}{(1+C'_i)}\right] \tag{8.38)}$$

The conditions placed on the summation terms mean that self-influence coefficients are not included in this algorithm. Furthermore, for any element–element interaction, only one of the summation terms is operative. If the dominant interaction between two elements is absorption, the A_{ij} term is used, the corresponding B_{ij} term being zero (this is then equivalent to the Lachance–Traill alpha coefficient expression). If, however, the dominant effect is enhancement, A_{ik} is set to zero and the B_{ik} expression is used to produce an improved mathematical expression for the inter-element interaction.

(*c*) *De Jongh delta function.* As most intensity data for quantitative analysis is ratioed against a suitable monitor standard, De Jongh (1973) described an empirical correction expressed in the form of *differences* in the concentration of an element i in the sample (C_i) and that in a reference standard (C_i^*) of similar composition. If I_i and I_i^* are the respective x-ray intensities, de Jongh's delta function has the form

$$C_i = C_i^*\cdot\frac{I_i}{I_i^*}[1+d_{ii}\Delta C'_i+d_{ij}\Delta C'_j+d_{ik}\Delta C'_k\ldots] \tag{8.39}$$

where $\Delta C'_i$ is the difference in concentration between unknown and standard, and d_{ii} the corresponding influence coefficient.

If δ_{ij} is defined as the difference $(d_{ij}-d_{ii})$, then the above equation may be rewritten

$$C_i = C_i^* \cdot \frac{I_i}{I_i^*}\left[1+\sum_{j\neq 1} \delta_{ij}(C'_j - C_j^*)\right] \qquad (8.40)$$

This is the de Jongh delta function.

(*d*) *General requirements and status of influence coefficient procedures*. The equations set out above illustrate the diverse effort that has been put into develop empirical algorithms to model accurately absorption–enhancement effects. In practice, these algorithms require that an instrument is calibrated with at least as many reference standards as the number of elements to be determined so that all the appropriate influence coefficients may be calculated from the relevant set of simultaneous equations. It is indeed preferable to use at least twice this number of standards to create sufficient redundancy to ensure the accuracy of the calibration. This requirement has limited the applications of such procedures to the analysis of silicate samples, due to the lack of reliable standards to cover the entire calibration range.

Nevertheless, the Rasberry–Heinrich correction expression (though using the absorption term only) was used by Schroeder *et al.* (1980) to determine the major elements on heavy absorber glass discs. In comparison, Lindsay *et al.* (1982) undertook a critical evaluation of the Rasberry–Heinrich model for the analysis of silicate rocks. They concluded that although this algorithm was of use in the analysis of undiluted samples (e.g. metals), it was not as valuable in the correction of data from geological materials diluted with flux.

8.8.4 *Mass attenuation corrections based on Compton scatter*

Correction methods based on scattered radiation originate from the observation of Andermann and Kemp (1958) that the intensity of secondary fluorescent radiation is affected by absorption in the sample to the same degree as primary x-ray tube radiation scattered from the sample at an adjacent wavelength. The ratio between the two is therefore virtually independent of sample attenuation effects. This phenomenon represents a particularly attractive method of correction since it depends on simple measurements from the sample rather than detailed calculations based on mathematical models.

Procedures based on this principle were readily adapted for the correction of trace-element data in silicate rocks. However, practical application depends on further considerations.

(i) Hower (1959) showed that the ratio of mass attenuation coefficients of samples of similar matrix remained essentially constant over a range of energies providing that no significant absorption edge intervened. This condition is satisfied in silicate rocks at energies above the iron absorption edge at 7.11 keV, since none of the trace elements emitting higher-energy radiation is normally present at a sufficiently high concentration to cause a significant absorption edge effect. At energies below the iron absorption edge, mass attenuation coefficients change abruptly by an amount related to the fraction of iron present in the sample and, without further correction, this relationship breaks down.

(ii) In order to make use of Andermann and Kemp's correction, it is necessary to measure precisely the intensity of scattered tube radiation. Earlier work had used the rather low-intensity background measurement to a fluorescent line. However, Reynolds (1963, 1967) showed that much more precise measurements could be made by measuring the intensity of the Compton scatter peak formed from characteristic x-ray tube lines (Figure 8.19). The intensity of this line is inversely proportional to the mass attenuation coefficient of the sample. This observation is only strictly true if matrix effects are caused by absorption within the sample and if secondary fluorescence effects are negligible. This is a reasonable assumption in the determination of trace elements in silicate rocks from lines with energies above the iron absorption edge. Tertian and Claisse (1983, p.263) have shown that there is a small approximation inherent in this approach.

Thus, in schemes of analysis where all count rate data is normalized to a suitable monitor, then the concentration C of an element is given by:

$$C = C_{\text{mon}} \cdot \left(\frac{I}{I_{\text{mon}}}\right)_{K\alpha} \cdot \left(\frac{I_{\text{mon}}{}^*}{I^*}\right)_{\text{compton}} \qquad (8.41)$$

where I is the background corrected intensity of the appropriate fluorescent line of the element to be determined, I^* is the background corrected intensity of the Compton scatter peak, and the subscript 'mon' refers to symbols representing equivalent measurements of the monitor standard.

The term

$$\left(\frac{I_{\text{mon}}{}^*}{I^*}\right)_{\text{compton}}$$

thus corrects for mass attenuation differences between sample and monitor for all x-ray energies above the iron absorption edge, this term being effectively equal to $\bar{\mu}/\bar{\mu}_{\text{mon}}$.

This procedure has been used successfully in the analysis of trace elements in silicate rocks by several groups of workers: for example, Leoni and Saitta (1977) (16 trace elements in international reference materials using a silver tube); Nesbitt *et al.* (1976) (trace elements using a molybdenum tube); Sumartojo and Paris (1980) (Mo-tube excitation); Lee and McConchie (1982) (trace elements in low dilution glass discs using a Mo tube); Harvey and Atkin (1981) (Rb, Sr using a Rh tube); Galson *et al.* (1983) (U, Th using a Rh tube).

The extension of this correction procedure to energies below the iron absorption edge has been considered by Walker (1973), Nesbitt *et al.* (1976), Leoni and Saitta (1977) and others. This is achieved by multipying equation (8.41) by the ratio of observed intensity of iron in the monitor to that in the sample. This factor corrects for differences in mass attenuation below the absorption edge caused by differences in concentration of iron between sample and monitor. This correction is only appropriate down to the next significant absorption edge, that due to titanium.

Some workers, for example Pankhurst and O'Nions (1973), have used the Rayleigh peak scattered from the characteristic tube lines for mass attenuation corrections. This correction procedure is not now considered to be as reliable as measurements based on the Compton scatter line.

The obvious advantage of the Compton scatter correction procedure is that measurements are simple and rapid and do not require analysis of all major components of the sample. However, it is necessary to use Compton scatter radiation from a suitable x-ray tube, Mo, Rh or Ag anode tubes being preferred. Particular care is required in correcting scattered Compton intensities for the background component, estimation of which is difficult owing to variations in slope caused by the adjacent Rayleigh scatter peaks. The technique does not account for enhancement effects nor for primary beam absorption. Furthermore, samples must be infinitely thick with regard to the selected Compton scatter radiation. Additional corrections may be necessary if the concentration of any trace element exceeds several hundred ppm due to the breakdown of assumptions on which the Compton scatter method is based.

Table 8.10 Critical (maximum) thickness of thin films for selected x-ray lines

Element	Line	X-ray energy (keV)	X-ray tube	Critical thin film thickness (μm)
Na	$K\alpha$	1.041	Cr	0.087
Si	$K\alpha$	1.739	Cr	0.23
Ca	$K\alpha$	3.690	Cr	0.65
Fe	$K\alpha$	6.398	Rh	3.4
Rb	$K\alpha_1$	13.393	Rh	16.4
Nb	$K\alpha_1$	16.612	Rh	25.1
Rh	$K\alpha_1$	20.213	W	72
La	$K\alpha_1$	33.436	W	193
Eu	$K\alpha_1$	41.535	W	281

Calculations are based on a pellet of andesite USGS: AGV-1 of assumed effective density 2.5 kg m^{-3}.
Critical thin film thickness has been calculated from equation (8.43). To simplify calculation of primary beam attenuation effects, it has been assumed that the element is excited exclusively by the $K\alpha_1$ radiation from the listed x-ray tube.

8.8.5 *Thin film techniques*

One method that can be used in special circumstances to overcome matrix effects is to make samples so thin that absorption-enhancement effects are negligible. It can be shown (see for example Tertian and Claisse, 1982, p.279) that the intensity I_x of fluorescent radiation from a sample of thickness x in comparison with that from an infinitely thick sample I_∞ is given by

$$\frac{I_x}{I_\infty} = 1 - \exp(-\rho x \bar{\mu}_{\text{net}}) \tag{8.42}$$

where ρ is the sample density, and $\bar{\mu}_{\text{net}}$ is defined by equation (8.23).

This expression can be simplified, using the approximation $1 - e^{-k} = k$, to

$$\frac{I_x}{I_\infty} = \rho x \bar{\mu}_{\text{net}} \tag{8.43}$$

providing $\rho x \bar{\mu}_{\text{net}} < 0.1$. The error compared with the rigorous expression (equation 8.42) is less than 5% at this limit.

The normal criterion used to define a thin film is that $\rho x \bar{\mu}_{\text{net}}$ is less than 0.1. Under these conditions, measured intensities may be considered to be independent of matrix interference effects.

For a given sample of known density and mass attenuation, the limiting factor in satisfying this criterion is sample thickness (x). The value of x has been estimated for selected fluorescent x-ray lines in Table 8.10, using the composition of an andesite, USGS: AGV-1 as the basis for calculations. These data indicate that in the analysis of thin films prepared from powders the critical sample thickness for sodium $K\alpha$ is about 0.9 μm, increasing to 3.4 μm for Fe $K\alpha$ and 25 μm for Nb $K\alpha$.

It would be unusual to analyse powdered material in this way, not least because of the very small mass of material presented for analysis and concern about the particle size and mineralogical homogeneity. More often, thin films are used to determine elements that have undergone some form of preconcentration procedure. However, Govindaraju and Montanari (1978) determined six major elements in silicate rocks by a thin-film technique. Samples were fused with lithium tetraborate and then taken up into solution. Selected elements were then absorbed on to a strongly acidic cation exchange resin. After completion, the resin was dried in an oven and mounted as a monolayer of resin beads on to adhesive paper. These samples were analysed by XRF as thin films without the necessity for matrix correction.

Fryer (1977) developed the ion exchange procedure of Eby (1972) to preconcentrate the rare-earth elements for analysis by thin film XRF. After ion-exchange separation, the rare-earth element eluent was taken up into dilute hydrochloric acid solution and equilibrated with an ion exchange paper that had been cut to fit the XRF sample holder. After 24 hours, the solution was evaporated to dryness using an infra-red lamp and analysed by XRF. Counts data were corrected for background, inter-REE spectrum overlap interferences, the fraction of REE sample recovered and then quantified using single or multi-element standard filter papers.

Because of the small mass of sample with which the primary x-ray beam may interact, scattered x-ray tube intensities are potentially much lower than in a 'normal' sample. Therefore, peak-to-background ratios are higher in thin film samples, leading to potentially improved detection limits. For this reason careful consideration has to be paid to the substrate on which the sample is mounted as the major proportion of scattered background radiation may originate from this support and not from the sample itself.

8.9 Instrumentation for wavelength dispersive XRF analysis

A conventional x-ray fluorescence spectrometer is shown schematically in Figure 8.25. The principal components are (i) an x-ray tube that emits polychromatic radiation used to excite the sample, and (ii) an x-ray spectrometer used to detect the fluorescence x-ray spectrum excited within the sample. The spectrometer itself consists of several components, including various collimators to maintain adequate

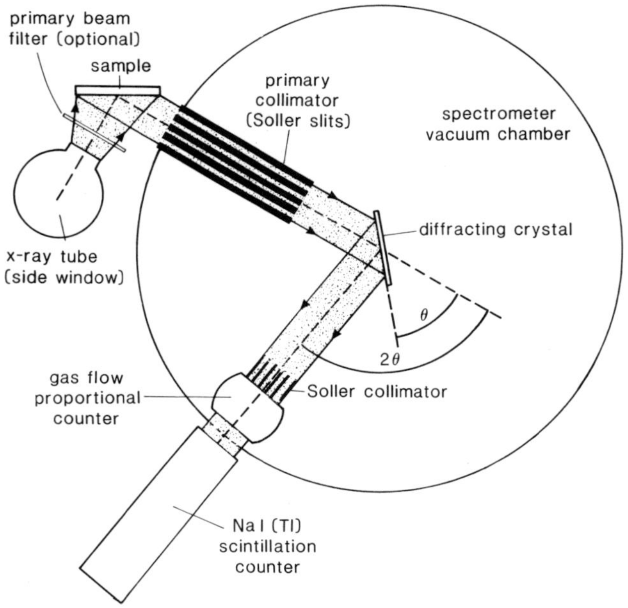

Figure 8.25 Schematic diagram of a wavelength dispersive x-ray fluorescence analyser showing the configuration of the x-ray tube, sample and spectrometer.

spectrum resolution, an x-ray diffracting crystral, an x-ray detector, normally a gas flow proportional counter or a scintillation counter, and a goniometer mechanism to maintain the correct angular relationship between these components. The analytical characteristics of these devices will now be considered.

8.9.1 *The x-ray tube*

The principles of operation of an x-ray tube in relation to its polychromatic excitation spectrum have been discussed in section 8.5. The parameters that are of interest to the analyst in developing an analytical program include (*a*) the choice of x-ray tube anode material, (*b*) the use of primary beam filters and (*c*) optimization of tube operating potential and current:

(*a*) *Choice of x-ray tube anode.* A variety of tube-types are available for analytical x-ray fluorescence analysis, including those with anodes made of Sc, Cr, Mo, Rh, Ag, Au, and W. In developing a scheme for routine multi-element analysis, criteria in the selection of anode material include:

(i) The elements to be determined must be efficiently excited by the characteristic lines of the x-ray tube

(ii) In extending the range of elements that may be excited efficiently, the emission of a significant continuum may be an important consideration

(iii) Choice of tube type may also be affected by the proximity of first- or multiple-order lines or Compton scatter lines originating from the tube that overlap fluorescent lines of interest.

For general-purpose schemes of analysis, the rhodium anode tube has achieved some prominence in recent years. The rhodium tube *K*-lines at 20 to 23 keV efficiently excite the heavier trace elements, including the Rb–Nb group of

Table 8.11 Excitation efficiency of selected x-ray tubes

Tube anode:			Cr	Mo	Rh	Au
Energy of principal tube line (keV)			5.4	17.5	20.2	68.8, 9.7
Analysed element	Line	Critical excitation energy (keV)	Sensitivity			
Zn	*Kα*	9.66	60	243	153	423
U	*Lα*	17.16	40	276	166	108
Mo	*Kα*	20.00	128	—	686	334
Sn	*Kα*	29.19	140	245	223	365

Data from Kikkert (1983).
Sensitivity is listed in units of thousands of counts per second per percent of element.

elements. Furthermore, there is a useful boost to the excitation of the light elements from the rhodium *L*-lines at 2.5 to 3.1 keV. For the elements important in silicate rock analysis, the rhodium tube lines cause few spectrum overlap interferences.

In schemes for the analysis of major elements on glass discs, chromium anode tubes are widely used, although excitation of the lightest elements is strongly influenced by the intensity of the low-energy continuum and hence the thickness of the beryllium end window. Tubes with scandium anodes have recently been introduced to boost the excitation of the light elements. It is possible to selectively excite ranges of trace elements by using a tube with an anode material that emits characteristic lines just above the respective absorption edges. Molybdenum, tungsten and gold tubes have a role to play in this respect, although none can be used to determine the full range of elements for which data are normally required in silicate rocks because of spectrum overlap interferences.

Data comparing the performances of four of the more commonly used x-ray tubes (having chromium, molybdenum, rhodium and gold anodes) are shown in Table 8.11, from Kikkert (1983). These results demonstrate the comparative efficiency in exciting the elements Zn, Mo, Sn and U. Note that the chromium tube does not have any characteristic lines at energies greater than the absorption edges of any of the lines analysed. The excitation of these elements arises from the tube continuum alone and is therefore relatively unfavourable compared with the other tubes. The high excitation efficiency of both the Rh tube for Mo *Kα* and the Au tube for Zn *Kα* arises from the proximity of characteristic tube lines at energies just above the critical excitation energies of these elements.

(*b*) *Primary beam filters.* Primary beam filters are metal foils placed in the primary beam between x-ray tube and sample. These filters may be used to modify the energy spectrum of tube radiation reaching the sample.

Such filters are not as widely used in wavelength dispersive XRF as in energy dispersive XRF analysis (section 9.7). The most important use of filters on the present instrumentation is in situations where it is necessary to determine an element that suffers a severe overlap interference from the characteristic line of the tube anode material itself. A case in point is the determination of Cr (or Mn) with a chromium tube; this is most convenient when this tube is used to analyse other major elements on glass discs. The measured Cr $K\alpha$ radiation then originates not only by fluorescence of the sample itself but also from Rayleigh scatter of x-ray tube lines from the sample into the spectrometer. Similarly, Mn $K\alpha$ radiation is interfered with by Cr $K\beta$ x-rays from the tube. In the determination of these elements, therefore, it is normal to insert an aluminium foil (200 μm thick) between tube and sample. The foil absorbs all low-energy radiation from the tube including the tube Cr $K\alpha$ and $K\beta$ lines. Excitation of the sample then occurs, with considerable loss in sensitivity, from the higher-energy tube continuum that can penetrate the aluminium foil when the tube is operated at say 60 kV.

(*c*) *Optimizing x-ray tube operating potential and current.* In analytical applications it is normal to optimize tube kV for the efficient excitation of the group of elements of interest. The tube current is then selected such that the energy dissipated by the tube is between 80 and 100% of maximum. The lower figure would be chosen out of consideration of conserving the useful life of the tube.

The tube kilovolt and milliamp settings cannot be varied entirely independently of one another and must be selected to operate within a design envelope (Figure 8.26). Apart from limitations on the maximum power output of the generator, the low kV cut-off is due to the economics of designing power supplies that will give exceptional high kV stability as well as having adequate low kV characteristics.

As a general rule of thumb, operating potentials are selected at about four to six times the critical excitation energy of the element x-ray line to be determined. As it is generally desirable to determine as many elements as possible without changing tube parameters, some compromise must be reached, generally in favour of adequate excitation of heavier elements within a group.

Excitation of individual elements is usually optimized by employing a *figure of merit* technique. The parameter recommended by Jenkins (1976, p.90; see section 8.11.2(*b*)) is

$$\text{Figure of Merit} = \sqrt{R_g} - \sqrt{R_b} \tag{8.44}$$

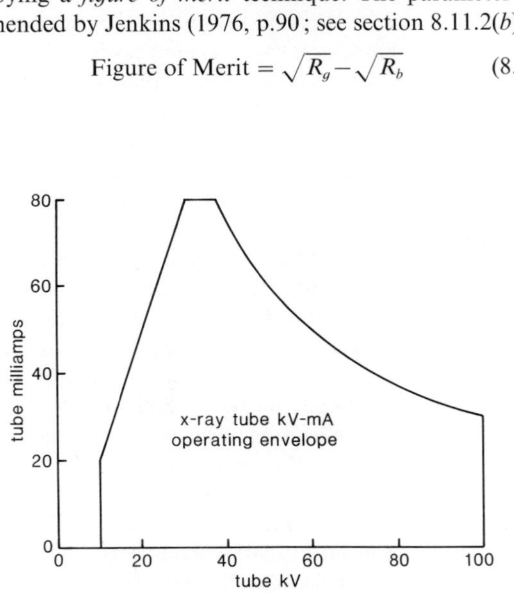

Figure 8.26 Example of the maximum kilovolt-milliamp power settings available on a 3 kW x-ray tube power supply.

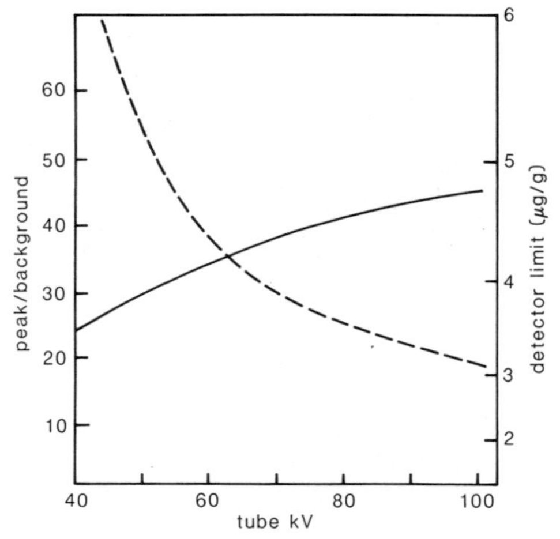

Figure 8.27 Optimization of tube kilovoltage in the excitation of Mo $K\alpha$ using an Rh tube. An improvement in both detection limit (dotted line) and peak-to-background ratio (solid line) is apparent as tube potential is increased up to 100 kV. (After Kikkert, 1983.)

where R_g is the count rate of the element x-ray peak, and R_b is the associated background count rate (see p. 279).

In general, the figure-of-merit parameter is calculated at individual kV settings over a range of operating values. The maximum in the figure-of-merit *v.* tube kV plot represents optimum counting conditions for the element. The error in determination under fixed counting times will be at a minimum.

In the case of silicate rocks, optimum operating conditions are usually found to be at tube potentials of 40 to 60 kV (to determine the major elements) or 80 to 100 kV (for the heavier trace elements). In this respect, data plotted in Figure 8.27, taken from Kikkert (1983), indicate that in the determination of molybdenum even the limit of 100 kV does not afford the optimum analytical setting when using a rhodium tube.

(*d*) *Tube contamination.* With the large amount of energy dissipated within a modern x-ray tube (3–5 kW), it is not surprising that metallic components within the tube envelope can become vaporized and contaminate the inside surfaces of the tube. During this process, contaminants can be deposited on the anode. During normal operation, these elements, as well as the anode material, will become excited, emitting characteristic x-ray lines that will appear in measured spectra due to Rayleigh scatter from the sample. The potential for such interference effects depends on both design and materials used for tube construction. However, it would not be unusual to observe contamination from some of the elements Fe, Cu, Ni, Zn and/or W (from the filament). The presence of tube contaminant may be detected by reflecting the tube spectrum off a low-atomic-number sample such as polyethylene, placed in the normal sample position. This interference must be corrected either by filtering the x-ray beam (see (*b*) above) or subtracting, from an observed x-ray count, that proportion which originates from tube contamination. Since the intensity of Rayleigh scatter is matrix-dependent, this correction can be uncertain. The severity of this problem is likely to increase with tube age.

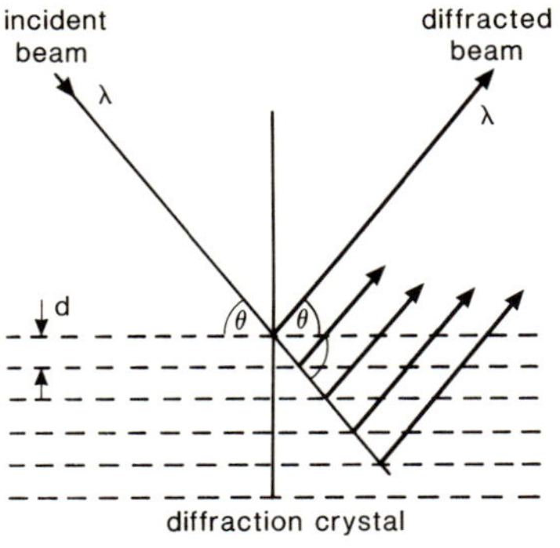

Figure 8.28 A representation of diffraction in a crystal, showing the incident beam diffracted from successive lattice layers in the crystal.

8.9.2 *Wavelength dispersive spectrometers*

The function of a wavelength dispersive spectrometer is to measure the intensity of x-rays of discrete energy in the fluorescent spectrum of a sample, using the phenomenon that certain crystalline materials are capable of strongly diffracting x-rays as a function of wavelength (Bragg diffraction). Such spectrometers are therefore known as wavelength dispersive devices in comparison with energy dispersive detectors, which simultaneously measure the full spectrum of x-ray energies emitted from a sample.

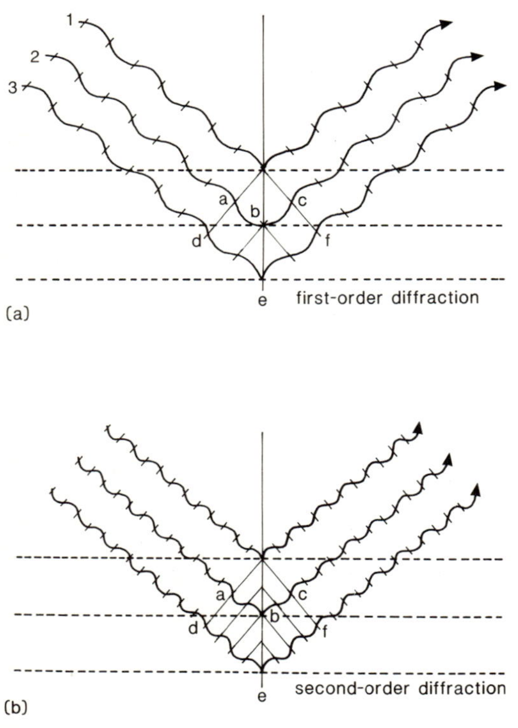

Figure 8.29 A representation of (*a*) first-order and (*b*) second-order x-ray diffraction for x-ray beams having identical angles of incidence. The pathlength difference between x-rays diffracted from adjacent atomic layers corresponds to an integral number of wavelengths (1 for first-order, 2 for second-order diffraction). Nodes corresponding to $\lambda/2$ are marked. (After Jenkins, 1976.)

A wavelength dispersive spectrometer consists of four principal components: a diffracting crystal, an x-ray detector which is normally a gas flow proportional counter or a scintillation counter, beam defining collimators, and the goniometer, a mechanism for maintaining the correct angular and spatial relationship between these components.

(*a*) *Bragg's Law*. The basis of wavelength spectrometers is the observation, formulated by Bragg, that when an x-ray beam is incident on certain crystalline substances, very strong diffractions occur when the wavelength (λ) of the beam and the angle of incidence (θ) with respect to the diffracting crystal satisfy the relationship

$$n\lambda = 2d \sin \theta \qquad (8.45)$$

where d is the lattice spacing between individual layers of atoms within the crystal, n is an integer ($=1,2,3,...$), and the geometric relationship between these parameter is as shown in Figure 8.28.

Note that the diffracted beam subtends the same angle of diffraction (θ) with respect to the crystal and that the incident beam is diffracted through a total angle of 2θ.

A schematic representation of first- and second-order Bragg diffraction is shown in Figure 8.29. Constructive interference will only occur in the diffracted beam when the difference in path length between photons diffracted from successive lattice layers is an integral number of wavelengths (1 λ for a first-order diffraction). Thus, compared to photon 1 in Figure 8.29*a*, the difference in path length of photon 2 is ($ab+bc$), i.e. one wavelength, and that of photon 3 is ($de+ef$), i.e. one wavelength relative to photon 2. If this condition is not met, then destructive interference occurs in photons diffracted from different layers in the crystal. Diffracted photons are therefore polarized in the plane perpendicular to the surface of the crystal. A second constructive interference will occur when the difference in path length between photons diffracted from successive layers is equal to two. This condition is satisfied for an x-ray beam having a wavelength of $\lambda/2$ compared with the first-order diffraction (λ), assuming the angle of incidence remains unchanged. Thus an important limitation of a wavelength-dispersive spectrometer is that it cannot distinguish between x-ray photons having wavelengths λ, $\lambda/2$, $\lambda/3$ etc. by diffraction alone. Pulse height analysis must be used to eliminate these interferences (see section 8.9.7 below).

In conventional instruments, the spectrometer take-off angle is fixed during manufacture and the spectrometer is designed to scan through the x-ray spectrum by rotation of the diffracting crystal about its axis. An x-ray beam, of angle of incidence θ, is diffracted through an angle 2θ. In order to maintain the spectrometer exit slit and counter in the diffracted beam, a mechanism must be provided to maintain a precise angular relationship between crystal and counter assembly, as illustrated in Figure 8.30. Because of this relationship, spectrometer movements and corresponding tables of x-ray lines are often quoted in values of 2θ.

(*b*) *Flat crystal spectrometer*. The most widely used design of spectrometer for analytical XRF instruments is based on diffraction from a flat crystal (Figure 8.31*a*). The exit slit and counter block assembly are normally mounted on an arm of fixed length, the other end of which is pivoted at the axis of rotation of the diffracting crystal. Rotation of this

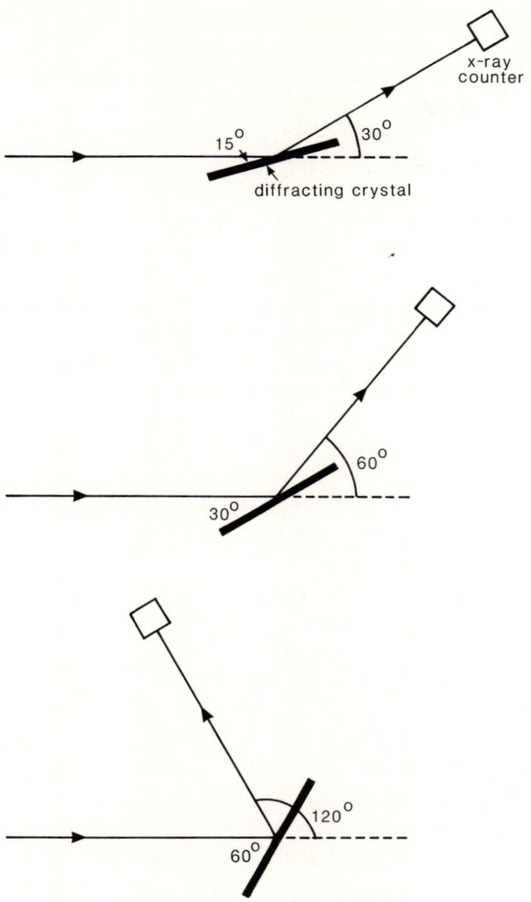

Figure 8.30 Movement of a flat crystal x-ray spectrometer, showing the angular relationship between incident beam, diffracting crystal and x-ray counter for 2θ angles of 30°, 60° and 120°.

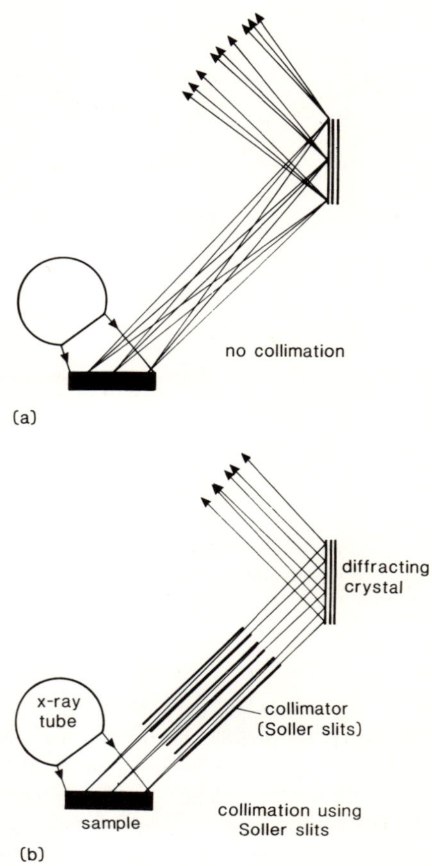

Figure 8.31 The effect of collimation on the angular divergence of the diffracted beam in a simple flat crystal x-ray spectrometer.

arm is geared in such a way that an angular rotation of θ in the diffracting crystal will cause an angular rotation of 2θ in the counter area.

The design of this spectrometer must compensate for the fact that, unlike microprobe applications, the source of fluorescent x-rays excited within the sample does not approximate to a point source. In fact, the area of sample excited by the tube is quite large (perhaps in excess of 2000 mm^2). Fully focusing spectrometer designs based on curved diffracting crystals are not, therefore, appropriate in this application (see section 10.10). Since the secondary x-ray source is comparatively large, the x-ray beam will subtend a range of angles at the diffracting crystal. In consequence, diffraction will occur over a range of wavelengths, since Bragg's equation will no longer be obeyed at any unique wavelength (see Figure 8.31*a*). Without further modification, spectrometers of this design would have inherently poor energy resolution. The difficulty is considerably reduced by employing a collimator between sample and crystal to limit the angle of divergence of x-rays accepted for diffraction (Figure 8.31*b*). The collimator usually consists of a set of parallel metal foil blades spaced, typically, 150 μm (fine) or 450 μm (coarse) apart, and called *Soller slits* after their originator. The device effectively acts as a set of multiple entrance slits for the spectrometer, and a compromise must be chosen between the improved resolution achieved with finer collimation and the higher beam intensities transmitted through coarser collimator slits.

(*c*) *Mosaic diffracting crystals.* A further phenomenon affects the resolution of all diffracting crystals, related to their microcrystalline surface structure. Although perfect crystals are expected to give optimum resolution, their diffracted beam intensities are disappointingly low due to the phenomenon of internal diffraction, shown schematically in Figure 8.32*a*. A perfect crystal structure may be represented by parallel blocks of material. Incident photons diffracted from blocks within the crystal can be internally diffracted a second time from the underside of surface blocks and so are lost from the diffracted beam. In order to reduce this effect, it is necessary to disrupt the microcrystalline surface texture of the crystal to give it an 'ideally imperfect' mosaic structure (Figure 8.32*b*). Photons diffracted from lower blocks are now unlikely to satisfy the Bragg equation for higher blocks that are slightly mis-orientated. By this process, the intensity of the diffracted beam is increased by perhaps more than an order of magnitude. However, a range of wavelengths will now satisfy Bragg's equation with respect to individual crystals in the surface mosaic layers. In consequence, the resolution in the diffracted beam is slightly degraded.

To counter both this, and residual effects due to the diver-

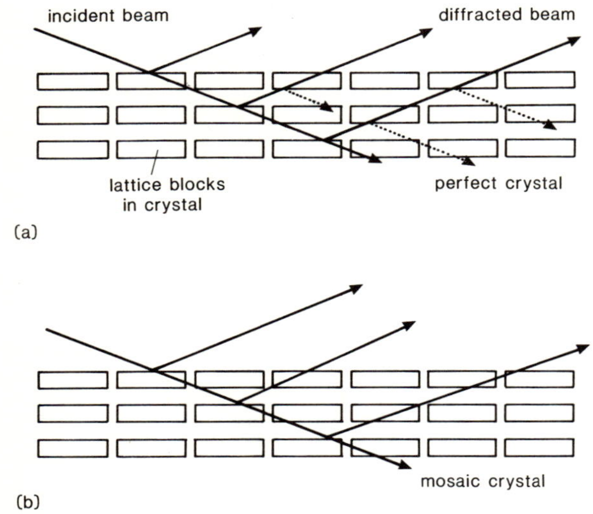

Figure 8.32 (*a*) Loss in intensity due to internal diffraction from a perfect crystal and (*b*) the improvement resulting from the use of an 'ideally imperfect' mosaic crystal. (After Reed, 1975, p. 75.)

gence of the incident beam discussed in (*b*) above, it is usually necessary to place a second set of Soller slits in front of the x-ray counter (Figure 8.25). The overall resolution of the spectrometer depends on choice of diffracting crystal, angle 2θ, spectrometer geometry and choice of collimation (fine or coarse). For standard instruments, full-width half-maximum resolution is likely to vary from about 10eV at 1 keV to over 500 eV at 20 keV under the compromise conditions normally adopted for quantitative analysis. This should be compared with the natural line width of fluorescent x-rays of between 0.1 to 1 eV, and further, with the equivalent resolution of an energy-discriminative detector which is likely to vary from 120 to 300 eV over the same energy range (see section 9.3).

8.9.3 *Diffracting crystals*

From an examination of Bragg's equation, it may be seen that crystals suitable for diffracting x-rays must have a lattice spacing $2d$ of the same order of magnitude as the wavelength of incident x-rays. Furthermore, there are practical limitations to the angular movement that can be designed into a practical instrument. Most spectrometers can measure 2θ diffraction ranges of typically 10° to 145°. Within these restrictions, it would be possible to select one crystalline material capable of diffracting the whole range of x-ray wavelengths from 1 to 20 eV (12–0.6 Å). However, this single crystal would not have sufficient angular dispersion (i.e. the degree of separation between adjacent x-ray lines) to allow analysis of the shorter wavelength peaks.

It is necessary therefore to select a range of diffracting crystals to cover the full range of wavelengths required for analysis. Selected crystals must have both an adequate angular dispersion and high reflectivity (i.e. high relative intensity in the diffracted beam).

Of the many crystals available (72 of which are listed, for example, by Bertin (1975, pp.981–999) only a few are now used for routine analytical work. The properties of these materials are summarized in Table 8.12. The range of elements for which individual crystals are normally used is indicated by data plotted in Figure 8.33. For the heavier elements (Ca to Nb *K*-lines), the standard crystal is lithium fluoride, cut in a specific orientation denoted by LiF(200). If line overlaps in this region dictate the use of a crystal having a higher angular dispersion, the LiF(220) may be substituted, although some reduction in diffracted beam intensity will then also occur. For the lighter elements, pentaerythritol (PET) and thallium acid phthalate (TAP) are generally used for the Si to Ti *K*-lines (PET) and F to Si *K*-lines (TAP) respectively. The TAP crystal has largely supplanted rubidium acid phthalate (RAP) and potassium acid phthalate (KAP) due to inferior reflected intensities of these materials (see, for example, Wybenga, 1978). Stearate and other pseudo-crystals (see below) are normally used in elec-

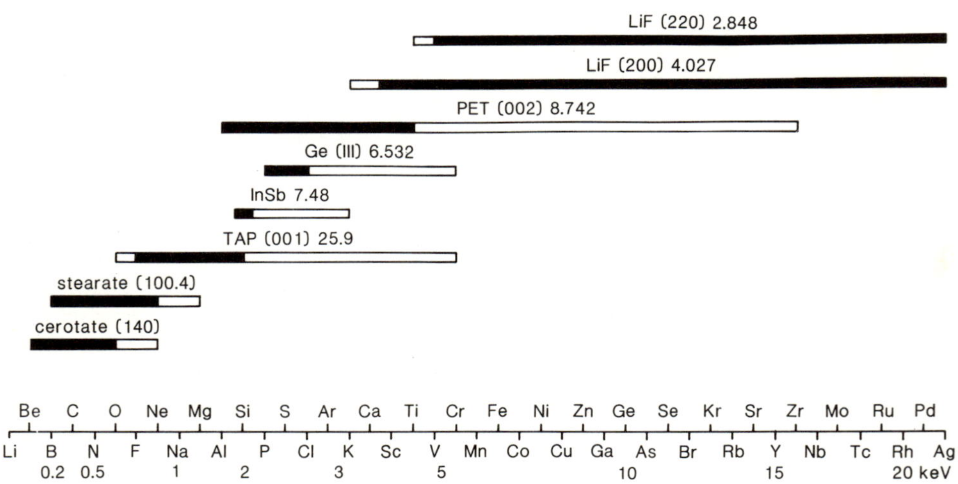

Figure 8.33 Diffraction range of the more common analysing crystals used in x-ray spectrometers as a function of x-ray energy. Most generally used ranges are shown in black. Lattice spacing ($2d$ in Å) and crystal orientation are indicated. The emission energy of element $K\alpha$ lines is marked.

Table 8.12 X-ray diffraction crystals and their properties

Diffracting crystal	Symbol and orientation	Crystal structure	2*d* spacing (Å)	Usable range (*K*α lines)	Comments
Lithium fluoride	LiF(220)	Cubic	2.848	Ti–U	Used in place of LiF(200) when higher angular dispersion is required but diffracted intensity is 40 to 80% of LiF(200).
Lithium fluoride	LiF(200)	Cubic	4.028	K–Ce	Standard general crystal for the *K*-lines of potassium and above. Combines high intensity with high dispersion.
Germanium	Ge(111)	Cubic	6.532	P–Pd	Preferred crystal for P (and S, Cl) *K*α lines as it eliminates the second-order diffraction interference from Ca *K*α.
Indium antimonide	InSb		7.48	Si	Preferred crystal for Si *K*α owing to high relectivity for this line (typically 2 to 3 times higher than PET).
Penta-erythritol (Tetrabis(hydroxymethyl)methane $C(CH_2OH)_4$	PET (002)	Tetragonal	8.742	Al–Zr	Preferred crystal for Al to K *K*α lines, partly due to absence of crystal fluorescence effects. PET has a relatively high coefficient of thermal expansion, which can cause drift errors.
Ammonium dihydrogen phosphate $NH_4H_2PO_4$	ADP (101)	Tetragonal	10.642	Mg–Rb	Formally favoured for Mg *K* but prone to crystal fluorescence effects due to P. Largely superseded by TAP.
Thallium acid phthalate (thallium hydrogen phthalate)	TAP(001)	Orthorhombic	25.75	O–Cr	Preferred crystal for F–Si *K*α having approximately twice the diffraction intensity of RAP.
Rubidium acid phthalate (rubidium hydrogen phthalate)	RAP(001)	Orthorhombic	26.12	O–Cr	Generally superseded by TAP
Potassium acid phthalate (potassium hydrogen phthalate)	KAP(001)	Orthohombic	26.63	O–V	Generally superseded by TAP or RAP. Potassium crystal fluorescence can cause high backgrounds.
Myristate (lead tetradecanoate)	MYR	Pseudo	80.5	B–Si	See stearate
Stearate (lead octadecanoate)	STR	Pseudo	100.4	B–Mg	Preferred crystal for *K*α lines of elements lighter than O, especially in electron microprobe applications.
Cerotate (lead hexacosanoate)	CER	Pseudo	137	Be–Ne	See stearate
Multi-layer (layered synthetic microstructure)	ML	Synthetic	40–200	B–Mg	A recent innovation. Multilayer crystals of optimized 2*d* spacing will probably take over for the detection of all elements lighter than A1.

Data based on Bertin (1975), pp.982–999.
Diffracting crystal name in brackets is the preferred chemical nomenclature.

tron microprobe analysis to determine the lightest elements (B to O *K*-lines). There are some exceptions to this general scheme. Indium antimonide gives exceptionally high diffracted beam intensities for the silicon *K*-line and is generally used to determine this element alone. Germanium (111) has the unusual property of suppressing second-order diffractions and is extremely useful in the determination of P *K*α which can suffer a serious second-order interference from Ca *K*α.

(*a*) *Miller indices*. The crystallographic notation used to describe the orientation of a plane cut through a crystal is known as the Miller index. This index defines the plane with reference to the appropriate unit cell of the crystal lattice. A detailed discussion of this convention is not necessary here. However, by way of example, the orientations of planes represented by the crystallographic notations LiF(200), LiF(220) and Ge(111) are shown in Figure 8.34.

(*b*) *Pseudo-crystals*. There is a limit to the magnitude of 2*d* spacings available in natural crystalline materials. In order to increase the available range and allow satisfactory diffraction of low-energy x-rays (B to N *K*-lines) pseudo-crystals have been developed, based chiefly on the lead salts of soap molecules. Most important among these include lead stearate, myristate and cerotate. These materials are artificially made by arranging layer upon layer of the appropriate molecule, as shown in Figure 8.35. The 2*d* spacing of the resultant pseudo-crystal is largely controlled by the length of soap molecule separating the layers of lead atoms. Diffracting crystals of this type are important in electron probe microanalysis, but are not generally used in XRF applications to silicate rocks, where critical excitation depth considerations normally limit low-atomic-number element analyses to sodium.

(*c*) *Multi-layered crystals*. A new category of diffracting materials has recently become available for the analysis of low-atomic-number elements. Multi-layer crystals are made

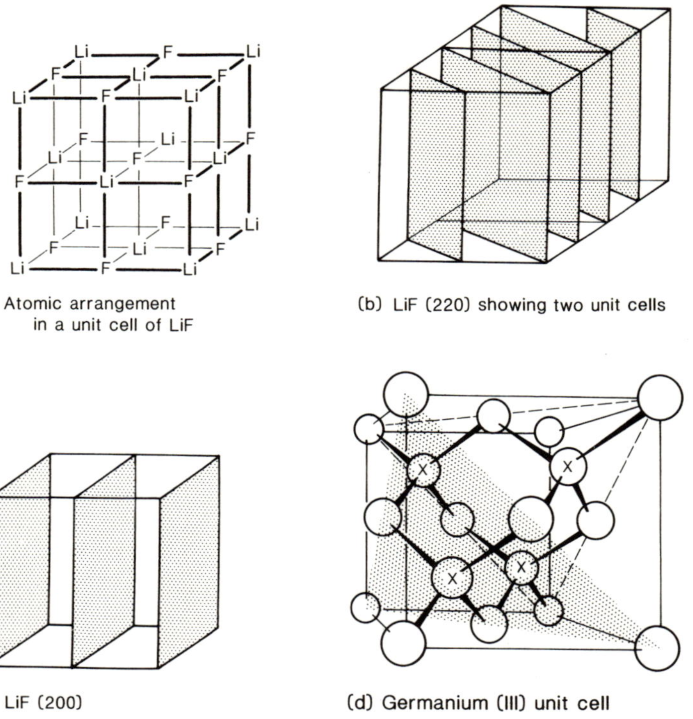

Figure 8.34 Structure of selected crystals used for x-ray diffraction: (*a*) ionic structure of an LiF crystal; (*b*) LiF (220), and (*c*) LiF (200) lattice planes; (*d*) germanium crystal showing the (111) lattice plane. Inter-lattice atoms (labelled *x*) cause second-order diffractions from this plane to be extinguished.

by coating with alternate layers of a high- and a low-atomic-number element such as tungsten. Diffraction occurs off the layers of tungsten. Manufacturing conditions can be controlled to vary the $2d$ spacing of such multi-layer devices, although a close tolerance on the periodicity of layers is required. Commercially available multi-layers normally have $2d$ spacings of between 40 and 120Å. The larger $2d$ spacings are of interest in electron probe microanalysis. The lower $2d$ spacing appear to have applications in the XRF determination of Na, Mg in silicate samples due to the higher reflected intensities in diffracted beams compared with conventional crystals.

(*d*) *Effect of temperature on diffraction crystals.* Perhaps the largest effect of temperature variations on x-ray spectrometers arises from the thermal expansion of crystal lattices, or more specifically, thermal expansion of the plane perpendicular to diffracting crystal lattice (all crystal planes do not usually expand at the same rate). Thermal expansion causes a change in the $2d$ lattice spacing and a consequential shift in the angle at which specific x-ray lines are diffracted. For this reason, most spectrometers are operated in a controlled environment with a tolerance on temperature of better than ± 2°C. Under these conditions, the only crystal with a sufficiently large linear coefficient of expansion to cause significant angular drift problems is penta-erythritol (PET) with a linear coefficient of expansion of 0.0013Å per °C for the 002 plane. In Table 8.13, the change in angle of the $K\alpha$ lines of the elements Si to Ti diffracted from a PET crystal is listed for each °C change in temperature. The calculated shift for the lighter elements, especially Al (0.054° 2θ) and Si (0.024° 2θ), is relatively large compared with the specified angular repeatability of 0.002° 2θ guaranteed on some commercial spectrometers. It is fortunate that in most analytical applications the PET crystal would be used only to determine the heavier elements (perhaps K to Ti $K\alpha$), where this problem is not as severe.

(*e*) *Crystal fluorescence.* The polychromatic x-ray beam incident on the diffracting crystal can excite the crystal material itself. A fraction of this fluorescent radiation is then detected along with diffracted x-ray lines and can in some circumstances cause a serious increase in background intensities. (Crystal materials that can be excited in this way include Tl in TAP, P in ADP, K in KAP, Rb in RAP.) When

Table 8.13 Effect of temperature on the angle of diffraction from a PET crystal

Element	2θ-angle of $K\alpha$ line	Change in 2θ angle per °C change in temperature
Al	145.12	0.054
Si	109.21	0.024
P	89.56	0.017
S	75.85	0.013
Cl	65.49	0.011
K	50.69	0.008
Ca	45.20	0.007
Sc	40.59	0.006
Ti	36.67	0.006

Data calculated assuming that the coefficient of thermal expansion of PET(002) is 0.0013Å per °C.

the preferred crystals in Table 8.12 are restricted to their normal operating range, this effect can normally be ignored. However, on older instruments where the KAP crystal was routinely used for determining the lighter elements (especially Mg *Kα*), fluorescence of K *Kα* in the crystal caused a significant increase in background. The careful selection of pulse height analysis settings was then essential to minimize this interference. In more modern instruments, the KAP crystal has been replaced first by the RAP and now by the TAP crystal, so that this effect is no longer prevalent.

8.9.4 *X-ray counters*

The x-ray counter is used to measure the intensity (usually in counts per second) of the diffracted x-ray beam. Counters used for x-ray spectrometry are designed so the size of the output pulse is proportional to the energy of the detected x-ray photon; this allows compensation for some interference effects by pulse height discrimination (see section 8.9.7). Two types are normally used: the gas proportional counter and the scintillation counter. These devices are commonly mounted in tandem along the axis of the diffracted beam (Figure 8.25). Both are required, since neither has adequate detection efficiency to cover the entire energy range from about 1 to 20 keV. The gas proportional counter is normally used to detect x-rays of energy up to 6 to 8 keV and the scintillation counter for x-rays above this range. At the changeover point, optimum performance is usually obtained by combining the signal from both detectors. At higher x-ray energies, the diffracted beam is arranged to pass through entrance and exit windows of the gas proportional counter into the scintillation counter mounted behind.

(*a*) *The gas proportional counter*. The gas proportional counter consists of a metal box with an entrance window and usually an exit window in the opposite side (Figure 8.36). The counter is filled with an ionizable gas, usually a mixture of argon–10% methane or argon–2.5% carbon dioxide. A thin wire, 10–40 μm in diameter, passes down the centre of

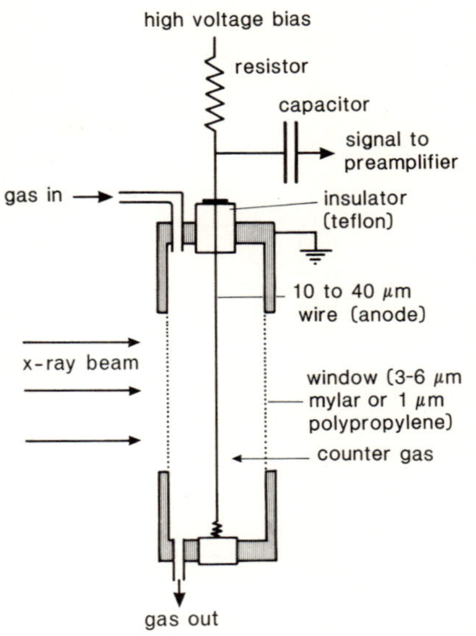

Figure 8.36 A gas flow proportional counter.

the counter. In normal operation a high-voltage bias of between 1 and 2 kV positive is applied to this wire, and with the counter body earthed, a potential gradient is set up inside the counter gas. To facilitate the formation of a uniform field the window material (1–6 μm mylar or polypropylene) must be aluminized and earthed to the body. X-rays that penetrate the entrance window cause ionization of the counter gas, and the resulting free electrons are swept rapidly to the wire anode under the influence of the electrostatic field. The complementary positively charged gas atoms drift rather more slowly in the opposite direction to earth. As electrons approach the wire anode, they experience a large electrostatic field that causes them to accelerate rapidly. Collisions between these electrons and other gas molecules causes further ionization and results in a cascade of electrons reaching the anode. This process is called *gas multiplication* and the magnitude of this effect depends both on the diameter of the wire and the magnitude of the applied voltage (the smaller the diameter of the wire, the greater the electrostatic field). Due to gas multiplication, the detected electronic signal is typically 10^3 to 10^5 times greater than the number of primary electrons originating from ionization of the counter gas by incident x-ray photons. Under the conditions described, the magnitude of the output pulse height is directly proportional to the x-ray energy.

Proportional counter characteristics are observed only over a relatively narrow range of applied voltages. The effect of varying applied voltage is shown in Figure 8.37. At very low voltages, no gas multiplication occurs and the counter effectively behaves as an ion chamber: only primary ions are detected. As the anode bias is increased beyond the onset of gas multiplication, the proportional region is reached. Under these conditions, the height of the output pulse is directly proportional to the incident x-ray energy. However, if the bias is increased even further, Geiger breakdown of the counter gas occurs. The initial ionization of the counter gas triggers a general discharge of the entire counter. At excessively high voltages, an arc discharge will occur as the

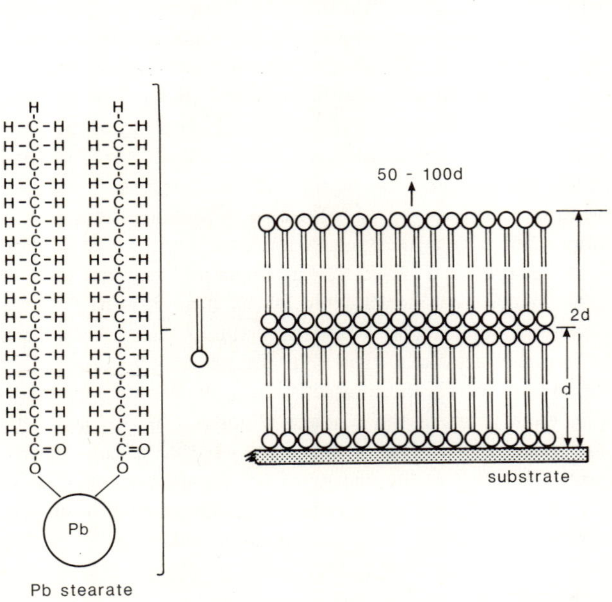

Figure 8.35 Structure of a lead stearate pseudo-crystal. (Bertin, 1975, p. 189.)

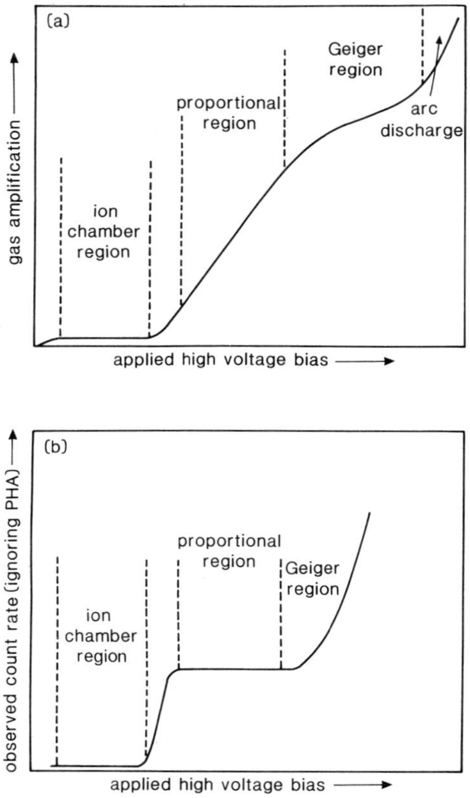

Figure 8.37 The effect of increasing the applied anode voltage on (*a*) the gas amplification factor and (*b*) the observed count rate (ignoring pulse height analysis settings) for a gas proportional counter. The rapid increase in observed count rate at the threshold of Geiger breakdown is indicated.

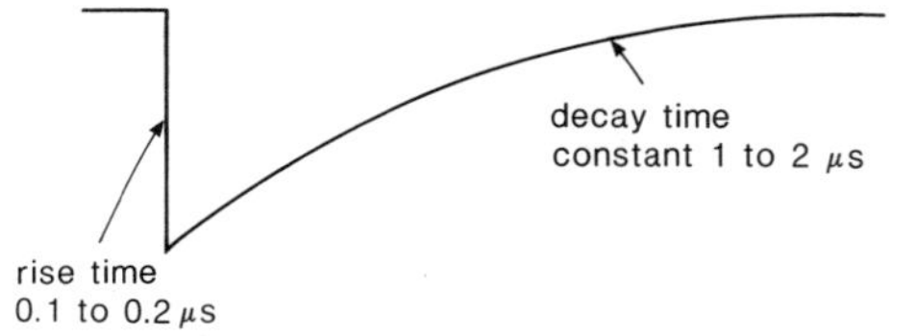

Figure 8.38 Output pulse from a gas flow proportional counter representing the detection of a single x-ray event. The height of this negative-going pulse is in the microvolt range.

counter insulation effectively breaks down. The onset of Geiger breakdown is heralded by a rapid increase in the apparent count rate. It is undesirable to operate for any length of time under these conditions as the counter rapidly becomes damaged.

During normal operation, positively charged gas atoms drift to the counter body to release their charge. However, at the moment of discharge, both ultraviolet radiation and secondary x-rays are generated as the positively charged ion is earthed. These radiations could themselves cause ionization of counter gas, initiating a new avalanche of electrons that would contribute to counter noise. To suppress this effect, a *quench gas* is added to the counter gas. In the case of argon gas counters, carbon dioxide or methane are the most commonly used quench gases; for krypton- or xenon-filled counters, halogen molecules are widely used.

The function of the quench gas is as follows, taking carbon dioxide as an example. Molecular interactions in the counter gas result in exchange of charge between argon ions and the more readily ionized carbon dioxide molecules. Quench gas molecules are therefore predominant carriers of positive charges in gas counters. When the latter reach the walls of the counter, the excess energy of the interaction is used up in dissociating the quench molecule. Furthermore, carbon dioxide is an efficient absorber of ultraviolet radiation. In this way, mechanisms that could initiate cascades of secondary electrons are suppressed.

The output pulse from a proportional counter is shown in Figure 8.38. It consists of a negative going pulse with a rapid 'rise time' (0.1 to 0.2 μs) corresponding to the high mobility of electrons. However, this signal has a long tail (time constant = 1–2 μs) corresponding to the much slower movement of positively charged gas molecules to the counter wall.

A problem affects the size of output pulses at high count rates. Since the majority of electrons formed by gas multiplication originate in the vicinity of the central wire, so too do the positively charged gas molecules. Immediately after the detection of an event, these positive ions will form a temporary coaxial sheath around the counter wire which then progressively drifts outwards to the counter wall. The primary electrons generated by a second x-ray event detected immediately after the first will, therefore, suffer an electrostatic shielding effect due to this sheath of positive ions which will cause a reduction in the gas multiplication factor. This phenomenon of *pulse height depression* is observed as count rates significantly exceed 10^4 cps. If uncorrected, pulse height depression can cause systematically low intensity measurements if, for example, the detected pulse drifts out of a pulse height analysis window set for lower count rates (see below). Modern spectrometers can compensate electrically for this effect. It is desirable to operate proportional counters at the lowest practical gas multiplication setting to minimize pulse height depression and dead time effects.

The principal interaction between incident x-ray and counter gas is the photoelectric effect. The energy of the x-ray is absorbed to create ionized electrons and corresponding positively charged gas molecules that generate the conventional counter signal. However, a second interaction is important. The critical ionization energy for a *K*-shell electron in argon is 3.202 keV. Any incident x-ray that exceeds this energy will be capable of exciting argon *K* x-ray radiation. The argon x-rays may themselves be absorbed within the outer gas. However, there is a finite probability that x-rays will escape completely from the counter (for example, back through the counter window). In this case, the total energy deposited in the counter is that of the incident x-ray photon less the energy of the escaped argon photon (Ar $K\alpha$ = 2.957 keV). Two pulse distributions will therefore appear in the counter output, the main pulse corresponding to the incident photon energy and a subsidiary escape peak pulse distribution at an energy equivalent to 2.957 keV below this main event (see Figures 8.46 and 8.47). The relative intensity of the escape peak will depend on the efficiency of the incident x-ray in fluorescing argon. This is greatest for x-rays which just exceed the critical excitation energy of argon (for example K $K\alpha$, 3.313 keV; Ca $K\alpha$, 3.691 keV). If follows that escape peaks are not observed at all for x-rays of energies below 3.202 keV (in argon).

The presence of this phenomenon requires some consideration when choosing pulse height analysis windows (to be discussed further in section 8.9.7), from the point of view of whether to include the escape peak in the window or not. It the two distributions are not fully resolved, the escape peak must be included with the main peak. If they are well resolved, it is probably better to include only the main peak within the pulse height window.

A further problem can arise from an increase in counter detection efficiency for x-rays at energies above as compared with below the argon absorption edge. A scan of the spectrum across this absorption edge (Figure 8.39) shows a jump in background counts. This particularly affects the measurements of potassium $K\alpha$ x-rays since analytical bias would result if one background count for this x-ray line were selected below, and the other above the absorption edge.

A further example of this phenomenon is in the electron probe microanalysis of uranium minerals. The *M*-lines of uranium must normally be used for determinations since the *L*-lines are not usually adequately excited by microprobe. However, the uranium $M\alpha$ (3.170 keV) and $M\beta$ (3.336 keV) lines straddle the argon absorption edge and although the relative intensity of $M\alpha$ to $M\beta$ is 100 : 60, it is prefereable to determine uranium from the latter due because of the enhanced detection efficiency using an argon flow counter (Bowles, 1978).

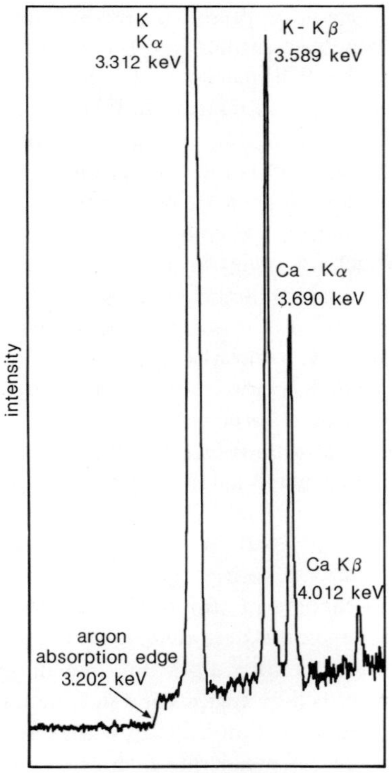

Figure 8.39 X-ray emission spectrum of a sanadine mineral, showing the step in background caused by the change in efficiency across the argon absorption edge of an argon gas flow proportional counter. The spectrum was recorded by electron probe excitation and recorded by scanning a wavelength dispersive spectrometer fitted with a PET crystal.

Escape peaks are observed from other counter gases, but for krypton and xenon, for example, they are offset to much higher energies (the critical excitation energies for the *K*-shells of Kr and Xe are 14.32 and 34.59 keV respectively—see Figure 8.41).

Ionization of the counter gas is a random process, the mean energy required to create a primary electron–ion pair being 26.4 eV in argon. Considerations similar to those for germanium detectors discussed in section 12.8 show that if an individual x-ray event is capable of creating a large number of electron–ion pairs in the detecting medium, then the statistical error in measuring that event will be relatively low and the resolution of the detector good (and vice versa). The mean number of primary electron–ion pairs generated by interaction of Mn $K\alpha$ radiation (for example, from a ^{55}Fe source) with various kinds of detectors are listed in Table 8.14, together with other physical characteristics of these devices. From these data it can be seen that gas proportional counters have inherently poor resolution characteristics compared, for example, with solid state Si(Li) or Ge detectors, because of the much larger mean energies required to create ionization in gas counters.

In this context it is necessary to emphasize the context in which the term resolution is used. The resolution of an energy dispersive spectrometer is controlled by the interaction characteristics of x-rays with Si(Li) or Ge as summarized in Table 8.14. However, the resolution of a wavelength dispersive spectrometer is mainly controlled by the angular dispersion characteristics of the diffracting crystal (see section 8.9.2). The proportional counter is only used on this instrument to detect the near-monochromatic diffracted beam. An estimate of the resolution of an entire crystal spectrometer is therefore included in Table 8.14 for comparison.

The resolution of a gas proportional counter defined as the full width of the counter pulse distribution at one half its maximum (FWHM) may be determined from:

$$\mathrm{FWHM} = 2.355\sqrt{F\varepsilon E} \qquad (8.46)$$

where F is the Fano factor, which for a gas proportional counter is not very well defined but is estimated to take the values 0.3 to 0.5 (Jenkins, 1976 p.66), ε is the mean energy required to create an electron–ion pair, and E_P is the photon energy.

The meaning and application of these terms is discussed further in section 12.8. It is often more convenient to express gas counter resolution as a percentage of the incident x-ray energy E, calculated from:

$$\mathrm{FWHM\,(in\,\%)} = \left[\frac{100 \times \mathrm{FWHM}}{E}\right] \qquad (8.47)$$

For an argon-filled proportional counter taking $F = 1$, $\varepsilon = 0.0264$ keV, this equation reduces to:

$$\mathrm{FWHM\,(in\,\%)} = \frac{38.3}{\sqrt{E}} \qquad (8.48)$$

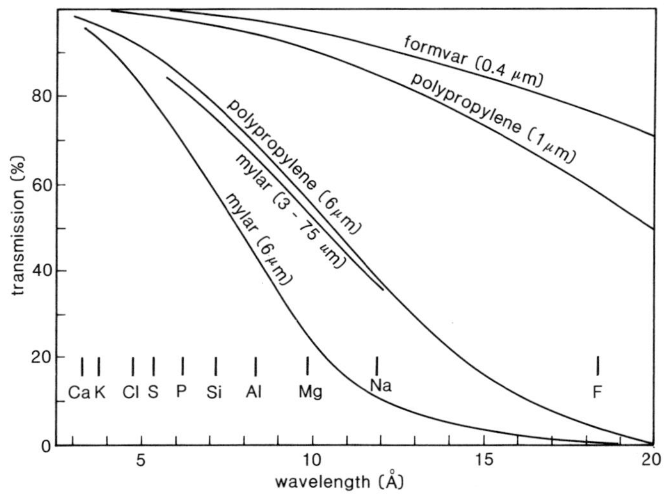

Figure 8.40 X-ray absorption properties of various materials used as window membranes for gas proportional counters as a function of energy. Data for mylar and polypropylene taken from Seaman (1979) and for formvar from Bertin (1975, p. 224).

Counter resolution is normally specified for the Mn *Kα* line (5.894 keV) (also emitted by a ^{55}Fe source), Fe *Kα* (6.398 keV) or Cu *Kα* (8.040 keV). The optimum FWHM (in %) resolutions for an argon proportional counter at these energies is thus 15.8% (Mn *Kα*), 15.1% (Fe *Kα*) and 13.5% (Cu *Kα*).

The most common reason why manufactured counters do not achieve this resolution specification concerns the quality of the anode wire. Any variation in the diameter or smoothness of this wire will significantly affect the local electrostatic field and therefore gas gain, causing degraded counter resolution. Specks of dirt and other extraneous particles settling on this wire will have the same effect.

The efficiency of gas proportional counters in detecting incident x-ray photons is limited at low energies by absorption in the counter entrance window and at high energies by the reduced probability of the photon interacting with the detector gas at all.

Severe absorption of x-rays of energy below 2 keV occurs in the counter entrance window. Careful consideration must be given both to the absorption properties of a membrane and its mechanical strength in containing counter gas at a pressure of about one atmosphere against a vacuum in the spectrometer chamber.

The standard window material for argon flow counters is 6 μm mylar (polyethylene tetraphthalate). However, severe attenuation of radiation lower than Si *Kα* occurs in this material (Figure 8.40). For instruments required to measure radiation down to Na *Kα* and F *Kα*, 1 μm polypropylene is normally used, though this material must be supported to withstand vacuum (usually by the secondary Soller slits attached to the counter assembly). Such thin counter windows usually have a limited life, typically of 3–8 months.

In electron probe microanalysis applications, even 1 μm polypropylene is too opaque to permit satisfactory detection of the lightest elements (B to O). Formvar (0.2 μm) or collodion must be used as window materials, supported for strength on a wire-mesh grid. The absorption properties of mylar and polypropylene have been investigated by Seaman (1979) and are compared in Figure 8.40.

A further difficulty in the detection of very low-energy x-rays (F *Kα* and below) is that they are unable to penetrate very far into the gas of a standard argon proportional counter. Charge collection from ionization events is not then efficient. One suggestion to reduce these difficulties is to operate the counter at reduced gas pressures. Low-energy x-rays are then able to penetrate further into the counter gas. As an alternative, Loch (1973) has described the use of a helium–12% carbon dioxide gas mixture in the determination for example of fluorine in apatite. A significant further advantage of this gas mixture is the elimination of a third-order phosphorus *Kα* escape peak which overlaps the F *Kα* pulse distribution if an argon gas counter is used.

Counters which use mylar (or thinner) entrance windows must be operated with a flow of gas passing through them to compensate for diffusion losses to the spectrometer vacuum system which would otherwise occur if the counter was completely sealed. Krypton- and xenon-filled proportional counters are not used to detect low-energy x-rays (below 3–8 keV) and so a non-porous beryllium window may be used. Such counters may then be sealed at the time of manufacture.

The pulse height distributions from gas proportional counters are sensitive both to gas pressure changes (due for example to changes in atmospheric conditions) and to temperature changes. It is desirable, therefore, to control the supply of counter gas using a pressure-regulating manostat such as that described by Jenkins and de Vries (1970).

The change in detection efficiency of several counter devices as a function of energy is shown in Figure 8.41. These data show that the detection efficiency of an argon flow counter falls off rapidly at energies above about 5 keV. In some microprobe applications, a krypton or xenon sealed counter would be used for the measurement of higher energy radiation. However, in many XRF applications, a sodium iodide scintillation detector is substituted instead.

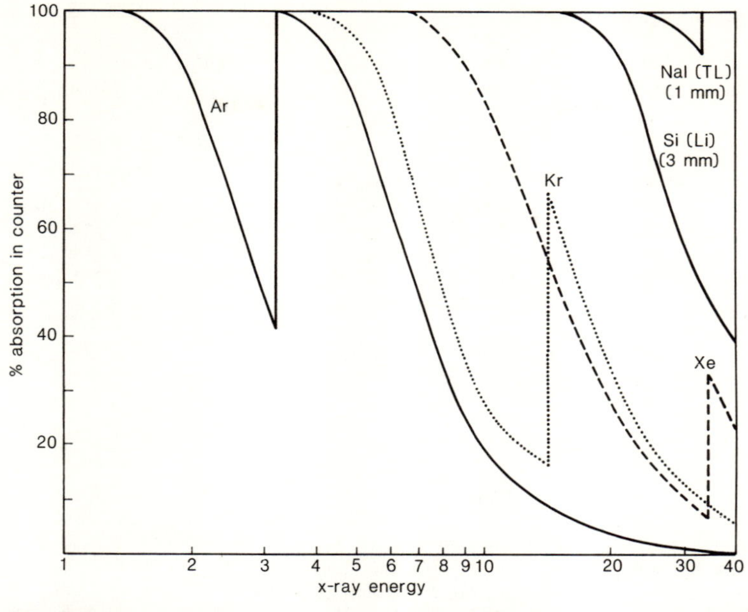

Figure 8.41 Calculated detection efficiencies of various x-ray counters as a function of photon energy. Data for argon, krypton and xenon gas proportional counters represent x-ray absorption within a path length of 20 mm. Thickness of NaI(Tl) scintillation detector had been taken as 1 mm and of the Si(Li) detector as 3 mm. Losses due to escape peaks have been ignored, nor has account been taken of absorption in the counter entrance window, data for which are plotted in Figure 8.40.

(*b*) *Sodium iodide scintillation detectors.* A scintillation detector consists of two components, a phosphor and a photomultiplier tube (Figure 8.42). The phosphor is a disc-shaped crystal of sodium iodide, perhaps 30 mm diameter and 1–3 mm thick. Thallium atoms are drifted into this crystal to improve its optical characteristics. The interaction of x-rays with the NaI(Tl) crystal causes this phosphor to scintillate. For each 50 eV or photon energy absorbed, an average of one light photon is emitted in the blue region of the visible spectrum (410 nm, 3 eV in energy). The surfaces of the NaI(Tl) crystal are aluminized so that as high a proportion as possible of these photons impinge on the photodynode (first stage) of the photomultiplier tube. Because of reflection losses and the inefficient conversion of scintillation photons into photodynode electrons, it takes on average 250 to 300 eV of absorbed x-ray energy to create one electron at the photocathode. This has unfavourable implications for the resolution of these detectors in comparison with other counting devices (Table 8.14).

The photomultiplier tube serves to amplify the electron signal generated at the photodynode by an overall factor of perhaps 10^6. To achieve this, the photomultiplier is designed with several electrodes (called dynodes) each of which is held at a progressively higher positive potential. Electrons emitted by the photodynode are accelerated by the potential gradient to the first dynode. On impact, each electron causes the emission of 2 to 4 secondary electrons which then progress through perhaps 9 or more further stages of multiplication at successive dynodes. The resultant photomultiplier output signal is collected at the anode. The magnitude of this signal is directly proportional to the energy of the incident x-ray photon.

Sodium iodide crystals must be hermetically sealed against moisture (NaI is hygroscopic) and surrounded by a reflective aluminized coating to prevent both the loss of scintillation photons and the admission of stray external light. These absorbers form a barrier on the front face of the detector preventing the efficient detection of radiation below about 3 to 4 keV.

The relative detection efficiency of a NaI(Tl) scintillation detector is compared with that of gas proportional counters in Figure 8.41. As mentioned earlier, at the crossover point

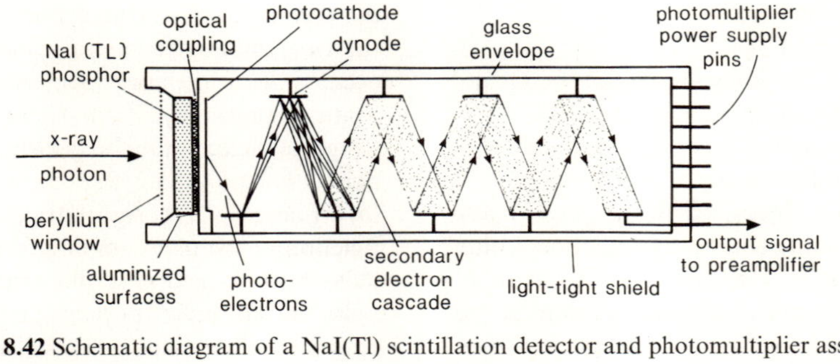

Figure 8.42 Schematic diagram of a NaI(Tl) scintillation detector and photomultiplier assembly.

Table 8.14 Resolution parameters for x-ray detectors

Detector	Atomic weight	Density (kg m^{-3})	Critical excitation energy (keV)	ε (eV)	Fano factor	N (Mn $K\alpha$)	Resolution at 5.89keV (eV)	Typical dead time (μs)
Gas proportional counter								
He	4.0	0.000179	—	27.8	—	212		
Ne	20.2	0.000900	0.867	27.4	—	215		
Ar	39.9	0.001874	3.202	26.4	0.3—0.5?	223	930	1.6
Kr	83.8	0.003708	14.32	22.8	—	258		
Xe	131.3	0.005841	34.59	20.8	—	283		
Scintillation detector								
NaI(Tl)	149.9	3.667	33.16	~300*1	—	20	3130	1.2
Crystal spectrometer							40*2	
Solid state detector								
Si(Li)	28.1	2.33	1.840	3.8*3	0.13	1550	160*4	100
Ge(Li)	72.6	5.35	11.102	2.98*3	0.13	1977	(150*4)	20*5

ε = mean energy to create an electron–ion pair.
N = mean number of electron–ion pairs produced by each Mn $K\alpha$ photon (5.89 keV).
Resolution = ideal full width half maximum of the Mn $K\alpha$ line at 5.89 keV.
Dead time is representative of detector coupled to its pulse processing electronics.
*1: Typical figure for average photon energy to general one electron at the photomultiplier photodynode. Precise data depend on the optical efficiency of scintillator.
*2: Typical resolution of an entire crystal spectrometer under routine analytical conditions.
*3: At 77 K (liquid nitrogen temperatures).
*4: Resolution figure calculated to include 100 eV of electronic noise.
*5: Typical data for gamma-ray applications. Data for comparison only as it would not be usual to use a Ge(Li) detector for low energy x-ray applications..

compared with the argon proportional counter, there is some advantage in measuring the sum of x-ray events detected on both devices. In XRF applications this mode is often used for the *K*-lines of Mn to As (6–11 keV). Above this energy, the scintillation counter is normally used alone. When counters are mounted in tandem, it is sometimes then desirable to flush the flow counter with air to reduce the absorption or scatter of higher energy x-rays by argon in the flow counter.

The rather poor energy resolution of a NaI(Tl) detector has already been commented on above (see also Table 8.14) and may be calculated by substituting the appropriate parameters into equation (8.46). Taking $\varepsilon = 300$ eV, and $F = 1$, then the resolution of a scintillation detector is given by

$$\text{FWHM (in \%)} = \frac{129}{\sqrt{E}} \qquad (8.49)$$

where E is the x-ray energy in keV.

The corresponding FWHM figures for Mn $K\alpha$, Fe $K\alpha$ and Cu $K\alpha$ are then 53%, 51% and 45.4% respectively.

Escape peak phenomena are observed in NaI(Tl) detectors due to excitation and loss on Na *K*-lines (1.04 keV), I *L*-lines (3.9–4.2 keV) and I *K*-lines (28.5–32.3 keV). None of these escape peaks is efficiently excited by radiation within the normal working range of this detector (6–20 keV), although the iodine *K*-line escape peak could be troublesome if determination of REE *K*-lines were attempted.

8.9.5 *Pulse processing electronics*

The function of pulse processing electronics is to amplify the very small charge generated by the x-ray detector (gas proportional or scintillation counter) and to convert it into a measured x-ray count rate. The principal components of such a system are shown in Figure 8.43. The brief function of each component will be described here; more detailed discussions are available (see, for example, Reed, 1975, pp. 113–131).

(*a*) *Preamplifier*. Detected x-ray photons appear at the preamplifier input as negative-going pulses with a rapid rise-time (~0.2 μs) and a longer tail (~2 μs) (see Figure 8.38). The function of the preamplifier is to amplify these pulses from a value in the μV range to one in the mV range. Added noise must be kept to a minimum since this too will be amplified by later electronic circuits. In order to achieve this, the preamplifier is normally located as close as possible to the x-ray counter to eliminate electronic pick-up that could occur in long interconnecting cables. The first stage amplification is normally based on an FET (field effect transistor), a high gain electronic component that contributes minimum noise.

(*b*) *Main amplifier*. Apart from further amplification, the function of the main ampifier is to 'shape' the preamplifier pulse both to suppress the distortion effect of electronic noise and to minimize pulse pile-up effects. This aspect of gaussian amplification is considered further in section 12.5.

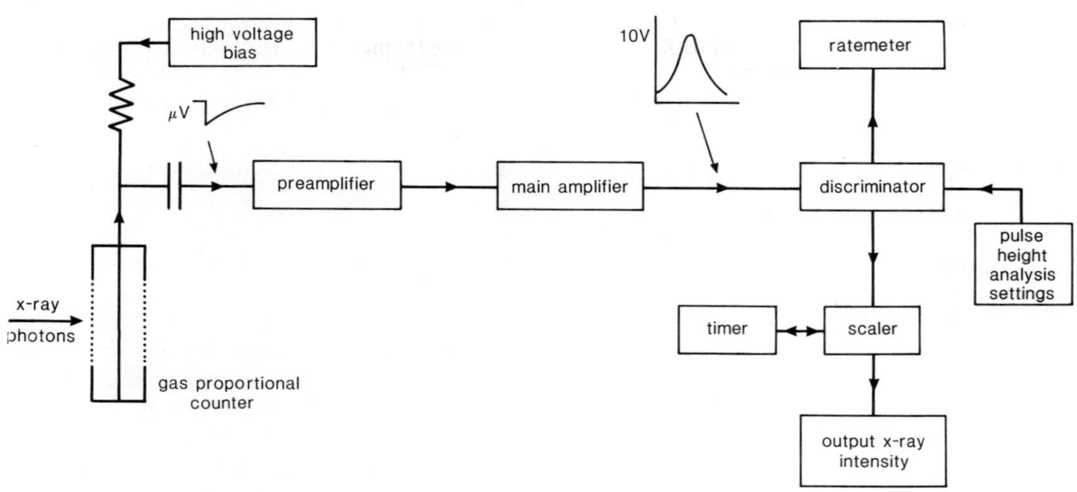

Figure 8.43 Block diagram of the electronic modules used in an x-ray spectrometer pulse processing chain.

The resultant output pulse (0–10 V) is semi-gaussian in shape, as shown in Figure 8.44.

(*c*) *Discriminator*. The discriminator compares the pulse output from the main amplifier with two user selectable threshold levels (Figure 8.44). Once an input pulse has exceeded a preset lower level threshold, the discriminator circuits are triggered. This causes a standard logic pulse to be created, representing one countable x-ray event, once the input has decayed back below the threshold level. The discriminator circuits are designed so that they cannot be reset to trigger further output pulses until the input has decayed below this threshold. Any x-ray event occurring immediately after the first within this threshold interval cannot therefore be recognized.

An important aspect of pulse height analysis is the ability to set a second upper discriminator threshold. If, having passed the lower threshold, an input also exceeds the upper level, output from the discriminator is effectively cancelled. The discriminator is not, however, reset until this rejected event decays below the lower threshold. In this way the discriminator can be set up to count specific pulse height distributions and avoid certain multiple-order interferences (see section 8.9.7).

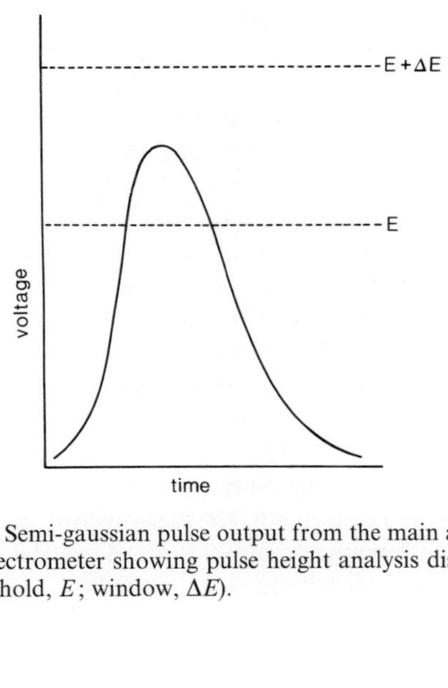

Figure 8.44 Semi-gaussian pulse output from the main amplifier of an x-ray spectrometer showing pulse height analysis discriminator levels (threshold, E; window, ΔE).

(*d*) *Scaler and timer*. These components are used to accumulate the number of countable events output by the discriminator under preset conditions.

In normal operation a preset time is entered into the timer which compares this with the number of 'clock' pulses generated by a high-frequency electronic oscillator. When the required number of clock pulses equivalent to the preset time has been accumulated, data acquisition to the scaler is stopped and the number of x-ray events accumulated there is inspected to determine the count rate.

The alternative mode of operation is to preset an integral number of counts in the scaler. The scaler and time are then started and data accumulated until the preset number of pulses have been stored from the discriminator. The time is then inspected to determine the time required for this measurement, this being related to x-ray count rate. The advantage of this method of operation is that all x-ray measurements are made to equivalent precision since the standard deviation of any x-ray measurement is equal to $\sqrt{N}$, the square root of the number of counts accumulated. This criterion cannot alwas be used, for example in samples with very high and very low emission intensities.

The discriminator and scaler/timer circuits behave as a single-channel analyser and may be compared with the operation of a multi-channel analyser described in section 12.5.

8.9.6 *Dead time corrections*

Dead time is introduced into the pulse processing electronics because of the finite length of time required to detect electronically and measure an x-ray event. During this time interval the detection system is unable to recognize any further x-ray events which are consequently lost from the accumulated count data. Dead times are categorized as being extendable (paralysable) or non-extendable (non-paralysable). Jenkins *et al.* (1981, p.230) distinguished the differ-

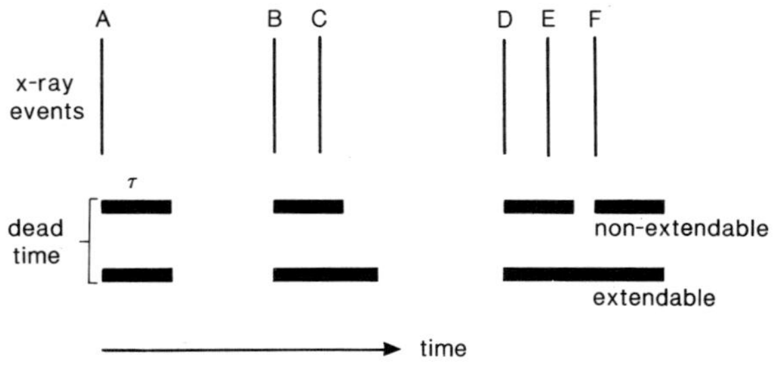

Figure 8.45 Schematic representation of extendable and non-extendable dead time. Lines A to F represent the occurrence of 5 individual x-ray events; the bars indicate the magnitude of the pulse processing dead time created by these events, assuming (i) non-extendable and (ii) non-extendable dead time. τ represents the dead time contribution for a single x-ray event. (After Jenkins *et al.*, 1981.)

ence as shown in Figure 8.45. If x-ray events are detected at times A, B, C,...,F and τ is the time required to process an individual pulse then dead time is classified as non-extendable if the detection of a second event (e.g. C and E) during the processing of the first (B and D) does not cause any further increase in pulse processing time. The offending event is simply ignored. Conversely, if the occurrence of this second event (e.g. C, E and F) does increase the pulse processing time of pulses B and D, dead time is classified as extendable.

If dead time is non-extendable, then the true count rate (R_T) is related to the measured count rate (R_M) as follows:

$$R_T = \frac{R_M}{1 - R_M \tau} \qquad (8.50)$$

where τ is the dead time.

If dead time is extendable, the following equation applies:

$$R_T = R_M \exp(R_T \tau) \qquad (8.51)$$

In x-ray counting systems, dead time originates in the counter itself. The sheath of positively charged ions surrounding the anode wire following the detection of one x-ray photon inhibits further gas multiplication for a period of about 0.5 μs (section 8.9.4). Any following photon detected within this time interval is therefore ignored, contributing to the non-extendable dead time of the spectrometer.

In most x-ray systems, it is possible to assume that dead time is non-extendable up to count rates of about $10^5\ s^{-1}$. The error-corrected count rates calculated according to this assumption are less than 2% compared with the full extendable correction formula. However, at the higher count rates attainable on modern XRF instruments (in excess of $7 \times 10^5\ s^{-1}$) it is necessary to take account of extendable contributions to dead time. Banerjee (1983) has reviewed errors in correcting for dead time losses at very high count rates and reports that the extendable correction described by equation (8.51) may be confidently applied to count rates of up to $7 \times 10^5\ s^{-1}$, the error being considerably less than 1%. Banerjee (1983) also describes a method of determining dead times in such circumstances.

In the simpler case of non-extendable dead time, it is possible to assume that x-ray tube current i (or if appropriate, the electron microprobe current) is directly proportional to R_T. The dead time (τ) may be determined from

$$\frac{R_M}{i} = k(1 - R_M \tau) \qquad (8.52)$$

A plot of R_M/i *v.* R_M for various values of i is a straight line of intercept k and gradient $-k\tau$ (k is a constant) from which τ may be calculated. Any deviation from linearity indicates a significant extendable contribution to dead time (Reed, 1975, p.130).

Typical counter dead times are about 1.6 μs for a gas proportional counter and electronics and 1.2 μs for a scintillation counter. Although dead time corrections are comparatively small and almost negligible for low count rates, this is not the case at higher count rates. If a value of $\tau = 1.6$ μs is assumed, the true count rate according to equation (8.50) for measured count rates of 10^4 and $10^5\ s^{-1}$ are calculated as 10 163 and 119 048 s^{-1}, differences of 1.6% and 19% respectively.

8.9.7 *Pulse height analysis*

Pulse height analysis uses the facility to set upper and lower discriminator threshold levels to filter the input signal distribution, as described in section 8.9.5. This offers the analyst two important benefits. The first is the ability to distinguish and eliminate multiple-order interferences on an x-ray line. The second concerns a reduction in background counts.

As well as diffracting an x-ray line of wavelength λ, a crystal will also diffract wavelengths $\lambda/2$, $\lambda/3$ etc. In the absence of a specific multiple-order interference, these wavelengths will normally represent part of the background continuum that is diffracted into the counter along with the first-order x-ray line of interest. Furthermore, a small component of the entire fluorescent spectrum emitted by the sample is scattered by the crystal into the x-ray counter. Since this scattered spectrum has not been diffracted, it causes a continuum of background pulses in the x-ray counter. By setting a pulse height analysis window, the background interference caused by this component is minimized. Jenkins (1976, p.75) estimated that under favourable circumstances, in particular in the analysis of fluorescent lines below about 11 keV, over 50% of the background can be eliminated in this way, and in the case of light elements (e.g. Mg or Al $K\alpha$), scattered background can be virtually eliminated.

The upper and lower threshold levels in the discriminator are normally set for pulse height analysis by adjusting a lower threshold potentiometer E and a window potentiometer ΔE. All pulses are accepted for analysis if they fall between the limits E and $(E+\Delta E)$. If the value of ΔE is set to say 0.1V and the value of E is swept (manually or by computer control) through the full range of output voltages (usually 0 to 10 V), then it is possible to measure an intensity–pulse height histogram of the main amplifier output. Typical pulse height distributions obtained in this way are shown in Figure 8.46.

In all pulse height measurements, it is necessary to set a lower threshold of about 0.2 V, since below this level, analytical signals are increasingly swamped by electronic noise. The pulse height distributions in Figure 8.46 are designed to

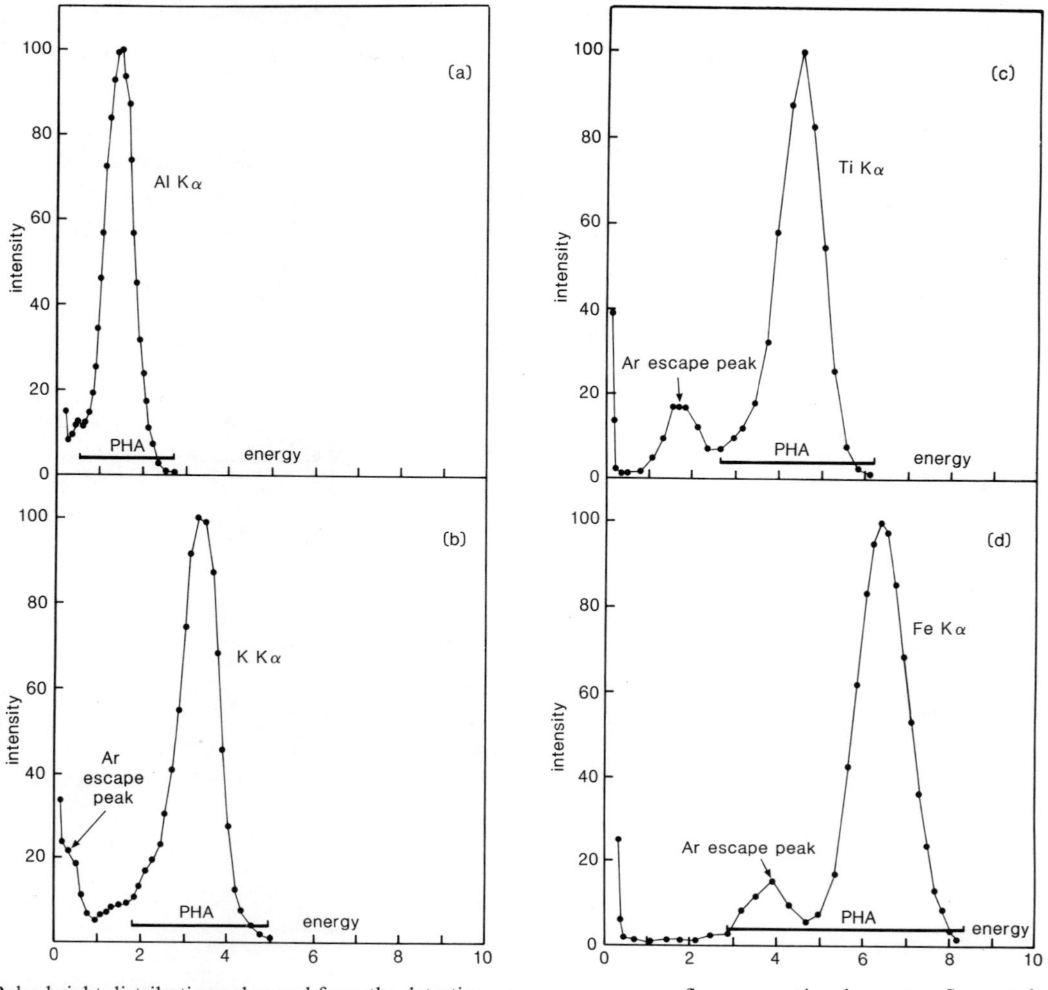

Figure 8.46 Pulse height distributions observed from the detection of (*a*) Al *Kα*, (*b*) K *Kα*, (*c*) Ti *Kα* and (*d*) Fe *Kα* x-ray photons by an argon gas flow proportional counter. Suggested pulse height analysis settings are indicated by the horizontal bars.

demonstrate aspects of pulse height selection using an argon gas flow proportional counter. The distribution shown in Figure 8.46*a* represents the spectrum observed from Al *Kα* x-rays indicating the proposed pulse height analysis window. The potassium *Kα* distribution (Figure 8.46*b*) shows the appearance of a prominent argon escape peak which cannot be fully resolved from the counter noise. This escape peak is, however, fully resolved in the case of Ti *Kα* x-rays (Figure 8.46*c*). Iron *Kα* is less efficient in exciting argon *Kα* radiation and the small escape peak distribution is now beginning to merge with the main peak (Figure 8.46*d*).

The pulse height analysis windows suggested for the determination of these lines are also indicated. However, the selection of precise upper and lower discriminator values must take into account any potential drift in the pulse height distribution due to changes in counter gas pressure, temperature and pulse height depression effects at high count rates.

The pulse distributions shown in Figure 8.47 represent situations that may be encountered in the analysis of silicate rocks where pulse height analysis can wholly or partially eliminate a multiple-order interference. These are spectrum interferences caused by the fact that Bragg diffraction is satisfied for $\lambda/2$, $\lambda/3$,... wavelengths as well as the first-order λ line. In the determination of F *Kα*, particularly in microprobe applications, a third-order interference from P *Kα* can be eliminated by pulse height analysis (Figure 8.47*a*). The situation is not as straightforward in the analysis of P *Kα* in the presence of a second-order Ca *Kα* interference (Figure 8.47*b*). The Ca escape peak distribution cannot be completely resolved from P *Kα* as investigated by Jenkins and Hurley (1968). The analyst has to decide whether to set a window to encompass both peaks and apply a correction or (if the counter resolution is good enough) to set a narrow peak about the P *Kα* peak alone. For measurements made with a germanium (III) crystal this interference is eliminated. A similar problem is encountered in the analysis of Mg *Kα* in the presence of K *Kα* radiation resulting from fluorescence of a KAP crystal. Again, partial overlap occurs between the potassium escape peak and Mg *Kα* distribution. Harvey *et al.* (1973) successfully minimized this interference by setting a narrow pulse height window similar to that shown in Figure 8.47*c*. Substitution of more modern RAP or TAP diffracting crystals eliminates this interference.

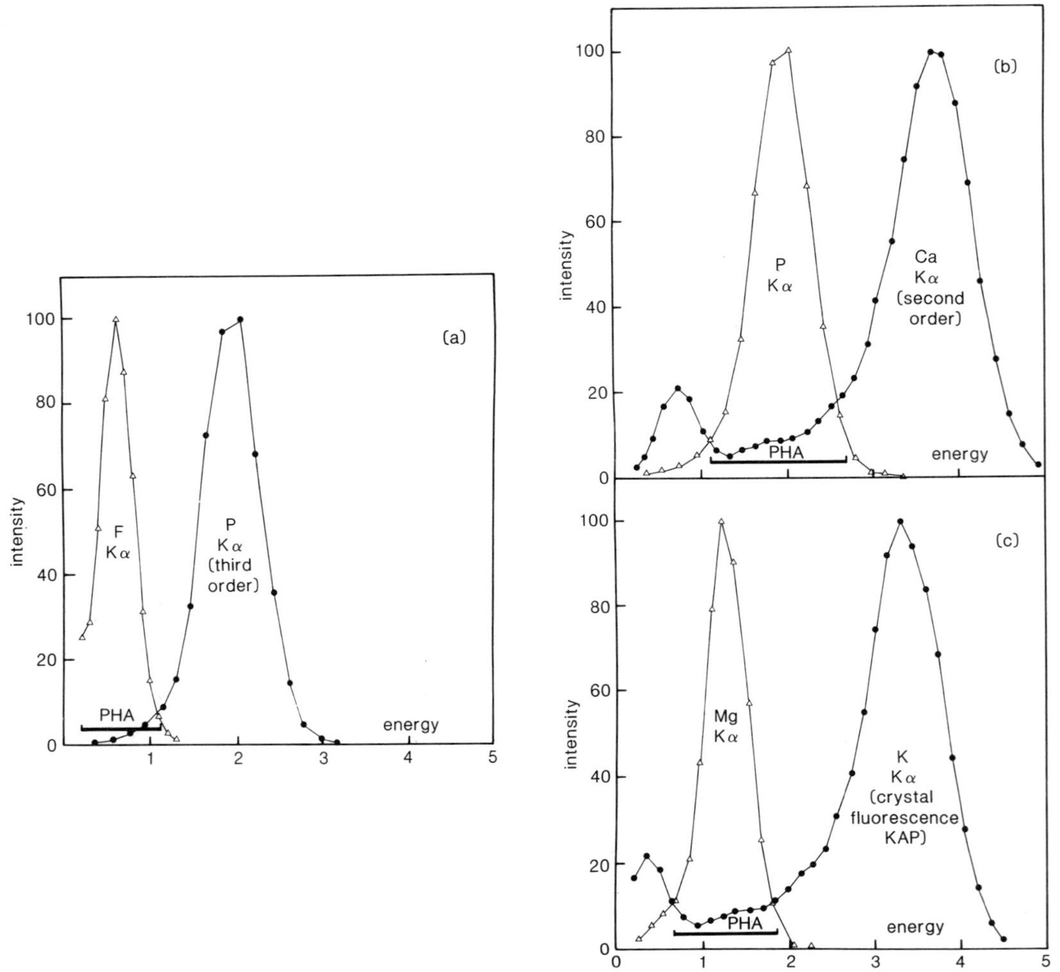

Figure 8.47 Pulse height analysis interferences: (*a*) P *Kα* third-order interference on F *Kα*; (*b*) Ca *Kα* second-order interference on P *Kα* (note the argon escape peak from Ca; (*c*) K *Kα* interference arising from the fluorescence of a KAP diffracting crystal used for the measurement of Mg *Kα*. In each case, the maximum intensity of individual pulse height distributions has been arbitrarily normalized in 100%. Suggested pulse height analysis settings are indicated by horizontal bars.

8.9.8 *Spectrometer vacuum consideration*

The path length of x-rays from sample to diffracting crystal to counter is several hundred millimetres in most spectrometer designs and if operated in air, severe absorption of low-energy radiation will occur. It is therefore necessary to determine these elements with the spectrometer either evacuated or, as an alternative, flushed with helium. The former is routinely used in the analysis of solid samples, the latter if liquid samples are to be excited. To indicate the effectiveness of these measures, the half-layer thicknesses for air and helium have been calculated in Table 8.15 for selected x-ray energies. The half-layer thickness is the thickness of a medium required to reduce the intensity of radiation to one-half of its incident value. As can be seen in Table 8.15, severe attenuation of x-rays below 5 keV occurs with the spectrometer in air. This absorption is considerably reduced under a helium atmosphere. Evacuating the spectrometer chamber minimizes absorption and has a further advantange. The excitation of argon *K*-line x-rays, caused by the fluorescence of this element in air (0.93% by volume), and the general scattering effect on x-rays by air molecules, are also eliminated.

Table 8.15 Path length of x-rays in helium and air

Photon energy (keV)	Half layer thickness (mm)	
	In helium	In air
1	716	2
2	3200	12
5	41 800	165
10	$>10^5$	306
20	$>10^5$	7600
50	$>10^5$	28 200
100	$>10^5$	38 100

8.10 Experimental considerations

8.10.1 *Sample preparation*

The analytical characteristics of powder pellets and lithium borate glass fusion beads have already been discussed in sections 8.7.3–8.7.4. This section is concerned with the practical aspects of sample preparation.

(*a*) *Lithium borate fusion discs*. Sample and the selected flux and pre-dried and then weighed in the appropriate proportions into a crucible. On the basis of a preliminary loss on ignition measurement, excess sample is added to compensate for any loss of volatiles from the sample during fusion. Platinum–5% gold alloy crucibles are normally used for fusion as this material is not wetted by the molten flux. However, difficulties can occur due to reaction of some elements with the crucible alloy. Thus, in experimental petrology applications, several workers (e.g. Biggar, 1977; Johannes and Bode, 1978) have reported losses of iron into platinum alloys at elevated temperatures and pressures, though not for samples diluted with lithium borate flux. It is advisable to carry out fusion under oxidizing conditions and to avoid samples containing sulphide minerals and high concentrations of some elements (such as tin, copper etc.), all of which will attack the crucible. To eliminate some of these problems, one manufacturer has developed crucibles made from platinum and platinum–5% gold in which the alloy grains are stabilized with finely dispersed zirconia crystals. These materials show greater resistance to attack than standard alloys.

Samples are fused, either in a muffle furnace or gas burners at between 950°–1200°C (depending on melting point and viscosity of flux) for 10 to 20 minutes. Sulphide sulphur is normally lost during fusion (sulphate sulphur is retained). There is some evidence that after prolonged fusion at temperatures above 1100°C, small amounts of other volatile elements such as potassium may also be lost from the melt.

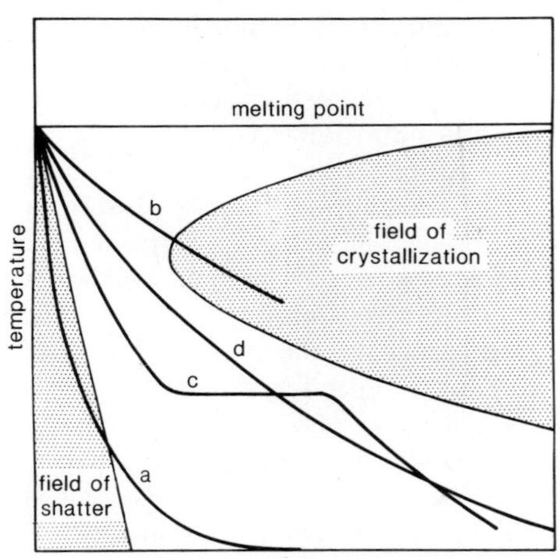

Figure 8.48 A representation of the cooling *v*. time diagram appropriate for the quenching of a lithium borate glass fusion disc. Shaded areas represent regions in which a disc will shatter, due either to the thermal shock caused by excessively rapid quenching or the strains arising from the formation of micro-crystals within the glass if cooling take place too slowly. (Adapted from Tertian and Claisse, 1982.)

It is essential to swirl the molten mixture several times during fusion to ensure complete dissolution and homogeneity of the sample.

Samples must then be cast as a flat disc in such a manner that the resultant glass does not subsequently devitrify. There are two difficulties, indicated schematically in Figure 8.48. This diagram represents the cooling-time graph of four notional sample annealing procedures. If the glass is chilled too rapidly (line *a*), strains are set up in the glass network, causing it to shatter. Conversely, if the glass is cooled too slowly (curve *b*), microcrystals separate out in the glass which on further cooling set up strains which again cause shattering. The art of successful bead making is to anneal the bead (lines *c* and *d*) so that the two shaded areas in Figure 8.48 are avoided. Unfortunately, the two fields of shatter vary with sample composition and flux type so that it is not possible to devise a universal scheme of sample preparation.

However, one widely used annealing procedure is that devised by Norrish and Hutton (1969) with subsequent modifications by Harvey *et al.* (1973). Molten samples are rapidly poured into a graphite, brass or duralumin platen which is held on a hot plate at about 200°C (Figure 8.49). An aluminium plunger also preheated to this temperature is then lowered on to the platen to squash the molten bead into a flat disc which is simultaneously quenched to form a glass. The glass disc is then annealed at 200°C on its platen for about 10 minutes before being allowed to cool to room temperature. The upper surface of the disc is analysed for major elements. The glass disc solidifies with a wrinkled texture using this technique. This does not seem to cause

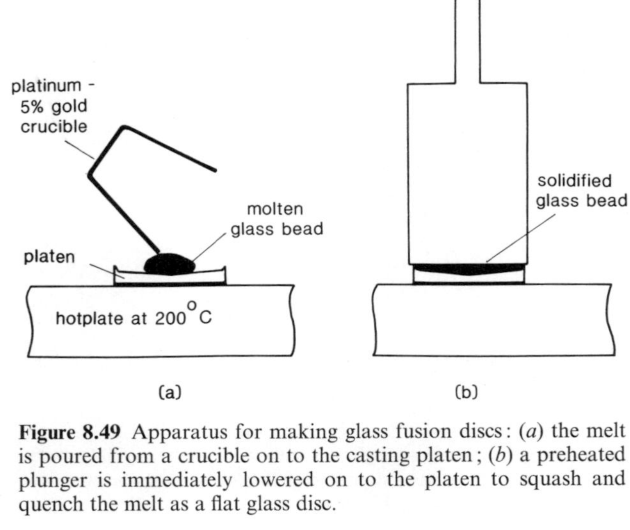

Figure 8.49 Apparatus for making glass fusion discs: (*a*) the melt is poured from a crucible on to the casting platen; (*b*) a preheated plunger is immediately lowered on to the platen to squash and quench the melt as a flat glass disc.

significant analytical errors in practice, especially if the XRF spectrometer is fitted with a sample spinner in the analysing position.

An alternative procedure used by some workers is to pour melts into platinum–gold casting dishes that have been preheated to the fusion temperature. The molten glass is normally solidified by blowing air over its surface and is then annealed in the casting dish. Using this procedure, the lower surface of the glass disc takes on the flatness and polish of the dish. Graphite casting dishes have also been used (Mahan and Leyden, 1982). Since the flux does not readily spread out in a graphite dish due to surface tension effects, it is necessary to use a relatively large mass of fusion mixture: whereas about 2.5 g of melt may be used to make a glass disc 31.5 mm diameter by about 1 mm thick using the plunger apparatus shown in Figure 8.49, about 10 to 12 g of melt is required to make an equivalent-diameter sample in a graphite casting dish. The advantage of graphite is that with its high heat capacity, a more ideal anneal curve (*d* in Figure 8.48) may be followed.

(*b*) *Powder pellets*. Powder pellets are prepared by mixing the rock with a suitable binder (to stop the sample crumbling) and compressing the mixture between tungsten carbide platens in a die at a suitable pressure. As can be seen in Figure 8.50, the measured intensity of fluorescent x-rays depends both on the particle size distribution in the powdered sample and the compression pressure used for making the pellet. The range 15 to 20 tonne inch^{-2} represents a reasonable compromise under which limiting constant intensity is obtained. For pellets prepared at lower pressures, there is a risk that discrepancies could occur due to small differences in packing densities.

It is necessary to use sufficient powder to ensure that the resultant pellet is infinitely thick with respect to x-rays in the 0–20 keV region. This condition is satisfied by the use of 6–7 g of sample to make a 31.5 mm diameter pellet of thickness 4 to 5 mm. Pellets prepared in this way offer two flat surfaces, each of which may analysed as (almost) independent subsamples.

Some workers prepare pellets using substantial proportions of binder (5–20% by weight of the pellet) including bakelite resin, cellulose powder and boric acid (for details, see for example, Fabbi, 1978, p.309). However, perhaps the most versatile procedure is to add a binder consisting of a

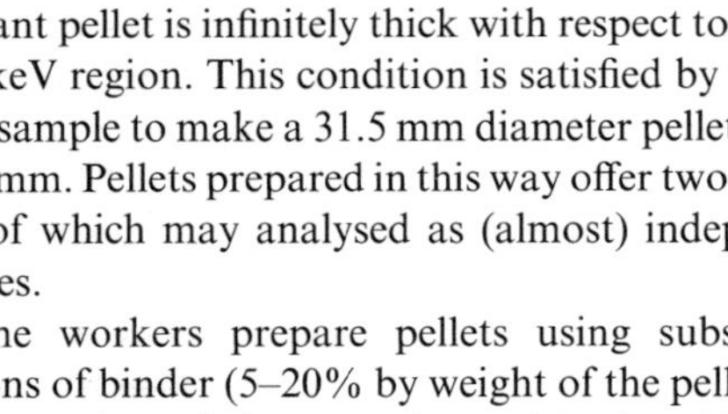

Figure 8.50 Intensity of Zn $K\alpha$ lines excited in samples prepared as pellets from powders varying in powder size from 30 to 300 μm and compressed at pressures ranging from 2.5 to 20 tons. (Togel, 1961; as presented by Tertian and Claisse, 1982.)

solution of polyvinylpyrrolidone–methyl cellulose solution in the ratio one drop per gram of rock powder (see, for example, Harvey and Atkin, 1982; Robinson and Bennett, 1981). The mixture is stirred into a paste, compressed into a pellet and then baked in an oven at 105°C to drive off excess solvent. The residual binder remaining in the pellet probably amounts to not more than 1% of the sample weight (Harvey and Atkin, 1982). This technique produces a durable pellet and overcomes some uncertainties in sample homogeneity when large proportions of binder are mixed with the sample.

This form of sample preparation is unusual in that it is not necessary to weigh accurately sample material. Furthermore, because the critical penetration depth of the heavier trace elements (for example, Rb to Nb) is about 1 to 1.5 mm (Table 8.6), analyses are representative of a significantly larger mass of sample (about 1.7 to 2.7 g) than is the case in optical emission techniques in which 1 g of sample is taken up into 100 mL solution.

8.10.2 *Spectrum analysis*

(*a*) *Spectrum interferences*. Although x-ray spectra are comparatively simple, and the choice of x-ray line for analysis straightforward, there are several sources of spectrum interference that must be considered and accounted for. These include:

(i) Direct overlap between first-order *K*-, *L*- or *M*-lines excited within the sample

(ii) Multiple-order interferences derived, in particular, from the overlap of second- and third-order major-element lines with trace-element lines

(iii) First-order and multiple-order overlaps from the characteristic lines of the x-ray tube and possibly their Compton scatter peaks

(iv) Additional interferences arising, for example, from the fluorescence of the lanthanum added to heavy absorber fluxes.

In extreme circumstances corresponding, for example, to multiple-order diffractions of tube lines overlapping trace elements approaching detection limit level, it may be necessary to consider up to the sixth-order overlap interference.

Table 8.16 lists the interferences that could be encountered in the analysis of major and trace elements in samples of 'normal' silicate composition. Interferences have been classified as those originating from the sample itself, and those artefacts caused by x-ray tube or lanthanum heavy absorber lines. Only interferences from the x-ray tube in use will be relevant to a particular scheme of analysis, and account must be taken of the operating potential. For example, if a tungsten tube is operated at less than 69.5 kV, the corresponding W *K*-line interferences will not be observed. Similarly, lanthanum heavy absorber *K*-lines are not excited unless the x-ray tube is operated at potentials above 38.9 kV.

The occurrence of significant tube or heavy absorber interferences is summarized in Table 8.17. These have been classified as those which may require correction (*) and those less severe interferences which either are not always excited or may readily be eliminated by pulse height analysis (●).

Table 8.16 X-ray fluorescence overlap interferences in silicate rock analysis

	Analytical line			Elemental interference						X-ray tube or heavy absorber interference					
	Line	keV	Å		Line	*n*	keV	Å	*I*		Line	*n*	keV	Å	*I*
F	$K\alpha_{1,2}$	0.667	18.32	Ca	$K\beta_{1,3}$	6	4.012	18.538	15	Ag	$L\beta_{2,15}$	5	3.347	18.517	25
				P	$K\alpha_{1,2}$	3	2.013	18.474	150	Mo	L_l	3	2.015	18.452	3
				Co	L_l	1	0.678	18.292	9	Rh	$L\gamma_{2,3}$	5	3.363	18.427	5
										Rh	$L\alpha_2$	4	2.692	18.422	10
										W	M_3–N_4	3	2.021	18.402	0.1
										W	$M\gamma$	3	2.035	18.276	1
										Sc	$K\alpha_{1,2}$	6	4.088	18.192	150
										La	L_1	6	4.124	18.036	2
Na	$K\alpha_{1,2}$	1.041	11.910	Zn	$L\beta_1$	1	1.034	11.983	26	Sc	$K\alpha_{1,2}$	4	4.088	12.128	150
										La	$M\gamma$	1	1.026	12.080	1
										La	$L\beta_3$	5	5.143	12.053	6
										La	L_l	4	4.124	12.024	2
										Ag	$L\beta_1$	3	3.150	11.804	42
										Au	$M\alpha_2$	2	2.118	11.708	100
										Au	$M\alpha_1$	2	2.123	11.680	100
Mg	$K\alpha_{1,2}$	1.253	9.890	Ca	$K\beta_{1,2}$	3	3.690	10.078	150	Mo	$L\beta_3$	2	2.473	10.027	3
				Ti	$K\beta_{1,3}$	4	4.931	10.056	20	Ag	$L\gamma_2$	3	3.743	9.936	3
				Ca	$SK\alpha_3$	3	3.711	10.020	2	Ag	$L\gamma_3$	3	3.749	9.919	2
				Zr	$L\gamma_{2,3}$	2	2.502	9.907	0.5	Mo	$L\beta_{2,15}$	2	2.518	9.846	1
				As	$L\alpha_{1,2}$	1	1.282	9.671	100	Rh	L_n	2	2.519	9.843	1
										La	$L\beta_1$	4	5.041	9.836	50
Al	$K\alpha_{1,2}$	1.486	8.340	Mn	$K\alpha_{1,2}$	4	5.894	8.413	150	Rh	$L\beta_6$	2	2.922	8.483	3
				Rb	L_l	1	1.482	8.364	3	W	L_l	5	7.386	8.391	3
				Ti	$K\alpha_{1,2}$	3	4.508	8.249	150	Ar	$K\alpha_{1,2}$	2	2.957	8.386	150
										Sc	$K\beta_{1,3}$	3	4.460	8.339	20
										Ag	$L\alpha_2$	2	2.978	8.326	10
										Ag	$L\alpha_1$	2	2.984	8.309	100
										Rh	$L\beta_{2,15}$	2	3.001	8.262	25
										La	L_n	3	4.524	8.220	1
Si	$K\alpha_{1,2}$	1.739	7.126	Rb	$L\beta_1$	1	1.752	7.075	45	Ag	$L\gamma_5$	2	3.428	7.233	0.1
				Y	L_n	1	1.761	7.041	1	La	$L\beta_3$	3	5.143	7.232	6
										Au	M_4–N_3	1	1.746	7.101	0.01
										Ag	$L\gamma_1$	2	3.519	7.045	10
P	$K\alpha_{1,2}$	2.013	6.158	Y	$L\beta_1$	1	1.995	6.211	45	Cr	$K\beta_{1,3}$	3	5.946	6.255	18
				Nb	L_n	1	1.996	6.211	1	W	$L\beta_2$	5	9.960	6.223	20
				Ca	$K\beta_{1,3}$	2	4.012	6.179	15	Mo	L_l	1	2.015	6.151	3
				Y	$L\beta_6$	1	2.034	6.094	3	La	$L\gamma_2$	3	6.059	6.138	1
				Zr	$L\alpha_2$	1	2.040	6.078	10	W	M_3–N_4	1	2.021	6.134	0.1
				Zr	$L\alpha_1$	1	2.042	6.071	100	La	$L\gamma_3$	3	6.073	6.123	1
										W	$M\gamma$	1	2.035	6.092	1
										Sc	$K\alpha_{1,2}$	2	4.088	6.064	150
S	$K\alpha_{1,2}$	2.307	5.373	Zr	$L\gamma_1$	1	2.302	5.384	1	Au	$L\beta_1$	5	11.440	5.418	50
				Nb	$L\beta_6$	1	2.312	5.361	3	Mo	$L\alpha_2$	1	2.289	5.414	10
				Nb	$L\beta_4$	1	2.319	5.346	3	Mo	$L\alpha_1$	1	2.293	5.407	100
										Au	$L\gamma_3$	6	13.807	5.387	2
										W	M_2–N_4	1	2.314	5.357	0.1
										Au	$L\beta_2$	5	11.583	5.351	20
										La	$L\alpha_2$	2	4.633	5.351	10
										Au	$L\beta_3$	5	11.608	5.339	6
										La	$L\alpha_1$	2	4.650	5.331	100
										W	$L\gamma_3$	5	11.672	5.310	2
Cl	$K\alpha_{1,2}$	2.621	4.729	Pb	M_3–N_4	1	2.629	4.715	5	Mo	$L\gamma_1$	1	2.623	4.726	1
										Ag	L_l	1	2.633	4.708	2
										Au	M_3–O_1	1	2.636	4.703	0.1

(Continued)

Table 8.16 (*Continued*)

	Analytical line			Elemental interference						X-ray tube or heavy absorber interference					
	Line	keV	Å		Line	n	keV	Å	I		Line	n	keV	Å	I
K	$K\alpha_{1,2}$	3.312	3.742	U	$M\beta$	1	3.336	3.716	60	Ag	$L\beta_1$	TC	3.13	3.96	
										Mo	$K\beta_3$	6	19.587	3.797	7
										Mo	$K\beta_1$	6	19.605	3.794	17
										W	$L\beta_3$	3	9.817	3.788	6
										W	$L\beta_{1,5}$	3	9.946	3.739	1
										W	$L\beta_2$	3	9.960	3.734	20
										Mo	$K\beta_2$	6	19.962	3.726	4
										Au	$L\gamma_1$	4	13.379	3.706	10
										Ag	$L\beta_{2,15}$	1	3.347	3.703	25
										Rh	$K\alpha_{1,2}$	6	20.165	3.688	150
										Rh	$L\gamma_{2,3}$	1	3.363	3.686	5
Ca	$K\alpha_{1,2}$	3.690	3.359							Ag	$K\alpha_2$	6	21.987	3.383	50
										Ag	$K\alpha_1$	6	22.159	3.356	100
										W	L_l	2	7.386	3.356	3
										Au	$L\beta_6$	3	11.158	3.333	0.1
										Au	$L\beta_4$	3	11.203	3.319	4
										Ag	$L\gamma_2$	1	3.743	3.312	3
Sc	$K\alpha_{1,2}$	4.088	3.032	Th	M_2–N_4	1	4.117	3.011	5	Rh	$K\alpha_1$	5	20.213	3.066	100
				La	L_l	1	4.124	3.006	2	Au	$L\beta_9$	3	12.145	3.061	0.01
				Cs	Ln	1	4.141	2.993	1						
Ti	$K\alpha_{1,2}$	4.508	2.750	Ba	$L\alpha_2$	1	4.450	2.786	10	Ag	$K\alpha_1$	5	22.159	2.797	100
				Sc	$K\beta_{1,3}$	1	4.460	2.780	20	Au	$L\gamma_1$	3	13.379	2.780	10
				Ba	$L\alpha_1$	1	4.465	2.776	100	La	L_n	1	4.524	2.740	1
										Rh	$K\beta_3$	5	22.695	2.731	8
										Au	$L\gamma_8$	3	13.624	2.730	0.1
										Rh	$K\beta_1$	5	22.720	2.728	16
La	$L\alpha_1$	4.650	2.666	Cs	$L\beta_1$	1	4.619	2.684	50	Au	$L\gamma_3$	3	13.807	2.693	2
				Nd	L_l	1	4.632	2.676	2	Rh	$L\beta_2$	5	23.169	2.675	4
				Cs	$L\beta_4$	1	4.649	2.667	5	Au	L_1–N_4	3	13.997	2.657	0.01
				Cs	$L\beta_3$	1	4.716	2.629	6	Au	$L\gamma_{11}$	3	14.017	2.653	0.01
Ba	$L\beta_1$	4.827	2.568	Ce	$L\alpha_2$	1	4.822	2.571	10	W	$L\beta_4$	2	9.524	2.603	4
				Ce	$L\alpha_1$	1	4.839	2.562	100	W	$L\beta_6$	2	9.610	2.580	0.1
										Au	$L\alpha_2$	2	9.626	2.575	10
										W	$L\beta_1$	2	9.671	2.564	50
										Au	$L\alpha_1$	2	9.712	2.552	100
V	$K\alpha_{1,2}$	4.949	2.504	Ba	$L\beta_3$	1	4.926	2.516	6	Mo	$K\beta_3$	4	19.587	2.531	7
				Ti	$K\beta_{1,3}$	1	4.931	2.514	20	Mo	$K\beta_1$	4	19.605	2.529	17
				Pr	L_n	1	4.935	2.512	1	W	$L\beta_3$	2	9.817	2.525	6
				Cs	$L\beta_{2,15}$	1	4.935	2.512	20	W	$L\beta_{15}$	2	9.946	2.493	1
				Ti	$K\beta_5$	1	4.961	2.499	0.02	W	$L\beta_2$	2	9.960	2.489	20
										Ag	$K\beta_3$	5	24.907	2.488	8
										Ag	$K\beta_1$	5	24.938	2.485	18
Pr	$L\alpha_1$	5.033	2.463	Ba	$L\beta_6$	1	4.993	2.483	0.1	Mo	$K\beta_2$	4	19.962	2.484	4
				Sm	L_1	1	4.994	2.482	2	Rh	$K\alpha_2$	4	20.070	2.471	50
				La	$L\beta_1$	1	5.041	2.459	50	Rh	$K\alpha_1$	4	20.213	2.453	100
				La	$L\beta_4$	1	5.061	2.449	5	W	$L\beta_7$	2	10.127	2.448	0.1
										Ag	$K\beta_2$	5	25.452	2.435	5
Ba	$L\beta_2$	5.156	2.404	La	$L\beta_3$	1	5.143	2.411	6	W	$L\beta_5$	2	10.199	2.431	0.1
				Nd	L_n	1	5.145	2.409	1	Au	L_n	2	10.307	2.405	1
				Eu	L_l	1	5.176	2.395	2						
Nd	$L\alpha_1$	5.229	2.370	Nd	$L\alpha_2$	1	5.207	2.381	10	Au	L_2–M_2	2	10.588	2.342	0.01
				Ba	$L\beta_7$	1	5.207	2.381	0.1						
Ce	$L\beta_1$	5.261	2.356	Ce	$L\beta_4$	1	5.276	2.350	5						
				Cs	$L\gamma_1$	1	5.279	2.348	5						
Cr	$K\alpha_{1,2}$	5.411	2.291	Ce	$L\beta_3$	1	5.364	2.311	6	La	$L\beta_{2,15}$	1	5.383	2.303	20
				V	$K\beta_{1,3}$	1	5.426	2.284	20	La	$L\beta_{10}$	1	5.413	2.290	0.01
										La	$L\beta_7$	1	5.449	2.275	0.1

Table 8.16 (*Continued*)

	Analytical line			Elemental interference						X-ray tube or heavy absorber interference					
	Line	keV	Å		Line	n	keV	Å	I		Line	n	keV	Å	I
Pr	$L\beta_1$	5.488	2.259	Ba	$L\gamma_1$	1	5.530	2.242	5	Au	L_1–M_1	2	10.926	2.269	0.01
				Cs	$L\gamma_2$	1	5.541	2.237	1	W	$L\gamma_5$	2	10.947	2.265	0.1
				Ta	L_l	1	5.546	2.235	2	Au	$L\beta_{17}$	2	10.990	2.256	0.01
				Cs	$L\gamma_3$	1	5.552	2.233	1	Ag	$K\alpha_2$	4	21.987	2.255	50
										Ag	$K\alpha_1$	4	22.159	2.238	100
										Au	$L\beta_4$	2	11.203	2.213	4
Nd	$L\beta_{1,4}$	5.721	2.167	Sm	$L\alpha_1$	1	5.635	2.200	100	W	$L\gamma_1$	2	11.284	2.197	10
				La	$L\gamma_1$	1	5.788	2.142	5	Rh	$K\beta_3$	4	22.695	2.185	8
				Ba	$L\gamma_2$	1	5.796	2.139	1	Rh	$K\beta_1$	4	22.720	2.182	16
				Ba	$L\gamma_3$	1	5.808	2.134	1	Au	$L\beta_1$	2	11.440	2.167	50
				Eu	$L\alpha_2$	1	5.816	2.132	10	W	$L\gamma_8$	2	11.466	2.162	0.1
										W	$L\gamma_6$	2	11.537	2.149	0.01
										Mo	$K\alpha_2$	3	17.371	2.141	50
										Rh	$K\beta_2$	4	23.169	2.140	4
										Au	$L\beta_2$	2	11.583	2.140	20
										W	$L\gamma_2$	2	11.606	2.136	1
										Au	$L\beta_3$	2	11.608	2.136	6
Mn	$K\alpha_{1,2}$	5.894	2.103	Nd	$L\beta_3$	1	5.828	2.127	6	Mo	$K\alpha_1$	3	17.476	2.128	100
				Pr	$L\beta_{2,15}$	1	5.849	2.119	20	W	$L\gamma_3$	2	11.672	2.124	2
				Cr	$K\beta_{1,3}$	1	5.946	2.085	18	Au	$L\beta_5$	2	11.914	2.081	1
Fe	$K\alpha_{1,2}$	6.398	1.937	Mn	$K\beta_{1,3}$	1	6.489	1.910	20	La	$K\beta_1$	6	37.795	1.968	21
										Ag	$K\beta_2$	4	25.452	1.948	5
										La	$K\beta_2$	6	38.723	1.921	7
										Au	$L\gamma_5$	2	12.972	1.911	0.1
Co	$K\alpha_{1,2}$	6.924	1.790	Nd	$L\gamma_3$	1	6.900	1.796	1	Au	$L\gamma_2$	2	13.707	1.809	1
				Fe	$K\beta_{1,3}$	1	7.057	1.757	20	Au	$L\gamma_3$	2	13.807	1.796	2
Ni	$K\alpha_{1,2}$	7.471	1.659	Co	$K\beta_{1,3}$	1	7.648	1.621	20	Ag	$K\alpha_2$	3	21.987	1.691	50
										Ag	$K\alpha_1$	3	22.159	1.678	100
										W	L_l	1	7.386	1.678	3
										La	$K\beta_3$	5	37.714	1.643	9
										La	$K\beta_1$	5	37.795	1.640	21
										Rh	$K\beta_3$	3	22.695	1.639	8
										Rh	$K\beta_1$	3	22.720	1.637	16
Cu	$K\alpha_{1,2}$	8.040	1.542	Ni	$K\beta_{1,3}$	1	8.263	1.500	20	W	$L\alpha_{1,2}$	TC	8.260	1.501	
Zn	$K\alpha_{1,2}$	8.630	1.436	Cu	$K\beta_{1,3}$	1	8.904	1.392	20	W	$L\alpha_1$	1	8.396	1.476	100
										Ag	$K\beta_2$	3	25.452	1.461	5
										Au	L_l	1	8.493	1.460	3
										Mo	$K\alpha_2$	2	17.371	1.427	50
										W	L_n	1	8.723	1.421	1
										Mo	$K\alpha_1$	2	17.476	1.419	100
Ga	$K\alpha_{1,2}$	9.241	1.341	Pb	L_l	1	9.183	1.350	3	Au	L_s	1	9.173	1.351	0.01
										W	$L\beta_1$	TC	9.491	1.306	
As	$K\alpha_1$	10.542	1.176	Ga	$K\beta_2$	1	10.356	1.196	0.3	Au	L_n	1	10.307	1.203	1
				Pb	$L\alpha_2$	1	10.448	1.186	10	Au	L_2–M_2	1	10.588	1.171	0.01
				Pb	$L\alpha_1$	1	10.550	1.175	100						
Pb	$L\beta_1$	12.612	0.983	Pb	$L\beta_3$	1	12.791	0.969	6	Ag	$K\beta_3$	2	24.907	0.995	8
										Ag	$K\beta_1$	2	24.938	0.994	18
										La	$K\beta_3$	3	37.714	0.986	9
										La	$K\beta_1$	3	37.795	0.984	21
										Ag	$K\beta_2$	2	25.452	0.974	5
Th	$L\alpha_1$	12.967	0.956	Th	$L\alpha_2$	1	12.807	0.968	10	La	$K\beta_2$	3	38.723	0.960	7
				Pb	$L\beta_5$	1	13.013	0.953	1	Au	$K\beta_3$	6	77.567	0.959	13
										Au	$L\gamma_5$	1	12.972	0.956	0.1
										Au	$K\beta_1$	6	77.971	0.954	27
										Au	L_2–N_3	1	13.184	0.940	0.01

(*Continued*)

Table 8.16 (*Continued*)

	Analytical line			Elemental interference						X-ray tube or heavy absorber interference					
	Line	keV	Å		Line	*n*	keV	Å	*I*		Line	*n*	keV	Å	*I*
Rb	$K\alpha_1$	13.393	0.926	Pb	$L\beta_{10}$	1	13.273	0.934	0.01	Au	$K\beta_2$	6	80.172	0.928	10
				Pb	$L\beta_9$	1	13.375	0.927	0.01	Au	$L\gamma_1$	1	13.379	0.927	10
				U	$L\alpha_2$	1	13.437	0.923	10	W	$K\beta_3$	5	66.940	0.926	12
										Au	$K\alpha_2$	5	66.978	0.925	50
										W	$K\beta_1$	5	67.233	0.922	26
U	$L\alpha_1$	13.612	0.911							Au	$L\gamma_8$	1	13.624	0.910	0.1
										Au	$L\gamma_2$	1	13.707	0.904	1
										Au	$K\alpha_1$	5	68.792	0.901	100
										W	$K\beta_2$	5	69.020	0.898	10
										Au	$L\gamma_3$	1	13.807	0.898	2
Sr	$K\alpha_1$	14.163	0.875	Pb	$L\gamma_5$	1	14.305	0.867	0.1	Au	L_1–N_4	1	13.997	0.886	0.01
				Th	L_n	1	14.507	0.854	1	Au	$L\gamma_{11}$	1	14.017	0.884	0.01
										Au	$L\gamma_4$	1	14.297	0.867	0.1
										W	$K\alpha_2$	4	57.972	0.855	50
Y	$K\alpha_1$	14.956	0.829	Pb	$L\gamma_1$	1	14.762	0.840	10	W	$K\alpha_1$	4	59.308	0.836	100
				Rb	$L\beta_3$	1	14.949	0.829	8						
				Rb	$K\beta_1$	1	14.959	0.829	16						
				Th	$L\beta_6$	1	14.973	0.828	0.1						
				Pb	$L\gamma_2$	1	15.099	0.821	1						
				Rb	$K\beta_2$	1	15.183	0.816	3						
				Pb	$L\gamma_3$	1	15.215	0.815	2						
Zr	$K\alpha_1$	15.772	0.786	Th	$L\beta_2$	1	15.621	0.794	20	Au	$K\beta_3$	5	77.567	0.799	13
				Th	$L\beta_4$	1	15.640	0.793	4	Au	$K\beta_1$	5	77.971	0.795	27
				Pb	$L\gamma_4$	1	15.775	0.786	0.1	Au	$K\beta_{2'}$	5	80.172	0.773	10
				Sr	$K\beta_3$	1	15.822	0.783	8						
				Sr	$K\beta_1$	1	15.833	0.783	16						
				Sr	$K\beta_2$	1	16.082	0.771	3						
Nb	$K\alpha_1$	16.612	0.746	Th	$L\beta_1$	1	16.199	0.765	50	La	$K\alpha_2$	2	33.028	0.751	50
				U	$L\beta_{15}$	1	16.383	0.757	1	La	$K\alpha_1$	2	33.436	0.741	100
				Th	$L\beta_3$	1	16.423	0.755	6	W	$K\beta_3$	4	66.940	0.741	12
				U	$L\beta_2$	1	16.425	0.755	20	Au	$K\alpha_2$	4	66.978	0.740	50
				U	$L\beta_4$	1	16.573	0.748	4	W	$K\beta_1$	4	67.233	0.738	26
				Y	$K\beta_3$	1	16.723	0.741	8						
				Y	$K\beta_1$	1	16.735	0.741	16						
				Y	$K\beta_2$	1	17.013	0.729	4						
Mo	$K\alpha_1$	17.476	0.709	U	$L\beta_5$	1	17.067	0.726	1	Au	$K\alpha_1$	4	68.792	0.721	100
				U	$L\beta_1$	1	17.217	0.720	50	W	$K\beta_{2''}$	4	69.020	0.718	10
				U	$L\beta_3$	1	17.452	0.710	6	W	$K\beta_{2'}$	4	69.089	0.718	10
				Zr	$K\beta_3$	1	17.651	0.702	9						
				Zr	$K\beta_1$	1	17.665	0.702	18						
				Zr	$K\beta_2$	1	17.967	0.690	4						

Analytical line: energy (keV) and wavelength (Å) of commonly used analyte lines are listed.
Element, tube and matrix interferences: principal spectrum overlap interferences listed together with order of diffraction (*n*), energy (kV) and relative intensity (*I*) of originating first-order line. Relative intensity is normalized to the intensity of the principal line in the *K*-, *L*- or M-series = 100.
Wavelength (Å) is that of the interference, taking into account order of diffraction.
A search has been made for major element, tube and lanthanum heavy absorber interferences corresponding to up to sixth-order diffractions.
TC = interference from scattered tube Compton.
Data from White and Johnson (1970).

The main features of the data in this table are as follows:

(i) A small interference from lanthanum in heavy absorber fluxes could be observed, in some circumstances, in the determination of the major elements Na, Mg, Al, Si, P, S, Ti, and Cr; the interferences on Na *Kα* (due to La *Mβ*) being particularly significant

(ii) The chromium tube, used extensively in major element analysis, is relatively free of interference except on P, Cr and Mn; in the determination of the latter two elements, an aluminium filter is normally used to remove characteristic tube lines (see section 8.9.1)

(iii) The rhodium tube is free of interferences in the determination of the heavy trace elements but may have some effect in the determination of the Mg, Al, K, Ti and Ni *Kα* lines

(iv) Gold, tungsten, silver and molybdenum may be used to excite selected trace elements, the molybdenum tube being particularly free of interferences on the heavier trace elements that have absorption edges below the energy of the Mo *K*-lines.

(*b*) *Interference corrections.* Many of the multiple-order interferences listed in Table 8.17 may be eliminated by careful use of pulse height analysis. However, several first-order interferences will remain, the most serious of which in silicate rocks affect the analysis of the following elements (interfering element in brackets)

Na $K\alpha$ (Zn $L\beta$)	P $K\alpha$ (Zr $L\alpha$)	S $K\alpha$ (Zr $L\gamma_l$)	Ti $K\alpha$ (Ba $L\alpha$)
V $K\alpha$ (Ti $K\beta$)	Cr $K\alpha$ (V $K\beta$)	Mn $K\alpha$ (Cr $K\beta$)	Fe $K\alpha$ (Mn $K\beta$)
Co $K\alpha$ (Fe $K\beta$)	Ni $K\alpha$ (Co $K\beta$)	Cu $K\alpha$ (Ni $K\beta$)	Zn $K\alpha$ (Cu $K\beta$)
Ga $K\alpha$ (Pb L_l)	As $K\alpha$ (Pb $L\alpha$)	Rb $K\alpha$ (U $L\alpha$)	Sr $K\alpha$ (Th L_n)
Y $K\alpha$ (Rb $K\beta$, Pb $L\gamma$)	Zr $K\alpha$ (Th $L\beta_4$, Sr $K\beta$)	Nb $K\alpha$ (U $L\beta$, Y $K\beta$)	
Mo $K\alpha$ (U $L\beta$, Zr $K\beta$)			

In some of the examples listed here, the inherent resolution of the spectrometer may be sufficient to eliminate the interference. However, in any general scheme of analysis, it will be necessary for an analyst to satisfy himself that either the relative overlap is sufficiently small that the interference may be safely ignored or that the appropriate correction is applied.

Table 8.17 Summary of spectral line overlap interferences from x-ray tube and flux

Element and line		X-ray tube							Flux
		Sc	Cr	Mo	Rh	Ag	W	Au	La
F	$K\alpha$	●		*	●	●	●		●
Na	$K\alpha$	●				*		*	*
Mg	$K\alpha$			*	*	*			●
Al	$K\alpha$	●			*	*	●		●
Si	$K\alpha$					*		*	●
P	$K\alpha$	*	*	*			*		●
S	$K\alpha$			*			*	●	*
Cl	$K\alpha$			*		*		*	
K	$K\alpha$			●	*	*	*	●	
Ca	$K\alpha$					*	*	●	
Sc	$K\alpha$	*			●			●	*
Ti	$K\alpha$	*			●	●		*	*
La	$L\alpha_1$				●			●	*
Ba	$L\beta_1$			●			*	*	
V	$K\alpha$			●		●	*	*	
Pr	$L\alpha$			●	●	●	*	*	*
Ba	$L\beta_2$				●	●	●	*	*
Nd	$L\beta_1$						●	*	*
Ce	$L\beta_1$		●					*	●
Cr	$K\alpha$		*			●	●	●	*
Pr	$L\beta_1$		*			●	●	*	*
Nd	$L\alpha_1$		*	*	●		*	*	
Mn	$K\alpha$		*	*	●		*	*	
Fe	$K\alpha$					●		●	●
Co	$K\alpha$							*	
Ni	$K\alpha$				*	*	*		●
Cu	$K\alpha$						*		
Zn	$K\alpha$			*		●	*	*	
Ga	$K\alpha_1$						●	*	
As	$K\alpha_1$							*	
Pb	$L\beta_1$					*			*
Th	$L\alpha_1$					●		*	*
Rb	$K\alpha_1$						●	*	
U	$L\alpha_1$						●	*	
Sr	$K\alpha_1$						●	*	
Y	$K\alpha_1$						●		
Zr	$K\alpha_1$							●	
Nb	$K\alpha_1$			*			●	●	*
Mo	$K\alpha_1$			*			●	●	

* = interference may be significant.
● = interference may not always be present (see text).
Element lines are listed in increasing order of energy.
In assessing application of a tube, account must also be taken of the excitation efficiency. For example, an Sc tube does not emit any characteristic lines capable of exciting lines below Ca $K\alpha$ in this table, nor a Cr tube above Ce $L\beta_1$.

The normal method of correcting overlap interferences of element B on element A, for example, is to excite a sample containing only element B. The intensity at the spectrometer angle of element A is then measured relative to the intensity of the principal line of this interfering element B. This ratio may then be used to subtract the relative intensity of the interference B on the count rate measured for A in composite samples.

To select spectrometer parameters to minimize errors resulting from overlap interferences, two methods are generally used.

(i) If the overlap between the two lines is not severe (for example, in Figure 8.51*a* where the tail of interfering element B overlaps the peak angle of element A), then a large reduction in the magnitude of this interference, perhaps by up to a factor of 10, can be gained by selecting spectrometer conditions that give higher resolution, as shown in Figure 8.51*b*. This may be achieved by substituting fine for coarse collimation and/or using a diffracting crystal of higher angular dispersion, for example LiF(220) in place of LiF(200). This strategy is normally accompanied by a reduction in measured intensities.

Thus, Norrish and Chappell (1977), recommend that both Rb $K\alpha$ and Nb $K\alpha$ should be measured using a LiF(220) crystal to minimize the overlap interference from U $L\alpha_1$, $L\alpha_2$ and Y$K\beta_{1,3}$ respectively.

(ii) The second situation is where the overlap interference is much more severe, the worst case being illustrated in Figure 8.51*c*. Here the overlap occurs at the most steeply sloping side of the interfering element peak. This situation often results in perhaps most serious introduction of systematic bias in the analysis of wavelength dispersive spectra. In routine operation, small discrepancies will occur in the precise angle at which the spectrometer settles (due to temperature effects, ageing, wear or mechanical stiffness). If the settling angle corresponds to the fairly flat top of a peak, little variation in count rate will be observed. However, if this angle corresponds to the flank of an x-ray line interference, as shown in Figure 8.52, large changes in the intensity correction ratio of the interfering element will occur with small changes in spectrometer settling angle. Precise interference correction will then be difficult to achieve. In this case, it is preferable to defocus the spectrometer by adopting conditions of lower resolution (compare Figures

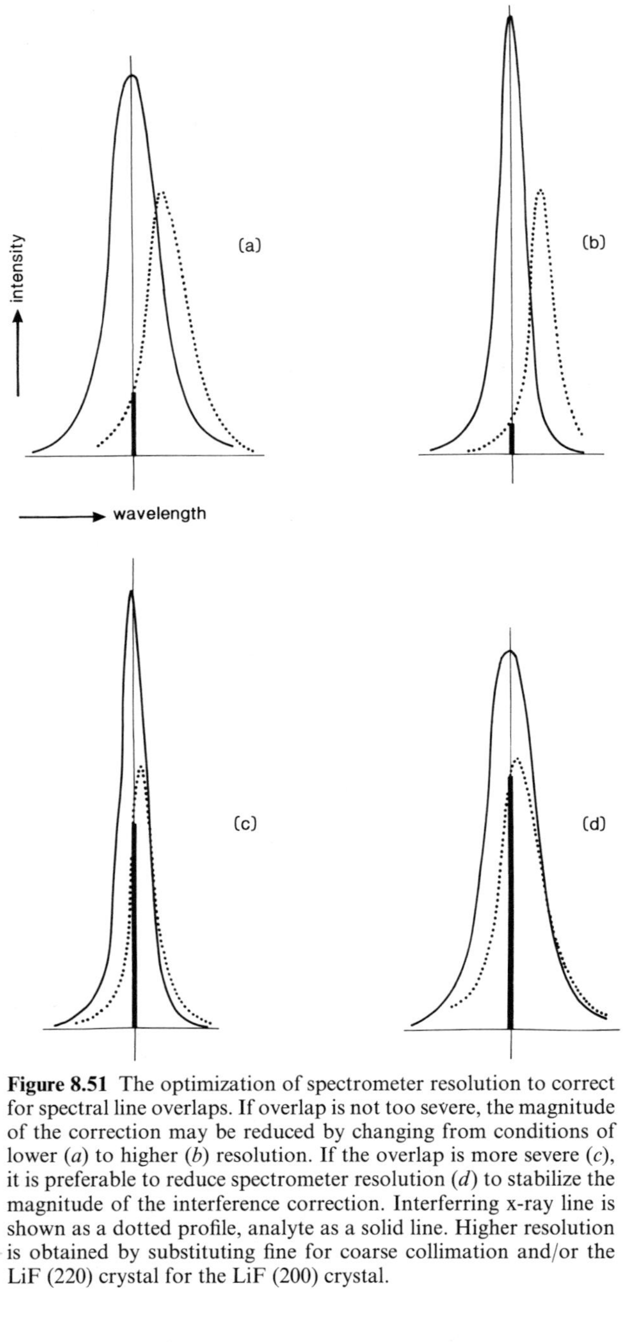

Figure 8.51 The optimization of spectrometer resolution to correct for spectral line overlaps. If overlap is not too severe, the magnitude of the correction may be reduced by changing from conditions of lower (*a*) to higher (*b*) resolution. If the overlap is more severe (*c*), it is preferable to reduce spectrometer resolution (*d*) to stabilize the magnitude of the interference correction. Interferring x-ray line is shown as a dotted profile, analyte as a solid line. Higher resolution is obtained by substituting fine for coarse collimation and/or the LiF (220) crystal for the LiF (200) crystal.

8.51*c* and *d*) to minimize the sensitivity of this overlap correction to spectrometer settling discrepancies.

Thus, Norrish and Chappell (1977) recommend determining Y $K\alpha$ and Zr $K\alpha$ with a lower resolution LiF(200) crystal in order to stabilize the interference correction from Rb $K\beta$ and Sr $K\beta$ respectively.

In fact, in geological samples, the Y $K\alpha$–Rb $K\beta$ interference represents a most serious analytical challenge, not least because the intensity of the Rb interference is sometimes equivalent to, or greater than, that of the Y $K\alpha$ line. For this reason, yttrium is one of the least accurately analysed of the Rb–Nb group of elements and determinations made in the presence of more than a few hundred ppm of Rb should be examined carefully for errors.

A novel form of interference correction was described by

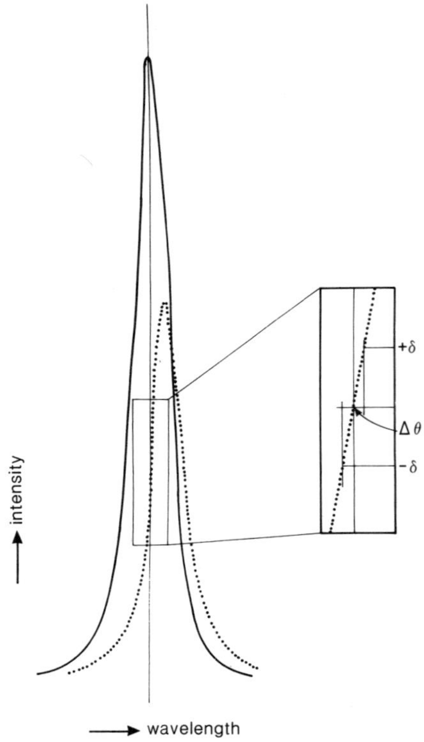

Figure 8.52 Sketch of a severe overlap interference showing the large effect on measured intensities that can be caused by a small discrepancy in the angular resetting of the spectrometer. The analyte and interfering lines are shown as solid and dotted profiles respectively. An error in spectrometer settling angle of $\Delta\theta$ causes the intensity of the interfering line to change by δ.

Kikkert (1983) and reproduced in Figure 8.53. This procedure involves selecting an angle for background measurement at a point where the intensity of the interfering element (in this case Sr $K\beta_1$) is equal to its intensity at the peak angle of the analyte (Zr $K\alpha$). The same criteria concerning high or low resolution conditions as discussed above still apply.

(*c*) *Background angles*. So far, the discussion of interferences has been restricted to those affecting analysed lines. However these considerations apply equally to interferences on background angles. In the analysis of metallic samples, background intensities may often be ignored. However, because of the relatively light matrix of silicate samples, scatter from the x-ray tube is relatively high and it is necessary to measure background intensities for each analysed line. Indeed in situations where the background slopes beneath an x-ray line, it is necessary to average the background counts from measurements made on either side of the peak. Galson *et al.* (1983) discussed the more rigorous procedures that must be followed for background correction in the determination of elements (e.g. U and Th) at ultra-trace levels involving linear interpolation of background measurements.

8.11 Routine operating conditions and statistical considerations

8.11.1 *Operating conditions*

An analytical program for routine silicate rock analysis is normally arranged to determine the major elements on glass

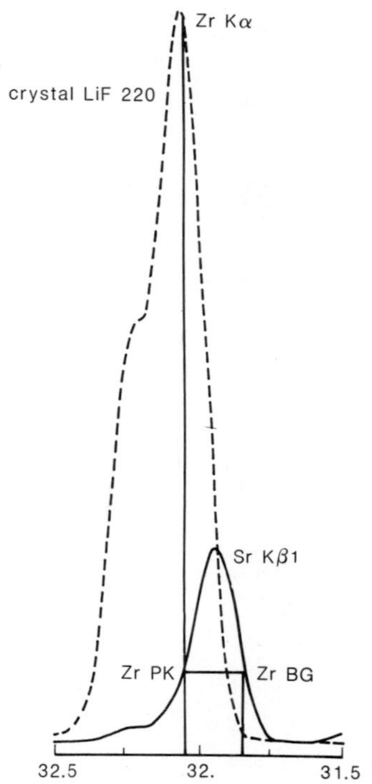

Figure 8.53 Compensation of spectral interferences (Sr $K\beta_1$ on Zr $K\alpha$) by careful choice of background angle as discussed in the text. (Kikkert, 1983.)

discs and the trace elements and possibly the heavier major elements on powder pellets. Since different operating conditions and often different x-ray tubes are required for each program, it is most convenient to operate the instruments under one set of conditions for sessions lasting weeks or even months to minimize drift and permit the instrument to 'settle down'.

The elements, x-ray lines and instrumental conditions under which determinations would normally be made are listed in Table 8.18. The line selected for analysis is normally the most intense ($K\alpha$ or $L\alpha$). However, barium must normally be determined from its $L\beta_1$ or $L\beta_2$ line to avoid the serious Ti $K\alpha$ overlap on Ba $L\alpha$. Cerium, if present in sufficient abundance, is normally determined from its $L\beta_1$ line (to avoid the Ce $L\alpha$–Ba $L\beta_1$ interference) and use of the Pb $L\beta_1$ line avoids the As $K\alpha$ overlap on Pb $L\alpha_{1,2}$.

The selection of diffracting crystal and x-ray counter used in determining these x-ray lines is also listed in Table 8.18. The choice between LiF(200) and LiF(220) in combination with coarse or fine collimators, is a question of fine analytical judgement, as discussed in section 8.10. The criteria used by Norrish and Chappell (1977) have been followed here.

8.11.2 *Analytical precision*

X-ray emission and detection, being a random process, is governed by Poisson counting statistics (see section 1.4). An estimate (s) of the true standard deviation (σ) is related to the total number of x-ray counts accumulated in a measurement (N) by

$$s = \sqrt{N}. \qquad (8.53)$$

As the measurement of the net peak intensity (N_p) is a result of subtracting a background count (N_b) from a gross peak plus background count (N_g), the standard deviation in a net line intensity (s_p) is given by

$$s_p = \sqrt{s_g{}^2 + s_b{}^2} = \sqrt{N_g + N_b} \qquad (8.54)$$

where s_g and s_b are standard deviations in the gross peak intensity and background intensity measurements respectively.

The relative standard deviation (or coefficient of variation) in a measurement is given by

$$\frac{s_p}{N_p} = \frac{\sqrt{N_g + N_b}}{N_g - N_b} \qquad (8.55)$$

(*b*) *Peak and background count times.* For elements present at levels approaching the detection limit, the contribution that the background count (N_b in equation (8.54)) makes to the relative standard deviation becomes equivalent to that of the gross peak count (N_g). It is essential under these conditions that peak and background data are accumulated for equivalent count times. However, as the peak increases in intensity, the contribution of N_b diminishes and it then becomes more efficient to spend longer accumulating count data on peak rather than the background.

It can be shown (see for example Bertin, 1975, p.479) that if R_g and R_b are the gross (peak + background) and background count rates respectively, then for minimum error in net peak measurements, the count time between peak (T_g) and background (T_b) should be apportioned according to

$$\frac{T_g}{T_b} = \sqrt{\frac{R_g}{R_b}} \qquad (8.56)$$

where the total time T available to make the measurement equals $T_g + T_b$.

From these equations (8.55) and (8.56), Jenkins and de Vries (1970, p.188) have derived the expression for relative standard deviation in a peak measurement:

$$\frac{s_p}{N_p} = \frac{1}{\sqrt{T}}\left[\frac{1}{\sqrt{R_g} - \sqrt{R_b}}\right] \qquad (8.57)$$

where the total time T is apportioned according to equation (8.56).

From these equations it may be seen that the relative standard deviation in a measurement approaches a minimum as $\sqrt{R_g} - \sqrt{R_b}$ tends to a maximum. Hence the recommendation to use this latter parameter as a figure of merit in section 8.9.

A consequence of equation (8.57) is that as the peak count rate (R_g) increases relative to the background count rate (R_b), the effect of the background measurement on the relative standard deviation decreases to a point where it can be ignored. Jenkins (1976, p.111) estimates that this point is normally reached at the ratio $R_g : R_b = 10:1$.

Unfortunately, in setting up a general scheme of analysis for silicate rocks, it is not often that advantage can be taken of either of these benefits. The composition of rocks is so variable that all the trace elements and many of the major elements will in some samples approach detection limit

Table 8.18 Operating parameters and detection limits for wavelength dispersive XRF

									Count times			Detection limits			
													Powder pellet		
Element	X-ray line	Energy (keV)	Wavelength (Å)	Diffracting crystal	Collimator	Tube kV	X-ray counter		Cr tube (s)	Mo tube (s)	W tube (s)	1 : 5 glass disc (ppm)	Routine analysis (ppm)	400 s count time per element (ppm)	
F	$K\alpha$	0.677	18.32	TAP		50	Ar1						5600		F
Na	$K\alpha$	1.041	11.910	TAP		50	Ar1		1892			2400	1600		Na_2O
Mg	$K\alpha$	1.253	9.890	TAP		50	Ar1		588			1200	1000		MgO
Al	$K\alpha$	1.486	8.340	TAP/PET		50	Ar1		211			600	492		Al_2O_3
Si	$K\alpha$	1.739	7.126	InSb/PET		50	Ar1/Ar2		200			1000	340		SiO_2
P	$K\alpha$	2.013	6.158	Ge/PET		50	Ar1/Ar2		170			140	270		P_2O_5
S	$K\alpha$	2.307	5.373	Ge/PET		50	Ar1/Ar2		30			160	20		S
Cl	$K\alpha$	2.621	4.729	Ge/PET		50	Ar1/Ar2						20		Cl
K	$K\alpha$	3.312	3.742	PET		50	Ar2/Kr		12			140	8		K_2O
Ca	$K\alpha$	3.690	3.359	LiF(200)		50	Ar2/Kr		10			140	12		CaO
Sc	$K\alpha$	4.088	3.032	LiF(200)		50	Ar2/Kr						8		Sc
Ti	$K\alpha$	4.508	2.750	LiF(200)		50	Ar2/Kr		7			240	6		TiO_2
La	$L\alpha_1$	4.650	2.666	LiF(220)	C	50	Ar2/Kr				111		15	2.6	La
Ba	$L\beta_1$	4.827	2.568	LiF(220)	C	50	Ar2/Kr		140*[1]		230		24	7.4	Ba
V	$K\alpha$	4.949	2.504	LiF(220)	F	50	Ar2/Kr		234*[1]				10	2	V
Pr	$L\alpha_1$	5.033	2.463	LiF(220)	C	50	Ar2/Kr							3	Pr
Ba	$L\beta_2$	5.156	2.404	LiF(220)	C	50	Ar2/Kr							16.2	Ba
Nd	$L\alpha_1$	5.229	2.370	LiF(220)		50	Ar2/Kr								Nd
Ce	$L\beta_1$	5.261	2.356	LiF(200)	F	50	Ar2/Kr				185		30	8	Ce
Cr	$K\alpha$	5.411	2.291	LiF(200)	C	50	Ar2/Kr				43		10	1.8	Cr
Pr	$L\beta_1$	5.488	2.259	LiF(220)		50	Ar2/Kr				404		30		Pr
Nd	$L\beta_1$	5.721	2.167	LiF(220)	C	50	Ar2/Kr				1678		36	6.2	Nd
Mn	$K\alpha$	5.894	2.103	LiF(200)	C	50	Ar2/Kr	Sc	171*[2]		138	240	13	2.3	MnO
Fe	$K\alpha$	6.398	1.937	LiF(200)	C	50	Ar2/Kr	Sc	26		380	260	14	2.3	Fe_2O_3
Co	$K\alpha$	6.924	1.790	LiF(200)	C	60	Ar2/Kr	Sc			724		6	1.2	Co
Ni	$K\alpha$	7.471	1.659	LiF(200)	C	60	Ar2/Kr	Sc	37*[1]	23	73		6	1	Ni
Cu	$K\alpha$	8.040	1.542	LiF(200)	C	60	Ar2/Kr	Sc	48*[1]	16			6	1	Cu
Zn	$K\alpha$	8.630	1.436	LiF(200)	C	60	Ar2/Xe	Sc	105*[1]	30			9	1	Zn
Ga	$K\alpha$	9.241	1.341	LIF(200)	C	100	Ar2/Xe	Sc		37			6	1.2	Ga
As	$K\alpha$	10.542	1.176	LiF(200)		100	Ar2/Xe	Sc					10		As
Pb	$L\beta_1$	12.612	0.983	LiF(200)	F	100	Xe	Sc		88			6	2.4	Pb
Th	$L\alpha_1$	12.967	0.956	LiF(200)	F	100	Xe	Sc		54			9	2.4	Th
Rb	$K\alpha_1$	13.393	0.926	LiF(220)	C	100	Xe	Sc		22			3	0.8	Rb
U	$L\alpha_1$	13.612	0.911	LiF(220)	F	100	Xe	Sc		42			6	2.8	U
Sr	$K\alpha_1$	14.163	0.875	LiF(200)	C	100	Xe	Sc		16			3	0.6	Sr
Y	$K\alpha_1$	14.956	0.829	LiF(200)	C	100	Xe	Sc		17			3	0.6	Y
Zr	$K\alpha_1$	15.772	0.786	LiF(200)	C	100		Sc		20	133		6	1.6	Zr
Nb	$K\alpha_1$	16.612	0.746	LiF(200)	F	100		Sc			101		6	1.8	Nb
Mo	$K\alpha_1$	17.476	0.709	LiF(200)		100		Sc					6		Mo

Diffracting crystals: most commonly used crystal is listed first. For elements above Ca, the selection of LiF(200) or LiF(220) follows the recommendations of Norrish and Chappell (1977). TAP: thallium acid phthalate, PET: pentaerythritol.
InSb: indium antimonide; Ge: germanium; LiF: lithium fluoride.
X-ray detectors:
Ar1: argon flow counter with an ultra-thin window (1 μm polypropylene or 0.2 μm collodion) must be used.
Ar2: argon flow counter with standard window (3 or 6 μm mylar).
Kr: krypton sealed counter.
Xe: xenon sealed counter.
Sc: sodium iodide scintillation counter.
Count times: *1: per 1000 ppm element for Ba, V, Ni, Cu and Zn.
Chromium tube: count times listed are those required to accumulate 100 000 counts for 1% element oxide calculated from the sensitivity data of Lee and McConchie (1982) for a 1:2 rock to flux glass disc excited at 50kV, 35mA. Mn*2 was excited with an aluminium primary beam filter, Ba was determined from the $L\alpha_1$ line.
Molybdenum and tungsten: count times listed are those required to accumulate 100 000 counts for 10 000 ppm of the element calculated from sensitivity data listed by (i) Lee and McConchie (1982) for 1:2 low dilution glass discs excited at 60 kV, 40 mA using a molybdenum tube, and (ii) Leake *et al.* (1969) for powder pellets excited at between 40 and 100kV (2kW generator) using a tungsten tube.
Some additional data for the Mo tube has been taken from Leake *et al.* (1969).
Detection limits have been recalculated as the six-sigma limit of determination for routine operating conditions using 'typical' count times. The glass disc data are based on parameters specified by Norrish and Hutton (1969) and the powder pellet data (routine analysis) on data from Robinson and Bennett (1981).
However, account has been taken of additional data presented by Leake *et al.* (1969), Nisbet *et al.* (1979), Hutton and Elliott (1980), Lee and McConchie (1982), and Vie le Sage *et al.* (1979). Data of Norrish and Chappell (1977), representative of 400s total count time per sample are listed in the last column. Samples were excited with gold or molybdenum tubes as appropriate.
Detection limit data are quoted as ppm major element oxide or ppm trace element as indicated.

levels. As a general strategy it is necessary to count both peak and backgrounds for equal count times.

Count times selected for analysis will vary from instrument to instrument depending on tube power, instrument geometry and detection efficiency. A compromise must be judged between obtaining the highest precision (from long count times) and the maximum sample throughout (short analysis times). As a general rule, it is suggested that one approach is to select count times so that each element would be determined to an equivalent precision if it were present at the same concentration level. On a modern instrument, this might be achieved by setting count times so that 100 000 counts would be accumulated if 2% of a major element oxide or 1000 ppm of a trace element were present in a sample. Alternatively precision targets may be selected as best fit local requirements.

Typical count times based on these criteria are listed in Table 8.18. It is not expected that these count times would be adopted without modification, and they should be taken as an indication of the relative sensitivity of individual elements. In many routine applications it would be unusual to assay an element for as long as some of the count times in Table 8.18. In such circumstances, it is likely that the light major elements will not be determined to as high a degree of precision as some of the heavier major elements. Manganese and iron measurements made with a chromium tube are likely to be somewhat degraded, due to unfavourable excitation.

(*c*) *Detection limits.* As the concentration of an element approaches detection limit level, so the gross peak count (N_g) approaches the background count (N_b). Under these conditions equation (8.54) approximates to

$$s_p = \sqrt{2N_b} \tag{8.58}$$

Using the criteria and approach discussed in section 1.5, detection limit signals in XRF analysis can be estimated from the total counts accumulated in a background measurement as follows.

Lower limit of detection $= 3s_p = 3\sqrt{2N_b} = 4.243\sqrt{N_b}$.
This is the smallest signal that can be qualitatively measured using three-sigma confidence criteria.

Limit of determination $= 6s_p = 6\sqrt{2N_b} = 8.485\sqrt{N_b}$.
This is the smallest signal that can be quantitatively measured using three-sigma confidence criteria.

Limit of quantitation $= 10s_p = 10\sqrt{2N_b} = 14.142\sqrt{N_b}$.
This is the smallest signal that can be quantitatively measured with sufficient confidence to be quoted for commercial or statutory use.

Since it has already been established that the magnitude of the background is dependent on sample matrix, it is desirable to select a silicate sample of 'representative' composition for detection limit measurements and to specify the material used for such calculations. For similar reasons, the count time (peak plus background) should be specified as well.

In some x-ray fluorescence applications, it is more convenient to calculate detection limits in terms of *count rates* rather than the total number of counts accumulated in an appropriate background measurement. Detection limit signals may then be recalculated in terms of element concentrations as follows.

If R_B is the count rate of the appropriate background signal, and T_B is the count rate for which the background signal was accumulated, then the total number of background counts accumulated is R_BT_B. From equation (8.58), the standard deviation in a peak minus background signal is thus $\sqrt{2R_BT_B}$.

In order to convert the detection limit signals into concentration, it is necessary to determine the analytical sensitivity S of an element (in counts per second per unit concentration). The detection limit signals listed above are then divided by $(S.T_B)$, where T_B is the count time for which the background measurement was recorded.

If a sample containing a concentration C of the element gives a count rate of R_P measured over a count time on the peak of T_P, then sensitivity

$$= \frac{P_P}{C} \text{ (counts per second per unit element \%)} \quad (8.59)$$

$$= \frac{R_PT_P}{C} \text{ (total count per unit element \%)} \quad (8.60)$$

The six-sigma 'limit of determination' is then given by:

$$\text{LoD} = 6\sqrt{2R_BT_B} \cdot \frac{C}{R_PT_P} \quad (8.61)$$

$$= 6\sqrt{\frac{2R_B}{T}} \cdot \frac{C}{R_P} \quad (8.62)$$

(assuming $T_P = T_B = T$, the count time on a peak or background measurement), or

$$= 12\sqrt{\frac{R_B}{T'}} \cdot \frac{C}{R_P} \quad (8.63)$$

(assuming $T_B = T_P$ and $T_B + T_P = T$, the total time for the peak and background measurement).

Limits of determination for various elements routinely analysed by XRF are listed in Table 8.18. Although not representative of any specific scheme of analysis (see footnotes to Table 8.18) these levels are the concentrations down to which it is generally possible to determine elements under routine operating conditions. If, however, substantially longer count times are used (perhaps 400 to 4800 s) and careful consideration is given to the subtraction of the background, then these limits may be significantly improved. To indicate the detection limits that may be achieved under special counting conditions, data from Norrish and Chappell (1977), recalculated as six-sigma detection limits, are listed in Table 8.18 for a total count time of 400 s per element using a 3 kW generator. If such conditions were used for routine analysis, total instrument time for measurement of the 21 elements for which data are listed would be 2.33 hours per sample, rather than the rate of two to four samples per hour for many routine trace-element applications. In the determination of low concentrations or uranium, thorium and potassium, Galson *et al.* (1983) reported six-sigma detection limits of 0.6 (U), 3.0 (Th) and 30 ppm (K), using count times of 4800, 480 and72 s respectively.

8.12 Performance in routine analysis

8.12.1 *Precision, accuracy and calibration procedures*

It is essential when evaluating the reliability of XRF results to distinguish between precision and accuracy (section 1.4). The precision of XRF data is usually extremely high and readily verified by the user from replicate analysis data. On the other hand, accuracy is much more difficult to assess and depends on the reliability of the attenuation–enhancement correction procedure and more especially on procedures used to calibrate instruments.

It is general practice to calibrate instruments against as many rock reference materials as possible, in part to minimize the effects of errors arising from discrepancies in individual 'standard' values. Compilations of reference material data such as those of Flanagan (1973), Abbey (1983) and Gladney *et al.* (1983) are useful in selecting suitable reference standards.

In order to minimize instrumental drift, all count data are generally normalized against corresponding measurements on a drift monitor. The drift monitor is counted every third sample, or with whatever decreased frequency is justified by the stability of the x-ray generator and instrumentation. As the entire calibration depends on the stability of response from the drift monitor, it is of concern to know how this is affected by the prolonged exposure to x-rays. Le Maitre and Haukka (1973) investigated this problem and observed some changes in the intensity of network-forming elements in monitors prepared as lithium tetraborate glass discs. Forté (1983) proposed procedures for preparing and using 'permanent' drift monitors and standards in the analysis of industrial raw materials. Monitor samples made up with epoxy-resin were found to degrade eventually after prolonged exposure to the x-ray beam. Reference pellets made up with aluminium powder were found to be more stable, a fact attributed to the high thermal conductivity of aluminium.

(*a*) *Major-element analysis.* Data for the routine determination of major elements on glass fusion discs are listed in Table 8.19. The data of Lee and McConchie for andesite USGS: AGV-1 were obtained from samples prepared in 1:2 dilution by fusing 1.5 g of sample with 3.0 g of ($Li_2B_4O_7$+ Li_2CO_3) and 0.2 g $LiNO_3$. The average and coefficient of variation on 10 analyses are listed in Table 8.18 and compared with the consensus data of Gladney *et al.* (1983). In general, the agreement between the two sets of data is exceptionally good, with the instrumental precision in several elements being less than 0.2% relative. However, it is interesting to note that the uncertainty in the 'consensus' data for several elements is an order of magnitude higher, despite the fact that AGV-1 must be one of the most extensively analysed reference materials available.

The data for reference material USGS: W-2 (diabase) in Table 8.19 (Schroeder *et al.*, 1980) was obtained on samples prepared as glass discs with heavy absorber flux in the ratio one part of sample to five of flux. The instrumental precisions are listed both for a single glass disc repeatedly analysed over a 10-day period and for six individual reference samples analysed within the same operating session. A reduction in precision in the 10-day data is particularly

Table 8.19 Major element analysis of glass discs by WD-XRF

	Andesite USGS: AGV-1				Diabase USGS: W-2				Andesite N-1005			
	Gladney *et al.* (1983)		Lee and McConchie (1982)		Abbey (1983)	Schroeder *et al.* (1980)			Harvey *et al.* (1973)			
	Expected (% oxide)	CV (%)	Analysed (% oxide)	CV (%)	Expected (% oxide)	Analysed (% oxide)	*A* CV (%)	*B* CV (%)	Analysed (% oxide)	*C* CV (%)	*D* CV (%)	
SiO_2	59.25	0.49	59.25	0.13	52.81	52.89	0.61	0.68	59.15	0.27	0.18	SiO_2
TiO_2	1.06	2.4	1.05	0.19	1.06	1.08	0.37	0.87	0.72	0.37	0.44	TiO_2
Al_2O_3	17.15	0.99	16.99	0.16	15.49	15.45	0.84	0.70	17.47	0.36	0.24	Al_2O_3
Fe_2O_3	6.76	2.8	6.81	0.16	10.86	10.94	0.30	0.49	5.73	0.19	0.42	Fe_2O_3
MnO	0.096	4.2	0.095	2.1	0.167	0.17	1.8	1.17	0.107	0.10	0.23	MnO
MgO	1.53	3.3	1.54	0.78	6.39	6.60	0.61	2.8	2.79	1.43	2.15	MgO
CaO	4.94	1.4	4.89	0.12	10.89	11.15	0.27	0.60	5.42	0.14	0.21	CaO
Na_2O	4.25	1.4	4.30	1.4	2.21	1.92	2.6	11.4				Na_2O
K_2O	2.90	1.7	2.95	0.14	0.63	0.60	0.83	(0)	2.28	0.18	0.24	K_2O
P_2O_5	0.48	3.1	0.49	0.1	0.14	0.15	6.7	4.6	0.244	0.08	0.23	P_2O_5
Total	98.42		98.37		100.65							Total

CV = coefficient of variation = 100.standard deviation/mean analysis (%).
(i) USGS: AGV-1. Analysed data (Lee and McConchie) was obtained on 1 : 2 low dilution glass discs excited with a chromium x-ray tube operated at 50 kV for count times of 10 to 100 s per element. Data represents mean and CV in 10 analyses. Expected data are the 'consensus' values of Gladney *et al.*; the CV was calculated from residual data remaining after discarding outliers.
(ii) USGS: W-2. Analysed data (Schroeder *et al.*) was obtained from samples prepared as glass discs by fusing in the ratio 1 : 5 with heavy absorber flux. A chromium tube was used to excite samples at 60 kV for count times of 20 to 100 s per element. Coefficient of variation *A* was calculated from the analysis of six separate glass discs. Coefficient of variation *B* was calculated from ten analyses of the same glass disc of USGS: W-1 carried out over ten working days. W-1 and W-2 have almost identical major element compositions. These data are compared with expected compositions for W-2 taken from Abbey (1983).
(iii) Andesite N-1005 (a Nottingham University in-house reference sample) was prepared as 1 : 5.4 glass discs with heavy absorber flux and excited with a chromium tube at 40 to 60 kV for 20 to 40 s per element. Coefficient of variation *C* represents the arithmetic mean of the standard deviation of five replicate measurements on six discs and is a measure of instrumental precision. Coefficient of variation *D* represents the standard deviation derived from the average analysis of each glass disc and therefore is a measure of additional errors caused by sample preparation.

apparent in the analysis of the light elements, Mg and Na, for this set of data.

In order to assess the influence on analytical precision of sample preparation in comparison with instrumental effects, Harvey *et al.* (1973) analysed data from a reference andesite (N-1005) in two ways. Instrumental precision was evaluated from a total of five measurements on each of six glass discs (prepared with heavy absorber flux), the arithmetic means of the standard deviation in the analysis of each disc is listed in Table 8.19. The variation between discs is based on the differences in the average analyses of each of the six replicate glass discs. These data (Table 8.19) suggest that for many elements the error in determinations on replicate samples is up to twice that averaged from single samples. This implies that it may be necessary to double instrumental precisions derived from measurements on single glass discs to allow for discrepancies attributable to sample preparation.

(*b*) *Trace-element analysis.* Data in Table 8.20 compare XRF results of routine trace-element programmes. The results were obtained from compressed powder pellets (Schroeder *et al.*, 1980; Leake *et al.*, 1969) and 1 : 2 low dilution glass discs (Lee and McConchie, 1982). The high degree of instrumental precision that can be achieved in routine trace-element analysis by XRF is demonstrated by the data for basalt CRPG-BR from Schroeder *et al.* Coefficients of variation are often in the range 1 to 2% relative in the average of 40 analyses obtained over a 10-day period. These results are compared with compiled data for CRPG-BR from Govindaraju and de la Roche (1977). Comparison between two sets of results reveals one of the difficulties in assessing the accuracy of XRF data. The relative standard deviation in many of the 'recommended' average analyses often exceeds 20% relative, far higher than the instrumental precision in the XRF results. It must be added that the errors in the 'recommended' values were calculated from all submitted data, without rejection of outliers, and indicate the large interlaboratory variability in data available to Govindaraju and de la Roche.

Comparison of standard deviations in the analytical results of Lee and McConchie (1982) and the 'consensus' data from Gladney *et al.* (1983) for trace elements in andesite USGS: AGV-1 (Table 8.20) shows a much closer agreement. In this case the compilers of consensus values (Gladney *et al.*, 1983) had much more data and were able to perform a more critical statistical assessment of the results.

The final two sets of data in Table 8.20 are designed to indicate the levels of precision that may be achieved in the analysis of a much wider range of trace elements. Account must be taken of the fact that the analysts (Leake *et al.*, 1969) used a comparatively low-power x-ray generator (2kW) and

Table 8.20 Trace element determinations on powder pellets by WD-XRF

	Basalt CRPG: Br				Andesite USGS: AGV-1				Galway Granite JW6012			Amphibolite BL3571			
	Govindaraju and de la Roche (1977)		Schroeder *et al.* (1980)		Gladney *et al.* (1983)		Lee and McConchie (1982)		Leake *et al.* (1969)			Leake *et al.* (1969)			
	Expected (ppm)	CV (%)	Analysed (ppm)	CV (%)	Expected (ppm)	CV (%)	Analysed (ppm)	CV (%)	Analysed (ppm)	*A* CV (%)	*B* CV (%)	Analysed (ppm)	*C* CV (%)	*D* CV (%)	
Sc									4.2	34	50	15.9	12.9	9.8	Sc
V	235	39	240.0	1.8	123	4.8	121	4.3	5.5	75	17	181.9	8.1	2.8	V
Cr	380	21	287.1	0.7	12	13			58.9	8.8	3.1	795.5	1.2	0.5	Cr
Co	50	24	56.2	2.8	15.1	4.0			(0.1)			55.2	22	29	Co
Ni	260	28	259.5	0.5	17	12	22	12.7	1.4	174	506	608.4	3.0	1.2	Ni
Cu	72	38	80.0	1.7	60	5.0	63	3.0	33	21	12.5	85.7	4.4	4.7	Cu
Zn	150	25	135.1	0.8	88	1.1	82	2.4	36.7	9.0	3.7	85.1	2.5	2.3	Zn
Ga	20	76			20	7.5			18.3	6.0	9.7	13.0	13.5	11.5	Ga
Rb	47	36	46.5	1.2	67	0.7	64	4.5	265.5	1.4	1.2	20.4	4.6	4.2	Rb
Sr	1320	22	1363.8	0.2	662	0.7	646	1.0	20.6	3.1	3.2	267.4	0.7	0.7	Sr
Y	30	40	31.2	1.3	21	14	26	13.0	32.1	4.4	3.6	9.5	7.0	3.9	Y
Zr	250	29	251.1	3.8	225	4.0	260	2.2	75.3	8.6	5.0	62.3	6.5	4.9	Zr
Nb	100	11	114.7	1.2	15	10			19.2	10.1	11.1	3.5	106	29	Nb
Ba	1050	20	1056.4	0.6	1221	0.7	1200	0.5	61.7	13.1	7.7	244.4	5.1	3.5	Ba
La	80	40			38	3.9			10.3	16	19.2	9.6	18.6	37.5	La
Ce	140	13			66	4.5			36.1	11.6	17.9	51.9	11.3	14.0	Ce
Nd					34	7.4			9.3	47	21	9.0	41	39	Nd
Sm					5.9	4.2			8.6	51	63				Sm
Pb	8	63			36	6.9	34	9.1	76.8	5.0	3.9	5.9	65	63	Pb
Th					6.50	2.8			62.1	6.2	6.1	6.6	73	24	Th
U					1.89	6.6			3.6	150	105				U

CV = coefficient of variation = 100 × standard deviation/mean analysis (%).
(i) CRPG: BR. Data of Schroeder *et al.* were obtained by exciting powder pellets at 60 kV with a tungsten tube for count times of 40 to 130 s per element line. Results are the average of 40 analyses carried out over a 10 day period. These data are compared with expected compositions compiled by Govindaraju and de la Roche; coefficients of variation were calculated from all submitted data, without rejection of outliers.
(ii) USGS: AGV-1. Analysed data (Lee and McConchie) were obtained by exciting a 1:2 low dilution glass disc with a chromium or molybdenum tube at 35 to 60 kV, using count times of 50 to 100 s per element. The coefficient of variation was calculated from 10 replicate analyses. Data are compared with 'consensus' values of Gladney *et al.*, the corresponding coefficients of variation being calculated from published analyses, after rejection of outliers.
(iii) JW6012 and BL3571 are two in-house reference samples analysed by Leake *et al.* (1969). Data represent the average of a total of 36 to 48 analyses of 12 pellets of these materials, together with the corresponding coefficients of variation (*A* and *C*). Samples were excited with molybdenum or tungsten tubes operated at 50 to 100 kV (2 kW generator). Count times were 30 to 100s. Coefficients of variation (*B* and *D*) were obtained from 9 to 30 replicate analyses of the same sample pellet.

early commercial XRF instrumentation. Relative standard deviations are listed for replicate analyses on 12 powder pellets of a Galway granite (JW A6012) and an amphibolite (BL 3571). Coefficients of variation are compared with equivalent data obtained by analysing the same pellet of each sample between 9 and 30 times.

In assessing this data it is clear that due to a combination of poor counting precisions, spectral overlaps and/or the inherently low abundance of particular elements in geological materials, it is not always possible to report XRF analyses of high accuracy for the elements Sc, Co, Mo, La, Ce, Nd, Sm and U.

(*c*) *Inter-laboratory agreement.* XRF spectrometers are standard instruments for routine major- and trace-element analyses in many geochemical laboratories. In view of the high degree of standardization in both instrumentation and operating procedures, it is relevant to enquire as to the agreement that can be obtained in results from different laboratories. Data in Table 8.21 compare the results for a granite AT 242 submitted for analysis to eight independent XRF laboratories. There are no recommended values for this sample, the test being the degree of agreement between submitted data, from which averages and standard deviation data have been calculated. Relative standard deviations generally lie in the range 5 to 30%. It is suggested that the significantly larger variation in yttrium data may in part be due to discrepancies in the correction of the spectrum interference from Rb discussed in section 8.10. Differences in the analyses of the REE and some of the lighter transition elements are particularly high. Conversely, there is relatively good agreement between data for the elements Zn, Rb, Sr, Zr, Nb, Pb and Th.

Whether the agreement shown in Table 8.20 is acceptable is a matter of judgement. However, it is interesting to note that the variations between these data are significantly greater than the analytical precision claimed for many indi-

Table 8.21 Interlaboratory comparison of XRF trace-element data for a granite, AT-242

	Laboratory number								Mean	s	CV
	1	2	3	4	5	6	7	8	(ppm)	(ppm)	(%)
Sc	6	—	—	—	—	—	—	—	6		
V	39	—	44	—	—	—	—	43	42	2.6	6.2
Cr	46	61	44	—	—	—	53	31	47	11.1	23.6
Co	—	5	11	—	—	—	<5	—	8		
Ni	19	17	24	17	—	17	32	15	20.1	5.9	29.3
Cu	2	1	—	11		3	7	11	5.8	4.5	77.6
Zn	31	31	39	37	—	30	34	30	33.1	3.6	10.8
Ga	—	17	—	17	—	17.5	—	17	17.1	0.25	1.5
Rb	214	205	230	216	225	210	209	205	214	9.1	4.3
Sr	256	247	300	262	268	251	248	238	259	19	7.3
Y	19	10	24	22	25	29	20	28	22.1	6.0	27.1
Zr	147	130	153	147	146	145	137	132	142	8.1	5.7
Nb	14	15	18	14	14	14	16	16	15.1	1.5	9.9
Ba	670	—	—	—	—	—	773	694	712	54	7.5
La	56	—	52	—	—	—	75	34	54.3	17	31.3
Ce	81	80	69	—	—	—	122	36	77.6	30.8	39.7
Nd	36	—	32	—	—	—	53	—	40.3	11.2	27.8
Sm	—	—	—	—	—	—	—	—			
Pb	32	32	—	34	—	35	31	26	31.7	3.1	9.8
Th	25	20	—	21	29	23	27	22	23.9	3.3	13.8
U	—	—	—	—	—	—	9	10	9.5		

Units of concentration = ppm rock.
CV = coefficient of variation = 100 × standard deviation/mean analysis (%).
s = standard deviation.
Data analysed by XRF.
Contributing laboratories include those at the Universities of Bergen, Bochum, Edinburgh, Leicester, London (Bedford College), Newfoundland, Nottingham and Open University, all of whom are gratefully acknowledged. The order of laboratories contributing data has been randomized.

vidual trace-element analyses, such as those listed in Table 8.20.

8.13 Concluding remarks

In summarizing the development and current status of wavlength dispersive XRF analysis, it is appropriate to quote Ron Jenkins *et al.* (1981, p.iii), who wrote of the technique. 'Like most other analytical methods, it [WD–XRF] has seen the birth pains of innovation and has struggled through a period of misunderstanding and over-exploitation before finally reaching the stable plateau of a reliable and well-accepted analytical technique'. This is certainly the case in applications of WD–XRF to silicate rock analysis.

Wavelength dispersive XRF is perhaps the analytical technique that has had the greatest influence on the development of a geochemistry during the last 20 years. The high precision of both major and selected trace element analyses are taken for granted these days. There are no other routine techniques capable of parts-per-million detection limits for selected trace elements where it is not even necessary to weigh the analysed aliquot. The capacity to analyse on a routine basis large numbers of samples has made possible new perspectives in planning geochemical sampling programs.

Because there have been few fundamental advances in instrumentation in recent years, WD–XRF has fallen slightly from the analytical limelight. This should not hide the fact that this technique often possesses the analytical specification that other techniques would seek to emulate. WD–XRF looks set to retain this analytical prominence in years to come.

9 Energy dispersive x-ray spectrometry

9.1 The development of energy dispersive XRF

An energy dispersive x-ray fluorescence spectrometer differs from a wavelength dispersive instrument in that, instead of being based on a Bragg diffracting crystal, the spectrometer uses a solid state 'lithium drifted silicon detector' (i.e. Si(Li)) to measure the fluorescence spectrum. Operating principles of the rest of the instrument and practice in analytical applications broadly follow those of the conventional wavelength dispersive technique.

Energy dispersive x-ray detectors became widely available in the early 1970s (Gedcke, 1972) and were developed as an offshoot of the rapidly expanding technology of silicon electronic devices. Energy dispersive (ED) spectrometers are distinguished by their potential to detect incident x-rays of all energies emitted from an excited sample. By comparison, the conventional wavelength spectrometer must be driven to the appropriate 2θ setting to detect x-rays of discrete wavelength. Early ED spectrometers were therefore unrivalled as qualitative instruments for the rapid and simultanteous identification of elements contributing to an x-ray spectrum, especially in electron microprobe or scanning electron microscope applications.

For quantitative analysis, a superficial comparison of the response characteristics of ED with WD spectrometers is discouraging. ED detectors have inherently poor resolution in the 0 to 10 keV region and are restricted to a much lower maximum count rate capacity than the WD spectrometer. This, together with the relatively unsophisticated algorithms available for spectrum analysis on early instruments, restricted the application of these devices for quantitative analysis. Several important developments have changed this situation. One was the introduction of high performance pulse processing amplifier designs (e.g. Kandiah *et al.*, 1968, 1975). The advanced performance of these amplifiers eliminated many of the count-rate-dependent gain and resolution instabilities that degraded the performance of early spectrometers. Another was the development of computer algorithms capable of successfully deconvoluting the severe overlap interferences present in many ED spectra (Schamber, 1973; Statham, 1977). Finally the application of 'pulsed' control over x-ray tube excitation systems, as implemented by Jaklevic *et al.* (1972), offers the advantage of increasing the count rate capability of ED spectrometers by a factor of about three.

Energy dispersive spectrometers are currently widely accepted in electron beam applications. Indeed, substituting such devices has rejuvenated the quantitative analysis capabilities of many older electron microprobe instruments (replacing worn-out WD spectrometers). ED detectors are standard bolt-on extras on new microprobe and scanning electron microscope systems. In quantitative x-ray fluorescence applications, ED detectors have not yet become widely accepted compared with conventional WD instrumentation, although current research indicates that the analytical performance in routine silicate rock applications is broadly comparable to that of WD systems for many major and trace elements.

The term 'energy dispersive' is a misnomer, since unlike wavelength dispersive spectrometers, the ED detector does not physically disperse the x-ray energy spectrum. 'Energy discrimination' would be a more appropriate description, but since 'dispersive' is now almost universally used in the literature, this term is perpetuated here.

In this section, the properties and performance of the ED detector are discussed, and related to applications in x-ray fluorescence analysis. Topics covered include secondary target excitation and the potential advantages in using polarized and novel excitation geometries.

9.2 The Si(Li) detector

An energy dispersive spectrometer consists of a lithium drifted silicon crystal mounted at the end of a copper cold finger in a vacuum housing (Figure 9.1). X-rays enter the spectrometer through a beryllium window. The cold finger is connected to an adjacent liquid nitrogen cryostat. Detected signals are fed to a preamplifier, the first stage of which (a field effect transistor) is mounted adjacent to the detector crystal. The characteristics of individual components are as follows.

9.2.1 *The detector crystal*

(*a*) *The lithium drifting process.* An energy dispersive detector comprises a crystal of highly refined silicon, usually manufactured in a cylindrical configuration 10 to 16 mm in diameter and 3 mm (sometimes 5 mm) deep (Figure 9.2). Silicon is a semiconductor material and its electronic properties relevant to photon detection closely follow those discussed in detail in section 12.4 for germanium crystals used for gamma-ray detection.

Like germanium, detector-grade silicon may be manufactured as an *n*-type or *p*-type semiconductor material, depending of the properties of impurity atoms. It is possible to refine silicon to extremely high purity suitable for the manufacture of so-called 'intrinsic' silicon detectors (e.g. Fink, 1981). However, these devices are not widely available. Conventional detectors are fabricated using the lithium drifting process similar to that described for germanium in section 12.4.

The starting material is *p*-type silicon, which contains very

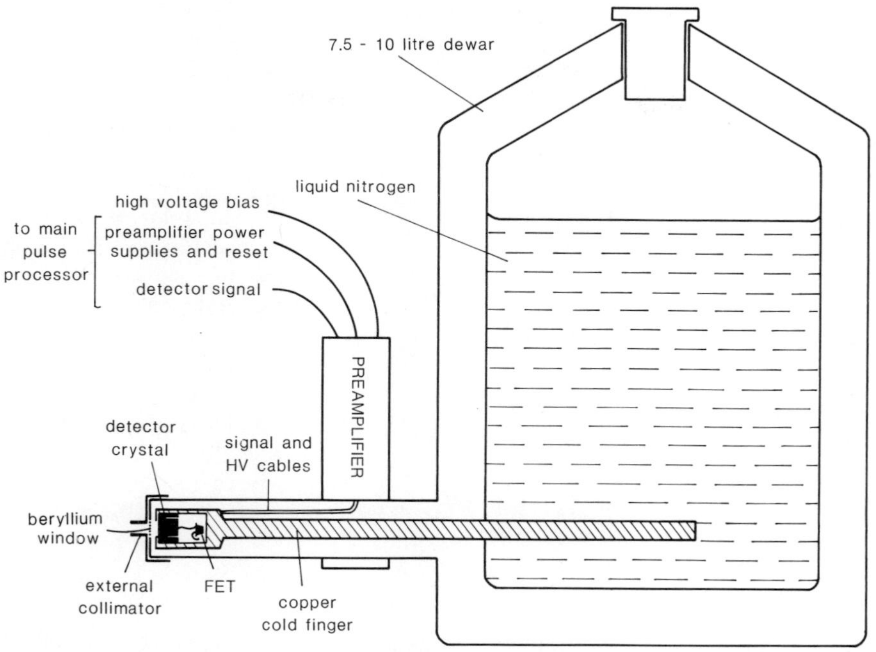

Figure 9.1 Overall diagram of an energy dispersive detector showing the arrangement of silicon crystal, cold finger, liquid nitrogen dewar and preamplifier.

small amounts of boron as impurity. Boron atoms have one too few electrons available to coordinate completely with four adjacent silicon atoms. These impurity atoms therefore form *acceptor levels* in the forbidden energy region between valence and conduction bands (see Figure 12.7) to which valence band electrons may be readily excited (section 12.4). The untreated silicon crystal is, therefore, unsuitable as a detection device. On application of a high voltage bias, the crystal would conduct due to the mobility of the positively charged 'holes' which remained in the valence band after electrons had been trapped in acceptor levels.

To compensate for this effect, lithium atoms are first diffused into the rear face of the crystal (the lithium diffusion layer, Figure 9.3). At elevated temperatures ($\sim 100°C$) and under the influence of an applied electrical bias, a fraction of these atoms are caused to 'drift' into the silicon crystal. Lithium atoms occupy interstitial lattice sites and seek out impurity atoms where they donate their readily ionized electron to neutralize the *p*-type properties of the impurity.

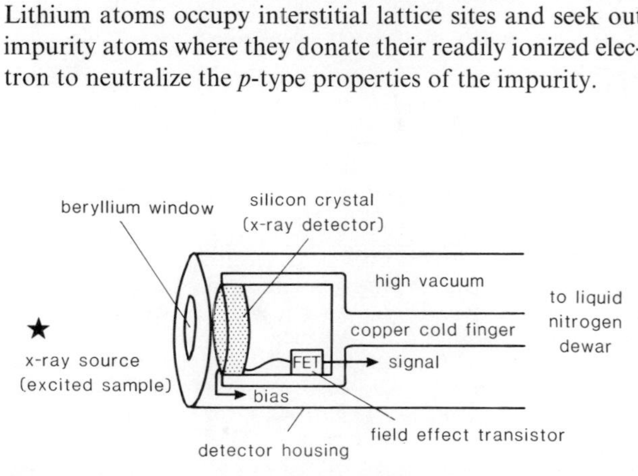

Figure 9.2 Schematic view of the mounting of the silicon crystal in an energy dispersive x-ray spectrometer.

The resultant crystal consists of three principal regions (Figure 9.3).

(i) The residual lithium diffusion layer (perhaps 300 μm thick) which forms the rear electrical contact to the crystal.

(ii) A much more extensive compensated region in which the number of charge carriers is greatly depleted due to the neutralization of *p*-type impurity atoms by the lithium drifting process—this is the active volume of the detector in which x-ray events are recognized.

(iii) A thin *dead layer* of silicon (perhaps 0.03 to 0.1 μm thick) on the front face of the detector in which the detrimental effect of impurity atoms is not fully compensated by the lithium drifting process. Some or all of the electrons that result from ionization of detector atoms by incoming x-rays within this dead layer are lost by charge trapping and recombination effects (section 12.4). The dead layer therefore comprises an undesirable additional absorber between x-ray source and the active volume of the crystal which has particularly important consequences for the measurement of low-energy x-rays (3 keV and below).

(*b*) *Critical penetration depth*. The attenuation properties of silicon are illustrated by data in Table 9.1, which lists the calculated critical penetration depth (defined as the thickness of silicon that will cause 99% attenuation of an incident x-ray beam) for selected $K\alpha$ photons. These data demonstrate the extremely high attenuation of phosphorus $K\alpha$ radiation, since P $K\alpha$ at 2.01 keV lies just above the silicon *K*-line absorption edge at 1.84 keV. It is expected, therefore, that severe charge collection problems will be encountered in the detection of this element (and S $K\alpha$, Cl $K\alpha$) because a significant proportion of these photons will be detected at the surface of the detector near (or in) the silicon dead layer.

Data in Table 9.1 also show that, due to the intervening

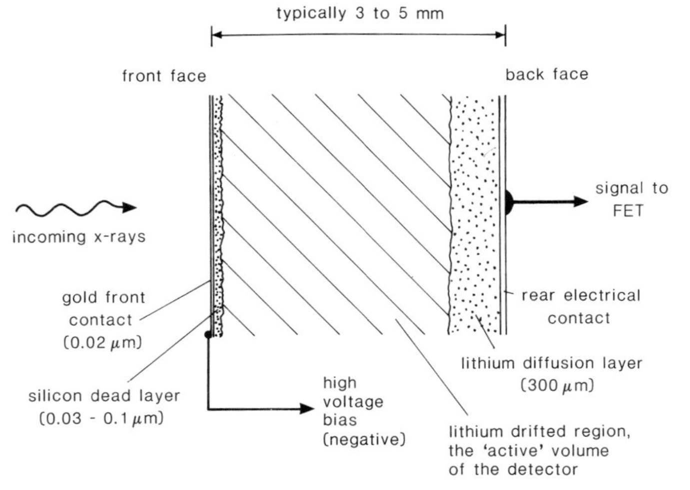

Figure 9.3 Cross-section through the silicon crystal of an ED detector. The dimensions of the boundary regions have arbitrarily increased in size to show their relationship to each other.

silicon absorption edge, the critical penetration depth of a light element such as Al *Kα* is very comparable to that of Ca *Kα*. At the other extreme, increasing proportions of radiation more energetic that Nb *Kα* (16.58 keV) will pass straight through a standard 3 mm thick Si(Li) detector, placing an upper limit to the useful detection range of these devices.

(*c*) *Gold front contact and high voltage bias.* A second electrical contact (in addition to the lithium diffusion layer, mentioned above) is required, this time on the front face of the detector crystal. This is usually formed by evaporating a very thin (0.02 μm) layer of gold on to the front surface. Statham (1981) reviewed data that suggested this layer is not normally uniform but may consist of areas completely free of gold surrounded by a filamentary network of gold. The front electrical contact is normally connected to a high voltage (negative) bias of between 500 and 1000 volts. The back contact (the lithium diffusion layer) is connected to the input stage of the signal preamplifier (a field effect transistor). X-rays entering the detector crystal cause ionization of silicon atoms. These electrons are promoted from the valence to the conduction band of the semiconductor. Under the influence of the high voltage bias, conduction band electrons are swept rapidly to the rear contact of the detector. The corresponding positive 'holes' migrate more slowly to the front contact. The resultant signal, detected by the preamplifier, is therefore a step function. The magnitude of each step is proportional to the number of ionized electrons caused by detection of an event which is itself proportional to the energy of the incident photon.

Table 9.1 Penetration of x-rays into a silicon detector

Element	Line	keV	Critical penetration depth (μm)
Na	*Kα*	1.04	15
Mg	*Kα*	1.25	24
Al	*Kα*	1.49	39
Si	*Kα*	1.74	61
P	*Kα*	2.01	7
Ca	*Kα*	3.69	37
Fe	*Kα*	6.40	170
Rb	*Kα*	13.37	1500
Nb	*Kα*	16.58	2800

Critical penetration depth is the distance into a silicon crystal over which 99% attenuation of an incident beam of photons occurs.

9.2.2 *Charge collection losses and guard rings*

Several designs of silicon crystal have been proposed for x-ray detection including grooved detectors (Figure 9.4) and

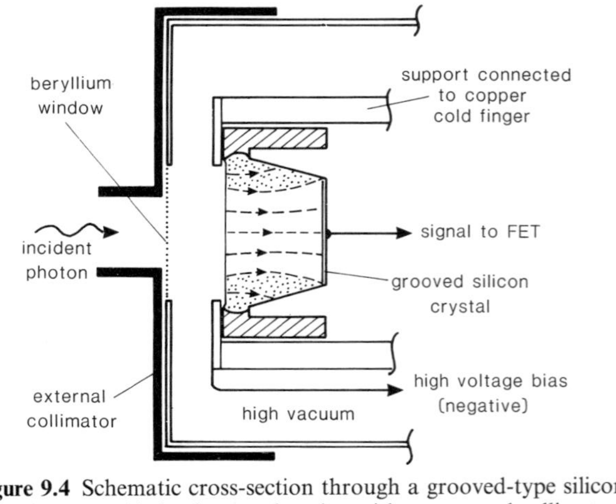

Figure 9.4 Schematic cross-section through a grooved-type silicon crystal mounted in its detector housing with an external collimator. Within the silicon crystal, arrows indicate the direction of electric fields set up when a high voltage bias is applied to the crystal. Shaded areas represent undrifted *p*-type silicon which is insensitive to x-ray detection, and the stippled areas represent regions of non-uniform electric field in which some or all of the charge resulting from x-ray detection may be lost. The detector possesses circular symmetry about its horizontal axis. (After Jaklevic and Goulding, 1978.)

'top hat' configurations (Figure 9.5*a*). The core of these crystals consists of lithium drifted silicon. However, the outer surfaces comprise uncompensated *p*-type silicon where the effects of impurity atoms have not been fully compensated by the lithium drifting process. The electrical fields created inside the detector crystal on application of a high voltage bias are indicated schematically in Figure 9.4 and 9.5. In general, electrons ionized within the crystal will follow the lines of force created by this applied bias.

As can be seen, x-ray photons that interact with the central region of the crystal will have their resultant electron charge efficiently collected at the rear contact. However, the electrical field becomes less uniform towards the edge of the crystal. The probability then becomes significant that some or all of the electron charge originating from an event detected there will be trapped in the uncompensated silicon region at the edge of the detector. This effect causes the loss of varying proportions of the total charge created by a photon and results in the apparent detection of an event that will randomly contribute to the background at energies below the full energy photopeak.

Goulding *et al.* (1972) estimated that a large proportion of the apparent continuum in an energy dispersive spectrum originates from this phenomenon. This is undesirable since peak-to-background ratios will be degraded, the very parameter in which ED detectors are at a disadvantage in comparison with WD spectrometers.

In order to achieve reasonable spectrometer performance, it is important, therefore, to minimize charge trapping at the edge of the detector crystal. The simplest way is to collimate the incoming x-ray beam so that events will primarily be detected in the central core of the crystal. The external collimator (Figure 9.4) should, therefore, be considered as an integral part of the spectrometer.

A more elegant way of minimizing peripheral charge losses was proposed by Goulding *et al.* (1972) using the concept of the *guard ring*. In its simplest form (Figure 9.5*b*) the rear electrical contact is divided into two sections, the central segment from which the analytical signal is taken, and an outer annulus. Both are maintained at ground potential, the function of the outer 'guard ring' electrode being to maintain a uniform electric field in the inner electrode from which the signal is measured. Goulding *et al.* (1972) claimed that by using this device, background due to poor charge collection may be reduced by a factor of between 2 and 10 depending on photon energy.

However, edge effects are still not completely eliminated. When an x-ray photon interacts with the crystal, Goulding *et al.* (1972) estimated that a charge cloud of diameter approximately 5 μm is formed (initially). Strong electrostatic repulsions occur within this cloud between negatively charged electrons (and also between positively charged 'holes'). The initial cloud therefore rapidly expands to about 100 μm before being swept by the applied bias to the appropriate electrode. Thus events that are detected within about 100 μm of the guard ring boundary may still have part of their charge lost from the signal collected at the central electrode.

To overcome this problem, Goulding *et al.* (1972) proposed a *double guard ring* configuration (Figure 9.5*c*). The outer guard ring us used to maintain a uniform field within the central regions of the crystal as before. The inner guard ring, however, has its signal gated in anticoincidence with the inner signal electrode. Any signals appearing simultaneously at the inner guard ring and signal electrode are rejected. Only those events entirely detected by the signal electrode contribute to the recorded spectrum. Although this system is claimed to result in a further significant reduction in background counts caused by peripheral charge losses, it has not found extensive application in commercial instruments used for the analysis of geological materials.

9.3 Detector configuration and characteristics

9.3.1 *Detector configuration*

Like germanium, Si(Li) crystals must be maintained at liquid nitrogen temperatures both to minimize electronic noise that

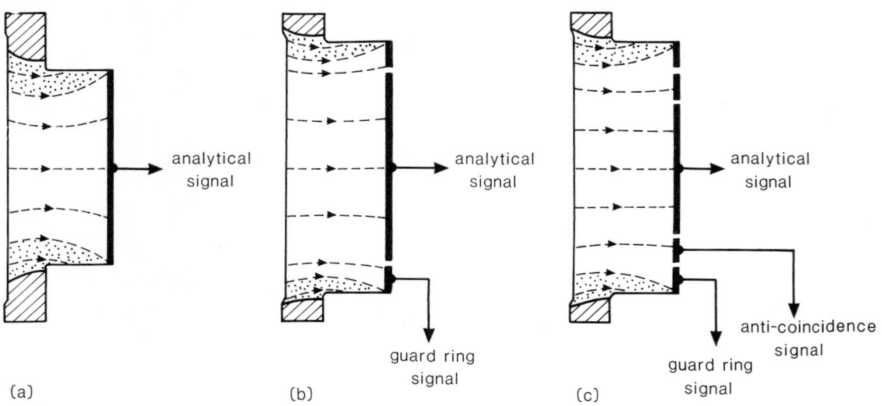

Figure 9.5 Silicon crystals for x-ray detection with the top-hat configuration: (*a*) standard crystal; (*b*) single guard-ring, the outer annulus of which is biased to maintain a uniform electric field in the central regions; (*c*) double guard-ring configuration in which an additional annulus on the rear contact is gated with the detector signal so that if an event is detected in the boundary region between central and inner guard-rings, it will be rejected. Arrows show schematic electric fields, shaded and stippled areas have the same representation as in Figure 9.4. (After Jaklevic and Goulding, 1978.)

affects the detection of very small electronic signals and to freeze the lithium atoms into the silicon matrix. The mobility of lithium in silicon is, in fact, less than that in germanium, so that temporary warm-up may not have such damaging effects. However, it is recommended that these detectors are never allowed to run dry of liquid nitrogen, since some degradation in performance is always likely to result.

The configuration of a typical Si(Li) spectrometer is shown in Figure 9.1. The crystal is mounted at one end of a copper cold finger, the other end of which enters the liquid nitrogen reservoir. The detector housing must be maintained under very high vacuum both to minimize conduction of heat from the surroundings to the detector and to prevent the deposition of contaminants on the crystal surfaces which would otherwise promote the breakdown of high voltage bias. The detector housing must also be light-tight to eliminate noise induced by light photons striking the detector crystal. A beryllium entrance window is provided to reduce attenuation of low energy x-ray photons. The attenuation characteristics of beryllium foil as a function of thickness are shown in Figure 8.14. Standard detectors are supplied with 8 μm windows to allow adequate sensitivity down to Na $K\alpha$ at 1.04 keV.

9.3.2 *Detector efficiency in comparison with germanium crystals*

The calculated efficiency of a 3 mm thick Si(Li) crystal as a function of photon energy is compared with that of a comparable planar germanium crystal in Figure 9.6. Three regions may be distinguished in the efficiency graph for the silicon crystal. At low energies (less than 5 keV), detection is largely limited by absorption in the beryllium window and silicon dead layer. At intermediate energies (5 to 20 keV), essentially all radiation entering the crystal is detected (neglecting charge collection losses). Above 20 keV, radiation increasingly passes straight through the crystal, causing a rapid fall-off in efficiency above about 30 keV.

By contrast, a 3 mm germanium crystal has a high detection efficiency up to about 80–100 keV. However, in the region from 30 keV down to the germanium *K*-absorption edge at 11.1 keV, increasing losses occur from the full energy photopeak, principally due to the increasing prominence of the Ge escape peak. Further difficulties are encountered in the detection of radiation below about 5 keV caused by the much higher attenuation of photons in the germanium dead layer in comparison with silicon. Due to incomplete charge collection effects, detection of events below about 5 keV is unsatisfactory since corresponding photopeaks have significant low-energy tails (see 9.2.2). Based on these characteristics, Si(Li) crystals are preferred for the detection of radiation between 0 and 20 to 30 keV, whereas germanium detectors are preferred for analytical measurements above 20 to 30 keV.

9.3.3 *Detector resolution*

The resolution of Si(Li) detectors follows closely the criteria relating to germanium crystals in section 12.8.

Briefly, the overall resolution of a Si(Li) detector depends on the magnitude of two components, a fundamental contribution from the detector crystal itself (related to statistical variations in the number of electron–hole pairs formed by an incident photon) and an additional effect due to electronic noise in the crystal and preamplifier circuits.

The overall resolution in terms of observed full width half maximum (FWHM) of a given photopeak may be calculated from:

$$\text{FWHM} = \sqrt{((\text{FWHM}_{\text{detector}})^2 + (\text{FWHM}_{\text{noise}})^2)} \quad (9.1)$$

Minimization of the electronic noise contribution is a major concern in x-ray analysis, since its contribution is

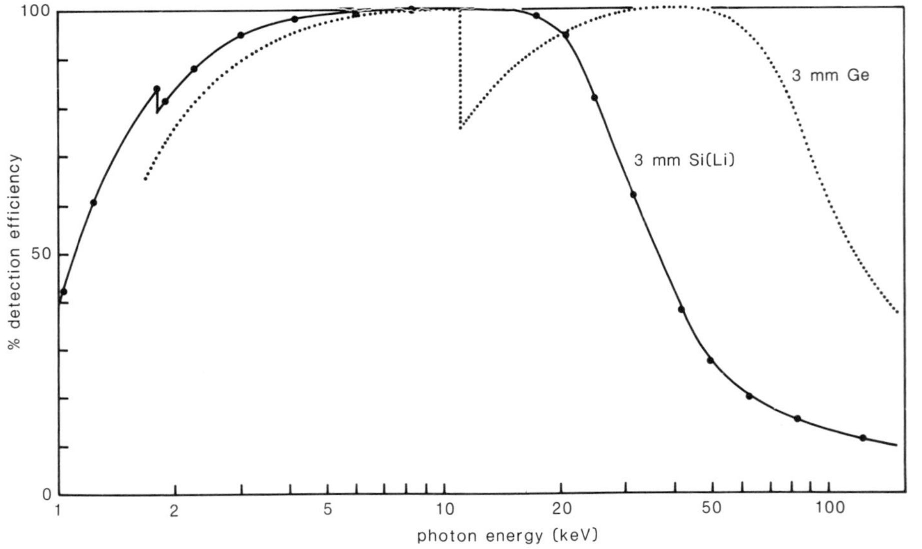

Figure 9.6 Calculated detection efficiencies of a Si(Li) and Ge detector, both 3 mm deep. It has been assumed that the silicon dead layer is 0.1 μm (relatively thick for a modern detector) and the germanium dead layer is 0.15 μm. Both detectors are assumed to have beryllium windows that are 8 μm thick.

much larger than statistical variations in the total charge deposited in the detector by lower energy photons. This noise component is reduced in two ways: first, by the selection of a low noise–high gain amplifier device, the field effect transistor (FET), as first stage in the preamplifier; second, by maintaining both detector crystal and FET close to liquid nitrogen temperatures to suppress electronic noise caused by thermal effects.

The detector term ($FWHM_{detector}$) is principally controlled by the average energy (ε) required to create an electron–hole pair in silicon. At 77 K, this parameter is 3.8 eV. The detector noise component is then given by

$$FWHM_{detector} = 2.344\sqrt{FE\varepsilon} \qquad (9.2)$$

where E is the photon energy, F is the Fano factor, which usually lies in the range 0.1 to 0.13 for modern silicon detectors (see section 12.8), and ε is the mean energy required to create an electron–hole pair, as above.

The overall resolution of a Si(Li) detector has been plotted as a function of photon energy in Figure 9.7 for a detector with a Fano factor of 0.1 and electronic noise contributions of (*a*) 80 eV (exceptionally low), (*b*) 100 eV (average) and (*c*) 120 eV (average for a larger detector). The equivalent resolutions at 5.9 keV (Mn $K\alpha$ line as obtained from an ^{55}Fe source) are 138, 151 and 165 eV respectively. These curves show that the resolution specification of a detector has a marked influence on performance towards lower-energy photons (below 5 keV) but has a progressively smaller effect at higher energies. Data in Table 9.2 show that an increase in the noise component from 80 to 120 eV causes an increase in detector resolution from 9.3% to 12.9% FWHM for a photon at 1 keV, but only 1.1 to 1.2% FWHM for a 20 keV photon.

Detector characteristics are compared with gas proportional counters and germanium detectors in Table 8.14. Because the mean energy to create an electron–hole pair in a germanium crystal (2.8 eV) is smaller than that in a silicon crystal (3.8 eV), the resolution of a germanium crystal is inherently better. However, advantage cannot be taken of this property in practice because of the charge collection problems at energies below 5 keV in a germanium crystal (section 9.3.2).

Table 9.2 Detector resolution as a function of energy

Noise	80 eV	100 eV	120 eV
FWHM at 1 keV	93 (9.3)	110 (11)	129 (12.9)
FWHM at 5.9 keV	138 (2.3)	151 (2.6)	165 (2.8)
FWHM at 10.0 keV	167 (1.7)	177 (1.8)	189 (1.9)
FWHM at 20.0 keV	222 (1.1)	230 (1.1)	240 (1.2)

Resolution calculations are based on equations (9.1) and (9.2), assuming the Fano factor = 0.1 and adding in the listed proportion of electronic noise.
FWHM data in keV.
Figures in brackets represent (FWHM × 100)/photon energy (%).

9.3.4 *Perspective on resolution compared with the crystal spectrometer*

The relatively poor resolution of Si(Li) detectors is often cited as one of the disadvantages of these spectrometers in comparison with wavelength dispersive instruments. It is important, therefore, to obtain some perspective on the variation in resolution of these two detector systems as a function of photon energy so that as assessment can be made of the analytical consequences.

Resolution data taken from Potts *et al.* (1985) for an XRF wavelength dispersive spectrometer fitted with TAP, PET and LiF crystals is compared with that of an XRF Si(Li) detector in Figure 9.8. The nominal resolution of the Si(Li) detector was 165 eV at 5.9 keV. The inherent resolution advantages of crystal spectrometers in the detection of low energy photons are clearly demonstrated by these data. However, it is not always appreciated that a cross-over occurs in the region at about 12–15 keV. For higher photon energies, resolution advantages lie with the Si(Li) detector.

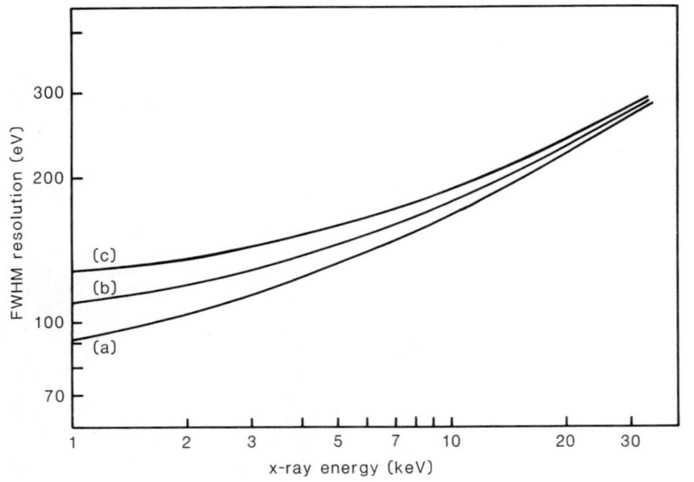

Figure 9.7 Variation with photon energy of the full-width half-maximum resolution of Si(Li) detectors of effective resolutions at 5.9 keV of 138, 151 and 165 eV (equivalent noise peaks = 80, 100 and 120 eV).

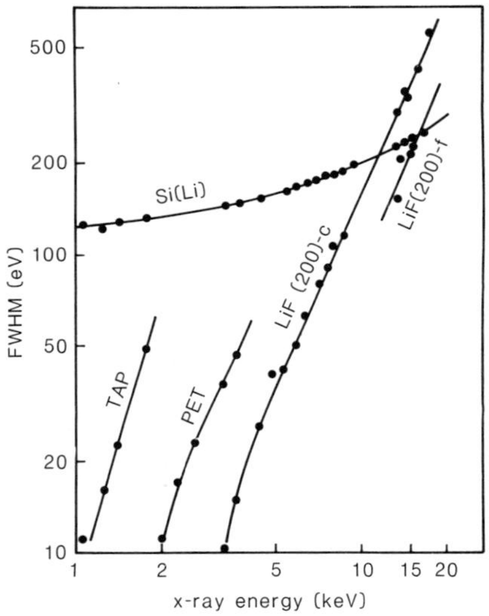

Figure 9.8 Comparison between the overall energy resolution of a WD-XRF crystal spectrometer (using TAP, PET and LiF diffracting crystals) and an XRF Si(Li) detector. (Potts *et al.*, 1985.)

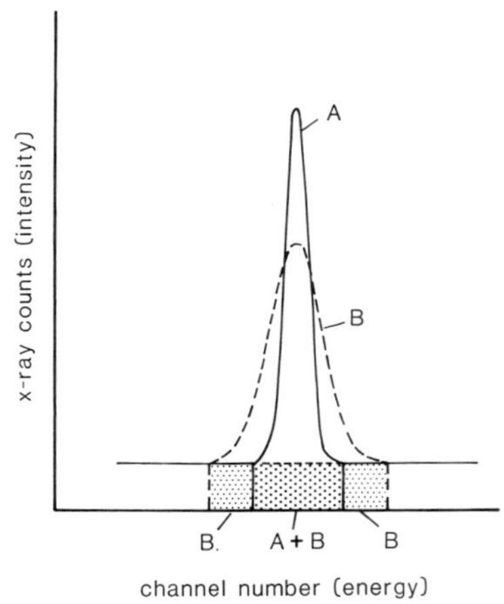

Figure 9.9 Exaggerated effect of an increase in FWHM of x-ray photopeak on the background contribution to that signal in a spectrum recorded with a Si(Li) detector.

The precise position of this cross-over depends on the specification of the Si(Li) crystal and the degree of angular dispersion selected for wavelength dispersive measurements (fine *v.* coarse collimation, LiF(220) *v.* LiF(200)).

The analytical consequences of this comparison are as follows.

(i) The resolution advantage unambiguously lies with crystal spectrometers for all measurements in the range 0 to 10 keV. This advantage is indeed considerable at the lower end of this range, particularly below 5 keV. Thus, on the basis of resolution criteria alone, a crystal spectrometer is the detector of choice both in electron microprobe applications and in the determination of major elements by XRF.

(ii) This advantage is not as clearly defined in the range 10 to 15 keV, and for photon energies significantly above 15 keV, the Si(Li) detector becomes the detector of choice, based again on resolution criteria alone. It is expected, therefore, that in the determination by XRF of the heavier trace elements (especially Rb, Sr, Y, Zr, Nb, Mo, Pb, Th and U) in silicate samples, the Si(Li) detector will be at no disadvantage in respect of resolution. Indeed, with the more advanced interference correction procedures available to analyse ED spectra, the Si(Li) detector may offer some advantages over conventional crystal spectrometry in this application.

9.3.5 *Resolution specification of ED detectors—analytical consequences*

Much importance is placed on resolution specification of ED detectors because this parameter summarizes the potential analytical capabilities of the spectrometer in two areas.

(i) The worse the resolution, the higher the contribution that background intensity will make in the measurement of an emission line intensity. This influences both precision (in terms of peak-to-background ratio) and detection limits that may be achieved. These effects are exaggerated in Figure 9.9. This diagram shows two peaks having identical areas. However, because the wings of one peak (*B*) extend further from the centroid than those of the other peak (*A*), the proportion of background that is effectively sampled with peak (*B*) is significantly larger. Data in Table 9.3 illustrate this effect by listing the increase in background contribution to selected photopeaks for detectors having resolutions of 151 eV and 165 eV at 5.9 keV compared with a premium-grade detector (resolution 138 eV at 5.9 keV). In calculating these data it has been assumed that the continuum intensity is constant with energy and that detected photopeaks can be considered as simple triangular functions having FWHM resolutions equivalent to the three sets of data plotted in Figure 9.7. The results show that the background area under

Table 9.3 Effect of change in detector resolution on the background under selected photopeaks

Analysed photopeak (keV)			FWHM resolution 138 eV at 5.9 eV	Increase in background contribution 151 eV at 5.9 eV (%)	Increase in background contribution 165 eV at 5.9 eV (%)
Na	*Kα*	1.04	93	19	39
Si	*Kα*	1.74	101	16	34
Ca	*Kα*	3.69	120	12	25
Fe	*Kα*	6.40	142	9	18
Rb	*Kα*	13.37	187	5	11
Mo	*Kα*	17.44	209	4	9

Calculations assume that gaussian photopeaks approximate to triangular functions, having the specified FWHM at 5.9 eV with effective noise components of 80, 100 and 120 eV, respectively (see equation (9.1)).
The increase in background represents the fractional increase in area under the base of the photopeak triangle compared with that for the detector with a FWHM of 138 eV at 5.9 eV, assuming constant continuum intensity.

the Na $K\alpha$ line at 1.04 keV increases by 19% and 39% when comparing detector resolutions changed from 138 eV at 5.9 keV to 151 eV and 165 eV respectively. However, for the Mo $K\alpha$ line at 17.44 keV, the corresponding values are much smaller, 4% and 9% respectively.

(ii) The second factor is the increase in the degree of overlap between adjacent x-ray lines as detector resolution becomes worse. The magnitude of this effect is illustrated by data in Table 9.4, again using the same three sets of detector resolution data and assuming that photopeaks approximate to triangular functions. Subject to the limitations of this model, these data list the percentage degree of overlap between selected pairs of adjacent x-ray photopeaks of interest in silicate rock analysis. The effect of resolution is clearly much less marked for higher energy photopeaks. Thus, the overlap for K $K\beta$ on Ca $K\alpha$ changes from 33 to 39 to 44% as nominal detector resolution icreases from 138 to 151 to 165 eV at 5.9 keV (assuming equal peak intensities). The corresponding data for the overlap between higher energy Mo $K\alpha$–Zr $K\beta$ interference at about 17.5 keV are 23% to 25% to 27%, an overall increase of only 4%. These data indicate that the resolution specification of an ED detector has an important effect on analytical performance in the 0–10 keV region of the spectrum. However, above 10 keV, the effect is much less critical.

9.4 Pulse processing electronics

The signal amplification and shaping circuits play an important role in controlling the performance of a Si(Li) x-ray detector. Their principal functions are as follows:

(i) To minimize the degree of electronic noise contributed to the detector signal—this is of critical importance if successful measurements are to be made down to photon energies of 1 keV or less

(ii) To minimize spectral distortions due to pulse pile-up effects that are likely to be serious at the long amplifier time-constants required to achieve (i)

(iii) To maintain an exceptionally high degree of stability both in spectrum gain and FWHM resolution over as wide a dynamic range of input count rate conditions as possible.

The individual electronic modules in the pulse processing chain of a Si(Li) detector are shown in Figure 9.10. Their functions are as follows.

9.4.1 *The preamplifier and pulsed optical feedback circuits*

The function of the preamplifier is to amplify any signal received from the detector. It is critical that additional electronic noise imparted on to this signal is minimized as this too will be amplified by subsequent circuits. Indeed, the effective signal-to-noise ratio is set by the noise imparted on the detector signal by the first-stage amplifier circuits. The signal is, therefore, fed to a low-noise field effect transistor (FET), mounted adjacent to the detector crystal and cooled to liquid nitrogen temperatures.

As has been indicated above (section 9.2.1(*c*)), the signal originating in the detector is a stepwise change in voltage, the magnitude of the step being proportional to the energy of the incident x-ray photon. An important consideration in the operation of the preamplifier is, therefore, the way in which the charge associated with this signal is restored to the quiescent state, since the detection of further x-ray photons would eventually bias input to the FET out of its linear operating range.

The conventional way of restoring the detector signals, as used on virtually all germanium gamma-ray detectors, is to conduct the charge back to the detector through an appropriate high value feedback resistor. However, noise considerations are so critical that this technique would cause an unacceptable increase in the noise of low-energy photon signals from the Si(Li) detector due to random fluctuations of electrons in the feedback resistor.

To minimize noise, an alternative charge restoration technique known as *pulsed optical feedback* is generally used. The operating principles of this system are shown in Figure 9.11. Each FET input signal is cumulatively stored on a feedback capacitance. When the voltage on this capacitance reaches a predetermined level, control logic circuits in the main amplifier momentarily switch on a light-emitting diode (LED) that is located adjacent to the FET. The LED has no direct electrical connection to the FET, but the light pulse causes the leakage current across the FET drain to gate to increase to such an extent that the feedback capacitance is

Table 9.4 Effect of change in detector resolution on the degree of overlap between selected photopeak interferences

Analysed peak		keV	Overlapping peak		keV	% Overlap FWHM: 138 eV at 5.9 keV	FWHM: 151 eV at 5.9 keV	FWHM: 165 eV at 5.9 keV
Mg	$K\alpha$	1.253	Na	$K\alpha$	1.041	0	0.4	3.6
Th	$M\alpha$	2.996	U	$M\beta$	3.145	12	18	23
Ca	$K\alpha$	3.690	K	$K\beta$	3.589	33	39	44
Fe	$K\alpha$	6.400	Mn	$K\beta$	6.489	47	51	54
Mo	$K\alpha$	17.44	Zr	$K\beta$	17.67	23	25	27

Data calculated assuming that photopeaks approximate to a triangular shape having a FWHM calculated according to equations (9.1) and (9.2).

Data represent the proportion of overlap between the listed interferences assuming each has an identical area.

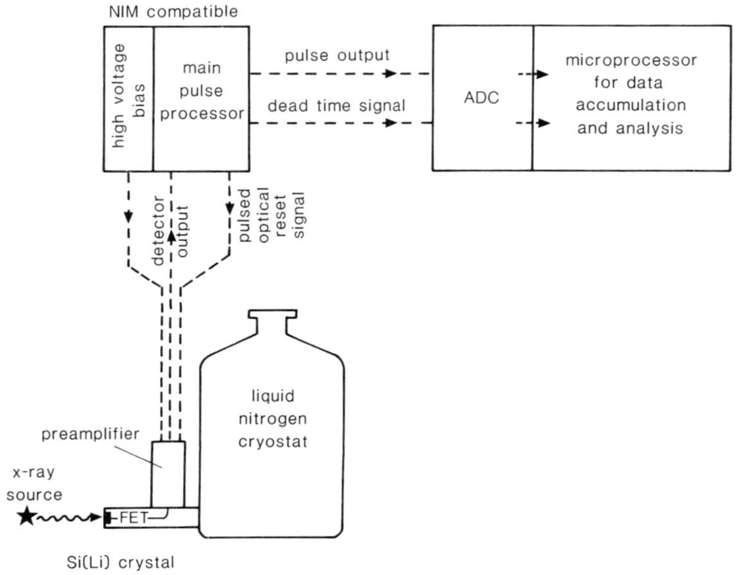

Figure 9.10 Electrical interconnections between detector and the pulse processing components of an ED spectrometer using pulse optical feedback charge restoration.

discharged and the FET input signal restored to its quiescent state. X-ray signals detected for a short interval after the LED has been pulsed must be ignored to allow photon-excited noise to die down. Between reset signals, x-ray events are accumulated on the feed-back capacitor as a ramp or staircase (Figure 9.12). Each step in the ramp corresponds to the detection of an x-ray photon, the magnitude of each step is proportional to the photon energy.

Other forms of charge restoration have been implemented on x-ray detectors, most notably dynamic charge restoration, discussed further by Jenkins *et al.* (1981, pp.172–177).

9.4.2 *The main pulse processor*

Electronic noise considerations are again fundamental to the design and operation of the main pulse processor. It is generally accepted that amplifiers for use with Si(Li) detectors must be operated at long time constants (in comparison with gamma-ray detectors, for example) to minimize electronic noise contributions. The way in which electronic noise in the signal generally varies with amplifier time constant is shown in Figure 9.13. A marked reduction is observed in signal noise as the amplifier time constant is increased. A minimum is observed at a time constant which varies with circuit design but is generally in the range 13 to 20 μs. This is a much longer amplifier time constant that would be selected on semi-gaussian amplifiers used in conjunction with germanium detectors, for example, where a value between 1 and 6 μs would be more usual (section 12.5).

Advantage may be taken of such long time constants only if careful consideration is given to the problem of pulse pile-up. This phenomenon occurs when a second x-ray photon is detected at a time interval shorter than the resolving time of the amplifier for a preceding photon (see section 12.5). Statham (1981) has estimated that if a standard semi-gaussian amplifier is operated in such a way that the time to peak of a signal is 20 μs, then a further 60 μs must elapse brfore the signal decays back to the baseline. Continuous pulse pile-up would then occur at evenly spaced signal input at a count rate as low as 3700 cps. More sophisticated semi-gaussian amplifier designs have been developed to cope with this problem, based on the principles outlined in section 12.5. However, a more revolutionary approach to the design and control of the entire pulse-processing chain was proposed by Kandiah *et al.* (1968, 1975) and manifest in the 'Harwell processor'. These concepts, based on the use of time variant filters, will be considered further here.

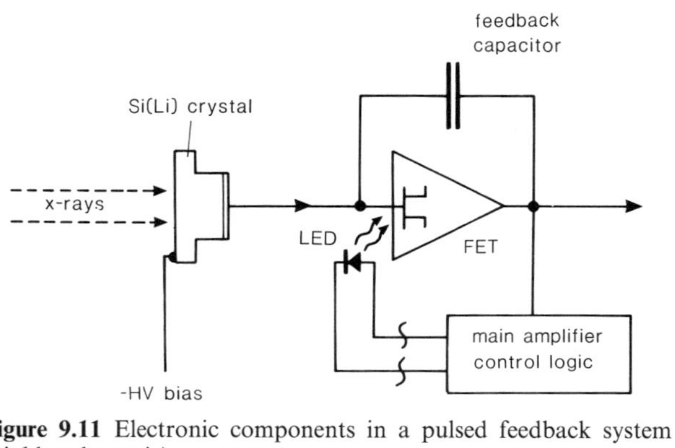

Figure 9.11 Electronic components in a pulsed feedback system (highly schematic).

9.4.3 *Amplifiers with time variant filters: the Harwell processor*

(*a*) *Circuit functions.* The pulse processor, designed by Kandiah *et al.* (1968, 1975) for use with Si(Li) detectors, is shown in block diagram form in Figure 9.14. The design

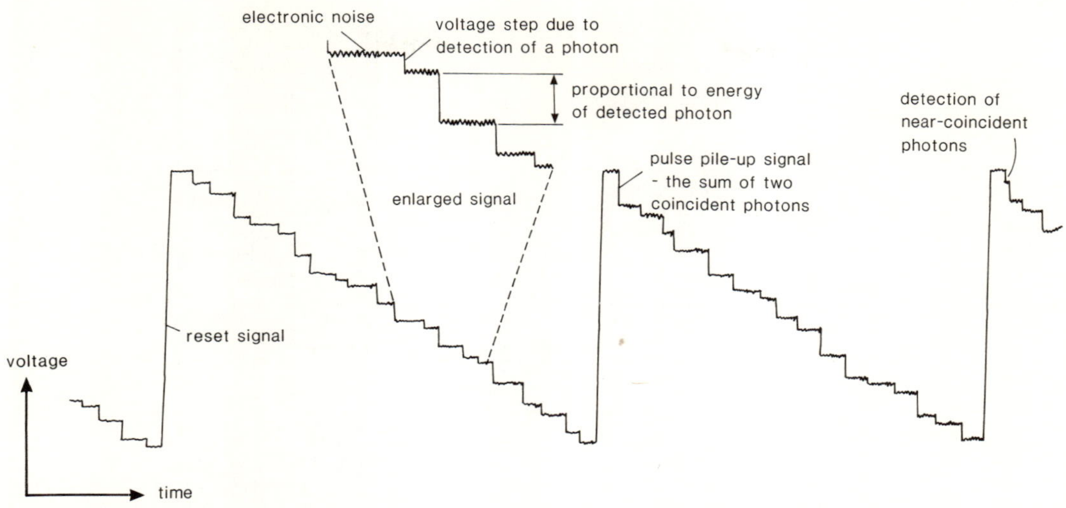

Figure 9.12 Schematic trace of the output from the preamplifier to the main pulse processor showing a staircase corresponding to the detection of successive x-ray events together with the pulsed optical feedback reset signals.

incorporates four 'channels', three for signal recognition and one main pulse amplification and shaping channel. The input signal is the ramped preamplifier output (Figure 9.12), sampled simultaneously at the input to each channel. Only if conditions imposed by channels A to C are satisfied is significant time expended in channel D in shaping and amplifying the signal for data accumulation. The conditions are as follows.

(i) Signal channel A acts as a *lower level discriminator* and is used to distinguish a new x-ray event from ambient electronic noise. It comprises an amplifier with a short time constant (1 to 2 μs). If an input signal exceeds the threshold set by channel A, control logic is triggered such that the input switch to channel D is enabled, initiating the full processing of the pulse.

(ii) Signal channel B acts as a *pulse pile-up recognition circuit*. If a second input signal is detected *after* initiation of pulse processing by channel A but *before* it has been completely integrated by channel D, the logic circuits that monitor channel B will initiate a restore cycle. Output of the first pulse is vetoed to avoid accumulation of a pile-up distortion in the output spectrum. It is most effective to split the detection of pile-up between two channels. One channel has a very short time constant of typically 100 ns, adequate to resolve pile-up involving photons of energy above about 3 keV; the second having a time constant of about 1 μs to permit resolution of pile-up down to energies below 1 keV. This circuit will resolve pile-up events providing they occur at intervals exceeding about 200 ns.

(iii) Signal channel C behaves as an *upper level discriminator*. Signals that would overload the main pulse processing channel D are recognized. If an overload signal is

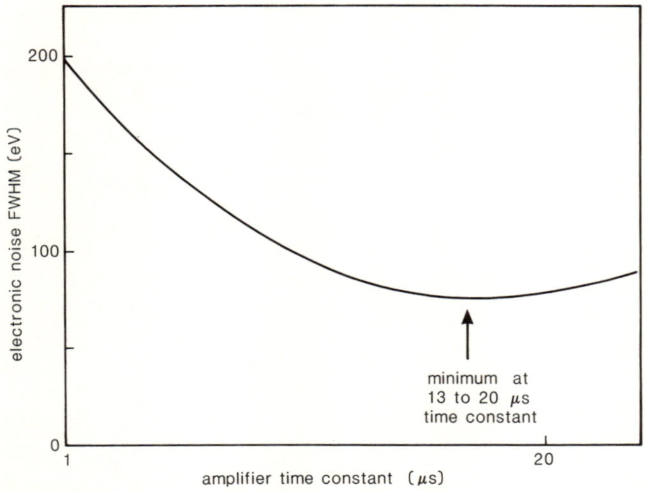

Figure 9.13 Effect of main amplifier ('pulse processor') time constant on the electronic noise in an x-ray signal showing a noise minimum at 13 to 20 μs, depending on manufacturer and circuit design.

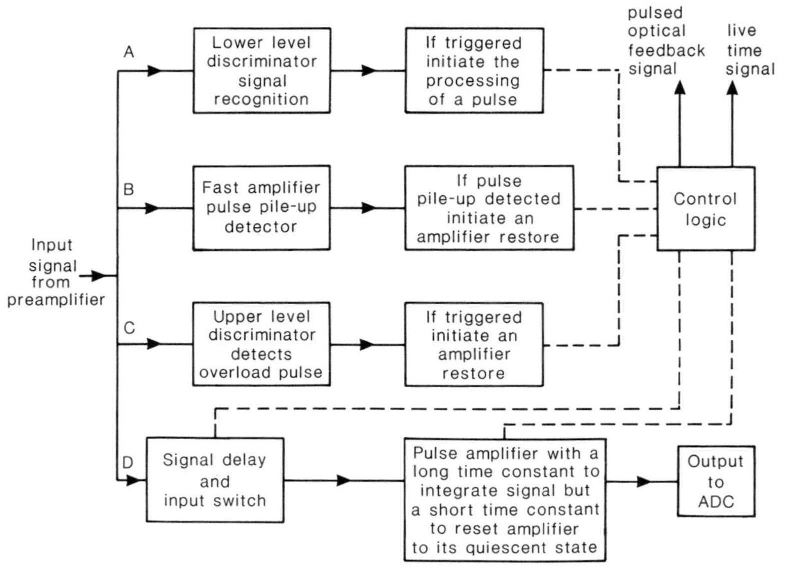

Figure 9.14 Block diagram showing operating functions of a Harwell pulse processor incorporating time variant filters. (After Kandiah *et al.*, 1968, 1975.)

detected, the control logic initiates a reset cycle to inhibit the further processing of the out of range event.

(iv) Signal channel D is the *main pulse amplification and integration channel.* The input to the amplifier is effectively grounded until arrival of a signal. The signal is first passed into a time delay line. This delays processing of the signal so that logic circuits can be triggered if appropriate by channels A–C and the input switch to the main amplifier circuit can close before arrival of the signal. The input signal is then amplified and integrated on a capacitor for a precise length of time (normally 20 μs time-to-peak). After integration, this peak voltage is held constant for sufficiently long to allow the ADC (analogue to digital converter) to measure its amplitude (Figure 9.15). The input switch is opened immediately after integration to prevent distortion of the signal by any succeding pulses. The time constant

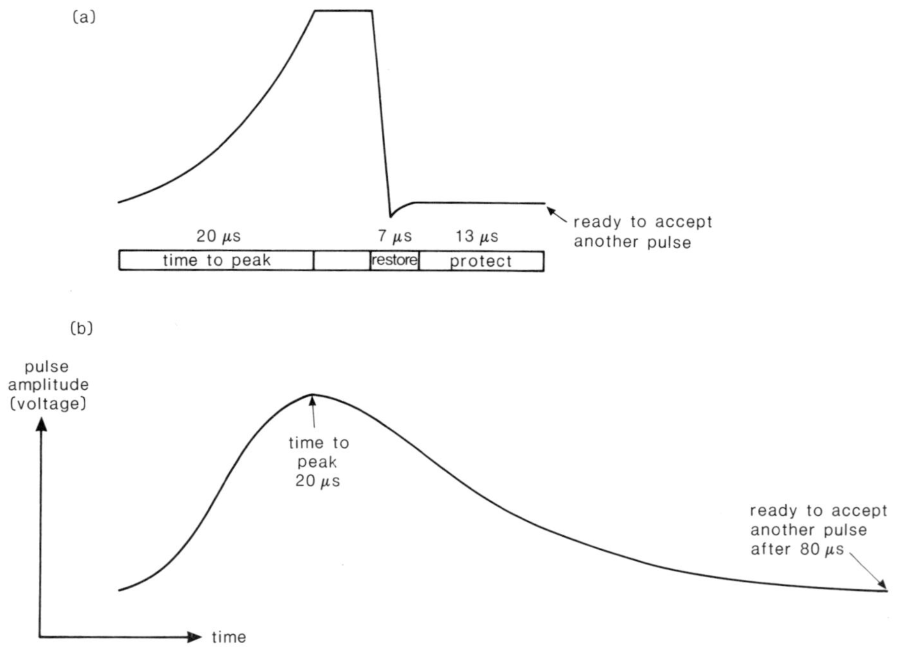

Figure 9.15 Diagram of pulse outputs from (*a*) Kandiah-type pulse processor with time variant filters; (*b*) conventional semi-gaussian amplifier operated at a time constant that gives an identical time-to-peak interval of 20 μs. (After Statham, 1981.)

characteristics of the amplifier are then switched from very long to very short so that the signal is rapidly discharged and the amplifier restored to its quiescent state to allow early acceptance of further pulses. This 'restore' interval (about 7 μs) must then be followed by a 'protect' interval (of perhaps 13 μs) to ensure that noise due to the switching process has decayed and will not distort subsequent measurements. The next pulse processing cycle cannot be initiated until the amplifier input has fallen below the channel A threshold. The output signal from this amplifier is compared with that from a simple semi-gaussian amplifier having an equivalent time-to-peak in Figure 9.15. The shorter time required to process a single pulse using the time variant filter system is apparent.

(*b*) *Zero stabilizer*. The Kandiah processor has a further refinement. At preset intervals (typically 600 cps), an oscillator triggers an electronic pulse which, if the main processing channel is not active, is amplified in the normal way. These 'strobe' pulses therefore accurately reflect the current electronic noise in the system and may be used to monitor and, if necessary, correct the dc level of the output signal. In this way, gain drift in recorded spectra caused either by changes in ambient conditions (e.g. temperature) or in input count rate can be virtually eliminated.

(*c*) *Count rate performance*. The count rate performance of a pulse processor after the Kandiah design is compared with that of conventional semi-gaussian amplifiers in Figure 9.16. The Kandiah pulse processor was operated at a time-to-peak of 20 μs, the gaussian amplifiers at time constants of 10 and 20 μs. At input count rates exceeding 3000 to 4000 cps, the Kandiah processor demonstrates significantly better performance due to its more efficient pulse pile-up and pulse restoration circuitry.

(*d*) *ADC and MCA*. Data from the main pulse processor are read by an *analogue to digital converter* (ADC) and accumulated in a *multi-channel analyser* (MCA). The design principles of these modules have been considered in section 12.5.3. Spectra are recorded on a histogram of channel number (i.e. photon energy) *v*. channel count (i.e. x-ray intensity). *Dead time* contributions from the pulse processor must be gated with additional dead time from the MCA for the accurate assessment of live time. Bloomfield *et al.* (1983) have critically investigated the reliability of dead-time corrections applied to ED counting systems, particularly those contributions from photons below 1500 eV that may go unrecognized on some systems.

9.5 Interaction of x-rays with the silicon detector

9.5.1 *Primary interactions*

Before considering the spectra of real samples, it is necessary to consider how mono-energetic x-rays interact with the Si(Li) detector. The three principal interactions involved, photoelectric absorption, Compton scatter and Rayleigh scatter, have been discussed elsewhere (sections 8.6 and 12.6). The relative probability (cross-section) of each of these interactions varies as a function of photon energy, data appropriate to silicon being plotted in Figure 9.17.

(*a*) *Photoelectric absorption*. This is the most probable interaction for photons of energy below about 50 keV. In this process, the entire energy of the photon is absorbed within the detector, usually by interaction with an inner (*K*-shell) electron of a silicon atom. This electron is ionized from the atom and rapidly loses its excess kinetic energy by collision with other silicon atoms causing further ionization. The original atom de-excites, by emitting either a characteristic Si

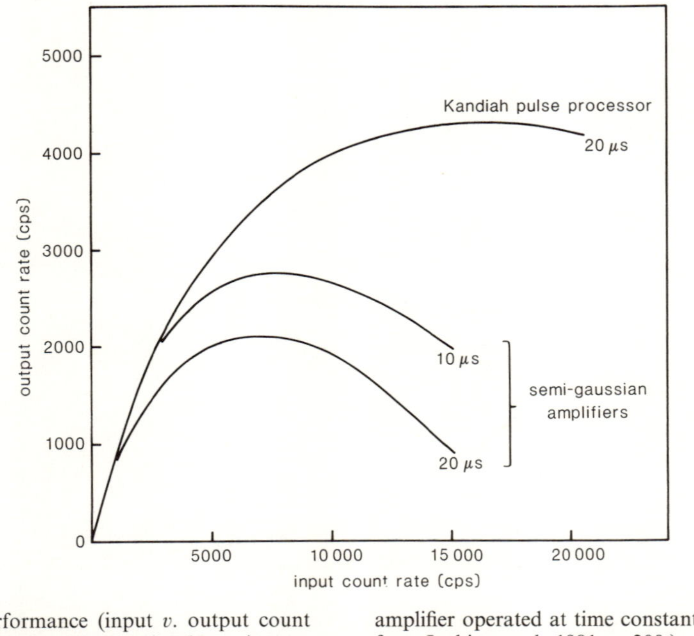

Figure 9.16 Pulse processor performance (input *v*. output count rates) of a Kandiah type pulse processor operated at 20 μs time-to-peak (taken from Link Systems) and a conventional semi-gaussian amplifier operated at time constants of 10 and 20 μs. (Data taken from Jenkins *et al.*, 1981, p. 200.)

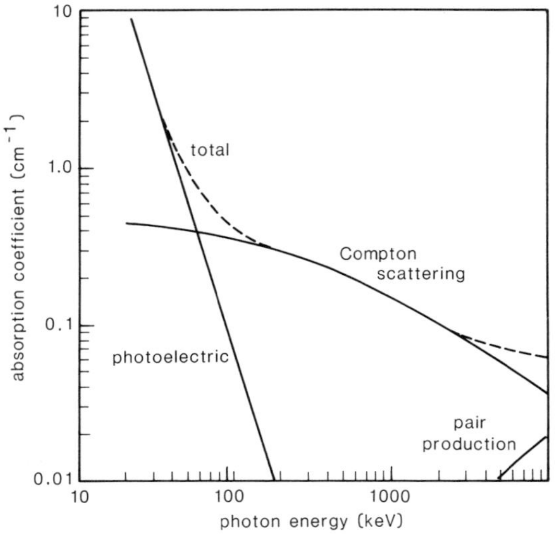

Figure 9.17 Linear absorption coefficient in silicon as a function of photon energy. (Grodstein, 1957.)

fluorescence x-ray or alternatively Auger electrons (section 8.6). Both have a high probability of being absorbed within the active volume of the detector. If this occurs, the full energy of the incident photon is detected. However, there is a significant probability that the Si $K\alpha$ x-ray will escape from the detector without further interaction. The energy of the incident photon less 1.74 keV, the energy of a Si $K\alpha$ x-ray, is then deposited in the detector. This event is, therefore, recorded as an escape peak, having an energy of 1.74 keV below the full energy peak. The magnitude of this escape peak varies as a function of incident photon energy, as shown in Figure 9.18. The highest probability is observed in the detection of P $K\alpha$ photons, the corresponding escape peak having a value of about 1.3% relative to the full energy peak. For photons above 10 keV, escape peak intensity is less than 0.1% of the full energy peak. Silicon escape peaks are not as prominent as those observed in germanium detectors, as discussed further in section 12.6. Photons below the energy of the Si K-shell absorption edge (1.8 keV) cannot excite Si K radiation and do not, therefore, generate escape peaks.

(*b*) *Compton scatter*. This occurs when part of the photon energy is absorbed, normally by ionization of an outer orbital electron of a silicon atom, the remainder being re-irradiated as a secondary x-ray. If this secondary x-ray escapes from the detector, only part of the incident photon energy is deposited in the detector. The event will then randomly contribute to the background spectrum at energies lower than the principal photopeak. Forunately, this interaction has a very low probability for the detection of 0–20 keV photons in comparison with the photoelectric effect (Figure 9.17).

(*c*) *Rayleigh scatter*. These interactions can also occur in the detector, but since incident photons are re-irradiated without transfer of energy, this process does not contribute to radiation detection.

9.5.2 *Spectrum artefacts*

In addition to the primary interactions listed above, several spectrum artefacts occur in spectra recorded on Si(Li) detectors and are particularly apparent in the spectrum of a monochromatic x-ray source such as ^{55}Fe (Figure 9.19).

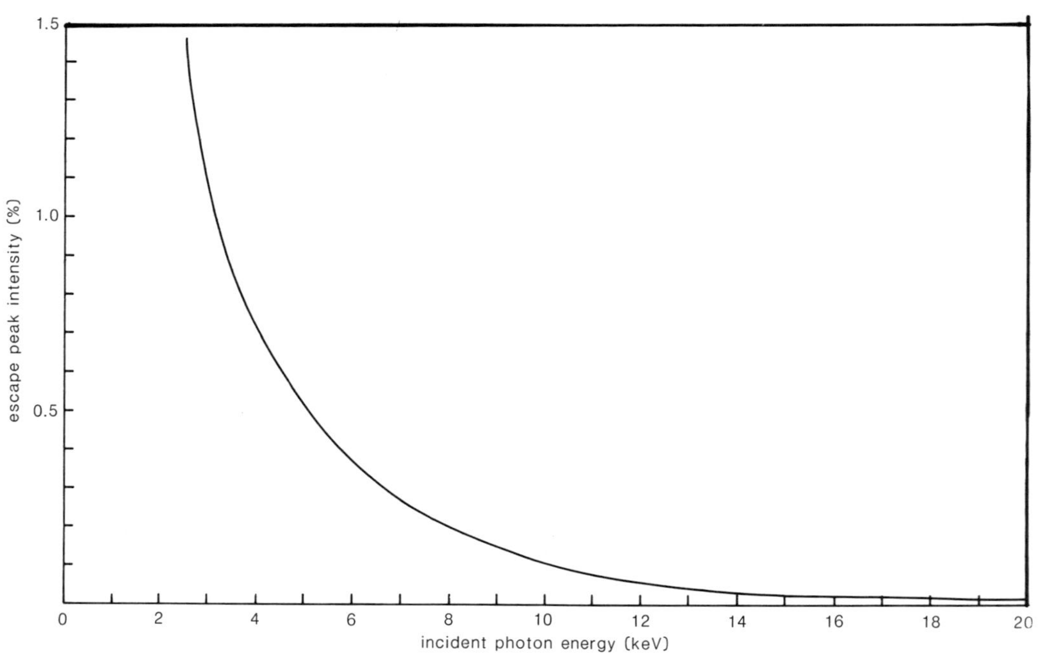

Figure 9.18 Intensity of silicon escape peak relative to that of the incident photon as a function of energy. (Data mainly from Woldseth, 1973, p. 2.12.)

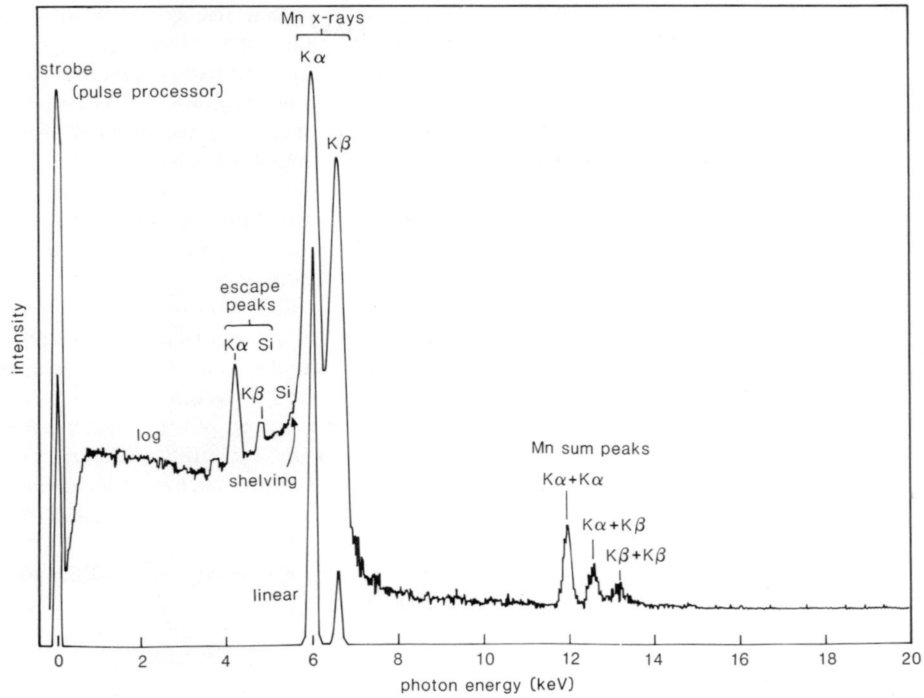

Figure 9.19 Spectrum of an ^{55}Fe source recorded on a Si(Li) detector plotted on (i) a log scale to demonstrate the low-intensity artefacts, and (ii) a linear scale to indicate relative intensities.

Artefacts are secondary effects arising from the design or construction of the detector crystal and the properties of pulse processing electronics.

A ^{55}Fe source emits monochromatic x-ray radiation corresponding to the manganese $K\alpha$ and $K\beta$ lines at 5.894 and 6.489 keV respectively. This source does not itself emit continuum radiation. The ^{55}Fe spectrum shown in Figure 9.19 shows several features, other than the full energy manganese photopeaks, common to all spectra recorded on Si(Li) detectors:

(i) *Escape peaks* are clearly visible at 4.155 keV (Mn $K\alpha$, 1.739) and 4.750 keV (Mn $K\beta$, 1.739). The Si $K\beta$ escape peak is not distinguished as it constitutes only 2% of the $K\alpha$ line.

(ii) The additional peak at 0 kV in Figure 9.19 is the pulse processor *zero strobe pulse*, which is conveniently offset to 0 keV. This apparent zero energy line is only observed in spectra recorded using a Kandiah-type pulse processor.

(iii) The Mn $K\alpha$, $K\beta$ lines are almost Gaussian in shape but show some tailing or *shelving* on the low energy side due to events that did not quite reach the full energy peak due to the charge collection problems already discussed in section 9.2.2.

(iv) Discrete pile-up peaks are observed at 11.788, 12.383 and 12.978 keV due to coincident detection of manganese ($K\alpha+K\alpha$, ($K\alpha+K\beta$) and ($K\beta+K\beta$) events, respectively. The resolution of these discrete pile-up peaks is not generally as good as that of a full energy photopeak: at the limit of pile-up resolution of the pulse processor, a small time lapse can occur between two pile-up events causing imperfect summing of the signals. The intensity of the pile-up is dependent on the resolving time of the pulse pile-up rejector (often 200 ns) as well as the input count rate.

(v) The background is enhanced at energies below the full energy peaks. The main cause of this phenomenon is the gross loss of charge from manganese photons that interact in the non-uniform electric fields near the edge of the detector crystal.

(vi) The relatively low, but significant background at energies higher than the principal x-ray lines arises from pulse pile-up of degraded full-energy events.

One further artefact may be apparent in ED spectra: a discrete Si $K\alpha$ line at 1.74 keV. This peak is associated with the detector dead layer: if photoelectric absorption of an incident photon occurs within this dead layer on the front face of the detector, then any ionization will almost certainly be trapped in this inactive region. However, there is a 50% probability that an associated Si $K\alpha$ x-ray fluoresced in this layer will penetrate deeper into the crystal and be detected as an independent event. Fortunately the probability of interaction in the dead layer is very low. However, this peak will sometimes be observed in spectra recorded on older detectors and is particularly noticeable in the analysis of metallic phases which are not themselves expected to contain silicon. These and other spectrum artefacts are reviewed in detail by Van Espen *et al.* (1980).

9.6 Comparison of ED and WD spectrometers

Having considered the basic response of an energy dispersive detector, it is appropriate to compare the salient analytical characteristics with those of a wavelength dispersive spectrometer.

Wavelength dispersive spectrometer	Energy dispersive spectrometer
Exceptional energy resolution below 10 keV	Energy resolution advantage above about 15 keV
Spectrometer behaves as a monochromator	All x-ray photons emitted from the sample may be detected
Maximum count rate at an individual wavelength is about 5 to 7×10^5 cps	Maximum count rate for entire spectrum is 3000–4000 cps (10 000 cps with a pulsed tube excitation source)
Very precise measurements of individual lines	Very rapid qualitative identification of x-ray lines present in spectra
Background requires separate spectrometer measurement	Background recorded simultaneously
Few serious spectrum overlap interferences in normal samples—corrected by simple ratio calculations	Many significant overlap interferences but sophisticated deconvolution procedures are available to correct these
Principal spectrum artefacts are caused by multiple-order line diffractions	Several spectrum artefacts, the most serious of which are escape and pulse pile-up peaks
Small solid angle of x-rays detected from sample (often less than 0.3°)	Large solid angle of x-rays detected from sample (often 5 to 10°)
High-power x-ray tube (3 to 5 kW) required as excitation source with water cooling	Low-power x-ray tube (50 W) is adequate for direct excitation, air cooling
Pulse processing consists of a simple semi-gaussian amplifier operated at a time constant of 0.5 to 2 μs. Data accumulated in a simple single channel scaler using pulse height discrimination	Pulse amplifiers must operate at long time constants (20 μs time-to-peak); sophisticated pulse amplification and pile-up rejection circuitry required: spectra accumulated in a 1024 channel multi-channel analyser
Precise mechanical alignment of spectrometer required with a sensitive servo-mechanical control system	No moving parts
Requires argon gas mixture for flow counters and cooling water services	Detector performance will probably be degraded if associated cryostat is ever allowed to run dry of liquid nitrogen
High capital cost	Moderate capital cost
Generally accepted for routine major and selected trace element determinations of the highest precision; qualitative analysis on older systems is inconvenient	The spectrometer of choice for exceptionally rapid qualitative analysis; quantitative performance viewed with unreasonable scepticism by some analysts in silicate rock applications

The comparison listed above shows that there are fundamental differences in the analytical characteristics of ED and WD spectrometers. To exploit the full potential of ED detectors in XRF analysis, it is therefore necessary to re-evaluate analytical strategy. Particular attention is required to mitigate the effects of inferior resolution at low energies and limited maximum count rate capability of ED detectors. Conversely, the much lower tube power required to excite ED spectra permits a more adventurous consideration of excitation geometries, including secondary target sources.

In the following sections, therefore, the capabilities of energy-dispersive spectrometers for quantitative analysis by x-ray fluoresence will be considered using the following modes of excitation:

(i) Direct tube excitation and its application in the routine analysis of major and trace elements on conventional glass fusion discs and pressed powder pellets (sections 9.7–9.9)

(ii) Secondary target excitation, particularly in the analysis of selected ranges of trace elements (section 9.10.2)

(iii) Polarized excitation geometries (section 9.10, 9.11)

(iv) Radioisotope excitation techniques (section 9.12).

9.7 Silicate rock analysis by ED-XRF using direct tube excitation

9.7.1 *Instrumental considerations*

(*a*) *Excitation geometry*. Direct tube excitation is the most flexible excitation technique in the analysis of geological samples for all the major elements as a group and all the trace elements as a group using an ED spectrometer. This flexibility results from the ability to excite a wide range of elements using both characteristic and continuum tube lines at operating potentials selected by the user. The excitation geometry is very similar to conventional WD-XRF instrumentation and is shown in Figure 9.20. The x-ray tube and detector are arranged to subtend an angle of about 90° at the sample to minimize Compton scatter of tube lines into the detector (Figure 8.18). Because of the large solid angle of x-rays that may be accepted for detection and the relatively low maximum count-rate capability, tube power requirements are considerably lower than in WD spectrometry. Whereas a 3000 W generator is standard equipment of WD spectrometer systems, ED-XRF will operate quite satisfactorily with a 50 W x-ray tube. Since power dissipation requirements are then so low, air cooling is adequate and generators may be constructed using solid-state circuitry which more readily satisfies specifications for high stability.

(*b*) *Transmission target tubes*. Spectrometers designed for direct tube excitation such as that in Figure 9.20 normally incorporate a conventional (low-power) side-window tube. However, a more specialized design has some restricted application. This is the transmission target tube (Figure 9.21). The transmission target tube operates in the conventional way, except that the target consists of a metal foil. Characteristic radiation is excited in this foil by the accelerated beam of electrons. Samples are excited by radiation that penetrates through the foil. The target, therefore, serves to filter its own photon output. In this way, much of the low energy bremsstrahlung is filtered from the useful x-ray beam by attenuation within the foil. In operation, a transmission target tube behaves as a source incorporating its own 'regenerative monochromatic filter' (see 9.7.3(*d*)), but without the flexibility for low-energy continuum exci-

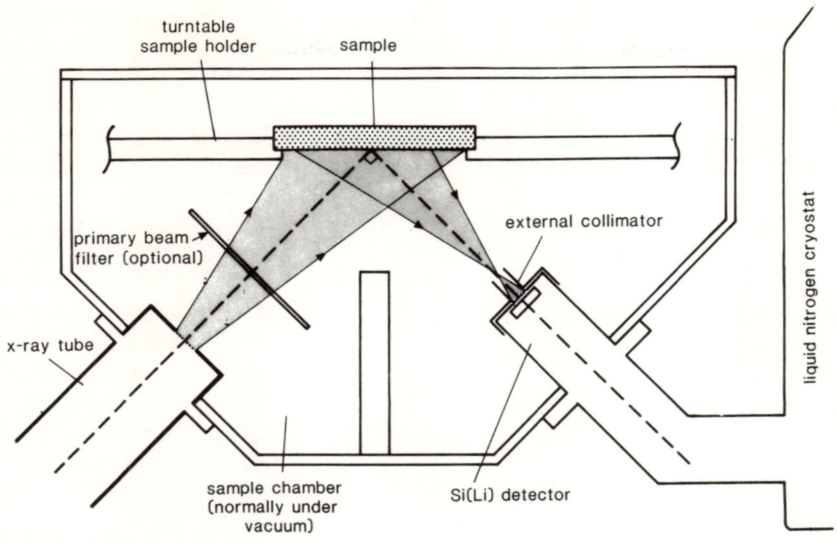

Figure 9.20 Overall view of excitation geometry in an ED–XRF spectrometer showing configuration of x-ray tube, sample and detector.

tiation. Owing to this lack of flexibility, transmission target tubes are only suitable in applications where analysis of a fixed range of elements is required.

(*c*) *Pulsed tube excitation.* A further refinement is possible with low-power x-ray tubes—that of pulsed operation. The tube is designed with an additional grid placed between filament and target. For convenience, this has been incorporated in the transmission tube shown in Figure 9.21, but could equally apply to a side-window tube. Power supplies are arranged so that an appropriate negative bias can be applied very rapidly to this grid. When this occurs, electrons emitted by the filament are deflected and no longer reach the target. The x-ray output is effectively switched off for the duration of the grid pulse. Removal of the grid bias causes the tube to revert to normal operation.

The benefit of this mode of operation is in increasing the output count-rate capacity of the detector by a factor of between two and three, as follows. As count rates increase above about 3000 cps, there is an increasing probability of pile-up in the pulse processor. Thus any event detected within a time interval of (typically) 20 μs of the first will cause the pile-up rejection circuits to inhibit an output from the amplifier. This causes a large increase in dead time and eventually the output count rate will begin to *decrease* with increasing input count rates as the pulse processor becomes saturated with pile-up events (Figure 9.16). In the pulsed tube mode of operation, the tube grid bias supply is controlled by the main pulse processor. As soon as an x-ray event is detected, the grid is pulsed, so switching off the excitation source. After amplification and integration of this signal, the pulse processor circuits cannot be reset to accept further photon events until the input signal level has fallen below the threshold set by the lower-level discriminator. The probability that this condition will be satisfied in pulsed tube operation is very much higher than in continuous (dc) tube excitation mode. In this way the contribution of extendable dead time (section 8.9) is significantly reduced, permitting a much higher rate of signal throughout. After a suitable delay interval following signal integration, the grid bias is removed from the x-ray tube and the cycle restarted.

This approach was described by Jaklevic *et al.* (1972, 1976) and the improved count rate performance has been considered by Sandborg and Russ (1977). Using an amplifier with a time-to-peak integration value of 20 μs, the rate of data accumulation may be increased by a factor of about three using this pulsed tube facility, typically from 30 000–40 000 cps (unpulsed) to 10 000 cps (pulsed).

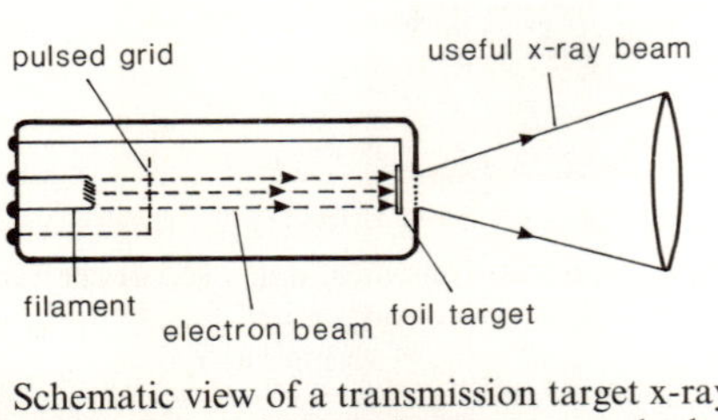

Figure 9.21 Schematic view of a transmission target x-ray tube. An additional grid has been added to demonstrate pulsed tube operation (see text).

9.7.2 *Excitation strategy*

The strategy for exciting samples in ED-XRF analysis is quite different from that used in conventional WD analysis. Excitation conditions must take into account both the more limited count-rate capability of ED detectors and the characteristic that all x-ray photons emitted from the sample, including scattered tube lines, have the potential of being detected. If the relatively high operating potentials used for WD analysis (40 to 60 kV for major, 60 to 100 kV for trace elements) were directly transferred to an ED system, a large proportion of the pulse processing capacity would be

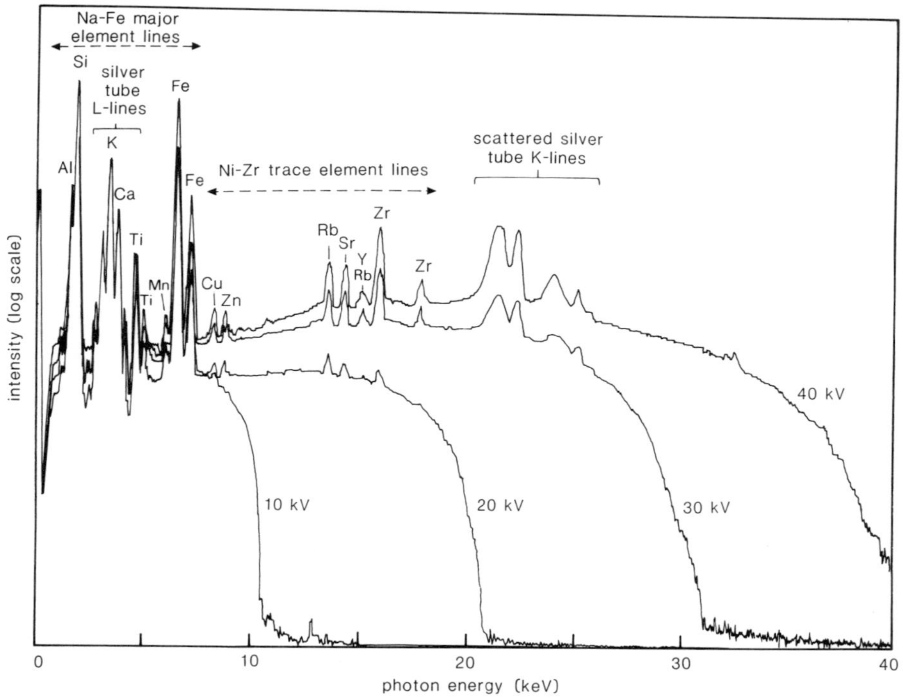

Figure 9.22 Spectra of a granodiorite USGS: GSP-1 (prepared as a powder pellet) excited by a silver tube operated at 10, 20, 30 and 40 kV. Intensity is plotted on a log intensity scale.

expended in accumulating unwanted tube lines scattered off the sample.

This problem is illustrated in Figure 9.22, which shows spectra recorded from a powder pellet of granodiorite USGS: GSP-1 excited at a range of potentials between 10 and 40 kV with a silver tube. Spectra are displayed on a log-intensity scale to exaggerate the contribution of scattered tube continuum to recorded spectra. Data plotted in Figure 9.23 show the proportion of time expended by the ED spectrometer in accumulating:

(i) Fluorescence x-ray lines of major elements (Na–Fe) excited in the sample

(ii) Fluorescence x-ray lines of the heavier trace elements (Cu–Zr) excited in the sample

(iii) Characteristic and Compton scatter tube lines scattered off the sample

(iv) Tube continuum radiation scattered off the sample.

These results illustrate the special considerations that must be given in exciting major and trace element spectra for ED-XRF analysis, namely:

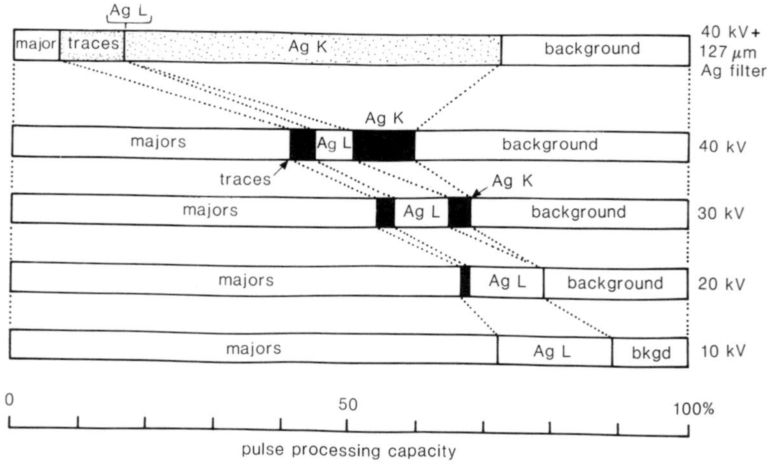

Figure 9.23 Pulse processor utilization factors for the spectra plotted in Figure 9.22, showing the proportion of total pulse processing capacity used in accumulating events due to (*a*) major element fluorescent x-rays: (*b*) trace element fluorescent x-rays (8–20 keV region); (*c*) silver tube *L*-lines; (*d*) silver tube *K*-lines; (*e*) background from scattered tube continuum lines. No upper level discriminator was set in recording these data. The upper bar plot shows equivalent data for the same sample excited at 40 kV with a silver filter in the primary beam. Although the overall proportion of detected continuum data is much larger under these conditions, so is the excitation of heavier trace elements.

(i) The major elements must be excited at about 10 kV, since higher tube potentials cause an unacceptable increase in unwanted tube continuum radiation scattered into the detector

(ii) Tube potentials in excess of 40 kV are necessary to excite adequately the heavier trace elements (Cu to Zr in Figure 9.22); however, unless other action is taken, analytical performance will be degraded by the large proportion of continuum radiation then scattered into the detector.

9.7.3 *Optimization conditions for routine major and trace analysis*

(*a*) *Selection of tube anode.* Selection of tube anode material follows similar considerations to those discussed for wavelength dispersive analysis in section 8.9. However, to take advantage of the characteristic of ED spectrometers to detect a range of x-ray photon energies in single spectra, it is desirable to select tubes that will efficiently excite the major elements as a group and the heavier trace elements as a group. Tubes with rhodium or silver anodes most closely satisfy this criterion in routine silicate applications.

(*b*) *Routine major element analysis.* Sample preparation requirements for major element determination, avoiding mineralogical effects, are identical to those discussed for wavelength dispersive analysis in section 8.10. It is, therefore, necessary to prepare samples as glass fusion discs. The important difference is that heavy absorber fluxes cannot be used since excitation of lanthanum *L*- and *M*-lines from this flux would obscure an important region of the major-element spectrum.

The requirement is, therefore, to excite effectively the elements sodium (*K*-absorption edge = 1.071 keV) to iron (*K*-absorption edge = 7.110 keV). A tube potential significantly greater than 7.11 keV is therefore required if all the major elements are to be excited as a group. Primary beam filters cannot be used to suppress tube continuum scattered into the detector, since it is the continuum component that is largely responsible for exciting the lightest major elements. Thus, although the use of cellulose (or paper) primary beam filters has been suggested by some workers to 'clean up' the low-energy tube continuum spectrum this practice cannot be followed if sodium and magnesium, in particular, are to be determined. A useful contribution to the excitation of the lightest elements is obtained from the characteristic *L*-lines at around 3 keV if silver or rhodium anode tubes are used as excitation source. However, experience suggests that this coupling is only effective in the excitation of phosphorus and, to a lesser extent, silicon. The factor controlling the excitation efficiency of aluminium, magnesium and sodium is the thickness of beryllium window fitted to the tube and therefore the intensity of low-energy continuum able to reach the sample. In this respect tubes with ultra-thin (50 μm) windows are far superior to general-purpose tubes having 200 μm beryllium windows.

Bearing these considerations in mind, the effect on the *proportion* of selected major element x-rays recorded in the spectra of an andesite, USGS: AGV-1, excited with a silver anode side-window x-ray tube operated at between 6 and 20 kV is shown in Figure 9.24. These data show that as, expected from absorption edge considerations, iron is not

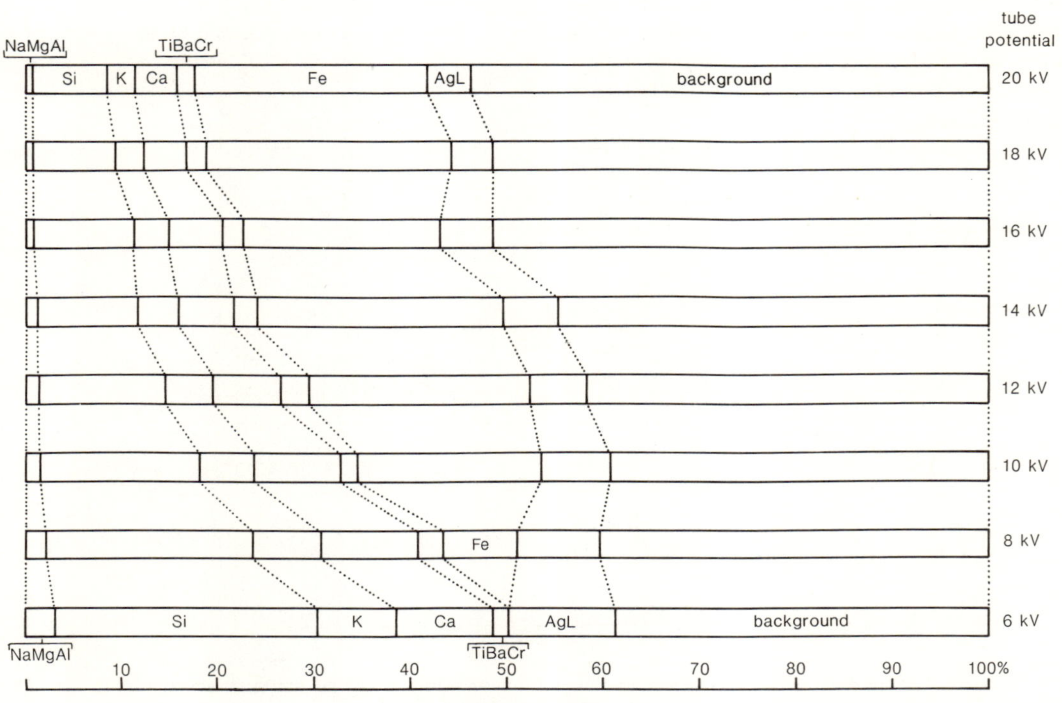

Figure 9.24 Pulse processor utilization for a glass fusion disc of andesite USGS: AGV-1 excited by a silver tube (no primary beam filter) operated at 6–20 kV. These bar plots show the proportion of pulse processor capacity expended in accumulating specified major element x-ray lines, scattered silver-tube *L*-lines and background counts derived mainly by scatter of tube continuum off the sample.

excited at all at 6 kV. However, above about 12 kV, iron is excited so efficiently by the silver tube that the accumulation of iron *K* x-rays begins to monopolize, to an unacceptable extent, a large proportion of the available pulse processing capacity. It is therefore suggested that a tube potential of between 10 and 12 kV offers the best balance between the excitation of iron and other major elements. Data in Figure 9.24 also demonstrate the large proportion of pulse processing capacity expended in recording scattered tube continuum events, significantly larger than that observed from a powder pellet because of the lower mean atomic number of lithium borate glass fusion discs. It is relevant to note that approximately 50% of all recorded events are unwanted scattered tube *L*-lines or continuum lines. The proportion of data accumulated at 10, 14 and 20kV in AGV-1 is as follows:

Tube potential	10 kV	14 kV	20 kV
Fe *K*-lines	19%	26%	24%
K-lines of other major elements	35%	24%	12%
Unwanted continuum and scattered tube lines	46%	50%	58%

The spectrum of andesite, USGS: AGV-1, excited with a silver tube under compromise conditions at 10 kV, is plotted in Figure 9.25.

(*c*) *Routine trace element analysis*. The trace elements Rb to Nb (*K*-lines) and Pb, Th and U (*L*-lines) in the range 11–18 keV are efficiently excited by the characteristic *K*-lines from a silver or rhodium anode x-ray tube (Rh–K: 20 to 23 keV; Ag–K: 22 to 25 keV). In order to excite these tube lines effectively, operating potentials in excess of 40 kV are required. However, as has been seen in the previous section, the penalty for operating under such conditions is that an increasing proportion of the pulse processor capacity is expended in analysing unwanted tube continuum scattered off the sample. The background under fluorescent lines of interest is also enhanced by scattered tube continuum radiation, seriously degrading detection limits. It is, therefore, essential to examine the improvement that may be achieved by modifying the energy distribution of tube radiation reaching the sample using metal foil primary beam filters.

(*d*) *Primary beam filters*. Filters for low-power ED tubes are made of metal foils, usually 10 to 150 μm thick, placed in the primary x-ray beam between tube and sample. They are particularly important in ED–XRF analysis because of the significant improvement that can be gained in peak-to-background ratios in the determination of the heavier trace elements. The influence of primary beam filters has been considered in some detail by Gedcke *et al.* (1977).

Selection of filter type depends on excitation source and the range of elements to be measured. The aim is normally to choose a filter that will:

(i) Transmit a large proportion of higher energy tube radiation (usually including the characteristic as well as higher-energy continuum radiation) that is capable of exciting the elements of interest

(ii) Attenuate the lower-energy tube continuum radiation that would otherwise be available for scatter off the sample and contribute to the background under the fluorescence lines of interest.

The effect of selected filters (150 μm Al, 50 μm Mo, 50 μm Ag and 127 μm Ag) on the spectrum recorded from a pelletted sample of granodiorite, USGS: GSP-1, excited with a silver anode x-ray tube operated at 40 kV, is shown in Figure 9.26. The behaviour of filters may be categorized as follows.

White filters are made of a material which has no absorp-

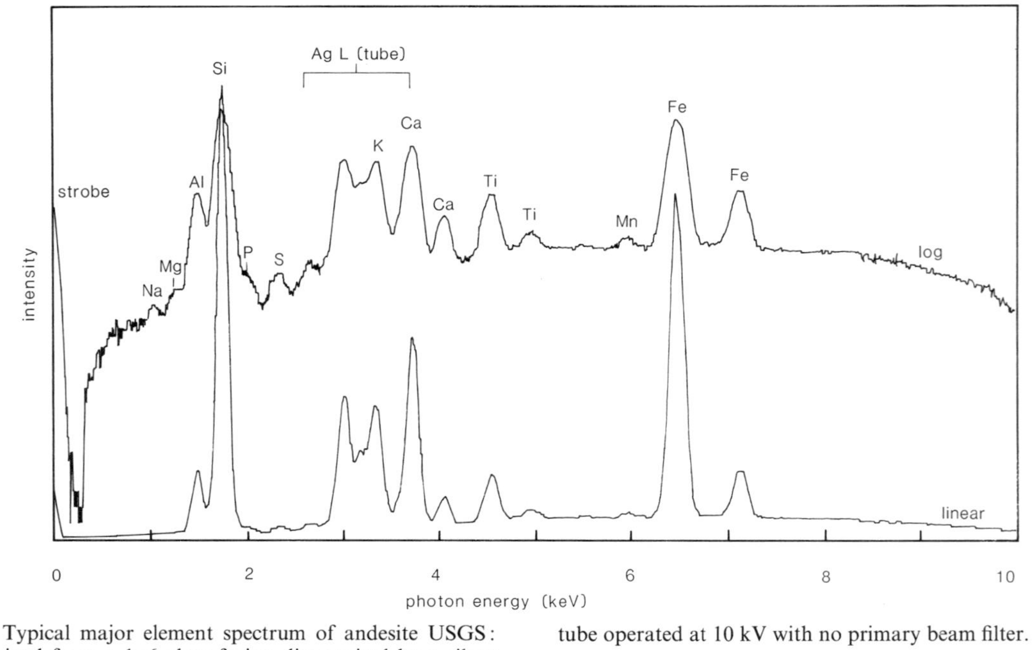

Figure 9.25 Typical major element spectrum of andesite USGS: AGV-1 obtained from a 1:6 glass fusion disc excited by a silver tube operated at 10 kV with no primary beam filter.

tion edge in the region of interest. X-ray energy from the tube is therefore continuously attenuated, such filters having the largest influence on the intensity of lower-energy tube photons. One use of these filters is to remove characteristic tube lines from recorded spectra. Thus in Figure 9.26*b*, the 50 μm aluminium filter effectively removes silver tube *L*-lines from the tube spectrum. These cannot then be scattered from the sample into the detector. At the same time all of

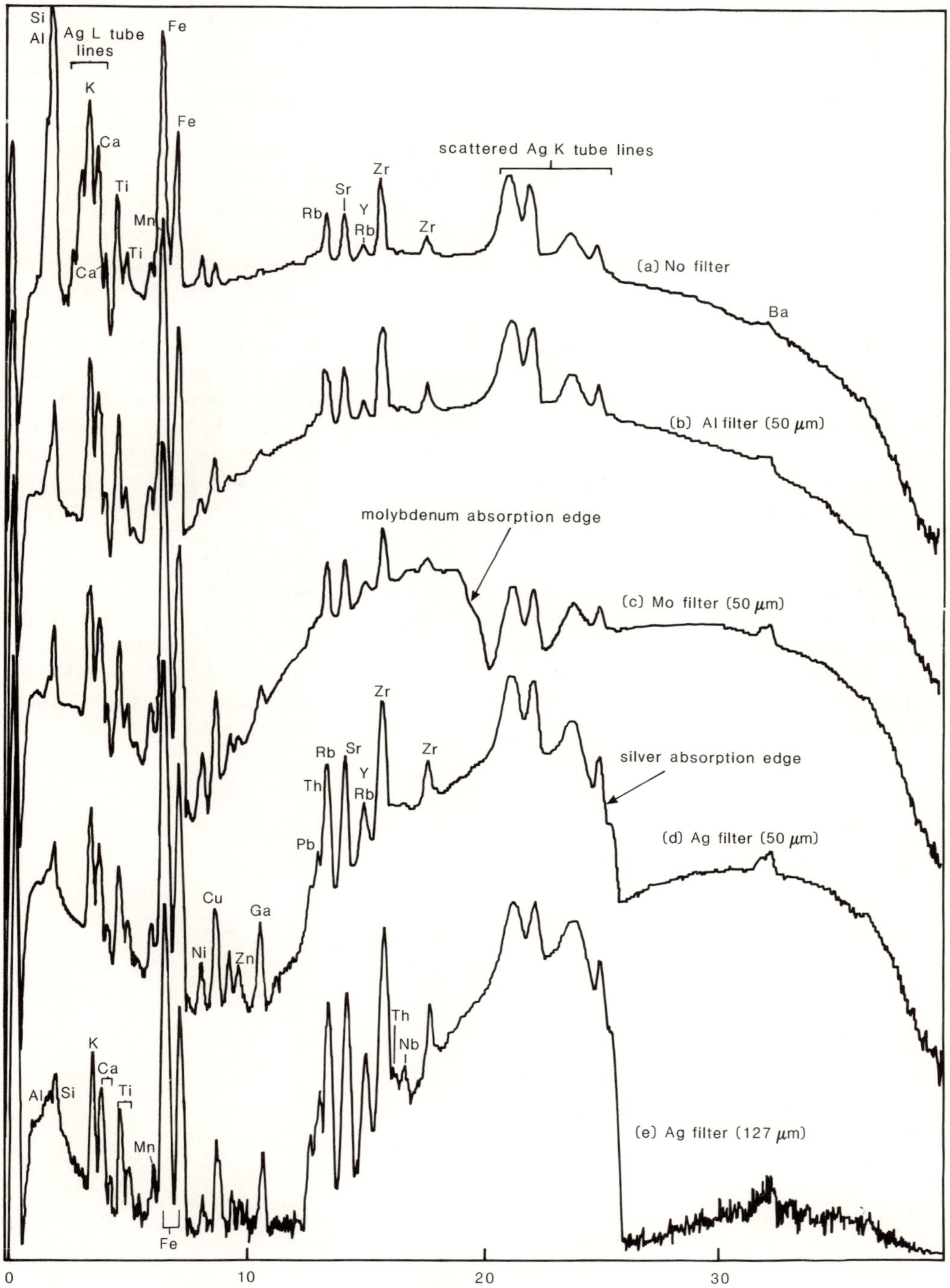

Figure 9.26 Effect of primary beam filters on x-ray spectra recorded from granodiorite USGS: GSP-1 (powder pellet) excited with a silver tube operated at 40 kV. Resultant spectra were recorded with (*a*) no filter (note universally high continuum); (*b*) 50 μm aluminium filter (note reduction in background below 10 keV and absence of scattered silver tube *L*-lines); (*c*) 50 μm molybdenum filter (note absorption edge at 20.0 keV and general reduction in background below about 8 keV); and (*d*) 50 μm and (*e*) 127 μm silver filter (note reduction in background below about 18 keV and absorption edge at 25.5 keV).

the low-energy tube continuum is attenuated severely, reducing the excitation of the lightest elements (Na–Si). An aluminium filter 50 μm thick has little influence on tube radiation above 15 kV, fluorescence spectra in this region being very similar to those recorded with an unfiltered tube spectrum. A further example of white radiation filters is the determination of manganese using a chromium tube in which a 250 μm aluminium filter is normally used to remove a scattered Cr $K\beta$ interference (section 8.9.1).

Absorption edge filters use the discontinuous attenuation across the characteristic absorption edge of the filter material in two ways: first, x-ray tube radiation at energies immediately *above* the absorption edge, and also some way below it, is significantly attenuated, and two windows are therefore available in spectra where scattered tube backgrounds are at low intensity; second, conversely, tube energies immediately *below* the absorption edge (and also some way above it) are efficiently transmitted, and elements with absorption edges immediately below these ranges are, therefore, efficiently excited. The effect of a 50 μm molybdenum filter on the spectrum of granodiorite, USGS:GSP-1 (Figure 9.26*c*) shows a region of low background in the range 5–10 keV. Tube continuum from about 12 keV up to the molybdenum absorption edge at 20.00 keV is effectively transmitted as indicated by the intensity of continuum scattered off the sample. Under these conditions favourable excitation of the Ti–Ga group of elements may be achieved if only a silver anode tube were available as excitation source.

A special case of the absorption edge filter is the so-called *regenerative monochromatic filter*. This filter is a foil of the same material as the x-ray tube anode. Hence the characteristic x-ray tube lines are efficiently transmitted, but higher-energy radiation is strongly absorbed. It is argued that this absorption will cause fluorescence of the silver foil, increasing the intensity of Ag K radiation available to excite the sample. Similar arguments should apply to the use of molybdenum filters. However, in view of the lack of any scattered Mo K lines (Figure 9.26c), it is thought that filter fluorescence has an insignificant effect on sample excitation compared with the much larger flux of primary tube photons.

The second important characteristic of the regenerative monochromatic filter is that tube continuum at some distance below the absorption edge is strongly attenuated. Under these conditions, therefore, optimum excitation of the elements Cu to Nb (8 to 17 keV) can be achieved with a silver tube (Figure 9.26*d*, *e*). Silver tube plus silver filter or rhodium tube plus rhodium filter is a powerful combination for the analysis of trace elements in the range Cu to Nb by ED–XRF (see Figure 9.23).

The effect of silver filters of different thicknesses, 50 μm and 127 μm, on the spectrum of granodiorite: USGS: GSP-1 excited with a silver tube at 45 kV is compared in Figure 9.27 with an unfiltered spectrum. Each spectrum has been normalized to approximately equal Zr $K\alpha$ photopeak areas. A marked reduction in background caused by scatter of tube continuum radiation is observed in the filtered spectra. The 127 μm silver filter was selected for the analysis of these heavier trace elements by Potts *et al.* (1984).

(*e*) *Optimizing tube potential for trace element analysis.* In the analysis of geological samples for trace elements, optimizing detection limits is generally the main consideration. If peak-to-background ratio is taken as a suitable criterion, the variation of this parameter with tube potential is shown in Figure 9.28 using a silver tube and 127 μm silver primary beam filter. These data follow a general trend indicating that optimum excitation is achieved at the upper end of the kV range. This may in part be caused by a corresponding shift in tube continuum intensity to higher ener-

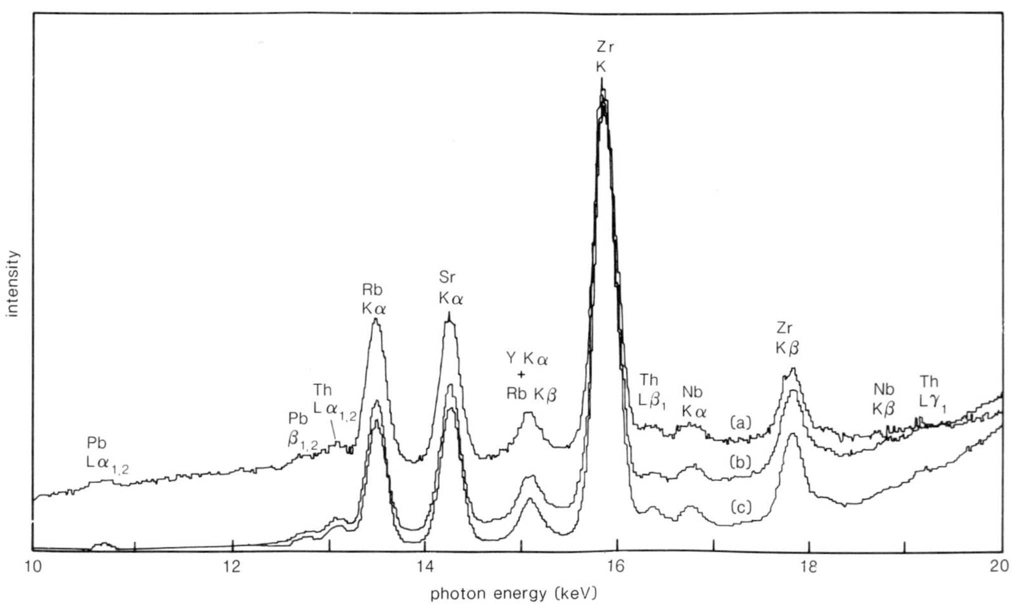

Figure 9.27 X-ray spectra (10–20 keV) of granodiorite USGS: GSP-1 excited by a silver tube operated at 45 kV using the following primary beam filters: (*a*) none, (*b*) 50 μm Ag, (*c*) 127 μm Ag. Each spectrum is plotted such that the Zr $K\alpha$ line is of approximately equal area, to demonstrate the effect of filters on suppressing the background. (After Potts *et al.*, 1984.)

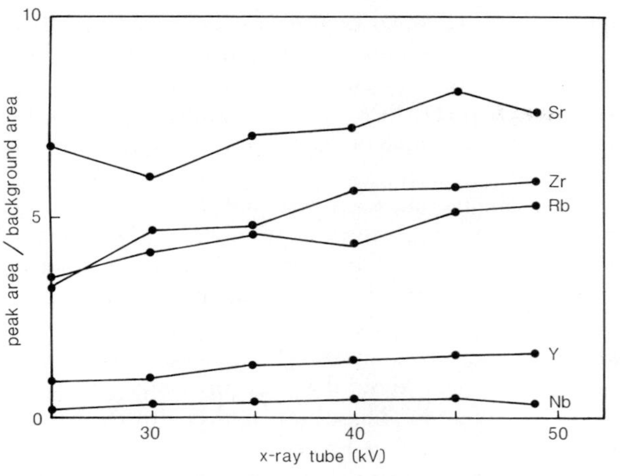

Figure 9.28 Variation in peak-to-background ratio as a function of x-ray tube kV for selected trace elements. Measurements were made on a powder pellet of USGS: GSP-1 (granodiorite) using a silver tube and 127 μm silver primary beam filter. Each spectrum was recorded at constant count time (200 s) and constant dead time (35%). (After Potts *et al.*, 1984.)

gies and in part be due to the more efficient excitation of silver tube *K*-lines.

It is therefore proposed that excitation of powder pellets with a silver (or rhodium) tube operated at more than 40 kV in conjunction with the appropriate regenerative monochromatic filter will permit the sensitive determination of the elements Ni, Cu, Zn, Ga, Rb, Sr, Y, Zr, Nb, Pb and Th. The effect on dead time and pulse processing capacity of the higher-energy scattered tube lines can be minimized by adjusting the pulse processor upper level discriminator threshold to about 20 keV. Unwanted higher-energy photons will then be recognized as out-of-range events and rejected before passing though the main amplifier channel of the 'Harwell processor' (section 9.4.3).

Determination of the trace elements that emit lines in the range 5 to 8 keV are not excited with sufficient sensitivity by direct tube excitation using silver or rhodium anode tubes. Of particular importance are the Cr *K*-lines, V *K*-lines and Ba *L*-lines, all of which are normally determined in routine WD–XRF programs. Recent work (Potts *et al.*, 1986) indicates that optimum excitation for these elements can be achieved by direct tube excitation using a cobalt anode x-ray tube. In order to suppress excess excitation of iron, it is necessary to filter tube radiation with 12.5 μm iron foil.

9.8 Spectrum analysis procedures

9.8.1 *General requirements*

As well as efficient excitation of the elements of interest, the reliability of spectrum deconvolution procedures in the accurate determination of photopeak areas is critical in quantitative ED–XRF analysis. Deconvolution of ED–XRF spectra is, in general, a much more demanding application than of spectra excited in electron microprobe analysis. The reason is that the ED–microprobe spectra of individual minerals often contain only a limited range of major elements and no 'trace' elements at analysable concentrations. In contrast, it is very unusual if XRF spectra do not contain photopeaks due to all the major elements (Na and heavier) and all trace elements present at concentrations greater than 2 to 10 ppm.

There are two important prerequisites to routine and reliable spectrum analysis in ED–XRF applications: one concerns the pulse processing electronics, and the other the peak-fitting algorithm.

(*a*) *Spectrum stability*. For any peak-fitting algorithm to work successfully, it is essential that drift in recorded spectra affecting both spectrum gain and FWHM peak resolution is minimized, with respect both to changes in input count rate and ambient temperature. In this respect pulse processors using the Kandiah design principles (section 9.4.3) have significant advantages in the protection afforded against drift in signal gain at high count rates. This is partially connected with the internal regulation afforded by the strobe pulse facility. In the real world, small drift effects will inevitably occur. It is necessary, therefore, that peak fitting algorithms have the flexibility to compensate for small changes in spectrum stability.

(*b*) *Peak-fitting algorithms*. Several peak-fitting algorithms have been sucessfully applied to the analysis of microprobe spectra, and will be considered in more detail in section 10.12. However, one of the few approaches that appears to be sufficiently robust to deal satisfactorily with the more complex interferences found in ED–XRF spectra is that based on digital filters and library spectra. This procedure is considered in more detail in the next section.

9.8.2 *Digital filter algorithms*

A digitized x-ray spectrum can be considered to incorporate three components having distinct characteristic frequencies:

(i) A high-frequency component caused by noise which appears as channel by channel fluctuations in spectrum intensity.

(ii) A low-frequency component representing the continuum background—to a first approximation, individual x-ray photopeaks can be considered to sit on the linear portion of a slowly curving background.

(iii) An intermediate frequency component corresponding to the x-ray photopeaks themselves.

A digital filter is an algebraic procedure used to separate the effect of (i) noise and (ii) continuum, so that (iii) photopeak areas may be unambiguously distinguished.

Digital filter procedures were first proposed by Schamber (1973) and have been further discussed by Schamber *et al.* (1977), Statham (1977) and McCarthy and Schamber (1981). The form of filter is a top-hat function (Figure 9.29). An envelope of adjacent data points is selected (15 in Figure 9.29). Each datum is multiplied by either +1 or −1 as indicated by the top-hat divison in Figure 9.29. The sum of this data array is then calculated, and the result, which represents the convoluted datum for the central point (asterisked in Figure 9.29) is passed over to the convoluted spectrum. The entire calculation is then repeated on the original data set with a one-channel increment on the top-hat array.

The effect of this digital filter on a simple gaussian peak is shown in Figure 9.30. The linear sloping background is

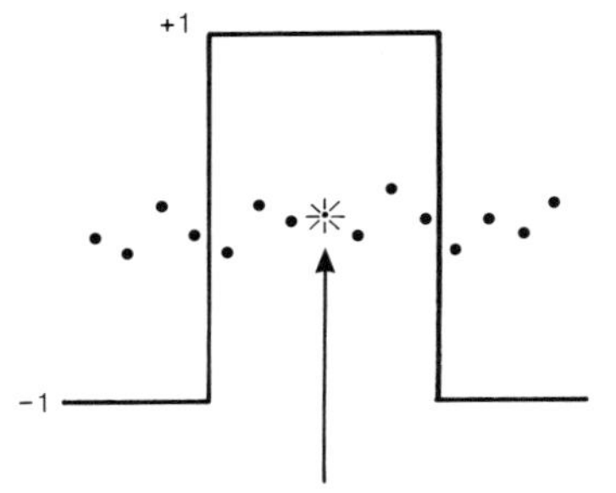

Figure 9.29 'Top-hat' digital filter as applied to a portion of an ED x-ray spectrum. The data points enclosed in the positive lobe of the filter are added, and from the sum is subtracted the sum of the data points in the negative lobes of the filter. The resultant is a 'filtered' count for the central (asterisked) channel. The filter function is then incremented by one channel and the cycle repeated.

entirely suppressed and the filtered spectrum is a smoothed function which approximates to the negative of the second derivative of the original data. The effect on an x-ray spectrum of coal ash is shown in Figure 9.31, taken from McCarthy and Schamber (1981). Background effects are entirely suppressed in the filtered spectrum, although it is no longer as simple to correlate visually the filtered photopeaks with the original x-ray lines.

Optimum dimensions of the filter have been investigated by Statham (1977) who considered that for spectra recorded at 20 eV per channel, the entire envelope should encompass 15 channels and the central envelope 7 channels, the latter being approximately equal to one FWHM.

The principle of the digital filter technique is that, having suppressed background and noise components, the intensities of photopeaks contributing to the digitally filtered sample spectrum may be quantified by the linear regression of library spectra. Each library spectrum consists of the spectrum of individual element *K*-, *L*- or *M*-lines, excited under similar conditions to the sample spectrum and convoluted by the same digital filter. A least-squares regression procedure is used to fit selected library spectra to the sample spectrum to determine the proportions of each that give the best data fit.

The technique has several advantages over some alternative procedures discussed in section 10.12. Variations in continuum intensity due to matrix between samples and standards are effectively suppressed. However, of greater importance is the fact that the procedure is not restricted to library spectra containing single photopeaks nor are any assumptions made about the shape or relative intensity of element photopeaks. Thus, the complete series of *K*-, *L*- or *M*-lines may be used as library spectra. In practical applications, this flexibility has considerable advantages, since a severe overlap affecting one photopeak in a multiple-peak library spectrum does not necessarily inhibit accurate fitting of the entire profile to a sample spectrum.

An important aspect of these procedures is that there should be adequate compensation for small discrepancies in both spectrum gain and FWHM resolution between library and sample spectra to ensure precise fitting. Practical aspects of preparing library spectra are considered further in the next section.

9.9 Routine analysis using direct tube excitation

9.9.1 *Schemes of analysis*

Several workers have investigated the analysis of geochemical materials by direct tube excitation procedures. Hebert and Street (1974) determined the major elements in rock samples prepared as glass discs by fusion in the ratio 1 : 10 with lithium metaborate flux. A highly specialized x-ray tube source was used which incorporated six selectable targets: magnesium (for the *K*-lines of O, F and Na), alu-

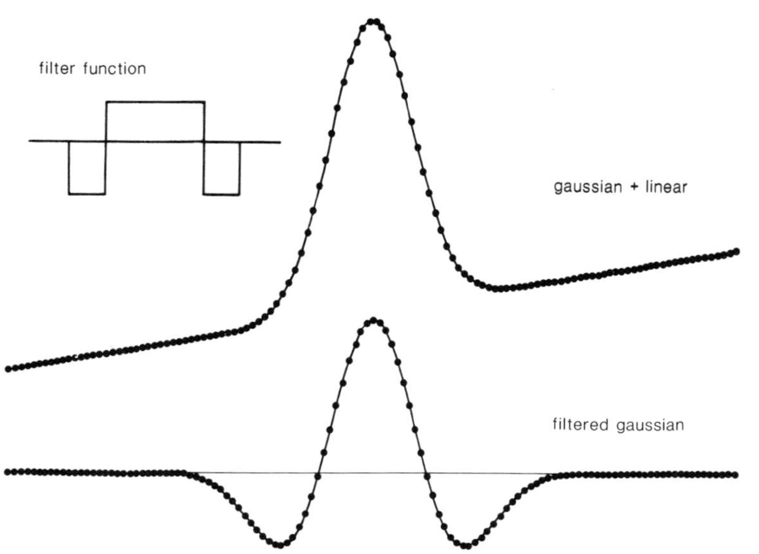

Figure 9.30 Effect of 'top-hat' digital filter on a Gaussian peak plus straight line background (upper spectrum) showing the convoluted result (lower spectrum). (McCarthy and Schamber, 1981.)

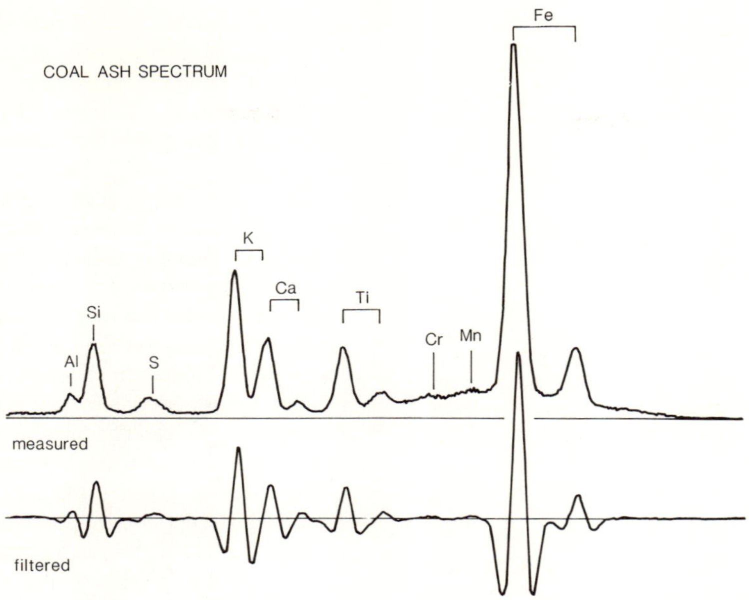

Figure 9.31 Effect of 'top-hat' digital filter on the x-ray spectrum of a coal ash sample. (McCarthy and Schamber, 1981.)

minium (O, Na, Mg), silicon (O, Na, Mg, Al), silver (Al, Si, P, S, Cl), titanium (Si Cl, K, Ca) and nickel (Cl, K, Ca, Sc, Ti, V, Cr, Cr, Mn, Fe). These target materials were bombarded with electrons accelerated to 7 kV (10 kV for Ni) and the resultant emission spectra filtered with a foil containing the same element as the target. Samples were, therefore, excited with radiation from each target in turn to optimize fluorescence of individual elements. In the analysis of selected USGS reference samples, standard deviations of better than 1% relative were generally observed at element concentrations exceeding 1% oxide. Reported analyses for granite, USGS: G-2 and basalt, USGS: BCR-1 are listed in Table 9.5. Analytical sensitivity corresponding to three standard deviations of the observed counting statistics for a five-minute analysis of samples prepared as glass discs varied from 0.09% for Na to 0.01% for Fe.

Giauque *et al.* (1977b) discussed a procedure for analysing trace elements in geochemical samples using a molybdenum transmission target tube and guard-ring type detector. the *K*-lines of the elements between Ti (4.5 keV) and Ba (32 keV) were analysed. Mahon and Leyden (1983) discussed the excitation of river sediments prepared as lithium tetraborate glass fusion beads using a pulsed silver x-ray tube and a 25 μm silver primary beam filter. The elements *K* (3.3 keV) to Zr (16 keV) were determined from spectra counted for 1000 s and excited at 30 kV. Potts *et al.* (1984) considered the excitation of conventional glass discs for ten major elements (Na to Fe) and pressed-powder pellets for ten selected trace elements using a pulsed silver tube and 500–800 s count times. Details of this procedure will now be considered.

9.9.2 *Analytical and instrumental requirements*

The procedure is based on the requirements to determine all XRF-analysable major elements (Na, Mg, Al, Si, P, K, Ca, Ti, Mn and Fe) on glass discs, and selected trace elements (Rb, Sr, Y, Zr, Nt, Pb, Th, Cu, Zn and Ga) on powder pellets. Sample preparation procedures, general principles of analysis and absorption correction algorithms are identical to those presented in Chapter 8. The specific requirements for ED instrumentation for the current procedures are:

(i) Sample excitation using a silver or rhodium anode side-window x-ray tube operated in pulsed mode (section 9.7.1 (*c*)).

(ii) A high-performance pulse amplifier incorporating time variant filters, a high degree of pulse pile-up protection, and internal stabilization against drift in signal gain, dc level or pulse resolution related to changes in detector count rate (section 9.4.3).

(iii) Spectrum deconvolution algorithms based on digital filtering and the fitting of library spectra (section 9.8).

9.9.3 *Detector precalibration*

In order to operate under the instrumental specification outlined above, there are several precalibration measurements which must be carried out to characterize the detector response.

(*a*) *FWHM variations.* The precise resolution of Si(Li) detectors increases slightly as a function of system count rate. It is, therefore, necessary to precalibrate this effect so that appropriate corrections may be applied to measured spectra to ensure precise match between sample and library spectra. This is most conveniently accomplished by recording a series of spectra of an element over a range of input count rates that exceeds those to be encountered during routine analysis. The precise FWHM is then recorded as a function of input count rate. In systems based on a pulse processor that generates an internal 'strobe' pulse, the area of the strobe peak can be measured as a convenient record of system count rate. Since the strobe pulse is generated at a precise frequency

Table 9.5 Energy dispersive XRF analysis of major elements in selected reference materials using direct tube excitation

	Basalt USGS: BCR-1						Granite USGS: G-2						NBS SRM-1645 River sediment		
	Gladney *et al.* (1983) Expected	CV	Hebert and Street (1974) Analysed	CV	Potts *et al.* (1984) Analysed	CV	Gladney *et al.* (1983) Expected	CV	Hebert and Street (1974) Analysed	CV	Potts *et al.* (1984) Analysed	CV	NBS Expected	Mahan and Leyden (1983) Analysed	
SiO_2	54.35	0.5	55.0	0.4	54.84	0.4	69.04	0.4	69.2	0.3	68.73	0.5			SiO_2
TiO_2	2.22	2.3	2.20	0.5	2.24	0.5	0.492	3.8	0.51	2.0	0.50	2.0			TiO_2
Al_2O_3	13.63	0.9	13.42	0.7	13.59	0.7	15.14	0.8	15.26	0.6	15.22	0.1			Al_2O_3
Fe_2O_3	13.46		13.40	0.08	13.31	0.4	2.68		2.67	0.5	2.70	0.8	16.2	13.4	Fe_2O_3
MnO	0.182	3.3			0.19	2.6	0.034	7.4			0.03		0.101	0.055	MnO
MgO	3.45	2.5	3.52	0.6	3.56	1.7	0.763	4.3	0.77	2.5	0.73	12.3			MgO
CaO	6.95	1.1	7.02	0.3	7.05	0.5	1.97	2.5	1.98	1.0	1.92	1.1	4.1	3.8	CaO
Na_2O	3.27	1.7	3.30	1.5	3.36	4.5	4.07	1.5	4.17	1.2	4.09	4.4			Na_2O
K_2O	1.69	2.4	1.68	1.2	1.66	1.2	4.49	1.6	4.50	0.4	4.50	0.7	1.4	1.3	K_2O
P_2O_5	0.37	2.7			0.36	2.8	0.14	3.6			0.13	15.4			P_2O_5
Total	99.52		99.54		100.16		98.82		99.06		98.55				Total

	Anorthosite GIT-IWG: AN-G		Basalt GIT-IWG: BE-N		Granite GIT-IWG: MA-N		Detection limits			
	Govindaraju (1980) Expected	Potts *et al.* (1984) Analysed	Govindaraju (1980) Expected	Potts *et al.* (1984) Analysed	Govindaraju (1980) Expected	Potts *et al.* (1984) Analysed	Potts *et al.* (1984) ED–XRF	Mahan and Leyden (1983) ED–XRF	Norrish and Hutton (1969) WD–XRF	
SiO_2	46.30	46.83	38.20	38.40	66.60	66.87	0.21		0.10	SiO_2
TiO_2	0.22	0.22	2.61	2.58	0.01	0.02	0.03	0.014	0.024	TiO_2
Al_2O_3	29.80	29.38	10.07	10.39	17.62	17.59	0.19		0.06	Al_2O_3
Fe_2O_3	3.36	3.25	12.84	12.61	0.47	0.25	0.025	0.004	0.026	Fe_2O_3
MnO	0.04	0.04	0.20	0.19	0.04	0.04	0.015	0.004	0.024	MnO
MgO	1.80	1.97	13.15	13.11	0.04	0.26	0.33		0.12	MgO
CaO	15.90	15.94	13.87	13.86	0.59	0.61	0.025	0.038	0.014	CaO
Na_2O	1.63	1.56	3.18	3.10	5.84	6.05	0.96		0.24	Na_2O
K_2O	0.13	0.06	1.39	1.34	3.18	3.17	0.04	0.057	0.014	K_2O
P_2O_5	0.01	0.07	1.05	1.14	1.39	1.52	0.05		0.014	P_2O_5
Total	99.19	99.32	96.56	96.72	95.78	96.38				Total

Units: % oxide.
CV = coefficient of variation = 100 × standard deviation/mean analysis (%).
The CV of Gladney *et al.* represents the standard deviation of compiled data after rejection of outliers.
The CV of Hebert and Street represents the standard deviation based on counting statistics.
The CV of Potts *et al.* represents the standard deviation of 12 replicate analyses carried out over a four-month period.
Analysis conditions
Hebert and Street: samples prepared as glass discs with lithium metaborate in the ratio 1:10 of flux and excited with a multiple target soft x-ray tube operated at between 7 and 10 kV. Average analysis time was 40 minutes per sample.
Potts *et al.*: samples were prepared as glass discs with lithium tetra/metaborate in the ratio 1:6 of flux and excited with a silver anode x-ray tube operated at between 7 and 10 kV. Analysis times were 500 live seconds per sample.
Mahan and Leyden: samples were prepared as glass discs with lithium tetraborate in the ratio 1:3 of flux and excited at 30 kV with a silver anode x-ray tube operated in pulsed mode with a 25 μm silver primary beam filter. Data was accumulated for 1000 s per sample.
Detection limits represent the six-sigma limit of determination.

(600 cps), but is only processed by the amplifier integration circuits when the amplifier is not recording a real x-ray event, the total area of the strobe peak recorded in any spectrum is inversely proportional to input count rate. This provides a useful measure by which to calculate the FWHM correction appropriate to individual spectra.

Elemental samples used for these resolution variation measurements include cobalt ($K\alpha_{1,2} = 6.924$ keV), copper ($K\alpha_{1,2} = 8.040$ keV) or zirconium ($K\alpha_{1,2} = 15.744$ keV), the choice depending on the energy range of spectra to be measured.

(*b*) *Gain drift.* Precise fitting of library spectra to sample spectra also demands that the gain of each should be precisely matched despite small changes in dc zero level and amplifier gain that inevitably occur. One technique for doing this is to record a reference spectrum at least once during each operating session, whether library standard or sample spectra are being recorded. A characteristic photopeak in the reference sample is fitted to determine the channel number of its centroid. The difference in channel number between the zero energy strobe and the reference photopeak of known energy then gives precise calibration of the spectrum gain during the current operating session. Reference samples containing cobalt, copper or zirconium can conveniently be used for gain calibration.

(*c*) *Library spectra.* The success of the digital filter deconvolution procedure rests on the quality of library spectra. These constitute sections of the spectra of individual elements that include all the *K*-, *L*- or *M*-series of x-ray lines of that elements. These spectral 'profiles' are filtered using the top-hat digital function and stored so that they can be subsequently recalled in digital form and fitted to recorded sample spectra. There are several requirements that must be satisfied:

(i) Library samples must not contain other elements that emit x-ray lines in the region of those of the element of interest.

(ii) Measurements should be made on samples that possess a similar matrix to unknown samples to minimize any discrepancies that could arise due to non-linear variations in continuum under the emission lines of interest.

(iii) Library spectra should be excited at the same tube potential and filter conditions as routine samples.

(iv) Library spectra should be recorded to high statistical precision, as any channel-by-channel errors will be carried over to the spectrum deconvolution algorithm.

In order to satisfy these requirements it is desirable to record library spectra for the major elements from high-purity standard reagents fused with lithium borate and cast as glass discs. Trace elements may be prepared as powder pellets by dilution in high-purity silica.

Some practical difficulties encountered in such procedures have been discussed by Potts *et al.* (1984). In particular, not all major element reagents could be fused with lithium borate and cast as stable glass discs. To these samples, high-purity silica was added to promote glass formation. Difficulties were still encountered in the preparation of sodium and magnesium library glass discs since it was considered inadvisable to add silica due to the proximity of the respective emission lines. Quenched melts of these samples were finally ground and prepared as powder pellets for recording library spectra.

Segments of the x-ray spectra recorded from a set of library standards are plotted in Figures 9.32 (0–10 keV) and 9.33 (10–20 keV). For clarity, only the region of each spectra covering the *K*- or *L*-lines peaks has been plotted. Data are included only for those elements that can be commonly detected in 'normal' silicate samples, and *L*-line profiles have been arbitrarily plotted at half the relative intensity of *K*-lines. From these data, it is possible to assess the severity of

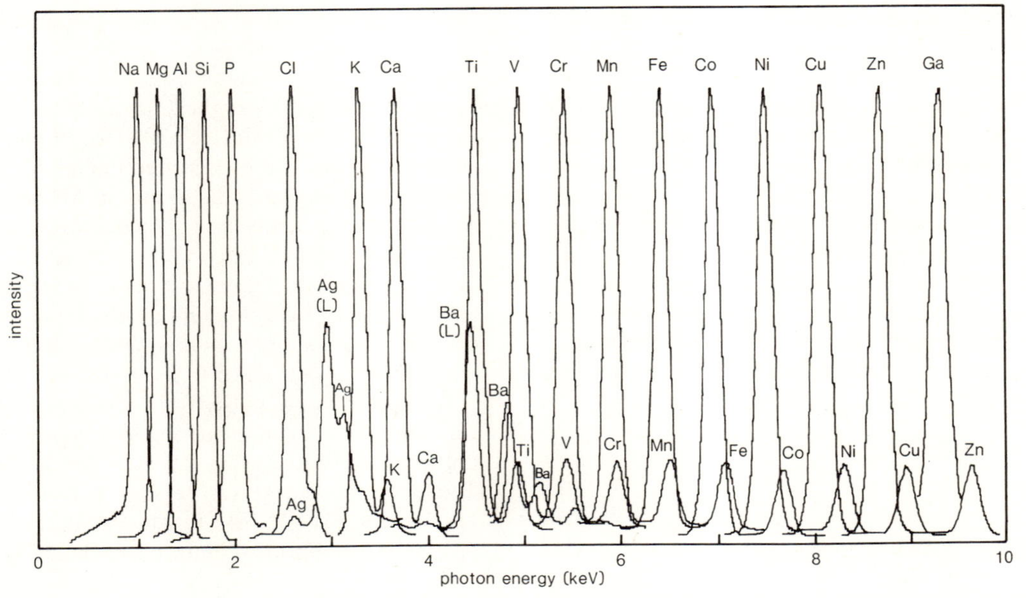

Figure 9.32 X-ray 'library' spectra (before the application of a digital filter) for the elements normally analysed from silicate rock spectra in the region from 0 to 10 keV. The *L*-lines of Ba and of Ag (scattered from a silver tube) have been included and are plotted at half the relative intensity. (After Potts *et al.*, 1984.)

L

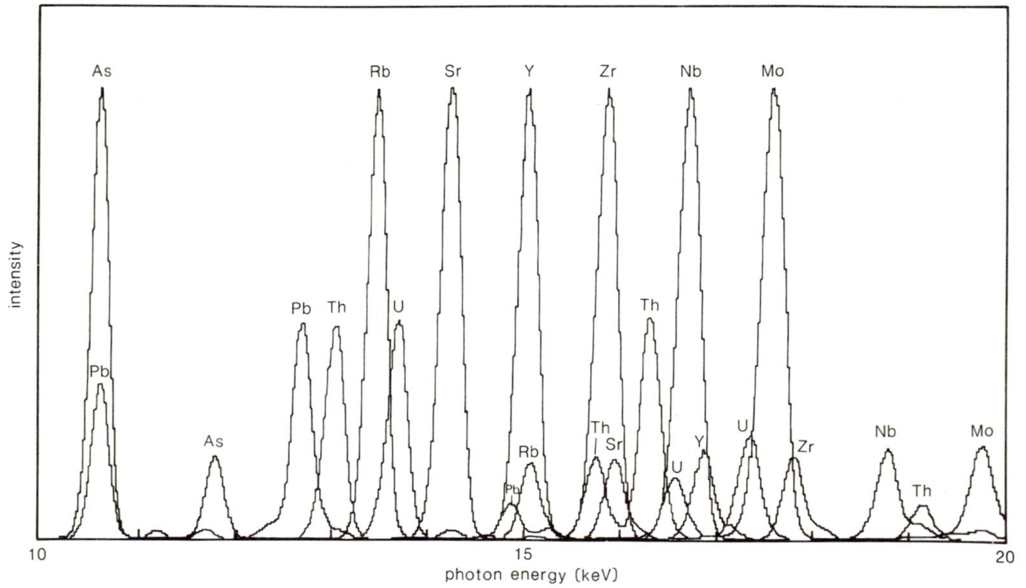

Figure 9.33 X-ray 'library' spectra (before the application of a digital filter) for the trace elements normally determined between 10 and 20 keV in the spectra of geological samples. The *L*-lines of Pb, Th and U have been arbitrarily plotted at half the intensity of the *K*-lines. (After Potts *et al.*, 1984.)

spectrum overlap interferences in ED–XRF spectra. As will be seen, of particular concern in the analysis of trace elements is the overlap interference of Rb $K\beta$ on Y $K\alpha$ and the Nb $K\alpha$ overlap on the only other prominent Y peak, the Y $K\beta$ profile (Figure 9.33).

A further requirement is the necessity of recording library spectra to represent x-ray tube lines (normally Ag *L*- or Rh *L*-lines) scattered off the sample. These must normally be fitted to spectra even though the resultant data are subsequently discarded. The particular difficulty is that the relative intensity of silver or rhodium scattered *L*-lines and any associated Compton scatter peak varies with sample composition. Whereas the dilution effect of lithium borate flux allows a representative spectrum to be recorded by scatter off a fusion disc containing the appropriate amount of silica, it is more difficult to measure a fully representative spectrum scattered of powder pellets where differences in matrix attenuation effects between samples are larger.

Van Dyck and van Grieken (1983) have described an algorithm that can account for variations in the relative intensity within line series by making use of the matrix information present in scatter peaks. The time required to complete this algorithm is significantly longer than for normal spectrum deconvolution but was claimed to be justified when the difference in matrix effects between standards and unknowns was significant.

9.9.4 *Routine operations*

(*a*) *Major elements.* Routine analysis for the major elements is carried out on glass fusion discs that are excited at 10 to 12 kV with a rhodium or silver anode side-window x-ray tube operated in pulsed mode. No primary beam filter is used since low-energy tube continuum is essential to efficiently excite sodium and magnesium. Spectra are recorded for about 500 live seconds with the tube current adjusted to give a dead time of 20 to 40% (pulsed tube mode). After spectrum deconvolution, all count data are normalized against a suitable monitor standard, for example, andesite USGS: AGV-1, which was routinely counted several times during an operating session so that drift in instrument sensitivity, particularly tube output, can be corrected. The following library spectra are normally fitted during the deconvolution procedure: Na, Mg, Al, Si, P, K, Ca, Ti, Ba *L*, (V), Cr, Mn, Fe, (Ni). In addition, it is desirable to fit profiles of Zr *L* and S *K* to account for those samples that may contain unusually high abundances of these elements. The Rh *L*- or Ag *L*-line profile must also be fitted to account for scatter from the tube off the sample. It is not possible to detect the *L*-lines of the light rare-earth elements in 'normal' silicate rocks due to both the dilution effect of the glass fusion disc and the inferior detector resolution in this region of the spectrum. Absorption-enhancement effects are best corrected by the procedure of Norrish and Hutton (1969) as discussed in section 8.8.

(*b*) *Trace elements.* Trace elements are analysed on powder pellets excited with a rhodium or silver tube operated at 40 kV or more using a 125 μm rhodium or silver primary beam filter. Spectra are counted for typically 800–1000 s at about 35% dead time in pulsed tube operation. After deconvolution, count data are normalized relative to a suitable drift monitor, for example, the granodiorite USGS: GSP-1. The following elements can be determined from spectra: Fe, Ni, Cu, Zn, Ga, Rb, Sr, Y, Zr, Nb, Pb *L*, Th *L*, and, if present at the upper end of their abundance range in silicate rocks, As, Mo and U *L*. In addition it is desirable to fit library spectra due to W *L* to account for those samples that have been ground in tungsten carbide milling equipment. Matrix effects can be corrected by an iterative mass attenuation calculation procedure following Holland and Brindle (1966) and Brown *et al.* (1973) (section 8.8). However, estimates must then be obtained of major element compositions by an additional analysis run in which powder pellets are excited at 10 kV using a silver or rhodium tube with no primary beam filter. Count times of 200 live seconds are adequate.

More efficient matrix correction can be carried out using the Compton scatter procedure (section 8.8), providing suitable characteristic lines from the x-ray tube are detected within the range of the recorded spectrum. Giauque *et al.* (1977*b*) used this technique in the analysis of geochemical samples using the Compton scatter peak from a molybdenum transmission target tube. Markowicz (1984) evaluated theoretical aspects of the efficiency of the Compton scattered radiation method in ED–XRF analysis. Alternative procedures include the use of the Lachance–Traill procedure (section 8.8) by Mahan and Leyden (1983) in the analysis of major and minor elements in river sediments. Samples were fused with lithium tetraborate and cast as glass discs. Application of the Lachance–Traill procedure was facilitated by the possibility of preparing synthetic standards using this fusion procedure.

A detailed assessment of the reliability of direct silver tube excitation coupled with the filtered least squares spectrum deconvolution procedures has been given by Potts *et al.* (1984). The effect of severe spectrum overlaps between, for example, the silver tube *L*-lines and the K *K*- and Ca *K*-lines or the Pb+Th and Rb, Sr, Y, Zr, Nb regions of the spectrum, was far smaller than otherwise might be expected, providing the procedures and precautions discussed above were followed. Indeed, one advantage of the digital filter–library spectra fitting procedure is that it is not usually necessary to carry out any specific overlap correction. Providing care is taken over the preparation and specification of library spectra, overlap interferences are automatically accounted for in the deconvolution algorithm.

The one intractable interference encountered in the above work was in the determination of yttrium in samples containing more than about 300 ppm Rb. The Rb $K\beta$ photopeak then begins to swamp the Y $K\alpha$ signal and if the sample also contains significant amounts of Nb, the Y $K\beta$ line is also distorted by the Nb $K\alpha$ overlap (Figure 9.33). There is then insufficient of the yttrium library spectrum free of interference to constrain the digital fitting procedure to allow accurate yttrium determinations. However it should be noted that samples of such composition would also present a significant analytical challenge in wavelength dispersive spectrometry.

9.9.5 *Precision, accuracy and detection limits*

Selected results for the analysis of samples for the major elements and selected trace elements are listed in Tables 9.5 and 9.6 respectively.

(*a*) *Major elements.* In Table 9.5, data of Hebert and Street (1974) are listed for the analysis of reference materials USGS: G-2 and BCR-1 on glass discs excited with the specialized multiple anode soft x-ray generator described in section 9.9.1. These results are compared with the data of Potts *et al.* (1984) for glass discs of the same reference materials excited with a silver anode x-ray tube operated in pulsed mode with no primary beam filter. Precision data for these samples were obtained from 12 determinations of the same glass discs obtained over a three-month period. These data indicate that coefficients of variation are generally better than 1 to 2% relative (one sigma), the most important exceptions being sodium and also phosphorus and magnesium at low concentration levels. Additional data are listed in Table 9.5 for the analysis of reference materials GIT-IWG: AN-G, BE-N and MA-N from Potts *et al.* (1984) to indicate the performance of the technique for samples over the full range of silicate compositions. The results of Mahan and Leyden (1983) for NBS river sediment SRM-1645 were obtained on a glass disc prepared with lithium tetraborate in the dilution 1 part to 3 of flux. The sample was excited at 30 kV with a pulsed silver tube using a 25 μm silver filter.

These data indicate typical results for major element determination by ED–XRF and show that all elements down to sodium can be measured, providing the appropriate excitation conditions are used.

ED detection limits, taken from Potts *et al.* (1984) (for samples prepared as 1:6 fusion discs with lithium meta/tetraborate flux, 500 s count times) and Mahan and Leyden (1983) (samples prepared as 1:3 glass fusion discs with lithium tetraborate, 1000 s count time) are compared with 'typical' data for wavelength dispersive XRF (for 1:5 fusion with heavy absorber flux) taken from Table 8.18. At the current state of development, WD-XRF detection limits for the light major elements are significantly lower than ED data. This difference is particularly serious for the lightest element, sodium, where ED–XRF data of 0.96% Na_2O must be compared with 0.24% Na_2O) by WD–XRF. However, data in Table 9.6 indicate that the detection limit performance of the two techniques becomes much more comparable when analysing the heavier major elements and may have the advantage in the determination of manganese and iron, elements that are not as effectively excited in routine WD–XRF programs that use a conventional chromium anode x-ray tube.

(*b*) *Trace elements.* Trace-element results for the analysis of selected reference materials taken from the work of Potts *et al.* (1984) and Giauque *et al.* (1977*b*) are listed in Table 9.6. Samples were analysed as compressed powder pellets excited under the following conditions: Potts *et al.* (1984)—pulsed silver tube operated at 45 kV using a 127 μm silver filter, 800s count time; Giauque *et al.* (1977)—molybdenum transmission tube (incorporating a 100 μm combined anode and window) operated at 42 and 48 kV using an additional external 100 μm molybdenum filter.

The data of Potts *et al.* (1984) indicate that the precision, based on replicate analyses of USGS: GSP-1 and AGV-1 over a four-month period, is typically in the range 1–3% (one sigma, relative), rising to about 10% as lower concentration levels are encountered. Analytical results for GIT-IWG AN-G and BE-N show that the technique is capable of analysing samples covering the full range of trace compositions. Giauque *et al.* (1977) reported that accuracies were expected to be typically 5% or better using 20-minute analysis periods. In the analysis of river sediments, Mahan and Leyden (1983) reported that coefficients of variation for eight quantifiable

Table 9.6 Direct tube excitation of selected trace elements by ED–XRF

	Granodiorite USGS: GSP-1			Andesite USGS: AGV-1			Granite CRPG: GA			Basalt CRPG: BR			Anorthosite GIT-IWG: AN-G		Basalt GIT-IWG: BE-N		Detection lines			
	Gladney *et al.* (1983) Expected	Potts *et al.* (1984) Analysed	CV	Gladney *et al.* (1983) Expected	Potts *et al.* (1984) Analysed	CV	Abbey (1983) Expected	Giauque *et al.* (1977*b*) Analysed	CV	Abbey (1983) Expected	Giauque *et al.* (1977*b*) Analysed	CV	Govindaraju (1980) Expected	Potts *et al.* (1984) Analysed	Govindaraju (1980) Expected	Potts *et al.* (1984) Analysed	ED-XRF Mahan and Leyden (1983)	ED-XRF Potts *et al.* (1984, 1985)	WD-XRF Table 8.18	
V							38	<160		240	175	4.9					24		10	V
Cr							12	<51		380	310	7.1					30		10	Cr
Co							5			50							14		6	Co
Ni							7	<8		260	232	2.6					10	19.4	6	Ni
Cu	34	31	9.0	60	58	8.3	16	20	20	72	80	6.9	19	21	72	70	12	9.9	6	Cu
Zn	103	105	3.0	88	89	3.4	80	72	4.9	150	151	4.0	20	21	120	114	4	11.5	9	Zn
Ga	22	21	8.1	20	20	6.9	16	15	6.7	20	17	8.8	18	20	17	16		7.8	6	Ga
Rb	254	250	1.0	67	68	1.3	175	180	0.6	47	43	2.3	1	(2)	47	48	8	5.6	3	Rb
Sr	234	236	0.8	662	675	0.7	310	313	0.6	1300	1326	0.3	76	79	1370	1385	18	3.5	3	Sr
Y	29	30	6.7	21	21	3.3	21	21	4.8	30	35	2.9	8	8	30	31	14	3.5	3	Y
Zr	530	495	0.4	225	233	0.7	150	123	4.5	250	259	4.8	15	21	265	276	12	4.0	6	Zr
Nb	26	25	3.2	15	15	2.7	10?			100?			2	(3)	100	103		4.4	6	Nb
Pb	54	54	9.4	36	36	3.9	30	32	3.1	8	6	33	2	7	4	6	10	7.7	6	Pb
Th	105	104	0.7	6.5	6.9	14.5	17			12			—	1	11	9		8.6	9	Th

All data in ppm rock.
CV = coefficient of variation = 100 × standard deviation/mean analysis (%).
Detection limits represent the six-sigma limit of determination.
Analysis conditions:
Potts *et al.* (1984): samples prepared as powder pellets and excited using a silver anode tube at 45 kV operated in pulsed mode with a 127 μm silver primary beam filter. Spectra were recorded for 800 live seconds. Coefficient of variation data are representative of replicate analyses (28 for GSP-1, 13 for AGV-1) carried out over a four-month period.
Giauque *et al.* (1977*b*): samples prepared as powder pellets and excited using a molybdenum transmission target tube operated at 42 kV (Ti-Zr) or 48 kV (Pd–Ba). In the latter case, an additional 100 μm Mo primary beam filter was used. Count times were 20 minutes per sample.
Mahan and Leyden (1983): detection limit data were derived for samples prepared as 1 : 3 glass discs using dilithium tetraborate as flux. Samples were excited using a silver anode x-ray tube operated at 30 kV in pulsed mode with a 25 μm silver filter. Spectra were counted for 1000 s.

major and minor elements ranged from 2 to 9% relative (one sigma) with improved accuracy being obtained at lower sample to flux ratios (1 : 10) rather than for the 1 : 3 glass discs also investigated. Samples were excited with a pulsed silver tube operated at 30 kV using a 25 μm silver filter, count times being 1000 s.

Detection limits taken from Potts *et al.* (1984, 1985) for powder pellets counted for 800 live seconds each and from Mahan and Leyden (1983) on glass discs counted for 1000 s are compared in Table 9.6 with WD–XRF data from Table 8.18, judged to be representative of routine WD–XRF trace operating conditions.

These data indicated that ED-XRF detection limits (six sigma) down to the 3.5 to 6 ppm level can be obtained for the elements Rb, Sr, Y, Zr and Nb. These data are fully comparable to the performance expected for WD–XRF systems. Indeed, this is so for most of the trace elements listed in Table 9.6.

Potts *et al.* (1985) have undertaken a detailed comparison of the performance of ED and WD spectrometers for major and trace elements in silicate rocks. The ED date were acquired by direct excitation using a silver anode x-ray tube. It was concluded that although detection limits for the lightest major elements by ED–XRF were inferior to routine WD–XRF data, analytical precisions at concentration levels encountered in normal silicate rocks were comparable. In the ED–XRF analysis of the heavier trace elements that are effectively excited by a silver tube, detection limits, instrumental precision and routine analytical performance were equivalent to results obtained by WD–XRF. It was further suggested that the accuracy of data, whichever technique were used, could be significantly influenced by uncertainties in the recommended values for international rock reference materials that were used to set up calibrations. The validity of these statements may be judged from WD–XRF data listed in Table 8.21, which compares the analysis of a granite (AT 242) from eight independent laboratories. The analysis from laboratory 4 was obtained on an energy dispersive spectrometer using the analysis procedures discussed in section 9.9.4 (*b*). Data are statistically indistinguishable from the remaining seven sets of wavelength dispersive results.

9.10 Indirect excitation methods

The analytical potential of ED–XRF analysis is not completely exploited by the use of conventional excitation geometries. In particular, a second category of instrument, in which samples are excited by radiation from secondary target materials, has found widespread application in the analysis of geological samples. In this section, the principles and practice of indirect excitation geometries are considered, and in particular, methods involving the polarization of x-rays from excitation sources.

9.10.1 *Polarized x-ray beams*

An unpolarized beam of x-rays can be considered to consist of two electrical vectors, at right angles, one oriented in the x–z plane and the other in the y–z direction (Figure 9.34). When such a beam is scattered off an amorphous surface, that component of the scattered beam viewed along the x axis at an angle of 90° to the incident beam has no electrical component in the x–z plane: it is plane polarized in the x–y direction (Figure 9.34). This polarization phenomenon applies to both Rayleigh and Compton scatter components of the beam.

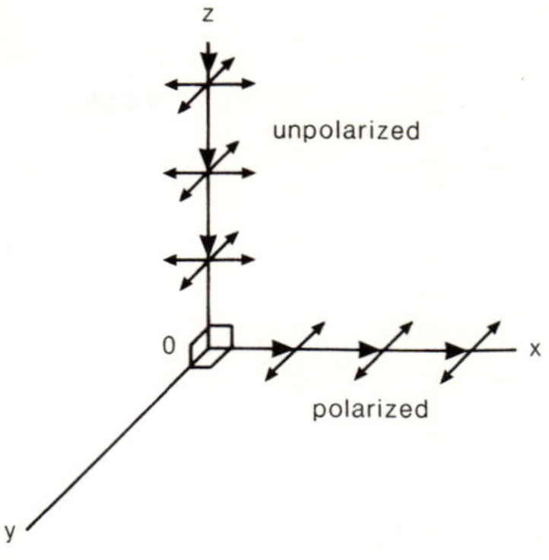

Figure 9.34 Polarization of scattered x-ray photons. The electric vectors in an unpolarized photon propagated along the z-axis are shown before and after scatter at the origin. The scattered beam view at 90° along the x-axis is polarized, having an electrical vector only in the x-y plane.

If, as in Figure 9.35, the scatter surface is a specimen and a detector is placed in the beam scattered at 90° (as in conventional direct-tube excitation geometries), then no apparent advantage is gained. Fluoresced (unpolarized) x-rays originating from the sample are measured together with polarized radiation scattered from the tube.

However, different considerations apply if the scatter surface is a secondary target arranged so that the sample is excited either by tube radiation scattered off this target or by characteristic radiation of the target itself (Figure 9.36). Provided that tube, secondary target, sample and detector are in an orthogonal geometry, radiation from the tube, polarized by scatter off the secondary target, cannot be scattered a second time into the detector. Two situations arise:

(i) The secondary target may be selected as a low atomic number material (such as boron carbide), which does not

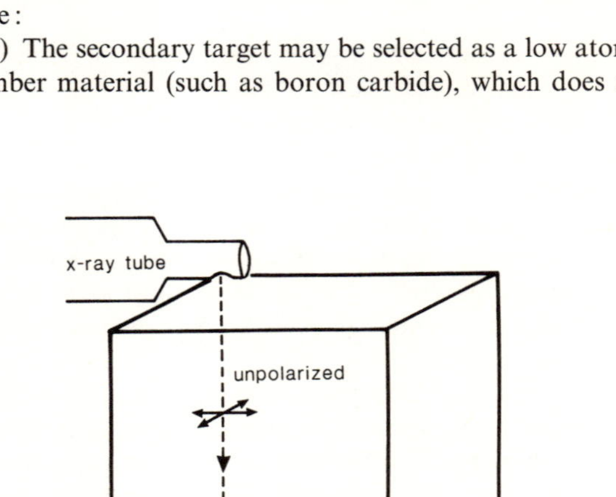

Figure 9.35 Conventional excitation geometry showing the electric vectors in the x-ray tube radiation before and after scatter from the sample. The cube is drawn to give perspective to this diagram.

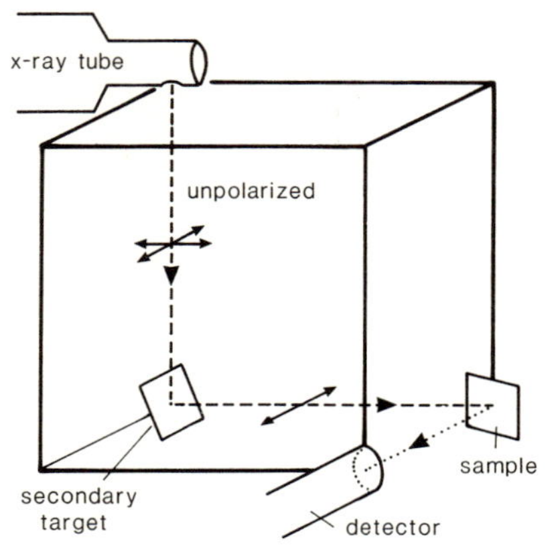

Figure 9.36 Secondary target excitation geometry showing x-ray tube–secondary tube–sample–detector in an orthogonal relationship to each other. Scattered tube lines from the secondary target are polarized, any characteristic lines generated in the secondary target are not.

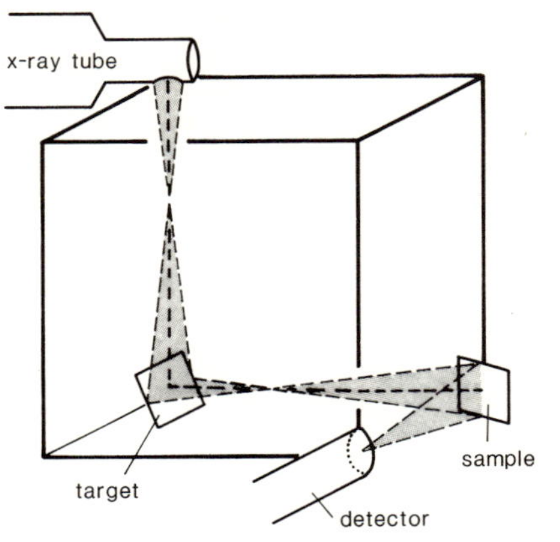

Figure 9.37 Secondary target excitation geometry demonstrating the solid angle effect. Due to the close spatial coupling of components, perfect 90° scatter cannot be achieved in practical systems.

emit characteristic x-rays but is simply designed to scatter polarized tube radiation on to the sample; this polarized tube radiation is then capable of exciting the sample, but under idealized circumstances cannot be scattered a second time into the detector.

(ii) The secondary target may be a metal in which characteristic x-rays are excited by the tube radiation. The target is then selected so that its characteristic x-rays will efficiently excite a range of elements in the sample that have absorption edges at immediately lower energies. In this situation, the sample is excited by polarized tube radiation scattered from the secondary target *and* by characteristic radiation from the secondary target. Only the latter can be scattered off the sample into a detector held in an orthogonal configuration in ideal circumstances.

In either case, a significant reduction in background scattered into the detector should be observed, permitting improved detection limits compared with direct tube excitation.

In analytical applications of these effects, several practical considerations must be taken into account.

(i) Each time an x-ray beam is scattered off a surface, the beam intensity (and hence its exciting power) is reduced by at least two orders of magnitude. Orthogonal excitation geometries must, therefore, be used in conjunction with energy dispersive detectors because of physical limitations in designing a generator with sufficient power to satisfy the more stringent collimation requirements of a wavelength dispersive spectrometer.

(ii) As excitation efficiency is at a premium, it is desirable to couple the components in an orthogonal excitation system as closely as possible. The large solid angle of x-rays collected by secondary target and sample then optimizes excitation efficiency.

(iii) However, the close-coupling requirements outlined in (ii) simultaneously increase the range of scatter angles about the ideal 90° orthogonal geometry (Figure 9.37). Ideal polarization cannot be achieved with practical spacings between tube–target–sample and detector components. In the design of orthogonal-geometry instruments, some compromise must be reached between excitation efficiency (and therefore tube power), which requires short path lengths, and a high degree of polarization (and therefore background suppression), which requires long path lengths.

9.10.2 *Excitation using secondary target radiation*

The principles of secondary target excitation outlined above indicate that secondary targets must be chosen such that the characteristic lines are at higher energies than the absorption edges of elements to be determined. Unfortunately, excitation efficiency is only satisfactory for a comparatively narrow range of elements. It is, therefore, necessary to select a range of target materials for a complete multi-element analysis program so that the groups of elements will be excited under optimum conditions. Providing orthogonal excitation conditions are used, a significant suppression of tube continuum scattered into the detector will be achieved. The technique retains the advantage of being able to use the Compton scatter method for matrix correction, using, however, characteristic secondary target rather than tube lines scattered off the sample.

One disadvantage of secondary target excitation is that the element that is most efficiently excited (that is the one having an absorption edge immediately below the secondary target radiation) is likely to suffer enhanced backgrounds due to Compton scatter of the unpolarized secondary target lines in the sample.

A second disadvantage arises in the analysis of elements lighter than potassium. Satisfactory excitation can only be achieved by selecting a secondary target composed of the element next higher in atomic number. An analytical programme using exclusively secondary targets then becomes impractical. In such schemes of analysis, direct tube excitation is normally reverted to for excitation of the lightest major elements.

These problems are demonstrated by detection limit data for secondary target excitation using Sn, Mo and Cu *K*-lines

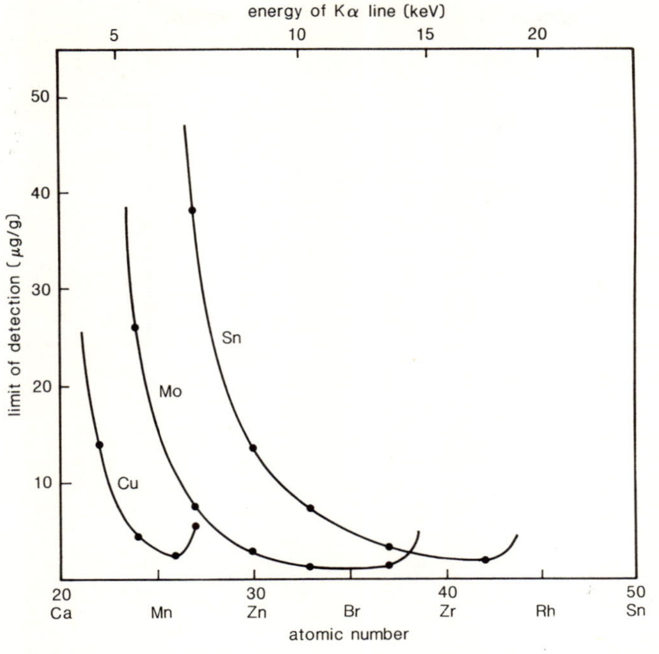

Figure 9.38 Detection limit as a function of secondary target excitation, using data from Spatz and Lieser (1977). Detection limit data are plotted here for elements excited by (i) the 25–29 keV radiation from a tin secondary target, (ii) the 17.4–20 keV lines from a molybdenum target, and (iii) the 8–8.9 keV lines from a copper target.

presented by Spatz and Lieser (1977, 1979) (Figure 9.38). A secondary target made of tin, emitting $K\alpha/K\beta$ x-rays in the range 25–29 keV, will effectively excite a relatively wide range of elements. In comparison, a copper secondary target emitting $K\alpha/K\beta$ x-rays in the range 8–9 keV is capable of optimum excitation of a much narrower range of elements (Cr–Co *K*- lines).

In order to reduce the proportion of tube radiation scattered off the target, Spatz and Lieser (1979) placed a filter between target and sample. The filter was made of the same material as the target. Secondary radiation was, therefore, transmitted efficiently but other scattered radiation was selectively attenuated (see regenerative monochromatic filters, section 9.7.3 (*d*)). Some attenuation of target x-rays will also occur in this filter material, so that careful optimization of filter thickness and tube operating potential is necessary. The optimum parameters used by Spatz and Lieser are listed in Table 9.7.

9.10.3 *Analytical applications of secondary target excitation*

The range of secondary target materials most commonly used for the excitation of geological samples for element lines in the range 1 to 20 keV is shown in Figure 9.39. In routine applications in which all major and trace elements are to be determined, a compromise might be reached by selecting silver and germanium secondary targets together with direct excitation using a rhodium anode tube for the lightest elements (Na–P). Johnson (1984) used, in addition, a gadolinium secondary target to excite efficiently barium *K*-lines in geological samples.

The precision of secondary target excitation for ten major elements and selected trace elements is demonstrated by data presented in Table 9.8. These results for a granite, USGS: G-2, prepared as a powder pellet, were obtained by a combination of secondary target excitation using silver and germanium fluorescence and direct rhodium tube excitation. Precisions are generally in the range 0.4 to 2% (coefficient of variation) except for those elements approaching detection limit levels. Detection limits derived from these data are listed in Table 9.8 and compared with data for direct tube excitation from Table 9.5. These latter data were derived for major elements determined on 1 : 6 glass fusion discs rather than powder pellets. There appears to be broad correlation between these two sets of results for the heavier trace elements which were obtained from comparable sample types.

Table 9.7 Optimum operating conditions for secondary target excitation

	Tungsten tube		
Secondary target	Operating potential (kV)	Anode current (mA)	Filter thickness (μm)
Cu	31	40	30*
Mo	38	45	60*
Sn	44	20	50
Ba	50	3	filter paper

Data from Spatz and Lieser (1979), representative of the instrument designed by them.
Filter thickness is for the x-ray beam filter placed between secondary target and sample to reduce the intensity of scattered tube radiation reaching the sample.
* thinnest filter available to the authors.
Ba : filter paper loaded with 0.01mmol cm^{-2} Ba^{2+}.

Table 9.8 Analysis of granite USGS:G-2 by secondary target ED-XRF analysis

	Secondary target and tube kV		Expected	Analysed (% oxide)	CV (% oxide)	Detection limit: Secondary target excitation (% oxide)	Detection limit: Direct tube excitation* (% oxide)
SiO_2	Rh direct	5 kV	69.04	69.81	0.27	0.02	0.21
TiO_2	Ge	15 kV	0.492	0.53	1.5	0.007	0.03
Al_2O_3	Rh direct	5 kV	15.14	16.05	0.43	0.06 .	0.19
Fe_2O_3	Ge	15 kV	2.68	2.66	0.80	0.002	0.025
MnO	Ge	15 kV	0.034	0.04	11.1	0.002	0.015
MgO	Rh direct	5 kV	0.763	0.84	6.3	0.12	0.33
CaO	Ge	15 kV	1.97	1.89	1.5	0.036	0.025
Na_2O	Rh direct	5 kV	4.07	4.13	7.0	0.90	0.96
K_2O	Ge	15 kV	4.49	4.27	0.80	0.024	0.04
P_2O_5	Rh direct	5 kV	0.14	0.10	15.2	0.03	0.05
TOTAL			98.82	100.32			
			ppm	ppm		ppm	ppm
Rb	Ag	40 kV	170	166	1.6	4.5	5.6
Sr	Ag	40 kV	478	465	0.9	4.2	3.5
Y	Ag	40 kV	11.4	10	12.0	5.7	3.5
Zr	Ag	40 kV	300	278	1.0	4.2	4.0
Nb	Ag	40 kV	13	12	4.7	4.2	4.4

Analysed data from Kevex Corporation application report: Geological applications, Kevex Corp., Foster City, California.
Expected compositions were taken from Gladney *et al.* (1983).
The sample was analysed as a powder pellet, prepared with chromatographic cellulose as binder and counted for 300 s (Ge and Ag secondary targets) or 200 s (Rh direct tube excitation).
Detection limit data are an estimate of the six-sigma limit of determination based on USGS: GSP-1.
Detection limit data for direct tube excitation are taken from Tables 9.5 and 9.6, the major-element data* being representative of a 1:6 glass fusion disc (500 s count time), the trace-element data for powder pellets counted for 800 s.

Further data obtained by secondary target excitation are listed in Table 9.9 for the following materials:

(i) Basalt USGS: BCR-1 prepared as a glass fusion disc diluted in the ratio one part of rock to 6 parts of dilithium tetraborate and excited by germanium secondary target and rhodium tube direct excitation (Kevex Corporation).

(ii) Andesite USGS: AGV-1 and basalt USGS: BCR-1 prepared as powder pellets using sulphur as a binder (Giauque *et al.*, 1977*a*) and excited by molybdenum and terbium secondary target radiation using a tungsten tube and guard-ring type detector.

(iii) Basalt GSJ: JB-1 prepared as a powder pellet using

Table 9.9 Secondary target excitation of selected reference materials by ED-XRF

	Basalt USGS: BCR-1 Gladney *et al.* (1983) Expected	Kevex Corporation Analysed	Kevex Corporation Excitation		Giauque *et al.* (1977*a*) Analysed	CV	Excitation	Andesite USGS: AGV-1 Gladney *et al.* (1983) Expected	Giauque *et al.* (1977) Analysed	CV	
SiO_2	54.35	54.44	Rh	direct 5 kV							SiO_2
TiO_2	2.22	2.13	Ge	20 kV	2.79	4.2	Mo 54 kV	1.06	1.28	1.9	TiO_2
Al_2O_3	13.63	14.86	Rh	direct 5 kV							Al_2O_3
Fe_2O_3	13.46	13.50	Ge	20 kV	13.18	2.4	Mo 54 kV	6.76	6.92	1.4	Fe_2O_3
MnO	0.182	0.18	Ge	20 kV	0.184	3.4	Mo 54 kV	0.096	0.101	3.7	MnO
MgO	3.45	3.23	Rh	direct 5 kV							MgO
CaO	6.95	6.59	Ge	20 kV							CaO
Na_2O	3.27										Na_2O
K_2O	1.69	1.65	Ge	20 kV							K_2O
P_2O_5	0.37	nd	Rh	direct 5 kV							P_2O_5
V	404				414	11.6	Mo 54 kV	123	<270		V
Cr	16				<175		Mo 54 kV	12	<140		Cr
Co	36.3				<120		Mo 54 kV	15.1	<115		Co

Table 9.9 (*Continued*)

	Basalt USGS: BCR-1						Andesite USGS: AGV-1			
	Gladney *et al.* (1983)	Kevex Corporation		Giauque *et al.* (1977*a*)			Gladney *et al.* (1983)	Giauque *et al.* (1977)		
	Expected	Analysed	Excitation	Analysed	CV	Excitation	Expected	Analysed	CV	
Ni	13			17	17.6	Mo 54 kV	17	16	18.7	Ni
Cu	19			23	24	Mo 54 kV	60	58	10.3	Cu
Zn	129			116	1.3	Mo 54 kV	88	84	2.4	Zn
Ga	22			20	12.5	Mo 54 kV	20	18	11.1	Ga
Rb	47.1			42	2.4	Mo 54 kV	67	64	1.6	Rb
Sr	330			317	1.9	Mo 54 kV	662	657	1.1	Sr
Y	39			33	29	Tb 80 kV	21	17	5.9	Y
Zr	191			181	2.8	Tb 80 kV	225	229	2.4	Zr
Nb	14			13	7.7	Tb 80 kV	15	15	10	Nb
Sn (*K*-line)	2.1			<4		Tb 80 kV	4.2	3	33.3	Sn
Ba (*K*-line)	678			680	3.2	Tb 80 kV	1221	1239	1.5	Ba
La (*K*-line)	25.0			<36		Tb 80 kV	38	33	27.3	La
Ce (*K*-line)	53.7			<70		Tb 80 kV	66	65	30.8	Ce
Pb (*L*-line)	13.56			10	10	Mo 54 kV	36	33	10.6	Pb
Th (*L*-line)	6.04			<8		Mo 54 kV	6.50	6	25	Th

	Basalt GSJ: JB-1				Diabase USGS: DNC-1			
	Abbey (1983)	Verbeke and Adams (1979)			Abbey (1983)	Johnson (1984)		
	Expected	Analysed	CV	Secondary target	Expected	Analysed	Secondary target	
SiO_2	52.60?	47.49	2.7	Ti				SiO_2
TiO_2	1.34	1.35	2.0	Mo				TiO_2
Al_2O_3	14.62	18.16	3.3	Ti				Al_2O_3
Fe_2O_3	9.01	8.32	1.0	Mo				Fe_2O_3
MnO	0.15	0.145	1.8	Mo				MnO
MgO	7.76	7.74	3.9	Ti				MgO
CaO	9.35	9.61	0.6	Mo				CaO
Na_2O	2.79							Na_2O
K_2O	1.42	1.44	25.0	Ti				K_2O
P_2O_5	0.26							P_2O_5
V	210	324	2.9	Mo				V
Cr	400	468	5.0	Mo				Cr
Co	39	<45						Co
Ni	135	(134)			250	248	Ge 35 kV	Ni
Cu	56	62	1.6	Mo	100?	104	Ge 35 kV	Cu
Zn	84	84	3.6	Mo	70	65	Ge 35 kV	Zn
Ga	17?	18	5.6	Mo				Ga
Rb	41	42	2.4	Mo	4.7?	3	Ag 40 kV	Rb
Sr	440	477	0.6	Mo	145?	146	Ag 40 kV	Sr
Y					18?	17	Ag 40 kV	Y
Zr					38?	41	Ag 40 kV	Zr
Nb					3.2?	3	Ag 40 kV	Nb
Sn								Sn
Ba (*K*-line)					120	100	Gd 60 kV	Ba
La								La
Ce								Ce
Pb (*L*-line)	11.5	(13)		Mo				Pb
Th (*L*-line)	9?	9	5.6	Mo				Th

Concentration units: majors: % oxide; traces: ppm element.
CV = coefficient of variation = 100 × standard deviation/mean analysis (%).

Analysis conditions

Kevex: samples were prepared as glass fusion discs using dilithium tetraborate mixed in the proportion 1 part of sample to 6.5 of flux using HBr as releasing agent. Spectra were counted for 100 s using Ge secondary target and 100 s for direct Rh excitation.

Giauque *et al.*: samples were prepared as powder pellets mixed in the ratio one part of sample to four of sulphur powder as binder. Secondary targets were excited with a tungsten tube. Radiation from the Mo secondary target was filtered with vanadium (250 μm) prefilter followed by an aluminium (150 μm) filter. A tantalum (100 μm) prefilter was used in conjunction with the terbium secondary target.

Verbeke and Adams: samples were prepared as powder pellets compressed in aluminium cups for support.

Johnson (1984): samples were analysed as powders packed into a counting cell, closed by a 6 μm mylar foil. Matrix effects were corrected by the Compton scatter method. Count times using Ge, Ag and Gd secondary targets were 200, 400 and 500 s respectively.

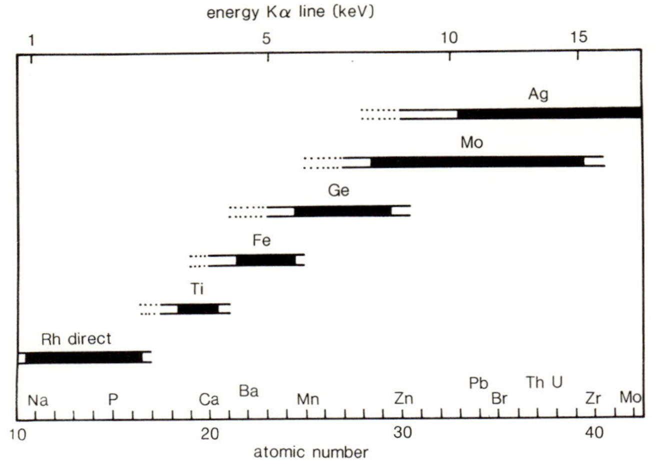

Figure 9.39 Ranges of elements that may be sensitively analysed using secondary excitation targets commonly used for geochemical applications. In some schemes of analysis, targets are limited to silver and germanium which, together with rhodium direct excitation, are adequate to cover the range of elements normally determined in geological samples. The solid bars show the range of optimum sensitivity.

an aluminium cup as support (Verbeke and Adams, 1979) and fluoresced with titanium or molybdenum secondary target radiation.

(iv) Diabase USGS: DNC-1 analysed as a powder packed into a sample holder having a 6 μm mylar end window (Johnson, 1984), samples being excited with germanium, silver and gadolinium secondary radiation.

Data listed in Table 9.9 indicate the range of results obtained in the secondary target excitation of geological samples. Of particular interest is the flexibility of these procedures to determine the *K*-lines of Sn, Ba, La and Ce using terbium secondary target excitation (Giauque *et al.*, 1977a) and Ba *K*-lines using a gadolinium secondary target (Johnson, 1984). These analyte lines cannot normally be excited efficiently by direct tube excitation techniques.

An elegant application of secondary target excitation for the rapid determination of Rb and Sr abundances in samples used in conjunction with isotope geochronology measurements was described by Updegrove (1982). Rock powders were simply poured into a mylar cup and excited with molybdenum secondary target radiation using 200 s count times. Very efficient fluorescence of Rb and Sr $K\alpha$ lines was achieved and variable powder packing and other sample attenuation effects were effectively corrected by normalizing net peak areas to the area of the Compton scatter molybdenum $K\alpha$ line generated in the sample as described in section 8.8. The resultant calibration line for Sr (Figure 9.40) shows excellent linearity, and the accuracy of analyses was reported to be within 2 to 3% of isotope dilution mass spectrometry calibrations.

9.10.4 *Excitation using tube radiation polarized by scatter off a secondary target*

If the conventional secondary target in orthogonal excitation geometry is replaced by a material which has a very high scatter efficiency but does not itself emit characteristic x-rays, then the sample is excited exclusively by scattered

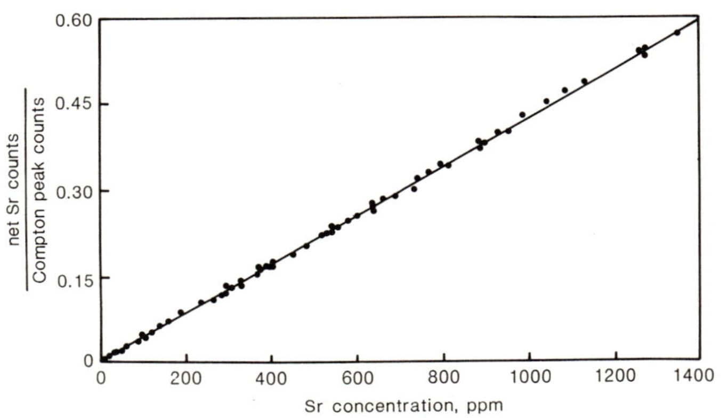

Figure 9.40 Calibration plot for strontium in geological samples using molybdenum secondary target excitation. The net Sr counts normalized to the area of the Compton scattered Mo $K\alpha$ peak are plotted against Sr concentration determined by isotope dilution mass spectrometry. (Updegrove, 1982.)

tube radiation. Since this radiation is polarized by the secondary target, only a small proportion will be scattered a second time into the detector. In theory, this technique has the potential of both retaining the flexibility of direct tube excitation and giving a significant reduction in tube radiation scattered into the detector.

Thus Dzubay *et al.* (1974) demonstrated significant improvements in peak to background ratios in the fluorescence of orchard leaf samples using a carbon rod as polarizer. Kauffman and Camp (1975) showed that very high polarized intensities could be achieved using boron carbide as target. Ryon (1977) established that in addition to geometric factors related to finite solid angles in the polarizing system, multiple scatter in the polarizing target or sample further limited the ultimate degree of polarization. Ryon observed optimum detection limits at about 85% polarization when the size of collimators was increased to allow the detection system to operate at higher count rates. Ryon and Zahrt (1979) discussed the design and operating characteristics of a circular polarizer which could produce a much higher intensity of polarized tube radiation. Furnas *et al.* (1982) described the application of toroidal monochromators in hybrid XRF systems.

The degree of background suppression achieved by such polarized excitation is illustrated in Figure 9.41, taken from Ryon (1977). This shows the spectra recorded from a pressed pellet of NBS orchard leaves excited with radiation from a molybdenum tube using (*a*) direct excitation with a molybdenum foil primary beam filter and (*b*) polarized excitation using the unfiltered tube spectrum. The improvement in peak to background ratios, especially below about 11 keV, is very apparent.

These instrumental developments have not yet led to a routine technique for geochemical analysis, but indicate the possible future improvements that could be made in ED–XRF analysis using polarized excitation sources. The ultimate advanced equivalent of polarized radiation sources is the synchrotron, on which some analytical applications have been investigated (Gilfrich *et al.* 1982; Deslattes, 1983). This, however, is never likely to achieve the status of a general laboratory-based technique.

9.11 Monochromatic polarized excitation using Bragg diffraction at $2\theta = 90°$

A special case of orthogonal geometry excitation is where the secondary target is replaced with a crystalline material selected such that its lattice spacing satisfies Bragg's Law at a 2θ angle of 90° with respect to a characteristic line from the x-ray tube as proposed by Aiginger *et al.* (1974). The instrumental arrangement is shown schematically in Figure 9.42. The advantages offered by this geometry are as follows:

(i) The intensity of the tube line which satisfies the Bragg relationship is increased, sometimes by a factor of 10 over that from amorphous polarizers (Wobrauschek and Aiginger, 1983).

(ii) Any tube radiation scattered off the diffracting crystal will be polarized and will not then be efficiently scattered into the detector.

However, there are some difficulties:

(i) There is the obvious problem of finding combinations of x-ray tube emission line and diffracting crystal material that satisfy this Bragg relationship—some of the more suitable combinations are listed in Table 9.10.

(ii) A related problem is the difficulty in finding any material with a sufficiently small lattice spacing to operate successfully by diffraction of x-ray tube lines more energetic than those of a molybdenum anode.

(iii) Interference effects can be observed due to fluorescence of the diffracting crystal material itself. The crystal then starts to behave as a secondary target and decreases the degree of polarization. Thus, Wobrauschek and Aiginger (1983) estimated that the degree of polarization in the Cu $K\alpha$ line from a copper anode tube, diffracted by a single crystal of copper (Table 9.10), was reduced to about 90% due to fluorescence of the copper crystal.

The most effective instrumental arrangement is based on curved diffracting crystals which satisfy Johansson fully-focusing criteria (see section 10.10). The polarized beam is then focused on to the sample. Minimum detection limits down to 0.3 μg/g Mn in aqueous solution have been reported (Aiginger and Wobrauschek, 1983). Application of the technique is likely to be limited to specific ranges of elements

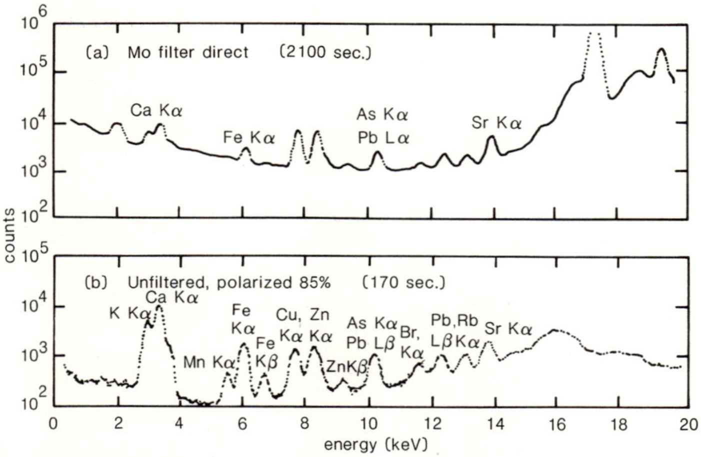

Figure 9.41 Comparison of x-ray spectra from a sample of NBS orchard leaves obtained by excitation with a molybdenum anode x-ray tube using (*a*) direct excitation using a molybdenum primary beam filter and (*b*) unfiltered polarized radiation. (Ryon, 1977.)

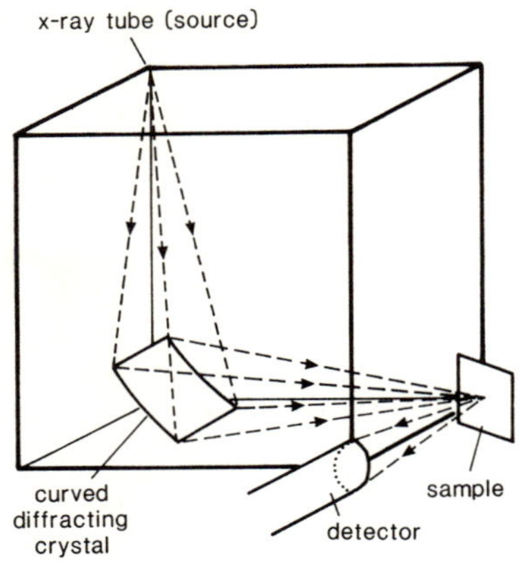

Figure 9.42 Excitation using Bragg diffraction of characteristic tube lines at $2\theta = 90°$. Efficient excitation using Cr, Cu or Mo tube lines may be achieved by selecting the appropriate diffracting crystal material.

that may be effectively excited by essentially monochromatic radiation in aqueous solution have been reported (Aiginger and Wobrauschek, 1983). Application of the technique is likely to be limited to specific ranges of elements that may be effectively excited by essentially monochromatic radiation.

9.12 Radioisotope excitation

An alternative means of exciting samples for ED–XRF analysis is by replacing the x-ray tube by a radioactive isotope source. Isotopes that decay by electron capture are normally selected since they intense x-ray emitters, and energetic beta emanations are absent. Beta particles would interact with the sample and cause unwanted bremsstrahlung. The sources most frequently selected in XRF applications are listed in Table 9.11.

One excitation configuration is shown in Figure 9.43. The isotope source is manufactured as an annulus with adequate shielding to ensure that primary radiation cannot enter the detector directly. Source, sample and detector are coupled as closely as possible to minimize the activity required to generate an adequate count rate.

Two of the most widely used sources are ^{109}Cd, which emits silver *K*-line radiation in the range 22–25 keV (plus a

Table 9.11 Radio isotope sources for x-ray fluorescence analysis

Isotope source	Decay mode	Half life	Energy and relative intensity per decay of x-ray emission
^{55}Fe	EC	2.7 y	Manganese x-rays:
			5.90 keV } (26%)
			6.49 keV }
^{109}Cd	EC	1.24 y	Silver x-rays:
			21.99 keV } (85.3%)
			22.16 keV }
			24.91 keV } (14.3%)
			24.94 keV }
			25.51 keV (2.4%)
			γ 88.04 keV (3.8%)
^{241}Am	alpha	433 y	Neptunium x-rays:
			13.76 keV } 28.0%
			13.94 keV }
			16.84 keV
			17.75 keV
			17.99 keV
			20.78 keV
			γ 26.36 keV (2.5%)
			γ 33.20 keV (0.11%)
			γ 43.42 keV (0.072%)
			γ 55.54 keV (0.017%)
			γ 59.54 keV (36.30%)
			γ 67.50 keV (0.00048%)
			γ 70.00 keV (0.023%)
			γ 75.80 keV (0.00011)
^{57}Co	EC	0.74 y	Iron x-rays:
			6.40 keV
			7.06 keV
			γ 14.41 keV (9.5%)
			γ 122.07 keV (85.6%)
			γ 136.43 keV (10.6%)

EC = electron capture, alpha = alpha decay, y = years.
Data from Erdtmann and Soyka (1979) and White and Johnson (1970).

weaker gamma line at 88 keV) and ^{241}Am, which emits a strong gamma line at 59.6 keV as well as weaker lines at 43.4, 33.2, 26.4 and 12–21 keV (Np x-rays) (Table 9.11).

In many respects the analytical potential of radioisotope excitation is analogous to secondary target excitation but without the flexibility of being able to change the source emission characteristics. Thus, both forms of excitation are essentially monochromatic, both cause some background

Table 9.10 Crystal–x-ray tube line combinations suitable for Bragg diffraction at $2\theta = 90°$

X-ray tube line (keV)			Diffracting crystal	d (Å)	Bragg angle (°2θ)
Cr	*Kα*	5.41	Mg (1,1,0)	1.605	91.111
Cr	*Kα*	5.41	Ta (2,0,0)	1.650	87.934
Cu	*Kα*	8.04	Cu (1,1,3)	1.090	90.055
Cu	*Kβ*	8.90*	LiF (2,0,0)	2.014	87.34
Mo	*Kα*	17.44	Mo (6,1,1)	0.510	88.260

d = crystal lattice spacing. * 2nd order.
Data from Wobrauschek and Aiginger (1983) and Ong and Randall (1978).

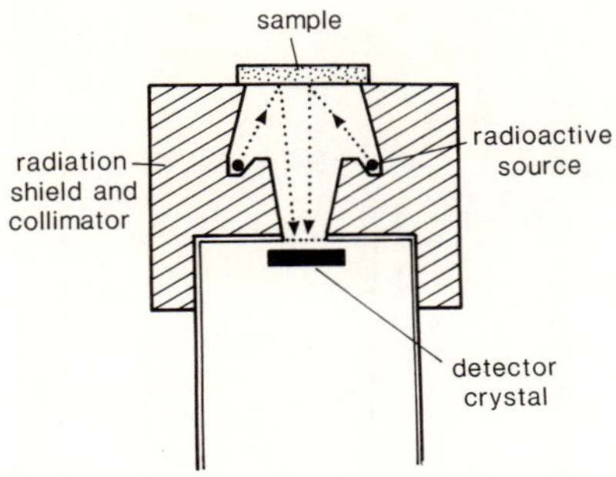

Figure 9.43 Excitation geometry using radioisotope sources. The source is arranged here as an annulus, and close spatial coupling is necessary between source, sample and detector if source activity is to be restricted to reasonable levels.

enhancement due to Rayleigh and Compton scatter within the sample, and both may be used effectively for restricted ranges of elements beyond which the excitation efficiency falls off to an unacceptable level.

As an illustration of these phenomena, the spectra of a silica gel sample spiked with various elements and excited by the silver lines of a ^{109}Cd source is shown in Figure 9.44, taken from Spatz and Lieser (1977). The Compton scatter peaks from the silver $K\alpha$, $K\beta$ lines emitted by this source form a prominent feature of this spectrum. Estimates of the detection limits that may be obtained using both a 5 mCi ^{109}Cd source and a 100 mCi ^{241}Am source are plotted in Figure 9.45, also taken from Spatz and Lieser (1977). This diagram has a similar form to that for secondary target detection limit data shown in Figure 9.38.

Radioisotope ED–XRF analysis has been used by Labrecque *et al.* (1982) to determine strontium and yttrium in laterites using molybdenum as internal standard. Samples (20 mg aliquots of powder) were excited using a 7 mCi ^{109}Cd source. Analytical precisions were about 10% (relative standard deviation) and detection limits for 1000 s counts were about 75 ppm for either element.

Excitation using the 59.5 keV line of a ^{241}Am source on the 122 keV line of ^{57}Co allows the direct analysis of *K*-lines of some rare-earth elements and Ba in the 30–40 keV region of the spectrum. A hyperpure germanium detector is required as conventional Si (Li) detectors do not have adequate efficiencies at these higher energies. Thus, Cotten and Hall (1976) determined light rare-earth elements and barium in low-grade ore materials (containing 0.1% REE) using a ^{57}Co excitation source. Laurer *et al.* (1982) analysed soil and sediment samples using a 40 mCi ^{57}Co source. Elements determined included Fe, Zr, Ba, La, Ce, Pr, Nd, Sm, Pb, Th and U. Ba to Sm were determined from *K*-lines in the 30 to 40 keV region of the spectrum and Pb, Th and U from *K*-lines between 70 and 100 keV. In order to confirm the reliability of calibrations, three reference samples were analysed and data for andesite USGS: AGV-1 and granodiorite USGS: GSP-1 counted for 23 511 s and 46 581 s respectively are listed in Table 9.12. For more usual count times of 1000 s using this excitation source, detection limits (six sigma) were generally in the range 20 to 40 ppm (Table 9.12).

In applications involving high-energy x-rays, it is important to give due consideration to the relationship between sample thickness and critical penetration depth (section 8.7).

9.13 Total reflection of primary beam

A novel technique that substantially reduces background in ED–XRF spectra was proposed by Wobrauschek and Aiginger (1975). The technique depends on the principle that if a collimated beam of x-rays impinges on an optically flat surface at a grazing angle of a few millirad of arc (5 to 20 minutes of a degree), total reflection will occur. Thus if a

Table 9.12 ED–XRF analysis of andesite USGS: AGV-1 and granodiorite USGS: GSP-1 using radioisotope excitation from a ^{57}Co source

	Andesite USGS: AGV-1			Granodiorite USGS: GSP-1			
	Abbey (1983) Expected	Laurer *et al* (1982) Analysed	CV	Abbey (1983) Expected	Laurer *et al* (1982) Analysed	CV	Detection limit
Fe_2O_3	6.78%	5.85	3.8	4.30	4.21	4.2	1.54
Zr	230	266	6.8	500	479	2.1	196
Ba	1200	1262	0.16	1300	1359	0.15	22
La	36	32	3.1	195	191	0.52	22
Ce	71	72	1.4	360	426	0.23	20
Pr	(7)	25	4.0	(50)	67	1.5	20
Nd	37?	45	2.2	190?	231	0.43	20
Sm	5.9	7	14.3	25?	29	3.4	16
Pb	33	44	2.3	54	67	1.5	46
Th	6.4	5	40	105	111	0.90	42
U	1.95	3	33	2.1	4	25	24
Count time		23571 s		46581 s			

All data in ppm except Fe_2O_3: % oxide (total).
Detection limits represent the six-sigma limit of determination appropriate for samples counted for 1000 s.
CV = coefficient of variation = 100 × standard deviation/mean analysis, based in this case on simple counting errors.

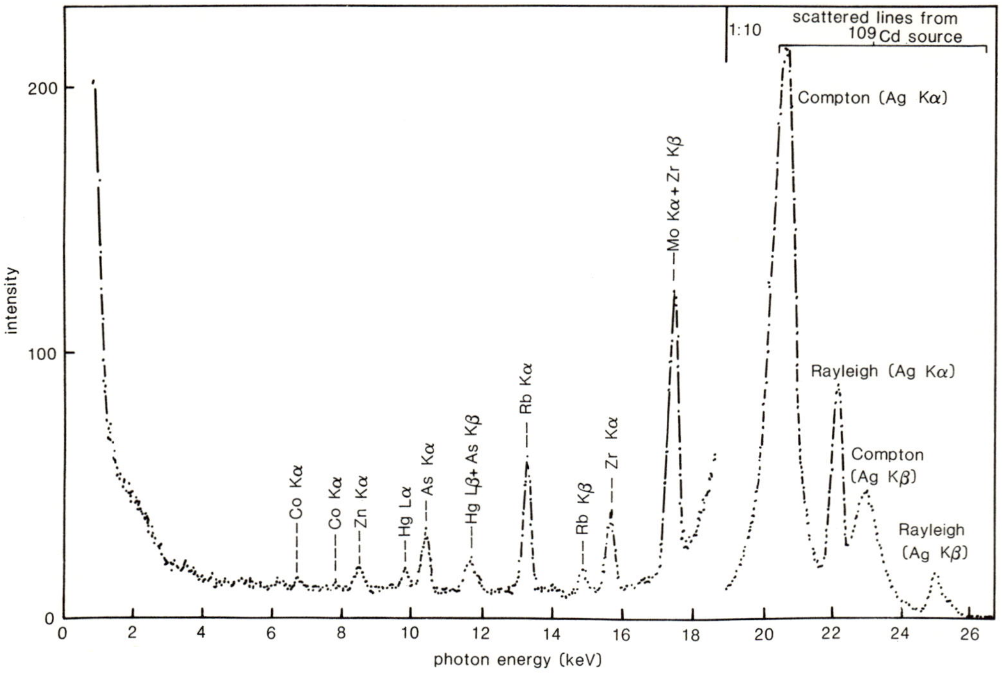

Figure 9.44 Spectrum of a sample containing 100 μg/g Ti, Cr, Co, Zn, Hg, As, Rb, Mo, Gd, Sb and Ba and excited using the silver *K*-lines of a ^{109}Cd radioisotope source. (Spatz and Lieser, 1977.)

small volume (1 to 5 μL) of samples that has been taken up into solution is evaporated on to this reflector, x-ray fluorescence of the sample film will occur, but the primary tube photons that do not interact with the sample will be totally reflected. These photons will not therefore contribute to the sample background spectrum. The experimental arrangement is shown in Figure 9.46. Quartz, germanium or gold substrates may be used as reflectors, and Wobrauschek and Aiginger (1980) and Knoth and Schwenck (1978, 1980) have discussed various theoretical and experimental aspects of the technique. Elemental abundances in the nanogram range may be detected, although control of

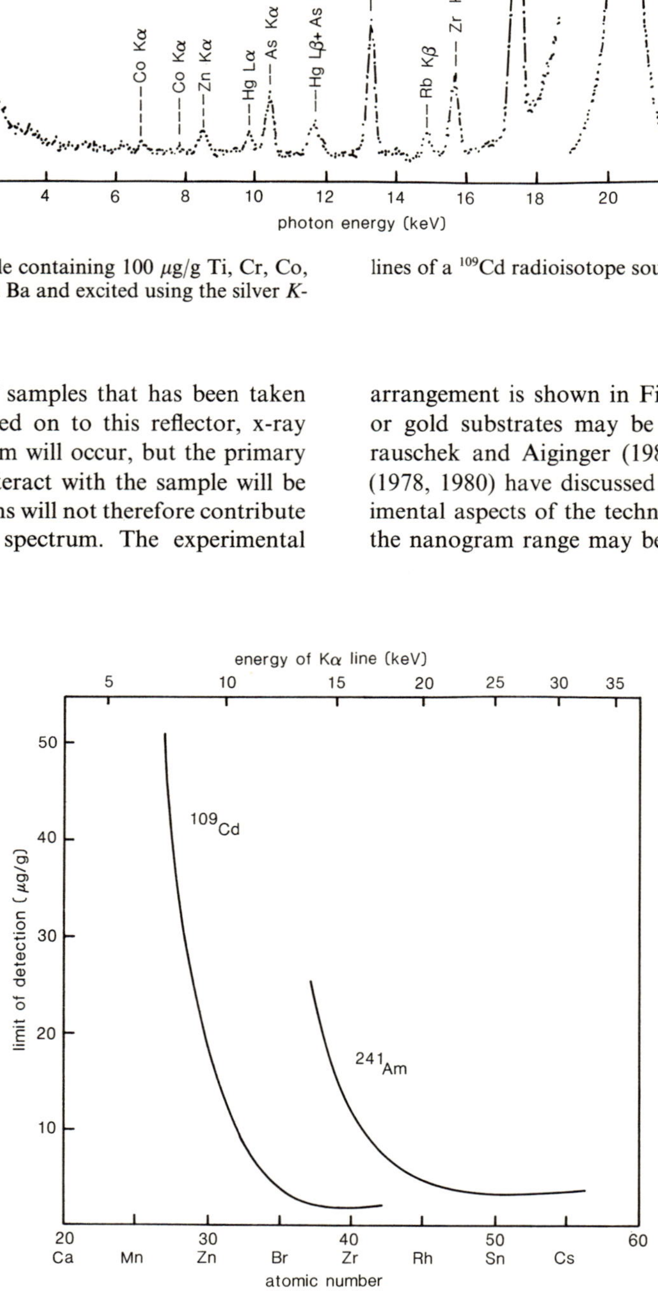

Figure 9.45 Detection limits of elements excited using the radioisotope sources. Data for the sources ^{109}Cd and ^{241}Am. (Data from Spatz and Lieser, 1977.)

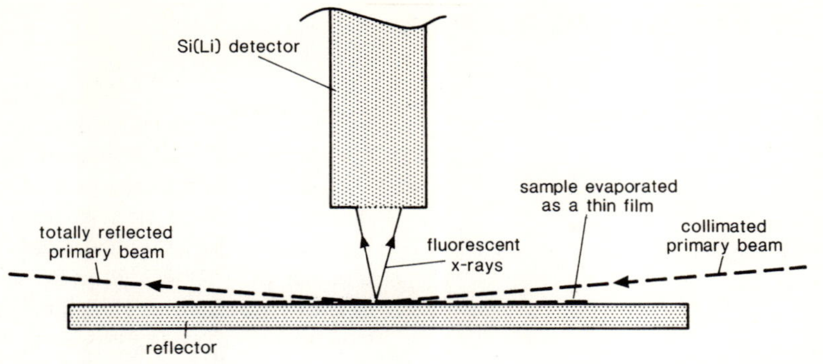

Figure 9.46 Schematic diagram of an excitation geometry based on total reflection of the primary x-ray beam.

the uniformity and thickness of the evaporated film are important practical considerations.

9.14 Concluding remarks

Energy dispersive XRF analysis is a technique that has not yet been fully exploited in applications to silicate rock analysis. Instrumentation is still benefiting from advances in the technology of silicon devices, microcircuits and fast microcomputer-based digital analysers. However, with the instrumentation currently available, traditional doubts about inferior energy resolution and poor count rate capability are no longer limitations, providing the appropriate excitation and spectrum deconvolution strategies are followed. Indeed, ED analysis procedures are now available that have established equivalence with WD–XRF in most aspects of the routine analysis of silicate rocks. Further developments in energy dispersive–XRF instrumentation are likely to enhance the standing of this technique as a routine geochemical tool.

10 Electron probe microanalysis

10.1 The development of microprobe techniques

Microprobe analysis techniques differ from most other techniques used to analyse geological materials in that they involve the excitation and chemical analysis of selected areas of diameter as small as a few microns on the surface of samples. Specimens must therefore be prepared as polished geological thin sections mounted on a glass slide backing, or alternatively as samples polished in a resin block. Microprobe techniques have, therefore, the ability to determine the composition of individual minerals in thin section or even concentration variations within a single grain.

The most widely used of these techniques is electron probe microanalysis. An electron beam, accelerated to a selected voltage, normally in the range 15 to 30 kV, is focused on the surface of a sample. Interactions between this primary electron beam and the sample cause a number of phenomena, including the generation of x-rays characteristic of the atoms of the excited sample. The intensity of these x-rays is measured using wavelength dispersive or energy dispersive spectrometers. After correction for matrix effects, count data are compared with data from minerals of standard composition in order to quantify the analysis. Due to the unavoidably higher intensity of the continuum formed by electron excitation, detection limits are of the order of 50 to 100 times poorer than those expected for bulk analysis by x-ray fluorescence.

The first experimental electron probe microanalysis instruments were developed in the early 1950s following the pioneering work of Castaing and Guinier (1949). However, it was not until the 1960s that commercial instrumentation became widely available. This stimulated a rapid expansion in this new field of microanalysis, since detailed mineral data, which had previously been virtually unobtainable, became available on a routine basis. A notable advance in the technique followed the work of Sweatman and Long (1969) who brought together the procedures necessary to ensure reliable quantitative analysis in mineralogical applications.

There is considerable similarity between the instrumentation used for electron probe microanalysis and that for scanning electron microscopy. The principal difference is that in the latter technique, the electron column is designed to allow measurement of high resolution backscattered electron images, x-ray analysis being of secondary importance. Indeed, modern microprobe instrumentation is normally designed as a hybrid between these instruments, allowing measurement of high-resolution scanning electron photomicrographs as well as incorporating a fully quantitative x-ray analysis capability. Such hybrid instruments are normally referred to as *scanning electron probe microanalysers*.

The original microprobe concept must now compete with a number of related techniques for selected area and surface analysis, which are normally referred to under their appropriate acronyms: SEM, TEM, ion probe (SIMS), PIXE, ESCA (XPS). The essential features of these techniques will first be summarized before returning to a more detailed consideration of electron probe microanalysis. The geological applications of these related techniques will be considered in Chapter 11.

10.2 Microbeam techniques

(*a*) *Electron probe microanalysis*. Polished surfaces of samples are excited with an electron beam accelerated to between 15 and 30 kV and focusable to a diameter of less than 1 μm. This electron probe is arranged to strike the sample at right angles to its surface. Characteristic x-rays generated within the sample are detected using wavelength dispersive or energy dispersive spectrometers for quantitative analysis of volumes a few microns across. An important design consideration is the take-off angle of the x-ray spectrometers; there are advantages if this is as large as possible. Instruments have been designed with 75° spectrometer take-off angles. In more recent hybrid instruments this is normally restricted to 40° to 52° to improve performance in the scanning electron microscope mode. Analyses can be obtained for all elements between boron and uranium in the Periodic Table, subject to sensitivity considerations.

(*b*) *Scanning electron microscope* (*SEM*). The design concepts of the SEM are similar to those of the electron probe microanalyser. However, the electron column and sample stage are designed to record high-resolution scanning electron microscope images using backscattered and secondary electron signals emitted when the electron probe is scanned across the surface of the sample. Samples are normally three-dimensional in shape (not necessarily polished), and the probe is often scanned across the surface of the sample at an average angle of incidence of about 45°. Image resolutions of between 3 and 7 nm may be obtained using probe accelerating voltages of between 5 and 20 kV. The measurement of x-ray emissions is normally a secondary design consideration, and fully quantitative analyses are often difficult to achieve, due to poor control over sample geometry and low spectrometer take-off angle (often 20-45°).

(*c*) *Transmission electron microscope* (*TEM*). Samples presented as thin foils (typically 0.15 μm thick for silicate materials) are excited with a beam of electrons accelerated to between 60 and 200 kV (or more in highly specialized instruments). The electron beam transmitted through the foil is focused to allow the recording of transmission electron photomicrographs with very high resolution (down to the 0.1–1 nm level). Alternatively, electron diffraction patterns may be measured to identify crystal structures with a spatial resolution of about 20 nm. The x-ray signal generated as the electrons are transmitted through the silicate foil may be recorded using energy dispersive detectors, allowing semi-quantitative analysis (data must usually be normalized to 100%) with a spatial resolution of perhaps 30 nm.

(*d*) *Ion probe micro-analyser*. This is a form of secondary ion mass spectrometer (SIMS). The surface of a polished sample is excited with a beam of ions (normally $^{16}O^-$) accelerated through 10 to 30 kV. The surface of the sample is sputtered away, generating secondary ions composed of material from the sample itself. These are collected by ion lenses for mass analysis. The ion probe is capable of analysing discrete isotope ratios in individual minerals (for example $^{87}Sr/^{86}Sr$ in calcite or $^{207}Pb/^{206}$ Pb in zircon) and can determine trace elements down to the ppm level (in comparison with about 100 ppm by electron probe). Spatial resolution for analysis can be as low as 1 μm, but is more often in the range 10 to 20 μm. The general usefulness of the ion probe depends on the successful correction of complex atomic and molecular interferences in the mass spectrum, and for this reason isotope ratio measurements are often limited to relatively simple mineral matrices.

(*e*) *Particle induced x-ray emission* (*PIXE*). This can be used for bulk analysis with a defocused beam or for selected area analysis with a beam focused to 1–10 μm. In the latter configuration the technique is sometimes known as a nuclear microprobe. Samples are normally excited with a beam of protons (hydrogen nuclei, H^+) accelerated to between 2 and 5 MeV (alpha particles can also be used). The characteristic x-ray spectrum excited from the sample by interaction of this beam is measured on an energy dispersive detector. Nuclear interactions also cause the emission of characteristic gamma radiation, which may be detected using a germanium detector. Potentially, the background x-ray continuum is much lower than for electron-probe-excited spectra, giving much improved detection limits (down to the ppm level at best). However, high technology instrumentation is required to generate the accelerated proton beam (a Van de Graaff generator or a cyclotron), reducing the widespread availability of PIXE to specialized installations. The technique has particular applications in the determination of the light elements from gamma rays emitted as a result of proton-induced nuclear transformations.

(*f*) *Electron spectroscopy for chemical analysis* (ESCA). ESCA, also known as *x-ray photoelectric spectroscopy* (XPS), is not a microanalysis technique, but rather a highly specialized surface analysis technique. Samples are excited with a very intense beam of low energy x-rays (usually Al $K\alpha$ or Mg $K\alpha$). The photoelectric effect causes electrons to be ejected from the outer orbitals of atoms in the sample. These electrons have a characteristic energy, equivalent to the energy of the incident photon *less* the ionization potential of the appropriate electron orbital. Excited electrons are collected and measured using an electron energy analyser. From these data it is possible to determine the binding energy of outer orbital electrons of specific elements in the sample. The binding energy for any element varies from mineral to mineral and is a function of both oxidation state and sometimes coordination number (degree of chemical bonding) of that element. ESCA data can therefore assist in the interpretation of ambiguous chemical composition data for minerals. The depth of sample analysed constitutes the first 5 to 10 nm of surface layer, and extreme care must be taken to minimize surface contamination.

10.3 Instrumentation for the electron probe microanalyser

10.3.1 *Instrument configuration*

The overall components of an electron microprobe are shown in Figure 10.1 and consist of:

(i) An electron gun and column to generate the electron beam and focus it onto a sample

(ii) A specimen stage and optical microscope to view and set up areas on the sample for analysis

(iii) One or more x-ray spectrometers (wavelength dispersive and/or energy dispersive) to record x-ray emission intensities for analysis

(iv) A vacuum system to maintain the electron and x-ray flight paths under high vacuum (10^{-5} to 10^{-6} Torr)

(v) Most modern analysers benefit from a dedicated computer to control and coordinate the movement and operation of stage and spectrometers, to record x-ray data and to calculate matrix corrections

(vi) An imaging facility: in most microprobes, the electron beam can be scanned in a 'raster' over the surface of a sample. The intensity of the backscattered or secondary electron emission signal is then displayed as a scan on a cathode ray tube so that the movement of electron beam and CRT spot are synchronized. High resolution scanning electron images may be viewed and recorded. This facility both permits the examination under high magnification of areas to be analysed, and allows individual phases to be distinguished on the basis of differences in their effective atomic weight (since this controls the intensity of the backscattered electron signal).

The instrumentation required to form and control the electron probe will first be considered before turning to interactions that occur between the electron beam and sample.

10.3.2 *The probe-forming system*

The probe-forming system consists of an electron gun in which electrons are released and accelerated through a preselected potential gradient in the range 15 to 30 kV and a series of electromagnetic lenses and collimators which together constitute the electron column. In operation, this column must deliver an electron beam which may be focused on to the surface of a sample to at least as small as 0.5 μm

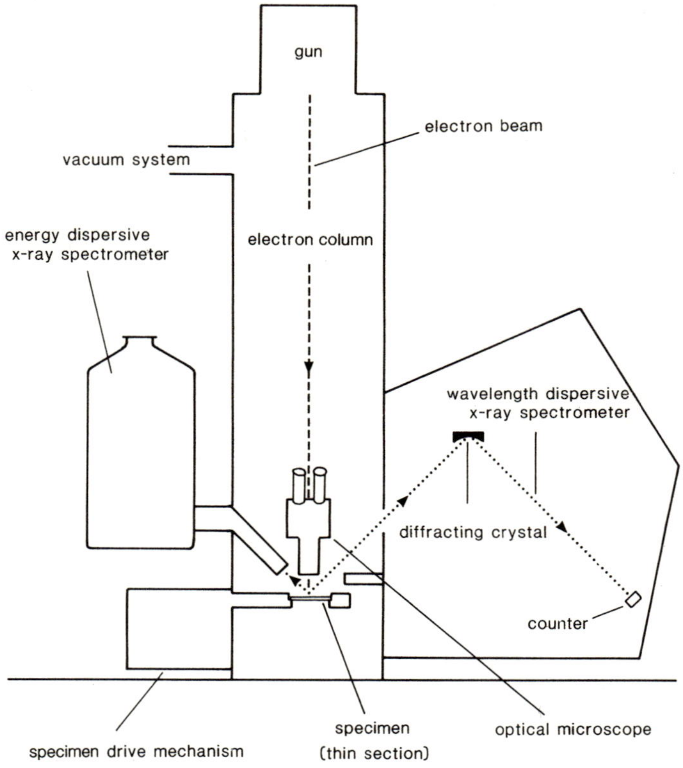

Figure 10.1 Overall configuration of an electron microprobe, showing relationship between the electron column, specimen and x-ray spectrometers. The electron beam is shown as a dashed line.

diameter (smaller in an SEM) to give a total current adjustable in the range 10^{-6} to 10^{-10} A (this may be as low as 10^{-12} A in SEM).

10.3.3 *The electron gun*

The source of electrons in the gun is normally a filament made of tungsten wire 100–150 μm in diameter and bent into the shape of an inverted 'V' (the so-called tungsten hairpin filament—see Figure 10.2). When an electric current of between 2 and 3 A is passed through this filament, the wire becomes incandescent, attaining temperatures of 2600–2700 K. Under these conditions, thermionic emission occurs, as electrons then have sufficient thermal energy to overcome the work function of the tungsten wire. If the filament is held at a large negative potential with respect to an anode (which is itself earthed to the rest of the instrument), then these emitted electrons will be rapidly accelerated to the operating potential of the gun.

It is, however, essential to restrict the area from which electrons are emitted to the very tip of the filament, since it is this source that is demagnified by the rest of the electron column. This is achieved with a grid (called the *Wehnelt cylinder*) placed between filament and anode in a configuration similar to an electrical triode valve (Figure 10.2) The grid is held at a small *negative* bias with respect to the filament, so that the whole of the filament except its tip is held in a field of negative potential. The electrostatic fields created by the grid suppress the emission of electrons except from a small area at the tip of the filament. These electrons are accelerated by the potential gradient between grid and anode, which themselves behave as a simple electrostatic lens. This causes the divergent beam of electrons emitted from the filament to be focused at a rather crude cross-over point of diameter 10 to 100 μm within the gun. It is this cross-over image that is subsequently demagnified by the rest of the electron column.

In operation the gun has to be adjusted rather sensitively for stable operation. The minimum filament current required to induce thermionic emission varies with the age of the filament since its diameter is progressively reduced by evaporation of tungsten during use. However, once this threshold has been exceeded, the self-regulatory action of the control circuit in which filament and grid are connected by a bias resistor (Figure 10.2) ensures that a stable beam current is emitted despite further increases in filament current (Figure 10.3). To preserve filament operating life, the filament current must be set on the plateau just above the knee of the saturation point. Adjustment to this point both ensures stable operation and minimizes the rate of evaporation of tungsten which would lead to premature failure of the filament. Filament life may be restricted to as little as 40 hours of operation. The critical factor is the quality of vacuum within the gun, and in particular the residual partial pressures of water vapour and oxygen (Bloomer, 1957). These gases become ionized by collision with energetic electrons in the gun. The heavy positively-charged ions so formed are accelerated by potential gradients in the opposite direction

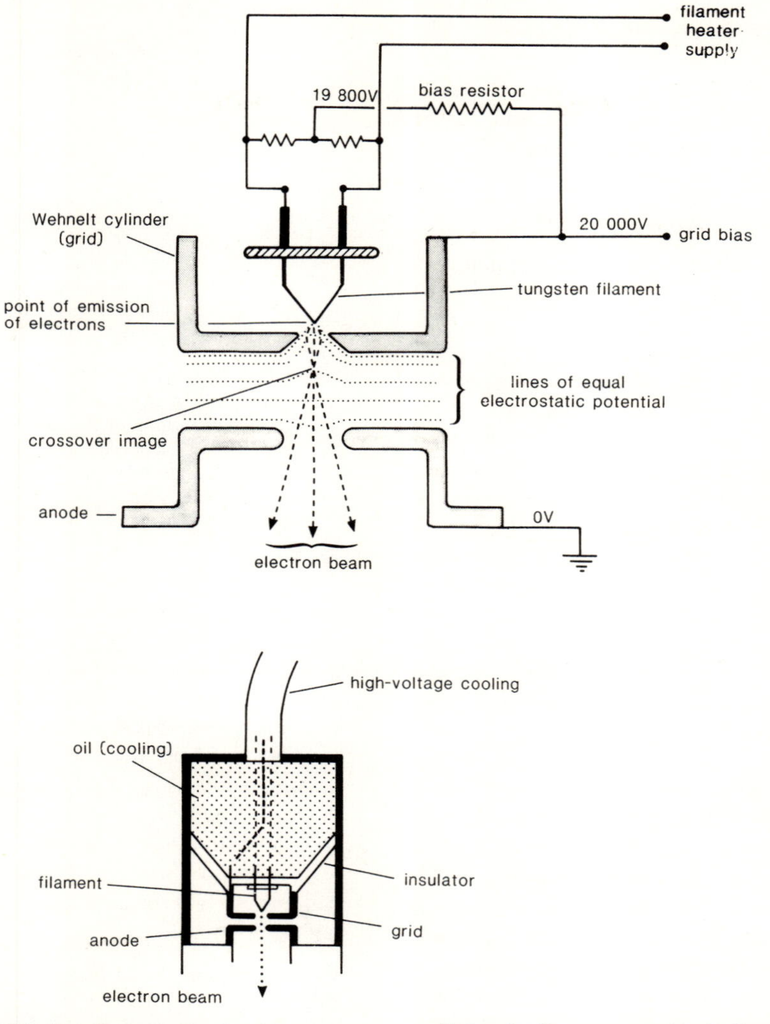

Figure 10.2 Diagram of a self-biasing electron gun, showing tungsten hairpin filament, Wehnelt cylinder and anode. The dotted lines contour areas of equal potential created when a negative bias is applied to the filament assembly. Indicated operating potentials are approximate. Inset diagram shows overall configuration of gun.

to electrons. On impact with the filament, tungsten is sputtered away ('*ion erosion*'). This eventually causes local overheating and premature filament failure.

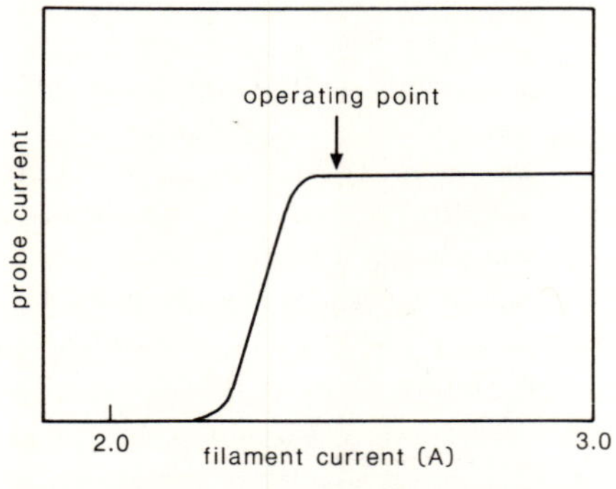

Figure 10.3 Effect of increasing the current passing through the filament on electron probe current. Electrons are not emitted from the filament until it attains a sufficiently high temperature to induce thermionic emission. The gun becomes saturated when increase in filament current causes no further increase in probe current. The normal operating point is indicated.

For these reasons, filament changing is the most frequent maintenance requirement on an electron microprobe. During this operation it is usually necessary to clean both grid and anode to remove cracked hydrocarbon (from diffusion pump oil) and evaporated tungsten deposits. The grid and anode are normally manufactured from chromium that must be given a highly polished finish to prevent the formation of corona discharges and consequent flashovers when the operating bias is applied.

For each operating kV, a unique position exists for the distance of the filament tip behind the grid so that the gun will operate at maximum brightness (measured in amps cm^{-2} $sterad^{-1}$). However, such accurate alignment is required only in the most demanding SEM work where ultimate image resolution is required. In microprobe work it is normal to position the filament further behind the grid where more stable emission characteristics and maximum operating life may be obtained.

Accurate alignment of the filament tip along the optical axis of the column is, however, essential for stable operation. Two orthogonal controls, which are adjustable externally, are normally provided to centre the filament. During the first thirty minutes of operation, some new filaments tend to wander and require careful attention to re-centring after that time.

The operating potential of the gun is measured by the potential difference between grid and anode, and not between filament and anode as might be otherwise expected.

10.3.4 *Electromagnetic lenses: general principles*

The diverging cone of electrons emitted from the gun must be focused and demagnified by a series of two or three electromagnetic lenses.

Each electromagnetic lens is constructed as a hollow cylinder containing a large number of turns of copper wire surrounded by a shroud of soft iron (Figure 10.4). The electron beam passes down the hollow axis of the cylinder. The iron shroud and copper windings must possess as near perfect circular symmetry as possible, to minimize astigmatic distortions (see section 10.4.4). Part of the iron shroud at the lens centre is cut away to form a gap (the pole pieces). Normally the gap is formed by machining a spacer made of a non-magnetic material such as brass to separate the pole pieces. The lens is excited by passing a current through the windings that may be adjusted by the user. This creates an intense electromagnetic field across the pole pieces that serves to focus the electron beam.

The mechanism of focusing works on the principle that when a negatively charged electron passes through a magnetic field, it suffers a force of deflection. The direction of this force is given by *Fleming's left-hand rule* (Figure 10.5). As each electron passes through the magnetic field, it suffers a radial force that causes the electron to be deflected in direction perpendicular to the axis of the lens. The combination of electron velocity and force of deflection induces a spiralling motion that results in the electron beam being brought into focus at a point on the axis of the lens (see

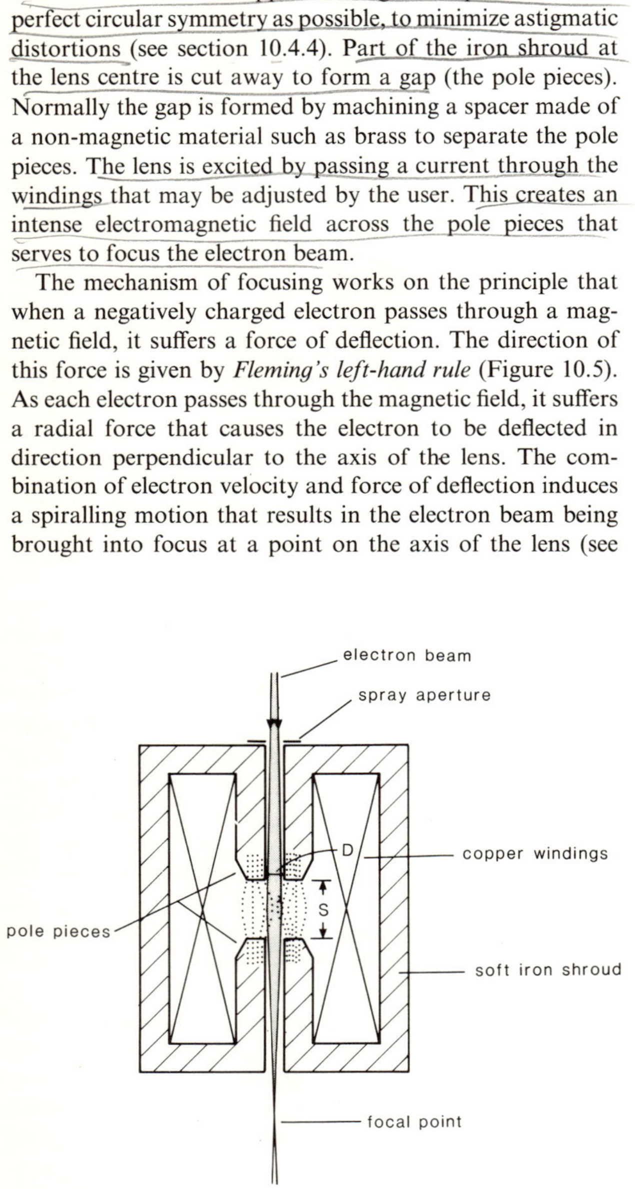

Figure 10.4 Electromagnetic lens showing deflection of the electron beam by the intense magnetic field (the lines of force of which are shown as dotted lines) created across the pole pieces; dimensions have been exaggerated for clarity.

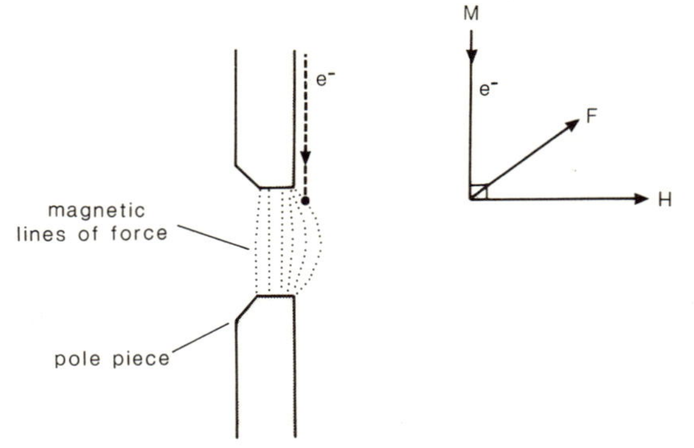

Figure 10.5 Fleming's left-hand rule as applied to an electron trajectory adjacent to the pole piece of an electromagnetic lens. *M* is the direction of motion of the electron, *H* is that component of the magnetic field created across the pole pieces which is perpendicular to the electron trajectory and *F* is the resultant force of deflection induced on the electron.

Figure 10.4). This electron–lens interaction is such that the velocity of the electron beam (corresponding to the nominal gun operating potential) is not significantly affected.

In a way analogous to an optical lens, the strength of an electromagnetic lens is measured by its focal length (f). The stronger the lens, the shorter the distance between focal point and pole pieces. In the case of an electromagnetic lens, the focal length is dependent on the lens bore (of diameter D), the distance between pole pieces (S), the number of turns of copper wire (n), the current passing through these windings (i) and the gun accelerating potential (V_0) according to the equation

$$f = \frac{k\,.(S+D)\,.\,V_0}{(ni)^2} \qquad (10.1)$$

where k is a constant which is, however, not quite independent of i or V_0. The parameters S, D and n are fixed by design. The strength of the lens during operation therefore increases according to the square of i and decreases with higher operating potentials.

10.3.5 *Lens aberration effects*

The beam of electrons is never brought into perfect focus by an electromagnetic lens. Several phenomena cause the cross-over to become slightly diffuse, the most serious being spherical aberration.

(*a*) *Spherical aberration*. This results because electrons passing through an electromagnetic lens at some distance from its optical axis are brought more strongly into focus than electrons closer to the lens axis (Figure 10.6). The point of electron focus is therefore in fact a '*disc of least confusion*', the diameter of which (d) is given by

$$d = \tfrac{1}{2}\,.\,C_s\alpha^3 \qquad (1.2)$$

where C_s is the spherical aberration constant of the lens and is fixed by design, and α is the semi-angle of convergence of the cone of electrons which are brought into focus.

(*b*) *Chromatic aberration*. This is a much smaller defocusing effect and is caused by variations in the energy of electrons

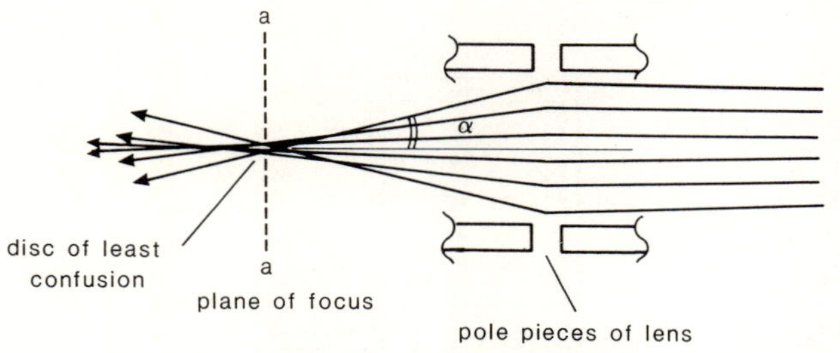

Figure 10.6 Spherical aberration: the electron beam is brought into focus as a beam of least confusion in plane *a–a*. α is the semi-angle of convergence.

emitted by the gun. Electrons ejected from the filament are emitted with random thermal motion equivalent to perhaps ±0.5 V. Noise on the high-voltage power supplies can, in addition, contribute up to ±1 V to the fluctuation in emission energies of electrons from the gun. Both these effects will cause small changes in the effective focal length of electromagnetic lenses since this parameter is dependent on V_0 (see equation (10.1)).

10.4 Electron column design

In conventional electron probe microanalysis, the smallest size of sample that may be successfully analysed is limited by the scatter of electrons within the sample (section 10.6), not just by probe diameter. There is, therefore, little analytical advantage to be gained if the electron beam is focused on the sample to a diameter much smaller than 0.5 μm. This can be achieved adequately using two electromagnetic lenses, as in the general column arrangement shown in Figure 10.7. However, the same design criteria do not apply in scanning electron microscope instruments designed to achieve the highest *image resolution*. Three electromagnetic lenses are often used in such instruments to generate the smallest spot size possible by minimizing spherical aberration.

By analogy with optical microscopy, the two electromagnetic lenses of a conventional microprobe analyser are normally called the condenser and objective (or final) lenses, and they function as follows.

10.4.1 *The condenser lens*

The function of the condenser lens is to form an intermediate demagnified image of the gun cross-over and to permit control of the total electron current (the probe current) reaching the sample. The typical simplified configuration is shown in Figure 10.8. The lens is protected at the gun end by a spray aperture, which prevents electrons striking and contaminating the bore of the lens. In order to understand the way in which the lens may control the probe current, it is necessary to consider the effect of changing its focal length in conjunction with an associated beam-limiting aperture.

With no current passing through the condenser lens windings, the divergent beam of electrons originating from the gun passes through the lens unfocused (Figure 10.8*a*). A

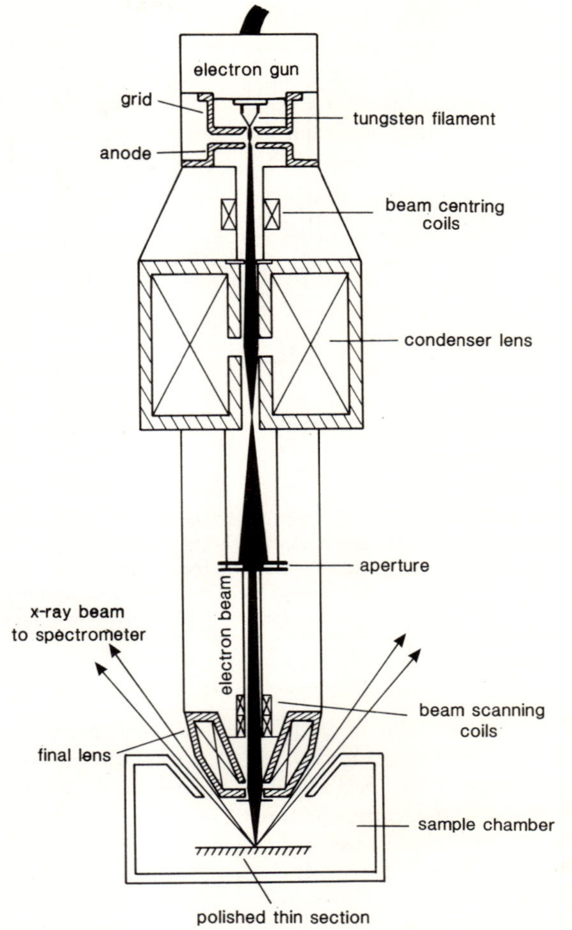

Figure 10.7 Schematic configuration of the electron column showing the gun, electromagnetic lenses and apertures. The x-ray beam, available for measurement by the x-ray spectrometers, is shown.

very small cone of this beam is transmitted through the beam-limiting aperture.

At low winding currents, the lens brings the electron beam weakly into focus until a point is reached where the beam is focused on the beam-limiting aperture itself. At this point the maximum current is transmitted down the column (Figure 10.8*b*).

As the condenser lens current is increased further, the focal length progressively shortens and a smaller and smaller cone of electrons is able to pass through the beam-limiting aperture. This condition represents the normal operating state for routine analysis. The effect on probe current of these changes in lens winding current is summarized in Figure 10.9. For routine microprobe analysis, it is usual to select a fixed value of the probe current and then to make small adjustments in the condenser winding current, as necessary, to compensate for any drift. Probe currents would normally be selected in the range 10–60 nA for analysis using wavelength spectrometers and between 1–6 nA for energy dispersive detection, depending on instrument design.

The accuracy of quantitative analysis depends, in part, on the stability of the probe current. This must normally be measured regularly throughout an operating session and

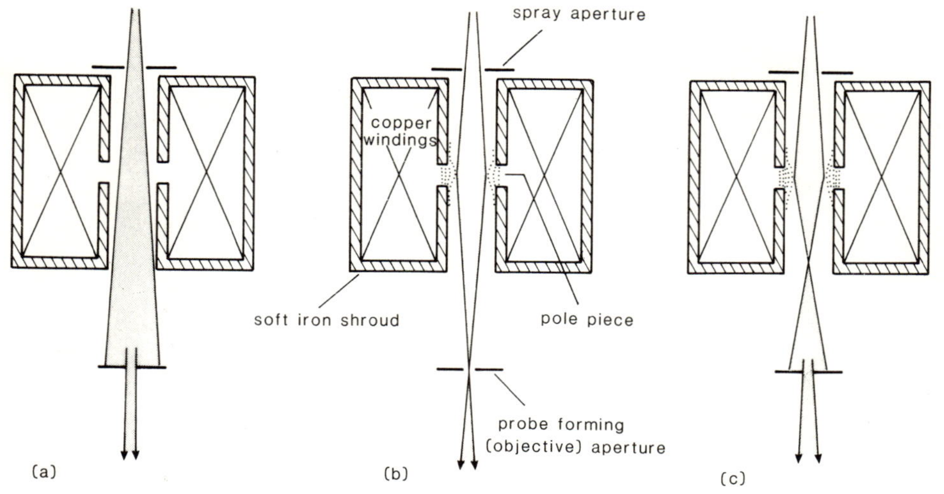

Figure 10.8 Effect of changing current through the windings of a condenser lens: (*a*) no current, electron beam is unfocused; (*b*) small current, lens is weakly excited with long focal length which in this diagram is shown to coincide with the condenser lens aperture; (*c*) larger current, electron beam is focused with shorter focal length. Under these conditions, a smaller cone of electrons is transmitted through the condenser lens aperture. In this way, the probe current reaching the sample may be adjusted. Dimensions have been exaggerated for clarity.

corrections for drift made if necessary. One instrumental method of compensating for drift is shown in Figure 10.10. The condenser lens beam-limiting aperture is in fact composed of two concentric apertures. The outer one defines the shape of the cone of electrons falling on the inner aperture. The fraction of electrons which strikes the inner aperture is used to monitor continuously the probe current. If this current changes, a signal is generated in a feedback circuit that causes the current passing through the condenser lens windings to change to compensate for this drift. High probe stability over long operating sessions can be achieved by this means.

In hybrid scanning electron microprobe analysers, the condenser lens is sometimes replaced by a tandem lens arrangement incorporating two pole pieces (Figure 10.11). This design represents a compact way of reducing spherical aberration, without the necessity of providing two independent lens assemblies.

10.4.2 *The objective lens*

The function of the objective lens is to demagnify further the intermediate gun image formed by the condenser lens and to focus it on to the surface of the sample. It is desirable that this lens should have as small a spherical aberration coefficient as possible so that optimum resolution is obtained in scanning electron imaging mode. However, this aim is not always compatible with a lens design that incorporates a

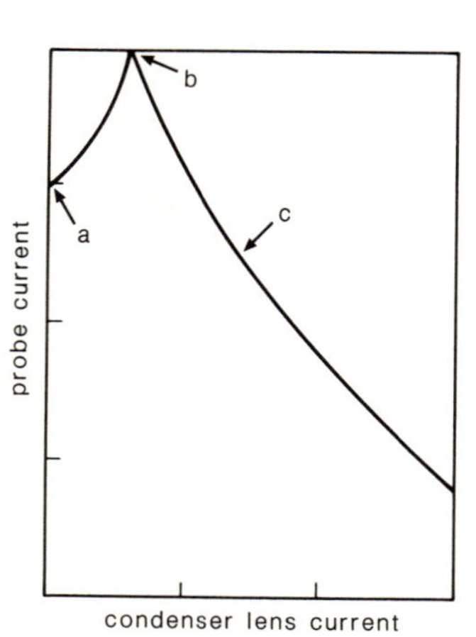

Figure 10.9 Change in probe current resulting from varying the condenser lens current. The points *a*, *b* and *c* correspond to the lens excitation conditions illustrated in Figure 10.8.

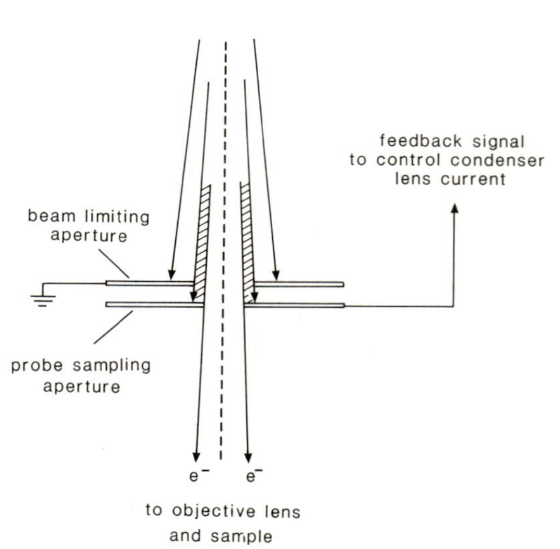

Figure 10.10 Concentric apertures used to monitor drift in probe current. The shaded portion of the electron beam falls on the probe sampling aperture and is used to create a signal which is used to control the condenser lens current to compensate for any long-term drift. For maximum effectiveness, this aperture assembly should be positioned close to the sample.

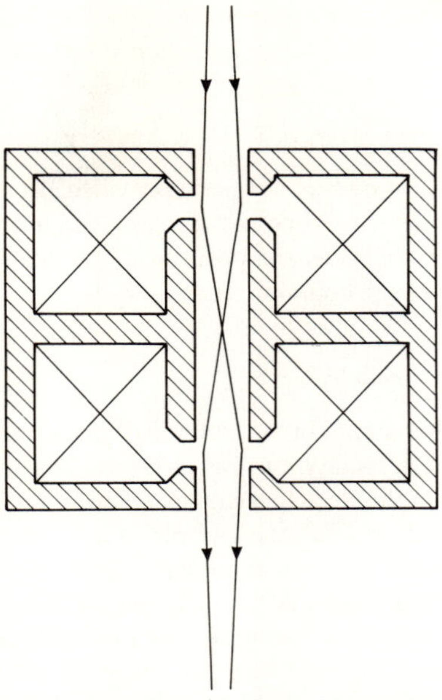

Figure 10.11 Schematic view of tandem condenser lens used as a compact way of reducing spherical aberration without having to provide two separate lenses.

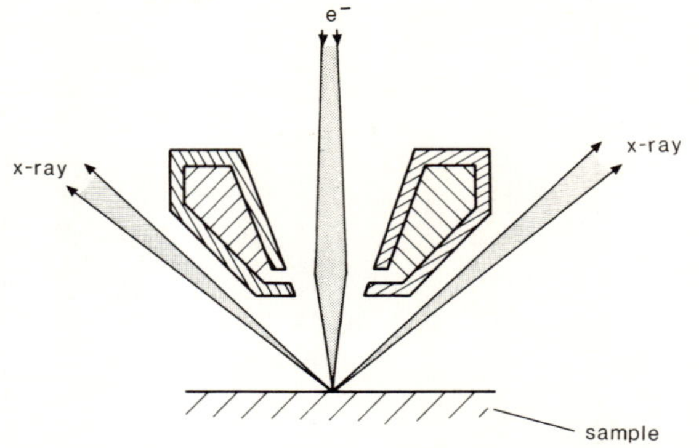

Figure 10.13 The mini-objective lens. This compact design offers low spherical aberration. Spectrometer take-off angles are restricted to a maximum of about 52°. Due to the large currents used to excite the lens, water cooling may be required to dissipate excess heat.

high x-ray take-off angle (the angle between the specimen surface and the entrance slit of the x-ray spectrometer). A high take-off angle is a desirable feature for x-ray analysis to minimize discrepancies due to imperfect sample polish or uncertainties in absorption corrections to count data.

One design of final lens, the so-called pin-hole lens, is shown in Figure 10.12. X-rays pass back through the bore of the lens before entering the spectrometer, permitting design of instruments with take-off angles as high as 75°. Modern hybrid scanning electron microprobe instruments often incorporated a *mini-lens* design as shown in Figure 10.13. This lens has a small number of turns of wire in its windings which carry a high current (n is small but i is large in equation (10.1)). Heat dissipation from these windings can be high, so that the design sometimes requires water-

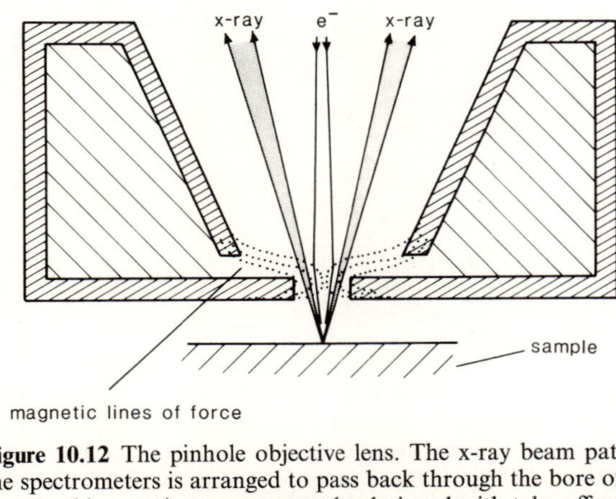

Figure 10.12 The pinhole objective lens. The x-ray beam path to the spectrometers is arranged to pass back through the bore of the lens. In this way, instruments can be designed with take-off angles as large as 75°. Lines of magnetic force created across the pole pieces are indicated.

cooling. However, the spherical aberration characteristics are lower than for some other lenses, allowing improved imaging capacity. Conversely, x-ray take-off angles must normally be restricted to between 40° and 52°. All final lens configurations suffer from the problems of finding room for the various optical lenses and prisms required adjacent to the sample to allow visual observation of the area to be analysed, by means of a binocular microscope.

10.4.3 *Spot size*

As is shown in equation (10.2), the minimum probe diameter at the point of focus of an electromagnetic lens depends, not only on the spherical aberration coefficient of that lens, but also on the semi-angle of convergence of the beam of electrons.

When the focal length of the condenser lens (and therefore also the probe current) is reduced (Figure 10.8), the semi-angle of the electron cone transmitted through the condenser lens aperture is also reduced. From this observation, it is expected that the smallest probe spot sizes will be observed at the lowest probe currents. The minimum probe diameter that may be selected depends on instrument design and the cleanliness and alignment of the electron-optical system. Data for a 'typical' column are plotted in Figure 10.14. This shows that under ideal conditions, the smallest probe diameter at 10^{-6} A is about 0.8 μm and that at 10^{-11} A is about 0.01 μm. At the currents normally selected for wavelength analysis, the minimum probe diameter, limited by instrument design, will be in the range 0.1 to 0.3 μm. Corresponding figures for energy dispersive analysis are 0.07 to 0.1 μm. These figures must not be confused with the minimum size of mineral grain on which an x-ray analysis can be performed. This is limited by scatter of the electron beam within the sample, which is controlled principally by the gun operating potential and the mean atomic number of the sample (section 10.6).

In scanning electron microscope applications, the smallest probe current, and therefore spot diameter, is limited by the minimum acceptable signal to noise ratio in the signal obtained from the secondary or backscattered electron detector.

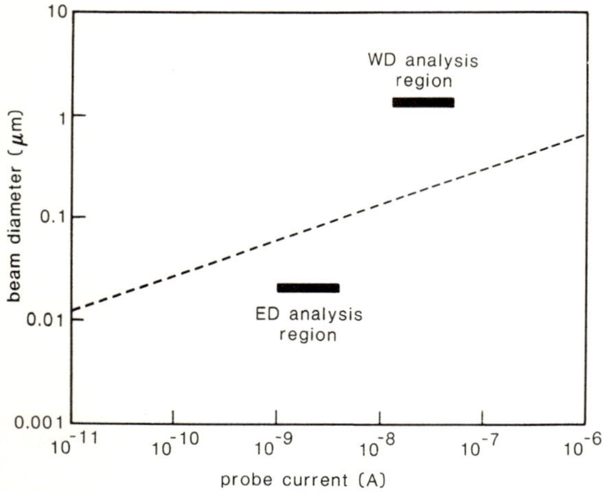

Figure 10.14 Minimum focused probe diameter on the sample as a function of probe current. Typical figures are plotted here; precise data depends on the design (and condition) of the electron column. The ranges for routine wavelength and energy dispersive analysis are indicated.

10.4.4 *Astigmatism*

Minimum spot size cannot always be achieved in routine operation, usually due to the effects of astigmatism. Astigmatism results from imperfections in the symmetry of magnetic fields formed in the column caused either by mechanical misalignment or contamination from dirt. In either case, the electron beam is distorted so that it can be focused on the sample only as an elliptical rather than a circular spot. It is possible to compensate for this effect by using a *stigmator coil*. To be most effective, the stigmator must usually be positioned close to the sample. Stigmators can be designed with varying numbers of compensation coils; the one in Figure 10.15 has eight. These coils are wired so that alternate coils are connected in series in such a way that they create opposing magnetic fields when a current is passed through their windings. The magnitude and direction of the current through each set of coils can be varied independently so that the size and orientation of the net magnetic force required to correct for astigmatism can be optimized.

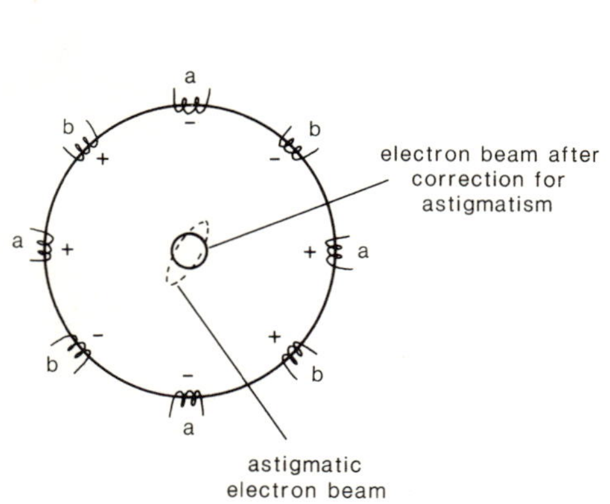

Figure 10.15 The stigmator assembly showing a design with eight poles. Alternate coils are connected in series but wound in opposite directions. By adjusting the size and direction of the current through the two sets of coils, it is possible to compensate for astigmatism of the electron beam apparent as elliptical distortion (shown as a dotted line).

10.4.5 *Beam centring coils*

A further set of coils is normally provided between the gun and condenser lens to correct for any residual mechanical misalignment between the position of the filament tip and optical axis of the column (see Figure 10.7).

10.4.6 *Beam scanning*

The heart of a scanning electron microscope and an important additional facility on the electron microprobe is the ability to scan the electron beam in a controlled manner over a pre-selected area of the sample. The electron beam is usually scanned over a rectangular area of the sample, line by line, in a pattern called a *raster* (Figure 10.16). Deflection of the electron beam is achieved using the electromagnetic fields formed by a set of scan coils. The scan coils normally consist of four individual electromagnetic coils, alternate coils being connected in series in such a way that when an electric current is passed through them, they generate opposing electromagnetic fields (Figure 10.16). These pairs of coils are independently controlled so that the electron beam is caused to scan each line at a predetermined speed, to fly back to the beginning of the next line and repeat the process until the required area has been covered. The circuits that control the movement of the electron beam are also used to trigger in synchronization the movement of a spot on a cathode ray tube (CRT) display screen. The intensity of this display CRT spot is modulated by a signal selected to give information about the sample surface. Signals that could be selected for display include the intensity of backscattered or secondary electron emissions, or the ratemeter output of an x-ray spectrometer set for a specified element.

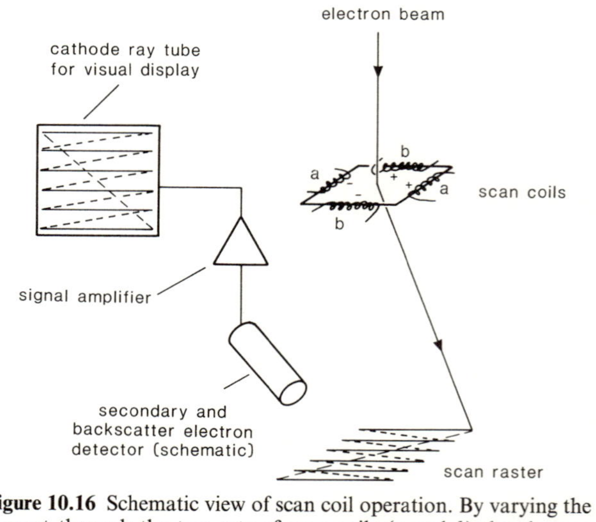

Figure 10.16 Schematic view of scan coil operation. By varying the current through the two sets of scan coils (*a* and *b*) the electron beam can be scanned, line by line, in a raster over an area of the sample. Flyback is shown as a dotted line. The corresponding secondary or backscatter electron signal is amplified and used to modulate the intensity of a CRT display screen which is scanned in synchronization with the electron beam.

Table 10.1 Length of line scanned on sample at different magnifications

Magnification factor	Length of line scanned on sample
×50	4 mm
×100	2 mm
×200	1 mm
×500	400 μm
×1000	100 μm
×5000	40 μm
×10000	20 μm
×20000	10 μm
×50000	5 μm

For a given magnification setting, these data list the distance across the sample traversed by the electron beam in completing one line of a raster scan, assuming that the corresponding image data is displayed on a 200 mm-wide CRT screen.

Each of these signals varies according to the surface composition of the sample at the point excited by the electron beam. A composite display image may, therefore, be built up to permit visual interpretation of compositional, structural or mineralogical features of the sample.

The magnification of the displayed image depends on the dimensions of the CRT display screen (a fixed parameter) in relation to the size of the raster over which the electron beam is scanned. As an indication of the variation of electron beam raster size with magnification, the apparent lengths of an individual scanned line at various magnifications assuming that the width of the CRT is about 200 mm (about 8 inches), are listed in Table 10.1. The maximum usable magnification is defined by the image resolution of the instrument (the limiting distance separating two features that can still be distinguished from one another with a specified confidence). Resolution is determined by probe diameter in the case of secondary electron signals and by the size of the excited volume in the case of x-ray mapping images. In general, a decrease in image resolution is observed in the order: secondary electrons < backscattered electrons < x-ray image.

10.4.7 *Distinguishing electric currents*

Several different electric currents are often specified in descriptions of electron microprobe operating procedures. To avoid confusion, the sources and uses of these currents and the circuits which they drive are distinguished as follows:

(i) *Filament current*: The current used to heat the tungsten filament to 2600–2700 K to allow thermionic emission of electrons.

(ii) *Beam current*: The total current emitted by the filament. The beam current may be selected by the user by adjusting the value of the EHT feedback resistor (Figure 10.2). The larger the beam current, the larger the area of the filament tip from which thermionic emission occurs, the larger will be the probe diameter.

(iii) *Lens current* is the current passing through the windings of an electromagnetic lens. An increase in the lens current causes an increase in the strength of the electromagnetic field strength which reduces the effective focal length.

(iv) *Probe current* is the total current delivered to the specimen. This quantity must be maintained at a constant value throughout an analytical operating session. The probe current represents the residual fraction of the original beam current that has not gone to earth by striking intermediate apertures within the electron column.

(v) *Specimen current*: Some of the probe current striking the sample is backscattered out again. The residual fraction that comes to 'rest' within the sample and must be leaked to earth to prevent the sample from charging-up is the specimen current. For a constant probe current, the magnitude of the specimen current will vary from sample to sample inversely according to the fraction of backscattered electrons. This in turn depends on the mean atomic number of the sample.

10.5 Vacuum requirements

The electron column and specimen chamber must be maintained under conditions of high vacuum for several reasons:

(i) The tungsten filament has a short operating life under poor vacuum conditions due to ion erosion (see section 10.3.3)

(ii) A high voltage could not be maintained between grid and anode in the gun under poor vacuum due to low-pressure discharge effects

(iii) Heavy absorption and scatter of the probe beam electrons will occur unless the column is maintained at high vacuum

(iv) Low-energy x-rays (e.g. Na Kα) suffer severe attenuation in air.

In order to reach a satisfactory operating vacuum of 10^{-5} to 10^{-6} Torr, two stages of pumping are required, normally provided by a mechanical rotary pump backing an oil diffusion pump.

10.5.1 *Rotary pump*

The rotary pump consists of an eccentric (A) that is rotated by electric motor within an air-tight housing (B), partially filled with vacuum pump oil (Figure 10.17). A vane (C, D)

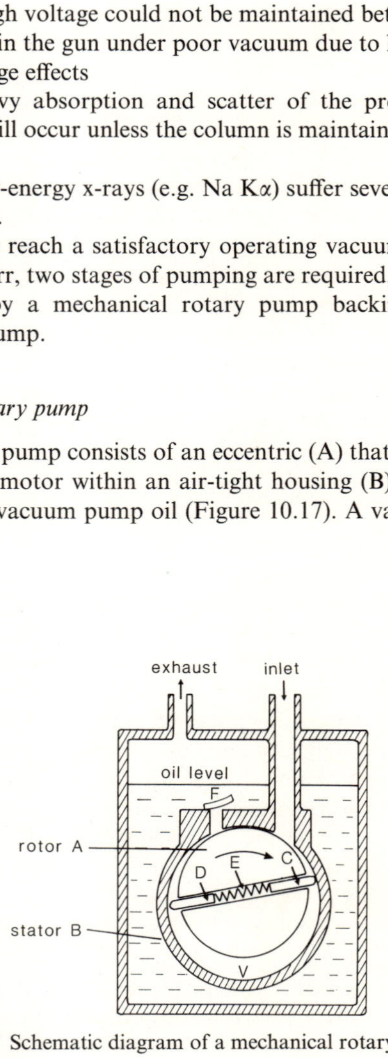

Figure 10.17 Schematic diagram of a mechanical rotary pump.

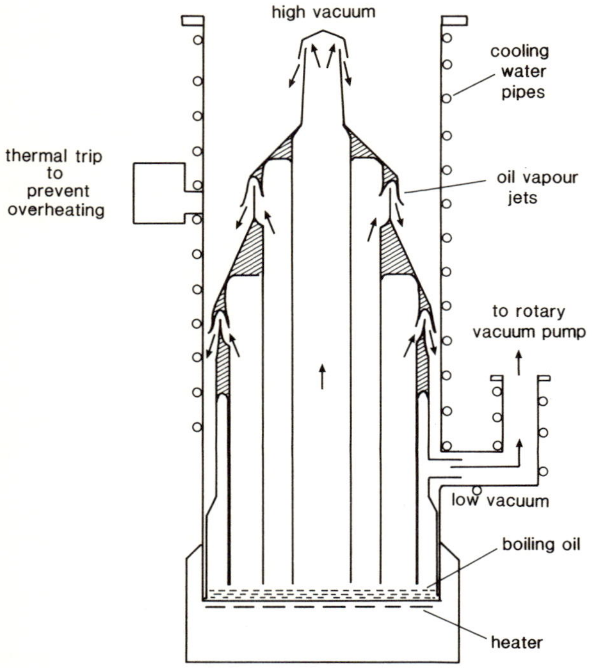

Figure 10.18 Schematic diagram of an oil diffusion pump. Arrows indicate the movement of boiling oil vapour.

rests against the edge of the rotating eccentric and forms a gas-tight seal. Air or other gases from the vacuum system are sucked into the pump (V) by the rotation of the eccentric, compressed, and expelled through a non-return valve at the other side. In practice, this type of pump can achieve a working vacuum of about 10^{-2} to 10^{-3} Torr. The limit is reached when compressed gases no longer have sufficient pressure to open the exhaust valve. Mechanical pumps depend on the phenomenon of viscous drag for efficient pumping action and are not appropriate for lower pressures. On many instruments, oil diffusion pumps are used to achieve a working vacuum below 10^{-3} Torr.

10.5.2 *Oil diffusion pump*

Oil diffusion pumps consist of a tubular barrel in which highly refined diffusion pump oil is boiled by an electric heating element. The boiling oil vapour is directed up a funnel-shaped baffle (the 'Christmas tree') and discharged as jets of vapour directed downwards against the water cooled walls of the pump (Figure 10.18). Gas molecules from the vacuum system are entrained in these jets of oil vapour so creating a high vacuum above the pump. The oil vapour is cooled by the walls of the pump, condenses and falls back to the reservoir. Gas molecules are pumped away by a rotary pump connected to the low-vacuum side of the diffusion pump. In this way the diffusion pump can maintain a high vacuum in excess of 10^{-6} Torr against a rotary pump (backing) pressure of between 10^{-2} and 10^{-3} Torr. The backing pressure must not be allowed to rise significantly above about 5×10^{-1} Torr, otherwise the pump effectively stalls and oil vapour is swept back into the electron column. Similarly, the pump must not be allowed to evacuate the electron column before this has first been pumped out to less than 10^{-2} Torr, as under these conditions, boiling diffusion pump oil is swept out of the pump into the rotary pump system. The operation of vacuum systems requires careful control and reliable interlocks especially for routine sample changing.

The oil used in a diffusion pump must possess some special properties. It must have a very low vapour pressure and a high resistance to both thermal degradation and oxidation effects. Silicone-based oils are widely used on SEMs. However, for microprobe applications, where the deposition of small quantities of silica on the surface of geological minerals is highly undesirable, oils based on polyphenyl ethers are normally substituted. One such oil (Santovac 5) has a boiling point of 295°C, and a vapour pressure of 4×10^{-4} mbar at 150°C.

More recent developments have included the replacement of oil diffusion pumps by turbomolecular pumps. These consist of a set of turbine blades which are rotated at very high speed (10 000 to 60 000 rpm). Gas molecules which diffuse on to the blades of the turbine are impelled into the backing system and pumped away. The vacuum created by these pumps is much cleaner, as it eliminates contamination by boiling oil.

10.6 Interactions between the electron beam and sample: the excited volume

The capabilities and limitations of electron probe microanalysis are all connected with interactions that occur within the volume of about 300 μm^3 where the electron beam and sample interact.

Electrons from the primary beam penetrate the surface of the sample and collide with its atoms, transferring energy in the process. A small fraction of the collisions are so violent that primary electrons are scattered back out of the sample, and are unable to cause further excitation. The majority of electrons, however, follow trajectories within the sample, lose energy by collision and eventually come to 'rest'. The electric charge so deposited in the sample must be conducted to earth and may be measured as the specimen current. Non-conducting samples must be coated with a thin (25 nm) layer of carbon to prevent charging of the specimen surface.

The region within the sample in which these collisions occur is known as the *excited volume*. The shape and size of this region are of some interest since it represents the source from which analytical signals originate.

The range of electrons within a sample (defined as the depth below the surface reached by the electrons) depends on the incident energy of the electron beam E_0 (in keV) and the density of the sample ρ (in kg m^{-3}) and is given approximately by:

$$x = \frac{k \, . \, E_0^{+n}}{\rho} \simeq \frac{0.1 \, . \, E_0^{1.5}}{\rho} \qquad (10.3)$$

(see Read 1973, p.57) where x is in the range in μm, and k is a constant.

Thus for a typical silicate mineral of density 3 kgm^{-3}, the penetration depth of electrons accelerated to 15, 20 and 30 kV would be approximately 2, 3 and 5.5 μm, respectively.

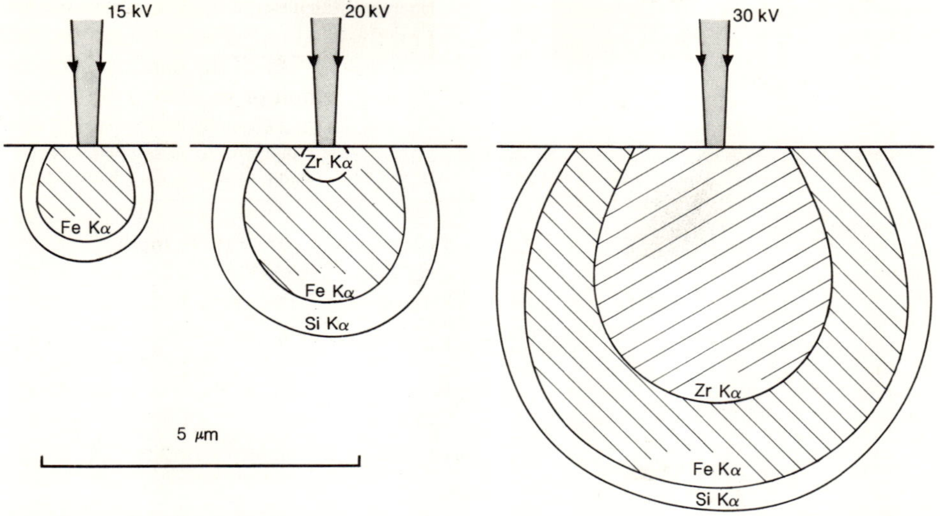

Figure 10.19 Schematic diagram of the size and shape of the excited volume within a 'typical' silicate mineral of density 3 kg m^{-3} for probe beams excited at 15, 20 and 30 kV. Regions from which Si *Kα*, Fe *Kα* and Zr *Kα* x-rays are emitted are delineated.

Not only do electrons penetrate into the sample, they are also scattered by collision. The distance across the excited volume may be estimated from

$$y = \frac{0.077 \,.\, E_0^{1.5}}{\rho} \tag{10.4}$$

(see Read, 1975, p.217).

The resultant excited volume typically has a truncated pear shape, and its relative size for three accelerating voltages, 15, 20 and 30 kV is shown in Figure 10.19.

It is possible to attempt to calculate the trajectories of individual electrons using Monte Carlo computer simulation techniques, as shown by Curgenven and Duncumb (1971). Data in Figure 10.20 are reproduced from their work. Figure 10.20 (*a*) shows the simulated trajectories of five electrons accelerated to 20 kV and penetrating copper metal. The generally erratic path caused by multiple collision with atoms is apparent. Figures 10.20 (*b*) and (*c*) show electrons of the same energy penetrating a sample of aluminium (*b*) together with the calculated points of origin of Al*Kα* x-rays excited along individual electron tracks (*c*). Such data are very valuable in giving a visual indication of the region within a sample from which x-rays originate.

Since electrons lose their incident energy progressively by multiple collision, elements may only be excited whilst the mean electron energy remains above the appropriate critical excitation energy. The volume from which x-rays of a given energy originate in a composite sample will, therefore, decrease with increasing fluorescence photon energy (see section 10.8).

The data in Figure 10.19 show the size of the excited volume for a typical silicate mineral of density 3 kg m^{-3}. However, because of the inverse dependence of penetration depth on sample density, these sizes would be almost halved for an opaque mineral such as rutile, of density up to 5.5 kgm^{-3}.

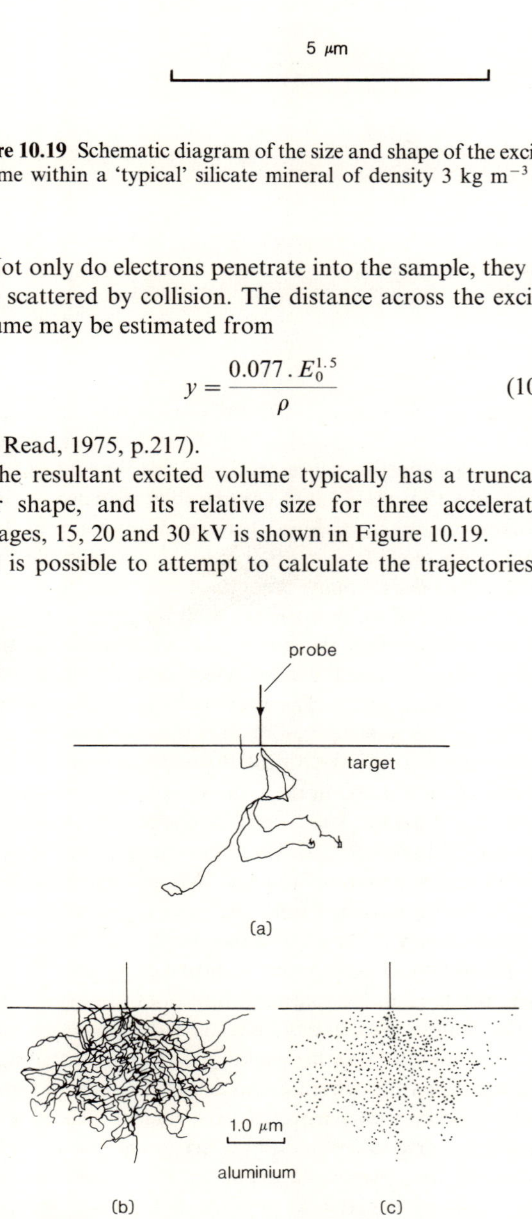

Figure 10.20 Monte Carlo simulations of interaction between primary beam electrons and sample: (*a*), the trajectories of five individual electrons penetrating a copper target at 20 kV; (*b*), the trajectories of many electrons penetrating an aluminium target at 20 kV; (*c*), the computed distributions of atoms emitting Al *Kα* radiation as a result of the interactions plotted in (*b*). (Curgenven and Duncumb, 1971.)

10.7 Phenomena within the excited volume

The transfer of energy that accompanies collisions between primary beam electrons and sample atoms causes a number of phenomena, including the generation of heat, back-scattered electrons, secondary electrons, cathodo-luminescence and x-rays (characteristic and continuum). These will now be considered in turn.

10.7.1 *Heat*

With the electron beam focused to a spot of 1 μm or less, a considerable amount of electrical energy is concentrated

Table 10.2 Temperature rise (°C) induced in selected minerals by an electron beam at specified probe diameters

	15 kV						30 kV					
	50 nA WD analysis			3 nA ED analysis			30 nA WD analysis			1 nA ED analysis		
Mineral (κ)	1 μm	5 μm	10 μm	1 μm	5 μm	10 μm	1 μm	5 μm	10 μm	1 μm	5 μm	10 μm
Obsidian glass (0.014)	260	52	26	15	3	2	309	62	31	10	2	1
Zircon (0.042)	86	17	9	5	1	<1	103	21	10	3	<1	<1
Quartz (0.10)	36	7	3	2	<1	<1	43	9	4	1	<1	<1
Calcite (0.05)	72	14	7	4	1	<1	86	17	9	3	<1	<1
Iron metal (0.80)	4	<1	<1	<1	<1	<1	5	1	<1	<1	<1	<1
Epoxy resin (0.002)	(1800)	(360)	(180)	108	22	11	(2160)	(432)	(216)	72	14	7

Listed data are an estimate of the temperature rise in the excited volume, induced by the electron beam under the stated conditions. Values in brackets for epoxy resin note the fact that this polymer will chemically decompose long before the listed temperature is reached.
Thermal conductivity (κ) in units of watts cm^{-1} K^{-1} is listed under sample type; data from Weast (1973).

into a small volume of the sample (approaching 8×10^7 W cm^{-2}). It is not, therefore, surprising that this energy causes a rise in temperature which, if the specimen has a low thermal conductivity, can be considerable.

Castaing (1951) estimated that the rise in temperature (ΔT °C) resulting from an electron beam of diameter d (μm) impinging on a sample of thermal conductivity k (W cm^{-1} K^{-1}) is given by:

$$\Delta T = 4.8 \,.\left(\frac{E_0 i}{kd}\right) \qquad (10.5)$$

where E_0 is the incident energy of the electron beam (in keV) of current i (μA).

As an indication of this heating effect, the expected temperature rise for various materials analysed with an electron beam focused to diameters of 1, 5 and 10 μm is listed in Table 10.2. These data indicate that the temperature rise in minerals of low thermal conductivity (such as obsidian glass) can be large—sufficient indeed to cause decomposition and/or structural alterations in some materials. Data for epoxy resin are included in Table 10.2 for interest. This material would, of course, decompose well before the very high equilibrium temperatures listed are reached. The rise in temperature in metallic phases such as iron is almost negligible.

10.7.2 *Thermal decomposition of minerals*

Minerals most subject to thermal decomposition effects are those containing structural water (or carbon dioxide), and glass and crystalline minerals containing high concentration of potassium and sodium. Two effects may be distinguished.

(i) In minerals containing structural water, for example scapolite or zeolites and other clay minerals, the heating effect of the probe appears to cause the structural water to boil off. The effective concentration of the remaining matrix elements within the excited volume then increases, causing a corresponding increase in their apparent x-ray emission signal. Similar effects are observed when carbon dioxide is driven off from carbonate minerals.

(ii) The loss of sodium and potassium signals from certain minerals is also often attributed to volatilization, but this is almost certainly not the case. More complex interactions occur in these minerals. The electron beam appears to make the ions of these elements mobile within the sample lattice, causing them to diffuse out of the excited volume. Again, this causes an increase in the x-ray signal due to the remaining matrix elements. The variation of such signals with time is shown in Figure 10.21. This phenomenon has been discussed in some detail by Siivola (1969) and by Craw (1981), the latter work relating to the analysis of iron-bearing phyllosilicates. Craw reported that oxidation of Fe^{3+} to Fe^{2+} in this mineral could be induced by probe excitation.

The mechanism for ion migration does not appear to be simple. Kane (1973, p.250) reviewed the phenomenon of Na^+ ion migration within glass and ceramic samples, and reported that although an initial fall in the sodium signal is often observed, some workers have then measured a substantial increase in the sodium x-ray intensity. It has been postulated that prolonged exposure to the electron beam causes a local negative space charge to form just below the surface of a sample and that this eventually attracts positively-charged sodium ions back into the excited volume. Graham *et al.* (1984) reported the migration of sodium, potassium, magnesium and calcium ions in meta-autunities ((Na,K,H_3O)UO_2PO_4.3.25H_2O) and related natural and synthetic materials. The effect was attributed to the

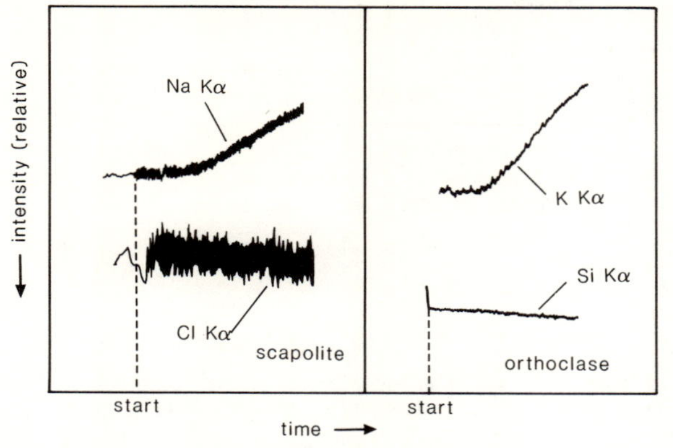

Figure 10.21 Electron-beam-induced decomposition of scapolite and orthoclase at 20 kV. After a short period of stability (about 30 seconds) the x-ray ratemeter output for sodium (in scapolite) and potassium (in orthoclase) shows a progressive reduction in intensity. The corresponding count rates for chlorine and silica show a complementary increase in intensity. These measurements were made with a probe diameter of 5 μm. X-ray intensity increases towards the bottom of each diagram.

formation of a space charge within the sample which caused the rapid migration (almost instantaneous in the case of sodium) of these cations. The effect was exacerbated because these ions were held in interlayer sites within the mineral structure.

(iii) The decomposition effects described above are dependent on the power per unit area dissipated within the sample, and may be reduced by defocusing the electron beam or selecting a lower probe current. Substituting an energy dispersive detector, with its lower probe current requirement, for wavelength dispersive spectrometers is one means of achieving this aim.

For accurate and reliable analyses, decomposition effects must not be observed in the course of an analysis. Using sequential wavelength dispersive spectrometers for results of high precision, excitation periods could extend to 12–15 minutes per sample.

It is not possible to generalize on the degree of stability for individual minerals. However, to give some indication of the general trend, data in Figure 10.22 show the time that elapsed before the onset of decomposition was identified in selected minerals (Potts and Tindle, unpublished data). Decomposition was correlated with changes in the ratemeter signal set to the Na and K *Kα* lines and by changes in the specimen current. The lines represent the elapse time for a given probe diameter before which decomposition effects were observed. The probe was operated at 30 nA, 20 kV. If analysis conditions (probe diameter and probe current) are set in the field to the top left side of the stability line appropriate for the mineral in question, decomposition effects are unlikely to be observed. These data indicate the poor stability of scapolite (structural water and carbon dioxide) and albite (high sodium) and the moderate stability for biotite and orthoclase (both high potassium) under the conditions indicated. Conversely, some crystalline minerals (olivine, garnet and opaques) are stable under virtually all reasonable operating conditions.

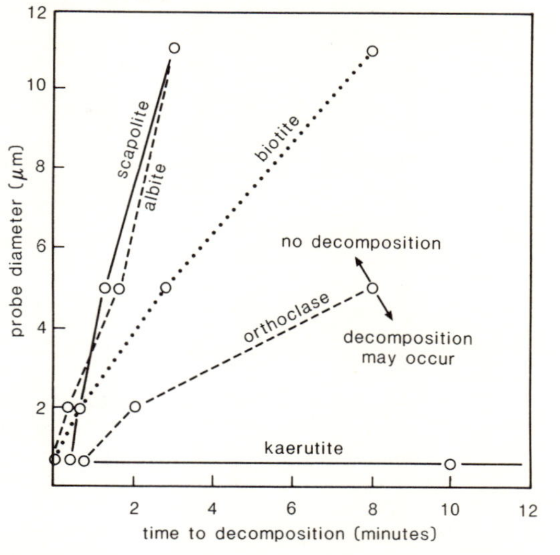

Figure 10.22 Stability of minerals to electron-beam-induced decomposition as a function of probe diameter. This diagram plots the time elapsed before the onset of decomposition for selected minerals at specified probe diameters. If measurements are made with the probe diameter and count time adjusted in the field to the upper left of the appropriate line, no decomposition is likely to be observed. All measurements were made at 20 kV, 30 nA. (Potts and Tindle, unpublished data.)

10.7.3 *Contamination effects*

The vacuum systems of all electron columns that are pumped by diffusion pumps contain organic oil vapour at small partial pressures. This oil, together with residual water vapour, is absorbed on to the surface of samples. Interaction with the electron beam causes the hydrocarbon molecules to be cracked, liberating hydrogen and leaving a residue of carbon. Campbell and Gibbons (1966) estimated that in normal electron optical systems, the rate of contamination varies between 0.1 and 1.0 nm per second. Carbon contamination is a particular problem in the analysis of the light elements (such as boron to oxygen and especially in the analysis of carbon itself) due to the additional absorption suffered by the very low energy characteristic x-rays. Rates of contamination can be minimized by placing a copper cold finger, cooled to liquid nitrogen temperatures, adjacent to the sample surface. Volatile vapours are then condensed on the cold finger and not on the sample. Ong (1966) discussed the interference effects in the determination of carbon by microprobe and described the construction of a cold plate designed to minimize contamination effects. Using this device, the rate of carbon build-up was found to decrease by a factor of about 6 with the probe operated at 30 kV and 18 at 10 kV.

10.7.4 *Cathodoluminescence*

When certain minerals are bombarded by electrons, they emit visible light, a phenomenon known as cathodoluminescence. The mechanism for cathodoluminescence depends on the mineral in question having some semiconductor properties (see, for example, section 12.4). This

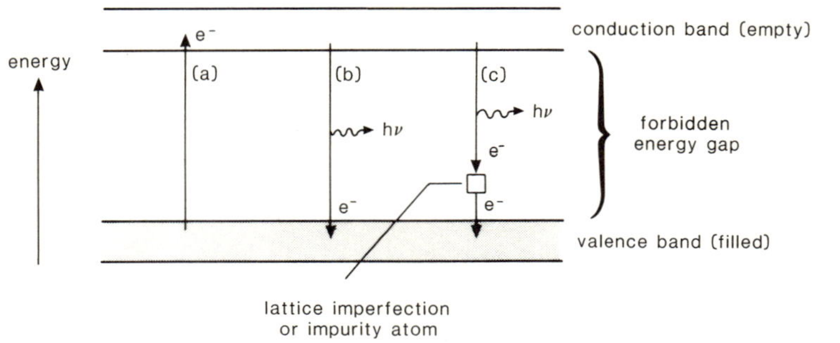

Figure 10.23 Cathodoluminescence. An electron excited from valence to conduction band in a crystal material (*a*) may cause luminescence if light of suitable wavelength is emitted when the electron falls either directly back to the valence band (*b*) or via a suitable trapping centre (*c*). The trapping centre may be a crystal lattice defect or an impurity atom which makes available intermediate energy levels within the forbidden energy region. (After Kniseley and Laabs, 1973, p. 373.)

means that electrons within the mineral crystal lattice can occupy two bands of energy: the ground state (or valence band) and a higher energy conduction band. The valence band is normally fully occupied with electrons, the conduction band is normally empty. These two energy bands are separated by a forbidden energy region. Interactions between the primary electron beam and sample, considered in terms of its crystal lattice properties, can cause electrons to be excited from valence to conduction bands (Figure 10.23). After a short half-life in this excited state, such electrons return to the ground state, a process that involves the release of energy. In some minerals the energy difference between these bands is such that this energy is emitted as electromagnetic radiation in the visible region of the spectrum (i.e. cathodoluminescence). Minerals such as scheelite ($CaWO_4$) and some uranium minerals are thought to fall into this category (Long and Agrell, 1965).

Often, however, the electron falls back to an intermediate 'trapping' centre lying at an energy level within the forbidden energy region. Such trapping centres are caused by imperfections in the structure of the crystal lattice or by the presence of impurity ions (for example, manganese). In the latter case, transition from conduction band to trapping centre can cause excitation of the impurity ion, which subsequently decays by emitting its characteristic uv–visible lines. As these transitions involve outer orbital electrons, the energy levels of which are continuously perturbed by thermal motions within the crystal lattice, the observed emissions appear as broad band spectra rather than the sharp lines normally associated with atomic transitions (as in optical emission spectrometry, for example). However in the case of rare-earth element impurity atoms, the appropriate electron orbitals are shielded by outer electrons so that there is some possibility of observing much higher resolution optical spectra in cathodoluminescence caused by rare-earth element activators (Kniseley and Laabs, 1973).

Cathodoluminescence phenomena are normally only observed if the light, emitted in association with the appropriate electron transition, falls within the uv–visible region of the spectrum. The phenomenon can be used very effectively in determining the distribution of manganese and hence revealing the structure of limestone or dolomite samples at concentration levels below those detectable by electron microprobe x-ray analysis. In this application, the intensity and colour of cathodoluminescence emissions depends not only on the concentration of manganese as activator, but also on the presence of iron, which inhibits the emission of light. Sommer (1972 *a,b*) described how shifts in the wavelength of the manganese cathodoluminescence band spectra can be used to distinguish changes in coordination, symmetry, bond length and composition of this ion in carbonate matrices from various geological settings.

The cathodoluminescence colours of several minerals are shown in Table 10.3. A practical use of this phenomenon is the pale blue glow observed when the electron beam is focused on a glass slide sputter-coated with a thin layer of aluminium. The size and shape of this emission can be used to measure spot size and identify astigmatism in the electron beam when checking the alignment of the electron column.

10.7.5 *Backscattered and secondary electrons*

Three electron signals result from the interaction of an electron beam with a bulk sample: a backscatter electron, sec-

Table 10.3 Cathodoluminescence properties of selected minerals

Mineral	Cathodoluminescence colour	Comments
Calcite	Orange-red	Mn^{2+} or Pb activator
Calcite	Blue-black	Fe extinguisher
Dolomite	Bluish	
Enstatite	Bright red	Mn^{2+} activator
Enstatite	Bright blue	Mn absent
Hibonite	Bright red-orange	High Al, low Mg, Ti
Hibonite	Bright blue	Low Al, high Mg, Ti
Kilchoanite	Orange-red	
Oldhamite	Yellow	
Plagioclase	Blue	
SiO_2	Orange	
Sinoite	Green	
Spurrite	Green	
Tilleyite	Red	
Uranium minerals	Blue	
Wollastonite	Green	

Data from Kniseley and Laabs (1973), Long and Agrell (1965) and Keil (1973).

ondary electron and specimen current signal. All may be used in conjunction with beam scanning to generate an image of the sample surface which incorporates information on composition and/or topography.

(*a*) *Backscatter electron signal.* Backscattered electrons are primary electrons that undergo violent collisions with atoms of the sample and are scattered back out of its surface. They show a continuum of energies ranging from zero up to the incident beam energy (Figure 10.24) and are emitted from typically the top one-third of the excited volume. Below this region, electrons suffer increasing absorption within the sample and do not have sufficient energy to re-emerge from the surface.

The proportion of electrons that undergo backscatter shows a marked dependence on the mean atomic number of the sample. The backscatter electron intensity from carbon is about 8% whereas that from uranium is 50%. This phenomenon has important consequences in quantitative microprobe analysis, since the proportion of primary beam electrons available to excite the sample is inversely affected by this backscatter phenomenon.

The backscatter electron signal is used extensively in scanning electron microscopy to give compositional data on a sample based in the contrast in intensity from adjacent minerals having different mean atomic numbers. Modern electron detectors can distinguish contrast in this signal for differences as low as about 0.1 to 0.3 of an atomic number unit.

(*b*) *Secondary electrons.* Secondary electrons are generated by the ionization of atoms of the sample by interaction with the primary beam and have relatively low energies (0–50 eV) (Figure 10.24). In consequence, these electrons cannot escape from deep within the sample and are ejected from the surface from a relatively localized region round the area of impact of the primary beam. The secondary electron signal therefore has the potential of offering a higher image resolution than that obtainable from backscattered electrons which originate from a larger sample volume. Traditional detectors were capable of detecting a much larger solid angle of low-energy secondary electrons than backscattered electrons. This signal had, therefore, more favourable signal to noise ratios and was less sensitive to topographic details

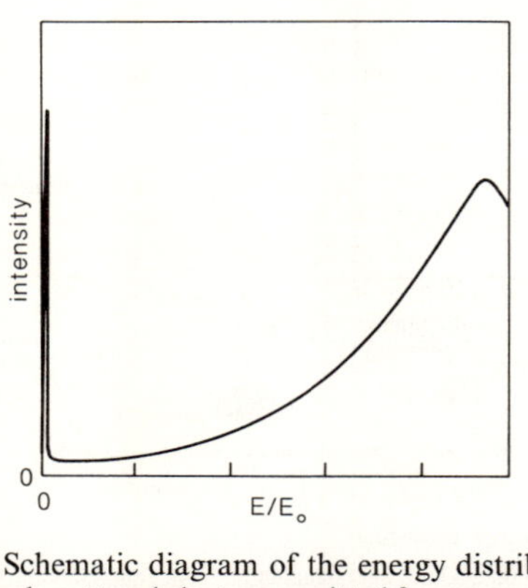

Figure 10.24 Schematic diagram of the energy distribution of secondary and backscattered electrons emitted from a sample. Intensity of electrons is plotted as a function of the ratio energy E to the primary electron energy E_0. The narrow peak at low energies corresponds to secondary electron emissions. (Heinrich, 1981.)

than the backscattered electron image. However, with recent developments in high-efficiency backscatter electron detectors, this advantage no longer exists.

(*c*) *Specimen current.* The specimen current represents the complementary signal to the backscattered plus secondary electron signals, representing those primary-beam electrons that come to rest within the sample. It is necessary to earth this current, usually through a current-sensitive amplifier, otherwise, under high-vacuum conditions, the sample would rapidly charge up to the operating potential of the gun. The local and transient electrostatic charges so formed would cause the electron beam to be deflected in an erratic manner over the sample surface. Non-conducting materials must therefore be coated with a 25 nm layer of carbon (section 10.8). During a quantitative analysis, the specimen current may be usefully monitored to check for drift in probe current or to indicate the onset of specimen damage.

10.7.6 *Electron detectors*

(*a*) *Scintillation detectors.* Traditional electron detectors are based on the design of Everhard and Thornley (1960) (Figure 10.25). This is a scintillation detector that takes advantage of the low-noise amplification that can be obtained from a scintillator–photomultiplier tube combination. The detector consists of a thin layer of phosphor which is coated on to one end of a perspex or quartz light pipe. The other end of this pipe is optically coupled to a photomultiplier tube (Figure 10.25). The phosphor is positioned close to the sample, and a bias of 10 to 15 kV must be applied to it, since secondary electrons have such a low energy that they cannot excite the phosphor. The applied bias imparts sufficient extra energy to these electrons to effect efficient scintillation. A grid is placed between scintillator and sample to control the collection of electrons. If a bias of about +200 to +250 V is applied to this grid, low-energy secondary electrons emitted over a large solid angle are swept towards the scintillator. Conversely, if a bias of about −50 V is applied, secondary electrons are repelled and only those higher-energy backscattered electrons emitted in the direction of the scintillator are detected. The intensity of light generated within the phosphor is directly proportional to the number of electrons impinging on it. These photons are optically coupled via the light pipe to the photocathode of the photomultiplier, where this signal is amplified and made available to modulate the intensity of a CRT display. A large solid angle of secondary electrons may be efficiently collected in this way, so that signal to noise ratios in amplified signals are low. For this reason, the Everhard–Thornley detector has been widely used in many SEM and microprobe applications for a number of years. However, backscattered electrons are too energetic to be deflected on to the scintillator in the same way as secondary electrons, giving a somewhat inferior signal. In cramped instrumental configurations within a sample chamber, the Everhard–Thornley detector may be mounted relatively far from the sample without excessive degradation in performance as regards secondary electrons.

(*b*) *Silicon surface barrier detector.* The backscattered electron signal may be effectively measured using a silicon surface barrier detector. This consists of a wafer of silicon,

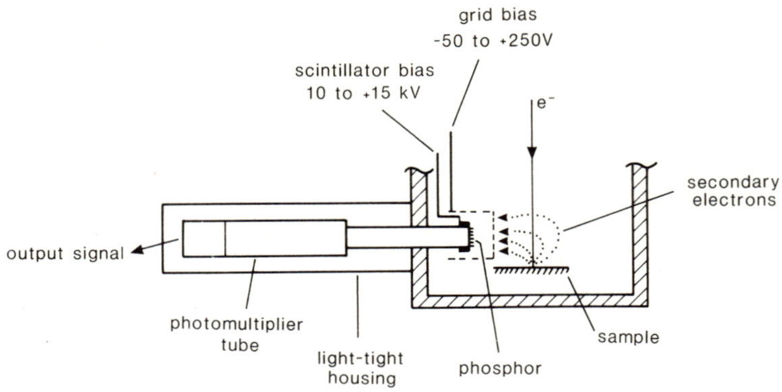

Figure 10.25 A secondary electron scintillation detector following the design of Everhart and Thornley. Low-energy secondary electrons (trajectories shown by dotted lines) are collected by applying a suitable bias (−50 to +250 V) to the grid. These electrons must be further accelerated by the 10 to 15 kV bias in order to give them sufficient energy to scintillate the phosphor.

normally mounted in front of the sample so that as large a solid angle of secondary electrons as possible may be collected. In some applications, the silicon wafer is split into two to provide additional flexibility in signal handling (see below) (Figure 10.26). Low-energy secondary electrons do not have sufficient energy to activate this device. However, energetic backscattered electrons deposit their energy within the silicon wafer, which is electronically controlled to behave as a reverse bias diode. Because the detector is designed to collect a large solid angle of electrons, it gives a satisfactory signal even at very low probe currents (say 10^{-11} A), conditions under which the smallest beam diameters can be achieved (section 10.4.3). Using a split detector arrangement (Figure 10.26), it is possible to distinguish compositional and topographic details in a sample. The signal from each half of the detector disc is independently amplified. If the sum of these signals is displayed, images will give mainly compositional details of the sample based on variations in mean atomic number. Conversely, if the signals are subtracted, topographic features in the sample will be emphasized. This mode of operation is normally useful for examining relatively smooth sample surfaces, the secondary electron signal being preferred for more 'rugged' samples.

(*c*) *The Robinson detector.*An alternative backscatter electron detector is available, following the designs of Robinson (1975). Like the Everhard–Thornley detector, the Robinson detector is a scintillation device, making use of the high-gain, low-noise capabilities of the photomultiplier tube. However, this detector overcomes the deficient backscatter electron detection inherent in the Everhard–Thornley design since it is positioned in front of the sample surface and, therefore, collects a large solid angle of backscattered electrons (Figure 10.27). The Robinson detector does not require a high voltage bias and the backscattered beam is detected over a relatively large area of scintillator. There has been a vigorous debate in the literature between the proponents of the Robinson detector (Robinson and Nickel, 1979, 1983) and those of the semiconductor device (Hall and Lloyd, 1981, 1983). The main conclusions are that in mineralogical applications, an efficient backscattered electron detector, whichever

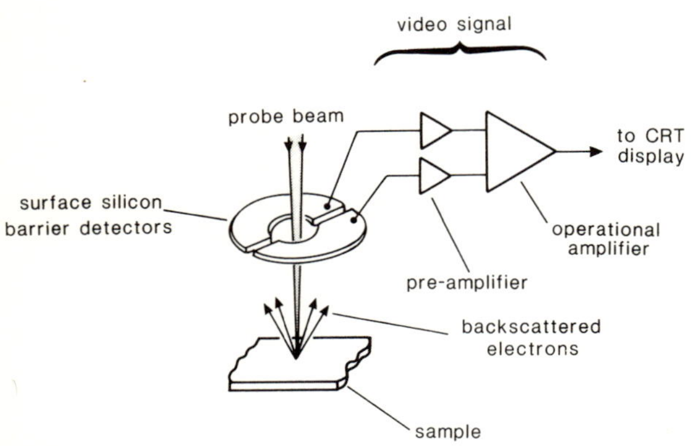

Figure 10.26 Backscattered electron detector consisting of two hemicircular wafers of silicon. Unlike secondary electrons, only those electrons emitted in the direction of the silicon wafers can be detected. If the signal from both silicon wafers is summed, compositional information (based on the mean atomic number of the sample) is displayed. If the difference between these two signals is displayed, topographic information is obtained (after Jeol).

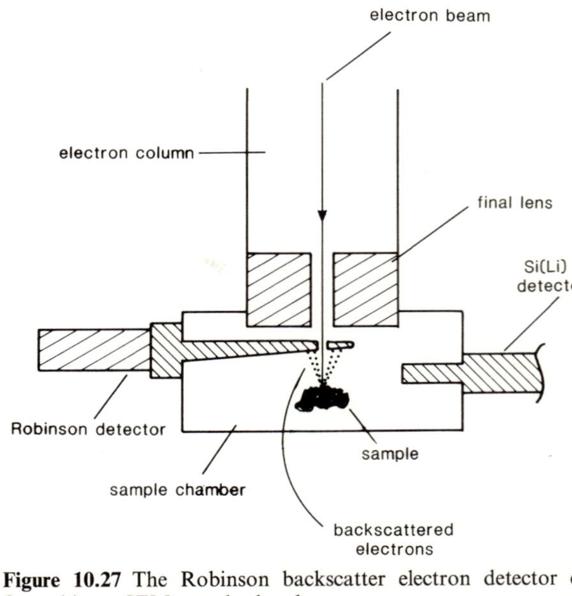

Figure 10.27 The Robinson backscatter electron detector configured in an SEM sample chamber.

design is adopted, is much more useful than a secondary electron detector, since the former are far more sensitive to small differences in the composition of minerals based on mean atomic number. A good backscattered electron detector can detect differences in atomic number of as low as 0.1 units, this corresponding to the difference in composition between Fe_7S_8 and Fe_9S_{10}, for example (Robinson and Nickel, 1979). One difficulty with semiconductor detectors is that they will not always operate satisfactorily at TV scan rates, frame periods of 1 to 2 s being required (Hall and Lloyd, 1981). A corresponding debate has occurred over the relative merits of SEM analysis with the sample chamber maintained at low vacuum (about 0.1 Torr) as opposed to conventional high-vacuum operation ($>10^{-4}$ Torr). At low vacuum, samples need not be coated with a conducting film, the low pressure of gas being adequate to conduct the charge away. Under high-vacuum conditions a conducting film (normally of carbon, 5 to 10 nm thick) is required. For atomic number contrast studies, low-vacuum conditions appear to offer some operating convenience. However, in all quantitative microanalysis (as opposed to SEM) applications, high-vacuum operation with samples coated with carbon is essential.

10.8 X-ray production

Interaction of the electron beam with the sample also leads to the emission of x-rays, this being the phenomenon used to perform quantitative chemical analysis.

10.8.1 *Characteristic x-rays*

Within the excited volume, collisions between the primary electron beam and atoms of the sample cause an exchange of energy during which the ionization of orbital electrons can occur. If an inner orbital electron is ionized, the unstable excited atom rapidly rearranges itself, and during the ensuing electron transitions, characteristic x-rays and Auger electrons are emitted. The mechanism is similar to the appropriate x-ray fluorescence-induced excitation models discussed in section 8.3.

10.8.2 *Continuum x-rays*

The emission of characteristic x-rays by the mechanism described in section 10.8.1 is a relatively inefficient process. Most interactions occur between the primary electron beam and electrons in outer orbitals of atoms within the sample. As a result of the deceleration of primary beam electrons under the influence of the electric fields created by these outer orbital electrons, a continuous x-ray spectrum (the *bremsstrahlung* or breaking spectrum) is generated. This continuous spectrum extends in energy from zero up to the incident energy of the electron beam. In this respect the electron microprobe could be considered to behave as a giant x-ray tube with the sample as the target (compare section 8.5). However, whereas the x-ray continuum in an XRF spectrum originates from the rather inefficient scatter of the x-ray tube spectrum off the sample, the electron microprobe continuum is generated *in situ*, together with the characteristic x-ray lines of the sample. In consequence, detection limits which may be achieved by electron excitation are up to two orders of magnitude worse than those by XRF analysis. The most sensitive microprobe determinations are therefore made down to the 50 to 100 ppm level rather than the 1–10 ppm range typical of XRF analysis.

10.8.3 *Excitation efficiency of microprobe compared with XRF*

In view of similarity in excitation mechanisms, it is of interest to compare the characteristics of electron microprobe excitation with those of XRF analysis.

(i) In terms of power required to excite the sample, electron excitation is a very much more efficient process, the typical power delivered to the sample being of the order of 10^{-3} W in comparison with the 3000 W power rating of many XRF x-ray tubes.

(ii) Even taking into account the enhanced continuum and relatively poor detection limits, electron excitation does have advantages in the analysis of the light elements (below silicon in the Periodic Table). Whereas sodium is the lightest element routinely determined in geological samples in x-ray fluorescence analysis, many electron probe microanalysers have the potential to determine elements down to boron ($K\alpha = 0.183$ keV), at least qualitatively. The limitation in microprobe analysis is not so much the excitation efficiency (as in XRF) but the degree of absorption of these very soft x-rays in the window of the detecting device (proportional counter or ED detector). Fluorine may usually be determined routinely at % levels by electron probe excitation without difficulty.

(iii) In x-ray fluorescence excitation, the primary tube photons with the highest probability of exciting an element are those with an energy that just exceeds the critical excitation energy of the appropriate electron orbital. Higher energy photons, however, show a decreasing probability

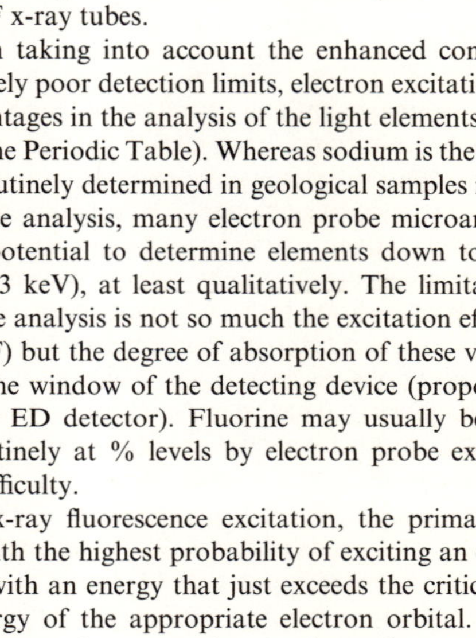

Figure 10.28 Excitation efficiency of electrons, photons and protons. The number of x-ray photons emitted from a bulk specimen is plotted as a function of the energy of the exciting quanta. (Birks *et al.*, 1964.)

of excitation. The opposite trend is observed in electron excitation, as shown in Figure 10.28. For primary beam electrons that exceed the critical excitation energy of the orbital in question, the probability of exciting an x-ray line increases with increasing incident electron energy according to the relationship

$$I \propto \left(\frac{E_0}{E_c} - 1\right)^{1.67} \tag{10.6}$$

where I is the characteristic x-ray intensity, E_c is the critical excitation energy, and E_0 is the energy of the incident electron beam.

The observed x-ray intensity, therefore, increases strongly in accord with the overvoltage ratio E_0/E_c.

10.8.4 *Optimum excitation conditions*

From equation (10.6), optimum excitation should be achieved at the maximum gun operating potential. However, there are opposing considerations.

(i) A significant increase in the size of the excited volume occurs as the operating potential is increased (Figure 10.19). Of particular relevance to analytical applications is the size of the volume from which x-rays of a particular energy are emitted. Each collision between primary beam electron and atom causes some transfer of energy, so that a point is reached along the electron track length where the mean energy of the electron has fallen below the critical excitation energy of a particular x-ray line. The depth at which this point is reached (x) may be estimated from

$$x = 0.077.\left(\frac{E_0^{1.5} - E_c^{1.5}}{\rho}\right) \tag{10.7}$$

where E_0 is the incident electron energy (in keV), E_c is the critical excitation energy (in keV), and ρ is the mineral density (kg m^{-3}).

To illustrate this concept, the volumes from which Si, Ca, Fe and Zr K x-rays are emitted are shown for accelerating potentials of 15, 20 and 30 kV in Figure 10.19. Note that Zr K x-rays are not excited at all at 15 kV and that the volume from which they originate at 20 kV is very small because beam electrons rapidly fall below the relevant critical excitation energy of 17.996 keV.

(ii) A second factor relevant to the selection of operating potential is the intensity of continuum and its variation with gun kV. As discussed in section 8.5, the intensity of the continuum spectrum (I) as a function of energy (E) is given by Kramers' equation

$$I \propto i.Z.(E_0 - E)/E \tag{10.8}$$

where E_0 is the incident electron energy (keV), E is the photon energy (keV), i is the probe current, and Z is the mean atomic number of the target (sample).

Two observations may be derived from this equation: (i) heavy minerals will show a higher continuum intensity than light minerals, and (ii) the continuum intensity at any energy is increased by selecting a higher gun operating potential.

Optimum detection limits are therefore obtained by selecting the lowest gun operating potential (to minimize background intensities) compatible with adequately exciting the elements of interest (to maximize sensitivities).

(iii) A third factor that must also be taken into account is uncertainties in the absorption matrix correction (discussed further in section 10.9). The magnitude of this correction depends on the size of the excited volume. Uncertainties in absorption corrections when comparing standard and unknown x-ray counts are reduced at lower, rather than higher, gun operating potentials.

Taking all these factors into account, the compromise conditions normally selected for the analysis of silicate minerals are a gun potential of 15 to 20 kV for the determination of all elements up to Fe $K\alpha$ (critical excitation energy = 7.11 keV). For the analysis of heavy minerals containing the heavy rare-earth elements, gold or the platinum group elements (Pt, Ir or Os) gun potentials of between 20 and 30 kV are commonly used. An extension of the arguments presented above would suggest that high precision in the analysis of light elements (for example Na to S K lines) should be obtained by exciting samples at between 5 and 10 kV. However, Reed (1972) has warned of the additional uncertainties that may then affect the results due to variable attenuation of primary beam electrons in the carbon coat deposited on the surface of samples (see section 10.8.6 (*f*)).

10.8.5 *X-ray resolution*

Although the extent of the excited volume delineated by the range of primary beam electrons has been demonstrated in Figure 10.19, this is not the only process that must be taken into account when considering the excitation of characteristic x-ray lines.

(*a*) *Continuum excitation*. Continuum x-rays, generated within the excited volume, can penetrate further into the sample, and those with energies exceeding the critical excitation energy of atoms within the sample can induce x-ray fluorescence by the photoelectric effect. The probability that this will occur varies approximately with Z^4 (Z being the mean atomic number of the sample), and data in Figure 10.29 show the proportion of observed fluorescent x-rays which results from continuum excitation in pure elements. The contribution is almost negligible for the K-lines of elements below potassium or the L-lines below zirconium in the Periodic Table. However for heavier elements, this excitation mechanism makes a significant contribution to observed x-ray intensities.

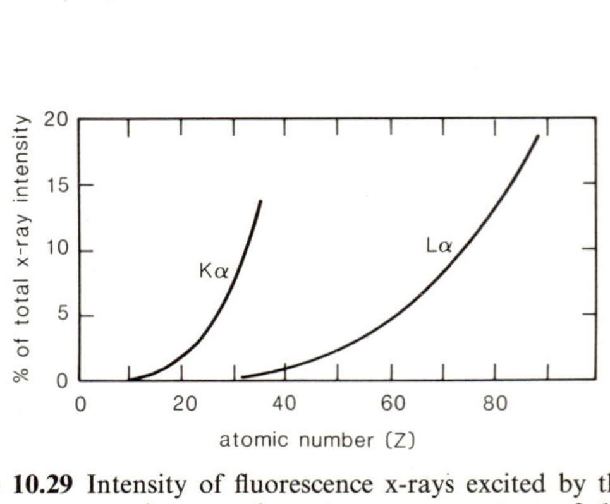

Figure 10.29 Intensity of fluorescence x-rays excited by the continuous spectrum in pure elements as a percentage of the total characteristic x-ray intensity. (Reed, 1973.)

(*b*) *Secondary fluorescence.* Characteristic x-rays generated within the excited volume can also penetrate further into the sample, where they are capable of fluorescing lighter elements having a lower critical excitation energy. The x-ray signal due to this second element is thus enhanced. The probability of this occurring depends both on how close the primary x-ray energy is to the critical excitation energy of the second atom and on its fluorescence yield (Figure 10.37). The effect is likely to be significant for specific pairs of heavier elements such as Ni *Kα* and Fe *Kα*.

(*c*) *Effective x-ray resolution.* The importance of continuum and secondary fluorescence in practical situations is that the source of characteristic x-rays within the sample is somewhat larger than the electron excited volume. This has important consequences in the interpretation of data in two circumstances: first, in the measurement of concentration gradients close to mineral grain boundaries, and second, in assessing the analysis of small minerals (less than 30 μm across). In each case, the recorded spectra are likely to include x-ray lines that are not present in the mineral under study but which are excited by continuum or secondary fluorescence of the matrix beyond the grain boundary.

In order to test this effect, Reed and Long (1963) carried out spot analyses for iron across an artificially created phase boundary between this metal and nickel. These elements were selected because the secondary fluorescence of Fe *Kα* by Ni *Kα* was likely to be significant. Their results are reproduced in Figure 10.30. These data show that as the point of analysis traverses the boundary, the Fe *Kα* intensity drops sharply over an interval of about 5 μm corresponding to the movement of the excited volume from one phase to the other. However, even with the excited volume entirely penetrating the nickel phase (point (*a*) in Figure 10.30), the apparent concentration of Fe *Kα* is still between 4 and 5%, entirely due to secondary fluorescence by Ni *Kα*/*Kβ*. A more disturbing observation is that a significant Fe signal (equivalent to about 0.3% Fe) is still observed in the iron-free nickel phase when the probe is positioned as far as 30 μm from the phase boundary.

A further difficulty will arise if the phase boundary is not perpendicular to the surface of the sample. In Figure 10.31 an Ni–Fe–Ni sandwich is shown to demonstrate this effect. Based on Reed and Long's work, hypothetical signals due to Fe *Kα* and Ni *Kα* are plotted and indicate the following.

(i) The fall-off in both Fe *Kα* and Ni *Kα* signal intensity is apparently displaced relative to the surface exposure of the phase boundary.

(ii) The nickel phase appears to contain iron due to the secondary fluorescence of the adjacent iron phase by Ni *Kα*/*Kβ*, and the iron phase appears to contain nickel, due to the continuum excitation of the adjacent nickel phase by bremsstrahlung radiation excited in iron, an effect that is likely to be much less significant.

(iii) If the iron sandwich is less than about 50 μm thick, its apparent nickel content will never reach zero.

(iv) Measurements at the phase boundary make it appear that iron is diffused into the nickel phase and vice versa, rather than there being a sharp discontinuity in elemental concentrations.

It must be emphasized that these observations are not based on direct experimental measurement but on an extrapolation of the data of Reed and Long. These phenomena illustrate the considerations that must be taken into account in interpreting apparent differences in concentration derived from x-ray measurements at or near phase boundaries.

Reed and Long (1963) discussed mathematical correction procedures that may be used in some circumstances to correct for fluorescence effects. The examples cited above are expected to represent the worst case. In mineralogical applications the secondary fluorescent of Ti or Ca *Kα* by Fe *Kα* in an adjacent mineral is perhaps a more appropriate example. It must be conclude that concentration data from

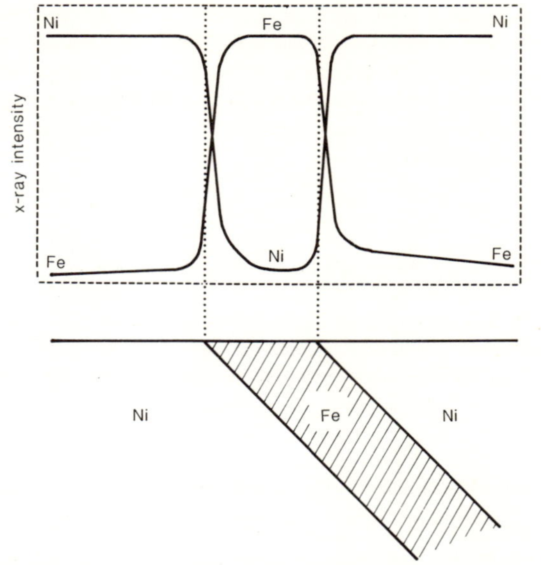

Figure 10.31 Apparent content of iron and nickel across a nickel–iron–nickel sandwich in which the iron layer is inclined to the surface. This diagram shows the apparent asymmetric distribution which may be observed in iron and nickel x-ray signals as the electron beam is scanned across the inclined phase boundary. Excitation of iron in the nickel phase is caused by secondary fluorescence of Fe *Kα* by Ni *Kα* and of nickel in the iron phase, by the continuum fluorescence of Ni *Kα* across the phase boundary (hypothetical data).

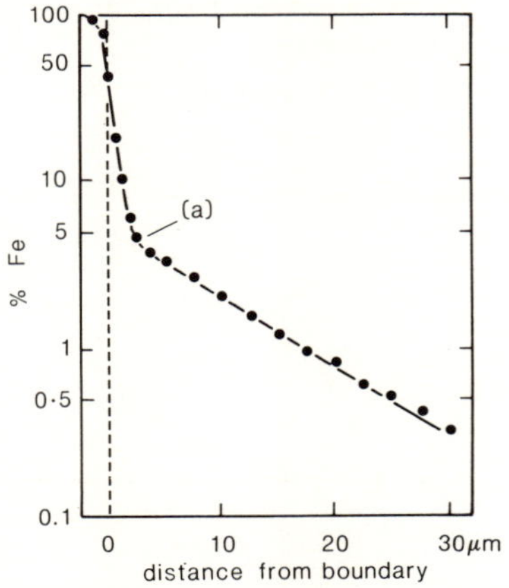

Figure 10.30 Apparent iron content of pure nickel as a function of the distance from a boundary with pure iron (Reed and Long, 1963.)

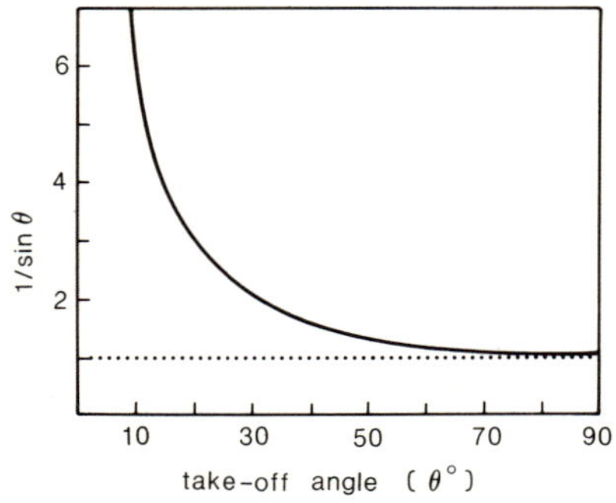

Figure 10.32 Effect of x-ray take-off angle (θ) on the relative magnitude of the x-ray absorption effect as measured by $1/\sin\theta$.

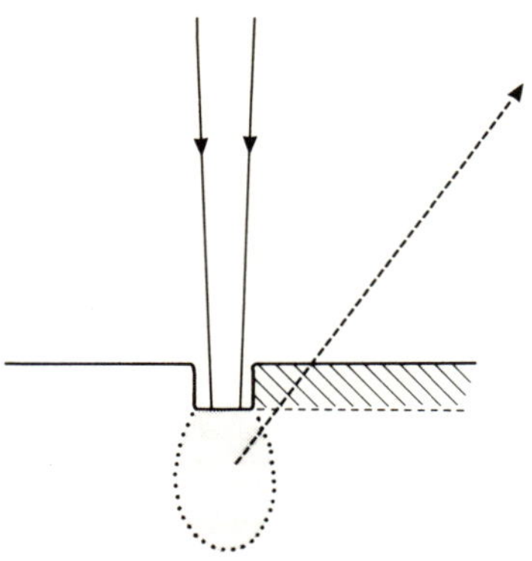

Figure 10.33 Errors in absorption correction due to polishing deficiency. The probe is focused on a 1 μm scratch. The dotted line indicates the x-ray take-off path to the spectrometer. The shaded area represents an additional thickness of sample through which x-rays must pass, absorption in which is not accounted for in ZAF correction procedures. (After Sweatman and Long, 1969.)

a mineral of uncertain depth obtained within say 30 μm of a grain boundary must be interpreted with caution.

10.8.6 *Other phenomena affecting the quantification of x-ray intensities*

(*a*) *Take-off angle.* The angle between the surface of the sample and the spectrometer entrance slits (or collimator) is known as the take-off angle. It is desirable to make this angle as large as possible for the following reasons:

(i) One of the largest matrix interference effects is caused by absorption of x-rays within the sample; the uncertainty in this correction is reduced at high take-off angles since mean x-ray path length within the sample is then minimized

(ii) The influence of scratches and other surface polish irregularities is minimized at high spectrometer take-off angles (section 10.8.6 (*b*)).

Take-off angle is normally fixed by instrument design and is often dictated by objective lens design (section 10.4.2). The largest take-off angle available on a commercial instrument is 75°. In hybrid scanning electron microprobe instruments, take-off angles are normally fixed at 40°, 45° or 52°. As the path length within the sample over which absorption of x-rays can occur varies according to $1/\sin\theta$, the relative effect of take-off angle on the magnitude of the absorption correction may be judged from data in Figure 10.32. These data show that the effective path length within the sample doubles as take-off angle is reduced from the hypothetical 90° down to 30°. Below 30°, a rapid increase in path length occurs.

(*b*) *Polishing irregularities.* If the electron beam is inadvertently focused on a scratch in the polished surface (Figure 10.33) additional (and unaccounted) absorption of measured x-ray intensities may occur. The severity of this effect will depend on instrument take-off angle. Sweatman and Long (1969) estimated that the error in analysing Mg $K\alpha$ at 15 kV in a pyroxene of composition 50% MgO, 50% SiO_2 amounts to 10% relative if the scratch is 0.5 μm deep for a spectrometer take-off angle of 20°. Conversely, instruments with take-off angles in the range 50 to 75° are much less sensitive to surface irregularities.

(*c*) *Specimen tilt.* If a specimen whose surface is assumed to be normal to the electron beam is in fact tilted so that the actual take-off angle is no longer identical to its nominal value, an error will occur in the calculated absorption correction. Sweatman and Long (1969) showed that the magnitude of this error decreases with increasing take-off angle (Table 10.4). At low take-off angles (20° to 40°), the error can be appreciable, as is seen in their data for the determination of Mg $K\alpha$ in an olivine, assuming specimen tilts of

Table 10.4 Relative error (%) in the analysis of Mg and Si in olivine due to specimen tilt and error in assumed operating potential

	Spectrometer take-off angle				
	20°	40°	52.5°	75°	
Tilt = 1°	2.1	1.2	0.7	0.2	Error in analysis
Tilt = 2°	4.3	2.4	1.3	0.4	of Mg due to indicated
Tilt = 5°	10.6	5.9	3.3	1.0	specimen tilt at 20kV
ΔkV = 0.5	1.3	1.0	—	0.8	Error in absorption correction
ΔkV = 1.0	2.5	2.0	—	1.6	for Si due to indicated
ΔkV = 2.0	5.0	4.0	—	3.3	error in assumed gun kV

Data from Sweatman and Long (1969).
Assumed sample composition of olivine: MgO: 48.1%; SiO_2: 40.2%; FeO: 11.5%.
Standards: MgO, SiO_2, Fe metal.

1°, 2° and 5°. Tilts of this order can be obtained from the rounding effects during over-polishing of hard minerals.
(*d*) *Specimen height*. If the height of the specimen in relation to the spectrometer entrance slits is not accurately reproduced to within a micrometre or so, discrepancies in measurements will occur, because this has a defocusing effect on wavelength dispersive spectrometers. As is discussed further in section 10.10, the specimen surface (x-ray source), diffracting crystal and counter entrance slits must be accurately maintained on the Rowland's circle of a wavelength dispersive spectrometer to avoid defocusing effects. One of the subsidiary uses of the optical microscope fitted to microprobe instruments is to ensure accurate and reproducible alignment of samples. This is achieved by the requirement to bring all analysed surfaces into sharp focus using high-magnification optics (> × 500) prior to analysis.
(*e*) *Errors in nominal gun kV*. Both the absorption and the generation correction factor (section 10.9) are sensitive to errors between the assumed and actual gun operating potential. This error is again dependent on spectrometer take-off angle. Data presented by Sweatman and Long (1969) (Table 10.4) show that the error in the absorption correction for Si *Kα* in an olivine can reach several per cent for discrepancies in the assumed gun accelerating potential of only 1 kV. The accuracy of the gun kV meter can be checked by making x-ray measurements of the elements shown in Table 10.5, as described by Sweatman and Long (1969). The observed count rate of the specified *Kα* line is plotted as a function of the kV meter reading (Figure 10.34). These data are extrapolated to determine the appearance threshold of characteristic x-ray counts. This value should correspond to the precise energy listed in Table 10.5. On instruments fitted with energy dispersive detectors, the gun kV may be more readily monitored by observing the continuum spectrum emitted from a sample. The true gun kV is indicated by the point at which the continuum spectrum merges into the low intensity random summing background, at this so-called Duane–Hunt limit.
(*f*) *Effect of carbon coating*. The surfaces of all non-conducting samples must be coated with a thin layer of carbon to ensure that the electron charge deposited in the sample during analysis is conducted uniformly to earth. A traditional carbon coater is shown in Figure 10.35. The polished sample is first cleaned to remove all traces of oil contamination and any polishing debris. It is then placed face upwards in a bell jar evacuated to less than 10^{-5} Torr. A dc

Table 10.5 Elements and x-ray lines suitable for calibrating the gun accelerating voltage

Element	Line	Excitation potential (kV)
Ti	*Kα*	4.964
Cu	*Kα*	8.979
Zn	*Kα*	9.659
Rb	*Kα*	15.200
Y	*Kα*	17.035
Mo	*Kα*	20.001
Ag	*Kα*	25.512
Sn	*Kα*	29.190

Based on data of Sweatman and Long (1969).
Excitation potentials from White and Johnson (1970).
It may be necessary to consider analysing second order diffraction lines to measure the *Kα* lines of Rb–Sn.

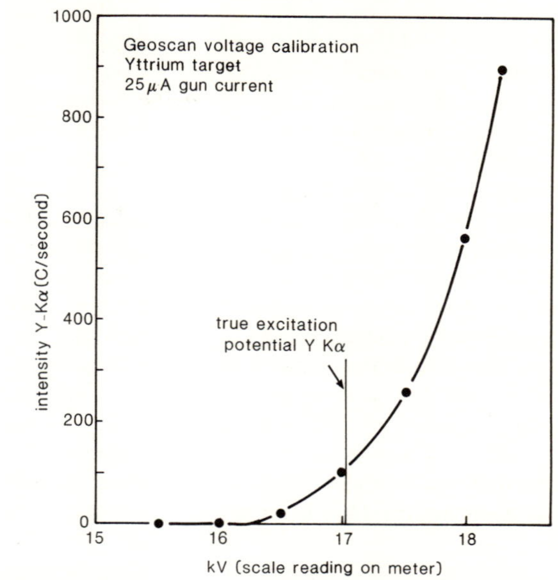

Figure 10.34 Measurement of the true gun accelerating potential from measurements of the intensity of x-rays from an element (in this case yttrium) of known critical excitation potential (17.035 keV). In this example, the kV meter reads 16.3 kV for a true potential of 17.04. (Sweatman and Long, 1969.)

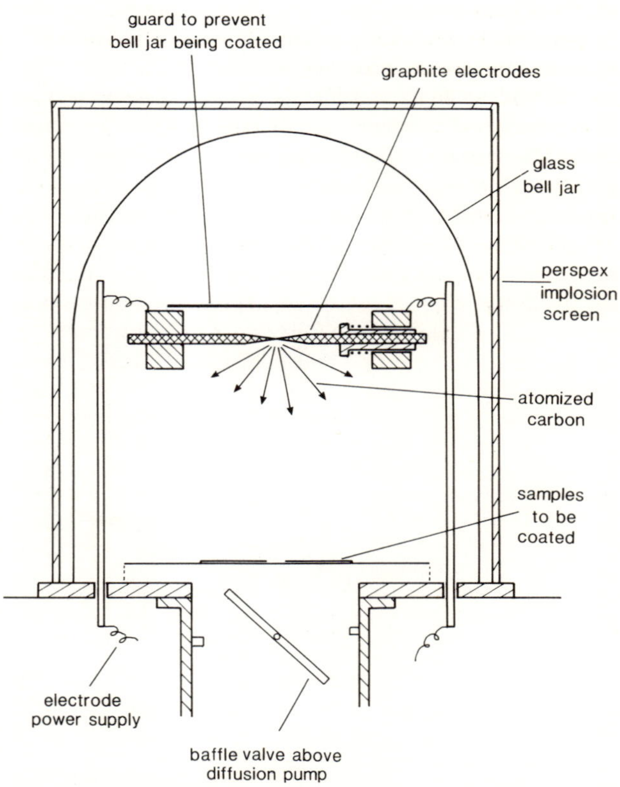

Figure 10.35 Schematic diagram of a carbon coater showing the vacuum chamber, carbon electrodes and sample.

current of up to 100 A at either 10 or 30 V is then passed through a pair of carbon electrodes. These electrodes are normally 5 to 6 mm in diameter and have pointed ends, spring loaded to maintain electrical contact (sometimes one electrode has a pointed end, the other a flat end). When the dc current is passed through this contact, the ends of the electrodes, which represent the region of highest electrical resistance, heat up and rapidly become incandescent. This causes carbon to become vaporized. All surfaces within line of sight of the electrode tips are then coated with a thin film of carbon. The current is maintained through the electrodes until either the required thickness of carbon has been deposited on the sample or the electrode tips have become so blunt that they can no longer be maintained at a sufficiently high temperature to continue vaporizing carbon. Some modern carbon coaters use a carbon filament 1 to 2 mm in diameter in place of two thicker, pointed electrodes so that more uniform rates of carbon emission can be achieved.

Any coating on the surface of a sample represents an additional layer in which absorption of x-rays can occur. Variation in carbon thickness can therefore affect the accuracy of analysis. Kerrick *et al.* (1973) studied the effect of carbon coats of thickness 20–170 nm (200–1700Å) and report that differences in film thickness of 20 nm can cause discrepancies in the measured count rate of F $K\alpha$ of up to 4%. Springer (1974) emphasized that in addition to the absorption of the x-rays of low-atomic-number elements, it is of equal importance to consider the absorption of *primary beam electrons* in the carbon coat. At accelerating potentials of 20 kV or more, differences in coating thickness of 20 nm cause errors of less than 1%. However, in the range 15 to 20 kV, errors may be up to 2%, a figure which may be substantially exceeded if accelerating potentials of 10 kV or less are used, as emphasized by Reed (1972).

Providing standard minerals are coated at the same time as samples, differences in the thickness of carbon coats should be negligible. This is, however, rather inconvenient, and Kerrick *et al.* (1973) have shown that the coating thickness can be effectively monitored by observing the interference colours developed on a surface of a highly polished brass block, coated at the same time as the samples. Distinctive colour changes are observed (Table 10.6) permitting reproducible coating to an uncertainty within ± 2.5 nm. For routine analysis, a carbon film thickness of 25 ± 2.5 nm is normally considered satisfactory.

Table 10.6 Interference colours formed by carbon films deposited on a polished brass surface

Thickness of carbon coat (approximate) nm	Colour
15	Orange
20	Indigo red
25	Blue
30	Bluish green
35	Greenish blue
40	Pale green
45	Silver gold

Data from Kerrick *et al.* (1973).

10.9 Matrix correction procedures

As in x-ray fluorescence analysis, the observed x-ray intensity of an element determined by electron microprobe is not linearly proportional to its concentration but is affected by the concentrations of all the other elements present in the sample.

A considerable amount of work has been undertaken to model mathematically the correction procedures required to account for these discrepancies. One widely used category of procedure is the *ZAF correction*, so called because it accounts for the influence of atomic number (Z), x-ray absorption (A) and secondary fluorescence (F) effects. These procedures will now be summarized; fuller treatments are available in Sweatman and Long (1969) and Reed (1975). A detailed comparison between the accuracy of computer programs then available for quantitative electron microprobe analysis has been undertaken by Beaman and Isasi (1970).

In the full ZAF correction procedure, the measured x-ray intensity is multiplied by a number of factors modelling the influence of the effects listed above. These are derived as follows.

10.9.1 *The atomic number effect or generation factor*

The atomic number effect comprises two factors which describe (*a*) the proportion of incident electrons backscattered out of the sample and therefore unavailable for further x-ray excitation, and (*b*) the way in which the rate of energy transfer from primary beam electrons is affected by the atomic composition of the sample.

(*a*) *Backscatter electron correction.* A proportion of electrons from the incident electron beam is scattered back out of the sample and these electrons are, therefore, no longer available for x-ray generation. This fraction shows a marked dependence on the mean atomic number of the sample, heavier elements being more efficient at backscattering electrons.

The proportion of backscattered electrons is measured experimentally as the backscatter coefficient η. However, the overall correction factor must also take into account the energy spectrum of backscattered electrons, since the higher the energy of such an electron, the greater would have been its potential for x-ray generation had it remained within the sample. This potential for x-ray excitation varies according to the over-potential ratio E_0/E_c, where E_0 is the incident electron beam energy and E_c the critical excitation energy of the x-ray line in question.

The overall effect of this phenomenon on x-ray intensity is measured by the *backscatter ratio R* (that is the fraction of x-rays actually generated within the sample compared with the intensity that would have been generated in the absence of backscatter). The backscatter ratio is given by

$$R = 1 - \eta_x \qquad (10.9)$$

where η_x is the *backscatter coefficient* corrected for the energy spectrum effect mentioned above.

In the literature, values of R are tabulated as functions of both atomic number (Z) and E_0/E_c derived mainly from experimental measurements. As virtually all of these experimental data have been obtained from pure elements, it must be assumed that for compound samples, the effective backscatter ratio R may be calculated from

$$R = \sum_i C_i R_i \tag{10.10}$$

where C_i is the concentration of element i having an effective backscatter ratio R_i.

(*b*) *Stopping power correction*. When the electron beam enters the sample, energy is transferred mainly by interaction of primary beam electrons with orbital electrons of atoms of the sample. As low-atomic-number elements have a larger proportion of orbital electrons per unit volume than do high-atomic-number elements, a larger proportion of the primary beam energy will be transferred to atoms of the former, leading to a correspondingly greater probability of x-ray excitation. The stopping power of the light elements is, therefore, greater than that of the heavy elements.The magnitude of this effect is quantified by the stopping power factor of an element S, defined as the fractional loss of energy of a primary beam electron (dE) per unit path length (dx) normalized to sample density:

$$S = -\frac{1}{\rho} \cdot \frac{\mathrm{d}E}{\mathrm{d}x}. \tag{10.11}$$

S thus has units of eV cm^2 g^{-1} and may be calculated from the Beth expression (see Long, 1977):

$$S = \mathrm{const.} \frac{Z}{A} \cdot \frac{1}{E} \cdot \left[\ln \left\{ 1.166 \left(\frac{E}{J} \right) \right\} \right] \tag{10.12}$$

where Z is the atomic number, A is the atomic weight, J is the mean ionization potential averaged over all electron shells in the target atom, and E is the mean primary electron energy usually taken as

$$E = \frac{E_0 - E_c}{2} \tag{10.13}$$

where E_0 is the incident electron beam energy, and E_c is the critical ionization energy of the x-ray line under consideration.

In compound samples it is assumed that the effective stopping power S may be calculated from

$$\bar{S} = \sum_i C_i S_i. \tag{10.14}$$

(*c*) *Overall generation factor*. The overall atomic number or generation correction factor must take into account the two opposing factors described above. These may be qualitatively described as follows. The stopping power of silicon in a higher mean atomic number matrix (e.g. fayalite, Fe_2SiO_4) is greater than silicon in a lower mean atomic number matrix (e.g. forsterite, Mg_2SiO_4) since in the former case a larger proportion of electron energy is absorbed by silicon atoms leading to a higher relative Si $K\alpha$ intensity. However, since the mean atomic number of fayalite is greater than that of forsterite, a larger proportion of incident electrons will be backscattered out of the fayalite, tending to reduce the excitation of Si $K\alpha$ in this sample.

As it is usual to compare the measured x-ray intensity with that from a standard mineral, corrections for backscatter (R) and stopping power (S) effects must be applied to both sample and standard data. The overall correction is then

$$C_{\mathrm{CORR}} = C_{\mathrm{APP}} \cdot \frac{R_{\mathrm{STD}}}{R_{\mathrm{UNK}}} \cdot \frac{S_{\mathrm{UNK}}}{S_{\mathrm{STD}}} \tag{10.15}$$

where C_{CORR} and C_{APP} refer to the corrected and apparent concentration of the element respectively.

10.9.2 *The absorption correction*

An absorption correction is required to compensate for the fact that x-rays, generated at distance z below the surface of the sample, will suffer attenuation before emerging for detection by an x-ray spectrometer. The degree of attenuation in part depends on the spectrometer take-off angle, since x-rays emitted in the direction of the spectrometer will travel a distance z. $(1/\sin\theta)$ within the sample before emerging, and partly on the mean mass attenuation coefficient of the sample $(\bar{\mu}/\rho)$, where

$$\left(\frac{\bar{\mu}}{\rho}\right) = \sum_i C_i \cdot \left(\frac{\mu}{\rho}\right)_i. \tag{10.16}$$

(See sections 8.6, 8.8 for a discussion of mass attenuation coefficients.)

The factor by which the x-ray beam is reduced in intensity is given by

$$\exp\left[-\left(\frac{\bar{\mu}}{\rho}\right) \rho z \cdot \frac{1}{\sin\theta} \right] \tag{10.17}$$

In practice, the attenuation correction term must take into account the fact that the probability of x-ray generation varies with depth within the excited volume. The function that describes this depth distribution is the so-called $\phi(\rho z)$ function, formulated as a function of density (ρ) as well as depth (z) so that it is generally applicable to all materials. The typical shape of the $\phi(\rho z)$ function is shown in Figure 10.36. As the operating kV is increased, this function rapidly expands along the z axis in accordance with increased electron penetration. However at constant potential, the depth distribution (normalized to density) varies only slightly with sample composition. The overall difference in x-ray attenuation between sample and standard is then controlled by differences in the mean mass attenuation coefficient.

Thus, to determine the overall attenuation of x-rays within a sample, the attenuation factor (equation (10.17)) must be integrated with respect to the x-ray generation depth distribution function $\phi(\rho z)$ for all values of z:

$$f(\chi) = \int_{\rho z = 0}^{\infty} \{\phi(\rho z) \cdot \exp[-\chi \rho z]\} \cdot \mathrm{d}(\rho z) \tag{10.18}$$

where for convenience

$$\chi = \left[\left(\frac{\bar{\mu}}{\rho}\right) \cdot \frac{1}{\sin\theta} \right] \tag{10.19}$$

and $f(\chi)$ represents the fraction of x-rays attenuated within the sample.

The absorption correction factor, F_A (the factor by which the observed x-ray intensity must be increased to compensate

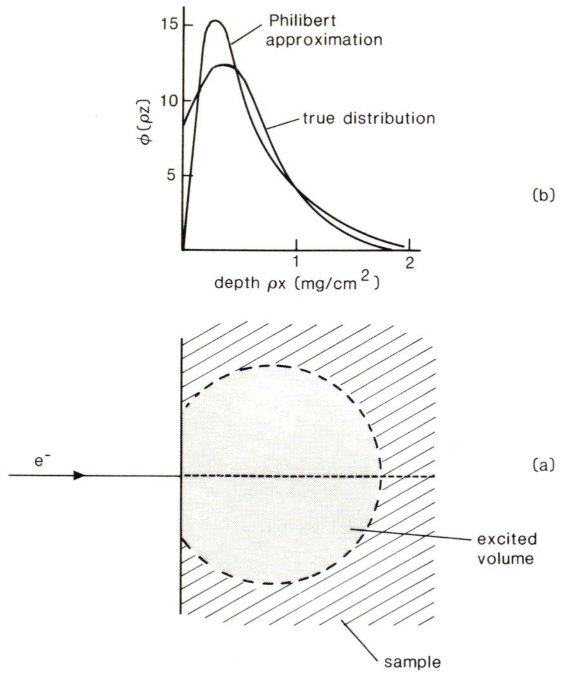

Figure 10.36 The distribution ($\phi(\rho z)$) of electrons within the excited volume as a function of depth: (*a*), the schematic excited volume; (*b*), the corresponding electron distribution function ($\phi(\rho z)$) as a function of normalized depth into the sample (ρx). (After Reed, 1975.)

for attenuation within the sample) is then given by

$$F_A = \frac{1}{f(\chi)}. \qquad (10.20)$$

One widely used approximation to calculate this absorption correction factor was devised by Philibert (1963) and is given by

$$F_A = \frac{1+h}{(1+\chi/\sigma)(1+h[1+\chi/\sigma])} \qquad (10.21)$$

where $h = 1.2\,A/Z^2$, Z is the atomic number, A is the atomic weight

$$\sigma = \left[\frac{4.5\times 10^5}{E_0^{1.65} - E_c^{1.65}}\right] \qquad (10.22)$$

and for a compound sample,

$$h = \sum_i C_i h_i \qquad (10.23)$$

E_0 is the incident electron beam energy, E_c is the critical excitation energy of x-ray line under consideration, and

$$\chi = \left(\frac{\bar{\mu}}{\rho}\right)\cdot\frac{1}{\sin\theta} \quad \text{as before.}$$

(See Reed, 1975, pp.244 *ff*. for a more complete review).

The degree of agreement between the Philibert approximation for the depth distribution function and observed values is indicated in Figure 10.34. Mass attenuation coefficients widely used in these calculations are taken from the data of Heinrich (1966) as modified by Springer and Nolan (1976).

The overall correction for x-ray attenuation in a sample ($F_{A(UNK)}$) in comparison with that occurring in the standard ($F_{A(STD)}$) is given by:

$$C_{CORR} = C_{APP}\cdot\frac{F_{A(STD)}}{F_{A(UNK)}}.$$

10.9.3 *The fluorescence correction*

Although important in the analysis of some metallic alloy phases, the fluorescence correction does not make a large contribution in the analysis of silicate minerals, rarely exceeding 5% relative in magnitude.

However, this correction is normally incorporated in ZAF programs and must take into account two phenomena: the fluorescence of one element caused by a second that emits a characteristic x-ray of energy above the absorption edge of the first. An example would be calcium *K*-lines (critical excitation energy: 4.037 keV) fluoresced by Ti $K\alpha$ (4.508 keV). The second and potentially more serious problem is the fluorescence of an element by that component of the x-ray continuum spectrum which exceeds the energy of the absorption edge.

(*a*) *Fluorescence by characteristic x-rays.*If I_F is the intensity of x-rays of an element generated by secondary fluorescence and I_P is the intensity generated by primary interaction with the electron beam, then a correction for secondary fluorescence (F_{SF}) must be applied to the observed intensity data as follows:

$$F_{SF} = \frac{I_P}{I_P+I_F} = \frac{1}{1+I_F/I_P} \qquad (10.24)$$

The calculation of the intensity ratio I_F/I_P must take into account:

(i) The probability that an x-ray from element B will excite the characteristic radiation from element A

(ii) The intensity distribution of radiation from element B within the sample

(iii) The probability that either of these x-rays will suffer absorption within the sample.

The full derivation of a fluorescence correction has been given by Reed (1965) and further reviewed by Reed (1975). It has the form

$$\frac{I_F}{I_P} = 0.5C_B\cdot\frac{(\bar{\mu}/\rho)_B^A}{(\bar{\mu}/\rho)_B}\cdot\frac{r_A-1}{r_A} \times \omega_{K(B)}\cdot\frac{A_A}{A_B}\cdot\left(\frac{U_B-1}{U_A-1}\right)^{1.67} \times\left[\frac{\ln(1+u)}{u}+\frac{\ln(1+v)}{v}\right] \qquad (10.25)$$

where the characteristic x-ray from element B excites element A. C_B is the weight fraction of element B, $(\bar{\mu}/\rho)^A{}_B$ is the bulk mass attenuation coefficient of pure element A, and $(\mu/\rho)_B$ is the bulk mass attenuation coefficient of the sample, both with respect to the energy of fluorescence x-rays from element B, $(r_A-1)/r_A$ is the absorption jump ratio of the $K\alpha$ line from element A (section 8.6), $\omega_{K(B)}$ is the fluorescence yield of *K*-line radiation from element B (section 8.6), A is the atomic weight, U is the over-potential ratio E_0/E_c,

$$u = \left[\frac{(\bar{\mu}/\rho)_A}{(\bar{\mu}/\rho)_B}\right] \cdot \frac{1}{\sin\theta} \cdot \qquad (10.26)$$

Subscripts A and B refer to bulk mass attenuation coefficients at the energies of the characteristic radiation from elements A and B respectively, and

$$v = \frac{\sigma}{(\bar{\mu}/\rho)_B} \qquad (10.27)$$

where σ is the Lenard coefficient and represents the electron mass attenuation coefficient used to describe the attenuation of an electron beam of initial intensity i_0 which is transmitted through a thin film of thickness x in a material of density ρ according to the equation

$$i = i_0 \,.\, \exp(-\sigma\rho x). \qquad (10.28)$$

The full derivation of this complex expression can be obtained from the references cited above.

Fluorescence yield data ω is quite well known for the *K*-line series, not as accurately known for *L*-lines and largely unavailable for *M*-lines. Reed (1973) has suggested that an approximate calculation can be used to test if the magnitude of the characteristic fluorescence correction is significant, namely

$$F_{SF} = \frac{1}{1 + 0.5 C_B \omega_B} \qquad (10.29)$$

where the symbols have the same meaning as in equations (10.24) and (10.25).

As can be seen from equation (10.25), the magnitude of the characteristic fluorescence correction depends on the fluorescence yield. From Figure 10.37, it can be seen that this parameter falls rapidly in magnitude with atomic number so that corrections are negligible for the *K*-lines of elements lighter than about Ca or the *L*-lines of elements lighter than about La. *M*-line fluorescence effects can probably be ignored.

(*b*) *Fluorescence by continuum radiation.*The approach for estimating the continuum fluorescence correction is similar to that outlined above. Calculations must take into account:

(i) The distribution of intensities in the continuum spectrum of energy exceeding the critical excitation energy of the element line of interest as determined by Kramers' equation (section 10.8.4).

(ii) The probability that this radiation will excite the line of interest

(iii) The probability that continuum and fluoresced radiation will be attenuated within the sample.

One widely used simplification is that of Springer (1967) where the ratio of continuum excited x-ray intensity (I_F) to that excited by primary interaction with the electron beam (I_P) is given by:

$$\frac{I_F}{I_P} = 4.34 \times 10^{-6} \,.\, \frac{r_A - 1}{r_A} \,.\, A \,.\, \bar{Z} \,.\, E_C \,.\, \frac{(\mu/\rho)_K^A}{(\mu/\rho)_K} \,.\, h(uU_o) \qquad (10.30)$$

where A is the mean atomic weight, $\bar{Z}$ is the mean atomic number given by $Z = \Sigma_i\ C_i Z_i$, E_c is the critical excitation energy of the x-ray line from element A, $(\mu/\rho)_K^A$ is the mass attenuation coefficient of pure element A at a point just on the high-energy side of the absorption edge of element A, $(\mu/\rho)_K$ is the same parameter average for the sample and $h(u, U_o)$ is the function describing the absorption of continuum radiation within the sample.

The factor F_{CF}, which must be used to correct the observed x-ray intensity for continuum fluorescence, is thus

$$F_{CF} = \frac{1}{1 + I_F/I_P} \qquad (10.31)$$

where I_F/I_P is determined from equation (10.30) above. The magnitude of the continuum fluorescence effect in pure elements is shown in Figure 10.29. These data indicate that continuum fluorescence corrections can be effectively ignored for elements lighter than K (*K*-line series) or Zr (*L*-

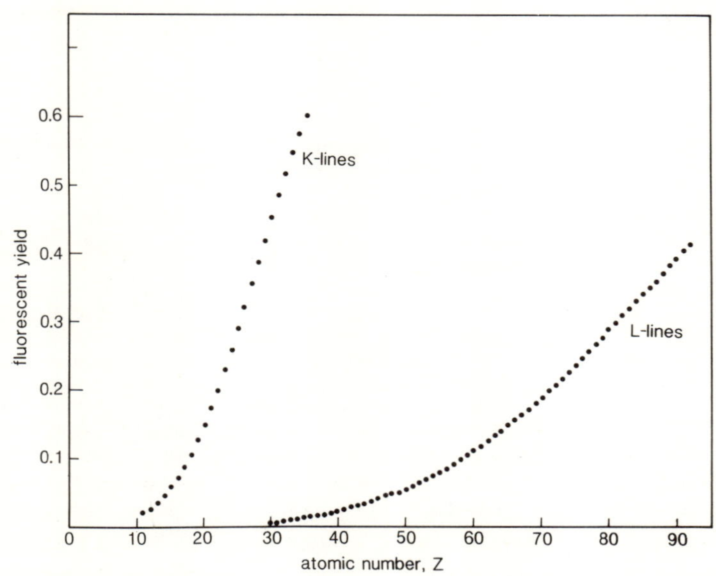

Figure 10.37 Fluorescence yield for *K*- and *L*-line x-rays as a function of atomic number. (Data from Reed, 1975.)

M*

line series) but that for the heavier elements the correction becomes more significant.

(*c*) *Errors in fluorescence corrections.*In all the calculations set out above, the assumption is made that the analysed sample is uniform in composition. However, as discussed in section 10.8.5, secondary and continuum fluorescence operates over a much longer range than direct electron excitation. Indeed, data in Figure 10.30 indicate that corrections may, in some circumstances, be required for effects occurring up to 30 μm from the point of impact of the electron beam.

In the practical analysis of geological samples, it is often unreasonable to expect mineral homogeneity on this scale due to the small size of mineral grain being analysed, the proximity of a grain boundary or the uncertain nature of mineral structure beneath the surface of the sample. It is therefore fortunate that the magnitude of the fluorescence correction, when applied to silicate minerals, is often negligible and rarely exceeds 5% in magnitude.

10.9.4 *The full ZAF correction*

The full correction formula, incorporating each of the terms described above, is

$$C_{COR} = C_{APP} \cdot \frac{R_0}{R_1} \cdot \frac{S_1}{S_0} \cdot \frac{f(\chi)_0}{f(\chi)_1} \cdot \frac{F_{SF1}}{F_{SF0}} \cdot \frac{F_{CF1}}{F_{CF0}} \quad (10.32)$$

where subscripts 0 and 1 refer to the mineral standard and unknown sample, respectively, and the individual factors correct for effects due to backscattered electrons (R) and stopping power of atoms (S)

$f(\chi)$: absorption of fluorescence x-rays within the sample, F_{SF}: secondary fluorescence by characteristic radiation, F_{CF}: secondary fluorescence by continuum radiation.

This equation may be summarized as

$$C_{APP} = C_{CORR} \cdot \frac{ZAF_{UNK}}{ZAF_{STD}} \quad (10.33)$$

where

$$ZAF = \frac{S}{R} \cdot \frac{1}{f(\chi)} \cdot F_{SF} \cdot F_{CF}$$

As each of these correction factors depends on the sample composition (represented by mean atomic number or bulk mass attenuation coefficient), it is necessary to use an iterative procedure. The apparent (uncorrected) concentrations are used to estimate correction factors which are themselves used to calculate improved estimates of the 'true' concentration. This cycle of calculations is continued until the differences between successive iterations are statistically insignificant. About five such iterations may be necessary.

10.9.5 *Relative magnitude of correction effects*

The relative magnitudes of individual correction factors may be judged from data in Table 10.7, taken from Sweatman and Long (1969). Correction factors for three minerals, an olivine, garnet and sphene are itemized and demonstrate the following.

(i) The atomic number or generation correction is independent of take-off angle, being a measure of the backscatter and stopping power interaction of electron beam with sample

(ii) The absorption correction can be large (up to 26% for Si $K\alpha$ in olivine) and shows, as expected, a marked dependence on take- off angle

(iii) The fluorescence correction is usually small (the maximum here being 2.5% on Ca $K\alpha$ in sphene). Contrary to the trend observed in the absorption correction, the fluorescence correction shows a decrease with decreasing take-off angle due to the enhanced absorption of secondary fluoresced x-rays excited at depth within the sample

(iv) The overall correction factor in the examples given here is in general less than 25% with factors appropriate to the 75° take-off angle being usually significantly smaller than those for 40°.

10.9.6 *Other forms of matrix correction—alpha coefficients*

The full ZAF correction procedure has been described above because it gives an insight into the physical processes that affect the generation and attenuation of x-rays within the sample. However, an alternative approach is to adopt the

Table 10.7 Examples of ZAF correction factors for spectrometers with 40° and 75° take-off angles

Mineral	Element	Oxide concentration	Atomic number (generation) correction	Absorption correction 40°	Absorption correction 75°	Fluorescence correction 40°	Fluorescence correction 75°	Overall correction 40°	Overall correction 75°
Olivine	Mg	48.1	0.9846	1.1730	1.1240	0.9980	0.9978	1.1526	1.1043
	Si	40.2	0.9886	1.2615	1.1806	1.0000	1.0000	1.2471	1.1671
	Fe	11.5	1.2197	0.9987	0.9991	1.0000	1.0000	1.2181	1.2186
Garnet	Al	22.0	0.9573	1.3063	1.2152	0.9960	0.9956	1.2455	1.1582
	Si	37.9	0.9449	1.2903	1.2009	0.9999	0.9999	1.2191	1.1346
	Fe	31.3	1.1587	0.9992	0.9995	1.0000	1.0000	1.1578	1.1581
Sphene	Si	30.0	0.9515	1.1010	1.0704	0.9988	0.9986	1.0463	1.0171
	Ca	27.5	0.9717	0.9922	0.9947	0.9765	0.9743	0.9415	0.9417
	Ti	39.1	1.1368	1.0437	1.0292	1.0000	1.0000	1.1865	1.1699

Data from Sweatman and Long (1969).
Data are calculated for analyses carried out at 15kV using the following calibration standards: jadeite (Na), MgO, Al_2O_3, SiO_2, wollastonite (Ca), iron metal, titanium metal.

empirical influence coefficient procedure similar to the techniques described for x-ray fluorescence analysis in section 8.8. The intensity of x-ray emission of element A is assumed to be influenced by the concentrations of all the other elements (B, C, D...) in the sample. The effect of these elements may be corrected from

$$C_{CORR} = C_{APP} \cdot \frac{1}{\alpha_A} \cdot (\alpha_A C_A + \alpha_B C_B + \alpha_C C_C + \ldots) \tag{10.34}$$

where C_{CORR} is the corrected concentration of element A, C_{APP} is the apparent concentration of element A, α is the alpha coefficient representing the influence of element A, B, C,... on element A, C is the respective concentration.

The alpha coefficients quantify the influence of a particular element on the one being determined and are assumed to be constant for the selected operating kV and spectrometer take-off angle. The most widely used application of this procedure for the microprobe analysis of silicate and oxide materials was described by Bence and Albee (1968). They discussed the determination of individual alpha coefficients for a probe operated at 15 kV with a 52° spectrometer take-off angle. The concentration terms in equation (10.34) were treated as element oxides and the procedure was applied to minerals such as amphiboles and micas using simple silicate minerals as standards.

The assumption that alpha coefficients are independent of concentration is not strictly true if the fluorescence correction is significant. However, it has already been shown above that this effect is usually either small or negligible in silicate minerals. Alpha correction procedures have proved successful in practice, giving data equivalent to that from the ZAF correction.

10.10 X-ray spectrometers

The principles of design and operation of both wavelength dispersive and energy dispersive spectrometers fitted to electron microprobes follows closely the criteria set out in Chapters 8 and 9. It is proposed here to contrast the differences in the more specialized electron microprobe application.

10.10.1 *Wavelength dispersive spectrometers*

Design considerations for the type of spectrometer fitted to an electron microprobe differ from x-ray fluorescence applications. In XRF analysis, a broad area of the sample is excited, making it desirable to use a flat crystal spectrometer. These require a high degree of collimation to maintain adequate spectrometer resolution. However, the origin of x-rays in the electron microprobe is almost a point source. Furthermore, there is some merit in being able to measure as large a solid angle of radiation as practicable so that the instrument will operate satisfactorily at relatively low probe currents. This would limit heating and therefore decomposition effects in samples, or alternatively permit operation at smaller probe diameters (section 10.4.3). There is a further important design criterion. Background intensities in electron excited spectra are inherently higher than XRF spectra, causing a significant deterioration in detection limits. It is advantageous, therefore, to make measurements using a spectrometer with good resolution since then the contribution of the background signal is minimized.

These requirements are, in general, best satisfied by the fully focusing spectrometer design which will now be considered further.

10.10.2 *The fully focusing x-ray spectrometer*

(*a*) *Description of mechanism.* The fully focusing x-ray spectrometer, shown schematically in Figure 10.38, consists of a diffracting crystal, exit slits, x-ray counter and a goniometer mechanism to maintain the correct angular and spatial relationship between these and the x-ray source. Unlike the XRF flat crystal spectrometer, the x-ray beam is not collimated, since the origin is effectively a point source. The instrument is designed to focus the diffracted beam as a line on the exit slits in front of the proportional counter. The

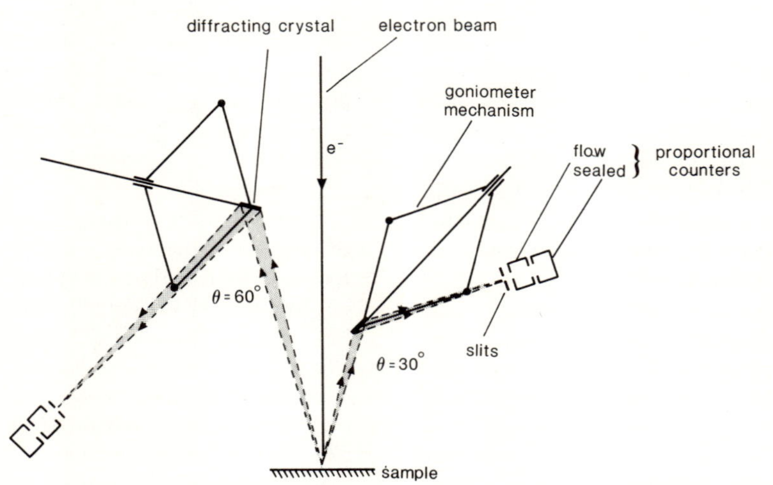

Figure 10.38 Fully-focusing wavelength dispersive spectrometer showing a parallelogram mechanism used to maintain diffracting crystal and proportional counters at the correct angular and spatial relationship with one another.

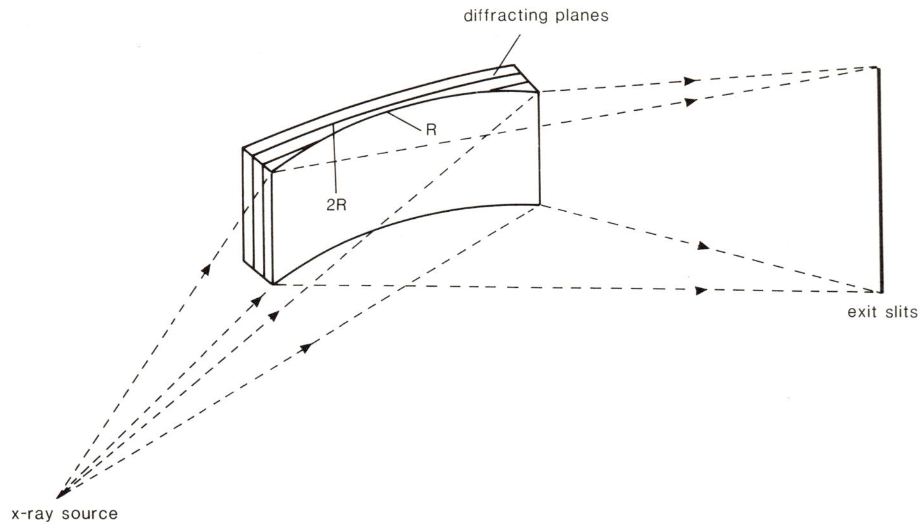

Figure 10.39 Johansson diffracting crystal geometry used on fully-focusing x-ray spectrometers. The diffracting planes of the crystal are bent to a radius $2R$ and the crystal surface is ground to a radius R (where R is the radius of the Rowland's circle). Using this type of crystal, the point x-ray source is focused as a line of twice the height of the diffracting crystal.

physical width of this focused line depends on spectrometer characteristics, but may be about 0.3 mm.

In order to achieve fully focused conditions some stringent geometric requirements must be met:

(i) The source, surface of the diffracting crystal and exit slit must all lie on the circumference of an imaginary circle of constant diameter called the Rowland's circle.

(ii) The distance between source and diffracting crystal must equal that between crystal and exit slit at all spectrometer angles.

(iii) The lattice planes of the diffracting crystal must be bent to a radius equal to twice that of the Rowland's circle, and moreover the crystal surface must itself then be ground to the radius of the Rowland's circle (Figure 10.39).

If condition (iii) is satisfied, the spectrometer is said to conform to *Johansson* diffracting geometry.

A fourth possibility is as follows. Simple geometric considerations show that the diverging beam diffracted by such a spectrometer is focused on the exit slits as a line of height twice that of the diffracting crystal. Small edge aberrations will then be present due to the angular discrepancy between x-rays striking the centre of the crystal and those incident at the top and bottom edges. Strictly, the crystal should be curved about a second circle, the plane of which is perpendicular to the plane of the Rowland's circle. The diffracting crystal could then have a toroidal shape. The practical advantages in terms of improved resolution do not justify the technical complexity of this arrangement.

One dynamic arrangement for satisfying the above conditions at all spectrometer angles is shown in Figure 10.40. The crystal is arranged to move in a linear manner with respect to the source in order to maintain constant take- off angle. The angle of the crystal is varied as it moves, and the position of the counter slits changes to maintain the required angular and spatial arrangement with respect to the crystal. Over the full range of movement of the spectrometer, the counter slits move along one lobe of a geometric curve known as an *epicycloid*. The distance from source to detector (x) is then described by the equation

$$x = R \, . \sin 2\theta \qquad (10.35)$$

where R is the diameter of the Rowland's circle, and θ is the spectrometer Bragg angle.

During spectrometer movement, the diameter of the Rowland's circle remains constant but its imaginary position is allowed to move (Figure 10.40).

(*b*) *Some practical considerations*. Some crystals can be readily manufactured in the bent and ground configuration shown in Figure 10.39, notably the LiF crystal. Other crystalline materials are more fragile and cannot be ground satisfactorily. Layered pseudocrystals (lead stearate, for example) cannot by their very nature be ground after manufacture. Such materials are usually mounted on a substrate bent to a diameter of twice the Rowland's circle. Since the entire surface of the crystal no longer then lies on the Rowland's circle, fully focusing characteristics cannot be maintained (Figure 10.41). The diffracted beam is no longer focused as a sharp line. The small reductions in the resolving power of the spectrometer caused by this effect do not seriously degrade analytical performance. This arrangement, in which the diffracting crystal is bent but not ground, is called the *Johann geometry*.

A spectrometer defocusing effect will also be observed if the surface of the sample is not accurately aligned on the Rowland's circle. The magnitude of this effect will depend on the width of the counter slits; the narrower these slits, the more sensitive will the spectrometer be to defocusing effects. As mentioned in section 10.8.6, an important function of the optical microscope facility on microprobes is to permit accurate alignment of specimen surface on to the plane of the Rowland's circle.

In many schemes of analysis, it is desirable to optimize the spectrometer angle for maximum count rate on individual elements no more than once at the beginning of each oper-

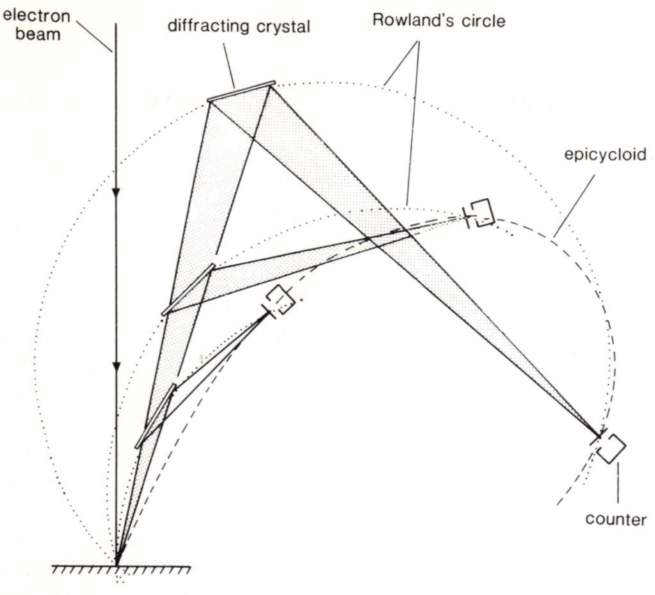

Figure 10.40 Movement of a linear fully-focusing wavelength dispersive spectrometer at angles of 15, 30 and 60° θ. The radius of the Rowland's circle passing through source, crystal and exit slits, remains constant but its centre moves. For simplicity the diffracting crystal has been drawn flat instead of curved. The dashed line represents one lobe of an epicycloid (see text) and indicates the path followed by the counter slits during spectrometer movement.

ating session. Throughout the operating session, it is then essential that no significant drift occurs in the count rate response of the spectrometer. It is not a trivial problem for any mechanical device to satisfy the precise repeatability specification required to achieve this aim, and certainly not for an instrument with complex bearings and linkages that must operate under high vacuum. The specification for many servo-controlled spectrometer systems is for an angular repeatability of better than $\pm 0.002°\ \theta$ (relative). In order to achieve this, it is essential that the drive mechanism incorporates a facility for overcoming *backlash*. Backlash is a phenomenon affecting the movement of mechanical devices in which small discrepancies occur in the settling position despite reproducible operation of the drive mechanism. Backlash is most commonly associated with stiff or worn bearings or resistance to movement due to inflexible x-ray counter cables. One way of reducing the effects of backlash is to arrange that the spectrometer always approaches a settling angle from the same direction.

The angular repeatability necessary to satisfy reproducible counting requirements varies with spectrometer angle and design. Data in Table 10.8 shows the angular movement permissible on one system to maintain the observed count rate for selected elements within $100 \pm 0.5\%$ of the maximum. At low spectrometer angles ($\theta = 15°$), the required tolerance approaches the instrument specification ($\pm 0.002°\ \theta$).

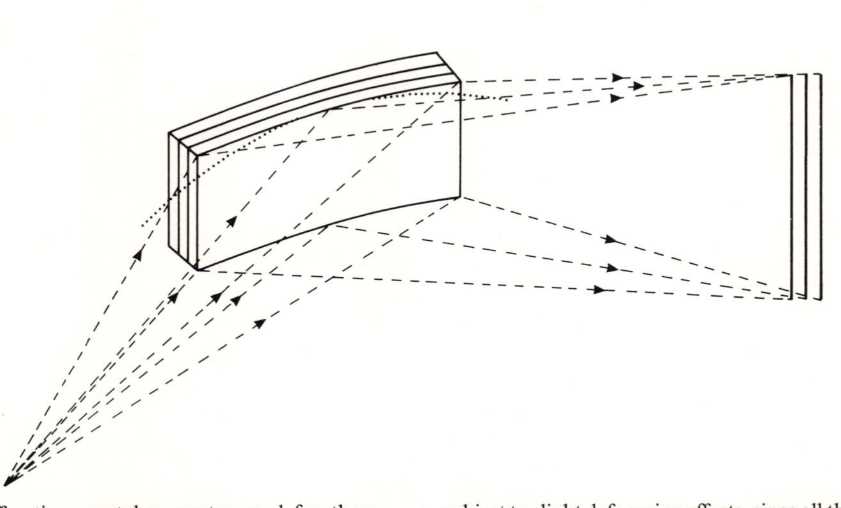

Figure 10.41 Johann diffraction crystal geometry used for those crystals that can be bent to a radius of twice the Rowland's circle but which cannot effectively be ground. The diffracted beam is then subject to slight defocusing effects, since all the surfaces of the crystal do not then exactly lie on the Rowland's circle (cf. Figure 10.39).

Table 10.8 Specification for the angular repeatability of a wavelength dispersive spectrometer

Spectrometer Bragg angle ($\theta°$)	θ range for maximum intensity
15	$\pm$0.0025
20	$\pm$0.0035
25	$\pm$0.0045
30	$\pm$0.0055
35	$\pm$0.0065

θ range represents the spread about the listed spectrometer angle within which the intensity of an emission line is greater than 99.5% of its maximum. Under routine analytical conditions, this tolerance represents, therefore, the maximum permissible discrepancy in spectrometer angular repeatability.
Data were obtained on a Cambridge M9 microprobe using a fully focusing wavelength spectrometer with the exit slits removed. The manufacturer's specification for this instrument is an angular repeatability of $\pm 0.0017°\theta$.
(P.J. Potts and A.G. Tindle, unpublished data.)

It is unfortunate that Si $K\alpha$ and Al $K\alpha$, elements for which particularly high precision is desirable in silicate mineral analysis, are both measured at spectrometer angles of less than 20° θ using TAP or RAP crystals.

In order to tolerate small spectrometer alignment errors without inducing unacceptable errors in x-ray measurements, it is desirable to operate a fully focused spectrometer with coarse, not fine, counter slits or even with the slits removed completely. Fine collimation is then reserved only for situations where it is essential to resolve spectral line overlap interferences.

A problem not encountered in XRF analysis but which can cause a random and undesirable increase in background counts in microprobe applications is that of backscattered electrons. Under certain circumstances, a small fraction of electrons, backscattered from the sample, can enter the x-ray counter directly.

This effect is most likely to occur at low spectrometer angles (counter close to the sample) and at high accelerating voltages (especially above 20 kV). An electron trap is, therefore, normally incorporated into spectrometer designs. This takes the form of a strong permanent magnet positioned in front of the counter slits (Figure 10.42). When backscattered electrons enter this magnetic field, they are deflected away from the counter entrance windows.

(*c*) *Choice of diffracting crystals*. The selection of diffracting crystals for use on a microprobe spectrometer is often more restricted than those used in XRF analysis, and would normally include:

(i) LiF(200) for the *K*-lines of all x-rays above about Cr $K\alpha$, (the higher-resolution LiF(220) crystal is not normally used for routine silicate mineral analysis but would be of value in specialized applications where serious spectrum overlap interferences are expected, for example, in the analysis of rare-earth element minerals).

(ii) PET(002), normally used for intermediate x-ray energies (P $K\alpha$ to V $K\alpha$).

(iii) TAP(001) (or as second choice RAP, with lower diffracted intensities) normally used for the light major elements (F $K\alpha$ to Si $K\alpha$).

(iv) Lead stearate pseudocrystals, with which most microprobe spectrometers are fitted for the analysis of the light elements B *K* to N *K*, an application which is not often encountered in routine analysis of silicate minerals.

(v) More specialized crystals such as InSb (Si $K\alpha$, only) or Ge(111) (to eliminate the second-order interference of Ca $K\alpha$ on P $K\alpha$)—not widespread in electron microprobe analysis.

(vi) Synthetic multi-layer crystals (artificially made from layers of tungsten and carbon, sputtered under controlled conditions on to a substrate—see section 8.9). These are likely to have a significant impact on wavelength dispersive microprobe analysis in the detection of the elements magnesium and lighter due to the higher diffracted beam intensities compared with existing crystals. If all the elements in the range boron to magnesium are to be detected, at least two synthetic multilayer crystals will be required, one with a 2*d* spacing of 45–50 Å for the elements oxygen to magnesium and a second with a 2*d* spacing of about 90 to 120 Å for lighter elements.

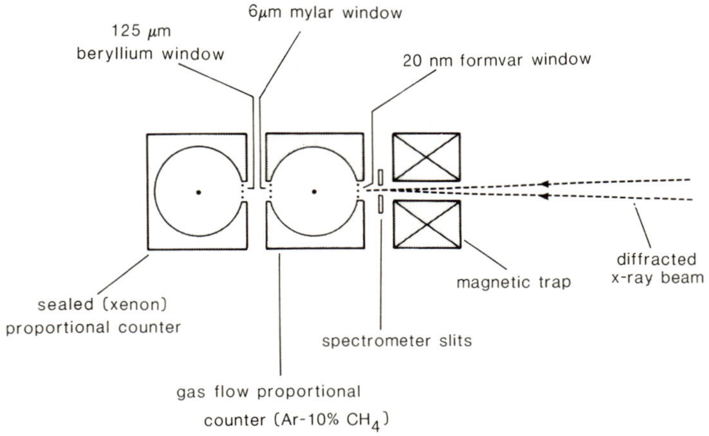

Figure 10.42 Arrangement of x-ray detectors on a microprobe spectrometer, showing magnetic trap, exit slit, and tandem arrangement of flow and sealed proportional counters.

(*d*) *Proportional counters*. Electron microprobe measurements are normally restricted to x-ray lines in the 0–10 keV range since those in the 10–20 keV range cannot normally be excited with sufficient efficiency. In a consequence, microprobe spectrometers are usually fitted with two gas proportional counters mounted in tandem (Figure 10.42). The first is a flow counter containing an argon–10% methane gas mixture to cover the energy range 0–4 keV (B to Ca $K\alpha$) and the second a xenon sealed counter for the 4–10 keV energy range (Ti to Zn $K\alpha$). These two counters are normally mounted in tandem, the flow counter having both entrance and exit windows. When it is necessary to switch over to the use of the sealed counter, the flow counter is normally first flushed with air. This minimizes attenuation of higher energy x-ray photons that must first traverse the flow counter before entering the sealed counter. The scintillation detector, standard on XRF spectrometers, is not normally used in microprobe applications.

Since microprobes are capable of detecting very low-energy x-rays from the light elements, the standard 1 μm polypropylene window, used on flow counters in XRF applications, is often replaced by an ultra-thin window consisting of a 20 nm film of collodion or formvar mounted on a nickel wire support grid (section 8.9).

(*e*) *Wavelength shifts*. X-ray emissions are normally associated with transitions between inner electron orbitals, which are shielded by outer electrons from the perturbation effects of chemical bonding. This assumption breaks down for the lighter elements. Significant changes in the energy and, therefore, wavelength of some *K*-line emissions can be detected due to changes in the *L*-orbital shell energy levels caused by chemical bonding.

This phenomenon does not normally affect x-ray fluorescence measurements, since these light elements are normally determined on glass fusion discs, where the mineralogical structure of the sample has been broken down.

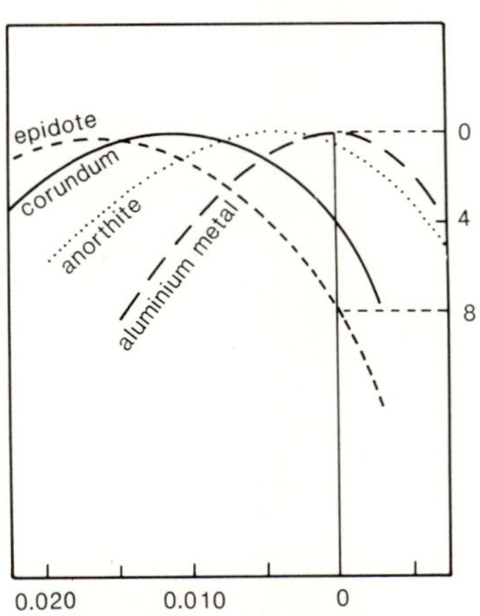

Figure 10.43 Wavelength shift in microprobe analysis of aluminium $K\alpha$. Intensity data are plotted for a wavelength spectrometer that has been stepped through the Al $K\alpha$ for measurements made on the minerals epidote, corundum, anorthite and aluminium metal. If aluminium metal were used as standard, the error in intensity measurement on epidote would be 8% due to this wavelength shift effect. (After Sweatman and Long, 1969.)

In microprobe applications, Sweatman and Long (1969) considered the problem of wavelength shift in some detail, and data for the Al $K\alpha$ peak are shown in Figure 10.43. These graphs show the count rates observed when a microprobe spectrometer is stepped through small angles about the Al $K\alpha$ maximum measured on aluminium metal, corundum (Al_2O_3), anorthite ($CaAl_2Si_2O_8$) and epidote ($Ca_2Fe^{III}Al_2Si_3O_{12}OH$). If the spectrometer angle was originally optimized by measurements on aluminium metal, a count rate error of up to 8% would be observed in the determination of aluminium in epidote. However, these authors concluded that if the spectrometer slits were removed to decrease resolution slightly and light element *oxides* rather than the metals were used for calibration (that is SiO_2, Al_2O_3 and MgO), discrepancies due to wavelength shift in the analysis of oxide and silicate minerals are likely to be negligible. Wavelength shifts are not likely to be observed if metals of elements heavier than Ti are used to calibrate the microprobe, although some workers might then be concerned at the magnitude of the matrix correction difference between metal standard and silicate/oxide mineral. It should, however, be noted that the $K\beta$-lines of the intermediate atomic number elements will be more sensitive than $K\alpha$-lines to wavelength shift, as the transitions that give rise to the former involve the outer *M*-shell orbital electrons. Advantage can be taken of wavelength shifts in the studies of valency and chemical bonding effects. Thus Urch and Wood (1978) investigated the way in which the manganese $K\beta_{1,3}$, $K\beta'$ emission profile varies with valency of manganese. From these measurements, manganese with a valency of 2 (in rhodonite and rhodochrosite) could be distinguished from manganese with a valency of 3 (manganite) and 4 (pyrolucite). These studies were carried out on a wavelength dispersive XRF spectrometer using tungsten tube excitation and fine Soller slits for collimation.

10.10.3 *Semi-focusing spectrometers*

Fully focusing spectrometers of the type described above are expensive to manufacture and sometimes temperamental in operation. In less demanding applications, semi-focusing spectrometers are often substituted (Figure 10.44). These consist of a much simpler mechanism, often with the crystal pivoting about a fixed point and the counters held on an arm of fixed length. Rotation of the counter arm is controlled by a 2:1 gear mechanism relative to the crystal. Since the spectrometer operates in a semi-focused state, counter slits must be wide (if fitted at all) and steps taken to prevent x-rays from the source passing directly into the counter.

10.10.4 *Spectrum interference effects*

The first- and multiple-order line overlap interferences encountered in the operation of wavelength dispersive spectrometers have been discussed in section 8.9. Microprobe spectra do not of course contain the characteristic tube line interferences listed in Table 8.16. Indeed, the x-ray spectra from individual minerals, which generally contain a limited number of elements, are often much simpler than XRF spectra observed from bulk materials. Nevertheless, overlap interferences in microprobe spectra can affect analytical data

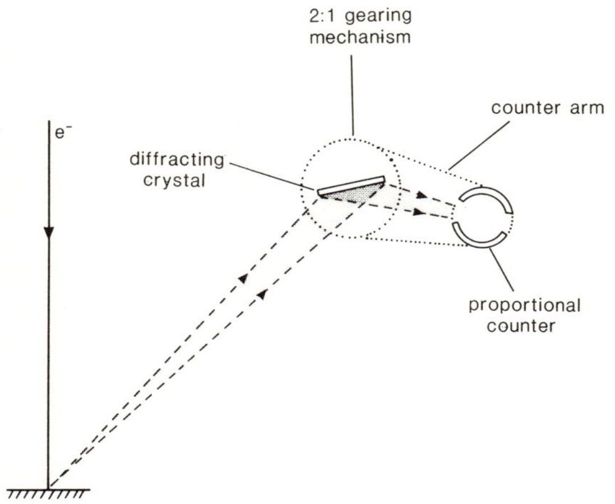

Figure 10.44 The semi-focusing wavelength spectrometer in which the diffracting crystal rotates about a fixed point and the counter is located on an arm of fixed length. The diffracting crystal is normally curved but because the spectrometer is then focused at only one wavelength it is desirable to discard the exit slits and to reduce the distance between crystal and counter to optimize x-ray count rates.

in some circumstances and the magnitude of some of the most significant interference effects encountered on one instrument are shown in Table 10.9. One of the most serious of these is represented by the presence of a third-order P *Ka* line that overlaps F *Kα* to the extent that the presence of 10% phosphorus (P_2O_5) results in apparent fluorine concentration of 2%. This interference seriously affects the analysis of fluorine in apatites, for example. Similarly, the third-order Ba *Lα* interference overlaps the Al *Kα* line so that it is possible to determine significant apparent concentration of aluminium in a mineral such as barite ($BaSO_4$), even though aluminium may be absent.

The precise magnitude of these interferences depends on the slit width selected for analysis, and for multiple-order diffractions, the selection of pulse height analysis conditions.

10.11 Calibration and routine operation

In XRF analysis, all x-ray count data is usually normalized to that from a suitable drift monitor, full calibration with reference to elemental standards being carried out infrequently. This is not normally the case in microprobe analysis, where it is usual to recalibrate instruments on a daily basis with reference to primary mineral standards. Count intensities are not usually normalized to a monitor. Individual calibrations are then sensitive to changes in the absolute response of the instrument; drift in probe current and the mechanical repeatability of the wavelength dispersive spectrometers then become critical.

The daily calibration procedure would depend on the design of instrumentation, including the degree of automation and the aim of the analytical program. Data for one scheme designed for routine silicate mineral analysis of high precision is summarized in Table 10.10. If a daily calibration is required, together with close control over calibration drift, a routine operating scheme would be as follows.

(i) Switch on the instrument to the normal operating mode for at least 30 minutes to allow probe current, gun kV, proportional counter gas flows and high-voltage bias supplies to stabilize.

(ii) Adjust the diameter of the probe beam to the appropriate value (5 to 10 μm for routine analysis to avoid decomposition effects). Set the probe current to the appropriate value. Optimize the proportional counter high-voltage bias to standardize the pulse height discriminator response. The diameter of the probe may be observed from the cathodo-

Table 10.9 Spectral overlap interferences on a wavelength dispersive spectrometer

Element/line analysed		Interfering element/line			Apparent concentration of element analysed due to overlap interferences	Examples of minerals affected
F	*Kα*	P	*Kα*	(3rd order)	2.0	Apatite
Al	*Kα*	Ba	*Lα*	(3rd order)	0.3	Ba minerals
Al	*Kα*	Ti	*Kα*	(3rd order)	0.007	Rutile, Ti minerals
P	*Kα*	Ca	*Kα*	(2nd order	0.012	Apatite, biotite
Ti	*Kα*	Ba	*Lα*	(1st order)	0.55	Ba minerals
Mn	*Kα*	Cr	*Kα*	(1st order)	0.04	Chromite
Fe	*Kα*	Mn	*Kα*	(1st order)	0.02	Manganese minerals
Sr	*Lα*	Si	*Kα*	(1st order)	0.008	Plagioclase
Sr	*Lα*	Cr	*Kα*	(3rd order)	0.1	(Chromite)
Ba	*Lα*	Ti	*Kα*	(1st order)	0.15	Rutile, Ti minerals
K	*Kα*	negative background (argon absorption edge)			—	Silicates

Magnitude of interference will vary with instrument design. The data listed above were obtained on a Cambridge M9 microprobe fitted with fully focusing spectrometers with the exit slits removed. Samples were excited at 20kV and no pulse height analysis window was set.
All concentrations quoted in weight % oxide (Fe as Fe_2O_3) except F (% element).
The magnitude of the interference is the apparent weight % of the element oxide, assuming that the interfering element were present at the 10% oxide level.
Measurements were made with the following diffracting crystals: TAP (F, Al), PET (P, K, Ti, Sr, Ba), and LiF (Mn, Fe).

Table 10.10 Instrumental conditions for routine electron probe microanalysis

Element	Line	keV	Mineral standard	% element in standard	Spectrometer Crystal	Spectrometer Counter	Count times: Standard 100 000 counts	Count times: Sample 50 000 counts	Detection limit: Wavelength dispersive (% oxide)	Detection limit: Energy dispersive (% oxide)	
F	$K\alpha$	0.677	Synthetic LiF	48.7	TAP	FLO1	200	486	0.24*		F
Na	$K\alpha$	1.041	Jadeite	11.2	TAP	FLO1	42	24	0.07	0.5	Na_2O
Mg	$K\alpha$	1.253	Olivine	29.6	TAP	FLO1	9	12	0.07	0.5	MgO
Al	$K\alpha$	1.486	Jadeite	52.9	TAP	FLO1	4	9	0.06	0.4	Al_2O_2
Si	$K\alpha$	1.739	Wollastonite	18.9	TAP	FLO1	12	12	0.05	0.4	SiO_2
P	$K\alpha$	2.013	Apatite	18.5	PET	FLO2	24	22	0.06	0.7	P_2O_3
S	$K\alpha$	2.307	Pyrite	36.5	PET	FLO2	10	19	0.04	0.2	S
Cl	$K\alpha$	2.621	Synthetic KCl	47.6	PET	FLO2	7	16	0.04	0.2	Cl
K	$K\alpha$	3.312	Synthetic KCl	52.4	PET	FLO2	21	10	0.04	0.2	K_2O
Ca	$K\alpha$	3.690	Wollastonite	34.2	PET	FLO2	6	10	0.03	0.2	CaO
Ti	$K\alpha$	4.508	Rutile	59.9	PET	FLO2	4	10	0.05	0.25	TiO_2
Cr	$K\alpha$	5.411	Metal	100	LiF	SEA	10	48	0.06	0.3	Cr
Mn	$K\alpha$	5.894	Metal	100	LiF	SEA	10	48	0.06	0.3	MnO
Fe	$K\alpha$	6.398	Fayalite	53.0	LiF	SEA	21	55	0.06	0.5	Fe_2O_3
Ni	$K\alpha$	7.471	Metal	100	LiF	SEA	14	67	0.12	0.35	Ni
Cu	$K\alpha$	8.040	Metal	100	LiF	SEA	18	85	0.18	0.5	Cu
Zn	$K\alpha$	8.630	Metal	100	LiF	SEA	21	100	0.20	0.6	Zn
Ba	$L\alpha$	4.465	Barite	63.2	PET	FLO2	11	32	0.11	1.0	Ba
Sr	$L\alpha$	1.806	Celestite	47.7	TAP	FLO1	32	76	0.07	1.0	Sr

Diffracting crystals:
TAP: thallium hydrogen (acid) phthalate
PET: penta-erythritol
LiF: lithium fluoride (200).
Gas proportional counters:
FLO1: argon flow counter with ultra-thin entrance window essential (e.g. 20 nm formvar).
FLO2: argon flow counter with standard 6 μm mylar window adequate.
SEA: xenon sealed counter with beryllium entrance window.
Count times (s) were determined on samples excited at 20 kV, 30 nA on a Cambridge M9 microprobe using the following criteria:
Standards: time required to accumulate 100 000 counts on specified mineral.
Samples: time required to accumulate 50 000 counts on a sample containing 10% of the element. These data are, therefore, indicative of the relative sensitivity of elements.
All count time appropriate for a FLO1 counter.
Detection limits are estimates of the six-sigma limit of determination. Wavelength dispersive data are appropriate for the sample count times listed. Energy dispersive data are compiled from the work of Reed and Ware (1975), Dunham and Wilkinson (1978) and Tindle and Potts (unpublished data) and are representative of spectra counted for 100 to 200 s (depending on instrument) at 15 kV. Precise data for light elements may be affected by the thickness of the beryllium window fitted to the ED detector. Detection limits for the heavier major elements could be improved by increasing the gun operating potential.
* 100 s count time.

luminescence emission from an aluminized glass slide or a suitable mineral (benitoite, $BaTi(SiO_3)_3$ is reported to generate an effective cathodoluminescence emission), or alternatively by measuring the burn marks formed in epoxy resin.

(iii) Select fresh areas on the surfaces of calibration mineral standards, avoiding those affected by polishing or other surface defects. If necessary, re-optimize spectrometer angle for each element to be determined. Measure the standard count rate at peak and background spectrometer angles. A minimum of four such measurements should be carried out with recorded data screened to eliminate erroneous values before averaging. Mean count rate data must be corrected for drift in the probe current, dead time effects in the counters, background count rates and ZAF factors appropriate to the mineral standard. It is suggested that the count time chosen for these measurements should allow the accumulation of 100 000 counts at each peak angle, as listed in Table 10.10.

(iv) Analyse as an unknown a secondary mineral standard of composition similar to that of the samples to be determined during the current operating session. Assess the accuracy of the calibration by comparing the analysed results with the accepted elemental composition, recalibrating individual elements if necessary.

(v) The microprobe is now ready for the analysis of unknown minerals. It is desirable to reset count times, an appropriate approach being to choose the time for 50 000 counts to be accumulated if the element were present at 10% concentration (see Table 10.10). Higher or lower criteria of precision may be selected as appropriate. The secondary mineral standard should be reanalysed at suitable intervals to check for calibration drift.

The scheme of analysis outlined above has been derived for an instrument that possesses a degree of computer automation in controlling the movement of spectrometers and sample stage and is capable of carrying out an on-line calibration and ZAF correction. Other procedures may be appropriate for instruments having different facilities.

10.12 Energy dispersive spectrometers

The design, construction and response of energy dispersive detectors is fully discussed in Chapter 9. This section is restricted to those specialized aspects related to electron microprobe applications.

10.12.1 *Instrumental requirements*

(*a*) *Analytical characteristics.* Energy dispersive detectors are an invaluable accessory on an electron probe microanalyser because of their ability to perform extremely rapid qualitative analysis. In particular, it is often possible to identify within a few seconds those opaque minerals that cannot readily be recognized using the optical microscope. However, it was in microprobe analysis that the quantitative capabilities of energy dispersive detectors were first generally accepted, and the vitality of many older wavelength dispersive instruments has been restored by the installation of an ED spectrometer for quantitative analysis. The current status of such instrumentation has been reviewed by Statham (1983).

Energy dispersive spectrometers have the advantage that they can be positioned relatively close to the sample so that a much larger solid angle of x-rays may be detected than would be possible with a wavelength dispersive spectrometer. In view of this, probe currents may fall by approximately a factor of 10. Heating effects within the sample are correspondingly reduced (Table 10.2), making ED probe analysis more attractive for the analysis of phases susceptible to decomposition. This advantage must be set against the limited count rate capability of these detectors, about 3000 cps in normal operation. Some specific points that affect the operation of ED detectors fitted to electron beam instruments are considered in the next sections.

(*b*) *Collimation.* Unlike wavelength dispersive spectrometers, there are no focusing/defocusing considerations in x-ray measurements using an ED detector. However, there are collimation requirements to ensure both that radiation from the sample itself, and not that scattered off the walls of the sample chamber, is measured, and to restrict the measured beam to the central sensitive volume of the detector (section 9.2). Reed and Ware (1975) considered the optimum size of collimator in relation to the diameter of the Si(Li) crystal (Figure 10.45). In the optimized configuration shown in this diagram, the detector 'sees' a full solid angle of radiation only from a point source corresponding to the excited volume within the sample. This is surrounded by a penumbra region in which emitted radiation is increasingly collimated. As such an arrangement is very sensitive to the exact mechanical alignment of the detector, it is essential to make the collimator diameter somewhat larger. This is also essential if the detector is to be used for x-ray measurement whilst the x-ray beam is being scanned across the sample. It may then be necessary to consider whether a collimator designed for spot analysis is clipping the x-ray emission at the extremes of beam deflection. For these reasons, the configuration shown in Figure 10.45 represents the limiting minimum diameter of collimation.

(*c*) *Backscattered electrons.* Backscattered electrons affect spectra recorded using ED detectors in two ways:

(i) Backscattered electrons can excite other materials within the specimen chamber causing spurious x-ray emissions—it is not unusual to observe extraneous x-rays due to Cu and Zn in recorded spectra from brass material in the vicinity of the sample.

(ii) Energetic backscattered electrons can also penetrate the beryllium window of the detector, contributing to a significant increase in continuum counts.

Neither of these problems is likely to be important at accelerating voltages of 20 kV and below. However at 30 kV, it may be desirable to place an aluminized mylar film (6 μm) in front of the detector in order to absorb a significant proportion of these electrons before they can strike the detector crystal (Statham, 1981). Some sacrifice in the detection of low-energy x-rays will then have to be made.

(*d*) *Oil contamination on detector window.* A potential source of long-term calibration drift, especially in the detection of low-energy x-rays, is the risk of condensation of oil on the beryllium window of the detector. This occurs because the detector crystal must be cooled to liquid-nitrogen temperatures by a copper cold finger which extends down the

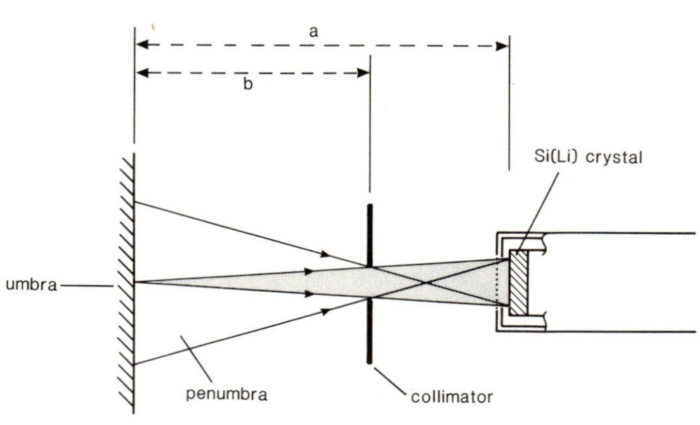

Figure 10.45 Collimator geometry for a Si(Li) detector showing the minimum collimator diameter necessary to ensure that the complete solid angle of x-rays from the source can enter the detector crystal. It is, however, desirable to select a collimator diameter significantly larger than this to ensure that x-ray defocusing effects are not observed when the electron beam is deflected over the surface of the sample. (After Reed and Ware, 1975.)

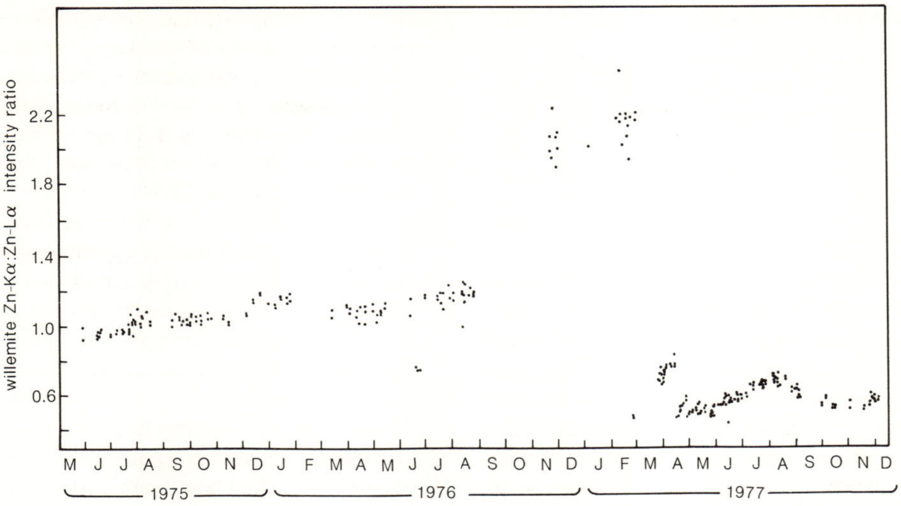

Figure 10.46 Change in the intensity ratio of Zn *Kα* : Zn *Lα* measured on a willemite mineral sample over a period of about $2\frac{1}{2}$ years. The sharp discontinuity (Sept.–Nov. 1976) coincided with serious vacuum problems. (Smith, 1981.)

snout of the detector. The beryllium window, situated close to the detector crystal is, therefore, often the coldest exposed surface within the vacuum system. Oil vapour, present at very low partial pressures, can then condense on to this surface, causing a progressive increase in the attenuation of low-energy x-rays. In a revealing study, Smith (1981) monitored the build-up of oil contamination by measuring the x-ray emission from the mineral willemite (Zn_2SiO_4). Changes in the observed intensity ratio of Zn *Kα* (8.63 keV) to Zn *Lα* (1.01 keV), recorded over a period of two and a half years, are plotted in Figure 10.46. Smith identified a progressive build-up of contamination which was exacerbated during a vacuum failure. Contamination could be virtually eliminated by substituting a polyphenol hydrocarbon diffusion pump oil for the original silicone-based oil (the former has a lower mass attenuation coefficient) and by placing miniature lamps having a total heat dissipation of about 3 W adjacent to the beryllium window.

(*e*) *Beam blanking.* As is discussed in section 9.7, the count rate capability of an ED detector used in XRF analysis can be dramatically increased by operating the x-ray tube in pulsed mode. The same mode of operation can be achieved in ED microprobe analysis by blanking the electron beam (Statham *et al.*, 1974; Pawley *et al.*, 1977; Short and Gleason, 1982). As soon as the pulse processor recognizes an x-ray event from the detector, a signal is generated which causes a voltage bias (typically 30–50 V) to be applied to a set of electrodes placed within the electron column. These electrodes are usually positioned between gun and condenser lens. On application of the bias, the electron beam is deflected sideways so that it strikes a spray aperture and is no longer transmitted down the column. The excitation source is effectively cut off and the probability that dead time will be extended by subsequent x-ray events is greatly reduced (section 9.7). When measurement of the x-ray event has been completed by the pulse processing electronics, the bias is removed from the beam blanking electrodes and the electron beam restored. The pulse processor must then be inhibited for a further short time until the electron beam has re-stabilized.

The main advantage of beam blanking is that an increase in the pulse processor output count rate by at least a factor of 3 can be achieved. Furthermore, the mean current delivered to the sample is significantly reduced (perhaps by 30%) in comparison with continuous excitation. The success of the technique depends on the design of control logic and voltage switching devices which must be capable of blanking the beam within a time interval of, say, 10 ns. Careful timing is also required to ensure that measurements are not made until the electron beam has recentralized on the sample following deflection. Because of the additional complexity of instrumentation, beam blanking is not yet widely implemented on ED microprobes.

10.12.2 *Spectrum deconvolution*

In some respects, the analysis of ED spectra from electron microprobes is less challenging than equivalent XRF spectra. As has already been mentioned (section 9.8), spectrum overlap interferences are often much less severe because of the more restricted range of elements present in many discrete minerals.

However, because of its higher intensity, successful background removal is an important aspect of microprobe ED spectrum analysis. Statham (1981) estimated that if ignored, the continuum component of a peak can contribute up to 2% by weight to the apparent concentration of an element.

The digital filtering procedures described in section 9.8 are fully compatible with, and are widely used in, ED microprobe analysis. However, alternative approaches have been successfully used in microprobe applications and selected procedures are considered further below.

(*a*) *Background subtraction by continuum modelling.* One technique in affecting background correction is to model the curvature of the continuum using a suitable algorithm, normally derived from Kramers' equation, which relates the intensity of the continuum (I) to photon energy (E) as follows:

$$I = k\bar{Z}\left[\frac{E_0 - E}{E}\right] \quad (10.36)$$

where $\bar{Z}$ is the mean atomic number of the sample (target) given by

$$\bar{Z} = \sum_i C_i Z_i$$

C_i is the weight fraction of element i having an atomic number of Z_i, E_0 is the incident beam energy in keV (equivalent to the gun operating potential), and k is a constant.

There have been several investigations of the accuracy of Kramers' formula for modelling microprobe continuum spectra, including a recent study by Brunetto and Riveros (1984). Practical applications have been described by Smith *et al.* (1975) and further summarized by Smith and Reed (1981). These developments are in part dependent on the proposal by Rao-Sahib and Wittry (1972, 1974) that inaccuracies in the simple Kramers' equation can be overcome if an expression of the following form is used:

$$\mathrm{I} = \mathrm{k}\bar{Z}^n\left[\frac{E_0 - E}{E}\right]^x \quad (10.37)$$

where the exponents n and x have the form

$$n = E.(0.0739\text{–}0.0051.\ln(\bar{Z})) + p$$
$$p = 1.6561\text{–}0.1150.\ln(\bar{Z})$$
$$x = a\text{–}0.00145.\bar{Z}/\bar{E}$$

and a is a constant at a given operating kV having the value 1.76 at 15 kV.

The practical application of this background calculation technique, as described by Smith *et al.* (1975), is as follows.

(i) The experimental spectrum is recorded from a sample of some material, such as diamond, which is free of characteristic x-ray lines, in order to determine the basic shape of the continuum.

(ii) The recorded intensities from this spectrum are corrected for electron backscatter (electrons scattered back out of the sample and which can therefore no longer contribute to continuum generation) and self-absorption effects (absorption of continuum x-rays within the sample). Mathematical procedures for carrying out this operation are similar to those described in section 10.9.

(iii) Corrected continuum intensities are then normalized with respect to that part of the modified Kramers' expression which depends on the atomic number of the sample, namely:

$$\bar{Z}^n\left[\frac{E_0 - E}{E}\right]^x \quad (10.38)$$

where n and x are as defined above.
The resultant normalized curve represents the hypothetical continuum spectrum of the hydrogen atom ($Z = 1$).

(iv) The shape of the continuum spectrum for any sample may then be calculated by multiplying the stored intensities by the expression in equation (10.38) using the appropriate value of $\bar{Z}$ for the sample and re-correcting for relevant backscattered electron and self-absorption effects.
The excellent agreement between continuum intensities calculated in this way and an experimentally recorded spectrum may be judged from data shown in Figure 10.47. The technique possesses the advantages of modelling both curvature and absorption edges in the continuum spectrum. Uncertainties can arise at low energies (<3 keV) due to background distortion caused by poor charge collection within the detector and by uncertainties in absorption correction coefficients (Statham, 1981).

Following application of the algorithms described above, areas of x-ray photopeaks essentially free of any background component must be calculated. The individual photopeaks are normally assumed to be of gaussian shape. It is then usually necessary to pre-calibrate the variation in gaussian FWHM as a function of energy for the detector in use, and also the intensity of the element $K\beta$ line as a function of the $K\alpha$ peak. A least-squares fit is then performed on specified element $K\alpha$ lines using an iterative stripping procedure, the associated $K\beta$ line being removed by calculation of its relative intensity. These and other spectrum quantification procedures have been reviewed by Statham (1976).

10.12.3 *ED calibration procedures*

ED calibrations on the microprobe are generally more stable than their WD counterparts, mainly because the spectrometer response is more reproducible. It is, therefore, quite common to perform a 'once and for all' calibration by recording the spectra of selected mineral standards under preset conditions. Small day-to-day variations in detector response are corrected by normalizing the intensity of all count data relative to a suitable drift monitor, cobalt metal or alternatively willemite (Zn_2SiO_4) having been selected by some workers. At the beginning of each operating session and periodically thereafter, the spectrum of this drift monitor is recorded so that the current response can be normalized

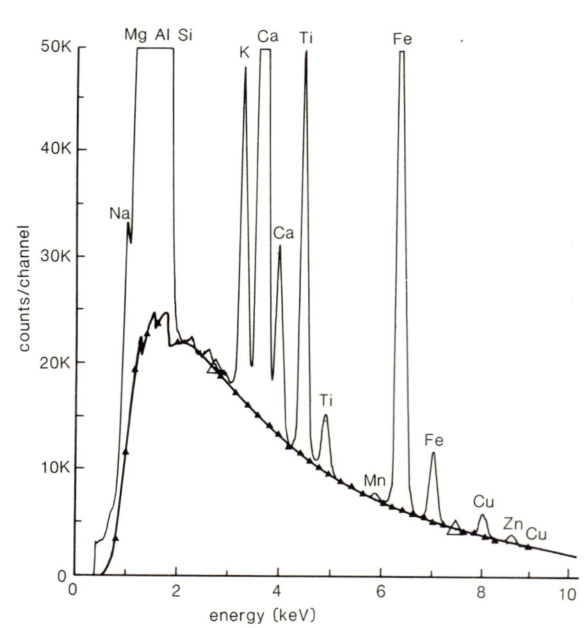

Figure 10.47 ED spectrum of Kakanui hornblende with the fitted x-ray continuum background. The larger open triangles represent the background measurements from which the continuum curve was calculated. (Mykelbust *et al.*, 1981.)

back to the conditions prevailing when the calibration was first set up. Drift monitor spectra may also be used to pre-calibrate the spectrum gain, an essential aspect of library deconvolution procedures (section 9.8).

The length of time for which any calibration is valid depends both on changes in detector response (due, for example, to ageing effects that may alter the detector electronic noise levels or resolution) and due to external factors such as discrepancies in the alignment of the detector if it were removed from the column for maintenance or to remove oil contamination.

10.12.4 *Qualitative analysis using an ED detector*

There is in general no difficulty in identifying the major components of an unknown mineral from the intense x-ray signature of its constituents. However, uncertainty can arise in attempting to identify some trace components, where the pattern of x-ray lines may be considerably more difficult to recognize and where confusion can arise between 'real' x-ray photopeaks and detector artefacts such as sum and escape peaks. To assist in qualitative interpretation, many of the x-ray photopeaks and artefacts found in ED spectra recorded between 0.7 and 20 keV are plotted for the elements heavier than fluorine in Figure 10.48.

A close examination of these data shows that some minerals contain combinations of elements which are difficult to distinguish in the ED spectra. Such minerals include lead sulphides/sulphates (Pb $M\alpha = 2.34$ keV, S $K\alpha = 2.31$ keV), molybdenum sulphides (Mo $L\alpha = 2.29$ keV), minerals containing potassium and uranium (K $K\alpha = 3.31$ keV, U $M\alpha = 3.17$ keV, U $M\beta = 3.34$ keV) and minerals containing silicon and strontium (Si $K\alpha = 1.74$ keV, Sr $L\alpha = 1.81$ keV, Sr $L\beta = 1.84$ keV).

Some assistance in identifying these mineral types can be obtained by examining the 10 to 20 keV region of the spec-

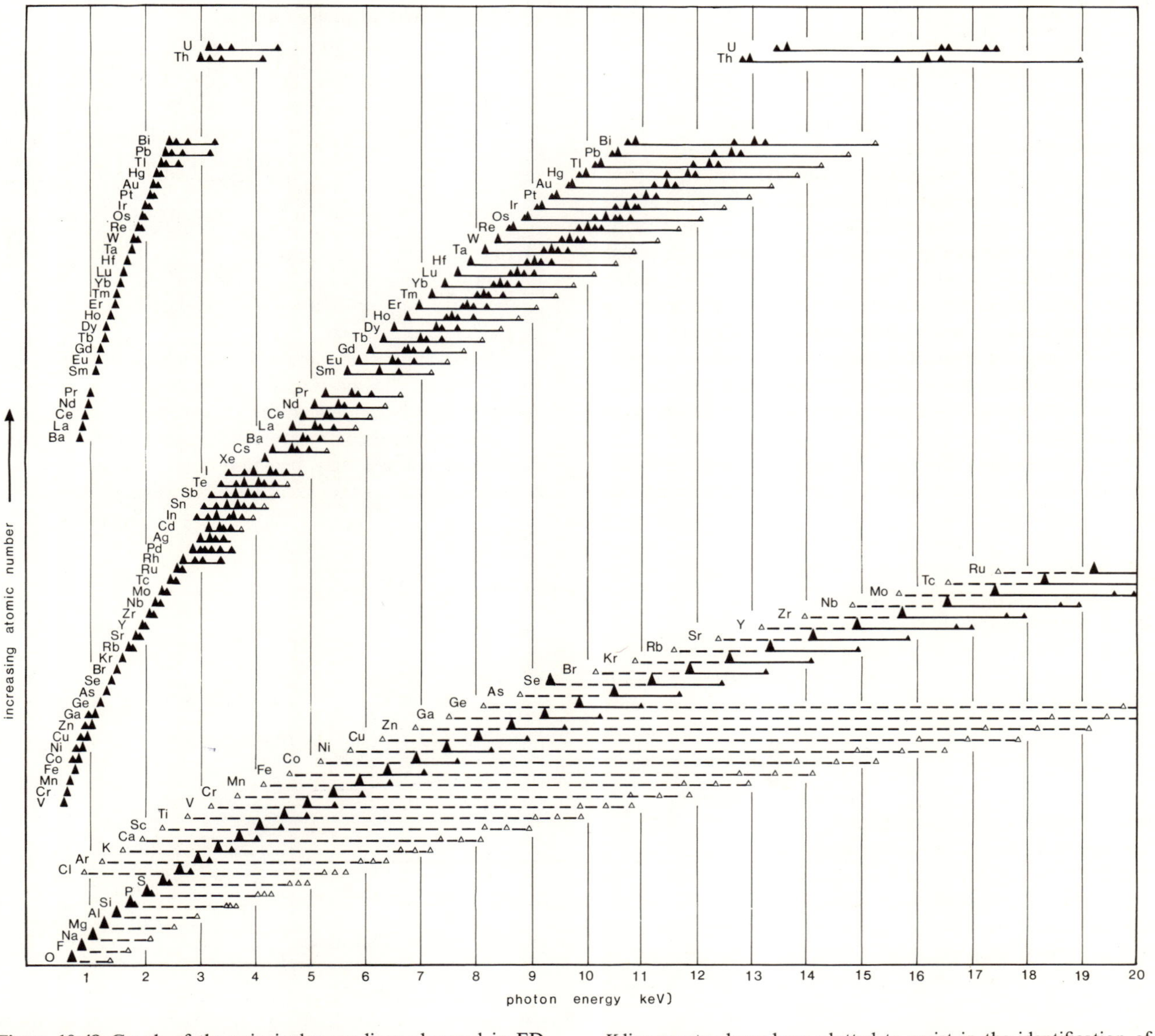

Figure 10.48 Graph of the principal x-ray lines observed in ED spectra in the range 0.6–20 keV. Sum peaks and escape peaks of *K*-line spectra have been plotted to assist in the identification of interferences.

trum for the presence of corresponding *K*- (Sr, Mo) or *L*- (Pb, U) lines. These higher-energy photopeaks are not usually used for quantitative analysis, due to the poor efficiency with which they are excited.

10.13 Sample preparation requirements

The surface polish requirements of samples to be analysed by microprobe is a critical and sometimes overlooked aspect of quantitative microanalysis. Two potential problems have already been discussed: namely, enhanced attenuation of x-rays if the electron beam is focused on a scratch as shallow as 0.5–1 μm, and the effect of specimen tilt, as will be observed if the beam is focused on the rounded edge of an overpolished mineral (section 10.8.6).

One semi-automated method of preparing polished thin sections from conventional geological sections is shown in Figure 10.49. Conventional sections are first prepared but are ground to a thickness of about 40 μm instead of the more conventional 30 μm. Several such sections, all of similar hardness and identical thickness, may then be mounted on a vacuum chuck and polished on paper or felt laps on a rotating table, as shown in Figure 10.49. The laps are impregnated with diamond polishing paste and it is usually necessary to reduce the grade of this paste successively from 13 μm to 6 μm to 3 μm to 1 μm to $\frac{1}{4}$ μm grit size to achieve a satisfactory finish. Scratches can only be avoided by scrupulously cleaning the semi-polished surfaces between grades and by using separate laps at each polishing stage. Careful judgement is required to produce a surface of high polish and yet to avoid relief due to prolonged polishing. Softer minerals erode more rapidly than harder ones, causing a three-dimensional component in the nominally flat surface.

Small individual mineral grains are more conveniently mounted in epoxy resin using a suitable mould. When set, a flat surface is formed by grinding on a diamond impregnated surfacing wheel before polishing as described above. Naney (1984) has described a simple jig to control the polishing of circular glass slides. The jig is designed to control the amount of material removed by grinding and to prevent loss of portions of the section through the application of differential pressure during polishing.

Having prepared a polished surface, a carbon film must be deposited on those samples which do not conduct electricity as described in section 10.8.6. Failure of the carbon coat due to flaking or cracking is almost always due to residual oil contamination adhering to the sample surface from polishing lubricants. In extreme circumstances, where oil has penetrated cracks in the sample, it may be necessary to place the section in an evacuated chamber (e.g. a carbon coater) to suck oil out of cracks before cleaning with a solvent and re-coating.

10.14 Microprobe mineral standards

10.14.1 *General requirements*

Minerals suitable for use as microprobe calibration standards must satisfy some demanding requirements. Suitable materials must be:

(i) Stable in air and under vacuum
(ii) Capable of sustaining a high polish
(iii) Not decomposed when excited by the electron beam
(iv) Homogeneous in all analysable elements down to the sub-micron level
(v) Available in sufficiently large quantities so that a representative fraction may be independently analysed, often by wet chemical procedures.

Few naturally-occurring minerals completely satisfy each of the last three conditions and there is often agonizing uncertainty over whether an independent bulk chemical analysis has been carried out on homogeneous and mineralogically pure material.

Minerals suitable as standard materials are usually selected to match the matrix of the sample materials to be analysed. In this way, uncertainties in the accuracy of matrix corrections due to compositional differences between standards and samples may be minimized. Instrumental discrepancies, such as those caused by wavelength shifts, are

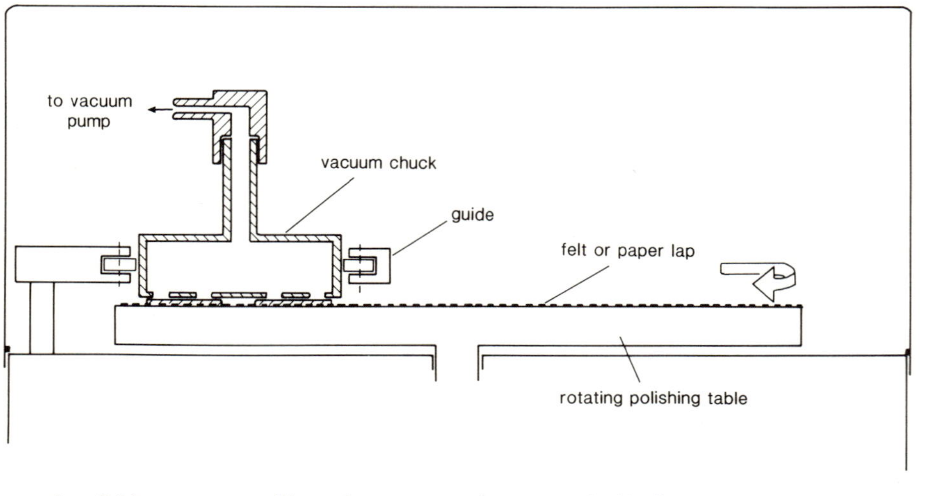

Figure 10.49 Semi-automatic polishing apparatus: thin sections are mounted on a vacuum chuck and polished on paper or felt laps impregnated with diamond paste and mounted on a rotating table.

Table 10.11 Minerals suitable for microprobe calibration standards

Element	Mineral
F	Fluorite CaF_2
	Synthetic LiF
Na	Jadeite $NaAlSi_2O_6$
	Albite $NaAlSi_3O_8$
Mg	Olivine (Mg, Fe)SiO_4
	Periclase MgO
Al	Jadeite $NaAlSi_2O_6$
	Corundum Al_2O_3
Si	Wollastonite $CaSiO_3$
	Quartz SiO_2
P	Apatite $Ca_5(PO_4)_3(OH,F,Cl)$
S	Pyrite FeS_2
Cl	Synthetic KCl
	Scapolite $(Na,Ca)_4Al_3(Al,Si)_3\,Si_6O_{24}$ (Cl,F,OH,CO_3,SO_4)
K	Orthoclase $KAlSi_3O_8$
Ca	Wollastonite $CaSiO_3$
Ti	Rutile TiO_2
	Titanium metal
V	Vanadium metal
Cr	Chromium metal
	Chromite $FeCr_2O_4$
Mn	Manganese metal
	Rhodonite (Mn,Fe,Ca)SiO_3
	Manganosite MnO
Fe	Fayalite Fe_2SiO_4
	Iron metal
	Hematite Fe_2O_3
Co	Cobalt metal
Ni	Nickel metal
Cu	Copper metal
Zn	Zinc metal
	Willemite Zn_2SiO_4
Sr	Celestite $SrSO_4$
Ba	Barite $BaSO_4$

also eliminated. Within these limitations, a remarkably narrow range of materials is normally used for the primary calibration of geological materials, as listed in Table 10.11.

10.14.2 *Reliability of mineral standards*

(*a*) *Homogeneity*. If a proposed mineral standard is not homogeneous over its entire surface area, any independent bulk chemical analysis is clearly meaningless. The established method of testing for homogeneity was described by Boyd *et al.* (1969), as follows. X-ray emission during microprobe analysis is a random process, so that statistical errors in measured intensities are described by Poisson counting statistics. The standard deviation (s) in any x-ray measurement, assuming errors are caused by random processes only, may then be determined from

$$s = \sqrt{N} \tag{10.39}$$

where N is the total number of counts detected during the measurement. There is therefore a 99.74% probability that any individual measurement will fall within the range $N \pm 3\sqrt{N}$ (see section 1.4).

Boyd *et al.* therefore proposed that count measurements for a particular element should be made for a representative number of points over the surface of a mineral, and the experimental standard deviation s', calculated from the set of count data for individual points. If the mineral is to be considered homogeneous, the ratio $s'/\sqrt{N}$ should have a value less than 3 (N is the average total count data representative of any one analysis).

Subsequently, Potts *et al.* (1983) proposed that the so-called K- factor, previously used as a precision indicator in XRF analysis, could also be used to test the homogeneity of mineral standards. The K-factor is calculated from complete analyses rather than raw count data and is determined from

$$K = \frac{s'}{\sqrt{C}} \tag{10.40}$$

where s' is the experimentally measured standard deviation of a specified element calculated from a set of analyses obtained over the surface of the mineral, and C is the mean concentration of that element.

This indicator can only be made universally applicable if the count times are adjusted so that the same number of counts would be emitted from each element where present at the same concentration level. Thus, if 50 000 is selected as the total counts that would be accumulated if an element were present at the 10% concentration level, the K factor is expected to take a value less than 0.042, providing the element is homogeneously distributed. Larger factors indicate increasing degrees of inhomogeneity.

(*b*) *Accuracy of standard mineral compositions.*The uncertainty that may exist in the assumption that an independent bulk chemical analysis of a potential mineral standard has been carried out on a mineralogically homogeneous sample has already been highlighted above. One way of circumventing this difficulty is to make use of an internally 'consistent' mineral calibration by the use of near end-members of a solid solution mineral series together with a judicial assumption of stoichiometry.

Thus, as discussed by Fredricksson (1966) and Long (1977), forsterite (Mg_2SiO_4), a potential Mg standard, will normally contain small amounts of iron. Fayalite (Fe_2SiO_4), at the other end of this solid solution series, is a potential Fe standard. It is normally expected that this mineral will contain small concentrations of magnesium. Neither mineral is likely to contain unanalysable components such as structural water or carbon dioxide, and both are expected to maintain a high degree of stoichiometry. Samples of these minerals may then be cross-calibrated by analysing the minor amounts of magnesium in fayalite against forsterite, assuming stoichiometric Mg_2SiO_4, and vice versa for iron. In this way, it is possible to analyse the minor levels of Fe and Mg in the respective minerals with some certainty even if the precise calibration concentrations are subject to small errors. Assuming that each mineral is stoichiometric, the concentration of magnesium in forsterite and iron in fayalite can be revised to take into account minor concentrations of the complementary element. In this way independent chemical analysis of some mineral standards may be avoided.

Walter (1966) attempted to circumvent this problem by preparing microprobe silicate standards from co-precipitated gels based on an ammonium-stabilized aqueous form of silica gel. Although of some value, results indicated that heterogeneities existed in these materials on a micrometre scale.

(*c*) *Reference mineral standards.* The final arbiter on the accuracy of a microprobe calibration is the reliability with which internationally available reference minerals can be analysed. At the time of writing the only comprehensive reference set of minerals is that prepared by the Smithsonian Institution (Jarosewich *et al.*, 1980). The compositions of these materials are listed in Table 10.12. The homogeneity of each of these minerals has been fully investigated and independent chemical analyses sought. It would be a valuable contribution to the harmonization of electron microprobe data if all silicate minerals analyses were referred back to this data set.

10.15 Routine analytical performance

10.15.1 *Wavelength dispersive microprobe analysis*

Typical results for the analysis of silicate minerals by wavelength dispersive electron microprobe are indicated by data in Table 10.13. Data for the DSDP basaltic glass (Potts *et al.* 1983) were obtained using the calibration standards and conditions set out in Table 10.10. These microprobe results are compared with independently derived analyses compiled by Staudigel (1979). The analysed values were obtained after routine calibration in which this sample was analysed as an unknown to 'prove' the quality of the calibration. The averages of 54 analyses are listed, derived from those calibrations considered to be acceptable. From this point of view, the precision errors in these results may be better than might be encountered in routine operation, although, since the data were obtained over a period of several months, the averages must realistically indicate day-to-day variations in instrument response.

Interlaboratory agreement is indicated by the comparison of data presented by Jarosewich *et al.* (1979). Three laboratories (at the US Geological Survey, Smithsonian Institution and Massachusetts Institute of Technology) performed replicate analyses on a glass sample using two methods of calibration. First, each laboratory used the same mineral (the Kanakui hornblende, Table 10.12) as calibration standard. Secondly, analyses were calibrated using the laboratories' preferred method. It is expected, therefore, that the first set of data ((a) in Table 10.13) will give an indication of instrumental errors, since each set of results is calibrated relative to the same material. The second set of data ((b) in Table 10.13) reflects discrepancies that arise from different calibration procedures. Jarosewich *et al.* (1979) identified small discrepancies in data for Mg, Na, K and P between individual laboratories in analyses normalized to the reference mineral, and in data for Na in the preferred calibration results. The latter errors were associated with the mobility of this element in the glassy matrix using near-focused electron beams.

Regrettably, there are few other reports of interlaboratory collaborative microprobe studies in the literature. It is therefore not simple to assess the accuracy of results, unless data are presented for independently analysed minerals such as those listed in Table 10.12.

10.15.2 *Energy dispersive analysis: comparison with wavelength dispersive results*

A detailed comparison between the performance of ED and WD spectrometers fitted to equivalent microprobe instruments, was carried out by Dunham and Wilkinson (1978). Results obtained by these authors for three silicate minerals, an albite, an orthopyroxene and a diopside/jadeite glass are listed in Table 10.14. The average analysis by ED-microprobe is compared with WD microprobe data. The authors concluded that ED microprobe analysis was at least as fast as the WD instrumentation used in this work and represented an accurate and convenient method of analysis providing more than about 1% (by weight) of an element was present. Amounts below this and down to about 0.15 weight % could be determined by ED microprobe but with decreasing accuracy and precision.

The precision of ED microprobe data in the analysis of serpentine has been investigated by Wicks and Plant (1983). The data presented by these workers in the analysis of a kaersutite mineral recorded at 20 kV, 10 nA (100 s count time) with an ED-microprobe are listed in Table 10.14. It is interesting to note that for elements present at equivalent concentration levels, the precision error in these ED data is equivalent to that in the WD data listed in Table 10.13.

10.15.3 *Detection limits*

In general, detection limits obtained by ED electron microprobe compare unfavourably with those by WD microprobe analysis. There are two principal reasons for this. The first is that, unlike XRF analysis, microprobe studies are almost exclusively confined to analysis of photopeaks between 0 and 10 keV, the range in which the ED detector is at an inherent disadvantage in terms of resolution in comparison with wavelength spectrometers. The second reason compounds these difficulties, since in comparison with XRF analysis, the continuum in electron excited spectra makes up a larger proportion of the analytical signal. Data in Table 10.10 compares the detection limit performance of these two spectrometer systems. These data show that in the determination of the major elements, ED detection limits, calculated as the six-sigma limit of determination, range down to about 0.2 weight % oxide for a 100 to 200 s count time (depending on instrumentation). For many elements, comparable detection limits using a wavelength dispersive spectrometer are almost an order of magnitude lower. The principal technical limitation to these data is that detection limits for the lighter elements (especially F, Na and Mg) are affected by absorption in the 6 to 8 μm beryllium window of an ED spectrometer and to a much lesser extent, the window of a WD proportional counter. For the heavier elements, detection limits using either detector are controlled by the more restricted excitation efficiency when the electron gun is operated at 15 to 20 kV.

Modern windowless energy dispersive detectors substantially overcome the light element attenuation problem. In these designs it is usually possible to interchange a standard beryllium window with an ultra-thin 20 nm formvar window or no window at all (windowless ED detectors). For routine operation, it is necessary to interlock reliable window selection with the microprobe vacuum control

Table 10.12 Chemical analysis of reference minerals prepared by the Smithsonian institution

Mineral	Source	USNM	SiO_2	Al_2O_3	Fe_2O_3	FeO	MgO	CaO	Na_2O	K_2O	TiO_2	P_2O_5	MnO	Cr_2O_3	H_2O^+	Total
Anorthite	Great Sitkin Island	137041	44.00	36.03	—	0.62	<0.02	19.09	0.53	0.03	0.03	—	—	—	—	100.33
Anorthoclase	Kanakui	133868	66.44	20.12	—	0.20	—	0.87	9.31	2.35	—	—	—	—	<0.05	99.29
Apatite	Durango	104021	0.34	0.07	0.06	0.00	0.01	54.02	0.23	0.01	—	40.78	0.01	—	0.01	99.94
Augite	Kanakui	122142	50.73	7.86	3.69	3.45	16.65	15.82	1.27	0.00	0.74	—	0.13	—	0.04	100.38
Benitoite	San Benito County	86539	43.75	—	—	—	—	—	—	—	19.35	—	—	—	—	100.15
Chromite	Tiebaghi Mine	117075	—	9.92	—	13.04	15.20	—	—	—	—	—	0.11	60.5	—	98.77
Corundum	Synthetic	657S	—	99.99	—	—	—	—	—	—	—	—	—	—	—	99.99
Diopside	Natural Bridge	117733	54.87	0.11	—	0.24	18.30	25.63	0.34	—	—	—	0.04	—	—	99.53
Fayalite	Rockport	85276	29.22	—	1.32	66.36	—	—	—	—	0.04	—	2.14	—	0.1	99.18
Gahnite	Brazil	145883	—	55.32	—	1.97	—	—	—	—	—	—	0.34	—	—	100.13
Garnet	Roberts Victor Mine	87375	39.47	22.27	2.77	13.76	6.55	14.39	—	—	0.39	—	0.59	—	<0.01	100.19
Garnet	Roberts Victor Mine	110752	40.16	22.70	2.17	9.36	7.17	18.12	—	—	0.35	—	0.19	—	<0.01	100.22
Glass (basaltic)	Juan de Fuca Ridge	111240/52 VG-2	50.81	14.06	2.23	9.83	6.71	11.12	2.62	0.19	1.85	0.20	0.22	—	0.02	99.86
Glass (basaltic)	Makaopuhi Lava lake	113498/1 VG-A99	50.94	12.49	1.87	11.62	5.08	9.30	2.66	0.82	4.06	0.38	0.15	—	0.02	99.39
Glass (basaltic)	Indian Ocean	113716	51.52	15.39	1.12	8.12	8.21	11.31	2.48	0.09	1.30	0.12	0.17	—	0.18	100.07
Glass (rhyolitic)	Yellowstone National Park	72854 VG-568	76.71	12.06	0.48	0.80	<0.1	0.50	3.75	4.89	0.12	<0.01	0.03	—	0.12	99.56
Glass (tektite)	Synthetic	2213	75.75	11.34	0.64	4.32	1.51	2.66	1.06	1.88	0.50	0.00	0.11	—	0.10	99.88
Hornblende	Arenal Volcano	111356	41.46	15.47	5.60	6.43	14.24	11.55	1.91	0.21	1.41	<0.01	0.15	—	1.21	99.64
Hornblende	Kakanui	143965	40.37	14.90	3.30	7.95	12.80	10.30	2.60	2.05	4.72	0.00	0.09	—	0.94	100.02
Hypersthene	Johnstown meteorite	746	54.09	1.23	—	15.22	26.79	1.52	<0.05	<0.05	0.16	—	0.49	0.75	0.00	100.25
Ilmenite	Ilmen Mountains	96189	—	—	11.6	36.1	0.31	—	—	—	45.7	—	4.77	—	—	99.40
Magnetite	Minas Gerais	114887	—	—	67.5	30.2	—	—	—	—	—	—	—	—	—	98.16
Microcline		143966	64.24	18.30	0.00	0.04	0.03	0.02	1.30	15.14	0.01	—	0.04	—	—	99.12
Olivine	San Carlos	111312/444	40.81	—	—	9.55	49.42	<0.05	—	—	—	0.00	0.14	—	—	100.29
Olivine	Springwater meteorite	2566	38.95	—	—	16.62	43.58	—	—	—	—	—	0.30	0.02	<0.05	99.47
Omphacite	Roberts Victor Mine	110607	55.42	8.89	1.35	3.41	11.57	13.75	5.00	0.15	0.37	—	0.10	—	0.02	100.03
Osumilite	Nain, Labrador	143967	60.20	22.60	—	6.38	5.83	<0.03	0.39	4.00	0.18	—	—	—	0.02	99.60
Plagioclase	Lake County	115900	51.25	30.91	0.34	0.15	0.14	13.64	3.45	0.18	0.05	—	0.01	—	0.05	100.17
Pyrope	Kanakui	143968	41.46	23.73	—	10.68	18.51	5.17	—	—	0.47	—	0.28	—	<0.01	100.30
Quartz	Hot Springs	R17701	99.99	—	—	—	—	—	—	—	—	—	—	—	—	99.99
Scapolite	Brazil	R6600-1	49.78	25.05	—	0.17	—	13.58	5.20	0.94	—	—	—	—	0.21	99.86

Data from Jarosewich *et al.* (1980). See original reference for preparation details and sources of analyses.
Units, weight % oxide.

Table 10.13 Analysis of selected minerals by wavelength dispersive electron probe microanalysis

	DSDP basaltic glass				Kanakui hornblende									
	Staudigel (1979)	Potts *et al.* (1983)		Jarosewich *et al.* (1980)	Jarosewich *et al.* (1979)									
	Expected	Analysed	CV	Chemical analysis		Laboratory 1 Analysed	CV		Laboratory 2 Analysed	CV		Laboratory 3 Analysed	CV	
SiO_2	51.52	51.53	0.47	40.37	(a)	40.35	0.42	(a)	41.09	1.2	(a)	40.42	0.69	SiO_2
					(b)	40.29	0.84	(b)	40.51	0.42	(b)	40.73	0.59	
TiO_2	1.30	1.32	1.9	4.36	(a)	4.34	0.69	(a)	4.34	2.8	(a)	4.32	4.2	TiO_2
					(b)	4.94	1.6	(b)	4.51	2.0	(b)	4.74	3.2	
Al_2O_3	15.39	15.01	0.65	14.90	(a)	14.68	0.75	(a)	15.02	1.5	(a)	14.93	2.7	Al_2O_3
					(b)	14.45	0.55	(b)	15.02	1.3	(b)	14.48	3.0	
FeO	9.12	9.05	0.74	10.92	(a)	10.99	1.4	(a)	10.88	2.4	(a)	10.51	1.3	FeO
					(b)	10.37	2.6	(b)	10.96	1.3	(b)	10.50	4.2	
MnO	0.17	0.18	7.2	0.09	(a)	0.04	25	(a)	—	—	(a)	0.09	—	MnO
					(b)	0.11	27	(b)	0.12	25	(b)	0.07	57	
MgO	8.12	7.96	1.0	12.80	(a)	12.59	1.0	(a)	12.75	3.4	(a)	12.99	2.2	MgO
					(b)	12.98	2.0	(b)	12.82	1.3	(b)	13.13	4.4	
CaO	11.31	11.15	0.68	10.30	(a)	10.13	1.5	(a)	10.72	2.4	(a)	10.66	4.6	CaO
					(b)	9.92	1.4	(b)	10.30	1.1	(b)	10.20	3.1	
Na_2O	2.48	2.76	1.6	2.60	(a)	2.71	4.4	(a)	2.58	3.1	(a)	2.68	4.9	Na_2O
					(b)	2.87	3.1	(b)	2.62	1.5	(b)	2.68	4.5	
K_2O	0.09	0.07	7.1	2.05	(a)	2.03	0.99	(a)	1.98	4.0	(a)	1.89	6.3	K_2O
					(b)	2.05	0.98	(b)	2.03	2.0	(b)	1.91	2.1	
P_2O_5				0	(a)	0.12	—	(a)	0.07	29	(a)	0.11	8	P_2O_5
					(b)	0.12	—				(b)	0.02	—	
Total	99.50	99.03		98.39	(a)	97.98		(a)	99.43		(a)	98.60		Total
					(b)	98.10		(b)	98.89		(b)	98.46		
n		54			(a)	5		(a)	7		(a)	10		*n*
					(b)	10		(b)	7		(b)	5		

CV = coefficient of variation = 100 × standard deviation/mean analysis (%).
n = number of replicate analyses from which mean and CV data were calculated.
Jaresowich *et al.* (1979): contributing laboratories Smithsonian Institution, Massachusetts Institute of Technology, US Geological Survey (order of contributing laboratories randomized).
Data set (a) was calibrated against the same mineral, Kakanui hornblende as reference standard.
Data set (b) was calibrated using the laboratory's preferred method.
All data listed as weight % oxide.

system, since formvar cannot maintain detector vacuum against atmospheric pressure and operation with no window causes the detector crystal itself to become susceptible to surface damage from oil and other contaminants in the microprobe vacuum system. The use of these devices has been reviewed by Russ and Sandborg (1981).

10.15.4 *The electron microprobe—overall performance*

In some respects the electron microprobe can be considered to approach the frontier of routine and reliable operation. Unless instrumentation is supported by high-quality maintenance and supervision, analytical results can easily become unacceptable. Two areas probably critically affect the success of operation of these instruments.

(i) The electron column is a finely tuned device and to operate successfully must be maintained free of contamination, under high vacuum and precisely aligned. A demanding specification is placed on gun and lens power supplies, both in terms of noise level and long-term stability. The ultimate goal would be the ability to operate for an eight-hour shift with negligible drift in electron optical parameters.

(ii) Fully focusing x-ray spectrometers are not ideally suited to highly repetitive operation for prolonged periods. It is essential to be able to reposition these instruments on to x-ray peaks with an extremely high degree of reproducibility. However, spectrometer bearings cannot be adequately oiled (most oils evaporate under prolonged vacuum conditions). Moreover, the electrical cables connecting preamplifier with signal counting electronics are prone to become stiff with age as the plastic insulation hardens under vacuum. Stiff bearings and hardened cables both restrict the free and repeatable movement of spectrometers and cause backlash, which is probably the most common cause of loss in analytical precision between routine maintenance. In this respect, ED detectors have some operating advantages.

10.15.5 *Reporting microprobe results: formula units*

As in XRF analysis, it is normal to determine oxygen in silicate minerals by difference: that is, for the purposes of the ZAF correction, the difference between 100% and the sum of the analysed elemental components is assumed to be oxygen.

Table 10.14 Energy dispersive microprobe analysis of selected minerals

	Albite Dunham and Wilkinson (1978)			Orthopyroxene Dunham and Wilkinson (1978)			Diopside/jadeite glass Dunham and Wilkinson (1978)				Kaersutite Wicks and Plant (1983)			
	Expected	ED Analysed	WD Analysed	Expected	ED Analysed	WD Analysed	Expected	ED Analysed	CV	WD Analysed	Expected	ED Analysed	CV	
SiO_2	68.17	68.29	68.89	55.1	55.28	54.24	56.81	56.83	0.40	57.22	40.42	40.35	0.64	SiO_2
TiO_2				0.1	0.20	0.14					4.43	4.41	1.4	TiO_2
Al_2O_3	19.62	19.57	20.34	1.7	1.79	1.84	8.44	8.91	1.2	8.63	13.90	13.87	0.72	Al_2O_3
FeO	0.08	nd	0.03	12.7	12.42	13.24					11.21	11.19	1.1	FeO
MnO				0.3	0.26	0.25					0.10			MnO
MgO				29.0	29.92	29.73	12.39	11.94	1.3	12.05	12.95	12.87	1.4	MgO
CaO	0.08	nd	nd	0.5	0.48	0.52	17.23	17.14	0.93	16.83	10.28	10.28	0.88	CaO
Na_2O	11.59	11.94	11.79				5.13	5.18	2.1	5.24	2.80	2.80	4.3	Na_2O
K_2O	0.28	0.20	0.26								2.05	2.05	1.5	K_2O
Cr_2O_3				0.2	0.25	0.20								Cr_2O_3
H_2O^+											0.96			H_2O^+
F											0.15			F
total	99.82	100.00	101.31	99.6	100.60	100.16	100.00	100.00		99.97	99.25	97.82		total
n		30	5		30	14		30		12		35		*n*

CV = coefficient of variation = 100 × standard deviation/mean analysis (%).
n = number of replicate analyses from which mean and CV were calculated.
nd = not detected. Units: weight % oxide.
Kaersutite: expected composition from Mason (1968).
Microprobe analyses carried out using an energy dispersive spectrometer by Wicks and Plant (1983), with samples excited at 20 kV, 10 nA (measured on kaersutite) for 100 s count times using a 5 μm diameter beam. The ED spectrometer take-off angle was 38.5°.
Dunham and Wilkinson (1978): Energy dispersive analyses were carried out at 15 kV using a probe current of 3 nA (measured on cobalt metal) for count times of 100 s.
The ED spectrometer take-off angle was 75°. Wavelength dispersive analyses were carried out at 15 kV, 15 nA (measured on olivine) using four counts of 20 s for each element.
Expected compositions of minerals are taken from the following sources: albite (Deer *et al*, 1963, Table 13, analysis 8); orthopyroxene (Howie and Smith, 1966, Table 5); diopside/jadeite glass (F.R. Boyd, pers. comm.).

After correction, element data is normally recalculated as the oxide and the sum calculated, data being accepted if it falls within the limits 100 ± 1%. However there are several circumstances where such a stringent criterion is inappropriate.

(*a*) *Structural water.* Some minerals (for example amphiboles) contain water as part of their chemical structure so that analysed oxide concentrations can legitimately add up to totals as low as perhaps 85 %.

(*b*) *Summation of* F^-, Cl^- *and* S^{2-}. Silicate minerals containing significant concentrations of fluoride, chloride or sulphide sulphur require further consideration. These ions substitute for O^{2-} in the crystal lattice. Thus, if the total composition is calculated by adding the element oxide concentrations to $F^- + Cl^-$ and S^{2-}, the sum will exceed 100%. A correction to this total is required to subtract the oxygen concentration displaced by these anions.

For singly charged anions (F^- or Cl^-), the oxygen equivalent concentration (taking F^- as an example) is given by:

$$\frac{1}{2}\left[(\text{concentration of } F^-) \times \frac{(\text{atomic weight of } O^{2-})}{(\text{atomic weight of } F^-)}\right] \quad (10.41)$$

For doubly charged anions (e.g. S^{2-}), the oxygen equivalent concentration is calculated from:

$$\left[(\text{concentration of } S^{2-}) \times \frac{(\text{atomic weight of } O^{2-})}{(\text{atomic weight of } S^{2-})}\right] \quad (10.42)$$

If sulphur is present in the form of sulphate ($SO_4{}^{2-}$) it is normally expressed as SO_3 and summed without further correction.

(*c*) *Oxidation state of iron.* As in other x-ray methods, iron is determined by electron microprobe as the elemental concentration regardless of oxidation state. However, in performing the mineral analysis summation, it is necessary to recalculate iron as either $Fe^{II}O$ or $Fe^{III}{}_2O_3$ as appropriate. For some minerals, oxidation state is known from crystallographic considerations. However, others can contain varying proportions of iron in both oxidation states. An appropriate estimate of the proportions of each may be calculated as follows.

Assuming all elements have been analysed (accurately !) and the mineral contains no structural water, then the concentration of oxygen, [O], which must be apportioned to iron is:

$$[O] = 100 - \left\{\sum_{\text{not Fe}} [\text{element oxides}] + [Fe]\right\} \quad (10.43)$$

where square brackets indicate concentration. If x is the fraction of iron present as Fe^{II}, then the concentration of Fe^{II} is given by

$$[Fe^{II}] = x[Fe] \quad (10.44)$$

and this combines with the following weight fraction of oxygen to form $Fe^{II}O$:

$$x[Fe]\frac{AW_{(O)}}{AW_{(Fe)}} \quad (10.45)$$

where AW is the atomic weight of the indicated element.

Using similar arguments, the concentration of Fe^{III} is $(1-x)$. (Fe), and this combines with the following weight fraction of oxygen to form $Fe^{III}{}_2O_3$:

$$\frac{3}{2}(1-x)[Fe]\frac{AW_{(O)}}{AW_{(Fe)}}. \quad (10.46)$$

As the sum of equations (10.45) and (10.46) must equal [O], the value of x may be obtained as

$$x = 3-2\frac{[O]}{[Fe]}\frac{AW_{(Fe)}}{AW_{(O)}}. \quad (10.47)$$

Although superficially an attractive approach, the practical application is severely limited, since errors accrue from each microprobe determination and any estimate of Fe^{II}/Fe^{III} content will only be reliable in the unlikely event that these errors exactly cancel out. A more usual approach depends on the known crystallographic structure and ion site occupancy of a mineral. For example, if total Fe must be apportioned between Fe^{II} and Fe^{III} in a given mineral that contains only one type of 2^+ ion site, and if the only other 2^+ ion is magnesium, then the fraction of Fe^{II} complementing Mg^{II} in this site may be calculated from the magnesium microprobe analysis and the known chemical formula of unit cell of the mineral. Approaches of this kind have been described by Gidskehaug (1975), for example. Finger (1972) has investigated the uncertainty in the calculation of the ferric iron contents from microprobe analyses.

An alternative approach based on analytical measurements has been proposed by Albee and Choodos (1969). They described a semi-quantitative method for the determination of both Fe^{2+}/Fe^{3+} and Mn^{2+}/Mn^{3+} ratios in oxides and silicates based on the analysis of the respective $L\beta$ and $L\alpha$ lines. The intensity of these lines varies in proportion to the ratio of oxidation states. However, there are some practical difficulties in making precise measurements since these emission lines occur at very low energies, in the 0.7 keV region of the spectrum.

Legkova *et al.* (1982), following a similar procedure, showed that it was possible to measure the Fe $L\beta_1$ and Fe $L\alpha_{1,2}$ line intensity ratio for the determination of Fe^{2+} and Fe^{3+} in amphiboles. Microprobe results were in acceptable agreement with chemical analyses.

(*d*) *Formula unit calculations.* Valuable geological information may be obtained by expressing the analysed microprobe composition of a mineral in terms of its crystal formula units. This gives data on the proportions of each element occupying individual crystal sites in the mineral. To demonstrate the way in which formula units are calculated from microprobe data, the stages involved for an olivine of simplified chemical formula $(Mg, Fe)_2SiO_4$ are shown in Table 10.15; the individual steps are as follows:

(i) Column A represents the microprobe analysis expressed in units of weight % element oxide

(ii) From these data the molecular proportion of each oxide is calculated by dividing these data by the appropriate molecular weight (*B* in column 3)

(iii) The atomic proportion of oxygen in each oxide is then calculated by multiplying this number by the number of oxygen atoms for each cation in the oxide formula (2 for SiO_2, 1 for FeO and MgO etc) ($B.x$ in column 5)

(iv) The data for the atomic proportion of oxygens is then normalized so that its sum for all elements adds up to the theoretical number of oxygen atoms in the ideal chemical formula (4 in this case) (*C* in column 6)

(v) These data are then divided by the number of oxygens to each element atom in the simple oxide (2 for SiO_2, 1 for FeO and MgO, etc.), the result being the number of cations in the chemical formula of the mineral (C/x in column 7).

A more detailed example of this procedure is given by Deer *et al.* (1976, p.515).

(*e*) *Minerals routinely encountered in microprobe analysis.* Formula unit calculations are normally incorporated as part

Table 10.15 Formula unit calculation for a microprobe analysis of an olivine

	A	MW	$B = A/\text{MW}$	x	$(B.x)$	$C = (B.x).4/T$	C/x
Oxide	Analysed microprobe composition	Molecular weight of oxide		Oxide ratio	Atomic proportion of oxygen from each oxide	Number of cations normalized to (O) = 4	Number of cations in formula unit
SiO_2	40.9%	60.09	0.6806	2	1.3613	2.0017	Si = 1.0008
FeO	9.6	71.85	0.1336	1	0.1336	0.1964	Fe^{II} = 0.1964
MgO	49.4	40.31	1.2255	1	1.2255	1.8020	Mg = 1.8020
Total	99.9				$T = 2.7203$		

x = ratio of oxygen to element atoms in stoichiometric oxide.
$T = \Sigma B.x$.
General formula of olivine = $(Mg, Fe^{II})_2SiO_4$.
Calculated formula from microprobe analysis = $(Mg_{1.8020}, Fe_{0.1964})Si_{1.0008})_4$.

of the ZAF correction procedure. However, relevant calculations depend on the correct identification of the mineral type and specification of the appropriate number of oxygens in unit chemical formula. In order to give an idea of the range of mineral types studied by electron probe microanalysis, selected minerals (predominantly silicate minerals, with some opaques) are listed in Table 10.16, together with chemical formula, from which unit cell oxygen data may be derived. Kretz (1983) has highlighted the 'disturbing' lack of uniformity in the choice of symbols in use as abbreviations for rock-forming minerals. The symbols proposed by Kretz to unify mineral abbreviations are also listed in Table 10.16.

10.16 Analysis of non-silicate minerals: uranium, thorium and rare-earth elements

10.16.1 *Scope of the problem*

The analysis by electron microprobe of silicate rocks in thin section does not concern only an examination of silicate

Table 10.16 Mineral names and chemical formulae

Amphibole group: a group of silicate minerals having the general formula
$A_{0-1}B_2Y_5Z_8O_{22}(OH,F,Cl)_2$

where A = Ca, Na
B = Ca, Fe^{II}, Li, Mg, Mn^{II}, Na
Y = Al, Cr, Fe^{II}, Fe^{III}, Mg, Mn^{II}, Ti
Z = Al, Si, Ti
including actinolite, glaucophane, grunerite, kaersutite, pargasite, riebeckite and tremolite.

Feldspar group: a group of silicate minerals having the general formula XZ_4O_8
where X = Ba, Ca, K, Na, NH_4, Sr
Z = Al, B, Si
including albite*, andesine*, anorthite*, anorthoclase, bytownite*, labradorite*, microcline, oligoclase*, and orthoclase. The asterisked minerals constitute the plagioclase subgroup, comprising a solid solution series of general formula $(Na, Ca)Al(Al,Si)Si_2O_8$.

Garnet group: a group of silicate minerals having the general formula $A_3B_2(SiO_4)_3$
where A = Ca, Fe^{II}, Mg, Mn^{II}
B = Al, Cr^{III}, Fe^{III}, Mn^{III}, Ti, V^{III}, Zr
Si can be partially replaced by Ti and Al
including almandine and pyrope.

Mica group: a group of minerals having the general formula $XY_{2-3}Z_4O_{10}(OH, F)_2$
where X = Ba, Ca, H_3O, K, Na
Y = Al, Cr, Fe^{II}, Fe^{III}, Li, Mg, Mn^{II}, V^{III}, Zn
Z = Al, Be, Fe^{III}, Si
including biotite, glauconite, lepidolite, muscovite.

Olivine group: a group of silicate minerals having the general formula A_2SiO_4
where A = Fe^{II}, Mg, Mn^{II}, Ni
including fayalite, forsterite

Pyroxene group: a group of silicate minerals having the general formula ABZ_2O_6
where A = Ca, Fe^{II}, Li, Mg, Na
B = Al, Cr^{III}, Fe^{II}, Fe^{III}, Mg, Mn^{II}
Z = Al, Si
including acmite, augite, diopside, enstatite, hedenbergite, jadeite, omphacite, pigeonite, spodumene.

Spinel group: a group of silicate minerals having the general formula AB_2O_4
where A = Co, Cu, Fe^{II}, Ge, Mg, Mn^{II}, Ni, Ti, Zn
B = Al, Cr^{III}, Fe^{II}, Fe^{III}, Mg, Mn^{III}, V^{III}
including chromite, magnetite, spinel, ulvospinel.

Mineral	Chemical formula	Symbol
Acmite	$NaFe^{III}Si_2O_6$ (aegirine)	Acm
Actinolite	$Ca_2(Mg,Fe^{II})_5Si_8O_{22}(OH,F)_2$	Act
Aegirine	= acmite	—
Albite	$NaAlSi_3O_8$	Ab
Allanite	$(Ca,Ce)_2(Fe^{II},Fe^{III})Al_2O.OH(Si_2O_7)SiO_4)$	Aln
Almandine	$Fe^{II}{}_3Al_2(SiO_4)_3$	Alm
Amphibole	(see amphibole group)	—
Anatase	TiO_2	Ant
Andalusite	Al_2OSiO_4	And
Andesine	$((NaSi)_{0.7-0.5}(CaAl)_{0.3-0.5})AlSi_2O_8$	—
Anglesite	$PbSO_4$	—
Anhydrite	$CaSO_4$	Anh

(*Continued*)

Table 10.16 (*Continued*)

Mineral	Chemical formula	Symbol
Ankerite	$Ca(Fe^{II},Mg,Mn)(CO_3)_2$	Ank
Anorthite	$CaAl_2Si_2O_8$	An
Anorthoclase	$(Na,K)AlSi_3O_8$	—
Apatite	$Ca_5(PO_4)_3(OH,F,Cl)$	Ap
Aragonite	$CaCO_3$	Arg
Augite	$(Ca,Na,Mg,Fe^{II},Mn,Fe^{III},Ti,Al)_2(Si,Al)_2O_6$	Aug
Barite	$BaSO_4$	Brt
Benitoite	$BaTi(SiO_3)_3$	—
Biotite	$K(Mg,Fe^{II})_3(Al,Fe^{III})Si_3O_{10}(OH,F)_2$	Bt
Bytownite	feldspar group	—
Calcite	$CaCO_3$	Cal
Celestite	$SrSO_4$	Cls
Chlorite	$(Mg,Al,Fe)_{12}(Si,Al)_8O_{20}(OH)_{16}$	Chl
Chromite	$Fe^{II}Cr_2O_4$	Chr
Clinochlore	$(Mg,Fe^{II})_5Al(Si_3Al)O_{10}(OH)_8$	—
Cordierite	$Mg_2Al_4Si_5O_{18}$	Crd
Corundum	Al_2O_3	Crn
Diaspore	$AlO(OH)$	Dsp
Diamond	C	—
Diopside	$CaMgSi_2O_6$	Di
Dioptase	$CuSiO_2(OH)_2$	—
Dolomite	$CaMg(CO_3)_2$	Dol
Enstatite	$Mg_2Si_2O_6$	En
Epidote	$Ca_2(Al,Fe^{III})_3(SiO_4)_3(OH)$	Ep
Fayalite	$Fe^{II}{}_2SiO_4$	Fa
Feldspar	see feldspar group	—
Fluorite	CaF_2	Fl
Forsterite	Mg_2SiO_4	Fo
Galena	PbS	Gn
Garnet	see garnet group	Grt
Gibbsite	$Al(OH)_3$	Gbs
Glauconite	$(K,Na,Ca)_{1.2-2.0}(Fe^{III},Al,Fe^{II},Mg)_4(Si_{7-7.6}Al_{1.0-0.4}O_{20})(OH)_4.nH_2O$	Glt
Glaucophane	$Na_2Mg_3Al_2(Si_8O_{22})(OH)_2$	Gln
Goethite	$FeO(OH)$	Gt
Grossularite	$Ca_3Al_2(SiO_4)_3$	Grs
Grunerite	$(Fe^{II},Mg)_7Si_8O_{22}(OH)_2$	Gru
Gypsum	$CaSO_4.2H_2O$	Gp
Halite	NaCl	Hl
Hedenbergite	$CaFe^{II}Si_2O_6$	Hd
Hematite	Fe_2O_3	Hem
Hibonite	$(Ca,Ce)(Al,Ti,Mg)_{12}O_{18}$	—
Hornblende	see hornblende group	Hbl
Hypersthene	$(Mg,Fe^{II})_2Si_2O_6$	—
Ilmenite	$Fe^{II}TiO_3$	Ilm
Jadeite	$NaAl(Si_2O_6)$	Jd
Kaersutite	$Ca_2(Na,K)(Mg,Fe^{II},Fe^{III})_4Ti(Si_6Al_2O_{22})(O,OH,F)_2$	Krs
Kaolinite	$Al_4(Si_4O_{10})(OH)_8$	Kln
Kilchoanite	$Ca_3Si_2O_7$	—
Kyanite	Al_2SiO_5	Ky
Labradorite	see plagioclase	—
Lawsonite	$CaAl_2(OH)_2(Si_2O_7).H_2O$	Lws
Lepidolite	$K_2(Li,Al)_{5-6}(Si_{6-7}Al_{2-1}O_{20})(OH,F)_4$	Lpd
Leucite	$KAlSi_2O_6$	Lct
Lizardite	$Mg_3Si_2O_5(OH)_4$	Lz
Magnesite	$MgCO_3$	Mgs
Magnetite	$Fe^{II}Fe^{III}{}_2O_4$	Mag
Manganite	$MnO(OH)$	—
Manganosite	MnO	—
Marcasite	FeS_2	—
Mica	see mica group	—

Table 10.16 (*Continued*)

Mineral	Chemical formula	Symbol
Microcline	$KAlSi_3O_8$	Mc
Microlite	$(NaCa)_2Ta_2O_6(O,OH,F)$	—
Molybdenite	MoS_2	Mo
Monazite	$(Ce,La,Nd,Th)PO_4$	Mnz
Monticellite	$CaMgSiO_4$	Mtc
Montmorillonite	$(0.5Ca,Na)_{0.7}(Al,Mg,Fe)_4((Si,Al)_8O_{20})(OH)_4.nH_2O$	Mnt
Muscovite	$K_2Al_4(Si_6Al_2O_{20})\ (OH,F)_4$	Ms
Natrolite	$Na_2Al_2Si_3O_{10}.2H_2O$	Ntr
Nepheline	$Na_3(Na,K)(Al_4Si_4O_{16})$	Ne
Oldhamite	$(Ca,Mn)S$	—
Oligoclase	see plagioclase	—
Olivine	$(Mg,Fe)_2(SiO_4)$ see olivine group	Ol
Omphacite	see plagioclase	Omp
Orthoclase	$KAlSi_3O_8$ see feldspar group	Or
Orthopyroxene	$(Mg,Fe)SiO_3$ see pyroxene group	Opx
Pargasite	$NaCa_2(Mg,Fe^{II})_4Al(Si_6Al_2)O_{22}(OH)_2$	Prg
Pentlandite	$(Fe,Ni)_9S_8$	Pn
Periclase	MgO	Per
Perovskite	$CaTiO_3$	Prv
Phlogopite	$KMg_3AlSi_3O_{10}(OH,F)_2$	Phl
Pigeonite	$(Mg,Fe^{II},Ca)(Mg,Fe^{II})Si_2O_6$	Pgt
Plagioclase	see feldspar group	Pl
Prehnite	$Ca_2Al_2Si_3O_{10}(OH)_2$	Prh
Pumpellyite	$Ca_2MgAl_2(SiO_4)(Si_2O_7)(OH)_2.H_2O$	Pmp
Pyrite	FeS_2	Py
Pyrochlore	$(Na,Ca)_2Nb_2O_6(OH,F)$	—
Pyrolusite	MnO_2	—
Pyrope	$Mg_3Al_2(SiO_4)_3$ see garnet group	Prp
Pyroxene	see pyroxene group	—
Pyrrhotite	$Fe_{1-0.8}S$	Po
Quartz	SiO_2	Qtz
Rhodochrosite	$MnCO_3$	Rds
Rhodonite	$(Mn,Fe^{II},Ca)(SiO_3)$	Rdn
Riebeckite	$Na_2Fe^{II}{}_3Fe^{III}{}_2(Si_8O_{22})(OH,F)_2$	Rbk
Ringwoodite	$(Mg,Fe^{II})_2SiO_4$	—
Rutile	TiO_2	Rt
Scapolite	$(Na,Ca,K)_4(Al_3(Al,Si)_3Si_6O_{24})(Cl,CO_3,SO_4,OH)$	Scp
Scheelite	$CaWO_4$	—
Serpentine	$Mg_3(Si_2O_5)(OH)_4$	Srp
Siderite	$Fe^{II}CO_3$	Sd
Sillimanite	Al_2OSiO_4	Sil
Sinoite	Si_2N_2O	—
Spencerite	$Zn_4(PO_4)_2(OH)_2.3H_2O$	—
Sphalerite	ZnS	Sp
Sphene	$CaTi(SiO_4)(O,OH,F)$ (titanite)	Spn
Spinel	$MgAl_2O_4$	Spl
Spodumene	$LiAlSi_2O_6$	Spd
Spurrite	$Ca_5(SiO_4)_2(CO_3)$	—
Staurolite	$(Fe^{II},Mg)_2(Al,Fe^{III})_9O_6(SiO_4)_4(O,OH)_2$	St
Stilbite	$NaCa_2Al_5Si_{13}O_{36}.14H_2O$	Stb
Strontianite	$SrCO_3$	Str
Sylvite	KCl	—
Talc	$Mg_3Si_4O_{10}(OH)_2$	Tlc
Thorianite	ThO_2	—
Thorite	$ThSiO_4$	—
Tilleyite	$Ca_5Si_2O_7(CO_3)_2$	—
Topaz	$Al_2SiO_4(OH,F)_2$	Toz
Tourmaline	$Na(Mg,Fe,Mn,Li,Al)_3Al_6\ Si_6O_{18}\ (BO_3)_3(OH,F)_4$	Tur
Tremolite	$Ca_2Mg_5Si_8O_{22}(OH)_2$	Tr

(*Continued*)

Table 10.16 *(Continued)*

Mineral	Chemical formula	Symbol
Tridymite	SiO_2	Trd
Ulvospinel	$TiFe^{II}{}_2O_4$	Usp
Uraninite	UO_2	—
Vivianite	$Fe^{II}{}_3(PO_4)_2.8H_2O$	—
Willemite	Zn_2SiO_4	—
Wollastonite	$CaSiO_3$	Wo
Xenotime	$Y PO_4$	—
Zircon	$ZrSiO_4$	Zrn

Data from Weast (1973), Deer *et al.* (1976) and Fleischer (1980).
Symbol abbreviations are taken from Kretz (1983).

minerals. Opaque phases are present in many samples constituting a range of oxide, sulphide and phosphate phases that can contain high concentrations of elements normally present in bulk samples at trace levels. More common examples of such minerals would include zircon ($ZrSiO_4$), ilmenite ($FeTiO_3$), magnetite (Fe_3O_4), pyrite (FeS_2), molybdenite (MoS_2) and rutile (TiO_2), as well as a host of less common phases.

The analysis of x-ray spectra emitted by these non-silicate minerals is often not a trivial matter, as the normal overlap and multiple-order line interferences appropriate for silicate materials (see Table 8.16) requires extensive revision. Further difficulties arise from the small size of these minerals (sometimes less than 10 μm across), it being important to be aware of the ambiguities in the identification of trace elements in such small phases that can arise from the secondary fluorescence of surrounding matrix elements (section 10.8.5).

As an example of the additional considerations applying to minor phases, it is proposed here to discuss the microprobe analysis of uranium, thorium and rare-earth element minerals.

There are significant analytical difficulties in the reliable determination of these elements in minerals such as monazite $(Ce,La,Nd,Th)PO_4$, allanite $(Ca,Ce)_2(Fe^{II},Fe^{III})Al_2O$ $OH(Si_2O_7)(SiO_4)$, apatite $Ca_5(PO_4)_3(OH,F,Cl)$, and xenotime $Y(PO_4)$. Not the least is the scarcity of suitable calibration standards.

Bowles (1978) discussed the problems of determining U and Th at the 1 to 15 weight % level in an oxide matrix such as uraninite (UO_2). It is not entirely satisfactory to use U or Th metals as calibration standards because U rapidly oxidizes in air and because of the large matrix discrepancy between metal and oxide phases. Furthermore, Bowles described the difficulties in analysing very old uranium minerals (uraninites 2000 to 3000 million years old) in which natural radioactive decay causes serious mineral inhomogeneity effects. These arise both from radioactive daughter products (alpha emitters), which disrupt the mineral structure, and the inhomogeneous distribution of the final decay product, lead, which is often present as galena (PbS) due to secondary reactions involving sulphur.

10.16.2 *Synthesized rare-earth mineral standards*

One of the few sets of rare-earth materials widely available was prepared by Drake and Weill (1972), who made up four REE microprobe standards using a synthesized glass matrix of composition $SiO_2:33\%$, $Al_2O_3:37\%$ and $CaO:30\%$. The appropriate weight of selected rare-earth element oxides, sufficient to give a final concentration of about 4% of each oxide, was mixed with the glass and homogenized by fusing to 1525°C under an atmosphere of argon. These materials have been successfully used to calibrate or standardize a large proportion of reported microprobe REE mineral analyses.

Styles and Young (1983) reported some inhomogeneity effects in one of these standards if beam diameters of less than 5 μm were used, although the average bulk composition agreed with the expected value.

Smellie (1972) experimented with the preparation of major element glass samples and reported losses of sodium and potassium when fusing charges under high vacuum at 1700°C using a molybdenum strip heater. Based on this experience, Smellie *et al.* (1978) prepared synthetic U and Th standards using a base glass comprising SiO_2 (42%), Al_2O_3 (31%) and CaO (26%) to which uranium and thorium were added to give a range of final compositions of between 1 and 15% U_3O_6 or ThO_2. Some local inhomogeneity effects were observed in these materials after fusion. In particular, concentration gradients were observed in the rim of glass, 20–30 μm thick, which had been in contact with the molybdenum heating boat. However, the bulk material was reported to be homogeneous.

10.16.3 *Spectrum overlap interferences*

Serious spectrum overlap interferences are encountered in the analysis of REE minerals, both between lines of the rare-earth elements themselves and between these and lines of other constituent elements which can include Ca, Ti, Mn and Fe, for example. As an indication of the complexity of REE mineral spectra, the *L*-line series spectra, superimposed with the intensity of each series normalized to allow for the natural relative abundance are plotted in Figure 10.50. This diagram represents very much the worst case, since spectral intensities are representative of a mineral that has a horizontal chondrite normalized plot. Many minerals are substantially more concentrated in either the light, or less frequently the heavy, rare-earth elements. An illustration of the REE spectra of a light REE enriched mineral (monazite) and a heavy REE enriched mineral (xenotime) is given in Figure 10.51.

Smith and Reed (1982) have quantified the degree of overlap interference in the spectra recorded on an ED detector by measuring the percentage of the measured $L\alpha$ intensity (integrated over a fixed number of channels representing $1.2 \times$ FWHM of the $L\alpha$ photopeak width) that originated from an interfering element. Their data are reproduced in Table 10.17 and indicate that, for example, 58% of the observed intensity in the Pr $L\alpha$ photopeak originates from an overlap interference from the La L-lines. Although derived for ED spectra, these data indicate the overall degree of interference in rare-earth element mineral spectra.

Amli and Griffin (1975) examined the analysis of such minerals by wavelength dispersive microprobe and reported that interferences were particularly severe in the analysis of the $L\alpha$ lines of Pr, Eu, Gd and especially Ho. Exley (1980) recommended using the $L\beta_1$ lines of Pr, Nd, Sm, Gd and Dy to avoid both these problems and also the Mn $K\beta_1$ overlap with Dy $L\alpha$. Exley found that the analysis of Eu was impossible, whichever line was used, at the low levels of this element encountered in allanite, zircon, sphene and apatite.

As well as direct overlap interferences, a second problem affects the wavelength dispersive analysis of REE minerals, that of accurate background subtraction. As the concentrations of some or all of these elements are likely to approach detection limit levels, accurate measurement of background under each measured line becomes the limiting factor in reliable analysis. As can be seen in the wavelength scan through the spectrum of a monazite (Figure 10.52) there are few angles at which accurate background measurements can be made adjacent to individual REE L-lines. An elegant solution to this problem has been proposed by Smith and Reed (1981) adapting the background calculation techniques described for the analysis of ED spectra in section 10.12.2. An empirical background shape was recorded by exciting a continuum spectrum in diamond and normalizing it to account for attenuation effects as described previously. The actual background intensity in the spectrum of the REE mineral was then sampled at a few points removed from the REE L-lines. The background at the precise wavelength of individual REE L-lines could then be calculated by fitting the pre-recorded continuum data to the measured background intensities with appropriate corrections for atomic number and fluorescence effects. In this way, it was not necessary to carry out direct background measurements for individual REE lines even if it had been possible to specify interference-free background angles. The continuum, calculated in this way, is shown fitted to the REE spectrum of a monazite in Figure 10.52.

Interferences must also be taken into account in the analysis of uranium and thorium minerals using wavelength spectrometers. The problem of the argon absorption edge splitting the U $M\alpha$ and $M\beta$ lines, for measurements made with proportional counters filled with that gas, has already been mentioned (section 8.9). In consequence, the $M\beta$ line may be detected more efficiently than the $M\alpha$ line and is, therefore, preferred for analysis. However, Bowles (1978) suggested that it is useful to determine uranium from the $L\alpha$ line in spite of the lower excitation efficiency. This is in part because the ZAF correction procedures for M-lines are not completely characterized but also because the absorption and fluorescence corrections for the $L\alpha$ line are considerably smaller. However, a WD spectrometer is capable of only partially resolving the U $L\alpha_1$ and $L\alpha_2$ lines and Bowles suggests that the relative intensity of the $L\alpha_1/L\alpha_2$ lines varies with mineral composition. In these circumstances, more reliable data are obtained by measuring the area of a combined $(L\alpha_1 + L\alpha_2)$ peak on an energy dispersive detector.

(*c*) *Detection limits.* Typical detection limits in the determination of the REEs by WD and ED microprobe are shown in Table 10.18. The wavelength dispersive data are largely governed by the presence of spectrum overlap interferences on individual $L\alpha$ or $L\beta$ lines, whereas the ED detection limits are restricted mainly by the significantly larger contribution from the continuum beneath the L-line series using the digital filtering procedures discussed in section 9.7.

Typical analyses of REE minerals are shown in Table 10.19, taken from the data of Exley (1980) for allanite, Exley and Smith (1982) for apatite and Styles and Young (1983) for monazite. Geological interpretations of such data are often based on the shape of the chondrite-normalized abundance diagram from individual minerals (Figure 10.53). Data such as these can be used to interpret differences in REE patterns which are relevant to fractionation processes.

Table 10.17 Overlap correction coefficients for the determination of $L\alpha$ lines of rare-earth elements by wavelength dispersive microprobe

Over-lapped element	Z	Overlapping element ΔZ											
		−8	−7	−6	−5	−4	−3	−2	−1	+1	+2	+3	+4
La	57									8	1	5	—
Ce	58								13	8	1	—	1
Pr	59							58	10	8	—	4	1
Nd	60						18	42	10	—	1	4	2
Sm	62				2	17	25	30	—	6	1	4	2
Eu	63			7	2	19	33	—	8	6	1	3	3
Gd	64		3	7	2	18	—	10	8	5	1	3	3
Tb	65	1	3	6	4	—	43	6	8	5	0	2	4
Dy	66	1	3	4	—	12	49	2	8	5	0	1	4
Ho	67	2	5	—	8	12	47	1	8	4	0	1	5
Er	68	3	—	0	12	12	41	1	8	4	0	0	
Tm	69	—	7	0	15	14	31	0	7	4	0		
Yb	70	2	6	0	17	20	21	0	7	5			
Lu	71	3	3	0	18	29	10	0	6				

Data from Smith and Reed (1982).
The missing element is promethium ($Z = 61$) which has no naturally-occurring isotope.
Tabulated data list the apparent concentration of the element listed in the left-hand column when a sample containing 100% of the interfering element is excited at 15 kV. Thus, for example, praesodymium ($Z = 59$) suffers an interference from lanthanum ($Z = 57$, $\Delta Z = -2$) so that a sample containing 100% La would also appear to contain 58% Pr.

N

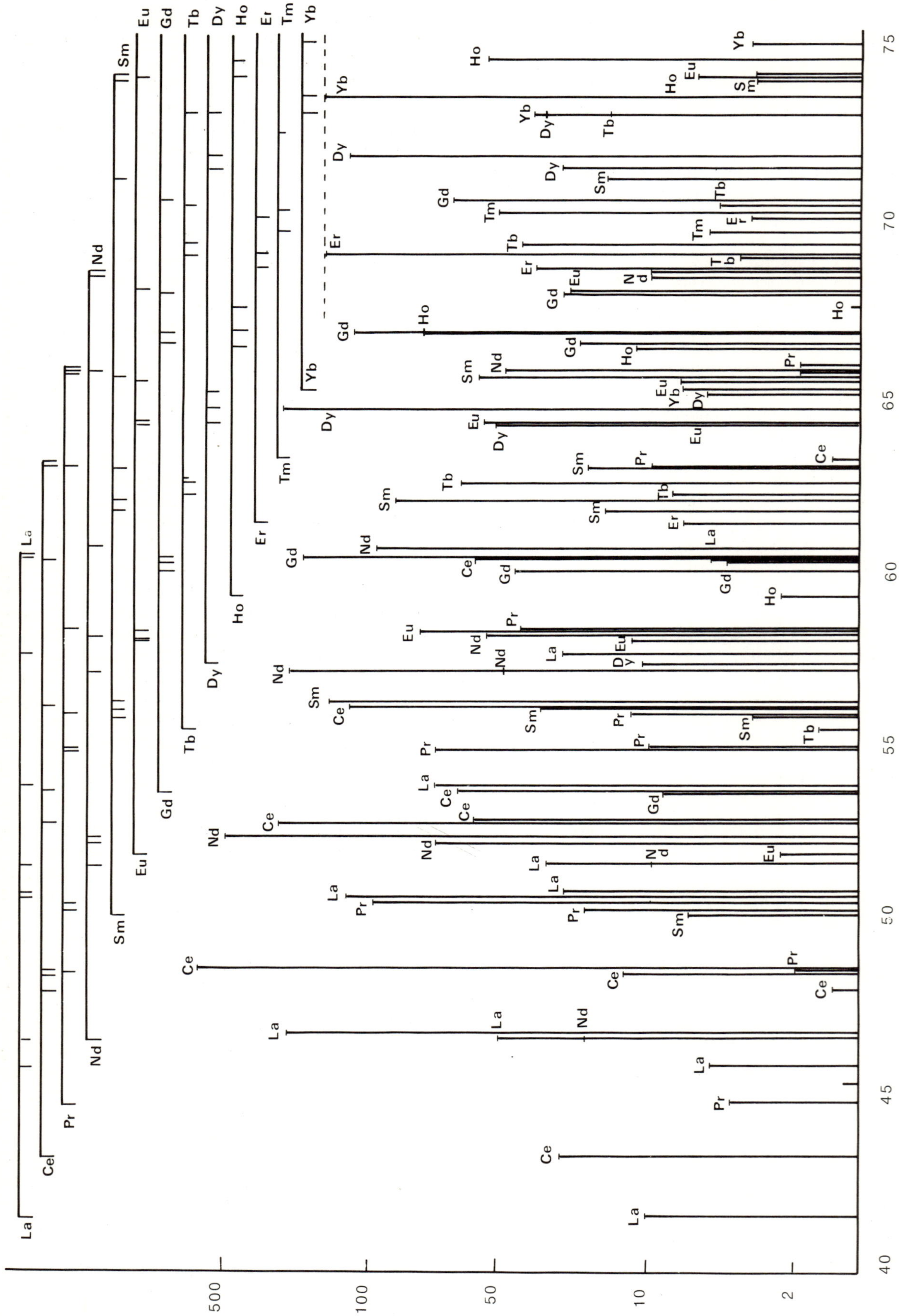

La
Ce
Pr
Nd
Sm
Eu
Gd
Tb
Dy
Ho
Er
Tm
Yb
500
100
50
10
2
40
45
50
55
60
65
70
75

(a)

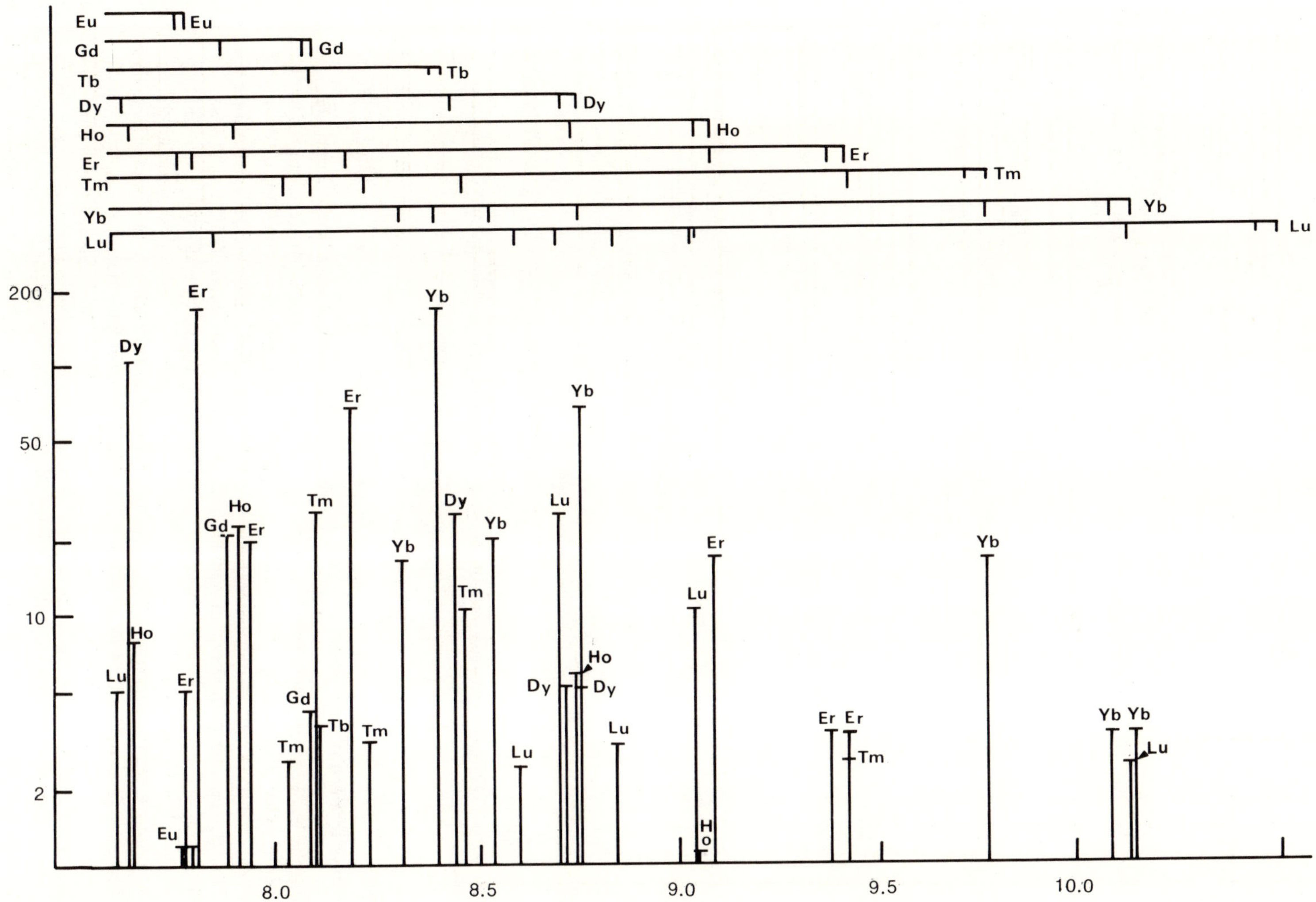

(b)

Figure 10.50 Rare-earth element L-lines superimposed on a logarithmic intensity plot to indicate the degree of overlap that may be encountered in x-ray analysis of REE-bearing minerals.
Data have been plotted so that the REE $L\alpha_1$ line is assumed to have an intensity of 2000 × chondrite abundance, with subsidiary line intensities scaled accordingly. Natural relative abundance of REEs has therefore been taken into account, spectral intensities representing a mineral that has a flat chondrite normalized plot.

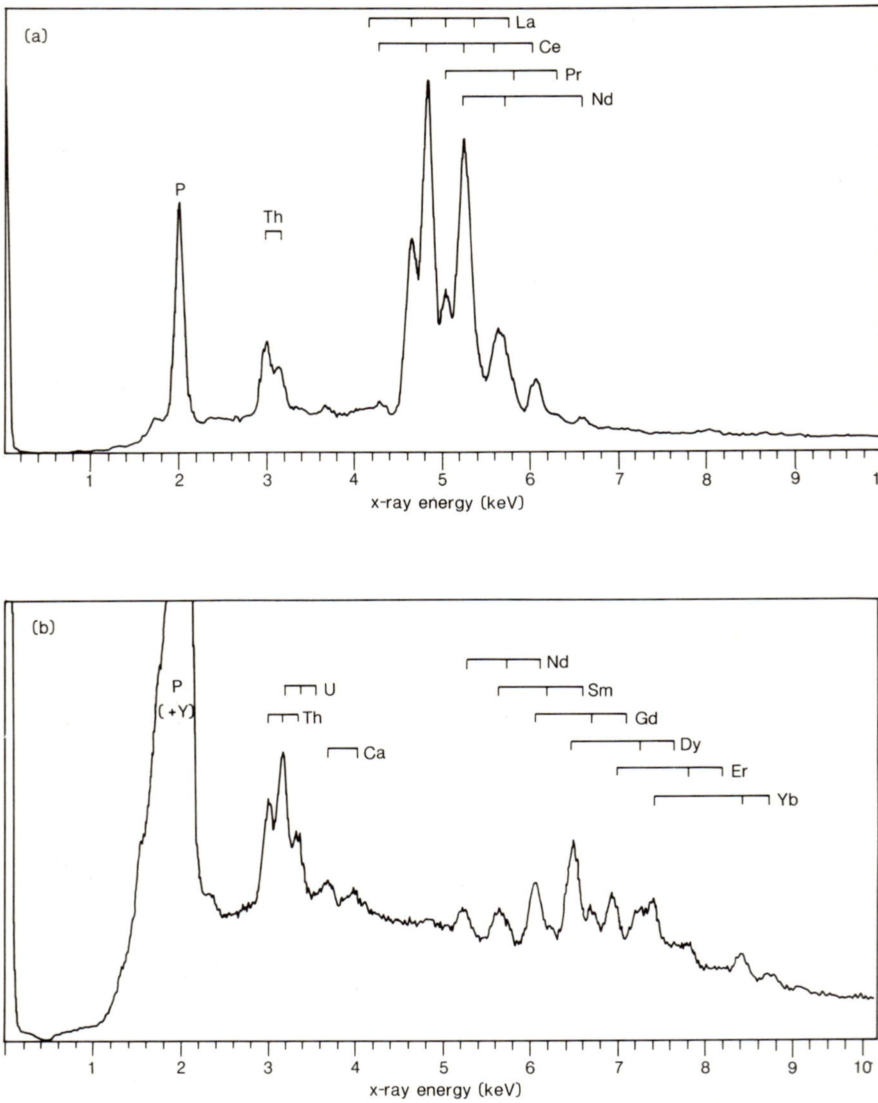

Figure 10.51 Energy dispersive spectrum of the REE-bearing minerals (*a*) monazite and (*b*) xenotime. (Courtesy of A. G. Tindle, Open University.)

10.17 Bulk rock analysis by electron microprobe

The electron microprobe has been used successfully in the bulk analyses of rock powders that can be fused to form homogeneous glass beads. This is an interesting application, since quantitative analysis can be performed on sample masses as low as 10 to 20 mg, far too small a mass for conventional analysis by XRF, for example.

Thus Nichols (1974) and Brown (1977) used iridium or molybdenum strip heaters to fuse 10–20 mg samples of finely powdered sample at temperatures of 1450–1650°C (andesites, basalts) or 1650–2000°C (syenite, granodiorite or granite) for up to about 30 s. Schimann and Smith (1980) constructed a special optical imaging furnace for fusing powdered samples. Foland and Wagner (1981) described the fusion of samples wrapped in platinum foil and heated for up to one minute inside a conventional AA graphite tube furnace to temperatures of about 1700°C.

Several points emerge from these studies:

(i) It is necessary to preheat samples to about 1000°C to prevent the explosive decomposition of volatile components (H_2O, CO_2, S, Cl and partly P) that would otherwise occur at higher fusion temperatures; the loss of Na and K is usually prevented by restricting fusion times.

(ii) Iridium is preferred to platinum foil for strip heaters because of the smaller proportion of Fe absorbed from the sample into iridium metal.

(iii) In some rock types, crystal intergrowths of Fe–Ti oxides, for example occur on fusion, causing inhomogeneity effects; bubbles formed in viscous melts (of granites, for example) can also cause sample preparation problems.

(iv) It is usually necessary to defocus the electron beam to about 20 μm to prevent losses of alkali element signals during analysis and to minimize the influence of small-scale sample inhomogeneity effects.

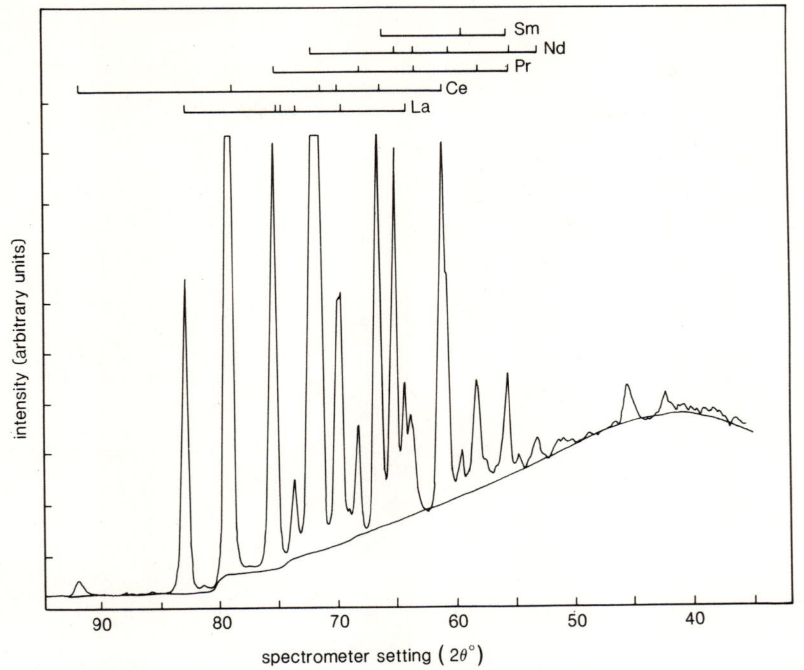

Figure 10.52 Wavelength dispersive spectrum of a monazite mineral scanned using a LiF (200) crystal showing calculated continuum. (Smith and Reed, 1982.)

Table 10.18 Detection limits of rare-earth elements by wavelength dispersive and energy dispersive microprobe analysis

		Wavelength dispersive		Energy dispersive
Element	Line	Exley (1980) 20 kV, 50 nA	Exley and Smith (1982) 25 kV, 4 μA	Tindle (pers. comm.) 30 kV, 3 nA
La_2O_3	$L\alpha$	0.16	0.023	1.4
Ce_2O_3	$L\alpha$	0.21	0.023	1.4
Pr_2O_3	$L\beta$	0.21		1.1
Nd_2O_3	$L\beta$	0.26	0.063	1.0
Sm_2O_3	$L\beta$	0.17	0.070	0.96
Eu_2O_3				1.1
Gd_2O_3	$L\beta$	0.15		1.0
Tb_2O_3				1.0
Dy_2O_3	$L\beta$	0.20		1.1
Ho_2O_3				1.0
Er_2O_3	$L\alpha$		0.034	1.0
Tm_2O_3				0.96
Yb_2O_3	$L\alpha$	0.12	0.038	0.93
Lu_2O_3				1.0
Y_2O_3	$L\alpha$	0.08		1.8

Concentration units: weight % oxide.
Detection limits represent estimates of the six-sigma limit of determination.
Count times:
Exley (1980): 60 s on peak (La,Ce,Na); 100 s for Y and other REEs plus two additional background measurements.
Tindle: 200 s for whole spectrum. Rare-earth elements determined from the *L*-profiles, Y from the *K*-profile using a filtered least-squares spectrum deconvolution procedure. La to Gd data representative of an allanite, Tb to Lu representative of a xenotime mineral analysis.

Table 10.19 Rare-earth mineral analysis by electron microprobe

Element	Allanite (Exley, 1980)		Apatite (Exley and Smith, 1982)	Monazite (Styles and Young, 1983)
	Wavelength dispersive 20 kV, 50 nA		Wavelength dispersive 25 kV, 4 μA	Energy dispersive 15 kV, 5 nA, 100 s count time
	Allanite 1	Allanite 2		
SiO_2	31.54	37.21	1.03	
Al_2O_3	13.32	23.45		
TiO_2	1.51	0.04		
Fe_2O_3	4.76	6.34		
MgO	0.25	0.23		
FeO	12.58	4.98		
MnO	0.06	1.27		
CaO	10.65	18.20	46.3	
P_2O_5			37.2	28.87
F			1.80	
Cl			0.58	
SrO	0.01	0.04	4.405	
ThO_2	0.33	0.20		1.28
Y_2O_3	0.40	3.20	0.039	
La_2O_3	6.29	0.17	0.797	16.63
Ce_2O_3	12.53	0.24	1.494	38.49
Pr_2O_3	1.06	nd		3.68
Nd_2O_3	2.94	0.08	0.515	9.69
Sm_2O_3	0.34	0.06	0.044	0.41
Gd_2O_3	0.15	0.51	0.025	
Dy_2O_3	nd	0.43		
Er_2O_3	nd	0.05		
Yb_2O_3	nd	0.24		
Total	98.72	96.94	94.76	99.05

Concentration data in % oxide.
Fe^{II} and Fe^{III} abundances in allanite were apportioned by Exley (1980) on ion charge balance criteria.
nd = not detected.

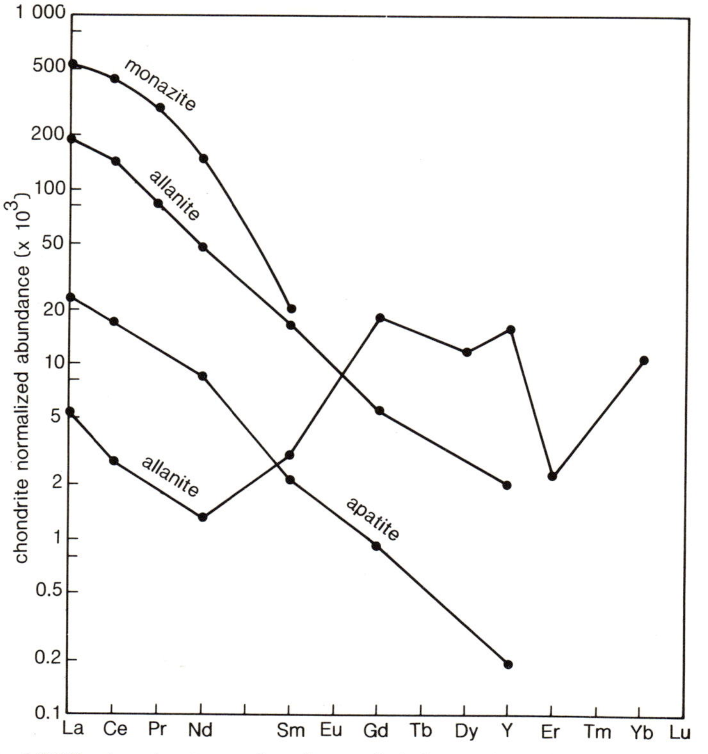

Figure 10.53 Chondrite normalized REE mineral patterns plotted from analyses listed in Table 10.19. Yttrium is plotted in place of holmium, using a chondrite normalizing factor of 2.0 ppm.

Notwithstanding these reservations, good agreement has been obtained by several workers in the analysis of glass beads prepared from reference materials. To illustrate the scope of such analyses, two sets of data for basalt USGS BCR-1, taken from Brown (1977) and Foland and Wagner (1981), are listed in Table 10.20.

10.18 The SEM as a microprobe

Scanning electron microscopes are primarily designed to record high-resolution backscattered and secondary electron images from samples that are normally three-dimensional, not polished flat. However, x-ray spectrometers are often fitted to these instruments (both ED and WD) as bolt-on extras, normally to permit rapid qualitative analysis of samples for mineral recognition, for example. Such spectrometers may be used for semi-quantitative analysis as well, and in this section, the limitations of this mode of operation are considered further.

10.18.1 *Geometric considerations*

One important limitation is in maintaining reproducible beam–sample– detector geometry. The specimen stage of an SEM is considerably more flexible than that of a microprobe, and is designed to permit continuous orientation of a sample, both in a rotational and angular sense. It is common to position the surface of a sample at an angle of about 45° to the electron beam with a similar take-off angle used for the x-ray spectrometer. However, as this plane may not pass directly through the detector, it is necessary to invoke three-dimensional geometry to calculate the true spectrometer take-off angle (ψ) required for accurate absorption and fluorescence correction calculations (Statham, 1981), as follows:

$$\operatorname{cosec}\psi = (\sin T . \cos E . \cos A + \cos T . \sin E)^{-1} \quad (10.48)$$

where T is the angle of tilt of the electron beam, relative to normal incidence, E is the detector elevation angle relative to a plane perpendicular to the electron beam, and A is the detector azimuthal angle relative to the direction of tilt. These angular relationships are shown in Figure 10.54.

Positional reproducibility of sample and standard minerals is probably the limiting factor in quantitative SEM analysis, particularly as such instruments are not normally fitted with an optical microscope to facilitate adjustment of the analysed surface to a reference plane. Some success may be achieved by careful centring of the gun filament (to prevent sideways drift of the electron beam) and by focusing samples in SEM imaging mode by adjusting the mechanical z-axis control of the stage rather than altering the final lens current. In routine applications, it is more common to apply some sort of normalization to the analytical results. It is not then possible to use the summation to 100% as a criterion for the reliability of data.

10.18.2 *Quantitative analysis using peak to background ratios*

One method of minimizing the effect of the uncertainties described above is to base analyses on the peak to local background ratio rather than the net peak intensities alone.

Table 10.20 Bulk analysis of basalt USGS: BCR-1 by electron microprobe after fusion of the powder

Element	Abbey (1983) Expected	Brown (1977) Analysed	CV $n = 20$	Foland and Wagner (1981) Analysed	CV $n = 5$
SiO_2	54.91	53.00	1.5	55.7	0.9
TiO_2	2.28	2.32	19.6	2.30	3.7
Al_2O_3	13.82	13.34	10.0	13.7	2.0
FeO	12.17	12.36	9.3	12.1	3.5
MnO	0.18	0.19	5.4	0.22	30
MgO	3.50	3.58	9.6	3.66	4.7
CaO	7.02	7.19	4.1	7.13	1.5
Na_2O	3.32	3.17	3.5	3.26	8.1
K_2O	1.71	1.65	5.7	1.69	5.6
P_2O_5	0.36	0.38	12.3		
total	99.27	97.18		99.76	

Data listed in % weight oxide.
CV = coefficient of variation = 100 × standard deviation/mean analysis (%).
Expected data for BCR-1 is taken from Abbey (1983) but with compositions corrected for loss of H_2O and CO_2.
Brown *et al* (1977): samples were fused on a molybdenum strip heater at 1450° to 1650°C for about 20 s after a 15 s preheat to 900°C to drive off H_2O and CO_2.
Glass beads were analysed by microprobe at 15 kV, 20 nA using a beam defocused to about 20 μm.
Foland and Wager (1981): samples were wrapped in platinum foil (0.03 mm) thick) and heated in the graphite furnace (of an atomic absorption instrument) at about 1700°C for about 1 minute after preheating to 150°C to remove absorbed water.
Microprobe analyses were carried out at 15 kV, 30 nA using a 20 μm defocused beam.
CV data of Foland and Wagner were obtained from 19 analyses of basalt USGS: BHVO-1 analysed under similar conditions. Data are listed here since the concentration levels in BHVO-1 are similar to those in BCR-1.

The reason is as follows. The most serious errors in SEM analyses arise either due to uncertainties in the precise excitation geometry or because the sample is not a flat polished surface. The calculation of absorption and fluorescence corrections are then subject to inaccuracies due to uncertainties as to the precise x-ray path length within samples. These uncertainties can be reduced significantly by basing determinations on peak to local background ratios since the magnitude of the absorption and fluorescence correction that must be applied to this ratio is significantly smaller than that required to correct net peak areas.

The apparent concentration of an element (C_{APP}) is then given by

$$C_{APP} = \frac{(P/B)_{SAMPLE}}{(P/B)_{STANDARD}} C_{STD} \qquad (10.49)$$

where P/B is the peak to local background ratio of the element.

Apparent concentrations must then be corrected for ZAF effects in the normal way.

This quantification technique has been discussed in some detail by Statham (1979*a*,*b*) and has been successfully applied to the analysis of both bulk SEM samples and particulate material suspended on a suitable substrate.

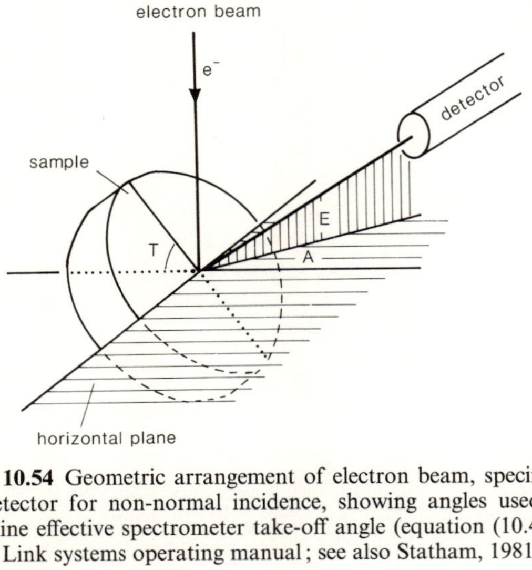

Figure 10.54 Geometric arrangement of electron beam, specimen and detector for non-normal incidence, showing angles used to determine effective spectrometer take-off angle (equation (10.48)). (From Link systems operating manual; see also Statham, 1981.)

10.18.3 *Secondary fluorescence excitation using an SEM*

One advantage offered by the large sample chamber fitted to most SEMs is the possibility of adapting it for excitation by secondary fluorescence. The electron beam is used to excite a metal foil or a metal target (Figure 10.55). Characteristic x-rays excited in this target cause a sample to fluoresce, and its spectrum is measured using an ED detector.

The main advantage of this mode of excitation is the considerable improvement in peak to background ratios observed in comparison with direct electron excitation. Thus, Gould and Healy (1975) were able to detect nickel in stainless steels excited using a gold target in an SEM, whereas this element was not visually apparent in a spectrum recorded by direct electron excitation. Linneman and Reimer (1978) compared in detail spectra of samples excited by direct electron and secondary x-rays and observed large improvements in the detectability of arsenic doped into silicon, for example, by secondary fluorescence, an application also examined by Eckert (1982), who reported arsenic detection limits of better than 1 ppm.

In spite of a large improvement in the detection limits, this excitation geometry suffers the disadvantage of a considerable loss in spatial resolution. Indeed, the technique is not really suitable for micro-analysis in the conventional sense since the analysed area is defined either by the size of the sample itself or the degree of collimation applied to the detector.

10.19 Concluding remarks

Electron microprobe instrumentation has seen dramatic improvements over the past 25 years. These days it is easy to forget that the earliest instruments had doubtful long-term stability and lacked any degree of automation either to move the sample stage and spectrometers or to compute calibrations or matrix effects. To perform routine analyses, users required a degree of endurance and robotic dexterity

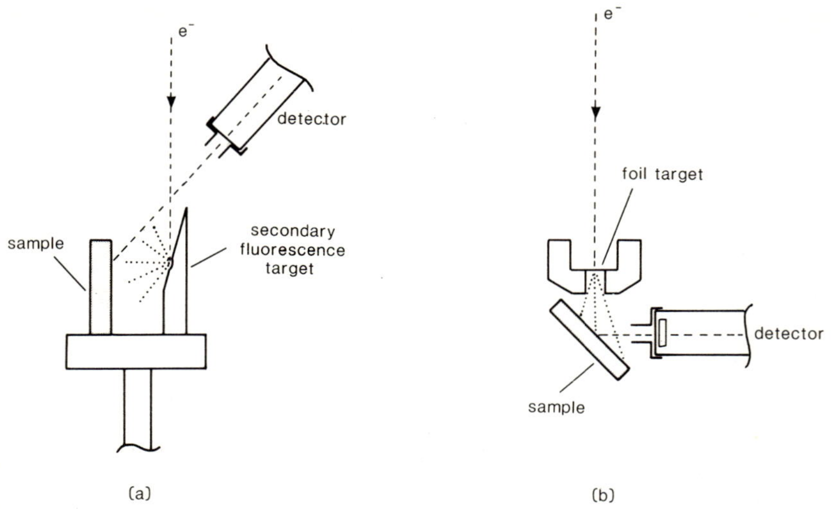

Figure 10.55 Secondary fluorescence excitation in an SEM: (*a*), after Gould and Healey (1975), the secondary fluorescence target and sample mounted on a standard SEM stub; (*b*), from Linnemann and Reimer (1978), in which the target is a metal foil mounted below the final lens of the SEM.

unheard of in modern laboratories. This is a far cry from modern instruments that incorporate a high level of automation such that users, having specified the minerals for analysis, may leave the rest of the analysis sequence to the coordinating computer. Indeed the latest instruments incorporate self-diagnostic error-checking routines that permit fault messages to be transmitted and instruments to be interrogated via a telephone link to a remote terminal in a maintenance engineer's office.

This high technology tends to mask the fact that microprobes afford a unique facility for the chemical analysis of individual minerals, information that was largely unavailable on a routine basis to the geologist before their introduction.

11 Other microbeam and surface analysis techniques

11.1 Introduction

The electron microprobe is by far the most widely used instrument for the analysis of selected areas on the surface of geological samples. However, there are other more specialized techniques that have been applied in more specialized circumstances to geological samples. It is proposed here to describe in outline five of these techniques.

(*a*) *Ion probe*. This is a technique in which a focused beam of ions (normally $^{16}O^-$) is used to sputter material from the surface of samples. Secondary ions so formed are collected and analysed by mass spectrometry. Trace elements down to the ppm level as well as classical isotope ratios may be determined in this way. However, applications are often limited by severe interferences in the mass spectrum unless simple mineral matrices are selected for analysis. Examples of applications include rare-earth element mineral analysis, the analysis of trace elements in selected matrices (e.g. nickel in olivines) and rubidium–strontium and lead–uranium isotope ratio measurements in calcite and zircon matrices respectively.

(*b*) *Laser probe*. Principles of this instrumentation are similar to the ion microprobe but selected areas on samples are excited by a power laser. Practical applications have been much more restricted than for the ion probe, although laser excitation has been particularly useful for fluid inclusion studies. In the latter case, laser excitation has been coupled with spark source optical emission detection.

(*c*) *Particle induced x-ray emission* (*PIXE*). Samples are excited with a beam of energetic nuclear particles, normally protons but sometimes alpha particles. Interaction with selected areas on the surface of samples causes the emission of characteristic x-rays. However, the beam can also initiate nuclear transformations, particularly involving selected lighter elements, permitting sensitive detection and analysis from the accompanying gamma-ray emissions. Detection limits from excited areas down to 5–10 μm in diameter are in the ppm range, significantly lower than can be achieved by electron microprobe excitation.

(*d*) *Transmission electron microscope*. This instrument is normally used either in biological applications for ultra-high-resolution transmission electron photomicrograph studies of thin slices of cell and tissue material, or in metallurgical studies, including, for example, the investigation of defect structures in alloy materials. Subject to some limitations, equivalent studies may be carried out on geological samples. However, samples must be prepared as thin foils (typically 150nm thick). As well as photomicrograph data, quantitative measurements may be derived in two ways. First, crystal structure data may be obtained from the patterns resulting from electron diffraction within the sample. Second, compositional data may be obtained from characteristic x-ray emissions excited by the electron beam as it is transmitted through the foil. Due to uncertainties in the precise foil thickness, it is usual to normalize such analyses to 100%.

(*e*) *Electron spectroscopy for chemical analysis* (*ESCA*). This technique is used to analyse the properties of the surfaces of samples. Low-energy radiation from an x-ray tube (normally Mg $K\alpha$ or Al $K\alpha$) is used to excite the 5 to 10 nm surface layer of a sample. These photon radiations interact by the photoelectric effect causing the emission by ionization of orbital electrons. By measuring the energy spectrum of these electron emissions, information on coordination number, oxidation states and chemical bonding effects may be obtained. The technique has found limited application in elucidating 'uncertain' mineral structures.

In general, all these techniques require expensive and highly specialized instrumentation and their availability is restricted to a limited number of laboratories.

11.2 The ion probe

11.2.1 *Principles of analysis*

The ion probe, like the electron microprobe, is used to analyse selected areas on the surface of a sample. However, the principles of excitation and detection are quite different. The probe itself is formed from a focused beam of heavy ions, $^{16}O^-$ being the most widely used species. The ion beam sputters away the surface of the sample on which it is focused, fragments of material being ejected as a complex mixture of ionic and molecular species. The ionic species are collected by electrostatic plates and are analysed and detected by mass-spectrometry procedures. The ion probe therefore belongs to a group of instruments known as secondary ion mass spectrometers (SIMS), and offers the unique advantage of being capable of determining isotope ratios of specific elements in individual minerals. Furthermore, the ion probe is well able to determine the concentration of some elements down to the ppm level—far beyond the capabilities of the conventional electron microprobe where detection limits are generally no better than 100 ppm. Instrumental effects and geochemical applications have been reviewed in detail by Shimizu and Hart (1982).

11.2.2 *Instrumentation and specimen requirements*

The overall instrumental arrangement of an ion probe is shown schematically in Figure 11.1. The ion probe itself comprises a beam of ions accelerated to between 15 and 30 keV and delivered to the sample at a beam current normally of 10–50 nA. The choice of ion depends on application, as discussed below, and although $^{16}O^-$ is the most widely-used species, Ar^+, $O_2{}^+$, and Cs^+ also have more specialized uses. The source of gaseous ions is the duoplasmatron in which the gas (oxygen or argon) is leaked into a hollow cathode at a pressure of about 10 Pa (0.1 Torr). An arc discharge is

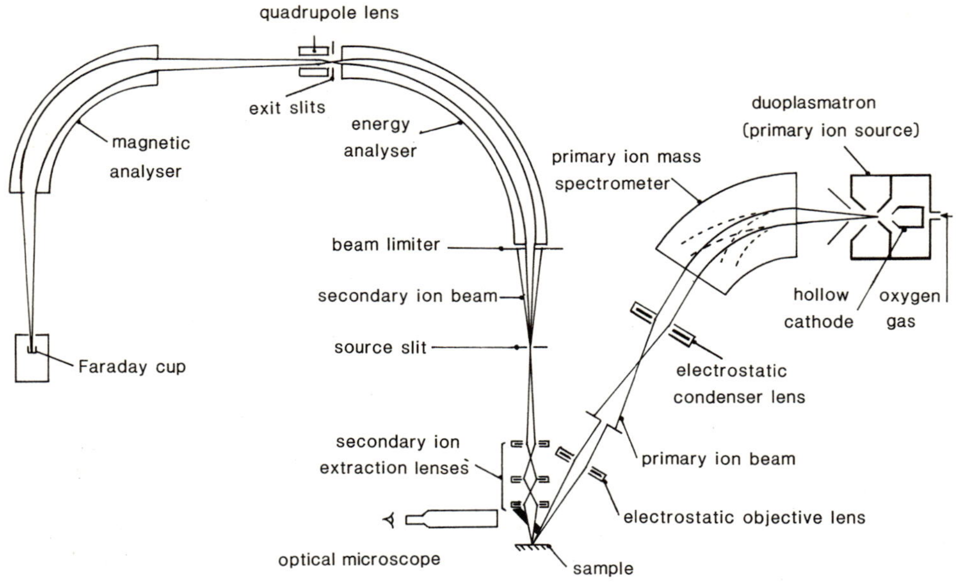

Figure 11.1 The ion probe showing schematic arrangement of primary ion source and secondary ion mass spectrometer. The components illustrated here may not all be found on the same instrument. The primary ion source with mass analyser is based on the design described by Banner and Stimpson (1974), whereas the secondary ion mass spectrometer is based on the design described by Compston *et al.* (1982).

created within the hollow cathode and the resultant ions are accelerated across a potential gradient (15 to 30 kV). On some instruments it is desirable for this primary beam of ions to be mass-filtered before exciting the sample, to remove contaminants. A mass spectrometer of low mass resolution (about 100 atomic mass units) is required to achieve this, and will adequately remove both $^{18}O^-$ (0.204% relative abundance) and hydrogen and carbon contaminants (from the vacuum system) to leave an isotopically pure $^{16}O^-$ ion beam. Any interferences in the secondary ion beam emitted from the sample are then restricted to those originating from ^{16}O.

The resultant ion beam is focused on to the surface of the sample using electrostatic lenses. It is not normally possible to focus this beam to the same degree as an electron probe, the minimum ion probe diameter normally being in the range 1–5 μm, although many reported analyses are undertaken with a 10–20 μm beam size.

Geological samples are normally analysed as polished thin sections or blocks which must be coated with a conducting layer of carbon (or alternatively gold) 20–30 nm thick. This conducting layer does not serve the same function as the carbon coating in electron probe excitation, since within a few seconds of ion beam excitation the surface film over the analysed area is completely eroded away. The main function of this conducting layer in ion probe analysis is to ensure that the sample offers a surface of equal potential with respect to the secondary ion extraction electrode to facilitate stable collection of secondary ion signals. There is not, therefore, the same concern about absorption of signals in the conducting coat as is the case in electron probe microanalysis, providing the sample has been sputtered for a short time before analysis to remove the surface coating and allow the secondary ion emission signal to attain equilibrium.

11.2.3 *Interaction of the ion beam with the sample*

An understanding of the excitation mechanism in ion probes stems mainly from the work of Andersen (1969, 1970). Heavy ions cannot penetrate into the sample as deeply as electrons, but instead deposit their energy within the surface layers of the sample, perhaps up to 10 nm deep. A large amount of energy is concentrated in this excited region, corresponding to about 10^7–10^8 W cm^{-2}. The kinetic energy transferred from the ion beam to the sample material is in fact so large that condensed plasma conditions are created within this excited volume causing localized temperatures equivalent to 6000–15 000 K. Under these circumstances, molecular and atomic fragments are formed within the plasma region as uncharged, singly-charged and multiply-charged species. As summarized by Robinson (1973), some of these ions will escape from the plasma, depending both on the work function of the surface barrier through which they must pass and their kinetic energy. The probability that a positively-charged ion of low kinetic energy will recombine with an electron as it passes through the barrier region is greater than for an ion of higher kinetic energy. As ion beam excitation continues, the surface layers are progressively eroded away at a rate estimated to be a few nanometres per second. Prolonged excitation will therefore pit the sample surface. Advantage may be taken of this phenomenon in some applications where it is necessary to measure variations in concentration with depth within the surface layers of a sample. In homogeneous samples, the intensity of emitted ion species is likely to change if the eroded hole is allowed to become too deep.

Neutral species eroded from the sample cannot be analysed by mass spectrometry. Ion species are collected and analysed by a medium- to high-resolution mass spectrometer. The intensity of individual ion species depends on the secondary ion yield of the element in question. This in turn is affected by the ion beam used to bombard the sample. Oxygen ($^{16}O^-$) is the most widely used primary ion, since many elements have high secondary ion yields with respect to this ion. One reason for this is that oxygen ions are electronegative, and they therefore reduce the availability of electrons within the plasma region that could otherwise recombine with positively charged ions from the sample, so reducing their yield. Although much of the early development work was carried out with argon primary ion beams, this gas is no longer widely used because the corresponding ion beam does not suppress the recombination of secondary ions with free electrons to the same degree as oxygen. The principal application of Ar^+ ion beam excitation is when oxygen isotopes are themselves to be determined within the sample. A comparison between the $^{27}Al^+$ ion intensity from aluminium metal excited by $^{40}Ar^+$ and $^{16}O_2^+$ ion beams is shown in Figure 11.2. An exponential reduction is observed in the signal excited by the $^{40}Ar^+$ beam as the ability to continue extracting Al^+ ions is progressively reduced.

As an indication of sensitivity, the relative positive secondary ion yields of selected elements after bombardment by oxygen (O^-) are plotted as a function of atomic number in Figure 11.3. These data show that elements that possess low ionization potentials (for example, the alkali and alkaline earth metals) have high secondary ion yields in contrast with elements such as O, C, F, S, Cl, P, As and Au, which are not efficiently excited since they possess large ionization potentials. However these latter elements also possess high electron affinities, that is a strong tendency to accept an electron to form a negatively-charged ion. If the species that forms the ion probe beam is replaced by an element with a high potential to donate electrons, these elements may be determined with enhanced sensitivity. An ion beam consisting of Cs^+ is the most widely-used means of achieving this objective, and Deline *et al.* (1980) have described the application of such beams to the determination of carbon and oxygen in electronic materials.

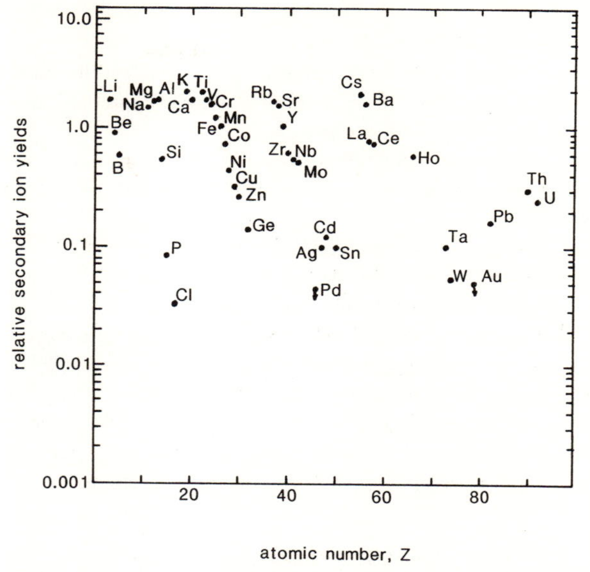

Figure 11.3 Relative yields of positive secondary ions of different elements in a refractory oxide matrix. (McHugh, 1975.) Precise data will vary with instrument design.

11.2.4 *Interferences in the secondary ion beam*

Samples must be excited under very high vacuum (normally better than 10^{-7} Torr). However, all vacuum systems inevitably contain residual water vapour and hydrocarbon molecules at low partial pressures. Although these contaminants can be reduced by inserting a cold finger maintained at liquid nitrogen temperatures in the vicinity of the sample, residual vapours will react with the secondary ion beam. The molecular ionic species then formed can cause interferences in the mass spectrum, the most troublesome normally being metallic hydride ions, which are formed by reaction with the breakdown products of water vapour.

The ion species emitted from the sample will therefore

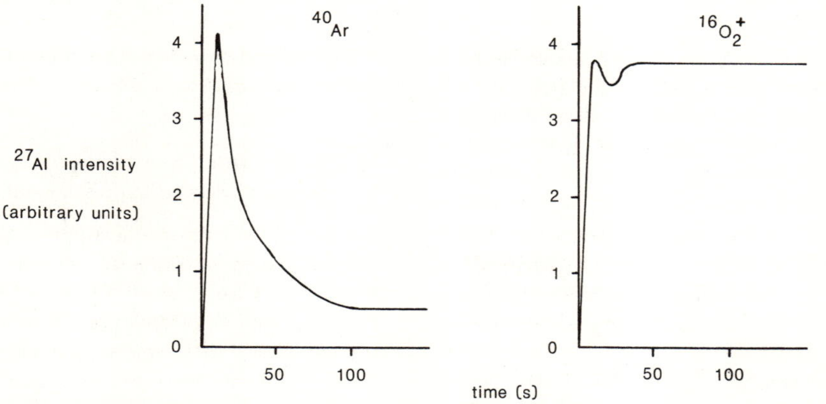

Figure 11.2 Variation in the $^{27}Al^+$ signal with time for aluminium metal samples excited with (*a*) $^{40}Ar^+$ and (*b*) $^{16}O_2^+$ primary ion beams. The emission of this positively-charged ion is stabilized by using a probe made up of electronegative gas ions (oxygen). (Andersen, 1969.)

consist of:
(i) Monatomic ions (usually the desired signal)
(ii) Element oxide ions (resulting from the recombination with O^- ions from the primary ion beam or element oxide fragments from the sample)
(iii) Molecular ionic species resulting from sputtered molecular fragments or recombination reactions within the plasma region
(iv) Molecular ionic species formed by reaction of any of the above with water or hydrocarbon fragments present as contaminants within the vacuum system.

It is important to note that in direct contrast with electron microprobe analysis, the ion probe signal is essentially free of a background component and the technique is therefore theoretically capable of low detection limits. The extent to which this can be realized depends on how successfully the ion interference effects (ii) to (iv) can be overcome.

11.2.5 *Correction of mass interferences*

Interferences in the secondary ion mass spectrum may be eliminated in two ways, by energy filtering and by selecting the appropriate mass spectrometer resolution.

(*a*) *Energy filtering*. Most secondary ions are emitted from the sample with kinetic energies of less than 20 eV, there being a rapid fall-off in the number of ions with higher energies. However, if this fall-off is examined more closely, it is observed that rate of decrease with kinetic energy of monatomic ions (the desired signal) is considerably lower than that of molecular ions (Herzog *et al.*, 1967), as can be seen in Figure 11.4. If, therefore, the ions permitted to enter the mass spectrometer are restricted to those of kinetic energy in excess of say 150 eV, interferences from molecular species can be considerably reduced and in some cases virtually eliminated (Shimizu *et al.*, 1978). The disadvantage of this technique is that energy filtering is accompanied by a very large reduction in the intensity (often by a factor of 100) of the signal entering the mass analyser, to an extent that may not be acceptable for trace element analysis.

(*b*) *Mass analyser resolution*. Residual interferences in the mass spectrum are due to:
(i) The presence of isobaric isotopes within the sample, that is ions of different elements having identical nominal atomic weights
(ii) Molecular species having the same nominal atomic weight as the isotope of interest.

In fact, as is discussed in section 16.13, there are small mass differences between these species due to the mass defect effect. Many interferences can, therefore, be resolved if a spectrometer with sufficiently high mass resolution is employed. However, it is then usually necessary to make some compromise, since the higher the mass resolution, the lower the ion beam intensity transmitted through the mass analyser. In this context, resolution is normally defined as atomic mass divided by the width of the mass peak (in atomic mass units) at 10% of its height, and is represented by $M/\Delta M$.

Two strategies are usually adopted, the choice being affected by the specification of the instrumentation available. Mass spectrometers with resolutions of about 200 ($M/\Delta M$) are capable of resolving adjacent integral mass peaks and give a flat-topped stable signal. Mass defect effects cannot be resolved, and in many cases it is necessary to calculate the intensity of overlap corrections from the ratio of the interfering element to that at an adjacent interference-free mass peak.

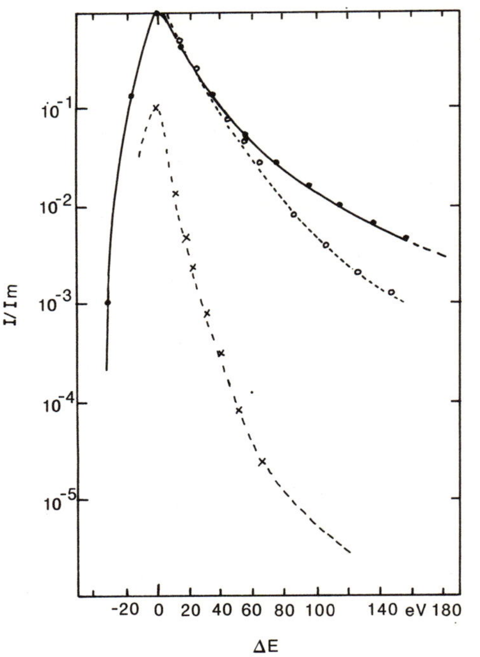

Figure 11.4 Distribution of the kinetic energy of secondary ion species. The intensity of the selected secondary ion (normalized to the maximum) is plotted as a function of excess energy relative to the ion extraction acceleration potential. Solid circles and lines represent data for mass 28^+ (the ion species $^{28}Si^+$) in olivine; crosses and the dotted line represent mass 44^+ (predominantly the molecular species $^{28}Si^{16}O^+$) in olivine; open circles represent mass 28^+ ($^{28}Si^+$) in silicon metal. Note that the intensity of the molecular species rapidly falls with increasing kinetic energy. (Shimizu, 1978.)

Mass spectrometers having resolutions of several thousands ($M/\Delta M$) are capable of resolving many (though not all) overlap interferences at identical nominal mass numbers. However, such performance can be achieved only by reducing the width of the mass analyser entrance slits. This can cause a serious reduction in transmitted ion intensities, as shown in Figure 11.5.

The different strategies are illustrated by the applications of ion probe analysis described below.

11.2.6 *Quantitative analysis by ion probe: matrix interferences, local thermal equilibrium model, secondary ion yields*

As in electron microprobe analysis, ion probe data are subject to matrix interference effects, that is, the secondary ion yield of an element varies with sample composition. However, unlike electron probe data, the theoretical basis for correcting these interferences is not fully understood. The only procedure to find widespread acceptance attempts to model the plasma conditions that exist in the region of interaction between ion beam and sample in terms of *local thermal equilibrium* (Andersen and Hinthorne, 1973). This

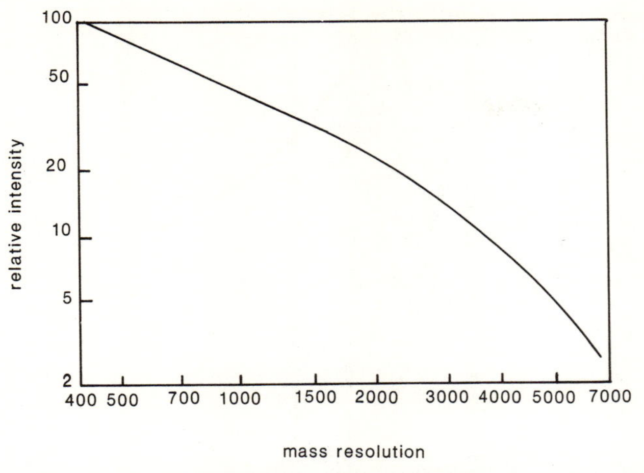

Figure 11.5 Variation in intensity of the secondary ion beam as a function of selected resolution. Higher mass resolutions, obtained by reducing the width of the spectrometer entrance slits, cause a progressive reduction in detected intensity. (Reed *et al.*, 1976.)

model is partially successful in accounting for interelement effects on secondary ion yields, and its mathematical derivation has been reviewed by Reed (1981) and Beusen *et al.* (1983).

From a practical point of view, it is essential to eliminate the effect of instrumental drift in measurements, so that satisfactory analytical precision may be achieved in determinations that may, in some cases, exceed 30 minutes per element. The most convenient way of satisfying this aim is to record the intensity of the appropriate mass number relative to a matrix element of known composition. Data are then accumulated by successive cycles of counts on the mass peaks of analysed and matrix element until the appropriate precision has been attained.

The atomic concentration of the element to be analysed (C_A) relative to that of the matrix element (C_B) is then given by

$$\frac{C_A}{C_B} = \frac{n_A}{n_B}\frac{a_B}{a_A}\frac{\varepsilon_B}{\varepsilon_A} \qquad (11.1)$$

where n is the ion count intensity at the appropriate mass peak, a is the natural isotopic abundance, and ε is the secondary ion yield: that is the fraction of atoms of the element sputtered in ionic form.

Providing non-radiogenic isotopes are being analysed, the value of the natural isotopic abundance can be looked up in tables. However, the secondary ion yield, being matrix dependent, will vary from sample to sample. Thus Hendry *et al.* (1981) found that the ion yield of ^{63}Cu relative to ^{54}Fe depended mainly on the FeO content of the source mineral. The ratio $\varepsilon_{Cu}/\varepsilon_{Fe}$ varied from 0.2 at 2.5% FeO to 0.6 at 20% FeO. Reed (1983) investigated the secondary ion yields of the rare-earth elements and found significant variations in the REE monoxide to REE atomic ion intensity ratios that correlated quite closely with the REE-monoxide bond energy.

The only practical solution to this matrix effect is to calibrate the secondary ion yields of elements of interest using 'standard' minerals of composition similar to the samples being analysed. In view of the scarcity of well characterized materials, it is often necessary to synthesize glasses of suitable composition (Hendry *et al.* 1981), precalibrate suitable minerals by independent electron microprobe analysis (Reed and Enright, 1981) or (exceptionally) to prepare reference samples by ion implantation techniques (Wilson and Long, 1983).

Some specific applications of ion probes in the determination of trace elements have been reviewed by Reed (1980) and selected examples will now be considered.

11.2.7 *Geological applications*

(*a*) *Determination of copper in mafic minerals.* As an example of quantitative analysis by ion probe, Hendry *et al.* (1980) determined copper in biotites, amphiboles and magnetites at concentration levels in the range 6 to 880 ppm. The isotope determined, ^{63}Cu (69.09% natural abundance), suffered an interference from an unspecified molecular peak which could only be partially resolved using a moderately high mass resolution ($M/\Delta M = 2000$). The ion probe consisted of a 15 nA beam of $^{16}O^-$ ions focused to a 15 μm beam diameter. Statistical detection limits were estimated to be about 1ppm copper for a total count time of 20 minutes, using a curve-fitting procedure to subtract the adjacent molecular interference (Figure 11.6). This should be compared with an electron microprobe detection limit of about 100 ppm for copper. All ^{63}Cu count data were normalized to the intensity of the ^{54}Fe peak to eliminate instrumental drift, and secondary ion yield matrix effects were calibrated using synthetic glasses containing known concentrations of Cu and Fe and covering a wide range of silicate compositions.

(*b*) *Rare-earth element analysis by ion probe.* A particular area of interest is the analysis of the rare-earth elements in minerals such as biotite and apatite at concentrations below the capability of the electron microprobe. Reed (1981, 1983, 1984) and Reed and Smith (1985) have discussed analytical problems with particular reference to overlap interferences and ion yield variations.

A particular problem in the determination of these elements is the presence of REE monoxide peaks in the mass spectra, which overlap the REE ion peaks of gadolinium and above. These can clearly be seen in an ion probe spectrum of an allanite mineral grain (Figure 11.7). Reed (1981) suggested that the six lightest rare-earth elements (La, Ce, Pr, Nd, Sm and Eu) may be determined from atomic peaks, but that monoxide peaks are preferred for the remainder. However, the monoxides of Gd and Tb coincide with atomic Yb and Lu and are, therefore, difficult to analyse satisfactorily by secondary ion mass spectrometry. In the analysis of apatite and merrillite, Reed *et al.* (1983) were unable to determine Tb, Yb and Lu due to excessive mass interference effects, and reported that molecular interferences from clusters of major element atoms could occur at any mass number and were difficult to predict. These workers therefore analysed individual elements at low mass resolution ($M/\Delta M = 400$) and applied corrections based on the relative intensity of molecular and isobaric interferences identified and precalibrated at high mass resolution ($M/\Delta M = 6000$). Chondrite normalized abundances in the range 30 to 200 times were measured from merrillite and apatite.

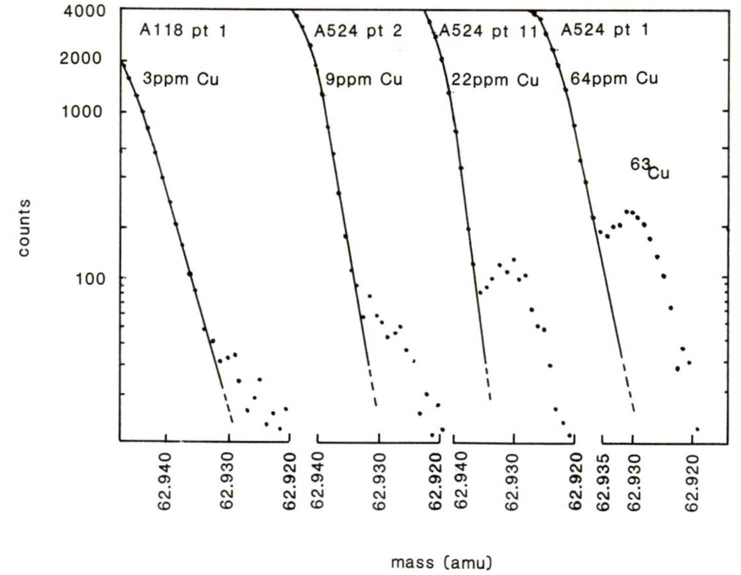

Figure 11.6 Scans through the mass-63 peaks for amphiboles of varying copper content showing the ^{63}Cu peak overlapped by a more intense molecular peak interference. (Hendry *et al.*, 1981.)

(*c*) *The determination of* ^{87}Sr / ^{86}Sr *isotope ratios by ion probe.* Exley and Jones (1983) and Exley (1983) have discussed the determination of strontium isotope ratios in a simple carbonate matrix where significant interferences in mass spectra were likely to originate from the elements Ca, Mg and O only. Samples were excited with a mass analysed beam of $^{16}O^-$ ions, at 20 kV, 10 nA, focused to a spot size of about 10 μm. Secondary ion measurements were undertaken at low mass resolution ($M/\Delta M = 200$), sufficient just to separate integral mass peaks. The mass peak due to ^{87}Sr suffered a mass overlap from ^{87}Rb (27.85% isotopic abundance), an interference not normally encountered in the determination of Rb/Sr ratios in bulk samples since mass analysis is there preceded by a chemical separation. The relative intensity of this interference could be determined by measuring the ratio relative to the ^{85}Rb mass peak (72.15% isotopic abundance). However, each of the mass numbers for which determination was required (86, 87 and 88) suffered interferences from molecular species due to Ca^{2+} and/or $CaMgO^+$ (Table 11.1). The magnitude of these interferences could conveniently be estimated by monitoring the species $(^{40}Ca)^{2+}$ and $(^{40}Ca^{25}Mg^{16}O)^+$ at mass numbers 80 and 81, respectively. A further difficulty was that the measured ratio ^{87}Sr/^{86}Sr must be corrected for small matrix-dependent differences in the secondary ion yields of these isotopes. It was necessary, therefore, to measure the isotopic ratio ^{88}Sr/^{86}Sr in each sample and compare data with the natural (and constant) value of 0.1194. Exley (1983) reported the isotope analysis for 10 μm spots selected on calcite grains in a hydrothermally altered basalt from the Isle of Skye, NW Scotland and of very small (<35 μm) calcite grains in a veined mantle lherzolite from South Africa to exemplify the potential application of this technique for strontium isotope ratio measurement on a micro scale. Exley and Jones (1983) applied the technique to calcite grains in a zoned kimberlite dyke from Kimberley, South Africa and used data to evaluate the role of hydrothermal alteration in determining the ^{87}Sr/^{86}Sr composition of kimberlite and the scale of crustal contamination processes. Selected data for the ion probe determination of ^{87}Sr/^{86}Sr ratios in three reference samples taken from Exley (1983) are compared with bulk solid source measurements on the same materials in Table 11.2.

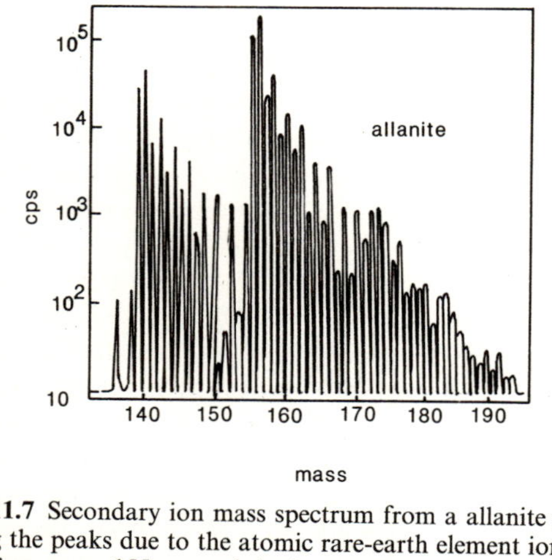

Figure 11.7 Secondary ion mass spectrum from a allanite mineral showing the peaks due to the atomic rare-earth element ions overlapped from mass 155 upwards by intense REE-oxide lines. (Reed, 1984.)

(*d*) *Lead isotope ratio measurements.* Lead isotope ratio determinations for individual minerals in thin section represent an important application of ion probe microanalysis. The desired signal is usually the ratio of ^{207}Pb to ^{206}Pb as these isotopes are the end members of the radioactive decay chains of ^{235}U and ^{238}U, respectively. By measurement of this isotope ratio, together with the abundance of the uranium parent, estimates of the age of the mineral can be made (section 17.5). One of the minerals most widely studied in

Table 11.1 Mass interferences in the ion probe determination of strontium isotope ratios in calcite/dolomite

Mass	Peaks present in calcite/dolomite	Intensity (cps)	Application in peak stripping	Other possible peaks considered negligible
80	$^{40}Ca_2$	80 000	Corrects 88 for $^{48}Ca^{40}Ca$	$^{24}Mg^{56}Fe$
	$^{40}Ca^{24}Mg^{16}O$	3100 (1%MgO)	and $^{44}Ca_2$	
81	$^{40}Ca^{25}Mg^{16}O$	400 (1%MgO)	Corrects 80, 86, 88	$^{24}Mg^{57}Fe$
			for CaMgO species	$^{26}Mg^{55}Mn$
85	^{85}Rb		check for ^{87}Rb	$^{42}Ca^{43}Ca$
			(could be ignored)	
	$^{44}Ca^{25}Mg^{16}O$	9 (1%MgO)		
86	^{86}Sr	3.5 (ppm^{-1})		$^{42}Ca^{44}Ca$,$^{43}Ca_2$
	$^{44}Ca^{26}Mg^{16}O$	9 (1%MgO)		$^{54}Fe^{16}O_2$
87	^{87}Sr	2.5 (ppm^{-1})		$^{55}Mn^{16}O_2$
	^{87}Rb			^{87}Rb
				$^{43}Ca^{44}Ca$
88	^{88}Sr	30 (ppm^{-1})		$^{56}Fe^{16}O_2$
	$^{48}Ca^{40}Ca$,$^{44}Ca_2$	180		
	$^{48}Ca^{24}Mg^{16}O$	6 (1%MgO)		

Data from Exley (1983).

lead isotope ratio measurements by ion probe is zircon (Zr,Hf)SiO_4. This mineral offers several advantages:

(i) Significant concentrations of uranium can be incorporated into zircon, a mineral commonly found in several rock types.

(ii) The mineral is relatively resistant to the loss or gain of U or Pb after crystallization adding to the confidence with which analytical results may be interpreted.

(iii) The matrix is relatively simple, permitting reliable prediction and correction of isobaric interferences in mass spectra.

Nevertheless, Hinton and Long (1979) reported that up to 200 potential molecular interferences could occur in the mass range of interest (202 to 210 atomic mass units) the most important being due to the species Zr_2O^+, Zr_2OH^+, $HfO_2{}^+$, $ZrSi^+$, $HfSi^+$ (Figure 11.8). Furthermore as zircon can contain significant concentrations of the rare-earth elements and yttrium, the $Y_2O_3{}^+$ interference at mass number 210, which is unresolved from Zr_2Si^+, must also be taken into account.

Table 11.2 Comparison between ion probe measurements of $^{47}Sr/^{86}Sr$ with independent bulk thermal ionization source mass spectrometry data

	Calcite vein in altered basalt	Magnesium calcite in carbonatite A	Magnesium calcite in carbonatite B
Individual ion probe analyses ($^{16}O_2^-$, 20 kV) 10 μm probe diameter)	0.7273±27	0.7054±12	0.7056±11
	0.7263±40	0.7059±11	0.7062±9
	0.7277±36		0.7051±9
	0.7288±50		
	0.7256±30		
	0.7255±25		
Weighted mean	0.7267±15	0.7056±9	0.7057±6
Bulk data by solid source mass spectrometry	0.72667±8	0.70588±9	0.70542±9
	0.72680±8	0.70587±9	0.70534±7

Data from Exley (1983).
Individual ion probe measurements are based on an average of 100 scans (minimum: 50, maximum: 150).
Error limits are two-sigma mean, quoted data being the uncertainty in the last quoted figure(s)).

The presence of hydride interferences is also of concern as these cannot normally be resolved from lead ion peaks even at the highest instrument resolutions. The magnitude of such interferences is strongly dependent both on the partial pressure of water vapour remaining in the sample chamber and on the presence of hydrated alteration products in the mineral itself. In order to minimize these problems Hart *et al.* (1981) cleaned each spot before analysis using a 200 μm defocused $^{16}O^-$ ion beam for two minutes. The beam diameter was then reduced to 100 μm and an aperture inserted in the secondary ion optics so that only the central 65 μm region was analysed. Under these conditions, the hydride interference of $^{206}PbH^+$ on $^{207}Pb^+$ was reported to be negligible in the analysis of galenas.

Analytical strategy adopted for the measurement of lead isotope ratios is very dependent on the specification of the mass spectrometer available for analysis. Thus Hinthorne *et al.* (1979) reported that a mass resolution of 1250 is sufficient to separate ^{206}Pb from Zr_2O, but that 4800 is required to separate ^{207}Pb from HfO_2 and 5450 to separate ^{207}Pb from HfO_2H. They suggested that when the concentration of lead is sufficiently high (~5000 ppm Pb), then any loss in sensivity incurred by operating at high resolution is not critical, but the magnitude of the interferences is likely to be negligible compared with the intensity of the lead peaks. High reslution is not then required. Conversely, when lead is present at low abundances, the reduction in sensitivity in high-resolution mode cannot be tolerated. They therefore considered that operation in low-resolution mode ($M/\Delta M = 250$) represented the best general method for separating interferences of lead in the secondary ion spectrum of zircons. Hinthorne *et al.* measured under these

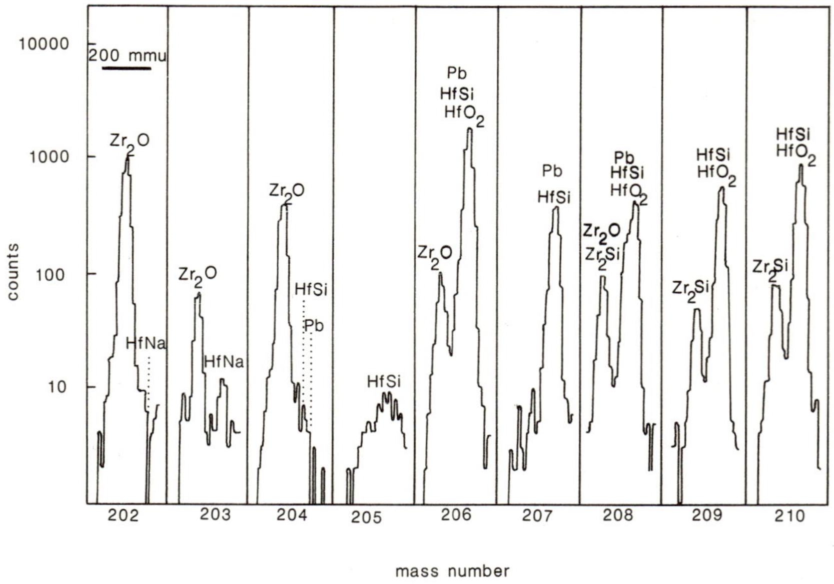

Figure 11.8 Mass spectrum of zircon in the mass range 202–210 showing lead atomic species overlapped by various Zr–O–Hf–Si molecular species. Data were accumulated by multiple-pass step scanning for a count period of about 5 hours at a mass resolution of about 1000. (Hinton and Long, 1979.)

conditions nine mass peaks in the range 202 to 210 atomic mass units corresponding to combinations of nine ion species (Zr_2O, Zr_2OH, HfSi, ^{204}Pb, HfO_2, ^{206}Pb, HfO_2H, ^{207}Pb and ^{208}Pb) (Table 11.3). By properly ordering the corrections, interferences on lead mass peaks could be subtracted.

In contrast, Hinton and Long (1979) operated their ion probe under conditions that permitted substantial resolution of many of the isobaric interferences. The magnitude of residual interferences may be judged from data in Table 11.3 A mass resolution of about 3200 was the highest that could be selected without suffering an unacceptable loss in sensitivity. Many of the interferences were then resolved, residual

Table 11.3 Mass interferences in the determination of lead isotope ratios in zircon by ion probe

Mass number	Pb		Zr_2O	Zr_2OH	HfSi	HfO_2	HfO_2H	Zr_2Si
202			9700	8	—			
203			690	19	—			
204	^{204}Pb	5	4400	1	24			
205			4	9	88			
206	^{206}Pb	17000	1100		130	60		
207	^{207}Pb	3800		2	74		—	
208	^{208}Pb	3100	88		170	1700		1200
209				—	11	6100	—	580
210					6	8900	—	920

Data from Hinton and Long (1979) and Hinthorne *et al* (1979).
Data represent the integrated counts for a 3-hour run for a zircon with 450ppm Pb and about 1% Hf taken from Hinton and Long; instrument conditions: $^{16}O^-$ beam operated at about 15 nA and normally defocused to 25 μm. The spectrometer resolution was set to $M/\Delta M = 3200$.
The presence of additional interferences reported by Hinthorne *et al.* is indicated by a dash. Their mass spectrometer was operated at low resolution ($M/\Delta M = 250$) and excited using a mass analysed $^{16}O^-$ beam.

overlap still being observed between HfO_2 and ^{208}Pb. This was removed by deconvoluting this doublet using the peak shape defined by ^{206}Pb. Other minor interferences were dealt with in this way. Detection limits for ^{204}Pb for measurements made over a three-hour run were about 0.2 ppm lead (three-sigma error on background).

Froude *et al.* (1983) reported extremely high zircon ages (4100–4200 My) using a new instrument which could operate at high mass resolution ($M/\Delta M = 6\,500$). Their mass spectrometer, which incorporated wide entrance slits, was physically large, permitting the transmission of high intensities of secondary ions (Compston *et al.*, 1982). Under these resolution conditions, almost all molecular interferences could be resolved except lead hydride overlaps on atomic lead lines.

Ion probe lead isotope data must be corrected for the presence of common (or non-radiogenic) lead using the ^{204}Pb mass number and related to the analysed composition of the parent ^{238}U, ^{235}U (or ^{232}Th if the $^{208}Pb/^{232}Th$ decay sequence is to be used). Details of these procedures are discussed in section 17.5. It is normally assumed that mass fractionation between various lead isotopes (that is the variation in secondary ion yields of specific isotopes) is negligible. However, Hart *et al.* (1981) reported that the sputtering region was inhomogeneous with respect to isotope ratio and that careful alignment of the primary beam spot with the secondary ion optics was important to minimize this effect. Fractionation effects could be as high as 0.5% per atomic mass unit. In their study of lead isotope ratios of galena crystals, Hart *et al.* (1981) compared some ion probe determinations with 'absolute' values from Stacey *et al.* (1969) measured by triple filament thermal ionization source mass spectrometry. These results are compared in Table 11.4 to indicate the precision of analytical data that may be obtained by ion probe. Hart *et al.* (1981) considered that with further attention to instru-

Table 11.4 Lead isotope ratios in a galena (Broken Hill) determined by ion probe

Isotope ratio	Ion probe analyses			Bulk mass spectrometry	Difference
	Analysis 1	Analysis 2	Average		
$^{208}Pb/^{206}Pb$	2.2274 ± 0.02%	2.303 ± 0.06%	2.2288	2.2287	0.004%
$^{207}Pb/^{206}Pb$	0.9606 ± 0.02%	0.9611 ± 0.06%	0.9608	0.9619	−0.11%
$^{206}Pb/^{204}Pb$	16.003 ± 0.04%	16.018 ± 0.09%	16.010	16.007	0.019%
$^{207}Pb/^{204}Pb$	15.372	15.395	15.384	15.397	−0.084%

Data from Hart *et al* (1981).
Ion probe data obtained using a $^{16}O^-$ probe operated at 50 nA using the 65 μm region of a 100 μm probe beam. Data were acquired for 30 to 60 minutes per sample to obtain the desired precision, and mass fractionation effects were corrected by analysing NBS lead reference materials SRM 981 and 982. The mass fractionation correction applied to the data presented here was 0.25% per mass unit.
Ion probe data are compared with triple filament lead determinations on bulk samples by Stacey *et al.* (1969).

mental parameters, these precision data could be improved further to give ion probe analyses on galenas that were comparable to data from state-of-the-art mass spectrometry.

11.3 The laser microprobe

The laser microprobe is used to excite selected areas on the surface of polished samples using the photon output from a power laser. On discharge, the laser is capable of delivering up to 10 joules of energy, focused on to an area usually 50 to 100 μm in diameter. The discharge pulse lasts a very short time (about 15 ns) and, therefore, represents a very large transient power input of between 10^9 and 10^{11} W/cm^2. Under these conditions, extremely high transient temperatures are attained (perhaps 10 000–15 000 K), forming a localized plasma in the excitation region. This causes the ejection of a plume of material from the sample containing both neutral and charged atomic and molecular species. The elemental composition of this plume may be determined by a variety of techniques including mass spectrometry, atomic absorption and optical emission measurements.

The precise mechanism of excitation depends on the type of laser selected, since this determines the power density, the duration of individual pulses and the pulse repetition rate, Bingham and Salter (1976) concluded that a Q-switched laser was ideal for materials analysis, since it produced a short-duration pulse of high intensity. This caused material to be removed early from the sample so that the energy of the pulse could be used for efficient ionization of evaporated material. The particular advantages of this mode of excitation were summarized by Bingham and Salter (1976) as follows:

(i) Short laser pulses prevent an excessive increase in plasma plume temperature that would favour the formation of multiply-charged ion species

(ii) Short laser pulses also avoid radiation from the plasma causing local evaporation of material around the plasma crater, which could cause preferential enhancement of signals due to volatile elements

(iii) Q-switched lasers have sufficiently high power densities (10^9–10^{11} W/cm^2) to avoid the discrimination effects against elements with high boiling points that can be observed when lasers of lower power density (e.g. 10^8 W/cm^2) are used.

Using a Q-switched laser, Harding-Barlow *et al.* (1973) considered that the plasma plume was approximately spherical in shape, several millimetres across, and if unaffected by electrostatic fields, would remain static for 1–20 μs. This plume is shown schematically in Figure 11.9. The material initially ejected forms a highly active plasma consisting mainly of ionic species. Later material may be ejected as atoms, which are in any case formed in the plume by recombination reactions as the plasma begins to decay. Laser-excited plasmas have been used as the source for both optical emission and atomic absorption signals (Keil and Snetsinger, 1973). Laser ablation has also been used as a technique for sampling solid materials for ICP–atomic emission spectroscopy (Thompson and Walsh, 1983, p.176) and ICP–MS (Gray, 1985). However, perhaps the most useful application of the plume is as a source of ions for analysis by mass spectrometry (e.g. Kovalev *et al.*, 1978; Bingham and Salter, 1976). A schematic layout of the laser excitation arrangement used by Bingham and Salter is shown in Figure 11.10. This source may be conveniently coupled to a mass spectrometer of the Mattauch–Herzog design (as used for spark source mass spectrometry measurements, see Chapter 19) to permit simultaneous detection of the entire mass spectrum on a photographic plate. In spite of some reported

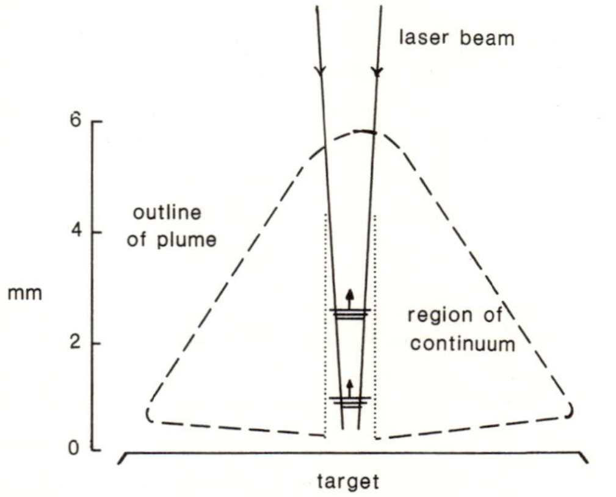

Figure 11.9 Schematic representation of laser excitation, showing the plume of material ejected from the sample. (Piepmeier and Malmstadt, 1969, as reproduced by Harding-Barlow *et al.*, 1973.)

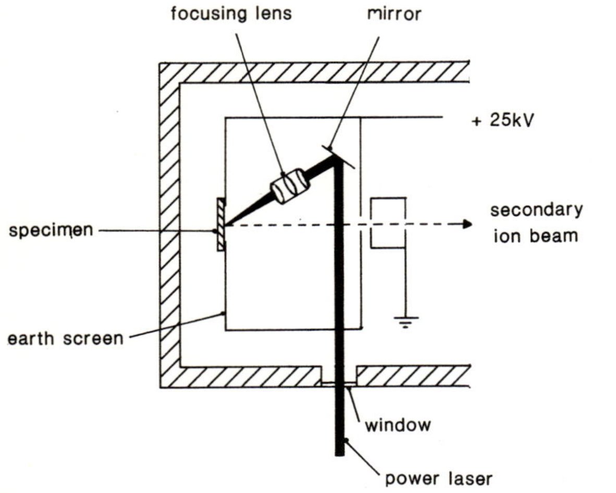

Figure 11.10 Schematic diagram of the laser microprobe based on the design of Bingham and Salter (1976) showing the arrangement of sample excitation only.

applications to geological samples (see Kiel and Snetsinger's review (1973) and Bingham and Salter (1976), for example) this technique has not yet had an important impact on the analysis of silicate minerals. A significant limitation is that analysed minerals must be opaque, since insufficient laser power is coupled into transparent minerals.

There has, however, been some interest in selectively analysing individual fluid inclusions using laser excitation (Tsui and Holland, 1975; Ishkov and Reyf, 1980). Ishkov and Reyf (1980) used a commercially available instrument in which a laser was used to evaporate fluid inclusion material. The resultant plume of material was passed between two electrodes held approximately 2 mm apart at a potential of 3 to 3.5 kV. Under these conditions a spark discharge was formed and the optical emission was recorded on an optical spectrograph. Analysis of the appropriate emission lines permitted the following minimum amounts to be detected: Be 3, B 30, Cu 70, Zn 500, Fe 300, Al 300 and Mn 400, all in pg units, on fluid inclusions of volume 10^{-8} mL. Laser input power (about 1 J) was not large enough to vaporize the host matrix, thus avoiding matrix interference effects.

Ishkov and Reyf (1980) used a novel calibration technique, there being no natural fluid inclusion reference materials. A drop of the appropriate standard solution was mixed with 'epoxide' resin on a glass slide. The resultant emulsion polymerized to form discrete liquid–gas inclusions within the resin matrix. The composition of these inclusions could be checked from freezing-point measurements if necessary.

11.4 Particle-induced x-ray emission (PIXE)

11.4.1 *Introduction*

PIXE is a technique for the multi-element analysis of solid samples excited with a beam of nuclear particles—normally protons or sometimes alpha particles or neon ions. The proton is equivalent to the nucleus of the hydrogen atom ($^1H^+$). In order to excite samples, a proton beam is usually accelerated to an energy equivalent to 2 to 3 MeV (sometimes up to 5 MeV). This beam can be collimated and if necessary focused by a series of quadrupole lenses to form a beam several millimetres across for the bulk analysis of specimens, or, alternatively, focused down to a minimum diameter of about 1 μm for microanalysis. In this latter configuration, the instrumentation is sometimes known as a *nuclear microprobe*. The general configuration of beam, sample and x-ray and gamma-ray detectors used to measure induced radiations is shown in Figure 11.11. PIXE instrumentation is available only at a limited number of laboratories since highly specialized (and expensive) equipment such as a Van de Graaff generator or a cyclotron is required to generate the proton beam.

Geochemical applications of this technique have been reviewed by Clark *et al.* (1975) and Kullerud *et al.* (1979). Basutçu *et al.* (1983) have described the analysis of light elements using a beam of singly-charged alpha particles of energy 3.5 MeV. In the literature, PIXE applications are

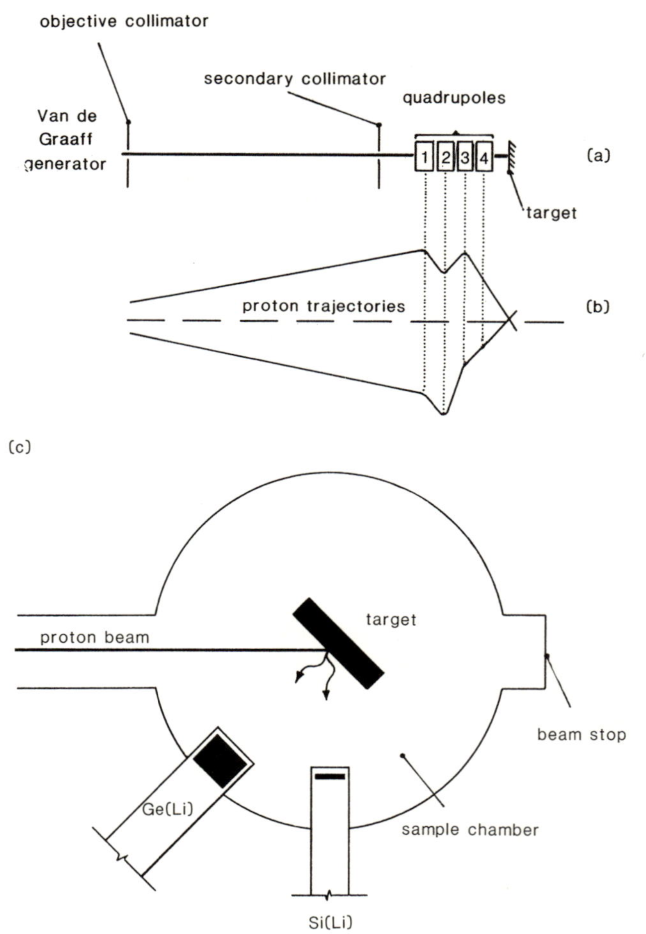

Figure 11.11 Schematic diagram of PIXE instrumentation showing: (*a*) components in the proton probe forming column, including collimators and quadrupole lenses; (*b*) calculated trajectories of protons in the horizontal plane (upper trace) and vertical plane (lower trace), both for proton paths starting 10 mm from the column axis with an angular divergence of 0.01 radians (based on Cookson *et al.*, 1972); (*c*) arrangement of the specimen chamber, showing proton beam, sample and detectors (based on Kullerud *et al.*, 1979).

more commonly reported for the analysis of biological and environmental samples, where material can readily be prepared in the form of a thin film, the preferred (though not exclusive) form of sample presentation for PIXE analysis.

11.4.2 *Excitation of characteristic x-rays*

A beam of protons can interact with a sample, causing ionization of orbital electrons, in a manner similar to that of an electron beam in electron microprobe analysis. Thus, ionization of inner-shell electrons causes electron transitions that result in the emission of characteristic x-rays. (The efficiency for excitation of characteristic x-ray spectra of proton beams is compared as a function of energy with that of electrons and x-ray photons in Figure 10.28.) As protons are about 1836 times heavier than electrons, they are deflected to a much smaller extent by interaction with target atoms. The corresponding intensity of continuum (bremsstrahlung) radiation is, therefore, much lower than that in electron excitation. Unfortunately, there are other sources of continuum radiation that contribute to proton excited spectra. One of these is bremsstrahlung generated by high-energy electrons ejected during collision between protons and target atoms. However, (see Clark *et al.*, 1975), the maximum energy (E) that can be imparted to an electron following head-on collision with a proton of energy E_0 is given by

$$E = 4\frac{m_{e^-}}{m_{H^+}}E_0 \qquad (11.2)$$

where m_{e^-} is the mass of an electron, and m_{H^+} is the mass of a proton.

Thus for a proton beam of energy 2 MeV, the maximum secondary electron energy is about 4.3 keV, placing an upper limit to the energy of the secondary electron induced continuum.

A second source of background is from Compton scatter of gamma radiation originating from proton-induced nuclear transformations within the sample, and these will be considered further in the next section.

The overall effect of these background contributions is considerably smaller than that encountered in electron excitation, giving PIXE analysis the potential of achieving significant reductions in detection limits compared with the electron probe microanalysis. The general ways in which continuum spectra are expected to change as a function of energy were compared by Kullerud *et al.* (1979) for proton and electron excitation. Their data for a tungsten target excited by an 80 keV electron beam and a proton beam are shown in Figure 11.12. Kullerud *et al.* estimated that the intensity of unwanted bremsstrahlung was reduced by a factor of 10^6 if protons rather than electrons were used for the excitation process.

To illustrate a typical PIXE spectrum, the x-ray spectrum of a sample of coal ash recorded by Kullerud *et al.* (1979) over the range of 0 to 20 keV using a proton beam accelerated to 4 MeV is shown in Figure 11.13. In this diagram, the authors have estimated the continuum contribution from secondary electron bremsstrahlung and nuclear gamma background (Compton scattering). As can be seen, the secondary electron bremsstrahlung was estimated to be negligible above about 18keV under these conditions of excitation.

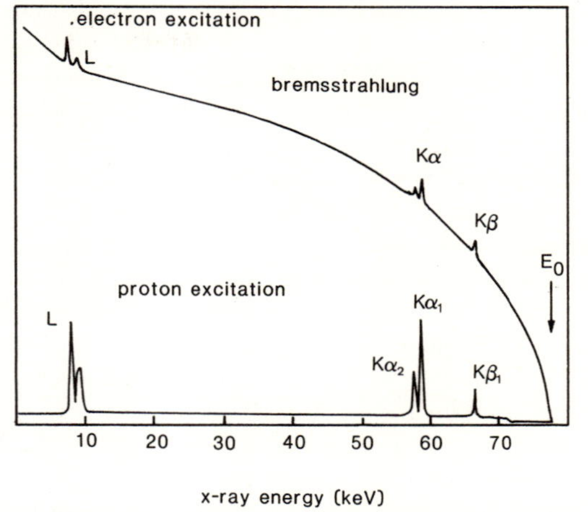

Figure 11.12 Comparison of continuum spectra generated by electron and proton excitation. The electron spectrum originated from a tungsten target excited by an 80 kV electron beam. The proton spectrum was observed from the same target excited with a proton beam of several MeV. (Kullerud *et al.*, 1979.)

11.4.3 *Proton-induced nuclear reactions*

The proton has sufficient energy to induce two kinds of nuclear reactions in target nuclei, both of which result in the emission of gamma radiation that may be used for quantitiative analysis.

(*a*) *Coulombic excitation.* Coulombic excitation results from the interaction of the bombarding proton with the coulombic (electrostatic) field surrounding the nucleus of a target atom. The result is that the nucleus is itself excited from the ground state to a discrete nuclear excited state (section 12.2). The excited state is normally highly unstable and relaxes back to the ground state, emitting a gamma-ray of energy equal to the difference in energy between the two nuclear

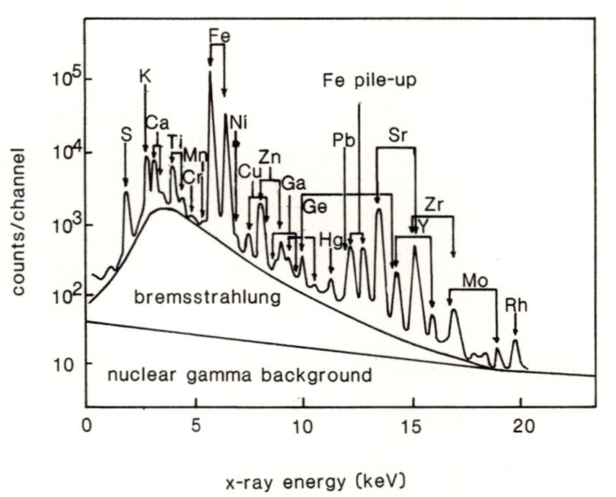

Figure 11.13 PIXE spectrum of coal ash excited with 4 MeV protons showing relative background contributions of secondary-electron bremsstrahlung and nuclear gamma-ray scattering. (Kullerud *et al.*, 1979.)

levels. The probability that this interaction will occur (i.e. the cross-section) depends in part on the availability of suitable low-energy nuclear excited states in the target atom. The isotopes which best satisfy this requirement and which can therefore be analysed from suitable gamma-radiation emitted during coulombic excitation, are listed in Table 11.5.

(*b*) *Nuclear resonance reactions*. Proton bombardment of samples can also induce nuclear transformations in which the proton is captured (or scattered) by target nuclei. The nucleus is left in an excited state, and decays by emitting characteristic gamma radiation, accompanied sometimes by alpha emission. These nuclear transformations may be classified as follows:

(p, p′, γ): proton scattering with subsequent gamma emission from the excited target nucleus

(p, γ): proton capture leading to the emission of prompt gamma-rays from the target nucleus

(p, α, γ): proton capture after which the compound nucleus decays by alpha emission, accompanied by characteristic prompt gamma-radiation.

The unusual characteristic of these nuclear reactions is that they have low cross-sections except for narrow ranges of energy of the bombarding proton beam. At these *resonant energies*, the probability of proton capture by a particular isotope can be high. PIXE excitation can thus be made specific for an individual element by tuning the proton beam energy to a value required for resonance excitation. The isotopes that may readily be determined in this way are listed in Table 11.6.

Overall, therefore, two signals are available in PIXE analysis: (i) a proton-excited x-ray spectrum having a reduced continum compared with electron excited spectra (an Si(Li) detector is used to measure this signal), and (ii) a gamma-ray spectrum excited from two sources: coulombic interaction of the proton beam with target nuclei and nuclear transformations resulting from proton capture or scattering from target atoms at resonance energies of the nuclear beam. Gamma spectra are normally measured using a suitable

Table 11.5 Isotopes that may be determined by Coulomb excitation in PIXE analysis

Isotope	Gamma energy (keV)	Limit of determination: Proton (ppm)	Limit of determination: Alpha (ppm)
^{75}As	199	6	1500
^{55}Mn	126	30	120
^{181}Ta	301	30	32 100
	136	90	600
^{56}Fe	847	300	54 000
^{47}Ti	159	600	2100
^{48}Ti	983	600	90 000
^{197}Au	279	600	33 000
^{80}Se	666	600	59 100
^{77}Se	239	900	5400

Data from Clarke *et al* (1975).
Detection limits are estimates of the six-sigma limit of determination for interference-free excitation of a thick target by a 4 MeV proton or alpha particle beam, based on 0.1 count per μC; detected by Ge(Li) at 100 mm from the target.

Table 11.6 Isotopes that may be determined by nuclear resonance reactions in PIXE analysis

Nuclear reaction	Resonance proton energy (MeV)	Gamma emission energy (keV)	Detection limit (ppm)
^{23}Na (p, p′γ) ^{23}Na	1.458	439	1.7
^{7}Li (p, p′γ) ^{7}Li	1.030	478	2.1
^{19}F (p, αγ) ^{16}O	1.375	6130, 7120	2.9
^{31}P (p, p′γ) ^{31}P	3.730	1266	3.3
^{32}S (p, p′γ) ^{32}S	3.379	2237	11
^{35}Cl (p, p′γ) ^{25}Cl	2.721	1220, 1763	19
^{27}Al (p, p′γ) ^{27}Al	2.727	1013	70
^{24}Mg (p, p′γ) ^{24}Mg	2.930	1370	86

Data from Clarke *et al* (1975).
Rock powder was mixed with 10% graphite and prepared as a thick (1 mm) target by pelletting.
Targets excited by a proton beam of the specified (resonance) energy.
Detection limits are estimates of the six-sigma limit of determination based on 1000 μC of integrated charge.

germanium gamma-ray detector although Clark *et al.* (1975) suggested that, notwithstanding the loss in resolution, a large-volume NaI(Tl) detector is of value in measuring higher-energy gamma radiation from resonant proton-capture reactions. In these circumstances, spectrum overlap effects are likely to be minimal, since conditions may be adjusted to ensure that only one nuclide is significantly excited. As a result of the nuclear transformations described above, the sample after analysis is left in a slightly radioactive state.

There is an overall consensus that for the excitation of x-ray spectra, proton beam energies of 2 to 3 MeV should be used (Russell *et al.*, 1981; Johansson and Johansson, 1976). At lower energies, *K*-shell ionization cross-sections fall off to an unacceptable extent. At higher beam energies, background intensities, due both to secondary electron bremsstrahlung and Compton scatter from nuclear reactions, increase to an unacceptable extent.

11.4.4 *Analytical results*

As a demonstration of the abilities of PIXE analysis in geological applications, Clark *et al.* (1975) reported the analysis of a basalt USGS: BCR-1 using granite USGS: G-2 as standard. Powdered materials were mixed with 10% graphite to conduct the proton charge to earth, and were analysed as pellets. Analytical results are listed according to the excitation mechanism in Table 11.7. In general, the sensitivities for several of the trace elements that may be determined by resonant proton capture reactions appear to be significantly better than other instrumental techniques based on x-ray/gamma-ray emission measurements.

11.4.5 *Excitation with alpha particles*

An alternative excitation method has been described by Basutçu *et al.* (1983) in which samples are excited by a beam of singly-charged alpha particles (corresponding to $^4He^+$). These alpha particles were accelerated by a Van de Graaff

Table 11.7 PIXE analysis of basalt USGS : BCR-1

Element	Abbey (1983) Expected	PIXE analysed data (Clarke *et al.*, 1975)	
		Proton resonance excitation	
Li	14	12.4±0.8	
F	500	491±9	
Na_2O	3.30	3.32±0.04%	
MgO	3.48	3.42±0.07%	
Al_2O_3	13.72	13.45±0.4%	
SiO_2	54.53	50.91±0.6%	
P_2O_5	0.36	0.32±0.07%	
S	400?	384±10	
Cl	58	43.2±2.5	
		Characteristic x-ray excitation	*Coulomb excitation*
K_2O	1.70	1.33±0.63%	
CaO	6.97	6.11±0.42%	
TiO_2	2.26	2.07±0.03%	1.88±0.47%
V	420	390±4	334±72
Cr	15	18.5±0.5	
MnO	0.18	0.179±0.002%	0.160±0.031%
Fe_2O_3	13.41	13.07±0.03%	12.10±2.2%
Cu	16	17.1±0.6	
Zn	125	118±2	106±28
Ga	22	19.5±2.2	
As	0.8?	0.70±0.02	
Rb	47	45.2±1.2	
Sr	330	332±0.4	
Y	40	38.2±2.1	
Zr	185	188±2.4	
Nb	19?	15.4±2.0	
Ag	0.034?	0.032±0.003	
Sb	0.6	0.66±0.06	
Ba	680	705±48	

Samples mixed with 10% graphite and compressed as powder pellets. Units: major elements % oxide, trace elements ppm.
Samples were excited with 4 MeV protons and count data were calibrated against corresponding data in granite USGS : G-2.

generator to an energy of 3.5 MeV, and could be collimated and focused by electromagnetic lenses down to a minimum spot size of 1 mm. Alpha-induced nuclear transformations within the sample caused the emission of gamma radiation

Table 11.8 Nuclear data and detection limits for excitation using singly-charged alpha particles of energy 3.5MeV

Nuclear reaction			Gamma energy (keV)	Detection limit (ppm)
^{7}Li	$(\alpha, \alpha'\gamma)$	^{7}Li	478	0.4
^{9}Be	(α, n)	^{12}C	3417	1
^{10}B	(α, p)	^{13}C	169	30
^{18}O	(α, n)	^{21}Ne	350	1800
^{19}F	(α, n)	^{22}Ne	1275	10
^{23}Na	$(\alpha, \alpha'\gamma)$	^{23}Na	440	60

Data from Basutçu *et al.* (1983).
Detection limits are 'lower detection limits'.

which was detected on a Ge(Li) detector. Principal applications of the technique were in the analysis of the light elements, Li, Be, B, O, F and Na in geological samples. The corresponding nuclear reactions and estimates of detection limits are listed in Table 11.8. Direct applications included the determination of lithium in synthetic micas, and lithium and fluorine in international reference materials. As it is not a conventional activation method, samples do not become radioactive after irradiation.

11.5 Electron spectroscopy for chemical analysis (ESCA)

This technique, also known as x-ray photoelectric spectroscopy (XPS), is a highly specialized method of analysing the surface layers of solid samples, 5 to 10 nm thick. Samples are excited with low-energy photons from an x-ray tube. The energy spectrum of photoelectrons emitted from outer electron orbitals as a result of this excitation process is measured. These data can give two types of information in geological applications: first, they can identify the oxidation and co-ordination state of elements in specific minerals with a view to elucidating their chemical structure; second, the concentration of elements adsorbed on the surface of minerals can be determined to give information on surface adsorption and reaction processes in geochemistry.

The arrangement of instrumentation is shown schematically in Figure 11.14. Solid samples are placed in a

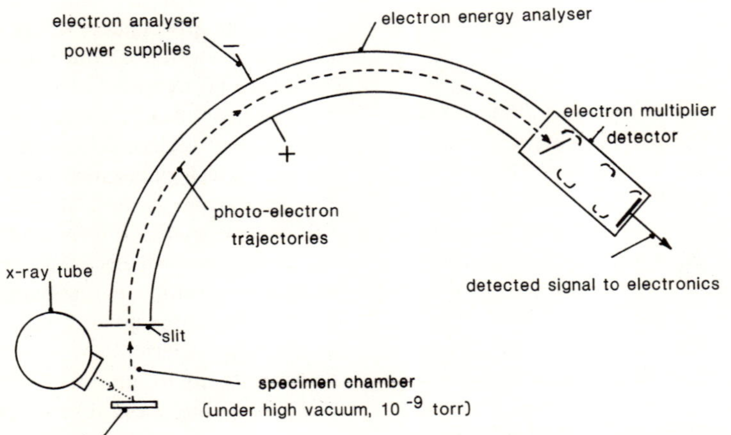

Figure 11.14 Schematic diagram of ESCA instrumentation showing overall arrangement of x-ray tube (usually employing Al or Mg anode), sample and electron energy analyser. (Adapted from Bertin, 1975, p. 894.)

special chamber under very high vacuum (up to 10^{-9} Torr) and excited with a large flux of low-energy x-ray radiation, Al $K\alpha$ (1486 eV) or Mg $K\alpha$ (1253 eV) being the most widely used. These x-rays interact with the sample in several ways (photoelectric effect leading to the emission of characteristic x-rays or Auger electrons, Compton scatter, Rayleigh scatter, section 8.6). The important mechanism here is photoelectronic absorption. In this process the energy of the incident photon is completely absorbed by an orbital electron, which is ionized from the atom with kinetic energy E_{KE} given by

$$E_{KE} = 1486 - E_B \quad (11.3)$$

where 1486 is the energy (in eV) of the incident photon (assumed here to be Al $K\alpha$), and E_B is the binding energy of the electron (critical ionization energy) in eV.

Thus, photoelectric absorption leads to the emission of a spectrum of electrons, each having an energy characteristic of the electron orbital from which it was ejected. This spectrum is measured using an electron energy spectrometer which operates on the principle that the degree of deflection induced in an electron by an electrostatic (or electromagnetic) field is related to its kinetic energy. Thus, the flight path of electrons is arranged to pass between two curved plates to which an appropriate voltage is applied (Figure 11.14). The electron beam is deflected; the lower the electron energy, the higher the degree of deflection. Under a given set of operating conditions, only electrons of a given energy are transmitted through the electrostatic sector to the electron multiplier. If the voltage applied to the electrostatic plates is varied, a spectrum may be measured representing the intensity of photoelectrons as a function of kinetic energy. An example of such a spectrum measured by Huntress *et al.* (1972) on an Apollo-11 lunar breccia is shown in Figure 11.15. Each peak in such a spectrum can be identified with the ionization of a specific orbital electron.

The energy of the exciting radiation is in fact so low that the *K*-orbital electrons of most elements cannot be excited. ESCA spectra, therefore, display peaks that in general correspond to the excitation of outer orbital electrons from the *L*-, *M*-, or *N*-orbital shells. The kinetic energy of these ejected electrons is so low that they suffer heavy absorption within the sample. The intensity (I) of an electron emission peak varies with depth below the surface of a sample (d) as follows:

$$I = I_0[1 - \exp(-d/\Lambda)] \quad (11.4)$$

where I_0 is the emission intensity from an infinitely thick sample, and Λ is the mean free path or escape depth for primary photoelectrons, normally having values in the range 1 to 5 nm (see Bancroft *et al.*, 1979). This equation demonstrates that over 80% of the emitted intensity is observed from electrons originating in the first 2Λ surface layers.

Practical applications of ESCA data stem from the fact that detected electrons are ejected from outer electron orbitals, the energy levels of which are affected by chemical bonding effects and the oxidation state of the element.

Geochemical applications of this technique have been reviewed by Bancroft *et al.* (1979). ESCA analysis has been used to identify the oxidation state of vanadium (V(III), not (V(IV)) in vanadium-bearing aegirine minerals

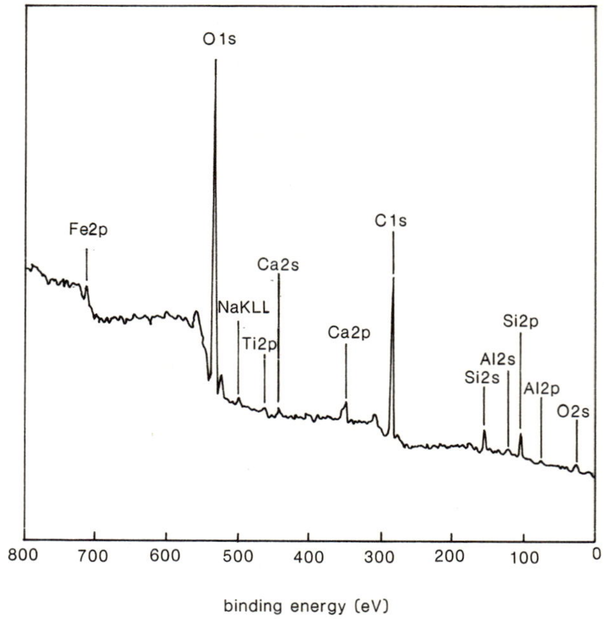

Figure 11.15 ESCA spectra of an Apollo-11 lunar breccia excited by Al $K\alpha$ radiation, showing the elements detected and the orbital levels from which photoelectrons were ejected. (Huntress and Wilson, 1972.)

($Na(Fe,V)Si_2O_6$) (Nakai *et al.*, 1976). Stucki *et al.* (1976) measured the binding energy of the Fe $2p_{3/2}$ orbital electron to identify Fe(II) and Fe(III) in reduced nontronites and biotites. Bancroft *et al.* (1979) also discuss other examples, including an investigation of the co-ordination number of Al and Mg in silicate materials and the adsorption and chemical reaction of ions at mineral surfaces.

11.6 Transmission electron microscope: the chemical analysis of thin foils

11.6.1 *Instrumentation and excitation*

The transmission electon microscope is normally used to record electron micrographs at high resolution from samples prepared as thin foils (less than about 0.15 μm thick for silicate minerals).

The sample is placed under high vacuum in a sample chamber and excited with a beam of electrons accelerated to at least 60 kV and often 100 kV or more (Figure 11.16). The electron beam, transmitted through the sample, is brought into focus on a fluorescent screen or photographic plate on which photomicrograph images may be recorded. The resolution of these transmitted images is limited by spherical aberration anomalies in the various electromagnetic lenses, but is normally in the range 0.5 to 1.0 nm under standard operating conditions. This is far better than can be achieved using the normal backscattered electron signal on a scanning electron microscope, where resolutions of between 3 and 10 nm are common, further improvements being limited by the scatter of the electron beam within the sample.

The electron beam passing through the foil is capable of exciting characteristic x-rays within the sample. However,

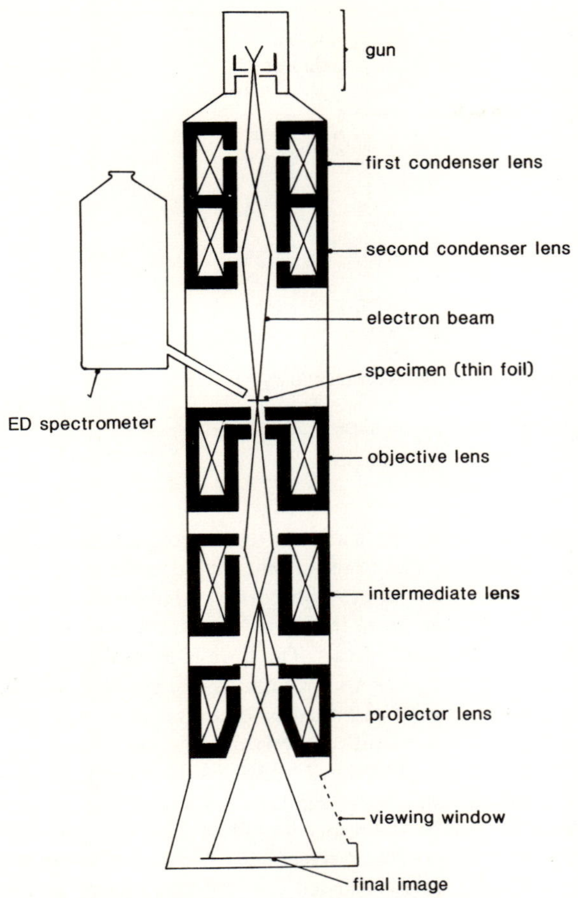

Figure 11.16 Schematic diagram of a transmission electron microscope, showing the electron beam and principal electromagnetic lenses.

the foil is so thin that appreciable scatter of the electron beam does not occur, in direct contrast to electron microprobe analysis (Figure 11.17). The size of the x-ray source region in TEM is largely controlled by the diameter of the electron beam. Chemical information is, therefore, normally obtained with a spatial resolution of less than about 20 nm.

Thus, in this mode of TEM operation, analyses of extremely small mineral structrues can be obtained, far smaller than could be attempted by conventional electron probe microanalysis. Set against this is the low intensity of the x-ray signal which can be compensated to some extent by placing a Si(Li) detector as close as possible to the sample. In spite of this, counting errors are normally large, and it is usually only possible to determine concentrations of elements above the 0.5 to 1 weight % level.

11.6.2 *Quantitative analysis by TEM*

As the x-ray source is effectively a thin film, x-ray absorption and fluorescence effects can, to a first approximation, be ignored. However, as it is not possible to control accurately the mass of sample analysed (ie the foil thickness), it is desirable to derive calibration procedures in terms of concentration *ratios*. Following the review by Champness *et al.* (1982), the concentration of an element C_X can be related to the concentration (C_{Si}) of silicon (say) in the sample according to

$$\frac{C_X}{C_{Si}} = k_{X,Si}\frac{I_X}{I_{Si}} \qquad (11.5)$$

where I is the x-ray intensity of the element; $k_{X,Si}$ is a calibration constant of element X relative to Si that is independent of sample thickness or composition.

In order to convert concentration ratios C_X/C_{Si} into weight per cent, it is necessary to assume that the weight fractions of all analysed components expressed as stoichiometric oxides sum of unity:

$$\sum_X C_X = 1. \qquad (11.6)$$

To indicate the results that may be obtained using this calibration procedure, some selected mineral analyses obtained by TEM are compared with wet chemical data in Table 11.9.

The accuracy and limitations of this form of analysis have been reviewed by Champness *et al.* (1982) who emphasized several points.

(i) A high degree of agreement is observed in the cali-

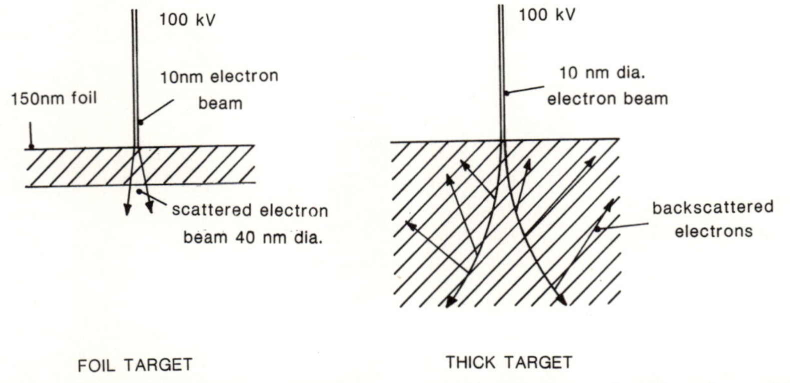

Figure 11.17 Electron beam scattering in a thin (150 nm) foil and a thick target. Due to the limited scatter, the spatial resolution of the electron beam transmitted through the foil (as in TEM analysis) is considerably better than that observed in a thick target (as in electron microprobe analysis).

Table 11.9 TEM analyses of silicate minerals

Element	Hornblende		Biotite		Actinolite	
	Expected	Analysed	Expected	Analysed	Expected	Analysed
Na_2O	2.8	3.0				
MgO	13.4	13.4	9.6	10.6	11.4	9.2
Al_2O_3	15.7	17.1	21.3	21.5		
SiO_2	43.2	42.2	39.9	38.9	53.5	54.7
K_2O	0.6	0.7	8.3	8.6		
CaO	10.4	10.1			23.0	23.0
MnO	0.2	0.2				
FeO	13.6	13.4	20.9	20.5	12.1	13.1

Data from Champness *et al.* (1982).
TEM data obtained by exciting mineral particulates at 100 kV using experimental $k_{X,\mathrm{Si}}$ values for calibration.
Expected values from P.W. Weiblen (Minnesota) using wet chemical techniques.

bration constant $k_{X,\mathrm{Si}}$ for measurements made on different instruments for elements heavier than Si ($K\alpha$). For lighter elements, the response appears to be detector-dependent, presumably affected by the degree of attenuation in the beryllium window of the Si(Li) spectrometer.

(ii) The maximum foil thickness of 0.15 μm is selected so that the error due to x-ray absorption within the sample is less than 5% for measurements made on orthopyroxene ($MgFeSi_2O_6$).

(iii) Some scatter of the electron beam does occur during transmission through the sample. Using an incident beam of diameter 7.5 nm, Champness *et al.* (1982) estimated that the increase in diameter due to electron scatter within the samples is less than 20 nm for a foil of thickness 0.15 μm.

Reports of the application of this technique to the analysis of natural minerals in ion-thinned rock sections have been limited. By far the most important application is in the analysis of minerals in the environment; asbestos fibres are the prime example of this. The fibre is mounted on a substrate of suitably low atomic number (for example, polished beryllium foil) and identified by quantitative analysis or by qualitative 'fingerprint' analysis using a Si(Li) detector to record the x-ray spectrum. To resolve some ambiguities from different minerals having similar x-ray signatures, it is sometimes necessary to resort to TEM electron diffraction measurements to obtain structural information. Champness *et al.* (1976) have discussed the identification of asbestos fibres using a TEM.

12 Neutron activation analysis

12.1 Introduction

Neutron activation analyis was first proposed as an analytical technique by von Hevesy and Levi in 1936. However, its successful application to geology (Gordon *et al.*, 1968; Hertogen and Gijbels, 1971) had to await the commercial development of high-resolution germanium detectors during the 1960s to allow satisfactory analysis of the complex gamma spectra obtained from irradiated silicate rock samples.

The technique is used predominantly for the analysis of specific trace elements down to detection limits in the ppm to ppb range, and is especially sensitive for the rare-earth elements (La, Ce, Nd, Sm, Eu, Tb, Yb, and Lu), Sc, Co, Cr, Cs, Hf, Ta, Th, U and some others.

Samples are 'activated' by irradiation with neutrons, usually in a nuclear reactor. Specific isotopes become radioactive by neutron capture-type nuclear reactions. After removal from the reactor, samples are allowed to decay ('cool') to permit unwanted short-lived activity (for example, from ^{24}Na) to diminish. Gamma-ray spectra from activated samples are then measured using solid-state germanium gamma-ray detectors, usually over the energy range 60 to 1600 keV. Specific isotopes may be identified by their gamma-rays of characteristic energy and quantitative determinations are made by comparing the areas of these photopeaks with those in the spectrum of a standard sample, irradiated and counted under identical conditions.

The two general procedures employed are classified as: (i) *instrumental neutron activation analysis* (INAA) which involves the irradiation and analysis of rock powders that undergo no chemical treatment in the course of the analysis (this procedure may, therefore, be considered as a non-destructive technique) and (ii) *radiochemical neutron activation analysis* (RNAA), which involves taking up irradiated samples into solution so that a chemical separation procedure may be carried out to isolate specific trace elements from the more active matrix. This latter procedure lowers detection limits by perhaps one to two orders of magnitude and significantly extends the range of elements that may be successfully analysed. However, such procedures are more time-consuming and involve specialized laboratory facilities for manipulation of radioactive material.

Neutron activation analysis has the advantage of being very sensitive for specified elements and virtually free of matrix interference effects (except overlap interferences in complex gamma spectra). The disadvantage is that the technique requires specialized irradiation facilities and for this reason is largely confined to establishments adjacent to nuclear reactors.

In the determination of the REEs, neutron activation analysis cannot achieve the high precision nor low detection limits claimed for isotope dilution mass spectrometry (e.g. Hooker *et al.*, 1975). Furthermore, recent developments in inductively coupled plasma emission spectrometry (e.g. Bolton *et al.*, 1983; Walsh *et al.*, 1981) indicate that sensitive REE determinations may be made by this technique provided sample preparation procedures include a chemical separation. Nevertheless, during the 1970s, INAA was the only technique available for the routine determination of REEs in rock samples and is still extensively used in this field. INAA is one of the few routine techniques available for the determination of other geochemically important trace elements including Ta and Hf (and Th), these elements being difficult to incorporate in schemes that involve solution chemistry.

In this chapter the theoretical and practical application of *instrumental* NAA is discussed in detail, this being the simplest and most widely used form of the technique. Other analytical derivations are then discussed more briefly.

Most counting systems are not purchased as integrated commercial packages but are sometimes put together as combinations of equipment from different manufacturers. A discussion is, therefore, included on individual instrumental components in the signal accumulation system with which a user must interact.

INAA is, by reputation, sometimes considered to be a technique of variable routine precision. It is felt that this is in part due to the 'do-it-yourself' way in which systems are put together by users and to the fact that calibrations are often set up against a single reference material. Data are presented to demonstrate that, provided care is taken in optimizing and controlling instrument parameters, precisions equivalent to other instrumental procedures used for trace analysis can be achieved.

12.2 The growth and decay of radioactivity

12.2.1 *Neutron sources—the nuclear reactor*

The most commonly used source of neutrons for activation analysis is the nuclear reactor. Most research reactors depend on the fission of ^{235}U. This isotope is naturally radioactive, with a half-life of 7.038×10^8 years, and will spontaneously decay by nuclear fission, in which the nucleus splits into isotopic fragments, releasing high-energy neutrons and energy. If *moderated* (i.e. slowed down) these neutrons can

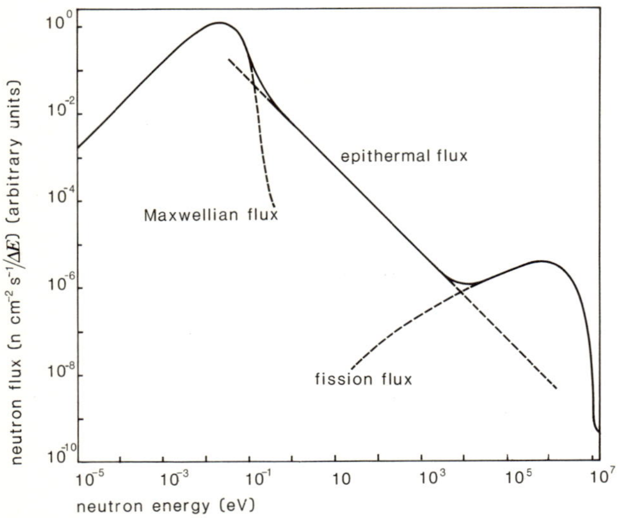

Figure 12.1 Neutron spectrum of a nuclear fission reactor. (Erdtmann, 1976.)

efficiently promote the further fission of ^{235}U by the reaction

$$^{235}U + {}^{1}n \rightarrow \text{fission products} + 2.2({}^{1}n) + \text{energy}.$$

On average, 2.2 additional neutrons are generated by this fission reaction. Under the appropriate conditions, these fission neutrons can promote a self-sustaining branching chain reaction. In a nuclear reactor this chain reaction is controlled by the selective absorption of neutrons with control rods so that a fission equilibrium is maintained (rate of emission of neutrons equals rate of absorption).

12.2.2 *Energy spectrum of neutrons*

Neutrons emitted during the fission of ^{235}U have energies in the range 0–25 MeV and are far too energetic to efficiently promote the activation of isotopes required for conventional neutron activation analysis. Their energies are moderated by interaction with a material of low atomic number in which efficient loss of kinetic energy occurs. Water is particularly effective for this purpose. The resultant energy spectrum of neutrons (Figure 12.1) may be subclassified as follows:

(*a*) *Thermal neutrons*. Thermal neutrons have very low energies, in the range 0 to 0.5 eV, but with a most probable energy of 0.025 eV at 20°C (corresponding to a mean neutron velocity of 2200 m s^{-1}). This energy is equivalent to that predicted by a Maxwellian distribution function for neutrons that have attained thermal equilibrium with their surroundings. Thermal neutrons efficiently promote (n,γ) reactions which are the basis of most neutron activation analysis measurements.

(*b*) *Epithermal neutrons*. Epithermal neutrons have energies in the range of 0.5eV to 10 keV and may be used to selectively activate particular isotopes in neutron activation analysis (see section 12.15). The threshold of 0.5 eV is chosen because of an unusual property of cadmium. A thin foil of this material (0.5–1mm thick) strongly absorbs neutrons with energies below this threshold (Figure 12.2), predominantly by the nuclear reaction

$$^{113}Cd(n,\gamma)\,^{114}Cd$$

where the product isotope ^{114}Cd is stable. Higher-energy neutrons are only weakly attenuated, as is apparent from the neutron capture cross-section diagram of ^{113}Cd (Figure 12.2). For this reason, this region is also known as the epi-cadmium neutron spectrum. *Resonance neutron capture* occurs when an isotope shows high neutron capture cross-sections at discrete neutron energies in the epithermal region (Figure 12.2). An appreciable degree of activation may then result from capture of resonance neutrons, permitting selective analysis of such isotopes in comparison with those that are only strongly activated by a thermal neutron flux. This extension of conventional thermal neutron activation analysis is known as *epithermal neutron activation analysis* (ENAA). Elements that may be preferentially activated in this way include U, Ta, Sb, Mo, Au, and Ir.

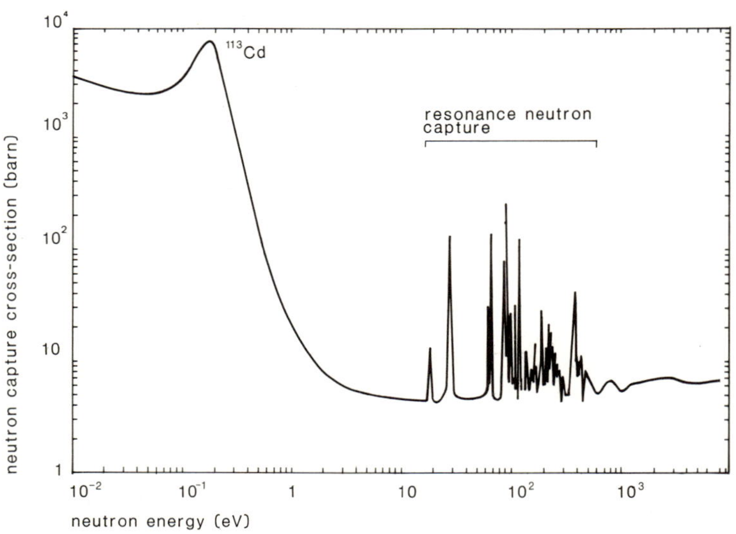

Figure 12.2 Variation in neutron capture cross-section of cadmium with neutron energy showing the high absorption of thermal neutrons of energies below about 0.5 eV. (Hughes and Schwartz, 1958; as reproduced by De Soete, 1972.)

(*c*) *Fast neutrons.* Fast neutrons have energies greater than 10 keV and principally promote (n,p) and (n,α) nuclear reactions. Although these reactions may be used for analysis, they are generally considered to be interferences in conventional INAA. The proportion of fast neutrons in a thermal neutron spectrum normally used for INAA is low, but varies according to the irradiation position relative to the reactor core.

12.2.3 *Neutron flux*

The 'concentration' of neutrons available for irradiation is called the neutron flux and is conventionally expressed in units of the number of neutrons passing through a square centimetre area per second. A minimum neutron flux of between 10^9 and 10^{10} n cm^{-2} s^{-1} is required for thermal activation analysis. Most small research reactors of total power 100–200 kW can generate fluxes of between 10^{11} and 10^{13} n cm^{-2} s^{-1}. Duration of irradiation varies with application but a reasonably long irradiation might require an *integrated neutron flux* (i.e. neutron flux × irradiation time) of 10^{17} to 10^{18} n cm^{-2}.

12.2.4 *Neutron capture cross-sections*

The reason why such a large integrated neutron dose is necessary lies in understanding the mechanism of neutron activation. This only occurs as a result of interaction between neutron and *nucleus* of an atom. A typical atom has a diameter of 10^{-10} m; however, its nucleus occupies only 10^{-14}m (Figure 12.3).

This very open structure means that the probability of a neutron (of diameter 10^{-18}m) interacting with the nucleus is low. The probability that such an interaction will lead to a nuclear transformation depends on the energy of the neutron and the nature of the target nucleus, and is referred to as the *neutron capture cross-section* of the isotope leading to a specified nuclear reaction. Such cross-sections are expressed in units of *barns*, where one barn equals $10^{-24}cm^2$. This

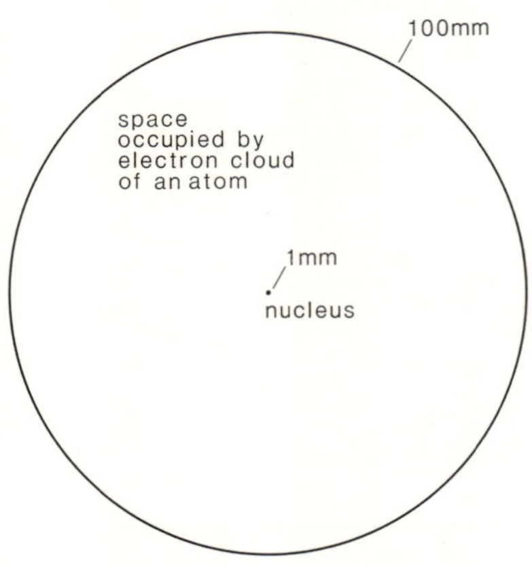

Figure 12.3 Schematic representation of the relative sizes of atom (10^{-10} m) and nucleus (10^{-14} m). Owing to the limitations in reproduction, the nucleus has been drawn 100 times too large; and the diameter of a neutron (10^{-18} m) would be 10 000 times smaller than this!

name is said to have been chosen by whimsical physicists in recognition of the very small size of this unit (Moore, 1966, p.798). (See also SI units, section 1.3.6.)

12.2.5 *Neutron activation reactions*

The nuclear reaction most likely to follow neutron capture of low-energy neutrons is the (n,γ) reaction. The compound nucleus formed by capture of neutron by nucleus usually has a very short half-life (10^{-9}–10^{-12} s) and loses its excess energy by the emission of gamma-rays whose energy is characteristic of differences between energy levels in the excited nucleus. These gamma-rays are called *prompt gammas* and they are generally emitted in the reactor during irradiation. A specialized form of analysis (*prompt gamma-ray NAA*), makes use of this radiation to determine specific elements, especially boron and some light elements (Peisach, 1981). Applications of such procedures in silicate rock analysis are discussed further in section 12.18. The product nuclide of an (n,γ) reaction has an *atomic number* that is unchanged but an *atomic weight* that has increased by one unit following capture of the neutron. This product nuclide is almost always radioactive and it is the measurement of gamma-rays emitted during this decay of this isotope after irradiation that forms the basis of neutron activation analysis.

A typical (n,γ) reaction is as follows:

		prompt gammas	
$^{140}Ce + {}^{1}n \rightarrow$	$(^{140}Ce\ {}^{1}n)^*$	$\rightarrow$	^{141}Ce
(natural isotopic abundance 88.5%)	(excited compound nucleus)		$T_{1/2} = 32.4d$ (radioactive product)

or

$$^{140}Ce\ (n, \gamma)\ ^{141}Ce.$$

Other types of nuclear reaction usually have low cross-sections to thermal neutrons and are normally considered to be interferences. Such reactions include:

^{54}Fe (n, p) ^{54}Mn (neutron capture followed by ejection of a proton from the nucleus)

^{235}U (n,f) ^{141}Ce (neutron-induced fission of ^{235}U yielding ^{141}Ce as fission product).

12.2.6 *Equation for the build-up of radioactivity*

The number of decays per second (activity A) induced in an isotope at the end of an irradiation is given by

$$A = N\sigma\phi\left[1 - \exp\left(-\frac{0.693T_{IRR}}{T_{1/2}}\right)\right] \qquad (12.1)$$

where N is the number of atoms of the target isotope in the irradiated sample, σ is the neutron capture cross-section, ϕ is the neutron flux, T_{IRR} is the duration of irradiation and $T_{1/2}$ is the half-life of the product isotopes.

$$N = \frac{m\theta}{AW} 6.023 \times 10^{23} \qquad (12.2)$$

where m is the mass of the *element* in the irradiated sample, θ is the fractional isotopic abundance of the target *isotope*,

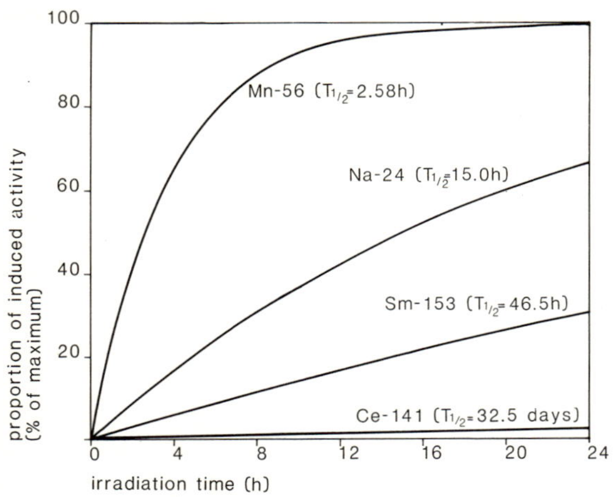

Figure 12.4 Build-up in activity induced in the isotopes ^{56}Mn, ^{24}Na, ^{153}Sm, ^{141}Ce as a function of irradiation time.

AW is the atomic weight of the element and 6.023×10^{23} is the Avogadro constant.

If components from both thermal (ϕ_{TH}) and epithermal (ϕ_{EPI}) fluxes are to be taken into account, $\sigma\phi$ (in equation (12.1)) is replaced by ($\sigma\phi_{TH} + \phi_{EPI}\, I$), where I is the epithermal neutron capture cross-section.

The exponential term in equation (12.1) represents the radioactive decay of the product isotope during irradiation in the reactor. The factor $0.693/T_{1/2}$ used in this expression is known as the *decay constant* of the isotope.

The factor

$$\left[1 - \exp\left(-\frac{0.693T_{IRR}}{T_{1/2}}\right)\right] \qquad (12.3)$$

is called the *saturation factor*. For irradiations that are long in comparison with isotope half-life ($T_{IRR} \gg T_{1/2}$), a point is reached where no significant increase in isotope activity results from further irradiation. Thus, over 99% of the maximum activity is induced for irradiations that exceed seven half-lives of a specified isotope. The build-up in activity for selected isotopes is shown in Figure 12.4. These data show that ^{56}Mn ($T_{1/2} = 2.58$h) is saturated after a 24-hour irradiation, but that the activities of ^{141}Ce ($T_{1/2} = 32.4$d) and ^{60}Co ($T_{1/2} = 5.27$y) (not plotted) have only attained 2.1% and 0.04% respectively of their potential saturation activities.

12.2.7 *Half-lives*

The half-life of an isotope is the time required for one-half of the activity of that isotope to decay, and is described by the function

$$A_T = A_0 \exp\left(-\frac{0.693T}{T_{1/2}}\right] \qquad (12.4)$$

where A_0 is the initial activity (decays per second), A_T is the activity after time T and $T_{1/2}$ is the isotope half-life.

The proportion of activity remaining after decay intervals corresponding to the first ten half-lives of an isotope is listed in Table 12.1. The relative rates of decay of five isotopes ranging from ^{56}Mn ($T_{1/2} = 2.58$h) to ^{60}Co($T_{1/2} = 5.27$y) are plotted in Figure 12.5 over a period of two months, assuming equal starting activities. Whereas ^{56}Mn has effectively decayed to negligible proportions during the first day, ^{60}Co retains 97.8% of its activity after 60 days.

Table 12.1 Proportion of radioactive isotope remaining after successive half-line decay intervals

Number of half-lives of decay	Residual activity (%)
0	100
1	50
2	25
3	12.5
4	6.3
5	3.1
6	1.6
7	0.78
8	0.39
9	0.20
10	0.10

12.3 Radioactive decay schemes

Isotopes formed by thermal neutron capture are almost invariably radioactive and decay by one, or sometimes by a combination, of the following routes.

12.3.1 *Beta-emission*

By far the largest number of isotopes formed by thermal neutron capture decay by beta-emission. The incorporation of a neutron upsets the balance in the nucleus between total numbers of protons and neutrons. Many isotopes return to a state of nuclear stability by converting a neutron to a

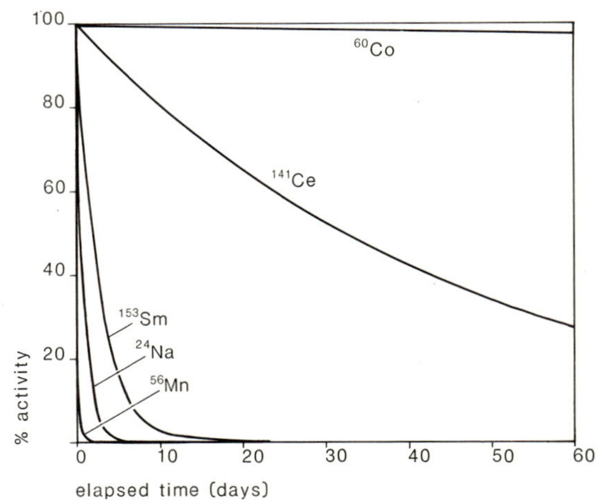

Figure 12.5 Relative rates of decay of the isotopes ^{56}Mn ($T_{1/2}$ = 2.58 h), ^{24}Na ($T_{1/2}$ = 15.0 h), ^{153}Sm ($T_{1/2}$ = 46.8 h), ^{141}Ce ($T_{1/2}$ = 32.4 d) and ^{60}Co ($T_{1/2}$ = 5.27 y) plotted on the assumption of equal initial activities.

proton with the simultaneous ejection of a beta-particle (electron) from the nucleus (Figure 12.6*a*). This transition results in a product isotope with the same atomic weight as the radioactive parent but with an atomic number which has increased by one unit.

The simplest form of beta-decay is shown in Figure 12.6*a*. ^{141}Ce ($T_{1/2}$ = 32.4d) decays to ^{141}Pr by two routes:

(i) 30% of ^{141}Ce nuclei decay with the ejection of 0.58 MeV beta particles directly to the ground state of ^{141}Pr.

(ii) 70% decay by emitting a 0.44 MeV beta particle to the first nuclear excited state of ^{141}Pr. This almost instantly decays to the ground state, emitting a 145 keV gamma-ray. Note however that only 48% (rather than 70%) of ^{141}Ce decays are associated with this gamma emission due to the *internal conversion* of a proportion of the 145 keV gammas (see 12.3.2).

A more complex scheme is shown in Figure 12.6*b* for the beta-decay of ^{153}Sm ($T_{1/2}$ = 46.8h) to ^{153}Eu (stable). Decay occurs by a range of routes in which beta-particles are ejected with energies of 0.63 MeV (34%), 0.70 MeV (44%), 0.71 MeV (0.6%) and 0.84 MeV (21%). Several excited states of the ^{153}Eu nucleus are involved (Figure 12.6*b*). Transitions from these excited levels of ^{153}Eu to the ground state are governed by selection rules which control the relative intensity of gamma emissions, the most intense of which occur at 103.2 keV (28.3%) and 69.7 keV(5.3%). The typical half-lives of intermediate excited states are between 20^{-11} and 10^{-13}s. Note that these 69.7 keV and 103.2 keV gamma rays (Figure 12.6*b*) are emitted *in cascade* (one very rapidly after the other). This is relevant to 'true' coincidence summing phenomena in radiation detectors to be discussed in section 12.8.1.

12.3.2 *Internal conversion*

Internal conversion is not a decay route as such, but a secondary process that can affect the proportion of gamma-rays emitted by an atom. Gamma photons, emitted by decay of the nucleus, can interact with orbital electrons of the atom or origin. The electrons with the highest probability of interaction in this manner are those from the *K*-shell. The energy of the gamma-ray is completely transferred to an orbital electron, causing its ionization from the atom with the excess appearing as kinetic energy. In consequence, the atom is left with a vacancy in the *K*-orbital shell which is filled by an *L*- or M- shell electron falling down the potential energy gradient. Simultaneously, $K\alpha$ and $K\beta$ x-rays are emitted in a mechanism similar to the Auger effect in x-ray fluorescence analysis (section 8.4).

In the case of ^{141}Ce, for example, internal conversion of 145 keV gammas results in the appearance of praesodymium x-rays (Figure 12.6*a*). This process is most likely to occur after the emission of gamma-rays that just exceed the ionization energy of the *K*-shell electrons. The probability decreases rapidly in the case of gamma radiation of much higher energy (>1000 keV).

Throughout this discussion, the distinction between gamma-rays and x-rays is not made on the basis of energy alone. Gamma-rays result from transitions between nuclear energy levels and are emitted from the nucleus: x-rays result from electron transitions between orbital energy levels in the atom. Some gamma-rays have sufficiently low energies to fall within what is conventionally known as the x-ray region of the spectrum.

12.3.3 *Electron capture*

Electron capture is the radioactive decay in which the nucleus 'captures' an orbital electron, almost always from the *K*-shell, resulting in the conversion of a proton to a neutron in the nucleus. Again the atomic weight is unchanged but the atomic number of the product isotope *decreases* by one unit. The product nucleus is usually formed in a nuclear excited state, emitting gamma radiation as it relaxes to the ground state. Isotopes that decay by electron capture always emit x-rays as the vacancy created by the capture of a *K*-shell electron is filled by higher orbital electrons. Thus ^{153}Gd ($T_{1/2}$ = 242d) decays by electron capture to ^{153}Eu via several intermediate nuclear excited states (Figure 12.6*c*). Characteristic gamma-rays are emitted following transitions between these excited states and the ground states together with the characteristic x-ray spectrum of ^{153}Eu. The isotopes ^{153}Gd and ^{153}Sm are unusual since their decay (one by electron capture, the other by beta decay) results in the formation of the same stable isotope (^{153}Eu) with the simultaneous emission of gamma photons of identical energies. However, the relative intensities of these gamma-emissions depend on the respective decay scheme transition probabilities.

12.3.4 *Isomeric transition*

When a transition between two nuclear excited states is strongly forbidden, one of these excited states may have a finite lifetime. Thus, neutron capture by the natural isotope ^{59}Co generates two radioactive products, ^{60}Co ($T_{1/2}$ = 5.27y) and ^{60m}Co ($T_{1/2}$ = 10.5m). The isotope ^{60m}Co corresponds to the first nuclear excited state of ^{60}Co. These two nuclides are called *mesomers* and the decay of ^{60m}Co by gamma emission to ^{60}Co is known as an isomeric transition. The latter nuclide (^{60}Co) then decays with a much longer half-life by beta emission as shown in Figure 12.6*d*.

12.3.5 *Positron emission*

A few isotopes formed by thermal neutron capture decay by emitting positrons. These result from the conversion of a proton to a neutron on the nucleus, and thus the atomic number of the product nucleus *decreases* by one unit. The nuclide ^{64}Cu ($T_{1/2}$ = 12.7h) follows a complex decay scheme (Figure 12.6*e*) with certain proportions of radioactive nuclei decaying by beta emission (40%), electron capture (40%) and positron emission (19%). The product nucleus after positron emission is ^{64}Ni. Positrons are unstable, and after losing their kinetic energy (0.66 MeV in this case), they spontaneously annihilate by interaction with an electron, emitting two gamma-rays each of energy 511 keV (see also section 12.6.3). Thus, positron emitters are always sources of 511 keV radiation due to annihilation events within the sample and surrounding matter.

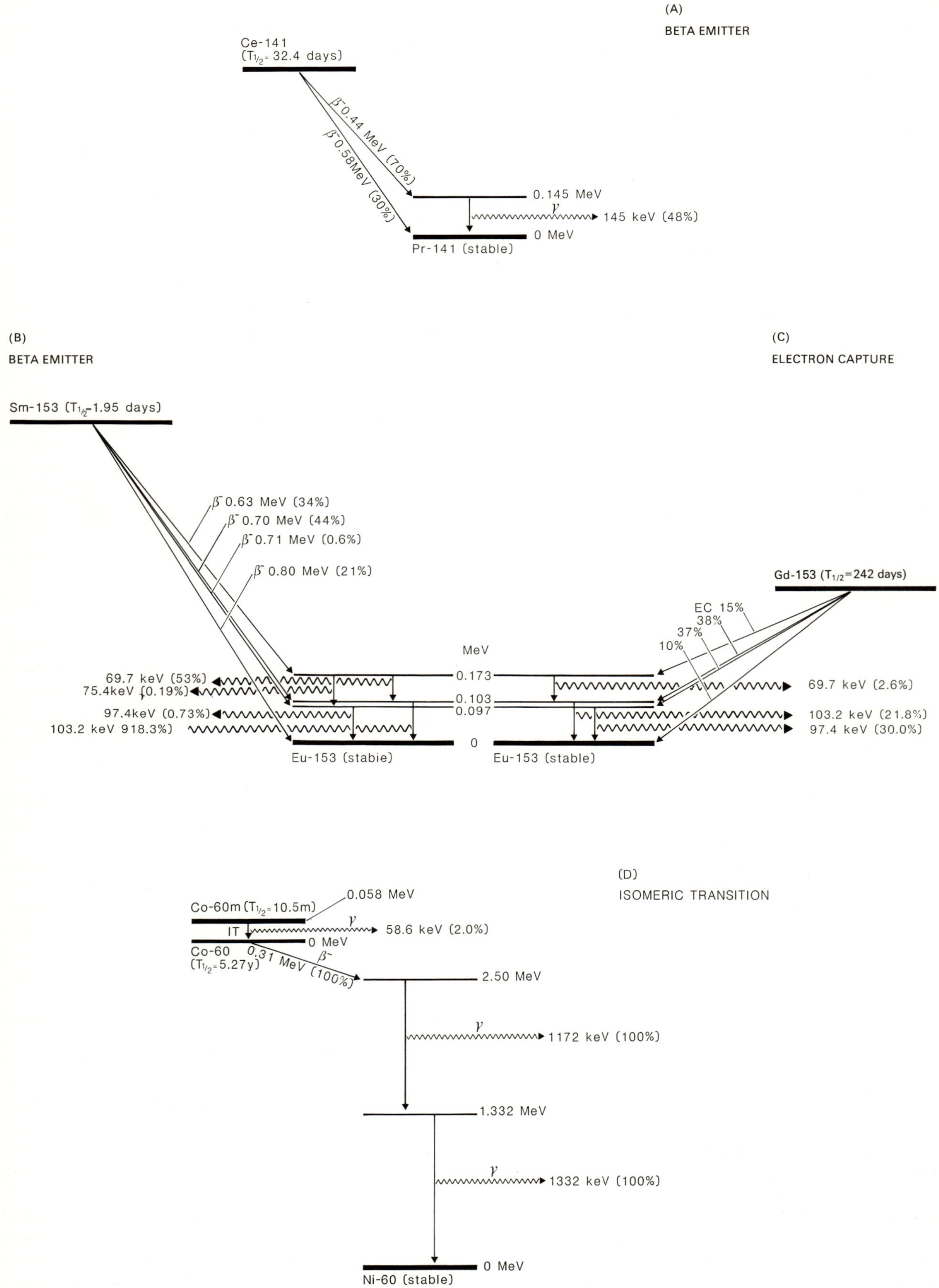
(A)
BETA EMITTER
Ce-141
(T½= 32.4 days)
β⁻ 0.44 MeV (70%)
β⁻ 0.58MeV (30%)
0.145 MeV
γ
145 keV (48%)
0 MeV
Pr-141 (stable)
(B)
BETA EMITTER
Sm-153 (T½=1.95 days)
β⁻ 0.63 MeV (34%)
β⁻ 0.70 MeV (44%)
β⁻ 0.71 MeV (0.6%)
β⁻ 0.80 MeV (21%)
69.7 keV (53%)
75.4keV (0.19%)
97.4keV (0.73%)
103.2 keV 918.3%)
Eu-153 (stabie)
MeV
0.173
0.103
0.097
0
(C)
ELECTRON CAPTURE
Gd-153 (T1/2=242 days)
EC 15%
38%
37%
10%
69.7 keV (2.6%)
103.2 keV (21.8%)
97.4 keV (30.0%)
Eu-153 (stable)
(D)
ISOMERIC TRANSITION
Co-60m (T½=10.5m)
0.058 MeV
γ
58.6 keV (2.0%)
IT
0 MeV
Co-60
(T½=5.27y)
0.31 MeV (100%)
β⁻
2.50 MeV
γ
1172 keV (100%)
1.332 MeV
γ
1332 keV (100%)
0 MeV
Ni-60 (stable)

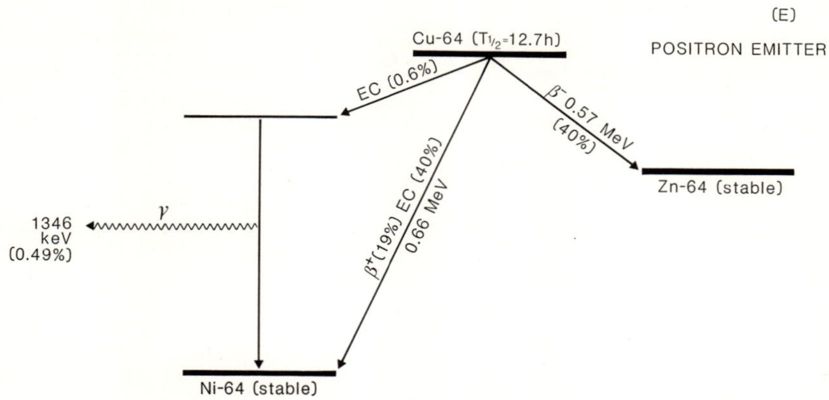

Figure 12.6 Schematic decay diagrams of (*a*) ^{141}Ce (beta-emitter); (*b*) ^{153}Sm (beta-emitter); (*c*) ^{153}Gd (electron capture); (*d*) ^{60m}Co (isomeric transition); (*e*) ^{64}Cu (positron-emitter).

12.3.6 *Bremsstrahlung*

Bremsstrahlung or 'braking radiation' is caused by the deceleration of energetic beta-particles within the strong electrostatic fields round a nucleus. The energy of this gamma radiation forms a background continuum from zero up to the maximum energy of emission of the beta-particle. Irradiated samples are therefore also sources of this continuum radiation.

12.3.7 *Compound decay schemes*

Occasionally, the isotope used for analytical measurements in gamma spectra does not correspond to the radioactive species formed by neutron capture in the nuclear reactor. Two common examples of this phenomenon are in the determination of uranium (from the activity of a neptunium isotope) and of thorium (from a protactinium isotope). In each case, the isotope formed by neutron capture is a short half-life isotope of the parent which rapidly decays by beta emission. The product of this decay is also radioactive and may be determined as a direct measure of its parent. The corresponding nuclear reactions are as follows:

Uranium:

$$^{238}\text{U}(\text{n},\gamma) \quad \underset{(T_{1/2}=23.5\text{m})}{^{239}\text{U}} \xrightarrow{\beta'} \underset{(T_{1/2}=2.36\text{d})}{^{239}\text{Np}}$$

Thorium:

$$^{232}\text{Th}(\text{n},\gamma) \quad \underset{(T_{1/2}=22.3\text{m})}{^{233}\text{Th}} \xrightarrow{\beta'} \underset{(T_{1/2}=27.4\text{d})}{^{233}\text{Pa}}$$

One day after irradiation, complete decay of these intermediate isotopes will have occurred and the concentration of uranium and thorium will be entirely represented by the activities of ^{239}Np and ^{233}Pa, respectively.

12.4 Instrumentation

Systems for measuring the gamma spectrum of an irradiated sample comprise three principal parts.

(i) The *detector* which in neutron activation work is invariably a germanium crystal—gamma photons entering the detector cause ionization of the germanium crystal, so generating a small electronic signal, the magnitude of which is proportional to the gamma photon energy

(ii) Electronic amplification networks, used to amplify this small signal, to maintain optimum signal-to-noise ratio and to minimize degradation of output signals at high count rates

(iii) A multi-channel analyser, to sort and store detected pulses and so accumulate a gamma spectrum as a histogram of the number of detected events as a function of photon energy.

12.4.1 *Semiconductor detectors*

A semiconductor is a regular crystalline material made up of atoms whose electrons can occupy two distinct bands of energies separated by a forbidden energy gap (Figure 12.7). The lower energy band is called the *valence band* and is occupied by orbital electrons which constitute the network of covalent bonds between the atoms of the crystalline material. At a temperature of 0 K, the valence band would be fully occupied by electrons and the higher *conduction band* completely empty. In this state the material would be an insulator. However, the energy required to excite electrons to the conduction band is relatively low in a semiconductor material. At normal temperatures, thermal energy is sufficient to raise a proportion of valence band electrons to this conduction band. This process leaves vacancies in the valence band which effectively behave as positive 'holes'. Under these conditions, an electric field (bias) placed across the semiconductor will cause the crystal to conduct as the conduction band electrons are efficiently swept in one direction and the positive holes swept rather more slowly in the other direction along the electrostatic gradient. If such a semiconductor device is to be used successfully to measure gamma radiation, it is necessary to minimize conduction cause by thermal excitation and to promote the efficient collection of all electrons resulting from gamma-induced ionization of semiconductor atoms.

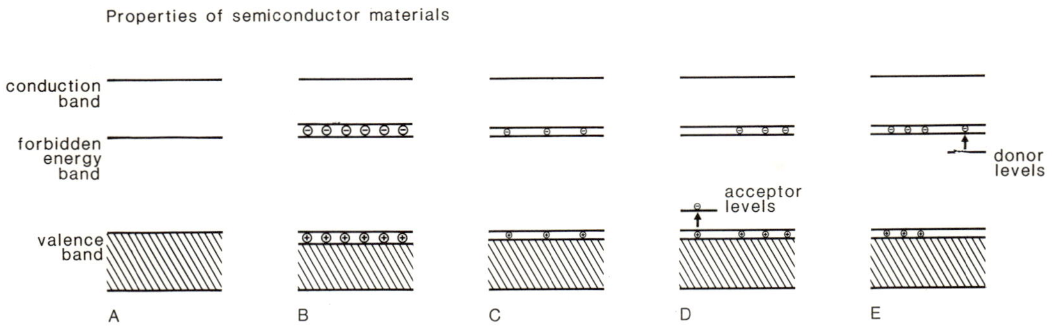

Figure 12.7 Schematic behaviour of a semiconductor crystal.
A: Perfect (intrinsic) semi-conductor at 0 K, the valence band is fully occupied by electrons, and the conduction band is empty, in this state the semiconductor cannot conduct
B: Semiconductor at room temperature; significant thermal excitation of electrons from valence to conduction band; in this state the semiconductor will conduct
C: Semiconductor at 77 K; vast reduction in thermal ionization to conduction band
D: Effect of 'acceptor' atom impurities in *p*-type semiconductor material
E: Effect of 'donor' atom impurities in *n*-type semiconductor material.
Shading indicates valence band fully occupied by electrons. Arrows indicate direction of ionization of electrons to or from impurity atoms.

12.4.2 *Germanium gamma detectors*

Germanium is chosen as the most suitable material for the construction of gamma radiation detectors because:

(i) Germanium is a semiconductor material which can be refined to an extremely high degree of purity

(ii) It has a relatively high atomic number and is therefore more effective at stopping high-energy gamma radiation than silicon, for example

(iii) The mean energy required to create an electron–hole pair (i.e. ionization of electrons from valence to conduction bands) in germanium is low.

The last point is particularly important. The magnitude of the energy gap between valence and conduction bands is 0.73 eV. However, because electron excitation occurs relatively inefficiently, the mean energy required to create one electron–hole pair is about 2.95 eV. Thus, a 1000-keV gamma-ray will create on average about 3.4×10^5 electrons (5.7×10^{-14} coulombs of charge). Because this is a relatively large number of electrons, the statistical spread in detecting gamma-rays due to ionization in the detector will be low, equivalent to about 1.5 keV at this gamma energy (see section 12.8.2). This compares favourably with gas ionization detectors and sodium iodide scintillation detectors where the equivalent electron–hole pair ionization energies are approximately 40 eV and 300 eV respectively. Germanium crystals thus have inherently good energy resolution characteristics in comparison with other devices. The detection characteristics of semi-conductor detectors and proportional counters are compared in Table 8.14.

12.4.3 *P- and n-type germanium*

Impurities in germanium crystals, even at very low concentrations, modify its semiconductor properties in one of two ways.

An impurity like arsenic has one extra electron over and above the four required to form covalent bonds with adjacent germanium atoms. The effect of such an impurity is to form localized energy levels in the forbidden region between conduction and valence bands from which this extra electron can readily be donated to the conduction band (Figure 12.7*e*). Germanium containing an excess of such *donor* atoms is called an *n-type* semiconductor, since it has an excess of electrons (*n*egative charges) in its conduction band.

Conversely, gallium for example has one too few electrons available to form the four covalent bonds with adjacent germanium atoms. This impurity causes *acceptor levels* to form in the forbidden energy region to which electrons from the valence band can readily be excited. Such a material will conduct due to the excess of *p*ositive holes in this valence band and is called *p-type* germanium (Figure 12.7*d*).

Both types of impurity have the undesirable consequence of causing the germanium to conduct under an applied electrostatic field even in the absence of ionizing radiation. Furthermore, impurity atoms provide local centres within the germanium crystal at which electrons can be trapped (*trapping centres*) or which can promote the *recombination* of electron–hole pairs. Both phenomena cause *incomplete charge collection* which shows itself as an asymmetric tailing effect on the low-energy side of a detected gamma photopeak (section 12.7).

12.4.4 *Lithium-drifted germanium detectors*

To compensate for the presence of impurity atoms, a process has been developed in which lithium atoms are drifted into the germanium crystal lattice.

Lithium will diffuse readily into a germanium crystal, occupying interstitial sites, at which its single valence electron acts as an *n*-type donor. The lithium drifting process is extensively used in the fabrication of coaxial germanium detectors (Figure 12.8). A cylinder of *p*-type germanium is coated with lithium and then baked at 350–450°C in an inert atmosphere. Lithium atoms rapidly diffuse into the surface of the crystal to form an *n*-type layer typically of thickness 300–1000 μm. Under the influence of an applied electric field, lithium atoms from this surface layer can be caused to migrate into the crystal lattice where they seek out

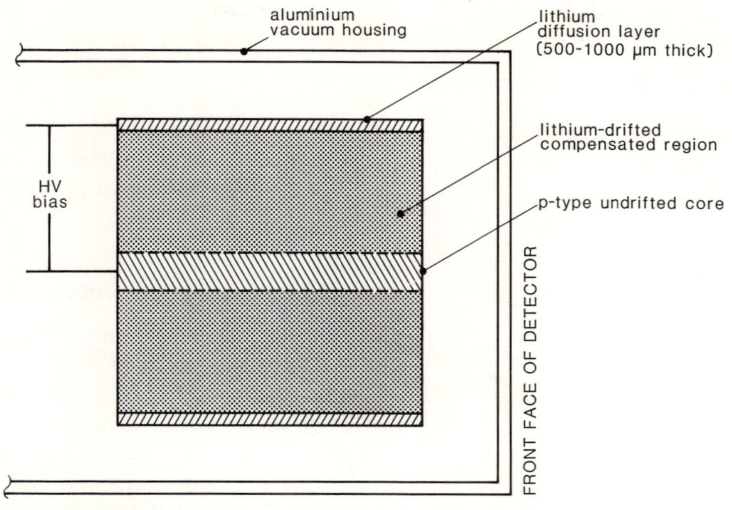

Figure 12.8 Schematic section through a lithium-drifted germanium detector with true coaxial geometry. Detector is a cylinder of germanium whose circular axis coincides with the *p*-type core.

impurity acceptor atoms and exactly compensate their *p*-type properties. Drifting is continued under carefully controlled conditions for sufficient time to create an extensive pseudo-intrinsic region in which the concentration of charge carriers (electron–hole pairs) is *depleted*. An undrifted *p*-type core remains (Figure 12.8). *Intrinsic* is a term applied to a semiconductor whose electrical properties are those characteristic of an ideal crystal. The schematic variation in lithium concentration from edge to core of such a device is shown in Figure 12.9.

In operation, a high-voltage bias (usually between 2 and 3 kV) is applied between the lithium diffusion layer and the *p*-type core. This bias causes efficient collection of any electrons ionized to the conduction band following the detection of a gamma photon in the depleted lithium-drifted region. The volume of coaxial germanium detector used in neutron activation work is usually between 30 and 100cm^3.

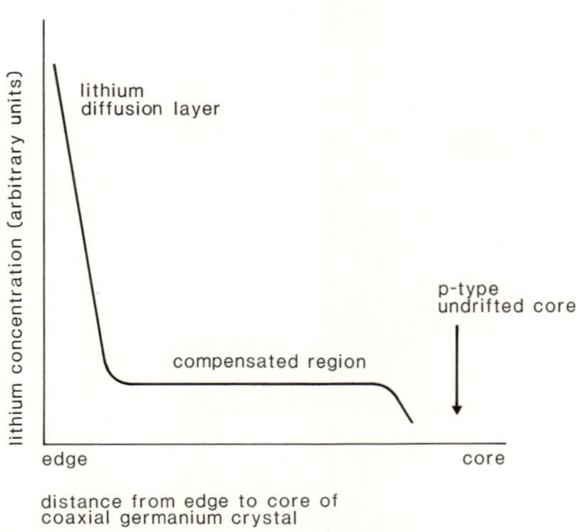

Figure 12.9 Schematic variation of lithium atom concentration between edge and core of a lithium-drifted germanium detector in coaxial geometry as shown in Figure 12.8.

12.4.5 *Hyperpure germanium detectors*

Modern developments in refining technology have resulted in the manufacture of germanium crystals of exceptionally low impurity concentrations, amounting to less than 10^{10} impurity atoms per cm^3 of germanium (1 atom per 10^{12} germanium atoms). Such material has a sufficiently low density of free charge carriers to allow it to be used as a gamma detector without the necessity for compensation by the lithium drifting process. The material is generally known as 'hyperpure' germanium, or less correctly, 'intrinsic' germanium.

Hyperpure germanium was originally used to manufacture relatively small planar detectors because it was difficult to refine large crystals of sufficiently uniform purity suitable for coaxial detectors. These difficulties have largely been overcome and large coaxial hyperpure germanium detectors are now generally available.

The particular advantage of such detectors is that they may be stored at room temperature with little danger of damage provided the detector bias is removed. Lithium-drifted detectors may suffer irreversible damage if they are ever allowed to warm up above liquid nitrogen temperatures.

12.4.6 *Planar detectors*

Whereas coaxial detectors with their large active volume are most appropriate for the detection of high-energy gamma radiation (300–2000 keV), small planar crystals are widely used to measure the lower energy region of the spectrum (60–300 keV). In this role they are known as *low-energy photon spectrometers* (LEPS detectors).

The construction of a planar hyperpure germanium detector is shown in Figure 12.10. It consists of a flat cylinder of germanium, typically 6 to 30 mm in diameter and 5 to 15 mm deep. Two electrical contacts are required on the front and back faces to apply the high voltage bias (usually 500–1500 V). The back contact is usually a 300–1000 μm layer of germanium into which lithium atoms have been diffused.

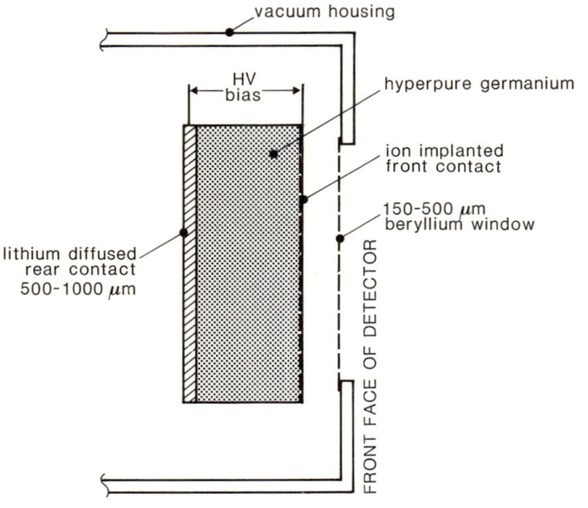

Figure 12.10 Schematic section through a planar hyperpure germanium detector. The detector is a flat cylinder of germanium whose circular axis lies perpendicular to the beryllium window.

In some detectors the front contact is a thin evaporated film of metallic gold or palladium. This metallic film can cause the absorption of low energy radiation passing through it. Fluorescence of characteristic Au or Pd X-rays originating from this layer may also be detected. To avoid these interferences, some detectors now employ an extremely thin (<0.3 μm) ion-implanted layer of boron as front contact. The vacuum housing in which the planar detector is mounted is normally fitted with a beryllium front window to minimize absorption of low-energy radiation, 150–500 μm being a typical range of thicknesses. The lower energy limit of detection is 3 to 5 keV and is restricted by absorption in the beryllium window, inefficient charge collection of low-energy gamma-ray events which do not fully penetrate the germanium crystal, and inappropriate time constants (1–6 μs) on gaussian amplifiers normally employed for gamma spectroscopy (see section 9.4 for comparison with silicon x-ray detectors).

12.4.7 *Liquid nitrogen and vacuum housing*

In operation a germanium detector must be cooled to liquid nitrogen temperatures (boiling point 77 K, -196°C) in order to minimize electronic noise created by the thermal excitation of electrons from valence to conduction bands.

Such low temperatures also serve to freeze lithium atoms in those detectors manufactured by the lithium-drifting process. This is necessary to avoid irreversible damage that can occur at higher temperatures due to the mobility of lithium atoms within the germanium crystal.

Germanium detectors must also be mounted in an ultrahigh vacuum environment to prevent surface contamination of the germanium crystal. Vacuum (10^{-6} to 10^{-8} Torr) is normally maintained by incorporating a molecular sieve within the vacuum housing to absorb extraneous gases. Surface contamination causes a *leakage current* to flow when the high-voltage bias is applied to the detector due to tracking across the surface of the crystal. This results in high electronic noise, poor detector resolution and ultimately the inability to maintain an adequate operating bias.

The typical configuration of a coaxial germanium spectrometer is shown in Figure 12.11. Detectors may be

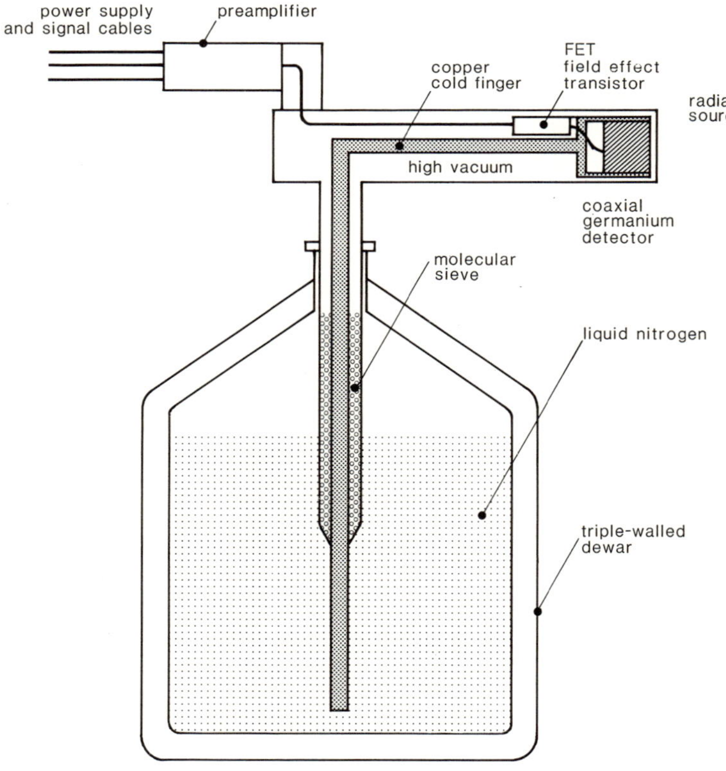

Figure 12.11 Schematic diagram of a typical gamma spectrometer with horizontally-pointing germanium crystal assembly.

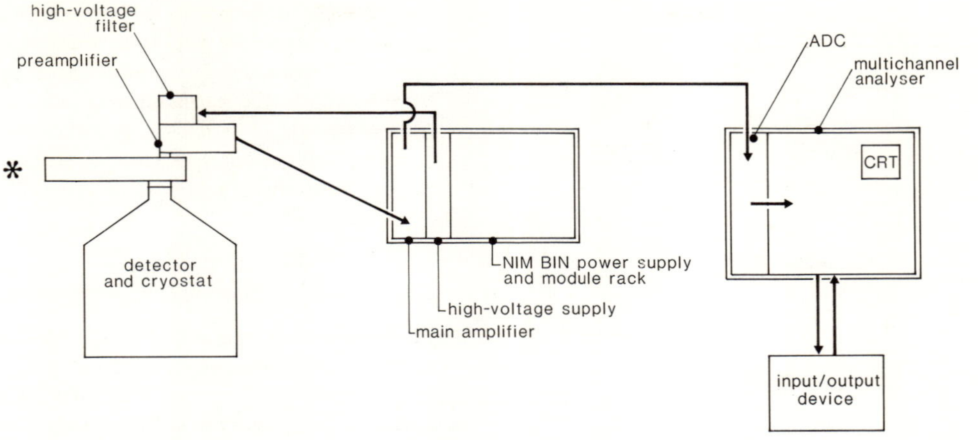

Figure 12.12 Schematic diagram of neutron activation counting system, showing cable interconnections between electronic modules.

mounted either horizontally or vertically depending on application and desired counting geometry.

12.5 Pulse-processing electronics

Pulse-processing electronics consists of preamplifier, main amplifier and multi-channel analyser and a typical configuration is shown in Figure 12.12.

12.5.1 *Preamplifier*

The function of the preamplifier is to collect and amplify the electronic charge created in the germanium detector by ionization due to a gamma-ray event. In operation a preamplifier must contribute an extremely low degree of electronic noise to the detected signal and maintain a high degree of gain stability. Variation in either of these parameters would be amplified by the rest of signal processing electronics.

The first stage of preamplification is a field effect transistor (FET), a charge-sensitive active electronic device which possesses inherently low noise characteristics. The FET is often mounted in the detector housing adjacent to the detector so that it too can be cooled to minimize electronic noise. In this way, interferences from stray signals that could be induced in long interconnecting cables between detector and FET are also eliminated. Other stages of voltage amplification follow this charge-sensitive first stage. The output signal is a step function with a very rapid rise time (typically 50 ns) but with a relatively long tail which decays with a time constant typically of 50 μs (Figure 12.13).

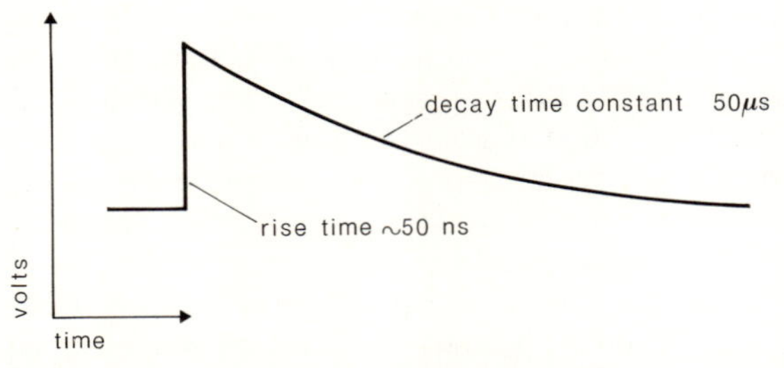

Figure 12.13 Typical preamplifier output pulse showing approximate rise and decay times.

12.5.2 *Main amplifier*

Because of the long decay time constant, pulses from the preamplifier are vulnerable to pile-up distortion effects where the rise of one pulse is distorted by the tail of the preceding one (Figure 12.14). The function of the main amplifier is to minimize this effect by employing pulse shaping and baseline restoration circuitry to shorten the effective length of individual pulses consistent with maintaining an adequate signal to noise ratio.

Main amplifier output pulses are near-gaussian in shape and are of an amplitude that is linearly related to the input step function from the preamplifier. More sophisticated units have additional circuitry to detect distorted pile-up events and to inhibit their conversion by the MCA. This enhances both gain and energy resolution stability at high count rates where pile-up events are most likely to occur. The control features in the main amplifier which must be optimized by the user are as follows.

(*a*) *Pole-zero cancellation.* Pole-zero cancellation is used to eliminate mismatch between preamplifier output and the first-stage differentiation network in the main amplifier. Mismatch manifests itself as either an overshoot or undershoot in the tail of the main amplifier pulse with respect to baseline. Succeeding pulses thus sit on a high or low but variable baseline causing pulse pile errors and poor energy resolution in recorded spectra.

(*b*) *Baseline restorer.* With increasing count-rate, it is possible for the dc baseline level to drift upwards causing a count rate dependent shift in amplifier gain. The baseline restorer circuit removes this effect. It is normally necessary to set a threshold level so that the restorer is triggered only by real pulses and not by electronic noise fluctuations.

(*c*) *Time constants.* Time constants are user-selectable at values normally between 1 and 6 μs in most NAA applications. The selected time constant controls the pulse-shap-

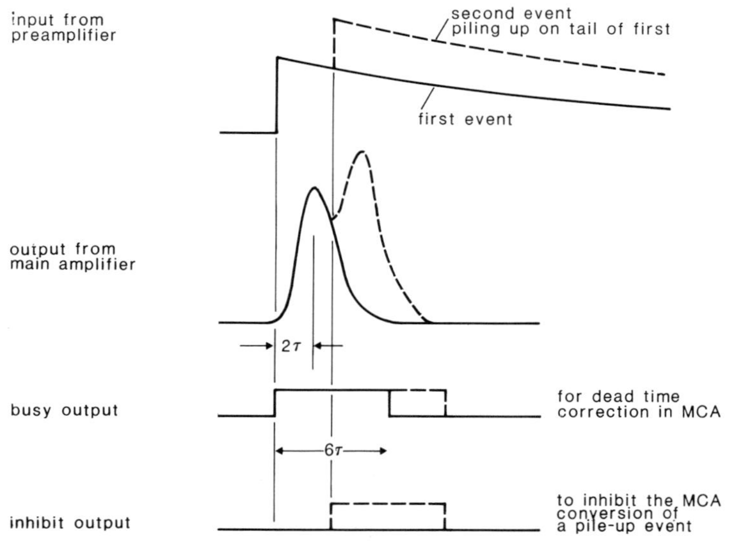

Figure 12.14 Pulse events in main amplifier incorporating pulse pile-up circuitry. (After E.G. & G. Ortec.)

ing electronic networks and therefore directly effects the length (in terms of time) of the gaussian output pulses. Long time constant settings improve the signal to noise ratio, and therefore energy resolution, of the amplifier. Short time constants reduce the probability of pulse pile-up effects, allowing the system to operate at high or variable count rates with minimum change in gain or resolution. Thus, resolution specifications of detectors are most favourably measured at 6 μs settings, but practical analytical systems usually operate at time constants between 1 and 3 μs. The change in energy resolution with time constant setting on a planar gamma spectrometer determined for the 121 keV photopeak of ^{57}Co is typically

Time constant	1	2	6 μs
121 keV FWHM	620	580	540 eV

(*d*) *Pulse pile-up.* Pulse pile-up occurs when two gamma events are detected simultaneously or in very close sequence. The pulse shaping electronics are then unable to distinguish individual events, and the main amplifier output pulse either represents the sum of the two energies (random summing) or a distorted combination of the two. In either case the output is unwanted. Pile-up effects usually become significant at count rates approaching 10 000 cps. Modern spectroscopy amplifiers can limit the effect of pile-up. The signals generated by one such circuit, designed to minimize this effect, are shown in Figure 12.14. Their significance is as follows. The pulse-shaping networks in the main amplifier convert input pulses from the preamplifier into near-gaussian output pulses with a rise time of typically twice the selected value of time constant (τ). Simultaneously the amplifier creates a busy signal (TTL logic '1') which extends from detection of input pulse for a time typically six times the selected time constant, to allow the baseline restoration circuits to stabilize the input and prevent interference effects on any following signals. The busy signal is fed to the ADC to be used for dead-time corrections (see section 12.5.4). If a second input pulse occurs within a very short time interval, the amplifier is unable to distinguish the two events and the output pulse represents the sum of both inputs. If, however, the second event occurs significantly after 2τ but before baseline restoration is completed at 6τ, the amplifier responds in two ways. First, the busy signal is extended by an appropriate amount to account for this second event. Second, an 'inhibit' logical signal is generated and fed to the multi-channel analyser to allow rejection of the distorted pile-up event. In this way the useful application of the amplifier can be extended to count rates of perhaps to 10^5 cps. It is interesting to compare this fairly simple pile-up circuit with the more sophisticated designs necessary in x-ray applications where amplifier time constants must be considerably longer to permit adequate reduction in signal noise (section 9.4).

12.5.3 *The multi-channel analyser*

The multi-channel analyser consists of three parts: an analogue-to-digital converter (ADC), a data scaler and a memory accumulator.

(*a*) *The analogue-to-digital converter* (*ADC*). The analogue-to-digital convertor (ADC) converts the analogue gaussian pulse from the main amplifier to a digitized signal which may then be stored as an event in an appropriate memory channel. This is usually accomplished by allowing a capacitor to charge up to the peak voltage of the input pulse (Figure 12.15). The capacitor is then discharged using a constant current source. The time to undertake this discharge is a direct measure of the input pulse voltage and is determined by counting the number of pulses generated by a high-frequency oscillator during this interval. The precision of this conversion depends on the rate at which the capacitor is discharged and the frequency of the oscillator. In order to minimize contributions to dead time which result from slow conversion times, it is usually necessary to use

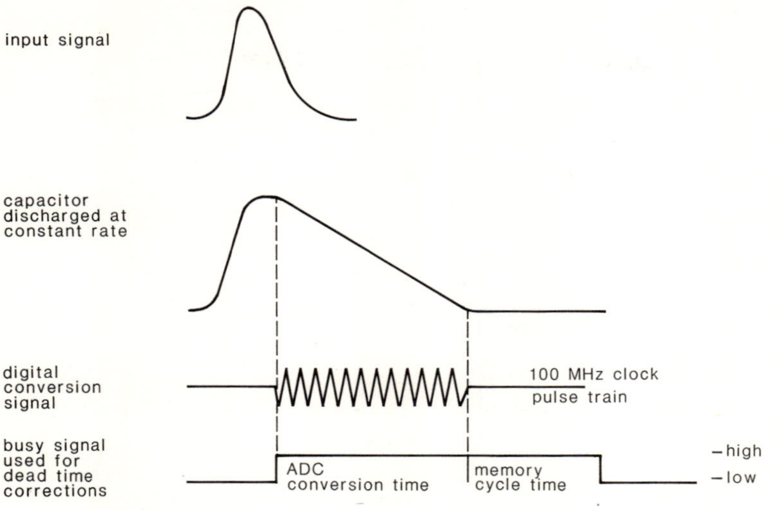

Figure 12.15 Multi-channel analyser conversion and dead-time signals.

high-frequency oscillators. For this reason most ADCs operate at a minimum of 50 MHz and more frequently 100 MHz. (*b*) *Memory accumulator*. Data from the ADC are accumulated in the memory which normally consists of 4096 individual channels, each of which can store an integer typically of up to 10^6 counts. Since the digitized signal from the ADC represents one gamma-ray energy, this datum is scaled to the appropriate memory channel so that the number accumulated at that address is caused to increase by one unit. A histogram of data accumulated in the MCA memory therefore represents a plot of gamma photon energy (channel number) versus intensity (memory counts) of the recorded spectrum.

Facilities are usually available to display the spectrum held in memory whilst it is being accumulated, allowing immediate visual appraisal of detected photopeaks. Older MCA systems were discrete instruments which used ferrite core memory, the contents of which could survive power failure. More modern systems are often integrated data acquisition and analysis instruments, either based on direct coupling of the MCA to a dedicated microprocessor, or with the memory accumulation function residing in the core of the microprocessor itself. The former can suffer from the relatively slow transfer of spectral data from MCA to computer; the latter can suffer from complete loss of accumulating spectrum data if the computer crashes for reasons independent of data acquisition (eg. corruption by an independent user program).

12.5.4 *Dead time*

Dead time is a phenomenon which results from the fact that pulse-processing electronics cannot carry out more than one operation at once. Thus when the ADC is performing a conversion on an individual pulse, the instrument is effectively unavailable for the acceptance of other pulses, so contributing to the dead time of the system. The true time for which the system was available to process pulses (the live time) is less than the real counting time. The difference is the dead time and expressed as a percentage:

$$\text{dead time} = \left(\frac{\text{real time} - \text{live time}}{\text{real time}}\right) \times 100 \qquad (12.5)$$

Since neutron activation analysis is a comparative technique, it is essential that spectra from individual samples are compared on the basis of equal live times. There are several ways of correcting for dead time. One widely used method, as described by Covell *et al.* (1960), involves extending the real time by an appropriate amount to account for dead time losses. This is achieved electronically by measuring the live time of a spectrum using a clock which is gated with the busy signal from the ADC (and main amplifier if appropriate). Each time the busy signal goes high, the clock is stopped (dead time incurred due to pulse pile-up, conversion or memory cycle). When the busy signal resets to low, the clock restarts. In this way the total time for which the spectrum was counted represents the total time for which the electronics were available to process pulses.

Most dead time in INAA counting systems is incurred in the ADC conversion and MCA memory cycle. When counting low-activity geological samples a week or more after irradiation, dead times rarely exceed 10–20%.

12.6 Interaction of gamma radiation with germanium detectors

Before discussing typical spectra recorded on neutron activation counting equipment, it is necessary to consider the artefacts introduced into spectra which result from the way in which high-energy radiation interacts with the atoms of a germanium detector. Artefacts are features in the spectrum that are caused by the detection and counting system and are not present in radiation emitted from the radioactive source. Gamma radiation causes ionization in matter by three principal processes: the photoelectric effect, Compton scattering and pair production. The relative cross-sections

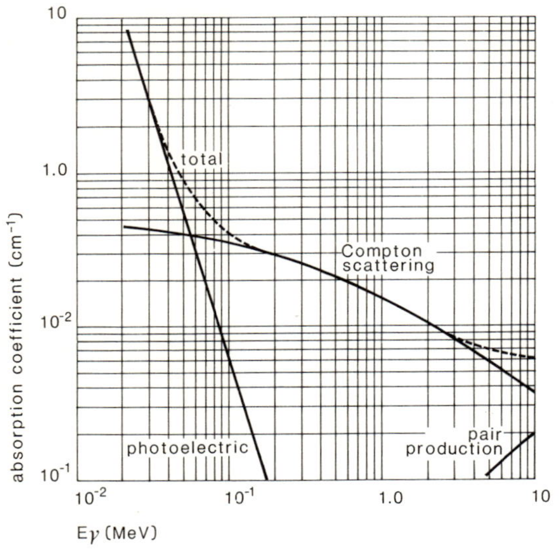

Figure 12.16 Linear absorption coefficient of gamma radiation in a germanium detector by the photoelectric effect, Compton scatter and pair production. (Grodstein, 1957, as reproduced by Adams and Dams, 1970.)

of each of these processes over the energy range under consideration here are shown in Figure 12.16. As can be seen, the predominant interaction in the range 100–4000 keV is Compton scattering. Below this energy, the photoelectric effect dominates. Above 5000 keV, interactions that induce pair production become significant.

12.6.1 *The photoelectric effect*

The photoelectric effect is the interaction between a gamma ray and an orbital electron in which complete transfer of energy to the electron occurs (Figure 12.17*a*). Part of this energy is used to overcome the ionization potential of the electron: the remainder appears as kinetic energy in the electron. This interaction is most likely to occur with orbital shells close to the nucleus (especially the *K*-shell). The vacancy left by the ionized electron is filled by an electron falling from the next higher shell, simultaneously emitting the characteristic *K* x-rays of germanium (see section 8.3). Thus, photoelectric interactions always yield characteristic x-rays of the detector material. Energies of both ionized electron and x-ray are usually absorbed by further ionization of detector atoms.

However, there is a possibility that the x-ray may escape from the detector without further interaction. The total energy then absorbed within the detector equals the full energy of the incident gamma photon less the energy of a germanium x-ray (12 keV for a Ge *K*-line). This event is called an escape peak, and is observed as a subsidiary peak at an energy 12 keV below the principal gamma photon line. However, as can be seen from Figure 12.18, the probability of escape peak losses become negligible for incident radiation above about 60 keV, due to the decreasing probability of photon interaction with the detector by the photoelectric effect.

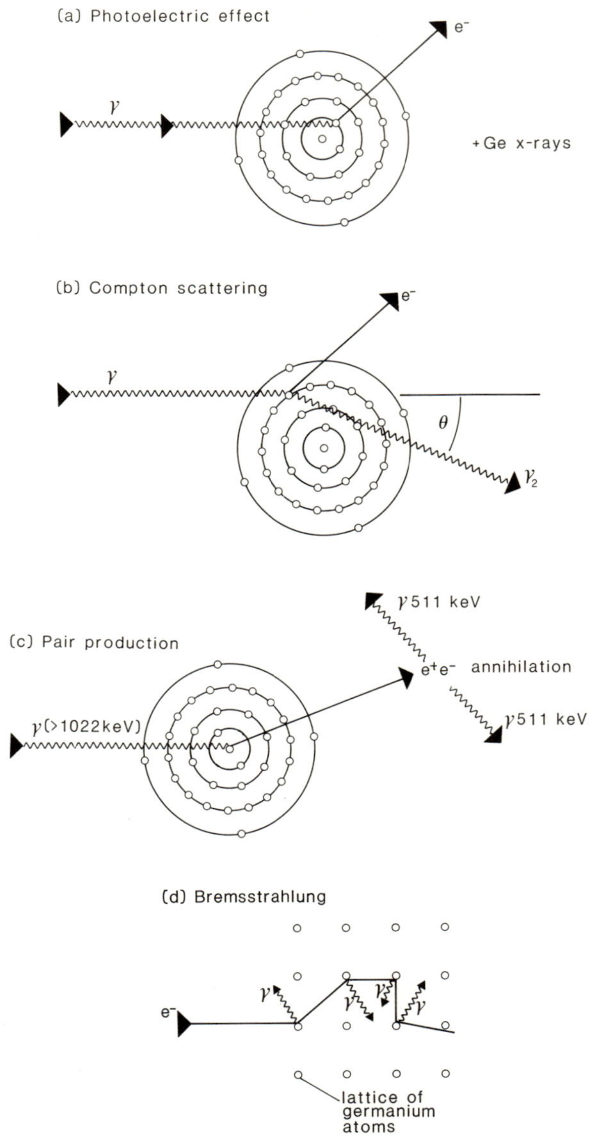

Figure 12.17 Interaction of gamma radiation with a germanium detector by: (*a*) photoelectric effect, (*b*) Compton scattering, (*c*) pair production, (*d*) bremsstrahlung effect.

12.6.2 *Compton scattering*

Compton scattering results from an *inelastic collision* between gamma-ray and (most probably) an outer orbital electron, in which only part of the gamma energy is transferred to the electron (Figure 12.17*b*). The remainder is re-irradiated as a lower energy gamma-ray emitted so that both total energy and momentum are preserved. Maximum transfer of energy results from a head-on interaction following which the secondary gamma-ray is emitted at 180° to the first. The secondary gamma photon can itself be absorbed by further Compton or photoelectric interactions. However, there is a probability that this gamma will itself escape from the detector. The detected event then corresponds to the full energy of the incident photon less that of the Compton escape gamma-ray. Compton scattering

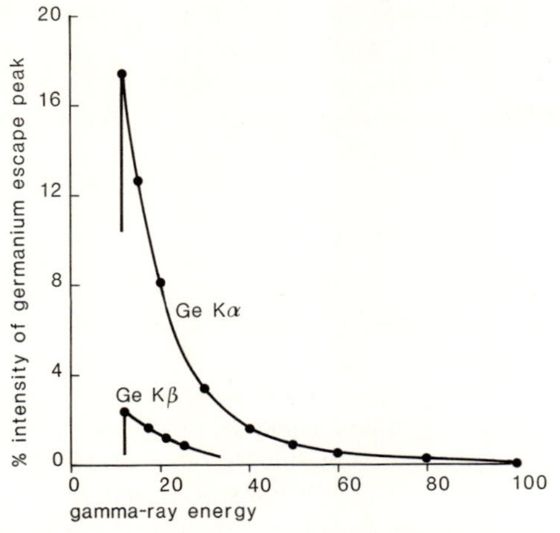

Figure 12.18 Relative Ge escape peak intensity as a function of incident gamma energy (keV). Intensity of escape peak = (escape peak area/incident photopeak area) × 100. Data from Fioratti and Piermattei (1971) ($K\alpha$) and Christensen (1979) ($K\beta$).

in the detector is the main cause of the high background continuum below the energies of principal gamma photopeaks recorded on germanium detectors.

12.6.3 *Pair production*

Pair production only occurs when incident gamma-rays exceed 1022 keV. There is then a small probability that interaction in the strong electromagnetic field surrounding the nucleus will result in complete transmutation of gamma photon energy into an electron–positron pair. Such particles lose their kinetic energy by further collision with atoms in the detector and then spontaneously annihilate to generate two 511 keV gamma-rays emitted at 180° to one another (Figure 12.17*c*). If one of these annhilation photons escapes from the detector, an event is detected at 511 keV below the parent photopeak (single annihilation escape peak). If both are lost, double annihilation escape peak is detected at 1022 keV below the parent gamma line. If the electron–positron pair escape the detector or if pair production occurs in matter surrounding the detector, annihilation radiation of energy 511 keV will be detected. Pair production is not a significant interaction for incident radiation below 2000 keV.

12.6.4 *Bremsstrahlung*

Bremsstrahlung continuum radiation is also created in the detector by the decceleration of energetic secondary electrons (Figure 12.17*d*) as described in section 12.3.6. If this bremsstrahlung radiation escapes from the detector, this process can randomly contribute to the continuum spectrum.

12.7 Typical spectrum

A typical spectrum from a monoenergetic gamma-ray source is shown in Figure 12.19. The source is ^{137}Cs, which decays by beta emission, generating a single gamma energy at 661.6 keV. The spectrum recorded on a coaxial detector shows several complex artefacts with only about 6% of the total detected events occurring in the full-energy photopeak. The features in this spectrum are as follows (compare Figure 12.19).

(i) Full energy peak at 661.6 keV (*a*) in which gamma events have been detected by photoelectric effect (with no escape of secondary Ge x-rays), and Compton scattering with complete detection of the scattered photon (*b*).

(ii) Degraded full energy events due to poor charge collection in the detector (see section 12.4.3 and also 9.5 for a more detailed account of charge trapping effects).

(iii) A broad Compton scattered continuum (*c*) of full energy peak events in which the secondary (scattered) gamma-ray escaped from the detector.

(iv) Compton edge (*d*) corresponding to head-on collision (180°) between gamma and electron. The position of the Compton edge varies with the energy (E) of the primary

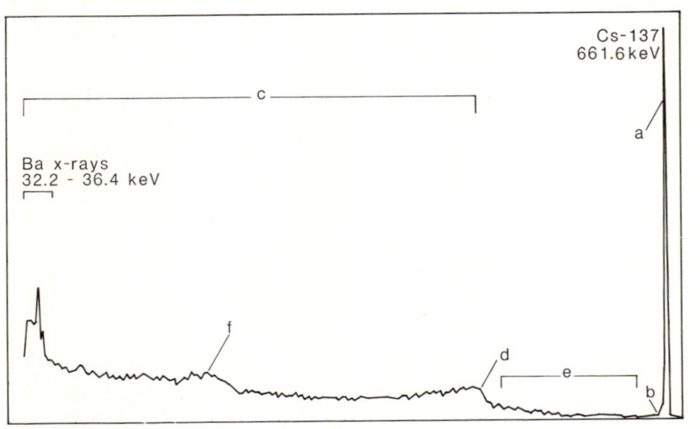

Figure 12.19 Gamma spectrum of a ^{137}Cs source recorded on a coaxial Ge(Li) detector demonstrating various spectrum artefacts referred to in the text (spectrum plot courtesy of E.G. & G. Ortec). For clarity the spectrum has not been extended up to the sum peak at 661.6 + 661.6 = 1323.2 keV. Barium x-rays are emitted following internal conversion of ^{137}Cs 661.6 keV gammas. (After Adams and Dams, 1970.)

photopeak, and is given approximately by

$$\text{Compton edge} = E - \frac{E}{1+0.004E} \qquad (12.6)$$

$$= 662 - 181 = 481 \text{ keV}$$

(E is in keV) (see also equation (8.10) with $\theta = 180°$).

(v) Background region where Compton scattering (at 180°) results in almost complete energy transfer to the secondary electron with a small probability of escape of the low-energy secondary gamma photon (*e*).

(vi) Broad backscatter peak centred at 200 keV (*f*) due to gamma-rays which are scattered through a large angle ($\sim 180°$) off matter outside the germanium crystal (e.g. the copper cold finger assembly). The scattered radiation then re-enters the germanium crystal and is detected. Figure 12.20 shows that there is a high probability that gamma photons of any energy will degrade to about 200keV when backscattered through 180°. It is these events that form the hump indicated in Figure 20.19.

(vii) Sum peak ($662+662 = 1324$ keV) due to the near-simultaneous detection of two gamma photons that have not been rejected as pulse pile-up events. As will be seen in section 12.9.5, irradiated rock spectra contain many individual photopeaks which sit on a continuum of superimposed Compton scatter peaks from the more intense gamma lines.

12.8 Detector characteristics

The quality of a germanium detector used to measure spectra such as those in Figure 12.19 is specified by three parameters: detector efficiency, FWHM resolution and signal to noise ratio. These are discussed individually below. An additional parameter which is critically important in successful routine analytical work is the long-term stability of the entire counting system, especially in terms of signal gain and photopeak resolution. Such phenomena usually have more to do with temperature control of the environment and quality of electrical power supplies rather than the specification of any individual component in the signal processing chain.

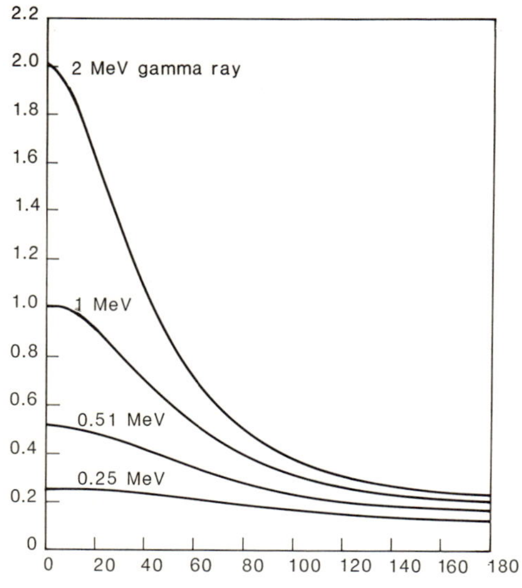

Figure 12.20 Energy of Compton scattered photon (MeV) as a function of scattering angle. (Zimmer, 1978.)

12.8.1 *Detector efficiency*

The absolute efficiency of a germanium detector at a given energy is the fraction of gamma events detected from a radioactive source of known gamma disintegration rate. This parameter is geometry-dependent, since changes in source–detector distance result in differences in the solid angle of gamma events entering the detector. Manufacturers normally specify only the efficiency of coaxial detectors (since planar detectors are usually selected on the basis of their energy resolution, rather than efficiency). To permit comparison between different detectors, resolution is usually quoted for the 1332 keV gamma of a ^{60}Co point source measured at 250 mm from the front face of the detector. The absolute efficiency so derived is then normally divided by 1.2×10^{-3}, the absolute efficiency of a 3×3 inch NaI(Tl) detector crystal under these conditions. Typical efficiencies of coaxial detectors depend on their volume but often range from 10 to 30% relative to a 3×3 inch NaI(Tl) scintillation counter.

Some spectrum analysis programs require accurate calibration of detector efficiency over the range of energies for which photopeaks are to be analysed. Calibration sources which may be used for this purpose are listed in Table 12.2. It has been found that above 130 to 150 keV, the efficiency of a coaxial detector (eff) is approximately dependent on gamma energy (E) according to the power function

$$\text{eff} \propto E^{1.4} \qquad (12.7)$$

Table 12.2 Isotopes for energy, full-width half maximum and efficiency calibration of germanium detectors

Isotope	$T_{1/2}$	γ energy (keV)	Intensity (per 100 decays)
^{55}Fe	2.7y	5.95	25.70
^{241}Am	433y	13.90	28.00
^{57}Co	270d	14.41	9.5
^{133}Ba	10.7y	53.17	2.20
^{241}Am	433y	59.54	36.3
^{133}Ba	10.7y	79.59	2.43
		81.01	32.75
^{109}Cd	1.241y	88.03	3.79
^{57}Co	270d	122.07	85.6
		136.43	10.6
^{133}Ba	10.7y	276.29	7.3
		302.71	18.62
		355.86	62.27
		383.70	8.84
^{22}Na	2.60y	511.0	179.8
^{137}Cs	30.1y	661.62	84.62
^{60}Co	5.2721y	1173.23	99.86
^{22}Na	2.60y	1274.54	99.94
^{60}Co	5.2721y	1332.51	99.98

Data from Erdtmann and Soyka (1979).
Isotopes selected on basis of commercial availability as standard calibration sources and long half-life.

and may be accurately fitted to the expression

$$\text{eff} = a+b(\ln E)+c(\ln E)^2 \qquad (12.8)$$

where *a*, *b* and *c* are constants (Adams and Dams, 1970, p.220).

Typical efficiency calibrations for coaxial and planar germanium detector are shown in Figure 12.21 and demonstrate the rapid fall-off in the efficiency of planar detectors to photon energies above 150–200 keV. One consequence of this fall-off is to reduce the probability that Compton scattered events derived from the partial detection of high-energy gamma photons will contribute to the background spectrum in the 50–200 keV region. Thus peak to background ratios of photopeaks in this 50 to 200 keV energy region should be improved on a planar detector, allowing potentially greater sensitivity.

A systematic error can arise in efficiency calibrations from certain standard radioactive sources that emit two (or more) gamma-rays *in cascade* (see section 12.3.1). This enhances the probability of *true coincidence summing* in the detector due to simultaneous detection of both events. Thus the 1172 and 1332 keV photons of ^{60}Co are emitted in cascade (Figure 12.6*d*). If true coincidence summing occurs, the (1172+1332) keV sum peak will be systematically more intense than would be predicted from the frequency of *random summing* (see section 12.5.2 (*d*)) hence robbing the full-energy peak at 1332 (and 1172) keV of counts. This effect is minimized by counting such sources at distances of at least 100 mm from the detector, since the two cascade gammas are emitted at random angular correlation with respect to each other and the probability of both being emitted in the direction of the detector at this distance is low.

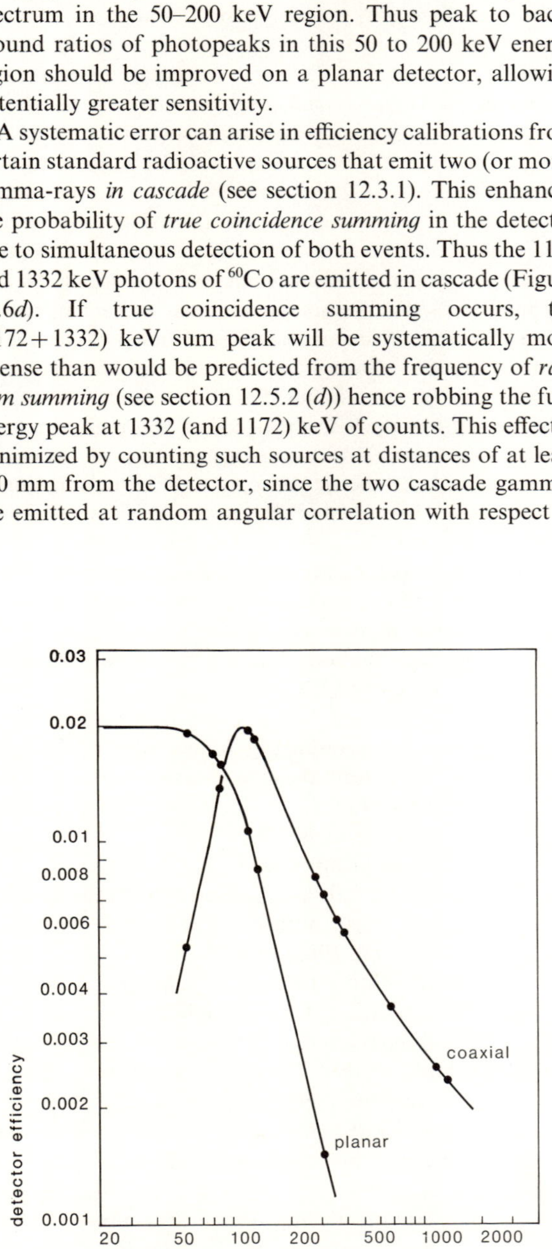

Figure 12.21 Efficiency of typical planar and coaxial germanium detectors as a function of photon energy (keV). The efficiency function of the planar detector has been arbitrarily increased by a factor of about $2\frac{1}{2}$ compared with original measurements so that the low-energy responses of the two detectors may be more readily compared.

12.8.2 *Detector resolution*

Detector resolution is defined as the full width at one-half the intensity maximum (FWHM) of a gamma photopeak at a given energy. The characteristic photopeaks normally used to define detector resolution are the 1332 keV gamma from ^{60}Co for coaxial detectors and the 122 keV gamma from ^{57}Co, and sometimes also the 5.9 keV x-ray from ^{55}Fe for planar (LEPS) detectors. As has already been noted, poor charge collection (section 12.4.3) results in an asymmetry in the low-energy tailing of a photopeak which does not significantly affect FWHM resolution. A more sensitive parameter which incorporates tailing effects is the FW1/10M (one-tenth of maximum) or the FW1/25M (one-twenty-fifth of maximum) but these values are not as frequently quoted.

The spread of energies over which monoenergetic gamma-rays are detected arises from two sources:

(i) The random nature of the ionization in the detector where each gamma-ray on average generates $E \times 1000/\varepsilon$ electron–hole pairs (*E* in keV)(section 12.4.2)

(ii) The electronic noise generated in the signal amplification and shaping circuits, especially in the coupling between the detector and the first stage of the preamplifier.

These parameters sum in quadrature, thus:

$$(\text{FWHM})^2_{\text{observed}} = (\text{FWHM})^2_{\text{noise}} + (\text{FWHM})^2_{\text{detector}} \qquad (12.9)$$

where

$$(\text{FWHM})_{\text{detector}} = 2.355\sqrt{FE\varepsilon} \qquad (12.10)$$

where *E* is the gamma energy *in MeV*, ε is the mean energy to generate an electron–hole pair = 2.95 eV at 80 K for germanium and *F* is the Fano factor $\simeq$ 0.13 for germanium at 80 K.

Experimental observations show that the process of ionization in the detector is not completely random and that the variation in the number of electron–hole pairs generated by gamma photons of a particular energy is *considerably less* than would be predicted theoretically. The Fano factor *F* is an empirical constant used to account for this observation.

The variation of resolution with energy predicted by equation (12.9) is plotted in Figure 12.22. Practical detectors do not always achieve these theoretical limits, and those used in activation analysis normally have FWHM resolutions in the range 1.8–2.2 keV at 1332 keV (coaxial detectors) or 530–580 eV at 122 keV (planar/LEPS detectors).

12.8.3 *Signal to noise ratio*

Signal to noise ratio of a detector is the proportion of full-energy events detected for a given photopeak in comparison with the total number of gamma events detected from a monoenergetic gamma-ray source. This figure can be as low as 6% for a coaxial detector. However, as it is difficult to measure the total number of counts detected, it is more usual to measure the *peak to Compton ratio* as a measure of this parameter, using the intensity of the 1332 keV photopeak from ^{60}Co and that of its corresponding Compton edge

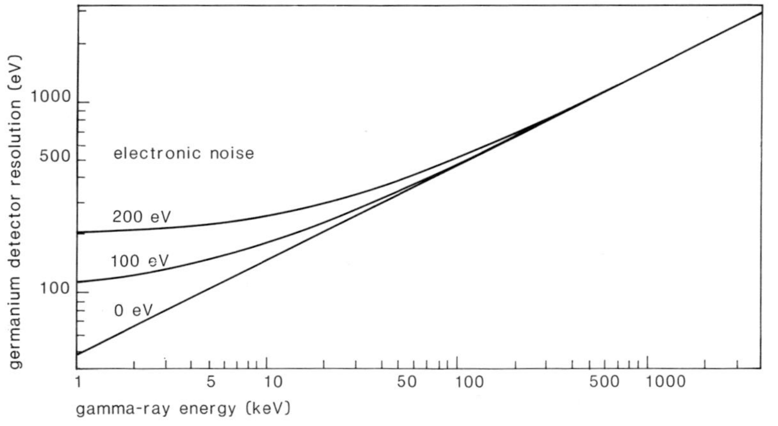

Figure 12.22 Resolution of a germanium detector as a function of gamma photon energy calculated for electronic noise components of 0, 100 and 200 eV.

(compare Figure 12.19). The height of the Compton edge is a measure of degraded gammas lost from the full-energy photopeak. This parameter is normally only determined for coaxial detectors, and typical values lie in the range 25:1 to 40:1.

12.9 Practical considerations—instrumental neutron activation analysis

A multitude of counting strategies have been developed for INAA depending on both the facilities and equipment available for analysis and elements for which the technique is to be optimized. Since many rocks, after irradiation, contain between 20 and 30 analysable isotopes, each of which has its characteristic sensitivity and half-life, any practical scheme of analysis must represent a compromise between optimum counting time for individual groups of isotopes and the time taken to complete a sample analysis. Differences of opinion exist on the best way to calibrate the technique; some workers prefer to employ rock reference materials, others to use single or multi-element solutions of high-purity element reagents which are absorbed on to various substrates and evaporated to dryness before irradiation.

Against this background, the following sections describe a general procedure for the instrumental analysis for 10 REE and other trace elements in rock samples. The method requires the minimum of sample preparation, a single irradiation and a counting strategy based on long-lived isotopes ($T_{1/2} > 2$ days) which allows samples to be analysed in a laboratory which is not necessarily adjacent to a nuclear reactor. More specialized techniques are described in later sections.

12.9.1 *Sample preparation and contamination*

Rock samples are normally analysed as powders milled to less than 200 mesh to ensure homogeneity in the analysed aliquot. Care must be taken in choosing the equipment used for crushing and milling. Grinding equipment lined with agate is preferred because the only significant contaminations imparted to the sample are silicon and perhaps iron. Tungsten carbide swing mills cause significant contamination by some elements, analysis for which may be required by INAA. The amount depends on the hardness of rock and length of time required to reduce it to the appropriate grain size. However, it is to be expected that most samples will be contaminated with hundreds of ppm of tungsten, tens of ppm of cobalt and a few ppm tantalum. This potential source of contamination must be taken into account if any of these elements are to be determined, since their respective background abundances in silicate rocks are normally very low.

Sample aliquots for irradiation are normally in the range 100–500 mg, and are limited by the space available in the reactor and the high activity induced in larger samples immediately after irradiation. An implicit assumption is that the elements to be analysed are homogeneously distributed in an aliquot of this size. Whereas this is probably the case for fine-grained rocks containing elements down to the ppm level, it is not necessarily the case at ppb concentrations or for certain elements that tend to form discrete minerals such as gold or iridium (see section 1.6).

Samples must be encapsulated before irradiation. Low-density polyethylene vials are most extensively used for this purpose and have the advantage of low blank levels (Kosta, 1982; Cornelius *et al.*, 1982). Polyethylene vials are simple to seal by heat welding to prevent subsequent escape of radioactive rock powder. However, polyethylene is degraded during irradiation, becoming brittle as its polymer cross-linking becomes damaged. Hence for irradiations exceeding 30 hours or in very high neutron fluxes, high-purity quartz or aluminium containers may be substituted. Some workers encapsulate samples by wrapping them in aluminium foil for irradiation (e.g. Steinnes, 1976). It is essential that such samples are subsequently transferred to a non-active container before counting to avoid the high blank activities which can occur in aluminium due to Fe, Co, Na, Ga, Cr, Hf, Sc and Cu (see, for example, Hollander, 1966). Blank concentrations of some elements found in quartz and polyethylene are listed in Table 12.3. A suggested design for a polyethylene sample vial is shown in Figure 12.23. The lid is heat-welded to the base, enclosing about 300 mg of rock

Table 12.3 Trace-element impurities in packaging materials

	Quartz (ppb)	Polyethylene (ppb)
Co	0.13	10–370
Cs	136	—
Cr	1.3	180–1500
Sb	0.1–118	~5
Zn	~0.3	300

From Cornelius *et al.* (1982) and references quoted therein.

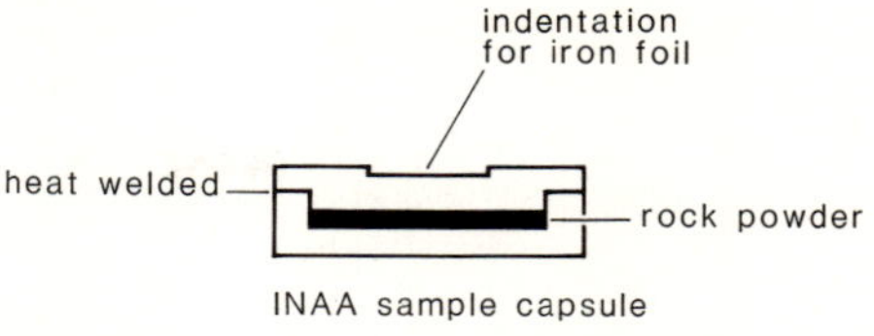

Figure 12.23 Cross-section of a vial suitable for the irradiation of powdered rock samples. The sample is contained as a flat disc of powder 1 mm deep and 16 mm in diameter and is held in position by light compression from the lid which is heat-welded to the base. The depression in the lid is for the positioning of an iron foil disc used as a neutron flux monitor during irradiation.

powder as a flat disc 16 mm in diameter and about 1 mm deep. After irradiation, individual samples and vials are assayed without further pre-treatment.

12.9.2 *Neutron flux monitors*

The spatial packaging of samples for irradiation depends on the reactor facility available. In a large reactor, and especially one in which samples are placed in a rotating irradiation facility (such as the 'lazy susan' device in the TRIGA reactor), it is usually possible to assume that neutron flux variations are negligible over the volume of the samples. However, in smaller research reactors, this is not necessarily the case, with samples in slightly different positions receiving different integrated fluxes in the course of irradiation. Failure to measure and correct such discrepancies can cause analytical errors of the order of 5–50% relative. When flux corrections are necessary, it is essential to know the spatial arrangement of samples and calibration standards relative to neutron flux monitors included in batches of samples prepared for irradiation.

The arrangement of one such batch suitable for a 25 mm irradiation tube is shown in Figure 12.24. Nine samples and two standards are encapsulated and packed in an outer polyethylene irradiation can with each vial separated by a small disc of pure iron foil (99.95% Fe) of diameter 8 mm and thickness 0.1 mm. After irradiation it is assumed that the activity induced in the iron foil is proportional to the neutron flux at that point in the irradiation can. Measurement of the iron activity in each foil then allows a correction factor to be calculated to determine the flux in each sample position relative to the standard sample (Figure 12.24).

Iron is particularly suitable for this measurement since it is fairly cheap, has a low capture cross-section and long half-life (45 days). The disadvantage is that it may rust in the reactor unless protected by a suitable lacquer.

Other workers have proposed the use of copper, gold or cobalt/aluminium foils (or wire) for measuring neutron flux corrections. Borley and Rogers (1979) employed several multi-element standards, spaced along the irradiation can so that neutron flux corrections could be applied by extrapolating the activity of the individual isotopes. This procedure has the advantage of producing more accurate corrections in situations where there is significant variation in the thermal, relative to epithermal, neutron flux along the length of an individual irradiation can. The activity induced in iron, with its low epithermal neutron capture cross-section, would not adequately correct variations in activity of isotopes having high epithermal neutron capture cross-sections in these circumstances.

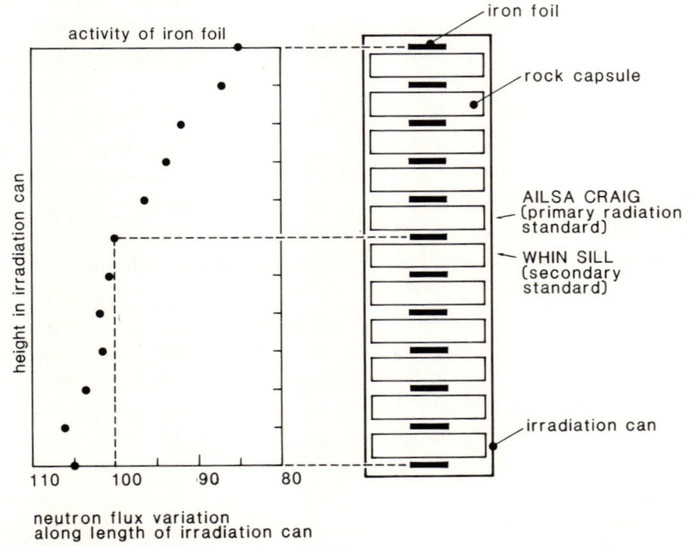

Figure 12.24 Schematic drawing of irradiation can containing 11 rock vials and 12 iron foils for neutron flux measurements. The graph above shows the activity induced in the iron foil after irradiation and indicates that the neutron flux varied by about 20% over the length of the irradiation can in this example.

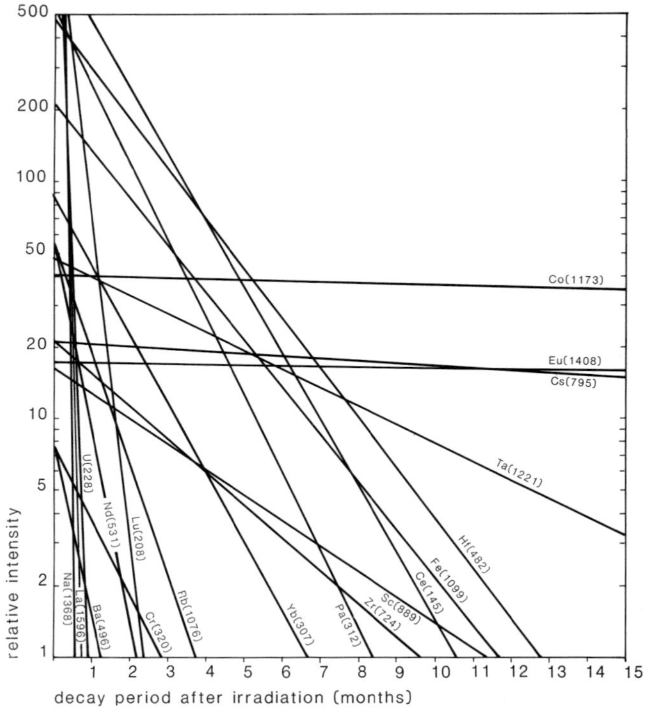

Figure 12.25 Schematic diagram of the relative rates of decay of isotope photopeaks which may be detected on a coaxial detector. Measurements were taken from a spectrum of an Ailsa Craig microgranite whose composition is as follows: La (60), Ce (161), Nd (95), Sm (25), Eu (2.0), Gd (28), Tb (4.9), Tm (2.8), Yb (18), Lu (2.7), Na_2O (6.89), Sc (0.1), Co (41), Fe_2O_3 (2.5%), Rb (155), Zr (834), Cs (2.9), Ba (50), Hf (27), Ta (7.3), Th (18.5), U (4.3). (Concentrations in ppm; photopeak energies (keV) in brackets.)

12.9.3 *Thermal irradiations and counting strategies*

Geological samples which are to be analysed for isotopes having half-lives exceeding 2 days are normally irradiated for at least 15 to 30 hours in a neutron flux of 5×10^{12} n cm^{-2} s^{-1}, or *pro rata*. Samples must then be allowed to cool for a further 5 days to allow the short-lived activity, especially that due to ^{24}Na ($T_{1/2} = 15h$) to decay. At this stage, most samples emit a complex gamma spectrum (see Figures 12.27, 12.28) originating from over 30 activated isotopes. The relative decay rates of these isotopes, each having its own characteristic half-life, is plotted in Figure 12.25 (coaxial spectrum) and Figure 12.26 (planar spectrum), each determined from a spectrum of microgranite from Ailsa Craig, SW Scotland.

It is clear from these figures that any counting strategy must be a compromise to allow the determination of groups of elements with similar half-lives at adequate sensitivity. Most schemes of analysis require at least two separate counts.

(i) *Count* 1: 7 to 10 days after irradiation for 1 to 3 hours per sample to determine the activity of the short-lived isotopes especially ^{140}La, ^{153}Sm, ^{239}Np (U).

(ii) *Count* 2: 2–4 weeks after irradiation to determine the longer lived isotopes from spectra counted for 8–16 hours each.

Some workers also count samples a third time:

(iii) *Count* 3: 8–10 weeks after irradiation for 10–24 hours per sample. This is especially beneficial in allowing the ^{233}Pa interference on ^{153}Gd to decay and for the preferential determination of those isotopes with half-lives longer than about 60 days, in particular, ^{46}Sc, ^{60}Co, ^{65}Zn, ^{95}Zr, ^{124}Sb, ^{134}Cs, ^{152}Eu, ^{160}Tb, ^{170}Tm, ^{182}Ta. However, for many elements the improvement in sensitivity and precision in comparison with data from Count 2 may not justify the additional utilization of the counting system.

12.9.4 *Choice of detectors : coaxial v. planar*

Table 12.4 lists the photopeaks available for the determination of those isotopes normally activated in silicate rocks. In preparing these data, it has been assumed that the photopeaks in the range 60–200 keV will be determined by preference on a planar (LEPS) detector and those in the range 200–1700 keV on a coaxial detector. Data in Table 12.4 show that for the determination of the REEs, a planar detector is preferred, except for the analysis of ^{140}La for which the 1596 keV peak is generally used (but see Henderson and Williams, 1981).

Many of the lighter elements may be determined most sensitively on a coaxial detector. The isotopes of Sc, Co, Fe, (Zn), Rb, Zr, Sb, and Cs fall into this category. Other isotopes including U, Th, Hf, and Ta emit gamma photons at several energies so that either detector may be used for determinations.

12.9.5 *Typical spectra*

Typical spectra of a microgranite from Ailsa Craig are shown in Figure 12.27 (coaxial spectrum) and Figure 12.28 (planar spectrum).

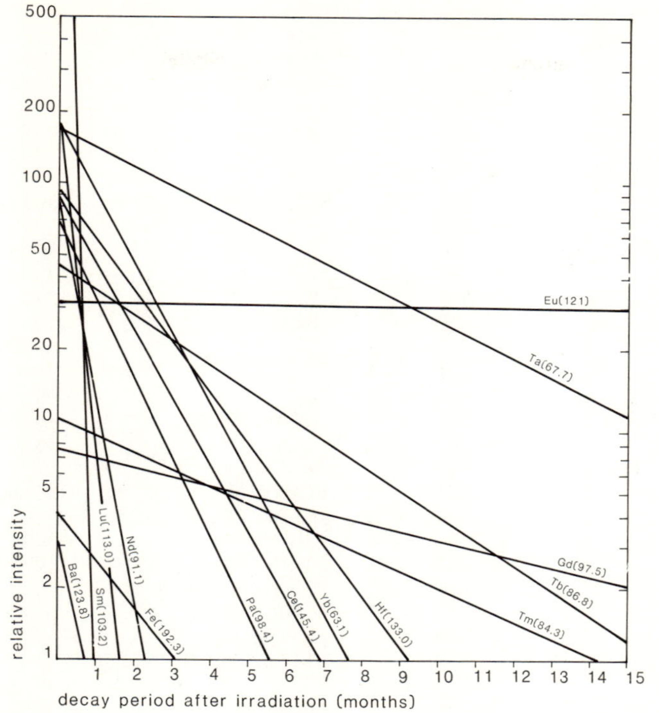

Figure 12.26 Schematic diagram of the relative rates of decay of isotope photopeaks which may be detected on a planar (LEPS) detector. Measurements were taken from a spectrum of an Ailsa Craig microgranite (composition indicated in Figure 2.25). Photopeak energies (keV) in brackets.

The coaxial spectrum demonstrates the problem of background enhancement due to Compton scattering. High-energy photopeaks for example due to ^{59}Fe and ^{60}Co generate Compton plateaux at lower energies on which other photopeaks of interest sit. Indeed, in many silicate rocks, detection limits for lower energy photopeaks are limited by Compton scatter of ^{59}Fe (1099, 1272 keV), ^{60}Co (1173, 1333 keV), ^{46}Sc (889, 1121 keV) and ^{51}Cr (320 keV). Thus ^{181}Hf may be determined from gamma peaks at 133.0 keV (43.0%) or 482.2 keV (86.0%). The improved efficiency of a coaxial detector at 482 keV should enhance the detectability of this photopeak in comparison with the less intense 133 keV line on a planar detector. In practice this is not always the case. The high Compton plateau under the coaxial detector photopeak often reduces the precision with which the 482 keV peak area may be determined.

12.10 Determination of photopeak areas

Gamma spectra are analysed first to determine the energies of photopeaks present in a spectrum (and therefore to identify parent isotopes) and, second, to calculate the areas of selected peaks. Several peak-search algorithms have been developed for identifying isotopes present in spectra. However, in the routine analysis of silicate rocks, photopeaks to be used for quantification and their interferences are normally pre-selected. This section will therefore describe the various methods of calculating areas.

12.10.1 *Total peak area calculation*

This is conceptually the simplest method for area determination. Channels on the high- and low-energy wings of the peak are identified at the position where the peak tail merges into the background (Figure 12.29). A straight line joining each represents the peak background. Peak area is then determined by summing data in all channels between upper and lower limits and subtracting the calculated background represented by the trapezoid area B in Figure 12.29.

$$\text{Gross peak area} = [P+B] \tag{12.11}$$

$$\text{Net peak area} = [P+B]-[B] \tag{12.12}$$

This technique has the advantage that it is a very simple algorithm to calculate but the disadvantage that it is only appropriate for relatively intense peaks with no overlap interferences and cannot be used on weak signals where statistical fluctuations in the limiting channels have a large influence in the position of the background.

12.10.2 *Covell's method*

Some attempts to overcome the limitations of the above procedure make use of the fact that the smallest contribution to the total peak area arises from the wings of a photopeak, and yet these are the channels with fewest counts which, therefore, contribute the largest uncertainty to the net area calculation. Thus Covell (1959) proposed a method which neglected these contributions. Peak area is computed by

Table 12.4 Details of isotopes and photopeaks for analysis by INAA

Product isotope	θ (%)	Target isotope	$T_{1/2}$	σ (barn)	I (barn)	A	Analytical photopeaks, LEPS detector	Analytical photopeaks COAX detector
^{24}Na	100	^{23}Na	15.03h	0.53	0.34	1.5		1368.55(100)
^{46}Sc	100	^{45}Sc	83.85d	26.5	13	1.1		889.26(99.98) [1120.52 (99.99)]
^{51}Cr	4.35	^{50}Cr	27.7d	15.9	7.7	1.1		320.07(9.83)
^{59}Fe	0.31	^{58}Fe	45.1d	1.15	1.7	3.3	142.65(1.03),192.34(3.11)	1099.22(56.5), 1291.56(43.2)
^{60}Co	100	^{59}Co	5.272y	37.2	75.5	4.4		1332.51(99.98),1173.23(99.86)
^{65}Zn	48.9	^{64}Zn	243.8d	0.78	1.8	4.8		[1115.52(50.75)]
^{88}Rb	72.2	^{85}Rb	18.6d	0.46	7.5	25		1076.63 (8.76)
^{95}Zr	17.4	^{94}Zr	64.4d	0.056	0.38	13		724.18 (44.2) 756.72(54.8)
^{99}Mo	23.8	^{98}Mo	2.76d	0.13	7.5	46	[140.51(90.9)]	739.47 (13.0)
^{122}Sb	57.3	^{121}Sb	2.70d	6.25	127	29		564.08(71.0)
^{124}Sb	42.7	^{123}Sb	60.2d	4.28	234	46		1691.04(50.0),[602.71(98.1)], 722.78(11.8)
^{134}Cs	100	^{133}Cs	2.062y	27	359	22		795.76(85.4) 604.66(97.6)
^{131}Ba	0.10	^{130}Ba	11.8d	11	263	31	[123.80(29.05)]	496.28(43.38)
^{140}La	99.9	^{139}La	40.3h	9.0	11.2	2.8		1596.20(95.5),328.75(18.5),815.80(22.32)
^{141}Ce	88.5	^{140}Ce	32.4d	0.57	0.43	1.7	145.45(48.0)	
^{147}Nd	17.2	^{146}Nd	11.1d	1.3	2.84	4.7	91.10(28.3)	531.00(13.5)
^{153}Sm	26.7	^{152}Sm	46.8h	206	3141	24	103.18(28.3),69.67(5.25)	
^{152}Eu	47.8	^{151}Eu	12.7y	5900	5564	2.1	121.78(29.24)	1408.08(21.21) 778.90 (13.0)
^{154}Eu	52.2	^{153}Eu	8.5y	390	1635	8.5	123.10(40.46)	1274.80(35.50)
^{153}Gd	0.2	^{152}Gd	241.6d	1100	3000	5.6	97.50(30.00)[103.20(21.8)]	
^{160}Tb	100	^{159}Tb	72.1d	25.5	532	29	86.80(13.40)	[876.37 (30.00)]
^{170}Tm	100	^{169}Tm	128.6d	103	1548	24	[84.26(10.0)]	
^{169}Yb	0.14	^{168}Yb	30.7d	3470	23040	12.2	63.12 (45.0),130.51(11.5)	
^{175}Yb	31.8	^{174}Yb	4.19d	65	37.7	1.3		396.32(6.2)
^{177}Lu	2.6	^{176}Lu	6.71d	2050	1069	1.2	112.95(6.60)	208.36(11.0)
^{181}Hf	35.1	^{180}Hf	42.5d	12.6	31.5	5.3	133.05(43.0),136.25(6.1)	482.16(86.0)
^{182}Ta	99.9	^{181}Ta	115.0d	21.0	710	38	67.75(41.3),100.11(14.07)	1221.42(27.4),1230.97(11.6),222.10(7.56)
^{187}W	28.6	^{186}W	23.9h	37.8	410	19	72.00(11.9)	479.57(23.4), 685.74 (29.2)
^{233}Pa	100	^{232}Th	27.4d	7.4	72.4	17	94.66(9.6),98.44(15.5)	311.90(33.7)
^{239}Np	99.3	^{238}U	2.36d	2.7	275	56	106.13(22.7)	277.60(14.10),228.19(10.72)

Product isotope is the radioactive isotope formed after neutron capture of the naturally-occurring target isotope.
θ is the natural abundance of the target isotope.
$T_{1/2}$ is the half-life.
σ is the thermal neutron capture cross-section.
I is the epithermal (resonance) neutron capture cross-section.
A is the advantage factor for epithermal irradiations (see section 12.15).
Neutron capture cross-section data taken from Erdtmann (1976).
Photopeak energies, intensities and half-lives taken from Erdtmann and Soyka (1981).
Analytical difficulties may be encountered in fitting the photopeaks denoted by square brackets.

integrating over a specified number of channels on either side of the centroid (not encompassing the entire peak) and subtracting a background obtained by drawing a straight line across the two limiting channels (Figure 12.29). The net area so defined then bears a constant relationship with the total peak area without the necessity of identifying the real peak baseline. These calculations result in areas which are determined relative to an artifically high background with some consequential loss in precision which becomes most serious as detection limits are approached.

12.10.3 *Wasson's method*

Wasson (1968) modified this algorithm so that net peak areas are calculated relative to a representative background (Figure 12.29). Sterlinski (1970) proposed a further modification where weighting factors are employed in area calculations so that those channels having the largest number of counts (and therefore the smallest statistical error) receive the highest weighting factor. Comparing these and other simple peak area algorithms, Baedecker (1971) favoured the Wasson method. However, Hertogen *et al.* (1974) propose that for optimum results different algorithms are appropriate for different peaks in the same spectrum.

12.10.4 *Peak-fitting methods : SAMPO*

The alternative approach for peak area determination is to choose a mathematical function to describe individual peak shapes and to use a least-squares minimization procedure to fit this function to appropriate regions of the spectrum. Such methods can successfully cope with partial overlap between adjacent photopeaks. Programs are, however, usually large and require pre-calibration of spectrum gain, peak resolution and sometimes detector efficiency.

Perhaps the best known of this class of algorithm is the peak-fitting program SAMPO (Routti and Prussin, 1969). This program has recently been revised for use in mini- and microcomputers (Koskelo *et al.*, 1981).

SAMPO assumes that peak shapes are basically gaussian

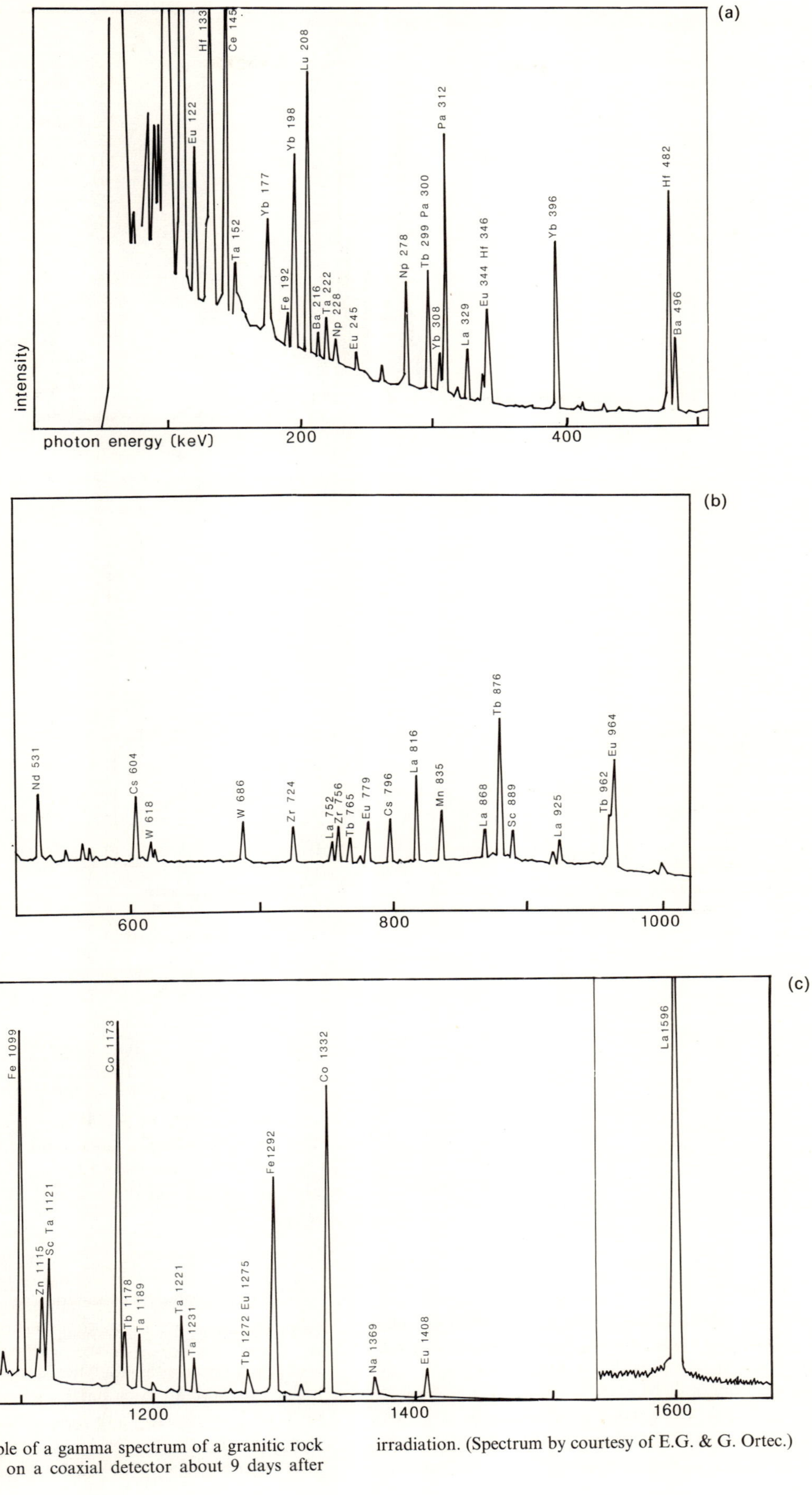

Figure 12.27 An example of a gamma spectrum of a granitic rock (Ailsa Craig) recorded on a coaxial detector about 9 days after irradiation. (Spectrum by courtesy of E.G. & G. Ortec.)

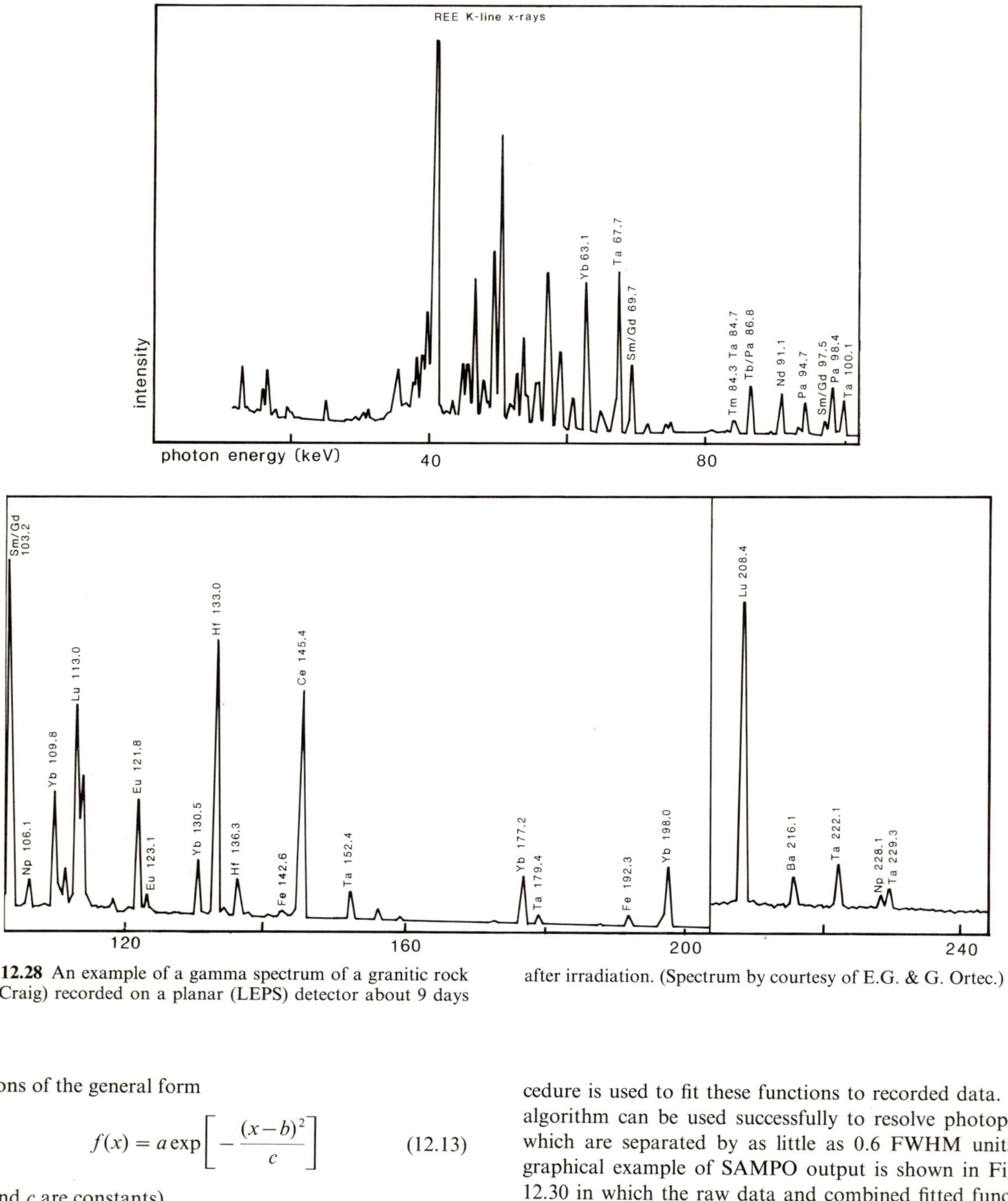

Figure 12.28 An example of a gamma spectrum of a granitic rock (Ailsa Craig) recorded on a planar (LEPS) detector about 9 days after irradiation. (Spectrum by courtesy of E.G. & G. Ortec.)

functions of the general form

$$f(x) = a \exp\left[-\frac{(x-b)^2}{c}\right] \tag{12.13}$$

(a, b and c are constants).

To allow for asymmetry in peak tails, these regions are modelled by simple exponential functions:

$$f'(x) = a' \exp\left[\pm\frac{(x-b')}{c'}\right] \tag{12.14}$$

contrived so that the function formed by combining equations (12.13) and 12.14) (and its first derivative) is continuous. A peak is then described by three parameters, the full-width half maximum of the gaussian peak and the channel difference from peak centroid to the start of each exponential tailing function. A linear or quadratic background is fitted to the continuum. A linear least squares minimization procedure is used to fit these functions to recorded data. This algorithm can be used successfully to resolve photopeaks which are separated by as little as 0.6 FWHM units. A graphical example of SAMPO output is shown in Figure 12.30 in which the raw data and combined fitted function are plotted for the ^{152}Eu 121.8 keV and ^{154}Eu 123.1 keV photopeaks, showing the success of fitting this overlap interference.

12.11 Other analytical considerations

12.11.1 *Calibration requirements*

INAA is unusual as an analytical technique in that it is quite normal to calibrate each batch of samples to be irradiated by including perhaps only one multi-element calibration standard (the *single comparator method*). The calibration graph is then the straight line drawn between the activity in

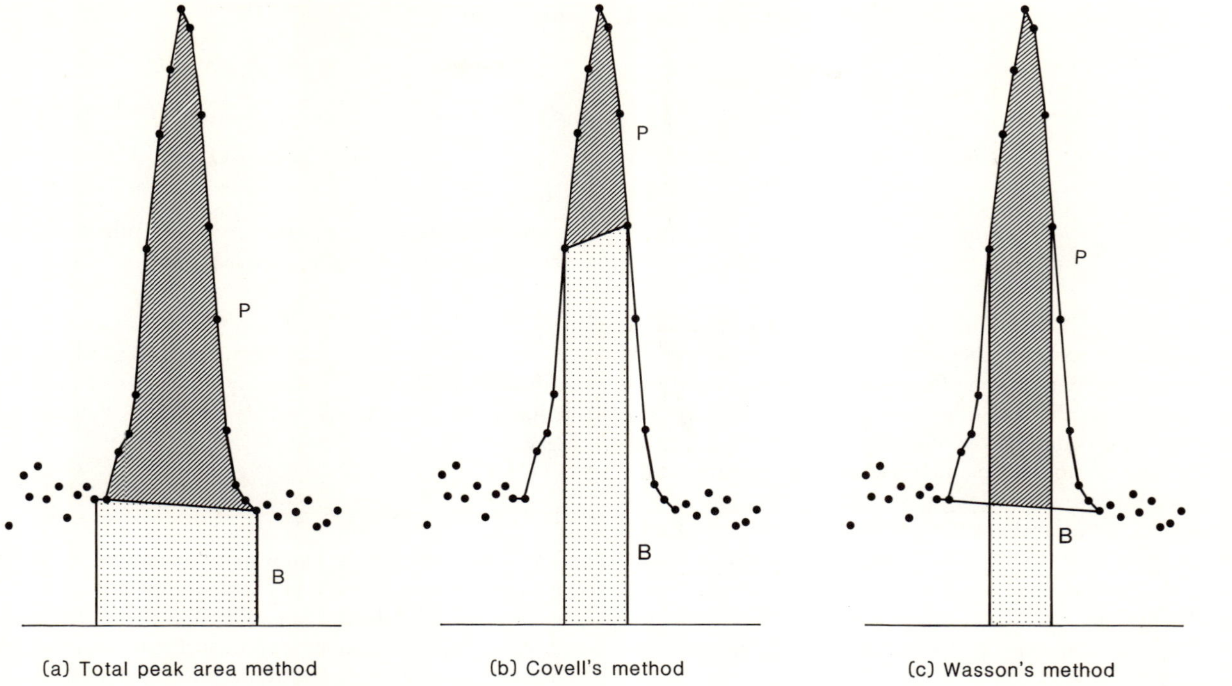

Figure 12.29 Methods of peak area determination. *P* represents the net peak area counts and *B* the net background counts derived from (*a*) total peak area calculation, (*b*) Covell's method and (*c*) Wasson's method.

the calibration standard and the origin. Whereas assumptions about the linearity of such calibrations are usually acceptable in INAA, this procedure does, of necessity, rely heavily on the confidence that may be placed in the composition of the selected calibration standard.

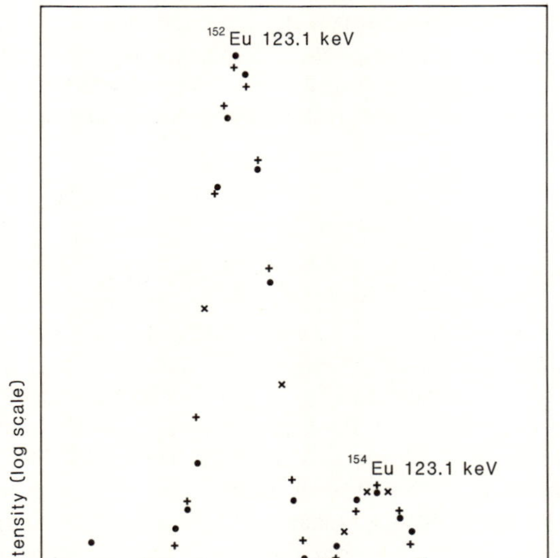

Figure 12.30 Example of output from the gaussian peak-fitting program SAMPO showing the fit of ^{152}Eu (121.8 keV) and ^{154}Eu (123.1 keV) in a sample where the activity due to ^{131}Ba (123.8 keV) as negligible. ●, raw data; +, computed (fitted) data; ×, coincidence of ● and +; –, computed background continuum.

A further difficulty arises when this multi-element standard is an international reference material. A significant proportion of the REE data on such standards has been contributed from INAA techniques, although data from alternative methods is increasingly becoming available. Thus, the danger of using such reference rocks is that their calibrations become circular, with compiled data incorporating the idiosyncratic errors of INAA.

Choice of calibration standards must also comply with the restrictions on accuracy imposed by attaining reproducible counting geometries (section 12.12.3) and the desirability of matching standard and unknown matrices.

Despite these reservations, many workers have selected individual rock reference materials for calibration and have selected data from compilations such as those published by Flanagan (1973) or Abbey (1983). The most popular reference rock has been USGS: BCR-1 (basalt), supplies of which are now exhausted, with AGV-1 (andesite), W-1 (diabase), and G-2 (granite) also being used extensively. Some workers have calibrated their own in-house rock standards (Staudigel, 1979; Perlman and Asario, 1969, Potts *et al.*, 1981). It is suggested that this approach does have some advantage from the view of long-term stability and reproducibility of calibration data, although accuracy depends on the care with which the in-house standard is itself calibrated.

Independent calibration methods use standard solutions prepared from high-purity chemical reagents. Normally the appropriate volume of a multi-element standard solution is pipetted on to a silica or magnesium oxide substrate held in an irradiation vial. The mixture is evaporated to dryness, sealed and irradiated with unknown samples. The method has the significant advantage that accuracies are independent of uncertainties in international reference materials.

Table 12.5 Element concentrations in INAA multi-element standard solution

La	Ce	Nd	Sm	Eu	Gd	Tb	Ho	Tm	Yb	Lu	Th
12.5	20.0	10.0	1.0	0.3	3.0	1.0	1.0	0.5	1.0	0.3	2.0

Units μg per 100 μL solution aliquot.
After Henderson and Williams (1981).

Furthermore, the precision with which the photopeaks of individual elements may be determined may be equalized by adjusting the composition of the standard solution. Serious photopeak interferences in the standard spectrum may also be avoided by omitting the offending element. On the debit side, standard and sample matrices no longer match, care must be taken over the long-term stability of standard solutions (stored as acid solutions to avoid adsorption of ions on to the container walls) and standard reagents must be carefully selected to avoid mixtures likely to react and cause precipitation. The composition of a REE standard solution recommended by Henderson and Williams (1981) is given in Table 12.5.

12.11.2 *Calibration equation*

Whichever calibration procedure is employed, unknown samples are determined in comparison with standard gamma spectra from the following equation:

$$C_u = C_s \frac{\text{area}_u}{\text{area}_s} \frac{W_s}{W_u} \text{NFF} \exp\left(-\frac{0.693T}{T_{1/2}}\right) \quad (12.15)$$

where C is the concentration (in ppm), area is the area in counts per second of appropriate isotope photopeak, W is the weight of irradiated sample, NFF is the neutron flux correction factor to correct for difference in neutron flux between standard and unknown samples, exp is the exponential decay term to correct for radioactive decay occurring in an isotope of half-life ($T_{1/2}$) for the time difference (T) between counting standard and unknown spectra and u,s are subscripts for parameters referring to the unknown sample (u) and standard (s) respectively.

12.11.3 *Precision*

The emission and therefore detection of gamma radiation from radioactive sources is a random process which may be described by the Poisson distribution function (see section 1.4). An estimate of the standard deviation (s) in any measurement is therefore given by

$$s = \sqrt{N} \quad (12.16)$$

where N is the total number of gamma-ray counts accumulated in a measurement.

Photopeak areas are calculated by subtracting the background area from the gross peak area (section 12.10, equation (12.12)).

If N_G is the total number of counts in the gross peak area, and N_P and N_B are the total counts in the net peak and background areas respectively then from equations (12.12) and (12.16):

$$s_G = \sqrt{N_G} = \sqrt{N_P + N_B} \quad \text{(gross peak area)} \quad (12.17)$$

$$s_B = \sqrt{N_B} \quad \text{(background area)} \quad (12.18)$$

As standard deviations sum in quadrature, the standard deviation on the net peak area is given by

$$s_N = \sqrt{s_G^2 + s_B^2} = \sqrt{N_P + 2N_B}. \quad (12.19)$$

Precision in peak area measurements is therefore significantly influenced by the magnitude of the background counts, particularly as signals approach detection limit levels.

12.11.4 *Detection limits*

The most influential philosophy on detection limits in gamma spectroscopy was put forward by Currie (1968). This is based on confidence limits placed in the identification of a peak sitting on a background which has itself an error in measurement whose standard deviation is described by

$$s_B = \sqrt{N_B} \text{ (see equation (12.16) above).}$$

Currie's terminology has been replaced here by modern equivalents, the derivation of which is discussed in section 1.5.

Detection limits in routine schemes of analysis are discussed further in section 12.13.

12.12 Interferences and systematic errors

INAA is remarkably free from analytical interferences, matrix effects being small. Probably the two most serious sources of error are in photopeak overlaps in gamma spectra and in maintaining reproducible sample geometry. These and other more fundamental interferences are discussed below.

12.12.1 *Photopeak interferences*

Photopeak interferences arise from the overlap of adjacent gamma peaks due to the energy broadening characteristics of the detection system (section 12.8.2). A full catalogue of gamma-rays emitted by radioactive nuclei has been published by Erdtmann and Soyka (1979), and from this Potts (1983) has abstracted those interferences most likely to be encountered in silicate rock analysis (Table 12.6). Not all these interferences will be significant in all spectra, depending on the relative abundances of elements in a sample. However, where an overlap does occur, there are three principal ways of overcoming the problem.

(i) Select an alternative interference-free gamma photopeak of the isotope: the 1121 keV peak of ^{46}Sc is severely overlapped by a ^{182}Ta peak, but the 889 keV ^{46}Sc peak is free of interference.

(ii) Choose to use a peak-fitting program rather than a simple total peak area algorithm so that overlapping peaks may be adequately fitted.

Table 12.6 Photopeak interferences in INAA

Analytical photopeak			Interfering photopeak		
	keV	Decay intensity (%)		keV	Decay intensity (%)
^{169}Yb	63.12	(45.0)	*^{181}Hf	63.20	(5.6)
^{182}Ta	67.75	(41.3)	^{182}Ta	67.20	(5.2)
^{153}Sm	69.67	(5.25)	*^{153}Gd	69.60	(2.6)
^{187}W	72.00	(11.9)	^{187}W	71.20	(1.1)
^{170}Tm	84.26	(10.0)	*^{182}Ta	84.68	(2.7)
^{160}Tb	86.80	(13.40)	*^{233}Pa	86.61	(1.8)
^{233}Pa	94.66	(9.60)	^{169}Yb	93.62	(2.4)
^{153}Gd	97.50	(30.0)	*^{153}Sm	97.43	(0.73)
^{233}Pa	98.44	(15.50)	^{239}Np	99.50	(15.0)
^{153}Sm	103.18	(28.3)	*^{153}Gd	103.20	(21.8)
			^{239}Np	103.70	(24.0)
			^{233}Pa	103.86	(0.59)
^{177}Lu	112.95	(6.60)	^{182}Ta	113.67	(1.9)
			^{175}Yb	113.80	(1.8)
^{181}Hf	133.05	(43.0)	^{131}Ba	133.61	(2.9)
^{141}Ce	145.45	(48.0)	^{175}Yb	144.86	(0.32)
			*^{233}Pa	145.40	(0.44)
^{177}Lu	208.36	(11.0)	^{239}Np	209.75	(3.2)
^{239}Np	228.19	(10.72)	^{182}Ta	229.32	(3.6)
^{51}Cr	320.07	(9.83)	^{147}Nd	319.40	(2.2)
^{175}Yb	396.32	(6.21)	^{233}Pa	398.50	(1.3)
^{187}W	479.57	(23.4)	^{154}Eu	478.30	(0.21)
^{122}Sb	564.08	(71.0)	^{134}Cs	563.26	(8.4)
			*^{152}Eu	563.97	(0.52)
^{134}Cs	604.66	(97.60)	^{124}Sb	602.71	(98)
^{95}Zr	756.72	(54.8)	*^{154}Eu	756.70	(4.1)
^{140}La	815.80	(22.32)	^{154}Eu	815.60	(0.53)
^{59}Fe	1099.22	(56.5)	^{160}Tb	1102.59	(0.54)
^{65}Zn	1115.52	(50.75)	^{182}Ta	1113.38	(0.40)
			*^{160}Tb	1115.12	(1.6)
^{46}Sc	1120.52	(99.99)	*^{182}Ta	1121.28	(35.0)
^{154}Eu	1274.80	(35.50)	^{182}Ta	1273.75	(0.67)
^{59}Fe	1291.56	(43.2)	^{182}Ta	1289.17	(1.4)
^{24}Na	1368.55	(100)	*^{124}Sb	1368.21	(2.6)
^{140}La	1596.20	(95.47)	^{154}Eu	1597.30	(1.7)

Data from Potts (1983).
* Interfering photopeaks cannot be resolved from the corresponding analytical photopeak.

(iii) Recount the sample after a suitable decay period after which the interfering activity has substantially disappeared. For example, in many spectra the ^{153}Gd 97.5 keV photopeak is swamped by the adjacent ^{233}Pa peak at 98.4 keV. This interference is greatly reduced if the sample is recounted several months after irradiation, due to the long half-life of ^{153}Gd (241.6d) in comparison with ^{233}Pa (27.4d).

There are inevitably isotopes for which all these remedies fail but for which quantitative analysis is desirable. It then becomes necessary to calculate the magnitude of the interference from the intensity of a neighbouring peak of the interfering isotope.

As an example, ^{160}Tb is most sensitively determined from its photopeak at 86.8 keV which is almost exactly overlapped by a ^{233}Pa interference at 86.6 keV. The relative contribution of this interference to the combined 86.8 keV peak can be calculated from the relative intensities of the adjacent ^{233}Pa 94.8 and 98.4 keV peaks. On one planar detector system, this interference is given by

$$\text{area } ^{233}\text{Pa at 86.6 keV} = 0.283 \text{ (area Pa at 94.8 keV)} = 0.163 \text{ (area Pa at 98.4 keV)} \qquad (12.20)$$

(Potts *et al.*, 1981).

12.12.2 *Background radiation*

Background radiation may be detected from both cosmic rays and the disintegration of naturally-occurring radioactive isotopes present in the environment, especially in building materials. These isotopes include the following.

Isotope	Natural abundance	Half-life
^{40}K	0.0118%	1.25×10^{9}y
^{232}Th (and its daughter isotopes)	100%	1.40×10^{10}y
^{238}U (and its daughter isotopes)	99.27%	4.47×10^{9}y

The latter two isotopes have complex decay chains (see Chapter 13) and the most prominent gamma photons emitted from these daughter isotopes are listed in Table 12.7.

A significant reduction in background counts can be achieved by surrounding the detector with a thick annulus of lead (or enclosing the entire detector in a lead box). The thickness of this shield should be 100 to 120 mm, the limiting factor being the thickness beyond which no further background reduction is obtained due to increased probability

Table 12.7 Gamma photopeaks present in background spectra

Parent	Isotope	keV	%	keV	%
^{40}K	^{40}K	**1460.8**	**10.7%**		
^{232}Th series	^{228}Ac	338.4	12.0	794.8	4.8
		964.6	5.5	**968.9**	**17.5**
		1587.9	3.7		
	^{212}Pb	74.8	9.6	77.1	17.5
		87.2	6.3	238.6	43.1
	^{212}Bi	727.2	11.8		
	^{208}Tl	277.4	6.5	**510.7**	**22.5**
		583.1	**86.1**	860.5	12.0
		2614.5	**100**		
^{226}Ra series	^{226}Ra	185.9	3.3		
	^{214}Pb	10.8	13.6	74.8	6.3
		77.1	10.7	241.9	7.5
		295.2	**19.2**	**351.9**	**37.1**
	^{214}Bi	**609.3**	**46.1**	768.4	4.9
		1120.3	**15.0**	1238.1	5.9
		1765	**15.9**	2204	5.0
	^{210}Pb	10.8	24.3	46.5	4.0
External pair production	Annihilation radiation	**511.0**			

Photopeaks in boldface are normally the most easy to detect in background spectra.
^{226}Ra is a radioactive daughter of the parent isotope ^{238}U.

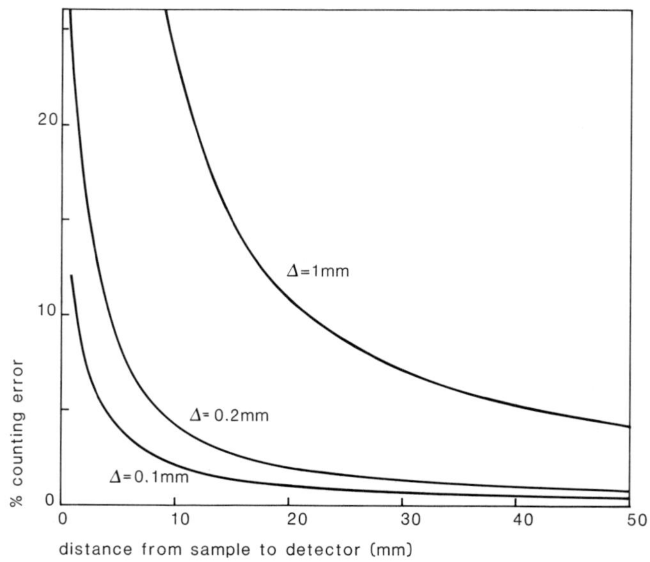

Figure 12.31 Error in counting caused by discrepancies (Δ) in repositioning samples in front of germanium detectors. Graphs are plotted for repositioning errors of 0.1, 0.2 and 1 mm at sample-to-detector distances of between 2 and 50 mm.

of interaction with high-energy cosmic radiation (Zimmer, 1978). Care must be taken to ensure that the lead itself does not contain significant radioactive impurities. Any radiation from a sample of energy greater than 88 keV is capable of fluorescing lead K x-rays at 72.8 to 74.9 keV (Pb $K\alpha$) and 84.4 to 87.4 keV (Pb $K\beta$). For this reason it is normal to line any surface of lead which may be viewed by the detector with a graded shield consisting of cadmium and then copper foil each about 1 mm thick (Crouch and Heath, 1963) to absorb fluorescence radiation.

12.12.3 *Sample–detector geometry*

Perhaps one of the critical experimental parameters in INAA is the reproducible positioning of samples in front of gamma detectors. Low-activity samples are often counted at distances of 10 mm or less from the front face of the detector. Sample positioning discrepancies then become critical.

If a sample can be considered as a point source and absorption of radiation in air ignored, then the measured activity varies with the inverse square of the distance (d) between source and detector. The error in activity due to a discrepancy Δ in positioning a sample is then given by

$$\%\ \text{error} = \left\{\left[\frac{1}{(d-\Delta)^2} - \frac{1}{d^2}\right] \Big/ \frac{1}{d^2}\right\} 100 \qquad (12.21)$$

This function has been plotted for repositioning errors of 0.1, 0.2 and 1 mm at source to detector distances of up to 50 mm in Figure 12.31. These data show, for example, that for a relatively small repositioning error of 0.2 mm, the counting error approaches 4% for source to detector distances of 10 mm. Indeed, very accurate sample positioning is required if samples are counted closer than about 20 mm from the detector.

Similar counting errors can arise from:

(i) Errors in repositioning samples perpendicular to the axis of the detector

(ii) Differences in the physical dimensions of samples, in particular small changes in the thickness of samples prepared to constant weight arising from differences in their effective density due to powder mesh size or the nature of sample.

These problems have been discussed by Hertogen and Gijbels (1971), Hoede and Das (1981) and Potts and Hussey (1983).

12.12.4 *Self-shielding of neutrons*

Self-shielding of neutrons occurs during irradiation when a sample contains high concentrations of an isotope which has a very high neutron capture cross-section. The effective concentration of neutrons within the sample is thus reduced, resulting in activities lower than expected. Virtually all the major elements in silicate rocks have very low neutron absorption cross-sections (often in the millibarn range). De Soete *et al.*, (1972, p.457) have reviewed this problem, and Gruber (1973) has presented data to estimate the magnitude of this effect in general activation analysis. In Table 12.8, the concentrations of those elements that would cause a 1% depression in neutron flux have been calculated from equations presented by De Soete *et al.* (1972, eqn. 10.12) for a 0.3 g sample irradiated as a finite sphere. These data should be treated as order of magnitude estimates, but agree approximately with some data reviewed by Laul (1979). It would be unusual to encounter concentrations in silicate rocks that approach these values; however, they are offered as a guide to when neutron self-shielding errors can be expected. A further error can arise, especially in epithermal irradiations, if the resonance integral of one isotope strongly overlaps that of another, a situation again unlikely to affect the analysis of silicate rocks (Steinnes, 1976). A more rig-

Table 12.8 Flux depressions caused by self-absorption of neutrons within sample

Element	Concentration to give 1% flux depression
Li	1000 ppm
B	150 ppm
Cl	1.1%
Sc	2.0%
Ti	9.0%
Cr	17%
Mn	4%
Fe	22%
Rh	0.7%
Cd	480 ppm
Sm	280 ppm
Eu	370 ppm
Gd	36 ppm

Data calculated for a sphere of effective radius 4 mm from the equation presented by De Soete *et al.* (1972) (their eqn. 10.12) assuming an effective sample density of 1.7 kg m^{-3} and using total thermal neutron cross-section data from Erdtmann (1976). Data presented above should be considered as order of magnitude estimates.

orous discussion of neutron self-shielding effects, especially in relation to sample particle size and elements such as gadolinium which may concentrate into minor phases, is given by Steinnes (1971).

12.12.5 *Self-absorption of x-rays and gamma rays*

X-rays and gamma-rays emitted from an irradiated sample have a finite probability of undergoing absorption within the sample. Croudace (1979) has shown that the mass absorption coefficients (section 8.6) vary between samples of basaltic and granitic composition to an increasing extent at energies below 100 keV. The magnitude of this discrepancy is plotted in Figure 12.32 for 1, 2 and 5 mm thick samples of assumed effective density 1.2 kg m^{-3} (i.e. lightly compressed powders). The ^{169}Yb peak at 63.1 keV is absorbed to an extent of about 1.5% more in a basalt than a granite when transmitted through a 1mm thickness of rock powder. At energies below 60 keV, this effect increases in severity. Thus there is some advantage in samples in the form of thin discs as in Figure 12.23.

12.12.6 *Second-order activation*

Second-order activation arises most seriously during long irradiations (several days) at high neutron fluxes ($>10^{13}$ n $cm^{-2}s^{-1}$). The neutron capture product of one isotope can decay during irradiation to a naturally occurring isotope which can itself undergo neutron capture. The activity after irradiation of the latter isotope is then enhanced. This interference is uncommon in silicate rock analysis, especially when measuring the longer-lived isotopes.

However, Kramar (1980) demonstrated an interference arising in the analysis of gadolinium from second-order activation products of europium. The competing reactions are :

$$\underset{(\theta = 0.2\%)}{^{152}\mathrm{Gd}}\,(n,\gamma)\,\underset{(T_{1/2} = 242\mathrm{d})}{^{153}\mathrm{Gd}} \qquad (12.22)$$

$$\underset{(\theta = 47.8\%)}{^{151}\mathrm{Eu}}\,(n,\gamma)\,\underset{(t_{1/2} = 9.3\mathrm{h})}{^{152m}\mathrm{Eu}} \xrightarrow{\beta^-} {^{152}\mathrm{Gd}}\,(n,\gamma)\,\underset{(T_{1/2} = 242\mathrm{d})}{^{153}\mathrm{Gd}} \qquad (12.23)$$

where θ is the natural isotopic abundance.

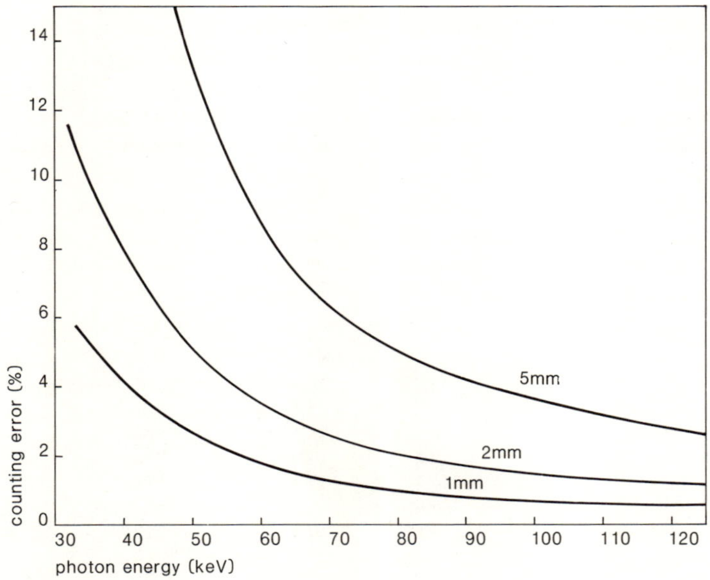

Figure 12.32 Estimates of the discrepancy in counting samples of basaltic and granitic composition due to self-absorption and gamma-rays. Plots are based on the mass absorption coefficient data of Croudace (1979) and represent the % difference in absorption of gamma radiation of specified energy when transmitted through granitic in comparison with basaltic rock powders of thickness 1, 2 and 5 mm. An effective powder density of 1.2 kg m^{-3} has been assumed.

Table 12.9 U fission interferences in neutron activation analysis

Element	Fission reaction	Half-life	% Fission yield
Zr	^{235}U (n,f) ^{95}Zr	64.4d	6.5
Mo	^{235}U (n,f) ^{99}Mo	2.76d	6.09
La	^{235}U (n,f) $^{140}Ba \rightarrow {}^{140}La$	40.3h	6.33
Ce	^{235}U (n,f) ^{141}Ce	32.4d	5.78
Nd	^{235}U (n,f) ^{147}Nd	11.1d	2.3
Sm	^{235}U (n,f) ^{153}Sm	1.95d	0.27

Data from Erdtmann and Soyka (1979).
Fission yield listed for ^{140}La is as for ^{140}Ba.
Fission yield is the percent of the listed isotope formed by thermal neutron fission of ^{235}U.

Data presented by Kramar indicated that 'fictitious' ^{153}Gd (derived from ^{151}Eu) amounts to 5% of the total ^{153}Gd activity after about 36 hours in a neutron flux of 10^{13} n cm^{-2}s^{-1} or after about 8 hours at 10^{14}n cm^{-2}s^{-1} for chondritic REE distribution patterns. Laul (1979) identified other second-order interferences affecting short-lived isotopes of Al, Mg, Na and Mn. The problem has been reviewed more generally by Kubota (1977) and Ricci and Dyer (1964).

12.12.7 *Fission products*

Almost all silicate samples contain uranium which, during neutron irradiation, can be induced to decay by fission. A wide range of fission isotopes is formed, and most are radioactive. A few of these isotopes are identical to neutron capture isotopes formed by activation of the elements of interest. In such circumstances, it is not possible to distinguish activity induced by conventional activation and that induced by neutron fission of uranium.

Ila *et al.* (1983) reviewed the analysis of uraniferous rocks by INAA with particular reference to interferences caused by uranium fission. In the determination of REEs, the elements directly affected are La, Ce, Nd and Sm; fission yields of these isotopes are listed in Table 12.9. Other elements that suffer fission interferences are zirconium (^{95}Zr) and molybdenum (^{99}Mo). Ila *et al.* (1983) also demonstrated that uranium fission can also cause an interference effect from radioactive fission isotopes that are not normally determined by INAA. This interference arises if these isotopes decay by the emission of a gamma photon that coincides in energy with that of a photopeak of interest. The isotopes that may cause such interefences are listed in Table 12.10.

Of particular concern is the uranium fission interference on ^{140}La. This isotope is not formed by direct fission of ^{235}U but via a radioactive intermediary, ^{140}Ba:

$$^{235}U(n,f)\ {}^{140}Ba \xrightarrow{\beta^-} {}^{140}La \qquad (12.24)$$
$$(T_{1/2} = 12.8d) \quad (t_{1/2} = 1.68d)$$

Ila *et al.* demonstrated graphically how activities of ^{140}Ba and ^{140}La varied during and after irradiation (Figure 12.33) and discussed mathematical correction procedures. Independently, Kennedy and Fowler (1983) addressed the same problem. These studies show that the interference factor (f) required to correct the observed ^{140}La activity for ^{235}U fission is given by

$$f = k\,(\exp(+\lambda_{La}T)\exp(-\lambda_{Ba}T) - 1) \qquad (12.25)$$

where λ_{La} and λ_{Ba} are the decay constants of ^{140}La and ^{140}Ba, respectively, T is the decay time since irradiation assuming that the irradiation time is short compared with the decay time and k is a constant that varies with irradiation conditions.

Kennedy and Fowler determined a value $k = 0.00282$ for

Table 12.10 Gamma photopeak interferences from ^{235}U fission isotopes

Fission isotope data					INAA isotope data		
Fission isotope	Half-life	Fission yield	Gamma energy	Intensity (%)	INAA isotope	Half-life	Gamma energy
^{72}Zn	1.94d	1.6×10^{-5}	144.8	86.0	^{141}Ce	32.4d	145.4
^{105}Rh	1.47d	0.93	319.2	19.6	^{51}Cr	27.7d	320.1
^{125}Sb	2.77y	0.030	427.9	29.6	—		
			463.5	10.0	—		
			600.8	18.4	^{134}Cs	2.06y	604.7
			636.2	11.2	—		
^{131m}Te	1.25d	0.19	773.7	38.1	^{152}Eu	12.7y	778.9
			793.8	13.8	^{134}Cs	2.06y	795.8
			852.2	20.6	—		
			1125.5	11.4	^{46}Sc	83.8d	1120.5
^{132}Te	3.25d	4.28	228.2	88.5	^{239}Np	2.36d	228.2
^{140}Ba	12.79d	6.33	537.4	19.9	^{147}Nd	11.1d	531.0
^{151}Pm	1.18d	0.42	340.1	22.4	^{152}Eu	12.7y	344.3
^{156}Eu	15.19d	0.013	88.9	8.9	^{160}Tb	72.1d	86.8
			811.8	10.3	^{140}La	40.3h	815.8
			1153.5	7.1	—		
			1230.7	8.9	^{182}Ta	115d	1231.0

Interferences from Ila *et al.* (1984).
Nuclear data from Erdtmann and Soyka (1979).
INAA isotopes have photopeaks used for analysis that suffer an overlap from a ^{235}U fission product.

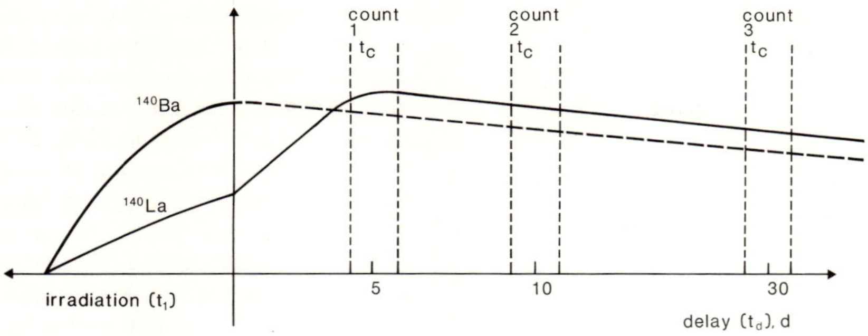

Figure 12.33 Build-up and decay in the activity of ^{140}Ba and ^{140}La during and following irradiation of ^{235}U. (Ila *et al.*, 1983.)

a 1-hour irradiation; Potts *et al.* (1985*b*) determined values of $k = 0.00347$ and 0.00732 for 3-day and 5-day sample irradiations, respectively.

Potts *et al.* also estimated the magnitude of the ^{140}Ba fission interference correction as a function of decay time since irradiation. Their data, listed in Table 12.11 as the apparent concentration of lanthanum induced by fission of a unit weight of uranium, show that the magnitude of this correction becomes unacceptably large if samples are counted after a delay significantly longer than 10 days following irradiation.

Several workers have estimated experimentally the magnitude of uranium fission interferences on elements determined by INAA, and selected data for corrections to ^{95}Zr, ^{99}Mo, ^{140}La, ^{141}Ce and ^{147}Nd activity are listed in Table 12.12.

If fission interference effects are measured experimentally, precautions must be taken in assessing the precise isotopic composition of uranium used for calibration measurements. Potts *et al.* (1985b) found that the high-purity reagent used for their measurements was significantly depleted in ^{235}U, having an isotopic composition of 0.325% ^{235}U rather than the natural abundance of 0.711%.

12.13 Routine schemes of analysis

A large number of reports has been published on the application of instrumental neutron activation analysis to routine

Table 12.11 Variation in the fission correction factor on lanthanum as a function of decay time after irradiation

Decay time after irradiation (days)	Fission interference factor (g La per g U)
6	0.056
8	0.12
10	0.26
12	0.54
14	1.1
16	2.3
18	4.7

Data from Potts *et al.* (1985*b*).
Factors appropriate for samples irradiated for 30 hours spread over 5 working days.

silicate rock analysis. To some extent counting strategy and the range of elements determined has depended on the choice of detector used for gamma spectrum measurements. The planar (LEPS) detector is favoured if reliable determinations of the rare-earth elements are particularly important, and the analytical characteristics of such devices have been described by, for example, Hertogen and Gijbels (1971), Henderson and Williams (1981), Hoede and Das (1981) and several others. In the pioneering application studies of Gordon *et al.* (1968) a coaxial Ge (Li) detector was used. However, it is clear from section 12.9.4 that if all gamma-activity data are to be obtained samples must be counted with both planar and coaxial germanium detectors. The performance characteristics of these devices have been compared by Baedecker *et al.* (1977), Schock (1977) and Potts *et al.* (1985*a*, 1981). Potts *et al.* (1985*a*) devised a counting strategy to measure all significant gamma photopeaks in irradiated rock spectra. Samples were counted simultaneously using both planar and coaxial detectors to cover the complete range of photon energies from about 60 keV to 2000 keV. Two series of counts were carried out: the first within about 10 days of irradiation for count times of 6000 s to 8000 s per sample; the second between 2 and 4 weeks of irradiation for count times of 6 to 10 hours per sample. The isotopes determined in this scheme of analysis, together with photopeaks fitted to spectra and interferences, are listed in Table 12.13. Also listed in this table are details of detection limits that may be obtained from individual photopeaks for 'typical' silicate matrices under routine counting conditions. Potts *et al.* (1985*a*) claimed that there were two advantages in their scheme of analysis in which several photopeaks were used (where available) to determine individual elements. First, the precision of determinations was improved by averaging data from several photopeaks and several individual gamma spectra. Second, this scheme offered sufficient analytical redundancy to permit the analyst to identify gross errors that might affect data from an individual photopeak, or even from an entire spectrum. The degree of reliablity that could be attributed to routine determinations was then increased.

In order to indicate the routine performance of INAA in multi-element silicate analysis, data for basalt USGS: BCR-1 from a variety of schemes are listed in Table 12.14. These INAA results are compared with the consensus values of Gladney *et al.* (1983) and the average of rare-earth element

Table 12.12 ^{235}U fission product correction factors

Fission isotope	Half-life	Erdtmann (1972) Thermal neutrons only	Erdtmann (1972) Including epithermal neutrons	Vobecky (1979) Thermal neutrons column	Vobecky (1979) Epithermal neutrons	Barnes and Gorton (1984) Thermal neutrons (Slowpoke)	Kennedy and Fowler (1983) Slowpoke 1 hour irradiation	Potts *et al* (1985b) Thermal irradiation 20 h over 3 days	Potts *et al* (1985b) Thermal irradiation 30 h over 5 days	Potts *et al* (1985b) Mean
^{95}Zr	64.4d	—	—	9.6	2.6	—	—	9.74±0.18	9.84±0.33	9.8
^{99}Mo	2.76d	0.9–3	0.3–1.0	1.4	0.13	—	—	—	—	—
^{140}La	40.3h	0.017	0.016	—	—	0.033 (7d)	0.032 (7d)	0.039 (7d)	0.083(7d)	—
^{141}Ce	32.4d	0.31	0.28	0.3	0.3	0.30	0.28	0.300±0.007	0.317±0.007	0.31
^{147}Nd	11.1d	—	—	0.4	0.2	0.26	0.23	0.238±0.004	0.245±0.003	0.24

Fission correction is listed as the apparent concentration of element per unit concentration of uranium (g per g of U).
(7d) indicates that the La correction is appropriate for a sample that has decayed for 7 days after irradiation before counting.
Precision errors are the standard deviation of 4 replicate determinations.
d = day, h = hour.

data evaluated from published isotope dilution mass spectrometry results taken from Table 17.17. These data indicate that the precision of neutron activation data is often better than 5% (coefficient of variation) although it sometimes exceeds 10% for less well activated elements.

As mentioned earlier, due to the limited space available in each sample batch prepared for irradiation and to the relatively long time spent counting individual samples, only one standard (the calibration standard) is often included in each batch of unknowns. There is then no independent check on the quality of data. In the scheme of analysis summarized in Table 12.13 it was considered important to include a 'quality assurance' sample in each irradiation batch. This sample was a silicate rock that was included in every irradiation batch and analysed as an unknown. From the resultant data, calibration drift could be readily identified and a realistic assessment made of long-term analytical precision. The results of the analysis of one such quality assurance sample (a basalt, Whin Sill) are listed in Table 12.15, data being calculated from 12 samples irradiated in separate batches over a period of several months. For many trace elements in this basalt analytical precision lies in the range 2 to 5% (one sigma).

Using instrumental NAA procedures, gadolinium cannot usually be determined reliably if more than a few ppm thorium are present unless samples are recounted several months after irradiation. Furthermore, data for thulium is not always reported unless the spectrum analysis program is capable of resolving the severe spectrum overlap with tantalum.

Since INAA data are often calibrated against an individual silicate reference material, it is possible that the occasional departure in results from earlier schemes of analysis may be due to the lack of reliable data for that element in the selected reference sample.

12.14 Chondrite normalized abundances

The relative fractionation trend of REEs may be assessed by plotting these elements as chondrite normalized abundances. This permits an independent indication of the self-consistency of REE analyses. In nature, the elements of even atomic number are more abundant than those of odd atomic number. The REEs comprise a homologous group of elements increasing in atomic number from 57 (La) to 71 (Lu). If the abundances of these elements are plotted against atomic number, a saw-tooth pattern is observed, as predicted by the Oddo–Harkins rule (Mason, 1966, p. 24). However, if REE values are divided by the corresponding elemental composition of chondritic meteorites, a smooth and regular variation is observed (Figure 12.34).

It is generally accepted that chondrites represent the primordial matter from which many silicate rocks are derived. Since the geochemical properties of the REEs depend on oxidation state (always 3+ in rocks, except Ce and Eu—see below) and ionic radius (which varies over a narrow range as a smooth function of atomic number—see Table 12.16), then the chondrite normalized REE abundance of a rock might reasonably be expected to represent the differentiation history of these elements from chondritic material.

Exceptions to a smooth REE pattern are cerium and

Table 12.13 Gamma photopeaks and detection limits for INAA using both planar and coaxial germanium detectors

Element/line		Planar detector		Coaxial detector		Detection limits (ppm)	
	keV	6000–8000 s, 7–10 days after irradiation	8–10 h, 2–4 weeks after irradiation	6000–8000 s, 7–10 days after irradiation	8–10 h, 2–4 weeks after irradiation	Planar detector ppm	Coaxial detector ppm
La	1596.2			YES Eu-1597.3 U-fission			*s*: 0.5
	328.7	YES U-fission		YES U-fission		*s*: 5	*s*: 1.5
	487.0			YES U-fission			*s*: 1.7
	815.8			YES U-fission			*s*: 2.2
Ce	145.4	YES Yb-144.9 Pa-145.4 U-fission	YES Yb-144.9 Pa-145.4 U-fission	YES Fe-142.6 Pa-145.4 U-fission	YES Fe-142.9 Pa-145.4 U-fission	*l*: 2.1	*l*: 1.5
Nd	91.1	YES U-fission	YES U-fission			*s*: 4.6	
	531.0			YES	YES		*s*: 15
Sm	103.2	YES Gd-103.2 Pa-103.9 Np-103.7				*s*: 0.1	
	69.7	YES Gd-69.6 Ta-69.1 (W-69.2)				*s*: 0.34	
Eu	121.8	YES (Np-120.7)	YES (Nd-120.5)			*l*: 0.05	
	123.1	YES Ba-123.8	YES Ba-123.8			*l*: 0.5	
	1408.1			YES	YES		*l*: 0.05
	778.9			YES (Mo-777.8)	YES		*l*: 0.17
Gd	97.5		YES (Sm-97.4) Pa-98.4			*l*: 4.6	
	103.2		YES Pa-103.9 (Sm-103.2)			*l*: 3.9	
Tb	86.8	YES Pa-86.6	YES Pa-86.6			*l*: 0.09	
	1177.9			YES	YES		*l*: 0.7
	876.4			YES Eu-873.2	YES Eu-873.2		*l*: 0.17
Ho	80.6	YES U-fission				*s*: 1.7	
Tm	84.3	YES Ta-84.7 Sm-83.4	YES Ta-84.7			*l*: 0.34	
Yb	63.1	YES Hf-63.2 Lu-61.3	YES Hf-63.2 Lu-61.3			*l*: 0.14	
	109.8	YES Ta-110.4 Pa?-111.4	YES Ta-110.4 YES Pa?111.4			*l*: 0.7	
	130.5	YES La-131.1	YES			*s*: 0.9	
	177.2	YES Ta-179.4	YES Ta-179.4	YES Ta-179.4	YES Ta-179.4	*l*: 0.7	*l*: 0.5
	198.0	YES Ta-198.4 Tb-197.0	YES Ta-198.4 Tb-197.0			*l*: 0.7	*l*: 0.34
	396.3			YES Pa-398.5 Tb-392.5			*s*: 0.34
Lu	113.0	YES Ta-113.7 Pa-114.5 Yb-113.8	YES Ta-113.7 Pa-114.5 Yb-113.8			*s*: 0.14	
	208.4	YES Np-209.8	YES	YES Np-209.8	YES	*s*: 0.10	*s*: 0.07
Ba	469.2			YES	YES		*l*: 60
	123.8	YES Eu-123.1	YES Eu-123.1			*s*: 140	
Co	1332.5			YES	YES		*l*: 0.14
	1173.2			YES (Cs-1167.8)	YES (Cs-1167.8)		*l*: 0.3
Cr	320.1	YES (Nd-319.4)	YES (Nd-319.4)	YES (Nd-319.4)	YES (Nd-319.4)	*l*: 12	*l*: 1.5
Cs	795.8			YES	YES		*l*: 0.24
	604.7			YES Sb-602.7	YES Sb-602.7		*l*: 0.17
Fe	1099.2			YES Tb-1102.6	YES Tb-1102.6		*l*: 0.03%
	1291.6			YES Ta-1289.2	YES Ta-1289.2		*l*: 0.01%
	192.3	YES	YES			*l*: 0.3%	*l*: 0.2%
	142.6	YES (Mo-140.5)	YES			*l*: 0.9%	*l*: 0.7%
Hf	133.0	YES (Ba-133.6) (W-134.2)	YES (Ba-133.6)			*l*: 0.22	
	136.2	YES (Hf-136.9)	YES (Hf-136.9)			*l*: 2.1	
	482.1			YES	YES		*l*: 0.17
Mo	739.5			YES			
Na	1368.6			YES Cs-1365.1 Sb-1368.2			*s*: 0.1%

(Continued)

Table 12.13 (*Continued*)

Element/line		Planar detector		Coaxial detector		Detection limits (ppm)	
	keV	6000–8000 s, 7–10 days after irradiation	8–10 h, 2–4 weeks after irradiation	6000–8000 s, 7–10 days after irradiation	8–10 h, 2–4 weeks after irradiation	Planar detector ppm	Coaxial detector ppm
Np (U)	106.1	YES				*s*: 0.5	
	277.6	YES Nd-275.4		YES Nd-275.4		*s*: 2.2	*s*: 1.9
				Yb-282.5			
	228.2	YES Ta-229.3		YES Ta-229.3		*s*: 2.9	*s*: 1.0
Pa(Th)	94.7	YES Yb-93.6	YES Yb-93.6			*l*: 0.9	
	98.4		YES Gd-97.5			*l*: 0.5	
		Sm-97.4					
	311.9	YES Yb-307.7	YES Yb-307.7	YES Yb-307.7	YES Yb-307.7	*l*: 0.7	*l*: 0.22
		Tb-309.6	Tb-309.6	Tb-309.6	Tb-309.6		
		(Np-315.9)		(Np-315.9)			
Rb	1076.6			YES	YES		*l*: 10
Sb	1691.0			YES	YES		
Sc	889.3			YES Eu-892.8	YES Eu-892.8		*l*: 0.03
Ta	67.7	YES (Ta-67.2)	YES (Ta-67.2)			*l*: 0.07	
	100.1	YES	YES			*l*: 0.14	
	152.4	YES	YES			*l*: 0.5	
	222.1	YES	YES	YES	YES	*l*: 0.9	*l*: 0.5
	1221.4			YES	YES		*l*: 0.15
	1231.0			YES	YES		*l*: 0.34
	179.4	YES Yb-177.2	YES Yb-177.2			*l*: 1.4	*l*: 1.2
	84.7	YES Tm-84.3	YES Tm-84.3			*l*: 0.7	
		Sm-83.4					
W	685.7			YES Nd-685.9			
	479.6			YES Cs-475.3			
				Eu-478.3			
	72.0	YES (W-71.2)					
Zr	724.2			YES Eu-723.3	YES Eu-723.3		*l*: 105
				Sb-722.8	Sb-722.8		
				U-fission	U-fission		
	756.7			YES Eu-756.7	YES Eu-756.7		*l*: 100
				U-fission	U-fission		

Data from Potts *et al.* (1985*a*) and Potts (1983).
Photopeaks to be fitted in a general silicate analysis program are listed together with potential spectrum interferences. Interferences in brackets may be trivial.
Detection limits are estimates of the six-sigma limit of determination *s*: means optimum data are obtained from short count spectra (i.e. those counted 7–10 days after irradiation); *l*: identifies optimum data from long count spectra (i.e. 2–4 weeks after irradiation).

europium. These elements are uncharacteristic, since both can exist in rocks in alternative oxidation states. Cerium can form the 4+ state in oxidizing environments and is often associated with rock–sea water interactions. Europium can be reduced to the 2+ state depending on the oxygen fugacity of the rock. Alternative lattice sites are accessible to Eu (2+) causing anomalous relative concentrations that are often associated with plagioclase fractionation. Thus cerium and europium anomalies (either positive or negative) can be found in chondrite normalized plots of some silicate samples.

These geochemical constraints on the REE have two analytical consequences. First, an examination of the chondrite normalized REE abundances of a sample is a sensitive method of assessing the relative analytical consistency of a determination. Second, there is considerable geochemical importance in the accurate determination of the elements (Sm and Gd) adjacent to Eu in order to assess the magnitude of any europium anomaly. It is unfortunate that INAA does not always provide sensitive gadolinium data (Th interference) unless special counting strategies are employed (see section 12.9.3). This disadvantage is, however, somewhat offset by the very sensitive determination that may be made for europium itself. In interpreting chondrite normalized REE data, it is important to use accurate and reliable average chondrite values as normalizing factors. In Table 12.16 two sets of chondrite REE abundances are listed, both being predominantly obtained by isotope dilution mass spectrometry (Chapter 17). Nakamura (1974) calculated the 'average of 10 ordinary chondrites' to which has been supplemented additional data for Pr, Tb, Ho and Tm (the elements that cannot be determined by isotope dilution) obtained by extrapolating the NAA data of Frey *et al.* (1968). More recently, Evensen *et al.* (1978) have made a very careful assessment of all available chondrite REE data and have calculated the 'average C1 chondrite' abundances also listed in Table 12.16, values which are now generally accepted for use as normalizing factors.

12.15 Epithermal *v.* thermal irradiations

Isotopes with high epithermal neutron capture cross-sections are preferentially activated when irradiated under a cadmium filter (section 12.2.2(*b*)). The improvement in acti-

Table 12.14 Instrumental neutron activation analysis of basalt USGS: BCR-1

Element	Gladney *et al.* (1983) consensus analysis		Average of 23 published isotope dilution mass spectrometry analysis		Potts *et al.* (1981)		Borley and Rogers (1979)		Henderson and Williams (1981)		Brunfelt *et al.* (1971)[1]	Nadkarni and Morrison (1978)[2]	van der Sloot and Zonderhuis (1979)[3]	Randle (1974)[4]		
	Expected	Uncertainty (±)	Mean	CV	Analysed	CV ($n = 2$)	Analysed	CV ($n = 10$)	Analysed	CV	Analysed	Analysed	Analysed	Analysed	CV	
La	25.0	0.08	24.72	3.6	24.2	—	25.1	6	26.5	6.2	23.7	23	26.4	23.7	2.5	La
Ce	53.7	0.8	53.54	2.0	53.2	1.3	56.2	6	56	3.6	53	43	49	46	4.3	Ce
Nd	28.7	0.6	28.90	1.8	28.5	5.0	27.0	7	27.4	4.4		32	38	33	12	Nd
Sm	6.58	0.17	6.646	2.4	6.75	3.1	6.77	4	6.81	4.7	6.52	6	7.40	7.5	6.7	Sm
Eu	1.96	0.05	1.964	2.2	1.97	0.7	1.95	3	1.98	4.3	1.94	2.0	1.79	1.82	5.5	Eu
Gd	6.68	0.13	6.722	2.3					6.5	11.5		5				Gd
Tb	1.05	0.09			1.08	1.3	1.11	7	1.11	3.2	0.96	1	1.16			Tb
Ho	1.25	0.14					1.27	7	1.2	16.7	1.20					Ho
Tm	0.590	0.035			0.57	3.8	0.60	14	0.56	9.8		0.6				Tm
Yb	3.39	0.08	3.41	2.4	3.36	8.5	3.46	5	3.38	2.5	3.21	3.1	3.35	3.3	6.1	Yb
Lu	0.512	0.025	0.528	6.5	0.53	—	0.53	10	0.49	7.1	0.535	0.54	0.58	0.60	15	Lu
Ba	678	16									580		880	520	12	Ba
Co	36.3	1.6			35.1	0.2					36.2	43	34.8	35	8.5	Co
Cr	16	4										14	15.2			Cr
Cs	0.97	0.13									0.97	1.1				Cs
Fe_2O_3	13.41	0.40									13.40	13.81	13.37			Fe_2O_3
Hf	4.9	0.3			4.76	1.8					4.72	4.8	4.71	5.1	9.8	Hf
Na_2O	3.28	0.11									3.26	3.33	3.19			Na_2O
Rb	47.1	0.6									50	44		38	18	Rb
Sb	0.62	0.10									0.60	0.68				Sb
Sc	32.8	1.7									30.7	31	31.5	33.1	2.7	Sc
Ta	0.79	0.09			0.79	0.9					0.74	1		0.93	11.8	Ta
Th	6.04	0.60			6.12	3.1			6.0	5.0	4.98	4.7	6.2	5.6	5.4	Th
U	1.71	0.16									1.68	1.75				U
W	0.40	0.09									0.38	0.44				W
Zr	191	5										274				Zr

Units: ppm rock (Fe_2O_3 and Na_2O in % oxide).
CV = coefficient of variation = 100 × standard deviation/mean analysis (%).
[1] *Talanta* **18,** 1197–1208; [2] *J. Radioanal. Chem.* **43,** 347–369; [3] *Geostand. Newslett.* **3,** 185–193; [4] *Chem. Geol.* **13,** 237–256.

Table 12.15 Routine analytical precision in the INAA analysis of a Whin Sill basalt quality assurance sample

Element	Mean analysis	CV
La	25.5	2.6
Ce	57.5	3.1
Nd	32.9	3.5
Sm	7.27	2.1
Eu	2.25	3.9
Gd	7.5	11.0
Tb	1.09	3.9
Tm	0.41	5.9
Yb	2.54	3.1
Lu	0.39	1.7
Co	47.4	3.1
Cs	1.03	9.5
Fe_2O_3(%)	12.62	3.1
Hf	4.93	3.0
Na_2O(%)	2.7	6.5
Rb	35.3	11.0
Ta	1.26	4.7
Th	3.05	4.3
U	0.90	26.5

Data from Potts *et al.* (1985*a*).
Concentration units: ppm.
CV = coefficient of variation
= 100 × standard deviation/mean analysis (%).
Data derived from 12 analyses of aliquots of the same sample irradiated in separate batches over a six-month period.

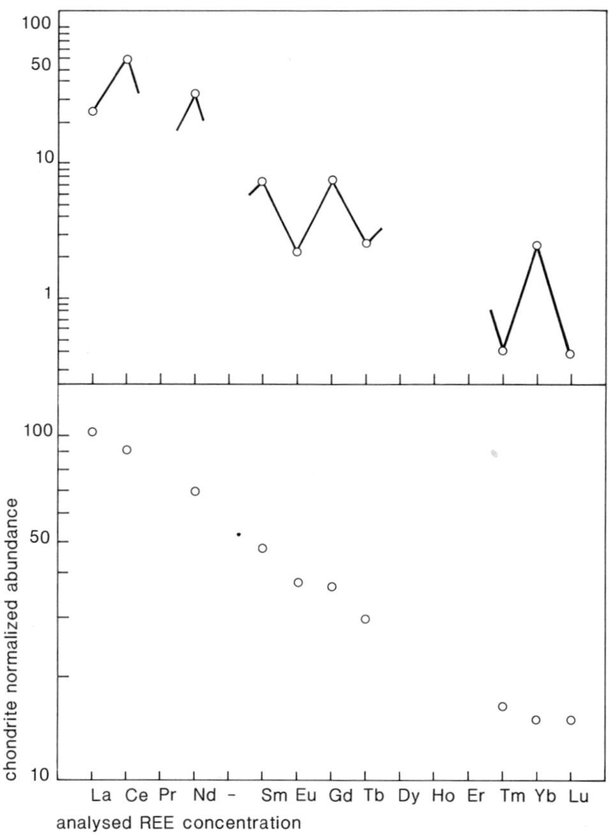

Figure 12.34 Analytical data for the REE in Whin Sill Basalt (Table 12.15) plotted against atomic number for (*a*) analysed concentrations and (*b*) chondrite normalized abundances. The upper diagram demonstrates the Oddo–Harkins rule where even- and odd-numbered elements show divergent natural abundances.

vation of an isotope by epithermal in comparison with thermal irradiation is measured by the *cadmium ratio* of the isotope (R_{Cd}) calculated from

$$R_{Cd} = \frac{\phi_{TH}\sigma + \phi_{EPI}(I+0.44\sigma)}{\phi_{EPI}(I+0.44\sigma)} \qquad (12.26)$$

where ϕ is the thermal (ϕ_{TH}) or epithermal (ϕ_{EPI}) neutron flux, σ is the thermal neutron cross-section, I is the epithermal neutron cross-section arising from resonance integrals only and 0.44 is the factor to account for neutron capture cross-section in the epithermal region arising from the so-called $1/v$ tailing effect, where neutron capture is inversely proportional to neutron velocity.

It is more convenient to determine the cadmium ratio of an element relative to that experimentally determined for gold under specified reactor conditions:

$$R_{Cd} = 1 + \left[\frac{\sigma(I^{(Au)}+0.44\sigma^{(Au)})}{\sigma^{(Au)}(I+0.44\sigma)}\right][R_{Cd}^{(Au)} - 1] \qquad (12.27)$$

where superscript (Au) now designates parameters for gold.

For nuclei that have *no* resonance integrals in the epithermal region ($I = 0$) and assuming a value $R_{Cd}{}^{(Au)} = 3.0$, then the cadmium ratio would equal 72 from equation (12.27). Thus Steinnes (1971) proposed the epithermal *advantage factor* for an isotope:

$$\text{advantage factor} = \frac{72}{R_{Cd}} \qquad (12.28)$$

in order to describe the improved sensitivity of epithermal irradiations in comparison with isotopes having no resonance integrals. This parameter is listed in Table 12.4 using

Table 12.16 REE ionic radii and chondrite abundances

	Ionic radius[1] (nm)	Chondrite abundances Nakamura[2] (μg/g)	Evensen *et al.*[3] (μg/g)
La	11.3	0.328	0.2446
Ce	10.9	0.865	0.6379
Pr	10.8	0.123	0.09637
Nd	10.6	0.630	0.4738
Sm	10.4	0.203	0.1540
Eu	10.3	0.0770	0.05802
Gd	10.2	0.276	0.2043
Tb	10.0	0.052	0.03745
Dy	9.9	0.343	0.2541
Ho	9.8	0.078	0.05670
Er	9.7	0.225	0.1660
Tm	9.6	0.034	0.02561
Yb	9.5	0.220	0.1651
Lu	9.4	0.0339	0.02539

Ionic radii quoted for 3+ oxidation state.
Ce(4+): ionic radius = 8.8 nm.
Eu(2+): ionic radius = 12.5 nm.
[1] Ionic radii data taken from Whittaker and Muntus (1970) for ions with a coordination number of 6.
[2] Chondrite abundances from Nakamura (1974) average of ten ordinary chondrites with additional values for Pr, Tb, Ho and Tm interpolated from the data of Frey *et al.* (1968).
[3] Chondrite abundances from Evensen *et al.* (1978) Cl chondrite average.

cross-section data from Erdtmann (1976). Many elements that contribute high activities to silicate rock spectra have low advantage factors, for example: ^{24}Na (1.5), ^{46}Sc (1.1), ^{51}Cr (1.1), ^{59}Fe (3.3), ^{60}Co (4.4). Since the activity of these isotopes is suppressed by epithermal irradiation, so is the Compton scatter contributed by their respective gamma photopeaks. Peak to background ratios of isotopes with high advantage factors will then be enhanced. Such isotopes (advantage factors in brackets) include ^{99}Mo (46), ^{124}Sb (46), ^{182}Ta (38) and ^{239}Np (U) (56).

An additional consideration is that the epithermal neutron flux is often two orders of magnitude lower than the thermal flux in many reactors, and so the activities of isotopes with middle-order advantage factors will not be as great as would otherwise appear. In practice, the balance point is usually between 10 and 20. Isotopes with advantage factors above this range are preferentially analysed by epithermal irradiation. Those with lower advantage factors are likely to be determined with lower precision by epithermal irradiation.

The analytical performance of epithermal irradiations has been discussed by Baedecker *et al.* (1977), Brunfelt *et al.* (1977) and Steinnes (1976). Parry (1980) has compared boron filters with the more conventional cadmium covers in the irradiation of geological samples and has suggested that the former are more suitable material for the analysis of some isotopes close to their detection limits. Bem and Ryan (1981) investigated the choice of boron shields for epithermal neutron irradiations and found significant improvements in the determination of As, Br, Cd, I, Mo, Sb, Sn, Th, W and U by their use, boron carbide being the preferred material.

12.16 Short-lived isotopes

The analysis of samples by neutron activation using such isotopes having half-lives in the range 1 minute to 3 hours has several advantages:

(i) Very short irradiation times are required, normally between 30 seconds and 30 minutes

(ii) Counting times are correspondingly short

(iii) In consequence, analytical costs are low, the potential for very high sample throughput leading to economies that are very attractive to commercial laboratories

(iv) The range of elements which may be analysed is extended (see Table 12.17)

(v) Quantitative data may be obtained to high precision from low flux reactors (10^{10}–10^{12} n $cm^{-2}s^{-1}$) so avoiding long irradiation times.

There are, however, several disadvantages. Such irradiations can only practically be carried out at a reactor site and preferably using a pneumatic 'rabbit' system to transport samples into and out of the reactor during the irradiation–cooling–counting–discard cycle. For isotopes with very short half-lives (less than 10 minutes) cyclic activation procedures have been proposed so that sufficient precision can be obtained from the combined spectrum recorded after successive irradiations (Amiel and Mantel, 1981). Care must be taken over counting dead time procedures which require special modification for samples whose activity changes significantly in the course of the count (Huysmans *et al.*, 1974).

Table 12.17 Isotopes which may be determined after short irradiations

Isotope	Half-life	Principal gamma photopeak (keV)	%
1–2 min irradiation, 20 min decay			
^{27}Mg	9.46 min	1014	(28.6)
^{28}Al	2.31 min	1779	(100)
^{49}Ca	8.7 min	3084	(91.7)
^{51}Ti	5.79 min	320.0	(95.0)
^{52}V	3.75 min	1434	(100)
^{56}Mn	2.58 h	1811	(27.2)
^{87m}Sr	2.81 h	388.4	(83.0)
^{139}Ba	1.42 h	165.8	(18.8)
^{165}Dy	2.33 h	94.7	(3.34)
^{239}U	23.54 min	74.7	(59.3)
5–10 min irradiation, 10–16 h decay			
^{24}Na	15.03 h	1368	(100)
^{42}K	12.36 h	1525	(17.9)
^{64}Cu	12.71 h	511	(38.6)
^{142}Pr	19.13 h	1576	(3.7)
^{152m}Eu	9.3 h	121.8	(7.2)
^{159}Gd	18.56 h	363.6	(10.3)
^{166}Ho	26.81 h	80.6	(6.2)
^{171}Er	7.52 h	308.3	(64.4)

min = minute, h = hour; photopeak energies in keV.
% decay intensities are given in parentheses.
Data from Hancock (1976) and Laul (1979).

Elements that are normally determined in silicate rocks from short-lived isotopes are listed in Table 12.17. Such procedures have been discussed, for example, by Hancock (1976), Duffield and Gilmore (1979) and Laul and Rancitelli (1977).

12.17 Radiochemical separation procedures

12.17.1 *General characteristics*

As has already been discussed, detection limits are constrained by the magnitude of the background continuum in gamma spectra which itself often arises from the high activity of a few isotopes (^{24}Na, ^{60}Co, ^{59}Fe, ^{45}Sc, and ^{50}Cr) (section 12.9.5). Epithermal irradiation selectively suppresses some of this background activity. However, this restriction may be effectively eliminated by using chemical procedures to separate the isotopes of interest from matrix activity.

Many such schemes have been devised and most involve multiple chemical separation steps which make a radiochemical separation an unattractive procedure unless the analytical data cannot be readily obtained by other means. Such situations might include the analysis of the REEs in samples where their chondrite normalized abundances fall significantly below a ratio of one to five (i.e. below the detection capability of INAA techniques) or in samples where these elements are present in a hostile environment, for example in apatites where the beta-activity induced in phosphorus (^{32}P: $T_{1/2} = 14.3$ days, pure beta-emitter) can interfere with the sensitive determination of REEs by instrumental techniques. In the former case (REE at low abundance), the cost effectiveness of a radiochemical spearation must be justified in comparison with more sensitive isotope dilution mass spectrometric techniques.

Most radiochemical separation procedures are carried out on samples which have already been irradiated. Some pre-irradiation procedures have been reported, e.g. Croudace (1980), Randa (1976), Kolesov (1976). *Post-irradiation* procedures have an important advantage in comparison with pre-concentration methods used in other forms of analysis: once a specimen has been irradiated, there is no further significant contribution to blank levels. Contamination by reagents is not important and in fact it is normal to add inactive aliquots of the elements to be analysed to irradiated samples to act as *carriers*. The function of carrier reagents is to minimize non-equilibrium effects that can influence the chemical separation of radioisotopes at very low concentrations (e.g. adsorption on the walls of containing vessels)—see de Soete *et al.* (1972) p.348.

The disadvantage of such radiochemical separation procedures is that they require specialized chemical manipulations that must often be carried out shortly after irradiation, when samples are highly active. A yield determination is almost always necessary to measure the fraction of radioisotope that has been separated into the final aliquot presented for counting (i.e. the *recovery efficiency* of the separation procedure for that isotope.)

Radiochemical separation procedures have been reviewed by Girardi and Pietra (1976). Three methods are summarized here, all relating to the recovery of the REEs: a traditional oxalate precipitation technique, a solvent extraction method for which 100% REE recoveries are claimed (yield determination unnecessary) and a method for the REE analysis of apatites.

12.17.2 *REE separation by oxalate precipitation*

Several workers have employed this technique; the adaptation summarized here was described by Duffield and Gilmore (1979). The irradiated sample (500 mg) and carrier (La^{3+}, 10 mg/g) are dissolved in 40% HF (10 mL). After evaporating to dryness at 150°C the procedure is repeated with more 40% HF (10 mL) and then a mixture of 40% HF (5 mL) and concentrated perchloric acid (5 mL) (to ensure complete removal of silica). The residue is taken up in concentrated hydrochloric acid, and saturated borax solution is added to complex any residual free fluoride. Insoluble hydroxides are then precipitated by adding ammonia solution. These are separated by centrifuge. The soluble fraction containing most of the matrix activity is discarded. The hydroxide precipitate is dissolved in HCl and the solution neutralized by adding ammonia and then a few drops of HCl solution until a clear solution is again obtained. The rare earths are then precipitated from this solution by boiling with saturated oxalic acid. The precipitate is separated, washed and dried, before gamma spectrometry to determine activities of individual rare-earth elements. Chemical yield is measured by re-irradiating the separated residue to determine the recovered concentration of lanthanum carrier. Yields usually exceed 85%.

12.17.3 *REE separation by solvent extraction*

Smet *et al.* (1978) described a radiochemical procedure for the determination of 24 trace elements in rocks and minerals. These authors' scheme is summarized in Figure 12.35. The procedure relevant to REE analysis alone is as follows: an irradiated sample to which has been added appropriate quantities of carriers, is dissolved in a PTFE-lined bomb

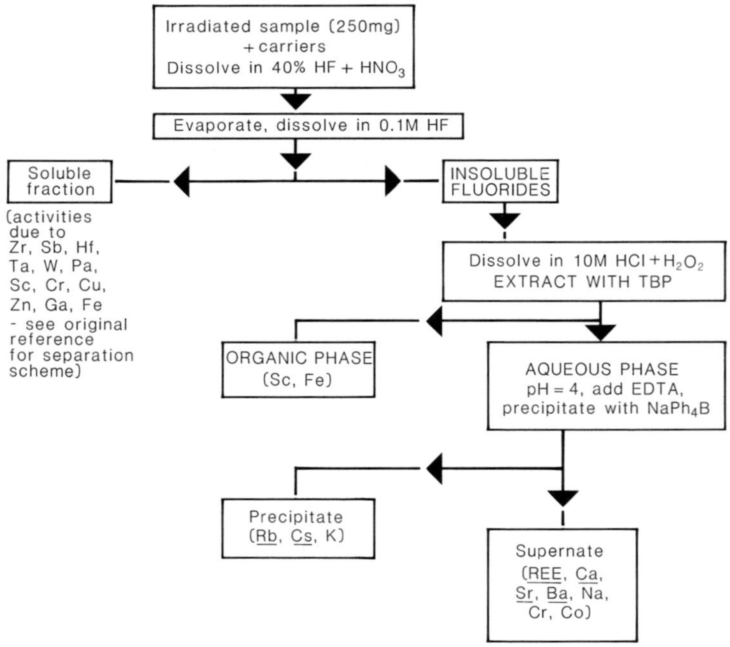

Figure 12.35 Abridged radiochemical separation procedure for REEs. (After Smet *et al.*, 1978.)

with a mixture of 40% HF (20 mL) and fuming nitric acid (1 mL) for 4 hours at 240°C. After cooling the solution is evaporated to dryness twice with 40% HF and the residue taken up with 0.1 M HF. The insoluble fluoride fraction is separated, washed and redissolved in 10 M HCl to which has been added hydrogen peroxide solution. This solution contains the REE fraction contaminated with Sc, Fe, Co, Cr, Rb, and Cs activities. These are removed as follows:

(i) Iron and scandium activities are removed by extraction into the organic solvent tri-*n*-butyl phosphate

(ii) Cobalt activity is removed from the residual aqueous phase by extraction into chloroform as the 1-nitroso-2-naphthol complex.

(iii) Rubidium, caesium and potassium activities are removed by precipitating these elements with sodium tetraphenyl boron in the presence of disodium ethylenediaminetetra-acetate solution to inhibit co-precipitation of the heavy REEs.

(iv) Chromium activity is removed by distillation as the volatile salt CrO_2Cl_2 in the presence of perchloric and hydrochloric acids

(v) The residue may then be counted to determine Ca, Sr, Ba and REEs.

The abridged scheme employed if ^{51}Cr, ^{59}Fe and ^{60}Co activities are acceptably low is shown in Figure 12.35. Recovery for REEs, Ca, Sr and Ba is quantitative as checked by yield determinations on the La, Sm and Yb carriers used in the original work. The technique has been successfully used in the analysis of USGS DTS-1 (dunite) at REE concentrations as low as 0.93 ppb Eu, and data obtained by Smet *et al.* (1978) following this procedure are listed in Table 12.18.

Table 12.18 Radiochemical neutron activation analysis of dunite USGS: DTS-1

Element	Gladney *et al.* (1983) Consensus values	Rey *et al.* (1970) Analysed values	Smet *et al.* (1978) Analysed values	Brunfelt *et al.* (1974) Analysed values
La	0.029±0.009	0.026±0.002	0.022±0.001	0.032
Ce	0.067±0.016	0.076±0.018	0.059±0.008	0.083
Pr	0.0063±0.0004	0.0058±0.0002		
Nd	0.030±0.004	<0.024 (3σ)	0.028±0.004	0.035
Sm	0.0048±0.0012	0.0050±0.0001	0.0041±0.0002	0.0061
Eu	0.0011±0.0003	0.0011±0.0001	0.00093±0.00006	0.0016
Gd	0.0038	<0.010 (3σ)		
Tb	0.0006±0.0002	<0.0003 (3σ)		0.0007
Dy	0.0034	0.0026±0.0006		
Ho	0.0013±0.0010	0.0026±0.0004		
Er	0.0039	<0.003 (3σ)		
Tm	0.00140±0.00035	0.0015±0.0001	0.0012±0.0006	
Yb	0.015±0.011	0.018±0.001	0.0108±0.0005	0.014
Lu	0.0026±0.0005	0.0031±0.0001	0.0022±0.0001	0.0025
Ba	2.35			0.70
Ca	1000±300		740±40	
Cs	—		0.0053±0.0003	0.0053
Cu	7.5±2.4			
Ga	0.15			
Hf	—		0.0036±0.0004	
In	0.0024±0.0003	0.0022±0.0001 0.0030±0.0004		
K	8±0.8	7.9±0.4 7.6±0.4		
Rb	0.058±0.010	<0.043 (3σ) <0.040 (3σ)	0.051±0.007	0.057
Sb	0.47±0.04		0.45±0.02	
Y	—	0.050±0.001		
Zn	48±11			

Units of concentration: ppm.
Analytical methods:
Rey *et al.* (1970) 0.5–1.0 g sample irradiated for 1 h in a thermal neutron flux of 7×10^{11} n cm^{-1}. Activities of longer-lived isotopes were enhanced by irradiating for 4–6 h at least 2 days before so that the activity of ^{24}Na would be reduced for the radiochemical procedure. After fusion with sodium peroxide in the presence of carriers and dissolution in water, rare-earth elements were separated by precipitation as hydroxides, then fluorides and finally by ion exchange. The isotopes ^{140}La, ^{143}Ce, ^{147}Nd, ^{153}Sm, ^{152m}Eu, ^{159}Gd, ^{160}Tb, ^{165}Dy, ^{166}Ho, ^{171}Er, ^{169}Yb, ^{175}Yb and ^{177}Lu were determined from gamma spectra; ^{142}Pr, ^{170}Tm and ^{99}Y by beta counting. An additional separation procedure was used to determine the remaining elements.
Smet *et al.* (1978) Radiochemical procedure described in the text; 95% confidence limits are listed.
Brunfelt *et al.* (1974): 0.2 g samples were irradiated in a thermal neutron flux of 1.7×10^{13} neutrons cm^{-2} s^{-1} and set aside for 5 days to permit short-lived activity to decay before radiochemical separation. After sodium peroxide/hydroxide fusion, solutions for REE analysis plus carriers were passed through an anion exchange column to remove interfering elements. The resulting solution was extracted with tri-*n*-butyl phosphate. The aqueous phase containing the REEs was separated into three fractions (Lu, Yb, Sr); (Sr, Tb, Ba, Eu, Sm, Gd) and (Nd, Ce, La) by passing through a second anion exchange column and eluting with a nitric acid–methanol mixed solvent solution. Chemical yields were determined by reactivation.

12.17.4 *REE separation from apatite minerals by anion exchange chromatography*

Roelandts *et al.* (1974) investigated the efficiency of strongly basic anion exchange resins (Dowex 1–X8, 200–400 mesh) in the separation of REE and other ions using mixed solvent methanol–nitric acid mixtures and reported distribution coefficients for various elements with respect to this resin. The following scheme was proposed for the analysis of apatites.

Samples are dissolved in nitric acid and, after evaporating to dryness, the residue is taken up in a 5% 7M nitric acid–95% methanol mixture. This solution is then passed through a 140 × 10 mm column of Dowex 1–X8 resin at a flow rate of 0.8 mL min^{-1}. The ions Na, Mg, Al, K, Sc, Mn, Fe pass straight through the column together with most calcium and phosphate ions. Elution of Ca is completed with 50 mL of the above mixed solvent. The REEs may then be eluted with a 45% 7 M nitric acid–55% methanol mixture and then water.

Apatites generally have relatively high REE content, and so NAA is not necessarily the only technique which may be used in the analysis of separated residues. Roelandts (1981) demonstrated the determination of the light REEs (La, Ca, Pr, and Nd) by x-ray fluorescence spectrometry using the above ion exchange procedure. Analytical results were compared with NAA data and yield recoveries were claimed to be reproducible in the range 95–100%. These and other ion exchange procedures are reviewed in Chapter 14.

Many other radiochemical separation schemes have been described in the literature for both groups of elements or individual isotopes. Perhaps the most important of these is the radiochemical determination of the platinum group elements which will be described, with other methods of analysing these elements, in Chapter 15.

12.17.5 *Analytical results*

The attraction of radiochemical neutron activation analysis lies in the exceptionally low backgrounds that may be achieved by removal of matrix activity. Table 12.18 compares analytical results for dunite USGS : DTS-1 obtained by three radiochemical procedures. These results indicate that detection limits in the ppb range may be achieved for selected rare-earth and other trace elements. Although some differences are apparent between sets of data, this must be set in the context that this sample can be analysed satisfactorily by few other techniques at these concentration levels. The absence of a reagent blank in radiochemical procedure once a sample has been irradiated is an obvious advantage in this application, particularly compared with the expensive clean-room facilities required to prepare samples for isotope dilution mass spectrometric analysis.

12.18 Prompt gamma neutron activation analysis

Prompt gamma neutron activation analysis is a rather specialized form of activation analysis involving the counting of gamma-rays emitting from the highly excited 'compound' nuclei formed at the moment of neutron capture. This compound nucleus has an extremely short half-life, of the order of 10^{-14} seconds, during which it must lose typically 4 to 8 MeV of energy, principally as gamma radiation. The product of this decay is occasionally a stable nucleus, but is more commonly radioactive, the activity of which may be measured after removal of the sample from the reactor by conventional neutron activation gamma spectrum analysis. However, since prompt gamma emission is effectively instantaneous with neutron capture, measurements of prompt

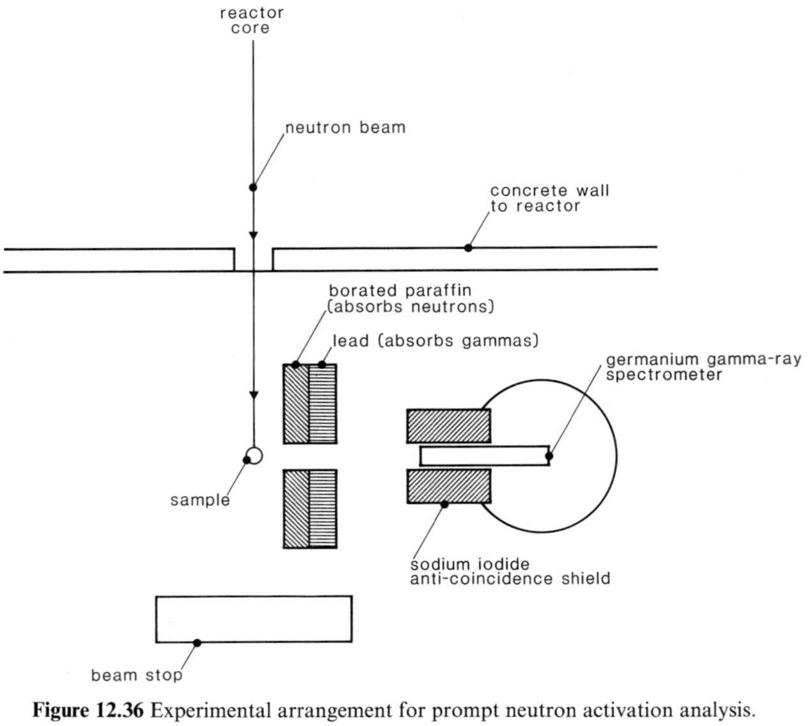

Figure 12.36 Experimental arrangement for prompt neutron activation analysis.

Table 12.19 Prompt gamma-ray analysis of boron, samarium and gadolinium in reference materials

Reference material	Boron (Higgins, 1984)			Samarium (Gladney *et al.*, 1985)			Gadolinium (Gladney *et al.*, 1985)		
	x	Analysed	keV	x	Analysed	keV	x	Analysed	keV
USGS: BHVO-1	—	2.1±0.3	478	6.1	5.6±0.5	334	6.0	6.4±0.6	182
					5.5±0.5	439		7.7±0.8	1185
GSP-1	—	0.9±0.1	478	25?	27±2	334	(15)	12.8±1.3	182
					26±2	439		11.7±1.6	1185
PCC-1	(6)	1.4±0.2	478	(0.008)	<0.023	334	—	<0.016	182
DTS-1	—	0.5±0.1	478	(0.004)	<0.014	334	—	<0.01	182

x = expected value from Abbey (1983).
keV = prompt gamma photopeak used for analysis.
All data in ppm.

gamma spectra can only be made whilst the sample is being irradiated.

This demands a rather specialized experimental arrangement, as shown in Figure 12.36. The sample is activated by a beam of neutrons extracted and collimated from the core of the reactor. The sample itself is normally mounted outside the reactor to facilitate the arrangement of counting equipment. The neutron beam is accompanied by an intense beam of gamma-radiation. Further analytical difficulties arise due to the scatter of the collimated beam of neutrons within the sample. These scattered neutrons can interact with surrounding shielding, causing additional, unwanted prompt gamma emissions. Prompt gamma-emission measurements are therefore made in the presence of a relatively intense gamma background that must be suppressed by the detecting system. Indeed, successful analysis of samples can only be made by mounting the germanium detector so that it views the sample at right angles to the neutron beam. In order to suppress unwanted gamma background, the germanium spectrometer must be surrounded by an annulus of sodium iodide designed to operate in anti-coincidence.

The highest neutron capture cross-sections are shown by the elements boron, cadmium, samarium and gadolinium. Cadmium, due to its very low natural abundance, cannot normally be determined in silicate rocks. However, the other three elements probably represent the most sensitive applications of this technique to geological samples. Determinations of boron are particularly valuable because of the greater sensitivity of prompt gamma analysis compared with more conventional techniques such as ICP–AES for example (Higgins, 1984). Possibilities also exist for the determination of H, V, Mn and Nd by prompt gamma emission (Gladney *et al.*, 1985).

Thus, Higgins (1984) determined boron in geological reference materials and discussed the problems of removing the interference from the sodium prompt gamma peak at 472 keV on boron at 478 keV. Some of Higgins' data is shown in Table 12.19. Detection limits were limited by blank levels of boron originating from boron-bearing materials close to the sample and including those within the detector itself and within the sample packaging material.

Reproducible blanks equivalent to 0.7 ppm B in a 2 g sample were recorded. With reference to sample containment, conventional polyethylene irradiation vials cannot be used for prompt gamma measurements since hydrogen is one of the elements that emits relatively intense prompt gamma radiation. Instead, samples must be contained within high-purity graphite or PTFE capsules for irradiation. Gladney *et al.* (1985) used prompt gamma activation to determine Sm and Gd in reference materials. In their experimental arrangement, the sample was placed within the reactor shielding and prompt gamma spectra measured externally with the Ge(Li) detector positioned some 6 m from the sample. The very long count time used to accumulate spectra (6–7 hours) made the technique most useful for analysing a small number of valuable materials (for example meteorites) rather than for routine batch analysis.

12.19 Concluding remarks

Neutron activation analysis has played a dominant role in the development of rare-earth element geochemistry over the last 20 years. Alternative techniques are now available, for instance ICP–AES, and isotope dilution mass spectrometry. However, for routine applications, each of these requires that samples are taken up into solution and then passed through ion exchange columns to separate rare-earth elements from matrix components. None can match the simplicity of sample preparation of INAA, requiring only that sample powders are sealed in vials.

Although known mainly for the analysis of the rare earths, INAA is a multi-element technique also capable of sensitive determinations of other important trace elements such as Ta, Hf and Th. These elements cannot normally be determined readily by other routine instrumental techniques. For this reason, neutron activation is likely to continue to retain an important role in routine geochemical analysis.

13 Nuclear techniques for the determination of uranium and thorium and their decay products

13.1 Techniques for uranium/thorium determination

Two naturally-occurring uranium isotopes (^{235}U, ^{238}U) and one thorium isotope (^{232}Th) occur in nature and are unusual in two respects. The first is that these isotopes are naturally radioactive, decaying mainly by alpha emission via a whole chain of daughter isotopes, the ultimate stable end products being ^{207}Pb, ^{206}Pb and ^{208}Pb respectively. The second unusual feature is that during neutron irradiation, they undergo nuclear fission, ^{235}U predominantly in a thermal neutron flux and ^{238}U, ^{232}Th in a fast neutron flux. These properties have led to the development of a number of techniques specific for uranium and thorium, including the following.

13.1.1 *Delayed neutron fission activation analysis*

Irradiation of a sample by neutrons in a nuclear reactor stimulates the fission of uranium (and thorium) nuclei within the sample. A small fraction of the fission products decays by the emission of neutrons. The half-life of neutron emission varies from a fraction of a second to just under a minute. Thermal neutrons will induce nuclear fission of ^{235}U. However, fission of ^{238}U and ^{232}Th can only be induced by fast neutron irradiation. The technique therefore involves irradiating a sample for about 60 seconds in a suitable neutron flux. The sample is then removed from the reactor using a rapid pneumatic rabbit system and, after a suitable decay period, the neutron emission count is measured using boron trifluoride neutron spectrometers. The technique is rapid and sensitive, and specific for uranium (normally with only a small interference from thorium). Minor interferences from oxygen and boron in the sample or sample container must be taken into account. Analytical details are discussed in section 13.3.

13.1.2 *Fission track analysis*

This technique is used for measuring the distribution of uranium (and thorium) over the surface of a polished section of a sample. The sample is prepared as a conventional polished thin section, normally 30 μm thick and mounted on a glass slide. A suitable 'solid state nuclear track detector' (SSNTD) (normally a polycarbonate plastic sheet, or sometimes mica) is mounted on the surface of the polished sample. Both are then irradiated by neutrons in a nuclear reactor. Neutron-induced fission of uranium (and thorium) nuclei within the sample causes the emission of energetic fission fragments. Fragments that penetrate the detector cause tracks to form within the first 10–20 μm of flight through the detector sheet before coming to rest. After irradiation the samples are allowed to decay (to permit safe handling). The SSNTD is then removed from the sample, and chemically etched to reveal the distribution of these fission tracks. The technique is therefore extensively used to locate uranium (and thorium) minerals as a prerequisite to further studies by optical microscopy and electron probe microanalysis. Quantitative data on the distribution of uranium can be obtained from fission detectors by counting the density of individual fission tracks and comparing data with a standard sample irradiated under identical conditions. Fission track analysis and other autoradiograph techniques are discussed in sections 13.4 and 13.5.

13.1.3 *Gamma spectrometry*

The radioactive decay chain of uranium and thorium includes daughter isotopes that emit gamma-rays of characteristic energy. In gamma spectrometry, the gamma spectra from natural (that is, unirradiated) samples are measured in order to determine the bulk uranium, thorium and potassium content of the sample. Spectrum measurements are normally carried out with large sodium iodide crystal detectors. The technique has two forms: first, a laboratory-based assay in which sample powders of about 300–400 g are counted in a special shielded enclosure designed to minimize interferences from background radiation; secondly, direct field measurements using portable detectors and simple data acquisition equipment. Measurements are often used to assess the geothermal potential of granite rocks arising from their radioactive element content. It is then convenient that all three radioactive components (U, Th, K) can be determined simultaneously by gamma spectrometry, as discussed further in section 13.6.

13.1.4 *Uranium and thorium series disequilibrium studies*

If a rock sample remains undisturbed, parent uranium and thorium isotopes attain a state of 'secular' equilibrium with their daughters. Under these conditions the rate of decay of parent and all daughter isotopes are identical (see section 13.8 for further considerations of secular equilibrium). However, various geochemical processes can disrupt this radioactive equilibrium. Samples that will show disequilibrium effects are those that have been subject to weathering or sedimentation, have taken part in chemical interactions in the marine environment or have participated in volcanic eruptions. The aim of thorium/uranium disequilibrium studies is therefore to measure the degree of disequilibrium between parent and daughter isotopes so that an estimate can be made of the time that has elapsed since secular equilibrium was disturbed. These studies therefore provide a basis for a geochronometer having a range that varies from a few days to a few million years, depending on which isotope pair is selected (or available) for analysis. The amount of radioactive daughter isotopes present in typical samples is so small that determinations can only be made by measurements of exceptionally high sensitivity based on the characteristic radioactive decay of individual isotopes. The most sensitive of these is alpha spectrometry (section 13.7), although direct beta-counting or sometimes gamma spectrometry must be used if the isotope does not decay by alpha-emission. Satisfactory alpha- or beta-measurements can only be made after extensive chemical treatment of the sample to isolate the isotope of interest from other radioactive and matrix components; procedures are discussed further in section 13.9.

13.2 The uranium–thorium decay chain

Details in the section above indicate that several techniques make use of a detailed knowledge of the mode of radioactive decay of ^{235}U, ^{238}U and ^{232}Th, and their daughter isotopes, the corresponding half-life of each and the interrelated decay sequence. These data are summarized in Figures 13.1, 13.2 and 13.3, which chart the decay sequence of the individual isotopes in each decay chain as a function of atomic number versus atomic weight. This permits the mode of decay to be readily identified as follows:

(i) *Alpha decay* is represented by a diagonal line and corresponds to a loss of two atomic number units and four atomic weight units from the decaying isotope

(ii) *Beta decay* is represented by a horizontal line in which the atomic number of the daughter isotope increases by one unit with no change in atomic weight

(iii) *Isomeric transition* (i.e. the decay of ^{234m}Pa to ^{234}Pa) does not involve change in atomic weight or atomic number.

Several daughter isotopes do not decay exclusively by one route; ^{227}Ac, for example (Figure 13.1) decays principally as a beta-emitter to ^{227}Th, but also with lower probability as an alpha-emitter to ^{223}Fr. An indication of the probability of such branched decay routes is given by the intensity of the decay lines in Figures 13.1 to 13.3, bold lines representing the most probable decay route.

There are several interesting aspects of these decay diagrams. The first is that individual daughter isotopes are unique to a particular decay chain. None appears in more than one decay series. Of particular relevance to geochemical applications is that several long- or very long-lived daughter isotopes are formed in the course of individual decay series. In particular: ^{231}Pa ($T_{1/2} = 3.2 \times 10^4$ y) (from ^{235}U) and ^{234}U ($T_{1/2} = 2.5 \times 10^5$ y), ^{230}Th ($T_{1/2} = 7.5 \times 10^4$ y), ^{226}Ra ($T_{1/2} =$ 1620y) and ^{210}Pb ($T_{1/2} =$ 22y) (all from ^{238}U). These are the isotopes that are of particular interest in uranium series disequilibrium measurements. A further consideration is that each decay sequence includes one isotope of radon, which is a gas at normal temperature and pressure. In particular, the isotope ^{222}Rn, formed in the decay chain from ^{238}U, has a relatively long half-life ($T_{1/2} = 3.8$ days). This causes some concern in gamma spectrometry measurements, in particular, where errors can arise due to a break in secular equilibrium resulting from diffusion of radon gas out of the sample.

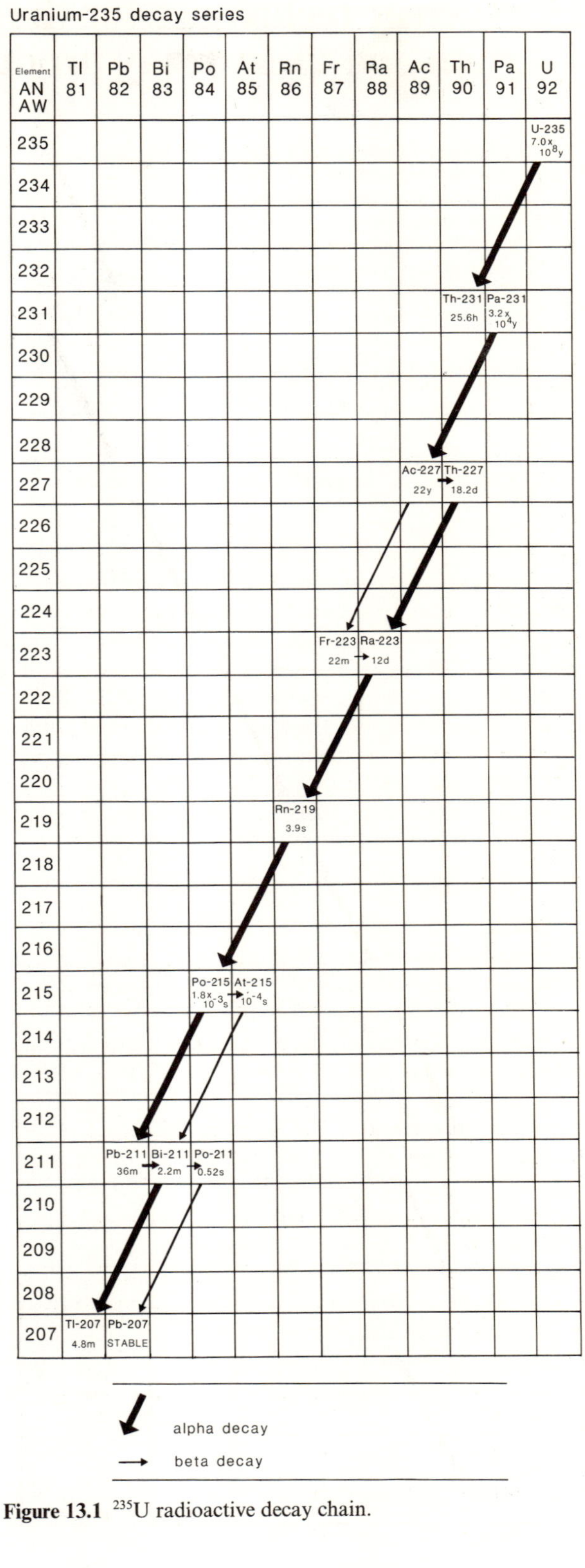

Figure 13.1 ^{235}U radioactive decay chain.

13.3 Delayed neutron fission activation analysis

(*a*) *Activation mechanism.* Delayed neutron fission activation analysis is a technique used specifically for the determination of uranium (and, less sensitively, thorium) in natu-

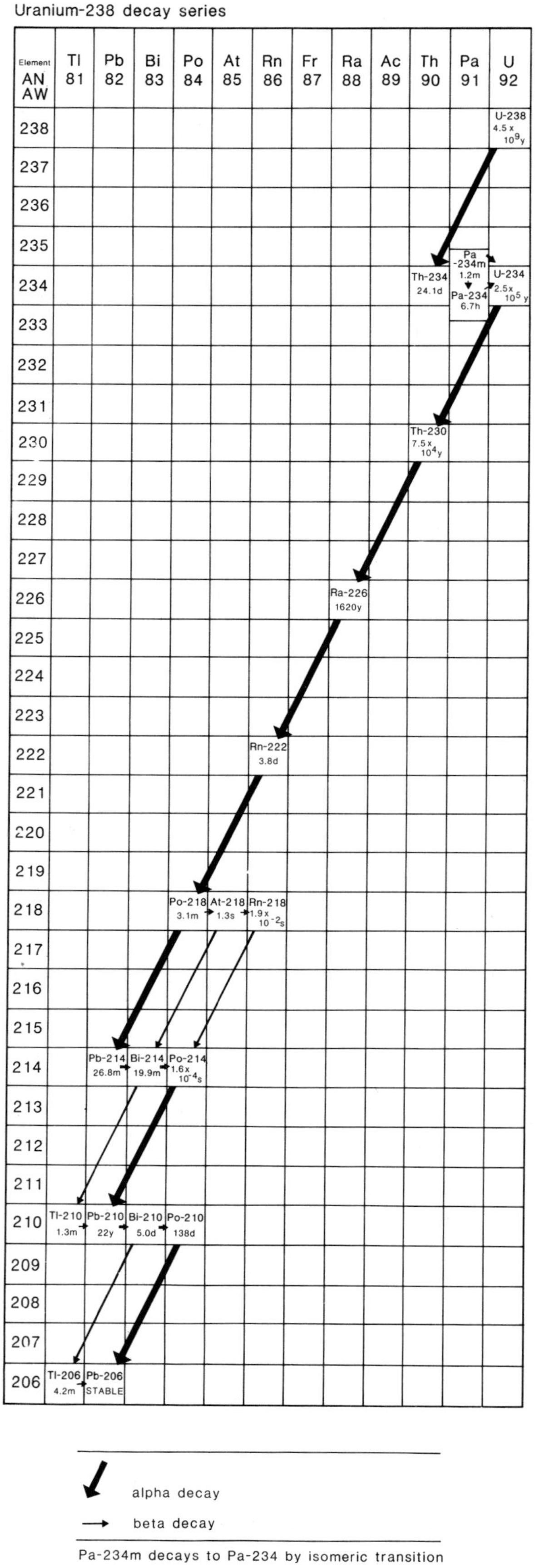

Figure 13.2 ^{238}U radioactive decay chain. (After Wilson, 1966.)

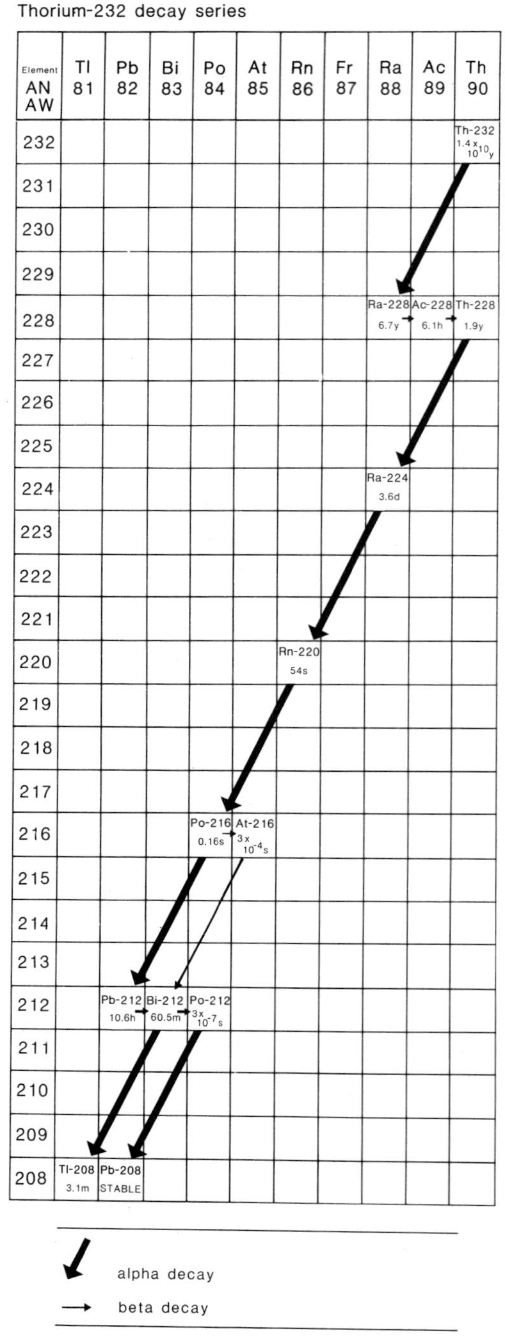

Figure 13.3 ^{232}Th radioactive decay chain. (After Wilson, 1966.)

ral samples. The technique is based on the phenomenon that when samples are irradiated by neutrons in a nuclear reactor, neutron interaction with the nuclei of ^{235}U, ^{238}U and ^{232}Th can induce fission of these isotopes. This process occurs in parallel with conventional neutron activation (i.e. the more conventional INAA (n, γ) reaction), which is observed as a competing activation mechanism. During the fission process, a neutron is incorporated into the nucleus of the uranium or thorium isotope forming a highly unstable compound nucleus. This splits into two more stable nuclear fragments in a process that is accompanied by the release of

Table 13.1 Half-life data for components that contribute to the delayed neutron emission spectra of uranium and thorium

Isotope	Number of delayed neutrons per fission	Neutron group index	Half-life (s)	Absolute group yield (%)
Thermal neutron irradiation				
^{235}U	0.0158	1	55.7	0.052
		2	22.7	0.346
		3	6.2	0.310
		4	2.3	0.624
		5	0.61	0.182
		6	0.23	0.066
Fast neutron irradiation				
^{235}U	0.0165	1	54.5	0.063
		2	21.8	0.351
		3	6.0	0.310
		4	2.2	0.672
		5	0.50	0.211
		6	0.18	0.043
^{238}U	0.0412	1	52.4	0.054
		2	21.6	0.564
		3	5.0	0.667
		4	1.9	1.599
		5	0.49	0.927
		6	0.17	0.309
^{232}Th	0.0496	1	56.0	0.169
		2	20.8	0.744
		3	5.7	0.769
		4	2.2	2.212
		5	0.57	0.853
		6	0.21	0.213

Data from Amiel (1962).

much energy. The fission products include isotopes (^{87}Br and ^{137}I, for example) that decay by emission of particles to products that are themselves unstable, decaying by the emission of neutrons (Ostle *et al.*, 1972). The precursors for delayed neutron emission can conveniently be classified into six groups having half-lives that range from about 0.2 s to about 55 s (Table 13.1). Analysis for uranium and thorium depends on the detection and quantification of these delayed neutrons.

Fission of ^{235}U can be effectively induced by a thermal neutron flux (that is, a very low-energy neutron flux, having a most probable energy of about 0.025 eV). However, fission of ^{238}U and ^{232}Th are threshold reactions and can only be induced by fast neutrons having energies in excess of 1 MeV. To determine both uranium and thorium, it is therefore necessary to carry out two irradiations, one in a thermal neutron flux for ^{235}U and a second with the sample surrounded by a cadmium foil. The cadmium foil absorbs the low-energy neutron spectrum (<0.4 eV) and therefore enhances the fast neutron component, permitting preferential fission of ^{232}Th. Data for the delayed neutron-emitting components observed after fission of ^{235}U (thermal neutron irradiation) and ^{235}U, ^{238}U and ^{232}Th (fast neutron irradiation) are given in Table 13.1. These results are obtained from analysis of data such as that in Figure 13.4. This shows the decay in emission of neutrons from a 1 mg sample of natural uranium as a function of time (in seconds) since the sample was irradiated. Analysis of these data shows that the compound decay curve (solid dots) is made up in this case of three distinct neutron-emitting components having half-lives of about 6, 21 and 55 seconds (Amiel, 1962).

(*b*) *Interferences*. Counting strategy for delayed neutron emission measurements must take into account potential interferences from two other isotopes formed during neutron irradiation that decay to excited-state neutron emitters. Oxygen and beryllium both undergo (n, p) reactions during neutron irradiation to form ^{17}N ($T_{1/2} = 4.2$ s) and ^{9}Li ($T_{1/2} = 0.17$ s), both of which decay by emitting neutrons. The activation reactions are:

$$^{17}O\ (n,\ p)\ ^{17}N \qquad (13.1)$$

$$^{9}Be\ (n,\ p)\ ^{9}Li \qquad (13.2)$$

The interference from ^{9}Li is negligible, since few samples have significant beryllium contents, and in any case a few seconds' delay after irradiation is sufficient to ensure the almost complete decay of ^{9}Li. The ^{17}N interference is poten-

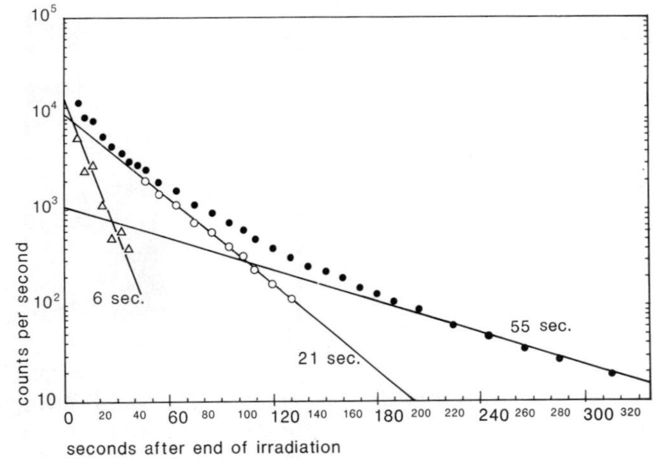

Figure 13.4 Decay of delayed neutron activity induced in a 1 mg sample of natural uranium after neutron irradiation. (Amiel, 1962.)

tially more troublesome, since oxygen is always a major component of silicate rocks (often >40% by weight) and the activation product has a significantly long half-life (4.2 s) compared with that of the uranium/thorium neutron emitters. It is, therefore, normally necessary to leave a sample to decay for a precisely-timed interval of about 20 to 25 seconds after irradiation, to permit substantial decay of ^{17}N. A small correction must then be applied for the proportion of neutron activity remaining due to ^{17}N. This is unlikely to exceed about 1 neutron per gram of oxygen emitted after the decay time indicated, and, although not significant for powdered rock samples, represents a more serious interference in the analysis of water samples (Ostle *et al.*, 1972).

Thermal neutron fluxes used to excite ^{235}U will also contain a small proportion of fast neutrons capable of inducing fission in ^{232}Th. The magnitude of this potential interference will then be influenced by the specific neutron spectrum used for irradiation. In the work of Ostle *et al.* (1972), the sensitivity of thorium was found to be 0.3% of that of natural uranium and was considered unlikely to cause a significant interference on a uranium determination.

A further consideration in experimental conditions is that neutron irradiation induces beta- and gamma-activity in a wide range of other isotopes present in the sample. Due to the rapid build-up in gamma activity, especially in ^{28}Al ($T_{1/2} = 2.3$ m), it is desirable to minimize irradiation times, both to reduce the radiological hazard and to prevent pile-up errors in the counting instrumentation. A typical analysis cycle would therefore be: 60 s irradiation, 20 s to 25 s decay out of the reactor, 60 s neutron counting. In order to achieve this cycle, an automatic pneumatic 'rabbit' sample transfer is required to transport samples into the reactor and then out again to the counting station. Analyses can only therefore be carried out adjacent to a reactor, at a rate equivalent to about one determination of uranium every 2–3 minutes.

(*c*) *Neutron detectors*. Neutrons are nuclear particles with no overall electronic charge and cannot be detected by conventional radiation counting equipment. Neutron counters are usually therefore based on the high probability of capture of neutrons by ^{10}B (or ^{3}He). In the case of ^{10}B, the neutron capture reaction leads to the formation of ^{7}Li with the accompanying emission of alpha particles:

$$^{10}B\ (n, \alpha)\ ^{7}Li \qquad (13.3)$$

Neutron counters used for delayed neutron measurements are normally made up as gas proportional counters (section 8.9) filled with boron-10 enriched boron trifluoride gas. A high voltage is applied to the anode wire within the counter. Both the energetic alpha-particles and recoil ^{7}Li ions released after neutron interaction with ^{10}B lose their kinetic energy by collision and ionization of the counter gas. Electrons released by this process drift towards the anode under the influence of the applied electrical voltage. In the vicinity of the anode, 'gas multiplication' occurs as electrons are accelerated by the large electric fields adjacent to the wire, and cause ionization of further counter gas molecules. Each neutron event is therefore detected as a voltage pulse, discharged at the anode of the counter. During delayed neutron counting the counter gas will also be subjected to intense gamma radiation emanating from the irradiated sample. Gamma-rays are rather inefficiently detected by a boron trifluoride neutron counter, and at moderate intensities the corresponding signals are easy to filter by setting a low-level discriminator on the counter output. However, during delayed neutron emission measurements, very high activities of gamma-rays may be encountered (from ^{28}Al, for example). Pile-up of corresponding counter signals can then interfere with neutron counting measurements, a problem now considered further.

(*d*) *Counting equipment*. Due to the very short cycle between irradiation and counting, it is necessary to carry out all delayed neutron measurements in the hall next to the reactor. Small but analytically significant neutron fluxes are normally detectable outside the reactor and can lead to unacceptably high background counts. Special consideration must be taken over the shielding arrangement of detectors in the counting position. One such arrangement was described by Amiel (1962). This consisted of a sample counting position surrounded by six $^{10}BF_3$ detectors embedded in a paraffin block 400 × 400 × 700 mm. This block was itself covered with an 0.8 mm sheet of cadmium and then by an additional 60 mm layer of borated paraffin wax shielding. Output from the six detectors was connected in parallel to maximize the delayed neutron emission signal and achieve 4π counting geometry as closely as possible. Even so, the counting efficiency of such a configuration is unlikely to exceed 10%. The successive layers of paraffin wax and cadmium sheet were designed to absorb as large a fraction of background reactor neutrons as possible. A similar arrangement using eight detectors is shown in Figure 13.5.

After irradiation the sample was automatically delivered

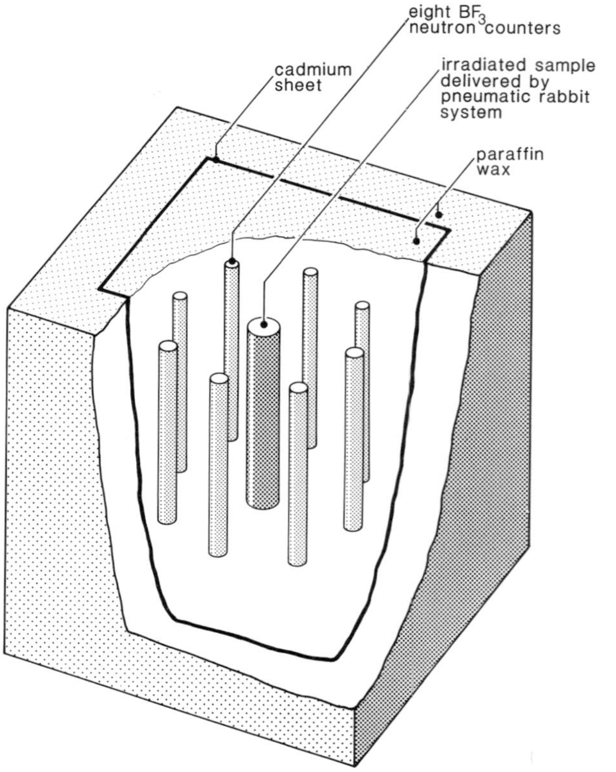

Figure 13.5 Delayed neutron counting assembly showing shielding and BF_3 neutron counters. (Adapted from Amiel, 1962.)

by a pneumatic rabbit system to the centre of the counting assembly, and the decay and counting cycles initiated automatically. A slightly different counting arrangement was used by Cumming (1974). The sample was mounted for counting in a 300 mm cube of polyethylene containing a 100 mm diameter core of lead. A ring of eight BF_3 neutron counters was arranged around the sample at a distance of about 200 mm from the sample. The lead ring significantly attenuated gamma-ray, but not neutron, emissions from the sample. Signals from each detector were connected in parallel, so ensuring efficient detection geometry. In order to minimize interference from fast reactor neutrons, the 300 mm cube of polythene was itself surrounded by a 1 mm sheet of cadmium, on the outside of which were placed 50 mm polyethylene and 200 mm paraffin blocks. These blocks moderated the energies of any fast reactor neutrons, causing them to be absorbed in the cadmium sheet. The signal from the counters was fed into a preamplifier and then into a discriminator and amplifier. Count data were accumulated in a scaler, programmed to count for a specified interval (64 s) after a precise timed decay interval (25 s).

(*e*) *Practical considerations.* Delayed neutron analysis is most commonly used to determine uranium only, after thermal neutron irradiation. If thorium is to be determined as well, a second irradiation under a cadmium cover is required. A typical procedure is as follows.

The sample (normally 0.2 to 1 g but sometimes up to 5 g) is sealed in a polythene vial and transferred via a pneumatic rabbit system into a reactor for 60 s irradiation. After irradiation the sample is transferred via the pneumatic tube to the counting assembly. Counting starts exactly 20 s (or 25 s) after irradiation. Neutron emissions are then detected over the following 60 s. The technique is normally calibrated with international reference materials. A small correction may be applied to data to subtract any residual neutron activity originating from the oxygen interference discussed above. Cumming (1974) found that the magnitude of this correction was 6.87 counts per gram of oxygen in the sample for a 25 s delay followed by a 64 s counting interval. Cumming also found it necessary to remove samples from the rabbit assembly vial prior to counting to avoid additional counts possibly caused by uranium contamination of the pneumatic transfer system. Other details covering the practical development of the technique have been given by Gale (1967).

If thorium is to be determined as well, the entire irradiation and counting sequence must be repeated, but this time with the sample covered with cadmium foil whilst in the reactor. This prevents thermal neutrons reaching the sample and therefore promotes activation of ^{232}Th with the fast component of the neutron spectrum. Gale (1967) showed that the concentrations of uranium [U] and thorium [Th] may be calculated from the respective thermal (I_{Th}) and cadmium shielded fast (I_F) neutron count measurements from the sample, as follows:

$$[U] = \frac{I_{Th}/r - I_F}{K(1/r - 1/R)} \qquad (13.4)$$

$$[Th] = \frac{I_F - I_{Th}/R}{k(1/r - 1/R)} \qquad (13.5)$$

where K is the calibration constant for uranium (counts per μg U), k is the calibration constant for thorium (counts per μg Th), R is the ratio between the unshielded to shielded count rate for uranium (i.e. R is the cadmium ratio for uranium) and r is the cadmium ratio for Th.

Determinations for uranium are mainly based on delayed-neutron emission from the isotope ^{235}U. If these data are to be converted into bulk uranium concentrations, assumptions must be made about the isotopic composition of standards and samples. In this respect it must be noted that many analytical reagents are supplied with non-natural uranium isotopic compositions depleted in ^{235}U. Sensitivity for thorium by delayed neutron activation is about 100 times less than uranium. Detection limits are often quoted of the order of 0.2 ppm U (and therefore about 2 ppm Th). Some results for selected rock reference materials are shown in Table 13.2 and indicate that two-sigma precisions are often in the range 5–10% relative at the low concentration levels (below 20 ppm) found in many silicate rocks.

13.4 Fission track analysis

(*a*) *Introduction.* Fission track analysis is primarily a technique for measuring the distribution of uranium over the polished surface of a geological thin section. Using the appropriate irradiation conditions, data for thorium can also be obtained. In the majority of applications, *qualitative* distribution data is all that is sought so that both uranium minerals may be located for subsequent microprobe analysis and any unusual secondary alteration effects that have mobilized uranium may be distinguished. However, in more demanding applications, quantitative data on the distribution of uranium between two phases can be obtained by counting and comparing the number of fission tracks formed in the appropriate detector with those produced from a uranium standard of known composition.

Fission is induced by irradiation with neutrons in a nuclear reactor. As outlined in section 13.3, only the isotope ^{235}U undergoes fission in a thermal neutron flux. A fast flux (comprising neutron energies exceeding 0.9 MeV) is required to induce fission of ^{238}U and ^{232}Th. Fission is measured using a suitable *solid-state nuclear track detector* (SSNTD) placed in contact with the polished surface of the sample. Since neutron-induced fission occurs only during irradiation, the thin section must be irradiated in the reactor with the detector *in situ*. After irradiation, samples are allowed to cool to permit decay of unwanted intense gamma activity. The detector is then removed from the sample and etched to reveal the tracks of fission fragments (Figure 13.6).

Two types of solid-state nuclear track detectors are widely used. The first is a natural muscovite mica (Price and Walker, 1963); the second a polycarbonate sheet manufactured under the trade name 'Lexan' (Fleischer *et al.*, 1965; Kleeman and Lovering, 1967). Tracks may readily be identified in a mica detector, although this material posesses a small background track density due to its natural uranium content. Lexan plastic scratches very easily, and tracks emitted perpendicular to the surface are not readily distinguished from scratch marks, but there is essentially no fission track background. Fission track autoradiography has been dis-

Table 13.2 Delayed neutron activation analysis of uranium

Reference material	Expected value Abbey (1983)	Delayed neutron activation data: Gladney *et al.* (1980 *a,b*)	Ledger *et al.* (1980*a,b*)	Conrad *et al.* (1982)		Reddy *et al.* (1982)
USGS:						
BHVO-1	0.4?	0.44±0.14				0.62
RGM-1	5.8?	5.93±0.40				6.0
BIR-1		0.02±0.04	<0.1			<0.2
DNC-1		0.10±0.04	<0.1			<0.2
W-2		0.61±0.08	0.42±0.04			0.66
CCRMP:						
SY-2	290	284±16	278±6	282±4	283±8	280±18
SY-3	650	705±144	557±6	625±12	631±22	730±50
CRPG:						
BR	3?		2.15±0.18			2.7
GA	4?		4.9±0.6			5.5
GH	18?		17.3±0.8			18.6
AN-G			<0.2			<0.2
BE-N	2.4		2.27±0.18			2.6
MA-N	12?		11.7±1.2			13.0
Mica-Fe	60?		75.9±2.2			88.6
Mica-Mg	0.5?		<0.2			<0.2

Data in ppm; ±error equals two standard deviations.

cussed in some detail by Wollenberg (1973), Fisher (1975), Bowie (1977) and Burnett and Woolum (1983).

(*b*) *Nuclear fission.* When neutron-induced fission occurs, massively charged fragments from the fission nucleus are ejected with energies in the range 85–90 MeV. With the detector lying on the surface of the sample, on average 50% of these fragments will enter the detector, causing damage to its structure by charge repulsion. The range of these fragments is about 0 to 15 μm, the threshold of zero arising because fission fragments can be attenuated within the sample itself before penetrating the detector. The effective range of fission fragments in most silicate and phosphate materials is quoted by Wollenberg (1973) to be of about 3 mg cm^{-2}. The damage caused by the passage of the fission fragment within the detector is revealed by etching. Natural mica is normally etched with hydrofluoric acid solution (40–48%) at room temperature for 3–4 hours. Lexan plastic sheet is etched with sodium hydroxide solution (6 M) for 1–2 minutes at 80°C. This latter process is somewhat critical since, if etching time is extended, the entire surface of the Lexan sheet is removed.

(*c*) *Fission track calculations.* The concentration of uranium present in the sample is proportional to the track density. The track density ρ (tracks cm^{-2}) following thermal neutron fission of ^{235}U is given by

$$\rho = k\phi\sigma \, \mathrm{Eff} \, N_U \qquad (13.6)$$

where ϕ is the integrated thermal neutron flux (neutrons cm^{-2}), σ is the thermal neutron fission cross section for ^{235}U (=580 barns), k is a geometric factor ($\sim$1), since one of the two fission fragments will enter the detector provided the latter lies on the surface of a flat polished sample, N_U is the number of nuclei of ^{235}U per cm^2 on the surface of the sample, the fission fragments of which are within range of the detector and Eff is an efficiency correction to allow for any fission tracks that do not register on the detector.

The number of ^{235}U atoms whose fission fragments are within the range of the detector is given by

$$N_U = \frac{[U]}{AW} I N_A R d \qquad (13.7)$$

where [U] is the weight fraction of uranium in the sample, I is the isotopic abundance fraction of ^{235}U, AW is the atomic weight of uranium, N_A is the Avogadro constant = 6.022×10^{23} mol^{-1}, R is the range of the fission fragment $\sim$3 mg cm^{-2} for silicate and phosphate minerals and d is the density of the appropriate mineral.

Similar equations are used to estimate the track density arising during the determination of thorium. However, since fission of thorium ^{232}Th only occurs in a fast neutron flux (>0.9 MeV), irradiations must be carried out with samples surrounded by 3 mm cadmium to absorb all thermal neutrons (<0.4 eV). In estimating track density of thorium, allowance must also be made for the simultaneous fission of ^{238}U. Equation (13.6) then becomes

$$\rho = k\phi \, \mathrm{Eff} \, (\sigma_{Th} N_{Th} + \sigma_U N_U) \qquad (13.8)$$

where ϕ now represents the integrated fast neutron flux, σ_{Th} and σ_U are now the fast neutron fission cross-sections for ^{232}Th ($\sim$0.3 barn) and ^{238}U ($\sim$1.1 barn) and N_{Th} and N_U are the concentrations of thorium and ^{238}U nuclei over the surface of the sample.

These equations suggest that it is possible to analyse uranium and thorium by 'absolute' fission track counting; that is, simply by irradiating the sample in a known neutron flux, counting the resultant fission tracks and calculating the

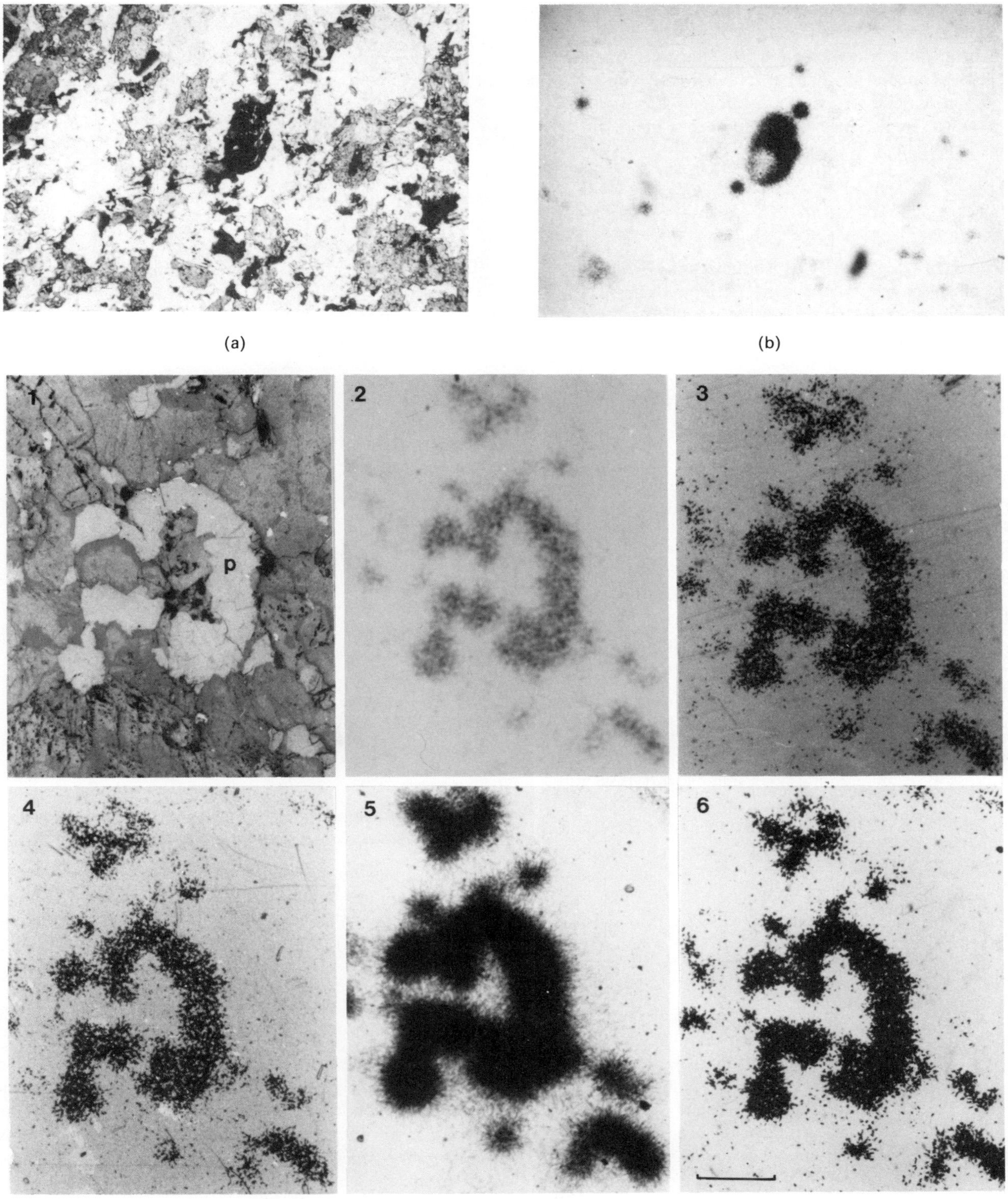

Figure 13.6 (*a*, *b*) Fission track autoradiography. (*a*) Photomicrograph of a large apatite grain in an amphibole–biotite hornfels from the contact zone of the Loch Doon granite, SW Scotland. (*b*) Fission track autoradiograph delineating the distribution of uranium in the oval-shaped apatite grain and in two smaller mineral grains, one a zircon and the other an allanite. The field of view is approximately 3 mm. (Sample by courtesy of A.G. Tindle, Open University.) (*c*) Alpha autoradiography of pitchblende (*P*) in a silicate gangue—comparison of autoradiographs recorded on photographic emulsion and cellulose nitrate detectors; (1) photomicrograph of rock section; (2) autoradiograph recorded on a photographic emulsion, 15 h exposure; (3) as (2) but recorded on a CA-80 cellulose nitrate sheet, etched for 6 h in 2.5M KOH at 25°C; (4) as (2) but recorded on a LR115 cellulose nitrate film, etched for 24 h in 2.5M KOH at 25°C; (5) emulsion autoradiograph exposed for 5 days; (6) as (3) but with the contrast enhanced by etching for 24 h. Scale bar 200 μm. (Basham and Easterbrook, 1977.)

thorium or uranium concentration from the appropriate formula. This is never the case. Quantitative analyses are almost invariably carried out by simultaneous irradiation of a glass or mineral standard of known uranium and thorium composition and comparing fission track densities with the unknown sample. The concentration of uranium, for example, in the sample is then given by

$$[\mathrm{U}]_{\text{sample}} = [\mathrm{U}]_{\text{standard}} \frac{\rho_{\text{sample}}}{\rho_{\text{standard}}} \times \frac{(R.D)_{\text{standard}}}{(R.D)_{\text{sample}}} K \quad (13.9)$$

where $(R.D)_{\text{standard}}/(R.D)_{\text{sample}}$ is almost unity $\pm$ a few percent for all silicate materials (see for example Fisher, 1975) and K is a correction factor to allow for any difference in neutron flux received between the sample and standard.

(*d*) *Fission track determination of uranium.* A typical procedure for determining uranium by fission track analysis is as follows: the sample is prepared as a polished thin section and mounted with its polished surface in contact with a suitable detector (often the Lexan polycarbonate plastic sheet). Normally several such sample–detector pairs are packed in an irradiation can, together with suitable uranium standards. The samples are irradiated in a thermal neutron flux for a suitable length of time. A minimum integrated neutron flux, that is, the product of the instantaneous neutron flux (neutrons $cm^{-2}\,s^{-1}$) and the irradiation time (s) of 10^{13} to 10^{14} is normally considered necessary. If uranium in the parts per billion range is to be determined, integrated fluxes as high as 10^{17} neutrons cm^{-2} may be used. After irradiation, samples and detectors are normally set on one side to permit the intense short half-life activity to decay. Detectors are removed and etched as described above. Fission track density is then determined, normally by optical microscope examination using a suitable grid. Fission track densities of sample and standard are compared as described above (equation (13.9)), applying a suitable correction for neutron flux differences between sample and standard.

(*e*) *Fission track determination of thorium.* This technique is far less successful in its application to thorium analysis because of the large uranium interference. It is necessary to first carry out a conventional fission track count for uranium as described as above. The sample is then re-irradiated with a fresh detector in a fast neutron flux, normally by selecting an irradiation position close to the reactor core and surrounding the irradiation can with cadmium foil to screen out the low-energy thermal neutrons. The sample must be accompanied by suitable separate uranium and thorium standards. After irradiation, samples are again allowed to 'cool'. The detectors are then removed and etched as described above. The resultant track density in the sample arises from fission of ^{232}Th and ^{238}U. From the previous analysis of U in the sample and the measured track density from the uranium standard, the contribution of ^{238}U can be calculated. The residual track density is due to ^{232}Th and may be compared with that due to the thorium standard for quantification.

However, the larger the proportion of uranium in the sample, the larger the uncertainty in counting precision of the thorium fission track density remaining after uranium fission correction. The relative error in thorium for typical irradiation and counting conditions is shown in Figure 13.7

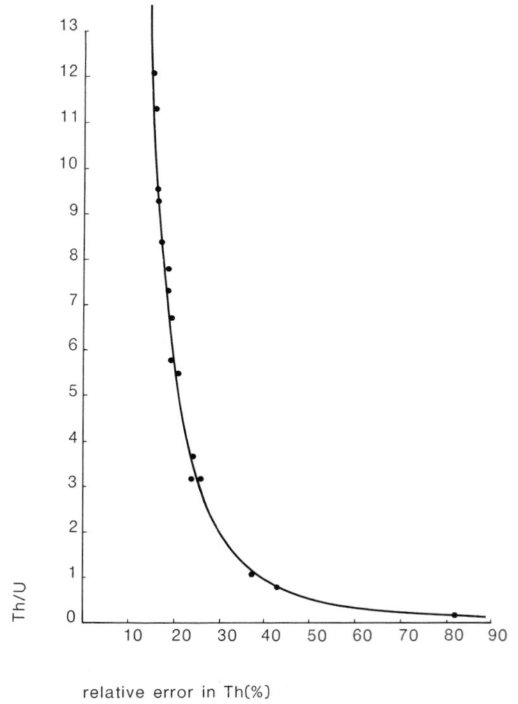

Figure 13.7 Relative error in thorium determination by fission track analysis as a function of the thorium ratio. (Wollenberg, 1972.)

as a function of both the Th and U content. For Th/U ratios less than about 3, errors in the determination of Th become unacceptably high. For this reason alternative procedures are usually employed for thorium analysis.

13.5 Other autoradiographic techniques for locating and analysing specific elements in thin section

Several other nuclear techniques apart from fission track analysis can be used to locate and analyse specific elements in thin section. The properties and characteristics of some of these are listed in Table 13.3 and will now be considered further.

13.5.1 *Spontaneous fission*

The naturally-occurring radioactive isotopes ^{235}U, ^{238}U and ^{232}Th have a very small probability of decaying by spontaneous fission (as opposed to neutron-induced fission, which occurs during neutron irradiation). However, the only isotope for which this phenomenon is at all significant is ^{238}U, and its decay constant has been variously estimated as 8.46×10^{17} and $7.03 \times 10^{17}\ y^{-1}$. The importance of this phenomenon is in offering the opportunity of measuring the age of a sample by counting the natural spontaneous fission track density in a suitable host mineral such as zircon. The host mineral is thus used as its own fission track detector.

One procedure is to prepare two polished sections of the same sample. In one sample, the spontaneous fission track density is counted after etching, but with no other pretreatment. The second sample is annealed by heat treatment to destroy any existing tracks prior to mounting as a thin

Table 13.3 A Summary of autoradiography techniques

Technique	Element distribution determined	Detector	Principle of technique	Reference
Fission track analysis	^{235}U (thermal neutron irradiation) $^{238}U + ^{232}Th$ (fast neutron irradiation)	(i) 'Lexan' polycarbonate sheet, or (ii) Natural mica	Fission fragments detected during neutron irradiation in a nuclear reactor	Wollenberg (1973) Fisher (1975) Bowie (1977) Burnett and Woolum (1983)
Spontaneous fission	^{238}U	Natural mineral (e.g. zircon) within sample to be analysed	Natural spontaneous fission of ^{238}U detected from fission tracks formed in host mineral — used as a dating technique	Hurford and Green (1982)
Alpha autoradiography	Uranium and thorium and daughter isotopes	(i) Cellulose nitrate sheet (ii) 'CR-39' polymer sheet (iii) Photographic emulsion	Natural alpha-decay of uranium thorium and daughter isotopes	Basham and Easterbrook (1977) Duane and Williams (1980)
Neutron activation-induced alpha autoradiography	^{10}B or ^{6}Li (interference: ^{17}O)	(i) Cellulose nitrate sheet (ii) 'CR-39' polymer sheet	Neutron activation-induced alpha-decay of specified isotopes detected during irradiation in a nuclear reactor	Burnett and Woolum (1983) Furst (1981)
Beta autoradiography	See Table 13.5	Photographic emulsion (Ilford K2 or K5)	Quantitative trace-element distribution data between minerals formed in an experimental charge. Data obtained by adding an artificial beta-emitting isotope to the charge before the experiment	Mysen and Sietz (1975) Mysen (1978)
Neutron activation-induced beta autoradiography	Various, especially REE, Au, Ir, As, Sb Ag, etc	As above	Distribution of minor mineral phases (especially REE, PGE, Au) in natural samples determined from beta-activity of isotopes measured after neutron irradiation of sample	Potts (1984) Potts and Prichard (1986)

section and polishing. This sample is then irradiated in a thermal neutron flux. Neutron-induced fission of uranium within zircon results in the formation of new fission tracks. These are subsequently revealed by etching so that the total uranium content of the mineral can be determined. From the analysed composition of uranium and the spontaneous fission track density, the age of the sample (T) can be estimated:

$$T = \frac{1}{\lambda_D} \ln \left[1 + \frac{\lambda_D \phi t \sigma I \rho_s}{\lambda_f \rho_f} \right] \tag{13.10}$$

where λ_D is the total decay constant of ^{238}U (1.55125×10^{-10} y^{-1}), λ_f is the spontaneous fission decay constant of ^{238}U, I is the isotopic ratio $^{235}U/^{238}U = 7.2527 \times 10^{-3}$, σ is the thermal neutron fission cross section of $^{235}U = 580.2$ barn, ϕt is the integrated thermal neutron flux, ρ_s is the spontaneous fission track density (of ^{238}U) and ρ_f is the thermal neutron irradiation induced fission track density (of ^{235}U).

The limitations of these procedures have been discussed in some detail by Hurford and Green (1982), who indicate that uncertainties in the spontaneous fission decay constant of ^{238}U and in the experimental measurement of the integrated neutron dose are important limitations to the accuracy of this technique. These uncertainties can be considerably reduced by simultaneous irradiation of a uranium-bearing standard mineral of known age. Practical application of the fission track dating technique has been described by Haggerty *et al.* (1983), for example, in relation to age determinations of zircon-containing kimberlites.

13.5.2 *Alpha-autoradiography for uranium and thorium*

It is not always convenient (or possible) to irradiate samples for the fission track determination of uranium (or thorium). An alternative procedure is to detect alpha-particles ejected during natural radioactive decay of uranium, thorium and daughter isotopes. The sample is therefore prepared as a polished thin section, and a suitable solid-state nuclear track detector placed in contact with the polished surface. This detector is exposed, often for several weeks, and is subsequently etched to reveal the distribution of alpha-emitting isotopes. In principle it is possible to identify the decaying nucleus by the number and energy (i.e. path length) of alpha-particles emitted. This is, however, impractical with geological samples. The abundance of isotopes can be measured by counting track densities and comparing with suitable standards. In practice, alpha-autoradiography is used mainly to locate uranium and thorium minerals in thin section for subsequent analysis by electron microprobe.

A variety of solid state nuclear track detectors have been used for alpha-autoradiography measurements. The characteristics of the silver halide/gelatine emulsion detection have been compared with cellulose nitrate sheets by Basham and Easterbrook (1977). Duane and Williams (1980) have described some practical applications in the use of cellulose nitrate detectors for the identification of textural features in uranium-bearing samples. An important development in alpha-autoradiography studies was the introduction of a thermosetting plastic detector, CR-39 (Cartwright *et al.*, 1978). CR-39 is a commercially-available plastic based on polymerization of the monomer oxydi-2,1-ethanediyl di-2-propenyl diester of carbonic acid. It shows a high sensitivity towards alpha-particles up to energies of 6 MeV as well as protons up to 1 MeV. Cartwright *et al.* claim that CR-39 has unrivalled sensitivity, resolution and optical properties as a nuclear track detector.

Natural samples are typically exposed for several weeks and particle tracks subsequently etched in NaOH or KOH solution (2.5 M). Etching conditions vary with the properties of the detector. KOH solution is reported to be more efficient than NaOH solution for cellulose nitrate sheets. Basham and Easterbrook (1977) preferred to use a room temperature etch for 2–24 hours, since this required less precise control than etching for 10–30 minutes at 60–80°C (Figure 13.6*c*).

Alpha-autoradiography is not considered to be as sensitive as fission track analysis in detecting relatively low abundances of uranium and thorium and cannot readily distinguish between these elements. However, the technique represents a relatively simple method of locating minerals of these elements in thin section without the necessity of neutron irradiation.

13.5.3 *Neutron-induced alpha-autoradiography for boron and lithium*

A small number of isotopes undergo (n, α) reactions on irradiation with neutrons: that is, a neutron is captured by the nucleus and the resultant metastable compound nucleus decays by ejecting a prompt alpha-particle. The distribution of these alpha-emitting nuclei can, therefore, be measured by irradiating a polished sample with a suitable alpha-sensitive detector in contact with the sample surface. The most suitable solid-state nuclear track detectors for this purpose are cellulose nitrate or the CR-39 polymer. After irradiation the detector is removed and etched to reveal the distribution of alpha-emitting nuclei. The technique is therefore the alpha-emitting equivalent of fission track analysis. The principal application of this technique is in measuring the distribution of boron and lithium in thin section. Both these elements undergo (n, α) reactions in a thermal neutron flux, the product isotopes being ^{7}Li (stable) and ^{3}H ($T_{1/2} = 12.3$ years) respectively. The principal interference if a thermal neutron is used for irradiation is the (n, α) reaction induced in ^{17}O (Table 13.4). Fast neutron irradiations are avoided to prevent more serious interferences from ^{18}O (n, α) ^{15}C.

In reviewing alpha-autoradiography measurements, Burnett and Woolum (1983) estimated that ultimate detection limits are about 0.2 ppm B and 1 ppm Li, relative track

Table 13.4 Neutron activation-induced alpha emitters

Target isotope	Nuclear reaction	Product isotope	Neutron capture cross-section	
			Thermal	Epithermal
^{10}B	(n, α)	^{7}Li	3837	1722
^{6}Li	(n, α)	^{3}H	940	425
^{17}O	(n, α)	^{14}C	0.235	0.105

Data from Erdtmann (1976).

production per unit weight of lithium to boron being 0.16. No simple means have been proved to distinguish alpha tracks from boron and lithium, and the technique has been applied to date in determining the distribution of boron in samples (e.g. molluscs) known to have a low lithium content (Furst *et al.*, 1976; Furst, 1981).

13.5.4 *Beta-autoradiography—applications in experimental petrology*

The application of beta-autoradiography to the measurement of trace-element distributions in experimental petrology has been described by Mysen and Seitz (1975) and Mysen (1978). These workers showed that if a small quantity of an artificial radioactive isotope that decays by beta-emission is added to an experimental charge, then after fusion and subsequent quenching, the distribution of that element can be determined by measuring the beta-activity of the various phases present in the quenched sample. The detector used for these studies was a silver halide gelatine emulsion 25 μm thick, mounted for convenience on a glass plate (Ilford K5 emulsion). Since these emulsions are sensitive to visible light, all manipulations must be carried out in a darkroom under red safety lights until the emulsion has been photographically developed and fixed. Initial examples (Mysen and Seitz, 1975) were given using pure beta-emitters ^{151}Sm and ^{14}C. However, Mysen (1978) subsequently showed that the range of applications could be extended to include other elements (e.g. ^{141}Ce, ^{171}Tm) that emitted x-rays and/or gamma-rays as well as beta-particles during decay. An alternative silver-gelatine-based emulsion with high detection efficiency to beta-particles but with low sensitivity to x-rays and gamma-rays was substituted in these studies (Ilford K2 emulsion, 25 μm thick). This technique has been applied to several isotopes and has been especially valuable in measuring partition coefficients of the rare-earth elements between various silicate phases in experimental charges. The isotopes selected for such studies include those listed in Table 13.5. These isotopes are all commercially available, decay by beta-emission, have reasonably long half-lives, and emit predominantly beta-particles with maximum energies in general less than 500 keV. This latter limitation is important, since higher energy beta-particles penetrate more deeply into the emulsion, activating several silver grains and forming diffuse tracks which become impossible to recognize after the development of the emulsion. Quantification of beta-distribution data is normally carried out by optical microscope examination of the beta-autoradiograph, track density being represented by the number of individual silver grains present per unit area. Holloway and Drake (1977) showed that it is also possible to measure track densities from the intensity of silver *L* x-rays excited by electron probe microanalysis of areas scanned on the autoradiograph. Benjamin *et al.* (1977) showed that electron microprobe reading of beta-autoradiograph track densities significantly extended the dynamic range of these emulsion detectors, since typically less than 5% of this range was available under optimum optical counting conditions. Electron probe measurement has, however, been shown to be subject to some errors due to volatilization of variable amounts of the organic emulsion at higher (microprobe) electron beam densities (Jones and Burnett, 1981).

Table 13.5 Beta-emitting isotopes with potential for experimental petrology studies

Isotope	Half-life	Beta-decay energies (MeV) and intensities (%)	Reference
^{3}H	12.3y	0.018 (100%)	
^{14}C	5760y	0.159 (100%)	1
^{33}P	25d	0.25 (100%)	
^{35}S	87.2d	0.167 (100%)	
^{46}Sc	84.0d	0.36 (100%) 1.48 (0.004%)	
^{60}Co	5.26y	0.31 (100%)	
^{63}Ni	120y	0.067 (100%)	
^{95}Nb	35d	0.16 (100%)	
^{103}Ru	40d	0.214 (89.4%) 0.112 (7%) 0.716 (3%)	
^{141}Ce	32.5d	0.44 (70%) 0.58 (30%)	2, 4
^{147}Pm	2.6y	0.225 (100%)	
^{151}Sm	93y	0.076 (100%)	1, 2, 3, 4
^{155}Eu	1.81y	0.140 (43%) 0.158 (32%) 0.247 (15%)	
^{169}Er	9.3d	0.34 (100%)	
^{171}Tm	1.92y	0.097 (98%) 0.03 (2%)	2, 3, 4
^{177}Lu	6.7d	0.50 (90%) 0.17 (7%) 0.38 (3%)	
^{181}Hf	45d	0.41 (92%) 0.40 (4%)	
^{182}Ta	115d	0.18 (38%) 0.44 (23%) 0.36 (20%)	
^{185}W	73d	0.43 (100%)	
^{203}Hg	47d	0.21 (100%)	

Isotopes selected as beta emitters with decay energies less than about 0.5 MeV.
y = year, d = day.
Isotope selection based on data of Mysen and Seitz (1975).
Nuclear data from Wilson (1966).
References: 1: Mysen and Seitz (1975)
2: Mysen (1978)
3: Harrison and Wood (1980)
4: Apted and Boettcher (1981).

13.5.5 *Neutron-activation-induced beta-autoradiography*

Irradiation of an entire thin section in a nuclear reactor induces activity in a wide range of isotopes, as is well known from neutron activation analysis studies (chapter 12). Many of these isotopes subsequently decay by beta-emission (a rather small number decay by electron capture or isomeric transition and cannot therefore be detected by this technique). Neutron-activation-induced beta-autoradiography is a technique for locating certain minor phase inclusions in thin sections for subsequent analysis by electron microprobe. Potts (1984) has shown that although many elements become activated by neutron irradiation, the ones that contribute most beta-activity are not the major silicate-mineral-forming elements, but rather elements normally present at whole rock abundances in the ppm or even ppb range. This technique is therefore valuable for locating minor phase inclusions of such elements. Important applications include the systematic location of mineral grains of rare-earth elements, platinum-group elements (due to their iridium and sometimes arsenic or antimony content) and gold. Potts and Prichard (1986) have shown that with careful optimization of emulsion exposure times, the distribution of gold in thin section can be unambiguously distinguished in the presence of silver, antimony and arsenic—elements that are also expected to contribute significant beta-activity to the sample. Samples are activated by irradiating polished thin sections in a neutron flux to give an integrated dose of about 10^{15} neutrons cm^{-2}.

After removal from the reactor, samples are allowed to decay for 10–12 days to allow the intense short-lived activity due to ^{24}Na ($T_{1/2} = 15$ h) in the glass mount to decay. Subsequently, beta-autoradiographs are recorded on Ilford K2 nuclear track emulsions using the sample handling and exposure techniques described by Mysen and Seitz (1975) for experimental petrology applications. It seems unlikely that quantitative distributions of isotopes can readily be measured by this technique, since activity is induced in such a wide range of isotopes and the energies of beta-emitters can be as high as 1 to 2 MeV, making it impossible to identify individual particle tracks in the emulsion (Potts, 1985). The principal application is in the systematic measurement of the distribution of minor phase inclusions, especially of minerals containing the elements mentioned above prior to microprobe determination.

13.6 Gamma spectrometry

13.6.1 *Introduction*

Gamma-ray spectrometry is a technique for determining the elements uranium, thorium and potassium in bulk samples. Instrumentation is normally laboratory-based but can be adapted for *in-situ* field-based measurements. The technique is traditionally associated with sodium iodide (NaI(Tl)) detectors. The photopeaks conventionally measured are the 1461 keV line of ^{40}K, the 1764 keV line of ^{214}Bi (a daughter of ^{238}U) and the 2614 keV line of ^{208}Tl (a daughter of ^{232}Th). In the late 1950s and early 1960s gamma-ray spectrometry was the principal technique used for laboratory-based determinations of K, U and Th. However, more recently this technique has been displaced by alternative techniques such as neutron activation analysis, delayed neutron fission and XRF spectrometry. Gamma-ray spectrometry is, however, still valuable for making direct measurements on suitable rock outcrops in the field so that the distributions of these elements can be assessed *in situ*. In this role the equipment offers several advantages, being simple, rugged and portable.

13.6.2 *Scintillation detectors*

The heart of a gamma-ray spectrometer is a sodium iodide crystal (Figure 13.8). This is grown as a single crystal normally in the form of a right-angled cylinder. The standard size for gamma spectrometry measurements is 3 inches (75 mm) long by 3 inches (75 mm) in diameter; larger crystals are available and offer higher efficiency to gamma-ray detection, though at greater expense. The crystal has a small concentration of thallium ions drifted into its lattice structure to give it the appropriate optical properties for gamma-ray detection and is then called a 'thallium drifted sodium iodide scintillation detector' or NaI(Tl) for short. Gamma radiation entering the detector interacts with the crystal structure, causing electrons to be raised to 'activator' excited states. These then 'relax' back to the ground state, simultaneously emitting light photons. The dominant half-life of this scintillation process is 230 ns (relatively long for some gamma detection applications with very high count rates) and comprises light photons emitted in the 370–450 nm wavelength range.

Sodium iodide is selected as the detector material because of its high scintillation light yield. The sodium iodide crystal is, however, relatively fragile from a mechanical point of view, and is hygroscopic and so must be hermetically sealed from moisture.

The scintillation signal is detected by optically coupling

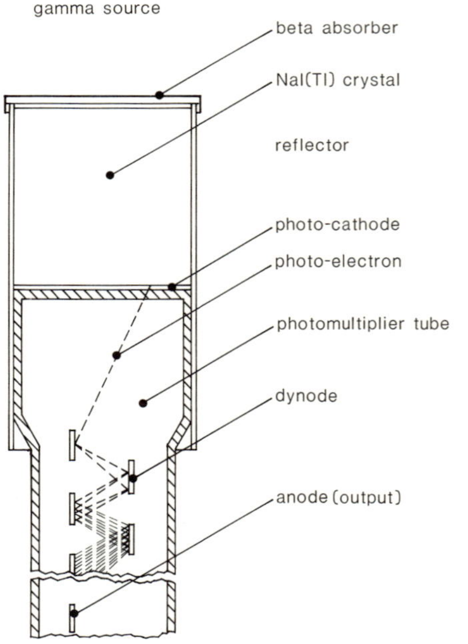

Figure 13.8 Gamma-ray spectrometer showing the sodium iodide scintillator crystal coupled to the photomultiplier tube.

the NaI(Tl) crystal to a photomultiplier tube. Photons impinging on the photocathode cause the emission of photoelectrons by the photoelectric effect. These photoelectrons are electrostatically focused on the first dynode in the photomultiplier tube, which is held at positive potential with respect to the photocathode. Each electron impacting on the first dynode causes the emission of 3 to 4 further electrons. These secondary electrons are successively focused, typically on nine further dynodes, each held at a successively higher positive potential. Due to the multiplication factor at each dynode stage, the overall electron signal at the final collector dynode is amplified by a factor, typically 10^6 to 10^7. The magnitude of this signal is proportional to the energy of the gamma photon deposited in the NaI(Tl) crystal. This signal is amplified and accumulated as part of the recorded gamma spectrum.

The efficiency of the NaI(Tl)–photomultiplier combination can be estimated from data quoted by Bertin (1975) as follows:

Fraction of scintillation photons produced in the crystal that reach the photocathode:	90%
Fraction of photons that reach the photocathode which give rise to photoelectric emission:	20–40%
Fraction of photoelectrons that reach the first dynode:	90%
Number of secondary electrons emitted for each electron incident on subsequent dynodes:	3–4
Overall signal gain using 9 to 14 dynodes:	10^6–10^7

Some properties of NaI(Tl) scintillation detectors compared with conventional Ge(Li) semiconductor detectors are listed in Table 13.6. This table shows that standard NaI single crystals (75 × 75 mm) have a significantly higher detection efficiency than conventional Ge(Li) semiconductor crystals. Thus NaI(Tl) detectors have a positive advantage in the detection of high-energy (> 1000 keV) gamma radiation in environmental samples. However, this advantage must be balanced against the fundamental limitations in the photopeak resolution of NaI(Tl) crystals compared with Ge(Li) detectors. Typical FWHM resolution of a standard coaxial Ge(Li) detector to the 1332 keV gamma from ^{60}Co is usually better than 2 keV. In comparison, the data for a NaI(Tl) detector is in the order of 120 keV.

Thus NaI(Tl) scintillation detectors can only successfully be used to analyse relatively simple gamma spectra, for example the natural activity due to U, Th and K. In such applications, the relatively high detection efficiency is a positive advantage. Furthermore, for field-based measurements the relative ruggedness of instrumentation, and the fact that unlike Ge(Li) detectors, liquid nitrogen is not required for successful operation, are also favourable aspects.

13.6.3 *U, Th, K gamma spectrum*

The gamma spectrum observed from natural samples originates from the radioactive decay of naturally-occurring ^{40}K ($T_{1/2} = 1.25 \times 10^9$ y) and from ^{238}U ($T_{1/2} = 4.468 \times 10^9$y and ^{232}Th ($T_{1/2} = 14.00 \times 10^9$y) and their daughter isotopes. Gamma activity from the decay chain of ^{235}U ($T_{1/2} = 0.7038 \times 10^9$ y) is not usually detected, since the parent isotope has a natural abundance of only 0.7204%.

The uranium and thorium decay chains have been considered in section 13.2, where it was shown that many of the daughter isotopes decay by alpha-emission. However alpha-decay is almost always accompanied by the simultaneous emission of characteristic gamma radiation. The most prominent gamma photons emitted by the daughter isotopes in the ^{238}U and ^{232}Th decay chains are listed in Table 13.7. These data have been restricted to those photopeaks having

Table 13.6 Some properties of NaI(Tl) scintillation and Ge(Li) semiconductor gamma-ray detectors compared

Property	NaI(Tl) scintillation detector	Germanium semiconductor detector
Typical size for gamma spectrum measurements (right-angled cylinder)	75 × 75 mm	30 × 30 mm
Density (g cm^{-3})	3.607	5.35
Energy required to produce one event	50 eV per light photon	2.98 eV per electron–hole pair
Typical resolution (FWHM) at 1332 keV	120 keV	1.9 keV
Detection efficiency (relative to NaI(Tl) = 100)	(100)	10–30
Properties	Crystal is hygroscopic	Detector must be cooled to liquid nitrogen temperatures
Signal amplification	Photomultiplier	Charge-sensitive field effect transistor
Detection process	Scintillation in NaI crystal	Ionization in Ge detector
Applications	Gamma spectrometry (U, Th, K) including field measurements	High-resolution gamma spectrometry e.g. neutron activation analysis

relative decay intensities greater than 4%. Corresponding gamma spectra are shown diagrammatically in Figure 13.9.

When selecting gamma photopeaks for analysis, account must be taken of the fact that it is desirable to analyse relatively large sample aliquots; 300–400 g is normal for laboratory-scale measurements. If low-energy photopeaks (< 500 keV) are selected for analysis, a significant attenuation will occur within the sample, causing low count rates that may vary with sample density. It is normally considered that significant attenuation will occur using photopeaks

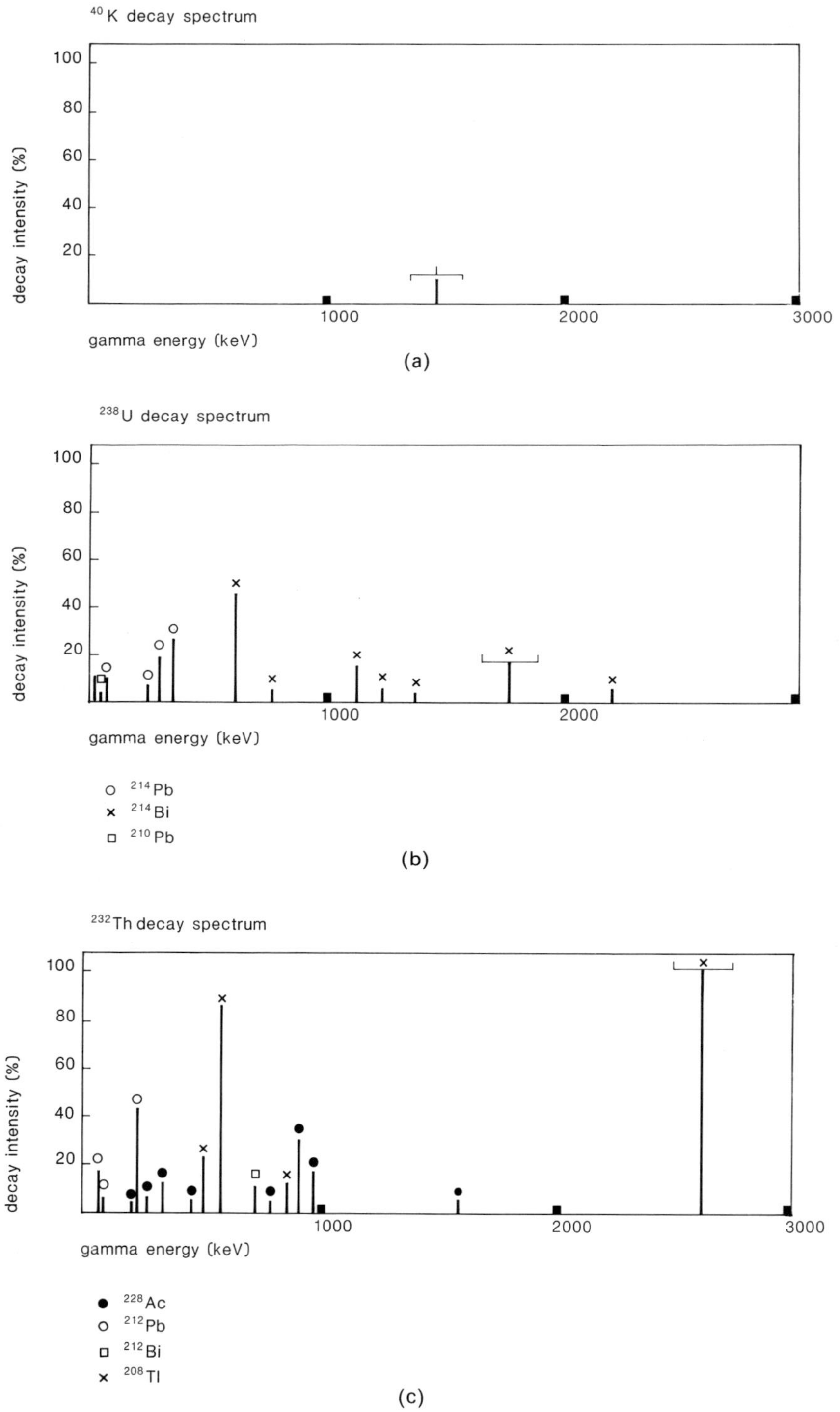

Figure 13.9 Schematic representation of gamma-ray spectrum emanating from ^{40}K and the decay series nuclides of ^{238}U and ^{232}Th. The windows normally set up for gamma-ray spectrometry measurements using a NaI(Tl) detector are marked.

Table 13.7 Gamma photopeaks originating from the uranium and thorium decay series and potassium

Decay series	Isotope	Gamma energy (keV)	% decay intensity
^{238}U	^{238}U	13.0	8.7
	^{234}Th	13.3	9.8
	^{234}U	13.6	10.4
	^{214}Pb	10.8	13.6
		74.8	6.3
		77.1	10.7
		241.9	7.5
		295.2	19.2
		351.9	37.1
	^{214}Bi	609.3	46.1
		768.4	4.9
		1120.3	15.0
		1238.1	5.9
		1377.7	4.0
		1764.5	15.9
		2204.1	5.0
^{232}Th	^{228}Ac	209.4	4.6
		338.4	12.0
		463.0	4.6
		794.8	4.8
		911.1	29.0
		964.6	5.5
		968.9	17.5
		1587	3.7
	^{212}Pb	74.8	9.6
		77.1	17.5
		87.2	6.3
		238.6	43.1
	^{212}Bi	727.2	11.8
	^{208}Tl	277.2	6.5
		510.7	22.5
		583.1	86.0
		860.5	12.0
		2614.5	100.0
K	^{40}K	1460.8	10.7

Data from Erdtmann and Soyka (1979).
Gamma photon intensities below about 4% have not been listed.

lower than 1000 keV in energy. This leaves little flexibility in the choice of gamma lines (see Table 13.7) available for analysis:

Uranium: 1764 keV (15.9%) photopeak from ^{214}Bi
Thorium: 2614 keV (100%) photopeak from ^{208}Tl
Potassium: 1461 keV (10.7%) photopeak (only choice).

In each case, values in brackets indicate the number of specified gamma photons emitted per 100 disintegrations of the isotope.

On the basis of relative decay intensities, the most sensitive determinations will be made for thorium; however, overall sensitivity must take into account detector efficiency, which decreases rapidly with photon energy. In converting gamma spectrum data into elemental analyses, several assumptions must be made:

(i) The uranium and thorium decay chains are in secular equilibrium (see section 13.8)

(ii) Natural isotopic abundances must be assumed for U and potassium measurements

(iii) The sample is homogeneous—this assumption is chiefly of concern for measurement made *in situ* in the field.

13.6.4 *Laboratory analysis procedure*

A typical arrangement of instrumentation for laboratory-based gamma spectrometry measurements is shown in Figure 13.10. The detector and photomultiplier assembly are held vertically inside a radiation shield (see also section 13.6.6). The sample itself is held in a counting capsule placed on the top surface of the detector crystal housing.

The weight of sample selected for analysis should be sufficient to give U, Th, and K count rates of at least ten times background. For 'typical' terrestrial samples, 350 g is sufficient. For some mineral separates, for example zircon or monazite, 5–10 g would be adequate for a U/Th determination.

Strictly, it is only necessary to crush samples to fulfil two criteria:

(i) To ensure that the analysed sample aliquot is homogeneous and representative of the bulk

(ii) To ensure that samples occupy identical counting geometry with respect to the detector so that all gamma-ray measurements will be reproducible.

These conditions are normally satisfied if samples are crushed rather coarsely, perhaps to 60–80 mesh. Furthermore it is preferable to use a sample counting container that is completely filled with sample rather than to weigh identical amounts of material into the capsule for analysis. Preparation to constant volume ensures that reproducible counting geometries are achieved. Since one of the daughter isotopes in both the uranium and thorium decay chains is

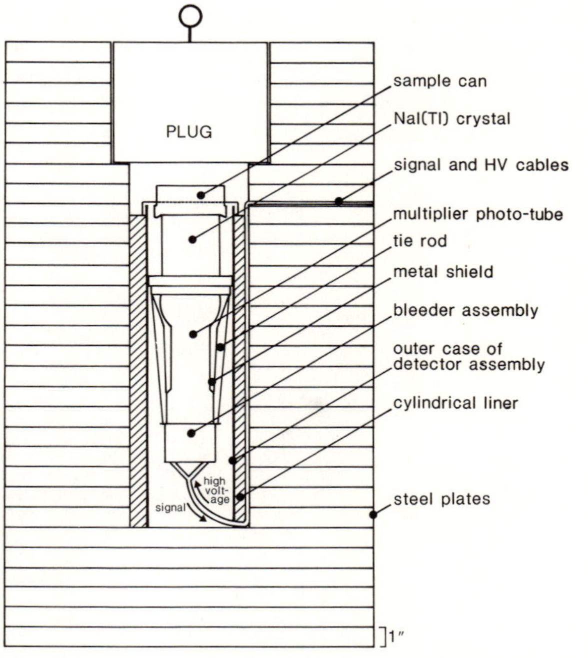

Figure 13.10 Gamma-ray spectrometry assembly for laboratory measurements. The NaI(Tl) detector and photomultiplier tube are shielded from ambient radiation in this design using steel plates. (Adams, 1964.)

Table 13.8 Analytical results by gamma spectrometry with different degrees of sample compaction

Sample weight (g)		300	350	404	448.5
K	(%)	4.04	4.01	3.84	3.75
U	(ppm)	5.9	4.8	4.5	4.4
Th	(ppm)	24.3	22.8	23.4	21.7

Potassium by flame photometry = 3.99%.
Data from Heier and Rogers (1963).

radon, a gas, there is some danger that, in crushing and sealing samples in the counting capsule, secular equilibrium will be broken, particularly since both ^{208}Tl and ^{214}Bi are daughters of their respective radon isotopes. Samples must therefore normally be left to stand for some time after sealing (up to one month) to allow radon equilibrium to be re-attained.

In order to demonstrate the effects of sample packing, Heier and Rogers (1963) analysed the same sample four times in a 350 g counting canister. In the first run (300 g sample) the canister was not completely full. In the following three runs (350 g, 404 g, 448.5 g) the sample was compacted into the canister to different extents. Their analytical data are shown in Table 13.8. These authors found that the largest variation was in the determination of potassium, which being analysed from the lowest-energy gamma photon, might be expected to most sensitive to sample self-absorption effects.

13.6.5 *Spectrum counting and analysis*

(*a*) *Count times.* Samples must be counted for sufficient time to achieve adequate statistical precision. Heier *et al.* (1965) set a target counting error of 5% (one sigma). Dolerites containing about 3 ppm Th, 1 ppm U and 0.7%K were counted for 12 hours to achieve this precision, but granophyres containing >10 ppm Th, 3 ppm U and 2%K were counted for 100 minutes.

(*b*) *Spectrum acquisition.* Spectrum data can be accumulated in a multichannel analyser prior to analysis (section 12.5). In view of the relatively low resolution of a NaI(Tl) crystal, a 256 channel analyser is probably satisfactory rather than the 2048 channel analysers normally specified for high-resolution gamma spectrometry with Ge(Li) detectors. Using such instrumentation, some of the simpler spectrum analysis procedures described in section 12.10 can be used for peak area determination. However, instrumentation can be simplified considerably (and almost invariably is for field-based measurements). The output from the scintillation detector is monitored and only those pulses that pass through one of three pre-selected 'windows' are accumulated in a corresponding single-channel scaler. The windows of each scaler are adjusted to correspond to one of the three gamma photopeaks selected for analysis. This arrangement is shown in Figure 13.11 and indicates the energy range of the three scaler windows in relation to the U/Th/K gamma spectrum.

(*c*) *Spectrum interferences.* The simplified diagram (Figure 13.11) shows that as well as the gamma photopeak data, interferences arising from background counts are also accumulated in each window as follows:

Th (^{208}Tl) 2614 keV: cosmic ray background
U (^{214}Bi) 1764 keV: cosmic ray background plus Compton scatter from 2614 keV (thorium) peak
K (^{40}K) 1461 keV: cosmic ray background plus Compton scatter from both the 2614 keV (thorium) and 1764 keV (uranium) peaks.

Instrument parameters used by Adams (1964) to set up these windows are shown in Table 13.9.

Other minor gamma-ray interferences can influence count rates in pre-selected windows. The most important of these are as follows:

2614 keV: U (^{214}Bi) interference at 2488 keV (1.6%)
U 1764 keV: Th (^{228}Ac) interference at 1587 keV (3.7%)
K 1461 keV: U (^{214}Bi) interference at 1378 keV (4.0%), 1408 keV (2.5%) and 1509 keV (2.2%), and Th (^{228}Ac) interference at 1587 keV (3.7 %).

(*d*) *Calculations of photopeak areas.* The areas of the Th, U and K components of the combined data accumulated in

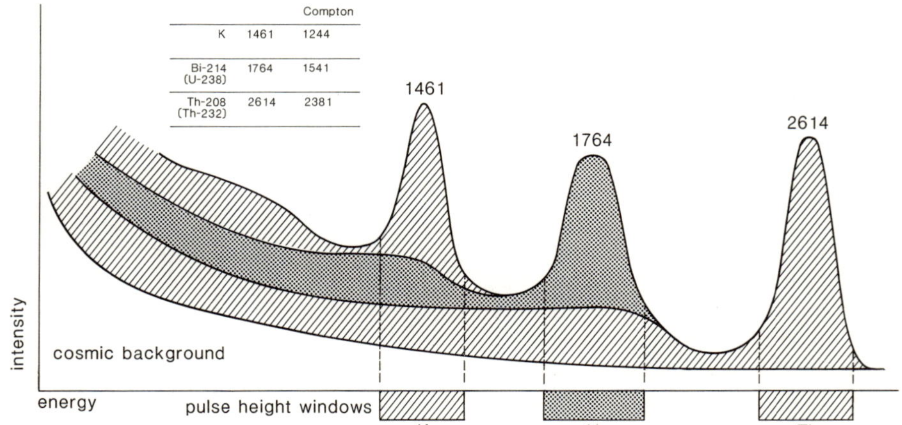

Figure 13.11 Schematic representation of a NaI(Tl) gamma spectrum showing the pulse height windows set up for the analysis of ^{40}K (1461 keV), ^{214}Bi (1764 keV) and ^{208}Tl (2614 keV). The contributions of individual isotopes to the background is indicated to demonstrate the ratio method of correcting for background interferences.

Table 13.9 Instrument settings used in the determination of K, U and Th by gamma spectrometry

	Potassium	Uranium	Thorium
Centre of window (MeV)	1.46	1.76	2.62
Width of window (MeV)	0.23	0.25	0.28
Background count rate (counts per minute)*	7.4	2.4	1.2
Weight of sample equivalent to total background in window (g)	1.7	0.57	1.1

* Sample: 350 g purified NaCl.
Data from Adams (1964).

Table 13.10 Gamma spectroscopy correction ratios and sensitivity factors

Correction ratio	R_1 (Th on U)	0.49 ± 0.07
	R_2 (Th on K)	0.63 ± 0.09
	R_3 (U on K)	0.99 ± 0.15
	R_4 (U on Th)	0.025 ± 0.004
Sensitivity factors	$(SENS)_{Th}$	8.1 ± 0.3
	$(SENS)_U$	20.0 ± 0.2
	$(SENS)_K$	210 ± 7

Data from Cassidy (1981).
Sensitivity factor is quoted in counts per minute per ppm (U, Th) or per % element (K).

each scaler window may be calculated by simple ratio as follows:

2614 keV window:

$$[AREA]_{Th} = [WINDOW]_{2614} - [BKGD]_{Th} \quad (13.11)$$

1764 keV window:

$$[AREA]_U = [WINDOW]_{1764} - R_1[AREA]_{Th} - [BKGD]_U \quad (13.12)$$

1461 keV window:

$$[AREA]_K = [WINDOW]_{1461} - R_2[AREA]_{Th} - R_3[AREA]_U - [BKGD]_K \quad (13.13)$$

where [AREA] is the corrected area (i.e. counts) due to the specified element, [WINDOW] represents the total number of counts accumulated within the specified window, [BKGD] represents the background contribution measured from a spectrum in which isotopes, other than the one being determined, are absent, R_1 is the fraction of thorium gamma-ray events that contribute to the uranium window at 1764 keV, R_2 is the fraction of thorium gamma-ray events that contribute to the potassium window at 1461 keV and R_3 is the fraction of uranium gamma-ray events that contribute to the potassium window at 1461 keV.

Having determined the true areas of each gamma photopeak, the concentration of the appropriate element may be calculated (assuming secular equilibrium) from formulae such as

$$\text{ppm}\,[Th] = [AREA]_{Th}/[SENS]_{Th} \quad (13.14)$$

where [SENS] is the sensitivity of thorium (counts per unit concentration) obtained from measurements on a suitable reference material.

In the most demanding applications, some workers include a fourth stripping ratio in the calculation of $[AREA]_{Th}$ to account for the small contribution of U (^{214}Bi, 2448 keV) on the thorium window at 2614 keV.

Correction ratios (R_1–R_4) can be measured by counting samples containing individual isotopes of Th, U and K. These samples should be formulated in the same geometry as silicate materials for analysis, since the magnitude of Compton scatter interference may otherwise vary. Correction and sensitivity factors derived by Cassidy (1981) for field measurements using a 75 × 75 mm sodium iodide crystal are shown in Table 13.10. It is expected that the precise magnitude of these parameters will change with detector size and sample configuration.

13.6.6 *Shielding*

The ultimate sensitivity of gamma spectrometry measurements depends on the degree of background suppression that can be achieved by shielding the detector. Background arises from:

(i) Cosmic ray interaction with the detector and surrounding material

(ii) Potassium, uranium and thorium in the environment surrounding the detector.

Building concrete contains significant proportions of potassium, and some forms of building blocks manufactured from pulverized fly ash (a waste product from coal-fired power stations) can contain measureable levels of radioactivity. Furthermore, all three elements are expected to be present in the adjacent subsoil/bedrock. Therefore, in contrast to neutron activation analysis, where environmental activity contributes only generally to the background continuum, in gamma spectrometry a direct contribution to the specific isotopes to be measured is expected from the natural background.

The detector assembly in Figure 13.10 is shown to be extensively shielded by steel plates. Unfortunately, as recognized by the original author, modern steel contains significant amounts of radioactive materials, notably ^{60}Co, and is therefore not suitable shielding material. The only steel suitable for shielding is that obtained from World War I battleships or similar historic sources, constructed before the advent of the atomic age. Lead is the traditional shielding material, but mercury has also been used. The effectiveness of various combinations of shielding is illustrated by data in Table 13.11.

These results show that the simple arrangement of surrounding the detector with a 100 mm barrier of lead is very effective in reducing external background. Further improvements are hard-won and expensive, involving the use of mercury and an anticoincidence arrangement. This involves surrounding the primary 75 × 75 mm detector with a 230 mm diameter outer NaI(Tl) collar coupled in such a way that any event detected in both detectors (a Compton scatter event) is rejected (see also section 12.18).

Table 13.11 Effect of various shielding materials on background count rate of 75 × 75 mm NaI(Tl) scintillation detector

Shield	Background count rate (counts per minute)		
	500 keV	1000 keV	2000 keV
No shielding	325	76	5.8
100 mm lead	4.0	1.7	0.36
100 mm mercury, anti-coincidence shield	1.9	0.3	0.04
200 mm lead, 300 feet concrete	1.0	0.2	0.02
100 mm lead, 100 mm mercury, anti-coincidence shield	0.16	0.05	0.01

Data attributed to J.H.Kaye, from Nielsen (1972).

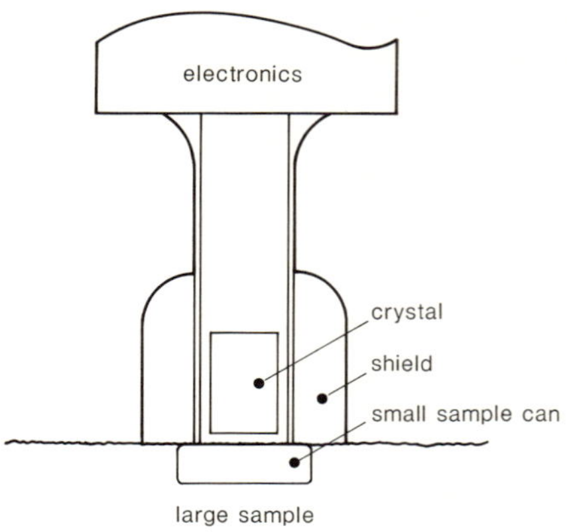

Figure 13.12 Configuration for field gamma-ray spectrometry measurements. (Adapted from Adams and Fryer, 1964.)

13.6.7 *Results*

Some analyses of international reference materials by gamma spectrometry for the elements K, U and Th are shown in Table 13.12. It is estimated that (six-sigma) detection limits are about 0.05%K, 0.2 ppm U and 0.5 ppm Th and that analytical precisions are typically about 5–10% (two-sigma).

13.6.8 *Field-based gamma spectrometry measurements*

One of the unusual aspects of gamma spectrometry is the possibility of carrying out analyses in the field using virtually the same instrumentation as for laboratory measurements. The spectrometer is placed on an outcrop and a gamma spectrum accumulated for a count time typically of 10–15 minutes. Spectra are analysed by counting gamma events that pass three discriminator windows centred on the photopeaks 1461 keV (^{40}K), 1764 keV (^{214}Bi) and 2614 keV (^{208}Tl) as described above.

(*a*) *Mass of sample analysed.* An important consideration in these measurements is the mass of sample for which measurements are representative. With the entrance window of the spectrometer placed on the surface of a flat rock outcrop, gamma radiation is accumulated from a 2π solid angle in front of the detector (Figure 13.12). Since attenuation of 1461 keV ^{40}K is greater than that of 2614 keV ^{208}Tl (Th) within the sample, the volume of outcrop sampled for potassium will be smaller than that for Th. Adams and Fryer (1964) estimate that with the configuration shown in Figure 13.12, approximately 50 kg of rock is sampled for 1461 keV ^{40}K radiation and approximately 80 kg for 2614 keV ^{208}Tl. Mahadavi (1964) analysed beach sands *in situ* by gamma spectrometry. The detector was buried 180 mm below the surface of the sand to achieve better counting geometry, and it was estimated that the volume of sand analysed in this way amounted to at least 80 L, which would have a dry weight of 140–160 kg.

The large volume of sample analysed by such field measurements places some constraints on the application and interpretation of data. It is necessary to make measurements on outcrops of at least a few square metres in area. Assumptions about the validity of secular equilibrium in the uranium and thorium decay chains are just as important in field measurements as in laboratory analyses (section 13.6.4). However, perhaps of equal importance is the necessity for the outcrop to be homogeneous in respect to its K, U and Th contents to a depth of up to 2 m.

(*b*) *Calibration of field instruments.* Reference samples must be measured in the same physical form as unknown samples. Two procedures widely used for field calibration are dis-

Table 13.12 Analyses of reference rocks for K, U, and Th by gamma spectrometry

			K (%)	U (ppm)	Th (ppm)
USGS:	W-1	Expected	0.53	0.58?	2.4
		Heier and Rogers (1963)	0.52	0.53	2.16–2.24
	AGV-1	Expected	2.42	1.95	6.4
		Belyaev and Sobornov (1981)	2.38 ± 0.03	2.0 ± 0.2	6.5 ± 0.3
	BHVO-1	Expected	0.45	0.4?	1.0
		Belyaev and Sobornov (1981)	0.33 ± 0.05	0.5 ± 0.1	1.0 ± 0.3
	GSP-1	Expected	4.57	2.1	105
		Belyaev and Sobornov (1981)	4.55 ± 0.05	2.2 ± 0.2	102.1 ± 0.6
CCRMP:	SY-2	Expected	3.72	290	380
		Belyaev and Sobornov (1981)		290 ± 3	382 ± 5

Expected data is taken from Abbey (1983).
Error limits presumed to represent ± one sigma.

cussed by Cassidy (1981):

(i) Spectrometers can be calibrated against large concrete blocks, doped with appropriate amounts of K, U and Th. In order to simulate a sample of infinite thickness, such blocks must have a diameter of a least 2 m and a thickness of 0.5 m (IAEA, 1976)

(ii) Alternatively, spectrometers can be calibrated in the field against a well-calibrated outcrop, hand specimens of which have been independently analysed in the laboratory. The outcrop must satisfy criteria for size and homogeneity at depth outlined above.

The results of one such calibration for thorium, taken from Lowder *et al.* (1964) are shown in Figure 13.13. Here, the spectrometer response (observed 2614 keV peak counts) is plotted as a function of independent laboratory analyses for thorium for a range of outcrop measurements.

Using such procedures, Cassidy (1981) compared field-based gamma spectrometry measurements with laboratory analyses of hand specimens from the same outcrops (U, Th by neutron activation analysis, K by flame photometry) and estimated that the standard deviation between field and laboratory measurements was ±2 ppm (Th), ±0.9 ppm (U) and $\pm 0.2\% K_2O$.

(*c*) *Secular equilibrium in the field.* For field measurements to be valid, secular equilibrium must be maintained within the decay chains of uranium and thorium. Cassidy (1981), summarizing earlier work, considers that secular equilibrium is normally maintained between ^{232}Th and its daughters in most geological environments. Uranium (^{238}U) and its decay products ^{234}U, ^{226}Ra, ^{222}Rn are somewhat more

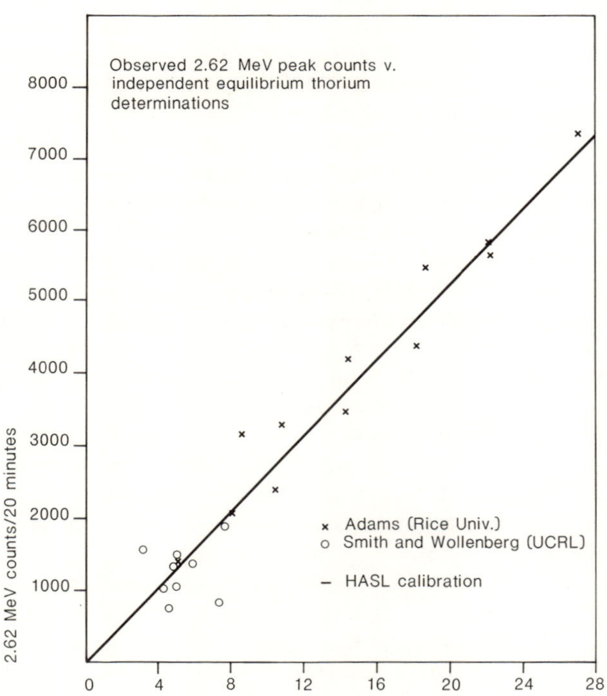

Figure 13.13 Field gamma-ray spectrometry calibration check showing count rate of the 2614 keV ^{208}Tl measured on an outcrop versus the thorium composition of hand specimens derived from independent laboratory measurements. (Lowder *et al.*, 1964.)

mobile, but can normally be considered to be in equilibrium in fresh unweathered rock.

(*d*) *Dose rate equivalents of field measurements.* It is sometimes desirable to convert field analyses into environmental radiation dose rates. Conversion factors taken from Adams and Lowder (1964, Appendix 3) are as follows:

$$1\ \mu R/h = 0.01\ \mu gray/h \equiv 0.58\%\ \text{potassium}$$
$$\equiv 1.32\ \text{ppm uranium}$$
$$\equiv 2.78\ \text{ppm thorium}$$

(assuming secular equilibrium in the decay chains of uranium and thorium).

(*f*) *Field mapping.* Despite the limitations in application and accuracy of field gamma-ray measurements outlined above, the outstanding advantage remains the capability of immediate field data. Thus in mapping the distributions of U, Th and K, decisions on sampling strategy do not have to be delayed until the outcome of complementary laboratory measurements on hand specimens is known. An example of such work in the contour data on the distribution of thorium in the Loch Doon granite complex, SW Scotland, based on gamma spectrometry measurements of Cassidy (1981). The distribution map (Figure 13.14) was obtained from 104 measurements on outcrops sampled at approximately 1 km grid intervals. The two centres of high thorium content are clearly revealed.

Although almost all measurements to date have been carried out using traditional NaI(Tl) scintillation spectrometers, it will be interesting to see what developments arise from survey work using the hand-held high-resolution hyperpure germanium detectors and portable multichannel analysers that are now becoming available.

13.7 Alpha spectrometry

13.7.1 *Alpha-decay*

Many of the naturally-occurring daughter isotopes in the ^{235}U, ^{238}U and ^{232}Th decay chains decay by alpha-emission. Alpha-emission involves the ejection of an alpha-particle from the nucleus of an unstable isotope. This alpha-particle is ejected with a discrete kinetic energy, usually in the range 4–7 MeV. The daughter nucleus is usually left in an excited state and spontaneously emits discrete gamma radiation whilst decaying to its ground state.

Alpha-particles have the structure of a helium nucleus ($^4He^{2+}$), consisting of two protons and two neutrons but no orbital electrons, and therefore a nuclear charge of +2. In nuclear terms they represent massive nuclear particles that interact with matter by causing a large degree of radiation damage over very short path lengths. Thus Moore (1966) estimated that the thickness of lead required to completely absorb 10 MeV alpha-particles is only 34 μm, and that of water, 130 μm. Alpha-particles ultimately come to rest as helium atoms, having picked up two electrons from surrounding matter. In terms of the external radiation hazard, alpha-emitters do not pose a serious problem to human beings, being unable to penetrate the outer dead layers of skin. However, they do pose a serious internal radiation

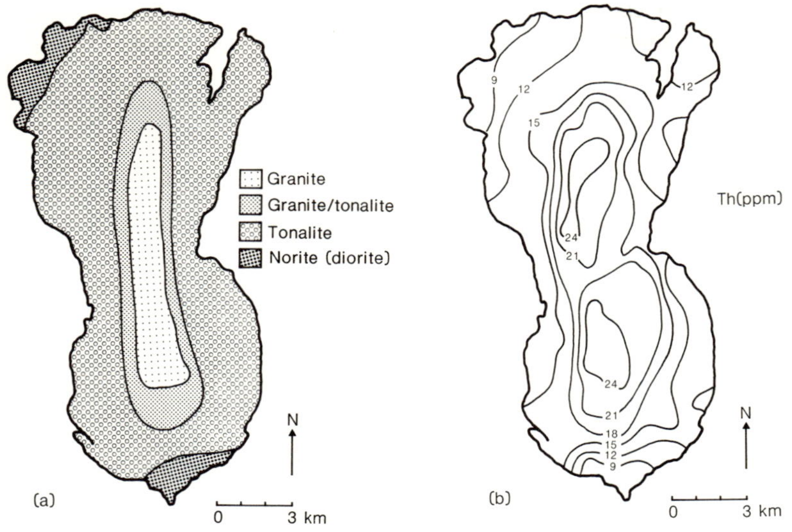

Figure 13.14 Thorium contour distribution (in ppm) of the Loch Doon intrusion, SW Scotland, determined from field gamma-ray spectrometry measurements from 104 sampling locations situated on an approximately regular grid spacing of 1 km. (Cassidy, 1981.)

hazard once ingested, since alpha-decay causes massive damage within individual cells of living tissue.

Alpha spectrometry is used to analyse samples for alpha-emitting isotopes based on the detection of the energy and intensity of discrete alpha-emissions. From the properties outlined above, it is apparent that considerable attenuation of alpha-particles will occur within the sample itself if attempts are made to analyse bulk materials. Extensive chemical pre-treatment is therefore necessary to remove matrix components and pre-concentrate the alpha-emitting residue. To minimize self-absorption effects, this residue is prepared for counting as a thin film coated on to a counting 'planchet'. Furthermore, measurements must be carried out in vacuum, since otherwise heavy absorption of alpha-particles would occur in the air gap between sample and detector. The range of 5 MeV alpha-particles in air at STP is estimated to be only about 35 mm.

One fundamental difference between alpha-emitting isotopes and those that decay by beta-emission must be emphasized. Beta-particles are not emitted from decaying nuclei at discrete energies. The beta spectrum from an isotope is observed as a continuum extending from a maximum energy (the value normally listed in tables of radioactive decay parameters) down to zero energy. It is not therefore readily possible to distinguish between a mixture of two isotopes by beta spectrum measurements. The reason for this continuum of decay energies is that beta-emission is accompanied by the ejection of a neutrino, and since the laws of conservation of momentum must be obeyed, the energy of either beta-particles or neutrinos can vary from zero up to a maximum characteristic of the nuclear transition.

Conversely, alpha-particles are emitted from a decaying nucleus at one or more discrete energies. Thus it is possible to identify, positively, alpha-emitting isotopes in a mixed sample from the characteristic peaks found in a high-resolution alpha spectrum.

A further advantage of alpha spectroscopy is that modern detectors are virtually insensitive to accompanying gamma- or beta-radiation. Measurements are therefore specific to alpha-emitting isotopes and since the natural alpha-background is very low and can readily be shielded, alpha spectrometry is a very sensitive technique. Indeed, for the limited number of naturally-occurring isotopes that can be analysed by this technique, determinations are more sensitive than mass spectrometry.

13.7.2 *Instrumentation*

A variety of instruments have been used for alpha-activity measurements including the following.

(i) *Windowless gas flow proportional counters.* The alpha-emitting sample is placed inside a proportional counter, which is flushed with argon–methane gas. Ionization within the counter caused by alpha-decay is detected as in a conventional proportional counter (section 8.9). Spectrum measurements are made with pulse-height analysis settings optimized for the decay energy in question.

(ii) *Scintillation counting*: when alpha particles impinge on zinc sulphide crystals (activated with silver, copper or manganese ions), the material scintillates. The zinc sulphide scintillator is therefore coupled to a photomultiplier tube to provide a means of alpha detection. However, due to poor resolution and counting losses in the thin window protecting the phosphor, such instruments are normally used for qualitative contamination monitoring rather than for alpha spectrometry.

(iii) *Silicon surface barrier detectors*: almost all alpha-spectral measurements are now carried out with silicon surface barrier detectors. These consist of a circular wafer of silicon about 1 mm thick (Figure 13.15). The front face of the detector is coated with a thin layer of gold (about 0.2 μm thick) to serve as one electrical contact. The rear surface is coated with aluminium (about 1.5 μm thick) to serve as a second electrical contact. The silicon wafer is itself normally composed of *n*-type silicon. In operation, a positive electrical

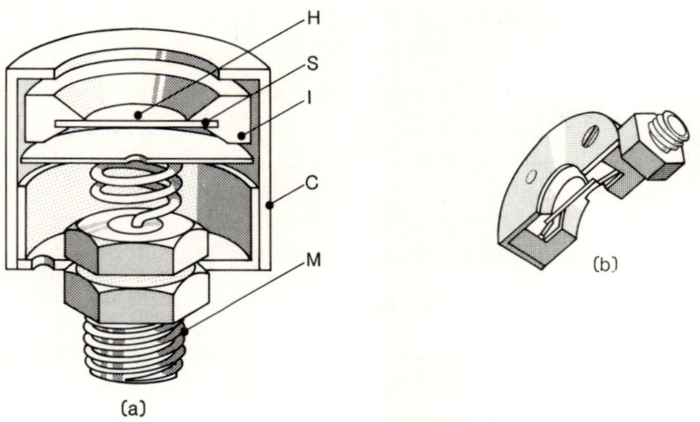

Figure 13.15 Silicon surface barrier detector for alpha-spectroscopy measurements. The silicon wafer (*S*) is mounted in a ceramic holder (*I*). The front face of the silicon wafer (*H*) is electrically grounded to the outer case (*C*); the rear contact is connected to the centre contact of the coaxial connection. (Ortec Inc.)

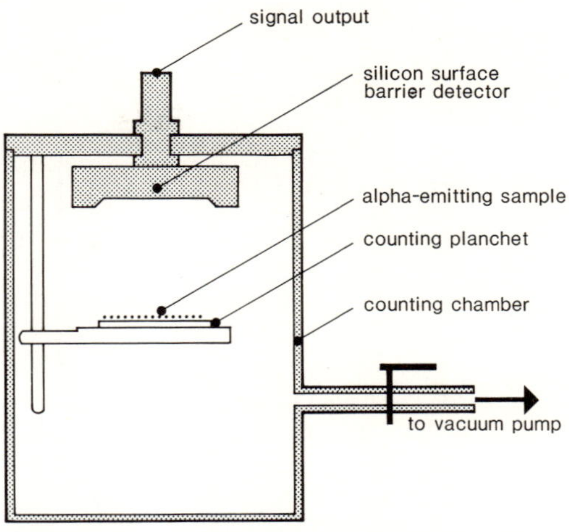

Figure 13.17 Counting configuration for laboratory alpha-spectrometry measurements.

bias is applied to the rear (aluminium) contact. Alpha-particles penetrating the gold entrance window cause ionization within the silicon wafer. The resultant electrons drift towards the positive electrode (the aluminium rear contact) and the corresponding positive 'holes' drift towards the gold front contact. The total charge deposited in the detector is directly

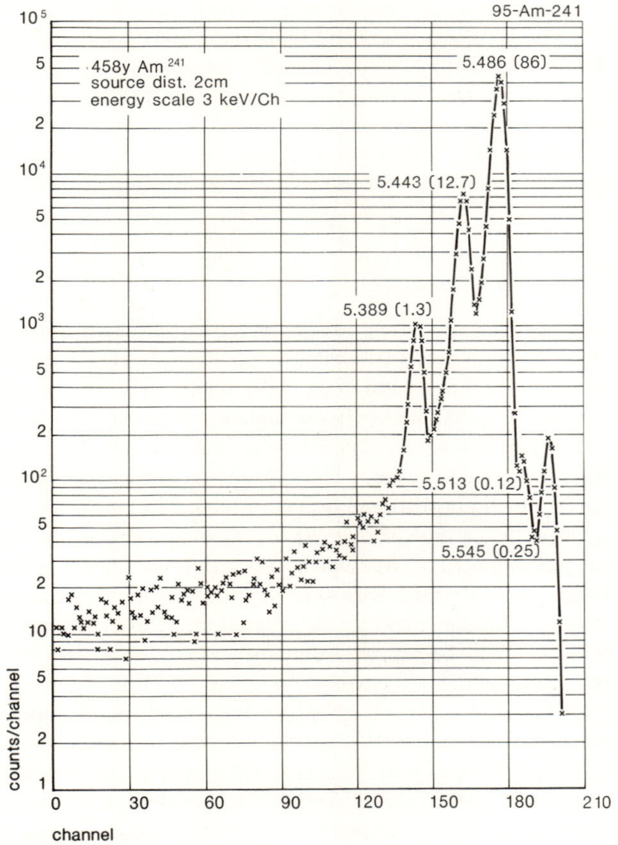

Figure 13.16 The alpha spectrum of ^{241}Am (5.0–5.6 MeV) recorded on a high-resolution silicon surface barrier detector. The energies (in MeV) and relative intensities of individual alpha-emissions are marked. (Chanda and Deal, 1970.)

proportional to the energy of the incident alpha-particle. This signal appears as a transient drop in the bias voltage applied to the detector, and is amplified and electronically processed for accumulation in a multichannel analyser. Compared with other forms of alpha-detection, surface barrier detectors offer several advantages, including:

(i) High-energy resolution (typically 15–20 keV FWHM for the 5.486 MeV alpha from ^{241}Am)

(ii) High stability and freedom from instrumental drift

(iii) A thin entrance window—inevitably some energy loss occurs in the thin gold entrance window and the silicon dead layer (~ 0.1 μm) lying beneath it, but for measurements on naturally-occurring alpha-emitters the effect on the relatively high alpha energies (4–7 MeV) is not important.

Compared with conventional silicon detectors used for x-ray and gamma-ray measurements, silicon surface barrier detectors do not require liquid nitrogen cooling, nor do they need a beryllium entrance window. Many commercial instruments are designed to operate at temperatures in the range -30 to $+25$°C.

A typical alpha spectrum from a ^{241}Am calibration source is shown in Figure 13.16. Five alpha-particle energies (labelled on the diagram) are clearly resolved. A typical counting configuration is shown in Figure 13.17. This shows the sample mounted on a platinum or stainless-steel planchet, positioned in front of a silicon surface barrier detector in an evacuated chamber. Alpha-spectrometer measurements are normally calibrated by isotope dilution techniques (section 17.5). This involves spiking samples prior to chemical separation with an artificial alpha-emitting isotope to permit recovery, detection efficiency and calibration corrections to be carried out simultaneously. The principal application of alpha spectrometry is in uranium/thorium series dating measurements, to be discussed in section 13.9.

13.8 Secular equilibrium with particular reference to uranium/thorium disequilibrium measurements

Secular equilibrium is a phenomenon observed when a radioactive parent isotope decays via one or more daughter

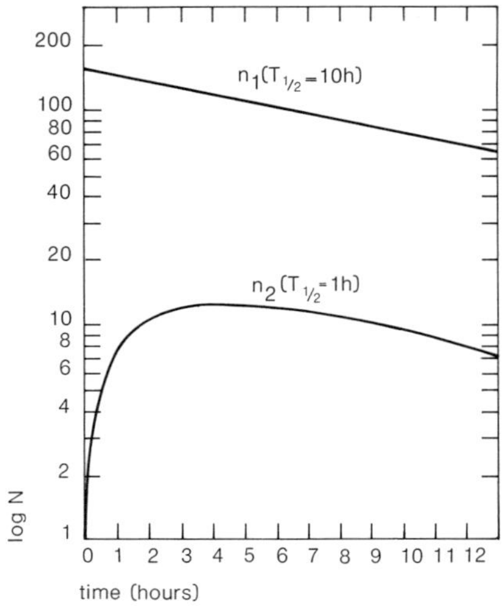

Figure 13.18 Change on the number of atoms of a radioactive parent isotope (N_1, of half-life 10 h) and its radioactive daughter (N_2, of half-life 1 h) as a function of time. Secular equilibrium is achieved after about 5 half-lives' decay of the daughter. Thereafter, the ratio of activities of parent to daughter remains constant. (Faure, 1977.)

isotopes to a stable product, and is only observed if the parent has a longer half-life than any of its intermediate daughters. Under these conditions a series of radioactive equilibria are set up so that the rate of decay of the parent isotope equals that of each of the daughters, which therefore equals the rate of formation of the stable product. The decay chain is then said to be in a state of secular equilibrium. This phenomenon was illustrated by Faure (1977) by considering the notional decay of a radioactive parent isotope having a half-life of 10 hours via a daughter isotope (half-life = 1 hour) to a stable product. The change in the number of atoms of parent (N_1) and daughter (N_2) with time are illustrated in Figure 13.18. During the first half-life of the daughter isotope, the proportion of its atoms formed by decay of the parent rapidly increases. However, after about five half-lives, the proportion of daughter isotope atoms effectively reaches a constant ratio with respect to the parent. Mathematically, this state of affairs is described by

$$N_1\lambda_1 = N_2\lambda_2 = \text{constant} \qquad (13.15)$$

where λ_1, λ_2 are the decay constants of parent and daughter respectively ($\lambda = 0.6931/T_{1/2}$; $T_{1/2}$ = half-life).

A full derivation of this equilibrium is given by Faure (1977, pp. 49-53).

If secular equilibrium is attained in a decay chain, the following conditions are met:

(i) The rate of decay of parent isotope equals that of each individual daughter which equals the rate of formation of the stable product

(ii) Decay of parent can therefore be considered to proceed directly to the stable product (the assumption made in the U–Th–Pb geochronometry measurements)

(iii) The ratio of radioactive decay of any two daughters will always be unity, that is:

$$\frac{\lambda_1 N_1}{\lambda_2 N_2} = 1.00 \qquad (13.16)$$

13.9 Uranium and thorium series disequilibrium

13.9.1 *Principles of analysis*

Uranium and thorium series disequilibrium measurements are based on situations where secular equilibrium within the appropriate decay chain has been broken by some geochemical phenomenon. Samples are then analysed so that the present content of a specified uranium or thorium decay series isotope can be compared with the concentration that would have been present had secular equilibrium not been broken. The difference permits an estimate to be made of the time that has elapsed since the original event occurred, and therefore offers the possibility of dating certain geochemical processes. Selected applications of uranium/thorium disequilibrium measurements are summarized in Table 13.13. This table lists the more common isotopes used for analysis, the mechanism that creates disequilibrium, and typical geological applications.

The useful time span of these geochemical chronometers depends on the half-life of the decay series isotope available for analysis and, as can be seen from Figure 13.19, varies from a few days to several million years. Clearly, the circumstances in which secular equilibrium can be broken are closely related to the geochemical properties of uranium, thorium and specified decay series isotopes. Here the more important of these geochemical properties relevant to disequilibrium applications will be reviewed, and then typical practical schemes of analysis will be examined.

13.9.2 *Geochemical properties of uranium, thorium and selected decay series isotopes*

(*a*) *Uranium and thorium isotopes in the marine environment.* Uranium is relatively soluble in water in oxidizing environments and is reported to exist as the uranyl ion UO_2^{2+} and $UO_2(CO_3)_3^{4-}$ complexes. In consequence the residence time of uranium in sea water is long, estimated to be of the order of 500 000 years (Faure, 1977, p.284). In contrast, thorium (and protactinium) are relatively insoluble in sea water, having a residence time of only 300 years, and are removed to bottom sediments, probably by adsorption on particulate material that slowly settles out.

Since uranium enters sea water predominantly by weathering, it might be expected that the decay activity of ^{238}U and its daughter ^{234}U would be equal. In fact the ratio $^{234}U/^{238}U$ in sea water is about 1.14. This discrepancy presumably arises because ^{234}U within terrestrial rocks is formed after alpha-decay of ^{238}U. During alpha-decay, a recoil is set up within the parent nucleus (gun and bullet effect). This recoil can disrupt the crystal lattice, forming micro-fractures that render ^{234}U more prone to hydrothermal weathering than ^{238}U. These mechanisms create the potential for disequilibrium measurements in the marine environment, as follows.

Table 13.13 Geological applications of uranium and thorium series disequilibrium studies

Isotope or ratio to be measured	Decay series	Timespan for applications	Mechanism for creating disequilibrium	Geological applications
			Methods based on the decay of unsupported intermediary daughter isotopes	
^{230}Th $^{230}Th/^{232}Th$	^{238}U	3500–350 000 y	Rapid precipitation of thorium (but not uranium) from sea water into marine sediments	Dating deep-sea sediments and measurements of sedimentation rates (often less than 0.1mm per year). Dating manganese nodules
^{231}Pa $^{231}Pa/^{232}Th$	^{235}U	1500–160 000 y	Protactinium is also insoluble in sea water and is rapidly incorporated into marine sediments (as is thorium)	Similar and in some respects complimentary to ^{230}Th and $^{230}Th/^{232}Th$ data outlined above. $^{231}Pa/^{30}Th$ is more susceptible to error (sediment mixing, analytical discrepancies) than use of ^{230}Th or ^{231}Pa alone (Ku, 1976)
^{210}Pb	^{238}U	1–100 y	^{222}Rn gas released to the atmosphere. After subsequent decay, ^{210}Pb falls out and is deposited in snow and shallow-water sediments	Dating permanent snowfields. Measuring depositional rates of near-shore and lake sediments (usually greater than 1 mm per year)
$^{228}Th/^{232}Th$	^{232}Th	30 d–10 y	^{228}Rn gas diffuses out of bottom sediments into water. After decay, ^{228}Th is precipitated from solution and re-deposited in top layers of sediments	Dating modern sediments (less than 10 years old). Accretion rates of freshwater manganese nodules
$^{234}U/^{238}U$	^{238}U	1200–120 000 y	Sea water ratio of $^{234}U/^{238}U$ is about 1.14. Uranium follows calcium into marine organisms where the subsequent decay of unsupported ^{234}U provides a means of dating samples assuming uranium is not remobilized	Dating corals, oolites, etc. formed in marine environments
^{228}Ra	^{232}Th	4 months–30 y	Radium is soluble in water and geochemically follows calcium into corals etc. Dating arises from the subsequent decay of unsupported ^{228}Ra in corals, etc.	Measurement of mean skeletal growth rates of fresh corals. Tracer to follow mixing of oceanic waters (Moore, 1969)
^{224}Ra ^{226}Ra	^{232}Th ^{238}U	4 h–18 d 75–8100 y	Radium isotopes, derived from the decay of respective thorium isotopes, dissolve in waters	Tracer for groundwater, estuarine water and nearshore sediments (Elsinger *et al.*, 1982; Michel *et al.*, 1981)
^{214}Pb	^{238}U	1 min–2 h	^{214}Pb is formed by radioactive decay of ^{222}Rn in groundwaters	Ground water tracer: mechanism for the precipitation of lead from waters (Hussain and Krishnaswami, 1982)
$^{234}Th/^{238}U$	^{238}U	1–120 d	^{234}Th, produced from decay of ^{238}U, is rapidly extracted from the water column by suspended particles and transported to underlying sediments	Tracer for short-term effects of biogenic and physical reworking of near shore sediments (Aller and Cochran, 1976)
			Methods based on the accumulation of decay products of uranium	
$^{230}Th/^{234}U$ $^{230}Th/^{238}U$ $^{231}Pa/^{235}U$	^{238}U ^{238}U ^{235}U	3500–350 000 y 3500–350 000 y 1500–160 000 y	Uranium is co-precipitated with calcium to form calcite and aragonite in natural waters essentially free of thorium. The decay products ^{230}Th and ^{231}Pa accumulated in such samples are therefore a function of time and uranium concentration	Dating of fossil bones, oolites, corals etc.
$^{230}Th/^{232}Th$ versus $^{238}U/^{232}Th$	^{238}U	3500–350 000 y	Fractionation of uranium and thorium in the upper mantle leads to disequilibrium between ^{238}U and daughter ^{230}Th. This forms a geochronometer capable of measuring the time that has elapsed since the magma crystallized	Age dating, normally by the internal isochron method (i.e. on mineral separates). Geochemical information on alteration history from discordant isochron data (e.g. Allegre and Condomines, 1976; Condomines and Allegre, 1980)
^{238}U ^{234}Th ^{230}Th		Parent 1200–120 000 y 3500–350 000 y	(i) Fractionation between uranium and thorium decay products in the upper mantle particularly as measured in	(i) Trace-element indicators of mechanisms and timing of events in magma chambers from the analysis of

(*Continued*)

Table 13.13 (*Continued*)

Isotope or ratio to be measured	Decay series	Timespan for applications	Mechanism for creating disequilibrium	Geological applications
^{226}Ra		75–8100 y	recently erupted volcanic rocks	recently-erupted lava flows (Bennett *et al.*, 1982; Capaldi *et al.*, 1976).
^{210}Pb		1–100 y	(ii) Re-equilibration of isotope ratios due to hydrothermal alteration, particularly involving the mobility of uranium and radium isotopes	(ii) Indicators of the geochemical stability of granitic rocks subjected to hydrothermal and other alteration effects (Schwarcz *et al.*, 1982)
^{210}Po		6d–2 y		
^{232}Th		Parent		
^{228}Ra		4 month–30 y		
^{228}Th		30 d–10 y		
^{235}U		Parent		
^{231}Pa		1500–160 000 y		

Applications date from Ku (1976) and references therein.
h = hour, d = day, y = year.

(*b*) $^{230}Th/^{232}Th$ *in marine sediments*. Since thorium is rapidly incorporated into marine sediments but uranium is not, ^{230}Th, originally formed by radioactive decay of ^{238}U, will be found in marine sediments unsupported by its parent ^{238}U. Providing the initial decay rate of ^{230}Th is known, then the proportion of this isotope that remains in analysed sediments offers the possibility of dating such samples. Furthermore, the variation in the rate of decay of ^{230}Th with depth in the sediment pile provides the opportunity of measuring the rate of sedimentation of a sample. Having regard to the half-life of ^{230}Th, this geochronometer is valuable for samples up to 350 000 years old.

(*c*) $^{230}Th/^{238}U$ *in corals and oolites*. From an examination of geochemical behaviour, the element uranium, but not thorium or protactinium, is readily deposited with calcium in calcite and aragonite to form corals and oolites, etc. Consider the decay series $^{238}U \rightarrow {}^{230}Th$. Since, on formation, a coral sample will contain uranium isotopes, but not those of thorium, the initial ratio of ^{230}Th to ^{238}U will be zero. However, as ^{238}U progressively decays, the proportion of ^{230}Th will gradually increase (Figure 13.20). After a decay of about five half-lives of ^{230}Th, secular equilibrium will be attained (providing the sample remains a closed system with respect to both uranium and thorium during this period) and the decay ratio of these isotopes will then approach unity. Taking the half-life of ^{230}Th as 7.52×10^4 years, this interval represents a period of about 350 000 years. If, therefore, a sample formed within the last 350 000 years is analysed for rates of decay of ^{238}U and ^{230}Th, an estimate of its age may be calculated from the degree of residual disequilibrium between these two isotopes.

(*d*) ^{210}Pb *in snow and shallow-water sediments*. The ^{238}U decay chain includes an isotope of the gas radon, ^{222}Rn ($T_{1/2} = 3.825$ days). The half-life of this isotope is sufficiently long to permit it to diffuse out of rocks and sediments into the atmosphere. The isotope decays via a series of short-half-life intermediates to ^{210}Pb ($T_{1/2} = 22$ years). This isotope is rapidly washed out of the atmosphere by precipitation (within a few days to one month) and incorporated in permanent snowfields and shallow-water sediments. Since ^{210}Pb

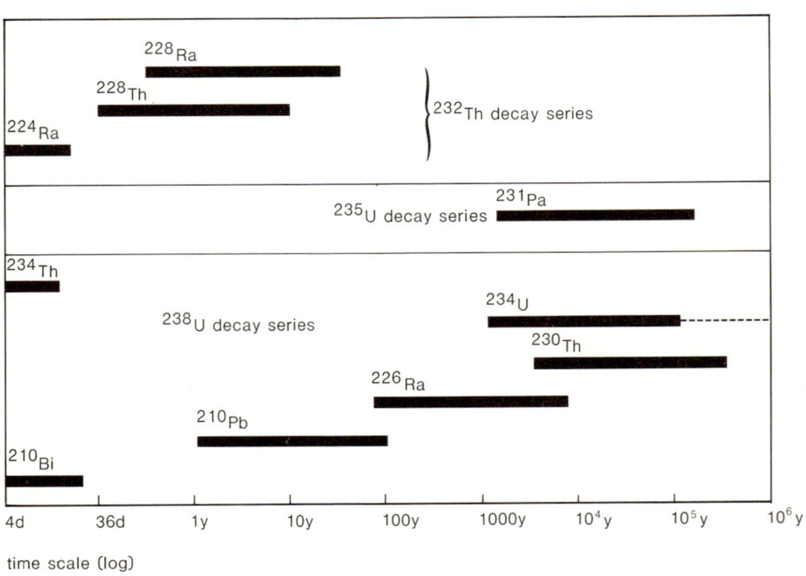

Figure 13.19 Time scales over which individual isotopes in the uranium and thorium decay series can be used for disequilibrium measurements.

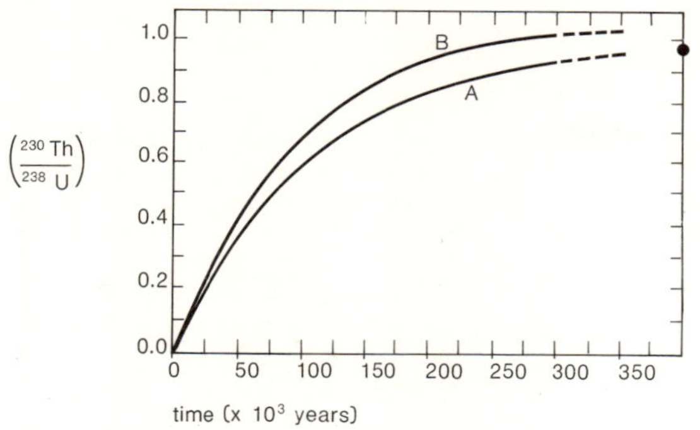

Figure 13.20 Growth of ^{230}Th (formed by radioactive decay of ^{234}U) towards equilibrium with ^{238}U in uranium-bearing minerals that initially contain no ^{230}Th. Curve *A*, the growth of ^{230}Th in a mineral in which ^{234}U is in secular equilibrium with ^{238}U. Curve *B*, the growth of ^{230}Th in a mineral having an initial activity $^{234}U/^{238}U$ of 1.15 (as in sea water). In this case ^{234}U is not initially in secular equilibrium with ^{238}U. (Faure, 1977.)

in these environments is completely unsupported by its parents, measurement of ^{210}Pb activity offers a means of dating such samples.

(*e*) *Application of disequilibrium studies in magmatic systems.* Application of disequilibrium studies has not had as great an impact on igneous systems, where in general timespans of interest substantially exceed one million years and for which alternate dating techniques exist. However, there has been some interest in dating mechanisms occurring within magma chambers by analysing samples derived from recent volcanic eruptions. During fractional crystallization or partial melting processes in a magma chamber, fractionation will occur between uranium and thorium so that these elements will be redistributed within minerals that recrystallize from the magma. Thus, the secular equilibrium between ^{238}U and its daughter ^{230}Th will be broken. Such magmatic processes are *not*, however, known to fractionate isotopes of the same element, ^{238}U and ^{234}U, for example. Some interest has been shown, therefore, in developing the ^{238}U to ^{230}Th geochronometer for magmatic systems. The theoretical basis for this has been reviewed by Allegre and Condomines (1976). This isotope pair forms a conventional isochron equation:

$$\left[\frac{^{230}Th}{^{232}Th}\right] = \left[\frac{^{230}Th}{^{232}Th}\right]_0 e^{-\lambda t} + \left[\frac{^{238}U}{^{232}Th}\right](1 - e^{-\lambda t}) \quad (13.17)$$

where λ is the decay constant of ^{230}Th, t is the time since crystallization and subscript 0 represents the initial ratio of $^{230}Th/^{232}Th$ at the time of crystallization.

In this equation ^{232}Th is selected as the normalizing element (i.e. natural thorium). Since the half-lives of ^{232}Th and ^{238}U are much longer than that of ^{230}Th, radioactive decay of these elements can be ignored.

The term

$$\left[\frac{^{230}Th}{^{232}Th}\right]_0 e^{-\lambda t}$$

represents the decay of ^{230}Th incorporated into the sample at the time of crystallization and unsupported by the presence of ^{234}U or ^{238}U parents.

The term

$$\left[\frac{^{238}U}{^{232}Th}\right](1 - e^{-\lambda t})$$

represents the ^{230}Th formed in the sample since crystallization from radioactive decay of its ^{238}U content.

The derivation of these equations is given in more detail by Faure (1977, pp.285–287). Allegre and Condomines (1976) have shown how isochron data can be obtained by plotting $^{230}Th/^{232}Th$ versus $^{238}U/^{232}Th$, and have interpretated such data for recent volcanic lavas from Costa Rica.

Other isotopes of shorter half-life, including ^{226}Ra, ^{210}Pb, ^{210}Po, ^{228}Ra, ^{228}Th and ^{231}Pa, have been used as tracers to elucidate geochemical mechanisms in magma chambers, for example in studies on recent volcanic lavas from Etna and Stromboli (Capaldi *et al.*, 1976) and Mount St Helens (Bennett *et al.*, 1982). Schwarcz *et al.* (1982) have used disequilibrium measurements in uranium-series isotopes as an indicator of the geological stability of granite plutons to hydrothermal and other alteration processes.

13.9.3 *Schemes of analysis*

Many chemical separation schemes of analysis have been employed in the determination of uranium and thorium series isotopes in a wide range of matrices (silicate rock, sediment, waters). The nuclear decay properties of the more commonly-determined isotopes are listed in Table 13.14. Details are also given in this table of the normal technique for detection (alpha spectrometry, beta-counting or gamma spectrometry). Indeed, the choice of detection technique largely dictates the complexity of preparative chemistry, varying from very simple for gamma spectrometry to a complete chemical separation for alpha spectrometry. The sample preparations associated with these techniques are as follows.

(*a*) *Alpha spectrometry.* This is the technique most commonly associated with the analysis of uranium and thorium-series isotopes, since many (though not all) decay by this mode. In order to carry out quantitative determinations, it is always necessary to carry out a chemical separation, not only to remove matrix elements, but also to chemically separate individual alpha-emitting isotopes. Such procedures must be designed to overcome the following limitations.

(i) The maximum sample thickness which can be tolerated is 100 $\mu g\ cm^{-2}$ (Cherry, 1964). In terms of a specific sample, this corresponds to a maximum thickness of uranium (as the uranyl nitrate hexahydrate, for example, of density 2.807 $g\ cm^{-3}$) of only 36 μm. If samples thicker than this are presented for analysis, significant attenuation of alpha-particles within the sample occurs, and instead of sharp alpha-peaks being recorded, spectra are measured with poor resolution characterized by prominent low-energy tails.

(ii) Not only must alpha-emitting isotopes be separated from the matrix elements, but it is necessary to isolate one element from another, because most elements decay by emitting alpha-particles in the energy range 4 to 6 MeV and serious spectral interferences would occur if uranium and thorium alpha-decay-series isotopes were counted simultaneously, for example. Set against these reservations are

Table 13.14 Radioactive decay schemes and methods of determination of selected uranium and thorium series isotopes

			Radioactive decay scheme							
			Alpha		Beta		Gamma			
Isotope	Decay series	Half-life	MeV	(% decay intensity)	MeV	(% decay intensity)	keV	(% decay intensity)	Method of analysis	Reference
^{238}U		4.468×10^9 y	4.2	(100%)					(i) Alpha spectrometry (ii) Epithermal NAA (iii) Gamma-counting of ^{234}Th daughter	 Goles and Seymour (1984) Kim and Burnett (1983)
^{235}U		0.7038×10^9 y	4.397 4.367 plus others	(57%) (18%)			185.7 143.8 205.3	(54%) (10.5%) (4.7%)	(i) Alpha spectrometry	
^{234}U	^{238}U	2.48×10^5 y	4.717 4.768	(28%) (72%)					(i) Alpha spectrometry	
^{231}Pa	^{235}U	3.248×10^4 y	4.666 to 5.046	(100%)					(i) Alpha spectrometry of ^{215}Po daughter	Schwarcz *et al.* (1982)
^{215}Po		1.8×10^{-3} s	7.360	(100%)						
^{234}Th	^{238}U	24.1 d			0.10 0.19	(35%) (65%)	92.8 63.3	(3.0%) (3.9%)	(i) Beta counting of ^{234m}Pa daughter; sample covered with plastic foil to absorb low-energy beta interference from ^{234}Th	Aller and Cochran (1976) Bhat *et al.* (1969)
^{234m}Pa		1.18 min			0.58 1.50 2.31	(1%) (9%) (90%)				
^{232}Th		14.0×10^9 y	3.948 4.007	(24%) (76%)					(i) Alpha spectrometry (ii) Epithermal NAA	 Goles and Seymour (1984)
^{230}Th	^{238}U	7.52×10^4 y	4.615 4.682	(24%) (76%)					(i) Alpha spectrometry	Allegre and Condomines (1976)
^{228}Th	^{232}Th	1.91 y	5.2 5.338 5.421	(1%) (28%) (71%)					(i) Alpha spectrometry	
^{228}Ra	^{232}Th	6.7 y			0.04	(100%)			(i) Gamma-counting of ^{228}Ac daughter (ii) Beta-counting of ^{228}Ac daughter	Michel *et al.* (1981) Koide and Bruland (1975)
^{228}Ac		6.13 h			1.18 1.76 2.10	(35%) (12%) (12%)	338.4 911.1 968.9 964.6	(12%) (29%) (17.5%) (5.4%)	(iii) Alpha-counting of ^{228}Th daughter	Koide and Bruland (1975)
^{226}Ra	^{238}U	1622 y	4.589 4.777	(5.7%) (94.3%)					(i) Gamma-counting of daughter isotopes ^{214}Pb and ^{214}Bi (ii) Alpha spectrometry of ^{222}Rn daughter extracted as a gas in a scintillation counting cell (iii) Alpha spectrometry	Michel *et al.* (1981) Kim and Burnett (1983) Schwarcz *et al.* (1982) Koide and Bruland (1975)
^{224}Ra	^{232}Th	3.6 d	5.445 5.681	(4.9%) (95%)					(i) Gamma-counting of ^{212}Pb daughter at 239 keV (ii) Alpha spectrometry	Elsinger *et al.* (1982)
^{210}Po	^{238}U	138.4 d	5.305	(100%)					(i) Alpha spectrometry	Koide and Bruland (1975)
^{214}Pb	^{238}U	26.8 min			0.59 0.65	(56%) (44%)	351.9 295.2 241.9	(37%) (19.2%) (7.5%)	(i) Beta-counting (ii) Gamma spectrometry	Hussain and Krishnaswami (1982)

^{212}Pb	^{232}Th	10.6 h	0.33 (80%) 0.57 (12%)	238.6 (43.1%) 77.1 (17.5%) 87.2 (6.3%)	(i) Beta-counting	
^{210}Pb	^{238}U	22.3 y	0.017 (85%) 0.063 (15%)	46.5 (4.0%)	(i) Gamma spectrometry	Kim and Burnett (1983) Gaggeler *et al.* (1976)
					(ii) Beta-counting of ^{210}Bi daughter	Koide and Bruland (1975) Craig *et al.* (1973) Somayajulu and Craig (1976)
^{210}Bi		5.01 d	1.17 (100%)			

min = minute, h = hour, d = day, y = year.
Radioactive decay data from Wilson (1966) and Erdtmann and Soyka (1979).

the advantages offered by alpha spectrometry. Alpha spectrometers are in general insensitive to gamma- and beta-radiation, and background counts are normally negligible (being probably chiefly caused by contamination introduced by the operator). The technique therefore offers exceptionally low detection limits towards the isotopes that can be determined in this way. Furthermore, specific isotopes can be positively identified from their corresponding alpha spectrum, and calibration can be carried out by isotope dilution techniques.

(*b*) *Beta-counting*. Not all isotopes in Table 13.14 decay by alpha-emission. For several, beta-counting has been adopted as an alternative. As has been mentioned above, individual isotopes cannot be identified positively from their beta-spectrum, and chemical separations are therefore essential to isolate the isotope of interest free from any radioelement contamination. Samples to be counted must again be in the form of thin films. The uniformity of this film is not as critical for the more penetrating beta-particles, provided the maximum decay energy exceeds about 0.5 MeV. However, this does pose a problem for two isotopes of interest, namely ^{234}Th ($\beta = 0.10$, 0.19 MeV) and ^{210}Pb ($\beta = 0.017$, 0.063 MeV). The beta-decay energies of these isotopes are so low that severe attenuation is likely to occur, making it impossible to carry out reproducible count measurements.

In the case of ^{234}Th ($T_{1/2} = 24.1$ days) this problem is solved by counting beta-activity of its immediate daughter ^{234m}Pa ($T_{1/2} = 1.18$ minutes). Since ^{234m}Pa has such a short half-life, secular equilibrium is rapidly achieved. Furthermore, the beta-decay energies of this isotope (2.31 MeV (90%), 1.50 MeV (9%), 0.58 MeV (1%) are suitable for beta-counting. In order to avoid a variable interference from the coexisting low-energy beta-decay of ^{234}Th, samples are normally covered with mylar foil (0.9 mg/cm^{-2}). This permits transmission of only the more energetic ^{234m}Pa beta particles (Craig *et al.*, 1973).

The isotope ^{210}Pb can be determined by gamma spectrometry (see below). However, a similar technique to that described for ^{234}Th can also be used. The immediate daughter of ^{210}Pb ($T_{1/2} = 22.3$ years) is ^{210}Bi ($T_{1/2} = 5.01$ days). Providing samples are left for 25 days or so to permit secular equilibrium to be established, ^{210}Pb can be assayed from the beta decay of ^{210}Bi ($\beta = 1.17$ MeV, 100 %). A mylar film is again employed to absorb the very low-energy beta interference from ^{210}Pb.

(*c*) *Gamma spectrometry*. This offers some advantages compared with alpha-and beta-measurements since sample preparation requirements are much less stringent. In some cases direct measurements on untreated powders are possible with no chemical separation, since photopeaks are more than adequately resolved in the corresponding gamma spectrum. Unfortunately, few of the isotopes in the uranium and thorium decay series emit prominent gamma photopeaks and measurements are usually made with far lower sensitivity than can be achieved by alpha-or beta-counting preceded by a chemical separation.

Thus Gaggeler *et al.* (1976) determined ^{210}Pb in lake sediments and air samples by gamma-ray measurement of ^{210}Pb ($\gamma = 46.5$ keV, 4.0 %). Sample aliquots of about 10 g were counted on a planar germanium gamma-ray spectrometer for about three days and detection limits of about 1.7 decays of ^{210}Pb per g of sample were reported with a background of 0.04 counts per minute. This was inferior to data that could be obtained by beta-counting. Kim and Burnett (1983) determined three isotopes in marine phosphorites by gamma-spectrometry:

(i) ^{238}U determined on the assumption that this isotope was in secular equilibrium with its daughter ^{234}Th ($T_{1/2} = 24.1$ d; $\gamma = 63.3$ keV, 3.9%; 92.8 keV, 3.0 %)

(ii) ^{226}Ra ($T_{1/2} = 1622$ y) determined on the assumption that this isotope was in secular equilibrium with two of its daughters: ^{214}Pb ($T_{1/2} = 26.8$ minutes; $\gamma = 242$ keV, 7.5%; 295.2 keV, 19.2% and 351.9 keV, 37.1%) and ^{214}Bi ($T_{1/2} = 19.9$ minutes); $\gamma = 609.3$ keV, 46%

(iii) ^{210}Pb ($T_{1/2} = 22.3$ years) determined directly from $\gamma = 46.5$ keV, 4.0%. These investigators used a 'well-type' intrinsic germanium detector to maximize gamma-counting efficiency. Analytical data compared well with independent analyses by alpha-spectrometry. The technique was particularly useful since the analysed samples contained relatively high uranium contents, permitting adequate precision to be achieved in the measurement of specified photopeaks.

13.9.4 *Chemical separation procedures for alpha spectrometry*

Several schemes for the chemical isolation of alpha-emitting isotopes have been described, including those of Ku (1965), Krishnaswami and Sarin (1976) and Rosholt (1957). Many of these procedures involve a large number of chemical manipulations to ensure purity of the fraction to be analysed. Furthermore, it is necessary to determine both the yield of the chemical separation (that is, the proportion of the total concentration of the isotope recovered in the fraction to be analysed) and the efficiency of the sample–detector configuration (to convert the apparent rate of decay to a total sample decay rate).

In order to satisfy the first of these requirements, an isotope dilution technique is normally used. For the determination of uranium and thorium isotopes, a spike containing ^{232}U ($T_{1/2} = 74$ years) in secular equilibrium with its daughter ^{228}Th ($T_{1/2} = 1.91$ years) is widely used. The isotope ^{232}U does not occur naturally and gives alpha-emissions with energies within the spectrum range of naturally-occurring uranium isotopes. Its daughter ^{228}Th is also an alpha-emitter and coincides with naturally occurring ^{228}Th from the ^{232}Th decay chain for which correction must be made. If it is necessary to determine accurately the ratio ^{228}Th/^{232}Th, as well as to quantify uranium isotope decay ratios, it is desirable to analyse two separate sample aliquots, one with no spike for accurate ratio measurements and one with the spike added for quantification of the uranium and thorium.

A typical scheme of analysis for the measurement of ^{238}U–^{232}Th disequilibrium schematics requires the determination of three isotopes: ^{238}U, ^{232}Th and ^{230}Th. In general, ^{230}Th must be analysed by alpha spectrometry after a chemical separation. There are, however, three methods widely used to determine ^{238}U and ^{232}Th: alpha spectrometry (for example, Ku, 1965; Capaldi *et al.*, 1976); thermal ionization mass spectrometry (e.g. Allegre and Condomine, 1976); and epithermal neutron activation analysis (for example, Goles and Seymour, 1984).

Principles covering the second two options have been covered elsewhere (Chapters 17 and 12 respectively).

(*a*) *Procedure for alpha spectrometry.* A typical alpha spectrometry procedure would be as follows. The mass of the sample taken for analysis is usually about 1 g, but in some cases may be as much as 10–20 g to improve sensitivity. In the analysis of sediments, samples are normally first ashed. Uranium and thorium are then extracted by a simple acid leach using hot 6 N HCl, for example (Aller and Cochran, 1976). For silicate rocks, full sample digestion is required, using 40%HF/nitric acid mixtures with silica removed by volatilization as the tetrafluoride. Any insoluble residue remaining after acid attack must be filtered and fused (e.g. with sodium hydroxide) to ensure complete dissolution of resistant materials that may contain significant concentrations of uranium and thorium.

An ion-exchange separation procedure follows, not only to quantitatively separate uranium and thorium isotopes, but also to remove activity due to ^{226}Ra, ^{231}Pa and ^{210}Po, all of which are alpha-emitters that cause serious overlap interferences in the spectra of U and Th isotopes.

One procedure based on the work of Harmon (1976) is to add the ^{232}U/^{228}Th spike to the combined sample solution from above, equilibrate and to remove some matrix elements by precipitating the hydroxides of Fe, Al, U, Th, Mg and Mn with ammonia. The soluble hydroxides are discarded. The insoluble fraction is redissolved in 9 M HCl and an initial separation made between uranium and thorium by passing the solution through a strong anion exchange resin (AG1-X8) in 9 M HCl (Kraus *et al.* (1956). Uranium is retained on the column, thorium passes through. Uranium is then eluted from the column with 0.1 M HCl. Any residual thorium contamination can be removed from this solution by extraction of an aqueous solution (at pH = 1.2) into a 0.2 M solution of thenoyltrifluoroacetone (4,4,4-trifluoro-1-(2-thienyl)butane-1-3-dione) dissolved in benzene (Poskanser and Foreman, 1961). Under these conditions thorium contamination in the aqueous phase is quantitatively extracted, whereas only a small fraction of uranium (5 %) is removed.

The thorium solution eluted from the first ion exchange is purified by passing this solution through a second column of strong anion exchange resin (AG1-X8) in 10% 5 M nitric acid–90% methanol medium. Any residual uranium contamination passes through the column, thorium is absorbed, and is subsequently eluted with 1 M nitric acid (Tera *et al.*, 1961). The last traces of radioactive contaminants can be removed from thorium by extraction of the aqueous solution at pH = 1.2 into a solution of thenoyl trifluoracetone in benzene. This time, the organic phase containing thorium is reserved for analysis.

As an alternative procedure, Allegre and Condomines (1976) followed a method originally devised by Goldberg and Koide (1958) for the isolation of thorium (^{230}Th). After dissolution with HF/HNO_3, this solution was taken up in HCl media and lanthanum chloride added. Ammonia solution was then added to precipitate insoluble hydroxides, including those of the lanthanum carrier and uranium and thorium. This precipitate was washed with HF. Thorium was quantitatively retained with insoluble LaF_3 in the residue and was purified by mesityl oxide ($(CH_3)_2C$:CH-

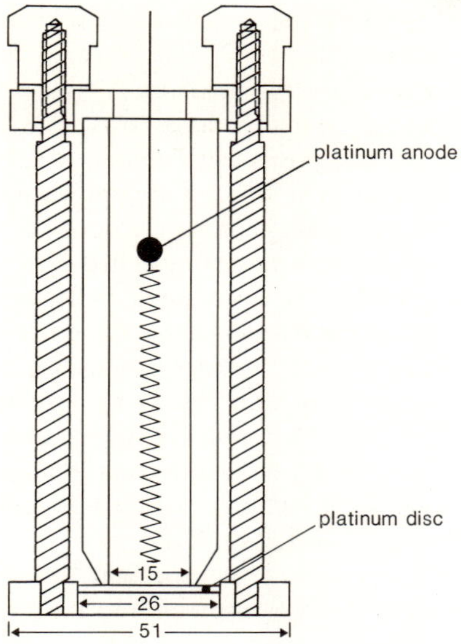

Figure 13.21 Electroplating cell for the deposition of thorium and plutonium isotopes on to counting planchettes. Dimensions in mm. (Krishnaswami and Sarin, 1976.)

CO.CH$_3$) and thenoyl trifluoroacetone/benzene extractions.

(*b*) *Plating*. Two techniques are widely used to prepare the sample as a thin film on a platinum or stainless steel counting planchet (this process is called plating). The simplest is to extract the uranium or thorium fraction into an organic phase (e.g. thenoyl trifluoroacetone/benzene solution) and to pipette this drop by drop on to a warmed planchet. Relatively rapid evaporation of the organic phase will occur, leaving an ultra-thin residue suitable for alpha counting. The procedure for extracting thorium into this organic phase has been described above. Uranium can also be extracted into a 0.2 M solution of thenoyl trifluoroacetone in benzene, simply by adjusting the pH of an aqueous solution of uranium to between pH 3 and 3.5 instead of pH 1.2 for thorium.

More sophistication can be achieved by electroplating radioactive residues directly on to the planchet using a cathodic deposition cell. Krishnaswami and Sarin (1976) described a cell suitable for the electrodeposition of thorium and plutonium for the determination of ^{239}Pu, ^{232}Th and ^{230}Th in marine suspended phases. Koide and Bruland (1975) used the same principle to plate radium (^{224}Ra, ^{226}Ra and ^{223}Ra spike) in the analysis of sea water and sediments. The apparatus used by Krishnaswami and Sarin (1976) is shown in Figure 13.21. A platinum counting disc is positioned at the bottom of a PTFE electroplating cell and forms the cathode. The anode is a platinum wire. In order to plate thorium isotopes, the purified sample was taken up in a mixture of 1 mL of 0.01 M HNO$_3$, 2 mL of 2 M NH$_4$Cl (at pH 2) and 2 mL of saturated ammonium oxalate. Alternatively, plutonium was plated from an electrolyte containing 2 M NH$_4$Cl at pH 2. Plating was started at a current of 0.8–1A, increased to about 1.2–1.4A and terminated when it had fallen to about 0.5 A after 30–40 minutes. Using a differently designed apparatus, Koide and Bruland (1975) electroplated radium isotopes from 0.05 M hydrochloric acid electrolyte solution. In this case, plating currents of 100–150 mA at 40 V were used.

Whichever method of plating is used, it is normal to 'flame' planchets before counting. This involves heating the planchet to dull red heat in a Bunsen flame. If this process is to be repeated, it is preferable to use platinum rather than stainless-steel planchets, to avoid surface oxidation. There are several reasons for this procedure: residual organic matter or volatile ammonium salts used during sample preparation are burnt off, but perhaps of greater importance, any radon gas or ^{210}Po contamination remaining in the plated sample is also volatilized during flaming, thus removing potential alpha spectrum interferences. Some contamination from ^{210}Bi ($T_{1/2}$ = 5.01 days) may also be present in purified fractions. This isotope is a beta-emitter and does not cause a serious spectrum interference. However, it decays to ^{210}Po (see above). Samples should therefore be flamed just prior to analysis to ensure complete volatilization of all ^{210}Po formed by radioactive decay of residual ^{210}Bi.

13.9.6 *Results*

Typical alpha spectra of the uranium and thorium fractions isolated from a rhyolite are shown in Figure 13.22. The position of potential interferences is marked. The uranium spectrum contains alpha emission from the ^{232}U spike. The thorium spectra shows alpha peaks due to the presence of ^{224}Ra, ^{220}Rn, and ^{216}Po contamination. These do not, however, significantly interfere with the determination of ^{232}Th, ^{230}Th or ^{228}Th.

In the determination of the ^{230}Th/^{232}Th ratio for ^{238}U–^{230}Th disequilibrium measurements on recent volcanic rocks, Allegre and Condomines (1976) estimated that the error computed from counting statistics (in alpha spectrometry) is generally 3–5%. Data for the analysis of uranium and thorium in some USGS reference rocks based on alpha spectrometry measurements is shown in Table 13.15, taken from Capaldi *et al.* (1976). Data are in general in good agreement with expected bulk compositions.

To illustrate the use of uranium/thorium decay series isotopes as tracer elements in volcanic systems, some isotope activity ratios taken from the work of Bennett *et al.* (1982) for pyroclastics from the Mount St Helens eruption are shown in Table 13.16. The authors interpreted these data as follows:

(i) If secular equilibrium were maintained throughout the uranium and thorium decay chains with no elemental fractionation, all activity ratios would equal unity

(ii) ^{234}U/^{238}U = 1.00: therefore no fractionation occurs between uranium isotopes during magnetic processes, as expected

(iii) ^{230}Th/^{232}Th = ^{238}U/^{232}Th $\approx$ 1.3: fractionation between uranium and thorium isotopes has not occurred during magmatogenesis or eruption

(iv) ^{226}Ra/^{230}Th = 1.4 (average): significant fractionation has occurred between radium and thorium during magma formation

(v) ^{228}Ra/^{232}Th $\sim$1.00. Consideration of the half-lives of ^{226}Ra ($T_{1/2}$ = 1622 years) and ^{228}Ra ($T_{1/2}$ = 6.7 years) indicates that the magma was formed more than 30 years ago to permit sufficient time for re-equilibration of ^{228}Ra and

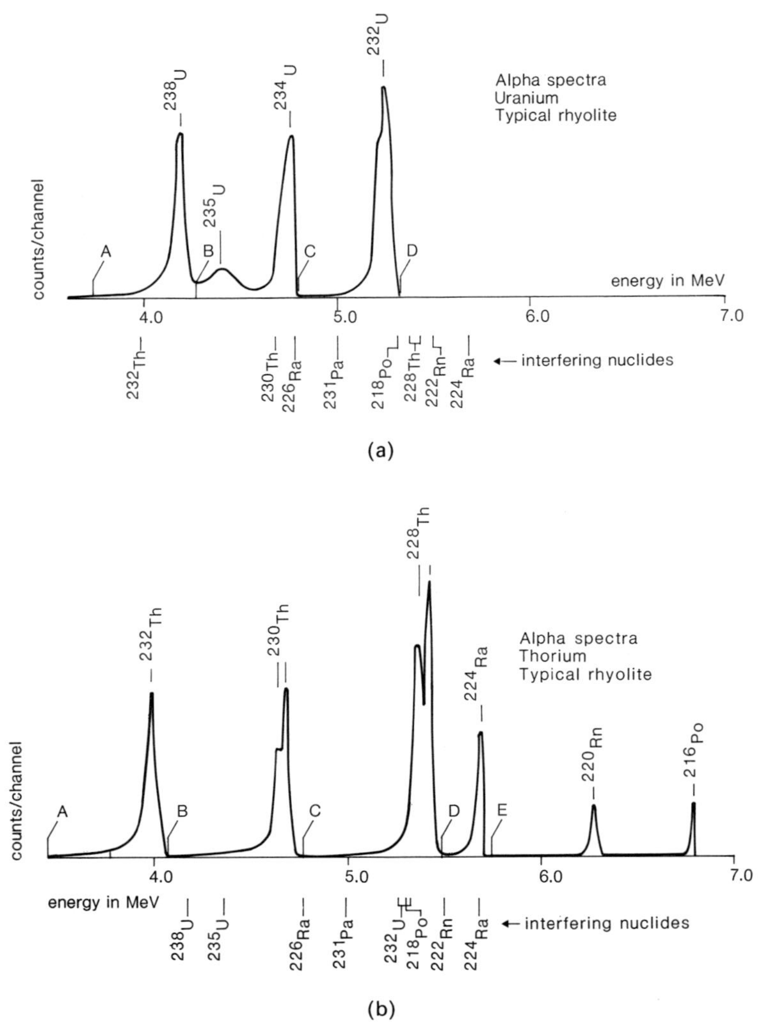

Figure 13.22 Alpha-spectra of (*a*) uranium and (*b*) thorium, after chemical separation from a rhyolite. The energy of potential interfering alpha-emissions is marked. (Harmon, 1976.)

^{232}Th compared with the anomalous ^{228}Ra/^{232}Th ratio (iv) above

(vi) ^{210}Po/^{210}Pb = 0.09: this fractionation appears entirely due to radioactive decay of ^{210}Pb ($T_{1/2}$ = 22.3 years) to ^{210}Po since eruption. Extrapolating data back to the date of eruption indicates that virtually all polonium is lost during the eruption by volatilization: an estimated total of 11–185TBq (300–5000 Ci) from the Mount St Helens eruption alone.

The accuracy of uranium series isotope ratio measurements has been reported by Harmon *et al.* (1979), who presented the results of an intercomparison in the determination of

Table 13.15 Alpha-spectrometry determinations of uranium and thorium in reference materials

Reference sample	Expected U	^{238}U	U (^{234}U equivalent)	U (^{230}Th equivalent)	Expected Th	^{232}Th
USGS: AGV-1	1.95	2.0 (±0.1)	2.3 (±0.1)	2.0 (±0.1)	6.4	6.6 (±0.2)
BCR-1	1.7	1.8 (±0.1)	1.7 (±0.1)	1.9 (±0.1)	6.1	6.4 (±0.2)
GSP-1	2.1	1.9 (±0.1)	1.9 (±0.1)	2.0 (±0.1)	105	102 (±3)

Data from Capaldi *et al.* (1976).
Error limits represent ± one standard deviation.
Uranium data were calculated from ^{234}U and ^{230}Th were derived on the assumption that these isotopes were in secular equilibrium with their ^{238}U parent.

Table 13.16 Uranium and thorium decay series isotope ratios from Mount St Helens pyroclastics

Isotope ratio	Mean activity ratio
$^{234}U/^{238}U$	1.0
$^{238}U/^{232}Th$	1.32
$^{230}Th/^{232}Th$	1.34
$^{226}Ra/^{230}Th$	1.41
$^{228}Th/^{232}Th$	1.02
$^{228}Ra/^{232}Th$	~1
$^{210}Pb/^{226}Ra$	1.00
$^{210}Po/^{210}Pb$	0.09

Data from Bennett *et al.* (1982).

$^{234}U/^{238}U$, $^{230}Th/^{234}U$ and uranium abundances in six natural carbonate reference materials. Fourteen laboratories in five countries participated in this study. Overall, the precision of $^{234}U/^{238}U$ measurements was significantly better than $^{230}Th/^{234}U$ or U abundance determinations. It was suggested that much of this variation was caused by systematic errors in calibration, spike activity, activity ratio determinations and spectral resolution.

Uranium/thorium decay isotope disequilibrium data such as those above give a unique insight into the magmatic behaviour of specific elements, a field of study in which further developments can be expected.

R

14 Ion exchange preconcentration procedures

14.1 Introduction

Ion exchange preconcentration techniques are used to extend the dynamic range of an analytical technique in two main sets of circumstances:

(i) The analysis for elements at concentration levels which would otherwise fall below the detection limit due to unfavourable signal to noise ratios

(ii) The analysis of trace elements where large and variable matrix interferences would otherwise preclude reliable analysis.

There are some groups of elements which can usually only be determined satisfactorily in silicate rocks by ion exchange or some other suitable preconcentration technique; for example gold and the platinum group elements (see also Chapter 15), and the rare-earth elements by spectroscopic methods including atomic absorption and ICP–atomic emission spectrometry. Conversely, some analytical techniques may only be successfully applied to the analysis of silicate rocks if matrix components are removed, usually by ion exchange, prior to analysis (for example, thermal ionization mass spectrometry, radiochemical neutron activation analysis, graphite furnace AAS).

The benefits of ion exchange preconcentration can be illustrated by the following two examples.

(i) In the determination of trace elements by radiochemical neutron activation analysis, detection limits are often restricted by the magnitude of the background gamma spectrum originating from isotopes of elements such as Na, Fe, Co, Sc and Cr. A dramatic improvement in signal to background ratios can be achieved if trace elements of interest are isolated from these gamma-active components in the fraction counted by germanium gamma-ray spectrometers (Chapter 12).

(ii) Isotope ratios cannot be determined by thermal ionization mass spectrometry unless the element to be determined is rigorously purified by ion exchange procedures. If other matrix or trace elements are present in the analysed aliquot, isobaric interferences are likely to be so severe and variable that reliable analysis is not possible.

Although ion exchange chromatography is the most widely used preconcentration technique, it is by no means the only procedure used for the separation of selected elements in silicate rock analysis. In a recent review, Leyden and Wegscheider (1981) considered the evaporation of solvents, electro-deposition, liquid–liquid extraction, surface adsorption and precipitation, as well as ion exchange (including the use of ion exchange impregnated materials and immobilizing reagents). All these techniques have found use in silicate rock analysis, some in highly specialized applications. In this chapter, the general aspects of ion exchange chromatography are considered together with applications to geochemical analysis. More specialized preconcentration techniques are considered under the appropriate chapter headings, as follows:

Fire assay and acid extraction (Chapter 15)

General radiochemical separation procedures (Chapter 12)

Isotope analysis by mass spectrometry for the ratios Rb/Sr, Nd/Sm and Pb/U/Th and selected elements (including REEs) by isotope dilution (Chapter 17).

14.2 Ion exchange techniques

14.2.1 *The technique*

Ion exchange chromatography involves the reversible exchange of ions between a solid phase (that is the ion exchange resin itself) and a mobile phase (in this case a solution of ions). The resin is normally packed into a suitable glass tube, the 'column', held vertically, and the sample solution poured ('loaded') on to the top of this column (Figure 14.1). The sample is then 'eluted' by washing its ionic components through the column using a suitable solvent at a controlled rate. Ions are separated one from another due to differences in their affinity towards the ion exchange resin (Figure 14.2). The more strongly an ion is attracted to the ion exchange resin, the larger the volume of eluent required to wash that ion out of the ion exchange column. In any specific scheme of analysis, columns must be pre-calibrated so that the element of interest may be eluted in a known fraction.

A typical column for the separation of REE for ICP–atomic emission analysis, for example, might be 20 mm in diameter by 100 mm long and contain 20 g of resin (slurry). Elution rates are typically 1 to 2 mL per minute and it may be necessary to pass up to 1000 mL of solvent through the column to elute rare-earth element ions, for example. Use of such large volumes of reagents inevitably causes a small contamination blank. This is not usually significant in techniques designed to determine bulk trace-element concentrations. However, it can become unacceptably large in precise isotope ratio determinations by mass spectrometry. This has led to the development of miniaturized ion exchange procedures which are discussed further in section 17.5.6.

14.2.2 *Ion exchange resins*

The ion exchange resin consists of a solid substrate to which functional groups are attached. In many modern resins, the substrate is made by polymerizing styrene with an appropriate proportion of divinylbenzene (DVB) (Figure 14.3). Styrene alone condenses as an essentially linear polymer. Divinylbenzene must be added to the polymerization mixture to induce cross-linking in the resultant resin. The degree of cross-linking may be varied by changing the proportion of divinylbenzene in the polymer reaction mixture. This

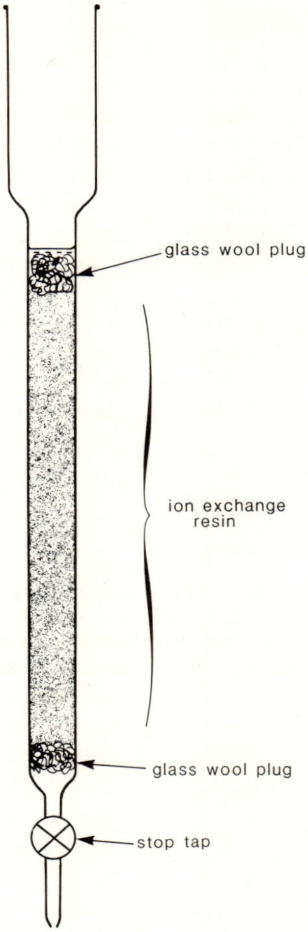

Figure 14.1 Diagram of an ion exchange column.

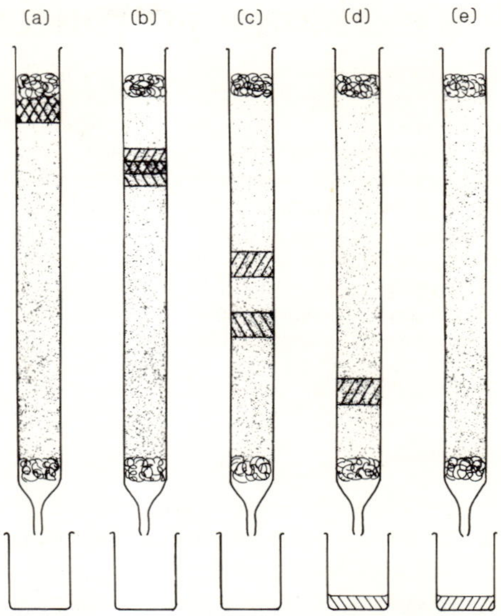

Figure 14.2 An idealized ion exchange separation. A solution containing two species of ions is loaded on to a column (*a*) and eluted successively with four fractions of solvent (*b–e*). After the second aliquot of solvent (*c*) the ions have been resolved into two distinct bands within the column. After the third addition of solvent (*d*), the first band of ions is eluted from the column. The second ionic species is eluted after addition of the fourth aliquot of solvent (*e*).

proportion can be varied in the range 1 to 20%. Most commercial resins contain 8% divinylbenzene. The advantage conferred by cross-linking is that the resin takes on a rigid and dense form which both resists swelling when mixed with water and has a large number of functional groups per unit volume of resin. Too high a degree of cross-linking restricts the pore size within the resin and so reduces the capacity for large ions to diffuse into and out of the resin.

Having formed the resin substrate, functional groups must be added to give the resin its ion exchange properties. A variety of functional groups are used, depending on the properties required of the resin. In the case of the strongly acidic cation exchange resin, the polystyrene substrate is reacted with fuming sulphuric acid. The result of this reaction is that one sulphonate group ($SO_3^-.H^+$) is added to each of the benzene rings. These groups are added at predominantly the meta- and para-positions (Figure 14.3).

The complete ion exchange resin therefore consists of styrene which has been polymerized with a specified proportion (normally 8%) of divinylbenzene. This substrate has an open porous structure (depending on the degree of cross-linking) and is reacted to incorporate a sulphonate group into each of the polymer-benzene rings. During the manufacturing process the polymer is condensed in spherical par-

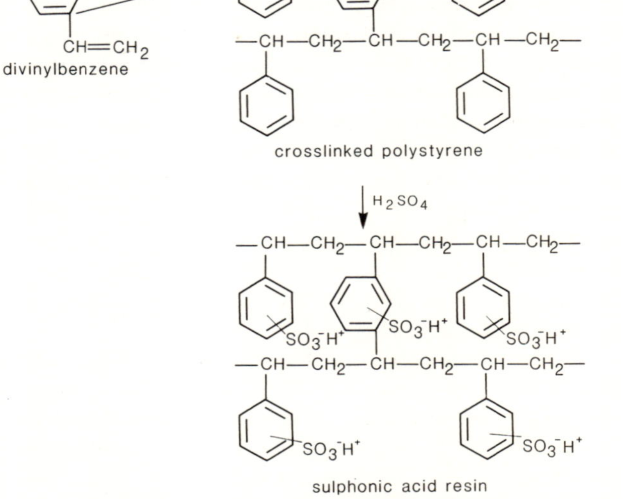

Figure 14.3 The synthesis of cross-linked polystyrene sulphonic acid resin by co-polymerization of styrene and divinylbenzene and subsequent sulphonation. (Grimshaw and Harland, 1975.)

Table 14.1 Mesh size and application of ion exchange resins

Mesh size	Applications
20–30	Suitable for high flow rates, in particular water purification and industrial applications
50–100 (fast flow) 100–200 (slow flow)	Suitable for analytical applications where solutions are to be eluted under the action of gravity
200–400	Except for the smallest columns, this mesh size is too fine to permit gravity elution. Suitable for applications where solutions are pumped through the column under pressure.

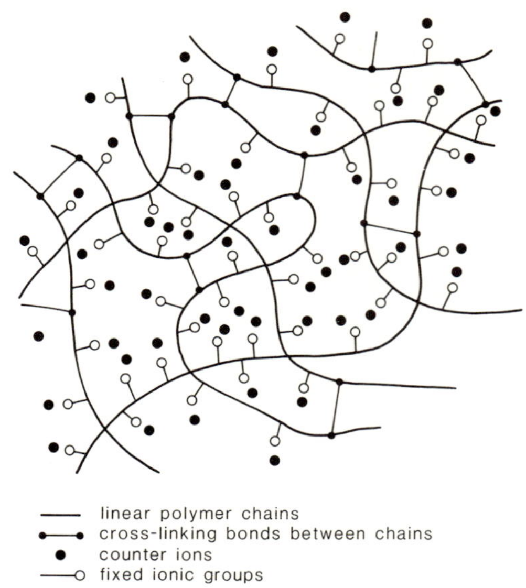

Figure 14.5 Representation of an ion exchange resin gel, based on a random three-dimensional structure of linear polymer chains that are cross-linked at intervals. Open circles represent fixed functional groups. Free solid circles represent counter ions. (Paterson, 1970.)

ticles, the size of which can be controlled ('pearl polymerization' for particular applications (Table 14.1).

Sulphonated polystyrene polymers have some desirable physical properties. They can withstand repeated cycles of drying and wetting and are stable on heating to temperatures of up to 120°C.

Although most resins are made from styrene, some cation exchange resins are manufactured from an alternative material, in particular, by condensing methacrylic acid with the appropriate proportion of divinyl benzene (Figure 14.4) in a manner similar to that described above. This resin substrate is widely used for weakly acidic cation exchange resins in which the functional group is the carboxylic group ($—COO^-.H^+$).

There are five principal categories of ion exchange resin: cation exchange resins, either strongly acidic or weakly acidic; anion exchange resins, strongly basic or weakly basic; and chelating resins. The specific character is controlled by the functional group and the most commonly encountered combinations of substrate, functional group and physical characteristics are listed in Table 14.2.

14.2.3 *Resin structure: macroporous and isoporous resins*

The conventional ion exchange resin is normally termed a 'gel' and has a rather heterogeneous cross-linked structure (Figure 14.5). The size of ion which may diffuse into the resin is restricted by the pore size and this is not usually larger than about 4 nm. However, by changing the conditions of polymerization, different substrate strutures can be formed. *Macroporous* resins are made by adding a solvent such as toluene to the polymerization mixture. The resultant resin has very much larger pore diameters than those in a conventional gel resin, often in the range 20–130 nm. The degree of cross-linking in macroporous resins is still heterogeneously distributed, the large pore spaces being separated by regions of relatively high cross-linking (Figure 14.6). The advantage of the macroporous resin is that ions of much higher molecular weight can diffuse into the resin. However, the exchange capacity is normally lower than the equivalent gel type. By further controls of the cross-linking reaction, isoporous ion exchange resins can be manufactured. These have a very much more uniform cross-linking structure and therefore pore size (Figure 14.6).

14.3 Characteristics of ion exchange resins

14.3.1 *Swelling*

Consider the behaviour of a dry polystyrene sulphonic acid resin on being mixed with water. The resin contains fixed functional groups ($—SO_3^-$) and mobile counter ions (H^+ if in the acid form, Na^+ if in the sodium form) to maintain electrical neutrality. Water molecules diffuse into the porous resin structure. Since water is a polar solvent, it is attracted

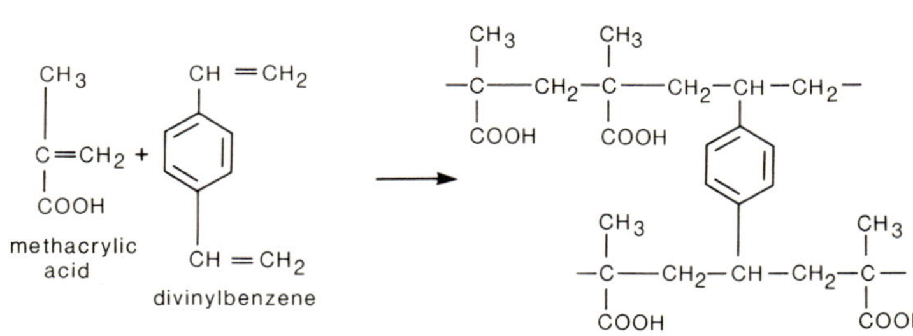

Figure 14.4 The synthesis of polymethacrylate ion exchange resin by co-polymerization of methacrylic acid and divinylbenzene. (Paterson, 1970.)

Table 14.2 Characteristics of ion exchange resins

Resin type	Functional group	Capacity (keq/m^3)	Maximum temperature	pH range	Typical resin matrix
Strongly acidic cation resin	$—SO_3^-.H^+$	1.9	120	0–14	Cross-linked polystyrene
Weakly acidic cation resin	$—COO^-.H^+$	4.1	100	4–14	or cross-linked poly-acrylic acid
Weakly basic anion resin	$—NH_2$, $—NHR$ or $—NR_1R_2$	1.2	70	0–9	Cross-linked polystyrene
Strongly basic anion resin	$—NR_3^+.Cl^-$	1.2	60	0–14	Cross-linked polystyrene
Chelating resins (conventional)	$—CH_2N(CH_2COO^-H^+)_2$	1.2	60		Cross-linked polystyrene

The capacity of a resin to absorb ions is expressed in kilo-equivalents per cubic metre.
$R = —CH_3$, $—CH_2CH_2OH$, etc.
Temperature data in °C.
Data from Grimshaw and Harland (1975).

to electronically charged groups in the resin, causing them to become 'solvated'. Each ionic group is then surrounded by a 'cloud' of water molecules. Effectively, a concentrated ionic solution is formed within the pores of the resin comprising the fixed solvated functional group ($—SO_3^-$) and the loosely-bound solvated counter ion (H^+). Intense osmotic forces are then set up between the bulk solvent and this concentrated ionic solution, inducing more water molecules to diffuse into the resin to reduce the strength of this ionic solution. This causes the resin to swell. The degree of swelling is opposed by the cross-linking which holds the pores together. However, Paterson (1970, p.23) estimates that the internal forces set up within the resin due to this osmotic action may be equivalent to a pressure of as much as 300 atmospheres. The degree of swelling is reduced by:

(i) Substitution of a less polar solvent such as methanol, since the degree of solvation of resin ions is then reduced

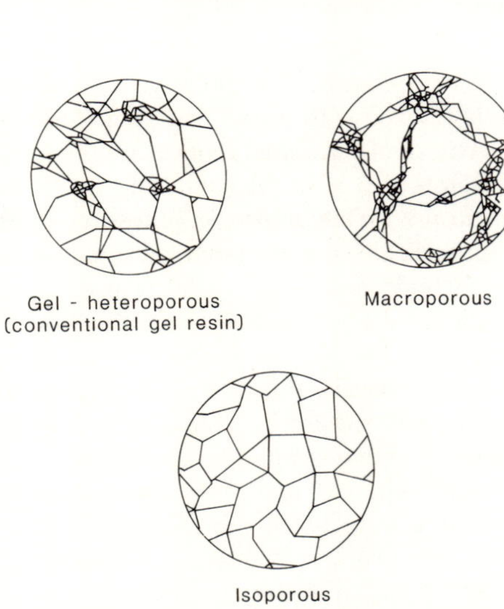

Figure 14.6 The structure of gel, macroporous and isoporous ion exchange resins. (Grimshaw and Harland, 1975; after T.R.E Kressman.)

(ii) An increase in the degree of resin cross-linking, since the pore spaces are then held more rigidly together

(iii) The addition of an external electrolyte to the solvating solution, since this then decreases the osmotic imbalance between bulk solvent and resin pore spaces.

14.3.2 *Ion exchange action: the Donnan potential*

As outlined above, in the dry resin, each fixed functional group (SO_3^-) is associated with a counter ion (say H^+) to preserve overall electronic neutrality. However, on immersion in water the counter ions become mobile, though still associated with the fixed functional group due to electrostatic attraction. Patterson (1970) envisaged the behaviour of ions in this situation to be as indicated in Figure 14.7. Some of the counter ions (H^+) will diffuse from the resin pore into the bulk solution. However, for each counter ion which diffuses out of the resin, an additional uncompensated negative charge remains which makes it more difficult for further counter ions to leave the resin. A dynamic equilibrium is rapidly set up so that the number of counter ions diffusing out equals the number diffusing back into the resin. A small electrial potential is thus set up at the resin surface, and this is called a Donnan membrane potential. Due to the dynamic equilibrium referred to above, opportunities exist for ions other than the original counter ions to diffuse back into the resin and associate with a fixed functional group. The proportion of such ions that participate in this ion exchange process depends on their affinity for the functional group compared with original counter ion. Ions of the same charge as the fixed functional group will show virtually no tendency to exchange, since they will experience a strong electrostatic repulsion on entering the pore space.

14.3.3 *Ion exchange equilibrium: selectivity coefficients*

The equilibrium for a polystyrene sulphonic acid resin (in acid form) and a solution containing the ion A^+ may there-

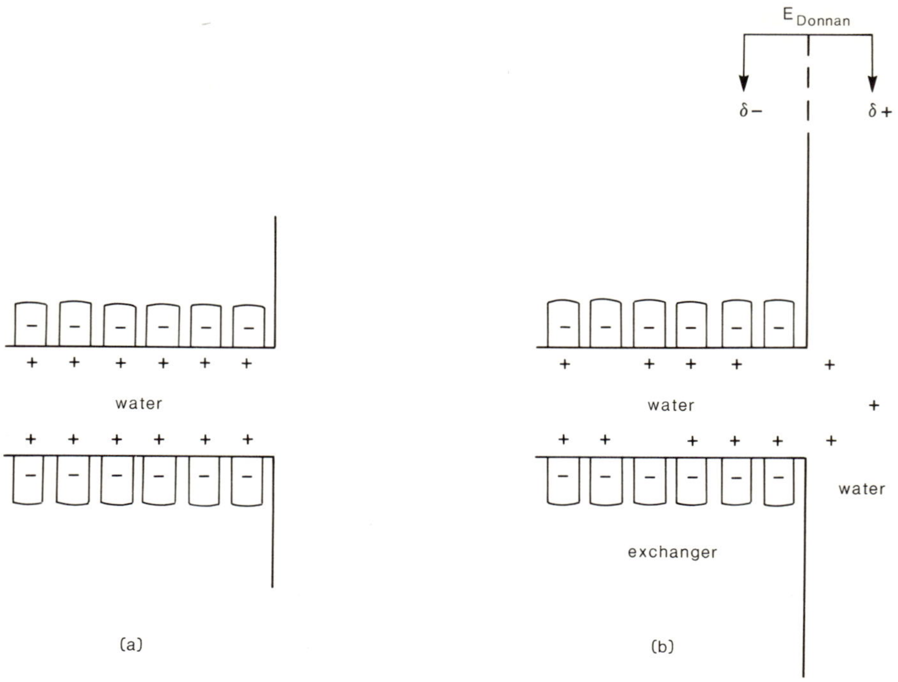

Figure 14.7 Schematic representation of the diffusion of counter ions out of an ion exchange resin on immersion in water showing the Donnan potential created at the resin boundary. (*a*) Idealized 'dry' resin; (*b*) resin after immersion in water. (Paterson, 1970.)

fore be represented by

$$\text{Resin SO}_3{}^-\text{H}^+ + \text{A}^+{}_{\text{aq}} \rightleftharpoons \text{Resin SO}_3{}^-\text{A}^+ + \text{H}^+{}_{\text{aq}} \quad (14.1)$$

where subscript (aq) indicates that the designated ion is solvated in aqueous solution.

From this may be derived an equilibrium constant, K:

$$K = \frac{\{\text{Resin SO}_3^-\text{A}^+\}\{\text{H}^+_{\text{aq}}\}}{\{\text{Resin SO}_3^-\text{H}^+\}\{\text{A}^+_{\text{aq}}\}} \quad (14.2)$$

where { } represents the activities of ions.

In terms of concentrations:

$$K = \frac{[\text{Resin SO}_3^-\ \text{A}^+][\text{H}^+_{\text{aq}}]}{[\text{Resin SO}_3^-\ \text{H}^+][\text{A}^+_{\text{aq}}]} \times \frac{\gamma_{\text{Resin SO}_3^-\text{A}^+}\,\gamma_{\text{H}^+_{aq}}}{\gamma_{\text{Resin SO}_3^-\text{H}^+}\,\gamma_{\text{A}^+_{aq}}} \quad (14.3)$$

where [ ] represents concentration and γ represents the corresponding activity coefficient.

In fact, the activities of ions in the rather concentrated solutions existing within the pores of an ion exchange resin are poorly modelled. To avoid these uncertainties, extensive use is made of a *selectivity coefficient*, K_{A}, based on the concentration terms in the above expression, as follows:

$$K^{\text{H}}_{\text{A}} = \frac{[\text{Resin SO}_3^-\text{A}^+][\text{H}^+_{\text{aq}}]}{[\text{Resin SO}_3^-\ \text{H}^+][\text{A}^+_{\text{aq}}]} \quad (14.3\text{a})$$

Thus if $K^{\text{H}}_{\text{A}} < 1$, the resin shows a preference for retaining H, and if $K^{\text{H}}_{\text{A}} > 1$, the resin shows a preference for exchanging with A^+.

Although the above equations have been formulated as applying to a strong acid cation exchange resin (in acid form), the form of the equation is the same for anion exchange resins. If, however, a participating ion possesses a multiple charge, this must be taken into account. Thus the selectivity coefficient for Fe^{3+} compared with Ca^{2+} may be formulated as follows:

$$K^{\text{Fe}}_{\text{Ca}} = \frac{[\text{Resin SO}_3^-\ \text{Fe}^{3+}]^2[\text{Ca}^{2+}_{\text{aq}}]^3}{[\text{Resin SO}_3^-\ \text{Ca}^{2+}]^3[\text{Fe}^{3+}_{\text{aq}}]^2} \quad (14.4)$$

where $[\text{Resin.SO}_3{}^-.\text{Fe}^{3+}]$ represents the total concentration of Fe^{3+} associated with the resin; this triply-charged ion would be expected to monopolize three $—\text{SO}_3{}^-$ functional groups.

The selectivity coefficient therefore represents a convenient experimental method for measuring the ion exchange properties of a resin.

14.3.4 *Ion selectivity*

The ion selectivity of an ion exchange resin depends on a number of factors.

(*a*) *Electroselectivity*. One of the dominant factors affecting the degree to which an ion is retained on a resin, particularly in the case of strong cationic or anionic resins, is the degree of electrostatic interaction between resin functional group and solvated ion. This depends both on the magnitude of the charge on the ion and the closest distance the ion can approach the functional group. Thus doubly-charged ions will, in general, be more strongly selected by ion exchange resins than singly-charged ions. For a series of similarly-charged ions, the selectivity will decrease with an increase in the size of the solvated ion. Thus for the alkali metal ions, the order of selectivity in relation to a strong cation resin is $\text{Cs}^+ > \text{K}^+ > \text{Na}^+ > \text{Li}^+$. This order is the reverse of what might be expected from the size of individual ions. However, the Li^+ ion has the smallest diameter and therefore the largest solvation shell of water molecules, since the electrostatic field created by this ion in aqueous solution is much greater than that created by Cs^+, for example.

(*b*) *Swelling factor*. A further contribution to ion selectivity arises from the swelling characteristics of a resin. Taking the alkali metals as an example again, a strong cation resin in the lithium form has a significantly larger swelling volume than when in the caesium form. This is again because lithium ions form large solvation clouds of water molecules that tend to expand pore volumes more than the rather more modest Cs^+ solvated ions. Since work must be done in expanding the resin, an exchange in which this work is released is energetically favourable. Thus, for example, swelling factor considerations favour the exchange of solvated Li^+ ions by solvated Cs^+ ions. In this case swelling considerations follow the same trend as electroselectivity factors. However, this is not always so.

(*c*) *The potential for chemical interaction between counter ion and resin*. In the discussion above, it has been assumed that the interaction between solvated ion and resin is purely electrostatic. However, some resins are specifically designed as chelating resins in which chemical bonding can occur between functional group and selected ions. Chelating resins will be considered in more detail in section 14.7. Chemical coordination is not usually a factor in the consideration of conventional cation or anion exchange resins. However it must be taken into account in some circumstances. For example, weakly acidic cation exchange resins use carboxylic acid as the functional group (i.e. Resin.COO^-. H^+). Some transition metal ions are capable of forming coordination complexes with this group, so altering the ion selectivity of the resin compared with that predicted on electroselectivity criteria alone.

(*d*) *Other factors*. These must also be considered in assessing the selectivity of ion exchange resins.

(i) The pore size may restrict diffusion into the resin of some particularly large ions

(ii) The nature of the co-ion affects counter ion selectivity to a small extent

(In consideration of the cation exchange chromatography of a sodium chloride solution, for example using a cation exchange resin (Resin.SO_3^-.H^+), Na^+ and H^+ represent *counter ions* and Cl^-, having the same charge as the functional group is referred to as a *co-ion*)

(iii) The degree of cross-linking affects the magnitude of the swelling factor

(iv) Temperature, pressure and ionic strength of solutions all influence resin selectivities

(v) Some unusually large metal complex anions have higher selectivity coefficients than would be expected on the basis of conventional electrostatic interactions with functional groups. There is some evidence that in these cases the complex anion experiences additional van der Waals' attraction with the resin backbone.

Taking into account all these factors, the approximate order of anion selectivity towards a strongly acidic cation exchange resin is as follows:

$Fe^{3+} > Al^{3+} > Ba^{2+} > Pb^{2+} > Sr^{2+} > Ca^{2+} > Ni^{2+} > Cd^{2+} > Cu^{2+} > Co^{2+} > Zn^{2+} > Mg^{2+} > Mn^{2+} > Be^{2+} > Tl^+ > Ag^+ > Cs^+ > Rb^+ > K^+ > NH_4^+ > Na^+ > H^+ > Li^+$.

Table 14.3 Ion selectivity factors for cation and anion exchange resins

Univalent cations relative to Li		Divalent cations relative to Ca		Univalent anions relative to Cl	
H	1.3	Mg	0.61	F	0.08
Na	2.0	Ca	1.00	Cl	1.00
K	2.9	Sr	1.25	Br	3.5
Rb	3.1	Ba	2.2	I	13.5
Cs	3.2	Zn	0.7	NO_3	3.0
Ag	8.5	Cd	0.8	ClO_4	23.0
Tl	12.5	Pb	1.9	OH	0.5
		Cu	0.75		
		Ni	0.75		

Cation exchange resin: 8% cross-linked polystyrene with sulphonic functional groups.
Anion exchange resin: 8% cross-linked polystyrene with quaternary ammonium functional groups.
Data from Walton and Reyes, 1973.

(These data are taken from Pecsok *et al.*, (1976, p.64) and BDH (1981), the latter data being representative of 0.1N solution of ions in the chloride form at 25°C.) The order of selectivity towards a strong anion exchange resin is expected to be:

citrate^{2-} > SO_4^{2-} > oxalate^{2-} > I^- > HSO_4^- > NO_3^- > Br^- > CN^- > HSO_3^- > NO_2^- > Cl^- > HCO_3^- > $H_2PO_4^-$ > formate$^-$ > acetate$^-$ > F^- $\simeq$ OH^-.

(Data taken from Pecsok *et al.*, (1976, p.64) and Patterson (1970, p.31)).

Although the data above indicate the order of affinity of ions for a resin, it must be emphasized that ion exchange is an equilibrium process and subject to the law of mass action. It is predicted that silver ions will readily displace hydrogen ions from a strong cation exchange resin:

$$[\text{Resin } SO_3^- \ H^+] + Ag^+_{aq} \rightleftharpoons [\text{Resin } SO_3^- \ Ag^+] + H^+_{aq}. \quad (14.5)$$

The hydrogen form can nevertheless always be regenerated by washing the resin with excess dilute acid:

$$[\text{Resin } SO_3^- \ Ag^+] + \text{excess } (H^+_{aq}) \rightleftharpoons [\text{Resin } SO_3^- \ H^+] + Ag^+_{aq}. \quad (14.6)$$

In order to give some numeric indication of the relative affinities of ions, selected values of the selectivity coefficient are listed in Table 14.3.

14.4 Some theoretical aspects of ion exchange

14.4.1 *The theoretical plate model*

The theory of ion migration down a column is based on the principle that when a solution is loaded on to a column, a dynamic equilibrium is rapidly set up such that ions in solution partition themselves between the stationary phase (the ion exchange resin) and the mobile phase (the solvent) according to the distribution law:

$$K = \frac{\text{CONC}_{\text{resin}}}{\text{CONC}_{\text{solution}}} \quad (14.7)$$

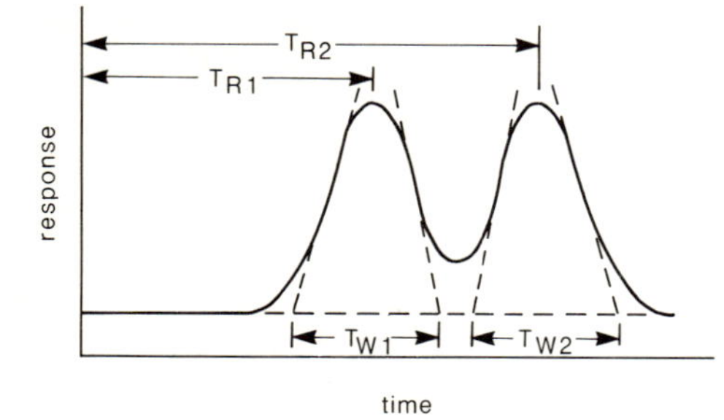

Figure 14.8 The time for elution of two ion species showing the retention time (T_R) and ion peak width (T_W). (After Pecsok *et al.*, 1976).

where CONC refers to the concentration of the ion on the resin or solvent as appropriate and K is the corresponding distribution coefficient.

As additional aliquots of solvent are added, ions are eluted and re-equilibrate with resin further down the column. In order to set this process into some sort of theoretical framework, Martin and Synge (1952) (the originators of modern chromatography) considered that a column could be divided up into a large number of segments called *theoretical plates*. It must be emphasized at the outset that these theoretical plates do not correspond to any physical reality but represent a fictitious concept which permits the efficiency of the chromatographic separation to be measured. A theoretical plate is defined as that length of column which will yield an effluent in equilibrium with the mean concentration over that length in the stationary phase (Ewing, 1975, p.358); that is, the length of resin over which the sample solution must pass to attain effective dynamic equilibrium with the ion exchange resin.

The number of theoretical plates in a column can be measured from the rate of elution of an ion. Consider a small aliquot of sample solution loaded on to the top of a column and then eluted with solvent at a suitable (constant) flow rate. If T_R is the time which must elapse before the ion of interest emerges at the base of the column (i.e. the ion retention time) and T_W is the width of the ion peak at the base of the column, that is the elapse time between first appearance of the ion and complete recovery of the ion at the base of the column (see Figure 14.8), then the number of theoretical plates (n) is given by:

$$n = \left(\frac{4T_R}{T_W}\right)^2. \tag{14.8}$$

These time parameters could equally be measured in terms of the corresponding volumes of solvent which must be passed down the column. If the column has a total length of L units, the height equivalent of one theoretical plate (HETP) is given by

$$\text{HETP} = \frac{L}{n} = L\left(\frac{T_W}{4T_R}\right)^2. \tag{14.9}$$

The height equivalent of a theoretical plate is a measure of the column ion exchange efficiency; the smaller the HETP, the more efficient the column. The effect of the theoretical plate model on a band of ions passing through a column can be judged from Figure 14.9. Here, the distribution of an ion is illustrated after equilibration with 3, 5, 7,...,13

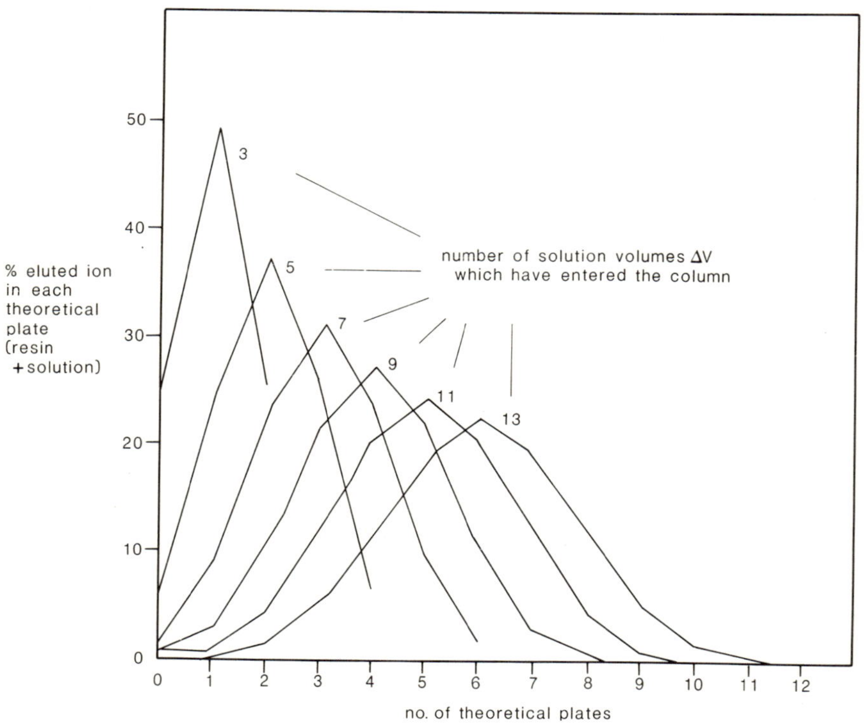

Figure 14.9 Distribution of ions within an ion exchange column after addition of 3, 5, 7, . . . , 13 solution volumes ΔV, assuming unit distribution coefficient. (Paterson, 1970.)

theoretical plates. By the 13th plate, the band of ions has adopted the form of a gaussian distribution when plotted as a function of distance along the column. In practice, columns used for chromatographic separations contain several hundreds to several thousands of theoretical plates.

14.4.2 *Column resolution*

In practical analytical applications, it is not the behaviour of ions of single elements which is of interest but rather the ability of a column to separate (resolve) into individual fractions the components of a multi-element solution. The degree of resolution between two ion species can be determined if their retention times (T_{R_1} and T_{R_2} and peak widths (T_{W_1}) and (T_{W_2}) are known (Figure 14.8). The resolution R is then given by:

$$R = 2\left(\frac{T_{R_2} - T_{R_1}}{T_{W_1} + T_{W_2}}\right) \tag{14.10}$$

The degree of separation achieved between two ions having resolution factors of 0.75, 1.00 and 1.50 is shown in Figure 14.10. It can be seen that a resolution of $R = 1.5$ is required to achieve adequate separation between the two species, provided they are loaded on to the column at equal concentration. Alternatively, the number of theoretical plates required to achieve a specified separation factor is shown in Figure 14.11. These data indicate that between 200 and 300 plates are required to obtain a separation factor of 1.5, for example.

In the development of ion exchange separation procedures, considerable effort must be expended in investigating conditions under which adequate separation is achieved between the ions of interest. It might be considered that the simplest way of achieving adequate separation was to increase the length of column, so increasing the number of theoretical plates and hence the resolving power. Indeed, if the length of a column is doubled, the retention time (and therefore the number of theoretical plates on the column) will also be increased by a factor of 2. However, equation (14.8) indicates that a corresponding increase in the width of ion bands emerging from the column will also occur (by a factor of $\sqrt{2}$) so that the overall improvement in resolution will be $\sqrt{2}$. Unfortunately there are several factors which oppose this improvement, so that at the limit, further increases in column length give no improvement in resolution. These factors are formulated in the van Deemter equation.

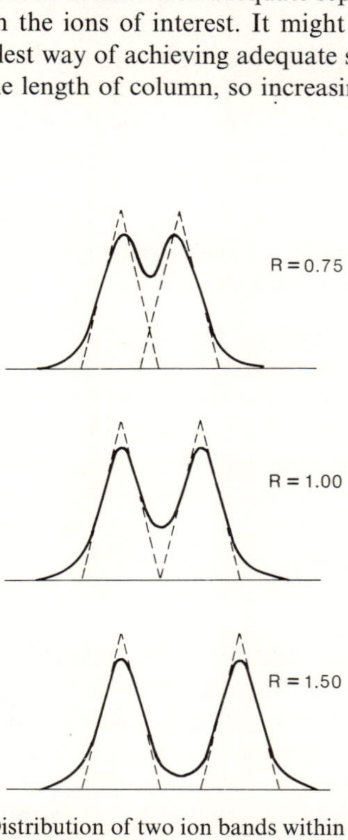

Figure 14.10 Distribution of two ion bands within an ion exchange column for resolution (separation) factors of 0.75, 1.00 and 1.50. The dashed lines show triangular approximations to individual peaks. (Ewing, 1975.)

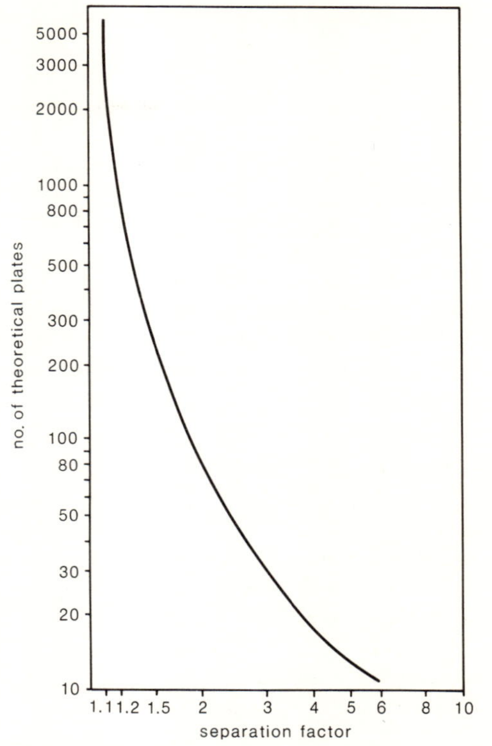

Figure 14.11 The minimum number of theoretical plates required to obtain a 99.9% separation of two solutes, originally loaded on to an ion exchange column at equal concentrations. The separation factor is the ratio of the selectivity coefficients of the two ions (Table 14.3). (Fritz and Schenk, 1969; as reproduced by Betteridge and Hallam, 1972.)

14.4.3 *The van Deemter equation*

The van Deemter equation describes the height equivalent of a theoretical plate (HETP), incorporating factors that allow for the fact that the experimental efficiency of an ion exchange column is never as high as would be expected from simple plate theory. The simplified form of the van Deemter equation is

$$\text{HETP} = A + \frac{B}{v} + Cv \tag{14.11}$$

where A, B and C incorporate several factors but can be regarded as constants for a particular column, and v is the linear velocity of the carrier (that is the flow rate of solvent) through the column.

R*

The non-ideal factors incorporated in this equation are as follows.

(*a*) *Eddy diffusion* (*A*). As ions pass down the column, they will follow random paths. The total apparent column length for each ion will therefore be slightly different, depending on how circuitous is its route down the column. The effect will be minimized for columns that are tightly packed with small particles so that near-laminar flow of solvent occurs. In practice this term is usually small.

(*b*) *Longitudinal diffusion* (*B*). As ions pass down the column, they have a natural tendency to diffuse into areas of lower concentration, inducing a greater spread in the ion concentration. Even with no solvent eluting through the column, this diffusion phenomenon will occur. As this term is inversely related to velocity, the phenomenon is likely to be significant only at very low rates of elution. As rates of diffusion in liquids are about 10^5 times lower than in gases, this term is likely to be much less important in solution than in gas chromatography.

(*c*) *Mass transfer effect* (*C*). This effect results from the small time delay which inevitably occurs in the establishment of equilibrium between solution and resin. At the leading edge of the ion zone the actual concentration ratio between ions partitioned between resin and solution will always be less than *K* (the equilibrium distribution coefficient); that at the trailing edge will always be greater than *K*. This effect will cause the band of ions to spread in both directions.

The general graphical form of the van Deemter equation is shown in Figure 14.12. This demonstrates that an optimum rate of elution exists for any particular column depending on the relative magnitude of the terms *A*, *B* and *C*. Very low elution rates are undesirable since the longitudinal diffusion term *B* then predominates. Fast elution rates are equally undesirable since the mass transfer effect then becomes large. The optimum elution rate is usually found by trial and error. The form of Figure 14.12 indicates that it is preferable to select flow rates slightly higher (rather than lower) than optimum.

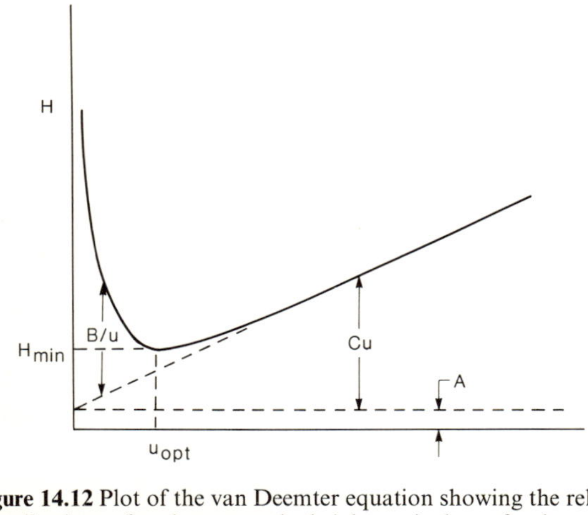

Figure 14.12 Plot of the van Deemter equation showing the relative contributions of each term to the height equivalent of a theoretical plate (*H*) as a function of the mobile phase velocity (*u*). (Pecsok *et al.*, 1976.)

14.5 Optimizing column separations

Data presented above show that several experimental parameters must be optimized in the development of ion exchange separation procedures once the most appropriate ion exchange resin has been chosen.

(*a*) *Diameter of column.* The larger the diameter of column (within reason), the larger the number of theoretical plates equivalents per unit length, the greater the potential separation factor. However, such improvements are limited by factors incorporated in the van Deemter equation as discussed above.

(*b*) *Length of column.* The greater the column length, the larger the number of theoretical plate equivalents present, and the better the potential ion separation factor. Again, the mass transfer effect as described in the van Deemter equation limits the improvement that can be achieved by excessive increases in column length.

(*c*) *Mesh size of resin particles.* Smaller particles permit more uniform packing and reduce the eddy diffusion effect described above. However, if a resin mesh size much smaller than 200 is selected, gravity elution of columns can no longer be carried out.

(*d*) *Composition of eluting solvent.* Large differences in the relative rates of elution of ions are observed when the composition of the eluting solvent is changed. Many schemes of ion separation use dilute acids as eluent. It is, however, common to change the acid strength of eluent in the course of an ion exchange separation ('gradient elution'). Changes in the polar nature of the eluent have a significant effect on ion selectivities, and mixed nitric acid–methanol solvents are popular for some schemes for rare-earth element separations, as described in the examples discussed below.

(*e*) *Eluent flow rate.* As has been discussed above, it is possible to optimize column resolution by selecting the appropriate flow rate (Figure 14.12). However, this parameter is probably much less sensitive in solution applications as compared with gas chromatography, for example.

14.6 Applications of ion exchange chromatography to rare-earth element separations

In order to demonstrate the analytical application of ion exchange chromatography, procedures devised for the group separation of rare-earth elements will now be considered. These procedures illustrate several aspects of ion exchange chromatography and have been used as preconcentration methods for several instrumental techniques.

14.6.1 *Rare-earth element separations using cation exchange chromatography*

There is widespread interest in the separation of rare-earth elements by ion exchange chromatography, mainly because these elements are often present at such low abundance that they cannot be determined without first removing matrix components. Ion exchange often affords the simplest and most rapid separation technique if determinations are to be made by graphite furnace atomic absorption, inductively coupled plasma–atomic emission spectrometry, neutron activation using a radiochemical separation, or mass spectrometry by isotope dilution. In the latter technique, 100% recoveries of REEs are not essential, as samples are 'spiked' prior to separation (see section 17.5). In radiochemical NAA using a post-irradiation separation, it is usual to add rare-earth element carriers to solutions of irradiated samples prior to ion exchange separation to improve the collection of small amounts of radioactive isotopes for subsequent counting. It is then necessary to determine the yield of carrier recovered during the separation so that any loss during the separation procedure can be corrected. The graphite furnace AA and ICP–atomic emission (and equivalent) applications are the most demanding, since the ion exchange procedure must be designed to ensure 100% recoveries of REE to permit confident quantification of data. The following two schemes satisfy this requirement.

14.6.2 *Rare-earth element separation by cation exchange chromatography using hydrochloric acid medium*

Walsh *et al.* (1981) investigated the separation of major and selected trace elements using a strong cation exchange resin (Dowex AG 50W–X8, 200–400 mesh) in a procedure designed for the determination of the rare-earth elements by ICP–AES. The ion exchange separation followed the earlier work of Strelow and Jackson (1974). Walsh *et al.* digested 0.5 g aliquots of rock powder using an HF/perchloric acid mixture. Silica was removed by fuming as silica tetrafluoride and the residue taken up in 25% HC1. Any undissolved residue (consisting of zircon and other acid-resistant minerals) was filtered off and fused with NaOH or KHF_2 at 800–850°C. After cooling the melt was dissolved in 25% HC1 and combined with the main fraction. Resultant sample solutions were diluted to give an acid strength of less than 10% HC1.

Chromatographic separations were carried out in glass columns of 20 mm ID containing 20 g of resin slurry which gave a settled resin height of 100 mm. The resin was preconditioned by passing 4N HCl solution and then 1N HCl solution through the column. The sample solution was then loaded on to the column and eluted with 400 mL 1.7N HCl. Tests indicated that all the major elements and selected trace elements were eluted from the column during this procedure. In Figure 14.13, the fractions of elements eluted in successive 100 mL aliquots of 1.7N HC1 eluent are plotted. Under the experimental conditions described here, the order of decreasing retention time on the column is estimated to be: REEs > Ba > Sr > Cr > Ca > Ni > Co > Fe > Mn > Mg > Cu > V > P > K > Na > Al > Ti. It is interesting to note that chromium appears to give two elution maxima (Figure 14.13), possibly indicating the presence of more than one chromium species in solution.

Following first elution with 400 mL 1.7N HCl, the rare-earth elements were quantitatively retained on the column together with all Ba and some Sr, Hf and Zr. These components were then eluted from the column with 500 mL of 4N HCl. This solution was filtered to remove any organic resin particles carried over into the effluent and then evaporated to dryness. The residue was taken up in 5 mL of 10% HCl for analysis by ICP–AES. Recoveries of REEs by this procedure were shown to be quantitative.

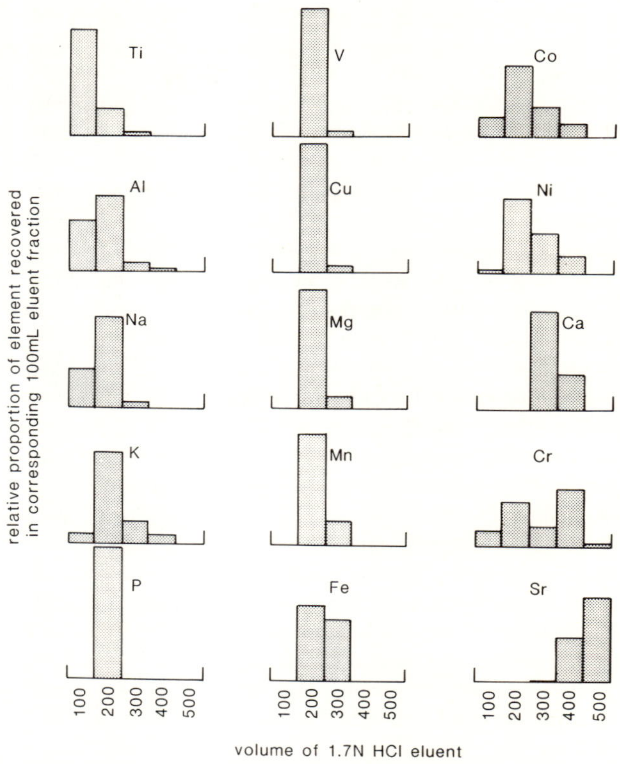

Figure 14.13 Relative proportion of selected elements eluted in successive 100 mL fractions of 1.7 N HCl from a Dowex AG 50W–X8 cation exchange column. (Data plotted from Walsh *et al.*, 1981.)

14.6.3 *Comparison of cation exchange separation of REEs using nitric and hydrochloric acid media*

A further study of the elution properties of cation exchange resins has been carried out by Crock *et al.* (1984), using both hydrochloric acid and nitric acid media. The aims of this study were to characterize more completely the behaviour of selected major and trace element ions to permit the quantitative recovery of the REEs and Y as a group. The authors used ion exchange separation procedures to take advantage of the following characteristics:

(i) To permit the REEs from a large sample to be concentrated in a small volume to allow maximum sensitivity, in their determination by ICP–AES

(ii) To reduce the total dissolved salt content in solutions prepared from samples which may have been fused with a flux

(iii) To remove elements which would otherwise cause spectral interferences in the ICP–AES method used for analysis.

The ion exchange resin used for this work was a strong cation resin (BioRad AG 50-X8), 100–200 mesh, prepared as a column 10 mm internal diameter by 200 mm long. Samples of up to 2 g weight were dissolved by acid digestion, silica being volatilized by fuming with HF. If insoluble accessory phases such as zircon, sphene, tourmaline, or monazite were present, samples were fused with lithium metaborate (0.2 g of sample, 0.6 g lithium metaborate) in a carbon crucible at 1050°C for 20 min). All samples were subsequently taken up in 1N nitric or hydrochloric acids (lithium metaborate residues would only dissolve satisfactorily in nitric acid).

Two parallel sets of experiments were carried out in which rock solutions were eluted in hydrochloric and nitric acid media using gradient elution techniques. That is, the concentration of successive aliquots of eluent was increased in acid strength. The schemes of elution were as below:

Nitric acid elutions

(i) Load 50 mL sample solution in 1N nitric acid on to a column that had previously been equilibrated with 1N nitric acid

(ii) Elute with 130 mL 2N nitric acid

(iii) Elute with 50 mL 6N nitric acid

(iv) Elute with 50 mL 8N nitric acid; this fraction was used for REE analysis

(v) Wash column for re-use by eluting with 100 mL 8N nitric acid.

Hydrochloric acid elutions

(i) Load 50 mL sample solution in 1N hydrochloric acid on to a column that has previously been equilibrated with 1N hydrochloric acid

(ii) Elute with 100 mL 2N hydrochloric acid

(iii) Elute with 160 mL 8N hydrochloric acid; this fraction was used for REE analysis

(iv) Wash column for re-use by eluting with 250 mL 8N hydrochloric acid.

In all cases flow rates through the column were adjusted to about 1 mL/min. The behaviour of selected major and trace elements during this elution procedure is shown in Figure 14.14. The features of these data are as follows.

(*a*) *Hydrochloric acid elution scheme.* Most major and selected trace elements are completely eluted by 100 mL 2N HCl. However, significant concentrations of Ca, Sr, Zr, and all Ba are carried over in the REE fraction. The light REEs (La and Ce) are not as effectively removed from the column as would be desired during final elution with 8N HCl.

(*b*) *Nitric acid elution scheme.* Most major and selected trace elements are effectively eluted by 130 mL 2N HNO_3, including Ba, Sr and Ca. However, a large proportion of Fe and some Zr is eluted in the REE fraction. The REEs are effectively eluted with 6N and 8N HNO_3 the light REEs being

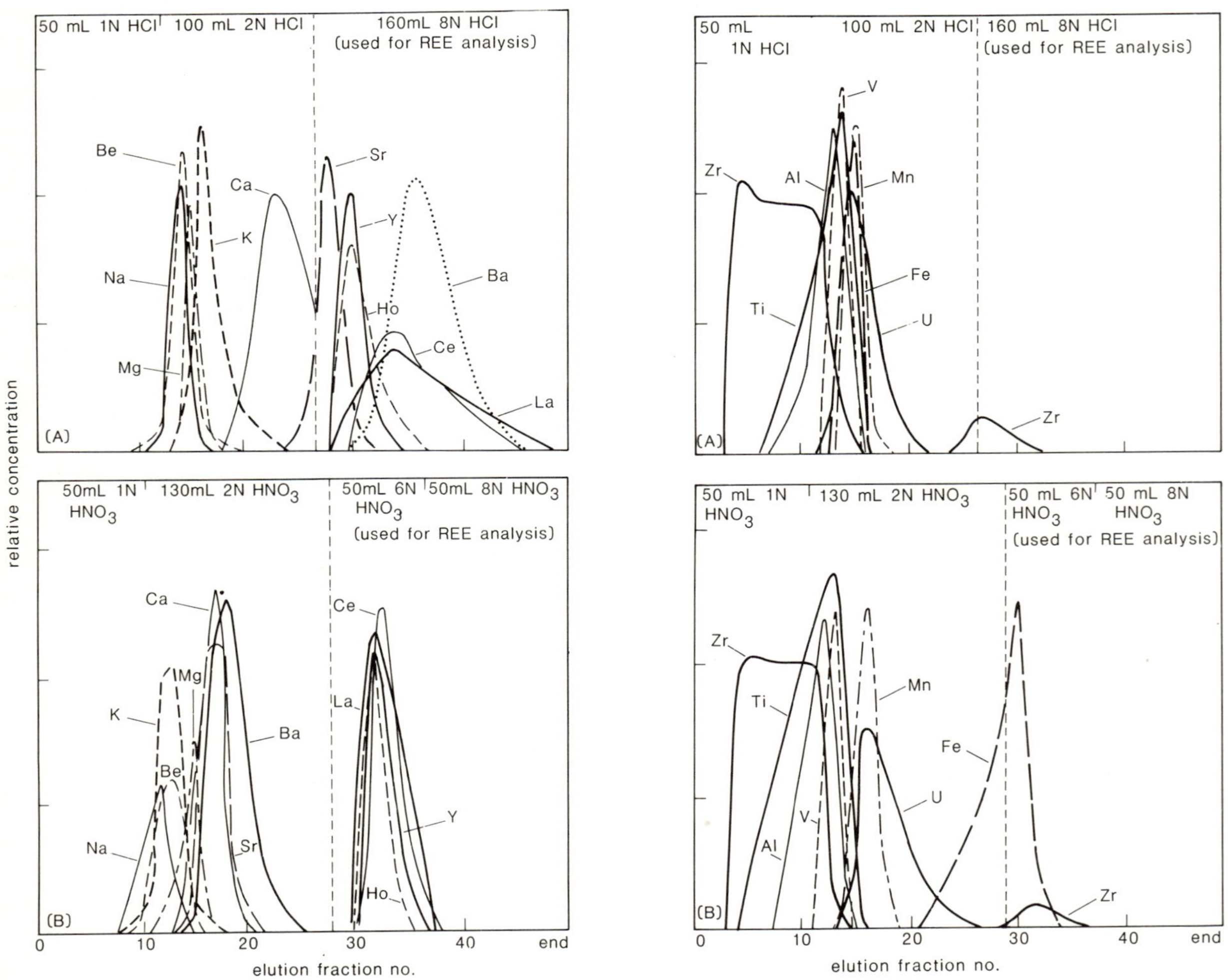

Figure 14.14 Elution of selected elements from a BioRad AG50-X8 cation exchange column using (*A*) hydrochloric acid and (*B*) nitric acid media. In each case gradient elutions were used (as indicated on the diagrams). Data show the relative concentrations of the appropriate element from an acid decomposition of BCR-1 found in successive fractions eluted from the column, each fraction having a volume of about 6 mL. (Crock *et al.*, 1984.)

removed as effectively as the heavy. Crock *et al.* suggest that the behaviour of iron can be explained as follows. In hydrochloric acid solution, iron is present as a chloro-iron (III) *anion* (e.g. $Fe(Cl)_4^-$) and is rapidly washed through the cation exchange column. However in dilute nitric acid iron is present as a hydrated Fe(III) *cation* and is retained on the cation exchange column sufficiently strongly to be eluted with the REE. As iron causes many serious spectral interferences in ICP–AES, a simple anion exchange separation was used by Crock *et al.* following the nitric acid elutions so that iron could be removed as the chloride complex from the REE fraction.

In assessing the two schemes of separation, Crock *et al.* preferred the cation separation of the REEs in nitric acid media, both because of the effective separation of REEs from Ca, Sr and Ba, and also because the REEs were more efficiently removed from the columns as a discrete group.

14.6.4 *Mixed solvent elution of REEs using an anion exchange resin*

There has been considerable interest in the separation of the rare-earth elements by anion exchange chromatography using *mixed solvent elution.* Two systems have been studied in some detail, the nitric acid–glacial acetic acid system (Korkisch and Arrhenius 1964) and the nitric acid–methanol system (Korkisch and Tera, 1961; Faris and Warton, 1962; Desai *et al.*, 1964; and Brunfelt and Steinnes, 1969). In dilute nitric acid solution alone, the REEs show no anion exchange properties at all. However, when dissolved in nitric acid mixtures containing excess methanol or glacial acetic acid, negatively charged REE ions are formed (presumably REE-methoxy ($REE(OCH_3)_x^{(x-3)}$) and REE-acetate ($REE(OOCCH_3)_x^{(x-3)}$) complexes). These ions are not stable in dilute aqueous solution, but these mixed solvent media offer an efficient method of separating the REEs.

(*a*) *The methanol–nitric acid system.* This has been studied in some detail by Roelandts *et al.* (1974) using eluent compositions ranging from 95% methanol–5% 7M nitric acid to 50% methanol–50% 7M nitric acid. None of the major elements investigated (Na^+, Mg^{2+}, Al^{3+}, K^+, Ca^{2+}, Mn^{2+}, Fe^{3+} and also Sc^{3+}) except Ca^{2+} showed any anion exchange properties. Barium was not studied. Roelandts *et al.* (1974) calculated the average separation factors of the REEs relative to calcium, and their data are listed in Table 14.4. These results indicate that the light rare-earth elements La–Nd are much more strongly retained on the anion column than the heavier REE elements. Brunfelt *et al.* (1974) devised a scheme for separating the REE into three fractions using methanol–nitric acid anion exchange chromatography: Lu and Yb were eluted with 15% 7.8M nitric acid–85% methanol; Tb–Sm were eluted with 15% 7.8M nitric acid–55% methanol and Nd–Ce and La were eluted with water).

Table 14.4 Average ion exchange separation factors of rare-earth element ions, relative to calcium, in nitric acid–methanol media

La	Ce	Nd	Sm	Eu	Gd	Tb	Dy	Ho–Lu
1700	1400	350	25	23	17	15	14	<10

Data from Roelandts *et al.* (1974).
The separation factor is calculated as K_d(REE)/K_d(calcium), where K_d is the ion exchange distribution coefficient and data has been averaged for elution mixtures ranging from 95% methanol—5% 7M nitric acid to 85% methanol—15% 7M nitric acid.

(*b*) *Glacial acetic acid–nitric acid/methanol–nitric acid elutions.* Perhaps the most popular use of mixed solvent anion exchange has been in the separation of the REEs for analysis by isotope dilution mass spectrometry. Hooker *et al.* (1975) investigated a two-stage separation technique using both the glacial acetic acid–nitric acid and methanol–nitric acid elution techniques. The three objectives to this method of development were:

(i) To remove all matrix elements and especially barium, to eliminate isobaric interferences from the REE mass spectrum

(ii) To keep the volumes of the reagents required for the procedure as low as possible to minimize reagent blank interferences

(iii) To split the rare-earth elements into three fractions, La, Ce–Gd and Dy–Yb; both for operational convenience and to eliminate some inter-REE isobaric mass interferences.

Since determinations were made by isotope dilution in which REE spike solutions were equilibrated with the rock sample during digestion, it was not essential that the ion exchange procedure should achieve quantitative recoveries. Rock powders (200 mg) were mixed with spike solutions and decomposed with HF/HNO_3. Reaction mixtures were then equilibrated with 6M HCl in a PTFE bomb at 130°C. After evaporating to dryness, the REEs were leached from the solid residue with an 80% glacial acetic acid–20% 5M nitric acid solution (5 mL) in a bomb at 140°C for 12 hours.

The first ion exchange separation was carried out on a strong anion exchange column (Bio-Rad AG-1X8, 200–400 mesh), prepared as a resin bed of 10 mm internal diameter ×50 mm long, containing about 2.5 g dry weight resin. This was washed with 1M nitric acid and pre-equilibrated for a few hours with 90% glacial acetic acid–10% nitric acid. The leached REE solution was then loaded on to this column and eluted with a total of 37 mL of the 90% glacial acetic acid–10% 5M nitric acid mixture at a flow rate of 0.1 to 0.3 mL/min. Hooker *et al.* estimate that 90% of the REEs are retained on this column but more than 99% of the Ba and other unwanted cations are removed in the eluent. The REEs are then eluted using 0.05M nitric acid solution (8 mL). If iron or barium contamination of this solution is suspected, the REE solution can be passed through the column a second time. Having separated the REEs as a group, Hooker *et al.* then further separated these elements into three groups using the same anion resin made up as a smaller column of about 5 mm diameter by 40 mm long, containing about 0.34 g (dry weight) of resin. This column was preconditioned with 3 mL of a 90% methanol–10% 5.25M nitric acid mixture. The rare-earth elements were taken up into about 1 mL of the same solution mixture and loaded on to the column. Columns were then eluted at a rate of about 0.1 mL/min as follows:

(i) 2 mL of 90% methanol/10% nitric acid solution to remove any residual Fe or Ti from the REE fraction–this effluent was discarded

(ii) 4.5 mL of same mixed solvent eluent to remove Dy to Yb

(iii) 16 mL of 90% methanol–10% 0.01M nitric acid to collect Gd to Ce plus about 50% La.

(iv) 1 mL of 1M nitric acid to remove residual La.

These solutions were then analysed by conventional thermal ionization mass spectrometry using procedures which are discussed in more detail in section 17.6.

Further use was made of the methanol–nitric acid elution procedure in anion exchange chromatography by Roelandts (1981) to determine the light rare-earth elements in apatite by x-ray fluorescence spectrometry (see section 12.17.4).

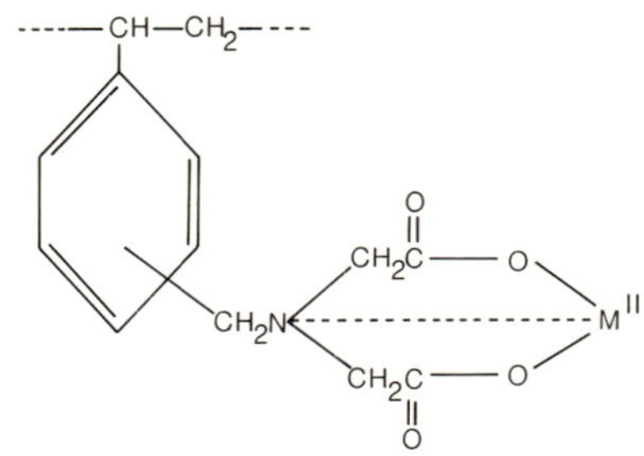

Iminodiacetate chelate - Dowex A-1 resin

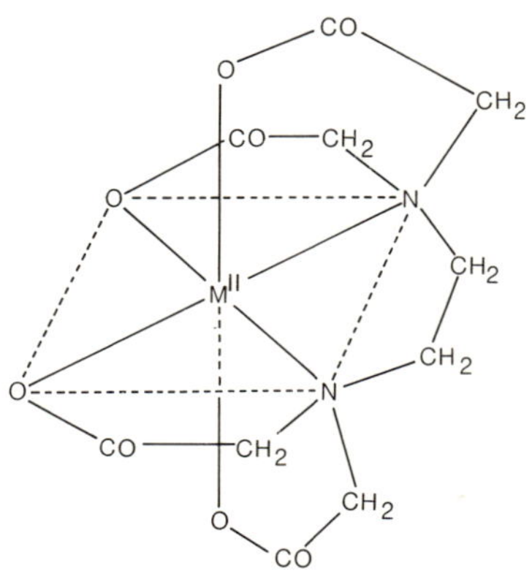

EDTA chelate (a suggested structure)

Figure 14.15 Structure of an iminodiacetate chelating ion exchange resin (Dowex A–1) compared with the chelating properties of EDTA. M represents a divalent metal ion. (Grimshaw and Harland, 1975.)

14.7 Chelating ion exchange resins

Chelating ion exchange resins do not work in the same way as conventional cationic or anionic resins. The selectivity of the latter is based mainly on the electrostatic interaction between solvated ions in solution and electronegative or electropositive functional groups attached to the resin. In contrast, the selectivity of a chelating ion exchange resins depends on the potential for complex formation between an ion in solution and the appropriate functional group.

14.7.1 *Chelating resins based on the iminodiacetate functional group*

One of the reagents which had an important impact on the development of rapid methods of analysis was EDTA (ethylenediamine tetra-acetic acid). This reagent will selectively complex particular ions in solution, especially those of transition metal elements (section 2.7). An analogous development in ion exchange chromatography was the synthesis of a cross-linked polystyrene resin in which half an EDTA molecule—the iminodiacetate group—was attached as the functional group on each benzene ring of the resin polymer. Not only does such a resin demonstrate some of the properties of a weakly acidic cation exchange resin (possessing two carboxylic groups within each functional site), but can also selectively complex ions in a manner analogous to that of EDTA (Figure 14.15). As might be expected from the properties of EDTA, chelating resins of this type are particularly suitable for complexing transition metal ions. However, possibly because alternative (instrumental) techniques are generally available for the determination of these elements in geological samples, preconcentration procedures using resins with the imino diacetate functional group have not found widespread application. Nevertheless, Hirose *et al.* (1978) described a technique for the preconcentration of rare-earth elements and selected heavy metals in river water using a chelating ion exchange resin (Chelex 100). A separation procedure was devised to obtain quantitative recoveries of the rare-earth elements Mn, Cu, and Zn, Co and Cd. The procedure was incorporated into a scheme of neutron activation analysis in which the elements of interest were preconcentated on the resin from natural river water. Sodium was then eluted from the column with ammonium acetate solution (0.1M) and the resin irradiated. Subsequently the rare-earth elements were eluted with hot sodium carbonate solution (1M) and the heavy elements with nitric acid (2M) so that they could be assayed separately.

14.7.2 *Other chelating resins*

Other more specific ion exchange resins have found geological application. *Srafion NMRR* is styrene-divinyl benzene copolymer to which is attached the guanidine functional group $—N{=}C(NH_2)_2$. This resin selectively complexes with ions with d^8 electronic configurations that are capable of co-ordination in a square planar configuration. It therefore has high specificity towards gold and the platinum group elements, and the use of this resin in preconcentration procedures in noble metal analysis is discussed in more detail in section 15.4.

A second resin which shows great promise in analytical applications is *poly(dithiocarbamate)*. This resin is synthesized from polyethylenimine (PE1–18) and polymethylene polyphenyl isocyanate, as described by Hackett and Siggia (1977). The functional group is the dithiocarbamate ligand:

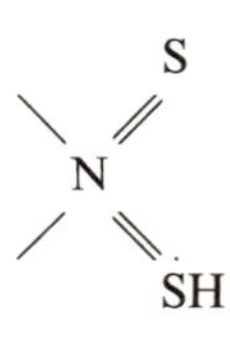

Both the sulphur and nitrogen atoms can participate in co-ordination complex formation. This resin has some very interesting properties for geological applications and these have been invesigated by Miyazaki and Barnes (1981). These workers demonstated that the resin quantitatively retains a wide range of heavier elements including the rare-earth elements, Cr(VI), V(IV), W(VI), Mo(VI) and Os. However the alkali and alkaline earth elements do not complex with this resin. The selectivity shown towards some elements depends on oxidation state: Cr(VI) and Fe(III) are retained by the resin, for example, whereas Cr (III) and Fe(II) are not. Mahanti and Barnes (1983) described the use of this resin to preconcentrate trace elements in coal for subsequent determination by ICP–AES. The pH of batches of sample solutions was adjusted to achieve optimum recovery of trace elements as follows: Cu, Fe, Mo, Ti, V, Ni, Co at pH 5; U, Th at pH 6; Ce and La at pH 8; Mn and Zn at pH 9. Following retention of selected ions on the column (and elution of interfering ions), the entire column was digested and taken up into solution with hydrogen peroxide (10%) and concentrated nitric acid. The final volume of this dissolved resin solution was 2 to 10 mL, so that overall concentration factor relative to the original sample solution was in the range 12 to 60. This procedure has several attractive features. Since it is not necessary to elute analytes from the resin (the resin itself being digested) uncertainties in the volume of eluent required to achieve 100% recovery are eliminated. Reagent blanks associated with large volumes of eluent are minimized. Analysis times are reduced. It seems likely that polydithiocarbamate has further potential applications in silicate rock analysis, particularly since the alkali and alkaline earth elements are not retained by this resin. Potential uses include the reduction in the total dissolved salt content of solutions, in particular of those samples which have been fused with lithium or sodium salts and are to be analysed by ICP–AES or ICP–MS techniques where clogging of the nebulizer/torch assembly could otherwise result. Further applications can be envisaged in radiochemical neutron activation procedures, particularly where it is necessary to remove sodium from analysed samples so that the gamma-ray background enhancement caused by ^{24}Na in samples counted within 2 weeks of irradiation can be minimized.

14.8 Other preconcentration procedures

14.8.1 *Absorption of REEs on lanthanum oxalate*

Radiochemical neutron activation analysis usually involves irradiating a rock powder in a nuclear reactor. After a suitable decay period, a group separation is carried out to remove unwanted (and analytically detrimental) gamma-ray activity which may be caused by the matrix and some minor elements. Examples of group separation schemes, often based on conventional ion exchange chromatography, have been discussed in section 12.17. An interesting variation with potential for wider analytical exploitation is reported by Csajka (1973), who found that when a rock power is irradiated and subsequently taken up into solution, all the rare-earth element activity could be rapidly and quantitatively removed from solution on to a prepared precipitate of lanthanum oxalate. The rock powder (5 to 10 mg) was irradiated and then digested by HF/perchloric acid attack, the final solution being taken up in dilute hydrochloric acid and diluted to 11 mL (0.05N HCl). This was then mixed with 150 mg of lanthanum oxalate precipitate. Quantitative recoveries of activated rare-earth element isotopes were reported on the lanthanum oxalate precipitate, the only other isotope detected being scandium. Separation factors of 10^4 were achieved from ^{56}Mn and ^{24}Na, isotopes which would otherwise considerably enhance background gamma-ray measurements for samples counted within 1 to 7 days of irradiation. Akillmali *et al.* (1979) studied absorption of activated REE ions on cerium oxalate in a similar role.

14.8.2 *Column cementation*

A novel preconcentration technique has been described by Welch and Ure (1980) and applied to the determination by spark source mass spectrometry of copper, lead and the noble metals by Fu *et al.* (1983) The principle of this technique is that when a solution is passed through a column of metallic aluminium powder, ions which are more electropositive than aluminium are reduced to the metallic state and deposited on the aluminium column by reactions such as, for example

$$3Cu^{2+} + 2Al^0 \rightarrow 3Cu^0 + 2Al^{3+} \qquad (14.12)$$

For determination by spark source mass spectrometry, the aluminium powder loaded with the elements of interest can be pressed into an electrode suitable for spark excitation without further treatment. The column (3 mm diameter × 5.3 mm long) was made up of 50 mg aluminium powder preconditioned by first eluting with dilute mercury (I) solution to remove the oxide coating on the aluminium powder by amalgamation. Optimum conditions for column cementation were reported by Fu *et al.* (1983) to be a solution pH of 3 and a column temperature of 70°C. It was necessary to add fluoride to analysed solutions to mask Fe(III) and to prevent the blocking of columns by precipitation of hydrated iron oxide. Under the conditions investigated, quantitative recoveries were obtained for the elements Cu, Pb, Pd, Rh, Ru, Au and Pt over a wide range of element concentrations. Preconcentration factors of more than 1000 could readily be obtained.

15 Gold and platinum group element analysis

15.1 Introduction

Throughout history, gold, silver and the platinum group elements have occupied a position of considerable social and economic importance. Indeed, the analytical chemistry of these elements extends back into pre-history, analyses being traditionally performed by fire assay techniques. Such techniques, which still form an important aspect of modern determinations of these elements, involve fusing together a sample and flux at a temperature of about 1000°C. The resultant melt separates into two immiscible liquids. The lighter glassy slag contains the alumino-silicates and base metals and is discarded, whereas the more dense 'collector' phase, usually composed of molten lead, sinks to the bottom. The platinum group elements are strongly partitioned into this collector phase from which they may be subsequently separated for analysis.

The elements of interest fall into two groups in the periodic table: (i) ruthenium, rhodium, and palladium, and (ii) osmium, iridium and platinum, which, together with silver and gold, are usually referred to as the noble metals. These elements possess some unique geochemical characteristics. They are very refractory, having high boiling points and a strong affinity for iron (siderophile). Some of their physical properties are listed in Table 15.1. They are normally present in silicate rocks at very low concentrations, normally in the parts per billion range (Crocket, 1979). Determination almost always requires some form of preconcentration. A further characteristic which must be taken into account in any scheme of analysis is that the noble metals often form discrete phases (such as alloys, sulphides, tellurides, arsenides, etc.). Unlike the rare-earth elements, for example, the noble metals do not appear to partition themselves extensively between silicate phases. It is not therefore a trivial task to select a representative sample mass for analysis, since the presence or absence of one noble metal mineral grain in the analysed aliquot could have a significant effect on the analysis particularly if small sample masses (<5 g) are chosen. The onus is therefore on the analyst to establish that his data are not affected by large sample inhomogeneity errors, a neglected aspect of noble-metal analysis which must call into question some of the published data on these elements undertaken on sample masses as low as 0.1 to 1 g. An inherent advantage of fire assay procedures is that they are suitable for sample masses in the range 5 to 100 g.

The analytical chemistry of the noble metals has been reviewed extensively by Beamish (1966), and Beamish and van Loon (1972). Procedures can normally be divided into two stages. The first is the preconcentration step in which these elements are separated from the host matrix. This traditionally involves a fire assay analysis (Haffty *et al.*, 1977; Robert *et al.*, 1971) although aqua regia extraction (Grimaldi and Schnepfa, 1968; Fryer and Kerrich, 1978), ion exchange separation (Nadkarni and Morrison, 1977; Parry, 1980*a*) and solvent extraction (Diamantatos and Verbeek, 1977) have also been used extensively.

The second stage is the determination of individual elements by a technique which must show high specific sensitivity. Some of the more commonly used techniques include flame atomic absorption (Sen Gupta, 1973) and electrothermal atomization–AAS (L'vov *et al.*, 1979; Page *et al.*, 1980; Haines and Robert, 1982). Inductively coupled plasma–mass spectrometry is likely to have some relevance to this area of analysis.

A large proportion of current analytical activity is related to exploration, mining and the economic importance of the

Table 15.1 Physical parameters of the noble metals

AN	Element	AW	mp. (°C)	bp. (°C)	σ (kg m^{-3})	Atomic radius (nm)	Chondrite abundance (ppm)
44	Ru	101.07	2310	3900	12.2	13.36	0.690
45	Rh	102.91	1966	3727	12.4	13.42	0.200
46	Pd	106.4	1552	3140	12.0	13.73	0.545
47	Ag	107.87	961.9	2212	10.5		
76	Os	190.2	3045	5027	22.5	13.50	0.514
77	Ir	192.22	2410	4130	22.4	13.55	0.540
78	Pt	195.09	1772	3827	21.45	13.85	1.020
79	Au	196.97	1064	2807	19.32		0.152

AN = atomic number, AW = atomic weight, mp. = melting point, bp. = boiling point.
σ = density.
Density data from Crabtree (1979).
Atomic radius data (12-fold coordination) from Westland (1981).
Chondrite abundance data from Naldrett (1981).
mp. and bp. data from Weast (1973).

noble metals. However, there is also a wider geochemical interest in these elements concerned with meteorite geochemistry (Ehmann *et al.*, 1970; Crocket, 1972), mantle–core equilibrium processes (Bird and Bassett, 1980; Arculus and Delano, 1981; Mitchell and Keays, 1981; Morgan *et al.*, 1981) and Cretaceous–Tertiary boundary layer extinctions (Alvarez *et al.*, 1980).

In this section, the methods of extracting the noble metals from samples for subsequent analysis will be considered, with particular emphasis on techniques capable of accepting samples larger than 5 g. Aspects of the methods of determination of these elements after preconcentration will not be considered in detail, since these are covered in the appropriate chapters.

15.2 Fire assay procedures

The principle of fire assay has been outlined above. The powdered sample (5 to 100 g) is mixed with a flux. On fusing in a fireclay crucible, the sample matrix dissolves in a borax slag, but the noble metals are strongly partitioned into an immiscible collector phase which settles to the bottom of the crucible. The most widely used collector phase for 'classical' fire assay is lead, and procedures based on this technique have been described in detail (Beamish, 1966; Haffty *et al.*, 1977). However, nickel sulphide as collector phase has also met with some success (Robert *et al.*, 1971; Hoffman *et al.*, 1978) and procedures based on the use of tin have been described (Faye and Moloughney, 1972; Moloughney and Faye, 1976). An important concern with all these procedures is the efficiency with which the noble metals are recovered in the collector phase. There is some evidence that complete recovery rarely occurs from a single fusion and that to ensure quantitative data it is often necessary to perform a second fire assay fusion on the glassy slag. These problems are discussed for both lead and nickel sulphide fire assay procedures below.

15.2.1 *The vocabulary of fire assay*

Fire assay is such an ancient art that a specialized terminology is still used today to describe various aspects of this procedure. Some of these terms, (especially 'cupellation' and 'scourification') are specific to lead fire assay analysis, others are more generally applicable. The more common of these terms are listed in Table 15.2.

15.2.2 *Lead fire assay*

Lead fire assay is the 'classical' method of analysis for gold and may normally be used to obtain quantitative recoveries of platinum, palladium, rhodium and silver. However iridium, osmium and ruthenium have only limited solubilities in molten lead and are largely collected because of their high densities. During cupellation of the resultant lead button (see below), some iridium and ruthenium and almost

Table 15.2 Specialized terms used in fire assay analysis

Assay ton: Weight of sample (in g) selected for analysis such that each milligram of noble metal contained in the sample is equivalent to a whole rock concentration of one ounce per ton.
An assay ton is equivalent to:
32.667 g of sample (ounces per long ton, 2240 lb).
29.167 g of sample (ounces per short ton, 2000 lb).

Button: the lead (or nickel sulphide) collector phase into which the noble metals are strongly partitioned. It takes its name from the button-shaped blob which solidifies at the bottom of the crucible after fire assay.

Charge: the mixture of sample and flux which is fused in the course of fire assay.

Collector: refers to the lead or nickel sulphide component of the flux into which the noble metals partition.

Cupellation: the procedure in lead fire assay analysis in which the lead button is heated to 900°–1300°C in an open bone-ash dish (a *cupel*) in an oxidizing atmosphere. Under these conditions, the lead matrix oxidizes and is absorbed by the bone-ash dish, leaving a noble-metal bead, called the *prill*.

Fire assay: a technique for separating gold and platinum-group elements from the matrix involving fusing the sample with a flux in a fireclay crucible. *'Classical'* fire assay normally refers to the lead fire assay technique.

Flux: the mixture of reagents added to the sample prior to fusion. The flux normally consists of sodium carbonate, sodium borate (borax) and either lead monoxide (litharge) and flour or nickel oxide (or nickel metal) and flowers of sulphur.

Gangue: the matrix material from which the noble metals must be separated.

Parting: the process in which the residue after fire assay (the prill) is taken up into solution prior to analysis of individual elements.

Prill: the noble metal bead remaining after the lead matrix of a button has been removed by cupellation.

Scourification: the procedure used in lead fire assay prior to cupellation by which the mass of lead in a button (or in two buttons which are to be combined) is reduced to about 30 g, suitable for cupellation. The button is placed in a flat open dish and heated to 950°C. The surface of molten lead rapidly oxidizes and, after a suitable period, the sample is allowed to cool and the oxidized layer removed and discarded.

Shot: globules of lead remaining in the silicate phase after fire assay fusion, signifying incomplete settling of the lead button.

Vanning: a method of qualitatively assessing the chemical and mineralogical composition of a sample in order to select the most appropriate flux composition for lead fire assay. Vanning consists of moistening a small amount of powdered sample and adding a few drops of 50% HCl. From the degree of effervescence, the carbonate content can be estimated; the odour of hydrogen sulphide signifies the presence of sulphide minerals and the residue can be examined to identify clay and some opaque minerals (see Haffty *et al.*, 1977).

Table 15.3 Flux components for lead fire assay

Na_2CO_3 (soda ash): Powerful alkali flux that readily dissolves alumino-silicates and sulphide minerals.

$Na_2B_4O_7$ (borax): Lowers the melting point for fusion and promotes alumino-silicate glass formation.

SiO_2 (silica): Added to samples deficient in silica to prevent the fireclay crucible being attacked by molten litharge.

PbO (litharge): Reduced to molten lead during fire assay fusion and 'collects' noble metals from the sample.

Flour: The traditional reducing agent; added to fire assay charges to reduce litharge to molten lead.

CaF_2 (calcium fluoride). Used in some recipes to increase the fluidity of melts that otherwise would be so viscous as to inhibit settling and separation of molten lead.

KNO_3 (nitre): An oxidizing agent with a low melting point (339°C). Used to oxidize sulphide minerals to sulphate to promote effective fire assay separation in some samples.

For further details, see Haffty *et al.* (1977).

all osmium are usually lost. More elaborate precautions must be taken if these elements are to be determined by this fire assay procedure. In order to assist in the quantitative recovery of elements from samples with very low abundance levels, it is necessary to add to the fire assay charge a small weight of either metallic silver (if gold, platinum or palladium are to be determined) or gold (if rhodium or iridium are to be determined) so that the metallic residue after cupellation (the prill) can be physically recovered. Indeed it is recommended that gold should always be added to the charge if quantitative recoveries of iridium at any concentration level are to be achieved (Grimaldi and Schnepfe, 1970).

(*a*) *Selection of flux composition.* The flux used for fire assay normally consists of sodium carbonate and sodium borate (to dissolve the sample and promote the formation of a glassy slag), lead monoxide (litharge), and flour (as a reducing agent). The latter two reagents react to form a 'mist' of molten lead droplets into which the noble metals are strongly partitioned. These lead particles coalesce and slowly sink through the melt under the action of gravity to form a molten lead button at the bottom of the fire assay crucible. As discussed by Haffty *et al.* (1977) other reagents are sometimes essential. Molten litharge has a strong affinity for silica, and if the sample itself has a low silica content, the walls of the fireclay crucible will be attacked instead. Silica must therefore be added as appropriate to compensate for any deficiency in the sample. Some samples melt to form viscous liquids which inhibit gravity separation of lead droplets. The addition of calcium fluoride to the flux causes a large increase in the fluidity of these melts. Samples which have high sulphide mineral contents can interfere with the lead collection process. Potassium nitrate (nitre) is added to such fluxes to ensure pre-oxidation of sulphide to sulphate to eliminate this interference. The functions and properties of these various flux components are summarized in Table 15.3.

The precise composition of the flux must be matched to the composition of individual samples. In particular it is necessary to know the silica content and the *reducing power* of a sample. The reducing power is the weight of lead button which 1 g of sample will form when fused with excess litharge but with no added reducing agent. *'Vanning'* is a qualitative procedure for estimating the elemental and mineralogical composition of a sample prior to selection of the most appropriate flux components.

These procedures have been described in detail by Haffty *et al.* (1977), and from this work, typical flux compositions for the analysis of 15 g aliquots of silicate and sulphide materials are shown in Table 15.4.

(*b*) *Lead fire assay procedure.* The sample and flux are mixed by rolling the powders together on a sheet of glazed paper or 'cellophane'. The mixture is poured into a fireclay crucible and silver or gold in the form of high-purity wire (or standard solution) is added to the charge. The crucible is sealed with a lid and placed into a muffle furnace at 950°C. The temperature of the furnace is increased to 1000–1200°C. As the temperature of the charge increases above 500–550°C

Table 15.4 Composition of fire assay flux for some silicate materials

	Weight (g)								Total
Sample type	Sample	Na_2CO_3	$Na_2B_4O_7$	SiO_2	PbO	Flour	CaF_2	KNO_3	weight of flux
Basalt	15	30	35	4 to 8	35	3.2	1	0	108 to 112
Dunite	15	30	35	4 to 8	35	3.2	0	0	107 to 111
Quartz	15	20	3	0	40	3.0	0	0	66
Rhyolite	15	20	3	1	50	2.8	1	0	78
Massive sulphides in gabbro	15	25	4	14	100	0	0	8.5	152

Data from Haffty *et al.* (1977).

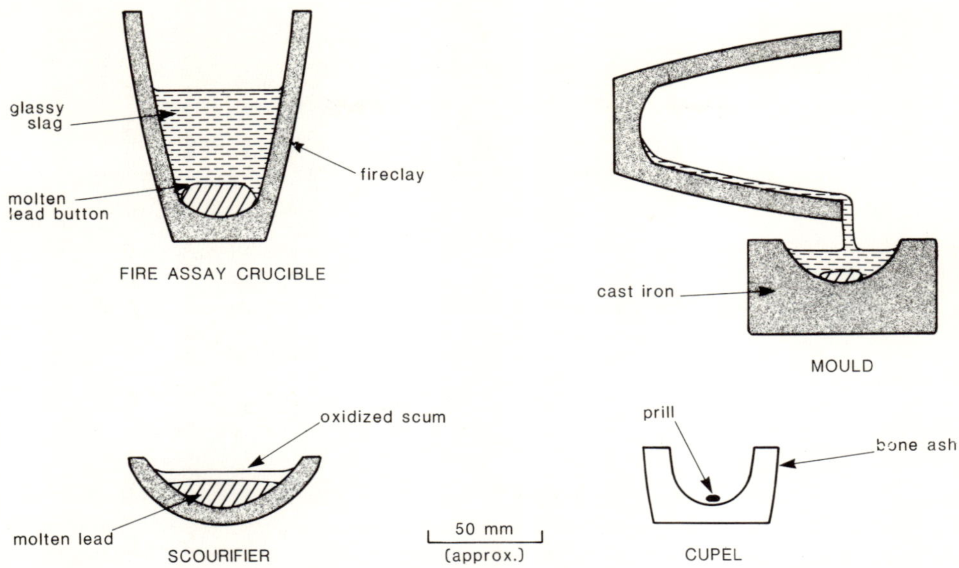

Figure 15.1 Fireclay crucible, bone-ash cupel, scourifying dish and iron mould used during lead fire assay analysis.

the litharge starts to react with flour or sulphides in the sample to form a mist of molten lead droplets which alloy with noble metals within the sample. It takes about one hour for the lead to coalesce as a molten blob at the bottom of the crucible. The mixture is then removed from the furnace and the melt poured into a mould made of cast iron. After solidifying and cooling, the lead button is physically separated from the glassy slag (Figure 15.1).

Test data of Bowditch (1973) (Table 15.5) show that significant amounts of some noble metals (Pt and Pd for example) can remain within the slag. To be sure of quantitative recovery, it is therefore advisable to crush the slag to a fine powder, mix with a further batch of litharge and flour and re-fuse in the original fireclay crucible. This second lead button is then combined with the first.

Before extraction of the noble metals, it is normally necessary to reduce the total amount of lead present in the two buttons by the process of *scourification*. The buttons are placed in an open fireclay dish (the 'scourifier') and heated in a furnace to 950–1000°C. Under these conditions the surface of the molten lead bead oxidizes (Figure 15.1), and after a suitable period, the mixture may be poured into a cast iron mould and the lead button containing the noble metals separated from the oxidized scum. The aim of this procedure is to reduce the total mass of the lead button to 25–30 g prior to cupellation.

Not all procedures require scourification, but almost all involve *cupellation*. During this process all the lead is removed from the button, leaving a metallic noble metal residue or *prill*. The lead button is placed in an open bone-ash dish (the 'cupel') and heated to a suitable temperature in the range 900–1300°. Under these conditions the lead is rapidly oxidized to molten PbO, most of which is absorbed by the bone-ash cupel (a small proportion, about 1.5%, is vaporized). During cupellation, almost all osmium is lost by volatilization as osmium tetroxide. Furthermore, it is estimated that perhaps 50% of the ruthenium and 20% of the iridium are partially oxidized and absorbed into the cupel (Bowditch, 1973).

If silver is added to the original fire assay charge and cupellation carried out at the relatively low temperature of about 960°C, a silver prill is recovered and may be analysed to determine the Au, Pt, Pd and Rh content of the sample. If, however, a high-temperature cupellation is carried out at about 1300°C, then all silver as well as any residual lead are completely removed from the prill which may then be weighed directly to determine the content of gold plus Pt, Pd and Rh without further analysis.

After cupellation, it is, however, usual to analyse the prill for individual elements. If a solution technique is to be used (e.g. graphite furnace atomic absorption), the prill would normally be dissolved in aqua regia (3 parts HCl + 1 part HNO_3). However, care must be taken to ensure complete dissolution, as some workers have reported that incomplete dissolution of gold occurs from a silver prill (if the ratio of silver to gold is less than 2.5, Van Wyk, 1980) or rhodium from a gold prill. To overcome these difficulties, Aruscavage *et al.* (1984) digested gold prills in aqua regia (4 mL), evaporated to dryness and heated the residue with a small volume of aqua regia (0.1 mL) in a sealed vial on a steam bath for 15 minutes. The residue was dissolved in 6N HCl (1 mL). Consistent rhodium data was observed using this dissolution technique. Part of the difficulty encountered by some workers appears to be the resistance to acid attack of intermetallic Rh–Ru–Ir alloys which can form in the prill.

(*c*) *Recovery of osmium and ruthenium by lead fire assay*. As outlined above, although osmium and ruthenium can be recovered quantitatively in the lead button after fire assay, these elements are lost during subsequent cupellation. Diamantatos (1977) proposed an alternative procedure in which the lead button was carefully cleaned of any slag residue and then dissolved in 70% perchloric acid to which glacial acetic acid had been added. When heated to 160–180°C, the perchloric acid mixture behaved as a non-oxidizing acid, dis-

Table 15.5 Recovery of noble metals after various preparation procedures

Technique	Sample type	Recovery step	% element recovered								Reference
			Ru	Rh	Pd	Ag	Os	Ir	Pt	Au	
Lead fire assay	Nickel sulphide concentrates	Recovered from slag									Bowditch (1973)
		First fusion:		(100)	83				90	(100)	
		Second fusion:			15				8		
		Total recovered:			98				98		
		Recovered after cupellation:	54	(100)	(98)		(0)	87	(98)	(100)	
Nickel sulphide fire assay	Massive sulphide:	One fusion:	99		97		99	98	97	96	Hoffman *et al.* (1978)
	gabbroic host rock:	One fusion:	97		99		95	99	94	95	
	ultramafic host rock:	One fusion:	95		96		97	100	96	96	
	Nickel sulphide concentrates	One fusion (worst case):	97	93	95				87	96	Bowditch (1973)
		One fusion (best case):	98	96	98				95	96	
	Radiotracers in SiO_2	First fusion:				90		90	88	86	Palmer *et al.* (1975)
		Second fusion:				7		10	10	12	
		Total recovered:				97		100	98	98	
Acid extraction (aqua regia)	Merensky reef ore	Worst case:		97	95				98	(100)	Palmer *et al.* (1971)
	Low-grade concentrate	Worst case:		89	96				98	(100)	Bowditch (1973)
	Nickel sulphide concentrate	Worst case:	46	67	96		0	7	88	90	
		Best case:	82	88	99		0	30	95	97	
Ion exchange Srafion NMRR			Ru(II)	Rh(II)	Pd(II)	Ag(I)	Os(IV)	Ir(IV)	Pt(IV)	Au(III)	
	Recovery efficiency of resin:		96.8	90.6	99.6	99.9	100	100	100	100	Nadkarni and Morrison (1977)

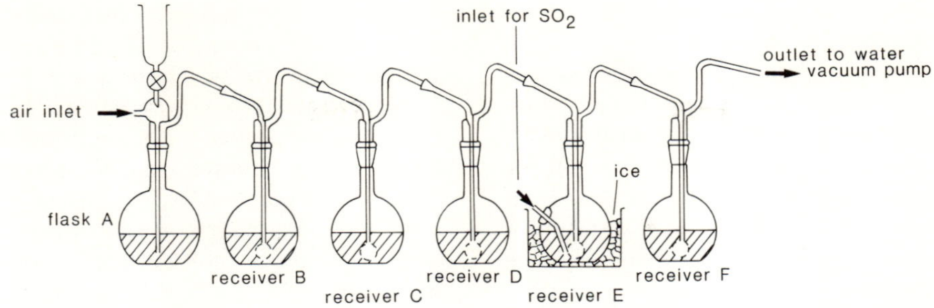

Figure 15.2 Distillation apparatus for the determination of osmium and ruthenium by a barium peroxide sinter procedure (Van Loon, 1980, pp. 147–150). Ruthenium distillate was collected in 6N HCl in B–D; osmium in 6N HCl+SO_2 in E, F.

solving completely the lead matrix. Some osmium was emitted as the volatile tetroxide and was absorbed by bubbling evolved gases through a solution of sodium hydroxide. The insoluble noble metal residue was then filtered off, fused with sodium peroxide and redissolved in water. This alkaline solution was combined with the previously vaporized osmium solution and poured into a still. Sulphuric acid was added to the still to acidify this mixture, followed by sodium bromate solution as an oxidizing agent. Ruthenium and osmium tetroxides were distilled off on heating to boiling. The former was absorbed in flasks containing 6M hydrochloric acid, the latter in 6M hydrochloric acid containing 10% thiourea (Figure 15.2). Both were determined spectrophotometrically by the colour intensity of their thiourea complexes.

(*d*) *Summary of lead fire assay procedures*. The application of lead fire assay to individual elements may be summarized as follows:

Gold: silver wire/solution can be added to charge to assist recovery

Palladium, rhodium, platinum: These elements can be recovered if either silver or gold are added to the charge to assist recovery of the prill

Iridium: Gold wire *must* be added to the charge both to eliminate losses on cupellation due to absorption of iridium oxide on cupel and to avoid errors due to low solubility of iridium as an alloy in a silver prill. Alternatively, Pt–13%Rh or Pt wire can be used as collector. Due to the uncertainty in quantitative recovery and the fact that determinations of Pt and Au are almost always required, iridium is rarely determined using a lead fire assay procedure.

Ruthenium: Partially lost as the tetroxide during cupellation; avoid cupellation step by dissolving lead button in perchloric acid and recover ruthenium by distillation (see (*c*) above)

Osmium: Almost completely lost during cupellation; as for ruthenium, dissolve lead button in perchloric acid and recovering osmium by distillation (see (*c*) above). Because of the specialized nature of these recovery procedures, osmium and ruthenium are not routinely determined by lead fire assay.

15.2.3 *Nickel sulphide fire assay*

The second fire assay procedure which has received considerable interest is that based on the use of nickel sulphide as collector (Robert *et al.*, 1971; Robert and van Wyk, 1975; Hoffman *et al.*, 1978). The advantage of the method is that gold and all the platinum group elements may be quantitatively recovered in the nickel sulphide button, sometimes after two fusions of the sample. The disadvantage is that physical methods of removing the nickel sulphide matrix from the button (such as scourification and cupellation) cannot be used in this procedure. Noble metals can only be extracted by dissolving the button by acid attack.

The flux normally consists of sodium carbonate, sodium borate, nickel oxide (or nickel metal), flowers of sulphur and sometimes also silica. A typical mixture would be:

sample	50 g
sodium carbonate	30 g
sodium borate	60 g
nickel oxide	20 g (or 16 g nickel metal powder)
	(or 32 g nickel carbonate)
flowers of sulphur	10 g
silica	0–10 g (as required)

The powdered mixture is poured into a fireclay crucible and ignited at about 1000°C for 1 to 1.5 hours. During fusion the sodium carbonate and borax dissolve the sample to form a dark glassy slag. The nickel oxide reacts with the sulphur to form a mist of molten nickel sulphide droplets which separate out under the influence of gravity and coalesce to form a molten blob at the bottom of the melt. The composition of this phase has been estimated to be Ni_3S_2 or Ni_6S_5 (Williamson and Savage, 1965). Gold and the platinium group elements are strongly partitioned into the nickel sulphide phase. After fusion, the melt is poured into a cast iron mould and, when cool, the nickel sulphide button is physically separated from the slag.

It is essential to pre-analyse samples to determine their nickel and sulphur content. The amount of these reagents in the flux is reduced to compensate for the intrinsic sample content. Silica must be added to samples low in this element to promote successful slag formation.

An important aspect of this technique is the purity of the nickel reagent used to form the fusion mixture. Robert *et al.* (1971) showed that reagent grade NiO was unsuitable since it contained individual noble metals in the range 0.3 to 2.7 ppm. Specially purified nickel carbonate was substituted instead. Hoffman *et al.* (1978) proposed the use of nickel metal of extremely low noble metal content (less than 1ppb of individual elements). This material had been purified by the carbonyl process in which metallic nickel is reacted to form nickel carbonyl ($Ni(CO)_4$) which, being volatile (its boiling point is 43°C) is distilled from impurities, recondensed and decomposed back to the metal.

The efficiency of the nickel sulphide procedure in recovering noble metals has been investigated by Robert *et al.* (1971), Palmer and Watterson (1971), Bowditch (1973), Dixon *et al.* (1975) and Hoffman *et al.* (1978). The effect of specific matrix elements on recovery efficiencies has been investigated by Robert and Van Wyk (1975). Some recovery efficiency data are listed in Table 15.5, and show that after a single fusion, the recovery of gold and ruthenium in the nickel sulphide button is normally near-quantitative. However, a significant proportion of Pt, Pd and Rh can remain in the slag (up to 10%). To be certain of quantitative recovery of all elements, it is therefore necessary to re-fuse the slag after separation of the first nickel sulphide button with a further batch of nickel (oxide) and sulphur. The second nickel sulphide button so formed is combined with the first.

The combined buttons are then normally crushed to a fine powder and dissolved in 12M hydrochloric acid with heating. The nickel sulphide matrix is slowly dissolved with the evolution of copious hydrogen sulphide, leaving a finely divided black insoluble residue comprising noble metal sulphides. These will not dissolve in HCl provided the solution remains saturated in hydrogen sulphide. This residue is therefore filtered off and can be dissolved in hydrochloric acid to which hydrogen peroxide has been added. Sometimes further filtration is required to remove any excess free sulphur carried over into the residue.

Individual elements may be determined by flame atomic absorption (Robert *et al.*, 1971). AAS with electrothermal atomization (Haines and Robert, 1982), or ICP optical emission analysis (Wemyss and Scott, 1978). Hoffman *et al.* (1978) omitted the final dissolution stage and irradiated the filter paper and noble metal sulphide residue to permit simultaneous determination by neutron activation analysis. In this way, Hoffman *et al.* avoided one of the difficulties which can affect solution techniques: Bowditch (1973) reported that difficulty can sometimes be encountered in complete dissolution of this sulphide residue and that small but significant losses of platinum can be observed.

15.3 Acid extraction of noble metals

The final preconcentration technique to be considered here in detail is simple acid leaching using aqua regia (1 part of nitric acid plus 3 parts of hydrochloric acid). The sample must be milled to as fine a grain size as possible. Aqua regia is then added and the mixture heated. The studies of Palmer *et al.* (1971) and Bowditch (1973) show that in many cases (near) quantitative recoveries of gold, platinum and palladium occur (see Table 15.5). The extraction of rhodium is not always complete. The recoveries of iridium and ruthenium are usually low and variable. Osmium is volatilized under the prevailing oxidizing conditions.

Addition of concentrated acid to a powdered sample can lead to vigorous or even violent reaction. In some cases it is therefore desirable to ignite powders at 600°C for one hour to expel volatile and organic matter and to oxidize sulphide minerals. The sample (25 g) is then heated on a steam bath with aqua regia (100 mL) overnight. The bulk of the sample remains undissolved, but noble metals are among the elements that are selectively dissolved. It is often desirable to remove nitrate ions from the extracted solution since these interfere with atomic absorption determination of many elements. This is achieved as follows. The aqua regia solution is evaporated to dryness and concentrated hydrochloric acid (50 mL) added. This mixture is evaporated to dryness and this HCl evaporation cycle repeated up to a total of six times. The final residue is taken up in dilute hydrochloric acid (1M) and the soluble fraction (containing the noble metals) filtered from the bulk insoluble residue.

Other elements dissolved by aqua regia attack include Cu, Ni, Zn and Fe. Variable proportions of these elements can cause interferences in atomic absorption determinations. A widely-used method of extracting the noble metals alone is by co-precipitation with tellurium. An acid solution of tellurium is added to the acid extraction solution. The tellurite in solution is reduced to the metallic state by the addition of stannous chloride solution. The resultant precipitate of noble metal tellurides is filtered and redissolved in aqua regia. Sighinolfi *et al.* (1984) suggest that some other trace elements can be co-precipitated with tellurium including Bi, Hg and Cu. Noble metal solutions purified in this way are most effectively analysed by atomic absorption spectrometry using electrothermal atomization (Sighinolfi *et al.*, 1984). Following the development of the acid extraction procedure using co-precipitation with tellurium by Grimaldi and Schnepfe (1968), further investigations by Palmer *et al.* (1971) and Bowditch (1973) showed that this technique may be used effectively to determine the elements gold, platinum and palladium, yielding data which are in general fully comparable with fire assay procedures. It is, however, necessary to qualify this statement. There is always the worry that in some samples, discrete platinum-group minerals may be completely occluded within the grains of the host matrix and may not then be effectively dissolved by aqua regia attack. An important prerequisite for obtaining reproducible data by this acid extraction technique may therefore be to ensure that all samples are uniformly milled to a grain size of less than 200–250 mesh.

15.4 Other methods of noble metal analysis

15.4.1 *Neutron activation analysis*

The largest volume of data published for the platinum group elements until recently was obtained by neutron activation analysis using a radiochemical separation. The technique has the advantage of high sensitivity, but in order to limit the activity encountered during the radiochemical manipulations, sample masses must normally be restricted to 100–500 mg, unacceptably small for some sample types. A typical procedure would involve adding platinum group element carriers to an irradiated sample and then fusing with NaOH–Na_2O in zirconium crucibles (not nickel, because of contamination problems). The digested residue is acidified and taken up into solution. In early schemes of analysis, the gamma-ray spectra of platinum group element isotopes often had to be recorded using NaI(Tl) detectors of poor energy resolution or even by measuring beta-activity with a Geiger counter. It was, therefore, necessary to go to extraordinary lengths to devise chemical procedures to separate individual elements from one another. Such techniques were, however, some of the first to enable determinations to be made in the ppb range in silicate rocks. (see, for example, Crocket *et al.*, 1968; Gijbels *et al.*, 1974; Gijbels, 1971; Crocket, 1972; Keays *et al.*, 1974).

Where a sample is irradiated before a group separation is carried out, a correction must be applied for an interference on ^{103}Ru ($T_{1/2}$ = 39.6 days) due to fission of uranium within the sample. Gijbels (1971) has considered this and other potential interferences and has estimated that the proportion of ^{103}Ru originating from uranium fission amounts to as much as 130 ppb apparent ruthenium per ppm uranium in the sample.

15.4.2 *Use of a specific chelating resin, Srafion NMRR*

An important simplification in these radiochemical procedures can be made by use of ion exchange chromatography. Investigations have shown that the chelating resin Srafion NMRR can quantitatively recover gold and the platinum group elements from solution (Koster and Schmuckler, 1967). The resin is specific to ions which form square planar complexes, and is capable, for example, of reducing the ions Ir^{4+} and Pt^{3+} to the square planar forms, Ir^{2+} and Pt^{+}, that can then be incorporated in to the resin (Nadkarni and Morrison, 1974). These workers report that quantitative recoveries (97–100%) of all noble metals can be achieved from solutions containing radioactive tracers of these elements in the presence of carriers, but that two passes are required through the column to obtain satisfactory recovery of ^{97}Ru and ^{194}Ir. Precise recovery data from the later work of Nadkarna and Morrison (1977) are listed in Table 15.5. These workers also investigated the recovery on the NMRR resin of a range of elements that might interfere with the gamma spectra of the platinum group elements. This work established that small quantities of ^{64}Cu and ^{51}Cr are also retained by this resin, causing significant background enhancement in the determination of some platinum-group element isotopes. Srafion NMRR has been incorporated into a simple radiochemical separation by Parry (1980*a*) that is capable of being carried out with a high degree of automation (Parry, 1980*b*).

The use of this platinum-group-element-specific resin is not only restricted to conventional radiochemical neutron activation analysis procedures. Green *et al.* (1970) have proposed a method of determining gold using resin-loaded papers and 'dangalation'. Gold was extracted from samples using an aqua regia attack. After dilution with water, filter papers loaded with Srafion NMRR resin were 'dangled' overnight in the acid extraction solution. Preliminary experiments showed that the recovery of gold in the resin-loaded paper was low and variable (10 to 30% after 16 hours). However, by spiking the sample solutions with a known amount of ^{195}Au tracer solution, the proportion of gold recovered could be determined by measuring the ^{195}Au take-up on the filter papers. The gold take-up originating from the sample itself was determined by neutron activation analysis of the irradiated filter paper, gold being analysed from its ^{198}Au isotope. However, the ^{195}Au spike is produced by cyclotron bombardment and not by neutron activation. The activity of the ^{195}Au spike was therefore unaffected by neutron irradiation and could furthermore be assayed from the same gamma spectrum used to determine ^{198}Au.

15.5 Noble metal analysis—comparisons of data

Earlier sections examined the effectiveness of fire assay and acid extraction preconcentration procedures without considering the determination stage. It is proposed here to compare results from schemes of analysis in common usage. A nickel sulphide fire assay procedure was developed by Hoffman *et al.* (1978), using neutron activation analysis as the determination stage. Details of isotope measured, count times and element detection limits are listed in Table 15.6.

Table 15.6 Gamma spectrum counting details and detection limits for nickel sulphide fire assay followed by neutron activation analysis

Element	Isotope	$T_{1/2}$	Irradiation time	Decay time	Count time (s)	Gamma-ray (keV)	Detection limit (ppb)
Ru	^{103}Ru	38.9d	16h	8d	5000	497	9
Rh	^{104m}Rh	4.41min	5min	1 min	200	51	3
Pd	^{109m}Pd	4.69min	5min	1 min	200	186	15
	^{109}Pd	13.5h	16h	4h	2500	88	150
Os	^{191}Os	14.6d	16h	8d	5000	129	6
Ir	^{192}Ir	74.2d	16h	8d	5000	317 468	0.3
Pt	^{199}Au	3.15d	16h	8d	5000	158	15
Au	^{198}Au	2.70d	16h	8d	5000	412	0.3

Scheme of analysis: Hoffman *et al.* (1978).
50 g sample, nickel sulphide fire assay. The residue after dissolving the button was filtered and irradiated without further treatment.
Blanks: Au, Ir: 0.1 ppb; Pt, Pd, Os, Ru, Rh < 1 ppb.
Limit of detection recalculated as concentration in whole rock equivalent to six standard deviations on the background.
s = second, min = minute, h = hour, d = day.

Table 15.7 Analytical conditions for graphite furnace atomic absorption analysis following lead fire assay or acid extraction preconcentration

Element	Reference	Wavelength (nm)	Graphite furnace temperature/hold time for:			Detection limit (ppb)	Blank (ng)
			Drying	Ashing	Atomization		
Rh	1	343.5	110°C/50 s	1250°C/50 s	2700°C/7 s	0.1	0.5
Pd	1	247.6	110°C/50 s	1100°C/50 s	2700°C/7 s	0.2	0.5
Ag	2	338.3	150°C/5 s	500°C/5 s	2400°C/5 s	5	
Pt	1	265.9	110°C/50 s	1600°C/50 s	2700°C/7 s	1.6	5
Au	2	242.8	150°C/5 s	900°C/5 s	2550°C/5 s	1	

Reference 1: Page *et al.* (1980). Lead fire assay procedure in which 2.5 mg of gold wire was added to the charge to assist in the collection of the platinum-group elements.
Detection limit data has been recalculated as the six-sigma limit of determination based on a 20 g sample.
Reference 2: Sighinolfi *et al.* (1984). Acid extraction procedure using aqua regia followed by co-precipitation with tellurium. An intermediate drying stage of 320°C held for 5 s was used in the graphite furnace heating program. Limit of determination data representative of a 5 g sample.

Noble metal residues remaining after dissolving the nickel sulphide button were first irradiated for five minutes, and after a one-minute decay period, gamma spectra counted on germanium detectors to permit determination of the short-lived isotopes ^{104m}Rh ($T_{1/2} = 4.41$ min) and ^{109m}Pd ($T_{1/2} = 4.69$ min). Samples were then re-irradiated for sixteen hours in a flux of 1×10^{11} n cm^{-2} s^{-1} to activate the longer-lived isotopes. All gamma spectra were counted on coaxial Ge(Li) detectors. Determination limits ranged from 0.3 ppb (Au, Ir) to 15 ppb (Pt, Pd) using 50 g sample aliquots. Results were reported for all six platinum group elements and gold by this technique.

Atomic absorption spectrometry with electrothermal atomization is capable of achieving very low detection limits, as may be judged from data in Table 15.7. Data in this table for the elements Rh, Pd and Pt are from Page *et al.* (1980), using a lead fire assay procedure on 20 g sample aliquots. Exceptional limits of detection in the range 0.1 ppb (Rh) to 1.6 ppb (Pt) were reported. Data in Table 15.7 for Ag and Au are from Sighinolfi *et al.* (1984) who used an acid (aqua regia) extraction on 5 g sample aliquots. Noble metals were co-precipitated with tellurium, and limits of determination of 5 ppb (Ag) and 1 ppb (Au) were obtained in this work.

Table 15.8 Comparison of detection limits between four instrumental methods of analysis

Element	Limit of determination (μg/l)			
	ETA–AAS	Flame–AAS	INAA	ICP–AES
Ru	3	500	10	125
Rh	0.5	100	1	25
Pd	0.2	100	4	30
Ir	2.5	750	0.1	160
Pt	1.6	1000	10	40

Data from Haines and Robert (1982).
ETA–AAS: Graphite furnace atomic absorption spectrometry.
Flame–AAS: conventional atomic absorption spectrometry.
INAA: instrumental neutron activation analysis.
ICP–AES: inductively coupled plasma atomic emission spectrometry.
Detection limits expressed as six-sigma limit of determination appropriate for solution analysis.

The high sensitivity of both graphite furnace atomic absorption spectroscopy and neutron activation analysis is confirmed by a comparison of detection limit data by Haines and Robert (1982) listed in Table 15.8. Flame atomic absorption and inductively coupled plasma–atomic emission spectrometry cannot achieve the sensitivities required in the analysis of silicate rocks at ppb levels, but are of value where ores and concentrates are to be analysed.

Accuracy and precision in this difficult area of analysis may be judged from comparative data in Table 15.9. Here data from various references are compared for the analysis of peridotite USGS: PCC-1 and a platinum group element flotation concentrate CCRMP: PTC-1. Analytical precision (one sigma) is quoted in brackets where available. The agreement between analyses submitted for some elements in PCC-1 is often fairly good, but spread over a larger range. As might be expected, the consistency of results for PTC-1 in the ppm range are far better than for PCC-1 at ppb levels.

In the present state of the art, it is clearly difficult to establish the accuracy of an individual technique by analysing a reference sample such as PCC-1. The uncertainties in 'expected' data averaged from the wide scatter of contributed data are uncomfortably large.

15.6 A note on the distribution of noble metals

The problem of inhomogeneous distribution of noble metals in samples and therefore the minimum sample mass which should be taken for analysis has already been mentioned (section 15.1). This problem has received little attention, and in general the mode of occurrence and distribution of the noble metals within silicate rocks is poorly understood, especially as their whole-rock abundance lies in the ppb range. It is proposed here therefore to draw analogies from studies of chromite samples where more work has been undertaken.

Grimaldi and Schnepfe (1969) studied the mode of occurrence of platinum, palladium and rhodium in chromite by

Table 15.9 Accuracy and precision in the noble metal analysis of two reference materials, USGS: PCC-1 (peridotite) and CCRMP: PTC-1 (flotation concentrate)

Sample	Technique	Analysed values (ppb)								Reference
		Ru	Rh	Pd	Ag	Os	Ir	Pt	Au	
PCC-1 (peridotite)	Compiled value	9.5?	1?	5?	10?	9?	6?	10?	0.7?	Abbey (1983)
	Lead fire assay/ ETA–AAS		1.1	4.4				5.7		Page *et al.* (1980)
	Radiochemical NAA			6.3				4.8		Ahmed *et al.* (1977)
	Ion exchange/NAA			<11			4.2 (2.2)	<20	1.0 (0.3)	Parry (1980*a*)
	Ion exchange/NAA			3.0 (0.1)		6.7 (0.4)	3.5 (0.4)	5.1 (1.5)	3.4 (1.4)	Nadkarni and Morrison (1974)
	Lead fire assay/ ETA–AAS		1.1 to 1.2	4.36 (0.56)				5.7 (0.7)		Aruscavage *et al.* (1984)
	Acid extraction/ ETA–AAS		1.4	6.2	16			20	<2	Sighinolfi *et al.* (1984)
	Mass spectrometry isotope dilution			2.9	3.5					Loss *et al.* (1983)
PTC-1 (flotation concentrate)	Certificated value	650?	620	12700	5800	240?	170?	3000	650	McAdam *et al.* (1973)
	Nickel sulphide fire assay/INAA	590 (50)	620 (20)	12800 (380)		340 (34)	200 (8)	2900 (340)	590 (60)	Hoffman *et al.* (1978)
	Ion exchange/NAA			11800 (1800)			200 (60)	1800 (1100)	390 (230)	Parry (1980*a*)
	Lead fire assay/AAS			12300 (290)	6000 (230)			2900 (170)	650 (50)	Moloughney (1980)

Data in brackets represent the standard deviation (one sigma).

decomposing various fractions of chromite and measuring their platinum-group element abundances. They concluded that rhodium occurred entirely within the chromite lattice, there being a direct correlation between the recovery of rhodium and the proportion of chromite dissolved. In contrast, palladium occurred almost entirely outside the chromite in the silicate matrix and platinum was distributed between the two.

This work is important in indicating the possibility of significant sampling errors due to the inhomogeneous distribution of individual noble metals. However, more recently, Prichard *et al.* (1986) have shown that chromites can contain a rich variety of discrete noble metal minerals and that a large proportion (if not all) of these elements

Table 15.10 Weight of sample requires for nickel sulphide fire assay as a function of platinum-group element content of sample

Estimated total PGE concentration (ppm)	Weight of sample taken for analysis (g)
>1000	5
500–1000	10
250–500	15
100–250	25
50–100	35
<50	50

Data from Robert *et al.* (1971).
Scheme designed for flame atomic absorption spectrometry.

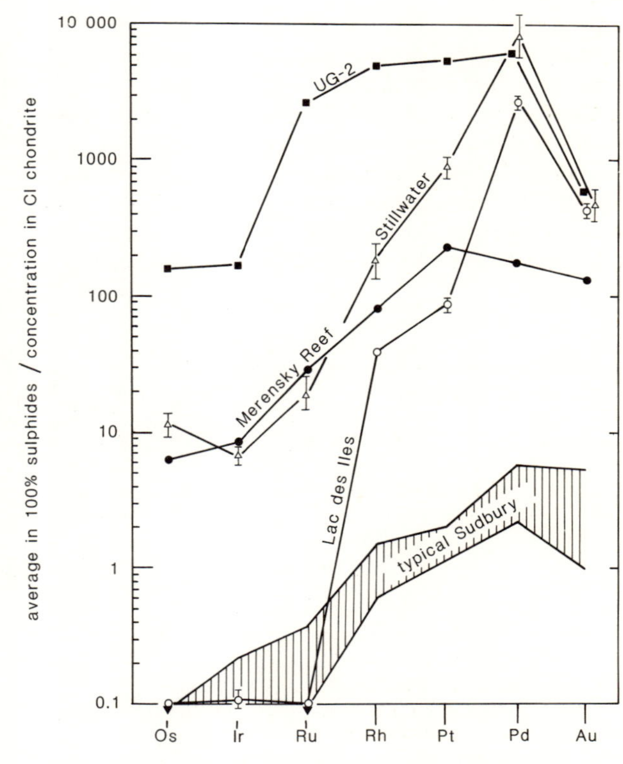

Figure 15.3 Chondrite-normalized platinum-group element diagrams of ore samples from selected deposits. Concentrations have been recalculated as though all PGEs were concentrated in the sulphide fractions of these ores. (From Naldrett, 1981.)

occur as very small alloy, sulphide, arsenic and antimony phases. These minerals can be occluded within both the chromite and the silicate phases. In an associated study, Potts (unpublished) has shown that in analysing gold and iridium in chromites by neutron activation analysis, errors of up to 30% (one sigma) can arise simply due to the inhomogeneous distribution of these elements between the 300 mg aliquots used for analysis.

Based on this experience, it is suggested that a *minimum* sample mass of 5 to 10 g should be selected for analysis unless it is specifically established that smaller sample aliquots can be used reliably.

In many fire assay procedures, larger weights than this are taken for analysis. As well as the homogeneity problem, it is necessary for some techniques to ensure that sufficient mass of platinum group element residue is recovered to permit adequate determination. In Table 15.10, the recommended sample weights are listed for a nickel sulphide fire assay followed by flame atomic absorption (Pt, Pd, Rh, Ru) and colorimetry (Ir, Os) taken from Robert *et al.* (1971). Although a weight of 5 g is considered adequate if the total platinum group element content of a sample exceeds 1000 ppm, 50 g is considered necessary for samples containing less than 50 ppm of these metals so that the recovered residue may be analysed with adequate precision.

15.7 Graphical presentation of platinum group element data

In a manner similar to that used for the rare-earth elements, platinum group element data may be presented as a graphical plot of abundance, normalized to that in C1 chondrite. Elements are plotted in decreasing order of refractivity, namely Os, Ir, Ru, Rh, Pt, Pd, and Au. There is some evidence that, as in the rare-earth elements, the distinctive trend obtained from platinum-group element chondrite normalized diagrams may be used to interpret and distinguish the geochemical chracteristics of the host rocks (Naldrett, 1981). As an illustration of this, the chondrite-normalized signatures of several platinum group element ore deposits are plotted in Figure 15.3, using data taken from Naldrett (1981). The appropriate chondrite normalizing factors are listed in Table 15.1.

16 Mass spectrometry: principles and instrumentation

16.1 Introduction

During the early 1970s, neutron activation analysis in the determination of rare-earth elements perhaps had the most significant influence on the advancement of geochemical thinking, but there can be little doubt that a decade later this accolade was firmly transferred to mass spectrometry. The reason for this is that mass spectrometry is the only technique capable of inferring the geochemical history of samples right back to the age of accretion of the Earth. This arises not only from the possibility of interpreting mass spectrometry isochron data in terms of the age of a sample, but also from the very precise isotope measurements of, for example, inert gases contained in some meteoritic samples which can give data on the nucleo-synthetic processes pertaining to the origin of the universe.

The importance of mass spectrometry as a technique stems from a combination of two analytical characteristics:

(i) The capability of determining the ratio of individual isotope abundances to very high precision

(ii) The very low background count rate present in most mass spectra, permitting exceptionally low ion beam detection limits. To attain this performance it is normally necessary to remove all isobaric interferences by careful chemical purification of the element to be determined.

Isobaric interferences are molecular and multiply-charged ion species having the same nominal mass to charge ratio as the isotope of interest. The presence of isobaric interferences in a mass spectrum seriously degrades the precision with which an isotope ratio measurement can be made. Furthermore, the magnitude of an interference can in some cases vary from sample to sample, due to matrix–dependent variations in the efficiency with which molecular and multiply-charged isobaric fragments are generated in the ion gun. In these cases it is essential to remove interferences, by chemically separating the element of interest from the matrix prior to analysis. For this reason the most sensitive forms of mass spectrometry (thermal ionization and gas source mass spectrometry) are often associated with extensive chemical pretreatment processes. As the ultimate detection limit capability is usually dependent not on instrumental factors (as in many other analytical techniques) but on the magnitude of the blank derived from reagent and environmental contamination, these chemical manipulations must be carried out under exacting conditions of cleanliness.

16.2 Mass spectrometry techniques in geology

There are five principal applications of mass spectrometry in the geological sciences.

16.2.1 *Spark source mass spectrometry*

Spark source mass spectrometry was the first mass spectrometric technique to be widely used in the analysis of geological samples. The sample is mixed with an electrically conducting binder (e.g. graphite) and pressed to form an electrode. Ionization is achieved by striking a radio-frequency electrical discharge between the sample electrode and a counter-electrode (an excitation technique similar to that used in spark source optical emission spectrometry, Chapter 6). Ions created in the discharge are sampled into a mass spectrometer. Because of the large range of kinetic energies of ions excited with this source, satisfactory mass resolution can only be achieved using a double-focusing mass spectrometer incorporating both magnetic and electrostatic sectors. The principal application of the technique is in the analysis of bulk samples for trace elements. In this respect the technique is potentially capable of sub-parts per million detection limits for a wide range of trace elements. This potential cannot always be realized, because of the presence of isobaric interferences in chemically untreated samples. The main difficulty with the technique relates to the large fluctuations which can occur in the rate of emission of ions from the spark source. Random effects can be minimized using a photographic plate detector and some form of internal standardization. However, instrumental precisions are rarely better than 5%, and a further problem arises from the fact that ion yields from the source are weakly matrix-dependent. Spark source mass spectrometry instrumentation is expensive in capital cost and requires sophisticated operation. In the hands of an expert, the technique is capable of generating excellent data, though it must be said that it is not widely implemented in many geological laboratories.

16.2.2 *Thermal ionization mass spectrometry*

This technique has made an important impact on progress in geochemistry over the last ten years. In thermal ionization mass spectrometry, the sample is loaded on to a tantalum or rhenium ribbon filament and ionized by electrically heating the filament to a controlled temperature. Successful determination can normally only be made by chemically separating elements prior to analysis to remove all potential isobaric interferences. A simple magnetic sector is usually adequate for mass dispersion. The technique is used in two distinct applications.

(i) To determine to very high precision the *relative isotopic abundances of the radiogenic elements*. These are the elements in which the abundance of one isotope slowly increases with time due to radioactive decay of a naturally–occurring parent isotope. The most commonly-determined elements which fall into this category include Rb/Sr, Nd/Sm and lead isotope ratios. This technique is most widely known as the method of determining the age of a rock. However, interpretations of ages can be affected by secondary geochemical processes following crystallization of the rock, and isotope ratio measurements are often just as important as tracers for these geochemical processes.

(ii) *Isotope dilution analysis* is the most commonly associated with thermal ionization mass spectrometry (although the technique can be used with the other forms of mass spectrometry as well). This technique is used to determine the bulk composition of selected trace elements. The rare-earth elements, for example, can be determined by isotope dilution, using an appropriate ion exchange separation. Very low detection limits can be achieved, controlled essentially by the procedural blank. Isotope dilution analysis involves 'spiking' the sample with an isotopically-enriched fraction of the element, and so cannot be used to determine elements which possess only one naturally-occurring isotope.

16.2.3 *Gas source mass spectrometry*

In the wider field of chemical analysis, gas source mass spectrometry is the most extensively used form of mass spectrometry. It is applied in the characterization of organic molecules in the organic/petrochemical industries. In the sphere of inorganic geochemistry, the technique is much more specialized, and is used almost exclusively to determine the stable isotope ratios of hydrogen, carbon, nitrogen, oxygen, sulphur and the inert gases. The isotope ratios of these elements do not vary due to radioactive decay, as described above. Rather, their isotope ratios can be changed in nature by various physical fractionation processes. For example, variations in oxygen isotope ratios in geological samples can be related to rock–water interactions and in some cases to palaeotemperatures. One exception to this general principle is the use of gas source mass spectrometers in the determination of radiogenic argon for traditional K–Ar age dating. Gas samples for analysis are normally released by a stepped ignition process and must be chemically purified prior to analysis to remove isobaric interferences. Since small, sometimes minute, amounts of gas are often released for analysis, expertise is required in vacuum-line technology and methods for reproducibly introducing sample and standard gases into the mass spectrometer for measurement.

16.2.4 *Secondary ion mass spectrometry*

Secondary ion mass spectrometry is most widely used in geological applications as a microbeam technique for the analysis of individual minerals in thin section, and as such is fully discussed in Chapter 11. In brief, the sample is bombarded with a focused beam of ions, $^{16}O^-$ being most widely used, although Ar^- and Cs^+ also have applications. Secondary ions sputtered from the surface of the sample are extracted into a mass spectrometer which must be designed as a double-focusing device due to the wide kinetic energy spread of ions in the source region. The technique is capable of significantly better detection limits than conventional electron probe microanalysis but suffers from serious isobaric interferences if minerals with complex matrices are analysed. Applications include the determination of trace elements in minerals, such as nickel at very low abundance in olivine, and the analysis of strontium and lead isotope ratios in isochron studies of calcites and zircons respectively.

16.2.5 *Inductively coupled plasma–mass spectrometry*

The latest application of mass spectrometry is to couple it to a radiofrequency inductively coupled argon plasma as an ion source. Plasma gases are extracted through an orifice into a differentially pumped chamber held at an intermediate vacuum of about 1 Torr. They are then passed into a quadrupole mass spectrometer for dispersion and measurement. The technique is therefore used for bulk analysis of trace elements in samples first taken up into solution (solid sampling variants using laser ablation are under development). The technique uses the characteristics of the plasma to ionize the sample efficiently, essentially free of matrix effects. Furthermore, the mass spectrum observed from the plasma is very simple, since ions from the plasma source are predominantly singly-charged, giving this technique important advantages over spark source mass spectrometry, for example. In practical applications, the mass spectrum is built up by repeatedly scanning over the required mass range (or peak hopping over the specified isotopes) so that the final spectrum is built up from 50 or more individual scans. ICP–MS is capable of isotope dilution analysis and the determination of relative isotope ratios, although at the present state of development the precisions required for isochron data cannot be achieved. The most important application of this technique is likely to be in the simultaneous multi-element determination of trace elements in geological samples first taken up into solution.

The mass spectrometer consists of four principal instrumental components:

(i) A device for ionizing the sample (the ion gun)

(ii) The mass analyser for dispersing ions according to their mass to charge ratio

(iii) An ion detector and readout system

(iv) A vacuum system capable of maintaining the ion flight tube at pressures better than 10^{-7} Torr.

16.3 The ion source

Of all the instrumental components of a mass spectrometer, the ion source is the most diverse, as can be seen from Table 16.1.

The characteristics of these sources are considered in the appropriate chapters. However, instrumentation for mass dispersion and detection obeys principles common to all forms of mass spectrometry, and these will now be considered.

16.4 The mass analyser

It has been known since the end of the last century that ions could be deflected by electric and magnetic fields. Indeed,

Table 16.1 Ion sources for mass spectrometry

Instrument	Ion source
Spark source mass spectrometer (Chapter 19)	Radiofrequency electrical discharge between two electrodes, one consisting of the bulk sample
Thermal ionization mass spectrometer (Chapter 17)	Chemically separated element is ionized from an incandescent tantalum or rhenium filament
Gas source mass spectrometer (Chapter 18)	Selected gas released by ignition purified and ionized in an electron impact source
Inductively coupled plasma mass spectrometer (Chapter 20)	Sample taken up into solution atomized and ionized in an argon plasma
Ion probe (secondary ion mass spectrometer) (Chapter 11)	Mineral sample in thin section sputtered and ionized by a focused primary beam of ions

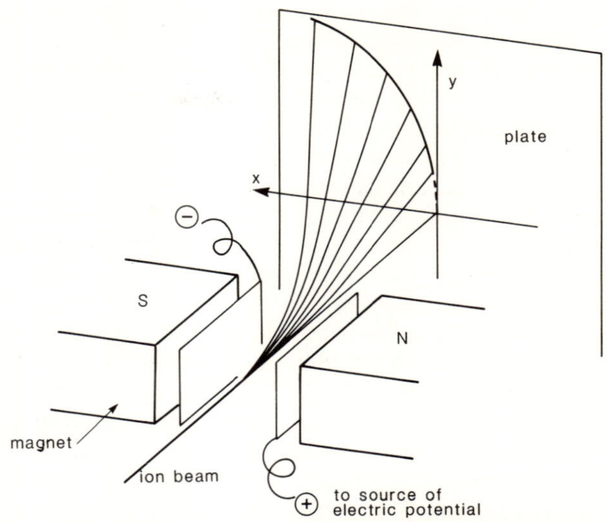

Figure 16.1 Parabolic mass spectrograph of Thompson using both magnetic and electrostatic dispersion. (Moore, 1966.)

an experimental mass spectrometer was developed in the first quarter of this century, one of the early proponents of the technique being J.J. Thompson. Thompson's instrument used both magnetic and electric fields to disperse ions. The combined effect of these fields was to deflect ions according to their mass to charge ratio in a parabolic arc. A photographic plate was used as a detector (Figure 16.1). This instrumentation was further developed by F.W. Aston, who used the technique extensively to measure the abundance of naturally-occurring isotopes. Modern mass spectrometers were derived through the work of A.J. Dempster, who in the early 1920s developed an instrument using a magnet as dispersing element and electronic detection of dispersed ions.

16.4.1 *The magnetic sector mass spectrometer*

An overall schematic view of a magnetic sector spectrometer is shown in Figure 16.2. Ions from the ion source are accelerated through a potential gradient, collimated by slits and passed into a strong magnetic field. Most types of ion sources generate three types of species, uncharged molecular and atomic fragments and positively and negatively-charged ions. Uncharged species cannot be measured by mass spectrometry at all. Instruments can be designed to measure both positively-charged and negatively-charged ions. In analytical applications, it is the positively-charged ions only that are determined (with the exception of ICP–MS; see section 20.7). Uncharged and negatively-charged ions ultimately collide with the walls of the instrument and are pumped away by the vacuum system.

A positively-charged ion entering a magnetic field is subject to a force which acts to deflect the ion. This force acts in a direction mutually perpendicular to both the direction of motion of the ion and the direction of the magnetic field (Figure 16.3). If the magnetic field were sufficiently extensive and homogeneous, the motion of the ion would describe a circular trajectory of radius r. The magnitude of this radius would depend on the balance between two forces. The first force is the centrifugal force due to the kinetic energy of the ion, which in the absence of the field would travel in a straight line. If the ion of mass m has been accelerated through a potential gradient (V) before entering the magnetic field, its kinetic energy will be given by

$$eV = \tfrac{1}{2}mv^2 \qquad (16.1)$$

where e is the charge on the ion and v its velocity.

The *centrifugal* force exerted on the ion travelling in a

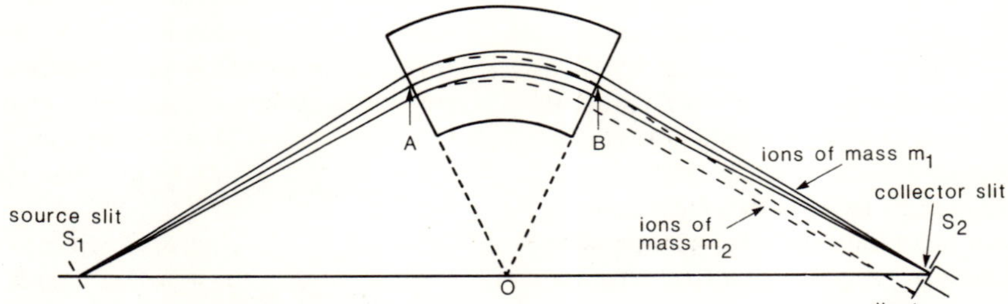

Fig. 16.2 A 60° magnetic sector showing trajectories of two ion beams, where mass of m_1 is greater than the mass of m_2. (Litzow and Spalding, 1973.)

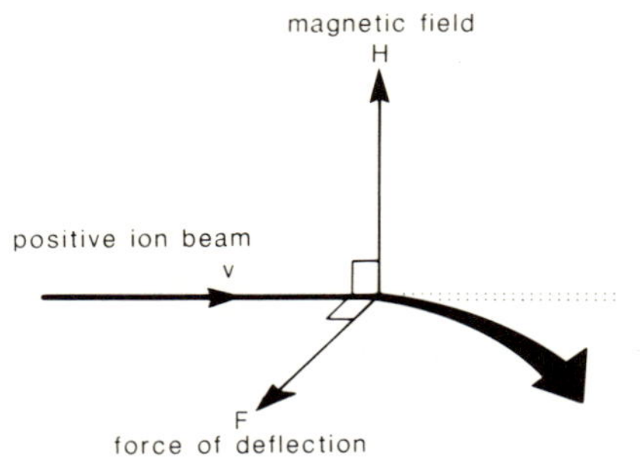

Figure 16.3 Force of deflection (F) caused by a magnetic field (H) acting on a mono-energetic positively-charged ion beam. If the magnetic field were sufficiently extensive and homogeneous, the ion beam would describe a circular trajectory.

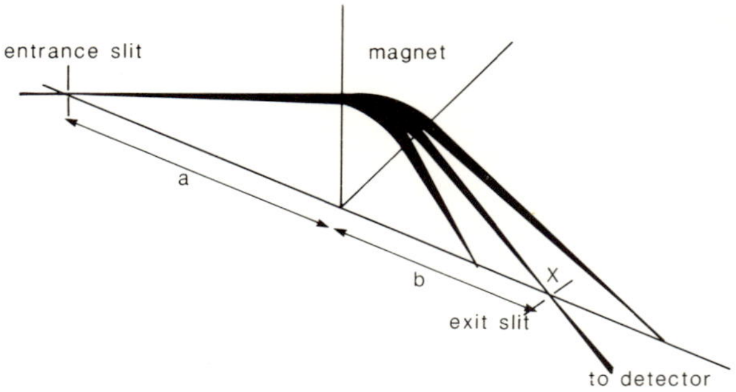

Figure 16.4 Focusing properties of a magnetic sector mass analyser. In idealized circumstances, the ion beam is brought into focus along a plane passing through the ion source slits and apex of the magnetic sector. (Beynon and Brenton, 1982.)

circle of radius r is

$$\frac{mv^2}{r} \tag{16.2}$$

The second force which must exactly balance this centrifugal force is the *magnetic centripetal* force, that is the force of deflection created by the magnetic field:

$$\text{force} = HeV \tag{16.3}$$

where H is the strength of the magnetic field.

For these two forces to balance, it can be shown from equations (16.1), (16.2) and (16.3) that

$$r = \frac{1}{H}\sqrt{\frac{2Vm}{e}} \tag{16.4}$$

or

$$\frac{m}{e} = \frac{H^2r^2}{2V}. \tag{16.5}$$

Equation (16.4) indicates that the degree of deflection of an ion of constant mass to charge ratio increases (r smaller) at higher field strengths or lower acceleration potentials in the ion source.

In practice, an ion beam is deflected through a magnetic sector having an apex angle typically of 60° (90° and 180° sectors are also used) (Figure 16.2). The radius of the ion flight path (r in equation (16.5)) is normally fixed by instrument design. The specific mass-to-charge ratio observed at a fixed exit slit depends therefore on the field strength of the magnet (H) or the accelerating potential (V) of the ion source. In order to scan a mass spectrum it is necessary to vary one or the other of these parameters. In most mass spectrometer designs it is normal to change the magnetic field strength. A few simple instruments are fitted with a permanent magnet as the dispersion element, it then being necessary to change the ion source accelerating potential to scan the mass spectrum.

Discussion so far has considered the behaviour of a single ion. In real instruments a small solid angle of ions enters the magnetic sector. Developments in the 1930s showed that the magnetic sector exerts a focusing influence on the divergent beam of source ions. Provided this ion beam enters and leaves the magnetic field at right angles, the plane of focus is as shown in Figure 16.4, lying on the line which passes through the entrance slit and the apex of the magnetic sector. In most instruments the deflected ion beam is constrained by a fixed exit slit across which the mass spectrum is scanned. Using such an arrangement, the resolution of the dispersed beam is often only moderate. One reason for this is that the flight path of the diverging source beam is such that not all components enter the magnetic sector perpendicular to the magnetic field. In order to overcome this difficulty, a modified magnetic sector geometry whose design is attributed to Cross (1951) has been adopted on some thermal ionization and gas source mass spectrometers. In this design the ion beam enters the magnetic sector at an angle of 26.5° and the source to magnetic sector and magnetic sector to collector slits are arranged to be equal to twice the radius of the ion beam flight path in the magnetic sector (Figure 16.5). One advantage claimed for this arrangement is that the mass dispersion (that is distance between adjacent mass peaks) is twice as large as that of a conventional arrangement having the same magnetic sector radius. Secondly, the design affords a focusing capability on that component of the ion beam which diverges in the z direction (i.e. perpendicular to the plane of the magnetic sector). This increases the proportion of ions transmitted through the magnetic sector and hence instrument sensitivity is improved compared with conventional designs. Sensitivity is further enhanced by the possibility of doubling the collector slit width in order to achieve a larger signal but the same mass resolution as a conventional instrument. Claimed ion transmission rates in thermal ionization mass spectrometry are one ion of uranium detected for every 500 atoms loaded on to the source.

By changing the angle at which the source beam enters the magnet sector it is possible to change the orientation of the focal plane. Multiple collector instruments must take into account this effect and allow for differences in the position of focal points of adjacent mass peaks.

16.5 Resolution

The resolution of a mass spectrometer is a measure of the ability of the instrument to distinguish between two adjacent mass peaks. Although several conventions are used, the one generally adopted is illustrated in Figure 16.6. Resolution is defined as $M/\Delta M$ and refers to two adjacent mass peaks of

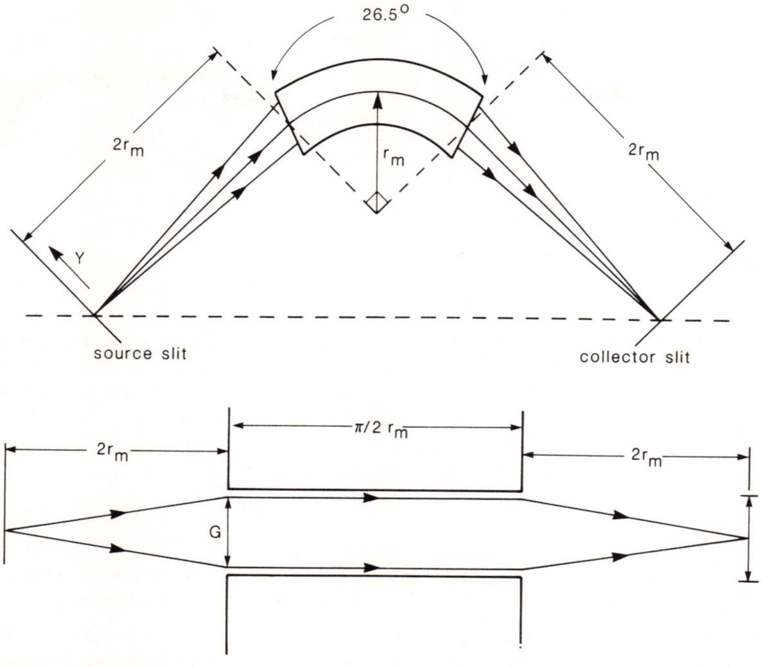

Figure 16.5 Extended geometry mass analyser, after Cross (1951). The ion beam enters the magnetic sector at an angle of 26.5°, and source–magnet and detector–magnet dimensions are arranged to be twice the radius of the magnetic sector flight tube. This design imparts a focusing action perpendicular to the plane of the magnetic sector. (V.G. Isotopes Newsletter.)

equal intensity of mean mass M such that the separation between their maxima ΔM is just sufficient to cause a valley at a relative intensity of 10% in their overlapping tails.

By this convention, an instrument with a resolution of better than 40 is required to separate adjacent unit masses when measuring the isotopes of argon, for example, and one of better than 200 to measure lead isotope ratios. In practice, instruments capable of achieving resolutions of at least twice these values would be used to permit greater analytical flexibility.

The criterion used above is that of separating adjacent mass peaks of nominal integral mass number. This satisfies the majority of requirements in gas source and thermal ionization mass spectrometric applications. However, in secondary ion mass spectrometry (ion probe analysis), it is not possible to chemically separate the elements of interest from matrix interferences. Some simplification in interference effects can be achieved by limiting applications to the analysis of minerals of relatively simple composition (e.g. calcite and zircon). However in some circumstances it is necessary to distinguish isotopes of nominally identical mass number on the basis of the mass defect effect (section 16.13). In order to do this, mass spectrometers having resolutions up to 5000–10 000 are required. Even at these high instrumental resolutions, it is not possible to resolve all interferences, in particular those arising from metal hydride ions (section 11.2).

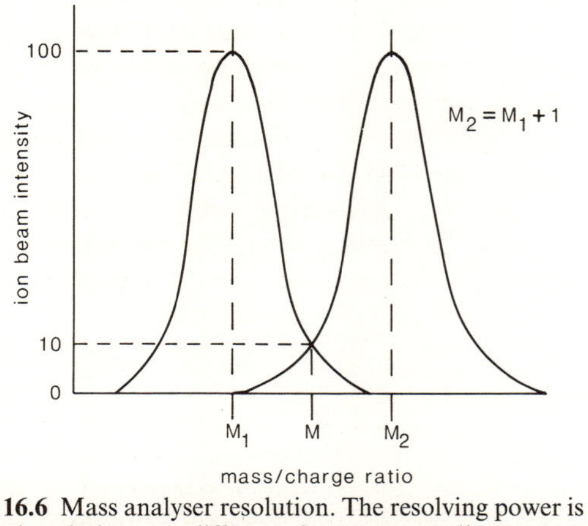

Figure 16.6 Mass analyser resolution. The resolving power is measured as the relative mass difference between two adjacent mass peaks of equal intensity, the tails of which overlap at an intensity of 10% peak height.

The single magnetic sector mass spectrometer gives adequate resolution performance in many applications. Thus a resolution of $M/\Delta M$ of 500 is adequate for most thermal ionization applications in which it is necessary to scan the mass range 2–700 amu. Conversely, a maximum resolution of about 100 is adequate for a gas source mass spectrometer designed to detect argon with mass of about 40. However, there are some applications where the performance of a single-focusing mass spectrometer is not adequate. These include both spark source mass spectrometry and secondary ion mass spectrometry (ion probe analysis). In both these applications, ions are generated in the source with a large spread of kinetic energies (v no longer constant in equation (16.2)). This spread is sufficient to degrade the resolution performance of a single magnetic sector instrument to an unacceptable extent. In such cases a double-focusing mass analyser is required, in which an

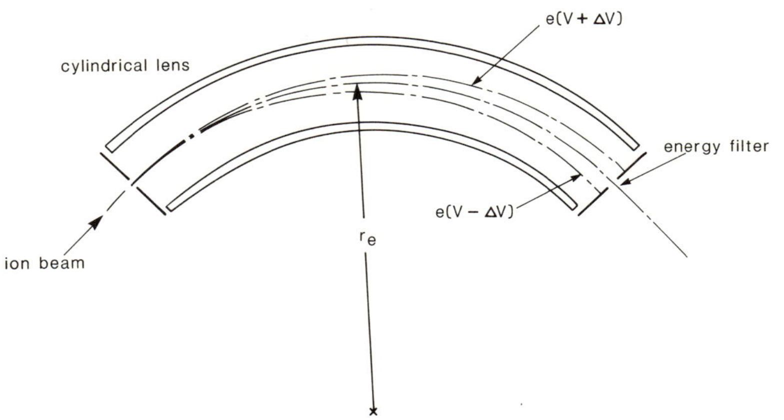

Figure 16.7 The electrostatic sector as an energy filter. Only ions of energy *V* describe a trajectory through the sector exit slits. Ions of other energies are not transmitted. (White, 1968.)

electrostatic sector is used in conjunction with the normal magnetic dispersion device.

16.6 Double-focusing mass spectrometer

The double-focusing mass spectrometer uses two dispersion elements, an electrostatic sector, used to filter the source ion beam on an energy basis, passing an essentially mono-energetic beam of ions into a magnetic sector for mass dispersion. This double-focusing arrangement is appropriate therefore in two circumstances:

(i) In spark source mass spectrometry where the large spread in energies of ions created in the spark source would otherwise degrade resolution to an unacceptable extent

(ii) In ion probe analysis where mass resolution of 1000–10 000 is required to permit separation of mass peaks of nominally identical integral mass number. The electrostatic sector is shown in Figure 16.7. It consists of two plates, one biased at positive and the other at negative electrical potential. The plates are held in a curved configuration with the ion flight path of mean radius r between them. On entering the electrostatic field created between these two plates, the ion beam is deflected along a circular path.

As with the magnetic sector mass analyser (section 16.4), the centrifugal force due to the kinetic energy of the ion beam entering the electrostatic sector is given by

$$\frac{mv^2}{r} \tag{16.6}$$

where m is the mass of the ion and v its velocity.

This must exactly balance the centripetal force created by electrostatic plates:

$$\frac{Ee}{d} \tag{16.7}$$

where E is the voltage difference between the electrostatic plates, d the distance between the plates and e the charge on the ion (E/d is therefore the field strength within the electrostatic sector). Since the kinetic energy (KE) of the ion entering the electrostatic field is given by

$$\tfrac{1}{2}mv^2 \tag{16.8}$$

the following expression for the radius of curvature of the ion beam can be determined:

$$r = \frac{Ee}{dmv^2} = \frac{2Ee}{d(\text{KE})} \tag{16.9}$$

The radius of deflection (r) of the ion beam therefore depends on the electrostatic field strength (E/d) and the kinetic energy (KE) of the ions entering the field, but *not* on the mass of the ion. Since r is fixed by instrument design, E/d can be adjusted so that the electrostatic sector transmits ions of the same kinetic energy irrespective of their mass. Ions of different kinetic energy do not describe an arc of the appropriate radius r, and are not transmitted through the sector exit slits. The electrostatic sector has some further focusing properties, in that the divergent solid angle of ions entering the sector is focused by the electrostatic field (Figure 16.8).

Two geometric arrangements of the electrostatic sector and magnetic sector are widely used in mass spectrometer design. The Nier–Johnson double-focusing mass spectrometer (Figure 16.9) uses a 90° electrostatic sector to bring an ion beam of filtered kinetic energy into focus on an intermediate slit. This ion beam is then transmitted into a magnetic sector (60° in the original Nier–Johnson design,

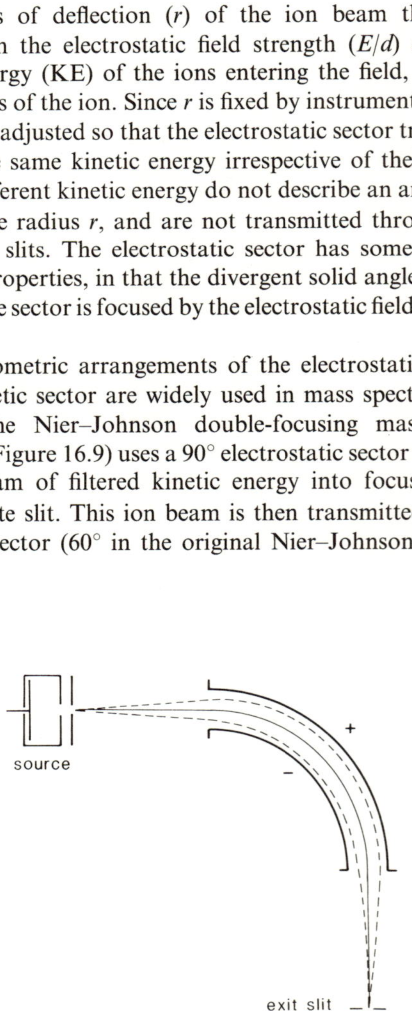

Figure 16.8 The focusing properties of an electrostatic sector towards a mono-energetic beam of ions. (Ewing, 1975.)

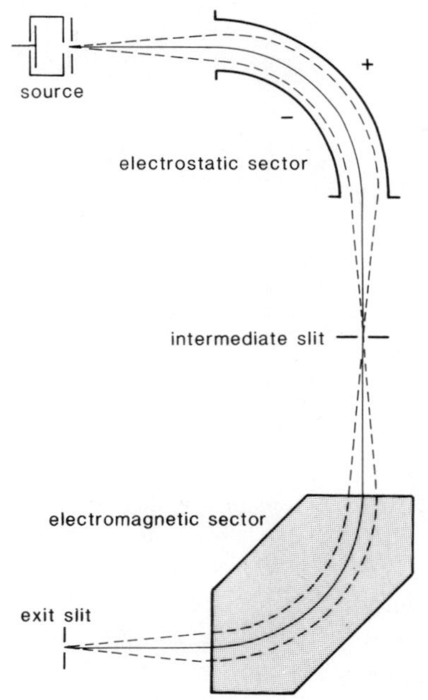

Figure 16.9 The Nier–Johnson double-focusing mass spectrometer incorporating both an electrostatic energy filter and a magnetic sector mass dispersion device. (Ewing, 1975.)

though 90° is now commonly preferred) where it undergoes mass dispersion. The ion beam transmitted through the exit slits is most suitably detected electronically (not photographically).

The second design is that of Mattauch and Herzog, an instrument widely used in spark source mass spectrometry where it allows the entire ion spectrum to be recorded simultaneously on a photographic plate (Figure 16.10). This is possible in the Mattauch–Herzog design by using an electrostatic sector with an angle of 31°50′. This has the property of deflecting the source ion beam to form a mono-energetic transmitted ion beam focused at infinity. This ion beam passes into a magnetic sector having an angle of 135°. Since the incoming beam is parallel, all ions enter the mass analyser at right angles to the magnetic field, thus eliminating aberration effects and permitting the entire spectrum of deflected ions to be brought into sharp focus on a photographic plate at the opposite face of the 135° sector.

Double-focusing mass spectrometers are potentially capable of achieving very high mass resolutions ($M/\Delta M$ in excess of 50 000). In practical situations, this cannot normally be achieved, since slit dimensions would have to be reduced to an unacceptably small width, leading to excessive signal amplification factors and electronic noise. The practical limit to resolution in ion probe operation using a Nier–Johnson mass analyser is 5000–10 000($M/\Delta M$). In spark source mass spectrometry, an instrument would probably be run at relatively low resolution (500 $M/\Delta M$) to obtain adequate signals for elements at trace concentrations. The double-focusing instrumentation is being used in this application largely to overcome serious resolution degradation that would result from the large kinetic energy spread of ions formed in the spark source.

16.7 Quadrupole mass spectrometer

The quadrupole mass spectrometer is a very compact mass dispersion device which was developed from ideas formulated in the 1950s by Paul *et al.* (1958). This instrument does not employ an electromagnet at all, but rather four rods which must be machined and aligned with a high degree of accuracy (Figure 16.11). Each pair of opposite rods is electrically connected, a positive dc voltage being applied to one pair and an equal, negative, dc voltage to the other pair. Superimposed on these dc voltages is a radiofrequency ac signal arranged so that the phase of this signal applied to one pair of rods is 180° different from that applied to the other pair. Under ideal circumstances, the rods should be hyperbolic in shape and the lines of equal potential then created are shown in Figure 16.12. In commercial instrumentation these rods are circular in cross-section to permit more reliable machining. The quadrupole mass filter is commonly used with inductively coupled argon plasma or gas chromatogram sources. Ions introduced into the instrument

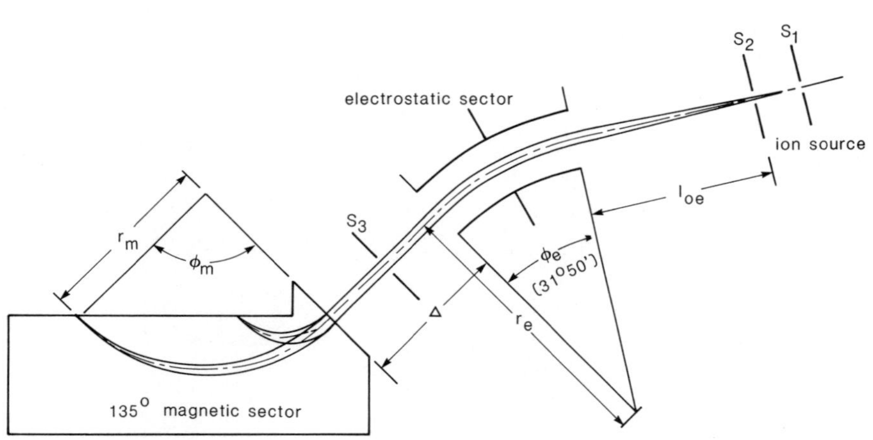

Figure 16.10 The Mattauch–Herzog double-focusing mass spectrograph used in spark source spectrometry. (White, 1968.)

s

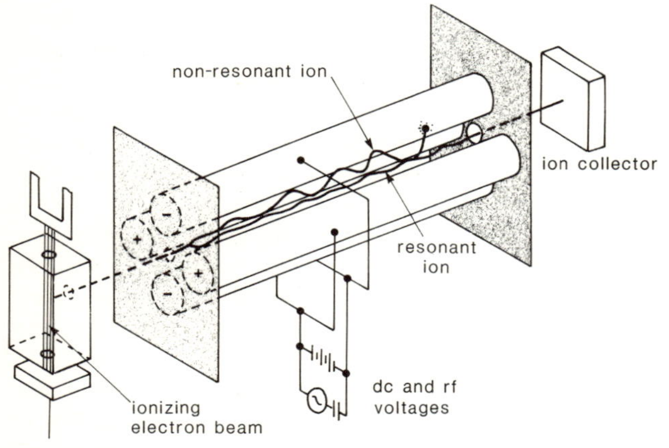

Figure 16.11 The quadrupole mass filter showing the flight path of ions through the field created between the four quadrupole rods. (Lichtman, 1964; as reproduced by Skoog and West, 1980.)

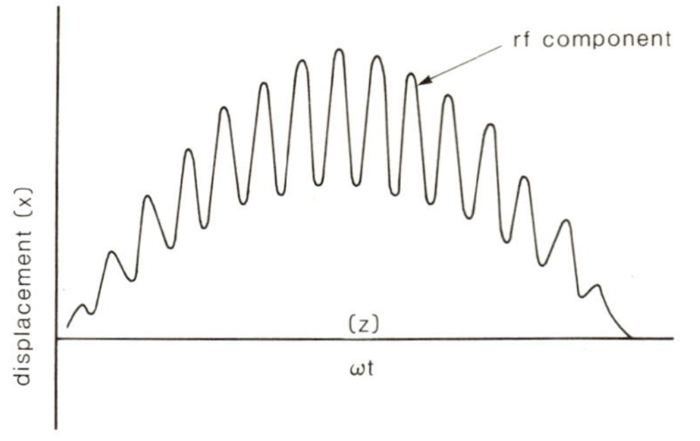

Figure 16.13 Trajectory of an ion in the x–z plane of a quadrupole mass filter. (White, 1968.)

are accelerated through a relatively low potential gradient (a few volts) and caused to drift along the axis of the quadrupole. Here they come under the influence of a varying combination of rf and dc fields which impart a complex motion to the ion beam (Figure 16.13). This motion includes a sinusoidal component in both x and y directions, the magnitude of which depends on the magnitude of the applied ac and dc fields, the frequency of the ac field, and the mass-to-charge ratio of the individual ion. For any set of operating conditions, there exists a unique mass-to-charge ratio which will satisfy the resonance requirements for these ions to be transmitted through the quadrupole to the exit slits. Other ions entering the quadrupole oscillate with frequencies which progressively increase in amplitude until they strike a quadrupole rod or are pumped away by the vacuum system. The ratio between the applied dc and ac fields is critical to successful operation. Theory predicts that there is a critical ratio between the applied dc voltage (V_{dc}) and the voltage amplitude of the ac signal (V_{ac}) of 0.168. For values of V_{dc}/V_{ac} which exceed 0.168, stable ion transmission by the quadrupole is not possible. For values of V_{dc}/V_{ac} which increase up to a limiting value of 0.168, mass resolution becomes increasingly better. In practice, there is only a limited range of voltage ratio values over which adequate resolution and stable ion trajectory can be obtained. One of the advantages of the quadrupole is the ease with which a trade-off can be made between high resolution (V_{dc}/V_{ac} tends to 0.168) and large signal (V_{dc}/V_{ac} lower than 0.168).

The degree of resolution attainable by the quadrupole mass filter depends on the number of oscillations induced in the motion of an ion as it passes down the quadrupole axis; the more the better. Resolution can be improved at the design stage, therefore, by increasing the length of pole pieces or the operating radiofrequency. The maximum resolution attainable on many instruments designed for ICP–MS is about $M/\Delta M = 500$.

Quadrupole mass spectrometers have different operating characteristics as compared with magnetic sector analysers:

(i) The quadrupole mass spectrometer may be rapidly scanned to cover, if desired, the entire mass range in a second or so; in ICP–MS applications, a spectrum is accumulated

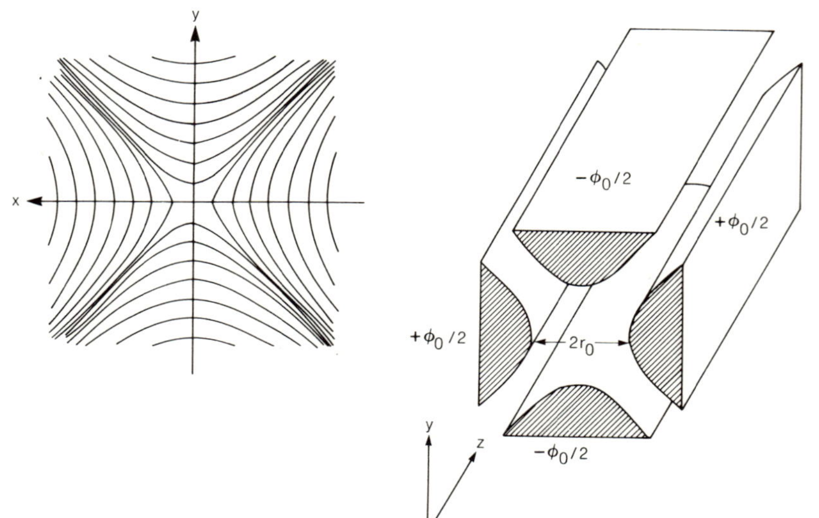

Figure 16.12 Equipotential lines of field created by ideal parabolic-shaped rods of a quadrupole mass filter. (Dawson, 1976.)

by superimposing data accumulated from many such scans. Magnetic sector analysers cannot be scanned anywhere near as rapidly.

(ii) The spectrum obtained by scanning a quadrupole mass filter is linear in mass, that from a magnetic sector analyser is exponential in mass.

(iii) The quadrupole is capable of accepting ions for dispersion from a source over a relative wide solid angle. The lateral motion of ions entering the quadrupole is rapidly swamped by the induced magnetic fields. Ions entering a magnetic sector analyser must be tightly collimated to preserve adequate resolution.

(iv) As referred to above, it is simple to adjust the trade-off between ion transmission intensity and resolution on a quadrupole mass spectrometer.

(v) The construction of a quadrupole is relatively compact and small.

(vi) The main disadvantage which prevents more widespread use of the quadrupole is that the precision of isotope ratio measurements is about one to two orders of magnitude worse than can be obtained on a magnetic sector instrument. At the present state of development, therefore, quadrupole filters are not suitable for high-precision isotope ratio measurements of the type traditionally obtained by thermal ionization mass spectrometry.

16.8 Ion detectors

The very earliest instruments used a fluorescent screen for the detection of ions. However on modern instruments, four devices are commonly used, depending on application. These are the photographic plate, the Faraday cup (or bucket), the electron multiplier (including the electron channel multiplier) and the Daly detector. The analytical characteristics of these detectors will now be considered.

16.8.1 *The photographic plate*

In modern applications, the use of photographic plate detection is restricted almost exclusively to spark source mass spectrometry using an instrument of the Mattauch–Herzog design. The detector itself consists of a glass plate, commonly 2″ wide by 10″ long, one surface of which is coated with a photographic emulsion. Ions impinging on the emulsion activate the formation of silver deposits, and after a suitable exposure period the resultant mass spectrum is revealed by photographic development and fixing of the plate. At any given mass, about 1000 ions (1.6×10^{-16} coulombs of charge) are required to give a perceptible image. The resultant spectrum is quantified using an optical densiometer to measure the absorbance of each mass line. A typical plate spectrum is shown in Figure 16.14. Conventional photographic emulsions cannot be used for ion detection since a large proportion of ions would be absorbed by the surface gelatin layers. Special emulsions have therefore been developed in which the photosensitive crystals are exposed near the surface of the emulsion. This makes these plates very sensitive to bruising during manipulation.

The principal advantage of photographic plate detection is that it permits simple simultaneous multi-element interpretation of mass spectra and provides a permanent visual record of an analysis. There are also some important disadvantages. At the practical level, mass spectrometers must be designed with special cartridges to permit loading and exposure of glass plates under vacuum. But, more fundamentally, the photographic plate has a rather limited dynamic range. For spark source mass spectrometry, Taylor and Gorton (1977) suggest that the dynamic range in mass intensities required to observe isotopes of interest in rock samples corresponds to a factor of about 2000. However, a photographic plate has a dynamic range of only about 50-fold. To detect and analyse the complete mass spectrum from a rock sample, it is therefore necessary to record typically 15 exposures ranging from 1 to 1500×10^{-9} coulombs of ion charge reaching the photographic plate for each exposure. This requires the exposure of between two and six photographic plates for each sample. Further difficulties arise from the fact that the sensitivity of individual photographic plates can vary significantly (Taylor, 1965) and that the exposure intensity of the mass lines within a plate is not a completely linear function of concentration. It is thus essential to calibrate each photographic plate individually; techniques for doing this in spark source mass spectrometry are discussed in Chapter 19. Perhaps the most important advantage of the photographic plate in spark source mass spectrometry is its ability to integrate mass spectrum intensities from what is an intrinsically noisy ion source.

16.8.2 *Faraday cup detection*

The Faraday cup is the simplest, most robust and least sensitive of the commonly-used electronic detection devices. It consists of three components: a beam-defining slit (the exit slit of the mass analyser), an electrode (the Faraday cup or bucket) and an electron repeller plate (Figure 16.15). The principle of operation is as follows. Each ion impinging on the Faraday cup deposits a charge e on this electrode. For an ion count rate of n ions per second, the corresponding charge deposited on the electrode is

$$\begin{aligned} ne &= 1.602 \times 10^{-19}.n \,\text{coulombs s}^{-1} \\ &= 1.602 \times 10^{-19}.n \,\text{amps} \end{aligned}$$

Figure 16.14 Mass spectrum recorded on a photographic plate. Ten exposures have been recorded, one displaced above the other to cover the range of signal intensities present. (Pecsok *et al.*, 1976.)

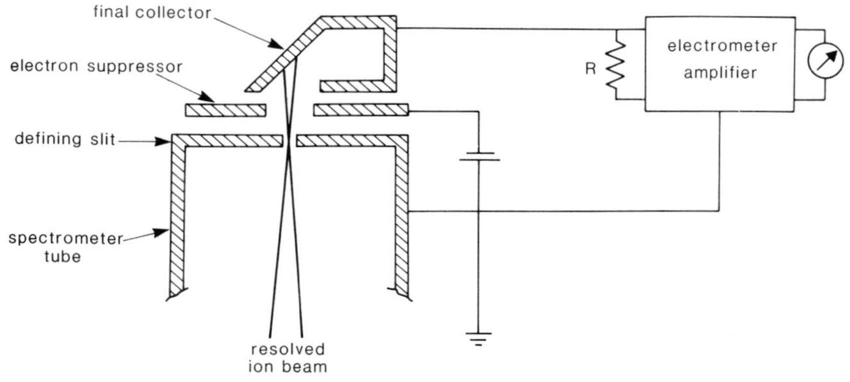

Figure 16.15 Faraday cage ion beam detector. (White, 1968.)

This charge is normally leaked to earth through a resistor having a very high value (in the range 10^8–10^{12} ohm). The ion count rate is proportional to the voltage drop across the resistor as measured with a sensitive voltmeter of high input impedance.

One of the potential difficulties with the Faraday collector is that ions impinging on the electrode cause, on impact, the emission of secondary electrons. These cause a loss of signal, the magnitude of this effect depending on the mass of the ion being detected as well as the accelerating potential in the ion source. The loss of secondary electrons from the Faraday cup would therefore cause a non-linear response of this detection device. In order to suppress this effect, an electron repeller plate is placed in front of the Faraday cup. This plate is held at a negative potential (50–90 volts) with respect to the Faraday electrode, which is itself designed in the shape of a cup or bucket to further suppress this effect.

The ultimate sensitivity of this detection arrangement is limited by pick-up of electronic noise in the signal cable connecting Faraday cup and amplifier and intrinsic electronic noise in the measurement resistor. The former is reduced by electronic screening and keeping all interconnecting signal cables as short as possible, the latter by introducing some smoothing in the output signal by operating the amplifier at time constants of one second or so. With these precautions, ion beam signals in the order of 10^{-15} amps can be detected (16 000 ions per second). Unlike the electron multiplier detector to be considered next, the Faraday cup has no internal gain in the detected signal except that which is provided by the operational amplifier.

16.8.3 *The electron multiplier detector*

In operation, the electron multiplier detector (Figure 16.16) is similar to the conventional photomultiplier (section 3.12, see also Figure 3.39). However, instead of being sensitive to light photons, the first-stage electrode of the electron multiplier is made from a material designed to optimize the yield of secondary electrons emitted when its surface is bombarded with ions. Beryllium/copper or silver/magnesium alloys are suitable for this process (Dalrymple and Lanphere, 1969, p. 75). These secondary electrons are then accelerated down a chain of dynodes. On impact with each dynode, each secondary electron induces the emission on average of two further electrons. The next dynode down the chain is held at an increasingly negative bias and shaped to create electrostatic fields which will both accelerate and focus the secondary electron beam on to its surface. A typical electron multiplier detector contains between 10 and 20 dynodes and therefore gives an overall internal signal gain of 2^n (n = number of dynode stages). Multiplication factors in the range 10^5–10^7 can readily be achieved. The last element in the electron multiplier is essentially a Faraday cup which collects the signal. This is then

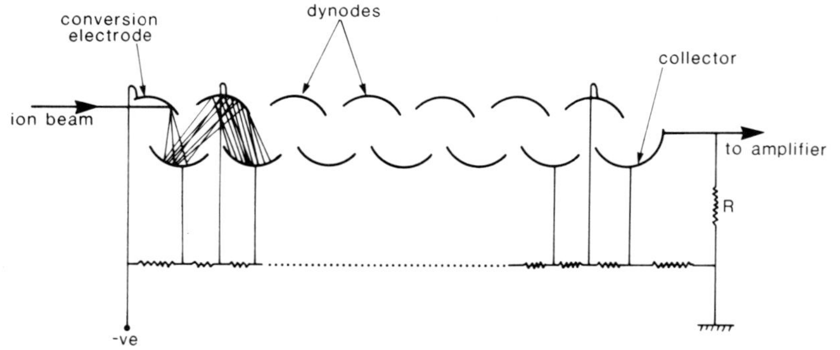

Figure 16.16 Electron multiplier ion beam detector. (Beynon and Brenton, 1982.)

fed to a high-value resistor and dc amplifier. Since the avalanche of electrons within the dynode multiplier occurs very rapidly, dead-time effects are very small and electron multipliers can be operated with amplifier time constants of 10^{-3}–10^{-5} s. In comparison with the conventional Faraday cup, the electron multiplier shows an improvement in signal sensitivity by a factor of 1000 or so, permitting ion beam currents below the 10^{-17} A level (160 ions per second) to be detected. Disadvantages of this device are that the conversion dynode is sensitive to both photons and electrons and adequate shielding must be provided in ICP–MS applications using a quadrupole mass filter, for example. Furthermore, the device suffers from the disadvantage that the secondary ion yield at the conversion dynode varies with ion mass, causing some non-linearity effects when the ratio of adjacent mass peaks is to be measured or when an ion with a variable isobaric interference is to be measured. However, these problems can be eliminated by operating the detector in an ion pulse counting mode. The measurement of the ion beam signal then depends on the rate of arrival of individual ion pulse signals and not on the precise current created by the ion beam.

16.8.3 *The electron channel multiplier*

The electron channel multiplier represents a continuous form of the electron multiplier detector. Instead of containing discrete dynodes, the electron channel multiplier consists of a glass tube, the inside surface of which is coated with a thin film of semiconducting material (Figure 16.17). This film has a high resistance, and a large electrical potential is applied across the tube. Ions striking the open end of the multiplier cause, on impact, the emission of secondary electrons. These electrons are accelerated by the applied potential further down the detector but rapidly collide with the semiconductor coated walls again, causing the emission of further secondary electrons. This process continues until the electron avalanche reaches the collector at the end of the detector tube. As in the electron multiplier, a large internal amplification is observed in the detected signal. In ICP quadrupole mass spectrometer applications, it is necessary to mount the detector off-axis to prevent line-of-sight photon interference from the ion source, and in these circumstances, very low background count rates can be observed, typically one count in 10 seconds (Dawson, 1976, p. 140).

16.8.4 *The Daly detector*

The Daly detector (Figure 16.18) is the most sensitive ion beam detector available in routine mass spectrometry and is used when the smallest signals are to be measured. It consists of an electrode (the Daly electrode) held at high negative potential (−30 kV). Ions transmitted through the exit slits of the mass analyser are accelerated into the electrode by the potential field, and on impact eject an average of eight secondary electrons per ion. These secondary electrons are scattered on to an adjacent scintillator which is coupled to a photomultiplier. The secondary electrons interact with the scintillator, causing the emission of photons which are detected and amplified by the photomultiplier. The Daly detector system is normally used in ion counting mode and can offer extremely low background count rates compared with other detectors. Background count rates as low as 0.05 cps can be achieved at a dead time of 5 ns. Ion beam currents as low as 10^{-18}–10^{-19} A can be detected in the most demanding applications.

Some mass spectrometers offer two ion beam detectors: the Faraday cup detector for the majority of applications and the Daly detector when ultimate performance is required. This configuration is particularly attractive since no moving parts are required. To switch from Faraday cup to Daly detector, the ion beam is deflected by switching on the −30 kV Daly operating bias (Figure 16.19).

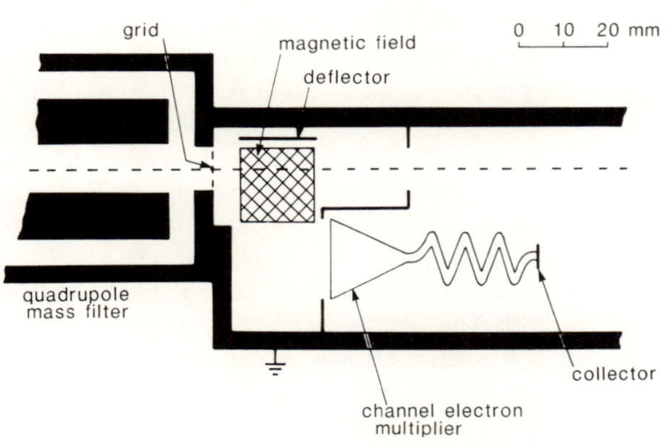

Figure 16.17 The channel electron multiplier fitted off-axis in a quadrupole mass spectrometer application. The ion beam is deflected into the multiplier by a positively biased deflector plate. (Benninghoven *et al.*, 1974; as reproduced by Dawson, 1976.)

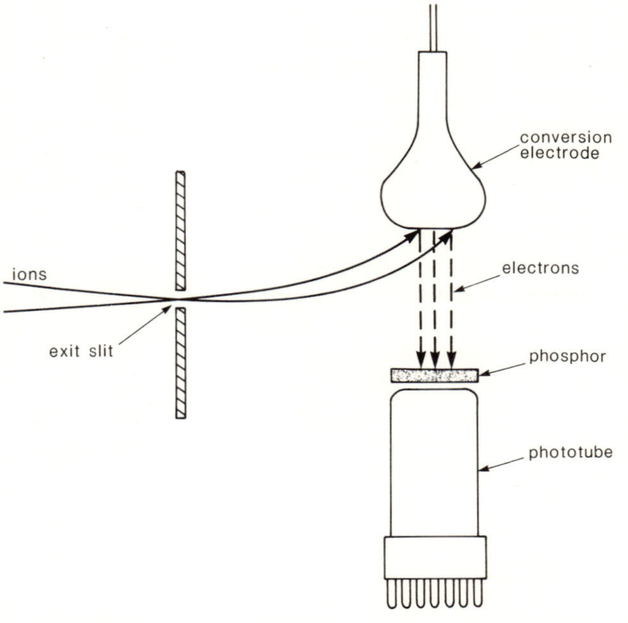

Figure 16.18 The Daly detector and scintillation counter. (Daly, 1960, as reproduced by White, 1968.)

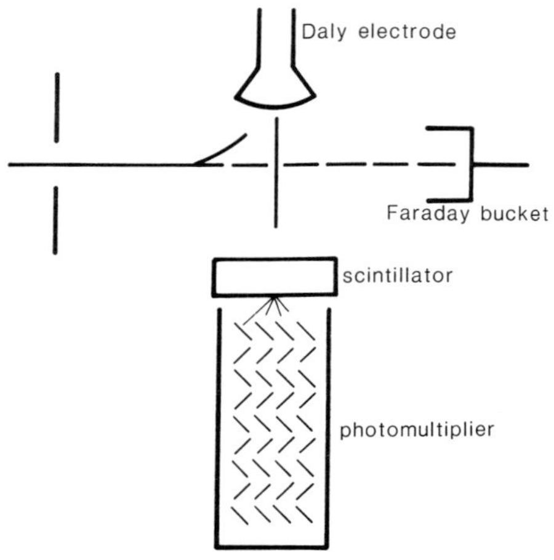

Figure 16.19 Changeover arrangement between Faraday cup and Daly detectors. Changeover is effected by applying a negative bias to the Daly conversion electrode. (V.G. Instruments.)

16.9 Vacuum requirements

It is implicit in the operation of a mass spectrometer that the entire ion flight path is pumped out to high vacuum. The main reason for this is that collisions between ions and residue gas molecules will cause some loss of kinetic energy from the ion so that it will no longer describe the appropriate flight path during deflection through the mass analyser (see also abundance sensitivity, section 16.10). In order to reduce the possibility of collision, the mean free path of ions within the vacuum system must be long compared with the ion flight path. The mean free path (l) of molecules of nitrogen (and therefore approximately for air) is given by

$$l = \frac{5 \times 10^{-5}}{p} \text{ m} \tag{16.10}$$

where p is the gas pressure (Torr) (Beynon and Brenton, 1982, p. 9).

From this equation it can be calculated that the mean free path varies with gas pressure as shown in Table 16.2. These data would indicate that an operating vacuum of at least 10^{-6} Torr is required to minimize the risk of collisions in the flight tube. In many applications an operating vacuum of better than 10^{-8} Torr is preferred (see section 16.11).

Some of the instrumentation for vacuum generation (rotary and oil-diffusion pumps) has already been discussed in section 10.5. Oil-diffusion pumps with rotary backing pumps are not generally considered capable of achieving a satisfactory operating vacuum and have been supplanted by titanium sublimation pumps or, more recently, by turbomolecular pumps. These can operate at higher rates and attain a higher ultimate vacuum. In order to reach vacuums of better than 10^{-6}–10^{-7} Torr within reasonable pumping times, it is usually necessary to bake-out the entire flight tube. At atmospheric pressure, gas molecules and in particular water molecules and organic solvents are adsorbed on to the surfaces of metals and other instrument components. On re-pumping, these molecules are released slowly, causing a long delay before an operating vacuum is reached. Heating the flight tube to at least 150°C causes more rapid release of these molecules and therefore higher pumping rates. Since vacuum requirements are much more rigorous in mass spectrometer than in microprobe applications, for example, vacuum grease, which itself has a significant vapour pressure, is avoided. Flanged components in the flight tube/detector assemblies are usually sealed with indium gaskets (synthetic viton gaskets would degrade at 150°C). Furthermore, it is necessary to avoid the use of some metals in the construction of mass spectrometers. Mercury has a vapour pressure of 10^{-3} Torr at room temperature, but cadmium and zinc also have finite vapour pressures (about 10^{-6} Torr at 120° or 180°C respectively) and would therefore inhibit pump-down during baking.

Table 16.2 Mean free path of nitrogen molecules as a function of vacuum pressure

Vacuum pressure (Torr)	10^{-5}	10^{-6}	10^{-7}	10^{-8}
Mean free path (m)	5	50	500	5000

16.10 Abundance sensitivity

Some instrumental factors which affect the mass resolution of a mass spectrometer have been discussed above. However, the usual definition of resolution refers specifically to the overlap between two adjacent mass peaks of equal intensity (section 16.5). Although adequate to describe general instrumental performance, this definition is not particularly useful in many practical analytical applications where it is necessry to make precise measurements of a small mass peak lying adjacent to an intense mass peak. It is not uncommon to find that the relative intensities of these peaks may differ by several orders of magnitude and it is then necessary to know to what extent an instrument is capable of resolving the tail of the major peak from measurements of the minor peak intensity. This phenomenon is incorporated in the definition of *abundance sensitivity*. The abundance sensitivity is normally defined in terms of a specified mass peak and is measured as the ratio of the intensity of the tail of a major peak (at the mass of the minor isotope) relative to the intensity of the major peak itself. Abundance sensitivity is therefore a measure of the resolution of an instrument, but measured in terms of a tail-to-peak overlap coefficient. In general, tail overlap arises from second-order effects. Figure 16.20 shows an instrument set up to determine a minor isotope peak in the presence of an adjacent peak of high intensity. At the selected instrument resolution, the overlap of the tail of the major peak with the minor peak should be negligible. That this is not so depends on several phenomena which involve collisions between major-element ions and residual gas molecules or the walls of the flight path. These collisions can result in the ion changing direction, changing its kinetic energy, or both. The net effect is that the expected flight path through the magnetic sector is changed so that the major element ion appears at the exit slit set for the minor ion of interest. This effect can be reduced by reducing the vacuum pressure or reducing the width of the mass analyser exit slit. Under ideal circumstances, analyser vacuums in the range 10^{-8}–10^{-9} Torr are preferred. The width of the mass analyser exit slit is specified largely to permit the detection

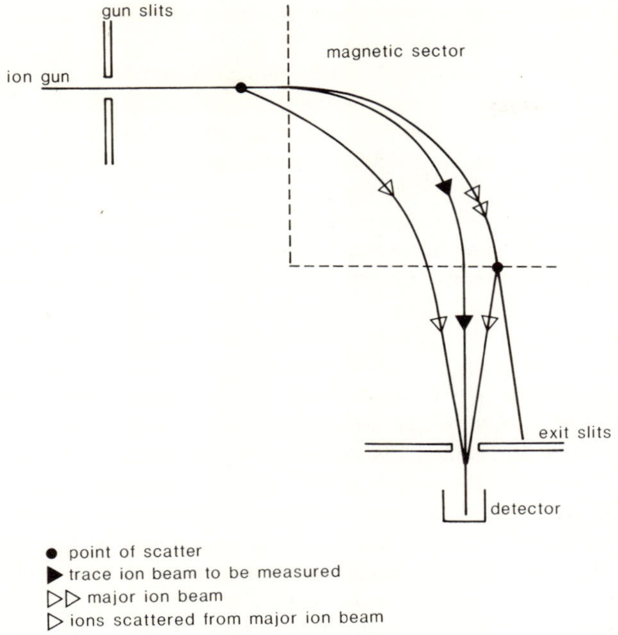

Figure 16.20 Scatter caused by collisions between the ion beam and residual gas molecules in the flight tube. A small fraction of the major ion beam is scattered into the exit slits optimized for measurement of the minor ion beam. (Adapted from Martin Elliott, V.G. Isotopes.)

of flat-topped peaks to facilitate accurate and reproducible measurement.

The most effective instrumental method of minimizing this tailing effect is to incorporate an electrostatic sector *after* the mass-dispersing magnetic sector. This eliminates ions which have suffered a significant loss in energy in the earlier part of the flight path. This arrangement should be contrasted with conventional double-focusing designs discussed in section 16.6, which use an electrostatic sector *before* the magnetic sector to obtain satisfactory resolution from sources where ions are emitted with a wide range of kinetic energies.

Typical values of abundance sensitivity for a modern thermal ionization mass spectrometer are 1 ppm for uranium and 0.5 ppm for strontium, both figures referring to the proportion of the uranium or strontium peak observed as a tail one atomic mass unit away.

16.11 Beam switching v. multiple collection

In almost all applications of thermal ionization and gas source mass spectrometry, instruments are used to measure not absolute peak intensities, but rather the ratio between selected mass intensities. In conventional mass spectrometry, such measurements are obtained in a repetitive cycle in which the ion beam is switched from one mass peak to the next, usually under computer control. In order to achieve this, a precise means of reproducibly changing the magnetic field is required to permit peak jumping over the specified mass range. Control of the magnetic field is possible by monitoring signals obtained from a Hall effect probe positioned in the magnetic sector. The Hall effect exploits the potential difference which is generated across a wire positioned in a magnetic field when an electric current is passed through the wire. This potential difference is created in a direction which tends to oppose the flow of the electric current and its magnitude is directly proportional to the magnetic field strength. By monitoring this potential difference, the magnetic field strength can be accurately controlled, permitting precise ion beam switching. Since the Hall effect signal also varies with temperature, it is also necessary to provide some form of temperature stabilization.

Ion beam switching is a widely-used method of measuring isotope ratios. However, the technique does have one drawback: any drift in the ion beam intensity between measurements will result in counting errors. Drift is likely to take two forms: a relatively slowly-changing long-term drift component in ion beam intensity due to changes in the emission characteristics of the ion gun, and more rapid random short-term fluctuations. The former can be accommodated by repetitively scanning over the appropriate mass peaks, an overall time-related drift correction being applied to the data. If two mass peaks A and B are to be measured, two measurement sequences ABBA, ABBA, ABBA . . . or alternatively ABABAB . . . may be used. These sequences are usually continued until a pre-specified precision in the averaged isotope ratio measurement has been achieved. A polynomial equation can be applied to account for the change in mass intensity over the entire run sequence, so that the accuracy of ratio measurement is not distorted by this intensity drift effect. Short-term fluctuations in ion beam intensity (that is, variations in intensity which occur within the time scale of an individual measurement) cannot be corrected by this counting procedure. It is normally necessary to screen intensity ratio output and reject the occasional measurement which appears to contain a gross analytical discrepancy.

Errors due to random fluctuations in ion gun output can, however, be largely overcome using a multiple-collector facility for data acquisition (Figure 16.21). Not only are

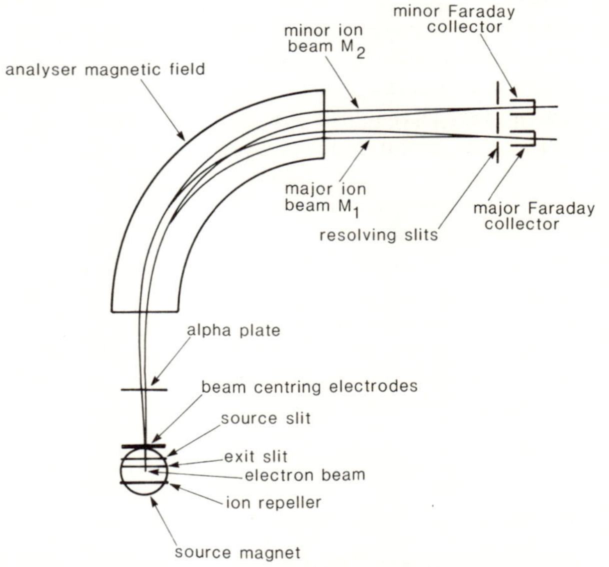

Figure 16.21 A gas source mass spectrometer designed for dual collection of a major and minor ion beam. (Adapted from V.G. Isotopes Newsletter.)

short-term fluctuations in beam intensity accounted for, but the time of analysis is considerably shortened. In a multiple-collector system, the single mass analyser exit slit and ion beam detector are replaced by two, three or more slits and detectors, the number depending on the ion ratio measurements of interest. Adjacent mass peaks can then be measured simultaneously without the necessity of switching the ion beam. This technique therefore has the potential of yielding isotope ratio data of significantly higher precision (or alternatively of meeting a pre-specified precision in a shorter analysis time) by dramatically reducing the influence of long-term and short-term drift on the precision of measurements.

Table 16.3 Physical properties of the proton, neutron and electron

Particle	Mass (amu)*	Mass (kg)	Charge
Proton	1.007825	1.6726×10^{-27}	+1
Neutron	1.008665	1.6749×10^{-27}	neutral
Electron	5.488×10^{-4}	9.1095×10^{-31}	−1

Mass data from McGlashan (1971).
* An atomic mass unit (amu) is defined as the mass relative to $^1/_{12}$ times the mass of a ^{12}C atom.

16.12 Isotopes and mass spectra: the structure of atoms and nuclear stability

The instrumental characteristics of mass spectrometers have been discussed above. Before turning to more specific geological applications, this section considers the nature and origin of mass spectra.

The atom is now known to be a complex structure, comprising several kinds of sub-atomic particles. For the purposes of mass spectrometry we need consider only three, the proton, neutron and electron, and we may use the simple concepts of atomic structure developed by Rutherford and Bohr in the 1920s. According to this simple model, the atom consists of a nucleus surrounded by a cloud of electrons (Figure 16.22); the nucleus contains certain proportions of protons and neutrons, and the physical properties of these particles are as shown in Table 16.3.

Each element in the Periodic Table has a unique number of protons on its nucleus. In the neutral atom, the positive charge associated with each proton must be balanced by a negative charge associated with an equal number of electrons occupying orbitals of discrete energy surrounding the nucleus. The atomic number of the element therefore characterizes the number of protons on the nucleus (hydrogen possesses one proton, helium two, lithium three and so on). However, for all elements (except hydrogen ^1H), protons on the nucleus are associated with a discrete number of neutrons. Since neutrons are electrically neutral, their presence is not balanced by any orbital electrons. The total number of protons plus neutrons on the nucleus is defined as the atomic weight of that element. (The atomic weight of (^1H) is one since its nucleus possesses only one proton; the atomic weight of helium (^4He) is four since its two protons are associated with two neutrons). These concepts are elucidated in more detail in section 3.4.

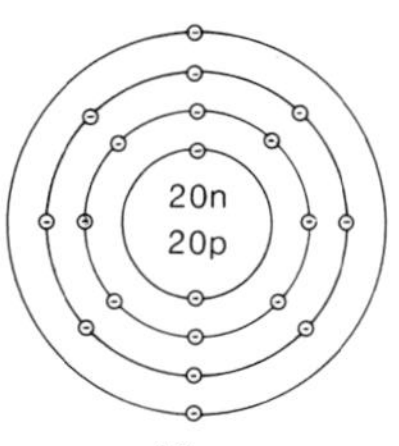

Figure 16.22 The simple Bohr–Rutherford model of a ^{40}Ca atom. The nucleus, comprising 20 protons and 20 neutrons, is surrounded by an electron cloud of 20 electrons, each occupying discrete electron orbitals. (Adapted from Dalrymple and Lanphere, 1969.)

This simple model is useful in understanding the overall structure of the atom but is incomplete in several respects. Firstly, since elements are composed of atoms whose nuclei contain discrete numbers of protons and neutrons, then the atomic weights might be expected to be integral whole numbers. In nature the atomic weights of some elements do not even approximate to whole numbers. For instance, the atomic weight of chlorine is 35.45. In the early 1900s it became clear that many elements could exist in nature in more than one form, that is, having the same number of protons on the nucleus (and therefore of orbital electrons) but with different numbers of neutrons.

In 1910 Frederick Soddy proposed the term 'isotope' to cover this phenomenon. Isotopes are therefore defined as different forms of the same element having the same number of protons on the nucleus, but different numbers of neutrons. Since the chemical properties of an atom are dependent on its outer electron orbital structure, the chemical properties of isotopes of a given element are virtually identical. The reason for non-integral atomic weights is now clear. Chlorine for instance exists in nature as two isotopes, both having an atomic number of 17, but one having an atomic weight of 35, the other 37. Since these isotopes occur in nature with relative abundances of 75.53% and 24.47% respectively, the mean atomic weight of natural chlorine is 35.45.

The first experimental evidence of the existence of isotopes is directly related to the development of mass spectrometry. This was the discovery by J. J. Thompson in 1914, using a prototype mass spectrograph, that neon existed as two naturally-occurring isotopes, of atomic weight 20 and 22 (a third neon isotope of AW 21 was subsequently discovered using more sensitive instrumentation). Following these first tentative steps, it was discovered by later workers (especially F. W. Aston) that a large proportion of elements exist in two or more naturally-occurring isotopic forms. Indeed the 83 naturally-occurring elements (including U and Th but not their daughter products) exist in about 287 naturally-occurring isotopic forms, making an average of over three isotopes per element.

Since the development of the nuclear reactor, many more (over 1200) isotopes have been discovered. These do not occur naturally, being unstable and decaying into daughter products with half-lives extending from a fraction of a second to several million years. Known nuclei, both stable and radioactive, are plotted in Figure 16.23. In this diagram the atomic number (Z) of each isotope is plotted as a function

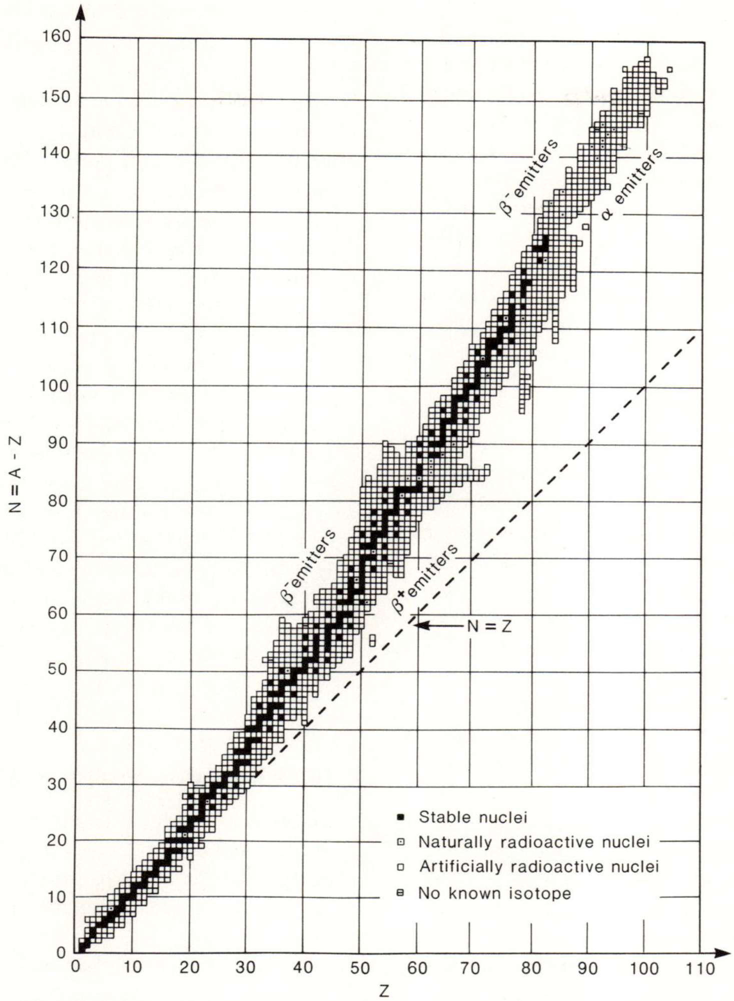

Figure 16.23 Chart of known nuclei plotted as a function of neutron number (*A–Z*) *v*. atomic number (*Z*). The dotted line represents nuclei possessing equal numbers of protons and neutrons. (Campbell, 1969, as reproduced by Dalrymple and Lanphere, 1969.)

of the number of neutrons on the nucleus. These data show that stable isotopes up to argon have approximately equal numbers of neutrons and protons on the nucleus. The stable isotopes of heavier elements possess larger numbers of neutrons than protons. The heaviest naturally-occurring stable isotope is bismuth-209. Uranium and thorium exist as naturally-occurring radioactive isotopes with half-lives in the range 10^8–10^{11} years. Two intermediate elements, technetium and promethium, do not possess any stable isotopes at all.

The trends indicated in Figure 16.23 also shed some light on the radioactive decay mechanism followed by unstable isotopes. This diagram indicates an approximate line of nuclear stability. Radioactive isotopes which fall above this line have a larger proportion of neutrons on the nucleus than that required to give stability. These isotopes tend to decay by beta-emission. In this decay mechanism, a beta-particle (an electron) is ejected from the nucleus, coinciding with the conversion of a neutron to a proton. The daughter isotope then approaches more closely the line of nuclear stability. Conversely, radioactive isotopes falling below the line of stability contain an excess of protons on the nucleus. They decay by one of several mechanisms. Some isotopes tend to decay as positron emitters; that is, a positively-charged electron is ejected from the nucleus, corresponding to the conversion of a nuclear proton to a neutron. Heavier isotopes decay by ejecting an alpha-particle (see below). Other isotopes decay by electron capture: the nucleus incorporates an orbital electron into its structure, causing a proton to be converted to a neutron. All these decay routes reduce the proportion of nuclear protons and enhance the stability of the daughter isotope.

Isotopes heavier than bismuth tend to decay by alpha-emission. An alpha-particle is a helium nucleus consisting of two protons and two neutrons ($^4He^{2+}$). After ejection of an alpha-particle, the daughter isotope has lost four units of atomic mass and two units of atomic charge. Many of the daughter isotopes of uranium and thorium decay by a combination of alpha-emission and beta-decay until they achieve a state of isotopic stability. However, there is an alternative route. In some heavy isotopes there is also a small probability of spontaneous fission. The heavy nucleus spontaneously splits into two lighter nuclei, attaining greater overall nuclear stability in the process.

16.13 Mass defect phenomena

Using the simple isotopic model for the structure of elements, it should then be a simple matter to predict the atomic weight of any isotope using the known weights of the proton and electron (i.e. the hydrogen atom) and the neutron. As an example of such calculations, Faure (1977, p. 16) considers the case of ^{27}Al. Taking the mass of hydrogen as 1.007825 amu and that of the neutron as 1.008665 amu, the atomic weight of ^{27}Al, with 13 protons and electrons and 14 neutrons, should be 27.223035 amu. However, the measured mass of ^{27}Al is less than this, namely 26.981541 amu. The reason for this difference is that some of the mass of the nuclear constituents is converted into binding energy. This binding energy confers on the nucleus its great stability. Without this mass-defect effect, the protons and neutrons which make up the nucleus would be no more stable in association within the ^{27}Al nucleus than as dissociated separate entities. The magnitude of the binding energy (E) can be calculated by substituting the mass defect (Δm) into Einstein's equation

$$E = \Delta mc^2 \tag{16.11}$$

where c is the velocity of light.

As a conversion factor it can be shown that 1 amu is equivalent to 931.58 MeV. The binding energy for ^{27}Al therefore corresponds to about 225 MeV or 225/27 = 8.33 MeV per nucleon.

The binding energy per nucleon of all the naturally-occurring isotopes is plotted in Figure 16.24. Except for the lightest elements, binding energies lie in the range 7–9 MeV/nucleon. The largest binding energy occurs for the nucleus of iron, this therefore being notionally the most stable element. The periodic maxima observed in the mass range 4–20 amu correspond to the isotopes ^{4}He, (^{8}Be), ^{12}C, ^{16}O and ^{20}Ne, all of which contain nuclei made up of multiple ^{4}He building blocks.

In some forms of mass spectrometry, it is important that instrumentation is capable of resolving ions of nominally identical mass on the basis of the mass defect effect. This is particularly the case in some schemes of ion probe analysis where serious mass interferences can occur due to the coincidence of atomic ions and molecular ion fragments. In the analysis of lead isotope ratios in zircons, for example, the molecular fragments $(^{180}Hf\,^{28}Si)^+$ and $(^{96}Zr_2\,^{16}O)^+$ both have the same nominal mass as common lead $^{208}Pb^+$. Using a low mass resolution mass spectrometer it is possible to calculate an overlap correction to remove these interferences. However, using a high resolution mass analyser, the ^{208}Pb peak can be resolved from its interferences due to the small differences in masses arising from the mass defect effect. Thus the difference in mass between ^{208}Pb and $^{208}(HfSi)$ is 0.05287 amu and between ^{208}Pb and $^{208}(Zr_2O)$ is 0.16529 amu. A mass spectrometer resolution ($M/\Delta M$) of 4000 and 300 respectively is required to separate these mass peaks. This performance is achieved at a significant reduction in count intensities.

16.14 Radioactive isotopes in nature

Stable isotopes have high binding energies per nucleon, as indicated in Figure 16.24. Radioactive isotopes have binding energies which fall below this line, the amount by which they do so dictating their stability and therefore radioactive half-life.

Naturally-occurring radioactive isotopes may be classified

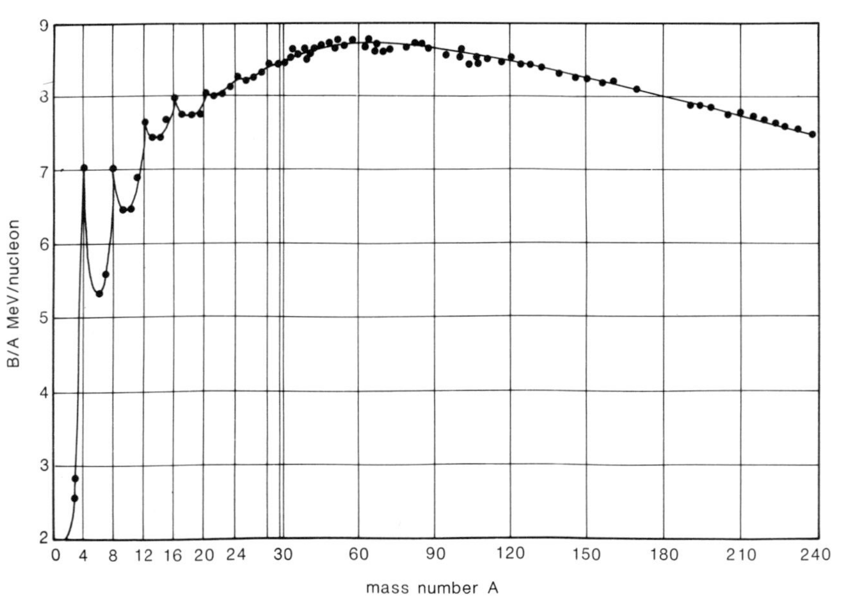

Figure 16.24 Average binding energy per nucleon plotted as a function of mass number (A) for naturally-occurring nuclides. The scale is expanded for mass numbers 0 to 30. (Moore, 1966.)

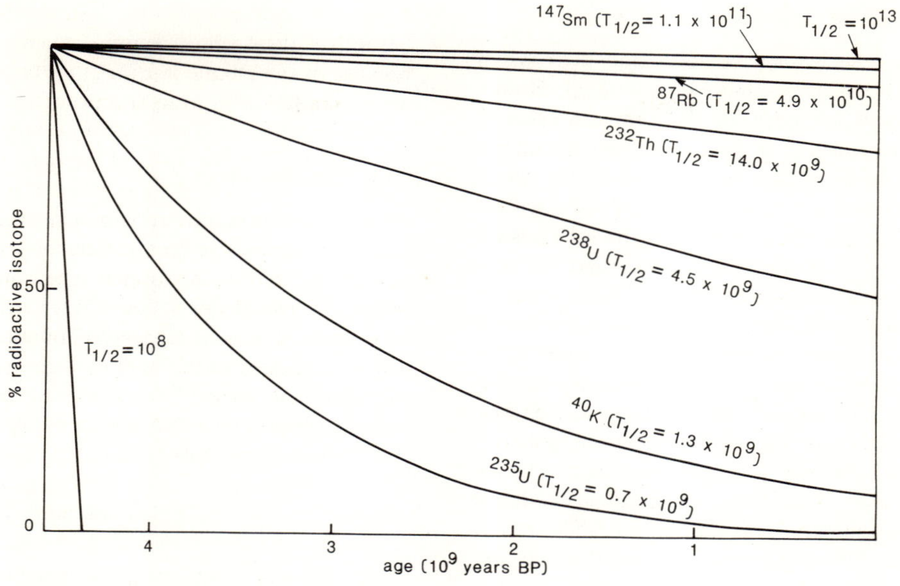

Figure 16.25 Radioactive decay of selected naturally-occurring isotopes over the last 4.57×10^9 years. Isotopes with half-lives less than about 10^8 years can be considered extinct, and those with half-lives in excess of about 10^{12} years are essentially stable.

into three groups of geological interest according to half-life.

16.14.1 *Isotopes with half-lives less than about* 10^5 *years*

These isotopes decayed so rapidly (on a geological time-scale) after the 'creation' of the universe that differences (if any) in the natural abundances of their stable daughter products in present-day rocks are expected to be immeasurably small.

16.14.2 *Isotopes with half-lives in the range* 10^6–10^8 *years*

These isotopes have now completely decayed in natural samples (Figure 16.25). However, their half-lives are sufficiently long that, in unusual circumstances, fractionation could have occurred between radioactive parent and stable daughter during the first hundred million years or so following the formation of the universe. Thus, in these exceptional circumstances, anomalies can be detected in the present-day isotopic ratios of radiogenic daughter elements. In practice, there are considerable technical difficulties in detecting the minute differences in isotope ratios in rock samples. However, several of these extinct radioactive isotopes have attracted some interest, for example ^{146}Sm which decays with a half-life of 7×10^7 years by alpha decay to ^{142}Nd (O'Nions, 1979; Lugmuir *et al.*, 1975). Some interest has also been shown in ^{129}I ($T_{1/2} = 1.7 \times 10^7$ y) and ^{244}Pu ($T_{1/2} = 8 \times 10^7$ y). The former decays by beta-emission to ^{129}Xe (Reynolds, 1960), the latter yields xenon as part of its spectrum of spontaneous fission products (Kuroda, 1960; Alexander *et al.*, 1971). In some mass spectrometer measurements of exceptional sensitivity, the former existence of ^{107}Pd ($T_{1/2} = 6.5 \times 10^6$ y) has been inferred from minute variations in radiogenic ^{107}Ag/stable ^{108}Ag ratios in meteorites (Kaiser *et al.*, 1981; Kelly *et al.*, 1978).

All radioactive isotopes with half-lives in this category are listed in Table 16.4. It must be re-emphasized that they are all extinct in natural samples (although ^{10}Be is continuously produced in minute quantities by nuclear reactions in nature). Furthermore, any residual isotope anomalies (if present) are likely to be very small and may be beyond instrumental capabilities. The principal geological interest in this category of isotope is in the analysis of meteoritic materials to elucidate nucleo-synthetic processes.

Table 16.4 Isotopes which decay with half-lives in the range 1 million to 80 million years

Isotope	Half-life (years)	Decay mode	Stable product	Relative abundance
^{10}Be	2.5×10^6	β^-	^{10}B	19.78
^{53}Mn	2×10^6	EC	^{53}Cr	9.55
^{93}Zr	1.5×10^6	β^-	^{93}Nb	100
^{97}Tc	2.6×10^6	EC	^{97}Mo	9.46
^{98}Tc	1.5×10^6	β^-	^{98}Ru	1.87
^{107}Pd	7×10^6	β^-	^{107}Ag	51.82
^{129}I	1.7×10^7	β^-	^{129}Xe	26.44
^{135}Cs	3×10^6	β^-	^{135}Ba	6.59
^{146}Sm	7×10^7	α	^{142}Nd	27.11
^{150}Gd	2.1×10^6	α	$^{146}Sm \rightarrow {}^{142}Nd$	see above
^{182}Hf	9×10^6	β^-	^{182}W	26.41
^{205}Pb	3×10^7	EC	^{205}Tl	70.50
^{210m}Bi	3×10^6	α	^{206}Pb	23.6
^{237}Np	2.1×10^6	α	^{209}Bi	100
^{244}Pu	8×10^7	α, SF	$^{232}Th \rightarrow {}^{208}Pb$	52.3

β^- = beta decay, EC = electron capture, α = alpha decay, SF = spontaneous fission.
Data from Weast (1973).

16.14.3 *Isotopes with half-lives above* 10^9 *years*

These isotopes have such long half-lives that their radioactive decay still persists in present-day samples. Radioactive isotopes which fall into this category are listed in Table 16.5. By no means all of these are of any geological interest. This can be demonstrated by plotting the radioactive decay of isotopes with selected half-lives over the last 4.57×10^9 years since the formation of the solar system (Figure 16.25). This diagram demonstrates that isotopes with half-lives significantly greater than 10^{12} years decay so slowly that they can be considered to be virtually stable over geological time. Conversely, isotopes with half-lives of less than about 10^8 years are now all extinct (see above). This leaves a narrow band of isotopes with half-lives in the range 10^9–10^{11} years. These are the naturally-occurring radioactive isotopes which have undergone significant decay over geological time and are likely to have imprinted significant variations in the isotopic abundances of their stable daughter elements. The isotopes of interest include ^{40}K, ^{87}Rb, ^{138}La, ^{147}Sm, ^{176}Lu, ^{187}Re, ^{232}Th, ^{235}U and ^{238}U. Several of these isotopes are of intense geological interest because of the unique information they can provide on the age of samples and geological processes.

Table 16.5 Naturally-occurring radioactive isotopes

Isotope	Natural abundance (%)	Half-life (year)	Mode of decay	Stable product
^{3}H	10^{-13}	12.33	β^-	^{3}He
^{14}C	$<10^{-10}$	5.73×10^3	β^-	^{14}N
^{40}K	0.01167	1.25×10^9	EC	^{40}Ar
			β^-	^{40}Ca
^{50}V	0.25	$>4 \times 10^{16}$	EC	^{50}Ti
			β^-	^{50}Cr
^{87}Rb	27.8346	4.88×10^{10}	β^-	^{87}Sr
^{113}Cd	12.3	9×10^{15}	β^-	^{113}In
^{115}In	95.7	5×10^{14}	β^-	^{115}Sn
^{123}Te	0.87	1.2×10^{13}	EC	^{123}Sb
^{138}La	0.09	1.05×10^{11}	EC	^{138}Ba
			β^-	^{138}Ce
^{142}Ce	11.1	$>5 \times 10^{16}$	α	^{138}Ba
^{144}Nd	23.9	2.1×10^{15}	α	^{140}Ce
^{147}Sm	15.0	1.06×10^{11}	α	^{143}Nd
^{148}Sm	11.2	8×10^{15}	α	^{144}Nd
^{149}Sm	13.8	$>1 \times 10^{16}$	α	^{145}Nd
^{152}Gd	0.20	1.1×10^{14}	α	^{148}Sm
^{156}Dy	0.06	2×10^{14}	α	^{152}Gd
^{176}Lu	2.6	3.53×10^{10}	β^-	^{176}Hf
			EC	^{176}Yb
^{174}Hf	0.17	2.0×10^{15}	α	^{170}Yb
^{180}Ta	0.012	$>1.6 \times 10^{13}$	EC	^{180}Hf
			β^-	^{180}W
^{187}Re	62.602	4.56×10^{10}	β^-	^{187}Os
^{190}Pt	0.013	7×10^{11}	α	^{186}Os
^{232}Th	100	14.00×10^9	α, β^-	^{208}Pb
^{235}U	0.7204	0.7038×10^9	α, β^-	^{207}Pb
^{231}Pa		3.248×10^4	dau	
^{238}U	99.2739	4.468×10^9	α, β^-	^{206}Pb
^{234}U	0.057	2.48×10^5	dau	
^{230}Th		7.25×10^4	dau	
^{226}Ra		1.622×10^3	dau	
^{210}Pb		22.26	dau	

Data from Weast (1972).
Where appropriate, half-lives are compatible with decay constants listed in Table 16.6. dau = daughter isotope.
β^- = beta decay; EC = electron capture; α = alpha decay.

16.15 Geochronology

Geochronology is the dating of a sample on the basis of the degree of radioactive decay of one or more of the naturally-occurring radioactive isotopes listed in the previous paragraph. A general discussion of radioactive decay has been considered in detail in Chapter 12 (neutron activation analysis). Of relevance to this subject is the application of the law of radioactive decay. This states that the rate of decay of a radioactive isotope ($-\mathrm{d}N/\mathrm{d}t$) is proportional to the number N of radioactive atoms remaining:

$$-\frac{\mathrm{d}N}{\mathrm{d}t} = N\lambda \qquad (16.12)$$

where λ is a constant called the decay constant (the minus sign indicates that the fraction of radioactive atoms $\mathrm{d}N$ is decreasing with time).

By integrating this expression and rearranging, it can be shown that the number of radioactive atoms N remaining after a decay interval of time t is given by

$$N = N_0 \exp(-\lambda t) \qquad (16.13)$$

where N_0 is the number of radioactive atoms originally present.

It can be shown that the decay constant λ is related to the half-life $T_{1/2}$ of the radioactive isotope by the expression

$$\lambda = \frac{\ln(2)}{T_{1/2}} = \frac{0.6931}{T_{1/2}}. \qquad (16.14)$$

To illustrate the application of the law of radioactive decay to geological problems, the simple decay of ^{87}Rb will be considered. This derivation is relevant, however, to any radioactive decay which occurs by a single stage.

The isotope ^{87}Rb is radioactive and decays, by beta-emission with a half-life of 4.88×10^{10} years, to ^{87}Sr. Equation (16.13) can therefore be written as

$$[^{87}\mathrm{Rb}] = [^{87}\mathrm{Rb}]_0 \exp(-\lambda t) \qquad (16.15)$$

or rearranging

$$[^{87}\mathrm{Rb}]_0 = [^{87}\mathrm{Rb}] \exp(+\lambda t) \qquad (16.16)$$

(note the change in exponent sign). $[^{87}\mathrm{Rb}]_0$ is taken as the initial number of atoms of this isotope in the rock sample when it crystallized, and $[^{87}\mathrm{Rb}]$ is the number of atoms remaining after a decay interval of time t. t is therefore the age of the sample measured in years before the present. Using this convention, the age of the solar system is dated as *c.* 4.57×10^9 years before the present, not as zero.

In general, small variations in $[^{87}\mathrm{Rb}]$ cannot be measured by mass spectrometry as precisely as the corresponding changes in the stable daughter isotope ^{87}Sr. Since each decay of a ^{87}Rb atom yields one atom of ^{87}Sr, the number of ^{87}Sr atoms formed over time t by radioactive decay of ^{87}Rb is given by

$$[^{87}\mathrm{Sr}]^* = [^{87}\mathrm{Rb}]_0 - [^{87}\mathrm{Rb}] \qquad (16.17)$$

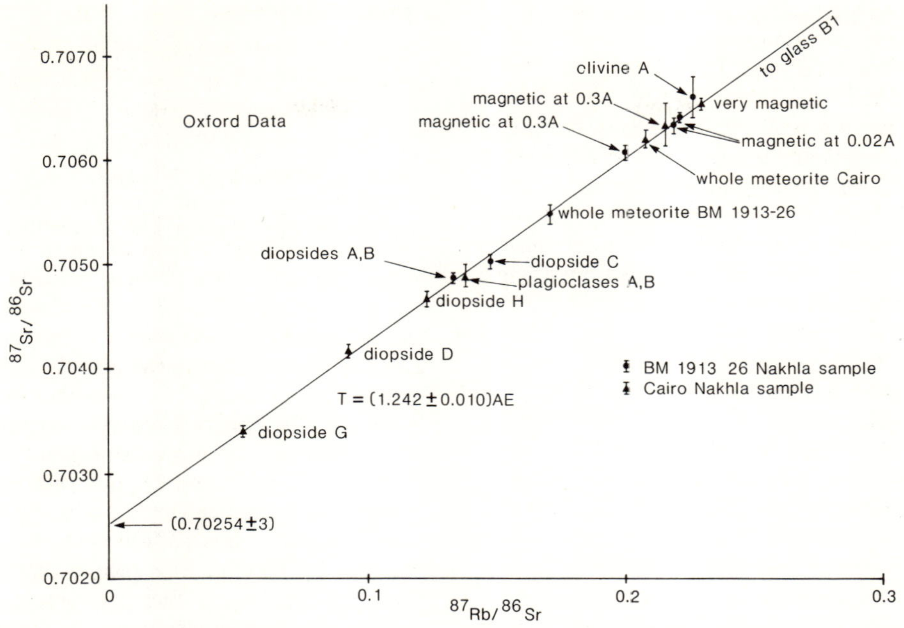

Figure 16.26 Example of an isochron diagram. Whole-sample and mineral separate data for $^{87}Sr/^{86}Sr$ and $^{87}Rb/^{86}Sr$ are plotted for the Nakhla achondritic meteorite. The 'age' derived from these data is $1.242 \pm 0.010 \times 10^9$ years before present, with an initial $^{87}Sr/^{86}Sr$ ratio of 0.70254 ± 3. (Gale *et al.*, 1975.)

Substituting $[^{87}Rb]_0$ from equation (16.16),

$$[^{87}Sr]^* = [^{87}Rb][\exp(+\lambda t) - 1] \qquad (16.18)$$

where $[^{87}Sr]^*$ accounts only for the radiogenic ^{87}Sr content of the sample. The total ^{87}Sr content must include that fraction of ^{87}Sr originally present before radioactive decay started, i.e. $[^{87}Sr]_0$.

The total $[^{87}Sr]$ content of the present-day sample is therefore

$$[^{87}Sr] = [^{87}Sr]_0 + [^{87}Rb][\exp(+\lambda t) - 1] \qquad (16.19)$$

It is extremely difficult to measure absolute concentrations of atoms by mass spectrometry, but much simpler to determine ratios of isotope abundances of the same element. Equation (16.19) is therefore normalized with respect to a non-radiogenic isotope of Sr, the concentration of which is therefore assumed to remain unchanged throughout geological time.

In the case of strontium, (^{86}Sr) is selected for normalization. Thus:

$$\left[\frac{^{87}Sr}{^{86}Sr}\right] = \left[\frac{^{87}Sr}{^{86}Sr}\right]_0 + \left[\frac{^{87}Rb}{^{86}Sr}\right][\exp(+\lambda t) - 1] \qquad (16.20)$$

This equation represents the general form of the age equation for the radioactive decay $^{87}Rb \rightarrow {}^{87}Sr$.

Equation (16.20) represents a straight line in coordinates $[^{87}Sr/^{86}Sr]$ *v.* $[^{87}Rb/^{86}Sr]$. The gradient of this line is $[\exp(+\lambda t) - 1]$ and the intercept $[^{87}Sr/^{86}Sr]_0$. In order to obtain data to construct this line, it is necessary to analyse a suite of samples by mass spectrometry to determine their $[^{87}Sr/^{86}Sr]$ ratios. The stable isotope ratio $[^{87}Rb/^{86}Sr]$ is normally determined by analysing the bulk elemental concentration of Rb and Sr and then calculating the $[^{87}Rb/^{86}Sr]$ ratio from the known natural isotopic abundance of these elements. If the suite of rocks all crystallized at the same time and possesses a sufficient range of Rb and Sr concentrations, then a plot of $[^{87}Sr/^{86}Sr]$ *v.* $[^{87}Rb/^{86}Sr]$ will form a straight line called an isochron (Figure 16.26). The age of the sample may then be calculated from the gradient of this line and the initial strontium ratio $[^{87}Sr/^{86}Sr]_0$ measured from the intercept.

Although derived for the decay of ^{87}Rb, this form of age equation may be applied to each of the radioactive decay schemes referred to above by substituting isotope ratios, as shown in Table 16.6.

The decay constant λ is normally simply related to the half-life of the radioactive isotope, as shown in equation (16.14). However, in the case of ^{40}K, for example, a branching decay occurs. ^{40}K decays by beta-emission to ^{40}Ca and by electron capture to ^{40}Ar. The decay constant used in the age equation must take into account the fact that only 11.2% of the ^{40}K atoms decay to ^{40}Ar and the majority,

Table 16.6 Isochron axis coordinates

Radiogenic isotope ratio	Radioactive isotope ratio
$^{87}Sr/^{86}Sr$	$^{87}Rb/^{86}Sr$
$^{40}Ar/^{36}Ar$	$^{40}K/^{36}Ar$
$^{138}Ce/^{142}Ce$	$^{138}La/^{142}Ce$
$^{143}Nd/^{144}Nd$	$^{147}Sm/^{144}Nd$
$^{176}Hf/^{177}Hf$	$^{176}Lu/^{177}Hf$
$^{187}Os/^{186}Os$	$^{187}Re/^{186}Os$
$^{206}Pb/^{204}Pb$	$^{238}U/^{204}Pb$
$^{207}Pb/^{204}Pb$	$^{235}U/^{204}Pb$
$^{208}Pt/^{204}Pb$	$^{232}Th/^{204}Pb$

88.8%, to ^{40}Ca. The total decay constant of 5.543×10^{-10} a^{-1} must therefore be partitioned into $\lambda_{\beta^-} = 4.962 \times 10^{-10}$ a^{-1} (for the decay to ^{40}Ca) and $\lambda_{EC} = 0.581 \times 10^{-10}$ a^{-1} (for the decay to ^{40}Ar).

As discussed above, it is possible to determine the age by analysing a suite of rocks all known to have crystallized at the same time. However, samples must be selected to cover as wide a range of rubidium and strontium compositions as possible to permit accurate regression of the isochron. An alternative approach is to construct isochrons from a single sample, provided it is possible to separate that sample into individual mineral components which themselves cover an adequate range of Rb/Sr compositions. The resultant data form an *internal* isochron (Figure 16.26).

The ages of samples derived from such studies are representative of the time which has elapsed since the isotopic composition of the sample was last homogenized, either by crystallization of the rock, or alternatively by a metamorphic event. Unfortunately, geological processes are not always as clear-cut as this, a wide scatter sometimes being encountered in isochron data. This generally occurs because one of the assumptions regarding age dating has been violated. These are as follows.

(i) The sample must have remained a closed system to the isotopes under study since crystallization. Unfortunately, temperature rises of even a few hundred degrees can cause diffusion of ions within a crystal lattice, causing partial or complete isotopic re-homogenization.

(ii) The rate of decay of the radioactive isotope remains constant. There is some evidence to suggest that small variations in the decay constant can occur in light elements such as ^{7}Be which decay by electron capture. In these unusual circumstances, radioactive decay appears to be slightly influenced by the chemical environment (Dalrymple and Lanphere, 1969, p. 23). There is no evidence that isotopes which decay by beta- or alpha-emission are affected by this phenomenon.

(iii) The isotopic composition of the daughter element has not been altered by fractionation processes other than radiogenic decay from the parent isotope.

(iv) The decay constant, and the isotopic ratio of a pair of stable isotopes of the radiogenic element are accurately known. The former is clearly essential in order to convert an isochron gradient into an age, and for some isotopes has been the subject of international agreement (Steiger and Jager, 1977). The latter is required to correct for instrumental mass fractionation effects in the mass spectrometer. In the past, different laboratories have adopted different conventions for these parameters, leading to the publication of ages and initial isotope ratios for identical samples which are internally consistent but disagree with data obtained from other laboratories. Where no international agreement exists, it is therefore essential to specify both the decay constant and the isotope ratio for mass fractionation correction used in the calculation of all quoted isochron data.

16.16 Geochronometers of geological importance

Specific details of the geochronometers applicable to geological samples are listed in Table 16.7. The more widely used ones are as follows.

16.16.1 *Carbon-14*

The isotope ^{14}C exists in nature as a result of a continuous nuclear reaction in the upper atmosphere. Cosmic-ray-produced neutrons react with atmospheric gases, resulting in the formation of ^{14}C. Several nuclear routes exist for this reaction involving oxygen, carbon and nitrogen. The latter is the most important, since the stable isotope ^{14}N will participate in an (n, p) reaction as follows:

$$^{1}n + ^{14}N \rightarrow ^{14}C + ^{1}H$$

or

$$^{14}N(n, p)^{14}C$$

where n represents a slow neutron captured by the ^{14}N nucleus and p a proton (a hydrogen nucleus) subsequently ejected. ^{14}C generated by this mechanism rapidly equilibrates in the atmosphere and hydrosphere. The present-day concentration is 13.56 decays per minute per gram of carbon. Living organisms such as plants, trees and shelled molluscs fix ^{14}C into their structure at a level initially in equilibrium with atmospheric ^{14}C. Once fixed, this incorporated ^{14}C decays, so that the age of the organism can be dated by measuring the proportion of ^{14}C remaining. Corrections must be made for the variations in the ^{14}C content of the atmosphere in recent geological time. The geological use of this chronometer is restricted to dating recent carbonate shells (less than about 30 000 years old) and possibly wood or other artefacts, engulfed for example in lava flows.

16.16.2 ^{40}K–^{40}Ar

This geochronometer was one of the first to be fully characterized, and is based on the natural radioactive decay of ^{40}K. This isotope decays by two routes with a total half-life of 1.25×10^{9} years: by electron capture to ^{40}Ar and by beta-decay to ^{40}Ca. The latter is the more probable decay route, accounting for an average 88.8% of the ^{40}K disintegrations. However, since ^{40}Ca is an abundant naturally-occurring isotope (96.94% of natural calcium), small variations in ^{40}Ca due to decay of ^{40}K can only be detected in special circumstances (Table 16.7). Decay to ^{40}Ar is therefore the most widely applicable decay scheme. The traditional method of analysing $^{40}Ar/^{40}K$ is to determine ^{40}Ar by isotope dilution using gas source mass spectrometry and to analyse bulk potassium by flame photometry or XRF on a separate sample aliquot. There is some advantage, however, in analysing ^{40}K as ^{39}Ar, produced when a sample is irradiated in a fast neutron flux. ^{40}Ar and ^{39}Ar can then be released and analysed simultaneously by gas mass spectrometry. Corrections must be made for competing routes for the production of ^{39}Ar from nuclear reactions involving Ca.

Furthermore, unknown samples can only be analysed against samples of known age, aliquots of which are irradiated simultaneously in the same neutron flux. The ^{40}K–^{40}Ar technique is extensively used to date silicate rocks and mineral separates. The technique can be prone to discrepancies due to premature loss of argon, since some minerals lose a proportion of their argon when heated to only a few hundred Celsius. Many metamorphic processes will therefore partially or completely reset the ^{40}K–^{40}Ar clock. The reliability of data obtained by the $^{40}Ar/^{39}Ar$ method can be tested by releasing argon for analysis by incremental heating of samples (stepped combustion). Variations in the $^{40}Ar/^{39}Ar$ ratio from beginning to end of the heating cycle indicate premature loss of ^{40}Ar from the sample. Because of this susceptibility to metamorphic effects, the ^{40}Ar dating technique is usually used for younger rather than older (ancient) samples, where the possibility of selecting fresh material is higher.

16.16.3 *^{87}Rb–^{87}Sr*

Rubidium–strontium isotope measurements became possible following the development of thermal ionization mass spectrometers of high sensitivity and the ability to determine bulk Rb and Sr concentrations in rocks at trace levels during the 1950s. This geochronometer arises from the beta-decay of ^{87}Rb to ^{87}Sr, the half-life of this decay being 4.88×10^{10} years. Rubidium and strontium are themselves widely dispersed in silicate rocks and minerals, making this technique relevant to a wide range of samples, particularly of igneous origin. However, these elements are also sensitive to metamorphic effects even when samples are heated to modest temperatures of a few hundred degrees. Ancient and altered samples are therefore sometimes difficult to date by this technique except to identify the last date at which isotopic re-homogenization due to metamorphism occurred.

16.16.4 *^{138}La–^{138}Ce*

^{138}La decays by beta-emission to ^{138}Ce and by electron capture to ^{138}Ba with a total half-life of 1.05×10^{11} years. Tanaka and Masuda (1982) have proposed that the ^{138}La–^{138}Ce pair forms a promising geochronometer, although the geochemical importance of this work is not yet established, particularly in the light of the widespread application of ^{147}Sm–^{143}Nd schematics.

16.16.5 *^{147}Sm–^{143}Nd*

^{147}Sm decays to ^{143}Nd by alpha-decay with a half-life of 1.06×10^{11} years. From the early 1970s there arose considerable interest in the analysis of these isotopes, a development made possible by improvements in thermal ionization mass spectrometry which permitted confident measurement of very small differences in isotope ratios. This interest was stimulated by several distinctive properties of Nd/Sm. These elements are much more resistant to metamorphic processes than are Rb/Sr or U/Th/Pb isotopes, and therefore are more likely to permit the analysis of the primary ages of metamorphic rocks. Furthermore, Nd and Sm often behave in a complementary manner to Rb and Sr permitting dating of samples which have far too low a concentration of Rb/Sr and vice versa. Indeed, there is considerable interest in schematic diagrams where $^{143}Nd/^{144}Nd$ isotope ratios for suites of samples are plotted against corresponding $^{87}Sr/^{86}Sr$ values in assessing the movement of trace elements in the upper mantle and the effects of contamination on the geochemistry of crustal rocks (Hawkesworth and van Calsteren, 1983). The development of $^{143}Nd/^{144}Nd$ isotope studies represents, therefore, one of the more important events in isotope geochemistry in recent years.

16.16.6 *^{176}Lu–^{176}Hf*

^{176}Lu decays by beta-emission to ^{176}Hf with a half-life of 3.53×10^{10} years. Hafnium isotope ratios have been measured in meteorites and some basalts (Patchett and Tatsumoto, 1980*a*,*b*), but this geochronometer has not yet been widely used.

16.16.7 *^{187}Re–^{187}Os*

^{187}Re is a beta-emitter with a half-life of 1.52×10^{11} years. Its daughter isotope is ^{187}Os. Rhenium is concentrated in molybdenite, and studies by Luck and Allegre (1982) indicate that all detectable osmium in molybdenite is radiogenic. This geochronometer is therefore specifically relevant to dating molybdenite in some sulphide deposits and also to certain meteorites. As such it has a specialized field of use.

16.16.8 *U, Th, Pb*

The naturally-occurring isotopes of uranium and thorium (^{238}U, ^{235}U and ^{232}Th) form complex decay chains which are discussed in more detail in section 13.2. As far as mass spectrometry is concerned, the important details are as follows:

^{238}U has a half-life of 4.468×10^9 years and decays by an alpha- and beta-emission chain to ^{206}Pb.

^{235}U (half-life of 0.7038×10^9 years) decays by a similar mechanism to ^{207}Pb.

^{232}Th (half-life of 14.00×10^9 years) decays to ^{208}Pb.

The only naturally-occurring stable isotope of lead is ^{204}Pb. Although for the purposes of this discussion ^{204}Pb can be considered completely stable, it is interesting to note that Faure (1977, p. 202) reports that this too is radioactive, decaying to ^{200}Hg with an extremely long half-life (1.4×10^{17} years). Decay and abundance data for these uranium and thorium isotopes are summarized in Table 16.8. A third isotope of uranium, ^{234}U, occurs in natural samples, and has a half-life of 2.48×10^5 years. This is an intermediary in the ^{238}U to ^{206}Pb decay chain.

16.16.9 *Secular equilibrium*

Age dating using uranium and thorium isotopes is normally based on the assumption that parent and daughter isotopes have attained a state of secular equilibrium. Secular equilibrium concerns the radioactive decay of a chain of isotopes from a long-lived parent isotope via one or more daughter isotopes having shorter half-lives. This situation is shown in Figure 13.18. The notional decay of an isotope with a half-life of 10 hours to a stable product via a short-lived daughter isotope of half-life 1 hour is plotted as a function of time. The concentration of the radioactive daughter is initially zero, but rapidly increases until a state is reached where the ratio of decay rates of parent and daughter is constant. If N_1 and N_2 are the number of atoms of parent and daughter

Table 16.7 Chronometers of geological importance

Radiogenic isotope	Stable reference isotope	Radioactive parent	Decay constant (year^{-1})	Isochron coordinates	Mass fractionation correction	Reference sample
^{14}C	none	^{14}N	1.209×10^{-4}			NBS 4990 (oxalic acid)
^{40}Ar (K–Ar method)	^{36}Ar	^{40}K	0.581×10^{-10}	$^{40}Ar/^{36}Ar$ *v.* $^{40}K/^{36}Ar$	$^{40}Ar/^{36}Ar$ = 295.5 in atmospheric argon (Steiger and Jager, 1977)	USGS P207 (muscovite)
^{40}Ar (^{40}Ar–^{39}Ar method)	^{36}Ar	^{40}K	0.581×10^{-10}	$^{40}Ar/^{36}Ar$ *v.* $^{39}Ar/^{36}Ar$	as above	Neutron flux monitor comprising a rock or mineral of known age and $^{40}Ar/K$ content
^{40}Ca	^{44}Ca	^{40}K	4.962×10^{-10}	$^{40}Ca/^{44}Ca$ *v.* $^{40}K/^{44}Ca$		Isotope dilution spike
^{87}Sr	^{86}Sr	^{87}Rb	1.42×10^{-11}	$^{87}Sr/^{86}Sr$ *v.* $^{87}Rb/^{86}Sr$	$^{85}Rb/^{87}Rb$ = 2.59265, $^{86}Sr/^{88}Sr$ = 0.11940 (Steiger and Jager, 1977)	Eimer and Amend $SrCO_3$ ($^{87}Sr/^{86}Sr$ = 0.7080) NBS SRM987 (strontium carbonate)
^{138}Ce	^{142}Ce	^{138}La	2.58×10^{-12}	$^{138}Ce/^{142}Ce$ *v.* $^{138}La/^{142}Ce$	$^{136}Ce/^{142}Ce$ = 0.01720 (Tanaka and Masuda, 1982)	Johnson and Matthey Ce reagent JMC 304: $^{138}Ce/^{142}Ce$ = 0.0228559 (Tanaka and Masuda, 1982)
^{143}Nd	^{144}Nd	^{147}Sm	6.54×10^{-12}	$^{143}Nd/^{144}Nd$ *v.* $^{147}Sm/^{144}Nd$	$^{146}Nd/^{144}Nd$ = 0.7219 (O'Nions *et al.*, 1977) $^{146}Nd/^{142}Nd$ = 0.636151 Wasserburg *et al.*, 1981)	USGS BCR-1 (basalt) $^{143}Nd/^{144}Nd$ = 0.51264 (Hawkesworth and van Calsteren, 1983) = 0.51184 (DePaolo and Wasserburg, 1976)
^{176}Hf	^{177}Hf	^{176}Lu	1.96×10^{-11}	$^{176}Hf/^{177}Hf$ *v.* $^{176}Lu/^{177}Hf$	$^{179}Hf/^{177}Hf$ = 0.7325 (Patchett and Tatsumoto, 1980)	Johnson and Matthey HfO_2, JMC 475: $^{176}Hf/^{177}Hf$ = 0.282195 ± 0.000015 (Patchett and Tatsumoto, 1980)
^{187}Os	^{186}Os	^{187}Re	1.53×10^{-11}	$^{187}Os/^{186}Os$ *v.* $^{187}Re/^{186}Os$	$^{186}Os/^{188}Os$ = 0.12035 (Luck and Allegre, 1983)	

Samples analysed	Age range applicable (year)	Reference	Detail of chemical separation and analysis
Wood, peat, organic materials, carbonates (shells), bone	50 to 50 000	Faure (1977) pp. 305–321 Ralph (1971)	Determination required of decay rate of ^{14}C in decays per minute per gram of carbon. ^{14}C decay detected by beta emission using Geiger–Müller counters with a high degree of background shielding. Sample usually analysed as CO_2 gas released by combustion with oxygen (for organic materials) or acid attack using phosphoric acid (carbonates)
Potassium-bearing minerals, especially feldspars, micas (e.g. biotite and muscovite) amphibole, pyroxene. Whole rock	10^4 to ancient	Dalrymple and Lanphere (1969)	^{40}Ar released by igniting the sample at about 1800°C and determined by isotope dilution gas source mass spectrometry using ^{38}Ar as spike. ^{40}K determined by x-ray fluorescence or flame photometry on separate sample aliquot
As above	As above	McDougall (1974) Dalrymple and Lanphere (1971)	Sample irradiated in a fast neutron flux to induce ^{39}K (n, p) ^{39}Ar. Radiogenic ^{40}Ar and ^{39}Ar (a measure of K) then released simultaneously by stepwise ignition and determined by gas source mass spectrometry. Determinations must be made relative to a sample of known age to calibrate irradiation conditions
Severely restricted to samples with high potassium and low natural calcium, for example micas in pegmatite, sylvite in evaporites		Faure (1977) pp. 188–190	Sample dissolution followed by ion exchange chromatography to isolate calcium. Isotope ratios determined by thermal ionization mass spectrometry
Rb-rich minerals: mica Potassium feldspar, clay minerals Sr-rich minerals: plagioclase, apatite carbonate minerals whole rock	10^7 to ancient	Moore *et al.* (1973) Sanz and Wasserburg (1969) Pankhurst and O'Nions (1973) Birck and Allegre (1973)	After sample digestion, strontium separated by ion exchange. $^{87}Sr/^{86}Sr$ ratio determined by thermal ionization mass spectrometry. Bulk rubidium and strontium can be determined either by thermal ionization mass spectrometry using an isotope dilution technique with ^{87}Rb and ^{84}Sr as spikes or by x-ray fluorescence analysis
Limited application at present to pyroxene, plagioclase and whole rock		Tanaka and Masuda (1982)	$^{138}Ce/^{142}Ce$ determined by thermal ionization mass spectrometry using ^{136}Ce–^{142}Ce for mass fractionation correction. La and Ce determined by isotope dilution using ^{138}La and ^{142}Ce as spikes
Whole rock, silicate minerals		O'Nions *et al.* (1979) Richard *et al.* (1976) Wasserburg *et al.* (1981) Hawkesworth and van Calsteren (1983)	After sample digestion, a two-stage ion exchange separation is required to separate Nd from matrix and adjacent REEs (especially Ce and Sm). $^{143}Nd/^{144}Nd$ then determined to very high precision by thermal ionization mass spectrometry. Bulk Sm and Nd are determined by the same technique using isotope dilution with ^{149}Sm and ^{145}Nd as spikes
Meteorites, gadolinite		Patchett and Tatsumoto (1980) McCulloch *et al.* (1976)	A three-stage ion exchange separation is required to isolate hafnium. The $^{176}Hf/^{177}Hf$ ratio and bulk Hf and Lu are determined by thermal ionization mass spectrometry, the latter using ^{180}Hf and ^{176}Lu as spikes
Molybdenites, copper sulphide ores, meteorites		Luck *et al.* (1980) Luck and Allegre (1982, 1983)	^{187}Os is determined by isotope dilution mass spectrometry using common Os as spike (common Os is not detected in molybdenites, only ^{187}Os by decay from ^{187}Re). The molybdenite is dissolved in HNO_3/HF and osmium plus spike distilled off as the tetraoxide prior to analysis. Due to inefficient ionization by a thermal source, osmium has been determined by ion probe, although ICP–mass spectrometry may prove more effective

(Continued)

Table 16.7 *(Continued)*

Radio-genic isotope	Stable reference isotope	Radio-active parent	Decay constant (year^{-1})	Isochron coordinates	Mass fractionation correction	Reference sample
^{206}Pb	^{204}Pb	^{238}U	1.55125×10^{-10}	$^{206}Pb/^{204}Pb$ *v.* $^{238}U/^{204}Pb$	NBS 981 (lead metal, natural)	Canyon Diablo troilite $^{206}Pb/^{204}Pb = 9.307$
^{207}Pb	^{204}Pb	^{235}U	9.8485×10^{-10}	$^{207}Pb/^{204}Pb$ *v.* $^{235}U/^{204}Pb$	NBS 982 (lead metal, equal	$^{207}Pb/^{204}Pb = 10.294$ $^{208}Pb/^{204}Pb = 29.476$
^{208}Pb	^{204}Pb	^{232}Th	4.9474×10^{-11}	$^{208}Pb/^{204}Pb$ *v.* $^{232}Th/^{204}Pb$	atom, 206/208) NBS 983 (lead	(Tatsumoto *et al.*, 1973)
207–206 combined isochron				$^{207}Pb/^{204}Pb$ *v.* $^{206}Pb/^{204}Pb$	metal, radiogenic 92%-206)	
Uranium–thorium series dating						

Decay constants:
^{14}C: Godwin (1962).
^{40}Ar, ^{87}Rb, ^{238}U, ^{235}U, ^{232}Th: Steiger and Jager (1977).
^{138}La: Tanaka and Masuda (1982).
^{147}Sm: Lugmair and Marti (1978).
^{176}Lu: Patchett and Tatsumoto (1980*b*).
^{187}Re: Luck and Allegre (1983).

and λ_1 and λ_2 their respective decay constants, then, following the establishment of secular equilibrium,

$$N_2 = \frac{\lambda_1}{\lambda_2}.N_1 \quad (16.21)$$

(the decay constant $\lambda = 0.6931/T_{1/2}$ where $T_{1/2}$ is the radioactive half-life). Under these conditions, therefore, the rate of decay of parent isotope is the same as the rate of decay of daughter isotope and also equals the rate of growth of the stable radiogenic isotope at the end of the decay chain. A further discussion of secular equilibrium in relation to uranium/thorium series dating is given in section 13.8.

Providing conditions of secular equilibrium have been attained, uranium and thorium afford three geochronometers represented by the alpha/beta decay of ^{238}U to ^{206}Pb, ^{235}U to ^{207}Pb and ^{232}Th to ^{208}Pb. The isochrons for these decay series have the same form as is described in equation (16.20). Thus, for ^{238}U decaying to ^{206}Pb:

$$\left[\frac{^{206}Pb}{^{204}Pb}\right] = \left[\frac{^{206}Pb}{^{204}Pb}\right]_0 + \left[\frac{^{238}U}{^{204}Pb}\right][\exp(\lambda t) - 1] \quad (16.22)$$

where $[^{206}Pb/^{204}Pb]$ is the analysed lead ratio, $[^{206}Pb/^{204}Pb]_0$ is the initial lead ratio representative of the time of crystallization of the sample, $[^{238}U/^{204}Pb]$ is the uranium:lead isotope ratio at the time of analysis, t the time which has elapsed since crystallization and λ the decay constant of ^{238}U.

Equivalent formulae can also be written for the decay of ^{235}U and ^{232}Th. However, it is possible to combine the two decay equations for ^{238}U and ^{235}U, unlike other radiogenic isotopes, into the form

$$\frac{\left[\frac{^{207}Pb}{^{204}Pb}\right] - \left[\frac{^{207}Pb}{^{204}Pb}\right]_0}{\left[\frac{^{206}Pb}{^{204}Pb}\right] - \left[\frac{^{206}Pb}{^{204}Pb}\right]_0} = \frac{[^{235}U][\exp(\lambda_2 t) - 1]}{[^{238}U][\exp(\lambda_1 t) - 1]} \quad (16.23)$$

where λ_2 now represents the decay constant of ^{235}U.

This equation may be simplified, since for all terrestrial samples the ratio ^{235}U to ^{238}U is a constant and equals 1:137.88 (Steiger and Jager, 1977). In this form (equation 16.23) the isochron is dependent only on the measurement

Table 16.8 Decay and abundance data for uranium and thorium isotopes

Radioactive parent	Decay series	Stable decay product	Present-day lead natural abundance (%)	Primeval lead abundance (%)
^{238}U	Uranium	^{206}Pb	23.6	18.585
^{235}U	Actinium	^{207}Pb	22.6	20.556
^{232}Th	Thorium	^{208}Pb	52.3	58.861
	Common lead	^{204}Pb	1.48	1.997

Primeval lead data from Tatsumoto *et al.* (1973) for troilite from the iron meteorite, Canyon Diablo.
Present-day lead data from Weast (1973).

Samples analysed	Age range applicable (year)	Reference	Detail of chemical separation and analysis
Zircon, galena	10^7 to ancient	Faure (1977) pp. 197–266 Krogh (1973) Korkisch and Arrhenius (1964) Tera and Wasserburg (1972*a*,*b*) Mahnes *et al.* (1978) Arden and Gale (1974) Interpretations: Stacey and Kramers (1975), Sun (1980)	After dissolution of the sample, Pb, U and Th are usually separated in three individual fractions by ion exchange. Lead isotope ratios are determined by thermal ionisation mass spectrometry using a silica gel/phosphoric acid filament loading technique. Bulk lead, uranium and thorium are determined by isotope dilution thermal ionization mass spectrometry using ^{205}Pb, ^{233}U and ^{230}Th as spikes
Sediments, corals, shells, igneous, volcanic rocks	10 to 10^6	See section 13.9	Chemical separation of selected radioactive daughter isotope of ^{238}U, ^{232}Th or ^{235}U followed by determination normally using alpha spectrometry

of lead isotope ratios and does not require the determination of either ^{238}U or ^{235}U. The equation may be further simplified by noting that

$$\left[\frac{^{207}\text{Pb}}{^{204}\text{Pb}}\right]^* = \left[\frac{^{207}\text{Pb}}{^{204}\text{Pb}}\right] - \left[\frac{^{207}\text{Pb}}{^{204}\text{Pb}}\right]_0 \qquad (16.24)$$

where

$$\left[\frac{^{207}\text{Pb}}{^{204}\text{Pb}}\right]^*$$

is the fraction of radiogenic ^{207}Pb formed from ^{235}U since crystallization of the sample.

Using a similar relationship for

$$\left[\frac{^{206}\text{Pb}}{^{204}\text{Pb}}\right]^*$$

equation (16.24) can be rewritten as

$$\left[\frac{^{207}\text{Pb}}{^{206}\text{Pb}}\right]^* = \left[\frac{^{235}\text{U}}{^{238}\text{U}}\right]\left[\frac{\exp(\lambda_2 t)-1}{\exp(\lambda_1 t)-1}\right] \qquad (16.25)$$

where $[^{207}\text{Pb}/^{206}\text{Pb}]^*$ is now the ratio of radiogenic ^{207}Pb to radiogenic ^{206}Pb. A more detailed account of this derivation is given by Faure (1977, pp. 203–204). This form of the equation requires the adoption of some suitable values for $[^{207}\text{Pb}/^{204}\text{Pb}]_0$ and $[^{206}\text{Pb}/^{204}\text{Pb}]_0$ so that the radiogenic component of these isotope ratios can be calculated. The precise ratio selected depends on the geochemical model adopted for interpretation of the lead isotope data. If a sample is assumed to be derived from material which has been unaltered since the formation of the Earth, then the initial $[^{207}\text{Pb}/^{204}\text{Pb}]_0$ and $[^{206}\text{Pb}/^{204}\text{Pb}]_0$ ratios required for ^{207}Pb–^{206}Pb dating are those representative of primeval lead. The best estimate for the composition of primeval lead has been obtained from isotope analysis of the mineral troilite (FeS) in an iron meteorite called 'Canyon Diablo'. This phase, troilite, contains significant concentrations of common lead but very low levels of uranium and thorium. Its lead isotopic composition will have remained very nearly constant since crystallization with a negligible radiogenic lead contribution. The initial ratios derived by Tatsumoto *et al.* (1973) from the analysis of troilite are

$$\left[\frac{^{206}\text{Pb}}{^{204}\text{Pb}}\right]_0 = 9.307 \left[\frac{^{207}\text{Pb}}{^{204}\text{Pb}}\right]_0 = 10.294$$

$$\left[\frac{^{208}\text{Pb}}{^{204}\text{Pb}}\right]_0 = 29.476$$

and these are the values used in a 'single-stage' lead evolution model to evaluate the ^{207}Pb–^{206}Pb isochron. This rather simple model assumes that terrestrial samples remained unaffected by geological processes over a period of about 4.57×10^9 years. Most terrestrial samples yield lead isotope date which do not fit this single-stage model. In order to overcome these difficulties, Stacey and Kramers (1975) postulated a two-stage evolution process for the interpretation of such lead isotope data.

From the above it is apparent that the interpretation of lead isotope data is far from complex than the more simple radiogenic isochron models discussed earlier. The technique is normally applied to mineral separates containing high levels of uranium and thorium and therefore large radiogenic lead anomalies. Such minerals include zircon, apatite, sphene and REEs, U, Th minerals including uraninite, thorianite, thorite, allanite, monazite and xenotime. Under ideal circumstances, four ages can be determined from the ^{238}U–^{206}Pb, ^{235}U–^{207}Pb and ^{232}Th–^{208}Pb decay series and by the ^{207}Pb–^{206}Pb method. Usually, in terrestrial samples, ages for a suite of samples derived by each of these isochrons do not agree. This is because uranium and lead are often lost from these minerals due to metamorphic alterations after crystallization. In general, it is considered that the ^{207}Pb–^{206}Pb data will generally come closest to the primary age of crystallization since this ratio is least sensitive to the recent loss of lead by thermal metamorphism or ion diffusion processes. Discordant ages are normally interpreted in respect of concordia diagrams obtained by plotting the evolution of radiogenic lead from variations in $^{206}Pb^*/^{238}U$ and $^{207}Pb^*/^{235}U$ ratios. These diagrams permit identification of an age at which major metamorphic events affected samples. For a detailed discussion of lead isotope modelling, see Faure (1977), Stacey and Kramers (1975) and Sun (1980).

Lead, uranium and thorium isotope ratios are determined by thermal ionization mass spectrometry and experimental procedures are discussed in more detail in Chapter 17.

16.16.10 *Uranium and thorium series disequilibrium*

A further method of dating samples using the uranium and thorium decay series is possible. The basis for U, Th and Pb dating outlined above is the assumption that secular equilibrium exists between U and Th and all the decay products. However, there are several ways in which this equilibrium can be broken by geological processes. One is weathering, since the geochemical properties of uranium and thorium are quite different. Thorium is preferentially incorporated into sediments, but uranium remains in solution. Then ^{238}U and its daughter ^{230}Th will then no longer be in secular equilibrium. The age at which the sediment was laid down can be found from the degree of disequilibrium remaining in the sample. Similar techniques can be applied to disequilibrium phenomena caused by volcanic eruptions. The time-scale over which events can be dated varies from a few days to about one million years, depending on the isotope pair selected or available for analysis. Isotope ratios must normally be determined by alpha spectrometry and are discussed further in Chapter 13.

17 Thermal ionization mass spectrometry

17.1 Introduction

Thermal ionization mass spectrometry is a technique which has been chiefly developed for the analysis of geological samples. The technique is used extensively for the isotope ratio measurements required for Rb–Sr, Nd–Sm and Pb–Th–U geochronology studies as well as the determination of rare-earth elements, and, less frequently, other selected elements by isotope dilution analysis.

Extensive chemical treatment of the sample, normally involving ion exchange separations, is required before analysis, since otherwise serious isobaric interferences are observed in mass spectra. After chemical separation, the element is loaded as a solution on to a metal filament and is then heated under vacuum to evaporate the solvent and precondition the sample to form an amorphous residue suitable for analysis.

The filament is mounted in the sample turret of the mass spectrometer (Figure 17.1) and is pumped down to high vacuum. The sample is atomized and ionized by heating the filament to the required temperature by passing an electric current through it. The sample must be in a suitable chemical form to ensure that a stable positive ion beam is maintained for the duration of the analysis. Ions are extracted into the mass analyser by applying a potential of several kilovolts (positive) to the filament relative to an anode plate. For some isotope ratio measurements (Nd–Sm for example) data must be acquired for several hours before the required precision is achieved. It is essential therefore that the thermal ionization source gives a highly stable rate of evaporation and ionization over the entire analysis period. This requires careful optimization of filament operating conditions. Contamination during sample preparation is of great concern, since detection limits are essentially restricted by the procedural blank. Strict conditions are therefore placed on the level of contamination which can be tolerated in laboratory environments and analytical reagents; this is discussed in more detail in section 1.7.

17.2 Ion production

17.2.1 *The Langmuir–Saha equation*

The proportion of positively charged ions (n^+) relative to neutral ions (n^0) emitted from a heated metal filament is given by the Langmuir–Saha equation:

$$\frac{n^+}{n^0} \propto \exp\left[\frac{e(W-\mathrm{IP})}{kT}\right] = \exp\left[\frac{11606(W-\mathrm{IP})}{T}\right] \qquad (17.1)$$

where W is the work function of the metal filament, IP is the ionization potential of the isotope, T is the temperature of the filament (K), k is the Boltzmann constant ($=1.38 \times 10^{-23}$ J/K) e is the electronic charge ($= 1.60 \times 10^{-19}$ coulomb).

The *work function* is the energy (i.e. voltage) which must be applied to the heated filament to cause free electrons within the filament to escape by overcoming their association with the bulk metal. The first ionization energy is the work (normally expressed in eV) which must be done on an atom to remove the most loosely-bound orbital electron.

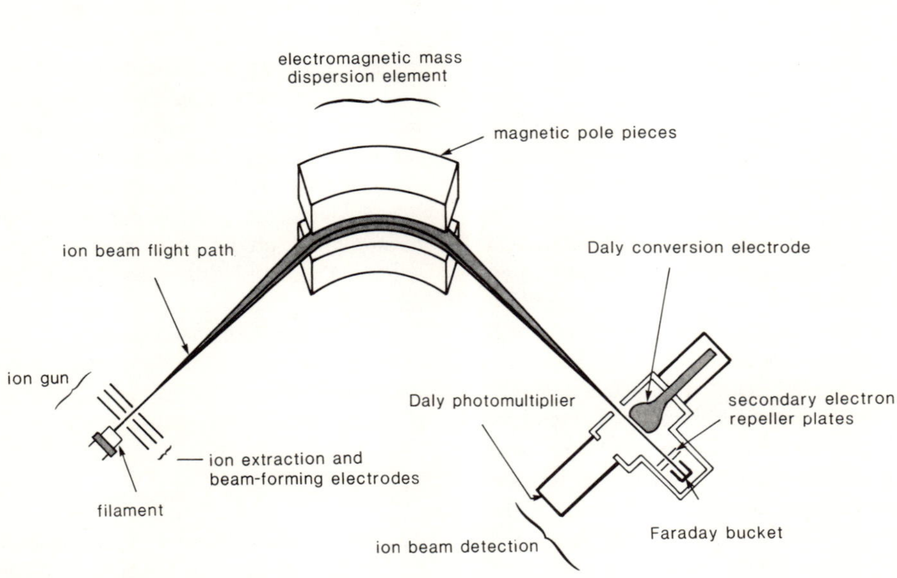

Figure 17.1 Overall schematic diagram of a thermal ionization mass spectrometer showing ion gun, magnetic sector and Daly Faraday cup detectors. (V.G. Isotopes Newsletter.)

Table 17.1 Work function and melting point of filament materials

Metal	Work function (V)	Melting point (°C)
Nickel	5.03	1453
Niobium	4.0	2468
Rhodium	4.80	1966
Palladium	4.99	1552
Tantalum	4.19	2996
Tungsten	4.52	3410
Rhenium	5.1	3180
Platinum	5.32	1772

There are several analytical characteristics which may be predicted from the Langmuir–Saha equation:

(i) In order to promote efficient production of positive ions, it is important that the term ($W-$IP) should be positive; the filament material must therefore be selected to have as high a work function as possible.

(ii) If W is greater than the ionization potential of the element to be determined, the filament must be run at as low a temperature as possible to maximize positive ion yield.

(iii) If the ionization potential (IP) is greater than the work function (W), the filament must be run at a relatively high temperature to maximize positive ion yield.

(iv) The exponential term becomes unacceptably small if the ionization potential of the element significantly exceeds 8 to 9 eV. A list of elements in this category is given in Table 17.14.

Since the second ionization potential of almost all elements exceeds 10–15 eV (Table 20.1), there is virtually no interference from doubly-charged ions from thermal ionization sources.

Metals which have been proposed as suitable filament materials are listed in Table 17.1. However, of these, only three have found extensive applications, namely rhenium, tantalum and tungsten. The first two in particular are in widespread use because of their high work functions, the operating flexibility afforded by the very high melting points and the resistance to chemical reaction with a sample.

17.2.2 *Single filament*

The simplest form of filament is the single ribbon filament (Figure 17.2). This consists of a strip of metal normally about 15 × 1 mm by about 10 μm thick, spot welded to suitable supporting legs. The sample is both evaporated and ionized off the filament at the appropriate operating temperature. This source is widely used for elements which have relatively low ionization potentials (Rb/Sr, for example). In such cases the heated metal ribbon has a higher affinity for retaining an electron from an evaporating atom that does the atom itself, and a beam of positively charged ions is efficiently formed. Rare-earth elements can be analysed satisfactorily on a single filament, particularly if the oxide species (such as NdO^+) are to be measured (Schuhmann and Philpotts, 1979). However, for those elements with higher ionization potentials a double or triple filament assembly is normally preferred.

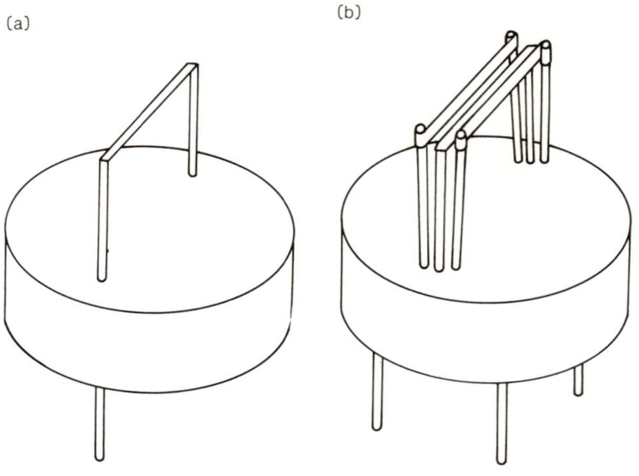

Figure 17.2 Schematic diagram of (*a*) single ribbon-type filament; (*b*) triple filament assembly. In the latter case, the sample is coated on the side filaments and the central filament is used to control ionization.

17.2.3 *Triple filament*

The triple filament consists of three metal ribbons (Figure 17.2), two side filaments on to which the sample is loaded and one central filament used for ionization. All three ribbons need not necessarily be made of the same material. In rare-earth element analysis, the side filaments are often constructed of tantalum and the central filament of rhenium. The importance of this configuration is that it permits independent control of sample evaporation (by optimizing the temperature of the side filaments) and ionization (by controlling the temperature of the central filament). The side filaments are therefore run at a sufficiently low temperature to promote stable evaporation of the element of interest over the entire analysis period. Conversely, the central filament is run hotter to achieve a more efficient degree of ionization than would be expected from a single filament. Thus, Thirlwall (1982) has shown how it is possible to load a combined rare-earth element sample on to the side arms of a triple filament and determine each element in turn by progressively increasing the temperature of the side filaments. The order in which these elements are vaporized from the filament is approximately Ba, Eu, Yb, Sm + Ce, Nd + La, Gd + Er + Dy. A further refinement in this application is that the temperature of the central filament can be adjusted to minimize interferences from rare-earth element oxide species. Details of this technique are discussed further in section 17.6.

The double filament is a simpler version of the triple filament, the sample being loaded on to one filament, and the temperature of the second element being controlled to promote efficient ionization.

17.2.4 *Ionization and sample matrix*

As outlined above, in some applications (where the element has a low ionization potential), efficient ionization is observed at a relatively low temperature. It is then necessary to load the sample in a chemical form which can be vaporized effectively at low operating temperatures. Thus, in the analysis of rubidium isotopes, samples are normally loaded in the chloride form, as chlorides are in general the most volatile of the common salts of inorganic acids. However, the opposite is true if the ionization potential of the element significantly exceeds the work function of the filament. Then the filament must be run at a relatively high temperature and it is necessary to load the sample in a form more resistant to rapid evaporation so that sufficient signal is maintained throughout the analysis run. Non-volatile salts cannot of course be analysed using a filament source.

In the early days of thermal ionization mass spectrometry, particular difficulty was observed in the efficient ionization of elements such as lead. An important advance was made in sample loading techniques by Cameron *et al.* (1969) and others. These workers found that a considerable increase in ionization efficiency could be achieved if the lead salt was loaded on to the filament together with a colloidal suspension of silica gel and phosphoric acid (see section 17.5.4 below). This technique is now widely used for lead and other elements with high ionization potentials.

17.2.5 *Mass fractionation effects*

Thermal ionization sources, involving the physical evaporation of samples from a heated filament, are subject to mass fractionation effects. These arise because the lighter isotopes of an element are slightly more volatile than heavier isotopes and will therefore be evaporated more rapidly at the beginning of a run. In order to correct for mass fractionation effects, the mass spectrum must be scanned repetitively and a time-dependent interpolation of isotope peak heights carried out, as discussed in section 17.3.

17.2.6 *Advantages of filament ionization*

Thermal ionization sources offer several advantages compared with other ionization sources in mass spectrometry. In particular, the spread of energies of ions emitted from a filament source is very small—perhaps 0.2 eV. As a consequence, the degradation in mass resolution due to kinetic energy effects in this source is also small. Thermal ionization mass spectrometers can therefore be designed with a single magnetic sector without the necessity of energy filtering and still retain adequate resolution for isotopic mass analysis. A further advantage is that since ionization from a filament is essentially a surface phenomenon, mass spectra are relatively free of interferences and background enhancements due to ionization of residual gases in the vacuum system.

17.2.7 *Instrument configuration*

The typical instrument (Figure 17.1) therefore consists of an ion source containing the filament and various collimating and alignment electrodes. Positive ions are extracted from the filament by accelerating them through a potential usually of 6–8 kV. Ions traverse a stainless-steel flight tube and are deflected by an electromagnetic mass analyser. The dispersed ion beam is focused on the exit slits. Detection is by Faraday cup, the Daly detector being a common alternative for low intensity ion beams. Since it is normally necessary to determine specific isotope ratios to very high precision (often better than 0.1 or even 0.01%), the electromagnet is normally programmed to scan selected mass peaks repeatedly and, after mass fractionation correction, to calculate the precision of data accumulated so far in the run. Analysis is terminated when the programmed precision (or time limit) has been attained. Instruments have traditionally employed single-collector detection although multiple-collection systems are under development and will permit an improvement in instrumental precision and/or analysis times.

As a prerequisite to successful thermal ionization mass spectrometry it is essential to chemically isolate the element for which precise isotope ratio measurements are to be made. In the following sections the more important procedures used for the preparation and analysis of samples for the determination of Rb–Sr, Nd–Sm and Pb–U–Th isotope ratios will be reviewed.

17.3 Rubidium–strontium isotope analysis

17.3.1 *Chemical separation*

Procedures for the preparation of samples for rubidium–strontium analysis are well established. The objective is to chemically isolate rubidium in one fraction and strontium in another fraction in a form suitable for thermal ionization mass spectrometry. Schemes are usually designed to use the minimum quantities of reagents to minimize procedural blanks. The sample is dissolved by acid attack using a Teflon pressure bomb. Rubidium and strontium are then separated both from matrix elements and from each other by anion exchange in hydrochloric acid medium.

For the analysis of whole rocks (as opposed to mineral separates), a typical procedure would be to digest 200–500 mg of powdered sample overnight at 180°C in a Teflon pressure bomb containing a mixture of 40% hydrofluoric acid and smaller volumes of 72% perchloric acid and/or concentrated nitric acid. The total volume of acid reagents would normally be limited to 5–10 mL. After cooling, small fractions of undissolved black residue are sometimes found. Gast (1962) found that such residues often contained high concentrations of Cr, Fe and Al and were generally considered to be undissolved spinel (chromite) or in some cases garnet. Some workers remove these fractions by filtration and discard them since their rubidium and strontium contents are expected to be negligible. Others would attempt to repeat the acid attack.

If rubidium and strontium bulk concentrations are to be determined by isotope dilution, the appropriate volumes of spike solutions (^{87}Rb : 99–99.5%, ^{86}Sr : 95–99.8%) must be added to the sample before acid attack to ensure complete isotopic equilibration with rubidium and strontium released from the rock powder.

The solution resulting from acid attack is then evaporated to dryness. In order to minimize contamination, a Teflon oven such as that designed by Sanz and Wasserburg (1969) is widely used (Figure 17.3). The residue is then taken up in the minimum volume of dilute hydrochloric acid for ion exchange separation. In some procedures an intermediate step is used in which the residue after evaporation is first taken up in dilute nitric acid which is itself evaporated to

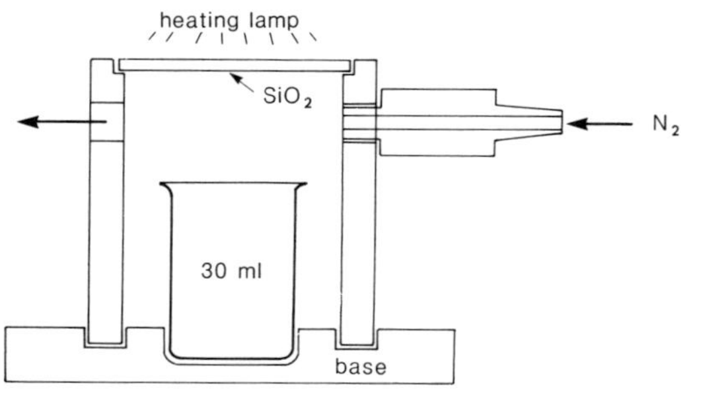

Figure 17.3 The Teflon evaporating oven. The sample solution is contained in a beaker placed in a Teflon chamber through which pure nitrogen is passed. The top of the chamber is covered with a quartz window above which an infra-red lamp is directed at the sample (in some designs, the quartz window is replaced by Teflon foil). The sample solution can thus be evaporated in an ultra-clean environment without risk of contamination. (Sanz and Wasserburg, 1969.)

dryness before being redissolved in hydrochloric acid. This ensures that any sparingly soluble perchlorates are completely redissolved, and that $HF/HClO_4$ residues which could cause ion exchange interferences are more completely expelled.

If the bulk concentration of rubidium and strontium is to be determined by XRF, a separation for strontium only is required, and it would then be unnecessary to use either Rb or Sr spiked additions since only the ratio $^{87}Sr/^{86}Sr$ need be determined by mass spectrometry. However, assuming Rb and Sr are to be determined by isotope dilution, procedures vary as to the appropriate ion exchange medium to employ. Most use a strong cation exchange resin AG 50W–X8, 200–400 mesh. The simplest procedure is to load the sample on to a suitable column dissolved in 1–2 mL 2.5M HCl and to elute with 45 mL 2.5M HCl. Rubidium is eluted in the 24–32 mL fractions, strontium is eluted with a further 10 mL of dilute hydrochloric acid (van Calsteren, pers. comm). Some variations on this procedure have been reported. For example, Gale *et al.* (1975) separated strontium by conventional ion exchange techniques but effected separation of rubidium salts by precipitation with sodium tetraphenylboron solution. Pankhurst and O'Nions (1973) split the digested sample into two fractions. Strontium was separated as above using an AG–50 ion exchange resin. However the rubidium was separated on a zirconium phosphate ion exchange column.

The eluted solutions are normally evaporated to a small volume and loaded on to a tantalum filament in the chloride form in the presence of phosphoric acid to promote stable ion emission.

17.3.2 *Mass spectrometry–strontium*

The mass peaks of interest are shown in Figure 17.4. About 0.1 to 1 mg of strontium is loaded on to the filament and run to give a stable ion beam of about 10^{-11} amps on ^{88}Sr with Faraday cup detection. The mass spectrum is scanned by repetitively counting the mass peaks listed in Table 17.2.

A typical spectrum is shown in Figure 17.5. Each sweep through the mass spectrum would involve 2–6 seconds' count integration on a selected mass peak followed by 2 seconds' or so switching delay to allow settling of the instrument before ion beam measurements are started on the next mass peak. Samples are analysed by acquiring many such cycles of mass data. Corrections must be applied for the following effects:

(i) Isobaric mass interference–mass 85 (^{85}Rb) can be monitored to detect if any correction is required for the ^{87}Rb interference on the ^{87}Sr mass peak.

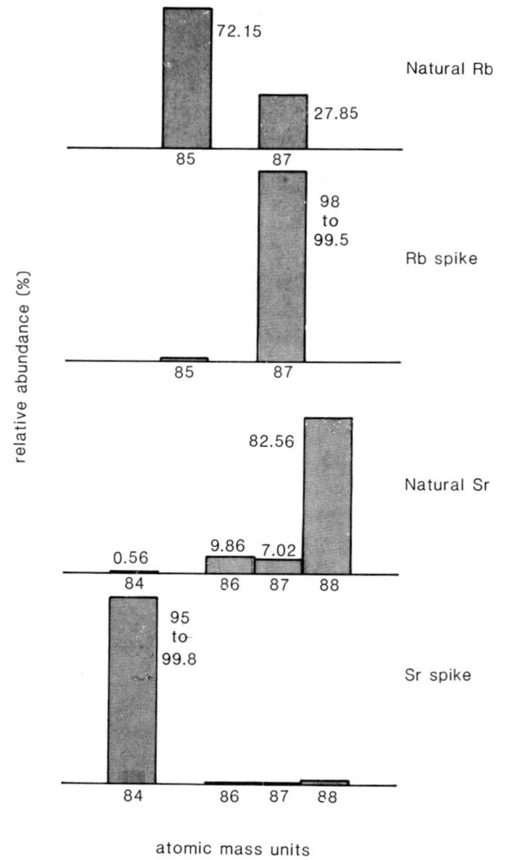

Figure 17.4 Schematic mass spectrum showing relative abundance of natural rubidium and strontium isotopes together with ^{87}Rb and ^{84}Sr spikes.

Table 17.2 Strontium mass peaks

Mass number	Isotope	Comments
88	^{88}Sr	Required for mass fractionation correction.
87	^{87}Sr	($+^{87}Rb$ if not completely removed): radiogenic isotope of strontium.
86.5	—	Background monitor
86	^{86}Sr	Required to normalize $^{87}Sr/^{86}Sr$ ratio.
85	^{85}Rb	Monitored at the beginning of a run to detect Rb contamination. Correction of ^{87}Rb rarely required if chemical separation has been undertaken satisfactorily.
84	^{84}Sr	Strontium spike isotope. Required if Sr is to be determined by isotope dilution.

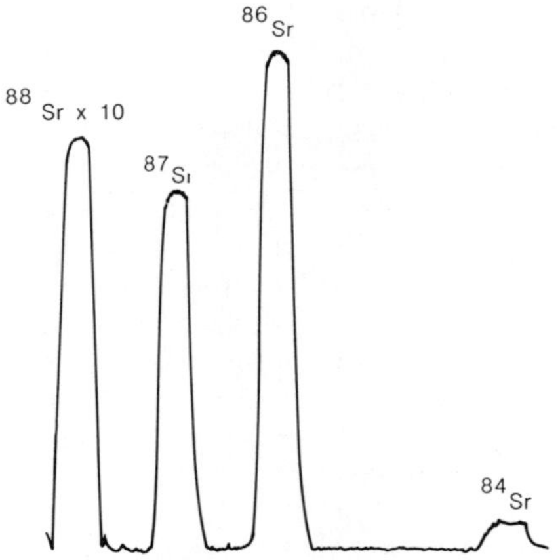

Figure 17.5 Mass spectrum of natural strontium. The heights of the mass peaks are proportional to the abundance of the respective isotopes. Note that the intensity of the ^{88}Sr has been reduced by a factor of one-tenth. (Faure, 1977.)

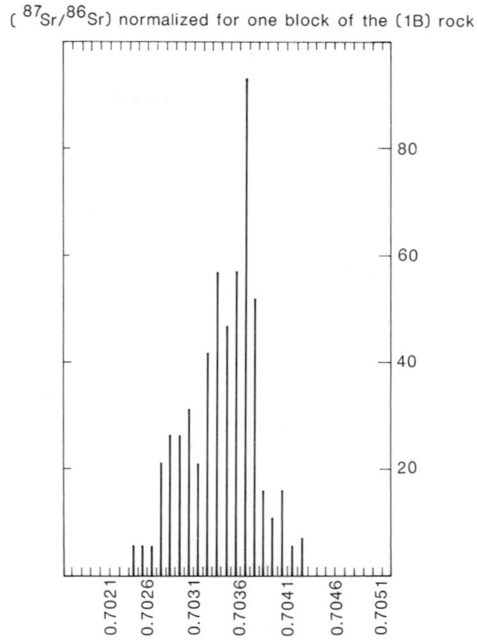

Figure 17.6 Histogram of ($^{87}Sr/^{88}Sr$) interpolated ratios measured during one block of data (24 scans of the strontium mass spectrum plotted as part of the scheme of data analysis of Birck and Allegre (1973).

(ii) Mass fractionation: a correction is usually applied to each cycle of data to correct for the higher volatility of lighter ions. The mass fractionation effect does not necessarily remain constant throughout a run, nor is it necessarily the same from one sample to the next, depending in part on changes in ionization conditions on the filament.

(iii) Variations in ion beam intensity with time as the sample is progressively burnt off the filament.

(iv) The occasional outliers; that is, the occasional anomalous mass peak measurement which is distorted by gross noise or some other non-random effect. The inclusion of such data in the final average could cause unacceptable bias in the mean ratio result.

By international agreement, mass fractionation effects are corrected by normalizing all data to a $^{86}Sr/^{88}Sr$ ratio of 0.1194 (Steiger and Jager, 1977). Various procedures have been developed for correction of the other effects. For example, Birck and Allegre (1973) divided mass count data into blocks, each consisting of 24 scans of selected mass peaks. The following corrections were applied to these data.

(i) A smoothing cubic spline function was used to model the variation in intensity of each mass peak with time within each 24-scan block. Fitted data were tested to see if any measurement lay outside the expected statistical errors. If a block contained any peak which did lie outside these error limits, data for that peak were discarded.

(ii) Spline function data were interpolated to permit calculation of $^{88}Sr/^{86}Sr$, $^{88}Sr/^{84}Sr$ and $^{87}Sr/^{86}Sr$ independent of time-dependent drift. These ratios were corrected for both mass fractionation and the presence of the ^{84}Sr spike. Detailed procedures for mass fractionation correction have been considered by Boelrijk (1968).

(iii) For each block of data, the $^{87}Sr/^{86}Sr$ and $^{86}Sr_{sample}/^{84}Sr_{spike}$ were plotted as a histogram (Figure 17.6) to test whether the distribution was normal. Values of these ratios lying outside two sigma from the mean were eliminated, and the mean and standard deviation of remaining ratios calculated.

(iv) Ratio data from each block were then classified according to precision into three categories, based on relative standard deviation, following the procedures of Papanastassiou and Wasserburg (1969). The three categories were (*a*) sigma $< 0.05\%$; (*b*) $0.05\% <$ sigma $< 0.07\%$; (*c*) $0.075\% <$ sigma $< 0.15\%$. Final results are averaged from data which fell into category (*a*) (or category ($a+b$) if insufficient data fell into category (*a*) alone).

In order to achieve adequate precision in the determination of the $^{87}Sr/^{86}Sr$ ratio, it may be necessary to average data from several hundred mass scans. Gale *et al.* (1975) required 250–400 mass scans to achieve a precision in the normalized $^{87}Sr/^{86}Sr$ of between 4 and 7×10^{-5} (two sigma). Using a 4-second integration period of six mass peaks measured during each scan with a 2.5-second delay after each peak jump, total analysis time for each $^{87}Sr/^{86}Sr$ measurement is 2 to 4 hours.

17.3.3 *Mass spectrometry—rubidium*

General procedures for the determination of ^{87}Rb by isotope dilution follow those for strontium above. The mass peaks of interest are 85 and 87 (Figure 17.4). However, one difficulty is the mass fractionation correction. Rubidium does not possess two stable isotopes from which the mass fractionation can be measured, since ^{87}Rb undergoes radioactive decay and its abundance relative to ^{85}Rb will vary from sample to sample. However an investigation of mass fractionation between rubidium isotopes by Pankhurst and O'Nions (1973) indicated that, under normal operating condition, this effect was small and generally much less than 0.5%. The $^{87}Rb/^{85}Rb$ ratio measured by these workers in

Table 17.3 Reagent blanks for Rb and Sr isotope analysis

Reagent	Sr (ng/mL)	Rb (ng/mL)
H_2O	0.00074	0.00016
HF	0.12	0.015
$HClO_4$	0.12	0.0028
HCl (6N)	0.0033	0.00043
column blank	0.4	0.0255

Data from Sanz and Wasserburg (1969).

Johnson and Matthey 'Specpure' (R) RbCl was 0.3829 ± 0.0013 (two sigma). Minster and Allegre (1976) measured this ratio at 0.3835 with an absolute spread of 0.11% and used this value (that is, $^{85}Rb/^{87}Rb = 2.6076$) for data correction. This ratio was found to be independent of both the quantity of Rb and of the proportion of Mg+K added to the filament.

17.3.4 *Precision and blank levels*

Based on the results reported by Pankhurst and O'Nions (1973), Arth *et al.* (1980) and Birck and Allegre (1973), typical analytical precision in the isotope dilution analysis of Rb is 1 to 1.5% and of Sr 0.4 to 1% (two sigma, based on a single determination). This gives an overall error in the $^{87}Rb/^{86}Sr$ ratio of typically ± 0.6 to $\pm 1.4\%$ (two sigma, Pankhurst and O'Nions, 1973). A much higher degree of precision must normally be achieved in the determination of $^{87}Sr/^{86}Sr$ ratio, of the order of ± 0.01 to 0.02% (two sigma).

Blank levels vary according to the purity of reagents, the amounts of these reagents used in the procedure and the laboratory environment. Sanz and Wasserburg (1969) reported the contamination levels shown in Table 17.3 in reagents specially purified for isotope analysis procedures.

Typical procedural blanks in the analysis of silicate rocks may be about 2 ng for Sr and 0.2 ng for Rb.

17.3.5 *Rb and Sr by XRF*

Clearly, a considerable amount of mass-spectrometer time is involved in the determination of ^{87}Rb and ^{86}Sr by isotope dilution. Savings in instrument time can be made by analysing bulk Rb and Sr by XRF and then calculating the required isotope ratio using the known isotopic abundances of these elements. The performance of XRF in Rb/Sr determinations has been evaluated by Pankhurst and O'Nions (1973), who concluded that, under the appropriate operating conditions, the absolute accuracy of XRF Rb/Sr ratios was better than $\pm 0.7\%$ and that the precision of individual Rb and Sr determinations was superior to isotope dilution data on the basis of a single determination, providing concentrations of Rb and Sr were in excess of 50 ppm.

17.4 Neodymium–samarium isotope analysis

In recent years the determination of Nd/Sm isotope ratios has attained considerable importance. However, there are some technical difficulties in obtaining satisfactory data. The first is that, because of the long half-life of ^{147}Sm (1.06×10^{11} years), differences in ^{143}Nd isotope abundance are small. It is only comparatively recently that the performance of mass spectrometers has advanced sufficiently to provide data of adequate precision. The second factor is that demanding ion exchange procedures are required, as it is necessary to separate neodymium from adjacent rare-earth elements, especially Ce and Sm, to avoid uncorrectable isobaric inteferences. Determinations are required for the isotope ratios $^{143}Nd/^{144}Nd$ which are measured as a function of the $^{147}Sm/^{144}Nd$ ratio, the latter being determined by isotope dilution mass spectrometry.

17.4.1 *Choice of spike*

The conventional enriched spike isotopes used to determine the rare-earth elements by isotope dilution mass spectrometry are ^{145}Nd and ^{149}Sm (Hooker *et al.*, 1975). However Wasserburg *et al.* (1981) argued that the isotopes ^{130}Nd and ^{147}Sm are preferable. ^{150}Nd was selected because it has a low abundance in nature; the small amounts of other Nd isotopes present in the spike will therefore cause only small changes in their relative abundance in spiked samples. Furthermore, Wasserburgh *et. al.* suggest that ^{149}Sm is not a universally suitable choice since levels of both this isotope and ^{150}Sm are variable in lunar and meteoric samples due to neutron capture effects. The spike of isotope ^{147}Sm does not suffer from this phenomenon.

17.4.2 *Chemical separation*

The chemical separation of the rare-earth elements from matrix is relatively straightforward, but the separation of Nd from adjacent REEs is much more demanding. Samples (150–250 mg) are normally brought into solution by acid attack using the Teflon bomb technique. In some rock types, it is particularly important to ensure complete dissolution of REE-bearing minor phase minerals (e.g. zircon, monazite etc.). O'Nions *et al.* (1977) used a development of the procedure for rare-earth element separation by Hooker *et al.* (1975) involving two ion exchange separations. The first anion exchange was designed to separate the rare-earth elements as a group, using a 90% glacial acetic acid–10% 5M nitric acid mixed solvent elution as described in section 14.6. In order to isolate neodymium for $^{143}Nd/^{144}Nd$ analysis, the REE fraction from the first separation was loaded on to a second anion exchange column and eluted with a mixed solvent consisting of acetic acid, methanol and nitric acid. Columns were thermostatically controlled to ± 0.1°C to ensure reproducible separations.

In contrast, Richard *et al.* (1976) performed a preliminary REE separation using a cation exchange column consisting of an AG50W–X8 (200–400 mesh) resin. Their procedure was as follows. After an HF–$HClO_4$ attack, the sample was evaporated to dryness to expel HF and Si. This residue was then redissolved in 4N HCl and loaded on to the column. All the major elements and some of the heavy REEs were eluted with 10 mL 4N HCl. The light REEs and Ba were collected in the subsequent 20 mL fraction. This was evaporated to dryness and redissolved in 0.3N HCl. Neodymium

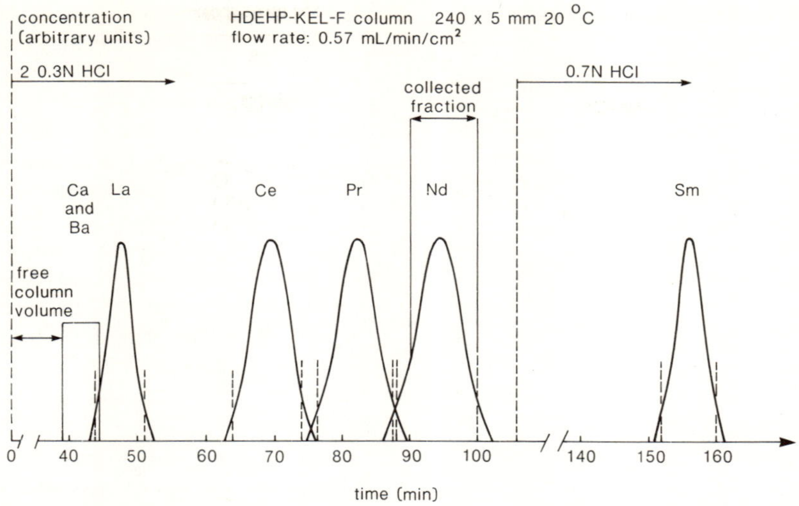

Figure 17.7 Elution curves showing the chemical separation of neodymium from other rare-earth elements. The ion exchange column consisted of di(2-ethylhexyl)orthophosphoric acid mixed with Teflon powder. The REE fraction from a previous ion exchange separation was loaded on the column and eluted with 0.3 N HCl. (Richard *et al.*, 1976.)

was separated from the solution using a second ion exchange column composed of di-(2-ethylhexyl)-orthophosphoric acid. After preconditioning the column, Nd was eluted with 0.3N HCl. A typical elution curve is shown in Figure 17.7.

In the procedure of De Paolo and Wasserburg (1976) the rare-earth elements were first separated from the major elements on a cation exchange column, using dilute hydrochloric acid as eluent. Nd and Sm were then separated on a second column using 0.2 M 2-methyl-lactic acid solution.

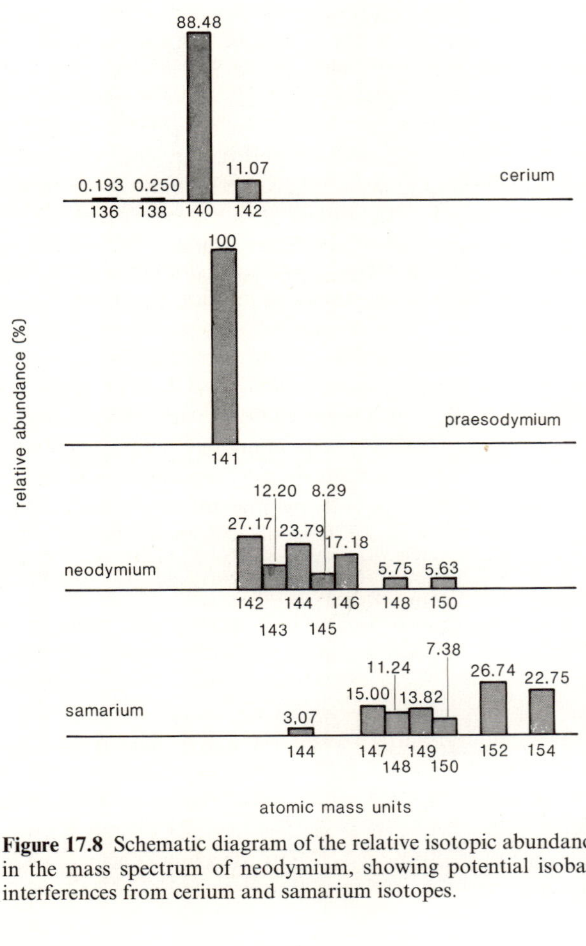

Figure 17.8 Schematic diagram of the relative isotopic abundances in the mass spectrum of neodymium, showing potential isobaric interferences from cerium and samarium isotopes.

17.4.2 *Mass spectrometry—neodymium isotopes*

Neodymium isotope ratios may be determined as the metal ion (Nd^+) or as the oxide ion (NdO^+). If the metal ion Nd^+ series is to be determined, the choice of peaks monitored (Figure 17.8) would depend on whether the spike is to be added to the same solution as is used for isotope ratio measurements or if isotope dilution is performed on a separate sample aliquot. However, isotope peaks of possible interest would include those shown in Table 17.4.

It is not suggested that all these isotopes would be measured in any one scheme of analysis. Thus O'Nions *et al.* (1977)

Table 17.4 Neodymium mass peaks

Mass number	Isotope	Comments
140	^{140}Ce	To monitor ^{142}Ce interference on ^{142}Nd
142	^{142}Nd	(+ minor ^{142}Ce if present): used for mass fractionation correction (De Paolo and Wasserburg, 1976)
142.5	—	Background measurement
143	^{143}Nd	For $^{143}Nd/^{144}Nd$ isotope ratio measurement
144	^{144}Nd	(+ minor ^{144}Sm): for $^{143}Nd/^{144}Nd$ isotope ratio measurement
145	^{145}Nd	Isotope dilution spike
146	^{146}Nd	Required for mass fractionation correction (O'Nions *et al.*, 1979)
146.5		Background measurement
147	^{147}Sm	To monitor interference on 144, 148 and 150 mass peaks
148	^{148}Nd	(+ ^{148}Sm)
150	^{150}Nd	(+ ^{150}Sm): isotope dilution spike (Wasserburg *et al.*, 1981)
152	^{152}Sm	To monitor interferences on 144, 148 and 150 mass peaks

Table 17.5 Mass overlaps for Nd isotope measurements

Mass no.	156	157	158	159	160	161	162	163
	$^{140}Ce^{16}O$	$^{140}Ce^{17}O$	$^{140}Ce^{18}O$					
		$^{141}Pr^{16}O$	$^{141}Pr^{17}O$	$^{141}Pr^{18}O$				
			$^{142}Ce^{16}O$	$^{142}Ce^{17}O$	$^{142}Ce^{18}O$			
			$^{142}Nd^{16}O$	$^{143}Nd^{17}O$	$^{142}Nd^{18}O$			
				$^{143}Nd^{16}O$	$^{143}Nd^{16}O$	$^{143}Nd^{18}O$		
					$^{144}Nd^{16}O$	$^{144}Nd^{17}O$	$^{144}Nd^{18}O$	$^{147}Sm^{16}O$
						$^{146}Nd^{16}O$	$^{145}Nd^{17}O$	$^{145}Nd^{18}O$
							$^{146}Nd^{16}O$	$^{146}Nd^{17}O$

Sm interference check: 163/162 = ($^{147}Sm^{16}O$/162)+($^{145}Nd^{18}O$/162)+($^{146}Nd^{17}O$/162). The sum of the second and third terms is 1.36×10^{-3}.
Data from Richard *et al.* (1976).

scanned mass peaks 143, 144 (neodymium isotope ratio measurement), 146, 146.5 (background) and 147 (to monitor Sm interference) in a scheme designed specifically to measure $^{143}Nd/^{144}Nd$ ratios. The 146 peak was monitored to permit mass fractionation correction. In comparison, McCulloch and Chappell (1982) scanned mass peaks 142, 146 (both for mass fractionation correction) 143, 144 (Nd ratio measurement), 150 (Nd spike), 140 (to monitor Ce interference) and 152 (to monitor Sm interference).

If neodymium is to be determined from the NdO^+ ion beam (e.g. Richard *et al.*, 1976), then a further interference must be considered. Oxygen itself exists as three stable isotopes, the approximate isotopic abundances of which are ^{16}O, 99.759%; ^{17}O, 0.037% and ^{18}O, 0.204%. The potential for isobaric intereferences is significantly increased, as can be judged from data in Table 17.5. Thus, for example, the isotopic abundance of $^{144}Nd^{16}O$ must be corrected for contributions from $^{142}Nd^{18}O$ and $^{143}Nd^{17}O$. These corrections can be carried out if the ratios $R_{18} = {}^{18}O/{}^{16}O$ and $R_{17} = {}^{17}O/{}^{16}O$ are accurately known. Most analysts have used values of these parameters originally determined by Nier (1950), namely $R_{18} = 0.002045$ and $R_{17} = 0.003708$. Wasserburg *et al.* (1981) recalculated these ratios based on the abundance ratios of $^{150}Nd^{18}O/^{150}Nd^{16}O$ and $^{150}Nd^{17}O/^{150}Nd^{16}O$ and obtained values of $R_{18} = 0.00211$ and $R_{17} = 0.00387$. These values were appropriate for correction of NdO^+ ion beam measurements, but may represent values only suitable for correction under the specific operating conditions/instrumentation used by Wasserburg *et al.* (1981).

Table 17.6 Neodymium oxide mass peaks

Mass number	Isotope	Comments
156	^{140}CeO	To monitor Ce interference on mass 158
157	^{141}PrO	To monitor Pr interference
158	^{142}NdO	For mass fractionation correction
159	^{143}NdO	For neodymium isotope
160	^{144}NdO	ratio measurements
159.5	—	Background
161	^{145}NdO	Nd spike isotope if specified
162	^{146}NdO	Required for mass fractionation correction
163	^{157}SmO	To monitor Sm interference on mass peaks 160, 164 and 166
166	^{150}NdO	Nd spike isotope if specified

Using correction ratios, a series of simultaneous equations can be set up to remove oxygen isotope interferences, namely:

$$^{142}Nd = {}^{158}NdO$$
$$^{143}Nd = {}^{159}NdO - {}^{142}NdR_{17}$$
$$^{144}Nd = {}^{160}NdO - {}^{142}NdR_{18} - {}^{143}NdR_{17} \text{ etc.}$$

The mass peaks which would normally be scanned would include those shown in Table 17.6.

17.4.4 *Mass fractionation correction*

(*a*) *Neodymium isotope ratios.* The lack of internationally agreed conventions for the correction of mass fractionation has caused endless confusion in neodymium isotope analysis. The requirements for a pair of isotopes suitable for such a correction have been considered in some detail by Wasserburg *et al.* (1981). These authors consider that the pair of isotopes selected should have as large a mass difference between them as possible and yet have ratio relatively close to unity to minimize uncertainties in the instrumental correction factor. This problem has been discussed further by Hawksworth and van Calsteren (1983). Different laboratories which use different conventions for this correction have published $^{143}Nd/^{144}Nd$ ratios which are in disagreement

Table 17.7 Isotopic composition of basalt USGS: BCR-1

Isotope ratio data			Isotopic abundance data		
Ratio	A	B	Isotope	A	B
$^{142}Nd/^{144}Nd$	1.141827	1.138305	^{142}Nd	27.17%	27.07%
$^{143}Nd/^{144}Nd$	0.51264	0.51184	^{143}Nd	12.20	12.172
$^{145}Nd/^{144}Nd$	0.348417	0.348956	^{144}Nd	23.79	23.78
$^{146}Nd/^{144}Nd$	0.7219	0.724134	^{145}Nd	8.290	8.299
$^{148}Nd/^{144}Nd$	0.241578	0.243075	^{146}Nd	17.18	17.22
$^{150}Nd/^{144}Nd$	0.236418	0.238619	^{148}Nd	5.780	5.781
			^{150}Nd	5.625	5.675

Data from Wasserburg *et al.* (1981).
A: data normalized against $^{146}Nd/^{144}Nd = 0.7219$ (O'Nions *et al.*, 1977).
B: data normalized against $^{146}Nd/^{142}Nd = 0.636151$ (Wasserburg *et al.*, 1981).
The $^{143}Nd/^{144}Nd$ ratio is specific to USGS: BCR-1; other samples will show small differences depending on their radiogenic ^{143}Nd content.

Table 17.8 Isotopic composition of samarium

Isotopic ratio data		Isotopic abundance data	
$^{144}Sm/^{154}Sm$	0.13516	^{144}Sm	3.075
$^{147}Sm/^{154}Sm$	0.65918	^{147}Sm	15.00
$^{148}Sm/^{154}Sm$	0.49419	^{148}Sm	11.24
$^{149}Sm/^{154}Sm$	0.60750	^{149}Sm	13.82
$^{150}Sm/^{154}Sm$	0.32440	^{150}Sm	7.380
$^{152}Sm/^{154}Sm$	1.17537	^{152}Sm	26.74
		^{154}Sm	22.75

Data from Wasserburg *et al.* (1981).

on identical samples. The two conventions in general use are as follows. O'Nions *et al.* (1977) corrected for mass fractionation using the value $^{146}Nd/^{144}Nd = 0.7219$, this being the average of a large number of measurements in their laboratory. Wasserburg *et al.* (1981) and De Paolo and Wasserburg (1976) adopted the value $^{146}Nd/^{142}Nd = 0.636151$, this being the isotope pair which they considered most closely fulfils the criteria set out above. The former mass fractionation correction ratio gives a $^{143}Nd/^{144}Nd$ isotope ratio in basalt USGS: BCR-1 of 0.51264, the latter gives a value of 0.51184.

The isotopic abundance of neodymium in USGS: BCR-1 according to these two conventions is as shown in Table 17.7.

(*b*) *Samarium isotope ratios.* The same confusion does not exist over the isotopic abundances of samarium, the most recent values being those of Wasserburg *et al.* (1981) (Table 17.8).

17.4.5 *Operating practice*

Neodymium isotope ratios may be measured adequately on single filaments by monitoring the NdO^+ ion beam. In the procedure of Richard *et al.* (1976), the neodymium residue was taken up in dilute hydrochloric acid and 200 ng loaded on to a single rhenium filament. This was preheated at 4.5 A for 2 minutes in a vacuum better than 7×10^{-7} Torr. After mounting in the mass spectrometer, stable ion beams were obtained by burning at 1280°–1300°C, giving a $^{158}NdO^+$ beam current of 2×10^{-12} A. Triple-filament operation is required for successful Nd^+ ion beam analysis. A typical scheme would involve loading neodymium solution dissolved in dilute hydrochloric acid on to the tantalum side filaments of a triple-filament assembly and using a rhenium centre filament to give a stable $^{144}Nd^+$ ion beam current in the range 10^{-11} to 10^{-12} A.

The appropriate mass peaks (see above) are scanned and data corrected for time-dependent drift and mass fractionation effects in a manner similar to that used in strontium ratio analysis (section 17.3). Repetitive scans of the mass spectrum are usually continued until the averaged $^{143}Nd/^{144}Nd$ ratio data reach the required precision. This is generally specified to be in the range 0.001–0.004% (two-sigma error in the ratio $^{143}Nd/^{144}Nd$). Depending on sample and analysis conditions, 800–1200 mass scans may be required before acceptable precision has been achieved (see, for example, Hawkesworth *et al.*, 1979). Isotope dilution determinations of ^{147}Sm and ^{144}Nd are not required to this standard of precision. There is, therefore, some merit in carrying out isotope dilution determinations in a separate run from the $^{143}Nd/^{144}Nd$ isotope ratio measurement, in order to limit the number of mass peaks monitored in the high-precision ratio run. If this procedure is followed, then four mass peaks must be monitored for $^{143}Nd/^{144}Nd$ isotope ratio measurement (^{143}Nd, ^{144}Nd, one mass peak for fractionation correction, e.g. ^{146}Nd, and one background mass peak). Each individual high-precision analysis for $^{143}Nd/^{144}Nd$ could then take 5–8 hours of instrument time. It might also be necessary to monitor mass peak 147, and possibly also 140, to check for interference from Sm and Ce.

The precision required in the isotope dilution determination of $^{147}Sm/^{144}Nd$ is less stringent, a two-sigma uncertainty of 0.05 to 0.1% being adequate in most circumstances. Isotope dilution procedures, particularly with respect to rare-earth element determinations, are discussed in section 17.6. Blank levels for Nd and Sm are usually very low and can often be ignored. For example, total procedural blanks of 30–60 pg Nd and 6–8 pg Sm have been reported (Lugmuir and Marti, 1978).

17.5 Lead, uranium and thorium isotope analysis

17.5.1 *Analysis of zircon*

Zircon was the first mineral to be used for Pb–U–Th isotope analysis and continues to be the most widely studied. The reason is that zircon is widely distributed in igneous rocks from which it can relatively easily be separated. Zircon is superficially resistant to alteration and, most importantly, readily incorporates uranium, but not lead, into its structure during crystallization. Most of the lead found in present-day samples is therefore radiogenic, making zircon an ideal medium for the investigation of the U–Pb decay sequence.

Zircon is one of the most resistant minerals and is very difficult to bring into solution. Early procedures used a borax fusion technique, but this was associated with undesirably high lead blanks; Krogh (1973) used an acid attack method, and, the key to this procedure was the development of a self-sealing Teflon bomb (Figure 17.9) which allowed

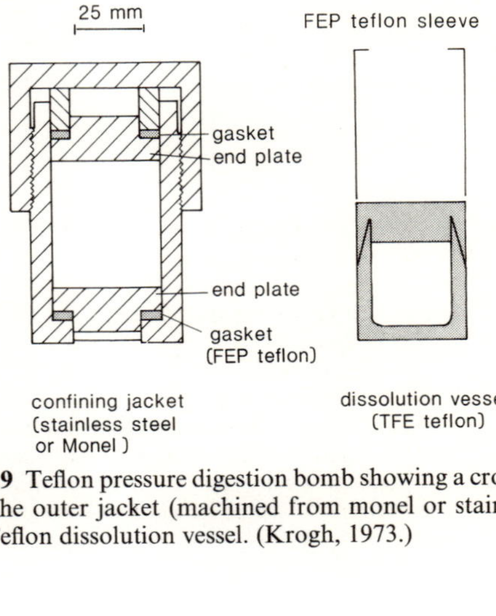

Figure 17.9 Teflon pressure digestion bomb showing a cross-section through the outer jacket (machined from monel or stainless steel) and the Teflon dissolution vessel. (Krogh, 1973.)

zircon to be attacked under pressure by 48% hydrofluoric acid solutions at temperatures up to 270°C. Particular features of this bomb were:

(i) A self-sealing V-shaped joint between the Teflon beaker and lid, to ensure quantitative retention of HF under pressure

(ii) An FEP Teflon sleeve round the Teflon beaker, to prevent contamination of the Teflon beaker from the outer metal jacket, which could cause problems during subsequent manipulation of the beaker in a clean environment

(iii) FEP Teflon gaskets to provide sufficient back pressure on the Teflon bomb after sealing

(iv) A screw-top monel (or stainless steel) outer jacket, which permits progressive adjustment in the closure pressure of the bomb and allows for shrinkage in the length of the Teflon inner beaker in continuous use. Experience showed that these beakers could shorten by 0.5 mm in the first six months of operation.

Tests on this bomb design indicated that some loss of HF occurred, amounting to about 1 g in 2.5 weeks at 220°C. The addition of FEP Teflon gaskets was reported to reduce losses by a factor of 5 to 6. No corresponding loss Pb or U from the sample inside the bomb was observed.

Krogh's (1973) separation procedure was therefore as follows. Zircon separates were pre-cleaned in warm 7N nitric acid, ultra-clean water and acetone. Hand-picked grains (50 mg in all) were added to the Teflon bomb together with 2 to 3 mL 48% HF, a few drops of nitric acid and the appropriate Pb and U spikes. The bomb was lightly sealed and heated in an oven at 220°C for one week. After cooling, the bomb was opened and 3.1N HCl added to give a total volume of 2 mL. The bomb was then resealed and heated overnight to redissolve any insoluble fluorides.

Lead and uranium are then separated from zirconium using a simple anion exchange column separation. The columns, consisting of 0.5 mL Dowex 1–X8 (200–400 mesh) resin in chloride form, were washed with 2 mL HCl (6.2N) and 2 mL H_2O and then preconditioned with 1 mL 3.1N HCl. After loading the sample solution, zirconium was eluted first with 3.1N HCl, U and Th being retained on the column. Lead was subsequently eluted with 1 mL 6.2N HCl, then uranium with 1 mL H_2O. The column elution curve is shown in Figure 17.10. Measured lead contamination levels are reported to be in the range 0.5 to 5 ng in contrast to 200 to 1000 ng obtained by the borax fusion method.

In order to undertake a complete analysis, Krogh divided sample solutions into two aliquots prior to chemical separation. The first aliquot was analysed for lead isotope ratios using the ion exchange procedure described above. The second was spiked with ^{208}Pb for the determination of lead concentration by isotope dilution analysis. ^{205}Pb would be the modern choice of lead spike (see 17.5.3). The analytical uncertainty in U–Pb zircon age determination by this method is about 3×10^6 year (two sigma). However, in many igneous samples, zircon separates are likely to show a range of ages due to remobilization of lead and uranium during subsequent alteration processes. The geological value of high-precision U–Pb zircon ages is therefore limited by the ability to select zircon minerals from a sample which have suffered minimal alteration. Not all zircon minerals in the same sample are affected by subsequent alteration processes to the same degree. In particular, the centres of large crystals are much less susceptible than the edges. Further work by Krogh (1982*a*,*b*) has shown that improved techniques of magnetic separation can, to a significant degree, overcome some discrepancies due to alteration effects. Krogh found that the least magnetic fraction of a group of zircon separates was the least likely to have suffered premature lead loss

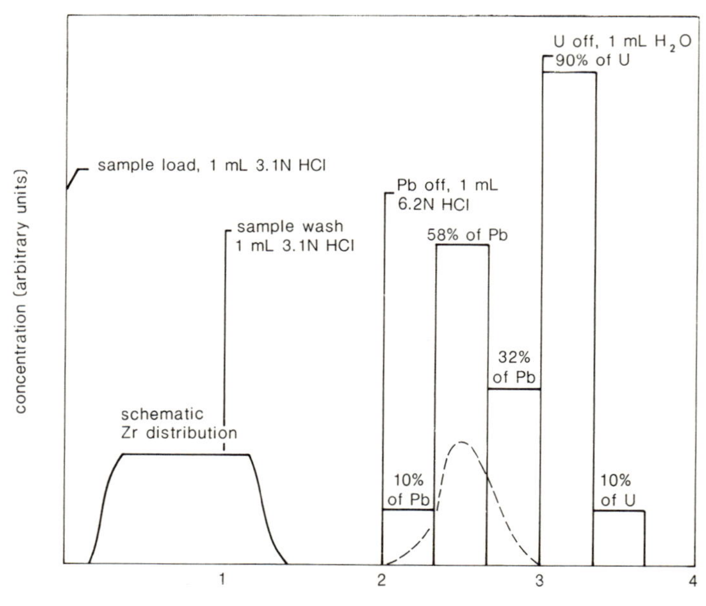

Figure 17.10 Ion exchange elution curves of zirconium, lead and uranium using a Dowex 1X–8 (200–400 mesh) ion exchange column. Samples were loaded on to the column and eluted with 3.1 N HCl (Zr elution), 6.2 N HCl (Pb elution), and H_2O (U elution), as used in the scheme of zircon analysis developed by Krogh (1973).

and was therefore likely to give the most concordant ages. Further developments using an air abrasion technique (Krogh, 1982*b*) permited the removal of outer altered layers of zircon, so eliminating discordant components of the sample. By this careful sample selection precedure, concordant ages of unprecedented accuracy could be achieved: that is, errors of $1–3 \times 10^6$ years in zircons 2700×10^6 years old.

Zircons may contain several hundred ppm of Th as well as U, and are therefore amenable to $^{232}Th/^{208}Pb$ studies as well as the more usual U–Pb age determinations (Allegre and Albarede, 1974), though such determinations are less often carried out.

17.5.2 *Analysis of silicate and meteoritic material*

Following the demonstration of the U–Th–Pb geochronometer in zircon analysis, there was considerable interest in extending these studies to silicate and meteoritic samples. This has led to the development of more sophisticated techniques to cope with the more variable nature of the matrix from which U, Th and Pb must then be separated. In particular in the study of meteoritic fragments, it is often necessary to perform an analysis on extremely small sample masses (down to the mg level). This requires miniaturization of the chemical separation procedure to reduce blank contamination levels. Several miniaturization procedures have been published, an exceptional example being that of Tera and Wasserburg (1975).

Several criteria influence the choice of chemical separation procedure compared with the relatively simple technique developed for zircons. Some of these criteria have been summarized by Manhes *et al.* (1984). In particular:

(i) Thallium gives an isobaric interference with ^{205}Pb (the normal lead spike, Figure 17.11). The ion exchange procedure must therefore efficiently separate Tl from the Pb fraction

(ii) Sufficiently intense lead ion beams are obtained by mounting samples on a filament in phosphoric acid medium. A small isobaric interference on lead from $BaPO_2^+$ can then occur. Barium must also therefore be efficiently separated from lead

(iii) Major elements, in particular the alkali metals, suppress the ionization of lead and must be removed from analysed fractions.

For *ion exchange separation* of lead, uranium and thorium, samples are usually first dissolved in a Teflon bomb using an $HF/HClO_4$ $(+HNO_3)$ attack. Digested residues are then taken up in HCl medium and separated on a strong anion exchange resin, AG1–X8 (200–400 mesh), using HCl/HBr mixtures for elution. Manhes *et al.* (1984) pre-

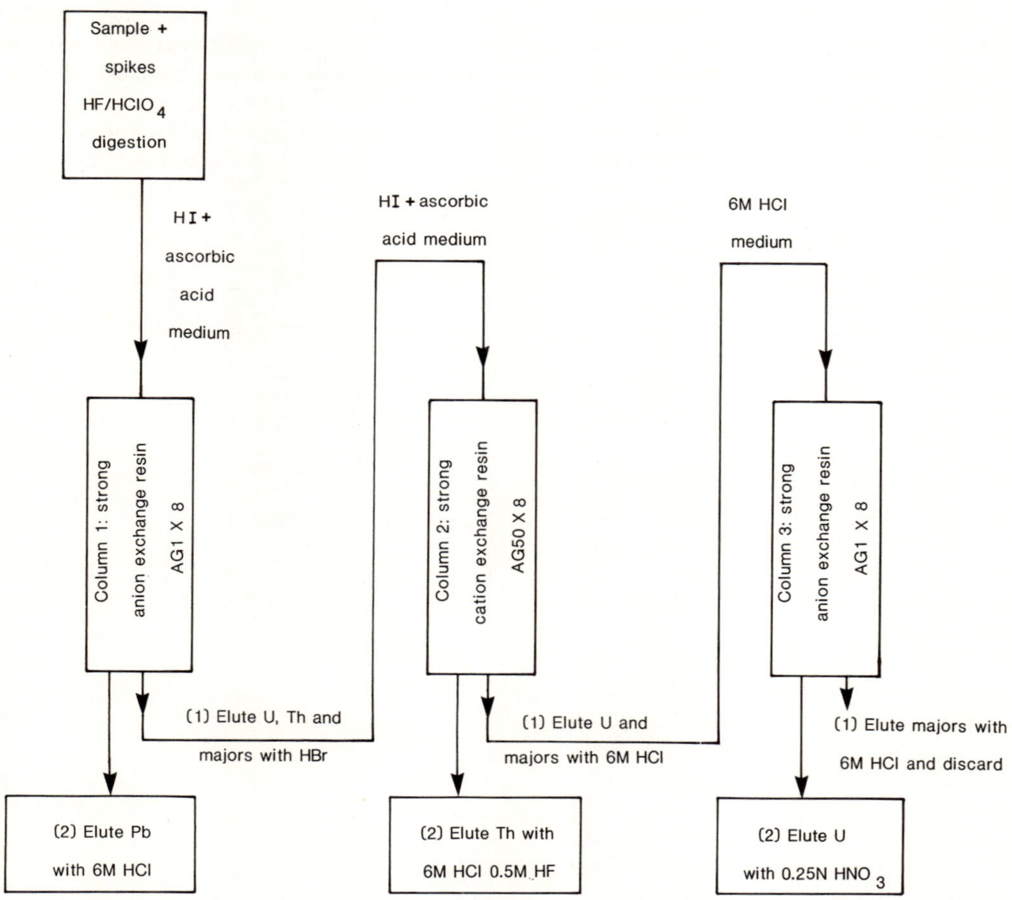

Figure 17.11 Flow chart showing the chemical separation scheme of Manhes *et al.* (1984) developed for the analysis of lead, thorium and uranium isotopes in silicate materials.

ferred HI for elution since this acid reduces the poisoning effect of residual $HClO_4$ on the anion exchange resin. In this case it was necessary to add ascorbic acid to HI solution to prevent oxidation of I^- by atmospheric oxygen or iron(III) in the sample. Lead is retained on the column and may be eluted with 6N HCl. It is normally necessary to repeat this procedure, using a second anion exchange column to ensure complete separation of lead from interfering elements. U and Th pass through this first column together with other major elements.

Oversby (1975) then separated uranium from the major elements by precipitating hydroxides, centrifuging and taking up the residue in 8N HCl (this procedure removes Br^-). This solution was then passed through an AG1–X8 (200–400 mesh) column. Major elements are eluted with 8N HCl, U and Fe^{3+} were retained on the column. Fe^{3+} was removed by eluting with a solution of $(NH_4)_2SO_4$ saturated with SO_2. This caused Fe^{3+} to be reduced to Fe^{2+} which could then be eluted, after which uranium could be eluted with 3N HBr.

By contrast, Manhes *et al.* (1984) used a different separation scheme, a simplified version of which is shown in Figure 17.11. After separating lead in the first anion exchange column using HI/ascorbic acid medium (as described above), the eluent was loaded on to a second, cation exchange resin (AG50–X8, 200–400 mesh). Uranium and the major elements are eluted with 6N HCl. Thorium was retained on the column and was subsequently eluted with a 6N HCl–0.5N HF mixture. The eluent from this second column separation was loaded on to a third AG1–X8 anion exchange column. Uranium was retained on this column, the major elements being removed with 6N HCl then 7.5N HNO_3. Uranium is finally eluted with 0.25N HNO_3 (Figure 17.11).

This by no means covers the variety of separation procedures used in Pb–U–Th isotope analysis, other schemes having been described, for example, by Tera and Wasserburg (1972, 1975) and Tilton (1973). Arden and Gale (1974) have described a novel method in which lead is separated from solution by electrodeposition on a copper-plated platinum cathode. This method has subsequently been developed to permit the separation of thallium for analysis by isotope dilution mass spectrometry (Arden, 1983*a*,*b*).

17.5.3 *Choice of spike*

Samples must be spiked in order to determine bulk lead, uranium and thorium concentrations by isotope dilution mass spectrometry. Uranium is determined using a spike enriched in ^{235}U.

Thorium is nominally mono-isotopic. However, the long-lived radioactive daughter of ^{238}U, namely ^{230}Th ($T_{1/2} = 7.52 \times 10^4$ years) is available for thorium determinations. In early lead work an enriched ^{208}Pb spike was normally used. However, since ^{208}Pb is a naturally-occurring isotope, whose ratio relative to ^{204}Pb is required as part of the isotope ratio determination, it used to be necessary to determine lead isotope ratios on one sample aliquot and to spike a separate aliquot with ^{208}Pb for the bulk lead determination. However, Krogh and Davis (1975) showed that both isotope dilution and isotope ratio measurements could be carried out simultaneously if samples were spiked with a long-half-life artificial isotope of lead, ^{205}Pb ($T_{1/2} = 3 \times 10^7$ years). Recently, Manhes *et al.* (1984) have shown that an artificial uranium isotope, ^{233}U ($T_{1/2} = 1.62 \times 10^5$ years) can be used for uranium isotope dilution.

17.5.4 *Mass spectrometry*

The development of thermal ionization mass spectrometry for Pb isotope ratio determinations was seriously hampered by the very low efficiency of ionization of lead from a heated metal filament. This necessitated the use of large samples for mass spectrometric analysis. Significant improvements were made in the 1950s and 1960s in increasing the ion yield from lead, culminating in the work of Cameron *et al.* (1969), who showed that stable ion beams could be obtained from 100 ng aliquots of Pb if samples were mixed with silica gel and phosphoric acid on the filament. This is now the standard method of analysing lead samples and makes an interesting analogy with matrix modification procedures used 10 years

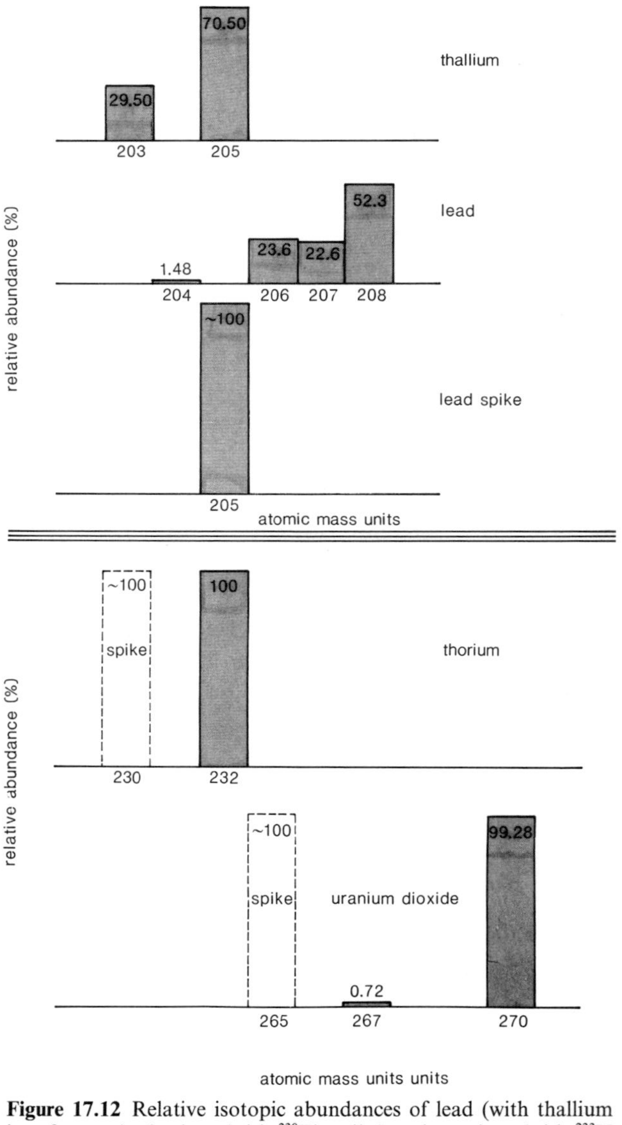

Figure 17.12 Relative isotopic abundances of lead (with thallium interferences), thorium (with ^{230}Th spike) and uranium (with ^{233}U spike). Uranium is measured as the UO_2^+ ion beam.

Table 17.9 Lead isotope mass peaks

Mass number	Isotope	Comments
204	^{204}Pb	Common lead
204.5	—	Background measurement
205	^{205}Pb	(Plus ^{205}Tl interference): spike for isotope dilution analysis of lead
206	^{206}Pb	Radiogenic lead from the ^{238}U decay chain
207	^{207}Pb	Radiogenic lead from the ^{235}U decay chain
208	^{208}Pb	Radiogenic lead from the ^{232}Th decay chain

later to improve volatilization efficiency of elements atomized in a graphite furnace (section 4.12). Lead can be analysed on a single rhenium filament, but beams of higher stability are generally considered to be obtained from triple-filament ionization. The mass peaks of relevance to Pb–U–Th analysis are shown in Figure 17.12. The mass numbers which would normally be counted in the determination of lead are shown in Table 17.9.

In addition, it may be desirable to monitor mass number 203 (^{203}Tl) to check for any interference on ^{205}Pb. An analogue display of the spectrum from mass numbers 194 to 212 is shown in Figure 17.13, taken from Tera and Wasserburg (1975). The presence and relative intensity of $BaPO_2^+$, Bi^+ and Tl^+ shown in this diagram depend on the success of the chemical separation procedure.

17.5.5 *Operating considerations*

A typical analysis procedure for lead would be as follows. Following the chemical separation, the sample is slurried with silica gel in phosphoric acid and loaded on to an outgassed rhenium filament which is then preheated to about 900°C under vacuum. The sample is then mounted in the mass spectrometer. Stable lead ion beams are usually obtained by running the filament at 1150–1200°C. Depending on instrumental conditions, 30 minutes' delay may be required before isotope ratio measurements to permit the beam to stabilize. Selected mass peaks are then scanned repetitively until the averaged lead ratio data has reached the specified precision. Typically, a minimum of 100 such scans may be required, giving a total analysis time of at least one hour. Mass fractionation between lead isotopes is normally measured by analysing isotope ratios from one of three lead isotope reference samples: NBS: SRM981 (natural lead metal), SRM982 lead metal equal atom 206/208, or SRM983 lead metal radiogenic (92% ^{206}Pb). These reference materials were themselves calibrated by determining the absolute lead isotope ratios relative to mixtures of isotopically pure ^{206}Pb and ^{208}Pb (Catanzaro *et al.*, 1968). Further work by Catanzaro (1967) has shown that these materials can be used to determine absolute isotope ratios in geochemical reference materials.

The corrections for mass fractionation measured from these reference samples usually amount to 0.09–0.12% per atomic mass unit under normal operating conditions.

17.5.6 *Blank levels*

Lead is widely distributed in the environment due to its use as an anti-knock agent in automobile fuels. The provision of ultra-clean laboratory facilities and the purification of reagents is therefore important in lead analysis. These aspects are considered in detail in section 1.7.

Blank levels depend very much on the mass of sample taken for analysis and therefore the volume of reagents required for the separation procedure. Some reagent blanks taken from Manhes *et al.* (1978) are listed in Table 17.10. A typical lead blank for a whole-rock determination in which 1 g of sample has been taken for analysis might be about 10–80 ng. However, for a miniaturized separation procedure using 1–50 mg of sample, procedural blanks could be as low as 0.02–0.1 ng Pb. In fact miniaturization of Pb separation procedures has now reached a high degree of sophistication. Manhes *et al.* (1984) report that it is possible to analyse 1 mg of silicate with a lead concentration of 0.1 ppm for

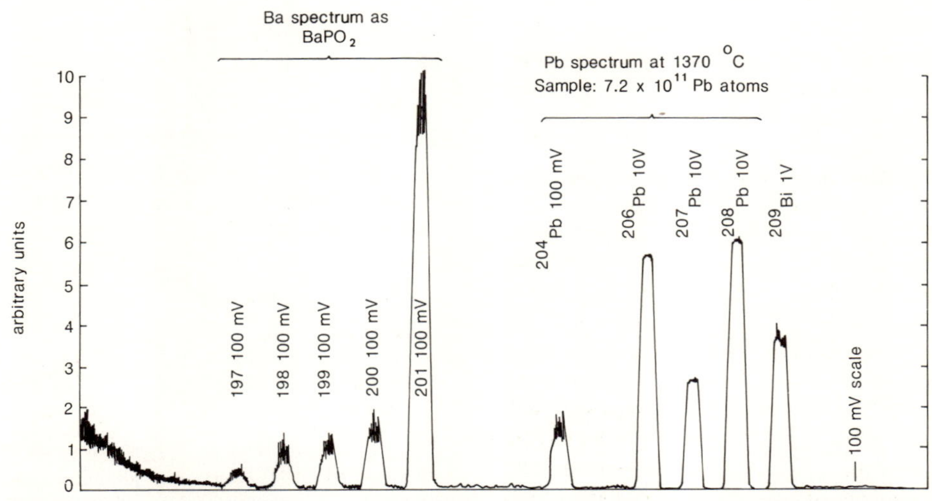

Figure 17.13 Mass spectrum of lead showing potential interferences from $BaPO_2^+$ and Bi^+. (Tera and Wasserburg, 1975.)

T

Table 17.10 Lead reagent blanks

Reagent	Purification procedure	Lead blank level (pg/g)
HBr (49%)	Sub-boiled twice in quartz	20
HF (50%)	Isothermally distilled in Teflon	50
H_2O	Sub-boiled in quartz	20
loading blank		15

Data from Manhes *et al.* (1978).

isotope ratios. The dissolution and separation procedure required 0.07 mL of reagents and 0.002 mL of anion exchange resin, the procedure introducing lead contamination of about 3 pg. The isotopic composition of such lead blanks is reported by Manhes *et al.* (1984) to be $^{206}Pb/^{204}Pb = 18.0$, $^{207}Pb/^{204}Pb = 15.55$ and $^{208}Pb/^{204}Pb = 37.4$.

17.5.7 *Analytical precision*

In an estimate of the precision of an isotopic analysis, it is desirable to use a rigorous error propagation procedure such as that of Nunes and Thurston (1980), which incorporates uncertainties due to:

(i) Statistical variation in measured isotopic data
(ii) Uncertainties in mass fractionation correction
(iii) Errors in sample and spike weighings
(iv) Estimated blanks.

Precisions in isotope dilution determinations of lead are usually reported to be in the range $\pm 0.1\%$ (two sigma).

17.5.8 *Mass spectrometry for the isotope dilution determination of uranium*

Uranium must be determined by isotope dilution using a ^{235}U (or ^{233}U) spike. Uranium isotope measurements are made using the UO_2^+ ion beam. To promote efficient ionization, samples are traditionally mixed with tantalum pentoxide (Ta_2O_5) and phosphoric acid on a tantalum filament. Stable ion beams are formed at 1350–1450°C. Mass fractionation is usually reported to be in the range 0.10–0.15% per atomic mass unit and can be estimated from the absolute isotopic composition of NBS: SRM960 (uranium metal). Gancarz and Wasserburg (1977) found that more stable and longer-lasting ion beams could be obtained from uranium samples mounted on rhenium 'V' filaments with silica gel and phosphoric acid. A stable ion beam of about 10^{-13} A could be obtained for four hours at ionization temperatures of 1400–1550°C from a filament loaded with 1 ng of uranium. The mass peaks of interest are shown in Table 17.11.

Since precisions in bulk uranium content by isotope dilution mass spectrometry are rarely quoted to better than 1–2%, no correction is necessary for oxygen isotope interferences (due to ^{17}O and ^{18}O) as was the case with NdO^+ measurements. Uranium blank levels are usually low, often much less than 1ng. In the analysis of zircons, Launcelot *et al.* (1983) found that it was possible to speed up instrumental analysis time by analysing both U and Pb for isotope dilution measurement on the same filament. Samples spiked with Pb and U were loaded on the same M-shaped rhenium filament together with silica gel. Lead was first run at temperatures of 1200–1400°C, and the UO_2^+ emission was subsequently measured by increasing filament temperatures to 1400–1550°C.

Table 17.11 Uranium isotope mass peaks

Mass number	Isotope	Comments
265	$^{233}UO_2^+$	Preferred spike for isotope dilution analysis
267	$^{235}UO_2^+$	Natural uranium isotope (traditional spike for isotope dilution analysis)
267.5	—	Background monitor
270	$^{238}UO_2^+$	Natural uranium isotope

17.5.9 *Mass spectrometry for the isotope dilution of thorium*

In the isotope dilution determination of thorium using a ^{230}Th spike, Manhes *et al.* (1984) loaded sample solutions prepared as sulphates on a tungsten filament mixed with colloidal graphite. Tungsten filaments which were degassed at 2050°C before loading the sample were preferred to rhenium, because of the lower observed thorium contamination levels. The Th^+ ion beam is monitored, and the mass peaks of interest are listed in Table 17.12.

Since only one natural isotope of thorium occurs (^{232}Th), mass fractionation cannot easily be measured. Based on equivalent data for U and Pb, it is normal to assume an arbitrary correction of about 0.1% per atomic mass unit. Blank levels are expected to be significantly less than 1ng, and precision in the isotope dilution measurement of Th is typically 1–2% (two sigma). Manhes *et al.* (1978) were able to determine uranium and thorium from the same rhenium triple filament. With the central filament held at 1940°C, the temperature of the lateral filaments was gradually increased until a stable uranium (UO_2^+) signal was observed. After uranium isotope measuremnts, the temperature of the central filament was increased to 2040°C and the lateral filaments further heated to ionize Th^+. Clearly this technique offers significant savings in instrument utilization time.

17.6 Isotope dilution

Isotope dilution mass spectrometry is one of the most precise and sensitive analytical techniques available for the bulk determination of trace elements. This section concerns the use of isotope dilution in conjunction with thermal ionization mass spectrometry.

Table 17.12 Thorium isotope mass peaks

Mass number	Isotope	Comments
230	^{230}Th	Spike for isotope dilution
231	—	Background monitor
232	^{230}Th	Natural isotope

17.6.1 *Principles*

The principle of the technique is simple. Many elements exist as two or more naturally-occurring isotopes. A known weight of rock sample is mixed with a spike consisting of a precisely known amount of the element enriched in one of those naturally-occurring isotopes. The sample is taken up into solution and a chemical separation procedure (normally ion exchange) is carried out to isolate the element of interest. Providing complete equilibration occurs between the element of interest dissolved out of the sample and the enriched spike, it is not necessary to achieve quantitative recovery during the separation procedure. The isolated element fraction is then analysed by thermal ionization mass spectrometry to determine the ratio between the spiked mass peak and an unspiked, naturally-occurring isotope. This situation is illustrated in Figure 17.14. In the example of strontium, four naturally-occurring isotopes exist: $^{84}Sr(0.56\%)$, $^{86}Sr(9.86\%)$, $^{87}Sr(6.98\%)$ and $^{88}Sr(82.5\%)$. A spike is available for strontium artificially enriched in the isotope ^{84}Sr. The degree of enrichment normally exceeds 99% (that is, the spike solution contains less than 1% total of $^{86}Sr+^{87}Sr+^{88}Sr$). The appropriate aliquot of ^{84}Sr spike is added to the sample, which is then digested and strontium chemically separated by a simple ion exchange procedure (see section 17.3). The ratio $^{88}Sr/^{84}Sr$ is then measured by mass spectrometry.

If $(^{88}Sr/^{84}Sr)_{measured}$ is the measured ratio between ^{88}Sr and ^{84}Sr in the analysed sample, I_{88} and I_{84} are the natural isotopic abundances of ^{88}Sr and ^{84}Sr (i.e. in an unspiked sample), and S_{88} and S_{84} are the known isotopic abundances of ^{88}Sr and ^{84}Sr in the enriched spike, then

$$\left[\frac{^{88}Sr}{^{84}Sr}\right]_{measured} = \frac{I_{88}+S_{88}}{I_{84}+S_{84}}. \tag{17.2}$$

From this may be determined the concentration of Sr in the sample, as follows.

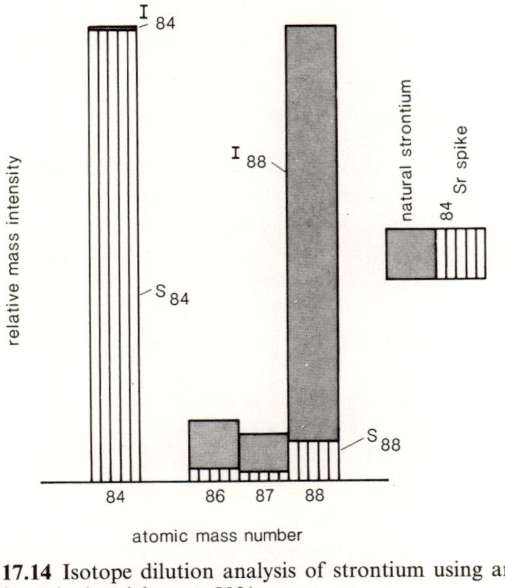

Figure 17.14 Isotope dilution analysis of strontium using an ^{84}Sr spike of nominal enrichment, 90%.

$[^{88}Sr/^{84}Sr]_{natural}$ is the known natural ratio of these isotopes (=148.01)

$[^{88}Sr/^{84}Sr]_{spike}$ is the pre-calibrated ratio of these isotopes in the spike

AW_{Sr} is the atomic weight of strontium (=87.6079amu)

W_{spike} is the weight of strontium spike added

W_{sample} is the weight of sample taken for analysis

F_{84} is the fraction of ^{84}Sr present in natural strontium (=0.0056)

$[^{84}Sr]_{spike}$ is the concentration (in ppm) of ^{84}Sr in the spike. The concentration (in ppm) of strontium in the sample $[Sr]_{sample}$ is then given by the isotope dilution equation

$$[Sr]_{sample} = [^{84}Sr]_{spike}\frac{AW_{Sr}W_{spike}}{F_{84}W_{sample}} \times \frac{\left[\frac{88}{^{84}Sr}\right]_{measured} - \left[\frac{^{88}Sr}{^{84}Sr}\right]_{spike}}{\left[\frac{88}{^{84}Sr}\right]_{natural} - \left[\frac{^{88}Sr}{^{84}Sr}\right]_{measured}}. \tag{17.3}$$

Measured isotope ratios must be corrected for mass fractionation (see section 17.2.5), and in the case of strontium this is achieved by normalizing measured data to a $^{86}Sr/^{88}Sr$ ratio of 0.1194 (Steiger and Jager, 1977). It may also be necessary to apply a mass-related correction to the data to account for any mass bias in the detector response (the magnitude of an electron multiplier signal, for example, is mass-dependent, section 16.8).

In order to minimize errors and to obtain optimum analytical sensitivity, the amount of spike added to the sample must be carefully controlled. Optimum sensitivity and accuracy are obtained when the proportion of spike is adjusted so that the measured ratio on the spiked sample is given by the following expression:

$$\left[\frac{^{88}Sr}{^{84}Sr}\right]_{measured} = \sqrt{\left[\left[\frac{^{88}Sr}{^{84}Sr}\right]_{sample} + \left[\frac{^{88}Sr}{^{84}Sr}\right]_{spike}\right]} \tag{17.4}$$

(see Heumann, 1982).

In practice, when highly-enriched spikes are used (e.g. more than 99% ^{84}Sr), the amount of spike added to the sample is adjusted so that the resultant ^{84}Sr mass peak is about equal to the ^{88}Sr mass peak. For optimum analysis, it is therefore necessary to know the approximate composition of the sample before spiking. Some indication of the order of magnitude of error introduced into an analysis as a result of inappropriate spiking is shown in Figure 17.15. This diagram has been calculated on the basis that the enriched abundance of the spike isotope is 99% and the natural abundance of that isotope in the sample is 1%. Assuming relative standard deviations of 5% for mass spectrometer ratio measurements, the variation in analytical error with sample to spike ratio is as shown in Figure 17.15. These data demonstrate that there is a broad area of minimum error about the optimum, but that if the spike to sample ratio deviates from this optimum by a factor of 10 or more, then errors in analytical measurements are magnified.

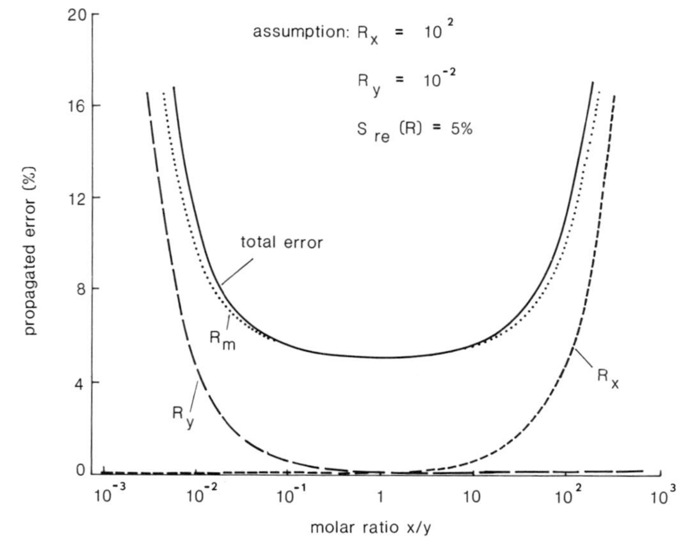

Figure 17.15 Propagated error resulting from variations in spike to spike ratio (x/y). Values have been calculated assuming 99% enrichment of the spike (R_y), 1% natural abundance for the isotope to be spiked (R_x) and 5% relative standard deviations in isotope ratio measurements. (Colby *et al.*, 1981, as reproduced by Heumann, 1982.)

17.6.2 *Scope of isotope dilution*

Isotope dilution analysis can be carried out provided an element exists as two or more naturally-occurring isotopes. If a mass fractionation correction is to be applied to the spiked sample, the element must possess three or more naturally-occurring isotopes, two of which are present in the analysed sample at a natural or known isotopic abundance. In consequence, there is a range of elements which cannot be analysed by isotope dilution since they are mono-isotopic. These are listed in Table 17.13, together with elements with two naturally-ocurring isotopes, one of which has a natural abundance of less than 1%. Analysis of these elements by isotope dilution may be subject to additional errors. Mono-isotopic elements can only be analysed if there exists a suitable radioactive isotope having a long half-life. Thus although thorium possesses only one naturally-occurring isotope, isotope dilution analysis is possible using ^{230}Th ($T_{1/2} = 7.52 \times 10^4$ years) a radioactive daughter product of the ^{238}U decay series.

There is a further limitation to isotope dilution analysis of particular relevance to thermal ionization mass spectrometry. The efficiency of thermal ionization is poor for elements having a first ionization potential of greater than 8–9 eV. These elements (Table 17.14) yield low ion beam currents from conventional heated filament sources. They include electronegative halides for which negative ion beam analysis is more appropriate, other non-metals, gold, and the platinum group elements, and the noble gases (for which gas-source mass spectrometer would in any case be more appropriate). A particularly important element in this category is osmium, isotopic determinations of which are required for the Re–Os geochronometer. This element cannot always be analysed satisfactorily on a mass spectrometer fitted with

Table 17.13 Elements which possess one naturally-occurring isotope or a minor isotope of relative abundance <1%

Isotope	Natural abundance (%)	Minor isotope and natural abundance	(%)
^{1}H	99.985	^{2}H	0.015
^{4}He	100	^{3}He	0.00013
^{9}Be	100		
^{14}N	99.63	^{15}N	0.37
^{16}O	99.759	^{17}O	0.037
		^{18}O	0.204
^{19}F	100		
^{23}Na	100		
^{27}Al	100		
^{31}P	100		
^{40}Ar	99.6	^{38}Ar	0.063
		^{36}Ar	0.337
^{45}Sc	100		
^{51}V	99.76	^{50}V	0.24
^{55}Mn	100		
^{59}Co	100		
^{75}As	100		
^{89}Y	100		
^{93}Nb	100		
^{103}Rh	100		
^{127}I	100		
^{133}Cs	100		
^{139}La	99.911	^{138}La	0.089
^{141}Pr	100		
^{159}Tb	100		
^{165}Ho	100		
^{169}Tm	100		
^{181}Ta	99.988	^{180}Ta	0.0123
^{197}Au	100		
^{209}Bi	100		
^{232}Th	100	(^{230}Th radioactive spike)	

Data from Weast (1973).

Table 17.14 Elements with first ionization energies greater than 8 eV and which are not efficiently ionized by thermal ionization mass spectrometry

Element	Ionization energy (eV)	Element	Ionization energy (eV)
Ar	15.755	N	14.53
As	9.81	Ne	21.559
Au	9.22	O	13.614
B	8.296	Os	8.5
Be	9.32	P	10.484
Br	11.84	Pd	8.33
C	11.256	Pt	9.0
Cd	8.991	S	10.357
Cl	13.01	Sb	8.639
F	17.418	Se	9.75
H	13.595	Si	8.149
He	24.481	Te	9.01
Hg	10.43	W	7.98
I	10.454	Xe	12.127
Ir	9	Zn	9.391
Kr	13.996		

Data from Weast (1973).

a thermal ionization source. On a more positive note, the elements for which isotope dilution analysis *is* possible are shown in Figure 17.16. Of these, the main geological interest is in two areas:

(i) The determination of radiogenic elements, especially Rb–Sr, Nd–Sm and Pb–Th–U to complement radiogenic isotope ratio measurements in geochronology studies (see previous sections).

(ii) In the analysis of the rare-earth elements by isotope dilution.

It should not be thought, however, that the rare earths are the only elements to which this technique has been applied. To give some indication of the scope of isotope dilution in geochemical applications, a selection of analysis schemes is summarized in Table 17.15, including details of spike, sample preparation and instrumental parameters.

17.6.3 *Rare-earth element analysis by isotope dilution*

Ten of the rare-earth elements can be analysed by isotope dilution mass spectrometry, namely La, Ce, Nd, Sm, Eu, Gd, Dy, Er, Yb, and Lu. Four elements (Pr, Tb, Ho and Tm) are mono-isotopic and cannot be analysed. Detection limits are very low, being essentially limited by the analytical blank and the magnitude of any isobaric interference. This technique is therefore one of the few reliable means of analysing samples containing very low rare-earth element abundances (down to chondrite levels and below). At higher concentration levels (that is, for samples containing more than about 10 times chondrite) the expense, in terms of both extended sample preparation and instrument utilization time, means that routine mass spectrometric analysis is less suitable than neutron activation analysis or ICP–AES. Nevertheless, isotope dilution mass spectrometry can still claim a precision which is unsurpassed in the analyses of such samples in comparison with these rival techniques. This performance is achieved at the expense of a detailed chemical separation procedure designed to separate the rare-earth elements from matrix elements and other elements capable of generating isobaric interferences (especially Ba and Ca).

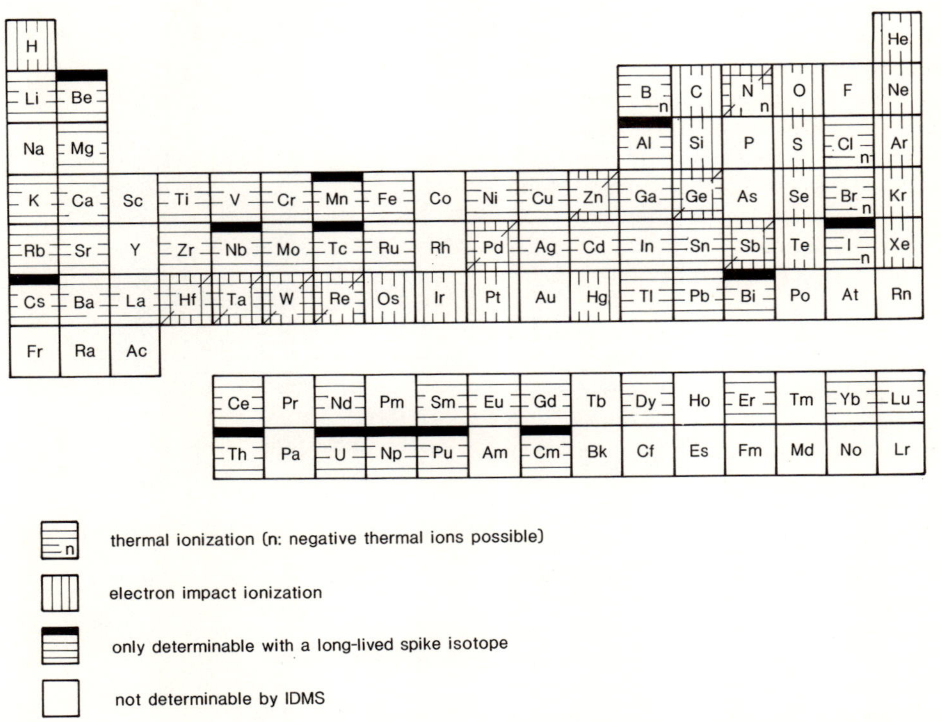

Figure 17.16 Elements that may be determined by isotope dilution mass spectrometer. (Heumann, 1982.) Refractory elements for which the ion yield is likely to be too small for conventional thermal ionization marked in crosshatch.

Table 17.15 Schemes of analysis for isotope dilution mass spectrometry

Element	Spike	Blank (ng)	Chemical separation	Filament/loading details	Ion beam (A)	Detector	Reference
K Rb Cs Ba	^{41}K ^{87}Rb ^{137}Cs ^{136}Ba	60 0.1 0.005 0.7	(1) HF–HNO_3 (2) HCl (3) Ion exchange separation	Single Ta filament, K, Rb, Cs, Ba loaded in HCl solution	5×10^{-15} to 4×10^{-13}	Daly ion multiplier	Verma (1981)
Sr	^{84}Sr	0.2		Sr loaded with H_3PO_4	10^{-11}	Faraday cup	Verma (1981)
Pd	^{102}Pd	1.2 ±0.2	(1) HF–$HClO_4$ (2) HCl (3) AG1-X8	Single Re filament silica gel technique	10^{-13}	Electron multiplier	de Laeter and Mermelengas (1978) Mermelengas *et al.* (1979)
Pd Te Ag Cd	^{102}Pd ^{124}Te ^{107}Ag ^{111}Cd	0.12 0.15 0.2	(1) aqua regia, HF HCl attack (2) HCl (3) AG1-X8 (4) AG50-X8 for Te	Single Re filament, Ag loaded in HCl, Te and Pd were electro-deposited on filament	10^{-15} $>10^{-14}$ (Cd)	Electron multiplier	Loss *et al.* (1983*a,b*) Loss *et al.* (1984)
Ba	^{136}Ba	300	(1) HF–$HClO_4$ (2) HCl (3) AG50W-X8	Re triple filament loaded in HCl solution			Schnetzler *et al.* (1967)
Ba	^{135}Ba	30±10	(1) HF–$HClO_4$ (2) HCl (3) Dowex 50	Re triple filament loaded in HCl solution	10^{-12}		de Laeter *et al.* (1969) de Laeter and Hosie (1977)
Tl	^{203}Tl	0.003	(1) HF–HNO_3 (bomb) (2) HCl (3) Electro-deposition	Re filament, loaded with silica gel and phosphoric acid	4×10^{-10} to 3×10^{-11}	Faraday cup	Arden (1983)
REE	*	10–100	(1) HF–$HClO_4$ (2) HCl (3) 50W-X8 eluted with HCl (4) Two fractions collected: (a) Gd–Lu (b) La–Eu	Re triple filament loaded in HCl solution	10^{-10} to 10^{-13}		Schnetzler *et al.* (1967)

REE + Ba	*	0.03 to 4.5	(1) HF–$HClO_4$ (2) HCl (3) 50W-X8, elute with HCl (4) 50W-X8, elute with HCl (5) Two fractions collected: (a) heavy REE (b) light REE	Single filament preferred for La, Gd, Dy; triple filament preferred for Ce, Eu, Yb, Lu			Masuda (1968) Masuda *et al.* (1973) Nakamura and Masuda (1973) Nakamura (1974)
REE	*	0.21 (Eu)	(1) HF–HNO_3 (bomb) (2) HCl then glacial acetic acid–HNO_3 (3) AG1-X8, mixed solvent elution (4) AG1-X8, mixed solvent elution (5) Fractions collected: (a) Dy, Er, Yb (b) La, Ce, Nd, Sm, Eu, Gd (c) La	Single Ta filament, samples loaded in HCl with H_3PO_4	10^{-12}	Faraday cup or Daly	Hooker *et al.* (1975)
REE	*	0.04 (Eu) 0.7 (Gd)	(1) HF–HNO_3 (bomb) (2) HCl then HNO_3 (3) AG1-X8, mixed solvent elution (4) AG1-X8, mixed solvent elution (5) AG1-X8, mixed solvent elution (to remove all Ba) (6) REEs collected as one fraction	Triple filament (Ta side filaments, Re central filament)	0.4×10^{-12} to 3×10^{-11}		Thirlwall (1982)

* For details of REE spikes see Table 17.16.
Chemical separation procedure gives (1) method of acid attack, (2) medium in which residue from (1) is redissolved, and (3) ion exchange details.
AG1-X8 is a strong basic anion exchange resin with 8% cross-linking.
AG50-X8 is a strong acidic cation exchange resin with 8% cross-linking.
Data for the isotope dilution analysis of radiogenic isotope pairs (Rb–Sr, Nd–Sm, U–Th–Pb) are given in Table 16.7.

Table 17.16 Isotope dilution analysis of rare-earth elements, instrumental parameters and interference corrections

Analysed mass peak	Interference	Mass peak monitored for interference correction	Magnitude of interference	Triple filament currents: Centre filament (A)	Triple filament currents: Side filament (A)	Beam size (10–12 A)	Comments
^{151}Eu ^{153}Eu (spike)	BaO BaO	$^{154}BaO + ^{154}Sm$	Usually trivial	4.2–4.6	1.1–1.4	1	BaO^+ emission suppressed by high centre filament temperature
^{171}Yb (spike) ^{173}Yb ^{174}Yb	LaO_2, GdO PrO_2, GdO CeO_2, GdO			4.9	1.6–1.8	1	Dioxide emission suppressed by high centre filament temperature GdO^+ emission inhibited by low side filament temperature
^{147}Sm ^{149}Sm (spike) ^{152}Sm	^{136}CeO, ^{136}BaO ^{152}Gd	$^{156}(CeO)$	CeO < 1%	4.5–4.9	1.9–2.1	5	Side filaments too cold for Gd^+ emission; BaO^+ emission suppressed by high central filament temperature
^{140}Ce (spike) ^{142}Ce	^{142}Nd	^{143}Nd ^{145}Nd	Requires correction	4.5–5.1	1.9–2.1	5	Measurement of ^{143}Nd plus ^{145}Nd allows mass fractionation correction of $^{142}Nd/^{145}Nd$ required for ^{142}Ce interference correction
^{143}Nd ^{145}Nd (spike) ^{146}Nd	^{130}BaO		Trivial	4.5–4.9	2.1–2.3	10	BaO^+ has largely been burnt off filament prior to Nd determination
^{138}La (spike) ^{139}La	^{138}Ce, ^{138}Ba	$^{137}Ba + ^{140}Ce$	Usually less than 5%	4.5–5.1	2.1–2.3	30	Precision of La determination is largely a function of Ba^+ beam stability
^{155}GdO (spike) ^{156}GdO ^{157}GdO	LaO_2, Yb CeO_2, Yb PrO_2, Yb		Usually trivial	2.9–3.1	2.8–3.1	0.4	GdO^+ measured since interferences on Gd^+ could neither be suppressed nor corrected. Dioxide emissions suppressed by high centre filament temperature. Yb^+ emission minimized by operating at low side filament temperature
^{167}Er (spike) ^{168}Er ^{170}Er	EuO SmO, (Yb) SmO, Yb	^{171}Yb	Yb: 10% of the 170 ion beam, 1% of the 168 ion beam	4.7–5.1	2.7–2.9	2	Oxide emission suppressed by high centre filament temperature. ^{171}Yb (measured for correction), may itself suffer interference from $^{171}(GdO)$
^{161}Dy (spike) ^{162}Dy ^{163}Dy	^{145}NdO ^{146}NdO, ^{162}Er	$^{166}(NdO) + ^{166}Er$ $^{167}(NdO)$	Requires correction	4.8–5.1	2.7–2.9	4	NdO^+ emission suppressed by high centre filament temperature
^{175}Lu ^{176}Lu (spike)	TbO Yb, ^{160}DyO GdO						Interferences could not be adequately corrected. Accurate analyses unlikely without chemical separation of Lu

Data from Thirlwall (1982).

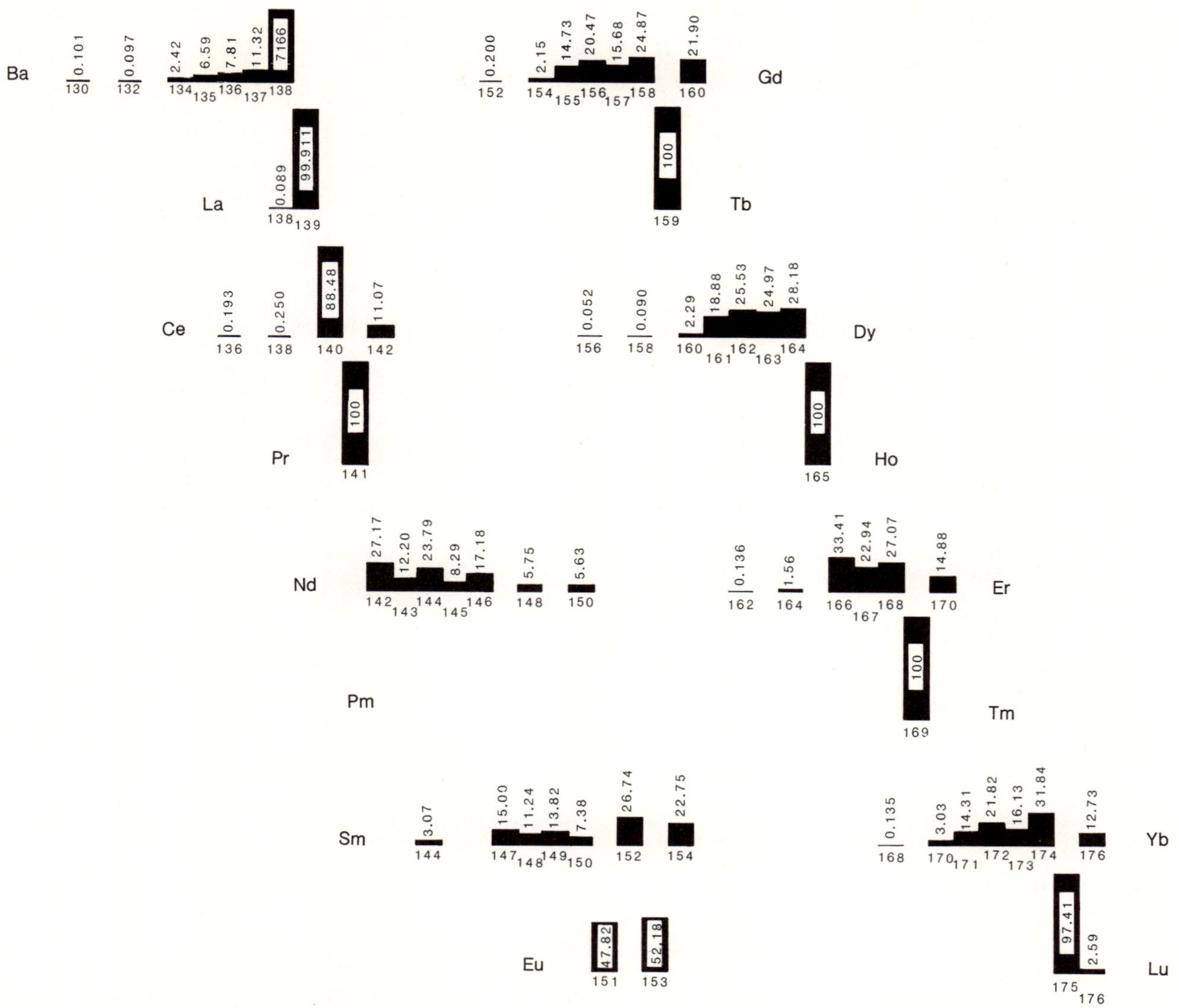

Figure 17.17 Relative isotopic abundances of the rare-earth elements. Interferences from oxide species, found at 16 and 32 atomic mass units above the plotted data, have been omitted for clarity.

Some schemes of analysis go further and separate the REEs into two or three fractions (for example La, Ce–Gd, Dy–Yb) to remove inter-REE interferences.

Ion exchange procedures for separating REEs have been reviewed in section 14.6. Korkisch and Arrhenius (1964) described a mixed solvent ion exchange procedure for separating uranium, thorium and the rare-earth elements, and this has been developed as a basis for isotope dilution analysis by Hooker *et al.*, (1975) and further refined in this application by Thirlwall (1982).

The method as developed by Thirlwall offers some insight into the reliability of the technique together with a full investigation of mass spectrometry interference effects. Thirlwall suggested that, although it is satisfactory for many samples, occasionally the digestion procedure of Hooker *et al.* does not ensure full equilibration between sample and spike, possibly due to the presence of minor REE phases such as zircon. To overcome this problem the following procedure was used.

(i) To 100–200mg of sample was added the spike solution (by weighing) together with 15M HNO_3 (1mL) and 40% HF (8mL). The mixture was evaporated overnight to volatilize silica.

(ii) Fresh HNO_3 and HF were added to the residue, which was heated in a Teflon bomb at 190°C for 48 hours to ensure complete sample dissolution

(iii) To ensure full equilibration with spike, HF was evaporated off and 15 M HNO_3 (3mL) added, the mixture being evaporated to dryness. The residue was then evaporated to dryness two more times with 8mL aliquots of 6M HCl and 3mL of 15M HNO_3 alternately before being redissolved in 2M HNO_3. If any undissolved material remained, it was filtered and re-digested in the Teflon bomb.

(iv) Two stages of anion exchange were undertaken to isolate and purify the REE fraction following the procedures of Hooker *et al.* (1975): (*a*) elution with 50mL 90% glacial acetic acid–10% 5M nitric acid to separate REEs from matrix elements; and (*b*) elution of the REE fraction from above through a second anion exchange column, using

Table 17.17 Isotope dilution mass spectrometric analysis of basalt USGS : BCR-1

	La	Ce	Nd	Sm	Eu	Gd	Dy	Er	Yb	Lu	Reference
1	25.23	53.76	29.042	6.6205	1.974	6.86	6.4	3.713	3.378		Thirlwall (1982)
2	25.36	53.58	29.037	6.6208	1.972	6.817		3.71	3.382		Thirlwall (1982)
3	24.8	53.7	29.2	6.67	1.89	6.75	6.35	3.67	3.47		Hooker *et al.* (1975)
4	23.5	53.2	29.4	6.67	1.98	6.77	6.31	3.67	3.50		Hooker *et al.* (1975)
5		52.3	28.9	7.11	1.92	6.54	6.12	3.48	3.42		O'Nions and Clarke (1972)
6	24.4	54.2	28.8	6.72	1.98	6.67	6.36	3.70	3.40	0.503	Nakamura (1974)
7		51.5	28.8	6.52	1.97	7.0	6.22	3.52	3.27	0.53	Philpotts *et al.* (1972)
8	23.8	53.7	29.2	6.56	1.93	6.63	6.31	3.67	3.34		Elderfield *et al.* (1981)
9	24.4	54.3	28.9	6.73	1.98	6.67	6.36	3.69	3.39	0.501	Nakamura and Masuda (1973)
10		52.9	27.5	6.57	1.943	6.80	6.36	3.48	3.44	0.548	Schnetzler and Nava (1971)
11		53.7	29.0	6.74	2.02	6.55	6.47	3.57	3.49		Shimizu (1975)
12	25.3	54.2	29.1	6.62	1.99	6.72	6.43	3.72	3.41		Sun and Nesbitt (1978)
13		51.4	29	6.77	1.93	7.01	6.00		3.29	0.529	Winzer *et al.* (1974)
14	26.1	54.9	28.8	6.74	1.97		6.20	3.71	3.68	0.59	Gast *et al.* (1970)
15		53.9	28.6	6.62	1.942	6.47	6.36	3.58	3.38	0.536	Philpotts and Schnetzler (1970)
16	25.1	53.6	28.9	6.70	2.00	6.76	6.37	3.68	3.36	0.500	Tanaka (1975)
17	23.2	51.8	29.9	6.22	2.00	6.67	6.42	3.46	3.40	0.52	Loubet *et al.* (1972)
18		53.1	28.6	6.66	2.06		6.51	3.65	3.40		Whitford and Arndt (1978)
19		53.44	28.38		1.87	6.64		3.54			Gill and Bridgwater (1979)
20		53.5	29.3	6.67	1.94	6.76	6.33	3.67	3.49		Hawkesworth and O'Nions (1977)
21		54.3	27.8	6.41	1.92	6.43	6.26	3.52	3.42	0.590	Arth and Hanson (1975)
22	25.5	54.4	29.1	6.59	2.00	6.79	6.46	3.71	3.38	0.499	Sun and Hanson (1975)
23		56.1	29.4	6.69	1.99	6.86	6.40	3.73	3.39	0.485	Simmons *et al.* (1980)
x	24.72	53.54	28.90	6.646	1.964	6.722	6.33	3.629	3.41	0.528	
σ	0.88	1.08	0.51	0.160	0.043	0.151	0.12	0.091	0.08	0.034	
CV	3.6	2.0	1.8	2.4	2.2	2.3	1.9	2.5	2.4	6.5	
	25.0	53.7	28.7	6.58	1.96	6.68	6.35	3.61	3.39	0.512	Gladney *et al.* (1983)
	0.08	0.8	0.6	0.17	0.05	0.13	0.12	0.09	0.08	0.025	± 2 sigma

x: mean of 23 ID–MS analyses.
σ: standard deviation.
CV: coefficient of variation (%) = $100\sigma/x$.
Data of Gladney *et al.* (1983) represents 'consensus' values derived from all published data, with standard deviation (two sigma).
Analytical data taken from the following sources: 1+2: CG **35**, 155–166; 3+4: CG **16**, 189–196; 5: EPSL **15**, 436–446; 6: GCA **38**, 757–775; 7: EPSL **17**, 13–18; 8: GCA **45**, 513–528; 9: EPSL **19**, 429–437; 10: EPSL **11**, 345–350; 11: EPSL **25**, 26–32; 12: CMP **65**, 301–325; 13: EPSL **23**, 439–444; 14: A **2**, 1143–1163; 15: A **2**, 1471–1486; 16: CMP **52**, 233–246; 17: EPSL **17**, 19–23; 18: EPSL **41**, 188–196; 19: JP **20**, 695–726; 20: JP **18**, 487–520; 21: CMP **37**, 161–174; 22: CMP **52**, 77–106; 23: PR **11**, 43–71; Gladney *et al.* (1983), GSNL **7**, 3–226.
Journal abbreviations:
CMP: *Contrib. Mineral. Petrol.*; GCA: *Geochim. Cosmochim. Acta*; EPSL: *Earth Planet. Sci. Lett.*; A: *Proc. Apollo 11 Lunar Sci. Conf.*; JP: *J. Petrol.*; GSNL: *Geostand. Newslett.*; CG: *Chem. Geol.*; PR: *Precamb. Res.*

15mL of the mixed solvent for elution. This was to minimize the barium contamination of the REE fraction.

No attempt was made in this procedure to separate the REEs into different fractions. Hooker *et al.* carried out such a separation, isolating three REE fractions (La, Ce–Gd, Dy–Yb) and used a simple tantalum single filament for mass spectrometry. Conversely, in Thirlwall's procedure, all the rare earths were loaded simultaneously from nitrate solution on to the tantalum side filament of a triple filament assembly. Programed heating of these side filaments was used to volatilize individual rare-earth elements in sequence and to reduce inter-element isobaric interferences. The first element volatilized was Eu. The temperature of the side filament was then increased, permitting subsequent analysis of elements in the order Yb, Sm, Ce, Nd, La, GdO, Er and Dy. The centre filament was kept at a high temperature throughout the analysis to ensure the decomposition of any rare-earth element oxides (except during the analysis of GdO). Potential isobaric interferences may be judged from the mass spectrum data in Figure 17.17. Details of Thirlwall's procedure, including REE spikes, instrumental parameters and interferences which required monitoring and correction are shown in Table 17.16. Gd was determined as GdO, since interferences from LaO^+, CeO^+ and PrO^+ on Gd^+ could neither be suppressed nor corrected satisfactorily. The temperature of the central filament was accordingly reduced during GdO^+ measurement to promote the formation of a stable oxide beam. Lutetium was not determined at all in this scheme of analysis, as it was considered that the mass interferences on ^{176}Lu from TbO^+ could not be adequately corrected. In some schemes of analysis, particularly earlier ones, determinations of lanthanum were not undertaken, due to the lack of a sufficiently enriched ^{138}La spike.

The generally-accepted international standard applied in

isotope dilution mass spectrometry to judge the accuracy of results is basalt USGS:BCR-1, used almost to the exclusion of any other reference material. Data for 23 ID-MS analyses of BCR-1 taken from the literature are listed in Table 17.17. Bearing in mind the degree of interlaboratory bias normally encountered in trace-element data, the consistency of these analyses is exceptional and testifies to the reliability of the isotope dilution technique and of those analysts who contributed these data. The spread of results for the elements Ce, Nd, Sm, Eu, Gd, Dy, Er and Yb is within $\pm 4\%$ (two sigma) of the mean. Data for La and Lu are slightly more variable (± 7 and 13% respectively).

It is suggested that there are three areas where errors could influence ID-MS REE determinations.

(i) Inhomogeneity errors in some sample types, due to the rather small sample mass normally taken for analysis (often 100mg)

(ii) Non-equilibration between sample solution and spike, due either to the difficulty in dissolving resistant minerals such as zircon or to insufficient chemical treatment

(iii) Errors in spike preparation: rare-earth element powders in particular are notorious for absorbing water and carbon dioxide from the atmosphere (up to 15–20% by weight). After ignition at 600–800°C, the resultant powder is not necessarily stoichiometric. There have also been reports that CeO_2 is particularly difficult to dissolve and retain in solution.

18 Gas source mass spectrometry

18.1 Geological applications

Geological applications of gas source mass spectrometry are centred on three specialized areas of analysis.

18.1.1 *The measurement of stable isotope ratios of the elements H, C, N, O, S*

Although stable in the sense that these elements are not radioactive, variations can occur in the isotopic abundances of these elements in geological samples. This arises because many physical and chemical processes in nature can induce mass fractionation between lighter and heavier isotopes of the same element. Fractionation can be caused, for example, by kinetic effects in chemical reactions, the rate of diffusion of molecular species through a porous matrix or the variations in the volatility of heavy and light water. The rate at which all these phenomena proceed is dependent to a small extent on the mass of isotopes participating. However, these effects are apparent only if the relative mass difference between isotopes of an element is large. Thus, at one extreme, there is a 100% difference between the mass of hydrogen (^{1}H) and deuterium (^{2}H), giving rise to a large potential mass fractionation effect. At the other end of the scale, the mass difference between ^{207}Pb and ^{208}Pb is only 0.5%. In natural samples, therefore, isotopic fractionation is detected only for elements lighter than about calcium and, of these, the elements which can be readily analysed as gaseous compounds are the ones which have received most attention. These include hydrogen (analysed as H_2), carbon (CO_2), nitrogen (N_2), oxygen (normally analysed as CO_2) and sulphur (SO_2). Applications include investigations of water–rock interactions, palaeotemperature measurements, ore genesis studies (S), the analysis of basaltic glasses to elucidate mantle processes, and meteorite studies.

18.1.2 *Noble gas analysis*

Considerable interest has been shown in the analysis of noble gases in silicate material to determine both the elemental abundance and isotopic composition. The elements of interest include helium, neon, argon, krypton and xenon. The latter two present considerable technical difficulties due to their very low abundances in terrestrial samples. These samples are of interest both for the estimation of the primordial noble gas composition of the Earth from the analysis of basaltic glasses, for example, and for the elucidation of nuclear synthetic processes involved in the formation of the solar system. In the latter field there is considerable activity in the analysis of meteoritic fragments.

18.1.3 *Potassium–argon and $^{40}Ar/^{39}Ar$ dating*

The isotope ^{40}K is a naturally-occurring radioactive species which decays, with a half-life of 1.25×10^9 years, to ^{40}Ar and ^{40}Ca. Following the development of sensitive gas source mass spectrometers in the 1950s there has been considerable interest in dating rocks by measuring the ^{40}Ar content. In the case of conventional potassium–argon dating this involves determining the potassium content of the sample by an alternative technique, normally XRF or flame photometry. In the $^{40}Ar/^{39}Ar$ technique, samples are irradiated in a nuclear reactor, inducing the production of ^{39}Ar by neutron interaction with ^{40}K. The age of the sample is then determined by measuring the $^{40}Ar/^{39}Ar$ ratio by gas source mass spectrometry using a suitable reference sample as neutron dose monitor. Independent determination of potassium on a separate sample aliquot is not then necessary. The advantage of this latter technique is that by using a stepwise heating technique to extract ^{40}Ar and ^{39}Ar from a sample, it is possible to investigate the alteration history of the rock and decide if an age determination is primary.

18.2 Instrumentation

The general principles of mass spectrometric instrumentation have been discussed in Chapter 16. Gas source mass spectrometers are normally based on a single magnetic sector and differ from other forms of mass spectrometer in two respects. First, the ion gun is almost always an electron impact source in instruments designed for geological applications. (Chemical ionization sources are extensively used in instruments designed for organic molecule analysis.) Secondly, instruments are designed with double or triple collection of the ion beam (depending on the application), permitting simultaneous detection of isotope ratios of interest.

18.2.1 *Electron impact ion source*

Gas samples introduced into an electron impact source are ionized by collision with a stream of electrons, as shown schematically in Figure 18.1. The electron beam is created by thermionic emission from a filament. Traditionally, the filament is made of tungsten wire about 200 μm in diameter that requires a current of between 3–4 A to sustain an emission temperature of 2200 K. However, carbon, especially from cracked hydrocarbon molecules, can react with such a filament, forming a coating of tungsten carbide. This causes variable emission current densities. To overcome such problems, rhenium can be used a filament material as it does not form carbide or oxide coatings at its normal operating temperature, nor does it become brittle like tungsten. Unfortunately, rhenium has a significantly higher vapour pressure and therefore a shorter operating life.

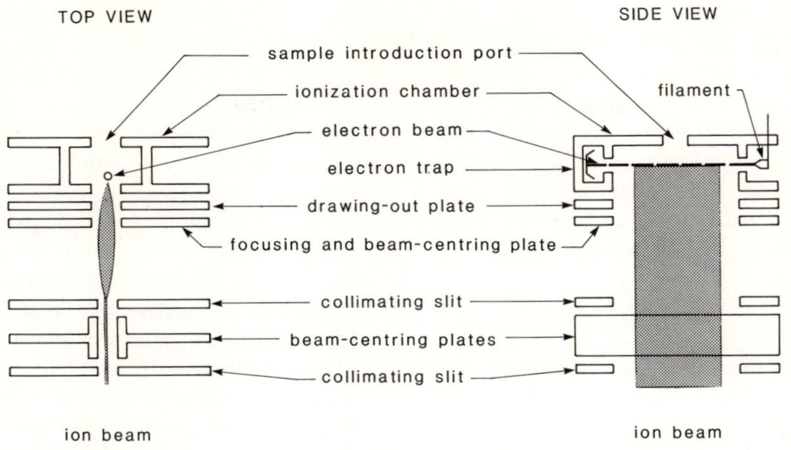

Figure 18.1 The electron impact ion source. (Dalrymple and Lanphere, 1969.)

The electrons emitted from the filament are collimated by slits and accelerated across an ionization chamber. The potential difference between filament and anode can normally be adjusted by the user between 0–100 V, 70 V being a normal setting. In order to minimize the energy spread of ions generated in the electron impact source, it is important to keep the electron beam as narrow as possible so that ions originate on an equipotential plane. This can be achieved by placing a small permanent magnet (50–300 gauss) across the gun so that any divergent electrons are deflected into helical paths away from the ionization region. It is desirable to surround the electron anode by an electron trap held at a potential of about 20 V higher than the ionization chamber to suppress secondary emission of electrons from the anode. The field strengths in the ionization region created by the drawing-out plate are kept small, perhaps 10 V cm^{-1}. All these factors help to minimize the energy spread of ions generated in the ion source.

Electron bombardment is a rather crude method of ionization. Ions so formed are given considerable vibrational and rotational energy during the ionization process. Sufficient energy transfer occurs to break up a fraction of molecular species into ionic fragments. Thus the ion spectrum of CO_2 will also contain CO^+ fragments, for example. Since the electron beam is not mono-energetic, and energy transfer during electron–molecule impact is not constant, a range of translational energies will be imparted to the resulting ion species. The maximum energy spread is generally considered to be in the range 0.5–3 eV. Once ions have been drawn out of the ionization chamber they are accelerated through a potential of several kilovolts. Since the energy spread of ions is small compared with this accelerating potential, the effect on degrading mass spectrometer resolution is small.

The efficiency of ionization depends in part on the voltage applied between filament and anode of the electron gun. The relative ion yield as a function of this gun accelerating potential is plotted for argon in Figure 18.2. Gases usually show a maximum ion yield in the range 50–100 V, and 70 V is frequently selected for the ionization of heavier gases.

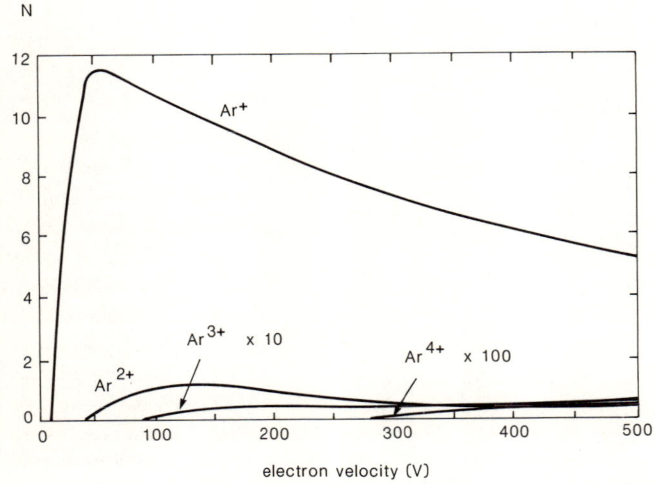

Figure 18.2 Efficiency of ionization of argon as a function of voltage applied to the electron beam in an electron impact ion gun source. Optimum potential for the production of Ar^+ ions is about 70 V. (Dalrymple and Lanphere, 1969.)

Chait (1972) considers that at typical electron currents of 100–200 μA within the source region, about one gas molecule in a million becomes ionized. Electron impact guns are capable of stable operation with ion beams in the range 10^{-7} to 10^{-14} A. Stability can be maintained by monitoring the electron gun anode current (the density of electrons reaching the anode) and altering the filament current to compensate for any detected drift. Stabilities in the order of one part in 10^4 are claimed using this feedback control.

18.2.2 *Dynamic and static sample introduction*

Conventional gas source mass spectrometers are operated in dynamic mode; that is, the sample gas is continuously introduced into the electron impact source through a suitable capillary leak. A very small fraction of gas becomes ionized, the rest is pumped to waste via the vacuum system. Dynamic sample introduction is normally used in conjunction with a change-over valve (Figure 18.3) which permits alternate measurements of sample and reference gas. The gas which is not being analysed is normally bled to waste to prevent differential isotopic fractionation. The best sensitivity which has been reported by this sample introduction is about 10^{-7} to 10^{-8} moles of sample (Pillinger, 1985). Since the sample must be introduced continuously from a reservoir of finite size, dynamic operation is satisfactory only for analysis of relatively large amounts of material. However, in some applications it is necessary to analyse very small gas volumes, for example in the measurement of noble gas concentrations in terrestrial samples or in the analyses of minute meteoritic fragments. These requirements were met by Reynolds (1956) who developed an instrument capable of noble-gas analysis in static vacuum mode. In this system, the mass spectrometer is isolated from the vacuum pumping system (with the exception of any getter pumps) before sample introduction. The sample is then permitted to expand from the appropriate reservoir into the mass spectrometer. Since the volume of the spectrometer will typically be at least ten times greater than that of the sample reservoir, almost complete transfer occurs once gas equilibration has been attained. Sensitivities using this technique are considerably higher than by dynamic operation because virtually the entire sample is available for ionization during analysis. Reynolds demonstrated that his instrument was capable of detecting about 10^7 atoms of xenon and argon. Pillinger (1984) has shown that sensitivities in stable isotope analysis can be 10^{-8} to 10^{-10} moles of sample or better, several orders of magnitude lower than for dynamic operation. There are, however, some serious limitations to static operation, firstly that continuous degradation of molecular gases ocurs due to interaction with the incandescent filament in the ion gun. Furthermore, adequate measurements can only be made if spectrometer backgrounds can be maintained at exceptionally low levels

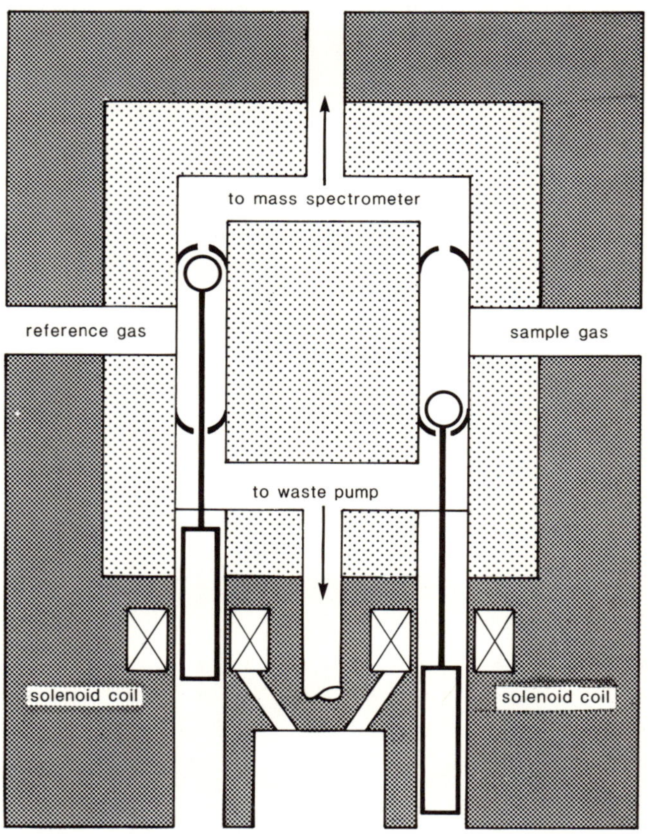

Figure 18.3 Change-over valve used in dynamic operation for the alternate introduction of sample and reference gas into a mass spectrometer. The gas which is not currently being analysed (the reference gas in the above diagram) is pumped to waste to prevent mass fractionation effects occurring in the sample reservoir. (VG isotopes technical brochure.)

and memory effects minimized. For this reason dynamic operation was for many years restricted to noble-gas analysis. However, more recently Gardiner *et al.* (1978) and Gardiner and Pillinger (1979) have demonstrated that static analysis of stable isotopes of carbon and nitrogen is possible. Filament degradation of CO_2 is avoided by converting carbon to deuteromethane (CD_4) prior to analysis (McNaughton *et al.*, 1983) (see section 18.5.2). Static operation has not yet been applied to hydrogen, oxygen or sulphur isotope measurements due to filament degradation or instrument background difficulties (Pillinger, 1984).

Table 18.1 Mass discrimination factors caused by molecular flow sample introduction

Gas	F	Gas	F
H_2	22%	He	12%
CO_2	1.1	Ne	2.5
N_2	1.8	Ar	1.3
SO_2	0.78	Kr	0.59
		Xe	0.38

F is the mass discrimination factor per atomic mass unit, calculated from $\sqrt{(m_2/m_1)}$.

18.2.3 *Mass fractionation corrections*

Mass fractionation is the phenomenon whereby the isotopic ratio measured on a mass spectrometer is not identical to that in the reservoir. Mass fractionation arises from several causes, including the characteristics of the sample input system, memory and vacuum system effects and the response of the ion detector (especially in the case of electron multiplier detectors). In relation to gas source mass spectrometry, the most important effects are as follows.

(*a*) *Sample introduction: dynamic operation.* In dynamic operation, a gas held in a reservoir is continuously leaked into the ion source of the mass spectrometer. Only a small fraction of the gas becomes ionized, the rest is pumped to waste. There are two types of gas leak used to introduce sample gases into the ion source. They are distinguished by their physical characteristic as *molecular flow* and *viscous flow*.

Molecules in gases have a translational motion so that they continually collide with one another. The mean free path between collisions is controlled mainly by gas pressure. *Molecular flow* is induced when the diameter of a sampling orifice is small compared with the mean free path of sample gas molecules. In mass spectrometer applications, molecular flow can be induced by puncturing four to six pinholes, each 5–20 μm in diameter, through a gold foil 250 μm thick (Roboz, 1968, p. 233). If the sample gas contains two isotopes held in a reservoir at partial pressures p_1 and p_2, then the number of molecules n_1 and n_2 of these isotopes passing through the molecular flow leak is given by

$$\frac{n_1}{n_2} = \frac{p_1}{p_2}\sqrt{\frac{m_2}{m_1}} \qquad (18.1)$$

where m_1 and m_2 are the masses of the isotopes.

From the form of equation (18.1) it is apparent that molecular flow causes mass fractionation, since the rate of flow depends on isotope mass. The 'ideal' mass discrimination factors (F) for several of the more common gas types are listed in Table 18.1.

As can be seen, very large mass fractionation effects are observed by molecular flow of the lighter gases. To minimize corresponding changes in the composition of gases left in the sampling reservoir, it is preferable to use large reservoir volumes, a remedy which is not always appropriate when only small sample masses are available for analysis. *Viscous flow* occurs when the diameter of the sampling orifice is large compared with the mean free path of gas molecules. Viscous flow conditions are met for example by injecting the gas into the ion source through a capillary tube 10 to 15 cm long and 0.1 mm diameter. Under these conditions, the flow rate (Q) through a capillary leak of a gas comprising a mixture of two components held at partial pressures p_1 and p_2 is given by

$$Q = \frac{k(p_1 - p_2)^2}{\eta} \qquad (18.2)$$

where k is a constant dependent on temperature and the geometry of the capillary and η is the gas viscosity (poise).

Under ideal viscous flow conditions, mass discrimination does not occur since there is no mass term in equation (18.2). The composition of the source reservoir will not then change with time as is the case in molecular flow. However, flow rate does vary with gas viscosity, which is itself an unpredictable function of gas composition. Furthermore, gas flow may become molecular (and therefore subject to some mass fractionation effects) on the ionization-chamber side of the viscous leak (Roboz, 1968, p. 233). In practical systems, therefore, capillary leak sample introduction gives much smaller, though not negligible, mass discrimination effects as compared with molecular flow. The principal disadvantage is the comparatively large sample mass required to undertake an analysis.

(*b*) *Sample introduction: static operation.* As outlined above, very small samples may be analysed with the mass spectrometer operated in static mode. Since the sample reservoir is normally at least ten times smaller than the mass spectrometer volume and the two chambers are permitted to attain equilibrium, mass fractionation does not occur during sample introduction. However, over the finite time of analysis, small but progressive changes do occur in measured isotope ratios. This appears to be mainly associated with the memory effect, that is, the gas to be analysed becomes contaminated by gaseous residues from samples previously analysed. These residues are absorbed on the internal surfaces of the spectrometer and progressively released during the subsequent analyses. Such contaminants can be removed by baking the spectrometer flight tube. However, it is not practical to do this between each sample run. It is normal, therefore, to monitor isotope peak heights (or ratios) as a function of time and to extrapolate these data back to the time of sample introduction (Figure 18.4). Usually a linear change in these data is observed as a function of time.

(*c*) *Detectors.* Mass discrimination effects can also be caused by the characteristics of the detector; in particular, when ions impinge on an electron multiplier detector, the sec-

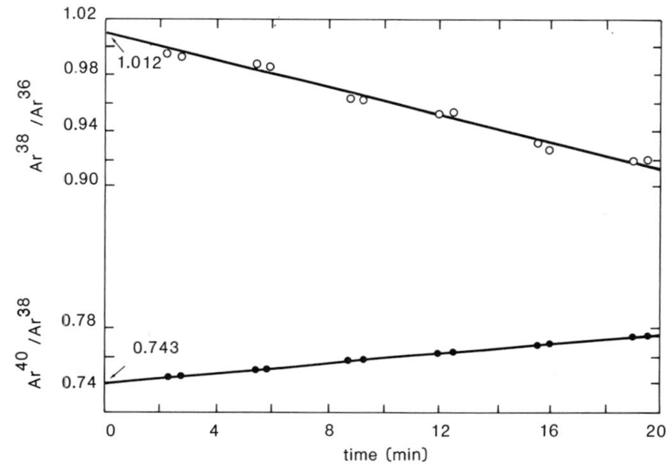

Figure 18.4 Memory effect in static gas source mass spectrometry analysis. The measured argon ratios change as a function of time due to 'memory' contamination from previous samples. Ratio data must be extrapolated back to the time of sample introduction to obtain the 'true' sample ratio. (Dalrymple and Lanphere, 1969.)

ondary electron yield varies slightly with ion mass. The resulting signal therefore shows a mass discrimination effect which requires correction. If such detectors are operated in ion counting mode, this problem is overcome, since it is the occurrence of an event rather than the magnitude of the detector signal which is taken as a measure of the ion beam intensity.

18.2.4 *Design of gas source mass spectrometers*

(*a*) *Nier type*. Most modern gas source mass spectrometers are developed from the designs of Nier (1947). The Nier-type mass spectrometer consists of a 60° (or 90°) wedge-shaped magnetic sector (Figure 18.5). The flight tube is fabricated from stainless steel and normally has a 150 mm (6-

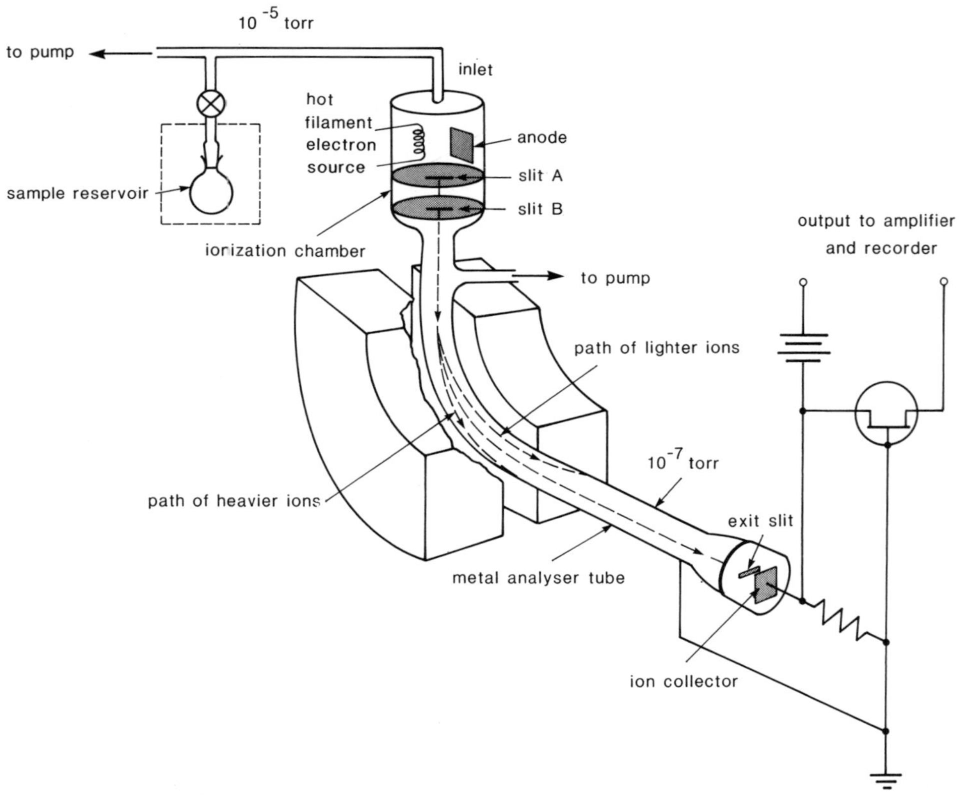

Figure 18.5 Schematic diagram of a 'Nier'-type gas source mass spectrometer. (Skoog and West, 1980.)

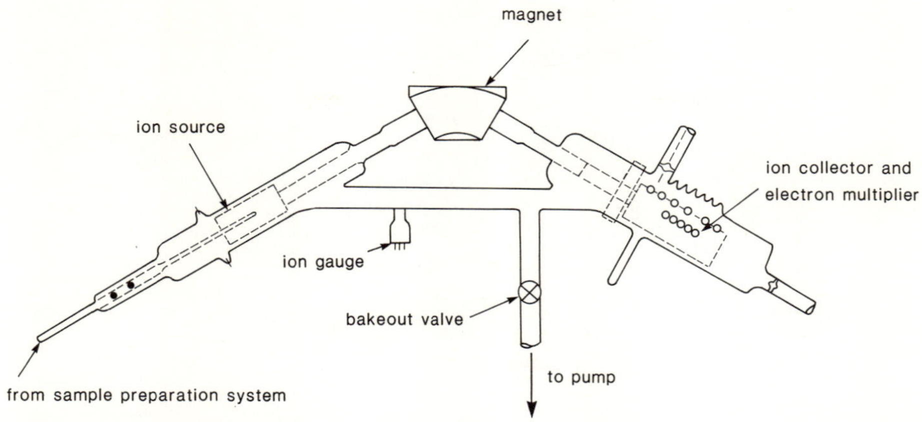

Figure 18.6 'Reynolds'-type gas source mass spectrometer. The flight tube is made of glass and the instrument designed for static operation. (Roboz, 1968.)

inch) radius. The flight tube can be baked and the instrument normally has a resolution in the range 150–200 ($M/\Delta M$). The ion source is the 'Nier-type' electron bombardment source and the instrument is normally designed for dynamic sample introduction and fitted with dual (or triple) collection of the dispersed ion beam.

(*b*) *Reynolds type*. The 'Reynolds-type' mass spectrometer, described by Reynolds (1956), is a development of the Nier type designed originally for noble-gas analysis. Features include the capability of operating in static mode (to enhance instrument sensitivities) and the possibility of baking the entire system to minimize the effects of contamination. Accordingly, the Reynolds instrument has a flight tube constructed of pyrex glass bakeable to 450°C, with a 60° magnetic sector and a $4\frac{1}{2}$ inch (114 mm) radius of curvature (Figure 18.6). Analyser pressures as low as 10^{-10} Torr can be achieved and usable resolution is about 135 (xenon) (Roboz, 1968, p. 258).

(*c*) *MS*-10. The MS-10 is a commercial instrument (AEI) which has been extensively used for many argon-dating studies. Deflection of ions in this instrument is achieved using a permanent magnet of strength 4–5 kilogauss. The ion beam is deflected through an angle of 180°, following the original designs of Dempster (Figure 18.7). Because of the small radius of curvature of the flight path (50 mm) resolution is only moderate (50–100 $M/\Delta M$). The mass spectrum must be scanned by changing the ion source accelerating voltage.

(*d*) *Clarke type*. The Clarke type mass spectrometer was originally designed for $^3He/^4He$ or tritium measurements. Two technical problems must be addressed for successful application in this area. The first is that the natural abundance of ^3He is of the order of 10^{-6} compared with that of ^4He, so that electron multiplier detection is essential for ^3He. The second is that in the analysis of ^3He, an instrument of relatively high resolution is required ($M/\Delta M$ 600–1000) so that a potentially serious interference from HD^+ can be

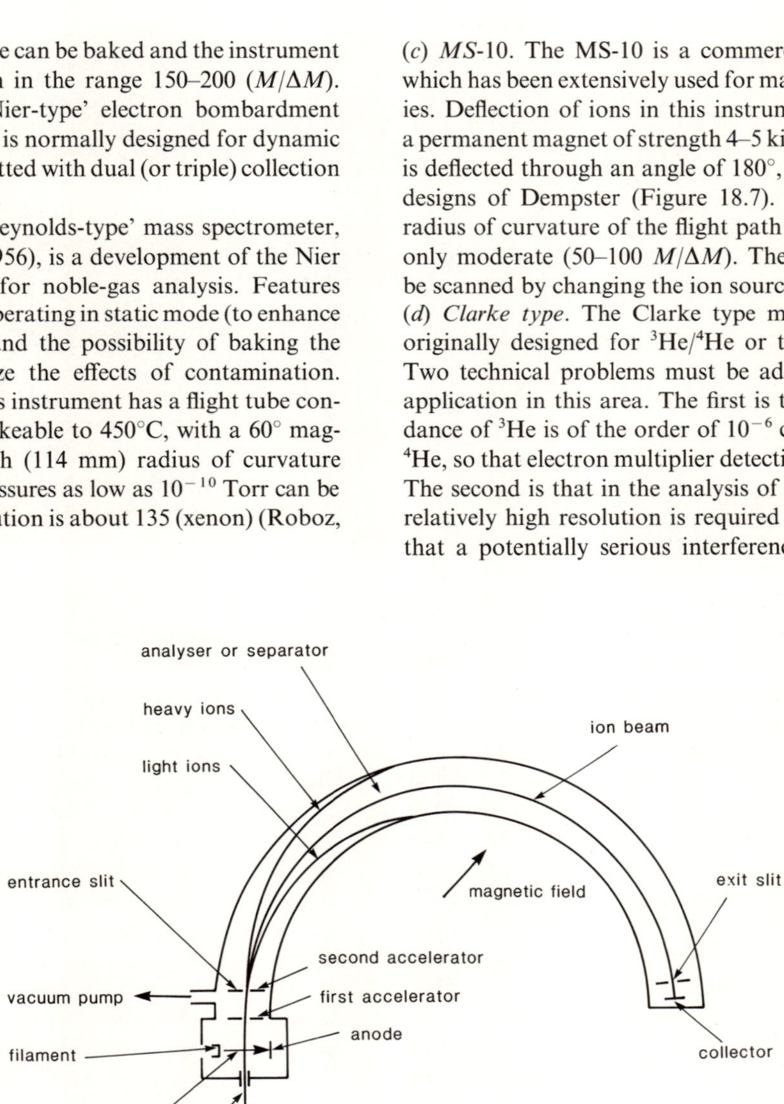

Figure 18.7 Single-deflection 180° mass spectrometer, a design originally proposed by Dempster. The magnetic field is perpendicular to the plane of the paper. (Pecsok *et al.*, 1968.)

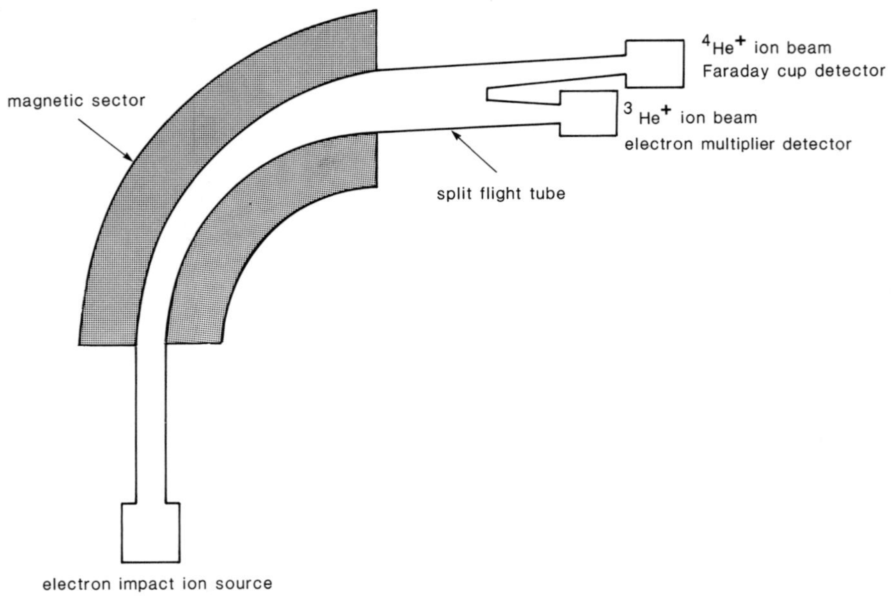

Figure 18.8 Schematic diagram of a 'Clarke'-type gas source mass spectrometer, showing split flight tube developed originally for the measurement of $^3He/^4He$ isotope ratios. (VG Isotopes.)

resolved. The Clarke-type instrument (Figure 18.8) is designed with a 90° magnetic sector and a split flight tube of radius 292 mm (^4He) and 254 mm (^3He). Electron multiplier detection is used for the minor, and Faraday cup for the major ion beam. The instrument is designed for static operation using long ion count times to ensure adequate precision in ^3He measurements.

18.3 The delta convention for reporting isotope data

Geological applications require the measurement of very small differences in stable isotope ratios between samples. Together with the discrepancies which could arise from uncertainties in sample preparation procedures and instrumental mass fractionation effects, this gives rise to the possibility of serious differences in the reported stable isotope data from different laboratories. It is fortunate, therefore, that at a relatively early stage in such studies, a universal convention was accepted for the reporting of stable isotope results. Two factors have contributed to this analytical harmony: (i) The international acceptance of a set of primary reference materials, relative to which all data are reported; and (ii) the convention proposed by Urey (1948) and formalized by McKinney *et al.* (1950), that all stable isotope results should be reported *relative* to the composition of the appropriate reference sample using a delta notation. Delta expresses the difference (either positive or negative) in the isotopic composition of the sample in parts per thousand ($^\circ/_{\circ\circ}$ or 'parts per mil') relative to the reference material as follows:

$$\delta = \frac{(H/L)_{sample} - (H/L)_{standard}}{(H/L)_{standard}} \times 1000 \qquad (18.3)$$

where (H/L) is the ratio of the selected heavy to light isotopic abundance.

The standards selected as reference materials and the isotopic ratios of interest are listed in Table 18.2. Specifically, the primary reference materials include:

(i) SMOW (standard mean ocean water) as defined by Craig (1961) for hydrogen and oxygen isotope ratios
(ii) Pee Dee belemnite (PDB) for carbon (a belemnite from the Pee Dee formation)
(iii) Canyon Diablo troilite for sulphur (Canyon Diablo is a primitive meteorite containing the mineral troilite, FeS)
(iv) Atmospheric gases (nitrogen and noble gases).

18.4 Hydrogen isotope analysis

Hydrogen is present in silicate rocks (such as basaltic glasses) mainly in the form of water. Samples are therefore ignited to extract structural water. H_2O as such is not suitable for analysis in a gas source mass spectrometer and must first be reduced to elemental hydrogen before measurement of the $^2H/^1H$ ratio (i.e. D/H). Considerable care must be taken to eliminate the effects of laboratory/atmospheric contamination. A typical procedure is as follows. The powdered sample (0.5–1 g, <200 mesh) is loaded into a tantalum tube (Craig and Lupton, 1976) or platinum crucible (Kyser and O'Neil, 1984) and heated at 100–150°C for up to 12 hours under vacuum to ensure complete removal of absorbed (atmospheric) water. The sample is then fused at 1200°C in a vacuum line for at least 20 minutes. The evolved gases are passed over copper oxide (CuO) heated to about 450–500°C to ensure complete oxidation of any molecular or organic hydrogen to water. Evolved gases are collected in a liquid nitrogen trap at −196°C and the more volatile gases (e.g. CO_2 and SO_2) are removed by differential heating. The residual water vapour is then volatilized and passed over uranium turnings heated to 750–850°C to reduce it to hydrogen, following a method originally proposed by Bige-

Table 18.2 Gas source mass spectrometry of stable isotopes: natural abundances and reference standards

Isotope	Natural abundance	Reference standard	Reference ratio	Reference
^{1}H	99.984	SMOW	$^{2}H/^{1}H = 0.0001557$	Craig (1961)
$^{2}H(D)$	0.01557	V-SMOW	$^{2}H/^{1}H = 0.00015595$	Gonfiantini (1978)
^{3}H	*			
^{12}C	98.888	Pee Dee	$^{13}C/^{12}C = 0.0112372$	
^{13}C	1.1112	formation		
^{14}C	*	belemnite		
		(PDB)		
^{14}N	99.634	Atmospheric	$^{15}N/^{14}N = 0.0036765$	
^{15}N	0.366	nitrogen		
^{16}O	99.759	SMOW	$^{18}O/^{16}O = 0.0020052$	Craig (1961)
^{17}O	0.037	V-SMOW	$^{18}O/^{16}O = 0.0020052$	Gonfiantini (1978)
^{18}O	0.204			
^{32}S	95.0	Canyon	$^{34}S/^{32}S = 0.0450451$	
^{33}S	0.76	Diablo		
^{34}S	4.22	troilite		
^{36}S	0.014			
Noble gases:				
^{3}He	0.00014	Atmospheric	$^{3}He/^{4}He = 0.00000140$	Lupton and Craig (1975)
^{4}He	(100)	helium		
^{20}Ne	90.48	Atmospheric	$^{21}Ne/^{20}Ne = (1)\ 0.002980$	(1) Bottomley
^{21}Ne	0.270	neon	(2) 0.002959	*et al.* (1984)
^{22}Ne	9.25		$^{22}Ne/^{20}Ne = (1)\ 0.10219$	(2) Eberhardt
			(2) 0.1020	*et al.* (1965)
^{36}Ar	0.337	Atmospheric	$^{36}Ar/^{40}Ar = 0.003384$	Steiger and Jager (1977)
^{38}Ar	0.063	argon		
^{40}Ar	99.60			
^{78}Kr	0.347	Atmospheric	$^{78}Kr/^{84}Kr = 0.006087$	
^{80}Kr	0.257	krypton	$^{80}Kr/^{84}Kr = 0.03960$	
^{82}Kr	11.523		$^{82}Kr/^{84}Kr = 0.20217$	Basford *et al.* (1973)
^{83}Kr	11.477		$^{83}Kr/^{84}Kr = 0.20136$	
^{84}Kr	56.998			
^{86}Kr	17.398		$^{86}Kr/^{84}Kr = 0.30524$	
^{124}Xe	0.0951	Atmospheric	$^{124}Xe/^{132}Xe = 0.003537$	
^{126}Xe	0.0887	xenon	$^{126}Xe/^{132}Xe = 0.003300$	
^{128}Xe	1.919		$^{128}Xe/^{132}Xe = 0.07136$	
^{129}Xe	26.43		$^{129}Xe/^{132}Xe = 0.983$	
^{130}Xe	4.071		$^{130}Xe/^{132}Xe = 0.1514$	Basford *et al.* (1973)
^{131}Xe	21.22		$^{131}Xe/^{132}Xe = 0.7890$	
^{132}Xe	26.89			
^{134}Xe	10.43		$^{134}Xe/^{132}Xe = 0.3879$	
^{136}Xe	8.857		$^{136}Xe/^{132}Xe = 0.3294$	

* ^{3}H and ^{14}C are radioactive isotopes which are continuously generated in minute quantities by cosmic-ray interaction with the upper atmosphere.
SMOW = standard mean ocean water (Craig, 1961).
V-SMOW = Vienna-SMOW, as prepared by the Atomic Energy Commission, Vienna (Gonfiantini, 1978).

leisen *et al.* (1952). Hydrogen is collected on charcoal held at −196°C. The volume of hydrogen released on heating the charcoal trap can be measured manometrically and is normally equivalent to the total water content of the sample. Kyser and O'Neil (1984) report that such water contents can be measured reproducibly to 0.02 weight %. The D/H ratio of this sample is measured by bleeding the gas into a double collector mass spectrometer. Isotope ratios are normally reported relative to standard mean ocean water (SMOW) (Table 18.2). Owing to the transfer of water vapour during sample preparation, it is desirable to heat the appropriate sections of the vacuum line to 100°C using heating tape to prevent gas losses due to surface adsorption.

The rate of release of water during ignition depends on the host mineralogy of the sample. The weight of water released from an Hawaiian tholeite during four heating stages between 25° and 1400°C, as reported by Kyser and O'Neil (1984), is given in Table 18.3.

The authors considered that water released below 200°C was adsorbed from the atmosphere on to the surface of the sample. Although most of the water was released between 600° and 1000°C, the δD value varied directly with temperature, suggesting that the release of water was controlled by diffusion out of the sample. There was no evidence that water released at higher temperatures resided in more than one mineral site within the basaltic glass.

Mass spectrometric ratios are measured by monitoring atomic masses 2 (H–H) and 3 (H–D). It is usually necessary to apply a correction to mass 3 to allow for both the presence of tritium (^{3}H) and instrument background. These corrections

Table 18.3 Water released from an Hawaiian tholeite during stepped combustion

Temperature interval (°C)	Water released (weight %)	δD (per mil)
25–200	0.08	−121
200–600	0.02	−89
600–1000	0.44	−57
1000–1400	0.04	+104
total wt.% above 200°C	0.50	−45

Data from Kyser and O'Neil (1984).

were estimated by Craig and Lupton (1976) to be 2% and 0.03% respectively. Analytical precision (two sigma) in δD measurements is normally around 0.5%.

Blank measurements recorded by Craig and Lupton (1976) were about 0.03 to 0.19 mL hydrogen (at standard temperature and pressure).

18.5 Carbon isotope analysis

Carbon exists in three naturally-occurring isotopic forms: ^{12}C, ^{13}C and ^{14}C. The latter is a radioactive isotope which is continuously generated in the upper atmosphere by cosmic-ray-induced nuclear reactions. There is considerable interest in determining ^{14}C in organic materials for age estimation. However, radioactive decay methods must be used, since mass spectrometry is not sensitive enough for such measurements. The stable isotopes which are analysed by mass spectrometry are $^{13}C/^{12}C$. Extraction techniques involve pyrolizing the sample to release carbon as CO_2 which, after purification, is measured by mass spectrometry. Much of the earlier work on carbon isotope analysis was inconclusive because the primary carbon isotope signature was obscured by surface contamination. The development of stepped combustion techniques has been very important in overcoming these difficulties. A typical sample preparation procedure for a basaltic glass would be as follows.

18.5.1 *Extraction procedure*

The sample is broken up into small chips and 100–700 mg are thoroughly cleaned to remove as much surface contamination as possible. For example, Mattey *et al.* (1984) washed samples with 0.3N HCl, then distilled water, then 1:1 methanol/toluene in an ultrasonic bath. The sample is then dried at 110°C for 24 hours to remove solvents and loaded into a pre-baked quartz tube and degassed under vacuum overnight. The sample is combusted in an atmosphere of oxygen to ensure complete oxidation of carbon to carbon dioxide. Stepped combustion is essential to permit quantitative removal of surface carbon contamination (fingerprints, organic matter, cotton waste).

Des Marais and Moore (1984) used a three-stage combustion in a study of mid-oceanic basaltic glasses, as follows:

(i) Combustion at 420–500°C for 40 minutes to remove surface carbon contamination—tests showed that fingerprints and cotton waste are removed under these conditions

(ii) Combustion at 505–740°C for 40 minutes to check for complete removal of surface contamination; unwanted carbon from calcite formed by weathering is also removed, as well as CO_2 from ruptured vesicles

(iii) 1190–1270°C for 90 minutes; magmatic carbon probably dissolved in the glass is removed at these temperature.

Mattey *et al.* (1984) followed a similar procedure and recorded the carbon yield recovered as the sample was heated in steps of 200°C up to a maximum of 1200°C. Their data for basalt glass from the Marianas Trough is shown in Figure 18.9. This diagram demonstrates a bimodal release of carbon:

(i) A light fraction of $\delta^{13}C_{PDB}$ about −25 per mil, evolved in the temperature range 200–600°C

(ii) An isotopically heavier fraction of $\delta^{13}C_{PDB}$ about −10 per mil, released in the temperature range 600–1200°C.

These workers proposed that the low-temperature fraction was superficial contamination, possibly present as some form of polymeric carbon; the higher-temperature carbon fraction was believed to be magmatic.

After release of gases by combustion, Des Marais and Moore (1984) passed them through an oven at 1200°C to ensure complete oxidation of CO to CO_2. Condensable gases were then trapped in a liquid nitrogen cold finger, and carbon dioxide was separated cryogenically from other gases by distillation at −155°C.

18.5.2 *Mass spectrometry*

Carbon dioxide is usually measured by dynamic gas source mass spectrometry using an instrument fitted with a double

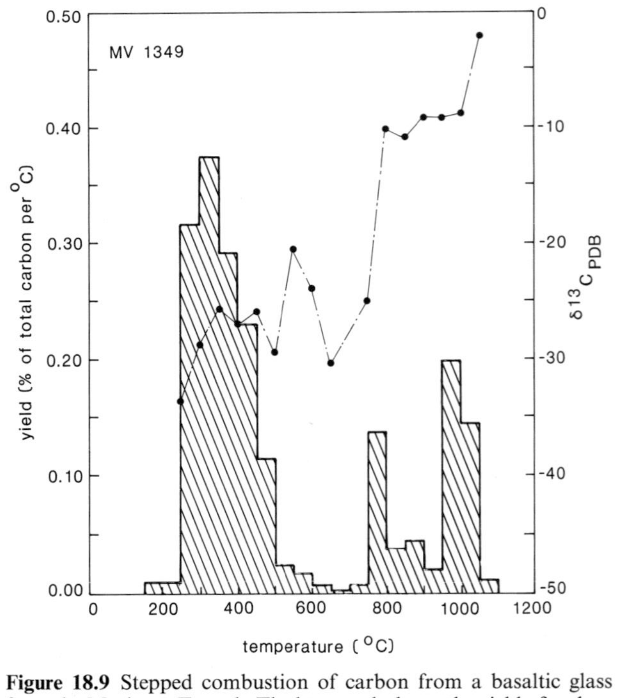

Figure 18.9 Stepped combustion of carbon from a basaltic glass from the Marianas Trough. The bar graph shows the yield of carbon for each 50° combustion step, the line graph shows the corresponding $\delta^{13}C_{PDB}$ measurement. (Mattey *et al.*, 1984.)

Table 18.4 Oxygen isotope interferences in the mass spectrometric determination of carbon dioxide

Mass number:	44	45	46
Ion species:	$^{12}C^{16}O^{16}O$	$^{13}C^{16}O^{16}O$	$^{12}C^{16}O^{18}O$
	(100)	(1.123)	0.204
		$^{12}C^{16}O^{17}O$	$^{13}C^{16}O^{17}O$
		(0.0417)	0.000417
			$^{12}C^{17}O^{17}O$
			(1.4×10^{-7})

Data in brackets represent the relative proportion of ionic species normalized to $^{44}(CO_2) = 100$.

collector. The mass numbers monitored are 44 ($^{12}CO_2$ and 45 ($^{13}CO_2$). For the most precise work, account must be taken of the ($^{12}C^{16}O^{17}O$) contribution to $^{13}C^{16}O_2$ at mass 45. The relative magnitude of these and interferences from other oxygen isotope ratios may be judged from data in Table 18.4.

In some applications it is desirable to determine simultaneously both $^{13}C/^{12}C$ and $^{18}O/^{16}O$ ratios from carbon dioxide gas extracted from a sample. A triple-collector gas source mass spectrometer can then be used to monitor mass numbers 44, 45 and 46, mass 46 being largely a measure of ^{18}O (Table 18.4).

Using the above procedures, carbon blanks of 0.7 μg carbon and precisions of 0.1% relative to the Pee Dee formation belemnite have ben reported by Des Marais and Moore (1984).

A significant advance in sensitivity of stable carbon isotope measurements has been reported by Wright *et al.* (1983). These workers have described a double-collector instrument capable of carbon isotope analysis in static mode. This instrument has been developed in particular for isotope ratio measurements on minute meteorite fragments. Samples as small as 10^{-8} g carbon can be analysed with a precision of the order of 0.16 per mil $\delta^{13}C$. Carbon dioxide itself suffers from fractionation effects in static mode, due to the decomposition of the gas by interaction with the electron gun filament. Wright *et al.* first converted the gas to deuterated methane (CD_4). This gas shows resistance to decomposition and is comparatively free from mass interference effects. In a companion paper, McNaughton *et al.* (1983) describes a method of synthesizing deuteromethane. Carbon dioxide from the combustion procedure described above is reacted with excess deuterium gas in the presence of a metallic nickel catalyst heated to 350–470°C:

$$CO_2 + 4D_2 \xrightarrow{\text{Ni catalyst}} CD_4 + 2D_2O \qquad (18.4)$$

Deuterated water formed by this reaction and excess deuterium are absorbed by a Zr/Al getter and residual CD_4 expanded into the mass spectrometer for analysis.

18.6 Nitrogen isotope analysis

Nitrogen exists as two stable isotopes, ^{14}N and ^{15}N, and the measurement of the isotope ratio poses some technical difficulties.

(i) Since nitrogen is a major component of the atmosphere, strict precautions are necessary to avoid contamination

(ii) Like carbon, nitrogen too is present in organic material which can contaminate samples; a stepped combustion method is essential if this nitrogen contamination is to be distinguished from the primary nitrogen signal

(iii) A particular difficulty with nitrogen is that this gas will react with some hot metal surfaces which must therefore be avoided in the operation of vacuum-line assemblies.

18.6.1 *Extraction procedures*

Procedures for the extraction of nitrogen, therefore, involve fusion in a quartz furnace at 1200°C. Des Marais (1983) used a three-stage stepped combustion procedure in the analysis of lunar rocks similar to that described for carbon extraction described above. Samples were heated as follows:

(i) 410–450°C (to remove terrestrial contamination)

(ii) 500–625°C (to check that no contamination remained)

(iii) 1220–1300°C (to release indigenous nitrogen for isotope ratio measurement).

This combustion procedure was carried out under an atmosphere of oxygen (15–40 Torr) to ensure that any carbon or sulphur evolved were fully oxidized to facilitate subsequent purification. Oxides of nitrogen formed during combustion were reduced back to nitrogen by passing over cupric oxide (CuO) heated to 350–450°C. Frick and Pepin (1981*b*) made use of the properties of cupric oxide, to provide and then to remove oxygen during the combustion process. Heated to 900°C, copper oxide released oxygen at a pressure of about 15 Torr for oxidative combustion. After ignition of the sample, the copper oxide furnace was allowed to cool to 450°C. At this temperature, oxygen in the evolved gas stream was re-absorbed. Condensable gases (CO_2, SO_2, H_2O) were frozen out using a liquid nitrogen cold finger. Frick and Pepin (1981*b*) found that unexpectedly large amounts of nitrogen could also be condensed out in these liquid nitrogen traps. This problem was reduced by miniaturizing the trap.

18.6.2 *Mass spectrometry*

Nitrogen ratios are measured using a double-collector gas source mass spectrometer which may be operated in static mode, since nitrogen is not decomposed significantly by the ion gun filament. The mass numbers monitored are 28 ($^{14}N_2$) and 29 ($^{14}N^{15}N$). Frick and Pepin (1981b) report that mass 29 suffers a significant interference from $^{14}N_2H^+$, and sometimes from $^{13}C^{16}O^+$. The hydride is formed in the ion source and is caused by hydrogen contamination in the purified nitrogen sample. This can be minimized by exposing the sample to a Ti–Zr getter for about one minute before expanding the gas for analysis into the mass spectrometer. The getter preferentially absorbs hydrogen but not nitrogen. The release pattern of nitrogen (and argon) measured by Frick and Pepin (1981*b*) during stepped combustive oxidation of an Allende meteorite matrix is shown in Figure 18.10. These authors suggest that gases evolved below 600°C may be released from lattice traps or may be structural nitrogen released from carbonaceous matter in the meteor-

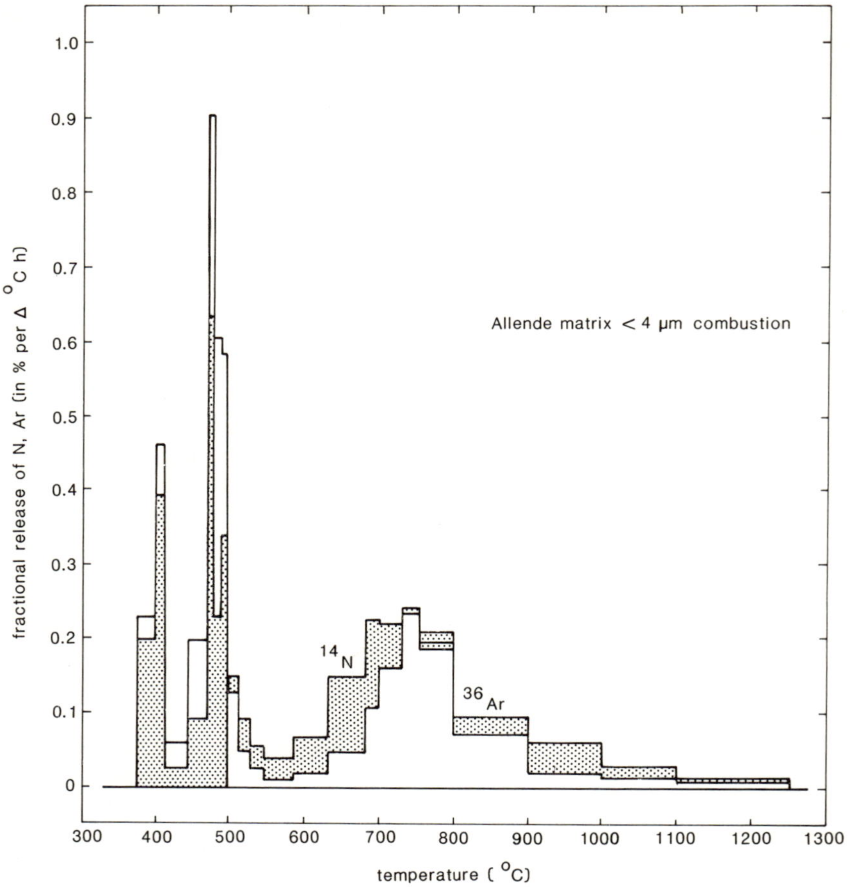

Figure 18.10 Stepped combustion of nitrogen and argon from Allende meteorite. The graph shows the percentage gas released per °C temperature change at various combustion temperatures between 400° and 1250°C. Gases released below 600°C were considered to be derived from a carbonaceous carrier, and those released above 600°C from a silicate host matrix within the meteorite lattice. (Frick and Pepin, 1981*b*).

ite. In an analysis of lunar rocks, Des Marais (1983) considered that nitrogen evolved below 626°C represented terrestrial contamination, and that released in the higher-temperature range (up to 1220–1300°C) was primary nitrogen.

Sakai *et al.* (1984) reported procedural blanks of about 0.06 μg nitrogen and isotope ratio precisions of the order of ±1% (delta values). Concentration measurements were estimated to be accurate to ±5%.

18.7 Oxygen isotope analysis

The extraction of oxygen is different from that of the other gases normally determined by gas source mass spectrometry in that oxygen cannot be released from the silicate matrix by simple ignition. Some of the early work involved reduction of samples by heating with carbon at 1600–2000°C. However, such procedures have been superseded by an oxidation reaction with fluorine or fluorine-related compounds. These follow the work of Clayton and Mayeda (1963), who used the reagents bromine pentafluoride (BrF_5) or bromine trifluoride (BrF_3) to attack the sample, and Taylor and Epstein (1962), who reacted samples with fluorine gas in the presence of HF.

Due to the high reactivity of these reagents and the fact that they will attack glass apparatus, special consideration must be given to the materials from which vacuum-line reaction vessels are constructed. The favoured materials are nickel, monel or copper. These are silver-soldered together. Valve seals are made from PTFE. There is a risk that the inside surfaces of apparatus could become oxidized during sample attack reactions. It is therefore essential to precondition the vacuum line with the fluorine reagent before extraction commences, to remove all trace of atmospheric contamination.

18.7.1 *Sample attack by bromine pentafluoride*

Perhaps the most widely-used procedure is that of Clayton and Mayeda (1963), who exploited the oxidizing properties of bromine pentafluoride. Taking $KAlSi_3O_8$ as the silicate matrix, the reaction sought is

$$KAlSi_3O_8 + 8BrF_5 \rightarrow KF + AlF_3 + 3SiF_4 + 4O_2 + 8BrF_3. \quad (18.5)$$

Bromine pentafluoride is a colourless liquid at room temperature (mp: −61.3°C, bp: 40.5°C) and requires great care in handling, as it will react explosively with some reagents, including water vapour. Reaction vessels must be made of pure nickel, since this is the only material known to withstand attack by this reagent at the elevated temperatures (up to 700°C) required for reaction.

The procedure of Clayton and Mayeda is as follows. The sample is loaded into the nickel reaction tube and outgassed at 400°C to remove all atmospheric oxygen and water vapour. This temperature is below the range at which hydroxyl minerals start to decompose. The metal vacuum line is then pretreated with BrF_5 to remove all traces of atmospheric oxygen. Reaction vessels are loaded with five times the quantity of BrF_5 calculated as necessary for stoichiometric reaction. The reaction tube is then isolated and heated for 12 hours at a temperature selected in accordance with the resistance of the minerals present:

450°C for quartz, micas and feldspar
600°C for magnetite, haematite and ilmenite
690°C for garnet and olivine.

After cooling the sample to room temperature, the volatile gaseous products are pumped off and the gases BrF_5, BrF_3, Br_2, SiF_4 and HF condensed in a liquid-nitrogen cold trap. Residual oxygen is pumped into a calibrated volume for determination of the extraction efficiency. The minimum extraction efficiency considered adequate for accurate oxygen isotope analysis is 97–99%. If smaller amounts are found to be extracted, there is a serious risk that isotopic fractionation will have occurred during the reaction procedure. The extracted gas will then be richer in ^{16}O than the original sample.

Oxygen is not normally determined directly by mass spectrometry since this gas is rather reactive and suffers too great a risk of contamination from the atmosphere. The extracted gas is therefore normally first converted to carbon dioxide by passing over a carbon rod (or filament) heated to 550–600°C. It is important that the rod is not heated to a higher temperature, since product gases will then contain appreciable quantities of carbon monoxide and the remaining carbon dioxide may again suffer from fractionation discrepancies due to isotope-related differences in reaction rates. Clayton and Mayeda reported oxygen blanks of usually less than 1 μmol which probably arose from oxygen reacting with the walls of the reaction vessel during sample loading. Mass spectrometric procedures follow those required for carbon isotope analysis (see above). Clayton and Mayeda (1963) also considered the use of bromine trifluoride as oxidizing agent and obtained data comparable to that obtained from BrF_5 extractions. The latter reagent was preferred in routine applications because it was easier to manipulate in the extraction line. A more recent development has been the substitution of chlorine trifluoride as extraction reagent.

18.7.2 *Oxygen extraction using fluorine*

An alternative oxygen extraction procedure described by Taylor and Epstein (1962) uses the reaction between the silicate sample and fluorine gas in the presence of anhydrous HF. The reaction mixture is heated at 450°C for at least 8 hours. For some resistant minerals, temperatures up to 520°C may be required. However, Taylor and Epstein report that at higher temperatures the nickel fluoride coating on the nickel reaction vessel became unstable, permitting fluorine gas to react with exposed metal. Quantitative reactions were reported with the minerals quartz, plagioclase, K-feldspar, muscovite, biotite, hornblende, clinopyroxene, orthopyroxene, spodumene, apatite and ilmenite. However, low recoveries were obtained from olivine, magnetite, epidote and garnet. It was suggested that one reason may have been the formation of non-volatile fluorides on the surfaces of sample particles of these minerals, which then inhibited further reaction. As in the earlier procedure, extracted oxygen was reacted with a heated carbon filament for conversion to carbon dioxide before subsequent mass spectrometric analysis.

18.7.3 *Application to hydrated minerals*

Haimson and Knauth (1983) have investigated the effectiveness of oxygen extraction procedures in the analysis of hydrated minerals (for example hydrated silicates (opal), clays and zeolites). In such samples it is not usually the isotope ratio of oxygen extracted from the entire sample (including that from hydrated phases) that is of geological interest, but specifically the isotope ratio of the silicate-bound oxygen only. The reason is that oxygen associated with hydroxyl radicals or 'structural' water is usually much more loosely bound than silicate oxygen and can suffer isotope exchange with environmental water. Haimson and Knauth (1983) described a stepwise fluoridation procedure based on the work of Taylor and Epstein (1962). The sample was reacted with fluorine in a nickel reaction vessel heated to 450°C for 12 hours. However, instead of adding 20 to 60% excess fluorine as would be required for a toal oxygen analysis, smaller fractions were reacted. After the first partial reaction, released oxygen was removed and analysed. Further fluorine was then added to the partially decomposed sample and a further extraction at 450°C carried out. In total 3 to 7 aliquots of oxygen were progressively extracted and isotopically analysed. Data showed that in the analysis of opal, the more loosely bound hydrous oxygen was first removed and that in natural crystallized samples, this had a distinct oxygen isotope ratio from the more strongly-bound silicate oxygen recovered in the latter extractions. By maintaining reaction temperatures at 450°C, undesirable side reactions were avoided. By removing water and exchangeable silicate oxygen in preliminary extractions, the isotopic composition of non-biogenic opals was found to remain constant in subsequent fractions to 0.2 per mil. Problems remained in the interpretation of data from biogenic opals.

18.7.4 *Analysis of carbonate samples*

Oxygen isotope ratios are also extensively measured on carbonate samples. Although it is relatively simple to release carbon dioxide from carbonate materials by acid decomposition, there are several factors which must be taken into account in choosing the most appropriate acid. McCrea (1950) has considered these factors in detail:

(i) Nitric acid and formic acid cannot be used, since both generate gaseous impurities which give mass interferences at 46 amu

(ii) Acid attack must yield soluble salts to permit rapid and complete decomposition of the sample (HF is therefore unsuitable)

(iii) The acid must be available in anhydrous form to prevent isotopic exchange between released CO_2 gas and water, e.g.

$$H_2{}^{18}O + \tfrac{1}{2}C^{16}O_2 \leftrightarrow H_2{}^{16}O + \tfrac{1}{2}C^{18}O_2. \quad (18.5)$$

(iv) No exchange must occur between liberated carbon dioxide and oxygen in any oxy-acid used for sample decomposition (sulphuric acid is therefore unsuitable).

After consideration of these factors, McCrea (1950) recommended anhydrous phosphoric acid for carbonate decomposition. This acid has the further advantage of a low vapour pressure which facilitates handling in a vacuum system. Data for oxygen isotope analysis by phosphoric acid decomposition was far superior to results obtained by a simple thermal decomposition. Early work indicated that serious errors could arise in the latter procedure due to isotopic exchange between carbon dioxide gas and residual calcium oxide. This discrepancy was found to be difficult to correct, since its magnitude was in part dependent on the particle size of the analysed sample.

Carbonates are therefore decomposed by reaction with anhydrous phosphoric acid at 25°C for between 12 hours and several days. The chemical reaction is as follows:

$$3CaCO_3 + 2H_3PO_4 \rightarrow Ca_3(PO_4)_2 + 3H_2O + 3CO_2 \quad (18.6)$$

This form of sample decomposition is not without its complications. These arise because some of the carbonate oxygen is released as water, some as carbon dioxide. The rates of these competing reactions vary slightly depending on the isotopic composition of oxygen. The net effect is that not only is the oxygen isotopic composition of carbon dioxide slightly different from that of the parent carbonate, but also the degree of this fractionation varies with sample type (e.g. limestone as opposed to dolomite). A small correction factor must therefore be applied to analysed carbon dioxide gas data to obtain the true isotopic compositon of the carbonate sample. This corection has been investigated for limestone and dolomite samples, for example, by Sharma and Clayton (1965) and Friedman and O'Neil (1977), and the correction factors normally used are as follows:

$$\alpha_{CO_2}\text{ (calcite, 25°C)} = ({}^{18}O/{}^{16}O)_{CO_2}/({}^{18}O/{}^{16}O)_{calcite} = 1.01020 \quad (18.7)$$

$$\alpha_{CO_2}\text{ (dolomite, 25°C)} = ({}^{18}O/{}^{16}O)_{CO_2}/({}^{18}O/{}^{16}O)_{dolomite} = 1.01103 \quad (18.8)$$

Sharma and Clayton (1965) investigated these isotopic fractionation factors associated with the acid liberation of CO_2 and showed that for carbonates which do not react appreciably with phosphoric acid (e.g. $MgCO_3$, $CoCO_3$, $NiCO_3$) a thermal decomposition plus bromine pentafluoride attack could be used successfully.

18.7.5 *Standardization*

Results are normally expressed relative to 'standard mean ocean water' (SMOW) as defined by Craig (1961). Standard mean ocean water is itself defined relative to a laboratory reference material NBS-1 (distilled water), originally distributed by the National Bureau of Standards:

$$({}^{18}O/{}^{16}O)\text{ (SMOW)} = 1.008\ ({}^{18}O/{}^{16}O)\text{ (NBS-1)} \quad (18.9)$$

More recently, Gonfiantini (1978) has described data for a replacement primary $^{18}O/^{16}O$ and $^2H/^1H$ water standard, V-SMOW (Vienna-SMOW). This was made up from distilled ocean water mixed with other waters so that the $^{18}O/^{16}O$ ratio is identical to the original SMOW defined by Craig (1961). The $^2H/^1H$ ratio of V-SMOW is slightly lower (by 0.2 per mil) than that of SMOW, although this difference is within the experimental uncertainty of most measurements. Gonfiatini (1978) also reports data on an additional water standard: SLAP (Standard Light Antarctic Precipitation). These are the internationally-recognized primary standards against which all laboratory measurements are reported.

18.8 Sulphur isotope analysis

Sulphur exists as four naturally-occurring isotopes, ^{32}S, ^{33}S, ^{34}S, and ^{36}S (Table 18.2), stable isotope ratios being determined for the pair $^{34}S/^{32}S$. Sulphur is normally extracted for analysis as sulphur dioxide by stepwise ignition of the sample in an atmosphere of oxygen. As in the case of carbon and nitrogen isotope analysis, Des Marais (1983) discarded sulphur dioxide evolved below 625°C to eliminate the possibility of contamination. However, unlike carbon and nitrogen, only very small fractions of sulphur were released at these lower temperatures.

Sulphur dioxide extracted from samples must be purified cryogenically to remove other gases (especially CO_2, N_2 and H_2O) before mass spectrometric determination. Dynamic operation is necessary since sulphur dioxide suffers degradation due to decomposition by the ion source in static mode. It is normally necessary to heat the spectrometer to 100°C during analysis to prevent the adsorption and condensation of gases on metal surfaces. The two mass peaks monitored are 64 ($^{32}S^{16}O_2$) and 66 ($^{34}S^{16}O_2$) plus minor amounts of ($^{32}S^{16}O^{18}O$), ($^{33}S^{16}O^{17}O$) and ($^{32}S^{17}O_2$).

A novel sample preparation procedure has been described by Ueda and Sakai (1983). This permits distinction between sulphate sulphur and sulphide sulphur in a sample. The rock powder was heated under vacuum with dehydrated phosphoric acid containing Sn^{2+} ions. The sample was first warmed to 90°C for 2 hours to remove contaminants. The temperature was then increased to 280°C over a one-hour period. At temperatures above 150°C, powdered silicates were rapidly decomposed with sulphide, sulphate and carbonate minerals being reduced to H_2S, SO_2 and CO_2 respectively. These gases may be trapped in a liquid nitrogen cold finger and separated by vacuum distillation. The hydrogen sulphide aliquot was oxidized to SO_2 by passing over heated Cu_2O. The volumes of the two SO_2 fractions (one from sulphide- and the other from sulphate-sulphur) and CO_2 were then measured to determine abundances and isotopic ratios of sulphide, sulphate and CO_2 in the sample. Sakai *et al.* (1984) used this technique to analyse ocean-floor basalts, although carbon blanks obtained by this acid-attack procedure were high compared with a conventional oxidative combustion. Hulston and Thode (1965*a,b*) attempted to analyse all four sulphur isotopes in an investigation of nuclear fractionation effects. However, the accuracy of these measurements was seriously restricted by corrections required to remove the effect of oxygen isotope interferences. These workers devised a technique whereby measurements were made by converting sulphur to SF_6 to eliminate these problems (^{19}F is mono-isotopic).

18.9 Noble gas analysis

The noble gases of interest in geological samples include helium (He), neon (Ne), argon (Ar), krypton (Kr) and xenon (Xe) (Table 18.2). Determinations are made both for the total abundance of the element in the sample as well as for stable isotope ratios.

18.9.1 *Separation procedures*

Noble gases are normally released by igniting samples at 1600–2000°C in a boat made of tantalum (e.g. Craig and Lupton, 1976), or zirconium oxide-lined molybdenum (e.g. Voshage and Feldman, 1978). It is desirable that this ignition is carried out in an atmosphere of oxygen so that other gases that are evolved simultaneously are converted to oxides (CO_2, SO_2, H_2O). These may subsequently be readily separated from the noble gases by condensing in a liquid nitrogen trap. In some applications it is desirable that a stepped combustion is used in which gases evolved at temperatures below 600°C are discarded as terrestrial contamination prior to evolution of primary igneous gases during the high-temperature ignition. Consideration of mass spectrometer interference effects (see Table 18.5) indicates that it is generally necessary to separate noble gases from one another prior to mass analysis. A typical procedure would be as follows.

The crushed sample (for example, 400–600 mg of a basaltic glass) is weighed into a suitable boat, mounted inside the ignition furnace and heated to 100–150°C under vacuum overnight (15 hours) to remove adsorbed gases. If desired, the sample can be ignited to about 600°C for several hours to remove trapped gases (atmospheric contamination). Samples are then ignited at 1800°C by resistive heating for about 20–30 minutes. Gases are passed over copper oxide (CuO) heated to 475°C to ensure complete oxidation of hydrogen, carbon and sulphur gases. Condensable gases are trapped in a liquid nitrogen cold finger at −196°C. Frick and Pepin (1981*a*) absorbed active gases (mainly CO_2 and SO_2) by 30-minute exposure to a Zi–Ti getter held at 800°C followed by a further clean-up using both a getter pump held at about 400°C and an incandescent rhodium filament to crack hydrocarbon molecules from the pumping system. If getters are not used, residual nitrogen can be removed from the noble-gas fraction by exposing gases to titanium metal held at about 800°C.

Separation of noble gases into individual components for mass analysis depends on the application, since they are not all always present in analysable quantities, and in some schemes of analysis coexisting nitrogen and or hydrogen must be measured. However, if the noble gases must be separated from one another, the following procedure can be used. A U-tube filled with activated charcoal and cooled to liquid-nitrogen temperatures will retain all Xe and Kr, more than 99.9% Ar and about 5% Ne (Lupton and Craig, 1975). These gases may subsequently be released by warming this charcoal trap to −140°C (to release Ar), −90°C (to release Kr) and +100°C (to release xenon) (Frick and Pepin, 1981*a*). If helium is to be separated from neon, gases must be passed through a charcoal trap cooled to 37K by a helium expansion refrigerator (Rison and Craig, 1983). Under these conditions, neon is quantitatively retained on the charcoal and helium passes straight through and may be expanded into the spectrometer for mass analysis. The neon fraction can subsequently be released by warming the trap to room temperature (Bottomley *et al.*, 1984). If precise isotope measurements are then required, it is desirable that this separation should be carried out, since Rison and Craig (1983) found that varying proportions of helium to neon in the analysed gas can affect measured helium isotope ratio measurements by up to 10%. Similarly Bottomley *et al.*, (1984) observed serious errors in neon isotope measurements if the analysed sample also contained helium.

18.9.2 *Mass spectrometry: interferences*

Isotope ratio measurements are normally carried out on a dual-collector gas source mass spectrometer operated in static mode. Some mass interferences are listed in Table 18.5.

Table 18.5 Mass spectrometer interferences in noble gas analysis

Isotope	Interference	Action	Comments
^{3}He	$^{3}H^+$, $^{3}(HD)^+$	Resolve using a mass spectrometer of adequate resolution	Helium isotope ratio measurements are reported to be variable, depending on the sample neon content (Rison and Craig, 1983). Separation from neon required
^{4}He			
^{20}Ne	$H_2{}^{18}O^+$, $^{40}Ar^{++}$	Monitor masses 18 and 40	Sample must be separated from argon
^{21}Ne	$^{20}NeH^+$	Monitor mass 23 ($^{22}NeH^+$)	
^{22}Ne	$CO_2{}^{++}$	Monitor mass 44	
^{36}Ar			
^{38}Ar			
^{40}Ar	$^{80}Kr^{++}$	Monitor mass 80	Sample must be separated from krypton if present at significant concentrations

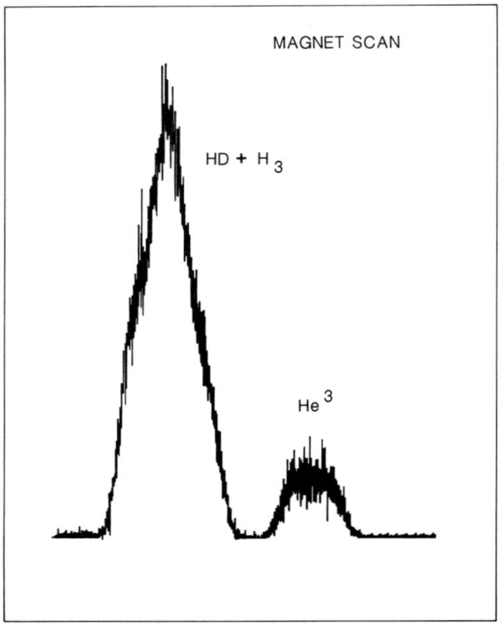

Figure 18.11 Magnetic scan through the $^3He^+$ ion beam (intensity 250 ions s^{-1}) showing the HD^+ and H_3^+ mass interference. (Lupton and Craig, 1975.)

Specifically, ^3He suffers interferences from both H^+_3 and HD^+. However these interferences can be adequately resolved using a mass spectrometer with a resolution $M/\Delta M$ of about 600. A mass spectrum scan through the ^3He isotope peak is shown in Figure 18.11. A titanium getter is desirable to pump hydrogen and deuterium out of the static volume of the mass spectrometer to minimize this interference. A Faraday cup is adequate to measure the ^4He ion beam, but an electron multiplier detector of higher sensitivity is required for the much weaker ^3He ion signal.

Neon suffers interferences from $^{20}(H_2{}^{18}O)^+$, $^{40}Ar^{2+}$, $^{20}NeH^+$ and $^{22}(CO_2)^{2+}$. Poreda and di Brozolo (1984) monitored these interferences at mass peaks 18, 23, 40 and 44. Under the extraction conditions used, the neon hydride interference was found to be negligible. The magnitude of the interferences due to H_2O, CO_2 and Ar did not vary significantly in the course of a run, and corrections were made. Podosek *et al.* (1971) carried out mass fractionation corrections for isotope ratio determinations of xenon assuming a linear discrimination and the value $^{129}Xe/^{132}Xe = 0.9833$. The resultant discrimination factor was 0.39% per mass unit, close to the $\sqrt{(m_1/m_2)}$ mass-dependent response (0.38%) expected of an electron multiplier detector.

18.10 Potassium–argon geochronometry

The naturally-occurring radioactive isotope ^{40}K decays by electron capture to ^{40}Ar (and by beta-emission to ^{40}Ca). The ^{40}Ar content of a rock or mineral is of great interest in measuring its age since crystallization. Argon itself exists as three naturally-occurring isotopes, the present-day percentages of which in the Earth's atmosphere are ^{36}Ar, 0.337%; ^{38}Ar, 0.063%; and ^{40}Ar 99.60%. Experimental procedures for the determination of ^{40}Ar must, therefore, take into account the following sources of argon in a sample:

(i) ^{40}Ar produced by radioactive decay of ^{40}K

(ii) Primordial argon incorporated into the mineral structure of the sample at the time of crystallization

(iii) Contamination of sample with argon from the atmosphere

(iv) Rehomogenization of argon within a sample due to subsequent metamorphic processes. This results both in argon being released from the sample and in atmospheric argon being incorporated into the crystal structure. Some minerals are more resistant than others to these effects.

The initial argon content is usually very small, often close to the blank level of the determination. Atmospheric contamination causes a significant contribution to the ^{40}Ar content of a sample and corrections must be made. These are calculated from the measured ^{36}Ar content of the sample and the known atmospheric $^{36}Ar/^{40}Ar$ ratio. However, perhaps of greatest importance is the recognition of alteration effects which could significantly influence the interpretation of model age data.

Two techniques are extensively used in argon geochronometry. The first is traditional potassium–argon isotope analysis in which ^{40}Ar is determined by isotope dilution using a gas source mass spectrometer and total potassium is determined on a separate sample aliquot by x-ray fluorescence or flame photometry. These data are then fitted to the standard age equation (section 16.15). However, in the 1960s an alternative and potentially more powerful technique was developed in which the sample is irradiated in a neutron flux to induce the nuclear reaction.

$$^{39}K\ (n,\ p)\ ^{39}Ar. \qquad (18.10)$$

The ^{39}Ar formed in the sample is thus a direct measure of its potassium content. Using a suitable reference material of known age, the ratio $^{40}Ar/^{39}Ar$ can be measured simultaneously by gas source mass spectrometry. If this procedure is used in conjunction with a stepped combustion extraction, the $^{40}Ar/^{39}Ar$ technique is capable of giving data on the thermal history of the sample, and therefore on the reliability of the age determination.

18.10.1 *Potassium–argon measurements*

In potassium–argon dating measurements, the gas source mass spectrometer is used to measure the concentration of ^{40}Ar in a sample by isotope dilution using ^{38}Ar as spike. A typical extraction line for argon is shown in Figure 18.12. The sample (typically 0.2–1 g) is weighed into a molybdenum boat and suspended in an induction furnace. The sample boat is connected to an extraction line and the entire vacuum system is baked at about 300°C to remove atmospheric argon and water vapour. The induction furnace portion of the extraction line is isolated from the vacuum pumping system and the sample heated to about 1800°C. Simultaneously, the seal on an ampoule containing the ^{38}Ar tracer is broken to ensure equilibration with sample gases. Released gases are collected on the charcoal cold finger which is cooled to liquid-nitrogen temperatures. Absorption

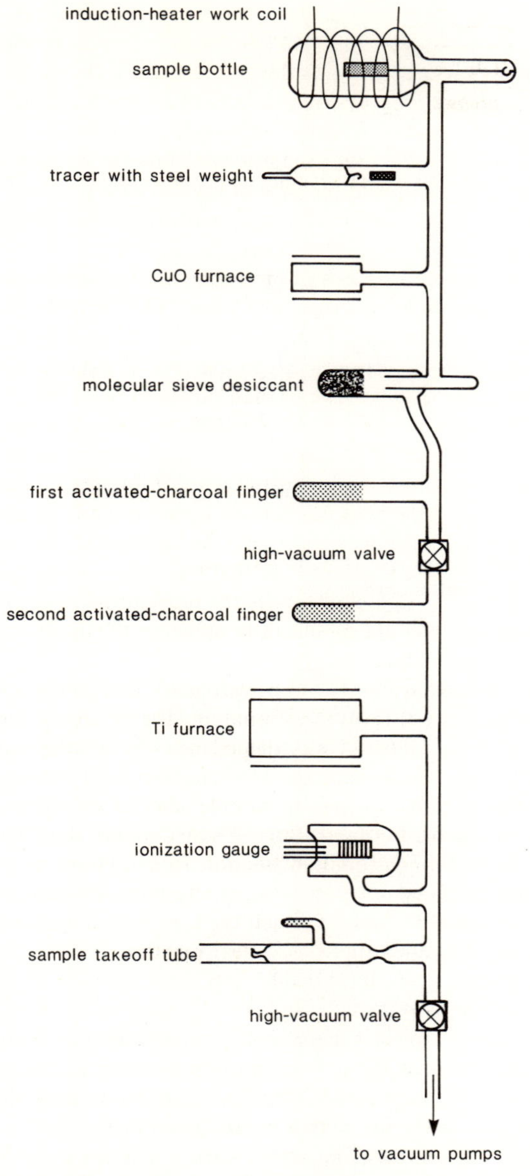

Figure 18.12 Schematic arrangement of an argon extraction line. (Dalrymple and Lanphere, 1969.)

of gases on this charcoal cold finger maintains a low vacuum in the induction furnace and prevents the formation of a radiofrequency-induced low-pressure discharge within the vacuum line. After ignition, the cold finger is warmed to room temperature and the released gases are passed over a Cu/CuO furnace heated to 450°–550°C to ensure complete oxidation of any residual hydrocarbons. Water is then adsorbed on a synthetic molecular sieve. Remaining gases are transferred into the second half of the extraction system where reactive gases (especially oxygen and nitrogen) are removed by reaction with a titanium metal getter heated to 800°C. Argon is condensed from the residual gas (which may also contain some helium and neon) on a liquid nitrogen cold finger and sealed in an ampoule for transfer to the mass spectrometer.

The mass peaks sought for analysis are:

^{40}Ar Radiogenic argon from ^{40}K decay plus atmospheric contamination
^{38}Ar Tracer plus atmospheric contamination
^{36}Ar Atmospheric contamination.

A small contribution from tracer gas may also enhance ^{40}Ar and ^{36}Ar. The relationship between these mass peaks is shown schematically in Figure 18.13. Mass analysis is carried out in the static mode with an appropriate correction for the memory effect (section 18.2.3). The ^{36}Ar peak is used to correct for the atmospheric contribution to the ^{40}Ar peak and the amount of radiogenic ^{40}Ar can then be determined by isotope dilution.

The potassium–argon method is extensively used but suffers one important disadvantage. It is difficult to establish if a sample has suffered premature loss of argon after crystallization except by carrying out a mineral separation on the bulk sample and performing individual determinations on each mineral fraction to see if the resulting data forms a concordant isochron diagram. For this reason the technique has to some extent been superseded by the ^{40}Ar/^{39}Ar dating technique since not only is much of this ambiguity overcome, but also isochron data can also usually be obtained by stepped ignition of a single sample.

18.10.2 *^{40}Ar/^{39}Ar dating technique*

The ^{40}Ar/^{39}Ar dating technique overcomes several of the shortcomings of the potassium–argon method. These advantages arise from the fact that the sample is irradiated in a nuclear reactor, during which a small fraction of ^{39}K is converted to ^{39}Ar by the nuclear (n,p) reaction outlined above. The radiogenic argon (^{40}Ar) to potassium (^{39}Ar) ratio

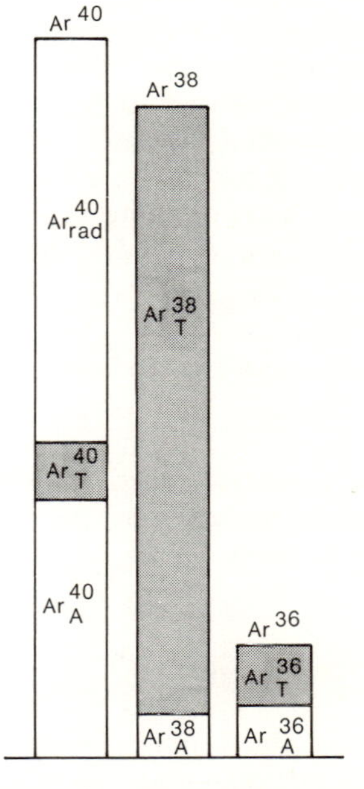

Figure 18.13 Isotope dilution and atmospheric contamination components in the analysis of radiogenic $^{40}Ar_{RAD}$. Ar_T is contributed from the $^{38}Ar_T$ spike and Ar_A represents atmospheric contamination. (Dalrymple and Lanphere, 1969.)

Table 18.6 $^{40}Ar/^{39}Ar$ dating: neutron activation interference products

Target isotope	Natural abundance	Nuclear reaction	Half-life	Second decay product	Neutron spectrum required	Cross-section (mb)	Comments
^{39}K	93.3	$(n, p)^{39}Ar$	269 y	—	Fast	20	The desired reaction
^{40}K	0.0118	$(n, p)^{40}Ar$	Stable		Slow	4400	Interference on radiogenic ^{40}Ar
^{42}Ca	0.65	$(n, \alpha)^{39}Ar$	269 y		Fast	2.7	Interference on ^{39}Ar produced from ^{39}K
^{40}Ca	96.94	$(n, n\alpha)^{36}Ar$	Stable		Fast		Interference on ^{36}Ar, measurement of which is required for correction of atmospheric contamination on ^{40}Ar
^{35}Cl	75.77	$(n, \gamma)^{36}Cl$	3.01×10^5 y	^{36}Ar	Slow	17000	As above
^{37}Cl	24.23	$(n, \gamma)^{38}Cl$	37.2 min	^{38}Ar	Slow	428	Interference on ^{38}Ar spike if used
^{79}Br	50.54	$(n, \gamma)^{80}Br$	17.4 min	^{80}Kr	Slow Fast	8500 14800	Interference as $^{80}K^{2+}$ on radiogenic $^{40}Ar^+$

Data from Erdtmann (1976).
mb = millibarns, y = years, min = minutes.
'Fast': the nuclear reaction is sustained by a fast neutron spectrum (>100 keV).
'Slow': implies a low-energy neutron spectrum comprising thermal (<0.5 eV) and epithermal (>0.5 eV) neutrons, except that cross-section data for ^{80}Br is tabulated for thermal (labelled slow) and epithermal (labelled fast) neutron fluxes.

can then be measured simultaneously by gas source mass spectrometry. The development of this procedure has been variously accredited to Merrihue (1965), Sigurgeirsson (1962) and Wanke and Konig (1959). The important development of this stepwise ignition of samples was introduced by Merrihue and Turner (1966).

The $^{40}Ar/^{39}Ar$ method has two potential advantages. Separate sample aliquots are no longer required for the determination of ^{40}Ar and potassium. Any uncertainly due to sample heterogeneity is thus removed. Further, since both potassium and argon are determined from a single mass spectrometric measurement, the analytical precision is potentially better than that of traditional potassium–argon measurements.

(*c*) *Interferences from competing nuclear reactions.* Neutron irradiation can cause some additional interference effects, which are summarized in Table 18.6. Neutron capture reactions involving calcium, potassium and chlorine are all capable of producing argon isotopes. These interferences have been considered in some detail by Tetley *et al.* (1980) and Turner (1971). The most serious arise from the reactions

$$^{40}K(n, p)^{40}Ar$$

$$^{40}Ca(n, n\alpha)^{36}Ar$$

$$^{42}Ca(n, \alpha)^{39}Ar.$$

Roddick (1983) considered other interferences, sometimes overlooked, due to neutron irradiation of chlorine, originating from the reactions

$$^{35}Cl(n, \gamma)^{36}Cl$$

$$^{37}Cl(n, \gamma)^{38}Cl.$$

Both reaction products are radioactive and decay with half-lives of 3.1×10^5 years and 37.3 minutes to ^{36}Ar and ^{38}Ar respectively. These interferences from chlorine are significant if irradiations are carried out in a reactor with a high thermal to epithermal neutron flux. Roddick (1983) also considered interferences caused by the production of ^{80}Kr by neutron capture of bromine. The resultant double-charge $^{80}Kr^{2+}$ ions create a small isobaric interference on $^{40}Ar^+$.

(*b*) *Optimization of irradiation conditions.* Irradiation conditions and their optimization have been considered in detail by Turner (1971). It is assumed that the minimum weight of sample required for precise mass spectrometric analysis is that which will yield a volume of ^{40}Ar of 10^{-6} cm^3 at STP (standard temperature, 0°C, and pressure, 1 atmosphere). Assuming a procedural blank of 10^{-8} cm^3, Turner's data for the sample mass which contains the minimum amount of ^{40}Ar are shown in Figure 18.14 as a function of both age and potassium content. In optimizing irradiation conditions, Turner (1971) discussed the integrated neutron flux required to minimize the interferences due to the reactions ^{40}Ca (n, nα) ^{36}Ar, ^{42}Ca (n,α) ^{39}Ar and ^{40}K (n,p) ^{40}Ar (listed in Table 18.6). His data are shown in Figure 18.15. This plots the integrated fast neutron flux as a function of sample age. The hatched area represents the field of optimum irradiation for varying potassium/calcium ratios. In the bottom right-hand quadrant, there is insufficient production of ^{39}Ar to permit accurate analysis. In the top left-hand quadrant, the interference from ^{40}Ar derived from ^{40}K and ^{36}Ar derived from ^{40}Ca becomes excessive.

Tetley *et al.* (1980) showed that the magnitude of the ^{40}K (n,p) ^{40}Ar correction was sensitive to the proportion of thermal as opposed to epithermal and fast neutrons used to irradiate samples. This interference could be minimized by irradiating samples in a cadmium cover (⩾0.5 mm thick). The cadmium cover has the effect of absorbing the thermal neutron spectrum (which promotes unwanted ^{40}K activation) but transmitting the higher-energy neutron spectrum which activates ^{39}K in the sample. Roddick (1983) considered that the chlorine interference could be suppressed in the same way.

(*c*) *Irradiation procedure.* A typical analysis scheme for $^{40}Ar/^{39}Ar$ dating would therefore be as follows. The appro-

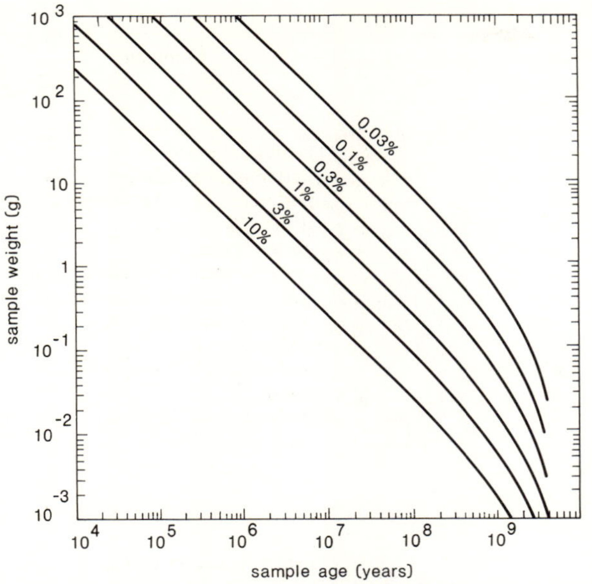

Figure 18.14 ^{40}Ar–^{39}Ar dating technique: sample size required to yield 10^{-6} cm^3 (STP) of radiogenic ^{40}Ar as a function of sample age. Lines are plotted for samples of varying potassium content. (Turner, 1971.)

priate sample weight (see Figure 18.14) is wrapped in high-purity aluminium foil, and sealed in air in a quartz irradiation vial. Several such vials are placed in an irradiation can. During irradiation it is necessary to monitor and correct for both the total integrated neutron flux and any neutron flux gradients along the length of the irradiation can whilst in the reactor. This is normally achieved by packing the unknown sample(s) together with one or more neu-

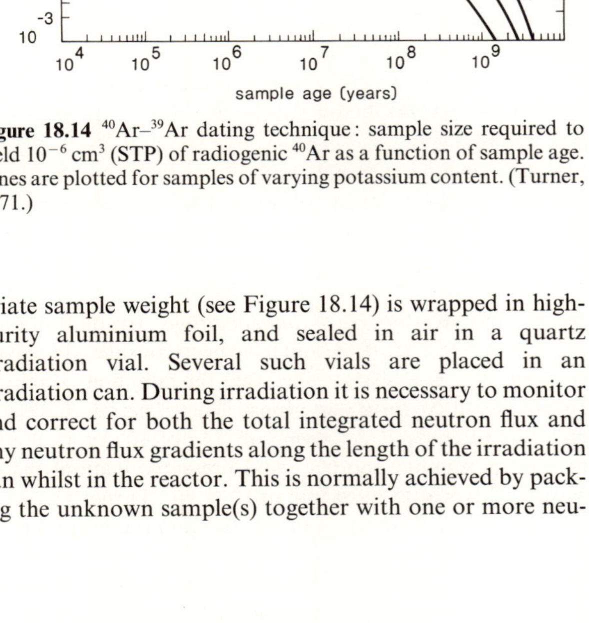

Figure 18.15 $^{40}Ar/^{39}Ar$ dating technique: optimum irradiation conditions. The shaded area shows the integrated neutron flux required to ensure adequate ^{39}Ar production from ^{39}K as a function of sample age. (Turner, 1971.)

Table 18.7 Example of correction ratios in $^{40}Ar/^{39}Ar$ dating

Correction	Magnitude
$(^{36}Ar/^{37}Ar)_{Ca}$	2.2×10^{-4}
$(^{39}Ar/^{37}Ar)_{Ca}$	6.33×10^{-4}
$(^{40}Ar/^{39}Ar)_{K}$	59×10^{-4}

Data from Dalrymple and Lanphere (1971)

tron flux monitors comprising pre-calibrated potassium–argon standards of known age. Roddick (1983) has described techniques for the high-precision intercalibration of $^{40}Ar/^{39}Ar$ standards. Corrections to data for interfering nuclear reactions can be undertaken either by simultaneously irradiating aliquots of K_2SO_4 and CaF_2 (e.g. Ozima *et al.*, 1979) or by using pre-calibrated correction factors appropriate for the specific irradiation position and reactor facility used to activate samples. As an example, correction ratios determined by Dalrymple and Lanphere (1971) for the production of ^{40}Ar (from K_2SO_4) and ^{36}Ar and ^{39}Ar from CaF_2 are listed in Table 18.7.

Samples, neutron monitors and interference correction samples (if appropriate) are then irradiated for the necessary length of time to give a fast neutron dose typically of 10^{17}–10^{19} neutrons cm^{-2} (see Figure 18.15). The irradiation canister can be wrapped in 0.5 mm cadmium foil if the irradiation positon has an unfavourably high thermal neutron component (Tetley *et al.*, 1980). Irradiation positions in the core of the reactor are favoured due to their enhanced fast-neutron spectrum. After removal from the reactor, samples are permitted to decay for several weeks/months to permit the intense short half-life gamma activity to decay.

(*d*) *Argon extraction.* Samples are then transferred to a suitable furnace for argon extraction and mass analysis by stepwise heating. The incremental heating technique, introduced by Merrihue and Turner (1966), offers considerable advantages over the $^{40}Ar/^{39}Ar$ dating method. As the temperature of the sample is progressively increased during the ignition cycle, it is expected that argon will first be released from the more loosely-bound crystal lattice sites. The more resistant minerals are expected to release their argon only at the highest temperatures. Assuming that the sample has remained a closed system since crystallization and contains no extraneous argon, the ratio of radiogenic ^{40}Ar to neutron-induced ^{39}Ar produced from ^{39}K during irradiation will remain constant. The resultant age spectrum obtained from the argon release pattern at each stage in the incremental heating cycle will then be constant. If, however, the sample has suffered disturbance during its history, leading to premature loss of radiogenic ^{40}Ar, the apparent age obtained at each incremental heating stage will vary. In such circumstances the age determined from a total sample fusion (as in conventional potassium–argon dating, for example) will not be meaningful. It may, however, be possible to infer the age as well as some of the geological history of thermal alteration effects from the spectrum of ages obtained during stepwise ignition using the $^{40}Ar/^{39}Ar$ dating method (Dalrymple and Lanphere, 1974).

After outgassing and baking the extraction line, samples

are normally heated for 30–45 minutes per stage at temperatures ranging from 400–500° to 1200–1400°C in steps of 100°C. After each step, released gases are purified and isolated for independent mass analysis.

(*e*) *Mass analysis and interference corrections.* Mass analysis is similar to that described for the conventional K–Ar technique. However, since individual aliquots of argon are normally much smaller than the gas released from a total fusion, procedural blank and atmospheric contamination corrections can be much more significant. Roddick *et al.* (1980) found it necessary to monitor masses 43 and 41 so that the presence of hydrocarbon contamination could be established. These peaks were found to be significant, particularly in low-temperature heating stages (<700°C) and could be removed by freezing out hydrocarbons on charcoal held at freezing methanol temperatures (−94°C) prior to admission of argon into the mass spectrometer. The corrections applied to mass spectrometer peak intensity data by these workers were as follows:

(i) A correction to the zero response of the electrometer due to a memory effect from varying beam intensity measurements

(ii) Linear extrapolation of ^{40}Ar peak height and isotopic ratios of ^{i}Ar/^{39}Ar to the time at which the mass spectrometer was equilibrated with the inlet section

(iii) Tailing effects under ^{39}Ar and ^{37}Ar peaks obtained by measuring zero levels on either side of the ^{36}Ar peak

(iv) Spectrometer mass discrimination effect (usually less than 0.25% per atomic mass unit) determined from a measurement of the known argon isotope ratios in air

(v) Radioactive decay during and since irradiation of ^{39}Ar ($\lambda = 7.05 \times 10^{-6}$ day^{-1}) and ^{37}Ar ($\lambda = 0.01975$ day^{-1}).

(vi) Interference corrections on ^{36}Ar, ^{39}Ar and ^{40}Ar caused by neutron irradiation of Ca and K in the sample

(vii) Correction for atmospheric contamination estimated from the corrected intensity of ^{36}Ar. Algebraic equations to correct the measured ^{40}Ar/^{39}Ar ratio for various interference effects have been considered by Dalrymple and Lanphere (1971). For samples which are sufficiently old and have a high K/Ca ratio, it may be adequate to correct for atmospheric contamination only as follows:

$$F = A - C_1 B. \qquad (18.11)$$

However, it is generally necessary to apply corrections for argon interferences caused by neutron irradiation of K and Ca:

$$F = \frac{A - C_1 B + C_1 C_2 D - C_3}{1 - C_4 D} \qquad (18.12)$$

where F is the ratio (radiogenic ^{40}Ar)/(^{39}Ar derived from ^{39}K), A is the measured ratio ^{40}Ar/^{39}Ar, B is the measured ratio ^{36}Ar/^{39}Ar, $C_1 = 295.5$, the ^{40}Ar/^{36}Ar ratio in the atmosphere, D is the ratio ^{37}Ar/^{39}Ar in the unknown sample after correcting for the decay of ^{37}Ar, C_2, C_3 and C_4 are the K, Ca interference correction factors

$$C_2 = (^{36}\mathrm{Ar}/^{37}\mathrm{Ar})_{\mathrm{Ca}}$$
$$C_3 = (^{40}\mathrm{Ar}/^{39}\mathrm{Ar})_{\mathrm{K}}$$
$$C_4 = (^{39}\mathrm{Ar}/^{37}\mathrm{Ar})_{\mathrm{Ca}}.$$

A summary of the mass measurements which would normally be performed includes some or all of the peaks listed in Table 18.8.

Table 18.8 Mass peaks for argon analysis

Mass number	Isotope	Comments
36	^{36}Ar	Required for atmospheric contamination correction on ^{40}Ar
37	^{37}Ar	Required for Ca, K-produced argon interference correction on ^{36}Ar, ^{39}Ar and ^{40}Ar
39	^{39}Ar	Derived from neutron irradiation of ^{39}K
40	^{40}Ar	Radiogenic argon plus atmospheric contamination
35.5		Background
36.5		Background
41, 43		Hydrocarbon contamination
80	^{80}Kr	Potential interference on ^{40}Ar (^{80}Kr is produced by neutron irradiation of bromine)

(*f*) *Age equation.* The apparent age of the samples can be calculated directed from the corrected ^{40}Ar/^{39}Ar isotope ratio as follows:

$$T = \frac{1}{\lambda} \ln_e(1 + FJ) \qquad (18.13)$$

where T is the age of the unknown sample, λ is the decay constant of ^{40}K (equals 5.543×10^{-10} year^{-1}), F is the corrected ^{40}Ar/^{39}Ar ratio (see above) for radiogenic ^{40}Ar and potassium-derived ^{39}Ar and J is a function of both the age of the monitor sample (T_{MON}) and the integrated fast neutron flux and is given by

$$J = \frac{\exp(\lambda T_{\mathrm{MON}}) - 1}{(^{40}\mathrm{Ar}_{\mathrm{rad}}/^{39}\mathrm{Ar}_{\mathrm{K}})_{\mathrm{MON}}}. \qquad (18.15)$$

For further details see Dalrymple and Lanphere (1971).

The potassium content of the uinknown sample may be found from

$$[\mathrm{K}] = \frac{\lambda\, ^{39}\mathrm{Ar}_{\mathrm{K}}}{\lambda_e J} \qquad (18.15)$$

where λ is the total decay constant of ^{40}K $= 5.543 \times 10^{-10}$ year^{-1} and λ_e is the partial decay constant of ^{40}K by electron capture to ^{40}Ar $= 0.581 \times 10^{-10}$ year^{-1}.

For further details see Kyser and Rison (1982).

The age spectrum obtained by an incremental heating experiment on a biotite reference material LP-6 is shown in Figure 18.16. This diagram shows both the proportion of argon released and the apparent age of the sample derived from individual gas fractions collected at the specified extraction temperature between 600°C and complete sample fusion. These data show that the largest proportion of argon is released above 1100°C and that the apparent age of this sample remains approximately constant for argon fractions extracted above 900°C.

The corresponding isochron is also shown. The axes of this isochron are $^{40}\mathrm{Ar}_{\mathrm{rad+atm}}/^{36}\mathrm{Ar}_{\mathrm{atm}}$ *v.* $^{39}\mathrm{Ar}_{\mathrm{K}}/^{36}\mathrm{Ar}_{\mathrm{atm}}$, where $^{36}\mathrm{Ar}_{\mathrm{atm}}$ represents the atmospheric and extraneous argon composition of the sample: $^{39}\mathrm{Ar}_{\mathrm{K}}$ is argon derived from

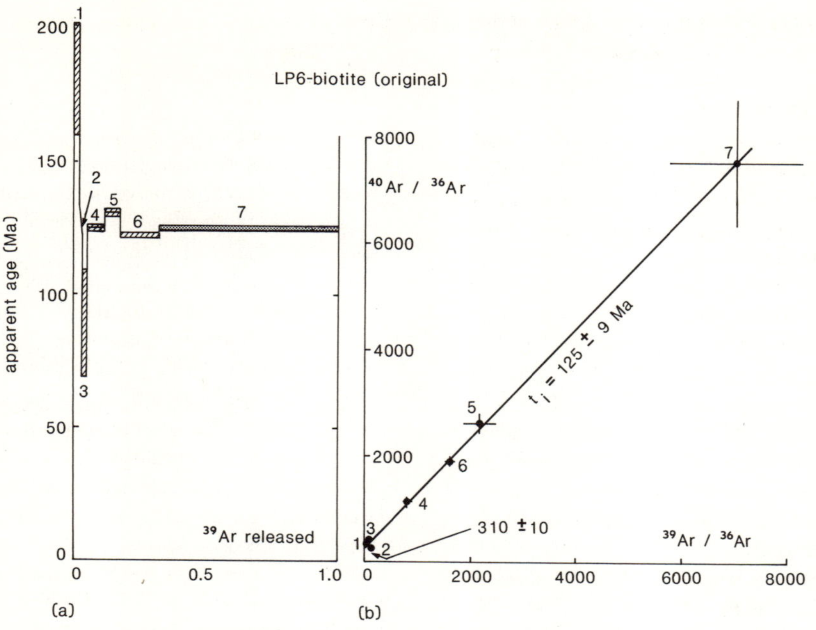

Figure 18.16 (*a*) Apparent age spectrum of LP-6 biotite plotted as a function of the fraction of ^{39}Ar released. The shaded area represents one-sigma errors. (*b*) Corresponding isochron; error bars correspond to one-sigma uncertainty. Data was acquired during ^{40}Ar–^{39}Ar dating using a stepped combustion procedure corresponding to the following extraction temperatures: (1) 600°, (2) 700°, (3) 800°, (4) 900°, (5) 1000°C, (6) 1100°C, (7) sample fusion. (Ozima *et al.*, 1979.)

neutron irradiation of ^{39}K, and $^{40}Ar_{rad+atm}$ is radiogenic ^{40}Ar plus atmospheric and extraneous components found in the sample. Use of these coordinates permits isochrons to be plotted without correcting for or making assumptions about the composition of atmospheric argon (Dalrymple and Lanphere, 1974). The age spectrum in Figure 18.16 is considered to be reasonably flat. Any major alteration of the sample which led to premature argon loss would be characterized by a non-linear age spectrum, data from each subsequent temperature increment forming a 'staircase' age pattern. Ozima *et al.* (1979) have demonstrated that the form of the age spectrum can be significantly disturbed by heating and pressure effects on the sample. Roddick (1978) has expressed caution about interpreting argon isochron data unless adequate corrections for atmospheric blanks are made.

It is estimated that the overall precision in ^{40}Ar/^{39}Ar measurements is about 0.2% (two sigma) for ^{39}Ar beams greater than 0.8×10^{-9} cm^3 (STP) and that the concentration of radiogenic argon can be determined to a precision of 1.5% (two sigma) for a single analysis (Roddick *et al.*, 1980). Dalrymple and Lanphere (1971) estimate that if the ratio K/Ca > 1, samples as young as about 10^5 years can be dated. Blank levels (including atmospheric contamination) have been compared by Roddick (1978) and are generally about $0.1\text{–}0.3 \times 10^{-10}$ cm^3 (STP) ^{36}Ar and $0.2\text{–}1 \times 10^{-8}$ cm^3 (STP) ^{40}Ar.

19 Spark source mass spectrometry

19.1 Introduction

Spark source mass spectrometry is one of the most sensitive techniques available for routine trace-element analysis of geological samples. Determinations down to the parts per billion level are possible with virtually no preparation of the sample other than crushing it to a fine particle size and compressing it with graphite to make a suitable electrode. Simultaneous determinations can be made typically on 30 trace elements with a potential throughput rate of about two samples per day. However, in spite of this impressive analytical specification, spark source mass spectrometers are operated in only a relatively small number of geochemical laboratories in the world. Reasons for this apparent unpopularity include the high capital cost of equipment and comparatively low sample throughput rate compared with other modern instrumental trace-element techniques (e.g. XRF, ICP–AES). A further potential drawback is that the very low detection limits cannot always be realized due to interference effects in the mass spectra of some samples. During analysis, a very small mass of sample powder (probably less than a milligram) is consumed. It is essential, therefore, that this should be fully representative of the bulk, and reservations must be expressed about whether this can always be achieved satisfactorily for some elements including zirconium and hafnium, for example, since these elements often occur in discrete zircon grains.

In spite of these reservations, spark source mass spectrometry has been mastered by a few laboratories where the technique has been shown to give high precision in trace-element analysis (typically 2–5%, one sigma). Applications of spark source mass spectrometry in geochemical analysis have been extensively developed by Taylor and co-workers, and it is appropriate here to acknowledge their contribution in refining this technique:

General analytical methodology (Taylor, 1965*b*)

Interpretation and processing of photoplate data (Taylor, 1971)

Element sensitivity, precision and accuracy (Taylor and Gorton, 1977)

Application to REE analysis (Taylor, 1979)

Analysis of tin and refined element sensitivity factors (Taylor and McLennan, 1983).

19.2. Instrumentation and ion production

The design principles of the spark source mass spectrometer are dictated by the characteristics of the ion source. The sample is ionized by passing a high-voltage radiofrequency spark across two electrodes, one of which consists of a briquette of sample mixed with graphite powder. This excitation process imparts a considerable range of kinetic energies to ions formed in the spark, up to about 1000 eV compared with about 0.2 eV for ions formed in a thermal ionization source. With such a wide range of energies, adequate resolution in the dispersed mass spectrum could not be achieved with a single magnetic sector mass analyser. Commercial instruments, therefore, use the Mattauch–Herzog geometry discussed in more detail in section 16.6. This design uses an electrostatic sector to select a mono-energetic beam of ions which are then dispersed by an electromagnetic analyser (Figure 16.10). The geometry is designed in such a way that the entire spectrum is focused on a flat plane suitable for detection using a photographic plate. The photographic plate is not an ideal detection device. It has a limited dynamic range for detection of only about 50 times instead of the 2000 times range desirable to cover the variation in trace-element concentrations found in silicate rocks. Multiple exposures of the plate at a range of ion beam currents are therefore necessary to complete a trace analysis of a sample. However, the photographic plate has one overall advantage in this application. It is an integrating detector, well suited to the detection of ion beams which may suffer transient variations in beam current of between 10^{-9} to 10^{-12} amps due to random fluctuations in ion emission from the spark source. Instruments fitted with such detectors are known as mass spectrographs rather than spectrometers in deference to the visual nature of the output signal.

19.2.1 *The spark source*

The radiofrequency electrical discharge is created between two electrodes, one of which contains the sample. Since silicate materials do not conduct, it is necessary to mix the powdered sample with graphite. Taylor (1965) prepared electrodes from equal weights of sample powder and graphite. 200 mg of the mixture (which also contained the internal standard, see below) was compressed at a pressure of 10 000 lb in^{-2} to form an electrode $20 \times 2 \times 2$ mm in a steel moulding die. Any surface contamination introduced from the steel die was removed by clamping the electrode in a wooden holder and trimming it into a circular section on a lathe. The rod was then broken in half to expose a fresh surface suitable for analysis. In view of the very small mass of sample excited in the spark source, considerable care must be taken in grinding and homogenization to ensure that the fraction ionized is representative of the bulk.

The electrodes are mounted in the source region of the instrument which is pumped down to a vacuum of about 10^{-7} Torr. The sample is excited by striking an electrical discharge between the electrodes. Two forms of discharge have been used: the radiofrequency spark discharge mentioned above and the triggered low-voltage dc arc discharge. Each has slightly different analytical characteristics (see below) and the the former appears to be more widely used in geological applications. The radiofrequency spark discharge uses a potential of between 20–200 kV oscillating at about 1 MHz. In order to limit heating effects in the electrodes, dissipated power may be controlled by adjusting the length of discharge pulses between 10–200 μs at a repetition frequency of 1–10^4 s^{-1}. The conditions selected by Taylor (1965) were as follows: spark potential, 25 kV; pulse length, 200 μs; repetition rate, 1000 cycles per second. Conversely, in the determination of ultratrace amounts of thorium in two reference materials after ion exchange separation, Victor and Rogers (1983) used a low-voltage arc discharge of about 5 A triggered for a duration of 140 μs every 33 ms. Parts of the instrument exposed to the ion beam are normally made of tantalum. This metal exists predominantly in one isotopic form ($^{181}Ta = 99.988\%$) so that contamination of the ion beam from the sample holder, for example, causes minimal interferences in the mass spectrum.

19.2.2 *Mechanism of ion production*

Ion production in a radiofrequency spark discharge has been reviewed by Adams (1982). The total energy dissipated in the discharge is extremely high, typically between 10^9–10^{22} W cm^{-2}, comparable, in fact, to that observed in a pulsed laser discharge (section 11.3). Adams, summarizing other published work, especially that of Ramendik (1980), suggests that a rapid series of mechanisms occurs during the spark discharge.

(i) *Atomization*: A weakly ionized plasma is formed near the surface of the electrode with a composition closely corresponding to that of the sample. The energy spread of atoms in this plasma is low and consumes only about 1% of the spark discharge energy.

(ii) *Ionization*: Nearly all the remaining energy is expended in ionizing atoms in this plasma, giving an average charge of 4–5 and a kinetic energy of 200–600 eV.

(iii) *Plasma expansion*: Following ionization, the plasma expands. During this process, intense recombination of ions occurs in the plasma gas. Most recombinations result in the formation of neutral species; only about 1% remain as ions with a charge of +1 or +2. The kinetic energy spread of these remaining ions increases to 2–3 keV.

The mechanisms and characteristics of a triggered dc arc are different from those of the spark outlined above. Although the dc arc gives better reproducibility, it is normally used with samples having good electrical conductivity. The ionization efficiency is lower and the proportion of more highly ionized species greater than the radiofrequency spark.

19.2.3 *Interferences in the spark source*

Under ideal conditions, the ion source would yield only singly-charged ions. However, spark source ionization is not a well-controlled process and several other species are formed that are capable of interfering with analysed masses. The nature and proportion of these species depends on the conditions selected for the electrical discharge. Some typical examples of mass overlap interferences are shown in Table 19.1. In general they can be categorized as follows.

Table 19.1 Interferences in the analysis of silicate rocks and minerals by spark source mass spectrometry

Element	Mass number	Isotopic abundance (%)	Interference
P	31	100	$^{62}Ni^{2+}$ in ultramafic rocks
S	32	95.0	$^{64}Zn^{2+}$, $^{64}Ni^{2+}$, $^{16}O_2^+$ in ultramafic rocks
Sc	45	100	$^{90}Zr^{2+}$, $^{135}Ba^{3+}$: granites and syenites
V	51	99.76	$^{204}Pb^{4+}$: granites and syenites
Cr	53	9.55	$^{159}Tb^{3+}$: granites and syenites
Co	59	100	$^{118}Sn^{2+}$: granites and syenites
Ga	69	60.4	$^{138}Ba^{2+}$
Ge	74	36.54	$^{148}Nd^{2+}$: syenites and granites with Nd > 100 ppm
Sr	88	82.56	$^{176}Lu^{2+}$: internal standard
Zr	91	11.23	$7\,^{13}C^+$, $CaAlC_2^+$ in plagioclase-rich rocks
Ag	107	51.82	$Si_2AlC_2^+$, $CaAlOC_2^+$
In	115	95.72	$CaAlC_2^+$ in plagioclase-rich rocks
Hf	179	13.75	$Si_2AlC_8^+$, $CaAlOC_8^+$
Pt	195	33.8	$15\,^{13}C^+$
Tl	203	29.5	$Si_2AlC_{10}^+$, $CaAlOC_{10}^+$
Pb	208	52.3	$16\,^{13}C^+$

Data from Taylor (1965*b*) and Taylor and Gorton (1977).

(i) *Multiply-charged ions that are formed in the high-energy discharge*. These are mainly 2^+ species in a spark (rf) discharge, but 3^+ species also exist in an arc (dc) discharge. Interferences from 2^+ species cannot occur on singly-charged odd-numbered isotope masses, making these more favourable for analysis.

(ii) *Molecular clusters from the graphite electrodes*. A whole series of carbon cluster ions is formed in the electrical discharge: $2C^+$, $3C^+$, $4C^+$, ... $16C^+$, etc. Those corresponding to clusters of ^{12}C atoms are sufficiently intense to make the corresponding mass number unusable for analysis. Providing a sufficiently high vacuum is maintained in the magnetic analyser ($<10^{-8}$ Torr), hydrocarbon interferences appear to be negligible (Taylor and Gorton, 1977).

(iii) *Other molecular fragments*. A whole range of molecular ions composed of fragments of the sample can also be observed in mass spectra. Taylor and Gorton (1977) report that plagioclase-rich samples (lunar highland samples) yield fragments of composition $CaAlC_{2n}^+$ which interfere with the determination of ^{91}Zr ($n = 1$), ^{115}In ($n = 2$), ^{139}La ($n = 3$) and ^{163}Dy ($n = 4$). These and other examples are listed in Table 19.1.

(iv) *Oxide and carbide formation*. A relatively small fraction of oxide and carbide ions is formed in the electrical discharge. However, correction for their presence can be important in some circumstances, particularly in the determination of Ba and the rare-earth elements (Taylor and Gorton, 1977). The proportion of these interferences as a percentage of the REE ion abundance is shown in Table 19.2, using data taken from Taylor and Gorton (1977). These data indicate that interferences from $(REE)O^+$ and $(REE)C^+$ are most significant. As an example, the interference on $^{159}Tb^+$ from $^{143}NdO^+$ and $^{135}BaC_2^+$ may be cor-

U

Table 19.2 Oxide and carbide interference corrections in the analysis of rare-earth elements and barium by spark source mass spectrometry

Interfering species	Average interference correction
$(REE)O^+$	1.5
$(REE)O_2^+$	0.02
$(REE)C_2^+$	0.6
BaO^+	0.05
BaO_2^+	0.002
BaC_2^+	0.2
BaC_3^+	0.05

Data from Taylor and Gorton (1977).
Interference correction is the proportion of the interfering ion expressed as % of the parent ion intensity (i.e. $(REE)^+$).

rected using data from Table 19.2 as follows:

$$({}^{159}Tb)_{correct} = ({}^{159}Tb)_{measured} - ({}^{143}Nd) \times \frac{1.5}{100} - ({}^{135}Ba)\frac{0.2}{100} \quad (19.1)$$

where the intensity of the indicated mass peak is shown in parentheses.

19.3 Internal standardization

If one or two elements are to be determined, it is possible to use isotope dilution techniques to standardize analyses (providing these elements are not mono-isotopic, section 17.6). However, this becomes impractical if 30 or so trace elements are to be determined simultaneously. In such circumstances, internal standardization procedures are used. An internal standard is an element added to the sample at a known concentration so that variations in analytical response can be corrected. All measured data, including those from the calibration run, are normalized against the intensity response of the internal standard during that particular run. Internal standardization is essential in spark source mass spectrometry because the spark conditions cannot be reproduced precisely from run to run (or sample to sample). Furthermore, the efficiency of detection can vary by a factor of 3 between different photoplates (Taylor, 1965). The criteria for selecting an element suitable for use as an internal standard in multi-element programs have been considered in detail by Taylor (1965), and may be summarized as follows:

(i) The internal standard should possess similar ionization characteristics in the electrical discharge to the elements sought and, in particular, have similar volatility and ionization potential

(ii) This element must be added to the sample in the same chemical form (i.e. as an oxide) as the elements sought

(iii) The internal standard should possess at least two isotopes with a natural abundance ratio of about 100 so that calibration of photoplates can be carried out over widely varying exposure settings

(iv) Isotopes of the internal standard should have odd atomic weights so that interferences from M^{2+} ions of the internal standard will be detected at fractional mass numbers, so avoiding isobaric interferences on trace elements; by analogous arguments, the mass of the internal standard should exceed 120 to avoid interferences of $2M^+$ ions on analysable elements

(v) The internal standard should not be an element of interest, and should be present at low natural abundances so that about 10 ppm will be sufficient both to swamp the occurrence of the element in a sample and to ensure adequate homogeneity on mixing.

Taking these factors into consideration, Taylor (1965) selected lutetium as the most appropriate internal standard element and added it at a concentration of 50 ppm to the graphite used for electrode making. Lutetium has two isotopes: ${}^{175}Lu$ (97.41% natural abundance) and ${}^{176}Lu$ (2.59%). The isotope ${}^{175}Lu$ was sufficient to swamp the signal due to natural lutetium in almost all samples, and insufficiently intense to normalize data in the shorter exposures. However, the weaker signal due to the isotope ${}^{176}Lu$ required correction for interferences from ${}^{176}Hf$, ${}^{176}Yb$ and natural lutetium as follows:

$$({}^{176}Lu)_{correct} = ({}^{176}Lu)_{measured} - 0.89({}^{171}Yb) - 0.19({}^{178}Hf) - 0.026({}^{169}Tm) \quad (19.2)$$

where parentheses again represent the intensities of the specified mass peaks. The (${}^{169}Tm$) term is used to correct for the presence of natural ${}^{176}Lu$ since thulium is commonly present in rocks at about the same concentration as lutetium. Subsequently, Taylor and Gorton (1977) recommended correction for natural ${}^{176}Lu$ on the assumption that the chondrite-normalized abundances of Yb and Lu are equal, since interferences on ${}^{169}Tm$ itself can cause inconsistencies in the correction outlined in equation (19.2).

19.4. Routine data acquisition

After the sample electrode has been mounted in the source holder and evacuated to operating pressure, the sample is ionized under the appropriate electrical discharge conditions. The transmitted ion beam is detected on a photoplate. However, a photoplate has a rather limited dynamic range for intensity measurements (times 50 compared with the intensity range of trace-element isotopes of about 2000 in silicate rocks). It is necessary, therefore, to carry out multiple exposures of the plate for varying integrated ion beam intensity. The total exposure is measured as the total charge of ions (in coulombs) reaching the plate. This charge is measured by intercepting about 50% of the ion beam on a monitor collector plate situated between electrostatic and magnetic sectors. The total current integrated as a function of time leaking to earth through this plate gives a measure of the total charge reaching the photoplate. In a multi-element scheme of analysis, it is necessary to record up to 15 individual exposures ranging in nominal charge from 1×10^{-12} coulombs to 3×10^{-6} coulombs (Taylor, 1965). The longest of these exposures takes over one hour. Exposures are recorded on 2×10 inch Ilford Q-2 plates, and 10 exposures can be recorded on each plate. The plate is removed from the instrument after exposure and developed photographically under standard conditions. The intensities of lines in mass spectra are then measured with a scanning microphotometer. Data to be analysed is therefore digitized from the microphotometer output as the photoplate exposure density at specified mass lines.

The principal drawback in this technique is that the total charge measured during a run is not a sufficiently accurate or reproducible measure of the actual charge reaching the photoplate. Taylor (1971) reports that differences of 100% between apparent and actual exposures are not uncommon. It is not, therefore, possible to compare directly data recorded in one photoplate exposure with those recorded in another: hence the role of the internal standard. Individual exposures must be internally quantified relative to the internal standard, as described in the next section. Normalized data can then be compared with pre-recorded calibration data to convert measured intensities into concentrations.

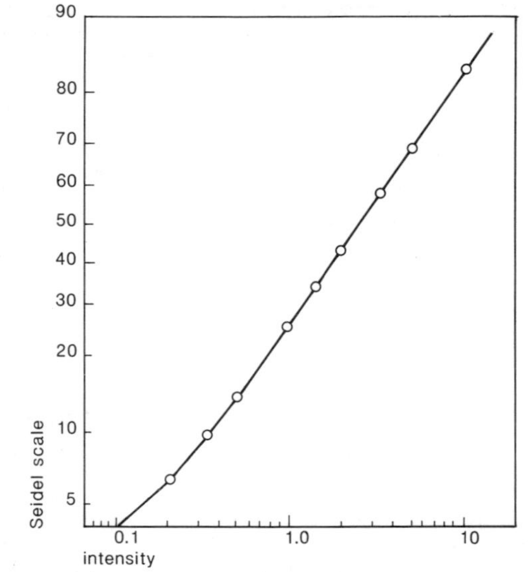

Figure 19.1 Photoplate intensity calibration curve. The intensity of the incident ion beam is plotted as a function of measured photoplate line transmittance (d). The latter has been recalculated as the reciprocal Seidel function $1/[(d_0/d)-1]$ (see text). (Taylor, 1971.)

19.5 Photoplate calibration and element sensitivities

Precise calibration of the photoplate response curve is necessary before observed line densities can be converted to ion beam intensities. This calibration must take into account two factors.

(a) *Seidel function.* The transmittance response of the photoplate is not a linear function of the intensity of the incident ion beam. The useful linear response range of the photoplate can be extended by measuring line density in terms of the empirical Seidel function:

$$\text{Seidel function} = \log\left[\frac{d_0}{d} - 1\right] \qquad (19.3)$$

where d_0 is the microphotometer galvanometer deflection on a clear region of the plate, normally selected at high mass number (d_0 is normally set to 100) and d is the galvanometer deflection on the line of interest (see Ahrens and Taylor, 1961).

A density v. ion beam intensity plot of an Ilford Q-2 photoplate is shown in Figure 19.1. Photoplate density is here plotted as the reciprocal of the Seidel function. The linear portion of this calibration line extends over a Seidel scale range of about 15 to 85 in this example.

(b) *Photoplate gamma.* The sensitivity of the photoplate transmittance v. ion beam intensity response curve is not identical from photoplate to photoplate: that is, the slope of the photoplate calibration curve (Figure 19.1) will vary from plate to plate within a batch. Photoplate sensitivity is measured by tan θ, where θ is the angle between the linear portion of the photoplate calibration plot and the abscissa (x axis). Tan θ is also known as the photoplate 'gamma'. It is not possible to calibrate a photoplate gamma from a set of mass spectra recorded over a range of exposure conditions. Difficulties in calibrating ion beam intensities precisely would introduce large errors into such data. Photoplate calibration measurements must, therefore, be made on elements having two or more isotopes of known relative intensity from data recorded in a single plate exposure.

Taylor and Gorton (1977) precisely measured the photoplate gamma from the relative intensity of isotopes of four elements, as follows: ^{177}Hf, ^{178}Hf and ^{179}Hf; ^{171}Yb, ^{172}Yb, ^{173}Yb and ^{174}Yb; ^{143}Nd and ^{146}Nd; and ^{134}Ba, ^{135}Ba and ^{137}Ba. Mass spectra were recorded over a range of exposures and intensity data from each isotope from individual exposures was plotted as a function of absolute isotopic abundance. Such data for the isotopes of Nd and Yb are plotted in Figure 19.2. From each set of isotope data, the value of gamma (tan θ) can be determined and a plate average calculated. If necessary, this technique can also be used to

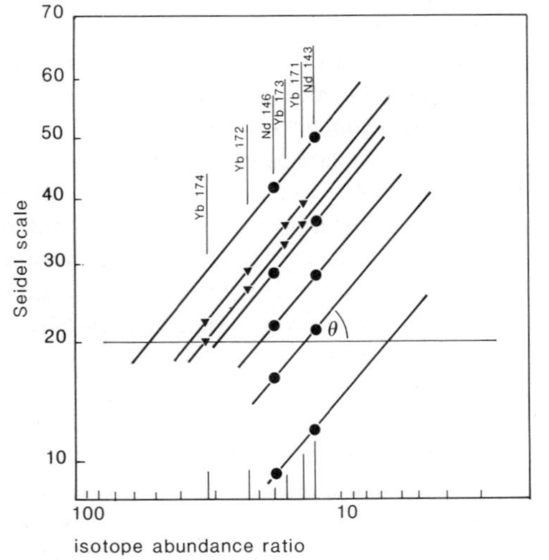

Figure 19.2 Determination of photoplate sensitivity factor or 'gamma'. Ion beam transmittance data for isotopes of Yb and Nd have been taken from several exposures of the same photoplate and their measured line intensities (recalculated to a Seidel scale) have been plotted as a function of isotope abundance ratio. The mean gradient of these calibration lines gives the photoplate gamma (tan θ). (Taylor, 1971.)

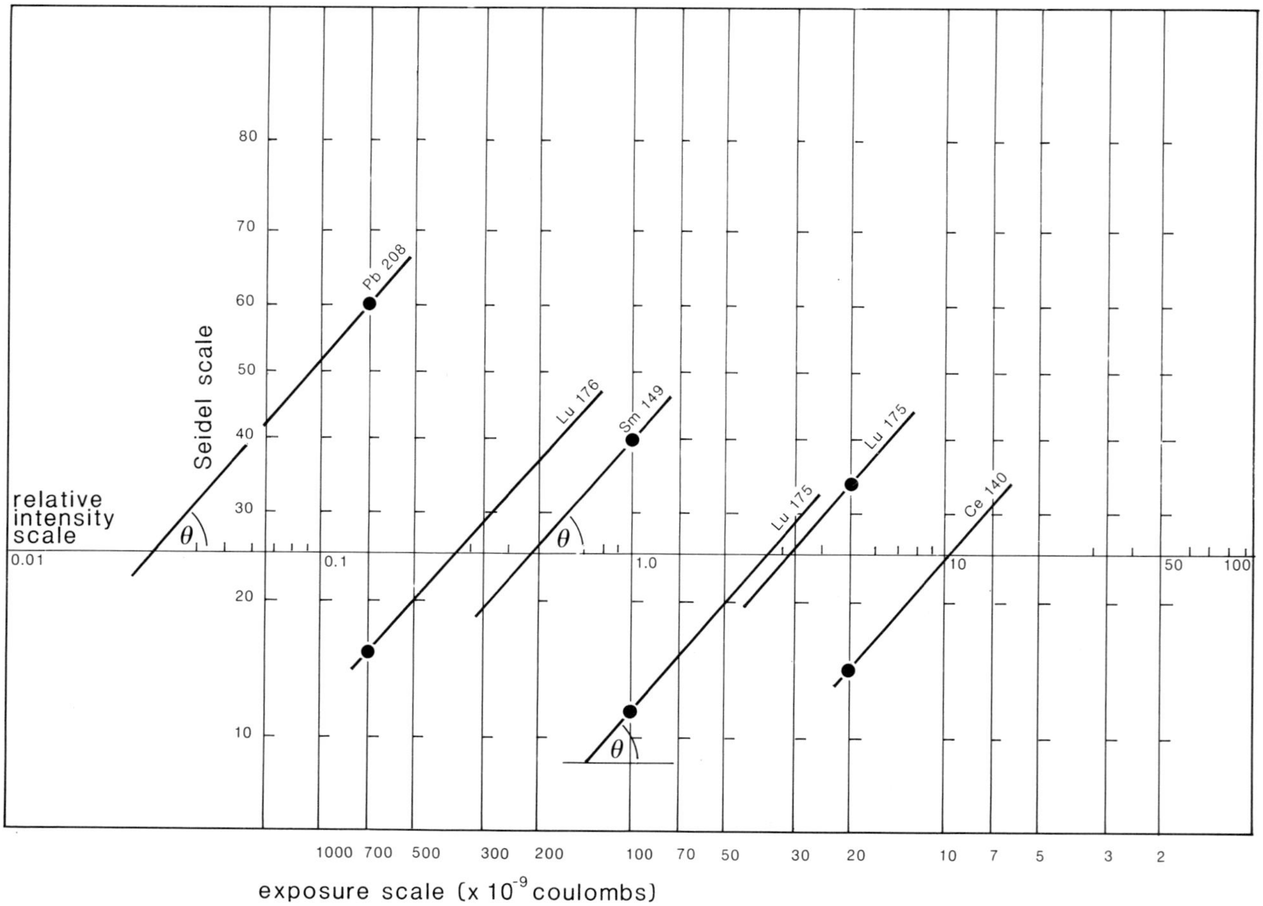

Figure 19.3 Graphical method of measuring line intensities relative to the lutetium internal standard using data from a single photoplate exposure. Measured line transmittance data are used together with the photoplate gamma (tan θ) to construct an intensity calibration line for each isotope. The intercept of this calibration line with a Seidel scale value of 25 gives the relative intensity. This value is ratioed against the internal standard intensity of ^{175}Lu or ^{176}Lu extrapolated in the same way from data recorded in the same plate exposure. Photoplate density (the vertical axis) is plotted as the reciprocal Seidel function $1/[(d_0/d)-1]$. (Taylor, 1971.)

calibrate the small variations in plate sensitivity with ion mass, although this refinement is not usually necessary.

(*c*) *Calculation of isotope ion beam intensities.* After calibration of the photoplate gamma, the ion beam intensity of any mass line can be calculated from its plate density relative to the corresponding density of an internal standard (lutetium) mass line, providing both measurements are made on the linear part of the plate intensity response curve. Taylor and Gorton (1977) suggest that, in their experience, plates with high gammas (>1.2) are readable over a range of 10–85% transmission, whereas those with low gammas (>0.9) are readable over a range of 25–90% transmission.

The data presented above describe the general principles of photoplate calibration. The actual procedure for calculating intensity ratios used by Taylor (1971) is shown in Figure 19.3. From a given plate exposure the line transmittance data d for each isotope line are recalculated as its reciprocal Seidel function. Data for the appropriate lutetium internal standard line (^{175}Lu for short exposures, ^{176}Lu for long exposures) is then used in conjunction with the pre-calibrated photoplate gamma (tan θ) to construct the appropriate density *v.* ion beam intensity calibration line for that exposure. The intercept of this line with a Seidel scale value of 25 is then read off (or calculated) as the lutetium ion beam intensity. The value of 25 was selected as the mid-point of the most reliable set of densiometer scale readings. Exactly the same procedure is then followed for each isotope to be analysed. Each time, the measured transmittance photoplate density is plotted as the reciprocal Seidel function and the photoplate gamma used to extrapolate the ion beam intensity at a Seidel scale value of 25. The intensity of ^{176}Lu is corrected for interferences as outlined above, and the ratio of the analyte ion beam intensity to that of the lutetium isotope calculated. Data in Figure 19.3 show the calibration of ^{140}Ce, ^{149}Sm and ^{208}Pb relative to the appropriate lutetium isotopes from three separate photoplate exposures corresponding to nominal ion beam exposures of 20, 100 and 700×10^{-9} coulombs.

The result of this plate calibration procedure is a set of ion beam intensities of analyte mass lines, each normalized to a lutetium isotope beam intensity recorded in the same exposure. If it is necessary to record 15 exposures at varying ion beam charges, density data for individual isotopes will fall within the linear part of the calibration curve in several

Table 19.3 Ion beam intensity for selected trace elements determined from successive plate exposures by spark source mass spectrometry

Isotope line	Exposure (coulombs × 10^{-9}) 500	300	200	100	70	30	20	10	7
U-238	1.28	1.03	1.15	1.02	0.99				
Th-232	3.90	3.43	3.51	2.82	2.83	3.30	2.88	2.59	
Eu-153	2.75	2.66	2.50	2.19	2.02				
Eu-151	2.75	2.32	2.38	2.01	2.29	2.78			
La-139					59.3	59.6	65.0	61.2	59.7

All data normalized to the appropriate lutetium internal standard ion beam intensity.
Data from Taylor (1971).

successive exposures. Intensity ratio data from individual isotopes can then be examined and outliers recognized and if appropriate discarded from the average. The consistency which can be achieved using these procedures can be judged from data presented in Table 19.3, which lists normalized intensity data for selected trace elements taken from the work of Taylor (1965).

(*d*) *Concentration calibration.* Concentration calibration curves can be set up either by analysing a set of prepared standards (quartz plus a weighed amount of trace element oxide (Taylor, 1965) or analysing a set of reliable international reference materials (Taylor, 1971). In either case the expected concentration of the element is plotted against the ion beam intensity of the element (corrected for interferences if necessary) normalized to that of the lutetium internal standard. A calibration curve for thorium obtained in this way is shown in Figure 19.4.

(*f*) *Element sensitivities.* The relative sensitivity of elements analysed by spark source mass spectrometry has been investigated by Taylor and Gorton (1977) and Taylor and McLennan (1983). These workers suggested that sensitivities

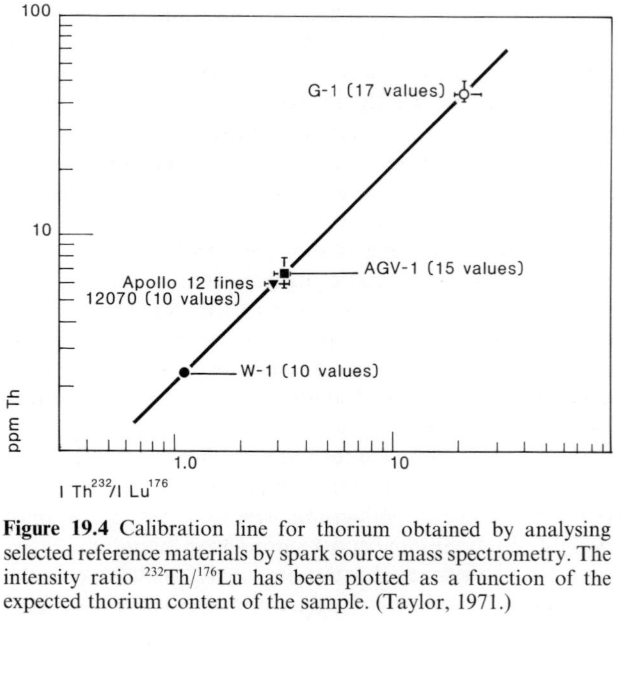

Figure 19.4 Calibration line for thorium obtained by analysing selected reference materials by spark source mass spectrometry. The intensity ratio ^{232}Th/^{176}Lu has been plotted as a function of the expected thorium content of the sample. (Taylor, 1971.)

Table 19.4 Excitation and sensitivity for trace elements determined by spark source mass spectrometry

Element	Ionization potential (eV)	Boiling point (K)	Relative sensitivity
Rb	4.18	961	0.089
Cs	3.89	951	0.149
Sr	5.70	1657	0.217
Ba	5.21	1913	0.358
Eu	5.67	1870	0.454
Y	6.38	3610	0.484
Sm	5.63	2051	0.496
Zr	6.84	4650	0.533
Nd	5.49	3341	0.548
Pr	5.42	3785	0.549
La	5.58	3730	0.612
Nb	6.88	5015	0.617
Ce	5.47	3699	0.619
Dy	5.93	2835	0.627
Tm	6.18	2060	0.695
Yb	6.25	1466	0.724
Tb	5.85	3314	0.734
Ho	6.02	2968	0.749
Gd	6.14	3539	0.753
Tl	6.11	1730	0.814
Er	6.10	3136	0.829
Lu	6.15	3668	1.00
Ru	7.37	4173	1.087
Pb	7.42	2013	1.230
Hf	7.0	4875	1.305
U	6.08	4091	1.646
Th	6.95	5063	2.203
Re	7.88	5900	2.532
Os	8.7	5300	3.198
Pt	9.0	4100	3.678
Ir	9.1	4403	3.720

Data from Taylor and McLennan (1983).
Sensitivity is relative to Lu = 1.00.

correlated with element ionization potential (as a measure of the ease of ion formation) and boiling point (as a measure of element volatility) (see Table 19.4). A linear relationship was demonstrated between element sensitivity, expressed as log (atomic weight/relative sensitivity factor) versus an empirical factor calculated as $\log[(\sqrt{\text{boiling point}}) \times (\text{ionization potential})^2]$ (Figure 19.5). These data appear to demonstrate that ease of ion formation and volatility are the two factors which control element sensitivities. This conclusion has recently been re-examined by Verlinden *et al.* (1985) who suggest that mechanisms in the thin surface layer of the electrode that is excited by the spark discharge control, to some extent, reactions in the spark gap.

19.6 Applications and results

The methodologies described above have largely followed the procedures of Taylor and co-workers using a radiofrequency arc discharge and lutetium as internal standard. However other methodologies have been described. In a pioneering paper, Nicholls *et al.* (1967) analysed 35 trace elements in silicate materials using similar instrumentation and radiofrequency spark excitation as described above. Rhenium was selected as internal standard and considerable

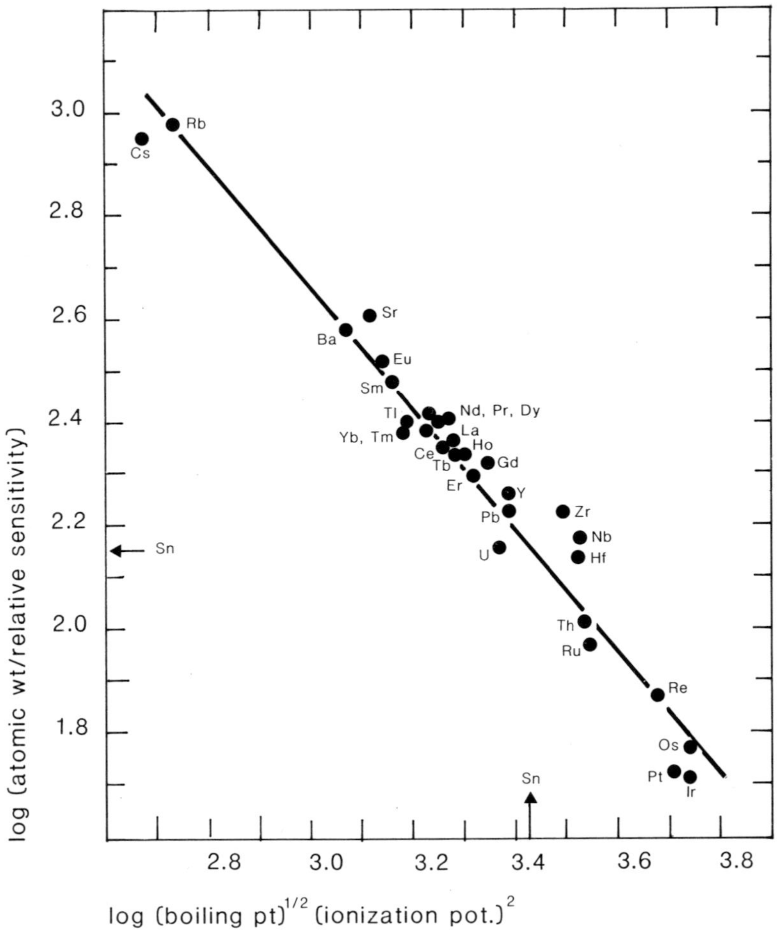

Figure 19.5 The relationship between element sensitivity, atomic weight, element boiling point and the first ionization potential for elements determined by spark source mass spectrography. Data determined by Taylor and McLennan (1983) using a pulsed 25 kV RF spark and powdered silicate samples. Element sensitivity increases up the *y*-axis.

attention was paid to homogenizing this with the sample prior to analysis. Rhenium was first fused with a silicate melt of low melting point in the fayalite–nepheline–alkali feldspar phase system. Aliquots of this mixture were then fused with the sample to ensure the highest degree of homogeneity in the distribution of the rhenium spike. Data for the analysis of USGS: W-1 (diabase) by this technique are shown in Table 19.5. The use of rhenium as an internal standard has subsequently been questioned, since its behaviour in the electrical discharge may not be fully representative of that of most trace elements analysed. For comparison, analyses of USGS: W-1 Taylor (1965) are also listed in Table 19.5.

Taylor and Gorton (1977) examined and refined the precision and accuracy of the technique, and Taylor (1979) investigated specifically the determination of rare-earth elements by spark source mass spectrometry. Particular attention was paid to the correction of interferences on elements (including the lutetium internal standard, as described above) and precise calibration of the photoplate-density function. The interferences which must be taken into account in rare-earth element analysis are listed in Table 19.6. The improvements in accuracy compared with earlier spark source mass analyses can be judged from the data for USGS: BCR-1 (basalt) in Table 19.5 which is listed together with additional spark source data from Roaldset (1975).

A different approach to spark source mass analysis was followed by Strelow and Jackson (1974) and Jackson and Strelow (1975). These workers determined rare-earth elements using an ion exchange separation procedure. The rare-earth element residue was dissolved in 2 mL of 1 M HNO_3 in 60% acetone, mixed with 0.2 g graphite and evaporated to dryness. After compression into electrodes, the sample was excited using a low-voltage discharge at a current of 2.5 A and repetition rate of 100 Hz. Under these excitation conditions, ionization and interference effects were quite different from those observed in the radiofrequency spark discharge described above. The doubly-charged rare-earth element spectrum was of almost equal intensity to that of singly-charged species, and could, therefore, be used for analysis. Interfering molecular species that are significant in radiofrequency discharges, including rare-earth oxide, BaO^+ and carbide species, were sought in the dc discharge but not found. Strelow and Jackson used tin as an internal standard and estimated detection limits of 0.00005 ppm for all rare-earth elements by this technique with blank levels of

Table 19.5 Analyses of rock reference materials by spark source mass spectrometry

	USGS: W-1 diabase			USGS: BCR-1 basalt			USGS: PCC-1 peridotite	
	Expected	Analysed		Expected	Analysed		Expected	Analysed
	Abbey (1983)	Nicholls *et al.* (1967)	Taylor (1965) (mean of 2)	Abbey (1983)	Taylor and Gorton (1977) (mean of 7)	Roaldset (1975)	Abbey (1983)	Strelow and Jackson (1974)
Sc	35?	43						
Cr	115?	98						
V	260?	260						
Co	47?	42						
Ni	76?	76						
Cu	110	110						
Zn	86	75						
Ga	16	16						
Ge	(1.4)	1.6						
As	1.9	2.38						
Rb	21	21	19					
Sr	190	175	188					
Y	25	26	29	40	34.7	50		0.04
Zr	105	90	95	185	180	—		—
Nb	9.5			19?	13.4	—		
Mo	0.57?	—	1.0					
In	0.065	0.068	—					
Sn	3.2	2.1	—					
Sb	1.0	0.85	0.89					
Cs	0.9	0.95	0.75					
Ba	160	145	198	680	725	—		
La	9.8?	11.9	13.5	27	24.6	26	(0.15)	0.05
Ce	(23)	17.7	18	53	53.6	45	(0.09)	0.12
Pr	(3.4)	4.15	2.5	7?	6.60	7.4	—	0.0099
Nd	15	12.5	10.7	26?	29.3	34	—	0.049
Sm	3.6?	—	3.3	6.5	6.44	7.0	(0.008)	0.0057
Eu	1.1	1.20	0.95	2.0	2.01	1.86	0.002	0.0017
Gd	(4)	3.82	3.0	6.6	6.25	5.4	—	0.0034
Tb	0.65	0.49	0.66	1.0	1.07	0.89	(0.001)	0.0012
Dy	4	3.89	2.6	7?	6.44	5.5	—	0.0065
Ho	0.69?	0.63	0.78	1.2?	1.32	—	—	0.0013
Er	2.4	2.08	1.8	3.5?	3.73	2.42	—	0.0047
Tm	0.30	0.28	0.31	0.6?	0.61	0.58	—	0.00077
Yb	2.1	—	1.58	3.4	3.69	—	(0.02)	0.0075
Lu	0.35?	0.20	—	0.5?	0.59	—	0.006?	0.0016
Hf	2.7?	0.93	2.0	5	4.5	—		
Tl	0.11	—	0.18					
Pb	7.8?	8.3	8	14	13.8	—		
Th	2.4	2.2	2.2	6.1	5.68	—		
U	0.58?	0.46		1.7	1.73	—		

All data in ppm.
Data of Strelow and Jackson obtained by ion exchange separation prior to analysis.
?: 'less reliable' data.
(): insufficient data available to calculate a 'usable' value; originators' values quoted.

a similar magnitude. Their data for USGS: PCC-1 (peridotite) are listed in Table 19.5. Due to the paucity of reliable international data on this material, comparison with the compilation data of Abbey (1983) should be made with care.

Victor and Rogers (1983) used a similar technique (ion exchange separation plus dc arc excitation) to analyse ultratrace amounts of thorium in reference materials NIM-D and NIM-P. The thorium residue after cation exchange separation was taken up in fuming nitric acid and mixed with 2 mL of standard uranium solution (500 ng U mL^{-1}) as internal standard. After evaporation to dryness, the residue was redissolved in hydrochloric acid and absorbed on 100 mg high-purity silver powder. After drying, the silver powder was compressed into 2 mm diameter electrodes in a nylon mould. These were mounted in the mass spectrometer, and volatilized using a dc arc of about 5 A triggered every 33 ms, with a duration of 140 μs. Data were calibrated against standard solutions and procedural blanks were less than 10 ng. Mean thorium analyses were: NIM-D, 0.15 ± 0.02 ppm; NIM-P, 0.36 ± 0.03 ppm.

19.7 Future developments

Many of the analytical results by spark source mass spectrometry reported in the literature were obtained on first-generation instrumentation. It is possible that more recent

Table 19.6 Interference in rare-earth element determinations by spark source mass spectrometry

Element	Mass number	Isotopic abundance (%)	Potential interference
La	139	99.91	$^{40}Ca_2AlO_2$ in plagioclase $^{40}CaAl^{48}TiC_2$ in Ti-rich samples
Ce	140	88.48	$^{28}Si_5$, Mg_2SiO_4, $Na^{35}Cl_2$ in chlorides
Pr	141	100	$^{47}Ti_3$ in rutile $Na_3{}^{37}Cl_2$ in chlorides
Nd	143	12.20	
	146	17.18	$^{26}MgC_{10}$ in olivine
Sm	147	15.00	$^{49}Ti_3$ in rutile
	149	13.82	
Eu	151	47.82	^{135}BaO
	153	52.18	^{137}BaO
Gd	158	24.87	$^{134}BaC_2$
Tb	159	100	$^{135}BaC_2$
Dy	161	18.88	$^{137}BaC_2$
	163	24.97	$^{139}LaC_2$
Ho	165	100	$^{141}PrC_2$
Er	166	33.41	$^{142}NdC_2$, $^{142}CeC_2$
	167	22.94	$^{143}NdC_2$
Tm	169	100	$^{12}C_4H$
Yb	171	14.31	$^{147}SmC_2$, Ca_2AlO_4 in anorthite, $^{135}BaC_3$
	172	21.82	$^{148}SmC_2$, $^{148}NdC_2$, $^{136}BaC_3$
	173	16.13	$^{149}SmC_2$, $^{137}BaC_3$
	174	31.84	$^{150}SmC_2$, $^{150}NdC_2$, $^{138}BaC_3$
Lu	175	97.41	
	176	2.59	^{176}Hf, ^{176}Yb
Y	89	100	$^{25}MgO_4$

Data from Taylor (1979).

developments could have some impact on the future application of this technique; such developments include the availability of fast solid state amplifiers better capable of giving a precise measurement of the ion beam charge used in any exposure. New electronic position-sensitive detectors show some promise in replacing photoplate detection and permitting more rapid computer-compatible data acquisition. Investigations of spark characteristics indicate that it may be possible to excite ions with a considerably narrower energy spread (−100 eV) than is observed in conventional spark excitation (2–3 keV). This could lead to some simplifications in mass analyser instrumentation. Examination of the ion production mechanisms in the plasma discharge also indicates that there is some delay in the emission of molecular ions compared with doubly-charged and singly-charged ions. Molecular interferences could be somewhat reduced if ion detection were synchronized in phase with the spark discharge. These and other developments have been reviewed by Adams (1982, 1983) and lead to the conclusion that a renaissance could occur in spark source mass spectrometry, particularly so when the attractive analytical properties of the spark source mass spectrometer are considered, namely:

(i) Almost uniform sensitivity for a wide range of trace elements

(ii) Detection limits in the range 0.1–0.001 ppm without chemical separation

(iii) Small, if not completely negligible, matrix effects

(iv) The capability of detecting all elements simultaneously.

20 Inductively coupled plasma–mass spectrometry

20.1 Introduction

The inductively coupled plasma–mass spectrometer combines two analytical facilities to produce an instrument with powerful potential in the field of multi-element trace analysis. The inductively coupled argon plasma is used, not as a source for optical emission measurements (as in ICP–atomic emission spectrometry, Chapter 5), but rather as a source of ions. The mass spectrum of this ion source is measured using a quadrupole mass spectrometer. This was made possible by the development of an interface whereby plasma gases can be physically sampled through an orifice, via a differential pumping unit and into the quadruple mass filter (Figure 20.1). The technique, therefore, combines the freedom from matrix interferences characteristic of the inductively coupled argon plasma with the very favourable signal-to-background ratios obtainable by mass spectrometry. Samples must first be taken up into solution and are then aspirated into the argon plasma in the conventional manner. Detection limits for applications in multi-element silicate rock analysis have not, at the time of writing, been fully characterized. However, they appear, in general, to be significantly lower than for ICP–AES. Nevertheless, it is not yet clear what position this technique will occupy in the hierarchy of analytical instrumentation. ICP–MS has a clear application in the analysis of waters and is likely to have a significant impact on routine trace geochemical analysis. This is so, in view of the potential of this technique to determine simultaneously a wide range of heavier trace elements, often down to ppm whole-rock levels, using a simple sample dissolution technique with no pre-concentration stage. Of particular interest in this respect is the application of ICP–MS in the determination of the rare-earth elements and a wide range of trace elements important in economic geology studies including As, Sb, Se, Te, Bi, Hg, Au, Ag, and the platinum group elements.

As a mass spectrometry technique, ICP–MS is capable of determining individual isotope ratios. There is, therefore, considerable potential in applying isotope dilution procedures for the calibration of selected elements. At the time of writing, the precision with which isotope ratios can be measured (about 1–3% two sigma) is not adequate to permit reliable determination of radiogenic isotope ratios (e.g. Rb–Sr, Nd–Sm, U–Th–Pb). The technique does not, therefore, rival conventional thermal ionization mass spectrometry in geochronological studies, although ICP–MS does show some potential in the specialized determination of osmium isotope ratios required for the Re–Os geochronometer (section 16.16).

In this rapidly developing field, the full impact of ICP–MS cannot yet be fully assessed. The following sections, therefore, describe in overview the instrumentation and potential of ICP–MS as one of the most promising techniques for trace-element analysis to become available in recent years.

20.2 Development of ICP–MS instrumentation: the plasma–mass spectrometer interface

Early experiments investigating the plasma as a source of ions for mass spectrometry were carried out by Gray (1975*a,b*). Gray used a dc capillary arc discharge plasma and established the potential of plasmas as ion sources in this application by demonstrating the excellent detection limits that could be achieved. However, interference effects were serious (Gray, 1978), largely due to the relatively low temperature and inefficient sample introduction using the dc capillary plasma source. The mainstream of development work subsequently involved the substitution of the inductively coupled argon plasma as ion source (Houk *et al.*, 1980; Date and Gray, 1981, 1983; Houk and Thompson, 1982). Douglas and French (1981) employed microwave excitation to create an argon plasma.

The inductively coupled argon plasma–mass spectrometer consists of three components, the relationship between which is shown in Figure 20.1:

(i) The argon plasma itself, comprising nebulizer, cloud chamber, torch, work coils, and associated power supplies. This technology is fully discussed in Chapter 5. The plasma torch is configured at 90° to conventional ICP–AES operation, so that the tail flame is directed along the horizontal axis at the sampling orifice.

(ii) A quadrupole mass spectrometer and associated data collection electronics which permit rapid scanning of selected mass ranges between 0–300 amu. Design and performance characteristics of the quadrupole mass filter have been considered in Chapter 16.

(iii) An interface unit which permits sampling of the plasma gases and transfer of the ion beam into the mass spectrometer proper. The design of this interface is critical to instrument performance and will be considered in more detail below.

20.2.1 *The plasma–mass spectrometer interface*

Mass spectrometers require a very high vacuum for successful operation (usually better than about 10^{-7} Torr). The sampling interface can, therefore, be considered to be a controlled leak, permitting plasma gases to enter the mass spectrometer for analysis. The design of this interface must take into account the interactions that occur when plasma gases at a pressure of one atmosphere and a temperature of 6000–7500 K expand through an orifice machined in a cool metal plate into a vacuum beyond. Early development instruments used rather small apertures of about 70 μm in diameter, and operated in the so-called *boundary layer*

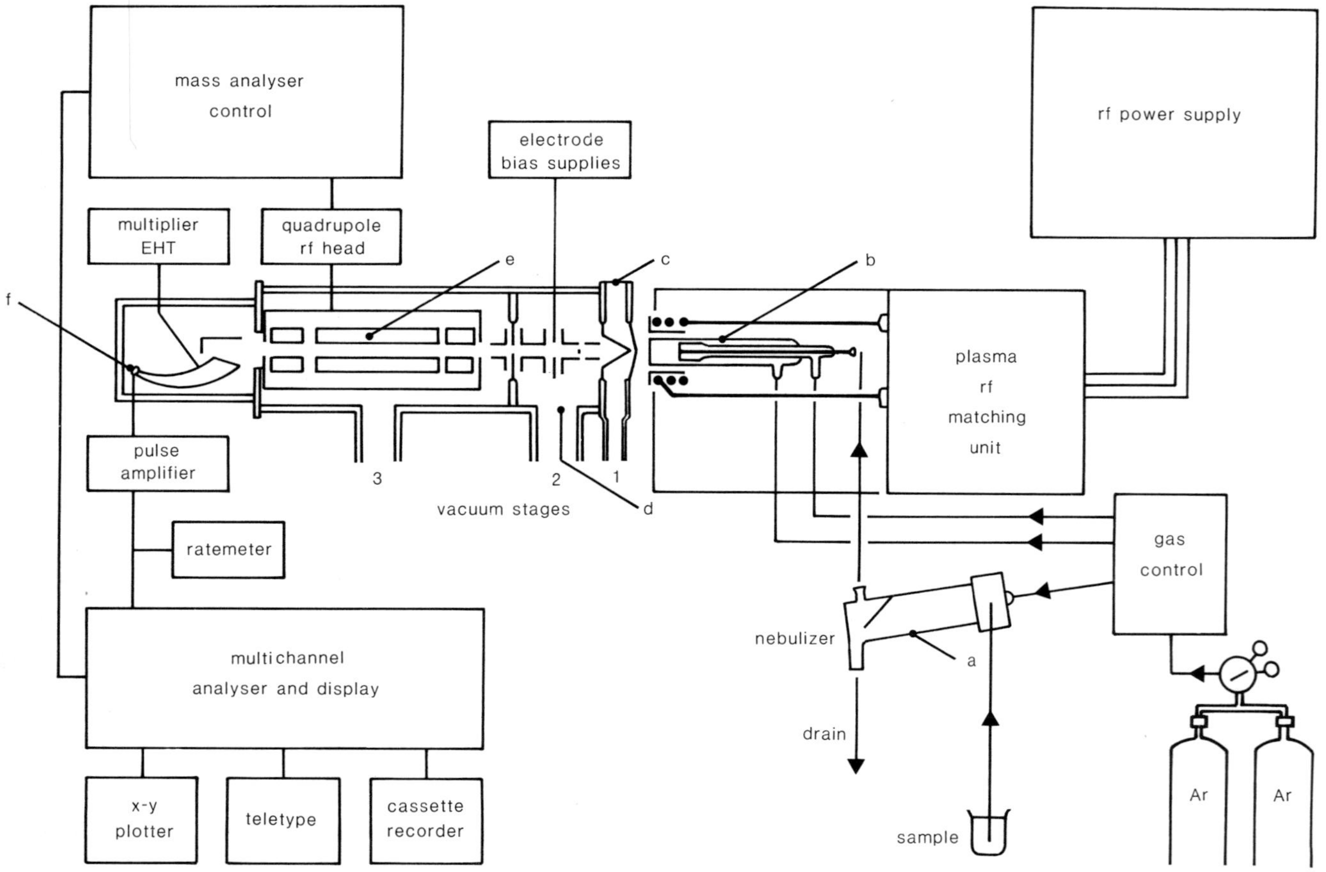

Figure 20.1 Schematic diagram of an inductively coupled plasma–mass spectrometer. The instrument consists of (*a*) a nebulizer; (*b*) a torch with radiofrequency work coils; (*c*) expansion stage held at intermediate vacuum (1 Torr) and incorporating the plasma sampling cone and skimmer (see Figure 10.4); (*d*) second intermediate vacuum stage incorporating the ion extraction electrodes and light baffle; (*e*) the quadrupole mass filter; (*f*) an electron channel multiplier ion detector; and (*g*) instrument control and data recording modules (Date and Gray, 1982*b*.)

sampling mode. This method has now been largely abandoned in favour of much larger apertures (400–800 μm in diameter) which operate in *continuum sampling mode*. The rationale behind these methods of operation is as follows.

(*a*) *Boundary layer sampling interface*. When a hot stream of gas impinges on a cool metal surface, a boundary layer of cool gas is formed next to the metal. In early instruments, plasma gases were sampled through an orifice directly into the mass spectrometer, held at high vacuum. Small apertures were therefore essential so that the spectrometer vacuum did not degrade excessively. However, these apertures were sufficiently small (50–80 μm) to cause a cool boundary layer of gas to form over the sampling aperture (Figure 20.2*a*). The kinetic energy of the plasma gas was insufficient to break directly through this layer of aerodynamically stagnant gas. Indeed, in early instrumentation, this 'breakthrough' situation was avoided because enlargement of the sampling aperture much above about 80 μm caused a *pinch discharge* to form in the aperture opening. This discharge arose from the compression of the free electron population of plasma gas as it passed through the aperture. Pinch discharges resulted in serious degradation in instrumental operation, firstly because of the large increase in noise caused by the presence of an intense photon source in the sampling aperture, and secondly because of the degradation in instrument mass resolution caused by the increase in kinetic energy of ions passing through this discharge.

Using small apertures, the plasma gases entering the instrument are those which have equilibrated with the cool boundary layer. At the lower temperatures of this equilibrated gas (perhaps 1000 K, Houk *et al.*, 1980) the energy spread of ions entering the mass spectrometer is also low (perhaps less than 1 eV), permitting operation of the mass analyser at high ion transmission rates with good mass resolution. Boundary layer sampling, therefore, offers excellent signal to background ratios, permitting detection limits to be achieved which are lower perhaps by a factor of 10 than in the alternative mode of operation, continuum sampling.

However, there is a serious disadvantage to operation with boundary layer sampling. This arises from the fact that plasma gases re-equilibrate in the boundary layer in front of the sampling aperture prior to analysis. It is estimated that during this 1–2 ms requilibration period, ions may suffer up to 10^6 collisions with adjacent plasma gas species (Houk *et al.*, 1980). This causes undesirable interactions to occur in the sampled beam which result in the enhanced formation

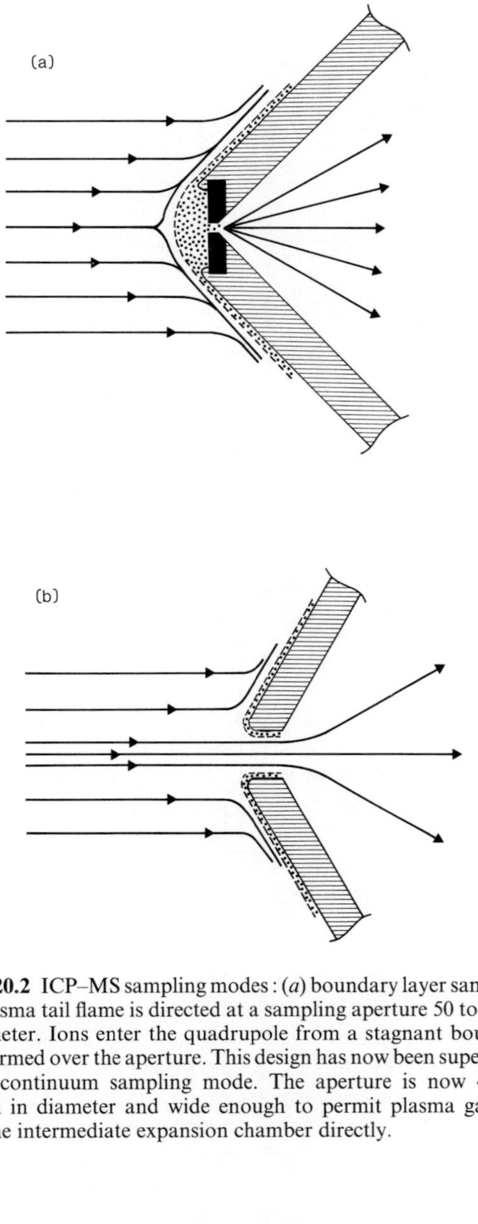

Figure 20.2 ICP–MS sampling modes: (*a*) boundary layer sampling. The plasma tail flame is directed at a sampling aperture 50 to 80 μm in diameter. Ions enter the quadrupole from a stagnant boundary layer formed over the aperture. This design has now been superseded by (*b*) continuum sampling mode. The aperture is now 400 to 800 μm in diameter and wide enough to permit plasma gases to enter the intermediate expansion chamber directly.

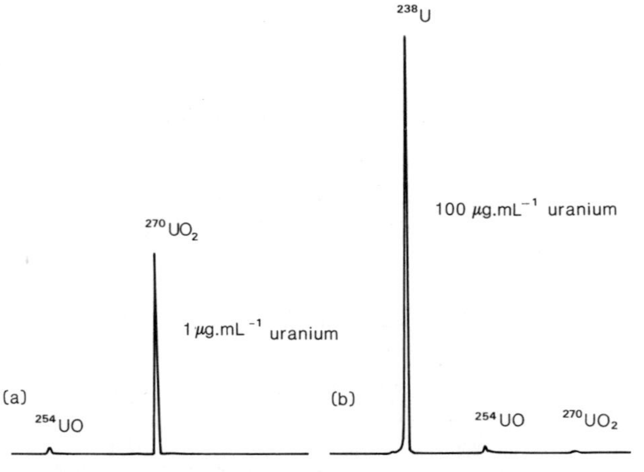

Figure 20.3 ICP–MS spectrum of uranium under (*a*) boundary layer and (*b*) continuum sampling modes. The boundary layer spectrum is dominated by the UO_2^+ peak resulting from recombinations within the boundary layer. This peak is virtually absent in the $^{238}U^+$ spectrum observed in continuum mode operation. (Date and Gray, 1983.)

of ion–molecular species. These not only add some complexity to the mass spectrum but also make boundary layer sampling sensitive to matrix effects. In particular, ionization suppression interferences can occur in the boundary layer due to interaction of ions with matrix components. Thus Houk *et al.* (1980) investigated the effect of increasing concentrations of Na^+ on the signal observed from $^{52}Cr^+$ and $^{59}Co^+$. Serious reduction in the ion beam signal intensity was observed at sodium concentrations exceeding 50 to 100 μg/mL due to boundary layer ionization suppression. Molecular interactions in the boundary layer also increase considerably the potential for ion oxide formation. Thus Date and Gray (1983) compared the spectrum of uranium in boundary layer and in continuum sampling modes and observed large signals due to $^{254}(UO)^+$ and $^{270}(UO_2)^+$ in the former mode of operation. These signals were virtually absent in continuum sampling mode, the $^{238}(U)^+$ signal dominating this spectrum (Figure 20.3). A further operational difficulty with boundary layer sampling is that the relatively cool gases have a tendency to cause salts to condense, with the eventual clogging of the sampling aperture, if solutions are analysed with salt contents much greater than about 10 μg/mL.

Thus, although boundary layer sampling is capable of good analytical performance using ideal dilute solutions, in practical applications a number of interference effects seriously limit the usefulness of this method of operation.

(*c*) *Continuum sampling operation.* Many of the disadvantages outlined above are eliminated if the sampling aperture is increased in diameter to permit plasma gases to punch their way through the boundary layer which otherwise covers the sampling orifice (Figure 20.2*b*). However, the serious problems which resulted from the formation of a pinch discharge in the sampling aperture (as outlined above) frustrated early attempts at such operation. These difficulties were overcome by the adaptation of molecular beam sampling technology to the argon plasma in a form of instrumentation described by Douglas and French (1981). The sampling aperture was enlarged to 400–800 μm. However, instead of sampling directly into a high-vacuum chamber, the region behind the sampling orifice was held at a relatively poor vacuum of about 1 Torr, maintained by a simple mechanical rotary pump (Figure 20.1). In this way the formation of a pinch discharge was avoided. Plasma gases passing through the aperture entered a small expansion chamber. The size of the ion beam then passing through to the mass spectrometer was defined by a skimmer having an aperture of about 1 mm in diameter (Figure 20.4). This skimmer is normally positioned about 100 mm behind the sampling aperture. However, the relative diameter of sampling and skimmer apertures and their relative configurations is currently an area of active design optimization.

With the sampled plasma gases no longer cooled in the boundary layer in front of the sampling aperture, many of the chemical interference effects disappear in continuum mode sampling. Thus the susceptibility to ionization sup-

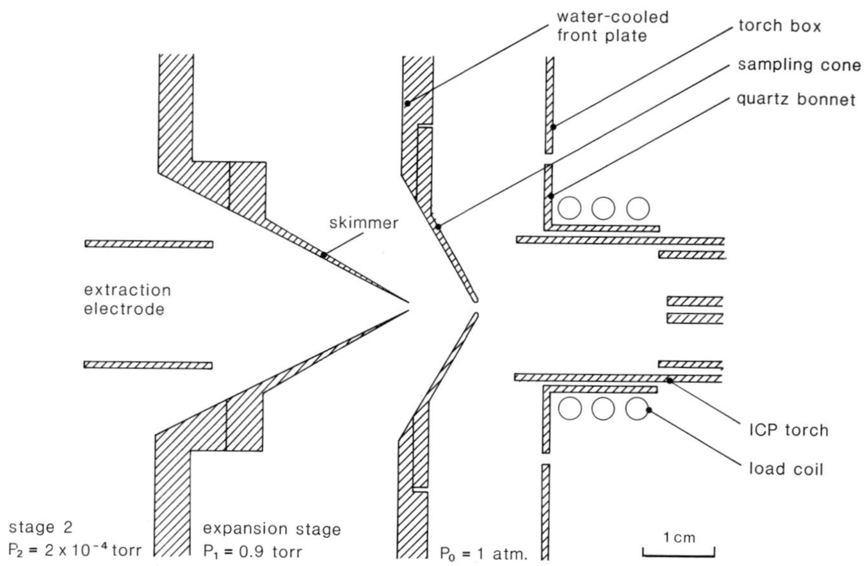

Figure 20.4 Configuration of sampling cone, expansion stage and skimmer cone designed for continuum sampling mode operation. (Date and Gray, 1982*b*.)

pression is significantly reduced. The potential for oxide formation is also minimized. Those elements which tend to form stable oxides are detected predominantly as the atomic ion. The problem with salt condensation within the aperture is greatly reduced, since aperture gases are much hotter. Little effect is observed using solutions of concentrations up to 1000 μg/mL. However, the mass resolution and background spectrum associated with continuum sampling operation are significantly worse than in boundary layer mode. The degraded resolution is because the energy spread of ions sampled in continuum mode (7eV) (Gray and Date, 1982) is significantly higher than that obtained from boundary layer mode. The significantly higher backgrounds may be due to the higher proportion of metastable argon atoms/ions which are swept into the mass spectrometer from the plasma. These decay within the mass analyser. Photons emitted during decay are picked up as noise by the ion detection system. For this reason instruments must be designed to avoid line-of-sight paths between aperture and ion detector to prevent photons from the intense plasma source reaching the detector.

20.3 The inductively coupled plasma as ion source

The mass spectrum observed from an inductively coupled plasma is remarkably simple compared with that observed, for example, in a spark source mass spectrometer. There are several reasons for this, associated with the nature of the plasma itself.

20.3.1 *Atomization*

One characteristic of the inductively coupled plasma is that it is a very efficient atomizer of samples. This is due to the very high temperatures experienced by the sample aerosol (about 8000 K) within the plasma fireball. This causes rapid desolvation, volatilization and atomization of samples. Thus molecular ion species, residues of incomplete atomization by the plasma, are almost undetectable, and even those atoms which form stable oxides (e.g. U, Th, REEs) are observed predominantly at atomic ion species in continuum mode operation.

20.3.2 *Ionization*

The second aspect of the ICP as an ion source is that samples are ionized very efficiently. This is well known from ICP–atomic emission spectrometry where, for many elements, the most sensitive lines are ion emission lines and not atomic emission lines. But perhaps equal in importance to the ionization efficiency is the fact that the ion spectrum is buffered by the first ionization energy of argon (15.76 eV). The argon plasma can only effectively ionize those elements that have an ionization energy significantly below this value. Data for the first and second ionization energies of the elements are shown in Table 20.1. These data demonstrate the following characteristics of the argon plasma:

(i) *Singly-charged ions* of almost all the elements are effectively excited by an argon plasma. The most notable exception is fluorine (and some noble gases). Elements such as Br, Cl, H, N, C, O have rather high first ionization energies (> 11 eV) and so will be ionized much less favourably.

(ii) Few elements have *second ionization energies* which fall below 15.76 eV and therefore interferences in the mass spectrum due to doubly-charged ions will normally be negligible. Of the elements which can form doubly-charged ions, barium has the lowest second ionization energy (10 eV). The others (including Ca, Sc, Sr, Ti, Y and REEs) all have second ionization energies above 11 eV and will be rather weakly excited

(iii) The *third ionization energy* is completely out of the

Table 20.1 First and second ionization potential of the elements

AN	Ionization potential I	II	AN	Ionization potential I	II
18 Ar	15.76	27.62	7 N	**14.53**	29.59
47 Ag	7.57	21.48	11 Na	5.14	47.29
13 Al	5.98	18.82	41 Nb	6.88	**14.32**
33 As	9.81	18.63	60 Nd	5.51	
79 Au	9.22	20.5	10 Ne	**21.56**	41.07
5 B	8.30	25.15	28 Ni	7.63	18.15
56 Ba	5.21	**10.00**	8 O	13.61	35.11
4 Be	9.32	18.21	76 Os	8.5	17
83 Bi	7.29	16.68	15 P	10.48	19.72
35 Br	11.84	21.6	82 Pb	7.42	**15.03**
6 C	11.26	24.38	46 Pd	8.33	19.42
20 Ca	6.11	**11.87**	59 Pr	5.46	
48 Cd	8.99	16.90	78 Pt	9.0	18.56
58 Ce	5.6	**12.3**	94 Pu	5.1	
17 Cl	13.01	23.8	88 Ra	5.28	**10.14**
27 Co	7.86	17.05	37 Rb	4.18	27.5
24 Cr	6.76	16.49	75 Re	7.87	16.6
55 Cs	3.89	25.1	45 Rh	7.46	18.07
29 Cu	7.72	20.29	86 Rn	10.75	
66 Dy	6.8		44 Ru	7.36	16.76
68 Er	6.08		16 S	10.36	23.4
63 Eu	5.67	**11.24**	51 Sb	8.64	16.5
9 F	**17.42**	34.98	21 Sc	6.54	**12.8**
26 Fe	7.87	16.18	34 Se	9.75	21.5
31 Ga	6	20.57	14 Si	8.15	16.34
64 Gd	6.16	**12**	62 Sm	5.6	**11.2**
32 Ge	7.88	15.93	50 Sn	7.34	**14.63**
1 H	13.59		38 Sr	5.69	**11.03**
2 He	**24.48**	54.40	73 Ta	7.88	16.2
72 Hf	7	**14.9**	65 Tb	5.98	
80 Hg	10.43	18.75	43 Tc	7.28	**15.26**
53 I	10.45	19.13	52 Te	9.01	18.6
49 In	5.79	18.86	90 Th	6.95	
77 Ir	9		22 Ti	6.82	**13.57**
19 K	4.34	31.81	81 Tl	6.11	20.42
36 Kr	13.99	24.56	69 Tm	5.81	
57 La	5.61	**11.43**	92 U	6.08	
3 Li	5.39	75.62	23 V	6.74	**14.65**
71 Lu		**14.7**	74 W	7.98	17.7
12 Mg	7.64	**15.03**	54 Xe	12.13	21.2
25 Mn	7.43	15.64	39 Y	6.38	**12.23**
42 Mo	7.10	16.15	70 Yb	6.2	**12.10**
			30 Zn	9.39	17.96
			40 Zr	6.84	**13.13**

Data from Weast (1973), Table E-67.
I, first ionization energy (eV); figures in bold indicate elements that *cannot* be ionized by an argon plasma.
II, second ionization energy (eV); figures in bold indicate elements that *can* form doubly-charged ions in an argon plasma.

range of excitation by an argon plasma and accordingly, triple-charged ions will be absent from mass spectra.

20.3.3 *Matrix interferences*

ICP–MS benefits from an inherent insensitivity to matrix interferences, as is observed in ICP sources used in atomic emission spectrometry. Potentially, the most significant interference is the suppression of ionization due to the presence of a more readily ionizable element. Data have already been presented to indicate that problems can occur due to chemical recombinations in boundary layer sampling, but that these effects are much smaller in continuum mode operation.

20.3.4 *Element sensitivity*

The variation in element sensitivity as a function of ionization energy for a wide range of elements is plotted in Figure 20.5 using data from Date and Gray (1982*b*). This plot shows that, for the majority of elements with first ionization energies below 10 eV, the instrument response is uniform to within a factor of about 3. Above 10 eV, sensitivity begins to fall off rapidly, the response for Br (11.84 eV) and Cl (13.01 eV) being between 1 and 1.5 orders of magnitude lower. The main constraint, therefore, controlling the relative intensity of adjacent elements (apart from concentration) will therefore be the relative isotopic abundance. This is in direct contrast to techniques such as ICP–AES, for example, where the relative sensitivity of adjacent elements in the Periodic Table often varies unpredictably.

20.3.5 *Interferences*

The consequence of the above is that ICP–MS spectra are simple, and interferences, where they occur, can be predicted with confidence. There are, in fact, only 211 mass lines in total contributed by all the elements in the Periodic Table (many *individual* elements have a larger number of discrete optical lines contributing to ICP–AES spectra). Furthermore, all elements, except indium, have at least one line free of overlap if background peaks and double-charged interferences are ignored (Table 20.2). Selection of mass numbers for analysis is, therefore, relatively simple and is usually based on the most abundant naturally-occurring isotope. Account must be taken of some background mass interferences which arise from five sources:

(i) The argon plasma gas (Ar^+, Ar_2^+)

(ii) The solvent in which the analysed species was dissolved—usually water (H_2O^+, H_3O^+, OH^+, ArH^+)

(iii) Air entrained in the plasma and impurity gases in the argon itself (N^+, O_2^+, NO_2^+)

(iv) Material eroded off the aperture/skimmer assembly (often isotopes of copper, nickel and molybdenum)

(v) Reagent contamination.

The background mass peaks reported by various workers are listed in Table 20.3. The presence and relative intensity of individual peaks is dependent on plasma–interface design. In particular, the intensity of interferences caused by air-entrained mass peaks is likely to be very low on modern instruments.

A further non-mass specific background enhancement appears to be caused mainly by the photon sensitivity of ion detectors used to measure the mass spectrum. As mentioned above, photons striking an electron channel multiplier are counted as part of the ion spectrum. These photons arise from two sources. Firstly, the plasma fireball itself is an intense light source. It is essential therefore that a light baffle is incorporated into the ion flight path to minimize the number of plasma photons which are scattered on to the ion detector (Figure 20.1). Second, metastable argon species enter the mass spectrometer along with other plasma gases. These excited species subsequently decay within the spectro-

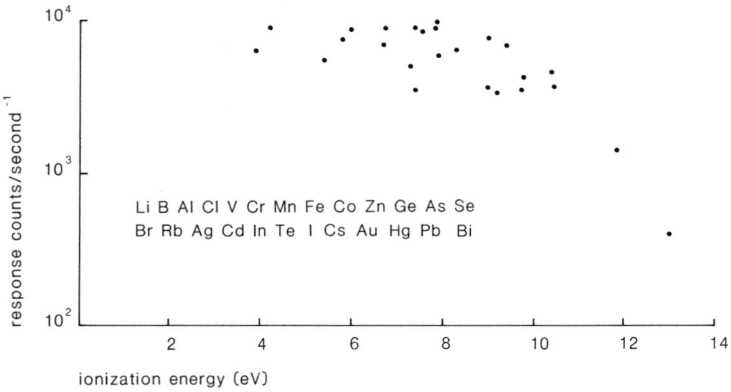

Figure 20.5 ICP–MS sensitivity data. The response of 1 μg/mL solutions of 25 elements in the mass range 7 to 209 amu is plotted as a function of ionization energy. (Date and Gray, 1982*b*.)

meter, emitting photons that contribute to the residual background. Since the proportion of metastable ions entering the sampling aperture is larger in continuum sampling mode than boundary layer operation, detection limits in the former mode are about an order of magnitude worse.

Table 20.2 Natural isotopic composition of the elements

	H	He	Li	Be	B	C	N	O
1	99.9							
2	0.01							
3	<0.01	10^{-4}						
4		100						
5								
6			7.4					
7			92.6					
8								
9				100				
10					19.8			
11					80.2			
12						98.9		
13						1.11		
14						<0.01	99.6	
15							0.37	
16								99.8
17								0.04
18								0.20

	F	Ne	Na	Mg	Al	Si	P	S
19	100							
20		90.9						
21		0.26						
22		8.8						
23			100					
24				78.7				
25				10.1				
26				11.2				
27					100			
28						92.2		
29						4.7		
30						3.1		
31							100	
32								95.0
33								0.76
34								4.2

20.3.6 *Optimizing plasma operation*

Criteria for optimizing plasma operating parameters (flow rates of argon to torch, forward power delivered to plasma, observation height, sample uptake rate) are different from ICP–atomic emission spectrometry. In the latter technique, compromise conditions are normally selected to maximize the optical emission signal from both atomic and ionic species. Optimum signals are often found at an observation height of 15–20 mm above the work coil. In ICP–MS it is

Table 20.2 (*Continued*)

	S	Cl	Ar	K	Ca	Sc	Ti
35		75.5					
36	0.01		0.34				
37		24.5					
38			0.06				
39				93.1			
40			99.6	10^{-3}	96.9		
41				6.9			
42					0.65		
43					0.14		
44					2.08		
45						100	
46					0.19		7.9
47							7.3
48					0.18		73.9
49							5.5

	Ti	V	Cr	Mn	Fe	Co	Ni
50	5.3	0.24	4.31				
51		99.8					
52			83.8				
53			9.6				
54			2.38		5.8		
55				100			
56					91.7		
57					2.19		
58					0.33		68.3
59						100	
60							26.1
61							1.13
62							3.59

necessary to maximize the total ion signal, and most studies to date have been carried out lower down in the plasma at observation heights of 5–100 mm. Under these conditions, the plasma gas is sampled at a temperature equivalent to about 7500 K after a residence time of about 1 ms in the plasma core (Gray and Date, 1982). Future investigations of optimum operating conditions are likely to be equally concerned about the suppression of background by minimizing the proportion of metastable argon ions sampled from the plasma.

Table 20.2 (*Continued*)

	Ni	Cu	Zn	Ga	Ge	As	Se
63		69.1					
64	0.90		48.9				
65		30.9					
66			27.8				
67			4.1				
68			18.6				
69				60.4			
70			0.62		20.5		
71				39.6			
72					27.4		
73					7.8		
74					36.5		0.9
75						100	
76					7.8		9.0
77							7.6

	Se	Br	Kr	Rb	Sr	Y	Zr
78	23.5		0.35				
79		50.5					
80	49.8		2.27				
81		49.5					
82	9.2		11.6				
83			11.6				
84			56.9		0.56		
85				72.2			
86			17.4		9.9		
87				27.8	7.0		
88					82.6		
89						100	
90							51.5
91							11.2

	Zr	Nb	Mo	Tc	Ru	Rh	Pd
92	17.1		15.8				
93		100					
94	17.4		9.0				
95			15.7				
96	2.8		16.5		5.5		
97			9.5				
98			23.8		1.9		
99					12.7		
100			9.6		12.6		
101					17.1		
102					31.6		1.0
103						100	11.0
104					18.6		
105							22.2

Table 20.2 (*Continued*)

	Pd	Ag	Cd	In	Sn	Sb	Te
106	27.3		1.2				
107		51.8					
108	26.7		0.9				
109		48.2					
110	11.8		12.4				
111			12.8				
112			24.1		1.0		
113			12.3	4.3			
114			28.8		0.66		
115				95.7	0.35		
116			7.6		14.3		
117					7.6		
118					24.0		
119					8.6		
120					32.8		0.09
121						57.3	
122					4.7		2.5
123						42.7	0.87

	Sn	Sb	Te	I	Xe	Cs	Ba
124	5.9		4.6		0.10		
125			7.0				
126			18.7		0.09		
127				100			
128			31.8		1.9		
129					26.4		
130			34.5		4.1		0.10
131					21.2		
132					26.9		0.10
133						100	
134					10.4		2.4
135							6.6

	Xe	Cs	Ba	La	Ce	Pr	Nd
136	8.9		7.8		0.19		
137			11.2				
138			71.7	0.09	0.25		
139				99.9			
140					88.5		
141						100	
142					11.1		27.1
143							12.2

	Nd	Pm	Sm	Eu	Gd	Tb	Dy
144	23.9		3.1				
145	8.3						
146	17.6						
147			15.0				
148	5.7		11.2				
149			13.8				
150	5.6		7.4				
151				47.8			
152			26.7		0.20		
153				52.2			
154			22.7		2.2		
155					14.7		
156					20.5		0.05
157					15.7		
158					24.9		0.09
159						100	
160					21.9		2.3
161							18.9

(*Continued*)

Table 20.2 (*Continued*)

	Dy	Ho	Er	Tm	Yb	Lu	Hf
162	25.5		0.14				
163	24.9						
164	28.2		1.6				
165		100					
166			33.4				
167			22.9				
168			27.1		0.14		
169				100			
170			14.9		3.0		
171					14.3		
172					21.8		
173					16.1		
174					31.8		0.18
175						97.4	
176					12.7	2.6	5.2
177							18.5
178							27.1
179							13.8

	Hf	Ta	W	Re	Os	Ir	Pt
180	35.2	0.01	0.14				
181		99.9					
182			26.4				
183			14.4				
184			30.6		0.02		
185				37.5			
186			28.4		1.6		
187				62.5	1.6		
188					13.3		
189					16.1		
190					26.4		0.01
191						37.4	
192					41.0		0.78
193						62.6	
194							32.9
195							33.8

	Pt	Au	Hg	Tl	Pb	Bi
196	25.3		0.15			
197		100				
198	7.21		10.0			
199			16.8			
200			23.1			
201			13.2			
202			29.8			
203				29.5		
204			6.9		1.5	
205				70.5		
206					23.6	
207					22.6	
208					52.3	
209						100

	Th	U
232	100	
233		
234		
235		0.72
236		
237		
238		99.3

Data adapted from Weast (1973).

Table 20.3 Background interferences in ICP–MS

Mass number	Reported interference	Coincident isotope		Alternative isotope	
14	N^+	^{14}N	(99.63)		
16	O^+	^{16}O	(99.8)		
17	OH^+	^{17}O	(0.04)	^{16}O	(99.8)
18	H_2O^+	^{18}O	(0.2)	^{16}O	(99.8)
19	H_3O^+	^{19}F	(100)	none	
20	Ar^{2+}	(^{20}Ne	(90.9))		
29	N_2H^+	^{29}Si	(4.7)	^{28}Si	(92.2)
30	$^{14}NO^+$	^{30}Si	(3.1)	^{28}Si	(92.2)
31	$^{15}NO^+$	^{31}P	(100)	none	
32	O_2^+	^{32}S	(95)	^{34}S	(4.2)
33	O_2H^+	^{33}S	(0.8)	^{34}S	(4.2)
36	$^{36}Ar^+$	none			
37	$^{36}ArH^+$	^{37}Cl	(24.5)	^{35}Cl	(74.5)
40	$^{40}Ar^+$	^{40}Ca	(97)	^{44}Ca	(2.1)
41	$^{40}ArH^+$	^{41}K	(6.9)	^{39}K	(93.1)
46	$^{14}NO_2^+$	^{46}Ti	(7.9)	^{49}Ti	(5.5)
47	$^{15}NO_2^+$	^{47}Ti	(7.3)	^{49}Ti	(5.5)
48	$N^{18}OO^+$	^{48}Ti	(73.94)	^{49}Ti	(5.5)
60	$(NO)_2^+$	^{60}Ni	(26.1)	^{58}Ni	(68.3)
80	Ar_2^+	^{80}Se	(50)	^{78}Se	(23)
81	$ArH.Ar^+$	^{81}Br	(49)	^{79}Br	(50)

Data from Houk *et al.* (1980), Date and Gray (1981), Douglas and French (1981) and Gray and Date (1982).
Identification of interfering isotopes follows conventions of original authors.
Alternative isotope is listed, but for some elements the magnitude of the interference may be too small to justify changing from the principal isotope.
Data in parentheses represent % natural abundance.

20.4 ICP–mass spectrometry instrumentation

An ICP–MS system is shown schematically in Figure 20.1. The principal components (nebulizer/plasma torch, sampling cone/expansion chamber, quadrupole mass filter/ion detector) have been discussed above. The ion beam, collimated by sampling the skimmer cones, then passes through a number of plates biased at an appropriate potential. The electrostatic fields created by these plates focus the ion beam in a form suitable for transmission through the mass filter. Any neutral species accompanying the plasma gases must be rapidly pumped away from the expansion chamber so that losses due to ion–molecular collisions are minimized. The quadrupole mass analyser possesses characteristics that make it particularly suitable for this application.

(i) It can be operated by repetitively sweeping through a selected mass range at a rapid rate. The maximum requirement is to scan the range 1–300 amu in about 1 second.

(ii) The quadrupole mass filter is tolerant of a wide angle of acceptance of the incoming ion beam and is not too sensitive to variations in kinetic energies of ions. Both these effects tend to be swamped by the large oscillating fields set up in the quadrupole.

(iii) It is simple to adjust the quadrupole between operation at high ion transmission rates with correspondingly moderate mass resolution, and vice versa. Adequate resolution is usually achieved if the instrument is set to separate adjacent integral mass peaks in the 1 to 300 amu range. Higher resolution is rarely required, since the proportion of doubly-charged species (which would cause an interference at half integral mass numbers) formed in an argon plasma is usually negligible (see section 20.3.2 above). In consequence, most instruments are operated at mass resolutions of 800 to 2000 ($M/\Delta M$). This is higher than would be required if transmitted peaks were an ideal triangular shape. However, in practice, ICP–MS spectrum peaks are usually characterized by small peak precursors and tails. These are caused by the spread of energies of ions accepted for measurement from the sampling aperture. The magnitude of these tails can be decreased by adjusting the mass filter to reduce the energy spread of ions reaching the detector. However, this is invariably accompanied by a significant reduction in ion transmission through the mass analyser and correspondingly smaller detected mass peak intensities. In practice, a satisfactory compromise must be made between mass resolution and ion beam intensitites.

In order to minimize discrepancies due to random fluctuations and short-scale noise, mass spectra must be accumulated by repetitively scanning a selected mass range. Spectra are therefore built up by rapidly scanning the quadrupole through the selected mass range using a sweep time of perhaps 0.1 to 1 s. Mass intensity data from each sweep are stored in a multi-channel analyser until 50 to 100 sets of superimposed data have been accumulated. The functions of the multi-channel analyser in recording and analysing gamma spectra have been considered in section 13.6. In ICP–MS applications, the multi-channel analyser is arranged to operate in *multi-channel scaling mode*. The signal which is used to ramp the quadrupole mass analyser over the selected mass range is also used to trigger the memory cycle pointer in the multi-channel analyser. The output count pulses from the ion detector are therefore added to the appropriate channel, selected in synchronization with the quadrupole mass sweep. If a mass spectrum is accumulated in this way in 2048 channels from 60 one-second scans of the mass range (i.e. one minute of data acquisition), then each MCA channel is available for data acquisition for slightly less than 0.5 ms per sweep or about 30 ms over the entire 60 mass scans. A typical channel-by-channel mass spectrum accumulated in this way is shown in Figure 20.6. Analyses are quantified by

Figure 20.6 ICP–MS multi-channel analyser output spectrum from NBS : SRM-981 lead reference material dissolved at 1 μg/mL in 1% HNO_3. Individual data are plotted as a histogram of channel counts (intensity) *v.* channel number (proportional to atomic mass units). (Date and Gray, 1983.)

integrating selected mass peaks over an appropriate channel range, then subtracting a background count which must take into account the presence of any residual counts in a blank solution (and the background) spectrum and comparing the resultant net area with that observed from a standard solution analysed under identical conditions.

20.5 Performance and applications

The two most exciting analytical characteristics of ICP–MS are the inherently low backgrounds and the simplicity and freedom from interferences observed in mass spectra. These characteristics, together with the uniform sensitivity towards elements with first ionization energies below 10 eV, give this technique a potentially low detection limit for a wide range of elements normally present in rock samples at trace levels (Table 20.4). In this respect the predicted detection limits achievable by ICP–MS are in the 1–5 ppm range for a wide range of elements determined from samples taken up into solution at a concentration level of 0.2 g rock to 100 mL solution with no element preconcentration or chemical separation. Improvements in these data are expected as instrumentation is further developed, and if it proves possible to run routinely solutions at the normal trace solution concentration of 1 g rock to 100 mL solution. Detection limit performance is compared with other instrumental solution techniques including AA and ICP–MS in Table 5.8. Particularly noteworthy is the ability of ICP–MS to determine combinations of elements from the same sample solution which previously demanded the use of more than one analytical technique. Such combinations would include elements of economic geochemical interest such as As, Sb, Se, Te, Cu, Ni, Zn, Pb, Bi, Au, Hg, etc.

A further unusual characteristic of this technique is the possibility of detecting all elements present within a selected mass range without having to specify precise instrument conditions for each. ICP–MS therefore joins a rather limited number of techniques (such as ED-XRF) which inherently possess this analytical characteristic. In routine trace-element analysis, steps will have to be taken to overcome two instrumental problems. One is the drift with time that is expected when a series of sample and/or standard solutions are to be analysed sequentially. The other is discrepancies which may arise due to mismatch between sample and standard solution composition and any limitation placed on the total salt content of sample solutions necessary to prevent clogging of the sampling aperture. In fact, matching of sample and standard solutions in terms of salt content and acid content is likely to be just as important in ICP–MS as it is in ICP–AES to ensure reproducible nebulization and atomization efficiencies in the plasma (section 5.11). At the current state of instrument development, such solutions must be restricted to total salt contents of $<1\%$ and total acid contents of 5%.

20.6 Internal standardization

One method of overcoming drift is to use an internal standardization procedure. An aliquot of a solution of known concentration is added to all sample and standard solutions. Analysed mass counts are then normalized with respect to the appropriate internal standard intensity. The technique

Table 20.4 Whole-rock detection limits by ICP–MS

Aluminium	Al	3	Iron	Fe	4	Samarium	Sm	4
Arsenic	As	60	Lanthanum	La	4	Scandium	Sc	6
Barium	Ba	2	Lead	Pb	2	Selenium	Se	100
Bismuth	Bi	2	Lithium	Li	8	Silver	Ag	2
Boron	B	4	Magnesium	Mg	2	Sodium	Na	4
Cadmium	Cd	2	Manganese	Mn	4	Strontium	Sr	2
Cerium	Ce	3	Mercury	Hg	4	Tellurium	Te	2
Caesium	Cs	2	Molybdenum	Mo	4	Terbium	Tb	4
Chromium	Cr	2	Neodymium	Nd	4	Thallium	Tl	2
Cobalt	Co	2	Nickel	Ni	4	Thorium	Th	4
Copper	Cu	2	Niobium	Nb	2	Thulium	Tm	1
Dysprosium	Dy	2	Osmium	Os	2	Tin	Sn	4
Erbium	Er	4	Palladium	Pd	1	Titanium	Ti	4
Europium	Eu	4	Phosphorus	P	600	Tungsten	W	4
Gadolinium	Gd	4	Platinum	Pt	2	Uranium	U	2
Germanium	Ge	4	Praesodymium	Pr	2	Vanadium	V	4
Gold	Au	1	Rhenium	Re	4	Ytterbium	Yb	2
Holmium	Ho	1	Rhodium	Rh	2	Yttrium	Y	2
Indium	In	2	Rubidium	Rb	2	Zinc	Zn	6
Iridium	Ir	4	Ruthenium	Ru	8	Zirconium	Zr	2

Data calculated as six-sigma detection limit for whole rock, assuming the sample has been dissolved in the proportion 0.2 g rock to 100 ml solution.
All data in ppm.
Data estimated from the ideal solution detection limit data of Date and Gray (1981), Cantle and Goddard (1983) and Sciex Ltd. (publicity leaflet). The original authors' data has been arbitrarily increased by a factor of two to allow for non-ideal response when analysing real rock solutions.

is particularly attractive in ICP–MS because the selection of the element to serve as internal standard is almost unrestricted on sensitivity criteria, and its addition is unlikely to cause significant mass spectrum interferences. Detailed discussions concerning the selection of internal standards are given in section 19.3 and are generally applicable here. In particular, the selected elements should have similar ionization characteristics in the plasma as the elements to be determined. Suitable internal standard solutions may contain several elements to satisfy these conditions for all the elements and mass ranges to be determined in an unknown solution.

20.7 Isotope dilution

The second procedure which can be used both for the correction of instrumental drift and for the quantification of individual elements is isotope dilution analysis. The principles behind this procedure are considered in more detail in section 17.6. Briefly, the element to be determined must possess more than one naturally-occurring isotope. An isotope dilution spike, containing an enriched fraction of one of those isotopes, is added at a known concentration to all solutions. Measurements are then made of the relative intensity of spike to natural isotope. Quantification is carried out from the isotopic abundance of the enriched spike, the measured isotope ratio and the known natural isotopic abundance of the element. Isotope dilution analysis in ICP–MS has several attractive features including the fact that quantification is carried out from one spectrum, so eliminating the effects of drift between sample runs. Set against this is the fact that enriched spikes are not available for all elements and tend to be expensive.

The above discussion has been addressed to the general

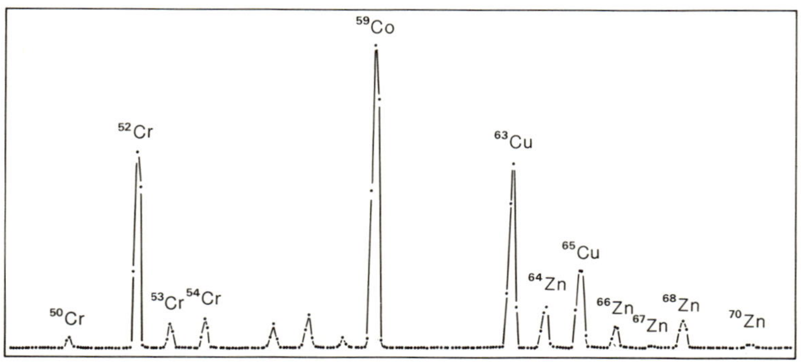

Figure 20.7 ICP–MS spectrum of a solution containing chromium, cobalt, copper and zinc all at 0.1 μg/mL in 1% nitric acid. This spectrum comprises 256 channels of output, the ^{59}Co peak containing 1192 counts. (Date and Gray, 1981.)

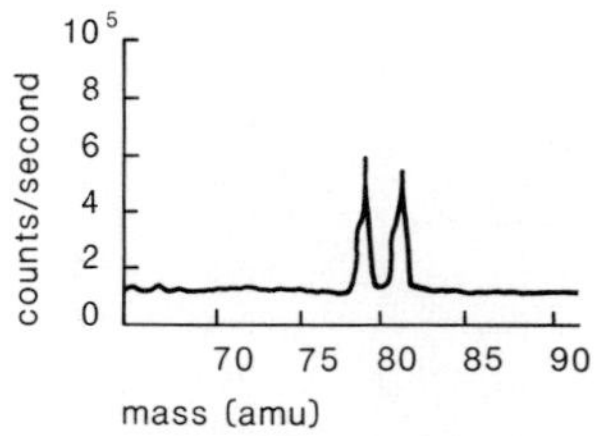

Figure 20.8 ICP–MS spectrum of bromide ions recorded in negative ion mode from a 10 ppm Br^- solution. (Douglas and French, 1981.)

Table 20.5 Lead isotope ratios in galena by ICP–MS

	$^{204}Pb/^{206}Pb$	$^{207}Pb/^{206}Pb$	$^{208}Pb/^{206}Pb$	Sample
Expected	0.05734	0.8837	2.137	
ICP–MS	0.05746	0.8872	2.144	Galena
RSD (%)	0.87	0.29	0.24	ORC 18
Error	0.21	0.40	0.33	
Expected	0.05515	0.8592	2.092	
ICP–MS	0.05523	0.8568	2.098	Galena
RSD (%)	1.03	0.34	0.28	ORC 17
Error	0.15	0.28	0.29	

Data from Date and Gray (1983).
Expected data was obtained by thermal ionization mass spectrometry.
RSD is the relative standard deviation of ICP–MS data based on counting statistics.
Error is the % difference between expected and ICP–MS data.

potential for trace-element analysis. There are, however some additional considerations for the determination of the lighter elements. The background spectrum is much more troublesome at low mass numbers and there is a serious interference on the principal mass peak of calcium (^{40}Ca) caused by $^{40}Ar^+$. A less abundant peak, ^{44}Ca (2.1% natural abundance), must be used instead. Fluorine can not be determined due to an H_3O^+ interference and the lack of an alternative isotope. However, this element cannot, in any case, be effectively excited in an argon plasma (section

Table 20.6 Rubidium isotope determinations by ICP–MS

Isotope	Relative abundance (%)	
	Certified value	Analysed value
^{85}Rb	72.17	71.14
^{87}Rb	27.83	28.86
		±1.8

Analysed data for reference material NBS 984.
Data from Houk and Thompson (1982).
Error represents two-sigma limits calculated from eight ICP–MS determinations.

20.3.2). Most of the remaining interferences listed in Table 20.3 superficially appear more serious than they probably are in practice, in that the principal (rather than the alternative) isotope can be used with appropriate correction factor. Of particular benefit to major-element analysis is the fact that ICP–MS calibrations are reported to be linear to over six orders of magnitude and limited more by the linearity of the counting electronics than the characteristics of the plasma.

The discussion above has covered in general terms quantitative analysis by ICP–MS. A survey of the literature indicates that at the time of writing (late 1984) only preliminary investigations of the application of this technique have been reported, and that these are restricted to simple solution analysis. Thus Date and Gray (1981) illustrated the simple mass spectrum observed from a solution containing 0.1 μg mL^{-1} of Cr, Co, Cu and Zn in 1% nitric acid solution (Figure 20.7). Douglas and French (1981) reported that chlorine and bromine could be determined (using a microwave excited argon plasma) from their corresponding negatively charged ions (Figure 20.8) offering some hope of compensating for the rather poor sensitivity of these elements in conventional positive ion analysis. Date and Gray (1982*a*) reported lead isotope ratios in a range of galenas and compared the results with conventional thermal ionization mass spectrometry. Their data is shown in Table 20.5. Precisions

Figure 20.9 ICP–MS spectrum of a mixed rare-earth element solution, each element present at a concentration of 10 μg/mL. This diagram shows the singly-charged part of the spectrum. Oxide peaks are very low in intensity in continuum sampling mode. (Date and Gray, 1982.)

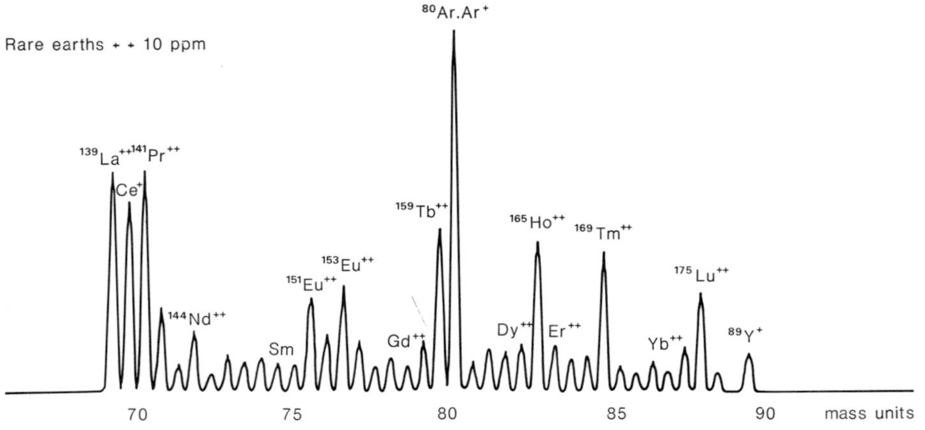

Figure 20.10 Rare-earth element spectrum as in Figure 20.9. This diagram shows the doubly-ionized region of the spectrum and includes the prominent Ar.Ar$^+$ interference. (Date and Gray, 1982.)

in the determination of these isotope ratios was of the order of 1% (one sigma) in comparison with precision errors of 0.1% expected for state-of-the-art thermal ionization mass spectrometric measurements. Houk and Thompson (1982) obtained similar orders of precision in the determination of Rb stable isotope ratios in NBS reference material 984 (see Table 20.6). Gray and Date (1982) demonstrated the determination of REE abundances by ICP–MS and their spectra for 10 ppm mixed REE solutions, as shown in Figure 20.9 (for singly-charged species) and Figure 20.10 (for doubly-charged ions). It is expected that rapid developments will occur in this field in the immediate future.

References

Chapter 1

Abbey, S. (1970) US Geological Survey standards—a critical study of published analytical data. *Can. Spectrosc.* **15,** 10–16.

Abbey, S. (1972) 'Standard samples' of silicate rocks and minerals—A review and compilation. Geol. Survey Canada, *Paper* 72–30.

Abbey, S. (1975) Studies in 'standard samples' of silicate rocks and minerals. Part 4: 1974 edition of 'usable' values. Geol. Survey Canada, *Paper* 74–41.

Abbey, S. (1977) Studies in 'standard samples' of silicate rocks and minerals. Part 5: 1977 edition of 'usable' values. Geol. Survey Canada, *Paper* 77–34.

Abbey, S. (1982) An evaluation of USGS III. *Geostand. Newslett.* **6,** 47–76.

Abbey, S. (1983) Studies in 'standard samples' of silicate rocks and minerals 1969–1982. Geol. Survey Canada, *Paper* 83–15.

Abbey, S. and R.M. Rousseau (1985) Pragmatism vs. rigour: a debate on the resolution of disparate analytical data on four Canadian iron formation reference samples. *Geostand. Newslett.* **9,** 1–16.

Ahrens, L.H. (1951) Spectrochemical analysis of some of the rarer elements in the granite and diabase samples. In: Fairburn and others (1951), *q.v.*, 53–57.

Ahrens, L.H. (1977) A story of two rocks. *Geostand. Newslett.* **1,** 157–161.

American Chemical Society, Committee on Environmental Improvements (1980). Guidelines for data acquisition and data quality evaluation in environmental chemistry. *Anal. Chem.* **52,** 2242–2249.

Andrews, D.F., P.J. Bickel, F.R. Hampel, P.J. Huber, W.H. Rogers and J.W. Turkey (1972) *Robust Estimates of Location.* Princeton University Press, Princeton.

Boumans, P.W.J.M. (1978) A tutorial review of some elementary concepts in the statistical evaluation of trace element measurements. *Spectrochim. Acta* **33B**, 625–634.

Chayes, F. (1951) Modal analyses of the granite and diabase test rocks. In: Fairburn and others (1951), *q.v.*, 59–68.

Chayes, F. (1969) A last look at G-1 and W-1. *Carnegie Inst. Yearbk.* **67**, 239–241.

Chayes, F. (1970) Another last look at G-1–W-1 *Int. Assoc. Math. Geol. J.* **2**, 207–209.

Christie, O.H.J. and K.N. Alfsen (1977) Data transformation as a means to obtain reliable consensus values for reference materials. *Geostand. Newslett.* **1,** 47–49.

Columbo, A. (1980) Another approach to the mode of a skewed set of data: the plot method. *Geostand. Newslett.* **4,** 239–242.

Coppola, P.P. and R.C.Hughes (1952) Polyethylene still for preparation of pure hydrofluoric acid. *Anal. Chem.* **24,** 768.

Cumming, G.L., J.S. Rollett, F.J.C. Rossotti and R.J. Whewell (1972) Statistical methods for the computation of stability constants. Part I: straight-line fitting of points with correlation errors. *J. Chem. Soc. (Dalton)* 2652–2658.

Curry, L.A. (1968) Limits for qualitative detection and quantitative determination. Application to radiochemistry. *Anal. Chem.* **40,** 586–593.

Dabeka, R.W., A. Mykytiuk, S.S. Berman and D.S. Russell (1976) Polypropylene for the sub-boiling distillation and storage of high-purity acids and water. *Anal. Chem.* **48,** 1203–1207.

Davis, J.C. (1973) *Statistics and Data Analysis in Geology.* John Wiley and Sons, New York.

Delaey, L. and O. Arkens (1981) The acronyms used in the world of spectroscopy, microscopy and diffractometry–I: compilation and classification. *Spectrochim. Acta* **36B,** 351–360.

Dixon, W.J. (1953) Processing data for outliers. *Biometrics* **9,** 74–89.

Dybczynski, R. (1980) Comparison of the effectiveness of various procedures for rejection of outlying results and assigning consensus values in interlaboratory programmes involving determination of trace elements or radionuclides. *Anal. Chim. Acta.* **117,** 53–70.

Dybczynski, R., A. Tugsavul and O. Suschny (1979) Soil-5, a new IAEA certified reference material for trace element determinations. *Geostand. Newslett.* **3,** 61–87.

Edelman, N. (1962) Mathematics and geology. *Geologiska Foren. Stockholm Forhand.* **84,** 343–350.

Ellis, P.J., I. Copelowitz and T.W. Steele (1977) Estimation of the mode by the dominant cluster method. *Geostand. Newslett.* **1,** 123–130.

Ellis, P.J., J.H. Venter and T.W. Steele (1983) Estimation of material variation in powdered materials. *Geostand. Newslett.* **7,** 261–272.

Fairburn, H.W. (1951) A cooperative investigation of precision and accuracy in chemical, spectrochemical and model analysis of silicate rocks. *US Geol. Survey Bull.* 980.

Ferguson, T.S. (1961) In: J. Neyman (ed.) *Fourth Berkeley Symposium on Mathematical Statistics and Probability*, University of California Press, Berkeley, 253–287.

Filby, R.H., S. Nguyen, C.A. Grimm, G.R. Markowski, V. Ekambaram, T. Tanaka and L. Grossmann (1985) Evaluation of geochemical standard reference materials for microanalysis. *Anal. Chem.* **57,** 551–555.

Flanagan, F.J. (1967) US Geological Survey silicate rock standards. *Geochim. Cosmochim. Acta* **31,** 289–308.

Flanagan, F.J. (1969) US Geological Survey standards - II. First compilation of data for the new USGS rocks. *Geochim. Cosmochim. Acta.* **33,** 81–120.

Flanagan, F.J. (1973) 1972 values for international geochemical reference samples. *Geochim. Cosmochim. Acta* **37,** 1189–1200.

Flanagan, F.J. (1976*a*) G-1 et W-1: Requiescat in pace! US Geol. Survey Prof. *Paper* **840,** 189–192.

Flanagan, F.J. (1976*b*) 1972 compilation of data on USGS standards. US Geol. Survey Prof. *Paper* **840,** 131–183.

Fleischer, M. (1965) Summary of new data on rock samples G-1 and W-1 1962–1965. *Geochim. Cosmochim. Acta* **29,** 1263–1283.

Fleischer, M. and R.E. Stevens (1962) Summary of new data on rock samples G-1 and W-1. *Geochim. Cosmochim. Acta* **26,** 525–543.

Gastwirth, J. (1966) On robust procedures. *J. Am. Stat. Assoc.* **61,** 929–948.

Gladney, E.S. C.E. Burns and I. Roelandts (1983) 1982 compilation of elemental concentrations in eleven United States Geological Survey rock standards. *Geostand. Newslett.* **7,** 3–226.

Govindaraju, K. (1980) Report (1980) on three GIT-IWG rock reference samples : anorthosite from Greenland, AN-G; basalte d'Essey-la-Cote, BE-N; granite de Beauvoir, MA-N. *Geostand. Newslett.* **4,** 49–138.

Govindaraju, K. (1984) 1984 compilation of working values and sample description for 170 international reference samples of mainly silicate rocks and minerals. *Geostand. Newslett.* Sp. Issue vol. 8.

Govindaraju, K. and H. de la Roche (1977) Rapport (1966–1976) sur les éłements en traces dans trois roches standardes géochimiques du CRPG: basalte BR et granites GA et GH. *Geostand. Newslett.* **1,** 67–100.

Griffiths, J.C. (1974) Problems of sampling in geoscience, In: M.P. Jones (1974) *q.v.* 1–11.

Grubb, F.E. (1950) Sample criteria for testing outlying observations. *Ann. Math. Stat.* **21,** 27–58.

Grubb, F.E. (1969) Procedures for detecting outlying observations in samples. *Technometrics* **11,** 1–21.

Gy, P.M. (1974) The sampling of broken ores : a review of principle and practice. In: M.P. Jones (1974) *q.v.*

Haber, A. and R.P. Runyon *General Statistics* (3rd edn.). Addison-Wesley, Reading, Mass.

Hayslett, H.T. (1981) *Statistics Made Simple.* Heinemann, London.

Heydorn, K. and E. Damsgaard (1982) Gains and losses of ultra-trace elements in polyethylene containers. *Talanta* **29,** 1019–1024.

Ingamells, C.O., J.C. Engels, and P. Switzer (1972) Effect of laboratory sampling error in geochemistry and geochronology. *Proc. 24th Int. Geol. Congress*, Section 10, 405–415.

Ingamells, C.O. and N.H. Suhr (1963) Chemical and spectrochemical analysis of standard silicate samples. *Geochim. Cosmochim. Acta* **27,** 897–910.

Jones, M.P. (ed.) (1974) *Geological, Mining and Metallurgical Sampling*. Inst. Mining Metallurgy, London.
Jones, M.P. and C.H.J. Beaven (1974) Sampling of non-Gaussian mineralogical distributions. In: M.P. Jones (1974) *q.v.*
Kaiser, H. (1978) Foundations for the critical discussion of analytical methods. *Spectrochim. Acta* **33B,** 551–576.
Kleeman, A.W. (1967) Sampling error in the chemical analysis of rocks. *J. Geol. Soc. Aust.* **14,** 43–47.
Kuehner, E.C., R. Alvarez, P.J. Paulsen and T.J. Murphy (1972) Production and analysis of special high purity acids purified by sub-boiling distillation. *Anal. Chem.* **44,** 2050–2056.
Lister, B. (1982) Evaluation of analytical data: a practical guide for geoanalysis. *Geostand. Newslett.* **6,** 175–205.
Lister, B. (1984) A note on robust estimates. *Geostand. Newslett.* **8,** 171–172.
Long, G.L. and J.D. Winefordner (1983) Limit of detection—a closer look at the IUPAC definition. *Anal. Chem.* **55,** 712A-724A.
Maas, R.P. and S.A. Dressing (1983) Purification of nitric acid at trace metal levels. *Anal. Chem.* **55,** 808–809.
Mattinson, J.M. (1972) Preparation of hydrofluoric, hydrochloric and nitric acids at ultra-low lead levels. *Anal. Chem.* **44,** 1715–1716.
McGlashan, M.L. (1971) *Physico-Chemical Quantities and Units*. Royal Institute of Chemistry, London.
Moody, J.R. (1982*a*) The sampling, handling and storage of materials for trace analysis. *Phil. Trans. Roy. Soc. London* **A308,** 669–680.
Moody, J.R. (1982*b*) NBS clean laboratories for trace element analysis. *Anal. Chem.* **54,** 1358A-1376A.
Moody, J.R. and R.M. Lindstrom (1977) Selection and cleaning of plastic containers for storage of trace element samples. *Anal. Chem.* **49,** 2264–2267.
Murphy, T.J. (1976) The role of the analytical blank in accurate trace analysis. *Proc. 7th Materials Res. Symp.*, US Govt. Printing Office, Washington, DC, 509–539.
Murphy, G.J. (1982) Some aspects of sampling in terms of mineral exploration and mine geology. In *Sampling and Analysis for the Mining Industry*, Inst. Min. Metall., London, 93–119.
Nicholls, G.D. (1974) Geochemical sampling problems in the analytical laboratory. In: M.P. Jones (1974) *q.v.*
Potts, P.J., O. Williams Thorpe and J.S. Watson (1981) Determination of the rare earth element abundances in 29 international rock standards by instrumental neutron activation analysis: a critical appraisal of calibration errors. *Chem. Geol.* **34,** 331–352.
Ratzlaff, K.L. (1979) Optimizing precision in standard addition measurement. *Anal. Chem.* **51,** 232–235.
Ridley, K.J., A. Turek and C. Riddle (1976) The variability of chemical analysis as a function of sample heterogeneity and the implications to the analysis of rock standards. *Geochim. Cosmochim. Acta* **40,** 1375–1379.
Sankar Das, M. (1979) Geometric means as probable values for compiled data on geochemical reference samples. *Geostand. Newslett.* **3,** 199–205.
Schlecht, W.G. and R.E. Stevens (1951) Results of chemical analysis of samples of granite and diabase. In: Fairburn and others (1951) *q.v.*, 7–24.
Skoog, D.A. and D.M. West (1982) *Fundamentals of Analytical Chemistry* (4th edn.). Holt-Saunders International, Philadelphia.
Steele, T.W., J. Levin and I. Copelowitz (1975) The preparation and certification of a reference sample of a precious-metal ore. Nat. Inst. Metall. Rept. (South Africa) No. 1696.
Stevens, R.E. and others (1960) Second report on a cooperative investigation of the composition of two silicate rocks. *US Geol. Survey Bull.* **1113**.
Thompson, M. (1982) Regression methods in the comparison of accuracy. *Analyst* (*London*) **107,** 1169–1180.
Thompson, G. and D.C. Bankston (1970) Samplings contamination for grinding and sieving determined by emission spectrometry. *Appl. Spectrosc.* **24,** 210–219.
Uriano, G.A. and C.C. Gravatt (1977) The role of reference materials and reference methods in chemical analysis. *CRC Critical Reviews in Analytical Chemistry* **6,** 361–411.
US Federal Standard No. 209b (1966) Clean room and work station requirements, controlled environments. Govt. Services Admin. Business Service Centre, Boston, Mass.
Vistelius, A.B. (1970) Statistical model of silicate analysis and results of investigation of G-1 and W-1 samples. *Int. Assoc. Math. Geol. J.* **2,** 1–14.
Vistelius, A.B. (1971) Some lessons of the G-1–W-1 investigation. *Int. Assoc. Math. Geol. J.* **3,** 323–326.
Weast, R.C. (editor-in-chief) (1973) *Handbook of Chemistry and Physics* (54*th edn*.). The Chemical Rubber Co., Cleveland, Ohio.
Wilson, A.D. (1964) The sampling of silicate rock powders for chemical analysis. *Analyst* (*London*) **89,** 18–30.
York, D. (1966) Least squares fitting of a straight line. *Can. J. Phys.* **44,** 1079–1086.
York, D. (1969) Least squares fitting of a straight line with correlated errors. *Earth Planet, Sci. Lett.* **5,** 320–324.
Zemany, P.D. (1978) Precision and accuracy. In: H.K. Herglotz and L.S. Birks, *X-ray spectrometry*, Marcel Dekker, New York, pp. 69–110.

Chapter 2

Abbey, S. (1983) Studies in 'standard samples' of silicate rocks and minerals 1969–1982. Geol. Survey Canada *Paper* 83–15.
Barker, J.F. and S. Chatten (1982) A technique for determining low concentrations of total carbonate in geological materials. *Chem. Geol.* **36.** 317–323.
Bernas, B. (1968) A new method for decomposition and comprehensive analysis of silicates by atomic absorption spectrometry. *Anal. Chem.* **42,** 1682–1686.
Buckley, D.E. and R.E. Cranston (1971) Atomic absorption analyses of 18 elements from a single decompositon of aluminosilicate. *Chem. Geol.* **7,** 273–284.
Carpenter, F.G. (1962). Use of gas chromatography for rapid determination of carbonate at low levels. *Anal. Chem.*, **34,** 66–67.
Cohen, R.S., P. Hemmes and J.H. Puffer (1975) *M*-benzene disulphonic acid (BDS) as a superior accompanying acid for routine silicate rock analysis. *Chem. Geol.* **16,** 307–309.
Cremer, M. and J. Schlocker (1976) Lithium borate decomposition of rocks, minerals and ores. *Am. Mineral.* **61,** 318–321.
Crock, J.G. F.E. Lichte and T.R. Wildeman (1984) The group separation of the rare earth elements and yttrium from geological materials by cation-exchange chromatography. *Chem. Geol.* **45,** 149–163.
Croudace, I.W. (1980) A possible error source in silicate wet-chemistry caused by insoluble fluorides. *Chem. Geol.* **31,** 153–155.
Culmo, R.F. (1972) The microdetermination of sulphur with an automatic elemental analyser. *Microchem. J.* **17,** 499–507.
Dean, J.A. (1960) *Flame Photometry*. McGraw-Hill, New York.
DeLong, S.E. and P. Lyman (1982) Water capture in rapid silicate analysis. *Chem. Geol.* **35,** 173–176.
Din, V.K. and G.C. Jones (1978) The determination of total carbon and combined water in silicates using a C,H,N elemental analyser. *Chem. Geol.* **23,** 347–352.
Ewing, G.W. (1975) *Instrumental Methods of Chemical Analysis*. McGraw-Hill, New York.
Fahey, J.J. (1946) Volumetric method for the determination of carbon dioxide. *US Geol. Survey Bull.* **950,** 139–141.
Fitton, J.G. and R.C.O. Gill (1970) The oxidation of ferrous iron in rocks during mechanical grinding. *Geochim. Cosmochim. Acta*, **34,** 518–524.
French, W.J. and S.J. Adams (1973) Polypropylene bottles in the decomposition of silicate materials. *Anal. Chim. Acta.* **62,** 324–328.
Fuge, R. (1967) The determination of chromium in rocks and minerals. *Chem. Geol.* **2,** 289–296.
Fuge, R. (1973) The automated photometric determination of vanadium in geological samples with 4-(2-pyridylazo) resorcinol. *Chem. Geol.* **12,** 61–66.
Fuge, R. (1976) The automated colorimetric determination of fluorine and chlorine in geological samples. *Chem. Geol.* **17,** 37–43.
Girardi, F. and E. Sabbioni (1968) Selective removal of radiosodium from neutron activated materials by retention on hydrated antimony pentoxide. *J. Radioanal. Chem.* **1,** 169–178.
Goldich, S.S. (1984) Determination of ferrous iron in silicate rocks. *Chem. Geol.* **42,** 343–347.
Govindaraju, K. and N. L'homel (1972) Direct and indirect atomic

absorption determination of silica in silicate rock samples. *Atomic Absorption Newslett.* **11,** 115–117.

Govindaraju, K., G. Mevelle and C. Chouard (1971) Laminating system for reducing borate fusion melts of silicates into glass sheets suitable for acid dissolution. *Meth. Phys. d'Analyse* **7,** 174–175.

Groves, A.W. (1951) *Silicate Analysis* (2nd edn.). Allen and Unwin, London, p.97.

Guern, F. Le, T.M. Gerlach and A. Nohl (1982) Field gas chromatograph analyses of gases from a glowing dome at Merapi Volcano, Java, Indonesia, 1977, 1978, 1979. *J. Volcano. Geotherm. Res.* **14,** 223–245.

Harmer, W.C.E. (1973) The effect of fluorine content on the determination of silica in quartzose rocks. *Mineral. Mag.* **39.** 112–113.

Haynes, S.J. and A.H. Clark (1972) A rapid method for the determination of chlorine in silicate rocks using ion selective electrodes. *Econ. Geol.* **67,** 378–382.

Hillebrand, W.F. (1919) The analysis of silicate and carbonate rocks. *US Geol. Survey Bull.* **700.**

Hopkins, D.M. (1977) An improved ion-selective electrode method for the rapid determination of fluorine in rocks and soils. *J. Res. US Geol. Survey* **5,** 589–593.

Ingamells, C.O. (1964) Rapid chemical analysis of silicate rocks. *Talanta* **11,** 665–666.

Ingamells, C.O. (1970) Lithium metaborate flux in silicate analysis. *Anal. Chim. Acta* **52,** 323–334.

Ingram, B.L. (1970) Determination of fluoride in silicate rocks without separation of aluminium using a specific ion electrode. *Anal. Chem.* **42,** 1825–1827.

Jeffery, P.G. (1975) *Chemical Methods of Rock Analysis*. Pergamon Press, Oxford.

Jeffery, P.G. and P.J. Kipping (1962) The determination of constituents of rocks and minerals by gas chromatography. Part I: The determination of carbon dioxide. *Analyst (London)* **87,** 379–382.

Jeffery, P.G. and A.D. Wilson (1960) A combined gravimetric and photometric procedure for determining silica in silicate rocks and minerals. *Analyst (London)* **85,** 478–486.

Kirschenbaum, H. (1983) The classical chemical analysis of silicate rocks—the old and the new. *US Geol. Survey Bull.* **1547.**

Kiss, E. (1977) Rapid potentiometric determination of the iron oxidation state in silicates. *Anal. Chim. Acta* **89,** 303–314.

Langmhyr, F.J. and P.R. Graff (1965) A contribution to the analytical chemistry of silicate rocks: a scheme of analysis for eleven main constituents based on decompositon by hydrofluoric acid. Norges Geologiske Undersokelse No. 230, Universitetsforlaget, Oslo.

Langmyhr, F.J. and K. Kringstad (1966) An investigation of the composition of the precipitates formed by the decompositon of silicate rocks in 38–40% hydrofluoric acid. *Anal. Chim. Acta* **35,** 131–135.

Langmyhr, F.J. and P.E. Paus (1968) The analysis of inorganic siliceous materials by atomic absorption spectrophotometry and the hydrofluoric acid decomposition technique. Part I: The analysis of silicate rocks. *Anal. Chim. Acta* **43,** 397–408.

Langmyhr, F.J. and S. Sveen (1965) Decomposability in hydrofluoric acid of the main and some minor and trace minerals of silicate rocks. *Anal. Chim. Acta* **32,** 1–7.

Maxwell, J.A. (1968) *Rock and mineral Analysis*. Wiley- Interscience, New York.

Medlin, J.H., N.H. Suhr and J.B. Bodkin (1969) Atomic absorption analysis of silicates employing $LiBO_2$ fusion. *Atomic Absorption Newslett.* **8,** 25–29.

Omang, S.H. (1969) A rapid fusion method for decomposition and comprehensive analysis of silicates by atomic absorption-spectrophotometry. *Anal. Chim. Acta* **46,** 225–230.

Peck, L.C. (1964) Systematic analysis of silicates. *US Geol. Survey Bull.* **1170.**

Penfield, S.L. (1894) On some methods for the determination of water. *Am. J. Sci.* (3rd ser.) **48,** 30–37.

Pratt, J.P. (1894) *Am. J. Sci.* (3rd Ser.) **48,** 150.

Riley, J.P. (1958*a*) The rapid analysis of silicate rocks and minerals. *Anal. Chim. Acta* **19,** 413–428.

Riley, J.P. (1958*b*) Simultaneous determination of water and carbon dioxide in rocks and minerals. *Analyst (London)* **83,** 42–49.

Riley, J.P. (1959) The use of continuous extraction for the removal of interfering elements in the determination of calcium and magnesium. *Anal. Chim. Acta* **21,** 317–323.

Robinson, J.W., (1982) *Undergraduate Instrumental analysis* (3rd edn.). Marcel Dekker, New York.

Schwarzenbach, G. (1957) *Complexiometric Titrations* (Translated by Harry Irving). Chapman and Hall, London.

Shapiro, L. (1967) Rapid analysis of rocks and minerals by a single-solution method. US Geol. Survey Prof. Paper **575–B.** B187–B191.

Shapiro, L. and W.W. Brannock (1955) Rapid determination of water in silicate rocks. *Anal. Chem.* **27,** 560–562.

Shapiro, L. and W.W. Brannock (1956) Rapid analysis of silicate rocks. *US Geol. Survey Bull.* **1036–C.**

Shapiro, L. and W.W. Brannock (1962) Rapid analysis of silicate, carbonate and phosphate rocks. *US Geol. Survey Bull.* **1144–A.**

Skoog, D.A. and D.M. West (1982) *Fundamentals of Analytical Chemistry*. Holt-Saunders, Philadelphia.

Stevens, R.E. and W.W. Niles (1960) Chemical analysis of the granite and the diabase. In: R.E. Stevens and others, 'Second report on a cooperative investigation of the composition of two silicate rocks', *US Geol. Survey Bull.* **1113,** 3–43.

Suhr, N.H. and C.O. Ingamells (1966) Solution technique for analysis of silicates. *Anal. Chem.* **38,** 730–734.

Theobald, L.S. (1942) Attack by fusion of pyrites and chromite. *Analyst (London)* **67,** 287–288.

Terashima, S. (1978) The rapid determination of total carbon and sulphur in geological materials by combustion and infra red absorption photometry. *Anal. Chim. Acta* **101,** 25–31.

Terashima, S. (1979) A direct method for the determination of carbonate and non-carbonate carbon in geological materials by infra-red spectrometry. *Anal. Chim. Acta* **108,** 113–118.

Troll, G., A. Farzaneh and K. Cammann (1977) Rapid determination of fluoride in mineral and rock samples using an ion selective electrode. *Chem. Geol.* **20,** 295–305.

Turek, A., C. Riddle, B.J. Cozens and N.W. Tetley (1976) Determination of chemical water in rock analysis by Karl Fischer titration. *Chem. Geol.* **17,** 261–267.

van Eenbergen, A. and E. Bruninx (1978) Losses of elements during sample decomposition in an acid-digestion bomb. *Anal. Chim. Acta* **98,** 405–406.

Van Loon, J.C. and C.M. Parissis (1969) Scheme of silicate analysis based on the lithium metaborate fusion followed by atomic absorption spectrophotometry. *Analyst (London)* **94,** 1057–1062.

Van Loon, J.C. (1980) *Analytical Atomic Absorption Spectroscopy, Selected Methods*. Academic Press, New York.

Washington, H.S. (1919) *Manual of the Chemical Analysis of Rocks*, (3rd edn.). John Wiley and Sons, New York.

Washington, H.S. (1932) *The Chemical Analysis of Rocks*. John Wiley and Sons, New York.

Walsh, J.N. (1980) The simultaneous determination of the major, minor and trace constituents of silicate rocks using inductively coupled plasma spectrometry. *Spectrochim. Acta* **35B,** 107–111.

Walsh, J.N. and R.A. Howie (1980) An evaluation of the performance of an inductively coupled plasma source spectrometer for the determination of the major and trace constituents of silicate rocks and minerals. *Mineral. Mag.* **43,** 967–974.

Walsh, J.N., F. Buckley and J. Barker (1981) The simultaneous determination of the rare earth elements in rocks using inductively coupled plasma source spectrometry. *Chem. Geol.* **33,** 141–153.

Weast, R.C. (editor-in-chief) (1973) *Handbook of Chemistry and Physics* (54th edn.). Chemical Rubber Publishing Co. Cleveland, Ohio.

Westland, A.D. and C.J. Kantipuly (1983) Potassium superoxide as a flux in ore and rock analysis. *Anal. Chim. Acta* **154,** 355–358.

Whipple, E.R. (1974) A study of Wilson's determination of ferrous iron in silicates. *Chem. Geol.* **14,** 223–238.

Wilson, A.D. (1955) Determination of ferrous iron in rocks and minerals. *Bull. Geol. Survey Great Britain* **9,** 56–58.

Chapter 3

Bauman, R.P. (1962) *Absorption Spectroscopy*. John Wiley and Sons, New York.

Cotton, F.A. and G. Wilkinson (1966) *Advanced Inorganic Chemistry*. Wiley-Interscience, New York.
Ebdon, L. (1982) *An Introduction to Atomic Absorption Spectroscopy*. Heyden, London.
Ewing, G.W. (1975) *Instrumental Methods of Chemical Analysis*. McGraw-Hill, New York.
Fastie, W.G. (1952) A small plane grating monochromator. *J. Opt. Soc. Am.* **42,** 641–647.
Meltzer, R.J. (1969) Spectrographs and monochromators. In: R. Kingslake (ed.), *Applied Optics and Optical Engineering*, Vol. V, Part II. Academic Press, New York, 47–84.
Moore, W.J. (1966) *Physical Chemistry*, Longman, London.
Price, W.J. (1979) *Spectrochemical Analysis by Atomic Absorption*. Heyden, London.
Richardson, D. (1969) Diffraction gratings. In: R. Kingslake (ed.) *Applied Optics and Optical Engineering*, Vol. V, Part II, Academic Press, New York. 17–46.
Ringbom, A. (1939) Accuracy of colorimetric determinations. *Z. Anal. Chem.* **115,** 332–342.
Robinson, J.W. (1982) *Undergraduate Instrumental Analysis*. Marcel Dekker, New York.
Rothman, L.D., S.R. Crouch and J.D. Ingle (1975) Theoretical and experimental investigation of factors affecting precision in molecular absorption spectrophotometry. *Anal. Chem.* **47,** 1226–1233.
Sharpe, M.R. (1984) Stray light in uv-vis spectrophotometers. *Anal. Chem.* **56,** 339A-356A.
Skoog, D.A. and D.M. West (1980) *Principles of Instrumental Analysis*. Saunders College, Philadelphia: Holt-Saunders Japan, Tokyo.
Skoog, D.A. and D.M. West (1982). *Fundamentals of Analytical Chemistry*. Saunders College, Philadelphia.
Sotera, J.J. and H.L. Kahn (1982) Background correction in AAS. *Am. Lab.*, November issue.
Thompson, M. and J.N. Walsh (1983) *A Handbook of Inductively Coupled Plasma Spectrometry*. Blackie, Glasgow and London.
Walker, S. and H. Straw (1966) *Spectroscopy, Vol. 1: Atomic Microwave and Radiofrequency Spectroscopy*. Science Paperbacks Chapman and Hall, London.
Walton, H.F. and J. Reyes (1973) *Modern Chemical Analysis and Instrumentation*. Marcel Dekker, New York.
Walsh, A. (1955) The application of atomic absorption spectra to chemical analysis. *Spectrochim. Acta* **7,** 108–117.

Chapter 4

Abbey, S. (1983) Studies in standard samples of silicate rocks and minerals 1969–1982. Geol. Survey Canada Paper 83–15.
Aldous, K.M., B.W. Bailey and J.M. Rankin (1972) Burning velocity of the premixed nitrous oxide-acetylene flame and its influence on burner design. *Anal. Chem.* **44,** 191–194.
Alkemade, C.J.T. and J.M.W. Milatz (1955) A double beam method of spectral selection with flames. *Appl. Sci. Res.* **B4,** 289–299.
Barnett, W.B. (1973) Electrodeless discharge lamps for atomic absorption. *Atomic Absorption Newslett.* **12,** 142–146.
Barredo F.B. and L.P. Diez (1980) Routine accurate determination of silica in silicate materials by atomic absorption spectrophotometry and subsequent computation. *Talanta* **27,** 69–70.
Barredo, F.B., P.A. Vasallo and L.P. Diez (1978) Determination of low amounts of strontium in geological materials by flameless atomic absorption spectrophotometry. *Chem. Geol.* **23,** 171–178.
Bernas, B. (1968) A new method for decompositon and comprehensive analysis of silicates by atomic absorption spectrometry. *Anal. Chem.* **40,** 1682–1686.
Bettinelli, M. (1983) Determination of trace metals in siliceous standard reference materials by electrothermal atomic absorption spectrometry after lithium tetraborate fusion. *Anal. Chim. Acta* **148,** 193–201.
Billings, G.L. and J.A.S. Adams (1964) Analysis of geological material by atomic absorption. *Atomic Absorption Newslett.* **23,** 65–71.
Boar, P.L. and L.K. Ingram (1970) The comprehensive analysis of coal ash and silicate rocks by atomic absorption spectrometry by a fusion technique. *Analyst (London)* **95,** 124–130.
Browner, R.F. and A.W. Boorn (1984) Sample introduction techniques for atomic spectroscopy. *Anal. Chem.* **56,** 875A–888A.
Buckley, D.E. and R.E. Cranston (1971) Atomic absorption analyses of 18 elements from a single decomposition of aluminosilicate. *Chem. Geol.* **7,** 273–284.
Capacho-Delgado, L. and S. Sprague (1965) *Atomic Absorption Newslett.* **4,** 363.
Casteldine, C.G., J.C. Robbins and N.A. Skerten (1983) A practical field-portable atomic absorption analyser. *International Labmate* **8,** (4).
Chakrabarti, C.L., S.Wu and P.C. Bertels (1983) Isothermal atomisation from a platform in graphite furnace atomic absorption spectrometry. *Spectrochim. Acta* **38B,** 1041–1060.
Cresser, M.S. and R.F. Browner (1980) A method for investigating size distributions of aqueous droplets in the range 0.5 to 1.0 μm produced by pneumatic nebulisers. *Spectrochim. Acta* **35B,** 73–79.
Delves, H.T. (1970) Microsampling method for the rapid determination of lead in blood by atomic absorption-spectrophotometry. *Analyst (London)* **95,** 431–438.
Dymott, T.C. (1985) *Atomic Absorption with Electrothermal Atomisation*. Pye Unicam, Cambridge.
Ebdon, L. (1982) *An Introduction to Atomic Absorption Spectroscopy*. Heyden, London.
Ediger, R.D. (1975) Atomic absorption analysis with the graphite furnace using matrix modification. *Atomic Absorption Newslett.* **14,** 127–130.
Ediger, R.D., G. Peterson and J.D. Kerber (1974) Application of the graphite furnace to saline water analysis. *Atomic Absorption Newslett.* **13,** 61–64.
Elson, C.M., J. Dostal, D.L. Hynes and C.A.R. de Albuquerque (1978) Silver, cadmium and lead contents of some rock reference samples. *Geostand. Newslett.* **2,** 121–123.
Ewing, G.W. (1975) *Instrumental Methods of Analysis*. McGraw-Hill, New York.
Fernandez, F.J. and D.C. Manning (1971) Determination of arsenic at sub-microgram levels by atomic absorption spectrophotometry. *Atomic Absorption Newslett.* **10,** 86–88.
Flanagan, F.J., R. Moore and P.J. Aruscavage (1982) Mercury in geological reference materials. *Geostand. Newslett.* **6,** 25–46.
Graydon, A.G. and H.G. Wolford (1960) *Flames, their Structure, Radiation and Temperature*. Chapman and Hall, London, p.304.
Gill, R.C.O. and B.I. Kronberg (1975) The precise determination of silicon in fluoborate solutions by atomic absoprtion spectrophotometry. *Atomic Absorption Newslett.* **14,** 157–159.
Govindaraju, K., R. Hermann, G. Mevelle and C. Chouard (1973) Solid sampling flame atomic absorption determination of caesium in silicate rock samples. *Atomic Absorption Newslett.* **12,** 73–76.
Govindaraju, K., J. Morel and N. L'Homel (1977) Solid sampling atomic absorption determination of silver in silicate rock reference samples. Application to a homogeneity study of silver in a one-ton two mica granite reference sample. *Geostand. Newslett.* **1,** 137–142.
Halls, D.J. (1977*a*) The formation of atoms in the air–acetylene flame. *Spectrochim. Acta* **32B,** 221–230.
Halls, D.J. (1977*b*) The efficiency of atomisation of sodium, potassium and lithium in the air–acetylene flame. *Spectrochim. Acta* **32B,** 397–412.
Holak, W. (1969) Gas sampling technique for arsenic determination by atomic absorption spectrophotometry. *Anal. Chem.* **41,** 1712–1713.
Horsky, S.J. and W.K. Fletcher (1981) Evaluation of a combined ion exchange–graphite furnace atomic absorption procedure for determination of rare earth elements in geological samples. *Chem. Geol.* **32,** 335–340.
Imai, N., S. Terashima and A. Ando (1984) Determination of selenium in twenty-eight geological reference materials by atomic absorption spectrometry. *Geostand. Newslett.* **8,** 39–41.
Ingamells, C.O. (1966) Absorptiometric methods in silicate analysis. *Anal. Chem.* **38,** 1228–1234.
Kahn, H.L. (1966) Topics in chemical instrumentation. XXV: Instrumentation for atomic absorption 2. *J. Chem. Educ.* **43,** A7 *et. seq.*, A103 *et seq.*
Kahn, H.L. (1968) Principles and practice of atomic absorption. In:

Advances in Chemistry Series **73,** 'Trace inorganics in water,' American Chemical Society, 183–228.

Kahn, H.L., G. Dulude, M. Conley and J.J. Sotera (1981) Determination of metals at high concentration by atomic absorption spectroscopy. *Canadian Res.* **14,** 27–31.

Kahn, H.L., G.E. Peterson and J.E. Schallis (1968) Atomic absorption microsampling with the sampling boat technique. *Atomic Absorption Newslett.* **7,** 35–39.

Kaiser, M.L., S.R. Koirtyohann, E.J. Hinderberger and H.E. Taylor (1981) Reduction of matrix interferences in furnace atomic absorption with the L'vov platform. *Spectrochim. Acta* **36B,** 773–783.

Khalighie, J., A.M. Ure and T.S. West (1979) An investigation of atom collection phenomena in the atomic absorption spectrometry of copper. *Anal. Chim. Acta* **107,** 191–200.

Khalighie, J., A.M. Ure and T.S. West (1980) Some observations of the mechanism of atomisation in atomic absorption spectrometry with atom trapping and electrothermal techniques. *Anal. Chim. Acta* **117,** 257–266.

Koirtyohann, S.R. and E.E. Pickett (1965) Background corrections in long path atomic absorption spectrometry. *Anal. Chem.* **37,** 601–603.

Koizumi, H. and K. Yasuda (1975) New Zeeman method for atomic absorption spectrophotometry. *Anal. Chem.* **47,** 1679–1682.

Koizumi, H. and K. Yasuda (1976*a*) Determination of lead, cadmium and zinc using the Zeeman effect in atomic absorption spectrometry. *Anal. Chem.* **48,** 1178–1182.

Koizumi, H. and K. Yasuda (1976*b*) A novel method for atomic absorption spectroscopy based on the analyte-Zeeman effect. *Spectrochim. Acta* **31B,** 523–535.

Kiss, E. (1982) Determination of silica in geological materials by atomic absorption spectrometry. *Anal. Chim. Acta* **140,** 197–204.

Langmyhr, F.J. and P.E. Paus (1968) The analysis of inorganic siliceous materials by atomic absorption spectrophotometry and the hydrofluoric acid decomposition technique. Part 1: the analysis of silicate rocks. *Anal. Chim. Acta* **43,** 397–408.

Langmyhr, F.J. and Y. Thomassen (1973) Atomic absorption spectrophotometric analysis by direct atomisation from the solid phase. I: The determination of rubidium and caesium in silicate rocks. *Z. Anal. Chem.* **264,** 122–127.

Lee, R.W. and N. Guven (1975) Chemical intereferences in atomic absorption spectrometric analysis of silicates in the fluoboric acid matrix. *Chem. Geol.* **16,** 53–58.

Lewis, B. and G. van Elbe (1943) Stability and structure of burner flames. *J. Chem. Phys.* **11,** 75–97.

Liddell, P.R. and K.G. Brodie (1980) Application of a modulated magnetic field to a graphite furnace in Zeeman effect atomic absorption spectrometry. *Anal. Chem.* **52,** 1256–1260.

Lundberg, E. and W. Frech (1981) Influence of instrumental response time on interference effects in graphite furnace atomic absorption spectrometry. *Anal. Chem.* **53,** 1437–1442.

L'vov, B.V. (1958) An investigation of atomic absorption spectra by complete vaporisation of a substance in a graphite cell. *Inzhener-Fiz. Zhur. Akad. Nauk. Belorus. SSR* (*Eng. Phys. J. USSR*) **2,** 44–52.

L'vov, B.V. (1961) Analytical use of atomic absorption spectra. *Spectrochim. Acta* **17,** 761–770.

L'vov, B.V. (1978) Electrothermal atomisation–the way towards absolute analysis. *Spectrochim. Acta* **33B,** 153–193.

L'vov, B.V., L.A. Pelieva and A.I. Sharnopolsky (1977) Decrease in the effect of the base during atomic absorption analysis of solutions in the tube furnaces by evaporation of samples from a graphite substrate. *Zh. Prikl. Spektrosk.* **27,** 395–399 (in Russian).

Manning, D.C. (1971) High sensitivity arsenic-selenium sampling system for atomic absorption spectroscopy. *Atomic Absorption Newslett.* **10,** 123–124.

Massmann, H. (1968) Comparison of atomic absorption and atomic fluorescence in graphite cuvettes. *Spectrochim. Acta* **23B,** 215–226.

Matousek, J.P. (1977) Aeropol deposition in furnace atomisation. *Talanta* **24,** 315–319.

Medlin, J.H. N.H. Suhr and J.B. Bodkin (1969) Atomic absorption analysis of silicates employing $LiBO_2$ fusion. *Atomic Absorption Newslett.* **8,** 25–29.

Morton, S.F.N. (1985) *Automated Multi-Element Analysis with the PU* 9000. Pye Unicam, Cambridge.

Oishi, K., Y. Arai, S. Mayama, S. Murayama and K. Fukuda (1980) An application of the Zeeman effect to atomic absorption spectrometry: development of spectral line sources stable in magnetic field. *Spectrochim. Acta* **35B**, 155–162.

Omang, S.H. (1969) A rapid fusion method for decomposing and comprehensive analysis of silicates by atomic absorption spectrometry. *Anal. Chim. Acta* **46,** 225–230.

Pecsok, R.L., L.D. Shields, T. Cairns and I.G. McWilliam (1976) *Modern Methods of Chemical Analysis.* John Wiley and Sons, New York.

Price, W.J. (1979) *Spectrochemical Analysis by Atomic Absorption.* Heyden, London.

Rann, C.S. and A.N. Hambly (1965) Distribution of atoms in atomic absorption flame. *Anal. Chem.* **37,** 879–884.

Rantala, R.T.T. and D.H. Loring (1975) Multi-element analysis of silicate rocks and marine sediments by atomic absorption spectrophotometry. *Atomic Absorption Newslett.* **14,** 117–120.

Rubeska, I. and V. Hlavinkova (1979) Determination of arsenic in rocks and soils by atomic absorption spectrophotometry using the MHS-1 automated hydride system. *Atomic Absorption Newslett.* **18,** 5–7.

Schnepfe, M.M. (1979) Germanium contents of USGS standard rocks by flameless atomic absorption. *Geostand. Newslett.* **3,** 93–96.

Scobbie, R. (1973) The determination of boron, silicon and tungsten by atomic absorption spectroscopy. *Technical Topics*, Varian Techtron, Springvale, Australia.

Sen Gupta, J.G. (1981). Determination of yttrium and rare earth elements in rocks by graphite furnace atomic absorption spectrometry. *Talanta* **28,** 31–36.

Sen Gupta, J.G. (1984) Determination of cerium in silicate rocks by electrothermal atomisation in a furnace lined with tantalum foil. *Talanta* **31,** 1053–1056.

Sen Gupta, J.G. (1985) Determination of the rare earths, yttrium and scandium, in silicate rocks and four new geological reference materials by electrothermal atomisation from graphite and tantalum surfaces. *Talanta* **32,** 1–6.

Shapiro, L. (1967) Rapid analysis of rocks and minerals by a single fusion method. US Geol. Survey Prof. Paper 575–B, B187–B191.

Shapiro, L. and W.W. Brannock (1962) Rapid analysis of silicate, carbonate and phosphate rocks. *US Geol. Survey Bull.* 1144–A.

Sighinolfi, G.P. (1973) Determination of thallium in geochemical reference samples by flameless atomic absorption spectroscopy. *Atomic Absorption Newslett.* **12,** 136–138.

Sighinolfi, G.P., C. Gorgoni and A.M. Santos (1980) Atomic absorption determination of ultratrace elements in geological materials by vapour, hydride-forming techniques. I: mercury. *Geostand. Newslett.* **4,** 223–227.

Skoog, D.A. and D.M. West (1980) *Principles of Instrumental Analysis.* Holt-Saunders International Edition, Philadelphia, Tokyo.

Slavin, W. (1982) Atomic absorption spectroscopy, the present and the future. *Anal. Chem.* **54,** 685A-694A.

Slavin, W. and D.C. Manning (1980) The L'vov platform for furnace atomic absorption analysis. *Spectrochim. Acta* **35B,** 701–714.

Slavin, W. and D.C. Manning (1982*a*) The graphite probe constant temperature furnace. *Spectrochim. Acta.* **37B,** 955–964.

Slavin, W. and D.C. Manning, (1982*b*) Graphite furnace interferences, a guide to the literature. *Prog. Analyt. Atom. Spectrosc.* **5,** 243–340.

Slavin, W., G.R. Carnrick and D.C. Manning (1984) Chloride interferences in graphite furnace atomic absorption spectrometry. *Anal. Chem.* **56,** 163–168.

Slavin, W., D.C. Manning and G.R. Carnrick (1981) The stabilised temperature platform furnace. *Atomic Spectrosc.* **2,** 137–145.

Slavin, W., G.R. Carnrick, D.C. Manning and E. Pruszkowska (1983) Recent experiences with the stabilized temperature platform furnace and Zeeman background correction. *Atomic Spectrosc.* **4,** 69–86.

Smith, D.D. (1983) Unpublished PhD thesis, Georgia Institute of Technology, Atlanta, Georgia.

Smith, S., R.G. Schleicher and G.M. Hieftje (1982) New atomic absorption background correction technique. Paper No. 422,

33rd Pittsburgh Conference on Analytical Chemistry and Applied Spectroscopy, Atlantic City, New Jersey.

Sotera, J.J. and H.L. Kahn (1982) Background correction in AAS. *Am. Lab.* **14,** 100–104.

Sotera, J.J., L.C. Cristiano, M.K. Conley and H.L. Kahn (1983) Reduction of matrix interferences in furnace atomic absorption spectrometry. *Anal. Chem.* **55,** 204–208.

Suhr, N.H. and C.O. Ingamells (1966) Solution technique for analysis of silicates. *Anal. Chem.* **38,** 730–734.

Trent, D. and W. Slavin (1964) Determination of the major elements in granitic and diabasic rocks by atomic absorption spectroscopy. *Atomic Absorption Newslett.* **19,** 17–22.

Van Loon, J.C. (1980) *Analytical Atomic Absorption Spectroscopy.* Academic Press, New York.

Van Loon, J.C. and C.M. Parissis (1969) Scheme of silicate analysis based on the lithium metaborate fusion followed by atomic absorption spectrophotometry. *Analyst (London)* **94,** 1057–1062.

Walsh, A. (1955) The application of atomic absorption spectra to chemical analysis. *Spectrochim. Acta* **7,** 108–117.

Walsh, A. (1980) Atomic absorption spectroscopy—stagnant or pregnant. *Spectrochim. Acta* **35B,** 643–652

Walsh, A. (1982) Atomic absorption and atomic fluorescence methods of analysis: their merits and limitations. *Phil. Trans. Roy. Soc. London* **A305,** 485–498.

Walsh, J.N. and R.A. Howie (1986) Recent developments in analytical methods: uses of inductively coupled plasma source spectrometry in applied geology and geochemistry. *Appl. Geochem.* **1.**

Warren, J. and D. Carter (1975) The determination of trace amounts of copper, vanadium, chromium, nickel, cobalt and barium in silicate rock using flame atomic absorption spectrometry. *Can. J. Spectrosc.* **20,** 1–5.

Watling, R.J. (1977) The use of a slotted quartz tube for the determination of arsenic, antimony, selenium and mercury. *Anal. Chim. Acta* **94,** 181–186.

West, M.H., J.F. Molina, C.L. Yuan, D.G. Davis and J.V. Chauvin (1979) Determination of metals in waters and organic materials by flameless atomic absorption spectrometry with a wire loop atomiser. *Anal. Chem.* **51,** 2370–2375.

West, T.S. and X.K. Williams (1969) Atomic absorption and fluorescence spectroscopy with a carbon filament atom reservoir. I: construction and operation of atom reservoir. *Anal. Chim. Acta* **45,** 27–41.

Willis, J.B. (1965) Nitrous oxide–acetylene flame in atomic absorption spectroscopy. *Nature (London)* **207,** 715–716.

Willis, J.B. (1967) Atomisation problems in atomic absorption spectroscopy. I: The operation of a typical nebuliser, spray chamber and burner system. *Spectrochim. Acta* **23A,** 811–830.

Yule, J.W. and G.A. Swanson (1969) A rapid method for decomposition and the analysis of silicates and carbonates by atomic absorption spectroscopy. *Atomic Absorption Newslett.* **8,** 30–33.

Chapter 5

Abbey, S. (1983) Studies in 'standard samples' of silicate rocks and minerals 1969–1982. Geol. Survey Canada, *Paper* 83–15.

Anderson, H., H. Kaiser and B. Meddings (1981) High precision (<0.5% RSD) in routine analysis by ICP using a high pressure (200 psig) cross-flow nebuliser. In: R.M. Barnes (ed.), *Developments in Atomic Plasma Spectrochemical Analysis.* Heyden, London, 251–277.

Anderson, T.A., A.R. Forster and M.L. Parsons (1982) ICP emission spectra: II, alkaline earth elements. *Appl. Spectrosc.* **36,** 504–509.

Babat, G.I. (1947) Electrodeless discharges and some allied problems. *J. Inst. Elect. Eng.* **94,** 27–37.

Babington, R.S. (1973) *Popular Sci. Mag.* 102.

Bankston, D.C. and N.S. Fisher (1977) Atomic emission spectrometer/spectrograph for the determination of barium in micro amounts of diatom ash. *Anal. Chem.* **49,** 1017–1023.

Bankston, D.C., S.E. Humphries and G. Thompson (1979) Major and minor oxide and trace element determination in silicate rocks by direct current plasma optical emission echelle spectrometry. *Anal. Chem.* **51,** 1218–1225.

Barnes, R.M. and R.G. Schleicher (1975) Computer simulation of RF induction-heated argon plasma discharges at atmospheric pressure for spectrochemical analysis—I preliminary investigations. *Spectrochim. Acta* **30B,** 109–134.

BDH (1983) *Laboratory Chemicals and Biochemicals.* BDH Chemicals, Poole, Dorset.

Blades, M.W. and G. Horlick (1981) The vertical and spatial characteristics of analyte emission in the inductively coupled plasma. *Spectrochim. Acta* **36B,** 861–880.

Bolton, A., J. Hwang and A.V. Voet (1983) The determination of scandium, yttrium and selected rare earth elements in geological materials by inductively coupled plasma optical emission spectrometry. *Spectrochim. Acta* **38B,** 165–174.

Borsier, M. and M. Garcia (1983) Analyse automatique d'échantillons géologiques par plasma ICP. *Spectrochim. Acta* **38B,** 123–127.

Boulos, M.I. and R.M. Barnes (1981) Induction plasma modelling: state of the art. In: R.M. Barnes (ed.), *Developments in Atomic Plasma Spectrochemical Analysis.* Heyden, London, 20–34.

Boumans, P.W.J.M. (1978) ICP atomic emission spectrometry: a multi-element analysis method for liquids and dissolved solids. *Sci. Industry* **12,** 1–6.

Boumans, P.W.J.M. (1980) *Line Coincidence Tables for Inductively Coupled Plasma Atomic Emission Spectroscopy.* Pergamon, Oxford/Elmsford, New York.

Boumans, P.W.J.M. (1982) Comment on a proposed excitation mechanism in argon ICPs. *Spectrochim. Acta* **37B,** 75–82.

Boumans, P.W.J.M. (1984) *Line Coincidence Tables for Inductively Coupled Plasma Atomic Emission Spectroscopy* (2nd edn.). Pergamon, Oxford (in two volumes).

Boumans, P.W.J.M. and F.J. de Boer (1975) Studies of an inductively-coupled high frequency argon plasma for optical emission spectrometry. II: Compromise conditions for simultaneous multi-element analysis. *Spectrochim. Acta* **30B,** 309–334.

Boumans, P.W.J.M. and F.J. de Boer (1976) Studies of a radiofrequency inductively coupled argon plasma for optical emission spectrometry. III: Interference effects under compromise conditions for simultaneous multi-element analysis. *Spectrochim. Acta* **31B,** 355–375.

Boumans, P.W.J.M. and M. Bosveld (1979) A tentative listing of the sensitivities and detection limits of the most sensitive ICP lines as derived from the fitting of experimental data for an argon ICP to the intensities tabulated for the NBS copper arc. *Spectrochim. Acta* **34B,** 59–72.

Brenner, I.B., E.A. Jones, A.E. Watson and T.W. Steele (1984) The application of a N_2–Ar medium power ICP and cation exchange chromatography for spectrometric determination of the rare earth elements in geological materials. *Chem. Geol.* **45,** 135–148.

Brenner, I.B., A.E. Watson, T.W. Steele, E.A. Jones and M. Goncalves (1981) Application of an argon-nitrogen inductively coupled radiofrequency plasma (ICP) to the analysis of geological and related materials for their rare earth contents. *Spectrochim. Acta* **36B,** 785–797.

Broekaert, J.A.C. and P.K. Hormann (1981) Separation of yttrium and rare earth elements from geological materials. *Anal. Chim. Acta* **124,** 421–425.

Broekaert, J.A.C., F. Leis and K. Laqua (1979*a*) Application of an inductively coupled plasma to the emission spectroscopic determination of rare earths in mineralogical samples. *Spectrochim. Acta* **34B,** 73–84.

Broekaert, J.A.C., F. Leis and K. Laqua (1979*b*) Some aspects of matrix effects caused by sodium tetraborate in the analysis of rare earth minerals with the aid of inductively coupled plasma atomic emission spectroscopy. *Spectrochim. Acta* **34B,** 167–175.

Browner, R.F. and A.W. Boorn (1984) Sample introduction techniques for atomic spectroscopy. *Anal. Chem.* **56,** 875A–888A.

Cantillo, A.Y., S.A. Sinex and G.R. Helz (1984) Elemental analysis of estuarine sediments by lithium metaborate fusion and direct current plasma emission spectrometry. *Anal. Chem.* **56,** 33–37.

Cantle, J. and P. Goddard (1983) Inductively coupled plasma source mass spectrometry for elemental and isotopic analysis. *Int. Labmate* **8,** issue 5.

Church, S.E. (1981) Multi-element analysis of fifty-four geochemical reference samples using inductively coupled plasma-atomic emission spectrometry. *Geostand. Newslett.* **5,** 133–160.

Crock, J.G. and F.E. Lichte (1982). Determination of rare earth elements in geological materials by inductively coupled argon plasma/atomic emission spectrometry. *Anal. Chem.* **54,** 1329–1332

Crock, J.G., F.E. Lichte and P.H. Briggs (1983) Determination of elements in National Bureau of Standards geological reference materials SRM 278 obsidian and SRM 688 basalt by inductively coupled argon plasma-atomic emission spectrometry. *Geostand. Newslett.* **7,** 335–340.

Crock, J.G., F.E. Lichte and T.R. Wildeman (1984) The group separation of the rare earth elements and yttrium from geologic materials by cation exchange chromatography. *Chem. Geol.* **45,** 149–163.

Date, A.R. and A.L. Gray (1981) Plasma source mass spectrometry using an inductively coupled plasma and a high resolution quadrupole mass filter. *Analyst* (*London*) **106,** 1255–1267.

Date, A.R. and A.L. Gray (1983) Progress in plasma source mass spectrometry. *Spectrochim. Acta* **38B,** 29–37.

Decker, R.J. (1980) Some analytical characteristics of a three-electrode dc argon plasma source for optical emission spectrometry. *Spectrochim. Acta* **35B,** 19–31.

Deming, S.N. and S.L. Morgan (1983) Teaching the fundamentals of experimental design. *Anal. Chim. Acta* **150,** 183–198.

Deming, S.N. and L.R. Parker (1978) A review of simplex optimisation in analytical chemistry. *CRC Crit. Rev. Anal. Chem.*, 187–202.

Ebdon, L. (1982) *An Introduction to Atomic Absorption Spectroscopy*. Heyden, London.

Ebdon, L., M.R. Cave and D.J. Mowthorpe (1980*a*) Simplex optimisation of inductively coupled plasmas. *Anal. Chim. Acta* **115,** 179–187.

Ebdon, L., D.J. Mowthorpe and M.R. Cave (1980*b*) A versatile torch for inductively coupled plasma spectrometry. *Anal. Chim. Acta* **115,** 171–178.

Fassel, V.A. (1977) Current and potential applications of inductively coupled plasma (ICP)–atomic emission spectroscopy (AES) in the exploration, mining and processing of materials. *Pure Appl. Chem.* **49,** 1533–1545.

Fassel, V.A. (1978) Quantitative elemental analyses by plasma emission spectroscopy. *Science* **202,** 183–191.

Fernando, L.A. (1982) Figures of merit for an ICP-echelle spectrometer system. *Spectrochim. Acta* **37B,** 859–868.

Floyd, M.A., V.A. Fassel and A.P. D'Silva (1980*a*) Computer-controlled scanning monochromator for the determination of 50 elements in geochemical and environmental samples by inductively coupled plasma-atomic emission spectrometry. *Anal Chem.* **52,** 2168–2173.

Floyd, M.A., V.A. Fassel, R.K. Winge, J.M. Katzenberger and A.P. D'Silva (1980*b*) Inductively coupled plasma-atomic emission spectroscopy: a computer controlled scanning monochromator system for the rapid sequential determination of the elements. *Anal. Chem.* **52,** 431–438.

Forster, A.R., T.A. Anderson and M.L. Parsons (1982) ICP spectra: I. Background emission. *Appl. Spectrosc.* **36,** 499–504.

Fry, R.C. and M.B. Denton (1977) High solids sample introduction for flame atomic-absorption analysis. *Anal. Chem.* **49,** 1413–1417.

Fry, R.C. and M.B. Denton (1979) Characteristics of a high solids nebuliser for flame atomic absorption spectrometry. *Appl. Spectrosc.* **33,** 393–399.

Garbarino, J.R. and H.E. Taylor (1980) A Babington-type nebuliser for use in the analysis of natural water samples by inductively coupled plasma spectrometry. *Appl. Spectrosc.* **34,** 584–590.

Greenfield, S. and D.T. Burns (1980) Comparison of argon-cooled and nitrogen-cooled plasma torches under optimised conditions based on the concept of intrinsic merit. *Anal. Chim. Acta* **113,** 205–220.

Greenfield, S., I.Ll. Jones and C.T. Berry (1964) High pressure plasmas as spectroscopic emission sources. *Analyst* (*London*) **89,** 713–720.

Greenfield, S., I.Ll. Jones, H.McD. McGeachin and P.B. Smith (1975*a*) Automatic multi-sample simultaneous multi-element analysis with a H.F. plasma torch and direct reading spectrometer. *Anal. Chim. Acta* **74,** 225–245.

Greenfield, S., H.McD. McGeachin and P.B. Smith (1975*b*) Plasma emission sources in analytical spectroscopy—I. *Talanta* **22,** 1–15.

Harrison, G.R. (1969) *MIT Wavelength Tables*. MIT Press, Cambridge, Mass.

Hieftje, G.M. (1983) Mini, micro and high efficiency torches for the ICP—toys or tools. *Spectrochim. Acta* **38B,** 1465–1481.

Human, H.G.C. and R.H. Scott (1976) The shapes of spectral lines emitted by an inductively coupled plasma. *Spectrochim. Acta* **31B,** 459–473.

Jeffery, P.G. (1975) *Chemical Methods of Rock Analysis*. Pergamon, Oxford.

Johnson and Matthey Chemicals (1978) *JMC Products for the Analyst*. Johnson and Matthey Chemicals Ltd., Royston, Herts.

Kalnicky, D.J., R.N. Kniseley and V.A. Fassel (1975) Inductively coupled plasma-optical emission spectroscopy. Excitation temperatures experienced by analyte species. *Spectrochim. Acta* **30B,** 511–525.

Kniseley, R.N., H. Amenson, C.C. Butler and V.A. Fassel (1974) Improved pneumatic nebuliser for use at low nebulising gas-flows. *Appl. Spectrosc.* **28,** 285–286.

Koirtyohann, S.R., J.S. Jones, C.P. Jester and D.A. Yates (1981) Use of spatial emission profiles and a nomenclature system as aids in interpreting matrix effects in the low-power argon inductively coupled plasma. *Spectrochim. Acta* **36B,** 49–59.

Langmyhr, F.J. and P.E. Paus (1968) The analysis of inorganic siliceous materials by atomic absorption spectrophotometry and the hydrofluoric acid decomposition technique. Part I: the analysis of silicate rocks. *Anal. Chim. Acta* **43,** 397–408.

Larson, G.F. and V.A. Fassel (1979) Line broadening and radiative recombination background interferences in inductively coupled plasma–atomic emission spectroscopy. *Appl. Spectrosc.* **33,** 592–599.

Leary, J.J., A.E. Brookes, A.F. Dorrzapf and D.W. Golightly (1982) An objective function for optimisation techniques in simultaneous multi-element analysis by inductively coupled plasma spectrometry. *Appl. Spectrosc.* **36,** 37–40.

McKinnon, P.J. and K.C. Giess (1981) A clog-free nebuliser for use in inductively coupled plasma-atomic emission spectroscopy. In: R.M. Barnes (ed.), *Developments in Atomic Plasma Spectrochemical Analysis*. Heyden, London.

McLaren, J.W., S.S. Berman, V.J. Boyko and D.S. Russell (1981) Simultaneous determination of major, minor and trace elements in marine sediments by inductively coupled plasma atomic emission spectrometry. *Anal. Chem.* **53,** 1802–1806.

McQuaker, N.R., P.D. Kluckner and G.N. Chang (1979) Calibration of an inductively coupled plasma-atomic emission spectrometer for the analysis of environmental materials. *Anal. Chem.* **51,** 888–895.

Meggers, W.F., C.H. Corliss and B.F. Scribner (1975) *Tables of Spectral Line Intensities*. Part I, arranged by element; Part II, arranged by wavelength. NBS Monograph 145, US Government Printing Off., Washington DC.

Meinhard, J.E. (1976) *ICP Inf. Newslett.* **2,** 163.

Mermet, J.M. (1975) Sur les mécanismes d'excitation des éléments introduits dans un plasma HF d'argon. *C. R. Acad. Sci. Paris* **281,** 273–275.

Michaud, E. and J.M. Mermet (1982) Iron spectrum in the 200–300 nm range emitted by an inductively coupled argon plasma. *Spectrochim. Acta* **37B,** 145–164.

Montaser, A., G.R. Huse, R.A. Wax, S.-K. Chan, D.W. Golightly, J.S. Kane and A.F. Dorrzapf (1984) Analytical performance of a low-gas-flow-torch optimised for inductively coupled plasma atomic emission spectrometry. *Anal. Chem.* **56,** 283–288.

Nelder, J.A. and R. Mead (1965) A simplex method for function minimisation. *Comput. J.* **7,** 308.

Nygaard, D.D., D.S. Chase and D.A. Leightly (undated) Choice of observation height for determination of alkali metals by ICP emission spectrometry. Instrumentation Laboratory Applications Note AID 165.

Olson, K.W., W.J. Haas and V.A. Fassel (1977) Multi-element detection limits and sample nebulisation efficiencies of an improved ultrasonic nebuliser and a conventional pneumatic nebuliser in inductively coupled plasma-atomic emission spectrometry. *Anal. Chem.* **49,** 632–637.

Parsons, M.L., A.R. Forster and D. Anderson (1980) *An Atlas of Spectral Interferences in ICP Spectroscopy*. Plenum, New York.

Ramsey, M.H., M. Thompson and B.J. Coles (1983) Modified con-

centric glass nebuliser for reduction of memory effects in inductively coupled plasma spectrometry. *Anal. Chem.* **54,** 1626–1629.
Reed, T.B. (1961) Induction-coupled plasma torch. *J. Appl. Phys.* **32,** 821–824.
Reed, T.B. (1962) *Internat. Sci. Technol.* June, 42.
Reed, T.B. (1963) Heat transfer intensity from induction plasma flames and oxy-hydrogen flames. *J. Appl. Phys.* **34,** 2266–2269.
Ripson, P.A.M., L. de Galan and J.W. de Ruiter (1982) An inductively coupled plasma using 1 L/min of argon. *Spectrochim. Acta* **37B,** 733–738.
Robin, J.P. (1982) ICP–AES at the beginning of the eighties. *Prog. Analyt. Atom. Spectrosc.* **5,** 79–110.
Schramel, P. (1983) Consideration of inductively coupled plasma spectroscopy for trace element analysis in the bio-medical and environmental fields. *Spectrochim. Acta* **38B,** 199–206.
Schutyser, P. and Jansenns, E. (1979) Evaluation of spray chambers for use in inductively coupled plasma-atomic emission spectrometry. *Spectrochim. Acta* **34B,** 443–449.
Scott, R.H., V.A. Fassel, R.N. Kniseley and D.E. Nixon (1974) Inductively coupled plasma–optical emission analytical spectrometry—a compact facility for trace analysis of solutions. *Anal. Chem.* **46,** 75–80.
Spendley, W., G.R. Hext and F.R. Himsworth (1962) *Technometrics* **4,** 441.
Suddenorf, R.F. and K.W. Boyer (1978) Nebuliser for analysis of high salt content samples with inductively coupled plasma emission spectrometry. *Anal. Chem.* **50,** 1769–1771.
Suddenorf, R.F. and K.W. Boyen (1981) New approaches to sample introduction with the inductively coupled plasma. In: R.M. Barnes (ed.), *Developments in Atomic Plasma Spectrochemical Analysis.* Heyden, London, 278–286.
Terblanche, S.P., K. Visser and P.B. Zeeman (1981) The modified sequential simplex method of optimisation as applied to an inductively coupled plasma source. *Spectrochim. Acta* **36B,** 293–297.
Thompson, M. and J.N. Walsh (1983) *A Handbook of Inductively Coupled Plasma Spectrometry.* Blackie, Glasgow and London.
Valente, S.E. and W.B. Schrenk (1970) Design and some emission characteristics of an economical dc-arc plasma jet excitation source for solution analysis. *Appl. Spectrosc.* **24,** 197.
Van Loon, J.C. (1980) *Analytical Atomic Absorption Spectroscopy.* Academic Press, New York.
Walsh, J.N. (1980) The simultaneous determination of the major, minor and trace constituents of silicate rocks using inductively coupled plasma spectrometry. *Spectrochim. Acta* **35B,** 107–111.
Walsh, J.N. (1982) Whole rock analysis by inductively coupled plasma spectrometry. In: *Sampling and Analysis for the Mining Industry,* Inst. Mining Metall., London, 79–91.
Walsh, J.N., F. Buckley and J. Barker (1981) The simultaneous determination of the rare earth elements in rocks using inductively coupled plasma source spectrometry. *Chem. Geol.* **33,** 141–153.
Walsh, J.N. and R.A. Howie (1980) An evaluation of the performance of an inductively coupled plasma source spectrometer for the determination of the major and trace constituents of silicate rocks and minerals. *Mineral. Mag. (London)* **43,** 967–984.
Walters, N.M., A. Strasheim and A.R. Oakes (1983) The influence of dispersion and stray light on the analysis of geological samples by inductively coupled plasma atomic emission spectroscopy (ICP–AES). *Spectrochim. Acta* **38B,** 959–965.
Watters, R.L. (1983) Practical limits of precision in inductively coupled plasma spectrometry. *Am. Lab.,* March, 16–25.
Wendt, R.H. and V.A. Fassel (1965) Induction-coupled plasma spectrometric excitation source. *Anal. Chem.* **37,** 920–922.
Winge, R.K., V.A. Fassel, V.J. Peterson and M.A. Floyd (1982) ICP emission spectrometry—on the selection of analytical lines, line coincidence tables and wavelength tables. *Appl. Spectrosc.* **36,** 210–221.
Winge, R.K., V.A. Fassel, V.J. Peterson and M.A. Floyd (1984) *Inductively Coupled Plasma-Atomic Emission Spectroscopy. An Atlas of Spectral Information.* Elsevier, Amsterdam.
Winge, R.K., V.J. Peterson and V.A. Fassel (1979) Inductively coupled plasma–atomic emission spectroscopy: prominent lines. *Appl. Spectrosc.* **33,** 206–219.
Yarbro, L.A. and S.N. Deming (1974) Selection and pre-processing of factors for simplex optimisation. *Anal. Chim. Acta* **73,** 391–398.

Chapter 6

Abbey, S. (1983) Studies in 'standard samples' of silicate rocks and minerals 1969–1982. Geological Survey Canada, *Paper* 83–15.
Ahrens, L.H. (1977) The story of two rocks. *Geostand. Newslett.* **1,** 157–161.
Avni, R., A. Harel and I.B. Brenner (1972) A new approach to the spectrochemical analysis of silicate rocks and minerals. *Appl. Spectrosc.* **26,** 641–645.
Boumans, P.W.J.M. and M. Bosveld (1979) A tentative listing of the sensitivities and detection limits of the most sensitive ICP lines as derived from the fitting of experimental data for an argon ICP to the intensities tabulated for the NBS copper arc. *Spectrochim. Acta* **34B,** 59–72.
Boumans, P.W.J.M. and F.J.M.J. Maessen (1969) Influences of the physical and chemical processes in the electrode cavity and the gaseous atmosphere in the arc on the emission characteristics of the dc arc for the spectrochemical analysis—I. Efficiency of particle transport from the electrode cavity to the excitation zone. *Spectrochim. Acta* **24B,** 585–610.
Bowditch, D.C. and A.E. Watson (1973) The commissioning of a direct-reading spectrometer for the analysis of silicate rocks by use of the graphite pellet technique. Rept. Nat. Inst. Metall. (South Africa) No 1571.
Brenner, I.B. and A. Harel (1976) Titanium and trace element data in USGS standard rocks SCo-1 and SGR-1. US Geol. Survey Prof. Paper 840, 95–98.
Date, A.R. (1984) British Geological Survey (pers. comm.).
de Pablo Galan, L. (1974) Direct reading emission spectroscopy analysis of geochemical samples. *Proc. 3rd Int. Geochem. Symp.,* Vancouver, 707–720.
Ewing, G.W. (1975) *Instrumental Methods of Chemical Analysis.* McGraw-Hill, New York.
Harrison, G.R. (1969) *MIT Wavelength Tables.* MIT Press, Cambridge, Mass.
Helz, A.W. (1964) A gas jet for dc arc spectroscopy. US Geol. Survey Prof. Paper 475-D, D176–D178.
Helz, A.W., F.G. Walthall and S. Berman (1969) Computer analysis of photographed optical emission spectra. *Appl. Spectrosc.* **23,** 508–518.
Holmes, K.A. (1982) British Geological Survey (pers. comm.).
Maessen, F.J.M.J., J.W. Elgersma and P.W.J.M. Boumans (1976) A systematic and rigorous statistical approach for establishing the accuracy of analytical results and its application to a comparison of alternative dc arc procedures for trace analysis of geological materials. *Spectrochim. Acta* **31B,** 179–199.
Meggers, W.F., C.H. Corliss and B.F. Scriber (1975) *Tables of Spectral Line Intensities.* Part I, arranged by element; Part II, arranged by wavelength. NBS monograph 145, US Government Printing Office, Washington DC.
Myers, A.T., R.G. Havens and P.J. Dunton (1961) A spectrochemical method for the semiquantitative analysis of rocks, minerals and ores. *US Geological Survey Bull.* **1084-I.**
Ordelman, J.E., H.A. Smit and A. Tolk (1966) A general method for quantitative spectrochemical analysis with the dc arc performed with solid samples in briquette form. *Spectrochim. Acta* **22,** 313–321.
Skoog, D.A. and D.M. West (1980) *Principles of Instrumental Analysis.* Holt Saunders International Edition, Philadelphia.
Stallwood, B.J. (1954) Air cooled electrodes for the spectrochemical analysis of powders. *Optical Soc. America J.* **44,** 171–176.
Stüwer, D. (1976) On-line evaluation of photoplates in spark source mass spectrometry. *Int. J. Mass Spectrom. Ion Phys.* **20,** 387–402.
Vanderborght, B. and R. van Grieken (1978) Automated evaluation of photographically recorded spark source mass spectra. *Anal. Chim. Acta* **103,** 223–235.
van Hoyle, E., F. Adams and R. Gijbels (1975) Critical evaluation of an automatic processing system for spark source mass spectrometry. *Bull. Soc. Chim. Belg.* **84,** 595–615.
Walthall, F.G., A.F. Dorrzapf and F.J. Flanagan (1976) Computerised spectrographic data for USGS standards. US Geol. Survey Prof. Paper 840, 99–115.
Watson, A.E. and G.M. Russell (1978) The analysis of geological samples for trace elements by direct-reading emission spectrometry. *Spectrochim. Acta* **33B,** 143–152.

Chapter 7

Abbey, S. (1983) Studies in 'standard samples' of silicate rocks and minerals 1969–1982. Geol. Survey Canada, *Paper* 83–15.

Aruscavage, P.J. and E.Y. Campbell (1983) An ion-selective electrode method for determination of chlorine in geological materials. *Talanta* **30,** 745–749.

Bixler, J.W. and L.S. Solomon (1984) Restoration of unresponsive fluoride ion selective electrodes. *Anal. Chem.* **56,** 3004–3005.

Bloxam, T.W. (1968) A rapid method for the determination of fluoride in silicate rocks. *Chem. Geol.* **3,** 89–94.

Bodkin, J.D. (1977) Determination of fluorine in silicates by use of an ion-selective electrode following fusion with lithium metaborate. *Analyst (London)* **102,** 409–413.

Boyle, D.R. (1982) The analysis of fluorine in geochemical exploration—reply. *J. Geochem. Explor.* **16,** 242–247.

Evans, K.L., J.G. Tarter and C.B. Moore (1981) Pyrohydrolytic-ion chromatographic determination of fluorine, chlorine and sulphur in geological samples. *Anal. Chem.* **53,** 925–928.

Farzaneh, A. and G. Troll (1977) Pyrohydrolysis for the rapid determination of small and large amounts of fluorine in fluorides, silicates and rocks using an ion-selective electrode. *Geochem. J.* **11,** 177–181.

Farzaneh, A. and G. Troll (1978) Pyrohydrolysis for the rapid determination of chlorine traces in silicate and non-silicate minerals and rocks. *Fresenius Z. Anal. Chem.* **292,** 293–295.

Frant, M.S. and J.W. Ross (1966) Electrode for sensing fluoride ion activity in solution. *Science* **154,** 1553–1555.

Haynes, S.J. and A.H. Clark (1972) A rapid method for the determination of chlorine in silicate rocks using ion-selective electrodes. *Econ. Geol.* **67,** 378–382.

Hopkins, D.M. (1977) An improved ion-selective electrode method for the rapid determination of fluorine in rocks and soils. *J. Res. US Geol. Survey* **5,** 589–593.

Ingram, B.L. (1970) Determination of fluoride in silicate rocks without separation of aluminium using a specific ion electrode. *Anal. Chem.* **42,** 1825–1827.

IUPAC (International Union of Pure and Applied Chemistry) (1976) Recommendations for nomenclature of ion-selective electrodes. *Int. Union Pure Appl. Chem., Anal. Chem. Div. Pure Appl. Chem.* **48,** 127–132.

Jagner, D. and V. Pavlova (1972) A standard addition titration method for the determination of fluorine in silicate rocks. *Anal. Chim. Acta* **60,** 153–158.

Nicholson, K. (1982) The analysis of fluorine in geochemical exploration—discussion. *J. Geochem. Explor.* **16,** 239–242.

Nicholson, K. (1983) Fluorine determination in geochemistry: errors in the electrode method of analysis. *Chem. Geol.* **38,** 1–22.

Nicholson, K. and E.J. Duff (1981) Fluoride determination in water: an optimum buffer system for use with the fluoride-selective electrode. *Anal. Lett.* **14,** 493–517.

Olade, M.A. (1976) Rapid and simultaneous determination of chlorine and fluorine in rocks and an application in detailed geochemical exploration. *J. Mining Geol.* **13,** 28–32.

Pohl, C.A. and E.L. Johnson (1980) Ion chromatography—the state of the art. *J. Chromat. Sci.* **18,** 442–452.

Shapiro, L. (1967) A simple and rapid indirect determination of fluorine in minerals and rocks. US Geol. Survey Prof. Paper 575D, 233–235.

Skoog, D.A. and D.M. West (1980) *Principles of Instrumental Analysis.* Saunders College, Philadelphia/Holt-Saunders, Japan, Tokyo.

Small, H., T.S. Stevens and W.C. Bauman (1975) Novel ion exchange chromatographic method using conductimetric detection. *Anal. Chem.* **47,** 1801–1809.

Stecher, O. (1983) Fluorine in twenty-two international reference rock samples and a compilation of fluorine values for the USGS reference samples. *Geostand. Newslett.* **7,** 283–287.

Terashima, S. (1979) Determination of total carbon and sulphur in forty-two geochemical reference samples by combustion and infra-red spectrometry. *Geostand. Newslett.* **3,** 195–198.

Troll, G. and A. Farzaneh (1980) Determination of fluorine, chlorine and water in eight USGS reference samples. *Geostand. Newslett.* **4,** 37–38.

Troll, G., A. Farzaneh and K. Cammann (1977) Rapid determination of fluoride in mineral and rock samples using an ion-selective electrode. *Chem. Geol.* **20,** 295–305.

Weast, R.C. (editor-in-chief) (1973) *Handbook of Chemistry and Physics* (54th edn.). Chemical Rubber Co., Cleveland, Ohio.

Wilson, S.A. and C.A. Gent (1983) Determination of chloride in geological samples by ion chromatography. *Anal. Chim. Acta* **148,** 299–303.

Chapter 8

Abbey, S. (1983) Studies in 'standard samples' of silicate rocks and minerals 1969–1982. Geol. Surv. Canada, *Paper* 83–15.

Andermann, G. and J.D. Allen. *Anal. Chem.* **33,** 1695.

Andermann, G. and J.W. Kemp (1958) Scattered x-rays as internal standards in x-ray emission spectroscopy. *Anal. Chem.* **30,** 1306–1309.

Bambynek, W., B. Crasemann, R.W. Fink, H.-U. Freund, H. Mark, C.D. Swift, R.E. Price and P. Venugopala Rao (1972) X-ray fluorescence yields, Auger and Coster-Kronig transition probabilities. *Rev. Mod. Phys.* **44,** 716–813.

Banerjee, P. (1983) A detailed evaluation of counting losses of detectors in wavelength-dispersion x-ray spectroscopy. *X-ray Spectrom.* **12,** 97–105.

Bennett, H. and G.J. Oliver (1976) Development of fluxes for the analysis of ceramic materials by x-ray fluorescence spectrometry. *Analyst (London)* **101,** 803–807.

Bertin, E.P. (1975) *Principles and Practice of X-ray Spectrometric Analysis* (2nd edn.). Plenum, New York.

Biggar, G.M. (1977) Some disadvantages of $Pt_{95}Au_5$ as a container for molten silicates. *Mineral. Mag.* **41,** 551–556.

Bowles, J.F.W. (1978) Quantitative microprobe analysis of uranium minerals. *Microscope* **26,** 55–67.

Brown, G.C., D.J. Hughes and J. Esson (1973) New XRF data retrieval techniques and their application to USGS standard rocks. *Chem. Geol.* **11,** 223–229.

Claisse, F. (1956) Accurate x-ray fluorescence analysis without internal standard. Preliminary Report No 327, Department of Mines, Quebec.

Claisse, F. (1957) Letter to the editor. *Norelco Reporter* **4,** 95–96.

Cremer, M. and J. Schlocker (1976) Lithium borate decomposition of rocks, minerals and ores. *Am. Mineral.* **61,** 318–321.

Criss, J.W. and L.S. Birks (1968) Calculation methods for fluorescent x-ray spectrometry. Empirical coefficients versus fundamental parameters. *Anal. Chem.* **40,** 1080–1086.

Eby, G.N. (1972) Determination of rare earth, yttrium and scandium abundances in rocks and minerals by an ion-exchange x-ray fluorescence procedure. *Anal. Chem.* **44,** 2137–2143.

Fabbi, B.P. (1978) Geology. In: H.K. Herglotz and L.S. Birks (eds.), *X-ray Spectrometry*. Marcel Dekker, New York, 297–353.

Flanagan, F.J. (1973) 1972 values for international geochemical reference samples. *Geochim. Cosmochim. Acta* **37,** 1189–1200.

Forte, M. (1983) Fabrication and use of permanent monitors and standards. *X-ray Spectrom.* **12,** 115–117.

Fryer, B.J. (1977) Rare earth evidence in iron formation for changing Precambrian oxidation states. *Geochim. Cosmochim. Acta* **41,** 361–367.

Galson, D.A., B.P. Atkin and P.K. Harvey (1983) The determination of low concentrations of U, Th and K by XRF spectrometry. *Chem. Geol.* **38,** 225–237.

Gladney, E.S., C.E. Burns and I. Roelandts (1983) 1982 compilation of elemental concentrations in eleven United States Geological Survey rock standards. *Geostand. Newslett.* **7,** 3–226.

Govindaraju, K. and R. Montanari (1978) Routine performance of a matrix correction-free fluorescence spectrometric method for rock analysis. *X-ray Spectrom.* **7,** 148–151.

Govindaraju, K. and H. de la Roche (1977) Rapport (1966–1976) sur les éléments en trace dans trois standards géochimiques du CRPG: Basalte BR et granites, GA et GH. *Geostand. Newslett.* **1,** 67–100.

Harvey, P.K. and B.P. Atkin (1981) The rapid determination of Rb, Sr and their ratios in geological materials by x-ray fluorescence spectrometry using a rhodium x-ray tube. *Chem. Geol.* **32,** 291–301.

Harvey, P.K. and B.P. Atkin (1982) Automated x-ray fluorescence

analysis. In: *Sampling and Analysis for the Mining Industry*, Inst. Mining Metall., London, 17–26.
Harvey, P.K., D.M. Taylor, R.D. Hendry and F. Bancroft (1973) An accurate fusion method for the analysis of rocks and chemically related materials by x-ray fluorescence spectrometry. *X-ray Spectrom.* **2,** 33–44.
Heinrich, K.J.F. (1966) X-ray absorption uncertainty. In: T.D. McKinley, K.J.F. Heinrich and D.B. Wittry (eds.), *Electron Microprobe*. John Wiley and Sons, New York, 296–377.
Holland, J.G. and D.W. Brindle (1966) A self-consistent mass absorption correction for silicate analysis by x-ray fluorescence. *Spectrochim. Acta* **22,** 2083–2093.
Hower, J. (1959) Matrix corrections in the x-ray spectrographic trace element analysis of rocks and minerals. *Am. Mineral.* **44,** 19–32.
Hutton, J.T. and S.M. Elliott (1980) An accurate XRF method for the analysis of geochemical exploration samples for major and trace elements using one glass disc. *Chem. Geol.* **29,** 1–11.
Jenkins, R. (1976) *An Introduction to X-ray Spectrometry.* Heyden, London.
Jenkins, R., R.W. Gould and D. Gedcke (1981) *Quantitative X-ray Spectrometry*. Marcel Dekker, New York.
Jenkins, R. and P.W. Hurley (1968) Escape peak interference in x-ray spectrochemical analysis. *Can. Spectrosc.* **13,** 35–39.
Jenkins, R. and J.L. de Vries (1970) *Practical X-ray Spectrometry* (2nd edn.). Macmillan, London.
Jenkins, R. and J.L. de Vries (1978) *Worked Examples in X-ray Analysis* (2nd edn.). Macmillan, London.
Johannes, W. and B. Bode (1978) Loss of iron to the Pt-container in melting experiments with basalts and a method to reduce it. *Contrib. Mineral. Petrol.* **67,** 221–225.
de Jongh, W.K. (1973) X-ray fluorescence analysis applying theoretical matrix corrections. Stainless steel. *X-ray Spectrom.* **2,** 151–158.
Kikkert, J. (1983) Practical geochemical analysis of samples of variable composition using x-ray fluorescence spectrometry. *Spectrochim. Acta* **38B,** 809–820.
Lachance, G.R. and R.J. Traill (1966) Practical solution to the matrix problem in x-ray analysis. *Can. Spectrosc.* **11,** 43–48.
Leake, B.E., G.L. Hendry, A. Kemp, A.G. Plant, P.K. Harvey, J.R. Wilson, J.S. Coats, J.W. Aucott, T. Lunel and R.J. Howarth (1969) The chemical analysis of rock powders by automatic x-ray fluorescence. *Chem. Geol.* **5,** 7–86.
Lee, R.F. and D.M. McConchie (1982) Comprehensive major and trace element analysis of geological material by X-ray fluorescence using low dilution fusions. *X-ray Spectrom.* **11,** 55–63.
Leoni, L. and M. Saitta (1977) Matrix effect corrections by Ag $K\alpha$ Compton scattered radiation in the analysis of rock samples for trace elements. *X-ray Spectrom.* **6,** 181–186.
Leroux, J. and T.P. Thinh (1977) *Revised Tables of X-ray Mass Attenuation Coefficients.* Corporation Scientifique Claisse, Quebec.
Lindsay, J.R., H.J. Rose and R.R. Larson (1982) An evaluation of the Rasberry-Heinrich model for the analysis of silicate rocks. *Applied Spectrosc.* **36,** 520–523.
Loch, G. (1973) Ein beitrag zur optimierung der röntgenspektralanalyse durch verwendung spezifischer zahlrohrgase. *X-ray Spectrom.* **2,** 125–128.
Lucas-Tooth, J. (1978) Electron excitation. In: H.K. Herglotz and L.S. Birks (eds.), *X-ray Spectrometry*. Marcel Dekker, New York, 111–122.
Mahan, K.I. and D.E. Leyden (1982) Techniques for the preparation of lithium tetraborate fused single and multi-element standards. *Adv. X-ray Anal.* **25,** 95–102.
le Maitre, R.W. and M.T. Haukka (1973) The effect of prolonged x-ray irradiation on lithium tetraborate glass discs as used in XRF analysis. *Geochim. Cosmochim. Acta* **37,** 708–710.
Moseley, H.G.J. (1913) The high frequency spectra of the elements. *Phil. Mag.* **26,** 1024–1034.
Moseley, H.G.J. (1914) The high frequency spectra of the elements II. *Phil. Mag.* **27,** 703–714.
Nesbitt, R.W., H. Mastins, G.W. Stolz and D.R. Bruce (1976) Matrix corrections in trace element analysis by x-ray fluorescence: an extension of the Compton scattering technique to long wavelengths. *Chem. Geol.* **18,** 203–213.
Nisbet, E.G., V.J. Dietrich and A. Esenwein (1979) Routine trace element determination in silicate minerals and rocks by x-ray fluorescence. *Fortschr. Miner.* **57,** 264–279.
Norrish, K. and B.W. Chappell (1977) X-ray fluorescence spectrometry. In: J. Zussman (ed.), *Physical Methods in Determinative Mineralogy* (2nd edn.). Academic Press, London, 201–272.
Norrish, K. and J.T. Hutton (1969) An accurate x-ray spectrographic method for the analysis of a wide range of geological samples. *Geochim. Cosmochim. Acta* **33,** 431–453.
Pankhurst, R.J. and R.K. O'Nions (1973) Determination of Rb/Sr and $^{87}Sr/^{86}Sr$ ratios of some standard rocks and evaluation of x-ray fluorescence spectrometry in Rb-Sr geochemistry. *Chem. Geol.* **12,** 127–136.
Rasberry, S.D. and K.F.J. Heinrich (1974) Calibration for inter-element effects in x-ray fluorescence analysis. *Anal. Chem.* **46,** 81–89.
Reed, S.J.B. (1975) *Electron Microprobe Analysis*. Cambridge University Press, Cambridge.
Reynolds, R.C. (1963) Matrix corrections in trace elements analysis by x-ray fluorescence: estimation of the mass absorption coefficient by Compton scattering. *Am. Mineral.* **48,** 1133–1143.
Reynolds, R.C. (1967) Estimation of mass absorption coefficients by Compton scattering: improvements and extensions of the method. *Am. Mineral.* **52,** 1493–1502.
Riviere, J.C. (1982) Surface specific analytical techniques. *Phil. Trans. Roy. Soc. Lond.* **A305,** 545–589.
Robinson, D. and M.C. Bennett (1981) XRF determinations of 19 trace elements in international geochemical reference samples. *Geostand. Newslett.* **5,** 175–181.
Robinson, J.L. and A.E. Davis (1977) Analysis of limestone survey samples by direct electron excitation x-ray spectrometry. *Inst. Geol. Sci. London Rpt.* **77,** 1–7.
Rose, H.J., I. Adler and F.J. Flanagan (1962) Use of La_2O_3 as a heavy absorber in x-ray fluorescence analysis of silicate rocks. US Geol. Survey Prof. Paper 450-B, 106–121.
Schroeder, B., G. Thompson, M. Sulanowska and J.N. Ludden (1980) Analysis of geological materials using an automated x-ray fluorescence system. *X-ray Spectrom.* **9,** 198–205.
Seaman, A. (1979) A comparison of the transmission properties of 1 μm and 6 μm polypropylene flow counter windows. *X-ray Spectrom.* **8,** 7–8.
Sumartojo, J. and M.W. Paris (1980) A method for measuring X-ray mass absorption coefficients of geological materials. *Chem. Geol.* **28,** 341–347.
Tertian, R. and F. Claisse (1982) *Principles of Quantitative X-ray Fluorescence Analysis.* Heyden, London.
Thomas, L. and M.T. Haukka (1978) XRF determination of trace and major elements using a single fused disc. *Chem. Geol.* **21,** 39–50.
Togel, K. (1961) Preparation technique for x-ray spectrometry. In: E.A.W. Muller (ed.), *Zerstorungsfreie Materialprufung*. R. Oldenburg, Munich.
Traill, R.J. and G.R. Lachance (1966) Practical solution to the matrix problem in x-ray analysis. II. Application to a multi-component alloy system. *Can. Spectrosc.* **11,** 63–71.
Turek, A., C. Riddle and T.E. Smith (1977) Determination of Rb and Sr by x-ray fluorescence in measurement of radiometric ages. *Can. Spectrosc.* **22,** 20–24.
Verdurmen, E.A.Th. (1977) Accuracy of x-ray fluorescence spectrometric determination of Rb and Sr concentrations in rock samples. *X-ray Spectrom.* **6,** 117–122.
Vie le Sage, R., J.P. Quisefit, R. Dejean de La Batie and J. Faucherre (1979) Utilisation du rayonnement primaire diffuse par l'échantillon pour une détermination rapide et précise des éléments traces dans les roches. *X-ray Spectrom.* **8,** 121–128.
Walker, D. (1973) Behaviour of x-ray mass absorption coefficients near absorption edges. Reynold's method revisited. *Am. Mineral.* **58,** 1069–1072.
White, E.W. and G.G. Johnson (1970) *X-ray Emission and Absorption Wavelengths and Two-Theta Tables.* ASTM Data series DS37A, American Society for Testing and Materials, Philadelphia.
Woldseth, R. (1973) *All You Ever Wanted to Know About X-ray Energy Spectrometry*. Kevex Corp., Burlingame, California.
Wybenga, F.T. (1978) The evaluation of thallium acid phthalate

crystal for x-ray fluorescence spectrometry. *X-ray Spectrom.* **7,** 33–37.

Chapter 9

Abbey, S. (1983) Studies in 'standard samples' of silicate rocks and minerals 1969–1982. Geological Survey of Canada, *Paper* 83–15.

Aiginger, H., P. Wobrauschek and C. Brauner (1974) Energy-dispersive fluorescence analysis using Bragg-reflected polarized x-rays. *Nucl. Instrum. Meth.* **120,** 541–542.

Bloomfield, D.J., G. Love and V.D. Scott (1983) Evaluation of dead-time corrections in EDS systems. *X-ray Spectrom.* **12,** 2–7.

Brown, G.C., D.J. Hughes and J. Esson (1973) New XRF data retrieval techniques and their application to USGS standard rocks. *Chem. Geol.* **11,** 223–229.

Cotten, D. and H.E. Hall (1976) X-ray rare earth analyses using an intrinsic germanium detector. *Proc 12th Rare Earth Research Conf.* **1,** 373–377.

Deslattes, R.D. (1983) X-ray fluorescence spectroscopy. *Nucl. Instrum. Meth.* **208,** 655–658.

Dzubay, T.G., B.V. Jarrett and J.M. Jaklevic (1974) Background reduction in x-ray fluorescence spectra using polarisation. *Nucl. Instrum. Meth.* **115,** 297–299.

Erdtmann, G. and W. Soyka (1979) *The Gamma-Rays of the Radionuclides.* Verlag Chemie, Weinheim.

Fink, R.W. (1981) Properties of silicon and germanium semiconductor detectors for x-ray spectra. In: K.F.J. Heinrich, D.E. Newbury, R.L. Myklebust and C.E. Fiori (eds.), *Energy Dispersive X-ray Spectrometry.* Nat. Bureau Stds Spec. Publ. **604,** 5–34.

Furnas, T.C., G.S. Kuntz and R.E. Furnas (1982) Toroidal monochromators in hybrid XRF system improve effectiveness of ED-XRF tenfold. *Adv. X-ray Anal.* **25,** 59–62.

Gedcke, D.A. (1972) The Si(Li) x-ray energy analysis system: operating principles and performance. *X-ray Spectrom.* **1,** 129–141.

Gedcke, D.A., E. Elad and P.B. Denee (1977) An intercomparison of trace element excitation methods for energy dispersive fluorescence analysers. *X-ray Spectrom.* **6,** 21–29.

Giauque, R.D., R.B. Garrett and L.Y. Goda (1977*a*) Determination of forty elements in geochemical samples and coal fly ash by x-ray fluorescence spectrometry. *Anal. Chem.* **49,** 1012–1017.

Giauque, R.D., R.B. Garrett and L.Y. Goda (1977*b*) Energy dispersive x-ray fluorescence spectrometry for determination of twenty-six trace and two major elements in geochemical specimens. *Anal. Chem.* **49,** 62–67.

Gilfrich, J.V., E.F. Skelton, D.J. Nagel, A.W. Webb, S.B. Qadri and J.P. Kirkland (1982) X-ray fluorescence analysis using synchrotron radiation. *Adv. X-ray Anal.* **26,** 313–323.

Gladney, E.S., C.E. Burns and I. Roelandts (1983) 1982 compilation of elemental concentrations in eleven United States Geological Survey rock standards. *Geostand. Newslett.* **7,** 3–226.

Goulding, F.S., J.M. Jaklevic and B.V. Jarrett (1972) Detector background and sensitivity of semi-conductor x-ray fluorescence spectrometers. *Adv. X-ray Anal.* **15,** 470–482.

Govindaraju, K. (1980) Report (1980) on three GIT-IWG rock reference samples: anorthosite from Greenland, AN-G; basalte d'Essey-la-Cote, BE-N; granite de Beauvoir, MA-N. *Geostand. Newslett.* **4,** 49–138.

Grodstein, G.W. (1957) *X-ray attenuation coefficients from 10 keV to 100 MeV.* Nat. Bureau Stds Circular **583.**

Hebert, A.J. and K. Street (1974) Non-dispersive soft x-ray fluorescence spectrometer for quantitative analysis of the major elements in rocks and minerals. *Anal. Chem.* **46,** 203–207.

Holland, J.G. and D.W. Brindle (1966) A self-consistent mass absorption correction for silicate analysis by x-ray fluorescence. *Spectrochim. Acta* **22,** 2083–2093.

Jaklevic, J.M. and F.S. Goulding (1978) Energy dispersion. In: H.K. Herglotz and L.S. Birks, *X-ray Spectrometry.* Marcel Dekker, New York, 17–57.

Jaklevic, J.M., F.S.Goulding and D.A. Landis (1972) High rate x-ray fluorescence analysis by pulsed excitation. *IEEE Trans. Nucl. Sci.* **3,** 392.

Jaklevic, J.M., D.A. Landis and F.S. Goulding (1976) Energy dispersive x-ray fluorescence spectrometry using pulsed x-ray excitation. *Adv. X-ray Anal.* **19,** 253–265.

Jenkins, R., R.W. Gould and D. Gedcke (1981) *Quantitative X-ray Spectrometry.* Marcel Dekker, New York, 172–177.

Johnson, R.G. (1984) Trace element analysis of silicates by means of energy-dispersive x-ray spectrometry. *X-ray Spectrom.* **13,** 64–68.

Kandiah, K., A.J. Smith and G. White (1975) A pulse processor for x-ray spectrometry with Si(Li) detectors. *IEEE Trans. Nucl. Sci.* **NS-22,** 2058–2063.

Kandiah, K., A. Stirling, D.L. Trotman and G. White (1968) A fast high resolution spectrometer for use with nuclear radiation detectors. *Int. Symp. Nuclear Electronics, Versailles* **1,** 69-1–69-22.

Kaufman, L. and D.C. Camp (1975) Polarised radiation for x-ray fluorescence analysis. *Adv. X-ray Anal.* **18,** 247–258.

Kevex Corporation. XRF applications report: geological applications. Kevex Corp., Foster City, California.

Knoth, J. and H. Schwenke (1978) An x-ray fluorescence spectrometer with totally reflecting sample support for trace analysis at the ppb level. *Fresnius Z. Anal. Chem.* **291,** 200–204.

Knoth, J. and H. Schwenke (1980) A new totally reflecting x-ray fluorescence spectrometer with detection limits below 10^{-11} g. *Fresnius Z. Anal. Chem.* **301,** 7–9.

Labreque, J.J., H. Schorin, P. Rosales and W.C. Parker (1982) The determination of strontium and yttrium in Venezuelan laterites from Cerro Impacto by both energy- and wavelength-dispersive x-ray fluorescence. *Chem. Geol.* **35,** 357–366.

Laurer, G.R., J. Furfaro, M. Carlos, W. Lei, R. Ballad and T.J. Kneip (1982) Energy dispersive analysis of actinides, lanthanides and other elements in soil and sediment samples. *Adv. X-ray Anal.* **25,** 201–208.

Link Systems. *Instruction Manual: 2010 pulse processor.* Link Systems Ltd, High Wycombe, Bucks.

Mahan, K.I. and D.E. Leyden (1983) Simultaneous determination of sixteen major and minor elements in river sediments by energy dispersive x-ray fluorescence spectrometry after fusion in lithium tetraborate glass. *Anal. Chim. Acta* **147,** 123–131.

Markowicz, A. (1984) Theoretical evaluation of the efficiency of Compton scattered radiation method in ED-XRF analysis. *X-ray Spectrom.* **13,** 166–169.

McCarthy, J.J. and F.H. Schamber (1981) Least squares fit with digital filter: a status report. In: K.F.J. Heinrich, D.E. Newbury, R.L. Myklebust and C.E. Fiori (eds.), *Energy Dispersive X-ray Spectrometry.* Nat. Bureau Stds Spec. Publ. **604,** 273–296.

Norrish, K. and J.T. Hutton (1969) An accurate x-ray spectrographic method for the analysis of a wide range of geological materials. *Geochim. Cosmochim. Acta* **33,** 431–453.

Ong, P.S. and J.N. Randall (1978) A focusing x-ray polariser for energy dispersive analysis. *X-ray Spectrom.* **7,** 241–248.

Potts, P.J., P.C. Webb and J.S. Watson (1984) Energy dispersive x-ray fluorescence analysis of silicate rocks for major and trace elements. *X-ray Spectrom.* **13,** 2–15.

Potts, P.J., P.C. Webb and J.S. Watson (1985) Energy dispersive x-ray fluorescence analysis of silicate rocks: comparisons with wavelength-dispersive performance. *Analyst (London)* **110,** 507–513.

Potts, P.J., P.C. Webb and J.S. Watson (1986) Selective excitation of Cr, V and Br in silicate rocks by energy dispersive x-ray fluorescence using a cobalt anode x-ray tube. *J. Anal. At. Spectrosc.* (in press).

Ryon, R.W. (1977) Polarised radiation produced by scatter for energy dispersive x-ray fluorescence trace analysis. *Adv. X-ray Anal.* **20,** 575–590.

Ryon, R.W. and J.D. Zahrt (1979) Improved x-ray fluorescence capabilities by excitation with high intensity polarised x-rays. *Adv. X-ray Anal.* **22,** 453–460.

Sandborg, A.O. and J.C. Russ (1977) Counting performance of pulsed-tube systems. *Adv. X-ray Anal.* **20,** 547–554.

Schamber, F.H. (1973) A new technique for deconvolution of complex x-ray energy spectra. *Proc. 8th Nat. Conf. on Electron Probe Analysis, New Orleans,* Paper 85.

Schamber, F.H. (1977) A modification of the linear least-squares fitting method which provides continuum suppression. In: T.G. Dzubay (ed.), *X-ray Fluorescence Analysis of Environmental Samples.* Ann Arbor Science Publ., 241–257.

Spatz, R. and K.H. Lieser (1977) Critical comparison of the measuring range and detection limits in energy dispersive x-ray fluorescence analysis with tube excitation (secondary target) and with radio-nuclide excitation (Cd-109 and Am-241) by means of bulk powder samples on silica gel basis as an example. *Fresenius Z. Anal. Chem.* **288,** 267–272.

Spatz, R. and K.H. Lieser (1979) Optimisation of a spectrometer for energy dispersive x-ray fluorescence analysis by x-ray tubes in combination with secondary targets for multielement determination. *X-ray Spectrom.* **8,** 110–113.

Statham, P.J. (1977) Deconvolution and background subtraction by least-squares fitting with pre-filtering of spectra. *Anal. Chem.* **49,** 2149–2154.

Statham, P.J. (1981) (*a*) The efficiency of Si(Li) detectors at very low photon energies. (*b*) Electronic techniques for pulse processing with solid state x-ray detectors. In: K.F.J. Heinrich, D.E. Newbury, R.L. Myklebust and C.E. Fiori (eds.), *Energy Dispersive X-ray Spectrometry*. Nat. Bureau Stds Spec. Publ. **604,** 127–139, 141–164.

Updegrove, W.C. (1982) XRF without standards: ratio methods at the USGS emphasise its speed and power. *Kevex Analyst* **2,** 5–7.

van Dyck, P. and R. van Grieken (1983) Automated matrix-correction of line ratios in energy dispersive x-ray fluorescence spectrum deconvolution. *X-ray Spectrom.* **12,** 111–114.

Van Espen, P., H. Nullens and F. Adams (1980) An in-depth study of energy dispersive x-ray spectra. *X-ray Spectrom.* **9,** 126–133.

Verbeke, P. and F. Adams (1979) Multi-element analysis of geological samples by energy dispersive x-ray fluorescence. *Anal. Chim. Acta* **109,** 85–95.

White, E.W. and G.G. Johnson (1970) *X-ray Emission and Absorption Wavelengths and Two-Theta Tables* (2nd edn.). Am. Soc. for Testing and Materials, ASTM Data Series DS 37A.

Wobrauschek, P. and H. Aiginger (1975) Total-reflection x-ray fluorescence spectrometric determination of elements in nanogram amounts. *Anal. Chem.* **47,** 852–855.

Wobrauschek, P. and H. Aiginger (1980) X-ray fluorescence in the ng region using total reflection of the primary beam. *Spectrochim. Acta* **35B,** 607–614.

Wobrauschek, P. and H. Aiginger (1983) X-ray fluorescence analysis with a linear polarised beam after Bragg reflection from a flat or a curved single crystal. *X-ray Spectrom.* **12,** 72–78.

Woldseth, R. (1973) *X-ray Energy Spectrometry*. Kevex Corp., Burlingame, California.

Chapter 10

Abbey, S. (1983) Studies in 'standard samples' of silicate rocks and minerals, 1969–1982. Geological Survey of Canada, *Paper* 83–15.

Albee, A.L. and A.A. Choodos (1970) Semiquantitative electron microprobe determination of Fe^{2+}/Fe^{3+} and Mn^{2+}/Mn^{3+} in oxides and silicates and its application to petrologic problems. *Am. Mineral.* **55,** 491–501.

Amli, R. and W.L. Griffin (1975) Microprobe analysis of REE minerals using empirical correction factors. *Am. Mineral.* **60,** 599–606.

Beaman, D.R. and J.A. Isasi (1970) A critical examination of computer programmes used in quantitative electron microprobe analysis. *Anal. Chem.* **42,** 1540–1568.

Bence, A.E. and A.L. Albee (1968) Empirical correction factors for the electron microanalysis of silicates and oxides. *J. Geol.* **76,** 382–403.

Birks, L.S., R.E. Seebold, A.P. Batt and J.S. Grosso (1964) Excitation of characteristic x-rays by photons, electrons and primary x-rays. *J. Appl. Phys.* **35,** 2578–2581.

Bloomer, R.N. (1957) The lives of electron microscope filaments. *Brit. J. Appl. Phys.* **8,** 83–85.

Bowles, J.F.W. (1978) Quantitative microprobe analysis of uranium minerals. *Microscope* **26,** 55–67.

Boyd, F.R., L.W. Finger and F. Chayes (1969) Computer reduction of electron probe data. *Carnegie Institute of Washington Yearbook* **67,** 210–215.

Brown, R.W. (1977) A sample fusion technique for whole rock analysis with the electron microprobe. *Geochim. Cosmochim. Acta* **41,** 435–438.

Brunetto, M.G. and J.A. Riveros (1984) A modification of Kramers' law for the x-ray continuum from thick targets. *X-ray Spectrom.* **13,** 60–63.

Campbell, A.J. and R. Gibbons (1966) Specimen contamination in the electron microprobe. In: T.D. McKinley, K.F.J. Heinrich and D.B. Wittry (eds.), *The Electron Microprobe*. John Wiley and Sons, New York, 75–82.

Castaing, R. and R. Guinier (1949) Application des sondes éléctroniques à l'analyse métallographique. *Proc. Conf. Electron Microscopy, Delft*, Martinus Nijhoff, The Hague, 60.

Castaing, R. (1951) Application des sondes éléctroniques à une méthode d'analyse ponctuelle chimique et cristallographique. PhD thesis, University of Paris.

Craw, D. (1981) Oxidation and microprobe-induced potassium mobility in iron-bearing phyllosilicates from the Otago schists, New Zealand. *Lithos* **14,** 49–57.

Curgenven, L. and P. Duncomb (1971) Simulation of electron trajectories in a solid target by a simple Monte Carlo technique. Tube Investment Research Laboratories Report No. 303.

Deer, W.A., R.A. Howie and J. Zussman (1963) *Rock Forming Minerals* (Vol. 4). Longman, London.

Deer, W.A., R.A. Howie and J. Zussman (1976) *An Introduction to the Rock Forming Minerals*. Longman, London.

Drake, M.J. and D.F. Weill (1972) New rare earth element standards for electron microprobe analysis. *Chem. Geol.* **10,** 179–181.

Dunham, A.C. and F.C.F. Wilkinson (1978) Accuracy, precision and detection limits of energy dispersive electron microprobe analysis of silicates. *X-ray Spectrom.* **7,** 50–56.

Eckert, R. (1982) An x-ray fluorescence set-up for trace analysis. *Beitr. Elektronenmikroskop. Direktabb. Oberfl.* **15,** 41–48.

Everhart, T.E. and R.F.M. Thornley (1960) Wide-band detector for microampere low-energy electron currents. *J. Sci. Instrum.* **37,** 246–248.

Exley, R.A. (1980) Microprobe studies of REE-rich accessory minerals: implications for Skye granite petrogenesis and REE mobility in hydrothermal systems. *Earth Planet. Sci. Lett.* **48,** 97–110.

Exley, R.A. and J.V. Smith (1982) The role of apatite in mantle enrichment processes and in the petrogenesis of some alkali basalt suites. *Geochim. Cosmochim. Acta* **46,** 1375–1384.

Finger, L.W. (1972) The uncertainty in the calculated ferric iron content of a microprobe analysis. *Carnegie Institution Yearbook* **71,** 600–603.

Fleischer, M. (1980) *Glossary of Mineral Species 1980*. Mineralogical Record, Tucson, Arizona.

Foland, K.A. and M.E. Wagner (1981) A simple method of preparing glasses from rock powders. *Am. Mineral.* **66,** 1086–1088.

Fredriksson, K. (1966) Standards and correction procedures for microprobe analysis of minerals. In: R. Castaing, P. Deschamps and J. Philibert (eds.). *X-ray Optics and Micro-analysis* (Proc. 4th Int. Congress, Orsay, 1965), Hermann, Paris, 305–309.

Gidskehaug, A. (1975) Method to determine the degree of non-stoichiometry of iron-titanium oxides. *Geophys. J. Roy. Astr. Soc.* **41,** 255–269.

Gould, R.W. and J.T. Healey (1975) Secondary fluorescent excitation in the scanning electron microscope: improved sensitivity of energy dispersive analysis. *Rev. Sci. Instrum.* **46,** 1427–1428.

Graham, J., C.R.M. Butt and R.B.W. Vigers (1984) Sub-surface charging, a source of error in microprobe analysis. *X-ray Spectrom.* **13,** 126–133.

Hall, M.G. and G.E. Lloyd (1981) The SEM examination of geological samples with a semiconductor back-scatter electron detector. *Am. Mineral.* **66,** 362–368.

Hall, M.G. and G.E. Lloyd (1983) The SEM examination of geological samples with a semiconductor back-scatter electron detector: reply. *Am. Mineral.* **68,** 834–844.

Heinrich, K.F.J. (1966) X-ray absorption uncertainty. In: T.D. McKinley, K.F.J. Heinrich and D.B. Wittry (eds.), *The Electron Microprobe*. John Wiley and Sons, New York, 296–377.

Heinrich, K.F.J. (1981) *Electron Beam X-ray Microanalysis*. Van Nostrand Reinhold, New York.

Howie, R.A. and J.V. Smith (1966) X-ray emission microanalysis of rock forming minerals. V: orthopyroxenes. *J. Geol.* **74,** 443–462.

Jarosewich, E., J.A. Nelen and J.A. Norberg (1980) Reference samples for electron microprobe analysis. *Geostand. Newslett.* **4,** 43–48.

Jarosewich, E., A.S. Parkes and L.B. Wiggins (1979) Microprobe analysis of four natural glasses and one mineral: an interlaboratory study of precision and accuracy. In: R.F. Fudali (ed.), *Smithsonian Contrib. Earth Sci.* **22,** 53–67.

Kane, W.T. (1973) Applications of the electron microprobe in ceramics and glass technology. In: C.A. Anderson (ed.), *Microprobe Analysis.* John Wiley and Sons, New York, 241–270.

Keil, K. (1973) Applications of the electron microprobe in geology. In: C.A. Andersen (ed.), *Microprobe Analysis.* John Wiley and Sons, New York, 189–239.

Kerrick, D.M., L.B. Eminhizer and J.F. Villaume (1973) The role of carbon film thickness in electron microprobe analysis. *Am. Mineral.* **58,** 920–925.

Kniseley, R.N. and F.C. Laabs (1973) Applications of cathodoluminescence in electron microprobe analysis. In: C.A. Andersen (ed.). *Microprobe Analysis.* John Wiley and Sons, New York, 371–382.

Kretz, R. (1983) Symbols for rock-forming minerals. *Am. Mineral.* **68,** 277–279.

Linnemann, B. and L. Reimer (1978) Comparison of x-ray elemental analysis by electron excitation and x-ray fluorescence. *Scanning* **1,** 109–117.

Legkova, G.V., V.G. Voitkevich and O.P. Sharkin (1982) The electron probe determination of the amounts of Fe^{2+} and Fe^{3+} in amphiboles. *Mineralogicheski Zhurnal* **4,** 90–93 (in Russian).

Long, J.V.P. (1977) Electron microprobe analysis. In: J. Zussman (ed.), *Physical Methods in Determinative Mineralogy.* Academic Press, London, 273–341.

Long, J.V.P. and S.O. Agrell (1965) The cathodoluminescence of minerals in thin section. *Mineral. Mag.* **34,** 318–326.

Mason, B. (1968) Kaersutite from San Carlos, Arizona, with comments on the paragenesis of this mineral. *Mineral. Mag.* **36,** 997–1002.

Myklebust, R.L., C.E. Fiori and K.F.J. Heinrich (1981) Spectral processing techniques in a quantitative energy dispersive x-ray microanalysis procedure (Frame C). In: K.F.J. Heinrich, D.E. Newbury, R.L. Myklebust and C.E. Fiori (eds.), *Energy Dispersive X-ray Spectrometry.* NBS Special Publication **604,** 365–389.

Naney, M.T. (1984) A grinding/polishing tool to aid thin section preparation of small samples. *Am. Mineral.* **69,** 404–405.

Nichols, I.A. (1974) A direct fusion method of preparing silicate rock glasses for energy-dispersive electron microprobe analysis. *Chem. Geol.* **14,** 151–157.

Ong, P.S. (1966) Reducing carbon contamination in an electron microprobe and measuring low energy backscattered electrons. In: R. Castaing, P. Deschamps and J. Philibert (eds.), *X-ray Optics and Microanalysis* (Proc. 4th Int. Congress, Orsay), Hermann, Paris, 181–192.

Pawley, J., P. Statham and T. Menzel (1977) Use of beam blanking and digital scan-stop to speed the microanalysis of particles. *Scanning Electron Microscopy* **1,** 297–306.

Philibert, J. (1963) A method for calculating the absorption correction in electron probe microanalysis. In: H.H. Pattee, V.E. Cosslett and A. Engstrom (eds.), *X-ray Optics and X-ray Microanalysis* (Proc. 3rd Int. Symp. Stanford, 1962), Academic Press, New York, 379–392.

Potts, P.J., A.G. Tindle and M.C. Isaacs (1983) On the precision of electron microprobe data: a new test for the homogeneity of mineral standards. *Am. Mineral.* **68,** 1237–1242.

Rao-Sahib, T.S. and D.B. Wittry (1972) The x-ray continuum from thick targets. In: G. Shinoda, K. Kohra and T. Ichinokawa (eds.), *Proc. 6th Int. Conf. X-ray Optics and Microanalysis, Osaka,* Tokyo University Press, Tokyo, 131–137.

Rao-Sahib, T.S. and D.B.Wittry (1974) X-ray continuum from thick elemental targets for 10–50 keV electrons. *J. Appl. Phys.* **45,** 5060–5068.

Reed, S.J.B. (1965) Characteristic fluorescence corrections in electron-probe micro-analysis. *Brit. J. Appl. Phys.* **16,** 913–926.

Reed, S.J.B. (1972) Electron microprobe analysis at low operating voltage: discussion. *Am. Mineral.* **57,** 1550–1551.

Reed, S.J.B. (1973) Principles of x-ray generation and quantitative analysis with the electron microprobe. In: C.A. Anderson (ed.), *Microprobe Analysis.* John Wiley and Sons, New York, 53–81.

Reed, S.J.B. (1975) *Electron Microprobe Analysis.* Cambridge University Press, Cambridge.

Reed, S.J.B. and J.V.P. Long (1963) Electron probe measurements near phase boundaries. In: H.P. Pattee, V.E. Cosslett and A. Engstrom (eds.), *X-ray Optics and X-ray Microanalysis* (Proc. 3rd Int. Symp. Stanford, 1962), Academic Press, New York, 317–327.

Reed, S.J.B. and N.G. Ware (1975) Quantitative electron microprobe analysis of silicates using energy dispersive x-ray spectrometry. *J. Petrol.* **16,** 499–519.

Robinson, B.W. and E.H. Nickel (1979) A useful new technique for mineralogy: the backscattered-electron/low vacuum mode of SEM operation. *Am. Mineral.* **64,** 1322–1328.

Robinson, B.W. and E.H. Nickel (1983) The SEM examination of geological samples with a semiconductor backscattered-electron detector: discussion. *Am. Mineral.* **68,** 840–842.

Robinson, V.N.E. (1975) Backscattered electron imaging. In: O. Johari and I. Corvin (eds.), *Scanning Electron Microscopy/1975,* IITRI, Chicago, 51–60.

Russ, J.C. and A.D. Sanborg (1981) Use of windowless detectors for energy dispersive light element x-ray analysis. In: K.J.F. Heinrich, D.E. Newbury, R.L. Myklebust and C.E. Fiori (eds.), *Energy Dispersive X-ray Spectroscopy,* NBS Special Publication **604,** 71–95.

Schimann, K. and D.G.W. Smith (1980) A new method for the optical fusion of whole-rock powders and their analysis by an electron microprobe technique. *Canadian Mineral.* **18,** 131–142.

Short, M.A. and T.G. Gleason (1982) Performance characteristics of a high resolution Si(Li) detector using a time variant amplifier and a pulsed source of x-rays. *Adv. X-ray Anal.* **25,** 45–48.

Siivola, J. (1969) On the evaporation of some alkali metals during the electron microprobe analysis. *Bull. Geol. Soc. Finland* **41,** 85–91.

Smellie, J.A.T. (1972) Preparation of glass standards for use in x-ray microanalysis. *Mineral. Mag.* **38,** 614–617.

Smellie, J.A.T., N. Cogger and J. Herrington (1978) Standards for quantitative microprobe determination of uranium and thorium with additional information on the chemical formulae of davidite and euxenite-polycrase. *Chem. Geol.* **22,** 1–10.

Smith, D.G.W. (1981) Window contamination of energy dispersive detectors in electron microprobe analysis. *X-ray Spectrom.* **10,** 78–81.

Smith, D.G.W., C.M. Gold and D.A. Tomlinson (1975) The atomic number dependence of the x-ray continuum intensity and the practical calculation of background in energy dispersive electron microprobe analysis. *X-ray Spectrom.* **4,** 149–156.

Smith, D.G.W. and S.J.B. Reed (1981) The calculation of background in wavelength dispersive electron microprobe analysis. *X-ray Spectrom.* **10,** 198–202.

Smith, D.G.W. and S.J.B. Reed (1982) Rare earth element determinations by energy-dispersive electron microprobe techniques. *Inst. Phys. Conf. Ser. No. 61* (*Proc. EMAG Conf., Cambridge 1981*), 159–162.

Sommer, S.E. (1972*a*) Cathodoluminescence of carbonates 1. Characterisation of cathodoluminescence from carbonate solid solutions. *Chem. Geol.* **9,** 257–273.

Sommer, S.E. (1972*b*) Cathodoluminescence of carbonates 2. Geological applications. *Chem. Geol.* **9,** 275–284.

Springer, G. (1967) The correction for 'continuous fluorescence' in electron probe microanalysis. *Neues Jahrb. Mineral. Abhandl.* **106,** 241–256.

Springer, G. (1974) The role of carbon film thickness in electron microprobe analysis: a comment. *Am. Mineral.* **59,** 1121–1122.

Springer, G. and B. Nolan (1976) Mathematical expression for evaluation of x-ray emission and critical absorption energies and of mass absorption coefficients. *Can J. Spectrosc.* **21,** 134–138.

Statham, P.J. (1976) A comparative study of techniques for quantitative analysis of the x-ray spectra obtained with a Si(Li) detector. *X-ray Spectrom.* **5,** 16–28.

Statham, P.J. (1979*a*) Measurement and use of peak-to-background ratios in x-ray analysis. *Mikrochim. Acta* Suppl. **8,** 229–242.

Statham, P.J. (1979*b*) A ZAF procedure for microprobe analysis

V*

based on measurement of peak-to-background ratios. In: D.E. Newbury (ed.), *Microbeam Analysis—1979*. San Francisco Press, San Francisco, 247–253.

Statham, P.J. (1981) X-ray microanalysis with Si(Li) detectors. *J. Microsc.* **123,** 1–23.

Statham, P.J. (1983) Prospects for improvement in EDX microanalysis. *J. Microsc.* **130,** 165–176.

Statham, P.J., J.V.P. Long, G. White and K. Kandiah (1974) Quantitative analysis with an energy dispersive detector using a pulsed electron probe and active signal filtering. *X-ray Spectrom.* **3,** 153–158.

Staudigel, H. (1979) Chemical analyses of interlaboratory standards. *Initial Repts., Deep Sea Drilling Project*, Vols. 51, 52, 53, Part II, US Govt. Printing Office, Washington DC, 1331–1333.

Styles, M.T. and B.R. Young (1983) Fluocerite and its alteration products from the Afu Hills, Nigeria. *Mineral. Mag.* **47,** 41–46.

Sweatman, T.R. and J.V.P. Long (1969) Quantitative electron-probe microanalysis of rock forming minerals. *J. Petrol.* **10,** 332–379.

Urch, D.S. and P.R. Wood (1978) The determination of the valency of manganese in minerals by x-ray fluorescence spectroscopy. *X-ray Spectrom.* **7,** 9–12.

Walter, L.S. (1966) A convenient method for the preparation of silicate standards for microprobe analysis. In: R. Castaing, P. Deschamps and J. Philibert (eds.), *X-ray Optics and Microanalysis* (*Proc. 4th Int. Congress, Orsay*), Hermann, Paris, 615–621.

Weast, R.C. (editor-in-chief) (1973) *Handbook of Chemistry and Physics* (54th edn.). Chemical Rubber Company Press, Cleveland, Ohio.

White, E.W. and G.G. Johnson (1970) *X-ray Emission and Absorption Wavelengths and Two-Theta Tables* (2nd edn.). ASTM Data Series DS 37A, American Society for Testing and Materials, Philadelphia.

Wicks, F.J. and A.G. Plant (1983) The accuracy and precision of routine energy-dispersive electron microprobe analysis of serpentine. *X-ray Spectrom.* **12,** 59-66

Chapter 11

Abbey, S. (1983) Studies in 'standard' samples of silicate rocks and minerals 1969–1982. Geological Survey Canada, Paper 83–15.

Andersen, C.A. (1969) Progress in analytic methods for the ion microprobe mass analyser. *Int. J. Mass Spectrom. Ion Phys.* **2,** 61–74.

Andersen, C.A. (1970) Analytic methods for the ion microprobe mass analyser II. *Int. J. Mass Spectrom. Ion Phys.* **3,** 413–428.

Andersen, C.A. and J.R. Hinthorne (1973) Thermodynamic approach to quantitative interpretation of sputtered ion mass spectra. *Anal. Chem.* **45,** 1421–1438.

Bancroft, G.M., J.R. Brown and W.S. Fyfe (1979) Advances in, and applications of, x-ray photoelectron spectroscopy (ESCA) in mineralogy and geochemistry. *Chem. Geol.* **25,** 227–243.

Banner, A.E. and B.P. Stimpson (1974) A combined ion probe/spark source analysis system. *Vacuum* **24,** 511–517.

Basutçu, M., J.N. Barrandon, M. Volfinger and J.-L. Robert (1983) A nuclear method for the analysis of light elements ($Z < 11$) in geological samples. *Chem. Geol.* **40,** 353–359.

Bertin, E.P. (1975) *Principles and Practice of X-ray Spectrometric Analysis* (2nd edn.). Plenum, New York.

Beusen, J.M., P. Surkyn, R. Gijbels and F. Adams (1983) Quantitative analysis of silicate minerals by secondary ion mass spectrometry and laser microprobe mass analysis—a comparative study. *Spectrochim. Acta* **38B,** 843–851.

Bingham, R.A. and P.L. Salter (1976) Analysis of solid materials by laser probe mass spectrometry. *Anal. Chem.* **48,** 1735–1740.

Champness, P.E., G. Cliff and G.W. Lorimer (1976) The identification of asbestos. *J. Microsc.* **108,** 231–249.

Champness, P.E., G. Cliff and G.W. Lorimer (1982) Quantitative analytical electron microscopy of metals and minerals. *Ultramicroscopy* **8,** 121–132.

Clark, P.J., G.F. Neal and R.O. Allen (1975) Quantitative multi-element analysis using high energy particle bombardment. *Anal. Chem.* **47,** 650–658.

Compston, W., I.S. Williams and S.W. Clement (1982) U-Pb ages within single zircons using a sensitive high mass-resolution ion microprobe. *Proc. 30th Amer. Soc. Mass Spectrometry Conf.* (*Honolulu*), Am. Soc. Mass Spectrom., 593–595.

Cookson, J.A., A.T.G. Ferguson and F.D. Pilling (1972) Proton microbeams, their production and use. *J. Radioanal. Chem.* **12,** 39–52.

Deline, V.R., R.J. Blattner and C.A. Evans (1980) Trace-level microanalysis of carbon and oxygen in electronic materials by Cs bombardment SIMS. In: D.B. Wittry (ed.), *Microbeam Analysis—1980*. San Francisco Press, San Francisco, 239–242.

Exley, R.A. (1983) Evaluation and application of the ion microprobe in the strontium isotope geochemistry of carbonates. *Earth Planet. Sci. Lett.* **65,** 303–310.

Exley, R.A. and A.P. Jones (1983) $^{87}Sr/^{86}Sr$ in kimberlitic carbonates by ion microprobe: hydrothermal alteration, crustal contamination and relation to carbonatite. *Contrib. Mineral. Petrol.* **83,** 288–292.

Froude, D.O., T.R. Ireland, P.D. Kinny, I.S. Williams, W. Compston, I.R. Williams and J.S. Myers (1983) Ion microprobe identification of 4100–4200 Myr-old terrestrial zircons. *Nature* **304,** 616–618.

Gray, A.L. (1985) Solid sample introduction by laser ablation for inductively coupled plasma source mass spectrometry. *Analyst* (*London*) **110,** 551–556.

Harding-Barlow, I., K.G. Snetsinger and K. Keil (1973) Laser microprobe instrumentation. In: C.A. Andersen, *Microprobe Analysis*. John Wiley and Sons, New York, 423–455.

Hart, S.R., N. Shimizu and D.A. Sverjensky (1981) Lead isotope zoning in galena: an ion microprobe study of galena crystal from the Buick Mine, Southeast Missouri. *Econ. Geol.* **76,** 1873–1878.

Hendry, D.A.F., A.R. Chivas, S.J.B. Reed and J.V.P. Long (1981) Geochemical evidence for magmatic fluids in porphyry copper mineralisation. Part II. Ion probe analysis of Cu contents of mafic minerals, Koloula igneous complex. *Contrib. Mineral. Petrol.* **78,** 404–412.

Herzog, R.F.K., W.P. Poschenreider, F.G. Rudenauer and F.G. Satkiewicz (1967) Recent results obtained with the sputtering ion source for solids. *Proc. 15th Ann. Conf. Mass Spectrometry and Allied Topics* (*Denver*), ASTM E-14, 301–307.

Hinthorne, J.R., C.A. Andersen, R.L. Conrad and J.F. Lovering (1979) Single-grain $^{207}Pb/^{206}Pb$ and U/Pb age determinations with a 10 μm spatial resolution using the ion microprobe mass analyser (IMMA). *Chem. Geol.* **25,** 271–303.

Hinton, R.W. and J.V.P. Long (1979) High resolution ion microprobe measurement of lead isotopes: variations within single zircons from Lac Seul, Northwestern Ontario. *Earth Planet. Sci. Lett.* **45,** 309–325.

Huntress, W.T. and L. Wilson (1972) An ESCA study of lunar and terrestrial materials. *Earth Planet. Sci. Lett.* **15,** 59–64.

Ishkov, Yu. M. and F.G. Reyf (1981) Laser spectral analysis of the liquid in individual inclusions. *Geochem. Int.* **1980,** 76–79.

Johansson, S.A.E. and T.B. Johansson (1976) Analytical application of particle-induced x-ray emission. *Nucl. Instrum. Methods* **137,** 473–516.

Keil, K. and K.G. Snetsinger (1973) Applications of the laser microprobe to geology. In: C.A. Andersen (ed.), *Microprobe Analysis*. John Wiley and Sons, New York, 457–476.

Kovalev, I.D., G.A. Maksimov, A.I. Suchkov and N.V. Larin (1978) Analytical capabilities of laser-probe mass spectrometry. *Int. J. Mass Spectrom. Ion Phys.* **27,** 101–137.

Kullerud, G., R.M. Steffen, P.C. Simms and F.A. Rickley (1979) Proton induced x-ray emission (PIXE)—a new tool in geochemistry. *Chem. Geol.* **25,** 245–256.

Lovering, J.F. (1973) Ion microprobe mass analyser (IMMA). Ultimate weapon for the geochemist. Comments. *Earth Sci. Geophys.* **3,** 153–163.

McHugh, J.A. (1975) Secondary ion mass spectrometry. In: A.W. Czanderna (ed.), *Methods of Surface Analysis*. Elsevier, Amsterdam, 223–278.

Nakai, I., H. Ogawa, Y. Sugitani, Y. Niwa and K. Nagashima (1976) X-ray photoelectron spectroscopic study of vanadium-bearing aegirines. *Mineral. J.* **8,** 129–134.

Piepmeier, E.H. and H.V. Malmstadt (1969) Q-switched laser energy

absorption in plume of an aluminium alloy. *Anal. Chem.* **41,** 700–707.
Reed, S.J.B. (1980) Trace element analysis with the ion probe. *Scanning* **3,** 119–127.
Reed, S.J.B. (1981) Secondary-ion spectra of rare earths. In: R.H. Geiss (ed.), *Microbeam Analysis—1981.* San Francisco Press, San Francisco, 87–90.
Reed, S.J.B. (1983) Secondary-ion yields of rare earths. *Int. J. Mass Spectrom. Ion Phys.* **54,** 31–40.
Reed, S.J.B. (1984) Secondary-ion mass spectrometry ion probe analysis for rare earths. In: O. Johari (ed.), *Scanning Electron Microscopy 1984*, 529–535.
Reed, S.J.B. and M.C. Enright (1981) Trace element distributions in meteorites determined by ion probe analysis. *Proc. Roy. Soc. London* **A374,** 195–205.
Reed, S.J.B., J.V.P. Long, J.N. Coles and D.M. Astill (1976) Ion microprobe trace element analysis with high mass resolution. *Int. J. Mass Spectrom. Ion Phys.* **22,** 333–338.
Reed, S.J.B. and D.G.W. Smith (1985) Ion probe determination of rare earth elements in merrillite and apatite in chondrites. *Earth Planet. Sci. Lett.* **72,** 238–244.
Reed, S.J.B., D.G.W. Smith and J.V.P. Long (1983) Rare earth elements in chondritic phosphates—implications for ^{244}Pu chronology. *Nature (London)* **306,** 172–173.
Robinson, C.F. (1973) Ion microprobe instrumentation. In: C.A. Andersen (ed.), *Microprobe Analysis.* John Wiley and Sons, New York, 423–455.
Russell, S.B., C.W. Schulte, S. Falq and J.L. Campbell (1981) Specimen backings for proton-induced x-ray emission analysis. *Anal. Chem.* **53,** 571–574.
Shimizu, N. (1978) Analysis of the zoned plagioclase of different magmatic environments: a preliminary ion-microprobe study. *Earth Planet. Sci. Lett.* **39,** 398–406.
Shimizu, N. and S.R. Hart (1982) Applications of the ion microprobe to geochemistry and cosmochemistry. *Ann. Rev. Earth Planet. Sci.* **10,** 483–526.
Shimizu, N., M.P. Semet and C.J. Allegre (1978) Geochemical applications of quantitative ion-microprobe analysis. *Geochim. Cosmochim. Acta* **42,** 1321–1334.
Stacey, J.S., M.H. Delevaux and T.J. Ulrych (1969) Some triple filament lead isotope measurements and an absolute growth curve for single-stage leads. *Earth Planet. Sci. Lett.* **6,** 15–25.
Stucki, J.W., C.B. Roth and W.E. Baitinger (1976) Analysis of iron-bearing clay minerals by electron spectroscopy for chemical analysis (ESCA). *Clays Clay Mineral.* **24,** 289–292.
Thompson, M. and J.N. Walsh (1983) *A Handbook of Inductively Coupled Plasma Spectrometry.* Blackie, Glasgow and London.
Tsui, T.E., H.D. Holland and K.G. Snetsinger (1975) Laser microprobe analysis of fluid inclusions. *Econ. Geol.* **70,** 1331.
Wilson, G.C. and J.V.P. Long (1983) The distribution of lithium in some Cornish minerals: ion microprobe measurements. *Mineral. Mag.* **47,** 191–199.

Chapter 12

Abbey, S. (1983) Studies of 'standard samples' of silicate rocks and minerals 1969–1982. Geological Survey Canada, *Paper* 83–15.
Adams, F. and R. Dams (1970) *Applied Gamma-ray Spectrometry* (2nd edn.; 1st edn., C.E. Crouthamel). Pergamon, Oxford.
Amiel, S. and M. Mantel (1981) Measurement of very short-lived nuclides. In: S. Amiel (ed.), *Nondestructive Activation Analysis.* Elsevier, Amsterdam, 53–70.
Baedecker, P.A. (1971) Digital methods of photopeak integration in activation analysis. In: A.O. Brunfelt and E. Steinnes (eds.), *Activation Analysis in Geochemistry and Cosmochemistry.* Universitetforlaget, Oslo, 175–182.
Baedecker, P.A., J.J. Rowe and E. Steinnes (1977) Application of epithermal neutron activation in multielement analysis of silicate rocks employing both coaxial Ge(Li) and low energy photon detector systems. *J. Radioanal. Chem.* **40,** 115–146.
Barnes, S.J. and M.P. Gorton (1984) Trace element analysis by neutron activation with a low flux reactor (Slowpoke II): results for international reference rocks. *Geostand. Newslett.* **8,** 17–23.
Bem, H. and D.E. Ryan (1981) Choice of boron shield material in epithermal neutron activation determinations. *Anal. Chim. Acta* **124,** 373–380.
Bolton, A., J. Hwang and A.V. Voet (1982) The determination of Sc, Y and selected REE in geological materials by inductively-coupled plasma optical emission spectroscopy. *Spectrochim. Acta* **38B,** 165–174.
Borley, G.D. and N. Rogers (1979) Comparison of rare-earth element data obtained by neutron activation analysis using international rock and multielement solution standards. *Geostand. Newslett.* **3,** 89–92.
Brunfelt, A.O., I. Roelandts and E. Steinnes (1977) Some new methods for the determination of rare-earth elements in geological materials using thermal and epithermal neutron activation. *J. Radioanal. Chem.* **38,** 451–459.
Christensen, L.H. (1979) Comparison between experimental and calculated relative escape peak intensities for an intrinsic Ge detector in the energy region 11–25 keV. *X-ray Spectrom.* **8,** 146–148.
Cornelius, R., J. Hoste and J. Versieck (1982) Potential interferences inherent in neutron activation analysis of trace elements in biological materials. *Talanta* **29,** 1029–1034.
Covell, D.F. (1959) Determination of gamma-ray abundance directly from the total absorption peak. *Anal. Chem.* **31,** 1785–1790.
Covell, D.F., M.M. Sandomire and M.S. Eichen (1960) *Anal. Chem.* **32,** 1086.
Crouch, D.F. and R.L. Heath (1963) Routine testing and calibration procedures for multichannel pulse analysers and gamma-ray spectrometers. US Atomic Energy Commission Report, IDO-16923.
Croudace, I.W. (1979) Errors in instrumental neutron activation analysis caused by matrix absorption. *Chem. Geol.* **25,** 175–177.
Croudace, I.W. (1980) The use of pre-irradiation group separations with neutron activation analysis for the determination of the rare earths in silicate rocks. *J. Radioanal. Chem.* **59,** 323–330.
Currie, L.A. (1968) Units for qualitative detection and quantitative determination. *Anal. Chem.* **40,** 586–593.
De Soete, D., R. Gijbels and J. Hoste (1972) *Neutron Activation Analysis.* Wiley-Interscience, London.
Duffield, J. and G.R. Gilmore (1979) An optimum method for the determination of rare earth elements by neutron activation analysis. *J. Radioanal. Chem.* **48,** 135–145.
Erdtmann, G. (1972) Interference in neutron activation analytical determinations by uranium fission. *J. Radioanal. Chem.* **10,** 137–138.
Erdtmann, G. (1976) *Neutron Activation Tables.* Verlag Chemie, Weinheim.
Erdtmann, G. and W. Soyka (1979) *The Gamma-Rays of the Radionuclides.* Verlag Chemie, Weinheim.
Evensen, N.M., P.J. Hamilton and R.K. O'Nions (1978) Rare-earth abundances in chondritic meteorites. *Geochim. Cosmochim. Acta* **42,** 1199–1212.
Fioratti, M.P. and S.R. Piermattei (1971) Calculation of the escape peak for Ge-Li and NaI radiation detectors. *Nucl. Instrum. Methods* **96,** 605–608.
Flanagan, F. J. (1973) 1972 values for international geochemical reference samples. *Geochim. Cosmochim. Acta* **37,** 1189–1200.
Frey, F.A., M.A. Haskin, J.A. Poetz and L.A. Haskin (1968) Rare earth abundances in some basic rocks. *J. Geophys. Res.* **73,** 6085–6098.
Girardi, F. and R. Pietra (1976) Multielement and automated radiochemical separation procedures for activation analysis. *Atomic Energy Rev.* **14,** 521–565.
Gladney, E.S., C.E. Burns and I. Roelandts (1983) 1982 compilation of elemental concentrations in eleven United States Geological Survey rock standards. *Geostand. Newslett.* **7,** 3–226.
Gladney, E.S., D.B. Curtis and D.R. Perrin (1985) Determination of selected rare earth elements in 37 international geochemical reference samples by instrumental thermal neutron capture prompt gamma-ray spectrometry. *Geostand. Newslett.* **9,** 25–30.
Gordon, G.E., K. Randle, G.G. Goles, J.B. Corliss, M.H. Beeson and S.S. Oxley (1968) Instrumental activation analysis of standard rocks with high-resolution gamma-ray detectors. *Geochim. Cosmochim. Acta* **32,** 369–396.
Grodstein, G.W. (1957) *X-ray attenuation coefficients from 10 keV to 100 MeV.* Nat. Bur. Stds Circular **583.**

Gruber, E. (1973) Neutronenselbstabsorption in der aktivierungsanalyse. *Z. Anal. Chem.* **263,** 194–202.
Hancock, R.G.V. (1976) Low flux multielement instrumental neutron activation analysis in archaeometry. *Anal. Chem.* **48,** 1443–1445.
Henderson, P. and C.T. Williams (1981) Application of intrinsic Ge detectors to the instrumental neutron activation analysis for rare earth elements in rocks and minerals. *J. Radioanal. Chem.* **67,** 445–452.
Hertogen, J. and R. Gijbels (1971) Instrumental neutron activation analysis of rocks with a low-energy photon detector. *Anal. Chim. Acta* **56,** 61–82.
Hertogen, J., J. DeDonder and R. Gijbels (1974) Experimental data on photopeak integration methods in activation analysis. *Nucl. Instrum. Methods* **115,** 197–212.
Higgins, M.D. (1984) Abundance of boron in international geochemical standards by prompt gamma neutron activation analysis. *Geostand. Newslett.* **8,** 31–34.
Hoede, D. and H.A. Das (1981) Accuracy and precision of instrumental neutron activation analysis based on gamma-spectrometry with a planar intrinsic Ge detector. *J. Radioanal. Chem.* **62,** 171–186.
Hollander, J.M. (1966) The impact of semiconductor detectors on gamma-ray and electron spectroscopy. *Nucl. Instrum. Methods* **43,** 65–109.
Hooker, P.J., R.K. O'Nions and R. Pankhurst (1975) Determination of rare-earth elements in USGS standard rocks by mixed-solvent ion exchange and mass spectrometric isotope dilution. *Chem. Geol.* **16,** 189–196.
Hughes, D.J. and R.B. Schwartz (1958) *Neutron Cross-Sections.* US Atomic Energy Commission Report, BNL-325.
Huysmans, K., R. Gijbels and J. Hoste (1974) Experimental comparison of multichannel analyser dead-time corrections for short-lived radionuclides. *J. Radioanal. Chem.* **20,** 51–69.
Ila, P., P. Jagam and G.K. Muecke (1983) Multielement analysis of uraniferous rocks by INAA: special reference to interferences due to uranium and fission of uranium. *J. Radioanal. Chem.* **79,** 215–232.
Kennedy, G. and A. Fowler (1983) Interference from uranium in neutron activation analysis of rare-earths in silicate rocks. *J. Radioanal. Chem.* **78,** 165–169.
Kolesov, G.M. (1976) Determination of rare-earth elements in rocks and meteorites by a radioactivation method. *J. Radioanal. Chem.* **30,** 553–560.
Koskelo, M.J., P.A. Aarnio and J.T. Routti (1981) Sampo 80: Minicomputer program for gamma spectrum analysis with nuclide identification. *Computer Phys. Comm.* **24,** 11–35.
Kosta, L. (1982) Contamination as a limiting parameter in trace analysis. *Talanta* **29,** 985–992.
Kramar, U. (1980) The importance of second order activation in the determination of trace elements in geological samples by instrumental neutron activation analysis. *Geochim. Cosmochim. Acta* **44,** 379–382.
Kubota, M. (1977) Interference by neutron induced second order nuclear reactions in activation analysis of rare earths. *J. Radioanal. Chem.* **36,** 565–576.
Laul, J.C. and L.A. Rancitelli (1977) Multielement analysis by sequential instrumental and radiochemical neutron activation. *J. Radioanal. Chem.* **38,** 461–475.
Laul, J.C. (1979) Neutron activation analysis of geological materials. *Atom. Energy Rev.* **17,** 603–695.
Mason, B. (1966) *Principles of Geochemistry* (3rd edn.). John Wiley and Sons, New York.
Moore, W.J. (1966) *Physical Chemistry* (4th edn.). Longmans, London.
Nakamura, N. (1974) Determination of REE, Ba, Fe, Mg, Na and K in carbonaceous and ordinary chondrites. *Geochim. Cosmochim. Acta* **38,** 757–775.
Parry, S.J. (1980) Detection limits in epithermal neutron activation analysis of geological material. *J. Radioanal. Chem.* **59,** 423–427.
Peisach, M. (1981) Prompt techniques. In: S. Amiel (ed.), *Activation Analysis.* Elsevier, Amsterdam, 93–111.
Perlman, I. and F. Asaro (1969) Pottery analysis by neutron activation. *Archaeometry* **11,** 21–52.
Potts, P.J. (1983) Gamma-ray interferences found in the instrumental neutron activation analysis of silicate rocks. *J. Radioanal. Chem.* **79,** 363–370.
Potts, P.J. and R. Hussey (1983) Effect of sample to detector geometry on the accuracy of instrumental neutron activation analysis and implications for the design of a simple automatic sample-changing wheel suitable for the routine counting of low activity geological samples. *J. Radioanal. Chem.* **78,** 339–346.
Potts, P.J., O. Williams Thorpe and J.S. Watson (1981) Determination of the rare earth earth element abundances in 29 international rock standards by instrumental neutron activation analysis: a critical appraisal of calibration errors. *Chem. Geol.* **34,** 331–352.
Potts, P.J., O. Williams Thorpe, M.C. Isaacs and D.W. Wright (1985*a*) High-precision instrumental neutron activation analysis of geological samples employing simultaneous counting with both planar and coaxial detectors. *Chem. Geol.* **48,** 145–155.
Potts, P.J., O. Williams Thorpe, M.C. Isaacs and N.W. Rogers (1985*b*) Instrumental neutron activation analysis of fourteen carbonate reference materials from the NBS and George Frederic Smith and eleven industrial geological samples from the Bureau of Analysed Samples. *Geostand. Newslett.* **9,** 173–179.
Randa, Z. (1976) Routine neutron activation determination of REE in rocks and similar materials by pre-activation group separation. *Radiochem. Radioanal. Lett.* **24,** 177–188.
Rey, P., H. Wakita and R.A. Schmitt (1970) Radiochemical neutron activation analyses of indium, cadmium, yttrium and the 14 rare earth elements in rocks. *Anal. Chim. Acta* **51,** 163–178.
Ricci, E. and F.F. Dyer (1964) Second-order interference in activation analysis. *Nucleonics* **22,** 45–50.
Roelandts, I. (1981) Determination of light rare earth elements in apatite by X-ray fluorescence spectrometry after anion exchange extraction. *Anal. Chem.* **53,** 676–680.
Roelandts, I., G. Duyckaerts and A.O. Brunfelt (1974) Anion-exchange isolation of rare-earth elements from apatite minerals in methanol-nitric acid medium. *Anal. Chim. Acta* **73,** 141–148.
Routti, J.T. and S.G. Prussin (1969) Photopeak method for the computer analysis of gamma-ray spectra from semiconductor detectors. *Nucl. Instrum. Methods* **76,** 115–124.
Schock, H.H. (1977) Comparison of a coaxial Ge(Li) and a planar Ge detector in instrumental neutron activation analysis of geologic materials. *J. Radioanal. Chem.* **36,** 557–564.
Smet, T., J. Hertogen, R. Gijbels and J. Hoste (1978) A group separation scheme for radiochemical neutron activation analysis for 24 trace elements in rocks and minerals. *Anal. Chim. Acta* **101,** 45–62.
Staudigel, H. (1979) Chemical analysis of interlaboratory standards. *Initial Reports, Deep Sea Drilling Project,* Vols. 51, 52, 53, Part 2. US Govt. Printing Office, Washington DC, 1331–1333.
Steinnes, E. (1971) Epithermal neutron activation analysis of geological material. In: A.O. Brunfelt and E. Steinnes (eds.), *Activation Analysis in Geochemistry and Cosmochemistry.* Universitetforlaget, Oslo, 113–128.
Steinnes, E. (1976) Simultaneous determination of U, Th, Mo, W, As and Sb in granitic rocks by epithermal neutron activation analysis. *Anal. Chem.* **48,** 1440–1443.
Sterlinski, S. (1970) Features of the modified Covell method for computation of total absorption peak areas in complex gamma-ray spectra. *Anal. Chem.* **42,** 151–155.
Vobecky, M. (1979) Nuclear interferences of uranium fission products in neutron activation analysis. *Radiochem. Radioanal. Lett.* **37,** 231–232.
von Hevesy, G. and H. Levi. (1936) The action of neutrons on the rare earth elements. Det. Kgl. Danske Videnskabernes Selskab Mathematisk-fysiske Meddelelser XIV, 5, as quoted by H.-J. Born (1976) *Proc. Int. Conf. Modern Trends in Activation Analysis* **1,** 2–16.
Walsh, J.N., F. Buckley and J. Barker (1981) The simultaneous determination of the rare earth elements in rocks using inductively coupled plasma source spectrometry. *Chem. Geol.* **33,** 141–153.
Wasson, J.T. (1968) Quoted in Baedecker, P.A. (1971) *q.v.*
Whittaker, E.J.W. and R. Muntus (1970) Ionic radii for use in geochemistry. *Geochim. Cosmochim. Acta* **34,** 945–956.
Zimmer, W.H. (1978) *What affects a gamma spectrum.* E.G & G Ortec systems application studies, PSD No. 9.

Chapter 13

Abbey, S. (1983) Studies in 'standard samples' of silicate rocks and minerals, 1969–1982. Geol. Survey Canada, Paper 83–15.

Adams, J.A.S. (1964) Laboratory γ-ray spectrometer for geochemical studies. In: J.A.S. Adams and W.M. Lowder (eds.) (1964, *q.v.*), 485–497.

Adams, J.A.S. and G.E. Fryer (1964) Portable γ-ray spectrometer for field determination of thorium, uranium and potassium. In: J.A.S. Adams and W.M. Lowder (eds.) (1964, *q.v.*), 577–596.

Adams, J.A.S. and W.M. Lowder (eds.) (1964) *The Natural Radiation Environment*. University of Chicago Press, Chicago.

Allegre, C.J. and M. Condomines (1976) Fine chronology of volcanic processes using ^{238}U-^{230}Th systematics. *Earth Planet. Sci. Lett.* **28,** 395–406.

Aller, R.C. and J.K. Cochran (1976) $^{234}Th/^{238}U$ disequilibrium in near shore sediment: particle reworking and diagenetic time scales. *Earth Planet. Sci. Lett.* **29,** 37–50.

Amiel, S. (1962) Analytical applications of delayed neutron emission in fissionable elements. *Anal. Chem.* **34,** 1683–1692.

Apted, M.J. and A.L. Boettcher (1981) Partitioning of rare earth elements between garnet and andesite melt: an autoradiographic study of P-T-X effects. *Geochim. Cosmochim. Acta* **45,** 827–837.

Basham, I.R. and G.D. Easterbrook (1977) Alpha particle autoradiography of geological specimens by use of cellulose nitrate detectors. *Inst. Min. Metall. Bull.* **86,** B96–B98.

Belyaev, Yu.I. and O.P. Sobornov (1981) Uranium, thorium and potassium in reference samples. *Geostand. Newslett.* **5,** 109–111.

Benjamin, T.M., N.T. Arndt and J.R. Holloway (1977) Instrumental techniques for β-track mapping. *Carnegie Institution Yearbook* **76,** 658–660.

Bennett, J.T., S. Krishnaswami, K.K. Turekian, W.G. Melson and C.A. Hopson (1982) The uranium and thorium decay series nuclides in Mt St. Helens effusives. *Earth Planet. Sci. Lett.* **60,** 61–69.

Bertin, E.P. (1975) *Principles and Practice of X-ray Spectrometric Analysis* (2nd edn.). Plenum, New York.

Bhat, S.G., S. Krishnaswamy, D. Lal Rama and W.S. Moore (1969) $^{234}Th/^{238}U$ ratios in the ocean. *Earth Planet. Sci. Lett.* **5,** 483–491.

Bowie, S.H.U. (1977) Radiographic techniques. In: J. Zussman (ed.), *Physical Methods in Determinative Mineralogy* (2nd edn.). Academic Press, London, 677–687.

Burnett, D.S. and D.S. Woolum (1983) In situ trace element microanalysis. *Ann. Rev. Earth Planet. Sci.* **11,** 329–358.

Capaldi, G., M. Cortini, P. Gasparini and R. Pece (1976) Short lived radioactive disequilibrium in freshly erupted volcanic rocks and their implications for the pre-eruptive history of a magma. *J. Geophys. Res.* **81,** 350–358.

Cartwright, B.G., E.K. Shirk and P.B. Price (1978) A nuclear track recording polymer of unique sensitivity and resolution. *Nucl. Instrum. Methods* **153,** 457–460.

Cassidy, J. (1981) Techniques of field gamma-ray spectrometry. *Min. Mag.* (*London*) **44,** 391–398.

Chanda, R.N. and R.A. Deal (1970) Catalogue of semiconductor alpha particle spectra. US Atomic Energy Commission publ. IN-1261.

Cherry, R.D. (1964) Alpha particle detection techniques applicable to the measurement of samples from the natural radiation environment. In: J.A.S Adams and W.M. Lowder (eds.) (1964, *q.v.*), 407–424.

Condomines, M. and C.J. Allegre (1980) Age and magmatic evolution of Stromboli volcano from ^{230}Th-^{238}U disequilibrium data. *Nature* (*London*) **288,** 354–357.

Conrad, C.P., M.W. Rowe and E.S. Gladney (1982) Comparative determination of uranium in silicates by delayed neutron activation analysis. *Geostand. Newslett.* **6,** 1–4.

Craig, H., S. Krishnaswami and B.L.K. Somayajulu (1973) ^{210}Pb-^{226}Ra: radioactive disequilibrium in the deep sea. *Earth Planet. Sci. Lett.* **17,** 295–305.

Cumming, G.L. (1974) Determination of uranium and thorium in meteorites by the delayed neutron method. *Chem. Geol.* **13,** 257–267.

Duane, M.J. and C.T. Williams (1980) Some applications of autoradiographs in textural analysis of uranium-bearing samples. *Econ. Geol.* **75,** 766–770.

Elsinger, R.J., P.T. King and W.S. Moore (1982) Radium-224 in natural waters measured by γ-ray spectrometry. *Anal. Chim. Acta* **144,** 277–281.

Erdtmann, G. (1976) *Neutron Activation Analysis Tables*. Verlag Chemie, Weinheim.

Erdtmann, G. and W. Soyka (1979) *The Gamma-Rays of the Radionuclides*. Verlag Chemie, Weinheim.

Faure, G. (1977) *Principles of Isotope Geology*. John Wiley and Sons, New York.

Fisher, D.E. (1975) Geoanalytical applications of particle tracks. *Earth Sci. Rev.* **11,** 291–335.

Fleisher, R.L., P.B. Price and R.M. Walker (1965) Tracks of charged particles in solids. *Science* **149,** 383–393.

Furst, M.J. (1981) Boron in siliceous materials as a paleosalinity indicator. *Geochim. Cosmochim. Acta* **45,** 1–13.

Furst, M., H.A. Lowenstam and D.S. Burnett (1976) Radiographic study of the distribution of boron in recent mollusc shells. *Geochim. Cosmochim. Acta* **40,** 1381–1386.

Gaggeler, H., H.R. von Gunten and U. Nyffeler (1976) Determination of ^{210}Pb in lake sediments and in air samples by direct gamma-ray measurement. *Earth Planet. Sci. Lett.* **33,** 119–121.

Gale, N.H. (1967) Development of delayed neutron technique as rapid and precise method for determination of uranium and thorium at trace levels in rocks and minerals, with applications to isotope geochronology. In: *Radioactive Dating and Methods of Low Level Counting*, International Atomic Energy Agency, Vienna, 431–452.

Gladney, E.S., D.R. Perrin, W.K. Hensley and M.E. Bunker (1980) Uranium content of twenty-five silicate reference materials. *Geostand. Newslett.* **4,** 243–246.

Goldberg, E.D. and M. Koide (1958) Ionium–thorium chronology in deep sea sediments of the Pacific. *Science* **128,** 1003.

Goles, G.G. and R.S. Seymour (1984) A novel use of disequilibrium ages in geothermal energy exploration. *Trans. 5th Int. Conf. Nuclear Methods in Environmental and Energy Research*, Puerto Rico.

Haggerty, S.E., E. Raber and C.W. Naeser (1983) Fission track dating of kimberlitic zircons. *Earth Planet. Sci. Lett.* **63,** 41–50.

Harmon, R.S. (1976) Unpublished PhD thesis, McMaster University, Hamilton, Ontario.

Harmon, R.S., T.-L. Ku, R.K. Matthews and P.L. Smart (1979) Limits of U-series analysis: Phase 1 results of the uranium series intercomparison project. *Geology* **7,** 405–409.

Harrison, W.J. and B.J. Wood (1980) An experimental investigation of the partitioning of REE between garnet and liquid with reference to the role of defect equilibria. *Contrib. Mineral. Petrol.* **72,** 145–155.

Heier, K.S., W. Compston and I. McDougall (1965) Thorium and uranium concentrations, and the isotopic composition of strontium in the differentiated Tasmanian dolerites. *Geochim. Cosmochim. Acta* **29,** 643–659.

Heier, K.S. and J.J.W. Rogers (1963) Radiometric determination of thorium, uranium and potassium in basalts and in two magmatic differentiation series. *Geochim. Cosmochim. Acta* **27,** 137–154.

Holloway, J.R. and M.J. Drake (1977) Quantitative microautoradiography by x-ray emission microanalysis. *Geochim. Cosmochim. Acta* **41,** 1395–1397.

Hurford, A.J. and P.F. Green (1982) A user's guide to fission track dating calibration. *Earth Planet. Sci. Lett.* **59,** 343–354.

Hussain, N. and S. Krishnaswami (1982) The behaviour of short-lived radiogenic lead isotopes (^{214}Pb and ^{212}Pb) in ground waters and laboratory leaching experiments. *Earth Planet. Sci. Lett.* **58,** 430–438.

IAEA (1976) Radiometric reporting methods and calibration in uranium exploration. Technical Reports Series **174,** International Atomic Energy Agency, Vienna.

Jones, J.H. and D.S. Burnett (1981) Quantitative radiography using Ag x-rays. *Nucl. Instrum. Methods* **180,** 625–633.

Kim, K.H. and W.C. Burnett (1983) γ-ray spectrometric determination of uranium series nuclides in marine phosphorites. *Anal. Chem.* **55,** 1796–1800.

Kleeman, J.D. and J.F. Lovering (1967) Uranium distribution in rocks by fission-track registration in Lexan plastic. *Science* **156,** 512–513.

Koide, M. and K.W. Bruland (1975) The electrodeposition and

determination of radium by isotopic dilution in sea water and sediments simultaneously with other natural radionuclides. *Anal. Chim. Acta* **75,** 1–19.

Kraus, K.A., G.E. Moore and F. Nelson (1956) Anion-exchange studies XXI. Th(IV) and U(IV) in hydrochloric acid. Separation of thorium, protactinium and uranium. *J. Am. Chem. Soc.* **78,** 2692–2695.

Krishnaswami, S. and M.M. Sarin (1976) The simultaneous determination of Th, Pa, Ra isotopes, ^{210}Pb, ^{55}Fe, ^{32}Si and ^{14}C in marine suspended phases. *Anal. Chim. Acta* **83,** 143–156.

Ku, T.-L. (1965) An evaluation of the ^{234}U/^{238}U method as a tool for dating pelagic sediments. *J. Geophys. Res.* **70,** 3457–3474.

Ku, T.-L. (1976) The uranium-series methods of age determination. *Ann. Rev. Earth Planet. Sci.* **4,** 347–379.

Ledger, E.B., T.T. Tieh and M.W. Rowe (1980*a*) Delayed neutron activation determination of uranium in thirteen French rock reference samples. *Geostand. Newslett.* **4,** 5–8.

Ledger, E.B., T.T. Tieh and M.W. Rowe (1980*b*) Delayed neutron activation determination of uranium in twelve rock reference standards. *Geostand. Newslett.* **4,** 153–155.

Lowder, W.M., W.J. Condon and H.L. Beck (1964) Field spectrometric investigations of environmental radiation in the USA. In: J.A.S. Adams and W.M. Lowder (eds.) (1964, *q.v.*), 597–616.

Mahadavi, A. (1964) The thorium, uranium and potassium contents of Atlantic and Gulf Coast beach sands. In: J.A.S. Adams and W.M. Lowder (eds.) (1964, *q.v.*), 87–114.

Michel, J., W.S. Moore and P.T. King (1981) Gamma-ray spectrometry for determination of radium-228 and radium-226 in natural waters. *Anal. Chem.* **53,** 1885–1889.

Moore, W.J. (1966) *Physical Chemistry*. Longman, London.

Moore, W.S. (1969) Oceanic concentrations of ^{228}radium. *Earth Planet. Sci. Lett.* **6,** 437–466.

Mysen, B.O. (1978) Experimental determination of rare earth element partitioning between hydrous silicate melt, amphibole and garnet peridotite minerals at upper mantle pressures and temperatures. *Geochim. Cosmochim. Acta* **42,** 1253–1263.

Mysen, B.O. and M.G. Seitz (1975) Trace element partitioning determined by beta track mapping: an experimental study using carbon and samarium as examples. *J. Geophys. Res.* **80,** 2627–2635.

Ostle, D., R.F. Coleman and T.K. Ball (1972) Neutron activation analysis as an aid to geochemical prospecting for uranium. In: S.H.U. Bowie, M. Davis and D. Ostle (eds.), *Uranium Prospecting Handbook*. Inst. Mining Metall., London, 95–109.

Poskanzer, A.M. and B.M. Foreman (1961) A summary of TTA extraction coefficients. *J. Inorg. Nucl. Chem.* **16,** 323–336.

Potts, P.J. (1984) Neutron activation induced beta autoradiography as a technique for locating minor phases in thin section: application to rare earth element and platinum group element mineral analysis. *Econ. Geol.* **79,** 738–747.

Potts, P.J. (1986) Neutron activation induced beta autoradiography as a technique for locating minor phases in thin section: emulsion response characteristics. In: R. Freer and P.F. Dennis (eds.), *Mass Transport in Silicate and Oxide Systems. Materials Science Forum* (TransTech Publications, Switzerland) **7,** 35–44.

Potts, P.J. and H.M. Prichard (1986) Mineralogical applications of beta autoradiography. In: M.J. Gallagher, C.R. Neary, R. Ixer and H.M. Prichard (eds.), *Metallogenesis in Basic and Ultrabasic Rocks*, Inst. Mining Metall., London.

Price, P.B. and R.M. Walker (1963) A simple method of measuring low uranium concentrations in natural crystals. *Appl. Phys. Lett.* **2,** 23–25.

Reddy, G.R., D.R. Pant and M. Sankar Das (1982) Uranium content of thirty-three rock reference samples determined by the delayed fission neutron counting method. *Geostand. Newslett.* **6,** 217–219.

Rosholt, J.N. (1957) Quantitative radiochemical methods for determination of the sources of natural radioactivity. *Anal. Chem.* **29,** 1398–1408.

Schwarcz, H.P., M. Gascoyne and D.C. Ford (1982) Uranium-series disequilibrium studies of granitic rocks. *Chem. Geol.* **36,** 87–102.

Somayajulu, B.L.K. and H. Craig (1976) Particulate and soluble ^{210}Pb activities in the deep sea. *Earth Planet. Sci. Lett.* **32,** 268–276.

Tera, F., J. Korkisch and F. Hecht (1961) Ion exchange in mixed solvents IV—the distribution of thorium between alcohol-nitric acid solutions and the strongly-basic anion exchanger Dowex-1, separation of thorium from uranium. *J. Inorg. Nucl. Chem.* **16,** 345–349.

Wilson, B.J. (ed.) (1966) *The Radiochemical Manual*. The Radiochemical Centre, Amersham, UK.

Wollenberg, H.A. (1973) Fission-track radiography of uranium and thorium in radioactive minerals. In: M.J. Jones (ed.), *Geochemical Exploration 1972*. Inst. Mining Metall., London, 347–358.

Chapter 14

Akillmali, K., B. Lumu and W. Mwamba (1979) Inorganic ion exchangers for the removal of scandium and rare earth elements in neutron activation analysis of geological samples. *Anal. Chem.* **51,** 165–166.

BDH (1981) *Ion Exchange Resins*. BDH Chemicals Ltd., Poole, Dorset.

Betteridge, D. and H.E. Hallam (1972) *Modern Analytical Methods*. The Chemical Society, London.

Brunfelt, A.O., I. Roelandts and E. Steinnes (1974) Determination of rubidium, caesium, barium and rare earth elements in ultramafic rocks by neutron activation analysis. *Analyst* (*London*) **99,** 277–284.

Brunfelt, A.O. and E. Steinnes (1969) Determination of lutetium, ytterbium and terbium in rocks by neutron activation and mixed solvent anion exchange chromatography. *Analyst* (*London*) **94,** 979–984.

Crock, J.G., F.E. Lichte and T.R. Wildeman (1984) The group separation of the rare earth elements and yttrium from geological materials by cation-exchange chromatography. *Chem. Geol.* **45,** 149–163.

Csajka, M. (1973) Fast radiochemical separation of rare earths from rock samples by retention on lanthanum oxalate. *Radiochem. Radioanal. Lett.* **13,** 151–159.

Desai, H.B., I.R. Krishnamoorthy and M. Sankar Das (1964) Determination of scandium, yttrium, samarium and lanthanum in standard silicate rocks G-1 and W-1 by neutron activation analysis. *Talanta* **11,** 1249.

Ewing, G.W. (1975) *Instrumental Methods of Chemical Analysis*. McGraw-Hill, New York.

Faris, J.P. and J.W.Warton (1962) Anion exchange resin separation of the rare earths, yttrium and scandium in nitric acid-methanol mixtures. *Anal. Chem.* **34,** 1077–1080.

Fritz, J.S. and G.H. Schenk (1969) *Quantitative Analytical Chemistry* (2nd edn.). Allyn and Bacon, Boston.

Fu, B., A.M. Ure and T.S. West (1983) Column cementation on aluminium powder as a preconcentration technique for trace element determinations by spark-source mass spectrometry. Part 1: Copper, lead, ruthenium and the noble metals. *Anal. Chim. Acta* **152,** 95–104.

Grimshaw, R.W. and C.E. Harland (1975) *Ion-exchange: Introduction to Theory and Practice*. The Chemical Society, London.

Hackett, D.S. and S. Siggia (1977) Selective concentration and determination of trace metals using poly(dithiocarbamate) chelating ion-exchange resins. In: G.W. Ewing (ed.), *Environmental Analysis*. Academic Press, New York, 253–265.

Hirose, A., K. Kobori and D. Ishii (1978) Determination of rare earth elements and heavy metals in river waters by preconcentration on Chelex 100 and neutron activation. *Anal. Chim. Acta* **97,** 303–310.

Hooker, P.J., R.K. O'Nions and R.J. Pankhurst (1975) Determination of rare earth elements in USGS standard rocks by mixed-solvent ion exchange and mass spectrometric isotope dilution. *Chem. Geol.* **16,** 189–196.

Korkisch, J. and G. Arrhenius (1964) Separation of uranium, thorium and rare earths in HNO_3-acetic acid medium. *Anal. Chem.* **36,** 850–854.

Korkisch, J. and F. Tera (1961) Separation of thorium by anion exchange. *Anal. Chem.* **33,** 1264–1266.

Leyden, D.E. and W. Wegscheider (1981) Preconcentration for trace element determination in aqueous samples. *Anal. Chem.* **53,** 1059A–1065A.

Mahanti, H.S. and R.M. Barnes (1983) Determination of trace

elements in coal and other energy-related materials by inductively coupled plasma emission spectrometry after collection on a poly-(dithiocarbamate) resin. *Anal. Chim. Acta* **149,** 395–400.

Miyazaki, A. and R.M. Barnes (1981) Complexation of some transition metals, rare earth elements and thorium with a poly-(dithiocarbamate) chelating resin. *Anal. Chem.* **53,** 299–304.

Paterson, R. (1970) *An Introduction to Ion Exchange*. Heyden, London.

Pecsok, R.L., L.D. Shields, T. Cairns and I.G. McWilliam (1976) *Modern Methods of Chemical Analysis* (2nd edn.). John Wiley and Sons, New York.

Roelandts, I. (1981) Determination of light rare earth elements in apatite by x-ray fluorescence spectrometry after anion exchange extraction. *Anal. Chem.* **53,** 676–680.

Roelandts, I., G. Duyckaerts and A.O. Brunfelt (1974) Anion-exchange isolation of rare earth elements from apatite minerals in methanol-nitric acid medium. *Anal. Chim. Acta* **73,** 141–148.

Strewlow, F.W.E. and P.F.S. Jackson (1974) Determination of trace and ultra trace quantities of rare earth elements by ion exchange mass spectrography. *Anal. Chem.* **46,** 1481–1486.

Walsh, J.N., F. Buckley and J. Barker (1981) The simultaneous determination of rare earth elements in rocks using inductively coupled plasma source spectrometry. *Chem. Geol.* **33,** 141–153.

Walton, H.F. and J. Reyes (1973) *Modern Chemical Analysis and Instrumentation*. Marcel Dekker, New York.

Welsh, K.H. and A.M. Ure (1980) Concentration procedures for trace element analysis by spark source mass spectrometry. *Anal. Proc.* **17,** 8–13.

Chapter 15

Abbey, S. (1983) Studies in 'standard samples' of silicate rocks and minerals 1969–1982. Canadian Geol. Survey, Paper 83–15.

Ahmed, I., S. Ahmed and D.F.C. Morris (1977) Determination of noble metals in geological materials by radiochemical neutron activation analysis. *Analyst (London)* **107,** 17–24.

Alvarez, L.W., W. Alvarez, F. Asaro and H.V. Michel (1980) Extra-terrestrial cause for the Cretaceous-Tertiary Extinction. *Science* **20B,** 1095–1108.

Arculus, R.J. and J.W. Delano (1981) Siderophile element abundances in the upper mantle: evidence for a sulphide signature and equilibrium with the core. *Geochim. Cosmochim. Acta* **45,** 1331–1343.

Aruscavage, P.J., F.O. Simon and R. Moore (1984) Flameless atomic absorption determination of platinum, palladium and rhodium in geological materials. *Geostand. Newslett.* **8,** 3–6.

Beamish, F.E. (1966) *The Analytical Chemistry of the Noble Metals*. Pergamon, London.

Beamish, F.E. and J.C. Van Loon (1972) *Recent Advances in the Analytical Chemistry of the Noble Metals*. Pergamon, Oxford.

Bird, J.M. and W.A. Bassett (1980) Evidence of a deep mantle history in terrestrial osmium-iridium-ruthenium alloys. *J. Geophys. Res.* **85,** 5461–5470.

Bowditch, D.C. (1973) A comparative study of three analytical procedures for the collection and determination of gold and platinoids in precious metal bearing ores. *Bull. Austral. Mining Develop. Labs (AMDEL)* **15,** 71–87.

Crabtree, R.H. (1979) The lattice parameters, densities and atomic volumes of the platinum metals. *J. Less Common Metals* **64,** P7–P9.

Crocket, J.H. (1972) Some aspects of the geochemistry of Ru, Os, Ir and Pt in iron meteorites. *Geochim. Cosmochim. Acta* **36,** 517–535.

Crocket, J.H. (1979) Platinum group elements in mafic and ultramafic rocks: a review. *Canadian Mineral.* **17,** 391–402, 889–890.

Crocket, J.H., R.R. Keays and S. Hsieh (1968) Determination of some precious metals by neutron activation analysis. *J. Radioanal. Chem.* **1,** 487–507.

Diamantatos, A. (1977) A new procedure for the recovery of ruthenium and osmium after lead collection. *Anal. Chim. Acta* **91,** 281–285.

Diamantatos, A. and A.A. Verbeek (1977) Method for the separation of platinum, palladium, rhodium, iridium and gold by solvent extraction. *Anal. Chim. Acta* **91,** 287–294.

Dixon, K., E.A. Jones, S. Rasmussen and R.V.D. Robert (1975) The efficiency of the fire assay procedure with nickel sulphide as the collector in the determination of platinum, silver, gold and iridium. *Rept. Nat. Inst. Metallurgy (South Africa)* **1714.**

Ehmann, W.D., P.A. Baedecker and D.M. McKown (1970) Gold and iridium in meteorites and some selected rocks. *Geochim. Cosmochim. Acta* **34,** 493–507.

Faye, G.H. and P.E. Mcloughney (1972) The tin collection scheme for the determination of platinum group metals, gold and silver. *Talanta* **19,** 269–284.

Fryer, B.J. and R. Kerrich (1978) Determination of precious metals at ppb levels in rocks by a combined wet chemical and flameless atomic absorption method. *At. Absorption Newslett.* **17,** 4–6.

Gijbels, R. (1971) Determination of noble metals by neutron activation analysis. *Talanta* **18,** 587–601.

Gijbels, R., H.T. Millard, G.A. Desborough and A.J. Bartel (1974) Osmium, ruthenium, iridium in silicates and chromites from the eastern Bushveld complex, South Africa. *Geochim. Cosmochim. Acta* **38,** 319–337

Green, T.E., S.L. Law and W.J. Campbell (1970) Use of selective ion exchange paper in X-ray spectrography and neutron activation; application to the determination of gold. *Anal. Chem.* **42,** 1749–1753.

Grimaldi, F.S. and M.M. Schnepfe (1968) Determination of palladium and platinum in rocks. US Geol. Survey Prof. Paper **600-B,** B99–B103.

Grimaldi, F.S. and M.M. Schnepfe (1969) Mode of occurrence of platinum, palladium and rhodium in chromitite. US Geol. Survey Prof. Paper **650-C,** C149–C151.

Grimaldi, F.S. and M.M. Schnepfe (1970) Determination of iridium in mafic rocks by atomic absorption. *Talanta* **17,** 617–621.

Haffty, J., L.B. Riley and W.D. Goss (1977) A manual on fire assaying and determination of the noble metals in geological materials. *US Geol. Survey Bull* **1445,** 58 pp.

Haines, J. and R.V.D. Robert (1982) The determination by atomic absorption spectrophotometry using electrothermal atomization of platinum, palladium, rhodium, ruthenium and iridium. Rept. Mintek (South Africa) **M34.**

Hoffman, E.L., A.J. Naldrett, J.C. Van Loon, R.G.V. Hancock and A. Manson (1978) The determination of all the platinum group elements and gold in rocks and ore by neutron activation analysis after preconcentration by a nickel sulphide fire assay technique on large samples. *Anal. Chim. Acta* **102,** 157–166.

Keays, R.R., R. Ganapathy, J.C. Laul, U. Krehenbuhl and J.W. Morgan (1974) The simultaneous determination of 20 trace elements in terrestrial, lunar and meteoritic material by radiochemical neutron activation analysis. *Anal. Chim. Acta* **72,** 1–29.

Koster, G. and G. Schmuckler (1967) Separation of noble metals from base metals by means of a new chelating resin. *Anal. Chim. Acta* **38,** 179–184.

Loss, R.D., K.J.R. Rosman and J.R. de Laeter (1983) Ag, Te and Pd in 17 geochemical reference materials by mass spectrometric isotope dilution analysis. *Geostand. Newslett.* **7,** 321–324.

L'vov, B.V., L.A. Pelieva, E.K. Mandrazhi and S.K. Kalinin (1979) Atomic absorption determination of platinum group elements in a HGA graphite furnace. *Industrial Lab.* **45,** 1348–1351 (Translated from *Zavodskaya Lab.* **45,** 1098–1101 (1979).)

McAdam, R.C., R. Sutarno and P.E. Moloughney (1973) Noble metals bearing sulphide concentrate PTC: its characterisation and preparation for use as a standard reference material. Technical Bull. TB176, Dept. Energy, Mines and Resources, Ottawa, Canada.

Mitchell, R.H. and R.R. Keays (1981) Abundance and distribution of gold, palladium and iridium in some spinel and garnet lherzolites: implications for the nature and origin of precious metal-rich intergranular components in the upper mantle. *Geochim. Cosmochim. Acta* **45,** 2425–2442.

Moloughney, P.E. (1980) A fire assay and wet chemical method for determination of palladium, platinum, gold and silver in ores and concentrates. *Geostand. Newslett.* **27,** 365–367.

Moloughney, P.E. and G.H. Faye (1976) A rapid fire assay atomic absorption method for the determination of platinum, palladium and gold in ores and concentrates. *Talanta* **23,** 377–381.

Morgan, J.W., G.A. Wandless, R.K. Petrie and A.J. Irving (1981)

Composition of the earth's upper mantle—1. Siderophile trace elements in ultramafic nodules. *Tectonophys.* **75,** 47–67.

Nadkarni, R.A. and G.H. Morrison (1974) Determination of the noble metals in geological materials by neutron activation analysis. *Anal. Chem.* **46,** 232–236.

Nadkarni, R.A. and G.H. Morrison (1977) Neutron activation determination of noble metals using a selective group separation scheme. *J. Radioanal. Chem.* **38,** 435–449.

Naldrett, A.J. (1981) Platinum group element deposits. In: L.J. Cabri (ed.), *Platinum Group Elements: Mineralogy, Geology, Recovery.* Canadian Institute of Mining and Metallurgy, Special Volume **23,** 197–231.

Page, N.J., J.S. Myers, J. Haffty, F.O. Simon and P.J. Aruscavage (1980) Platinum, palladium and rhodium in the Fiskenaesset complex, South Western Greenland. *Econ. Geol.* **75,** 907–915.

Palmer, I., G. Streichert and A. Wilson (1971) The acid extraction of noble metals from low grade concentrates and ores and the subsequent determination of platinum, palladium, rhodium and gold. Rept. Nat. Inst. Metallurgy (South Africa) **1218.**

Palmer, R. and J.I.W. Watterson (1971) The recovery of noble metals for analysis. A radiotracer investigation of losses. Rept. Nat. Inst. Metallurgy (South Africa) **1185.**

Parry, S.J. (1980*a*). Simultaneous determination of the noble metals in geological materials by radiochemical neutron activation analysis. *Analyst (London)* **105,** 1157–1162.

Parry, S.J. (1980*b*) Automated ion-exchange system for the radiochemical separation of the noble metals. *Analyst (London)* **105,** 816–820.

Prichard, H.M., C.R. Neary and P.J. Potts (1986) Platinum group minerals in the Shetland Ophiolite. In: M.J. Gallagher, C.R. Neary, R. Ixer and H.M. Prichard (eds.), *Metallogenesis in Basic and Ultra-basic Rocks.* Inst. Mining Metall., London, 395–414.

Robert, R.V.D. and E. van Wyk (1975) The effects of various matrix elements on the efficiency of the fire assay procedure using nickel sulphide as collector. Rept. Nat. Inst. Metallurgy (South Africa) **1705.**

Robert, R.V.D., E. van Wyk and R. Palmer (1971) Concentration of noble metals by a fire assay technique using nickel sulphide as collector. Rept. Nat. Inst. Metallurgy (South Africa) **1371.**

Sen Gupta, J.G. (1973) A review of the methods for the determination of platinum group metals, silver and gold by atomic absorption spectroscopy. *Miner. Sci. Engng* **5,** 207–218.

Sighinolfi, G.P., C. Gorgoni and A.H. Mohamed (1984) Comprehensive analysis of precious metals in some geological standards by flameless AA spectroscopy. *Geostand. Newslett.* **8,** 25–29.

Van Loon, J.C. (1980) *Analytical Atomic Absorption Spectroscopy: Selected Methods.* Academic Press, New York.

van Wyk, E. (1980) Laboratory method No. 78/29. The determination of platinum, palladium, rhodium and gold in ores and concentrates by the fusion technique with lead as the collector. Rept. Nat. Inst. Metallurgy (South Africa) **2068.**

Weast, R.C. (editor-in-chief) (1973) *Handbook of Chemistry and Physics* (54th edn.). Chemical Rubber Publishing Co., Cleveland, Ohio.

Wemyss, R.B. and R.H. Scott (1978) Simultaneous determination of platinum group metals and gold in ores and related plant materials by inductively coupled plasma-optical emission spectrometry. *Anal. Chem.* **50,** 1694–1697.

Westland, A.D. (1981) Inorganic chemistry of the platinum group elements. In: L.J. Cabri (ed.), *Platinum Group Elements: Mineralogy, Geology, Recovery.* Canadian Institute of Mining and Metallurgy, Special Volume **23,** 5–18.

Williamson, J.E. and J.A. Savage (1965) The determination of osmiridium in Witwatersrand ore. *J. S. Afr. Inst. Min. Metall.* **65,** 343–356.

Chapter 16

Alexander, E.C., R.S. Lewis, J.H. Reynolds and M.C. Michel (1971) Plutonium-244: confirmation as an extinct radioactivity. *Science* **172,** 837–840.

Arden, W. and N.H. Gale (1974) New electrochemical technique for the separation of lead at trace levels from natural silicates. *Anal. Chem.* **46,** 2–9.

Benninghoven, A., C. Plog and N. Treitz (1974) Measurements of relative secondary ion yields from oxidized tungsten (100) under bombardment by ions with different masses and energies. *Int. J. Mass Spectrom. Ion Phys.* **13,** 415–424.

Beynon, J.H. and A.G. Brenton (1982) *An Introduction to Mass Spectrometry*, University of Wales Press, Cardiff.

Birck, J.L. and C.J. Allegre (1973) ^{87}Rb-^{87}Sr systematics of Muntsche Tundra mafic pluton. *Earth Planet. Sci. Lett.* **20,** 266–274.

Campbell, J.A. (1969) *Chemical Systems—Energetics, Dynamics, Structure.* W.H. Freeman, San Francisco.

Cross, W.G. (1951) Two-directional focussing of charged particles with a sector-shaped uniform magnetic field. *Rev. Sci. Instrum.* **22,** 717–722.

Dalrymple, G.B. and M.A. Lanphere (1969) *Potassium-Argon Dating: Principles, Techniques and Applications to Geochronology.* W.H. Freeman, San Francisco.

Dalrymple, G.B. and M.A. Lanphere (1971) $^{40}Ar/^{39}Ar$ technique of K/Ar dating: a comparison with the conventional technique. *Earth Planet. Sci. Lett.* **12,** 300–308.

Daly, N.R. (1960) Scintillation type mass spectrometer ion detector. *Rev. Sci. Instrum.* **31,** 264–267.

Dawson, P.H. (ed.) (1976) *Quadruple Mass Spectrometry and Its Applications.* Elsevier, Amsterdam.

DePaolo, D.J. and G.J. Wasserburg (1976) Nd isotopic variations and petrogenetic models. *Geophys. Res. Lett.* **3,** 249–252.

Ewing, G.W. (1975) *Instrumental Methods of Chemical Analysis* (4th edn.). McGraw-Hill, New York.

Faure, G. (1977) *Principles of Isotope Geology.* John Wiley and Sons, New York.

Gale, N.H., J.W. Arden and R. Hutchinson (1975) The chronology of the Nakhla achondritic meteorite. *Earth Planet. Sci. Lett.* **26,** 195–206.

Godwin, H. (1962) Half life of radiocarbon. *Nature (London)* **195,** 984.

Hawkesworth, C.J. and P.W.C. van Calsteren (1983) Radiogenic isotopes—some geological applications. In: P. Henderson (ed.), *Rare Earth Element Geochemistry* (Developments in geochemistry, **2**). Elsevier, Amsterdam, 375–421.

Kaiser, T., D. Piepgras and G.J. Wasserburg (1981) A search for evidence of a fissionable nuclide in iron meteorites and implications on heat sources in planetary cores. *Earth Planet. Sci. Lett.* **52,** 239–250.

Kelly, W.R., F. Tera and G.J. Wasserburg (1978) Isotopic determination of silver in picomole quantities by surface ionisation mass spectrometry. *Anal. Chem.* **50,** 1279–1286.

Korkisch, J. and G. Arrhenius (1964) Separation of uranium, thorium and the rare earth elements by anion exchange. *Anal. Chem.* **36,** 850–854.

Krogh, T.E. (1973) A low contamination method for hydrothermal decomposition of zircon and extraction of U and Pb for isotopic age determinations. *Geochim. Cosmochim. Acta* **37,** 485–494.

Kuroda, P.K. (1960) Nuclear fission in the early history of the earth. *Nature (London)* **187,** 36–38.

Lichtman, D. (1964) *Res. Dev.* **15,** 52.

Litzow, M.R. and T.R. Spalding (1973) *Mass Spectrometry of Inorganic and Organometallic Compounds.* Elsevier, Amsterdam.

Luck, J.M. and C.J. Allegre (1982) The study of molybdenites through the ^{187}Re-^{187}Os chronometer. *Earth Planet. Sci. Lett.* **61,** 291–296.

Luck, J.M. and C.J. Allegre (1983) ^{187}Re-^{187}Os systematics in meteorites and cosmochemical consequences. *Nature (London)* **302,** 130–132.

Luck, J.M., J.L. Birck and C.J. Allegre (1980) ^{187}Re-^{187}Os systematics in meteorites: early chronology of the solar system and the age of the galaxy. *Nature (London)* **283,** 256–259.

Lugmair, G.W. and K. Marti (1978) Lunar initial $^{143}Nd/^{144}Nd$: differential evolution of lunar crust and mantle. *Earth Planet. Sci. Lett.* **39,** 349–357.

Lugmair, G.W., N.B. Scheinin and K. Marti (1975) Search for extinct ^{146}Sm. 1: the isotopic abundance of ^{142}Nd in the Juvinas meteorite. *Earth Planet. Sci. Lett.* **27,** 79–84.

Manhes, G., J.F. Minster and C.J. Allegre (1978) Comparative uranium-thorium-lead and rubidium-strontium study of the Saint Severin amphoterite: consequences for early solar system chronology. *Earth Planet. Sci. Lett.* **39,** 14–24.

McCulloch, M.J., J.R. de Laeter and K.J.R. Rosman (1976) The isotopic composition and elemental abundance of lutetium in meteorites and terrestrial samples and the ^{176}Lu cosmochronometer. *Earth Planet. Sci. Lett.* **28,** 308–322.

McDougall, I. (1974) The ^{40}Ar/^{39}Ar method of K-Ar age determination of rocks using HIFAIR reactor. *Atom. Energy Australia* **17,** 3–12.

McGlasham, M.L. (1971) *Physico-Chemical Quantities and Units* (2nd edn.). Royal Institute of Chemistry, London.

Moore, W.J. (1966) *Physical Chemistry* (4th edn.). Longman, London.

Moore, L.J., J.R. Moody, I.L. Barnes, J.W. Gramlich, T.J. Murphy, P.J. Paulsen and W.R. Shields (1973) Trace determination of rubidium and strontium in silicate glass standard reference materials. *Anal. Chem.* **45,** 2384–2387.

O'Nions, R.K., S.R. Carter, N.M. Evensen and P.J. Hamilton (1979) Geochemical and cosmochemical applications of Nd isotope analysis. *Ann. Rev. Earth Planet Sci.* **7,** 11–38.

O'Nions, R.K., P.J. Hamilton and N.M. Evensen (1977) Variations in ^{143}Nd/^{144}Nd and ^{87}Sr/^{86}Sr ratios in oceanic basalts. *Earth Planet. Sci. Lett.* **34,** 13–22.

Pankhurst, R.J. and R.K. O'Nions (1973) Determination of Rb, Sr and ^{87}Sr/^{86}Sr ratios of some standard rocks and evaluation of x-ray fluorescence spectrometry in Rb-Sr geochemistry. *Chem. Geol.* **12,** 127–136.

Patchett, P.J. and M. Tatsumoto (1980*a*) Lu-Hf total-rock isochron for the eucrite meteorites. *Nature* (*London*) **288,** 571–574.

Patchett, P.J. and M. Tatsumoto (1980*b*) A routine high precision method for Lu-Hf isotope geochemistry and chronology. *Contrib. Mineral. Petrol.* **75,** 263–267.

Paul, W., H.P. Reinhard and U. von Zahn (1958) The electric mass-filter as mass spectrometer and isotope separator. *Z. Phys.* **152,** 143–182.

Pecsok, R.L., L.D. Shields, T. Cairns and I.G. McWilliam (1976) *Modern Methods of Chemical Analysis* (2nd edn.). John Wiley and Sons, New York.

Ralph, E.K. (1971) Carbon-14 dating. In: H.N. Michael and E.K. Ralph (eds.), *Dating Techniques for the Archaeologist.* MIT Press, Cambridge, Mass., 1–48.

Reynolds, J.H. (1960) Determination of the age of the elements. *Phys. Rev. Lett.* **4,** 8–10.

Richard, P., N. Shimizu and C.J. Allegre (1976) ^{143}Nd/^{144}Nd, a natural tracer: an application to oceanic basalts. *Earth Planet. Sci. Lett.* **31,** 269–278.

Sanz, H.G. and G.J. Wasserburg (1969) Determination of an internal ^{87}Rb-^{87}Sr isochron for the Olivenza chondrite. *Earth Planet. Sci. Lett.* **6,** 335–345.

Skoog, D.A. and D.M. West (1980) *Principles of Instrumental Analysis* (2nd edn.). Saunders College, Philadelphia/Holt-Saunders, Japan, Tokyo.

Stacey, J.S. and J.D. Kramers (1975) Approximation of terrestrial lead isotope evolution by a two-stage model. *Earth Planet. Sci. Lett.* **26,** 207–221.

Steiger, R.H. and E. Jager (1977) Subcommission on geochronology: conventions on the use of decay constants in geo- and cosmochronology. *Earth Planet. Sci. Lett.* **36,** 359–362.

Sun, S.-S. (1980) Lead isotopic study of young volcanic rocks from mid-oceanic ridges, ocean islands and island arcs. *Phil. Trans. Roy. Soc. Lond.* **A297,** 409–445.

Tanaka, T. and A. Masuda (1982) The La-Ce geochronometer: a new dating method. *Nature* (*London*) **300,** 515–518.

Tatsumoto, M., R.J. Knight and C.J. Allegre (1973) Time differences in the formation of meteorites as determined from the ratio of lead-207 to lead-206. *Science* **180,** 1279–1283.

Taylor, S.R. (1965) Geochemical analysis by spark source mass spectrography. *Geochim. Cosmochim. Acta* **29,** 1243–1261.

Taylor, S.R. and M.P. Gorton (1977) Geochemical application of spark source mass spectrography—III. Element sensitivity, precision and accuracy. *Geochim. Cosmochim. Acta* **41,** 1375–1380.

Tera, F. and G.J. Wasserburg (1972*a*) U-Th-Pb systematics in Lunar highland samples from the Luna 20 and Apollo 16 missions. *Earth Planet. Sci. Lett.* **17,** 36–51.

Tera, F. and G.J. Wasserburg (1972*b*) U-Th-Pb analyses of soil from the Sea of Fertility. *Earth Planet. Sci. Lett.* **13,** 457–466.

Wasserburg, G.J., S.B. Jacobson, D.J. DePaolo, M.T. McCulloch and T. Wen (1981) Precise determination of Sm/Nd ratios, Sm and Nd isotopic abundances in standard solutions. *Geochim. Cosmochim. Acta* **45,** 2311–2323.

Weast, R.C. (editor-in-chief) (1973) *Handbook of Chemistry and Physics* (54th edn.). Chemical Rubber Publishing Co., Cleveland, Ohio.

White, F.A. (1968) *Mass Spectrometry in Science and Technology.* John Wiley and Sons, New York.

Chapter 17

Allegre, C.J. and F. Albarede (1974) ^{238}U/^{206}Pb-^{235}U/^{207}Pb-^{232}Th/^{208}Pb zircon geochronology in alpine and non-alpine environments. *Contrib. Mineral. Petrol.* **43,** 163–194.

Arden, J.W. (1983*a*) Electrochemical separation and isotopic determination of thallium at the nanogram level by surface ionisation mass spectrometry. *Anal. Chim. Acta* **148,** 211–223.

Arden, J.W. (1983*b*) Distribution of lead and thallium in the matrix of the Allende meteorite and the extent of terrestrial lead contamination in chondrites. *Earth Planet. Sci. Lett.* **62,** 395–406.

Arden, J.W. and N.H. Gale (1974) New electrochemical technique for the separation of lead at trace levels from natural silicates. *Anal. Chem.* **46,** 2–9.

Arth, J.G., F. Barker and T.W. Stern (1980) Geochronology of Archean gneisses in the Lake Helen area, south western Big Horn mountains, Wyoming. *Precambrian Res.* **11,** 11–22.

Birck, J.L. and C.J. Allegre (1973) ^{87}Rb/^{87}Sr systematics of Muntsche Tundra mafic pluton (Kola Peninsula, USSR). *Earth Planet. Sci. Lett.* **20,** 266–274.

Boelrijk, N.A.I.M. (1968) A general formula for double isotope dilution analysis. *Chem. Geol.* **3,** 323–325.

Cameron, A.E., D.H. Smith and R.L. Walker (1969) Mass spectrometry of nanogram-size samples of lead. *Anal. Chem.* **41,** 525–526.

Catanzaro, E.J. (1967) Absolute isotopic abundance ratios of three common lead reference samples. *Earth Planet. Sci. Lett.* **3,** 343–346.

Catanzaro, E.J., T.J. Murphy, W.R. Shields and E.L. Garner (1968) Absolute isotopic abundance ratios of common, equal-atom and radiogenic lead standards. *J. Res. Natl. Bur. Stand.* **A72,** 261–267.

Colby, B.N., A.E. Rosecrance and M.E. Colby (1981) Measurement parameter selection of quantitative isotope dilution gas chromatography/mass spectrometry. *Anal. Chem.* **53,** 1907–1911.

de Laeter, J.R., I.D. Abercrombie and R. Date (1969) Mass spectrometric isotope dilution analysis of barium in standard rocks. *Earth Planet. Sci. Lett.* **7,** 64–66.

de Laeter, J.R. and D.J. Hosie (1977) Mass spectrometric isotope dilution analysis of barium in geochemical reference samples. *Geostand. Newslett.* **1,** 143–145.

de Laeter, J.R. and N. Mermelengas (1978) Mass spectrometric isotope dilution determination of palladium in geochemical reference samples. *Geostand. Newslett.* **2,** 9–11.

DePaolo, D.J. and G.J. Wasserburg (1976) Nd isotopic variations and petrogenetic models. *Geophys. Res. Lett.* **3,** 249–252.

Faure, G. (1977) *Principles of Isotope Geology.* John Wiley and Sons, New York.

Gale, N.H., J.W. Arden and R. Hutchison (1975) The chronology of the Nakhla achondritic meteorite. *Earth Planet. Sci. Lett.* **26,** 195–206.

Gancarz, A.J. and G.J. Wasserburg (1977) Initial Pb of the Amitsoq gneiss, West Greenland and implications for the age of the Earth. *Geochim. Cosmochim. Acta* **41,** 1283–1301.

Gast, P.W. (1962) The isotopic composition of strontium and the age of stone meteorites, I. *Geochim. Cosmochim. Acta* **26,** 927.

Hawkesworth, C.J. and P.W.C. van Calsteren (1983). Radiogenic isotopes—some geological applications. In: P. Henderson (ed.), *Rare Earth Element Geochemistry* (Developments in Geochemistry, 2). Elsevier, Amsterdam, 375–421.

Hawkesworth, C.J., M.J. Norry, J.C. Roddick, P.E. Baker, P.W. Francis and R.S. Thorpe (1979) $^{143}Nd/^{144}Nd$, $^{87}Sr/^{86}Sr$ and incompatible element variations in calc-alkaline andesites and plateau lavas from South America. *Earth Planet. Sci. Lett.* **42,** 45–57.

Heumann, K.G. (1982) Isotopic analysis of inorganic and organic substances by mass spectrometry. *Int. J. Mass Spectrom. Ion Phys.* **45,** 87–110.

Hooker, P.J., R.K. O'Nions and R.J. Pankhurst (1975) Determination of rare earth elements in USGS standard rocks by mixed solvent ion exchange and mass spectrometric isotope dilution. *Chem. Geol.* **16,** 189–196.

Korkisch, J. and G. Arrhenius (1964) Separation of uranium, thorium and the rare earth elements by anion exchange. *Anal. Chem.* **36,** 850–854.

Krogh, T.E. (1973) A low-contamination method for hydrothermal decomposition of zircon and extraction of U and Pb for isotopic age determinations. *Geochim. Cosmochim. Acta* **37,** 485–494.

Krogh, T.E. (1982*a*) Improved accuracy of U-Pb zircon dating by selection of more concordant fractions using a high gradient magnetic separation technique. *Geochim. Cosmochim. Acta* **46,** 631–635.

Krogh, T.E. (1982*b*) Improved accuracy of U-Pb zircon ages by creation of more concordant systems using an air abrasion technique. *Geochim. Cosmochim. Acta* **46,** 637–649.

Krogh, T.E. and G.L. Davis (1975) The production and preparation of ^{205}Pb for use as a tracer for isotope dilution analysis. *Carnegie Institute Yearbook* **74,** 416–417.

Lancelot, J.R., A.M. Bouillier, H. Maluski and J. Ducrot (1983) Deformation and related radiochronology in a Late Pan-African myonitic shear zone, Adrar des Iforas (Mali). *Contrib. Mineral. Petrol.* **82,** 312–326.

Loss, R.D., K.J.R. Rosman and J.R. de Laeter (1983) Measurement of Ag, Te, Pd in geochemical reference materials by mass spectrometric isotope dilution analysis. *Talanta* **30,** 831–835.

Loss, R.D., K.J.R. Rosman and J.R. de Laeter (1983) Ag, Te, Pd in 17 geochemical reference materials by mass spectrometric isotope dilution analysis. *Geostand. Newslett.* **7,** 321–324.

Loss, R.D., K.J.R. Rosman and J.R. de Laeter (1984) Mass spectrometric isotope dilution analysis of palladium, silver, cadmium and tellurium in carbonaceous chondrites. *Geochim. Cosmochim. Acta* **48,** 1677–1681.

Lugmair, G.W. and K. Marti (1978) Lunar initial $^{143}Nd/^{144}Nd$ differential evolution of the lunar crust and mantle. *Earth Planet. Sci. Lett.* **39,** 349–357.

Manhes, G., C.J. Allegre and A. Provost (1984) U–Th–Pb systematics of the eucrite "Juvinas": precise age determination and evidence for exotic lead. *Geochim. Cosmochim. Acta* **48,** 2247–2264.

Manhes, G., J.F. Minster and C.J. Allegre (1978) Comparative uranium-thorium-lead and rubidium-strontium study of the Saint Severin amphoterite: consequences of early solar system chronology. *Earth Planet. Sci. Lett.* **39,** 14–24.

Masuda, A. (1968) Geochemistry of lanthanides in basalts of central Japan. *Earth Planet. Sci. Lett.* **4,** 284–292.

Masuda, A., N. Nakamura and T. Tanaka (1973) Five structures of mutually normalised rare earth patterns of chondrites. *Geochim. Cosmochim. Acta* **37,** 239–248.

McCulloch, M.T. and B.W. Chappell (1982) Nd isotopic characteristics of S- and I-type granites. *Earth Planet. Sci. Lett.* **58,** 51–64.

Mermelengas, N., J.R. de Laeter and K.J.R. Rosman (1979) New data on the abundance of palladium in meteorites. *Geochim. Cosmochim. Acta* **43,** 747–753.

Minster, J.F. and C.J. Allegre (1976) ^{87}Rb-^{87}Sr history of the Norton County enstatite achondrite. *Earth Planet. Sci. Lett.* **32,** 191–198.

Nakamura, N. (1974) Determination of REE, Ba, Fe, Mg, Na and K in carbonaceous and ordinary chondrites. *Geochim. Cosmochim. Acta* **38,** 757–775.

Nakamura, N. and A. Masuda (1973) Chondrites with peculiar rare earth patterns. *Earth Planet. Sci. Lett.* **19,** 429–437.

Nier, A.O. (1950) A redetermination of the relative abundances of the isotopes of carbon, nitrogen, oxygen, argon and potassium. *Phys. Rev.* **77,** 789–793.

Nunes, P.D. and P.C. Thurston (1980) Two hundred and twenty million years of Archean evolution: a zircon U-Pb age stratigraphic study of the Uchi-Confederation Lakes greenstone belt, northwestern Ontario. *Can. J. Earth Sci.* **17,** 710–721.

O'Nions, R.K., P.J. Hamilton and N.M. Evensen (1977) Variations in $^{143}Nd/^{144}Nd$ and $^{87}Sr/^{86}Sr$ ratios in oceanic basalts. *Earth Planet. Sci. Lett.* **34,** 13–22.

O'Nions, R.K., S.R. Carter, N.M. Evensen and P.J. Hamilton (1979) Geochemical and cosmochemical applications of Nd isotope analysis. *Ann. Rev. Earth Planet. Sci.* **7,** 11–38.

Oversby, V.M. (1975) Lead isotopic systematics and ages of Archaean intrusives in Kalgoorlie-Norseman area, Western Australia. *Geochim. Cosmochim. Acta* **39,** 1107–1125.

Pankhurst, R.J. and R.K. O'Nions (1973) Determination of Rb/Sr and $^{87}Sr/^{86}Sr$ ratios of some standard rocks and evaluation of x-ray fluorescence spectrometry in Rb-Sr geochemistry. *Chem. Geol.* **12,** 127–136.

Papanastassiou, D.A. and G.J. Wasserburg (1969) Initial strontium isotopic abundances and the resolution of small time differences in the formation of planetary objects. *Earth Planet. Sci. Lett.* **5,** 361–376.

Richard, P., N. Shimizu and C.J. Allegre (1976) $^{143}Nd/^{146}Nd$, a natural tracer: an application to oceanic basalts. *Earth Planet. Sci. Lett.* **31,** 269–278.

Sanz, H.G. and G.J. Wasserburg (1969) Determination of an internal $^{87}Rb/^{87}Sr$ isochron for the Olivenza chondrite. *Earth Planet. Sci. Lett.* **6,** 335–345.

Schnetzler, C.C., H.H. Thomas and J.A. Philpotts (1967) Determination of rare earth elements in rocks and minerals by mass spectrometric stable isotope dilution technique. *Anal. Chem.* **39,** 1888–1890.

Schnetzler, C.C., H.H. Thomas and J.A. Philpots (1967) The determination of barium in G-1 and W-1 by isotope dilution. *Geochim. Cosmochim. Acta* **31,** 95–96.

Schuhmann, S. and J.A. Philpotts (1979) Mass spectrometric stable isotope dilution analyses for lanthanides in geochemical materials. In: K.A. Gschneider and L. Eyring (eds.), *Handbook on the Physical Chemistry of Rare Earths.* Elsevier/North Holland Publishing, Amsterdam, **4,** 471–481.

Steiger, R.H. and E. Jager (1977) Subcommission of geochronology: convention on the use of decay constants in geo- and cosmochronology. *Earth Planet. Sci. Lett.* **36,** 359–362.

Tera, F. and G.J. Wasserburg (1972) U-Th-Pb systematics in lunar highland samples from the Lunar 20 and Apollo 16 missions. *Earth Planet. Sci. Lett.* **17,** 36–51.

Tera, F. and G.J. Wasserburg (1975) Precise isotopic analysis of lead in picomole and subpicomole quantities. *Anal. Chem.* **47,** 2214–2220.

Thirlwall, M.F. (1982) A triple filament method for rapid and precise analysis of rare earth elements by isotope dilution. *Chem. Geol.* **35,** 155–166.

Tilton, G.R. (1973) Isotopic lead ages of chondritic meteorites. *Earth Planet. Sci. Lett.* **19,** 321–329.

Verma, S.P. (1981) Mass spectrometric isotope dilution determination of K, Rb, Cs, Ba and Sr in five geochemical reference samples. *Geostand. Newslett.* **5,** 129–131.

Wasserburg, G.J., S.B. Jacobsen, D.J. DePaolo, M.T. McCulloch and T. Wen (1981) Precise determination of Sm/Nd ratios, Sm and Nd isotopic abundances in standard solutions. *Geochim. Cosmochim. Acta* **45,** 2311–2323.

Chapter 18

Basford, J.R., J.C. Dragon, R.O. Pepin, M.R. Coscio and V.R. Murphy (1973) Krypton and xenon in lunar fines. *Proc. 4th Lunar Sci. Conf.*, 1915–1955.

Bigeleisen, J., M.L. Pearlman and H.C. Prosser (1952) Conversion of hydrogenic materials to hydrogen for isotopic analysis. *Anal. Chem.* **24,** 1356–1357.

Bottomley, D.J., J.D. Ross and W.B. Clarke (1984) Helium and neon isotope geochemistry of some ground waters from the Canadian Precambrian Shield. *Geochim. Cosmochim. Acta* **48,** 1973–1985.

Chait, E.M. (1972) Ionisation sources in mass spectrometry. *Anal. Chem.* **44,** 77A–91A.

Clayton, R.N. and T.K. Mayeda (1963) The use of bromine pentafluoride in the extraction of oxygen from oxides and silicates for isotopic analysis. *Geochim. Cosmochim. Acta* **27,** 43–52.

Craig, H. (1961) Standard for reporting concentrations of deuterium and oxygen-18 in natural waters. *Science* **133,** 1833–1834.

Craig, H. and J.E. Lupton (1976) Primordial neon, helium and hydrogen in oceanic basalts. *Earth Planet. Sci. Lett.* **31,** 369–385.

Dalrymple, G.B. and M.A. Lanphere (1969) *Potassium–Argon Dating: Principles, Techniques and Applications to Geochronology.* W.H. Freeman, San Francisco.

Dalrymple, G.B. and M.A. Lanphere (1971) $^{40}Ar/^{39}Ar$ technique of K-Ar dating: a comparison with the conventional technique. *Earth Planet. Sci. Lett.* **12,** 300–308.

Dalrymple, G.B. and M.A. Lanphere (1974) $^{40}Ar/^{39}Ar$ age spectra of some undisturbed terrestrial samples. *Geochim. Cosmochim. Acta* **38,** 715–738.

Des Marais, D.J. (1983) Light element geochemistry and spallogenesis in lunar rocks. *Geochim. Cosmochim. Acta* **47,** 1769–1781.

Des Marais, D.J. and J.G. Moore (1984) Carbon and its isotopes in mid oceanic basaltic glasses. *Earth Planet. Sci. Lett.* **69,** 43–57.

Eberhardt, P., O. Eugster and K. Marti (1965) A redetermination of the isotopic composition of atmospheric neon. *Z. Naturforsch.* **20a,** 623–624.

Erdtmann, G. (1976) *Neutron Activation Tables.* Verlag Chemie, Weinheim.

Frick, U. and R.O. Pepin (1981*a*) On the distribution of noble gases in Allende: a differential oxidation study. *Earth Planet. Sci. Lett.* **56,** 45–63.

Frick, U. and R.O. Pepin (1981*b*) Microanalysis of nitrogen isotope abundances: association of nitrogen with noble gas carriers in Allende. *Earth Planet. Sci. Lett.* **56,** 64–81.

Friedman, I. and J.R. O'Neil (1977) Compilation of stable isotope fractionation factors of geochemical interest. In: M. Fleischer (ed.), *Data of Geochemistry*, US Geol. Surv. Prof. Paper **440KK.**

Gardiner, L.R., A.J.T. Jull and C.T. Pillinger (1978) Progress towards a direct measurement of $^{13}C/^{12}C$ ratios for hydrolysable carbon in lunar soil by static mass spectrometry. *Proc. 9th Lunar Planet. Sci. Conf.*, 2167–2193.

Gardiner, L.R. and C.T. Pillinger (1979) Static mass spectrometry for the determination of active gases. *Anal. Chem.* **51,** 1230–1236.

Gonfiantini, R. (1978) Standards for stable isotope measurements in natural compounds. *Nature* (*London*) **271,** 534–536.

Haimson, M. and L.P. Knauth (1983) Stepwise fluorination—a useful approach for the isotopic analysis of hydrous minerals. *Geochim. Cosmochim. Acta.* **47,** 1589–1595.

Hurlston, J.R. and H.G. Thode (1965*a*) Variations in the ^{33}S, ^{34}S and ^{36}S contents of meteorites and their relation to chemical and nuclear effects. *J. Geophys. Res.* **70,** 3475–3484.

Hurlston, J.R. and H.G. Thode (1965*b*) Cosmic ray produced ^{36}S and ^{33}S in the metal phase of iron meteorites. *J. Geophys. Res.* **70,** 4435–4442.

Kyser, T.K. and J.R. O'Neil (1984) Hydrogen isotope systematics of submarine basalts. *Geochim. Cosmochim. Acta.* **48,** 2123–2133.

Kyser, T.K. and W.R. Rison (1982) Rare gas isotope systematics of basic lavas and ultramafic xenoliths. *J. Geophys. Res.* **87,** 5611–5630.

Lupton, J.E. and H. Craig (1975) Excess ^3He in oceanic basalts: evidence for terrestrial primordial helium. *Earth Planet. Sci. Lett.* **26,** 133–139.

Mattey, D.P., R.H. Carr, I.P. Wright and C.T. Pillinger (1984) Carbon isotopes in submarine basalts. *Earth Planet. Sci. Lett.* **70,** 196–206.

McCrea, J.M. (1950) On the isotopic chemistry of carbonates and a paleotemperature scale. *J. Chem. Phys.* **18,** 849–857.

McKinney, C.R., J.M. McCrea, S. Epstein, H.A. Allen and H.C. Urey (1950) Improvements in mass spectrometers for the measurement of small differences in isotope abundance ratios. *Rev. Sci. Instrum.* **21,** 724–730.

McNaughton, N.J., P.I. Abell, I.P. Wright, A.E. Fallik and C.T. Pillinger (1983) Preparation of nanogram quantities of deuteromethane for stable carbon isotope analysis. *J. Phys. E: Sci. Instrum.* **16,** 505–511.

Merrihue, C. (1965) Trace element determinations and potassium-argon dating by mass spectrometry of neutron irradiated samples (abstract). *Trans. Am. Geophys. Union* **46,** 125.

Merrihue, C. and G. Turner (1966) Potassium-argon dating by activation with fast neutrons. *J. Geophys. Res.* **71,** 2852.

Nier, A.C.O. (1947) A mass spectrometer for isotope and gas analysis. *Rev. Sci. Instrum.* **18,** 398–411.

Ozima, M., I. Kaneoka and M. Yanagisawa (1979) Temperature and pressure effects on ^{40}Ar-^{39}Ar schematics. *Earth Planet. Sci. Lett.* **42,** 463–472.

Pecsok, R.L., L.D. Shields, T. Cairns and I.G. McWilliam (1968) *Modern Methods of Chemical Analysis* (2nd edn.). John Wiley and Sons, New York.

Pillinger, C.T. (1984) Light element stable isotopes in meteorites—from grams to picograms. *Geochim. Cosmochim. Acta.* **48,** 2739–2766.

Podosek, F.A., J.C. Huneke, D.S. Burnett and G.J. Wasserburg (1971) Isotopic composition of xenon and krypton in the lunar soil and in the lunar wind. *Earth Planet. Sci. Lett.* **10,** 199–216.

Poreda, R. and F.R. di Brozolo (1984) Neon isotope variations in Mid-Atlantic Ridge basalts. *Earth Planet. Sci. Lett.* **69,** 277–289.

Reynolds, J.H. (1956) High sensitivity mass spectrometer for noble gas analysis. *Rev. Sci. Instrum.* **27,** 928–934.

Rison, W. and H. Craig (1983) Helium isotopes and mantle volatiles in Loihi Seamount and Hawaiian Island basalts and xenoliths. *Earth Planet. Sci. Lett.* **66,** 407–426.

Roboz, J. (1968) *Introduction to Mass Spectrometry, Instrumentation and Techniques.* Wiley-Interscience, New York.

Roddick, J.C. (1978) The application of isochron diagrams in ^{40}Ar–^{39}Ar dating: a discussion. *Earth Planet. Sci. Lett.* **41,** 233–244.

Roddick, J.C. (1983) High precision intercalibration of ^{40}Ar-^{39}Ar standards. *Geochim. Cosmochim. Acta.* **47,** 887–898.

Roddick, J.C., R.A. Cliff and D.C. Rex (1980) The evolution of excess argon in alpine biotites—a ^{40}Ar-^{39}Ar analysis. *Earth Planet. Sci. Lett.* **48,** 185–208.

Sakai, H., D.J. Des Marais, A. Ueda and J.G. Moore (1984) Concentrations and isotope ratios of carbon, nitrogen and sulphur in ocean-floor basalts. *Geochim. Cosmochim. Acta.* **48,** 2433–2441.

Sharma, T. and R.N. Clayton (1965) Measurement of $^{18}O/^{16}O$ ratios of total oxygen of carbonates. *Geochim. Cosmochim. Acta.* **29,** 1347–1353.

Sigurgeirsson, T. (1962) *Dating Recent Basalt by the Potassium Argon Method.* Rept. Physical Lab., Univ. Iceland.

Skoog, D.A. and D.M. West (1980) *Principles of Instrumental Analysis* (2nd edn.). Saunders College/Holt-Saunders Japan, Philadelphia/Tokyo.

Steiger, R.H. and E. Jager (1977) Subcommission on geochronology: convention on the use of decay constants in geo- and cosmochronology. *Earth Planet. Sci. Lett.* **36,** 359–362.

Taylor, H.P. and S. Epstein (1962) Relationship between $^{18}O/^{16}O$ ratios in coexisting minerals of igneous and metamorphic rocks. Part 1, principles and experimental results. *Geol. Soc. Am. Bull.* **73,** 461–480.

Tetley, N., I. McDougall and H.R. Heydegger (1980) Thermal neutron interferences in the $^{40}Ar/^{39}Ar$ dating technique. *J. Geophys. Res.* **85** (B12), 7201–7205.

Turner, G. (1971) Argon 40-Argon 39 dating: the optimisation of irradiation parameters. *Earth Planet. Sci. Lett.* **10,** 227–234.

Ueda, A. and H. Sakai (1983) Simultaneous determinations of the concentrations and isotope ratio of sulphate-sulphur, sulphide-sulphur and carbonate-carbon in geological samples. *Geochem. J.* **17,** 185–196.

Urey, H.C. (1948) Oxygen isotopes in nature and the laboratory. *Science* **108,** 489–497.

Voshage, H. and H. Feldmann (1978) Investigations on cosmic-ray produced nuclides in iron meteorites. 1: The measurement and interpretation of rare gas concentrations. *Earth Planet. Sci. Lett.* **39,** 25–36.

Wanke, H. and H. Konig (1959) Eine neue methode zur Kalium-argon-alterbestimmung und ihre Anwendung auf Steinmeteorite. *Z. Naturforsch.* **14a,** 860.

Wright, I.P., N.J. McNaughton, A.E. Fallik, L.R. Gardiner and C.T. Pillinger (1983) A high precision mass spectrometer for stable carbon isotope analysis at the nanogram level. *J. Phys. E: Sci. Instrum.* **16,** 497–504.

Chapter 19

Abbey, S. (1983) Studies in 'standard samples' of silicate rocks and minerals, 1969–1982. Geol. Survey Canada, *Paper* 83–15.

Adams, F. (1982) Recent advances in analytical spark source mass spectrometry. *Phil. Trans. Roy. Soc. London* **A305,** 509–519.

Adams, F. (1983) Recent progress in mass spectrometry for inorganic analysis. *Spectrochim. Acta* **38B,** 1379–1393.

Ahrens, L.H. and S.R. Taylor (1961) *A Treatise on the DC Arc Analysis of Geological and Related Materials.* Addison-Wesley, Reading, Mass.

Jackson, P.F.S. and F.W.E Strelow (1975) The rare earth content of six international geological reference materials of South African origin. *Chem. Geol.* **15,** 303–307.

Nicholls, G.D., A.L. Graham, E. Williams and M. Wood (1967) Precision and accuracy in trace element analysis of geological materials using solid source spark mass spectrometry. *Anal. Chem.* **39,** 584–590.

Ramendik, G.I. (1980) A new model of vacuum discharge and the development of spark-source mass spectrometry. *Adv. Mass Spectrom.* **8,** 408–413.

Roaldset, E. (1975) Rare earth element distributions in some Pre-cambrian rocks and their phyllosilicates, Numedal, Norway. *Geochim. Cosmochim. Acta.* **39,** 455–469.

Strewlow, F.W.E. and P.F.S. Jackson (1974) Determination of trace and ultra-trace quantities of rare earth elements by ion exchange chromatography–mass spectrography. *Anal. Chem.* **46,** 1481–1486.

Taylor, S.R. (1965*a*) Geochemical application of spark source mass spectrography. *Nature* (*London*) **205,** 34–36.

Taylor, S.R. (1965*b*) Geochemical analysis by spark source mass spectroscopy. *Geochim. Cosmochim. Acta* **29,** 1243–1261.

Taylor, S.R. (1971) Geochemical application of spark source mass spectrography—II. Photoplate data processing. *Geochim. Cosmochim. Acta* **35,** 1187–1196.

Taylor, S.R. (1979) Trace element analysis of rare earth elements by spark source mass spectrometry. In: K.A. Gschneider and L. Eyring (eds.), *Handbook on the Physics and Chemistry of Rare Earths.* Elsevier-North Holland, Amsterdam, 359–376.

Taylor, S.R. and M.P. Gorton (1977) Geochemical application of spark source mass spectrography—III. Element sensitivity, precision and accuracy. *Geochim. Cosmochim. Acta* **41,** 1375–1380.

Taylor, S.R. and S.M. McLennan (1983) Geochemical application of spark source mass spectrography—IV. The crustal abundance of tin. *Chem. Geol.* **39,** 273–280.

Verlinden, J.A.A., K.M.E. Swenters and R.H.H. Gijbels (1985) Modification in the surface composition of sparked electrodes and its relation to relative sensitivity factors in spark source mass spectrometry. *Anal. Chem.* **57,** 131–136.

Victor, A.H. and Rogers, D.E.C. (1983) Determination of ultra-trace amounts of thorium in NIM-D and NIM-P standard geological reference materials using cation exchange chromatography—spark source mass spectrometry. *Geostand. Newslett.* **7,** 331–334.

Chapter 20

Cantle, J. and P. Goddard (1983) Inductively coupled plasma source mass spectrometry for elemental and isotopic analysis. *Int. Labmate* **8,** issue 5.

Date, A.R. and A.L. Gray (1981) Plasma source mass spectrometry using an inductively coupled plasma and a high resolution quadrupole mass filter. *Analyst* (*London*) **106,** 1255–1267.

Date, A.R. and A.L. Gray (1982*a*) Isotope ratio measurements on solution samples using a plasma ion source. *Int. J. Mass Spectrom. Ion Phys.* **48,** 357–360.

Date, A.R. and A.L. Gray (1982*b*) Progress in plasma source mass spectrometry. Poster presentation, 1st Biennial National Atomic Spectroscopy Symposium, Sheffield, July 1982.

Date, A.R. and A.L. Gray (1983) Progress in plasma source mass spectrometry. *Spectrochim. Acta* **388,** 29–37.

Douglas, D.J. and J.B. French (1981) Elemental analysis with a microwave-induced plasma/quadrupole mass spectrometer system. *Anal. Chem.* **53,** 37–41.

Gray, A.L. (1975*a*) Mass-spectrometric analysis of solutions using an atmospheric pressure ion source. *Analyst* (*London*) **100,** 289–299.

Gray, A.L. (1975*b*) Plasma sampling mass spectrometry for trace analysis of solutions. *Anal. Chem.* **47,** 600–601.

Gray, A.L. (1978) Isotope ratio determination on solutions with a plasma ion source. In: D. Price and J.F.J. Todd (eds.), *Dynamic Mass Spectrometry*, Vol. 5, Heyden, London, 106–113.

Gray, A.L. and A.R. Date (1982) Plasma source mass spectrometry of inorganic samples: recent developments of the technique. Paper presented at the 9th Int. Mass Spectrometry Conf., Vienna, Austria, 30th Aug–3rd Sept., 1982.

Houk, R.S., V.A. Fassel, G.D. Flesch, H.J. Svec, A.L. Gray and C.E. Taylor (1980) Inductively coupled argon plasma as an ion source for mass spectrometric determination of trace elements. *Anal. Chem.* **52,** 2283–2289.

Houk, R.S. and J.J. Thompson (1982) Elemental and isotopic analysis of solutions by mass spectrometry using a plasma ion source. *Am. Mineral.* **67,** 238–243.

Weast, R.C. (editor-in-chief) (1973) *Handbook of Chemistry and Physics* (54th edn.), Chemical Rubber Publishing Co., Cleveland, Ohio.

Index

Abbreviations

AAS	Atomic absorption spectrometry	MS	Mass spectrometry
DCP	Direct current plasma	NAA	Neutron activation analysis
ED–XRF	Energy dispersive x-ray fluorescence spectrometry	PGE	Platinum group elements
EPMA	Electron probe microanalysis	REE(s)	Rare-earth element(s)
ESCA	Electron spectroscopy for chemical analysis	SEM	Scanning electron microscope
ICP	Inductively-coupled (argon) plasma	SIMS	Secondary ion mass spectrometry
ICP–AES	Inductively-coupled plasma–atomic emission spectrometry	SSMS	Spark source mass spectrometry
		TEM	Transmission electron microscopy
ICP–MS	Inductively-coupled plasma–mass spectrometry	WD–XRF	Wavelength dispersive x-ray fluorescence spectrometry
INAA	Instrumental neutron activation analysis	XPS	X-ray photoelectron spectroscopy
ISE	Ion-selective electrodes	XRF	X-ray fluorescence spectrometry